DNA REPAIR AND MUTAGENESIS

DNA REPAIR AND MUTAGENESIS

ERROL C. FRIEDBERG
Department of Pathology, The University of Texas
Southwestern Medical Center, Dallas, Texas

GRAHAM C. WALKER
Department of Biology, Massachusetts Institute of Technology,
Cambridge, Massachusetts

WOLFRAM SIEDE
Department of Pathology, The University of Texas
Southwestern Medical Center, Dallas, Texas

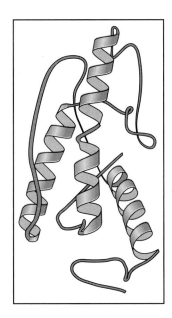

ASM Press
Washington, D.C.

Library of Congress Cataloging-in-Publication Data

Friedberg, Errol C.
    DNA repair and mutagenesis / Errol C. Friedberg, Graham C. Walker, Wolfram
Siede.
      p.   cm.
    Includes bibliographical references and index.
    ISBN 1-55581-088-8
    1. DNA repair. 2. Mutagenesis. I. Walker, Graham C. II. Siede, Wolfram.
III. Title.
    [DNLM: 1. DNA Damage. 2. DNA Repair. 3. Mutagenesis. QH 465 F899d
1955]
QH467.F753   1995
574.87'328—dc20
DNLM/DLC
for Library of Congress   94-43983
                                             CIP

10 9 8 7 6 5 4 3 2

Cover illustration: Proposed structure of the phage T4 PD-DNA glycosylase (endonuclease V or DenV protein). (*See Figure 4–24 on page 167.*)

For Jonathan, Gordon, and Lawrence:
the next generation

Contents

Preface

We have written *DNA Repair and Mutagenesis* to provide the scientific community with a thorough and state-of-the-art treatment of cellular responses to DNA damage. The general field of DNA repair has now become firmly integrated into numerous other areas of scientific inquiry, including the molecular basis of human health and disease. We therefore intend this book for students and scientists working in carcinogenesis, recombination, transcription and gene regulation, DNA replication, environmental studies, and biological evolution, in addition to those more directly involved in DNA repair and mutagenesis.

The first edition of this book was called *DNA Repair* and was published in 1985 by W. H. Freeman & Co. This new edition, *DNA Repair and Mutagenesis,* has been extensively revised and updated but does contain significant blocks of material essentially adapted verbatim from the first edition. We want to acknowledge generous permission from W. H. Freeman & Co. to reuse material from *DNA Repair.* Certainly a major feature of this new edition is the comprehensive coverage of mutagenesis and other aspects of DNA damage tolerance in five new chapters. Coverage of mutagenesis was not included in the 1985 edition, and we are pleased to be able to present this topic in a more complete text. We hope too that the revised treatment of DNA repair mechanisms and processes does appropriate justice to the enormous strides that this exciting field has witnessed in the decade since the writing of *DNA Repair.*

In very recent times, progress in the DNA repair and mutagenesis fields has been particularly rapid, and we feel compelled to apologize in anticipation for the inescapable fact that, the minimal time lapse between completion of the manuscript and the emergence of the published book notwithstanding, the literature may have already eclipsed some aspects of its content. We hope that such eclipses are mainly in matters of detail and have not rendered any fundamental concepts inaccurate or obsolete. To the extent that the book is current at the time of publication, we owe a tremendous debt of gratitude to colleages around the world, too numerous to mention by name, who have kept us well stocked with reprints and in many cases were generous enough to share information with us prior to its entry into the public domain. The book is well referenced through 1993 and contains numerous citations to work published in 1994, and even 1995. We are also enormously grateful to the individuals who provided us with original photographs and other illustrative materials.

We have continued the theme, developed in *DNA Repair*, of attempting to present the field of cellular responses to DNA damage with a historical perspective. We have refrained wherever possible from unadulterated dogma. While we are conscious of having presented viewpoints that are controversial and even conflicting, we hope that readers are challenged by such controversy and are not unduly confused or frustrated by different points of view or by our reluctance to give the definitive word, as it were. We also hope that readers, particularly students, will discover that this text is as much about how science advances and how ideas develop and attain full maturity as it is about DNA repair and mutagenesis itself. In the same historical vein, we realize that the names of many proteins and genes have changed over the years, sometimes for good reasons as more has been learned about them. In some instances we have elected to use current nomenclature, whereas in others we have retained the original nomenclature in deference to historic recognition and value. We apologize to those who might be offended by such license.

No work of this sort can be brought to fruition without special assistance. We owe an enormous debt of gratitude to many individuals for the help they have provided at every level of this labor. First and foremost, we have enjoyed scientific dialogue with an outstanding cadre of professional colleagues who have given unstintingly of their time, energy, and creative talents to review and discuss every chapter with us. In this respect we owe special and particular thanks to Jane and Lee Bardwell, Rick Bockrath, Vilhelm Bohr, Dirk Bootsma, Bryn Bridges, Tony Carr, James Cleaver, Priscilla Cooper, Rick Cunningham, Dick D'Ari, Pat Foster, Arthur Grollman, Larry Grossman, Jim Haber, Phil Hanawalt, Karla Henning, Jan Hoeijmakers, Peter Karran, Hannah Klein, Richard Kolodner, Tom Kunkel, Chris Lawrence, Alan Lehmann, Mike Lieber, Tomas Lindahl, John Little, Paul Lohman, Stephen Meyn, Kenneth Minton, Paul Modrich, Ethel Moustacchi, Eric Radany, Miroslav Radman, Hans Rahmsdorf, Leona Samson, Aziz Sancar, Gwen Sancar, Alain Sarasin, Roger Schultz, Mick Smerdon, Larry Thompson, Alan Tomkinson, Ben Van Houten, Bert van Zeeland, Harry Vrieling, Susan Wallace, Zhigang Wang, Felicity Watts, Steve West, Evelyn Witkin, Rick Wood, and Margaret Zdzienicka. We thank them all for their thoughtful criticisms, comments, and suggestions, which contributed immeasurably to the currency and the accuracy of the topics treated, not to mention our peace of mind concerning these issues. Particular thanks are due to Ken Minton and Karla Henning, who labored over much of the entire text with the most attentive detail. Ken Minton, Rick Wood, Susan Wallace, Roger Schultz, and Alan Tomkinson went so far as to rewrite some sections, for which we are especially grateful. If we have left anyone out, please forgive the transgression. Final responsibility rests with us, and we apologize for any inaccuracies and omissions that may remain at the time of publication. Readers are encouraged to inform us of such as they are discovered, for no other reason than the fact that we may be masochistic enough to contemplate revising this work some time in the future.

We acknowledge the outstanding talent, dedication, and commitment of Marty Burgin and her staff in the Medical Illustration Services at The University of Texas Southwestern Medical Center. This book is as much hers as it is ours. We are also enormously grateful to Patrick Fitzgerald of ASM Press for pursuing us as vigorously as he did to publish this work and for providing the strong sense of his personal commitment and that of his staff throughout the final stages. The extraordinary job of editing of the manuscript by Yvonne Strong merits special mention, as does Patrick Fitzgerald's recognition of the value of complete reference citations, despite the significant added expense of publishing these. This touch speaks highly of his interest in serving the academic community through ASM Press. Thanks are also due to Susan Birch, Senior Production Editor at ASM Press; to Susan Schmidler for the cover and book design; to Terese Winslow for the color illustration on the cover; and to Mary Boss, Marcie Abell, and the staff at Impressions for their outstanding coordination of the editing, typesetting, and proofing of the manuscript.

Each of us owes special thanks to particular individuals who provided special logistical support and help. E.C.F. thanks Mary Thurston, Glenda Westerfield, and Matthew Cavey for spending many hours laboring over the organization of the references for some of the chapters, particularly at a stage in the completion of the manuscript when the task of providing complete literature citations seemed well nigh impossible. Mary Thurston, Glenda Westerfield, and Theresa Peterson provided indispensable administrative and clerical assistance and sometimes also badly needed spiritual comfort to both E.C.F. and W.S. G.C.W. specially thanks Marianne White and Darlene McGurl for tracking down so many references and for invaluable technical assistance; his research group for their good-natured prodding and general interest; Evelyn Witkin, Cilla Cooper, and Dick D'Ari for their support and encouragement throughout this project; and, above all, Jan and Gordon for cheering every step of the marathon of working on this book. Finally, we collectively thank our families and friends, who suffered through this long labor, much of which was done at times of the week, day, and night that might otherwise have been spent in more leisurely and interactive, if not more intellectually productive, pursuits.

ERROL C. FRIEDBERG
GRAHAM C. WALKER
WOLFRAM SIEDE

October 1994

DNA Damage

Once it was recognized that DNA is the informationally active chemical component of essentially all genetic material (with the notable exception of RNA viruses), it was assumed that this macromolecule must be extraordinarily stable in order to maintain the high degree of fidelity required of a master blueprint. It has been something of a surprise to learn that the primary structure of DNA is in fact quite dynamic and subject to constant change. For example, gene transposition is a well-established phenomenon in prokaryotic and eukaryotic cells (116, 211, 212). In addition to these larger-scale changes, DNA is subject to alteration in the chemistry or sequence of individual nucleotides (45, 47, 210, 246, 249, 338, 366, 382, 436, 444). Many of these changes arise as a consequence of errors introduced during replication, recombination, and repair itself. Other base alterations arise from the inherent instability of specific chemical bonds that constitute the normal chemistry of nucleotides under physiological conditions of temperature and pH. Finally, the DNA of living cells reacts very easily with a variety of chemical compounds and a smaller number of physical agents, many of which are present in the environment. Some of these chemicals are products of the metabolism or decomposition of other living forms with which many organisms exist in intimate proximity. Others, particularly in recent decades, are man-made and contribute to the genetic insult faced by individuals living in highly industrialized communities. It should be appreciated, however, that outside of situations in which individuals are exposed to mutagens and carcinogens which are highly concentrated in the environment, in general the magnitude and real biological significance of this latter source of damage are difficult to assess and hence the issue of industrial genetic toxicity remains controversial.

Each of these modifications of the molecular structure of genetic material is appropriately considered to be *DNA damage*. Collectively, this multitude of modifications constitutes the substrate for the manifold cellular responses to such damage that are considered in this text. For convenience, DNA damage can be classified into two major classes, referred to as *spontaneous* and *environmental*. However, as we shall see, in some cases the actual chemical changes in DNA that occur "spontaneously" are indistinguishable from those brought about through interaction with certain environmental agents. In fact, the term "spontaneous" may sometimes merely imply that we have not as yet identified a particular environmental culprit. The latter category comprises an extensive list of known and potential agents, and it is not the intention of this text to treat each of these exhaustively. Rather, we discuss a number of specific physical and chemical agents for which there is substantial information concerning the mechanism of their interaction with DNA and the biological consequences of such interaction. While all of the primary components of DNA (bases, sugars, and phosphodiester linkages) are subject to damage, this chapter focuses largely on the nitrogenous bases, since these are the informational elements of the genetic blueprint encoded in DNA.

Spontaneous Alterations of and Damage to DNA

Mismatches

The chief source of DNA alterations arising during normal DNA metabolism is the mispairing of bases during DNA synthesis, resulting in *mismatches*. While mismatches can in theory arise from the synthetic events associated with a subset of nuclear DNA during repair and recombination, all of the bases in a genome must be replicated prior to cell division. Thus, semiconservative DNA synthesis probably accounts for the bulk of such alterations. A number of parameters are known to affect replicational fidelity in *Escherichia coli* (100, 154, 155, 221, 256, 258, 281, 324–326, 352), and a summary of these is provided in Table 1–1.

Without the intervention of any cellular factors, the difference in free energy for the stable pairing of a complementary relative to a noncomplementary base during DNA synthesis is only 2 to 3 kcal/mol (8 to 12 kJ/mol) (the equivalent of a single hydrogen bond) (256, 324). In the absence of other influences, this translates to a potential error frequency of 1 to 10% per nucleotide (256, 258, 324, 325). However, measurements of spontaneous mutation frequency show that the error frequency in newly replicated DNA in *E. coli* is in fact 6 to 9 orders of magnitude less than that predicted from the energetic considerations given above (256, 258, 324, 325). A significant contribution to this fidelity stems from the action of specific components of the replication machinery—the complex of DNA polymerases and accessory proteins that constitute the "replisome" required for normal semiconservative DNA synthesis in vivo (221). The combined effects of base selection and proofreading of newly inserted nucleotides by certain DNA polymerases, together with the enhancement of replication fidelity contributed by accessory factors, increase the accuracy of DNA synthesis by 3 to 6 orders of magnitude (256, 258, 324, 325, 352).

Not all of the determinants of semiconservative DNA replication that contribute to fidelity are known, but interesting clues are available from a variety of biological systems. Thus, for example, DNA polymerases that contain associated

Table 1–1 Mechanisms for maintaining genetic stability associated with DNA replication in *E. coli*

Mechanism	Cumulative error frequency
Base pairing	$\sim 10^{-1}$–10^{-2}
DNA polymerase actions (including base selection and $3' \rightarrow 5'$ proofreading exonuclease)	$\sim 10^{-5}$–10^{-6}
Accessory proteins (e.g., single-strand binding protein)	$\sim 10^{-7}$
Postreplicative mismatch correction	$\sim 10^{-10}$

Source: After Radman et al. (324).

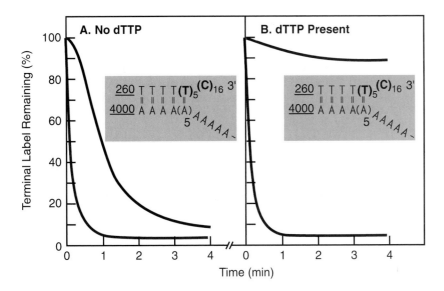

Figure 1–1 The 3′ → 5′ proofreading function of *E. coli* DNA polymerase I. When C is mispaired with A in a synthetic polymer, the mispaired nucleotide is excised both in the absence (A) and presence (B) of the appropriate triphosphate (dTTP). In the absence of the triphosphate terminal fraying of the polymer is associated with excision of some dTMP as well. (*Adapted from Kornberg [221] with permission.*)

3′ → 5′ exonucleases can use this catalytic function to "edit out" (proofread) 3′-terminal mispaired nucleotides (154, 221, 226, 256, 258, 324, 325). The phenomenon of proofreading by DNA polymerase-associated 3′ → 5′ exonuclease activity was first demonstrated with purified *E. coli* DNA polymerase I in vitro (40) (Fig. 1–1). The phenomenon has also been observed with the DNA polymerase of bacteriophage T4, an enzyme known to be required for the semiconservative replication of the DNA of this phage (221). Gene 43 of phage T4 is the structural gene that codes for the DNA polymerase core or holoenzyme (82), and mutations in this gene profoundly affect the overall mutation frequency of phage T4. Certain mutations in this gene (*mutators*) create a markedly enhanced mutation frequency in the phages (82, 330, 394, 395), while others significantly lower the mutation frequency (*antimutators*) (92) (Fig. 1–2). The enzyme from phages that behave as mutators exhibits reduced 3′ → 5′ exonuclease relative to polymerase activity, while DNA polymerases from the antimutators have relatively enhanced 3′ → 5′ exonuclease activity (150, 151, 291). These and other observations have led to models that explain incorporation errors in DNA synthesis as a consequence of the interplay of polymerase specificity, exonuclease proofreading activity, and the proneness of a given polymerase to extend a mismatched primer (62, 100, 137, 154–156, 226).

A 3′ → 5′ exonuclease can be associated with a polymerase as part of the same polypeptide or as a subunit. Whereas this situation is true for prokaryotic polymerases, purified eukaryotic polymerase preparations frequently lack 3′ → 5′ exonuclease activity (221). However, it has been suggested that the same proofreading exonuclease may function in concert with several different DNA polymerases (Fig. 1–3) (308).

For many years, measurements of replicative fidelity in vitro utilized as templates homopolymers or copolymers missing one of the four nucleotides, to quantitate the incorporation of a noncomplementary nucleotide (256). However, the polymers may be subject to artifactual effects such as complementary strand slippage and unnatural base-stacking interactions that confer conformations not present in natural DNA (256). Figure 1–3 shows an example of a different assay for testing proofreading efficiency and primer extension at a given site. Experiments of this kind have indicated the importance of neighboring base pairs for proofreading fidelity (154, 309).

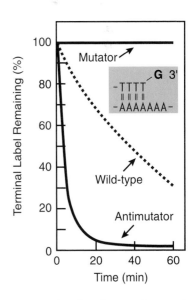

Figure 1–2 The 3′ → 5′ exonuclease activity of phage T4 mutator and antimutator DNA polymerases. The antimutator enzyme removes mismatched terminal nucleotides much more rapidly than does the wild-type enzyme, while the mutator enzyme does so much more slowly. (*Adapted from Kornberg [221] with permission.*)

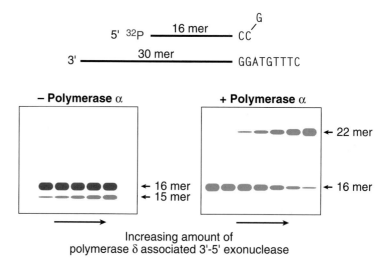

Figure 1–3 A mismatched terminal base is removed by a 3′ → 5′ exonuclease activity associated with mammalian polymerase δ (left). The single strand containing the mismatch is end labeled. Reaction products are separated in a denaturing polyacrylamide gel. Appearance of a 15-mer is indicative of the removal of the terminal mismatched base. Such an activity can enable another polymerase (polymerase α), which is normally not associated with proofreading activity, to resume DNA synthesis (right). Appearance of a 22-mer indicates primer extension. (*Adapted from Perrino and Loeb [308] with permission.*)

An alternative sensitive assay has been developed in which the error rate of DNA synthesis in vitro is quantitated by measuring the frequency of reversion of replicated ϕX174 DNA containing defined amber mutations to the wild type (1, 256, 383, 455) (Fig. 1–4). In this assay ϕX174 template DNA containing an amber mutation is replicated in vitro with a selected DNA polymerase of interest and the resulting product is used for the transfection of *E. coli* spheroplasts. The titer of the progeny phage is then measured on bacterial indicators either permissive or nonpermissive for the amber mutation. When deoxyribonucleoside (thio)triphosphates containing a sulfur atom in place of an oxygen on the phosphate are used as substrates for replication of the ϕX174 template DNA, the analog is incorporated as a thiomonophosphate at rates similar to those of the corresponding unmodified

Table 1–2 Effect of metals on the replicative fidelity of avian myeloblastosis virus DNA polymerase

Compound tested[a]	Maximal error frequency (ratio)[b]	Metal concn (mM)[c]
AgNO$_3$	1.85	0.03
BeCl$_2$	15.0	10.0
Cd(C$_3$H$_3$O$_2$)$_2$	2.22	0.24
CdCl$_2$	1.35	0.04
CoCl$_2$	8.37	4.00
CrCl$_2$	3.70	0.64
CrO$_3$	3.83	16.0
MnCl$_2$	3.75	10.0
NiCl$_2$	1.92	8.0
PbCl$_2$	1.48	4.0

[a] The metal ions shown in the table have all been implicated as mutagens and/or carcinogens.

[b] Fidelity of replication by avian myeloblastosis virus DNA polymerase was determined by measuring the frequency of misincorroration by using polymer templates. The maximal error frequency values quoted are the ratio of the highest error frequency during titration with a given compound to that determined without the compound added.

[c] The metal concentration that yielded the largest observed change in error rate is also shown.

Source: After Sirover and Loeb (384).

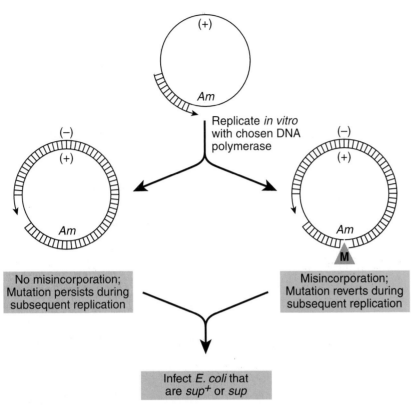

Figure 1–4 The fidelity of a DNA polymerase can be determined in vitro. φX174 template DNA (+ strand) containing an *amber* mutation (*Am*) is annealed to a primer and replicated with a given DNA polymerase. The replicated molecules are then transformed into *E. coli* strains that are either suppressors or not suppressors of the *amber* mutation. If replication resulted in an appropriate misincorporation opposite the site of the *amber* mutation (right), the phenotype reverts to wild-type and these molecules yield phage progeny on both permissive and nonpermissive indicator bacteria. If replication past the *amber* mutation was accurate (left), the *amber* phenotype persists and viral progeny will be obtained only on the permissive (suppressor) host. Measurement of the reversion frequency can thus be converted into an error rate for in vitro DNA synthesis. (*Adapted from Friedberg [131] with permission.*)

nucleoside triphosphate (227). However, the phosphorothioate diester bond is not hydrolyzed by the $3' \rightarrow 5'$ exonuclease of either *E. coli* DNA polymerase I or T4 DNA polymerase (Fig. 1–5) (227). Thus, reversion of amber mutations can be enhanced by incorporation of the analog as a mispaired base that cannot be edited out. By using this approach, it can be inferred that the proofreading functions of the *E. coli* and phage T4 DNA polymerases increase the fidelity of replication by factors of 20- and 500-fold respectively (227, 352).

An interesting hypothesis relating replicative fidelity to the use of RNA primers for DNA synthesis has been suggested (221). At or near the beginning of a DNA chain, proofreading and other possible fidelity-promoting mechanisms may not function as efficiently as during subsequent chain elongation. However, since the RNA primers are eventually excised, this provides a possible mechanism for eliminating the adjacent stretches of the newly synthesized genome that may contain an unusually high level of errors.

A number of parameters that affect the replicative fidelity of various DNA polymerases in vitro have been explored. Replacement of Mg^{2+} by divalent metal ions such as Mn^{2+}, Co^{2+}, Cd^{2+}, or Be^{2+} reduces the replicative fidelity of a number of DNA polymerases (384) (Table 1–2). Alterations in the relative proportions of the different deoxyribonucleoside triphosphates used as precursors for DNA synthesis (115, 228, 230), the addition of deoxyribonucleoside monophosphates

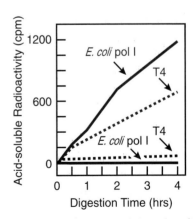

Figure 1–5 φX174 containing phosphorothioate (black lines) is resistant to digestion by the $3' \rightarrow 5'$ exonuclease functions of *E. coli* DNA polymerase I (pol I) and of phage T4 DNA polymerase. Unsubstituted DNA is readily degraded by these enzymes (red lines). (*Adapted from Kunkel et al. [227].*)

(228), and the presence of certain accessory proteins such as single-stranded DNA-binding protein (229) also affect the fidelity of DNA replication in the ϕX174 system. These accessory proteins may be particularly important for DNA polymerases which do not contain associated $3' \rightarrow 5'$ exonuclease activities.

While considering semiconservative DNA synthesis as a source of DNA alterations, it should be noted that under certain circumstances the replication of DNA occurs on parental template strands carrying unrepaired noninstructional or misinstructional base damage, which can act as miscoding lesions. We will revisit the issue of polymerase fidelity as an important factor in DNA damage-induced mutation in chapters 2, 11, and 12. Finally, it was pointed out above that the multitude of factors that determine replicative fidelity reduce the *misinsertion* frequency in newly replicated DNA to about 10^{-6} to 10^{-9} per nucleotide. However, the actual *mutation* frequency per nucleotide is even lower than this by about another 3 orders of magnitude (89). This reduction reflects the postreplicational repair of mispaired bases in DNA (mismatch repair), a topic that is discussed in detail in chapter 9.

A significant fraction of the book is devoted to a consideration of the biochemical mechanisms involved in the replacement of nucleotides following the enzymatic excision of damaged or mispaired bases from DNA. This replacement process does not result in the generation of entire new daughter DNA strands as in classical semiconservative DNA synthesis. This mode of synthesis is therefore referred to as *non-semiconservative DNA synthesis* or *repair synthesis* (310). In some instances, stretches of DNA as long as 3 kb or longer have been shown to be replaced via these mechanisms (see chapter 5) (70). Hence, this DNA synthetic mode also offers the potential for replicative infidelity. Alterations in nucleotide sequence can also occur during various modes of DNA rearrangement, including homologous and nonhomologous recombination, and these, too, constitute sources of mismatches in DNA (90).

Spontaneous Alterations in the Chemistry of DNA Bases

TAUTOMERIC SHIFTS
Each of the common bases in DNA can spontaneously undergo a transient rearrangement of bonding, termed a *tautomeric shift*, to form a structural isomer (*tautomer*) of the base (449). Formation of the tautomer of any base alters its base-pairing properties. For example, although the N atoms of the purine and pyrimidine rings are usually in the amino (NH_2) form, they can sometimes shift to the imino (NH) tautomeric form (449). When either cytosine or adenine is in the latter configuration, it can mispair with the other through two available hydrogen bonds (Fig. 1–6). Similarly, the oxygen on the C-6 atoms of guanine or thymine can shift from the usual keto (C=O) to the infrequent enol (C–OH) configuration, allowing for anomalous base pairing between them by three available hydrogen bonds (43) (Fig. 1–6). Thus, if any base in a template strand exists in its rare tautomeric form during DNA replication, misincorporation in the daughter strand can result. Several other mispairing schemes have been suggested (91), including switches of the base residue to the *syn* configuration (rotation by 180° around the glycosidic bond) or inclusion of a water molecule as a bridge. The contribution of all of these possible mispairing scenarios to the load of spontaneous DNA alterations is unknown.

DEAMINATION OF BASES
Three of the four bases normally present in DNA (cytosine, adenine, guanine; also 5-methylcytosine) contain exocyclic amino groups. The loss of these groups (*deamination*) occurs spontaneously in pH- and temperature-dependent reactions of DNA (245, 246, 366) and results in the conversion of the affected bases to uracil, hypoxanthine, xanthine, and thymine, respectively (Fig. 1–7). Some of these products of deamination can give rise to mutations, since during semiconservative synthesis of DNA they are miscoding lesions that result in altered base pairs in the genome (244) (Fig. 1–8).

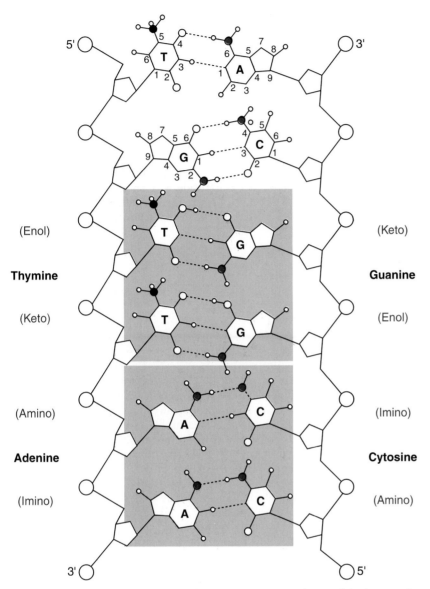

Figure 1–6 Anomalous base pairing involving rare tautomeric forms of the bases. When either thymine or guanine is in the rare enol form, they can pair with each other. Similarly, when adenine or cytosine is in the rare imino form, they can pair. (*Adapted from Drake [90] with permission.*)

Deamination of Cytosine

The deamination of cytosine to uracil is one of the ways in which the latter base, normally confined to RNA, can occur in DNA. It can in fact be argued that the use of T (i.e., methylated U) in DNA and not U (as in RNA) enables the cell to identify the deamination product of cytosine (i.e., uracil) as an inappropriate base. The formation of uracil in DNA by the deamination of cytosine is indeed biologically important. For example, strains of *E. coli* that are defective in the removal of uracil from DNA (see chapter 4) have an increased spontaneous mutation rate (98), and G·C → A·T base pair transitions have been observed at selected sites in such mutants (97).

Two chemical mechanisms have been proposed for the deamination of cytosine in solution at neutral pH (366, 368) (Fig. 1–9). One involves the direct attack at the 4-position of the pyrimidine ring by a hydroxyl ion. The other postulated

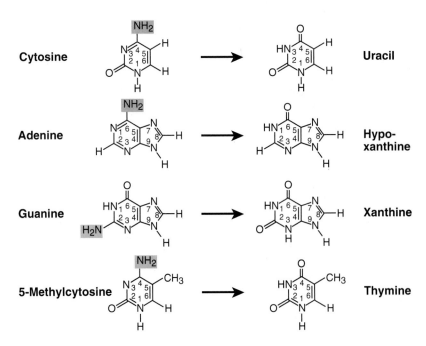

Figure 1–7 Products formed from the deamination of bases in DNA. (*Adapted from Friedberg [131] with permission.*)

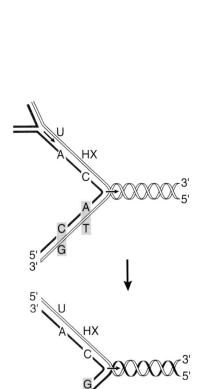

Figure 1–8 Deamination of cytosine to uracil (U) and of adenine to hypoxanthine (HX) can result in base pair transitions. The U and HX will code for A and G, respectively, during semiconservative DNA synthesis. The top panel of the figure shows a replicating DNA molecule in which U and HX have already mispaired. A second round of DNA replication is just beginning. As this second replication fork proceeds (lower panel), replication of the template strand containing the A and G results in *transition* (see chapter 2) of the G·C and A·T base pairs to A·T and G·C base pairs, respectively. (*Adapted from Friedberg [131] with permission.*)

pathway involves an addition-elimination reaction with the formation of dihydrocytosine as an intermediate. The hydrolytic deamination of cytosine in nucleotides and polynucleotides occurs at a measurable rate when these compounds are incubated at elevated temperatures in buffers at physiological ionic strength and pH (251). The rate constants for these reactions and their dependence on temperature have been determined. By extrapolation, the rate of deamination of cytosine in single stranded DNA at 37°C was calculated at $k = 2 \times 10^{-10}$/s (251) (Table 1–3), which translates into a half-life of an individual residue of about 200 years (246). This extrapolation has been confirmed in a genetic reversion assay similar to the one depicted in Fig. 1–4 (125).

What does this value tell us about the biological relevance of spontaneous deamination in living organisms? This question is difficult to answer quantitatively since the rate of spontaneous deamination of cytosine in duplex DNA in vitro is less than 1% of that in single-stranded DNA (251). Most of the genome of a living cell is presumably in a double-stranded configuration at any given time. Nonetheless, it is clear that the processes of replication, recombination, and transcription involve transient localized denaturation of DNA that could accelerate cytosine deaminations (245). Furthermore, duplex DNA undergoes spontaneous localized denaturation, or "breathing," that could increase the rate of the spontaneous deamination of cytosine (245). This rate is estimated to be 40-fold higher in *Saccharomyces cerevisiae* than in *E. coli*, possibly because the lower rate of eukaryotic transcription keeps DNA single stranded for a longer period (192). If cytosine deamination in vitro occurs by the addition-elimination mechanism discussed above, this could involve the formation of dihydrocytosine and dihydrouracil as intermediates (Fig. 1–9), either of which could represent distinct (albeit transient) forms of DNA damage not amenable to the same mechanism of repair that deals with uracil in DNA. However, if formed, these compounds may constitute substrates for other DNA repair processes (245).

The deamination of cytosine can also be enhanced by a number of chemical alterations and steric factors, by the formation of UV radiation-induced cyclobutane dimers (238, 414) (see below), by certain intercalating agents (286), or by the positioning of a mismatched or alkylated base opposite cytosine (126, 459). Deamination can also be promoted by reaction with nitrous acid (354) or sodium

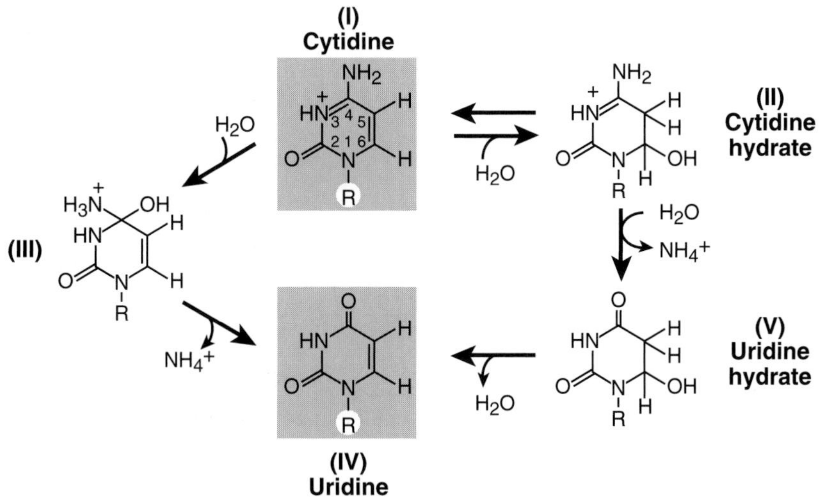

Figure 1–9 Proposed mechanisms for the hydrolytic deamination of cytidine to uridine. The path I → III → IV is analogous to the hydrolysis of an amide. It is called the *direct route* and involves the direct attack at the 4-position of the pyrimidine ring by a hydroxyl ion. Loss of ammonia yields uridine. The path I → II → V → IV is called the *addition-elimination* mechanism. It involves addition of water to the 5,6 double bond of protonated cytidine to yield cytidine hydrate (dihydrocytidine) (II). Further attack by water is followed by the loss of ammonia, yielding uridine hydrate (dihydrouridine) (V), which is dehydrated to uridine (IV). (*Adapted from Shapiro [366] with permission.*)

bisulfite (172). Nitrous acid reacts relatively nonspecifically since it also results in the deamination of adenine and guanine residues in DNA (245), and additionally it can promote the cross-linking of DNA strands (367). Nitrous acid attacks cytosine residues in double-stranded DNA almost as efficiently as in single-stranded DNA (354). In contrast, sodium bisulfite promotes the deamination of cytosine exclusively in single-stranded regions of DNA and is specific for cytosine residues under defined experimental conditions (172). Purines are not attacked, and although bisulfite-thymine adducts can be formed, they are transient and readily reversible (172). Bisulfite converts cytosine residues in DNA to uracil by an acid-catalyzed addition-elimination reaction with the intermediate formation of 5,6-dihydrocytosine-6-sulfonate and 5,6-dihydrouracil-6-sulfonate (172) (Fig. 1–10). At alkaline pH the latter compound is converted to uracil (Fig. 1–10). However, bisulfite-induced deamination of cytosine also occurs to a significant extent under physiological conditions (50). The appearance of CC → TT tandem double mutations suggests cooperative deamination in adjacent cytosine residues (51). Yet another mechanism for the conversion of cytosine to uracil in DNA is by exposure to strong alkali (426).

Table 1–3 Time needed for a single event at pH 7.4 at 37°C

Event	Time (h) for event in:[a]	
	Single-stranded DNA (2×10^6 bases)	Double-stranded DNA (1×10^6 bp)
Depurination	2.5	10
Depyrimidination	50.8	200
Deamination of C	2.8	700
Deamination of A	140	ND[b]

[a] Data computed from the rate constants for these events.
[b] ND, not determined.

Source: After Shapiro (366).

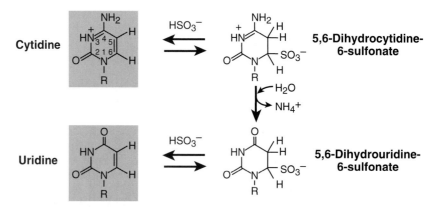

Figure 1–10 Mechanism of deamination of cytidine by bisulfite. Cytidine is converted to a sulfonated derivative (5,6-dihydrocytidine-6-sulfonate), which is then hydrolytically deaminated at acidic pH to yield a sulfonated uridine derivative (5,6-dihydrouridine-6-sulfonate). At alkaline pH this derivative is converted to uridine. (*Adapted from Shapiro [366] with permission.*)

Incorporation of Uracil into DNA during Semiconservative Replication

In addition to its origin by the deamination of cytidine, uracil can occur to a limited extent in DNA as a consequence of semiconservative synthesis. This phenomenon is observed in organisms that normally contain thymine as the principal pyrimidine. It is also observed in a rare class of bacterial viruses in which uracil replaces thymine completely (see below and chapter 4). With respect to the former, the presence of U·A rather than T·A base pairs is not expected to alter the replicational fidelity of DNA. However, subtle but biologically relevant parameters of the transcription of such DNA may be affected. Additionally, it is possible that significant replacement of T by U in DNA affects the recognition of substrate nucleotide sequences by various enzymes and/or regulatory DNA-binding proteins.

Small amounts of dUMP are normally incorporated into the DNA of *E. coli* (147, 178, 218, 219, 300, 422–424) and other prokaryotic cells such as *Bacillus subtilis* (263, 404) (Fig. 1–11). The extent of this incorporation is apparently directly related to the size of the intracellular dUTP pool, since the K_m values of *E. coli* DNA polymerase III for dUTP and TTP are not significantly different (372). In wild-type cells, incorporated uracil is excised very rapidly by an enzyme called uracil-DNA glycosylase (448) (see chapter 4); hence, this base is not normally

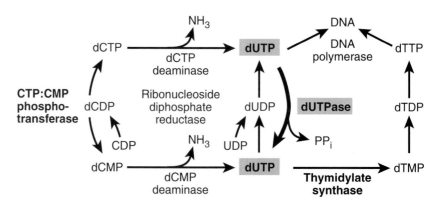

Figure 1–11 Uracil can be incorporated into DNA from dUTP during semiconservative DNA synthesis. The dUTP pool is generated both from dCTP and from dUDP. In wild-type cells the pool size of dUTP is small relative to that of dTTP, since most dUTP is degraded to dUMP by dUTPase. (*Adapted from Kornberg [221] with permission.*)

detected in DNA isolated from such cells (422). However, in mutants (designated *ung*) defective in the DNA glycosylase activity, dUMP is incorporated into DNA at a frequency of about 1 in 2,000 to 3,000 nucleotides (422). In double mutants (*ung dut*) defective in both uracil-DNA glycosylase and dUTPase activities (the latter mutation prevents the degradation of dUTP to dUMP and hence increases the size of the dUTP pool), the frequency of uracil in DNA can be as high as 0.5% of all bases (422). When such mutants are infected with bacteriophage T4, as much as 30% of the thymine in phage DNA can be replaced by uracil (244).

The incorporation of uracil into DNA has been observed in a number of viruses, including polyomavirus (41) and adenovirus (9), as well as in human lymphocytes in culture (159). Presumably most, if not all, organisms that synthesize TMP from dUMP incorporate small amounts of uracil into their DNA. In bacteria, dUMP is derived largely from dUTP; hence, a pool of the latter normally exists in these cells (Fig. 1–11). In mammalian cells, dUMP is generated mainly from dCMP and can be readily converted to the triphosphate form. In human lymphocytes, inhibition of the biosynthesis of TMP from dUMP increases the pool size of dUTP relative to TTP and hence promotes significant incorporation of uracil relative to thymine during DNA replication. For example, during the synthesis of TMP from dUMP by the enzyme thymidylate synthetase, methylenetetrahydrofolate contributes the methyl group to dUMP and is converted to dihydrofolate (221). Regeneration of tetrahydrofolate is essential for continued TMP synthesis by thymidylate synthetase. This regeneration is catalyzed by the enzyme *dihydrofolate reductase*, which is inhibited by the folate antagonist 4-amino-10-methylfolate (amethopterin; methotrexate) (221) (Fig. 1–12). Treatment of cells with methotrexate results in reduced utilization of dUMP by thymidylate synthetase, causing an increased pool size of this nucleotide, with an attendant drop in the concentration of TTP (195, 292). These relative pool size changes result in deregulation of other aspects of pyrimidine nucleotide metabolism, which augment this effect (159). As a net result, the intracellular dUMP level in human lymphocytes in culture is increased about 1,000-fold, and, despite the presence of dUTPase, the dUTP level also increases by at least 3 orders of magnitude (158, 159). The TTP pool size drops about 50-fold in these cells, resulting in a significant incorporation

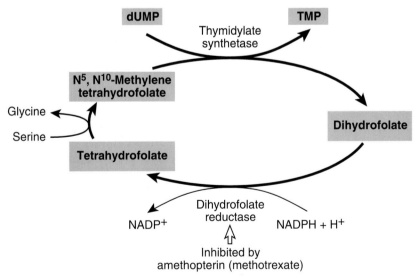

Figure 1–12 The formation of thymidylate from dUMP is catalyzed by the enzyme thymidylate synthetase. During this reaction 5,10-methylenetetrahydrofolate is converted into dihydrofolate and regeneration of tetrahydrofolate is catalyzed by dihydrofolate reductase. Inhibition of dihydrofolate reductase by amethopterin (methotrexate) results in reduced levels of tetrahydrofolate and hence reduced conversion of dUMP into TMP. (*Adapted from Kornberg [220].*)

of dUMP into DNA. In the presence of free uracil, which is an inhibitor of uracil-DNA glycosylase (see chapter 4), uracil accumulates as a stable component in DNA (159).

The related bacteriophages PBS1 and PBS2 normally contain uracil instead of thymine in their DNA (403). Following infection of their natural host *B. subtilis*, these phages induce a number of new enzyme activities, including dTMPase (202), dCTP deaminase (419), and dUMP kinase (202) activities. These activities significantly alter the normal host pathways of deoxynucleoside triphosphate biosynthesis, resulting in an increase in the dUTP pool size relative to TTP and thereby facilitating the incorporation of uracil instead of thymine during phage DNA synthesis (322) (see chapter 4).

Finally, certain bases that are not normally present in DNA but bear a strong structural resemblance to normal nitrogenous bases can be incorporated from the appropriate triphosphate precursor during DNA synthesis. These compounds are called *base analogs*, one of which is 5-bromouracil (5-BU), an analog of thymine. When DNA containing 5-BU is exposed to 313-nm radiation in the presence of cysteamine, it undergoes debromination to yield uracil in the DNA (252). Later chapters will deal with specific examples of the utility of 5-BU-substituted DNA in the study of the *repair* of DNA damage.

Deamination of Cytosine Analogs

Let us return to the topic of deamination of bases with exocyclic amino groups. Analogs of cytosine are encountered naturally in the DNA of some living forms. For example, 5-methylcytosine occurs as a small fraction of the total cytosine content of a number of prokaryotes including *E. coli*, as well as in all higher organisms (102, 245). Consider that the rate of heat-induced deamination of 5-methyl-dCMP is about four times as high as that of dCMP (245, 250) and that the rate of spontaneous deamination of cytosine in dCMP is approximately the same as in single-stranded DNA (245). Assuming that the same holds for 5-methylcytosine, the deamination of this base could account for as much as 10% of the total spontaneous deamination events occurring in the DNA of the average mammalian cell (246).

The deamination of 5-methylcytosine in DNA results in the formation of thymine (Fig. 1–7) and hence of T·G mispairs. Unlike the U·G mispair considered above, which arises from cytosine deamination, T·G mispairs are apparently less repairable lesions which result in mutational hot spots in *E. coli* (72). This is not too surprising, since in the former case only the recognition of an inappropriate base is required, whereas in the latter case, unless a DNA repair mechanism for the removal of thymine could specifically distinguish between the mismatched T·G and the correct T·A base pairs, all thymine in DNA would be potentially subject to removal. As discussed in chapter 9, such a repair mechanism does indeed exist.

In bacteriophages T2, T4, and T6, cytosine in DNA is completely replaced by 5-hydroxymethylcytosine, which is glucosylated to various extents in the different phages (221). The deamination of 5-hydroxymethylcytosine yields 5-hydroxymethyluracil. Interestingly, the enzyme in *E. coli* that catalyzes the removal of uracil from DNA (uracil-DNA glycosylase; see chapter 4) does not recognize the 5-hydroxymethylated derivative (132), nor is there evidence that phages T2, T4, and T6 encode different DNA glycosylases with the appropriate substrate specificity. This is consistent with the observation that at elevated temperatures phage T4 accumulates mutations due to G·C → A·T transitions at a significantly increased rate (18).

Deamination of Adenine and Guanine

Deamination of adenine and of guanine also occurs under physiological conditions in vitro, but at rates much lower than for cytosine deamination (245). At raised temperatures and at pH 7.4, the conversion of adenine to hypoxanthine in single-stranded DNA occurs at about 2% of the rate of the conversion of cytosine to uracil (245). Detailed studies on the acidic reactions have not been reported, but from the limited data available (199) it has been estimated that the deamination of adenine and guanine would occur at rates $\sim 10^{-4}$ that of the rate of loss of these bases from DNA (see the following section on depurination) (366). Adenine residues in DNA can be deaminated by nitrous acid at a rate similar to that of cytosine (354). Hypoxanthine in DNA is potentially mutagenic, since it can base pair with cytosine during DNA replication, giving rise to $A \cdot T \rightarrow G \cdot C$ transitions (245).

It is not clear whether misincorporation of dIMP (from dITP) instead of dGMP occurs during semiconservative DNA synthesis in vivo, analogous to the situation discussed above with respect to dUMP incorporation from existing dUTP pools. While IMP is a key metabolite in purine biosynthesis in *E. coli*, this organism (and probably others) is essentially devoid of a pool of ITP (245). Xanthine in DNA, arising from the deamination of guanine, is unable to pair stably with either cytosine or thymine (161) and thus may result in the arrest of DNA synthesis on templates containing this base.

Loss of Bases—Depurination and Depyrimidination

The loss of purines and pyrimidines from DNA has been most extensively studied at acidic pH. However, depurination and depyrimidination can also occur at appreciable rates at neutral or alkaline pH (161, 245, 246, 249, 257, 366). The rate of depurination and depyrimidination of single- and double-stranded DNA can be observed by incubating these substrates at various temperatures and pHs and measuring the rate of release of specifically labeled bases (250). Guanine is released from DNA about 1.5 times faster than is adenine at both acidic and neutral pH, but at alkaline pH dAMP is hydrolyzed more rapidly than dGMP (250). Figure 1–13 shows a pH-rate profile summarizing the behavior of the major deoxyribonucleosides of DNA in acid.

The chemical mechanism of hydrolytic DNA depurination at acidic pH is believed to be the same as that established for acid hydrolysis of deoxynucleosides, i.e., protonation of the base followed by direct cleavage of the glycosyl bond (250, 462). The mechanism of hydrolysis of nucleosides at neutral and at alkaline pHs (143, 426) is less well characterized (245). By extrapolation of Arrhenius plots derived by direct measurements at high temperatures, the rate of depurination of duplex DNA at physiological pH and ionic strength is calculated to be about $k = 3 \times 10^{-11}$/s at 37°C (250). In vivo, this value corresponds to the loss of approximately one purine per *E. coli* genome per generation, given a DNA doubling time of about 1 h (245) (Table 1–3). By the same argument, a thermophilic organism such as *Thermus thermophilus*, which grows optimally at temperatures around 85°C, may lose as many as 300 purines per genome per generation (245). For mammalian cells, in which genomes are much larger and replication times are longer,

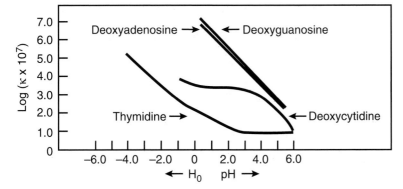

Figure 1–13 Dependence of the logarithms of the rate constants (κ) (reciprocal seconds) on pH and H_0 (a parameter used to indicate acidity of pH < 1) at 95°C, for deoxyribonucleoside hydrolysis. At acidic pHs, depurination occurs more rapidly than does depyrimidination. (*Adapted from Shapiro [366] with permission.*)

estimates of purine loss at the rate of 10,000 per cell generation have been made (245).

Pyrimidine nucleosides are considerably more stable than purine nucleosides with respect to the glycosylic linkage of the bases to deoxyribose. The mechanism of depyrimidination is the same as for depurination, but cytosine and thymine are lost at rates only 1/20 of that for adenine or guanine (248). This still translates into the loss of hundreds of pyrimidines per mammalian cell per generation. The influence, if any, of the packaging of DNA into nucleosomes and of the various levels of folding that chromatin assumes in the cell on the rate of spontaneous base loss has not been extensively explored.

The deoxyribose residues that are left at sites of base loss in DNA exist in equilibrium between the closed furanose form and the open aldehyde form (198). The 3' phosphodiester bonds associated with the latter are labile and can be hydrolyzed by a *β-elimination reaction* in which the pentose carbon β to the aldehyde is activated at alkaline pH and at elevated temperatures (198) (Fig. 1–14) (see chapter 4). The same reaction proceeds at a reduced rate at neutral pH. At physiological pH and temperature, a site of base loss in DNA has an average lifetime of ~400 h (247). The presence of Mg^{2+} and of primary amines promotes phosphodiester bond cleavage by β-elimination (405), reducing the average lifetime of sites of base loss to about 100 h (245). Polyamines further promote the rate of cleavage of the deoxyribose-phosphate backbone (245, 264). Nonetheless, it is apparent that the integrity of the DNA backbone would probably be maintained for several generations in the absence of DNA repair in *E. coli* (245, 264). In vitro this β-elimination reaction can be prevented by reducing the pentose aldehyde to an alcohol with a reducing agent such as sodium borohydride.

Oxidative Damage to DNA

PRIMARY REACTIONS

Besides deamination, spontaneous hydrolysis, and nonenzymatic methylation (see below), attack by *reactive oxygen species* must be considered a major source of spontaneous damage to DNA and also to other intracellular macromolecules such as proteins, lipids, and carbohydrates (4, 5, 36, 76, 167, 191, 197, 336, 350, 377). There are various intra- and extracellular sources of oxygen radicals (60, 350), and oxygen has been appropriately called the "sink" for electrons generated in various redox reactions of aerobic metabolism (336). The major intracellular source of oxygen radicals is probably leakage associated with the reduction of oxygen to water during mitochondrial respiration. These products are singlet oxygen, peroxide radicals ($\cdot O_2$), hydrogen peroxide (H_2O_2), and hydroxyl radicals ($\cdot OH$). Other intracellular processes resulting in the release of reactive oxygen include peroxisomal metabolism, the enzymatic synthesis of nitric oxide, and the metabolism of phagocytic leukocytes (16). Extracellular sources include radiation (especially ionizing radiation and near-UV light at 320 to 380 nm [425]), heat, various drugs, and redox cycling compounds. Indeed, as we will discuss below, most of the damaging effects of ionizing radiation are due to reactive radiolysis products of water (336, 436). Chemical agents that increase the amount of intracellular oxidative damage include known tumor promoters such as 12-*O*-tetradecanoylphorbol-13-acetate (451).

These radicals can also abstract electrons from residues of organic macromolecules according to the following general reaction:

$$RH_2 + \cdot OH \rightarrow \cdot RH + H_2O$$

and can initiate chain reactions that result in damage at considerable distances from the initial chemical event (350). A notable example of a chain reaction involves the peroxidation of unsaturated lipids initiated by reactive free radicals such as $\cdot OH$ (385). The half-life and diffusibility of the generated radicals are important parameters that can influence the potential for DNA damage. Peroxyl ($ROO\cdot$) and alkoxyl ($RO\cdot$) radicals are likely intermediates in many chain reactions, while per-

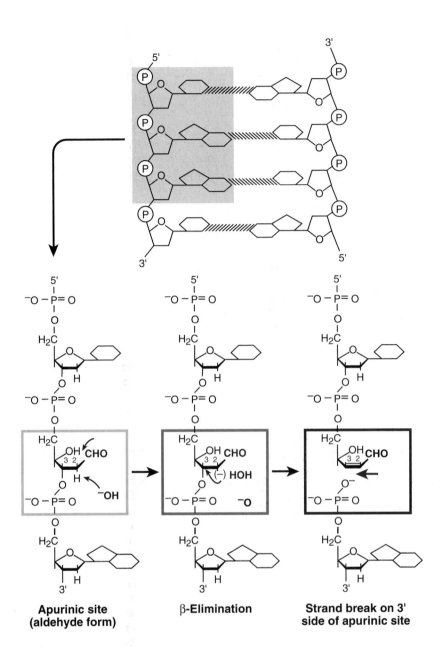

Figure 1–14 Mechanism of strand breakage in DNA by β-elimination. Deoxyribose residues at sites of base loss exist in equilibrium between the open (aldehyde) form (shown in the figure) and the closed (furanose) form (not shown). In the aldehyde form, 3′ phosphodiester bonds are readily hydrolyzed by a β-elimination reaction in which the pentose carbon beta to the aldehyde is activated at alkaline pH, as shown. (*Adapted from Friedberg [131] with permission.*)

Apurinic site (aldehyde form) | **β-Elimination** | **Strand break on 3′ side of apurinic site**

oxides (ROOH and H_2O_2) and hydroxylated (ROH) and carbonylated (HR=O) residues are likely end products (350). H_2O_2 is probably the most significant of these in terms of diffusion range. H_2O_2 by itself is relatively inert but can give rise to highly reactive ·OH radicals in a process catalyzed by transition metal ions, typically Fe^{2+}. This reaction is called the metal-catalyzed Haber-Weiss reaction, or the Fenton reaction (Fig. 1–15) (63).

Besides DNA repair, various other cellular defense mechanisms against damage by reactive oxygen are operative at different levels and merit brief consideration (336).

Compartmentalization of Sensitive Target Molecules

The compartmentalization of sensitive target molecules results in protection of genetic material in the nucleus from sources of active oxygen metabolism in mitochondria, chloroplasts, and peroxisomes. It has even been suggested that the transfer of certain vulnerable genes from mitochondria and chloroplast genomes

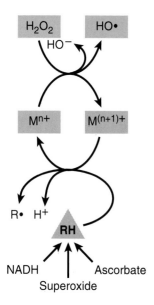

Figure 1–15 Metal-catalyzed formation of ·OH radicals (Fenton reaction). This reaction can function in a redox cycle in which a transition metal ion transfers electrons from donors such as NADH, superoxide, or ascorbate. (*Adapted from Riley [336] with permission.*)

into the nuclear genome reflects such a protective mechanism (336). The fact that nuclear DNA is surrounded by histones and polyamines and is organized into higher-order chromatin structures presumably affords additional protection against oxidative damage (254).

Elimination of Reactive Oxygen Species by Antioxidant Enzymes

Superoxide dismutase (268, 343) can eliminate reactive oxygen species by the following reaction:

$$2O_2\cdot\ +\ 2H^+ \rightarrow H_2O_2\ +\ O_2$$

Catalase can produce a similar antioxidant effect as follows:

$$2H_2O_2 \rightarrow 2H_2O\ +\ O_2$$

Disruption of Perpetuating Chain Reactions

Perpetuating chain reactions can be disrupted by reaction with cellular thiols or other antioxidants; this is illustrated by the reaction between hydrogen peroxide and glutathione (GSH) resulting in the formation of a glutathione dipeptide-adduct (GS-SG), catalyzed by glutathione peroxidase:

$$H_2O_2\ +\ 2GSH \rightarrow 2H_2O\ +\ GS\text{-}SG$$

Other radical scavengers include vitamin C, α-tocopherol, bilirubin, and urate (4, 6, 374).

Sequestration of Transition Metals

Another protective mechanism is the sequestration of transition metals otherwise available for Fenton-like reactions (e.g., the sequestration of Fe^{2+} in the form of ferritin).

Cellular Regulation

Potential cellular responses triggered by the presence of reactive oxygen species may include the inhibition of cell cycle progression, the initiation of apoptosis (programmed cell death), and the activation of degradative processes to replace damaged macromolecules. We will discuss such processes in greater detail in chapter 13.

NATURE OF RADICAL-INDUCED DNA LESIONS

There is good evidence that H_2O_2 and ·O_2 radicals do not react directly with DNA. The main source of damage by these compounds is believed to be the formation of ·OH radicals by Fenton-like reactions of H_2O_2 with iron-complexed DNA (87, 190, 191, 194). Such reactions can be generalized as follows:

$$DNA[Fe(II)]\ +\ H_2O_2 \rightarrow DNA[Fe(III)]\ +\ \cdot OH\ +\ OH^-$$

The carcinogenic effect of certain metal salts has been attributed to such radical-induced DNA alterations (84).

Radical attack on DNA can produce a variety of products (79, 84, 166, 191). Many of these have also been characterized by treating bases, nucleosides, or oligonucleotides in vitro with ionizing radiation (413, 436). Attack at the sugar residue leads to fragmentation, base loss, and strand breaks with a terminal sugar residue fragment (271, 332, 436). Examples of various types of radical attack on base residues are provided in Fig. 1–16. The ·OH radical can attack the double bond of thymine at C-5 or C-6 and, less frequently, can abstract hydrogen from the methyl group (200, 375). The hydroxythymine radical intermediate can react with O_2 to yield thymine glycol (81). This product has been extensively

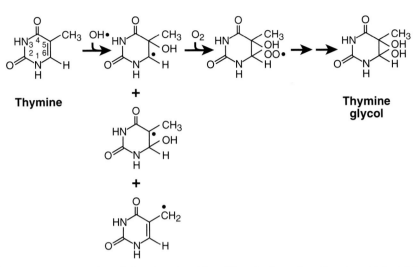

Figure 1–16 Formation of thymine glycol by ·OH radical attack at thymine. (*Adapted from Simic [375] with permission.*)

studied, mainly because it can be readily generated by the oxidation of DNA with osmium tetroxide (20, 130). Another well-studied oxidation product, 8-hydroxyguanine, resulting from the addition of ·OH to C-8 of guanine (201, 203), can be specifically generated by the action of methylene blue and light in the presence of oxygen (353). This guanine derivative is a mispairing lesion during DNA replication (52, 224, 371), whereas thymidine glycol appears to predominantly block replication (57, 58, 188).

The treatment of DNA with hydrogen peroxide or other free radical-generating systems (such as xanthine/xanthine oxidase) can result in the formation of an imidazole ring-opened derivative of guanine designated 2,6-diamino-4-hydroxy-5-formamidopyrimidine (FaPy) as an abundant lesion (see chapter 4) (11, 12, 36, 87). This lesion is prominent among the forms of base damage induced by H_2O_2 treatment of cultured mammalian cells, although a higher yield of 8-hydroxyguanine is observed under these particular conditions (84, 86). A scheme for the generation of this and the analogous adenine lesion is shown in Fig. 1–17. Note that each reaction proceeds through a carbinolamine-type intermediate.

We will discuss the repair of some of these lesions in chapter 4. Some of the DNA repair mechanisms that operate are specific for individual damaged bases. However, some forms of oxidative damage to bases are repaired by systems that

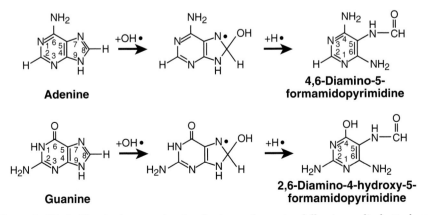

Figure 1–17 Imidazole ring opening in adenine and guanine following radical attack can yield the products shown. (*Adapted from Chetsanga et al. [54] with permission.*)

apparently recognize elements of DNA helix distortion rather than specific base damage (351). With respect to more general defense strategies against oxygen damage in living cells, studies with *E. coli* have revealed a complex coregulation of DNA repair and other protective enzymes. Two regulons controlled by the *soxRS* and *oxyR* gene products respond to $\cdot O_2$ or H_2O_2, respectively (77, 78, 397, 406). These proteins are redox-sensitive transcriptional activators. Enhanced $\cdot O_2$ levels cause a conformational change of SoxR protein, which in turn activates transcription of the *soxS* gene. Among other genes, *soxS* enhances transcription of the genes for superoxide dismutase (*sodA*) and the DNA repair enzyme endonuclease IV (*nfo*) (see chapter 4). The *oxyR* regulon responds to H_2O_2 and controls the expression of catalase (*katG*) and NADPH-dependent alkyl hydroperoxidase (*ahpFC*) (396).

IMPLICATIONS FOR AGING AND CANCER

DNA and other macromolecular damage induced by reactive oxygen species has attracted considerable interest in recent years and has generated extensive discussion on the relevance of spontaneous DNA damage to aging and to cancer in humans (4–6) (also see chapter 4). High-performance liquid chromatography (HPLC) and monoclonal antibodies have been used to quantitate intracellular levels of oxidative damage to DNA and to detect DNA repair-mediated release (in urine) of base products resulting from oxidative attack (2, 4, 5, 119, 335, 375). Thymine glycol and 8-hydroxydeoxyguanosine recovered in urine have been used as specific indicators of oxidative damage to DNA. Apparent correlations between the oxidative damage rate and the metabolic rate of organisms have also been the source of extensive comment (Fig. 1–18). The life span of an organism appears to be *inversely* correlated with its metabolic rate (74, 75) and thus with the rate of oxidative damage. Caloric restriction is certainly a proven means of prolonging the life span in rats and mice and of decreasing cancer rates, which increase dramatically with age (Fig. 1–18). All of these observations are consistent with the gradual accumulation of deleterious genetic alterations due to oxidative damage. As a consequence, one might expect an accumulation of certain poorly repaired forms of DNA damage with age (246). By using an assay that involves the labeling of damaged bases (postlabeling assay; below), a tissue-specific pattern of yet uncharacterized damage (termed I-compounds), which increases with age, has indeed been detected in rodent tissues (328). Perhaps the most dramatic confirma-

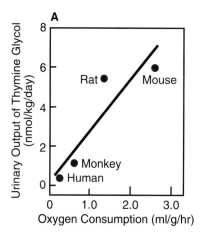
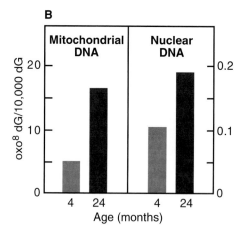
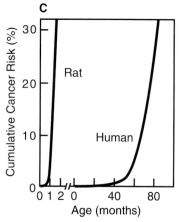

Figure 1–18 Correlations between levels of oxidative damage, aging, and cancer. (A) The metabolic rate of an organism correlates with the urinary output of thymine glycol, which serves as a marker for repair of oxidative DNA damage and appears to be inversely correlated with life span. (B) The level of oxidative DNA damage increases with age. Here, the level of 8-hydroxyguanine (oxo^8 dG) was measured in the DNA of rats of different ages. Note the different scales for mitochondrial and nuclear DNA. (C) Also, the cancer incidence increases dramatically with the relative age of an organism. (*Adapted from Ames [4, 6] with permission.*)

tion of the importance of cellular defense systems against reactive oxygen in aging is the recent observation that overexpression of superoxide dismutase and catalase can prolong the life span of *Drosophila melanogaster* (301). Besides their potential for causing DNA damage directly, many oxidants are also mitogens and thus may also contribute to cancer incidence by a different mechanism(s) (5, 48). These considerations indicate that antioxidants are excellent candidates for cancer-preventing agents (205).

CLASTOGENIC FACTORS

Another interesting aspect of the injurious effects of oxygen radicals is their potential to function as *clastogenic factors* (103), i.e., agents that cause chromosomal breakage. Most clastogenic factors derive from well-characterized exogenous sources of DNA damage such as ionizing radiation (see below). However, some clastogens may be generated in the course of cellular metabolism in mammalian cells. Thus, for example, the cocultivation of lymphocytes and plasma from patients suffering from the human disease ataxia telangiectasia (see chapter 14) with those of normal individuals results in a significant increase in chromosomal damage in the normal cells (365). Tissue culture medium from cultivated skin fibroblasts of ataxia telangiectasia patients also significantly increases chromosomal breakage in normal lymphocytes previously stimulated with phytohemagglutinin (365). Clastogenic activity has also been detected in the medium of cultures of fibroblasts isolated from patients with Bloom's syndrome (104) and Fanconi's anemia (103), other autosomal recessive human diseases that are characterized by chromosomal abnormalities and a high cancer incidence and in which DNA repair defects have been implicated (see chapter 14). Transferable clastogenic material has also been detected in the plasma of patients with chronic inflammatory diseases such as systemic lupus erythematosus, rheumatoid arthritis, chronic active hepatitis and ulcerative colitis (103, 108, 111); in patients infected with human immunodeficiency virus (103); and in certain strains of New Zealand Black mice with high frequencies of spontaneous chromosomal aberrations (109). Clastogenic factors can also be found in plasma from individuals exposed to high doses of ionizing radiation (180) (one study suggests that such an activity has persisted for 31 years in the plasma of atom bomb survivors! [302]) and can be generated by irradiating blood in vitro (355). Tumor promoters, as well as radical-generating chemicals, can also induce clastogenic factors (105, 106).

In general, the presence of radical-protective enzymes such as superoxide dismutase not only prevents the formation of clastogenic activity but also can reduce the effect of preformed factors (103, 106, 110, 111). Hence, it appears that the generation of reactive oxygen species somehow triggers the release of a diffusible clastogenic agent, which induces the same effect in other cells and therefore perpetuates the activity. Attempts to purify clastogenic factors indicate that they are not a single species of molecules but, rather, constitute a complex mixture of substances (103). Tumor necrosis factor alpha is one such component (103). Other suspected agents include inosine tri- and diphosphate (14) and 4-hydroxynonenal (107), a diffusible product of lipid peroxidation that can damage DNA.

Environmental Damage to DNA

Physical Agents That Damage DNA

IONIZING RADIATION

Ionizing radiation has been a source of naturally occurring physical damage to the DNA of living organisms on this planet since the beginning of biological evolution. However, at present various man-made therapeutic, diagnostic, or occupational sources of exposure to ionizing radiation are of far greater importance. Ionizing radiation can randomly cause damage to all cellular components and induces a variety of DNA lesions (46, 121, 153, 186, 241, 321, 336, 413, 436, 445, 446). The deposition of energy from ionizing radiation results in the formation of excited and ionized molecules. An important parameter of the rate of energy deposition by various kinds of ionizing radiation frequently stated in the literature is a quantity termed *linear energy transfer* (153). Linear energy transfer (in kiloelectron volts per micrometer) is defined as the average energy loss due to collisions per distance traversed by a particle.

Traditionally, radiation damage to DNA is ascribed to both *direct* and *indirect* effects (445). Direct effects result from the direct interaction of the radiation energy with DNA. Indirect effects result from the interaction of DNA with reactive species formed by the radiation. Since in living cells DNA exists in an environment containing numerous molecules, inorganic ions, and water, a large variety of potential sources of reactive species can exist as excited molecules, ion radicals, or free radicals, which are ultimately converted to chemically stable products by subsequent decay reactions. This multiplicity of reaction mechanisms and the even greater number of potential reactants create the possibility for a large spectrum of products of ionizing radiation in DNA. Correlations between specific lesions produced in DNA by ionizing radiation and biological end points such as cell killing and mutagenesis are not as well established as for some of the other forms of DNA damage discussed in this chapter.

The predominance of water in biological systems suggests that species formed by the radiolysis of water are major potential sources of indirect damage to DNA (336, 436, 445). It has been estimated that more than 80% of the energy of ionizing radiation deposited in cells results in the abstraction of electrons from water by the following reaction:

$$H_2O \rightarrow H_2O^+ + e^-$$

Several other reactive species (e.g., solvatized electrons, $\cdot OH$, $H\cdot$, H_2, H_2O_2) are formed in subsequent reactions. Of particular biological interest is the formation of relatively stable reactants such as $\cdot OH$, O_2^- and H_2O_2:

$$H_2O^+ + H_2O \rightarrow \cdot OH + H_3O^+$$
$$\cdot OH + \cdot OH \rightarrow H_2O_2$$
$$e_{aq} + O_2 \rightarrow O_2^-$$
$$2O_2^- + 2H^+ \rightarrow O_2 + H_2O_2$$

Note that the formation of superoxide radicals depends on the presence of oxygen. As discussed above, H_2O_2 can also give rise to $\cdot OH$ in Fenton-like reactions. Indeed, the majority of damage to DNA appears to be caused by hydroxyl radicals (a rough estimate is that ~65% of damage is due to $\cdot OH$ radical attack versus ~35% damage by direct ionization [445]). Many aspects of our previous discussion of DNA damage by reactive oxygen species are also applicable to damage by ionizing radiation and vice versa.

Base Damage

Damage to bases by ionizing radiation has been extensively studied in vitro by irradiation of free bases, nucleosides, oligonucleotides, or DNA in aqueous solution, under aerobic or anaerobic conditions, or in the solid state. The chemistry of lesion formation is fairly well understood and comprehensive accounts can be found in the literature (413, 436). Many products and short-lived intermediates have been characterized, especially for thymine. As described above (Fig. 1–16), $\cdot OH$ radicals typically attack the C-5 = C-6 double bond. It has been suggested that the *direct* action of ionizing radiation may lead to the ejection of an electron from the unsaturated C-5 or C-6 position and the resulting cation radical may further react with a hydroxyl ion (445). Thus, direct and indirect radiation effects may result in identical reactive intermediates.

Under aerobic conditions, further reaction with O_2 can result in various ring-saturated derivatives, particularly thymine glycol (5,6-dihydroxy-5,6-dihydrothymine) (Fig. 1–19). The saturated ring can undergo further fragmentation and give

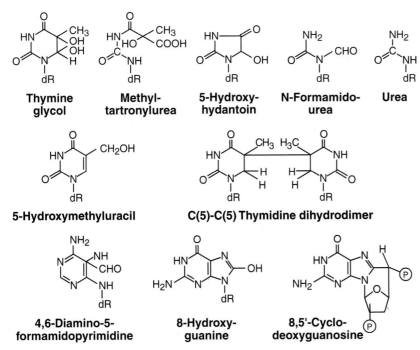

Figure 1–19 Examples of base damage induced by ionizing radiation.

rise to products such as methyltartronylurea, 5-hydroxyhydantoin (a five-ring derivative), *N*-formamidourea, and urea (413, 436) (Fig. 1–19). A different spectrum of ring-saturated thymine derivatives is found in oxygen-free solutions; the products 5-hydroxythymine and 6-hydroxy-5,6-dihydrothymine detected under these conditions may be of particular biological importance (37). 5-Hydroxymethyluracil is a less frequent lesion formed by attack with γ-rays or by radiolytic decay of [6-^{3}H]thymine in DNA (79, 412) (Fig. 1–19).

There have been fewer studies on the effects of ionizing radiation-mediated damage to cytosine and to the purines, although it has been established with respect to cytosine that the major site for attack by hydroxyl radicals is the C-5=C-6 double bond, as in the case of thymine (413, 436). We have already mentioned the biologically important purine derivatives 4,6-diamino-5-formamidopyrimidine, 2,6-diamino-4-hydroxy-5-formamidopyrimidine, and the mispairing lesion 8-hydroxyguanine (Fig. 1–19). Purine residues can also undergo cyclization, giving rise to products such as 8,5'-cyclodeoxyguanosine or 8,5'-cyclodeoxyadenosine (Fig. 1–19), which probably result in distortion of the normal double helical structure of DNA (43, 79, 85, 327, 413, 436). Table 1–4 illustrates the yield of various single base products found after γ-irradiation of human cells.

The possible physiological relevance of dimers of base residues formed within the same DNA strand or the covalent linkage of complementary strands by similar lesions (interstrand cross-links) has not been extensively explored (44, 94, 436). As discussed below, the latter type of lesion is typically associated with the attack of bifunctionally reacting chemical agents. An example of a thymine-derived dimer product is given in Fig. 1–19. Protein-DNA cross-links induced by ionizing radiation may also be important (136, 376, 389). A thymidine-tyrosine cross-link is the predominant species found after γ-irradiation in the presence of oxygen or after treatment with H_2O_2 in the presence of Fe(III) or Cu(II) (84, 293).

A phenomenon not usually encountered following treatment with other agents that generate reactive oxygen species is the formation of more than one damaged moiety in close proximity (186, 445). The explanation for this clustered damage or *locally multiply damaged sites* is that a single energy deposition event can result in several radical reactions in the immediate vicinity. The shape and length

Table 1–4 Yields of DNA base products formed in chromatin of cultured human cells following treatment with γ radiation (420 Gy)

Product	No. of molecules/10^5 DNA bases	Fold increase over background[a]
FaPyGuanine	34.4	13
8-Hydroxyguanine	23.3	3
5-Hydroxyhydantoin	23.2	2
Thymine glycol	10.2	6
FapyAdenine	10.0	3
8-Hydroxyadenine	5.5	2
2-Hydroxyadenine	4.9	2
5-Hydroxycytosine	4.7	2
5,6-Dihydroxycytosine	4.1	13
5-Hydroxymethyluracil	2.8	4
5-Hydroxyuracil	1.8	5

[a] Background level as determined with untreated cells.

Source: After Dizdaroglu (84).

of ionizing particle tracks in relation to the spatial dimensions of DNA in differently packaged structures are important in influencing the local density of damage (153, 186, 445). Ionizing radiation quite frequently induces regions of locally unpaired DNA strands which can be detected by digestion with an endonuclease that specifically recognizes such sites (e.g., *Aspergillus* S1 endonuclease) (8, 266). These S1-sensitive sites have been interpreted as the result of clustered base damage.

Sugar Damage and Strand Breaks

It has been estimated that damage to sugar residues in DNA may be by a factor of 2.7 less frequent than damage to bases (445). Nonetheless, such damage is of biological importance primarily because of the strand breakage that might result. Low-linear-energy-transfer radiation (X rays, γ rays) at 1.0 Gy can result in 600 to 1,000 single-strand breaks and 16 to 40 double-strand breaks, compared with 250 damaged thymine residues in DNA (445).

Damage to the DNA bases such as ring saturation can result in destabilization of the *N*-glycosylic bond and the formation of abasic deoxyribose residues (Fig. 1–20) (413, 436). The other lesions shown in Fig. 1–20 are the result of direct attack by ·OH radicals (413, 436). All of these lesions constitute alkali-labile sites and can be converted into strand breaks upon hot-alkali treatment. Irradiation in the presence of oxygen results in an increase in the number of alkali-labile sites by a factor of 4 (186).

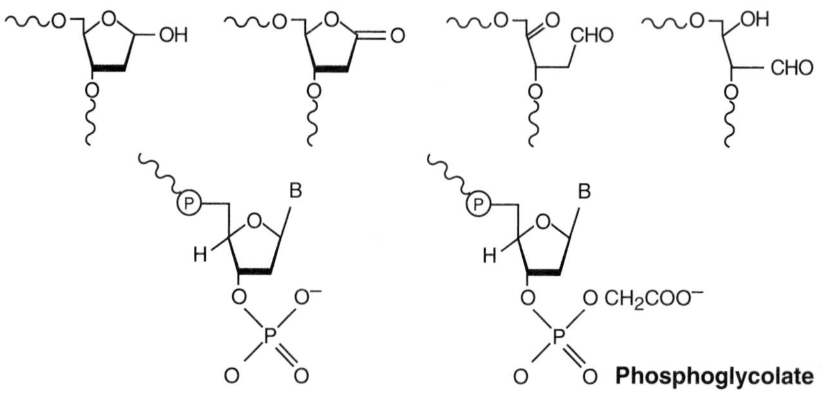

Figure 1–20 Examples of damage to the sugar-phosphate moieties induced by ionizing radiation.

Ionizing radiation also induces strand breakage directly, and most of the lethal effects of ionizing radiation can be attributed to these lesions, in particular to double-strand breaks, a striking example of clustered DNA damage (186, 189, 436, 445, 446). Single-strand breaks are initiated by radical formation at deoxyribose following the loss of a hydrogen atom. This can result from a direct ionization event or from abstraction by an ·OH radical. Subsequently the radical can react with oxygen and form a peroxy radical. Details of the sequence of reactions leading to strand breaks are not entirely clear. Additionally, it is not known whether double-strand breaks are caused by radical transfer to the opposite strand (e.g., the same initial event) (373) or whether they are the consequence of multiple radical attacks leading to independent single-strand breaks (444, 446). The presence of radiation-induced single-strand breaks in DNA can result in very localized denaturation in the vicinity of the break, thereby increasing the probability of free radical-mediated attack at that site as a result of loss of the protective effect of base stacking interactions. If single-strand breaks on opposite strands are sufficiently close, a double-strand break will result.

In considering the repairability of these lesions, it is important to note that the majority of strand breaks induced by ionizing radiation are characterized by unusual or damaged termini which preclude repair by a simple DNA ligation step (298, 436). Although the majority of the 5′ ends retain phosphate groups (28, 175), OH groups at 3′ ends are usually not available. Phosphate or phosphoglycolate groups are found instead (Fig. 1–20) (175, 176). The phosphate is also often entirely lost, and various fragmented sugar derivatives are left (298, 436). Frequently, the terminal base residue is missing, and many strand breaks are in fact single nucleotide gaps (298, 436).

On several occasions we have mentioned the availability of oxygen as a precondition for the "fixation" of radical damage. Indeed, differences in lesion spectrum and frequency are in general assumed to be the cause of increased radiosensitivity in the presence of enhanced concentrations of oxygen, a phenomenon termed the *oxygen effect* (122, 436, 446). It is worth repeating that cellular radioprotectors such as glutathione can efficiently counteract damage induced by radicals at various levels. Thus, the amount of radical formed at deoxyribose or base residues that is amenable to the formation of stable, deleterious DNA damage is determined by the competition of thiol groups with oxygen (436, 446), i.e.,

$$R\cdot \: + \: XSH \rightarrow RH \: + \: XS\cdot$$

as opposed to

$$R\cdot \: + \: O_2 \rightarrow RO_2$$

Diminishing cellular thiol levels as a therapeutic radiosensitization strategy is an area of highly active ongoing research (447).

In summary, random energy deposition by ionizing radiation causes a wide array of different DNA lesions. The relative importance of individual lesions for various genetic end points such as survival, mutagenesis, or chromosome aberrations is difficult to establish and varies from system to system (35, 36). The physiological relevance of a particular lesion can be inferred from the existence of an enzyme(s) that removes it (although it is the lesions that cannot be repaired that may in fact be the deleterious ones) (438). Such enzyme activities exist for FaPy (53), for thymine glycol (80) and other thymine derivatives such as urea (34), and for 8-hydroxyguanosine (79). Some of these enzymes have a broad substrate specificity. We will discuss examples of these activities in detail in chapter 4. The repair of double strand breaks requires more-complex repair reactions involving homologous DNA molecules. We will discuss these in chapters 11 and 12.

Protection of the human population from sources of ionizing radiation and careful evaluation of risk-versus-benefit ratio of exposure are areas of intense investigation and of obvious importance (134, 435). A fundamental problem arises from the fact that many of the conclusions discussed above are derived from studies involving high doses of radiation which are rarely, if ever, encountered in an organism's life. The validity of certain modes of extrapolation to low doses is debated, and more sensitive methods for the detection of low levels of damage are continuously being developed. Examples are the detection of base damage by specific antibodies (183, 184, 237, 439) and the chromosome-specific detection of double-strand breaks by pulsed-field gel electrophoresis (see chapter 12) (149), a method which might ultimately replace the less sensitive analysis of the elution characteristics of filter-bound DNA or of molecular weight distributions of DNA in neutral sucrose gradients.

Even in systems in which double-strand breaks have been established as the major source of lethal damage and a linear increase of the frequency of double strands with dose over a wide range of exposure levels has been demonstrated, survival curves are typically nonlinear (121, 445). It must therefore be concluded that cellular processes such as dose-dependent repair efficiencies are important parameters in determining lethal effects at various doses.

UV RADIATION

Historically, the investigation of UV radiation damage to DNA marks the beginning of the study of the repair of DNA damage, and the exposure of cells to UV radiation is probably the best studied and most extensively used model system for exploring the biological consequences of DNA damage and its repair and tolerance. One of its many attributes as a system for producing DNA damage is that UV radiation at 254 nm is readily available from an ordinary germicidal lamp and instrumentation for accurately measuring its intensity is commonplace. In addition, UV light in general is highly relevant biologically, because living organisms have had to contend with the genotoxic effects of solar UV radiation since the beginning of the evolution of life on this planet. Indeed, it has been suggested that unattenuated UV radiation during the pre-Phanarazoic aeons (from 3.5 billion until 0.5 billion years ago) might have precluded the development of terrestrial life as we know it (24).

The UV radiation spectrum has been subdivided into three wavelength bands designated UV-A (400 to 320 nm), UV-B (320 to 290 nm), and UV-C (290 to 100 nm) (Fig. 1–21). Solar UV radiation consists mainly of UV-A and UV-B, since penetration of the atmospheric ozone layer drops dramatically for wavelengths below 320 nm (Fig. 1–21). It has often been pointed out that most studies involving germicidal UV light of approximately 260 nm (which corresponds to the DNA absorption peak) are addressing a wavelength which has minor relevance for human health. On the other hand, one can argue that identical lesions in DNA are

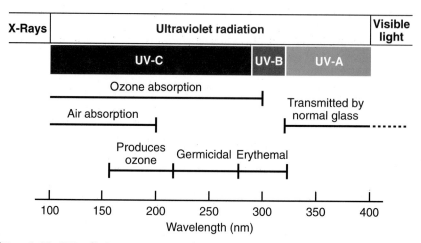

Figure 1–21 UV radiation spectrum. (*Adapted from Hu [182] with permission.*)

of relevance even at longer wavelengths of UV radiation but are simply more efficiently induced with UV-C radiation (32, 363). Clearly, the nature of a particular biological problem of interest determines the most appropriate wavelength for experimental use.

Cyclobutane Pyrimidine Dimers

When DNA is exposed to radiation at wavelengths approaching its absorption maximum (~260 nm), adjacent pyrimidines become covalently linked by the formation of a four-membered ring structure resulting from saturation of their respective 5,6 double bonds (255, 270, 359, 361, 362, 390, 432, 437, 442, 443). The structure formed by this photochemical cycloaddition is referred to as a *cyclobutane dipyrimidine* or *pyrimidine dimer*. These dimers can theoretically exist in 12 isomeric forms. However, only four of them, with the configurations *cis-syn*, *cis-anti*, *trans-syn*, and *trans-anti* occur in significant yields (208). Pyrimidine dimers are thought to exist predominantly in the *cis-syn* form in double-stranded B-form DNA (Fig. 1–22 and 1–23) (437). However, *trans-syn* dimers (Fig. 1–23) are generated to some extent, mainly in denatured DNA (304), and in vivo they can be found in single-stranded regions (possibly in highly transcribed genes) or in duplex DNA with special conformations such as B-DNA–Z-DNA junctions (407). Single-stranded DNA also permits the formation of dimers between nonadjacent pyrimidines (295).

The effect of these lesions on the local structure of DNA is an issue of considerable interest with regard to their recognition by repair enzymes or their ability to form correct base pairs if present in a replicating template strand. Additionally, significant perturbation of the normal DNA structure (e.g., extensive bending) may affect the binding of sequence-specific proteins and result in regulatory abnormalities. Traditionally, cyclobutane dimers have been considered to be bulky, helix-distorting lesions which are strictly noncoding and whose presence results in an obligatory arrest of replication, since no base can be inserted which can form hydrogen bonds of reasonable stability (49, 173, 460). However, more recent information derived from the thermodynamic and spectroscopic characterization of duplex decamers containing a single thymine dimer in a defined position indicates that the *cis-syn* lesion can be accommodated inside the double helical structure of B-DNA without much distortion (409) (Fig. 1–23). Correct hydrogen bond formation with adenine residues is apparently still possible, an assumption that has

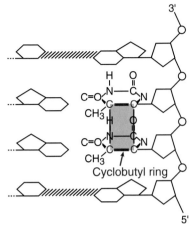

Figure 1–22 The cyclobutane pyrimidine dimer is formed in DNA by covalent interaction of two adjacent pyrimidines in the same polynucleotide chain. Saturation of their respective 5,6 double bonds results in the formation of a four-membered cyclobutyl ring (red area) linking the two pyrimidines. (*Adapted from Friedberg [131] with permission.*)

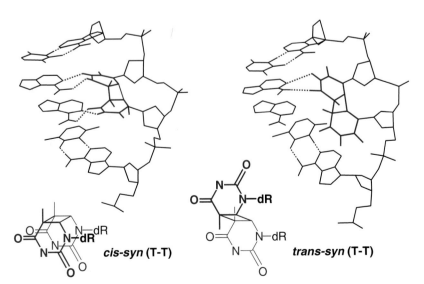

Figure 1–23 Structure of *cis-syn* and *trans-syn* cyclobutane thymine dimers and their proposed three-dimensional orientation in B-DNA. (*Adapted from Weinblum and Johns [452] and Taylor et al. [409] with permission.*)

been supported by in vitro DNA synthesis experiments (299, 411). The *trans-syn* dimer, however, results in a more severe pertubation, especially at its 5' side (Fig. 1–23). We will return to the issue of the coding properties of pyrimidine dimers in the context of mutagenesis and DNA damage tolerance in prokaryotes and eukaryotes (see chapters 10 to 12).

The amount of bending introduced by *cis-syn* dimers into the DNA helix is still unknown. Whereas in one study a bending of only 7° was measured (441), bending of as much as 30° has been detected in other studies based on molecular modeling, electron microscopy, and analysis of the anomalous electrophoretic behavior of dimer-containing oligonucleotides (185, 306) (see chapter 5).

Cyclobutane pyrimidine dimers are extraordinarily stable to extremes of pH and temperature (361). They survive total acid hydrolysis of DNA and can be separated from thymine in such hydrolysates by a variety of chromatographic techniques (133). Since thymine (unlike the other major bases) in DNA can be conveniently and specifically radiolabeled with ^{3}H or ^{14}C, the thymine-containing pyrimidine dimer content in a given sample of DNA can be directly measured by one or more of these techniques (see chapter 3).

The formation of pyrimidine dimers during irradiation of DNA is a reversible process, which can be represented as

$$Py + Py \overset{UV}{\rightleftharpoons} Py<>Py$$

Under normal conditions the equilibrium is shifted far to the right and so dimer formation is favored over dimer reversal (362). However, if *E. coli* DNA radiolabeled in thymine is continuously irradiated at 254 nm, the thymine-containing pyrimidine dimer content (thymine-thymine plus thymine-cytosine dimers) of the DNA does not increase beyond about 7% of the total thymine content (323). This steady state reflects a dynamic equilibrium in which the rates of dimer formation (which is pseudo zero order, to good approximation) and reversal (which is first order in dimer content) are equal (323) (see chapter 4).

The yield of cyclobutane pyrimidine dimers in DNA is influenced by nucleotide composition. It has been known for many years from measurements in bulk DNA that at moderately high doses of UV radiation the yield of C-C is significantly lower than that of C-T or T-T (362) (Table 1–5). More-recent studies indicate a ratio of T-T to C-T to T-C to C-C of 68:13:16:3 for irradiation of plasmid DNA with 254-nm light, and a similar trend has been observed in irradiated human cells (278, 421). Aside from these quantitative observations, it was tacitly assumed that thymine-containing pyrimidine dimers are randomly distributed in DNA and that all sites in which thymine is adjacent to another pyrimidine are sites in which di-

Table 1–5 Distribution of pyrimidine dimers in UV irradiated DNA

Source of DNA	Wavelength (nm)	Dose (J/m²)	Distribution of dimers (%)		
			C-C	C-T	T-T
Haemophilus influenzae (high A+T)	265	2×10^2	5	24	71
	280	4×10^3	3	19	78
E. coli (G+C ≈ A+T)	265	2×10^2	7	34	59
	280	4×10^3	6	26	68
M. luteus (high G+C)	265	2×10^2	26	55	19
	280	4×10^3	23	50	27

Source: After Setlow and Carrier (364).

merization can occur with equal probability. The advent of techniques for sequencing DNA has facilitated studies on the detailed distribution of a variety of forms of DNA damage. Pyrimidine dimers are an interesting and important case in point. The detailed distribution of cyclobutane pyrimidine dimers in a segment of DNA of known nucleotide sequence from the operator-promoter region of the *E. coli lacI* gene, present in a plasmid as a recombinant DNA insert, has been studied (157). A radiolabeled 117-bp segment of the *lacI* gene was exposed to various doses of UV radiation and then incubated with saturating amounts of an enzyme probe that specifically recognizes pyrimidine dimers in DNA and catalyzes the formation of a single-strand break (nick) in the DNA at each dimer site. (This enzyme is called pyrimidine dimer DNA-glycosylase/apurinic/apyrimidinic (AP) endonu-

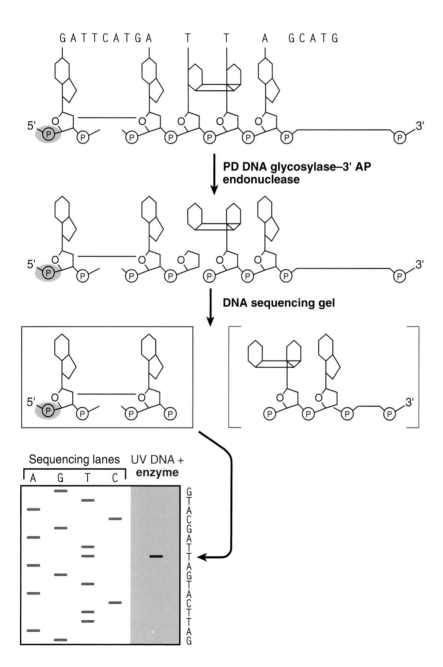

Figure 1–24 Determining the location of pyrimidine dimers in UV-irradiated DNA by using a dimer-specific enzyme probe. A double-stranded DNA fragment (only one strand is shown) is radiolabeled at the 5′ ends and incubated with saturating amounts of an enzyme (PD DNA glycosylase-3′ AP endonuclease) that specifically recognizes cyclobutane pyrimidine dimers in DNA. The enzyme cuts the 5′ glycosyl bond of the dimer and also the 3′ phosphodiester bond as shown. This procedure leaves stable 5′-end-labeled DNA fragments, whose lengths bear a precise relationship to the sites of pyrimidine dimers. The DNA is then loaded onto a denaturing polyacrylamide sequencing gel alongside (in separate lanes of the gel) standard DNA-sequencing reaction samples. [*Adapted from Gordon and Haseltine [157] with permission.*)

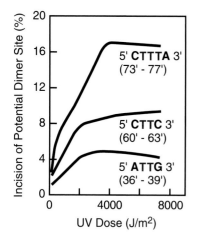

Figure 1–25 Dose response for cyclobutane thymine dimer formation in a DNA fragment of known nucleotide sequence. The different curves show the dose response for individual dimer sites, which are identified numerically by their precise location in the sequenced DNA (numbers in parentheses). Dimers were quantitated from the amount of radioactivity present in electrophoretic bands by using the technique shown in Fig. 1–24. (*Adapted from Gordon and Haseltine [157] with permission.*)

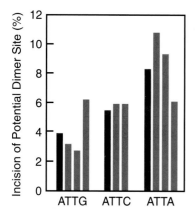

Figure 1–26 Influence of the DNA sequence on the formation of pyrimidine dimers in UV-irradiated DNA. Each grey bar represents the effect of nucleotides flanking the dimer sites shown, in different regions of the *E. coli lacI* gene. The red bar on the left of each set represents the mean of all determinations for the particular sequence, taken from all regions where that sequence occurred. The frequency of dimers was determined from the frequency of incision of UV-irradiated DNA by a dimer-specific enzyme probe (see Fig. 1–24). (*Adapted from Gordon and Haseltine [157] with permission.*)

clease and is discussed in detail in chapter 4.) Assuming that all sites of dimer formation are converted to nicks in DNA by the action of the enzyme probe, the precise location of dimers in the DNA sequence can be determined by denaturing the DNA to separate strands and comparing the electrophoretic distribution of each fragment with those generated from unirradiated DNA exposed to the Maxam-Gilbert sequencing technique (267) (Fig. 1–24). The frequency of a particular dimer is reflected by the amount of radioactivity associated with each fragment (represented as a band) in the gel. The effect of UV radiation on single-stranded DNA was examined by denaturing the 117 bp fragment and harvesting the labeled single strands prior to UV irradiation.

Such studies have shown that at high doses of UV light at each site at which dimerization is possible, the extent of dimer formation reaches a maximum which is unaffected by further irradiation (157). This reflects the establishment of a steady state between the formation of dimers and their monomerization by photoreversal. The dose at which the maximum is reached varies for different types of dimers. Thus the level of C-C plateaus at doses of approximately 500 J/m², whereas doses of 2,000 J/m² are required for T-T. This result is not surprising, because it is known that the quantum yields for formation differ for different pyrimidine dimers (362).

The yield of cyclobutane pyrimidine dimers in DNA is influenced by the sequence context. The steady-state level of dimer formation is also influenced by the nature of the nucleotides flanking potential dimer sites. For example, in the studies quoted in the previous section, different sites of potential T-T attained different steady-state levels that varied between 4 and 16% (Fig. 1–25) (157). In general, the equilibrium level of dimers was greater for TT sites flanked on both sides by A than for TT sites flanked on the 5' side by A and on the 3' side by G (Fig. 1–26) (157). However, this immediate nucleotide-flanking effect is not sufficient to account for all variations observed. Thus, the frequency of dimer formation in the sequence TGAAATTGTTAT was significantly higher than that observed in either of the sequences CTATATTGATTA and TGGAATTGTGAG (157).

In general, for most potential dimer sites, the dose response of dimer formation was the same in an *E. coli lacI* DNA fragment regardless of the secondary structure of the DNA (157). However, differences in the overall rate of pyrimidine dimer formation in single- and double-stranded DNA have been detected in other studies (270, 359, 361, 362, 390, 432, 443).

These experiments demonstrate that the formation of cyclobutane pyrimidine dimers in DNA is not a random phenomenon as was believed for many years. Furthermore, pyrimidine dimer yields for any given dose of UV radiation cannot be accurately extrapolated from measurements at any other dose, and they cannot be accurately extrapolated for any single dose from one site to another. These limitations notwithstanding, the cyclobutane pyrimidine dimer has served well as the classic "test lesion" for analysis of DNA repair, and much of the discussion in the remainder of this book centers around the cellular responses to this particular lesion in DNA.

The Pyrimidine-Pyrimidone (6-4) Photoproduct

Alkali-labile lesions at positions of cytosine (and, much less frequently, thymine) located 3' to pyrimidine nucleosides are also present in UV-irradiated DNA (253). These photoproducts, referred to as *pyrimidine-pyrimidone (6-4) lesions* or simply as *(6-4) lesions* (32, 123, 170, 279, 345) are apparently identical to a noncyclobutane type of dipyrimidine photoproduct (6,4',5',methylpyrimidin-2',one'-thymine) that was detected following the irradiation of thymine in frozen aqueous solution (433, 434). This photoproduct introduces a major distortion in the double-helical structure of DNA (410) (Fig. 1–27). These lesions are insensitive to certain enzyme probes specific for pyrimidine dimers or for apurinic DNA but can be detected by their lability to hot alkali (170, 253). The (6-4) products of TC, CC, and, less frequently, TT sequences are observed in UV-irradiated DNA, whereas that of CT is not. Cytosine methylation at C-5 apparently inhibits any (6-4) photoproduct formation (33, 152, 311). Irradiation of (6-4) products with 313-nm light leads to the formation of the Dewar isomer (408), a lesion which may be of considerable biological relevance (see chapter 11).

As is true for pyrimidine dimers, the number of (6-4) lesions formed is proportional to the incident UV radiation dose in the range 100 to 500 J/m² (33, 253). When the frequency of (6-4) photoproducts and that of pyrimidine dimers were compared at specific sites in the *E. coli lacI* gene (33) or in simian virus 40 DNA (30), the frequency of the former varied between 10- and 15-fold at different sites. The incidence of (6-4) TC lesions was generally greater than that of CC lesions (Fig. 1–28). TT lesions were detected only at very high UV doses (> 5 kJ/m²). At most sites in the DNA, (6-4) lesions occurred at a frequency severalfold lower than that of pyrimidine dimers. However, at some sites the lesion was detected at levels equal to or greater than that of dimers. A correlation exists between the frequency of nonsense mutations and of UV-induced damage at TC and CC sequences in the *E. coli lacI* gene. This correlation is somewhat better for (6-4) photoproducts than for pyrimidine dimers (32, 33, 124, 170, 171). The relative contributions of cyclobutane dimers and (6-4) photoproducts to mutagenicity and cytotoxicity of UV radiation have been investigated in various systems (279). We will discuss this topic in detail in several contexts later in the book.

Other Photoproducts in DNA

Spore photoproduct. The formation of other noncyclobutane dipyrimidines can be detected in irradiated solutions of free bases and of nucleosides. Figure 1–29 shows the structure of 5'-thyminyl-5,6-dihydrothymine, a major photoproduct produced in the UV-irradiated spores of *B. subtilis* (431). As much as 30% of the thymine in spore DNA can be converted to this product following exposure to very high doses of UV radiation (360, 390). The formation of this lesion appears to be related to the state of hydration of the DNA, since the so-called *spore photoproduct* can also be generated by UV irradiation of dehydrated DNA, i.e., DNA in the A form (390). The repair of this lesion is considered in chapter 3.

Complex lesions involving purines. A photoproduct found after irradiation of poly(dA) was identified as 8,8-adenine dehydrodimer (145, 319). Here, dimer formation is accomplished by a single 8,8 bond linking the imidazole rings. Another complex uncharacterized lesion involving adenine has been reported after UV irradiation of simian virus 40 DNA (30). The lesion is alkali labile and occurs specifically at ACA sequences. Other photoinduced complex lesions involving purines have been identified (95, 225). The fact that they are recognized by repair enzymes (95, 142) suggests their physiological significance. The formation of purine lesions is greatly enhanced if they are flanked by two or more contiguous pyrimidines on their 5' side (19).

Pyrimidine hydrates. Another type of photochemical reaction of a pyrimidine base is the addition of a molecule of water across the 5,6 double bond to form a 5,6-dihydro-6-hydroxy derivative designated *cytosine hydrate* (Fig. 1–30). The quantum yield for the formation of cytosine hydrates in UV-irradiated DNA and in UV-irradiated polymers is significantly greater in single-stranded than in duplex-DNA (117). Hydrates of cytosine, deoxycytosine, CMP, or dCMP are highly unstable; they readily dehydrate and revert to the parent form (117). However, their half life is dramatically increased in DNA, and it has been suggested that cytosine hydrate is the major nondimer photoproduct of cytosine. It can undergo hydrolytic deamination to yield uracil (27). The photoproduct 5-methylcytosine hydrate may undergo deamination to yield 5-thymine hydrate, which can convert to thymine upon dehydration (429). This could be a mechanism for UV radiation-induced mutation at methylated C residues.

Thymine glycols. Another lesion resulting from saturation of the 5,6-double bond of some pyrimidines is *thymine glycol* or *5,6-dihydroxydihydrothymine*. As discussed above, this lesion is one of the major forms of DNA base damage induced

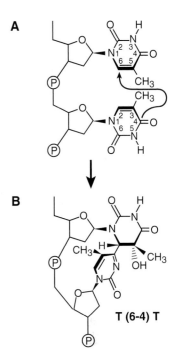

Figure 1–27 Structure of a thymine-thymine photoproduct produced by linkage between the C-6 position of one thymine and the C-4 position of the adjacent thymine. This covalent linkage is unstable to hot alkali. It can be assumed that formation of this lesion will result in a major distortion of the double helix. The plane of the 3' base moiety is shifted 90° relative to that of the 5' thymine. In DNA the 3' pyrimidine in these so-called (6-4) lesions is typically cytosine rather than thymine (see Fig. 1–28). (*Adapted from Taylor et al. [410] with permission.*)

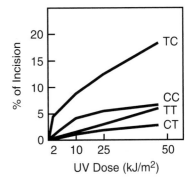

Figure 1–28 Percentage of incision at (6-4) products by hot alkali treatment of simian virus 40 DNA following treatment with increasing doses of 254-nm UV radiation. Average percentage of incisions at several different sites are shown for each dose and each class of dimer. (*Adapted from Bourre et al. [30] with permission.*)

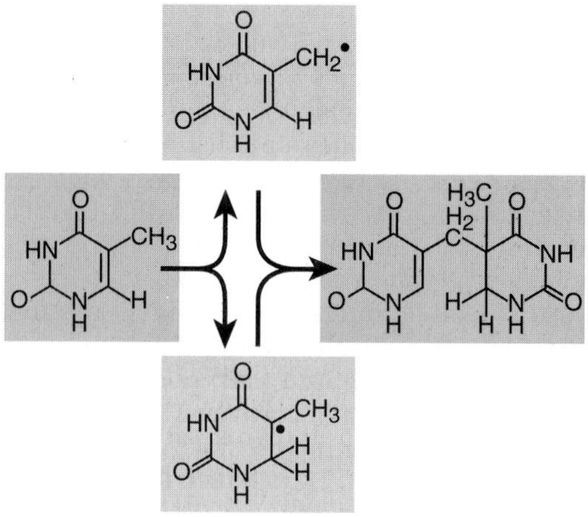

Thymine

5-Thyminyl-5,6-dihydrothymine

Figure 1–29 Formation of 5-thyminyl-5,6-dihydrothymine (spore photoproduct) by the addition of two different radicals of thymine generated by UV radiation. (*Adapted from Smith [388] with permission.*)

by ionizing radiation, but it can also result from UV radiation (81, 461). *E. coli* contains an enzyme activity that catalyzes the excision of this type of base damage from DNA (80). The properties of this enzyme are discussed in chapter 4.

DNA Cross-Links and Strand Breaks

UV radiation can result in the cross-linking of DNA to proteins (305). Cross-links between different duplex DNA molecules (DNA-DNA cross-links) have been also observed occasionally (265), mainly when DNA is irradiated in the dry state, in an extremely densely packed condition (such as in the heads of salmon sperm [387]), or in other special conformations (259). Irradiation of DNA at 254 nm can also result in breakage of the polynucleotide chain (305, 341). (Note that we are addressing strand breakage as an immediate UV radiation lesion and do not include *repair*-related single- or double-strand breaks in this context.) However, the amount of UV radiation required to reduce the molecular weight of *Streptococcus pneumoniae* DNA by 50% is about 100 times that required to reduce the transforming activity of the streptomycin resistance marker of that organism to the same extent (390). In addition, no chain breaks are detected in phage T7 DNA exposed to doses of UV that inactivated almost 100% of the phage population (390). Thus, there is no conclusive evidence that the formation of DNA strand breaks by 254-nm UV is of biological consequence. However, the frequency of strand breaks and DNA-protein cross-links is dramatically increased by irradiation at longer wavelengths (425).

Sensitized Photoreactions of DNA

The UV radiation-induced damage discussed thus far results principally from the direct absorption of photons by bases in DNA. DNA damage can also result from wavelengths in the electromagnetic spectrum which, though not absorbed directly by bases, are absorbed by other molecular species (*sensitizer molecules*) that then transfer energy to the bases in DNA. This phenomenon is referred to as *photosensitization* and can occur by a variety of possible reaction pathways (231, 316). Photosensitization reactions can be broadly classified into those in which the sensitizing molecule behaves as a true photocatalyst and those in which the sensitizing molecule is consumed (231).

The best example of a naturally occurring photosensitization process that is truly catalytic is called *enzymatic photoreactivation*. This is a ubiquitous biological reaction for the repair of cyclobutane-type *cis-syn* pyrimidine dimers in DNA and is discussed in detail in chapter 3. During this repair process, pyrimidine dimers

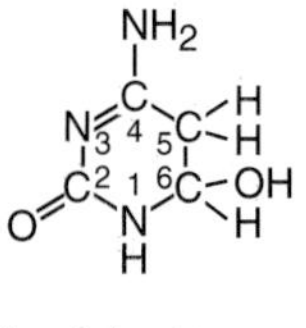

5,6-Dihydro-6-hydroxy-cytosine
Cytosine hydrate

Figure 1–30 Example of a monomeric pyrimidine base lesion caused by UV radiation.

are monomerized in the presence of near-UV light (300 to 600 nm) by an enzyme called photoreactivating enzyme or DNA photolyase.

Photosensitized reactions of interest with respect to DNA damage include those that result in the formation of thymine-thymine dimers in DNA by triplet excitation transfer. Energy can also be transferred to oxygen, resulting in highly reactive oxygen species, a reaction termed the *photodynamic effect* (316). Photosensitizers include not only exogenous drugs (such as 8-methoxypsoralen [144]; see below) but also endogenous agents such riboflavin or aromatic amino acids (316). Various ketones promote pyrimidine dimer formation in this fashion, a notable example being acetophenone (210, 231, 233). The lowest triplet energy state of acetophenone is slightly higher than that of thymine but lower than the triplet states of the other DNA bases (Fig. 1–31). Upon irradiation of DNA in the presence of acetophenone at wavelengths of ~300 nm, the triplet energy of the photosensitizer is transferred to thymine, facilitating the formation of thymine-thymine dimers (210, 231).

A particular advantage of the use of photosensitizers in photobiological research is that they promote the formation of thymine dimers at wavelengths of UV light that do not drive the reverse reaction (photoreversal). Thus, in contrast to the situation described above, in which irradiation of DNA at 254 nm results in a steady state when approximately 7% of the thymine in DNA is dimerized, irradiation at 300 nm in the presence of a molecular photosensitizer such as acetophenone can produce thymine-thymine dimer contents close to the theoretical maximum (232). Thus, photosensitizers such as acetophenone are extremely useful in the study of the repair of thymine dimers in DNA. In addition to the quantitative considerations just described, the use of photosensitizers permits the study of the repair of a homogeneous class of base damage, since at the longer wavelengths of UV radiation used for photosensitized reactions, other pyrimidine dimers, such as (6-4) products, and nondimer photoproducts produced at 254 nm are avoided (232, 428).

Effect of substitution by halogenated pyrimidines on the sensitivity of DNA to UV radiation. While on the general topic of photosensitized reactions in DNA, it is appropriate to consider some of the effects of the substitution of thymine by certain halogenated pyrimidines. Substitution of thymine in DNA by 5-BU is an effective means of perturbing the density of DNA, and this technique has contributed much to our current understanding of both semiconservative and non-semiconservative modes of DNA replication. In addition to altering the density of DNA, 5-BU imparts an increased sensitivity to irradiation at about 313 nm, resulting in polynucleotide strand breaks (*BU-photolysis*) and the production of alkali-labile sites (83, 187). The observed breaks are thought to arise from photochemical debromination followed by free radical attack on the sugar or sugar-phosphate backbone (187). The use of BU-substituted DNA for the detection and measurement of DNA repair synthesis by both the density shift and photolysis techniques is discussed in greater detail in chapter 5.

Chemical Agents That Damage DNA

Perhaps the earliest impetus to the study of the interactions of chemicals with DNA was the potential for using chemicals as lethal and injurious agents in warfare (39). A far more humane stimulus came from the field of cancer chemotherapy, which is predicated on the basic idea that damage to DNA can interfere with normal DNA synthesis, leading to the replicative arrest of rapidly dividing cell populations such as cancer cells. In more recent years, an increasing public awareness of environmental mutagens and carcinogens has led to an intense interest in studying the mechanisms by which genotoxic chemicals interact with and damage DNA. At present there is a wealth of literature on chemical damage to DNA. The goal in this text is to provide illustrative examples of some classes of chemicals whose interaction with DNA is well characterized and which have been widely used as models for the study of the repair of DNA damage. Biological consequences of specific lesions will be discussed in detail in the following chapters.

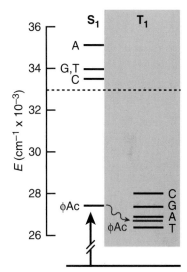

Figure 1–31 The absorption of a photon can promote an electron to one of several short-lived excited states termed *singlet states*, which are characterized by antiparallel electron spins. Return to the ground state by photon emission is accompanied by fluorescence. However, spin inversion results in the longer-lived *triplet state*, which can facilitate further reactions. The energy levels of the lowest excited singlet states (S_1) and lowest triplet states (T_1) of adenine (A), guanine (G), cytosine (C) and thymine (T), along with that of acetophenone (ϕAc), are shown. The lowest triplet energy state of ϕAc is slightly higher than that of thymine but lower than that of the other DNA bases. Thus, on irradiation of DNA at 300 nm, the triplet energy of ϕAc is transformed to thymine, thereby facilitating the formation of dimers between adjacent thymines. (*Adapted from Lamola [231] with permission.*)

ALKYLATING AGENTS

Alkylating agents are electrophilic compounds with affinity for nucleophilic centers in organic macromolecules (234, 260, 338, 342, 379, 381, 382). These include a wide variety of chemicals, many of which are proven or suspected carcinogens (235). However, intermediates of normal cellular metabolism such as S-adenosylmethionine have also been shown to alkylate DNA (246, 261) (see chapter 3).

These agents can be either monofunctional or bifunctional. The former have a single reactive group and thus interact covalently with single (although varied) nucleophilic centers in DNA. Some alkylating agents that cause DNA damage are bifunctional; i.e., they have two reactive groups, and each molecule is potentially able to react with two sites in DNA. Numerous potential reaction sites for alkylation have been identified in all four bases, although not all of them have equal reactivity. The sites of reaction in DNA for many monofunctional alkylating agents include the following (Fig. 1–32): in adenine, N^1, N^3, N^6, and N^7; in guanine, N^1, N^2, N^3, N^7, and O^6; in cytosine, N^3, N^4, and O^2; and in thymine, N^3, O^2, and O^4 (338, 380–382). In general, the ring nitrogens of the bases are more nucleophilic than the oxygens, with the N^7 position of guanine and the N^3 position of adenine being the most reactive (338, 380–382). Alkylation of oxygen in phosphodiester linkages results in the formation of phosphotriesters (380).

The reactivity of a given alkylating agent for particular chemical groups in DNA is roughly correlated with a constant (s) referred to as the *Swain-Scott constant* (402). Reagents with low s values tend to react more extensively with less-nucleophilic enters such as the O^6 position of guanine and the phosphodiester groups of the DNA backbone, whereas compounds with higher Swain-Scott constants tend to react with the more-nucleophilic nitrogen atoms such as the highly reactive N^7 nitrogen of guanine (338). Table 1–6 shows the relative extent of reaction of the bases with monofunctional alkylating agents frequently used experimentally as mutagens. It should also be noted that base modification by alkylation weakens the N-glycosylic bond. Hence, treatment with many alkylating agents leads to depurination/depyrimidination and the appearance of alkali-labile abasic sites (257).

The reaction of alkylating agents with base residues in DNA frequently appears to be nonrandom, and the relative distribution of alkylation damage may play a role in the generation of mutational hot spots (334). For example, the negative electrostatic potential of the N^7 position of guanine is enhanced if this guanine is flanked by other guanine residues and will preferentially undergo elec-

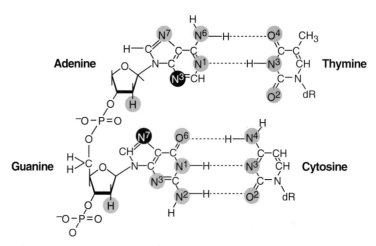

Figure 1–32 Nucleophilic centers in DNA that are the most reactive to alkylating agents. In general, the ring nitrogens of the bases are more reactive than the ring oxygens. Alkylations at phosphodiester linkages (to yield phosphotriesters), N^7 of guanine and N^3 of adenine, are the most frequently encountered. (*Adapted from Friedberg [131] with permission.*)

Table 1–6 Relative proportions of alkylated bases present in DNA after reaction with carcinogenic alkylating agents

Adduct	% of total alkylation after reaction with:		
	Dimethyl-nitrosamine, N-methyl-N-nitrosourea, 1,2-dimethylhydrazine	Methyl methane-sulfonate	Diethylnitrosamine, N-ethyl-N-nitrosourea
N^1-Alkyladenine	0.7	1.2	0.3
N^3-Alkyladenine	8	11	4
N^7-Alkyladenine	1.5	1.9	0.4
N^3-Alkylguanine	0.8	0.7	0.6
N^7-Alkylguanine	68	83	12
O^6-Alkylguanine	7.5	0.3	8
N^3-Alkylcytosine	0.5		0.2
O^2-Alkylcytosine	0.1		3
N^3-Alkylthymine	0.3		0.8
O^2-Alkylthymine	0.1		7
O^4-Alkylthymine	0.1–0.7		1–4
Alkylphosphates	12	1	53

Source: After Pegg (307).

trophilic attack (334). An increased negative charge of the alkylating moiety has the effect of diminishing this sequence preference (216).

The reaction of alkylating agents with specific sites in DNA may also be governed by steric effects. For example, when DNA is in the normal B (right-handed) helical configuration, both the O^6 and N^7 atoms of guanine are in the more accessible major groove, while the N^3 position of adenine lies in the relatively less accessible minor groove (see Fig. 4–7). However, portions of DNA molecules may assume a *left-handed* helical conformation called the Z form (221, 333). This conformation occurs in the alternating copolymer poly(dG-dC) at high salt concentrations (10, 73, 93, 440). In fact, this copolymer, when fully methylated at the 5 position of cytosine, undergoes a transition from B to Z at approximately physiological concentrations (21). A left-handed configuration also occurs in poly(dT-dG)·poly(dC-dA) (168). It was apparent as soon as the structure of Z-DNA was solved that certain atoms such as the O^6, N^7, and C-8 positions of guanine are much less sterically hindered in that structure than in the B form.

CROSS-LINKING AGENTS

As pointed out above, bifunctional alkylating agents can react with two different nucleophilic centers in DNA (234, 260, 338, 342, 379, 381, 382). If the two sites are on opposite polynucleotide strands, *interstrand* DNA cross-links result (Fig. 1–33 and 1–34). If these sites are situated on the same polynucleotide chain of a DNA duplex, the reaction product is referred to as an *intrastrand* cross-link (Fig. 1–34).

Interstrand DNA cross-links represent an important class of chemical damage to DNA, since they prevent DNA strand separation and hence can constitute complete blocks to DNA replication and transcription. It is precisely for this reason that a number of agents such as nitrous acid (148), mitomycin (29, 193), nitrogen mustard and sulfur mustard (56, 118, 217), various platinum derivatives [such as *cis*-platinum(II) diaminodichloride] (55, 99, 120, 339), and certain photoactivated psoralens (64, 144, 316) have been used extensively in cancer chemotherapy. In addition, UV radiation at about 254 nm (259, 265) and ionizing radiation (240, 436) can result in the formation of intermolecular DNA cross-links as minor products of DNA damage.

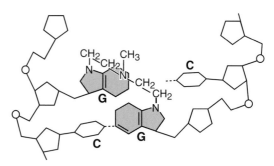

Figure 1–33 Schematic representation of the cross-linking of DNA by nitrogen mustard through the N^7 positions of two guanine moieties on opposite strands of the duplex. (*Adapted from Friedberg [131] with permission.*)

Nitrogen and sulfur mustard, mitomycin, and *cis*-platinum are well-studied examples (38). When bacterial DNA is reacted with concentrations of nitrogen mustard as low as 5×10^{-7} M, approximately 0.005% of the bases are alkylated (217). Studies with radiolabeled nitrogen mustard have shown that only a small fraction (approximately 4%) of the DNA-bound mustard molecules become effective interstrand cross-links between the N^7 positions of guanine moieties on opposite DNA strands (217) (Fig. 1–33).

The covalent interaction between the two strands constituted by cross-links in duplex DNA molecules facilitates their detection by a number of techniques, including gel-electrophoretic mobility, alkali elution and velocity, and special density-labeling procedures (338). For example, DNA can be labeled in one of the two strands by allowing replication in the presence of the thymidine analog [^{3}H]5-

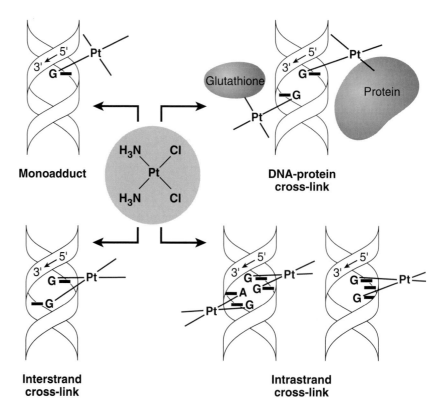

Figure 1–34 *cis*-Platinum can form monoadducts with DNA, interstrand cross-links, intrastrand cross-links, and protein-DNA cross-links, such as glutathione-DNA cross-links. (*Adapted from Eastman [99] with permission.*)

bromo-2′-deoxyuridine ([³H]BUdR). This compound is converted to the triphosphate form in living cells and is incorporated into DNA as BUdR monophosphate during semiconservative DNA synthesis (310) (Fig. 1–35). The analog imparts increased density to the regions (strands) of DNA in which it is present. Thus, if only one strand of a DNA duplex is density labeled, following denaturation of the DNA and sedimentation to equilibrium in a CsCl density gradient all fragments representing the unlabeled strand will have a normal (light) density and all those containing BUdR will have a heavy density. However, if some pieces of duplex DNA contain covalent intermolecular cross-links, these cross-links will prevent strand separation during denaturation and the DNA will sediment as molecules of intermediate density (17) (Fig. 1–35).

In addition, cross-links prevent complete separation of the two strands of the DNA duplex following exposure to denaturing agents, i.e., the DNA is rapidly renaturable, and this property can be measured by the differential binding affinity of single- and double-stranded DNA to hydroxylapatite. The reversibly denatured (cross-linked) DNA behaves as a duplex and can be separated from totally denatured (non-cross-linked) DNA by differential elution (135). Benzoylated-naphthylated DEAE-cellulose is another chromatographic matrix from which single- and double-stranded DNA can be differentially eluted (398).

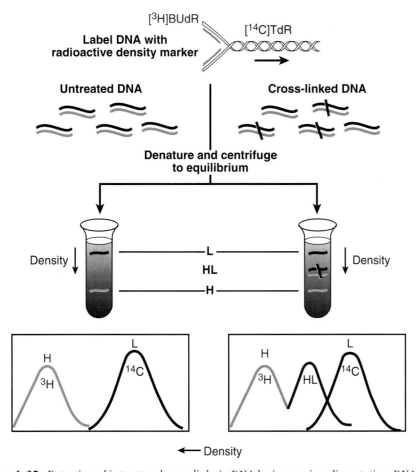

Figure 1–35 Detection of interstrand cross-links in DNA by isopycnic sedimentation. DNA uniformly labeled with [¹⁴C]thymidine ([¹⁴C]TdR) is replicated in the presence of [³H]BUdR to generate DNA of intermediate density in which one strand is light and the other is heavy. In the absence of cross-linking (left), denaturation of the DNA and sedimentation in alkaline cesium chloride yield ³H (heavy, H) and ¹⁴C (light, L) peaks of radioactivity. However, when the two DNA strands are cross-linked (right), DNA of intermediate density (HL) results. (*Adapted from Friedberg [131] with permission.*)

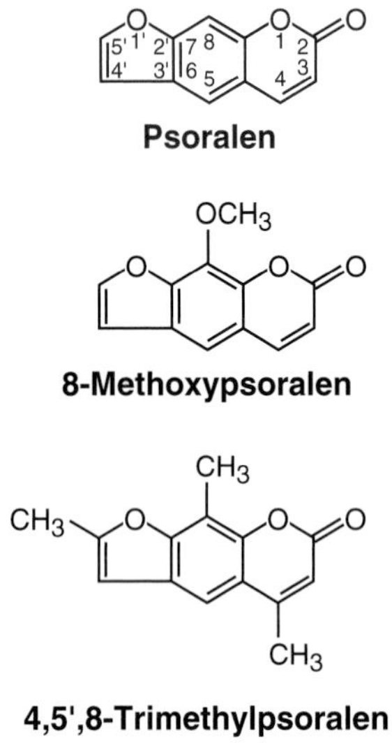

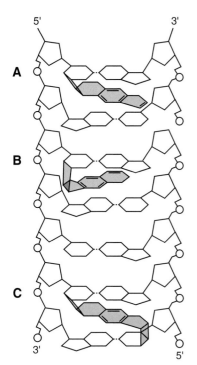

Psoralen

8-Methoxypsoralen

4,5',8-Trimethylpsoralen

Figure 1–36 Structures of psoralen and some psoralen derivatives.

Figure 1–37 Interaction of psoralen with DNA to form two types of monoadducts (A and B) or a diadduct (interstrand cross-link) (C). Two types of monoadducts can result because the 5,6 double bond of thymine can photoreact with psoralen at either its 3,4 double bond or its 4', 5' double bond (see Fig. 1–36 and 1–38). The formation of the cross-link requires independent UV absorption events at each reactive end. (*Adapted from Pathak et al. [303] with permission.*)

cis-Platinum (*cis*-diamminedichloroplatinum) is one of the most widely used chemotherapeutic agents in the treatment of a broad spectrum of tumors (120). The low intracellular chloride concentration induces loss of chloride from the compound, thereby converting the drug into a charged electrophilic agent (55, 99) (Fig. 1–34). Reaction with nucleophilic sites results in the formation of monoadducts, interstrand cross-links, and intrastrand cross-links (Fig. 1–34) (99, 317). The *trans*- isomer *trans*-platinum is therapeutically inactive. It may be important for this differential activity that *trans*-platinum is restricted in the types of intrastrand cross-links it can produce. Both compounds can also cause the formation of DNA-protein cross-links (99, 317).

We have mentioned the formation of DNA-protein cross-links on several occasions. After the exposure of living cells to a variety of DNA-damaging agents including UV radiation and ionizing radiation, the ease with which DNA can be extracted by deproteinizing procedures such as phenol-salt treatment is diminished (390). In addition, the rate of elution of DNA extracted from UV-irradiated human cells through nitrocellulose filters in alkali (alkaline elution of DNA; see chapter 7) is enhanced if the DNA is first incubated with proteases (215). These observations have been interpreted as evidence for the cross-linking of DNA to protein in vivo. This phenomenon has also been studied in vitro with purified DNA and proteins and with free bases and amino acids (369, 370). For example, the monofunctional alkylating agent β-propiolactone reacts with DNA principally at the N^7 position of guanine (31). Incubation of β-propiolactone with DNA and purified proteins results in DNA-protein cross-linking (296). A number of observations attest to the biological importance of DNA-protein cross-links (387, 390, 391). The study of the mechanism of formation of various DNA-protein interactions and their effects on DNA metabolism are important aspects of DNA damage that merit further exploration.

PSORALENS

Cross-links in DNA can also be formed during photosensitized reactions involving the photoaddition of certain planar compounds to the nitrogenous bases. Furocoumarins with planar tricyclic configurations (*psoralens*) (Fig. 1–36) can intercalate into DNA and, upon subsequent photoactivation by long-wavelength UV radiation, may form covalent adducts to the pyrimidines (144, 174, 303), principally by addition across the 5,6 double bond of thymine. Psoralens with a totally planar organization of the three aromatic rings (such as 8-methoxypsoralen) are able to react with pyrimidines at one or both ends to form monofunctional or difunctional adducts, respectively (Fig. 1–37). The latter join a pyrimidine in one DNA strand above the plane of the intercalated psoralen to an appropriately situated pyrimidine below in the other strand, thus cross-linking the two strands (144, 174, 303, 340). Such changes introduce significant helix distortion, kinking, and unwinding of DNA (306) (see Color Plate 6). This cross-linking reaction requires two independent UV absorption events. Some furocoumarins such as angelicin can form only monofunctional photoadducts, since the unreacted end is not appropriately juxtaposed with a pyrimidine in the native DNA helix, because of an angular arrangement of the three rings (144, 174, 303, 340) (Fig. 1–38).

The psoralen-plus-UV-light reaction is highly specific for native DNA, and the induced cross-links are stable to alkali. Psoralen monoadducts to pyrimidine bases are formed at ca. threefold-higher yield than are cross-links (65, 303, 378). However, the latter class of damage appears to be primarily responsible for biological inactivations induced by the treatment of bacteria, bacteriophages, and eukaryotic cells with photoactivated psoralen (13, 15, 38, 65, 144, 303, 378). For this reason, psoralen-induced cross-links have become a favored model for the study of the repair of interstrand DNA cross-links in a variety of biological systems (15, 38, 66, 67, 204, 262, 275, 346, 386). The ability of photoactivated psoralens to arrest DNA replication of dividing cells has also facilitated the use of these compounds in the treatment of the human disease psoriasis, a disease characterized by a marked proliferative disturbance of certain epithelial cells in the skin (213).

CHEMICALS THAT ARE METABOLIZED
TO ELECTROPHILIC REACTANTS

The study of chemical damage to DNA has led to the recognition that a variety of relatively nonpolar (and hence chemically unreactive) compounds undergo metabolic activation to more reactive forms, which, like typical alkylating agents, interact with nucleophilic centers in DNA (7, 69, 177). Many of these compounds are potent mutagens and carcinogens. Indeed, it was largely through their study as carcinogens that their metabolism by susceptible species was elucidated. For example, it was known for many years that some carcinogenic aromatic amines produce tumors at sites other than that of immediate administration (so-called *remote carcinogenesis*), such as the urinary bladder (331). The frequency of affliction of this particular organ led to the suggestion that the parent compounds were metabolized to water-soluble forms excreted in the urine (59). Another example was the observation that *N,N*-dimethyl-4-aminoazobenzene (butter yellow), a potent liver carcinogen in rats, did not itself bind to rat liver proteins, but a metabolite of this compound did (273).

It is now established that the metabolic activation of these compounds and of many other carcinogens, is effected by specific metabolizing enzymes in the affected cells (7, 69, 381, 427) (Fig. 1–39). The biological function of these enzyme systems is to protect the cell against cytotoxic effects by converting potentially toxic nonpolar chemicals to water-soluble excretable forms (7, 69, 427). While most of the products of these reactions do indeed enjoy this innocuous fate, some of them become activated to electrophilic forms that are particularly reactive with nucleophilic centers in organic macromolecules such as DNA (Fig. 1–39) (7, 69, 177, 381). Thus, while no longer directly cytotoxic, these agents have been converted to potent genotoxic forms. A well-studied example of activating enzymes is the inductive response of a series of membrane-bound proteins containing numerous monooxygenase activities (Fig. 1–39). Because of its strong absorbance at about 450 nm, this complex, in combination with one or more membrane-bound flavoprotein reductases is frequently referred to as the *cytochrome P-450 system* (68, 179, 320, 427).

These multicomponent enzyme systems require NADPH and atmospheric oxygen. An initial reaction sequence usually metabolizes hydrophobic nonpolar substrates to more polar oxygenated intermediates and products, which are sub-

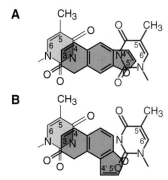

Figure 1–38 Projection of psoralen (A) and angelicin (B) molecules intercalated between two base pairs in DNA. In each case the thymines shown are on opposite strands of the DNA duplex. Note that angelicin cannot cross-link two DNA strands, because one end of the molecule is not appropriately juxtaposed with one of the thymines as a result of its angular configuration. (*Adapted from Musajo et al. [290] with permission.*)

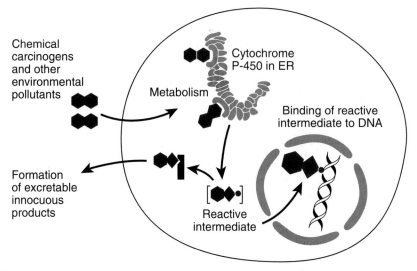

Figure 1–39 Scheme for the metabolic activation of nonpolar polycyclic chemicals by the cytochrome P-450 system in a typical mammalian cell to form reactive intermediates that bind to nucleophilic centers in DNA. ER, endoplasmic reticulum. (*Adapted from Nebert [294] with permission.*)

strates for secondary reactions with enzymes that catalyze the formation of conjugates (frequently esters) that are rapidly excreted from the cell and from the body (Fig. 1–39). Other examples of enzymes involved in activating genotoxic chemicals are microsomal and cytoplasmic glutathione-*S*-transferases, sulfotransferases, acetyltransferases, UDP-glucuronosyltransferases, and adenosylating as well as methylating enzymes (7). The following paragraphs detail specific examples of some of these metabolic activation reactions.

N-2-Acetyl-2-Aminofluorene

N-2-Acetyl-2-aminofluorene (AAF) belongs to a class of compounds known as the *aromatic amines,* many of which are associated with an increased incidence of cancer in humans. The first step in the metabolic activation of this compound (originally used as an insecticide) is the cytochrome P-450-catalyzed formation of an *N*-hydroxy derivative (22, 274) (Fig. 1–40). This intermediate (called a *proximate carcinogen*) is relatively unreactive with nucleic acids, but following formation of

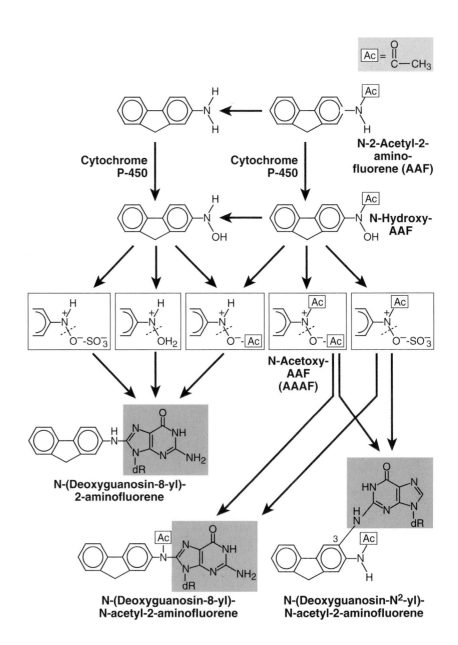

Figure 1–40 Metabolic activation of AAF proceeds through the formation of *N*-hydroxy intermediates before the formation of *N*-acetoxy-AAF and other esterified forms. These compounds are highly reactive with the C-8 (left and middle) and to a lesser extent the N^2 (right) positions of guanine in DNA. (*Adapted from Miller and Surh [274].*)

electrophilic metabolites (e.g., sulfate or acetate ester) by cytosolic enzymes, it becomes a highly reactive alkylating agent (22, 114, 274). Thus, for example, *N*-acetoxy-*N*-2-acetylaminofluorene (an *ultimate carcinogen*) reacts readily with guanine residues in nucleosides and nucleic acids at the C-8 and N^2 positions to yield *N*-(deoxyguanosin-8-yl)-*N*-acetyl-2-aminofluorene as the major component and *N*-(deoxyguanosin-8-yl)-2-aminofluorene as well as 3-(deoxyguanosin-N^2-yl)-*N*-acetyl-2-aminofluorene as minor components in DNA respectively (22, 223, 274, 454) (Fig. 1–40).

Normally the C-8 position of guanine in duplex DNA in the B conformation is relatively hindered sterically and is rather inaccessible to bulky adducts such as activated AAF (163). However, rotation of the base around the *N*-glycosylic bond from the *anti* to the *syn* conformation allows for attack at the C-8 position of guanine in B-DNA (163). During this conformational shift, the modified guanine residue can be displaced from its normal coplanar relationship to adjacent bases and its position can be subsumed by the fluorene (so-called *base displacement*) (163). When DNA is in the Z conformation, the C-8 position of guanine is highly accessible to reaction with activated AAF, because deoxyguanosine in Z-DNA exists in the *syn* conformation while base paired to deoxycytidine (440). Since AAF modification of deoxyguanosine in DNA also causes the rotation of the guanine from the *anti* to the *syn* conformation, it has been suggested that if stretches of DNA (especially alternating G·C base pairs) are modified by AAF, this might favor localized transitions to the Z form (163, 239, 347).

Benzo[*a*]pyrene

An association between cancer and hydrocarbon exposure goes back historically nearly 200 years, when Percival Pott, an English surgeon, observed the remarkable correlation between cancer of the skin of the scrotum and the occupation of chimney sweeping. Chimney sweeps in those times were exposed to very high levels of polycyclic aromatic hydrocarbons derived from the combustion of coal in the presence of air. Pott's observation provided the basis for the first preventive measures against environmental cancer, since, shortly after his publication in 1775, the Danish chimney sweepers' guild urged its members to take daily baths and by 1892 a lower incidence of scrotal cancer in Northern European than in English chimney sweeps was noted (42). It was not until some years later, however, that a very potent carcinogenic polycyclic aromatic hydrocarbon called *benzo[a]pyrene* was identified in and isolated from crude coal tar (206, 207, 313). Despite the subsequent identification of a large number of other carcinogens from industrial products, including coal tar, benzo[*a*]pyrene remains to this day one of the most highly carcinogenic compounds known. Exposure to coal tar is no longer the public health hazard it represented at the time of the industrial revolution in England. However, exposure to benzo[*a*]pyrene from sources such as cigarette smoke and automobile exhaust fumes is still highly prevalent in the environment (3).

Unmodified benzo[*a*]pyrene is an unreactive nonpolar compound with a planar configuration (Fig. 1–41). This configuration facilitates its intercalation between the hydrogen-bonded base pairs in duplex DNA, and this was initially thought to be a likely mode of DNA damage by this compound. It is now known that components of the P-450 system known as *arylhydrocarbon hydroxylases* can metabolize benzo[*a*]pyrene and other polycyclic aromatic hydrocarbons to phenols and dihydrodiols, which, together with their corresponding ester conjugates, are excretable (Fig. 1–41) (165, 242, 313, 358). However, some of the products of benzo[*a*]pyrene metabolism are electrophilic epoxides, and it is well established that the ultimate carcinogenic form of this hydrocarbon is an *anti* diol-epoxide called 7β,8α-diol-9α,10α-epoxy-7,8,9,10-tetrahydrobenzo[*a*]pyrene (165, 358) (Fig. 1–41).

Following a noncovalent interaction by intercalation into DNA (146), the C-10 position of the benzo[*a*]pyrene *anti* diol-epoxide (BPDE) binds predominantly to the exocyclic 2-amino position of guanine (71, 160, 358, 453) (Fig. 1–42). This product has been purified, and examination by nuclear magnetic resonance spec-

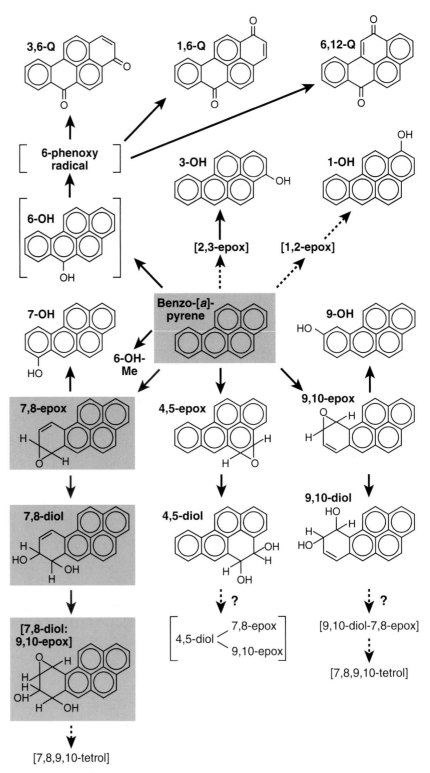

Figure 1–41 Metabolic products of the metabolism of benzo[a]pyrene by microsomal mixed-function oxygenases. Some of the products are electrophilic epoxides that have high reactivity for nucleophilic centers in DNA. The 7,8-diol-9,10-epoxide is thought to be the ultimate carcinogenic form of benzo[a]pyrene. (*Adapted from Selkirk et al. [358] with permission.*)

troscopy indicated only minimal perturbation of the B-DNA helix, with the benzo[a]pyrenediol residue being well aligned with the minor groove (71). BPDE interacts nonrandomly with DNA; long tracts of guanines are preferentially modified (26, 417).

Aflatoxins

Aflatoxins are among the most potent liver carcinogens known and represent an interesting example of DNA-damaging agents that have their origin as products of natural metabolism (162). Aflatoxins are mycotoxins produced by the fungi *Aspergillus flavus* and *Aspergillus parasiticus*, and human and animal exposure is usually a result of contamination of food, particularly peanuts, by these fungi (162). Chemically, the aflatoxins consist of a difurofuran ring system fused to a substituted coumarin moiety, with a methoxy group attached at the corresponding benzene ring (Fig. 1–43) (162). Among the aflatoxins of fungal origin, aflatoxin B_1 (Fig. 1–43) is the most potent hepatocarcinogen.

Aflatoxin B_1 is known to be oxidized by the mixed function oxygenases of the P-450 system present in the microsomal fraction of liver extracts, giving rise to aflatoxin B_1-8,9-epoxide as the major product (162, 181) (Fig. 1–44). The use of DNA substrates of known nucleotide sequence suggests that certain guanine residues in double-stranded DNA are preferentially attacked by this reactive electrophilic epoxide, in a manner predictable from a knowledge of adjacent nucleotides (23, 288). The major product is the N^7 adduct 8,9-dihydro-8-(N^7-guanyl)-9-hydroxyaflatoxin B_1 (112, 162) (Fig. 1–44). Mildly alkaline conditions can subsequently result in the formation of a formamidopyrimidine derivative containing an opened imidazole ring (Fig. 1–17) (162).

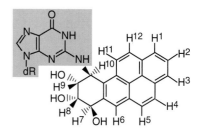

Figure 1–42 Adduct formation of the anti-benzo[a]pyrene dihydro-diolepoxide (7,8-diol-9,10-epoxide; see Fig. 1–41) (BPDE) with the exocyclic amino group of deoxyguanosine. (*Adapted from Cosman et al. [71].*)

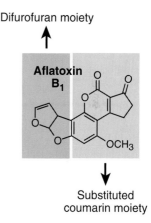

Figure 1–43 Chemical structure of aflatoxin B_1.

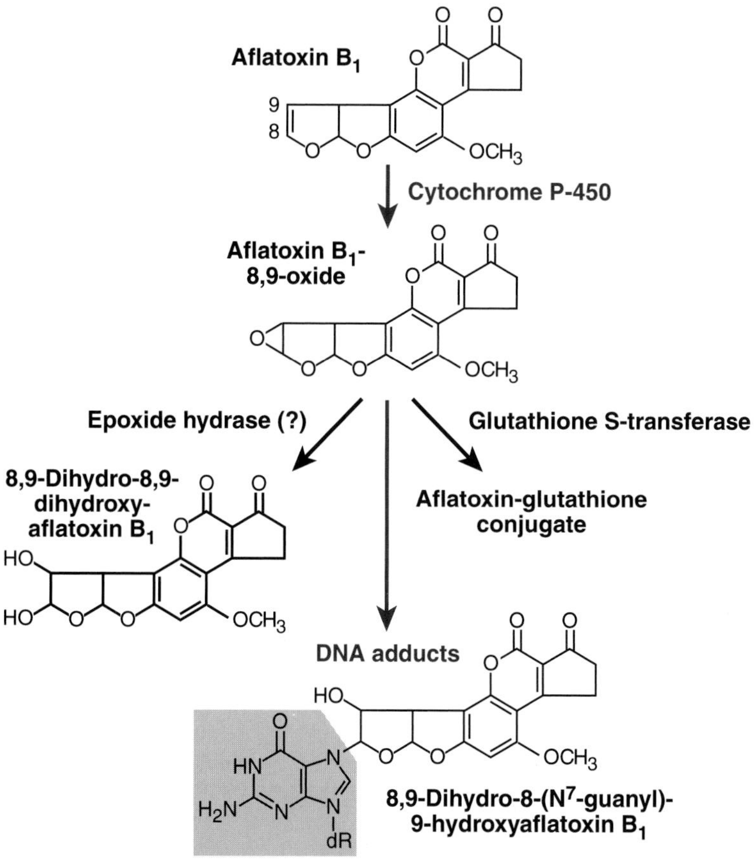

Aflatoxin B$_1$

Cytochrome P-450

Aflatoxin B$_1$-8,9-oxide

Epoxide hydrase (?)   Glutathione S-transferase

8,9-Dihydro-8,9-dihydroxy-aflatoxin B$_1$

Aflatoxin-glutathione conjugate

DNA adducts

8,9-Dihydro-8-(N^7-guanyl)-9-hydroxyaflatoxin B$_1$

Figure 1–44 Metabolic activation and detoxification pathways for aflatoxin B_1. (*Adapted from Groopman and Cain [162] with permission.*)

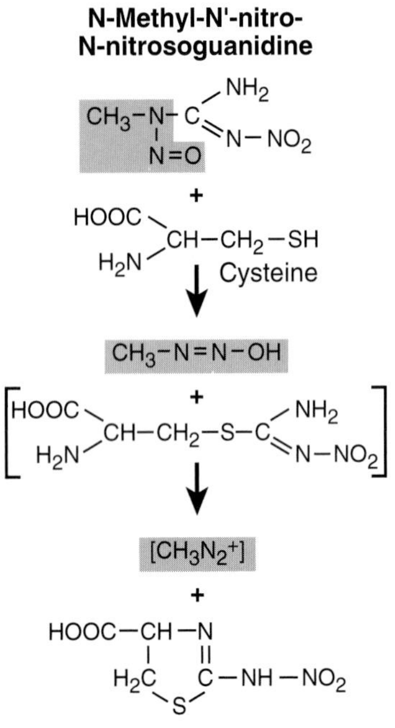

**N-Methyl-N'-nitro-
N-nitrosoguanidine**

Figure 1–45 Formation of a strong electrophilic agent by reaction of MNNG with a cysteine residue, typically part of glutathione. (*Adapted from Lawley [236] with permission.*)

N-Methyl-N'-Nitro-N-Nitrosoguanidine

Glutathione (γ-glutamylcysteinglycine) is a major intracellular sulfhydryl component. We have already mentioned its role as a scavenger of free radicals in cells. In its reduced state, glutathione is strongly nucleophilic and reacts with electrophiles. These conjugation reactions are catalyzed by various cytosolic or microsomal glutathione-S-transferases (7). Glutathione conjugation is in general considered to be a detoxification reaction (aflatoxin B_1 is an example [Fig. 1–44]), facilitating the excretion of xenobiotics, frequently after previous oxidation by P-450. However, an increasing number of studies suggest that glutathione conjugation can be potentially genotoxic when it involves the activation of carcinogenic agents such as dihaloalkane (e.g., ethylene dichloride), half-mustards, or N-nitroso compounds (nitrosamines) (164, 236, 381). An example of the activation of an N-nitroso compound by a thiol group is shown in Fig. 1–45. Reaction of N-methyl-N'-nitro-N-nitrosoguanidine (MNNG) with a cysteine residue as found in glutathione gives rise to a highly electrophilic methylating intermediate (236). The isolation of glutathione-deficient yeast mutants by virtue of their resistance to the toxic effects of MNNG attests to the physiological relevance of this activation pathway (209). We will discuss the products of the reaction of activated MNNG with DNA in later chapters.

4-Nitroquinoline 1-Oxide

DNA damage induced by *4-nitroquinoline 1-oxide* (4-NQO) is often referred to as "UV radiation-like" because this chemical produces bulky base damage of the type that, like cyclobutane pyrimidine dimers and (6-4) photoproducts, is repaired principally by the nucleotide excision repair system (see chapters 5, 6, and 8). Metabolic activation of and DNA adduct formation by 4-NQO are not entirely understood. 4-NQO is first converted to the proximate carcinogen 4-hydroxyaminoquinoline 1-oxide. This component apparently is further activated by a reaction with seryl-tRNA-synthetase in which 4-hydroxyaminoquinoline 1-oxide instead of tRNA is acylated by the seryl-AMP enzyme complex (381). This complex then introduces quinoline groups into DNA. Adducts at C-8 of guanine but also at the exocyclic N^2 of guanine and N^6 of adenine have been identified (140, 141) (Fig. 1–46). The N^2 adduct appears to be the major lesion that accounts for 50 to 80% of all quinoline base adducts, depending on the superhelicity of the target (272). 4-NQO treatment can also result in the formation of 8-hydroxyguanine (214) and leads to a significant amount of strand breakage, probably indicative of the formation of unstable adducts (141).

BASE ANALOGS

In general, the formal possibility exists that biologically relevant base damage does not originate from the direct reaction of chemical agents such as alkylating agents with DNA itself but, rather, from the incorporation of damaged deoxynucleotides during replication of undamaged templates (420). Such a mechanism has been observed in vitro. Apart from the incorporation of certain oxidized deoxynucleotides, the available data are not conclusive enough to assume its relevance as a mechanism in vivo (392).

Certain analogs of the four naturally occurring bases in DNA can be incorporated from the appropriate triphosphate substrates during DNA replication. These are mentioned here for the sake of completeness, since they are used experimentally in studies of DNA damage and of the cellular responses to such damage. Among the most extensively studied base analogs are the halogenated uracil derivatives, 5-bromouracil, 5-fluorouracil, and 5-iodouracil (Fig. 1–47), all of which are thymine analogs that can result in mutations when present in template DNA undergoing DNA replication (127-129, 393). The adenine analog, 2-aminopurine (Fig. 1–47) is also mutagenic (337, 348, 393).

DNA Damage and Chromatin Structure

This chapter has been devoted very largely to a consideration of damage to bases in DNA, without regard to higher levels of organization of DNA molecules in the genomes of living cells. However, in eukaryotes DNA is associated with both histone and nonhistone chromosomal proteins to form chromatin (221, 222, 415, 430). The relevance of this organization to DNA repair is discussed in detail in chapter 7. However, some repetition is warranted here to emphasize the importance of chromatin in the context of damage to DNA, especially by relatively inaccessible chemicals.

The association of DNA with histones results in their organization into repeating units called nucleosomes, which consist of a core of 140 bp of DNA wrapped around an octamer of histones: two each of the histones H2A, H2B, H3 and H4 (see Fig. 7–6 and 7–7) (221, 222, 415, 430). Variable (in different organisms) amounts of DNA ranging from 20 to 60 bp are referred to as linker DNA, which is loosely associated with another histone, H1 (221, 222, 415, 430). There is evidence that in living cells the nucleosomes are organized into further levels of folding to provide the structure of chromosomes. The highly compact organization of DNA in chromosomes accounts for the packaging of 5.3×10^9 bp of DNA, corresponding to a contour length of 180 cm, into 46 chromosomes whose total contour length is 200 mm in a human diploid cell (401).

DNA in linker regions is preferentially sensitive to digestion by nucleases such as micrococcal and *Staphylococcus aureus* nucleases. The frequency of damage per unit amount of DNA can be measured before and after selective removal of linker DNA, and thus the distribution of lesions among core and linker regions can be determined. Many studies have indicated that certain forms of damage are not random in their distribution in the nucleosome. Base damage produced by a variety of chemicals occurs selectively in linker regions (169, 243, 287, 357, 418).

In certain cases this may reflect an accessibility problem, but often the reactive groups in bases must be considered to be as accessible in nucleosomes as in naked DNA. However, protein-DNA contacts may also restrict conformational changes required for the formation of certain base adducts. This explanation has been invoked to explain the nonrandom distribution of damage induced by physical agents such as UV radiation. The distribution of the major species of damage, i.e., cyclobutane dimers and (6-4) photoproducts, has been found to be quite different. Whereas (6-4) products are formed preferentially in linker regions (280), there are no indications for a nonrandom distribution of cyclobutane dimers (280, 297, 458). This may be explained by the fact that (6-4) product formation requires more severe helix distortion than does the formation of cyclobutane dimers (409, 410). Consequently, a restriction of conformational changes by protein binding can be expected to have a more significant influence.

When damage formation within the nucleosomal core is investigated, a different picture emerges. To this end, core regions of UV-irradiated chromatin fibers were isolated and the extracted DNA was labeled at the 5′ end, denatured, and treated with T4 polymerase 3′ → 5′ exonuclease without any deoxynucleoside triphosphates added (138) (Fig. 1–48). Under these conditions, this enzyme will digest untreated DNA completely, starting from the unlabeled 3′ end. This activity is aborted at dimers (88), and the length of the resulting fragments corresponds to the position of a lesion in relation to the nucleosomal core border (88). The intensity of the electrophoretically separated bands reflects the photoproduct frequency at this particular position. The observed pattern indicates a periodicity of high susceptibility to photoproduct formation of 10.3-bp periodicity (138). This pattern seems to reflect the periodicity of maximum exposure of the DNA backbone in the nucleosome core structure and corresponds to the periodicity of DNase I susceptibility. In other words, the farther a DNA region is situated from the histone core, the more susceptible is this region to photoproduct formation. Initial experiments did not discriminate among different photoproducts. Subsequently, it was found that this pattern was largely abolished if cyclobutane dimers were selectively removed by DNA photolyase treatment (see chapter 3) (139). It must be concluded that the formation of the cyclobutane dimer, but not of the (6-4) product or other photoproducts, results in the observed periodicity. The (6-4) product appears to be formed more randomly and is less affected by histone binding, a result not anticipated from the previously discussed influence of protein

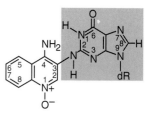

3-(Deoxyguanosin-N²-yl)-4-aminoquinoline-1-oxide

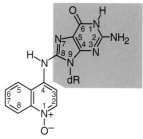

N-(Deoxyguanosin-C8-yl)-4-aminoquinoline-1-oxide

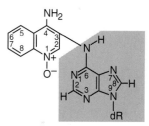

3-(Deoxyadenosin-N⁶-yl)-4-aminoquinoline-1-oxide

Figure 1–46 The major DNA adducts of 4-nitroquinoline 1-oxide (4-NQO). (*Adapted from Galiègue-Zouitina et al. [140, 141] with permission.*)

5-Bromo-(fluoro)-(iodo)uracil   **2-Aminopurine**

Figure 1–47 Some base analogs that can be incorporated into DNA.

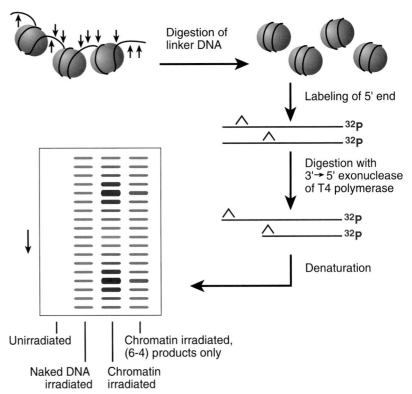

Figure 1–48 The nucleosome structure influences the probability of UV-induced cyclobutyl dimer formation. Nucleosome fibers are isolated, and linker DNA is digested with micrococcal nuclease. The remaining nucleosomal core DNA is 5′ labeled and subjected to digestion with the 3′ → 5′ exonuclease activity of T4 polymerase. The exonuclease digest arrests at photodimer positions, and the resulting mixture of end-labeled fragments is separated in a polyacrylamide gel. Untreated DNA is completely digested. The band intensity reflects the probability of photoproduct formation in a defined distance from the nucleosomal core border (however, it does not provide information about sequence preferences, since each nucleosomal core will contain a different DNA sequence). If the chromatin fibers and not the naked DNA were irradiated with UV, the photoproduct distribution shows a 10.3 bp periodicity. This periodicity is largely abolished if the cyclobutyl dimers had been selectively removed by treatment with photolyase before exonuclease digestion.

binding on the distribution of different photoproducts among core and linker regions.

While protein binding does affect photoproduct formation differentially, simple general rules cannot be derived. Consider the binding of transcription factors to promoter regions. Susceptibility to UV radiation-induced photoproducts has been studied in vivo in a regulated promoter under induced and uninduced conditions (i.e., with transcription factors bound or unbound) and compared with the pattern observed following irradiation of naked DNA (312). Transcription factor binding can indeed dramatically alter the frequency of photoproduct formation at certain potential dimer sites. It can create damage hot spots but can also render certain positions hypoactive (312). If the classes of photoproducts are compared, opposing effects are found quite frequently; e.g., creation of a cyclobutane dimer hotspot by protein binding will often result in the reduced formation of (6-4) photoproducts at the same site (312).

The influence of chromatin structure and protein binding on the concentration and distribution of various forms of DNA damage is an important consideration that merits further detailed study. As will be seen in later chapters, this phenomenon has significant bearing on the responses of living cells to damage, since there is evidence that sites of base damage in chromatin are not equally accessible to DNA repair enzymes. If identical rules govern the preferential formation of a particular species of damage and accessibility for repair enzmes (which can easily be imagined for linker regions), such damage will exhibit fast repair kinetics.

Detection and Measurement of Base Damage in DNA

The discussion of DNA damage in this chapter has included brief descriptions of selected forms of damage and their detection and quantification. This topic has not been extensively covered, partly because of space constraints but primarily because the essential focus of this book is the molecular biology of DNA damage and the responses that it elicits in living cells. Technical and methodologic descriptions are available from other sources in the literature (133). Nonetheless, the issues of the quantitation of DNA damage and the sensitivity of its detection are important. Their importance relates partly and quite obviously to the furthering of information on DNA damage and its repair in general. However, in addition it would be extremely valuable to establish accurate and meaningful quantitation of genotoxicity (i.e., the amount of DNA damage at any given time) in human populations in general and in selected populations after specific genotoxic exposure in particular. The following is intended as a brief overview of the different classes of methods, many of which will be discussed in more detail in conjunction with particular repair phenomena considered in later chapters.

Enzyme Probes for Measuring DNA Damage

The principle of the use of enzyme probes lies in the conversion of base damage into easily detectable physical DNA alterations, typically strand breaks. For example, uracil and hypoxanthine originating in DNA from the deamination of cytosine and adenine, respectively, as well as a number of other forms of base damage to DNA discussed in this chapter, can be detected by specific enzyme probes belonging to the class of DNA glycosylases, which recognize these bases in DNA and catalyze their excision by hydrolysis of the relevant N-glycosylic bonds (96, 245) (see chapter 4). In this way, sites of base damage are converted to sites of base loss, which in turn can be translated into DNA strand breaks either by virtue of their alkali lability or by a different class of enzymes that specifically recognize sites of base loss in DNA. The latter enzymes, called AP (apurinic/apyrimidinic) endonucleases, are also discussed in detail in chapter 4. Thus, the sensitivity for the detection of those forms of base damage that can be converted to DNA strand breaks is limited only by the sensitivity with which the latter can be detected and quantitated. By the judicious selection of appropriate DNA molecules for study, the detection of one strand break in some entire genomes is technically feasible. For example, one strand break in a form I (covalently closed circular) DNA molecule results in the relaxation of the molecule to an open circular (form II) configuration. Form I and II DNA molecules can be readily separated by electrophoresis in agarose gels (356) or by sedimentation (349) and can also be distinguished by visualization in the electron microscope (61). As described for the akali-labile (6-4) photoproduct (see Fig. 1–24), the precise site of damage and subsequent strand breakage following in vitro treatments of a defined DNA target can be determined by a comparison of the length of the resulting fragment with the corresponding bands in a sequencing gel (457). In vitro, arrest of polymerization processes or exonuclease digestion (Fig. 1–48) at damaged sites can also be used as a method for detection of damage for a broader range of adducts (88, 282, 283, 344, 457). In vivo, Southern hybridization (25, 416) or PCR based techniques (311) allow one to quantitate the frequency of strand breaks introduced by damage-recognizing enzymes in a given gene, even at a given site.

Other repair enzymes (discussed in chapters 5, 6, and 8) are even more useful, since they recognize a broad spectrum of DNA base damage (416). In addition, some nucleases not involved in DNA repair but more probably involved in DNA recombination, are specific for single stranded DNA and may even be able to recognize "denatured" regions created by interrupted hydrogen bonding in a single base pair. Examples are the so-called S1 nuclease from *Aspergillus orzyae* referred to earlier (269) and BAL 31 nuclease from *Alteromonas espejani* (450).

Postlabeling Methods for Detecting Base Damage

The classic approach to the detection of damaged bases in DNA typically includes the treatment of cells or animals with a radioactively labeled compound, isolation and hydrolysis of the radiolabeled DNA, and characterization of the base adducts by chemical methods. However, these methods are severely limited in sensitivity. Hence, high doses are usually needed, and such an approach is of course impossible for detecting low levels of base damage in human cells after occupational exposure to a DNA-damaging agent. Postlabeling and immunoassays have become the methods of choice for these and related investigations (314, 315, 400). For a postlabeling assay, a small amount of DNA is isolated and digested to completion with micrococcal nuclease and spleen phosphodiesterase, yielding deoxynucleoside 3'-monophosphates (Fig. 1–49). These are then labeled with ^{32}P at their 5' end with T4 polynucleotide kinase and separated by chromatographic methods (thin-layer chromatography, HPLC) (Fig. 1–49). The resulting pattern is compared with that produced by untreated samples. Defined adducts can serve as standards for identifying radioactive peaks or spots not found in untreated controls. Several refinements aimed at an enrichment of damaged nucleotides before labeling have increased the sensitivity of this method (315, 400). One important refinement is the inclusion of a nuclease P1 digest before labeling (329). This nuclease dephosphorylates deoxynucleoside 3'-monophosphates but frequently does not accept modified bases. Deoxynucleosides without 3' phosphates are not substrates for

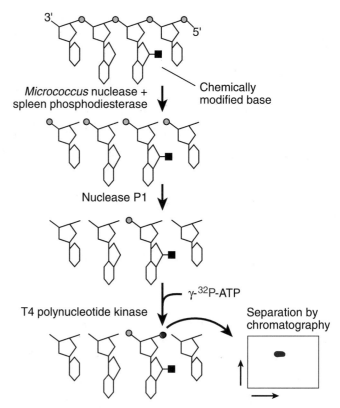

Figure 1–49 Scheme of the ^{32}P-postlabeling assay for the detection of base adducts. Isolated DNA is digested to completion with micrococcal nuclease and spleen phosphodiesterase. The resulting deoxynucleoside 3' monophosphates are further digested with *Penicillium citrinum* nuclease P1, yielding deoxyribonucleosides. However, many DNA base adducts (in particular bulky or aromatic adducts) are resistant to this cleavage. These are subject to the subsequent labeling reaction in which T4 polynucleotide kinase transfers a ^{32}P-labeled phosphate residue from γATP to the 5' OH group of 3' monophosphates, however, any dephosphorylated deoxyribonucleosides are not substrate for this enzyme. The labeled products are then separated by thin-layer chromatography and detected by autoradiography.

phosphorylation by T4 polynucleotide kinase. Consequently, base adducts which were not dephosphorylated in the previous step will be preferentially labeled (Fig. 1–49). The method allows the detection of 1 adduct per 10^{10} nucleotides. The identification of base adducts in peripheral lymphocytes and placentas of human smokers are among the impressive achievements realized by this technology (113, 196).

Specific Antibodies for Measuring DNA Damage

Damaged DNA or defined base adducts coupled to proteins can elicit an immune response in animals. Polyclonal and monoclonal antibodies raised against numerous forms of base damage have been characterized (289, 314, 456). UV radiation damage is an important example (101, 276, 277, 279, 285, 399), as evidenced by the fact that immundetection has facilitated the study of the specific repair of (6-4) photoproducts. Figure 1–50 shows the characterization of such an antiserum. Polyclonal antibodies were raised against DNA irradiated with 254-nm light [which produces cyclobutane dimers and (6-4) products] or 320-nm light in the presence of acetophenone (which produces exclusively cyclobutane dimers) (277). Antiserum raised against 254-nm-induced damage had a higher affinity for such damage than for that induced by 320-nm radiation, whereas antiserum raised against 320-nm-induced damage did not discriminate between them. This implies that (6-4) products are more immunogenic than cyclobutane dimers. The antiserum is in fact highly specific for (6-4) products (276, 277).

Apart from providing a specific and highly sensitive method of detecting base damage, other applications of the immunodetection of base damage are evident. Binding of a second fluorescent antibody directed to the adduct-specific antibody in fixed cells (indirect immunofluorescence) allows visualization of damaged sites in situ (314, 318), and flow-cytometric analysis can provide information on the distribution of damage among cell populations (284, 314).

Summary and Conclusions

This chapter has reviewed a number of mechanisms of damage to DNA. In living cells which have sustained such damage, a variety of cellular responses can ensue, many of which apparently do not operate in cells without damaged DNA. In general, we know the most about those forms of DNA damage for which specific cellular responses have been identified, particularly responses that eliminate the damage. Indeed, a historically fruitful approach to the discovery of new DNA repair mechanisms has been first to identify a particular form of base damage in DNA and then to ask whether living cells can remove that damage and, if so, how? Frequently, the discovery of a new repair enzyme or biochemical pathway for repair provides the incentive to characterize and quantitate the substrate (damaged DNA) in even greater detail. A particularly good example of this relationship between the study of DNA damage and of DNA repair is the cyclobutane dimer. Additionally, the discovery of enzymes that catalyze the removal of base damage produced by ionizing radiation has focused interest in defining these forms of damage more precisely.

DNA damage is an inescapable aspect of life in the biosphere. Of particular importance is damage caused by agents that have been present in the environment long enough to have provided selective pressures for the evolution of mechanisms for its repair and tolerance. It is likely that there are biological responses to all forms of such damage, and their continued exploration may be expected to uncover further examples of DNA repair and damage tolerance mechanisms. Non-repairable lesions in DNA are also very important, since these are most likely to be mutagenic and/or lethal. In this regard, the many synthetic products of modern technology merit special attention. Many of these may produce types of DNA damage that are recognized by evolutionarily long-established repair mechanisms. However, many may not, and thus a detailed understanding of their genotoxic potential is vital to the continued health of all living forms on this planet.

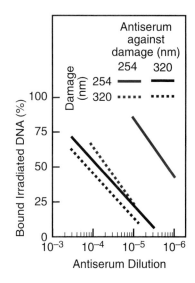

Figure 1–50 Characterization of an antiserum specific for (6-4) photoproducts. Polyclonal antibodies were raised against DNA irradiated with 254-nm or 320-nm UV light in the presence of acetophenone (and thus inducing cyclobutane dimers only). The antiserum concentration necessary to detect the same amount of binding of radioactively labeled DNA irradiated with 254- or 320-nm radiation is determined. The "anti-254-nm damage" serum (red lines) has a higher affinity for 254-nm damage than for 320-nm damage; the "anti-320-nm damage" serum (black lines) has the same (but lower) affinity for both. From this (and other experiments) it was concluded that the increase in affinity toward 254-nm-induced damage is due to the presence of (6-4) photoproduct-specific antibodies in the antiserum raised against DNA irradiated at 254 nm. (*Adapted from Mitchell et al. [277] with permission.*)

REFERENCES

1. **Abbotts, J., and L. A. Loeb.** 1985. On the fidelity of DNA replication: use of synthetic oligonucleotide-initiated reactions. *Biochim. Biophys. Acta* **824:**58–65.

2. **Adelman, R., R. L. Saul, and B. N. Ames.** 1988. Oxidative damage to DNA: relation to species metabolic rate and life span. *Proc. Natl. Acad. Sci. USA* **85:**2706–2708.

3. **Albert, R. E., and F. J. Burns.** 1977. Carcinogenic atmospheric pollutants and the nature of low-level risks, p. 289–292. *In* H. H. Hiatt, J. D. Watson, and J. A. Winston (ed.), *Origins of Human Cancer. A.* Cold Spring Harbor Laboratory, Cold Spring Harbor, N.Y.

4. **Ames, B. N.** 1989. Endogenous oxidative DNA damage, aging, and cancer. *Free Radical Res. Comm.* **7:**121–128.

5. **Ames, B. N., and L. S. Gold.** 1991. Endogenous mutagens and the causes of aging and cancer. *Mutat. Res.* **250:**3–16.

6. **Ames, B. N., M. K. Shigenaga, and T. M. Hagen.** 1993. Oxidants, antioxidants, and the degenerative diseases of aging. *Proc. Natl. Acad. Sci. USA* **90:**7915–7922.

7. **Anders, M. W., and W. Dekant.** 1994. Conjugation-dependent carcinogenicity and toxicity of foreign compounds. *Adv. Pharmacol.* **27:**1–519.

8. **Andrews, J., H. Martin-Bertram, and U. Hagen.** 1984. S1 nuclease-sensitive sites in yeast DNA: an assay for radiation-induced base damage. *Int. J. Radiat. Biol.* **45:**497–504.

9. **Ariga, H., and H. Shimojo.** 1979. Incorporation of uracil into the growing strand of adenovirus 12 DNA. *Biochem. Biophys. Res. Commun.* **87:**588–604.

10. **Arnott, S., R. Chandrasekaran, D. L. Birdsall, A. G. W. Leslie, and R. L. Ratliff.** 1980. Left-handed DNA helices. *Nature* (London) **283:**743–745.

11. **Aruoma, O. I., B. Halliwell, and M. Dizdaroglu.** 1989. Iron ion-dependent modification of bases in DNA by the superoxide radical generating system hypoxanthine/xanthine oxidase. *J. Biol. Chem.* **264:**13024–13028.

12. **Aruoma, O. I., B. Halliwell, E. Gajewski, and M. Dizdaroglu.** 1989. Damage to the bases in DNA induced by hydrogen peroxide and ferric ion chelates. *J. Biol. Chem.* **264:**20509–20512.

13. **Ashwood-Smith, M. J., and E. Grant.** 1977. Conversion of psoralen DNA monoadducts in *E. coli* to interstrand crosslinks by near UV light (320–360 nm). *Experientia* **33:**384–386.

14. **Auclair, C., A. Gouyette, A. Levy, and I. Emerit.** 1990. Clastogenic inosine nucleotides as components of the chromosome breakage factor in scleroderma patients. *Arch. Biochem. Biophys.* **278:**238–244.

15. **Averbeck, D.** 1989. Recent advances in psoralen phototoxicity mechanism. *Photochem. Photobiol.* **50:**859–882.

16. **Badwey, J. A., and M. L. Karnovsky.** 1986. Production of superoxide by phagocytic leukocytes: a paradigm for stimulus-response phenomena. *Curr. Top. Cell. Regul.* **28:**183–208.

17. **Ball, C. R., and J. J. Roberts.** 1971. Estimation of interstrand DNA cross-linking resulting from mustard gas alkylation of HeLa cells. *Chem.-Biol. Interact.* **4:**297–303.

18. **Baltz, R. H., P. M. Bingham, and J. W. Drake.** 1976. Heat mutagenesis in bacteriophage T4: the transition pathway. *Proc. Natl. Acad. Sci. USA* **73:**1269–1273.

19. **Becker, M. M., and Z. Wang.** 1989. Origin of ultraviolet damage in DNA. *J. Mol. Biol.* **210:**429–438.

20. **Beer, M., S. Stern, D. Carmalt, and K. H. Mohlenrich.** 1966. Determination of base sequences in nucleic acids with the electron microscope. V. The thymine-specific reactions of osmium tetroxide with deoxyribonucleic acid and its components. *Biochemistry* **5:**2283–2288.

21. **Behe, M., and G. Felsenfeld.** 1981. Effects of methylation on a synthetic polynucleotide: the B-Z transition in poly(dG- m5dC)·poly (dG-m5dC). *Proc. Natl. Acad. Sci. USA* **78:**1619–1623.

22. **Beland, F. A., and F. F. Kadlubar.** 1990. Metabolic activation and DNA adducts of aromatic amines and nitroaromatic hydrocarbons, p. 267–325. *In* C. S. Cooper and P. L. Grover (ed.), *Chemical Carcinogenesis and Mutagenesis I.* Springer-Verlag, KG, Berlin.

23. **Benasutti, M., S. Ejadi, M. D. Whitlow, and E. L. Loechler.** 1988. Mapping the binding site of aflatoxin B_1 in DNA: systematic analysis of the reactivity of aflatoxin B_1 with guanines in different DNA sequences. *Biochemistry* **27:**472–481.

24. **Berkner, L. V., and L. C. Marshall.** 1964. The history of oxygenic concentrations in the earth's atmosphere. *Discuss. Faraday Soc.* **37:**122–141.

25. **Bohr, V. A., C. A. Smith, D. S. Okomuto, and P. C. Hanawalt.** 1985. DNA repair in an active gene: removal of pyrimidine dimers from the DHFR gene of CHO cells is much more efficient than in the genome overall. *Cell* **40:**359–369.

26. **Boles, T. C., and M. E. Hogan.** 1986. High-resolution mapping of carcinogen binding sites on DNA. *Biochemistry* **25:**3039–3043.

27. **Boorstein, R. J., T. P. Hilbert, R. P. Cunningham, and G. W. Teebor.** 1990. Formation and stability of repairable pyrimidine photohydrates in DNA. *Biochemistry* **29:**10455–10460.

28. **Bopp, A., and U. Hagen.** 1970. End group determination in γ-irradiated DNA. *Biochim. Biophys. Acta* **209:**320–326.

29. **Borowy-Borowski, H., R. Lipman, and M. Tomasz.** 1990. Recognition between mitomycin C and specific DNA sequences for cross-link formation. *Biochemistry* **29:**2999–3004.

30. **Bourre, F., G. Renault, and A. Sarasin.** 1987. Sequence effect on alkali-sensitive sites in UV-irradiated SV40 DNA. *Nucleic Acids Res.* **15:**8861–8875.

31. **Boutwell, R. K., N. H. Colburn, and C. C. Muckerman.** 1969. *In vitro* reactions of β propiolactone. *Ann. N. Y. Acad. Sci.* **163:**751–763.

32. **Brash, D. E.** 1988. UV mutagenic photoproducts in *Escherichia coli* and human cells: a molecular genetics perspective on human skin cancer. *Photochem. Photobiol.* **48:**59–66.

33. **Brash, D. E., and W. A. Haseltine.** 1982. UV-induced mutation hotspots occur at DNA damage hotspots. *Nature* (London) **298:**189–192.

34. **Breimer, L., and T. Lindahl.** 1980. A DNA glycosylase from *Escherichia coli* that releases free urea from a polydeoxyribonucleotide containing fragments of base residues. *Nucleic Acids Res.* **8:**6199–6211.

35. **Breimer, L. H.** 1988. Ionizing radiation-induced mutagenesis. *Br. J. Cancer* **57:**6–18.

36. **Breimer, L. H.** 1990. Molecular mechanisms of oxygen radical carcinogenesis and mutagenesis: the role of DNA base damage. *Mol. Carcinogen.* **3:**188–197.

37. **Breimer, L. H., and T. Lindahl.** 1985. Thymine lesions produced by ionizing radiation in double-stranded DNA. *Biochemistry* **24:**4018–4022.

38. **Brendel, M., and A. Ruhland.** 1984. Relationships between functionality and genetic toxicology of selected DNA damaging agents. *Mutat. Res.* **133:**51–85.

39. **Brookes, P.** 1990. The early history of the biological alkylating agents. *Mutat. Res.* **233:**3–14.

40. **Brutlag, D., and A. Kornberg.** 1972. Enzymatic synthesis of deoxyribonucleic acid. XXXVI. A proofreading function for the 3′-5′ exonuclease activity in deoxyribonucleic acid polymerases. *J. Biol. Chem.* **247:**241–248.

41. **Brynolf, K., R. Eliasson, and P. Reichard.** 1978. Formation of Okazaki fragments in polyoma DNA synthesis caused by misincorporation of uracil. *Cell* **13:**573–580.

42. **Butlin, H. T.** 1892. Cancer of the scrotum in chimney-sweeps and others. II. Why foreign sweeps do not suffer from scrotal cancer. *Br. Med. J.* **1:**1341–1346.

43. **Cadet, J., and M. Berger.** 1985. Radiation-induced decomposition of the purine bases within DNA and related model compounds. *Int. J. Radiat. Biol.* **47:**127–143.

44. **Carmichael, P. L., M. Ni Shé, and D. H. Phillips.** 1992. Detection and characterization by ^{32}P-postlabelling of DNA adducts induced by a Fenton-type oxygen radical-generating system. *Carcinogenesis* **13:**1127–1135.

45. **Cerutti, P. A.** 1975. Repairable damage in DNA: overview, p. 3–12. *In* P. C. Hanawalt and R. B. Setlow (ed.), *Mechanisms for the Repair of DNA*, vol. I. Plenum Publishing Corp., New York.

46. **Cerutti, P. A.** 1976. DNA base damage induced by ionizing radiation, p. 375–401. *In* S. Y. Wang (ed.), *Photochemistry and Photobiology of Nucleic Acids*. Academic Press, Inc., New York.

47. **Cerutti, P. A.** 1978. Repairable damage in DNA, p. 1–14. *In* P. C. Hanawalt, E. C. Friedberg, and C. F. Fox (ed.), *DNA Repair Mechanisms*. Academic Press, Inc., New York.

48. **Cerutti, P. A., and B. F. Trump.** 1991. Inflammation and oxidative stress in carcinogenesis. *Cancer Cells* **3:**1–7.

49. **Chan, G. L., P. W. Doetsch, and W. A. Haseltine.** 1985. Cyclobutane pyrimidine dimers and (6-4) photoproducts block polymerization by DNA polymerase I. *Biochemistry* **24:**5723–5728.

50. **Chen, H., and B. R. Shaw.** 1993. Kinetics of bisulfite-induced cytosine deamination in single-stranded DNA. *Biochemistry* **32:**3535–3539.

51. **Chen, H., and B. R. Shaw.** 1994. Bisulfite induces tandem double CC → TT mutations in double-stranded DNA. 2. Kinetics of cytosine deamination. *Biochemistry* **33:**4121–4129.

52. **Cheng, K. C., D. S. Cahill, H. Kasai, S. Nushimura, and L. A. Loeb.** 1992. 8-Hydroxyguanine, an abundant form of oxidative damage, causes G → T and A → C substitutions. *J. Biol. Chem.* **267:**166–171.

53. **Chetsanga, C. J., and T. Lindahl.** 1979. Release of 7-methylguanine residues whose imidazole rings have been opened, from damaged DNA by a DNA glycosylase from *Escherichia coli*. *Nucleic Acids Res.* **6:**3673–3684.

54. **Chetsanga, C. J., M. Lozon, C. Makaroff, and L. Savage.** 1981. Purification and characterization of *Escherichia coli* formamidopyrimidine-DNA glycosylase that excises damaged 7-methylguanine from deoxyribonucleic acid. *Biochemistry* **20:**5201–5207.

55. **Chu, G.** 1994. Cellular responses to cisplatin. *J. Biol. Chem.* **269:**787–790.

56. **Chun, E. H. L., L. Gonzales, F. S. Lewis, J. Jones, and R. J. Rutman.** 1969. Differences in the *in vivo* alkylation and cross-linking of nitrogen mustard-sensitive and resistant lines of Lettré-Ehrlich ascites tumors. *Cancer Res.* **29:**1184–1194.

57. **Clark, J. M., and G. P. Beardsley.** 1987. Functional effects of *cis*-thymine glycol lesions on DNA synthesis in vitro. *Biochemistry* **26:**5398–5403.

58. **Clark, J. M., N. Pattabiraman, W. Jarvis, and G. P. Beardsley.** 1987. Modeling and molecular mechanical studies of the *cis*-thymine glycol radiation damage lesion in DNA. *Biochemistry* **26:**5404–5409.

59. **Clayson, D. B.** 1962. *Chemical Carcinogenesis*. Little, Brown & Co., Boston.

60. **Clayson, D. B., R. Mehta, and F. Iverson.** 1994. Oxidative DNA damage—the effects of certain genotoxic and operationally nongenotoxic carcinogens. *Mutat. Res.* **317:**25–42.

61. **Clayton, D. A.** 1981. Measurement of strand breaks in supercoiled DNA by electron microscopy, p. 419–424. *In* E. C. Friedberg and P. C. Hanawalt (ed.), *DNA Repair—A Laboratory Manual of Research Procedures*, vol. 1 B. Marcel Dekker, Inc., New York.

62. **Clayton, L. K., M. F. Goodman, E. W. Branscomb, and D. J. Galas.** 1979. Error induction and correction by mutant and wild type T4 DNA polymerases. Kinetic error discrimination mechanisms. *J. Biol. Chem.* **254:**1902–1912.

63. **Cohen, G.** 1985. The Fenton reaction, p. 55–64. *In* R. A. Greenwald (ed.), *CRC Handbook of Methods for Oxygen Radical Research*. CRC Press, Inc., Boca Raton, Fla.

64. **Cole, R.** 1970. Light-induced cross-linking of DNA in the presence of a furocoumarin (psoralen). Studies with phage γ, *Escherichia coli*, and mouse leukemia cells. *Biochim. Biophys. Acta.* **217:**30–39.

65. **Cole, R. S.** 1971. Psoralen monoadducts and interstrand cross-links in DNA. *Biochim. Biophys. Acta* **254:**30–39.

66. **Cole, R. S.** 1973. Repair of DNA containing interstrand cross-links in *Escherichia coli*: sequential excision and recombination. *Proc. Natl. Acad. Sci. USA* **70:**1064–1068.

67. **Cole, R. S., R. R. Sinden, G. H. Yoakum, and S. Broyles.** 1978. On the mechanism for repair of cross-linked DNA in *E. coli* treated with psoralen and light, p. 287–290. *In* P. C. Hanawalt, E. C. Friedberg, and C. F. Fox (ed.), *DNA Repair Mechanisms*. Academic Press, Inc., New York.

68. **Coon, M. J., X. Ding, S. J. Pernecky, and A. D. N. Vaz.** 1992. Cytochrome P450: progress and predictions. *FASEB J.* **6:**669–673.

69. **Cooper, C. S., and P. L. Grover.** 1990. *Chemical Carcinogenesis and Mutagenesis II*. Springer-Verlag, KG, Berlin.

70. **Cooper, P. K.** 1982. Characterization of long patch excision repair of DNA in ultraviolet-irradiated *Escherichia coli*: an inducible function under rec-lex control. *Mol. Gen. Genet.* **185:**189–197.

71. **Cosman, M., C. de los Santos, R. Fiala, B. E. Hingerty, S. B. Singh, V. Ibanez, L. A. Margulis, D. Live, N. E. Geacintov, S. Broyde, and D. J. Patel.** 1992. Solution conformation of the major adduct between the carcinogen (+)-*anti*-benzo[a]pyrene diol epoxide and DNA. *Proc. Natl. Acad. Sci. USA* **89:**1914–1918.

72. **Coulondre, C., J. H. Miller, P. J. Farabaugh, and W. Gilbert.** 1978. Molecular basis of base substitution hotspots in *Escherichia coli*. *Nature* (London) **274:**775–780.

73. **Crawford, J. L., F. J. Kolpak, A. H.-J. Wang, G. J. Quigley, J. H. van Boom, G. van der Marel, and A. Rich.** 1980. The tetramer d(CpGpCpG) crystallizes as a left-handed double helix. *Proc. Natl. Acad. Sci. USA* **77:**4016–4020.

74. **Cutler, R. G.** 1984. Antioxidants, aging, and longevity, p. 371–428. *In* W. A. Pryor (ed.), *Free Radicals in Biology*. Academic Press, Inc., New York.

75. **Cutler, R. G.** 1991. Antioxidants and aging. *Am. J. Clin. Nutr.* **53 (Suppl.):**373S–379S.

76. **Davies, K. J. A. (ed).** 1992. *Oxidative Damage and Repair*. Pergamon Press, New York.

77. **Demple, B.** 1991. Regulation of bacterial oxidative stress genes. *Annu. Rev. Genet.* **25:**315–338.

78. **Demple, B., and C. F. Amábile-Cuevas.** 1991. Redox redux: the control of oxidative stress response. *Cell* **67:**837–839.

79. **Demple, B., and J. D. Levin.** 1991. Repair systems for radical-damaged DNA, p. 119–154. *In* H. Sies (ed.), *Oxidative Stress: Oxidants and Antioxidants*. Academic Press, Ltd., London.

80. **Demple, B., and S. Linn.** 1980. DNA N-glycosylases and UV repair. *Nature* (London) **287:**203–208.

81. **Demple, B., and S. Linn.** 1982. 5,6-Saturated thymine lesions in DNA: production by ultraviolet light or hydrogen peroxide. *Nucleic Acids Res.* **10:**3781–3789.

82. **DeWaard, A., A. V. Paul, and I. R. Lehman.** 1965. The structural gene for deoxyribonucleic acid polymerase in bacteriophages T4 and T5. *Proc. Natl. Acad. Sci. USA* **54:**1241–1248.

83. **Dillehay, L. E., L. H. Thompson, and A. V. Carrano.** 1984. DNA-strand breaks associated with halogenated pyrimidine incorporation. *Mutat. Res.* **131:**129–136.

84. **Dizdaroglu, M.** 1992. Oxidative damage to DNA in mammalian chromatin. *Mutat. Res.* **275:**331–342.

85. **Dizdaroglu, M., M.-L. Dirksen, H. Jiang, and J. H. Robbins.** 1987. Ionizing-radiation-induced damage in the DNA of cultured human cells. Identification of 8,5-cyclo-2-deoxyguanosine. *Biochem. J.* **241:**929–932.

86. **Dizdaroglu, M., Z. Nackerdien, B.-C. Chao, E. Gajewski, and G. Rao.** 1991. Chemical nature of *in vivo* DNA base damage in hydrogen peroxide-treated mammalian cells. *Arch. Biochem. Biophys.* **285:**388–390.

87. **Dizdaroglu, M., G. Rao, B. Halliwell, and E. Gajewski.** 1991. Damage to the DNA bases in mammalian chromatin by hydrogen peroxide in the presence of ferric and cupric ions. *Arch. Biochem. Biophys.* **285:**317–324.

88. Doetsch, P. W., G. L. Chan, and W. A. Haseltine. 1985. T4 polymerase (3′-5′) exonuclease, an enzyme for the detection and quantitation of stable DNA lesions: the ultraviolet light example. *Nucleic Acids Res.* **13**:3285–3304.

89. Drake, J. W. 1969. Comparative rates of spontaneous mutation. *Nature* (London) **221**:1132.

90. Drake, J. W. 1970. *The Molecular Basis of Mutation.* Holden-Day Inc., San Francisco.

91. Drake, J. W. 1991. Spontaneous mutation. *Annu. Rev. Genet.* **25**:125–146.

92. Drake, J. W., and E. F. Allen. 1968. Antimutagenic DNA polymerases of bacteriophage T4. *Cold Spring Harbor Symp. Quant. Biol.* **33**:339–344.

93. Drew, H., T. Takano, S. Tanaka, K. Itakura, and R. E. Dickerson. 1980. High salt d(CpGpCpGp), a left-handed Z DNA double helix. *Nature* (London) **286**:567–573.

94. Duba, V. V., V. A. Pitkevich, N. G. Selyova, I. V. Petrova, and M. N. Myasnik. 1985. The formation of photoreactivable damage by direct excitation of DNA in X-irradiated *E. coli* cells. *Int. J. Radiat. Biol.* **47**:49–56.

95. Duker, N. J., and P. E. Gallagher. 1988. Purine photoproducts. *Photochem. Photobiol.* **48**:35–39.

96. Duncan, B. K. 1981. DNA glycosylases, p. 565–586. *In* P. D. Boyer (ed.), *The Enzymes.* XIV. Academic Press, Inc., New York.

97. Duncan, B. K., and J. Miller. 1980. Mutagenic deamination of cytosine residues in DNA. *Nature* (London) **287**:560–561.

98. Duncan, B. K., and B. Weiss. 1978. Uracil-DNA glycosylase mutants are mutators, p. 183–186. *In* P. C. Hanawalt, E. C. Friedberg, and C. F. Fox (ed.), *DNA Repair Mechanisms.* Academic Press, Inc., New York.

99. Eastman, A. 1987. The formation, isolation and characterization of DNA adducts produced by anticancer platinum complexes. *Pharmacol. Ther.* **34**:155–166.

100. Echols, H., and M. F. Goodman. 1991. Fidelity mechanisms in DNA replication. *Annu. Rev. Biochem.* **60**:477–512.

101. Eggset, G., G. Volden, and H. Krokan. 1987. Characterization of antibodies specific for UV-damaged DNA by ELISA. *Photochem. Photobiol.* **45**:485–491.

102. Ehrlich, M., X.-Y. Zhang, and N. M. Inamdar. 1990. Spontaneous deamination of cytosine and 5-methylcytosine residues in DNA and replacement of 5-methylcytosine residues with cytosine residues. *Mutat. Res.* **238**:277–286.

103. Emerit, I. 1994. Reactive oxygen species, chromosome mutation and cancer: possible role of clastogenic factors in carcinogenesis. *Free Radical Biol. Med.* **16**:99–109.

104. Emerit, I., and P. Cerutti. 1981. Clastogenic activity from Bloom syndrome fibroblast cultures. *Proc. Natl. Acad. Sci. USA* **78**:1868–1872.

105. Emerit, I., and P. Cerutti. 1982. Tumor promoter phorbol 12-myristate 13-acetate induces a clastogenic factor in human lymphocytes. *Proc. Natl. Acad. Sci. USA* **79**:7509–7513.

106. Emerit, I., S. H. Khan, and P. Cerutti. 1985. Treatment of lymphocyte cultures with hypoxanthine-xanthine oxidase system induces the formation of transferable clastogenic material. *Free Radical Biol. Med.* **1**:51–57.

107. Emerit, I., S. H. Khan, and H. Esterbauer. 1991. Hydroxynonenal, a component of clastogenic factors? *Free Radical Biol. Med.* **10**:371–377.

108. Emerit, I., A. Levy, and J. P. Camus. 1989. Monocyte-derived clastogenic factor in rheumatoid arthritis. *Free Radical Biol. Med.* **6**:245–250.

109. Emerit, I., A. Levy, and C. DeVaux Saint Cyr. 1980. Chromosome damaging agent of low molecular weight in the serum of New Zealand black mice. *Cytogenet. Cell Genet.* **26**:41–48.

110. Emerit, I., A. Levy, and A. M. Michelson. 1981. Effect of superoxide dismutase on the chromosomal instability of NZB mice. *Cytogenet. Cell Genet.* **30**:65–69.

111. Emerit, I., A. M. Michelson, A. Levy, J. P. Camus, and J. Emerit. 1980. Chromosome-breaking agent of low molecular weight in human systemic lupus erythematosus. Protector effect of superoxide dismutase. *Hum. Genet.* **55**:341–344.

112. Essigmann, J. M., R. G. Croy, A. M. Nadzan, W. F. Busby Jr., V. N. Reinhold, G. Buchi, and G. N. Wogan. 1977. Structural identification of the major DNA adduct formed by aflatoxin B1 *in vitro.* *Proc. Natl. Acad. Sci. USA* **74**:1870–1874.

113. Everson, R. B., E. Randerath, R. M. Santella, R. C. Cefalo, T. A. Avitts, and K. Randerath. 1986. Detection of smoking-related covalent DNA adducts in human placenta. *Science* **231**:54–57.

114. Falany, C. N., and T. W. Wilborn. 1990. Biochemistry of cytosolic sulfotransferases involved in bioactivation. *Adv. Pharmacol.* **27**:301–363.

115. Fersht, A., and J. W. Knill-Jones. 1981. DNA polymerase accuracy and spontaneous mutation rates: frequencies of purine·purine, purine·pyrimidine, and pyrimidine·pyrimidine mismatches during DNA replication. *Proc. Natl. Acad. Sci. USA* **78**:4251–4255.

116. Finnegan, D. J. 1990. Transposable elements and DNA transposition in eukaryotes. *Curr. Opin. Cell Biol.* **21**:471–477.

117. Fisher, G. J., and H. E. Johns. 1976. Pyrimidine hydrates. p. 169–294. *In* S. Y. Wang (ed.), *Photochemistry and Photobiology of Nucleic Acids,* vol. 1. Academic Press, Inc., New York.

118. Fox, M., and D. Scott. 1980. The genetic toxicology of nitrogen and sulphur mustard. *Mutat. Res.* **75**:131–168.

119. Fraga, C. G., M. K. Shigenaga, J.-W. Park, P. Degan, and B. N. Ames. 1990. Oxidative damage to DNA during aging: 8-hydroxy-2′-deoxyguanosine in rat organ DNA and urine. *Proc. Natl. Acad. Sci. USA* **87**:4533–4537.

120. Fram, R. J. 1992. Cisplatin and platinum analogues: recent advances. *Curr. Opin. Oncol.* **4**:1073–1079.

121. Frankenberg-Schwager, M. 1990. Induction, repair and biological relevance of radiation-induced DNA lesions in eukaryotic cells. *Radiat. Environ. Biophys.* **29**:273–292.

122. Frankenberg-Schwager, M., D. Frankenberg, D. Blöcher, and C. Adamczyk. 1979. The influence of oxygen on the survival and yield of DNA double-strand breaks in irradiated yeast cells. *Int. J. Radiat. Biol.* **36**:261–270.

123. Franklin, W. A., P. W. Doetsch, and W. A. Haseltine. 1985. Structural determination of the ultraviolet light-induced thymine-cytosine pyrimidine-pyrimidone (6-4) photoproduct. *Nucleic Acids Res.* **13**:5317–5325.

124. Franklin, W. A., and W. A. Haseltine. 1986. The role of the (6-4)photoproduct in ultraviolet light-induced transition mutations in *E. coli. Mutat. Res.* **165**:1–7.

125. Frederico, L. A., T. A. Kunkel, and B. R. Shaw. 1990. A sensitive genetic assay for the detection of cytosine deamination: determination of rate constants and activation energy. *Biochemistry* **29**:2532–2537.

126. Frederico, L. A., T. A. Kunkel, and B. R. Shaw. 1993. Cytosine deamination in mismatched base pairs. *Biochemistry* **32**:6523–6530.

127. Freese, E. 1959. The difference between spontaneous and base analogue induced mutation of phage T4. *Proc. Natl. Acad. Sci. USA* **45**:622–633.

128. Freese, E. 1959. On the molecular explanation of spontaneous and induced mutations. *Brookhaven Symp. Biol.* **12**:63–73.

129. Freese, E. 1959. The specific mutagenic effect of base analogues on phage T4. *J. Mol. Biol.* **1**:87–105.

130. Frenkel, K., M. S. Goldstein, N. J. Duker, and G. W. Teebor. 1981. Identification of the *cis* thymine glycol moiety in oxidized deoxyribonucleic acid. *Biochemistry* **20**:750–754.

131. Friedberg, E. C. 1985. *DNA Repair.* W. H. Freeman & Co., New York.

132. Friedberg, E. C., A. K. Ganesan, and K. Minton. 1975. *N*-Glycosidase activity in extracts of *Bacillus subtilis* and its inhibition after infection with bacteriophage PBS2. *J. Virol.* **16**:315–321.

133. Friedberg, E. C., and P. C. Hanawalt (ed). 1981. *DNA Repair—A Laboratory Manual of Research Procedures*, vol. 1A, 1B, 2, and 3. Marcel Dekker, Inc., New York.

134. Fry, R. J. M., and S. A. Fry. 1990. Health effects of ionizing radiation. *Med. Clin. North Am.* **74:**475–488.

135. Fujiwara, Y. 1983. Measurement of interstrand cross-links produced by mitomycin C, p. 143–160. *In* E. C. Friedberg and P. C. Hanawalt (ed.), *DNA Repair—A Laboratory Manual of Research Procedures*, vol. 2. Marcel Dekker, Inc., New York.

136. Gajewski, E., G. Rao, Z. Nackerdien, and M. Dizdaroglu. 1990. Modification of DNA bases in mammalian chromatin by radiation-generated free radicals. *Biochemistry* **29:**7876–7882.

137. Galas, D. J., and E. W. Branscomb. 1978. Enzymatic determinants of DNA polymerase accuracy. Theory of coliphage T4 polymerase mechanisms. *J. Mol. Biol.* **124:**653–687.

138. Gale, J. M., K. A. Nissen, and M. J. Smerdon. 1987. UV-induced formation of pyrimidine dimers in nucleosome core DNA is strongly modulated with a period of 10.3 bases. *Proc. Natl. Acad. Sci. USA* **84:**6644–6648.

139. Gale, J. M., and M. J. Smerdon. 1990. UV induced (6-4) photoproducts are distributed differently than cyclobutane dimers in nucleosomes. *Photochem. Photobiol.* **51:**411–417.

140. Galiègue-Zouitina, S., B. Bailleul, Y. Ginot, B. Perly, P. Vigny, and M.-H. Loucheux-Lefebvre. 1986. N^2-Guanyl and N^6-adenyl arylation of chicken erythrocyte DNA by the ultimate carcinogen of 4-nitroquinoline 1-oxide. *Cancer Res.* **46:**1858–1863.

141. Galiègue-Zouitina, S., B. Bailleul, and M. H. Loucheux-Lefebvre. 1985. Adducts from *in vivo* action of the carcinogen of 4-hydroxyaminoquinoline 1-oxide and from *in vitro* reaction of 4-acetoxyaminoquinoline 1-oxide with DNA and polynucleotides. *Cancer Res.* **45:**520–525.

142. Gallagher, P. E., and N. J. Duker. 1986. Detection of UV purine photoproducts in a defined sequence of human DNA. *Mol. Cell. Biol.* **6:**707–709.

143. Garrett, E. R., and P. J. Mehta. 1972. Solvolysis of adenine nucleosides. II. Effects of sugars and adenine substitutents on alkaline solvolysis. *J. Am. Chem. Soc.* **94:**8542–8547.

144. Gasparro, F. P. 1988. Psoralen-DNA interactions: thermodynamics and photochemistry, p. 5–36. *In* F. P. Gasparro (ed.), *Psoralen DNA Photobiology.* CRC Press, Inc., Boca Raton, Fla.

145. Gasparro, F. P., and J. R. Fresco. 1986. Ultraviolet-induced 8,8-adenine dehydrodimers in oligo- and polynucleotides. *Nucleic Acids Res.* **14:**4239–4251.

146. Geacintov, N. E. 1986. Is intercalation a critical factor in the covalent binding of mutagenic and tumorigenic polycyclic aromatic diol epoxides to DNA? *Carcinogenesis* **7:**759–766.

147. Geider, K. 1972. DNA synthesis in nucleotide-permeable *Escherichia coli* cells. The effects of nucleotide analogues on DNA synthesis. *Eur. J. Biochem.* **27:**554–563.

148. Geiduschek, E. 1961. "Reversible" DNA. *Proc. Natl. Acad. Sci. USA* **47:**950–955.

149. Geigl, E. M., and F. Eckardt-Schupp. 1991. The repair of double-strand breaks and S1 nuclease-sensitive sites can be monitored chromosome-specifically in *Saccharomyces cerevisiae* using pulsed-field gel electrophoresis. *Mol. Microbiol.* **5:**1615–1620.

150. Gillin, F. D., and N. G. Nossal. 1976. Control of mutation frequency by bacteriophage T4 DNA polymerase I. Accuracy of nucleotide selection by the L88 mutator, CB120 antimutator, and wild type phage T4 DNA polymerases. *J. Biol. Chem.* **251:**5225–5232.

151. Gillin, F. D., and N. G. Nossal. 1976. Control of mutation frequency by bacteriophage T4 DNA polymerase I. The CB120 antimutator DNA polymerase is defective in strand displacement. *J. Biol. Chem.* **251:**5219–5224.

152. Glickman, B. W., R. M. Schaaper, W. A. Haseltine, R. L. Dunn, and D. E. Brash. 1986. The C-C (6-4) UV photoproduct is mutagenic in *Escherichia coli. Proc. Natl. Acad. Sci. USA* **83:**6945–6949.

153. Goodhead, D. T. 1989. The initial damage produced by ionizing radiations. *Int. J. Radiat. Biol.* **56:**623–634.

154. Goodman, M. F. 1988. DNA replication fidelity: kinetics and thermodynamics. *Mutat. Res.* **200:**11–20.

155. Goodman, M. F., and E. W. Branscomb. 1986. DNA replication fidelity and base pairing mutagenesis, p. 191–232. *In* T. B. L. Kirkwood, R. F. Rosenberger, and D. J. Galas (ed.), *Accuracy in Molecular Processes: Its Control and Relevance to Living Systems.* Chapman & Hall, New York.

156. Goodman, M. F., W. C. Gore, N. Muzyczka, and M. J. Bessman. 1974. Studies on the biochemical basis of spontaneous mutations. III. Rate model for DNA polymerase-effected nucleotide misincorporation. *J. Mol. Biol.* **88:**423–435.

157. Gordon, L. K., and W. A. Haseltine. 1982. Quantitation of cyclobutane pyrimidine dimer formation in double- and single-stranded DNA fragments of defined sequence. *Radiat. Res.* **89:**99–112.

158. Goulian, M., B. Bleile, and B. Y. Tseng. 1980. The effect of methotrexate on levels of dUTP in animal cells. *J. Biol. Chem.* **255:**10630–10637.

159. Goulian, M., B. Bleile, and B. Y. Tseng. 1980. Methotrexate-induced misincorporation of uracil into DNA. *Proc. Natl. Acad. Sci. USA* **77:**1956–1960.

160. Gräslund, A., and B. Jernström. 1989. DNA-carcinogen interaction: covalent DNA-adducts of benzo(a)pyrene 7,8-dihydrodiol 9,10-epoxides studied by biochemical and biophysical techniques. *Q. Rev. Biophys.* **22:**1–37.

161. Greer, S., and S. Zamenhof. 1962. Studies on depurination of DNA by heat. *J. Mol. Biol.* **4:**123–141.

162. Groopman, J. D., and L. G. Cain. 1990. Interactions of fungal and plant toxins with DNA: aflatoxins, sterigmacystin, safrole, cycasin, and pyrrolizidine alkaloids, p. 373–407. *In* C. S. Cooper and P. L. Grover (ed.), *Chemical Carcinogenesis and Mutagenesis I.* Springer-Verlag, KG, Berlin.

163. Grunberger, D., and R. M. Santella. 1982. Alternative conformations of DNA modified by N-2-acetylaminofluorene, p. 155–168. *In* C. C. Harris and P. A. Cerutti (ed.), *Mechanisms of Chemical Carcinogenesis.* Alan R. Liss, Inc., New York.

164. Guengerich, F. P. 1994. Metabolism and genotoxicity of dihaloalkanes, p. 211–236. *In* M. W. Anders and W. Dekant (ed.), *Conjugation-Dependent Carcinogenicity and Toxicity of Foreign Compounds.* Academic Press, Inc., San Diego, Calif.

165. Hall, M., and P. L. Grover. 1990. Polycyclic aromatic hydrocarbons: metabolism, activation and tumour initiation, p. 327–372. *In* C. S. Cooper and P. L. Grover (ed.), *Chemical Carcinogenesis and Mutagenesis I.* Springer-Verlag, KG, Berlin.

166. Halliwell, B., and O. I. Aruoma. 1991. DNA damage by oxygen-derived species. Its mechanism and measurement in mammalian systems. *FEBS Lett.* **281:**9–19.

167. Halliwell, B., and J. M. C. Gutteridge. 1989. *Free Radicals in Biology and Medicine.* Clarendon Press, Oxford.

168. Hamada, H., and T. Kakunaga. 1982. Potential Z-DNA forming sequences are highly dispersed in the human genome. *Nature* (London) **298:**396–398.

169. Hanawalt, P. C., P. K. Cooper, A. K. Ganesan, and C. A. Smith. 1979. DNA repair in bacteria and mammalian cells. *Annu. Rev. Biochem.* **48:**783–836.

170. Haseltine, W. A. 1983. Site specificity of ultraviolet light induced mutagenesis, p. 3–22. *In* E. C. Friedberg and B. A. Bridges (ed.), *Cellular Responses to DNA Damage.* UCLA Symp. Mol. Cell. Biol. New Ser., vol. 11. Alan R. Liss, Inc., New York.

171. Haseltine, W. A. 1983. Ultraviolet light repair and mutagenesis revisited. *Cell* **33:**13–17.

172. Hayatsu, H. 1976. Bisulfite modification of nucleic acids and their constituents. *Prog. Nucleic Acids Res. Mol. Biol.* **16:**75–124.

173. Hayes, F. N., D. L. Williams, R. L. Ratliff, A. J. Varghese, and C. S. Rupert. 1971. Effect of a single thymine photodimer on the oligodeoxythymidilate-polydeoxyadenylate interaction. *J. Am. Chem. Soc.* **93**:4940–4942.

174. Hearst, J. E., S. T. Isaacs, D. Kanne, H. Rapoport, and K. Straub. 1984. The reaction of the psoralens with deoxyribonucleic acid. *Q. Rev. Biophys.* **17**:1–44.

175. Henner, W. D., S. M. Grunberg, and W. A. Haseltine. 1982. Sites and structure of γ radiation-induced DNA strand breaks. *J. Biol. Chem.* **257**:11750–11754.

176. Henner, W. D., L. O. Rodriguez, S. M. Hecht, and W. A. Haseltine. 1983. γ Ray induced deoxyribonucleic acid strand breaks. *J. Biol. Chem.* **258**:711–713.

177. Hiatt, H. H., J. D. Watson, and J. A. Winston (ed). 1977. *Origins of Human Cancer. Book B.* Cold Spring Harbor Laboratory, Cold Spring Harbor, N.Y.

178. Hochhauser, S. J., and B. Weiss. 1978. *Escherichia coli* mutants deficient in deoxyuridine triphosphatase. *J. Bacteriol.* **134**:157–166.

179. Hollenberg, P. F. 1992. Mechanisms of cytochrome P450 and peroxidase-catalyzed xenobiotic metabolism. *FASEB J.* **6**:686–694.

180. Hollowell, J. G., Jr., and L. G. Littlefield. 1968. Chromosome damage induced by plasma from irradiated patients. An indirect effect of X-ray. *Proc. Soc. Exp. Biol. Med.* **129**:240–244.

181. Hsieh, D. P. H., J. J. Wong, A. Wong, C. Michas, and B. H. Ruebner. 1977. Hepatic transformation of aflatoxin and its carcinogenicity, p. 697–707. *In* H. H. Hiatt, J. D. Watson, and J. A. Winston (ed.), *Origins of Human Cancer. Book B.* Cold Spring Harbor Laboratory, Cold Spring Harbor, N.Y.

182. Hu, H. 1990. Effects of ultraviolet radiation. *Med. Clin. North Am.* **74**:509–514.

183. Hubbard, K., H. Huang, M. F. Laspia, H. Ide, B. Erlanger, and S. S. Wallace. 1989. Immunochemical quantitation of thymine glycol in oxidized and X-irradiated DNA. *Radiat. Res.* **118**:257–268.

184. Hubbard, K., H. Ide, B. F. Erlanger, and S. S. Wallace. 1989. Characterization of antibodies to dihydrothymine, a radiolysis product of DNA. *Biochemistry* **28**:4382–4387.

185. Husain, I., J. Griffith, and A. Sancar. 1988. Thymine dimers bend DNA. *Proc. Natl. Acad. Sci. USA* **85**:2558–2562.

186. Hutchinson, F. 1985. Chemical changes induced in DNA by ionizing radiation. *Prog. Nucleic Acid Res.* **32**:115–154.

187. Hutchinson, F. 1973. The lesions produced by ultraviolet light in DNA containing 5-bromouracil. *Q. Rev. Biophys.* **6**:201–246.

188. Ide, H., Y. W. Kow, and S. S. Wallace. 1985. Thymine glycols and urea residues in M13 DNA constitute replicative blocks *in vitro*. *Nucleic Acids Res.* **13**:8035–8051.

189. Iliakis, G. 1991. The role of DNA double strand breaks in ionizing radiation-induced killing of eukaryotic cells. *BioEssays* **13**:641–648.

190. Imlay, J. A., S. M. Chin, and S. Linn. 1988. Toxic DNA damage by hydrogen peroxide through the Fenton reaction *in vivo* and *in vitro*. *Science* **240**:640–642.

191. Imlay, J. A., and S. Linn. 1988. DNA damage and oxygen radical toxicity. *Science* **240**:1302–1309.

192. Impellizzeri, K. J., B. Anderson, and P. M. J. Burgers. 1991. The spectrum of spontaneous mutations in a *Saccharomyces cerevisiae* uracil-DNA-glycosylase mutant limits the function of this enzyme to cytosine deamination repair. *J. Bacteriol.* **173**:6807–6810.

193. Iyer, V. N., and W. Szybalski. 1963. A molecular mechanism of mitomycin action: linking of complementary DNA strands. *Proc. Natl. Acad. Sci. USA* **50**:355–362.

194. Izatt, R. M., J. J. Christensen, and J. H. Rytting. 1971. Sites and thermodynamic quantities associated with proton and metal ion interaction with ribonucleic acid, deoxyribonucleic acid, and their constitutent bases, nucleosides, and nucleotides. *Chem. Rev.* **71**:439–481.

195. Jackson, R. C. 1978. The regulation of thymidylate biosynthesis in Novikoff hepatoma cells and the effects of amethopterin, 5-fluorodeoxyuridine, and 3-deazauridine. *J. Biol. Chem.* **253**:7440–7446.

196. Jahnke, G. D., C. L. Thompson, M. P. Walker, J. E. Gallagher, G. W. Lucier, and R. P. DiAugustine. 1990. Multiple DNA adducts in lymphocytes of smokers and nonsmokers determined by ^{32}P-postlabeling analysis. *Carcinogenesis* **11**:205–211.

197. Joenje, H. 1989. Genetic toxicology of oxygen. *Mutat. Res.* **219**:193–208.

198. Jones, A. S., A. M. Mian, and R. T. Walker. 1968. The alkaline degradation of deoxyribonucleic acid derivatives. *J. Chem. Soc. Sect. C* **1968**:2042.

199. Jordon, D. O. 1960. *The Chemistry of Nucleic Acids.* Butterworth's, Washington, D.C.

200. Jovanovic, S. V., and M. G. Simic. 1986. Mechanism of OH radical reaction with thymine and uracil derivatives. *J. Am. Chem. Soc.* **108**:5968–5972.

201. Jovanovic, S. V., and M. G. Simic. 1989. The DNA-guanyl radical: kinetics and mechanisms of generation and repair. *Biochim. Biophys. Acta* **1008**:39–44.

202. Kahan, F. M. 1963. Novel enzymes formed by *Bacillus subtilis* infected with bacteriophage. *Fed. Proc.* **22**:406.

203. Kasai, H., and S. Nishimura. 1991. Formation of 8-hydroxydeoxyguanosine in DNA by oxygen radicals and its biological significance, p. 99–116. *In* H. Sies (ed.), *Oxidative Stress: Oxidants and Antioxidants.* Academic Press, Ltd., London.

204. Kaye, J., C. A. Smith, and P. C. Hanawalt. 1980. DNA repair in human cells containing photoadducts of 8-methoxypsoralen or angelicin. *Cancer Res.* **40**:696–702.

205. Kelloff, G. J., C. W. Boone, W. F. Malone, and V. E. Steele. 1992. Chemoprevention clinical trials. *Mutat. Res.* **267**:291–295.

206. Kennaway, E. L. 1925. Experiments on cancer-producing substances. *Br. Med. J.* **2**:1–4.

207. Kennaway, E. L., and I. Huger. 1930. Carcinogenic substances and their fluorescence spectra. *Br. Med. J.* **1**:1044–1046.

208. Khattak, M. N., and S. Y. Wang. 1972. The photochemical mechanism of pyrimidine cyclobutyl dimerization. *Tetrahedron* **28**:945–957.

209. Kistler, M., K.-H. Summer, and F. Eckardt. 1986. Isolation of glutathione-deficient mutants of the yeast *Saccharomyces cerevisiae*. *Mutat. Res.* **173**:117–120.

210. Kittler, L., and G. Löber. 1977. Photochemistry of the nucleic acids. *Photochem. Photobiol. Rev.* **2**:39.

211. Kleckner, N. 1981. Transposable elements in prokaryotes. *Annu. Rev. Genet.* **15**:341–404.

212. Kleckner, N. 1990. Regulation of transposition in bacteria. *Annu. Rev. Cell Biol.* **6**:297–327.

213. Knobler, R. M., H. Hönigsmann, and R. L. Edelson. 1988. Psoralen phototherapies, p. 117–148. *In* F. P. Gasparro (ed.), *Psoralen DNA Photobiology.* CRC Press, Inc., Boca Raton, Fla.

214. Kohda, K., K. Taka, H. Kasai, S. Nishimura, and Y. Kawazoe. 1986. Formation of 8-hydroxyguanine residues in cellular DNA exposed to the carcinogen 4-nitroquinoline-1-oxide. *Biochem. Biophys. Res. Commun.* **139**:626–632.

215. Kohn, K. W., R. A. G. Ewig, L. C. Erickson, and L. A. Zwelling. 1981. Measurement of strand breaks and cross-links by alkaline elution, p. 379–401. *In* E. C. Friedberg and P. C. Hanawalt (ed.), *DNA Repair—A Laboratory Manual of Research Procedures*, vol. 1B. Marcel Dekker, Inc., New York.

216. Kohn, K. W., J. A. Hartley, and W. B. Mattes. 1987. Mechanisms of DNA sequence selective alkylation of guanine-*N*7 positions by nitrogen mustards. *Nucleic Acids Res.* **15**:10531–10549.

217. Kohn, K. W., C. L. Spears, and P. Doty. 1966. Inter-strand crosslinking of DNA by nitrogen mustard. *J. Mol. Biol.* **19**:266–288.

218. Konrad, E. B. 1977. Method for the isolation of *Escherichia coli* mutants with enhanced recombination between chromosomal duplications. *J. Bacteriol.* **130**:167–172.

219. Konrad, E. B., and I. R. Lehman. 1975. Novel mutants of *Escherichia coli* that accumulate very small DNA replicative intermediates. *Proc. Natl. Acad. Sci. USA* **72**:2150–2154.

220. **Kornberg, A.** 1980. *DNA Replication*. W. H. Freeman & Co., San Francisco.

221. **Kornberg, A., and T. Baker.** 1991. *DNA Replication*. W. H. Freeman & Co., New York.

222. **Kornberg, R. D.** 1977. Structure of chromatin. *Annu. Rev. Biochem.* **46:**931–954.

223. **Kriek, E.** 1972. Persistent binding of a new reaction product of the carcinogen N-hydroxy-2-acetylaminofluorene with guanine in rat liver DNA *in vivo*. *Cancer Res.* **32:**2042–2048.

224. **Kuchino, Y., F. Mori, H. Kasai, H. Inoue, S. Iwai, K. Miura, E. Ohtsuka, and S. Nishimura.** 1987. Misreading of DNA templates containing 8-hydroxydeoxyguanosine at the modified base and at adjacent residues. *Nature* (London) **327:**77–79.

225. **Kumar, S., N. D. Sharma, R. J. H. Davies, D. W. Phillipson, and J. A. McCloskey.** 1987. The isolation and characterization of a new type of dimeric adenine photoproduct in UV-irradiated deoxyadenylates. *Nucleic Acids Res.* **15:**1199–1216.

226. **Kunkel, T. A.** 1988. Exonucleolytic proofreading. *Cell* **53:**837–840.

227. **Kunkel, T. A., F. Eckstein, A. S. Mildvan, R. M. Koplitz, and L. A. Loeb.** 1981. Deoxynucleoside [1-thio]triphosphates prevent proofreading during *in vitro* DNA synthesis. *Proc. Natl. Acad. Sci. USA* **78:**6734–6738.

228. **Kunkel, T. A., and L. A. Loeb.** 1980. On the fidelity of DNA replication. The accuracy of *Escherichia coli* DNA polymerase I in copying natural DNA *in vitro*. *J. Biol. Chem.* **255:**9961–9966.

229. **Kunkel, T. A., R. R. Meyer, and L. A. Loeb.** 1979. Single-strand binding protein enhances fidelity of DNA synthesis *in vitro*. *Proc. Natl. Acad. Sci. USA* **76:**6331–6335.

230. **Kunkel, T. A., J. R. Silber, and L. A. Loeb.** 1982. The mutagenic effect of deoxynucleotide substrate inbalances during DNA synthesis with mammalian DNA polymerases. *Mutat. Res.* **94:**413–419.

231. **Lamola, A.** 1974. Fundamental aspects of the spectroscopy and photochemistry of organic compounds; electronic energy transfer in biologic systems; and photosensitization, p. 17–55. *In* M. A. Pathak, L. C. Harber, M. Seiji, and A. Kutika (ed.), *Sunlight and Man*. University of Tokyo Press, Tokyo.

232. **Lamola, A. A.** 1969. Specific formation of thymine dimers in DNA. *Photochem. Photobiol.* **9:**291–294.

233. **Lamola, A. A., and T. Yamane.** 1967. Sensitized photodimerization of thymine in DNA. *Proc. Natl. Acad. Sci. USA* **58:**443–446.

234. **Lawley, P. D.** 1966. Effects of some chemical mutagens and carcinogens on nucleic acids. *Prog. Nucleic Acid Res. Mol. Biol.* **5:**89–131.

235. **Lawley, P. D.** 1989. Mutagens as carcinogens: development of current concepts. *Mutat. Res.* **213:**3–25.

236. **Lawley, P. D.** 1990. *N*-Nitroso compounds, p. 409–469. *In* C. S. Cooper and P. L. Grover (ed.), *Chemical Carcinogenesis and Mutagenesis I*. Springer-Verlag, KG, Berlin.

237. **Leadon, S. A., and P. C. Hanawalt.** 1983. Monoclonal antibody to DNA containing thymine glycol. *Mutat. Res.* **112:**191–200.

238. **Lemaire, D. G. E., and B. P. Ruzsicska.** 1993. Kinetic analysis of the deamination reactions of cyclobutane dimers of thymidylyl-3′,5′-2′-deoxycytidine and 2′-deoxycytidylyl-3′,5′-thymidine. *Biochemistry* **32:**2525–2533.

239. **Leng, M., E. Sage, and P. Rio.** 1981. DNA chemically modified by N-acetoxy-N-2-acetylaminofluorene: nature of the adducts and conformation of the modified DNA, p. 31–34. *In* E. Seeberg and K. Kleppe (ed.), *Chromosome Damage and Repair*. Plenum Publishing Corp., New York.

240. **Lett, J. J., K. A. Stacey, and P. Alexander.** 1961. Crosslinking of dry deoxyribonucleic acids by electrons. *Radiat. Res.* **14:**349–362.

241. **Lett, J. T.** 1990. Damage to DNA and chromatin structure. *Prog. Nucleic Acid Res. Mol. Biol.* **39:**305–352.

242. **Levin, W., A. Y. H. Lu, D. Ryan, A. W. Wood, J. Kapitulnik, S. West, M.-T. Huang, A. H. Conney, D. R. Thakker, G. Holder, H. Yogi, and D. M. Jerina.** 1977. Properties of the liver microsomal monoxygenase system and epoxide hydrase: factors influenc-

ing the metabolism and mutagenicity of benzo(*a*)pyrene, p. 659–682. *In* H. H. Hiatt, J. D. Watson and J. A. Winston (ed.), *Origins of Human Cancer. B.* Cold Spring Harbor Laboratory, Cold Spring Harbor, N.Y.

243. **Lieberman, M. W., M. J. Smerdon, T. D. Tlsty, and F. B. Oleson.** 1979. The role of chromatin structure in DNA repair in human cells damaged with chemical carcinogens and ultraviolet radiation, p. 345–363. *In* P. Emmelot and E. Kriek (ed.), *Environmental Carcinogenesis*. Elsevier/North-Holland Biomedical Press, Amsterdam.

244. **Lindahl, T.** 1974. An *N*-glycosidase from *Escherichia coli* that releases free uracil from DNA containing deaminated cytosine residues. *Proc. Natl. Acad. Sci. USA* **71:**3649–3653.

245. **Lindahl, T.** 1979. DNA glycosylases, endonucleases for apurinic/apyrimidinic sites and base excision repair. *Prog. Nucleic Acid Res. Mol. Biol.* **22:**135–192.

246. **Lindahl, T.** 1993. Instability and decay of the primary structure of DNA. *Nature* (London) **362:**709–715.

247. **Lindahl, T., and A. Andersson.** 1972. Rate of chain breakage at apurinic sites in double-stranded DNA. *Biochemistry* **11:**3618–3623.

248. **Lindahl, T., and O. Karlström.** 1973. Heat-induced depyrimidination of DNA. *Biochemistry* **12:**5151–5154.

249. **Lindahl, T., and S. Ljungquist.** 1975. Apurinic and apyrimidinic sites in DNA, p. 31–38. *In* P. C. Hanawalt and R. B. Setlow (ed.), *Molecular Mechanisms for the Repair of DNA*. Plenum Publishing Corp., New York.

250. **Lindahl, T., and B. Nyberg.** 1972. Rate of depurination of native deoxyribonucleic acid. *Biochemistry* **11:**3610–3618.

251. **Lindahl, T., and B. Nyberg.** 1974. Heat-induced deamination of cytosine residues in DNA. *Biochemistry* **13:**3405–3410.

252. **Lion, M. B.** 1968. Search for a mechanism for the increased sensitivity of 5-bromouracil-substituted DNA to ultraviolet light. *Biochim. Biophys. Acta* **155:**505–520.

253. **Lippke, J. A., L. K. Gordon, D. E. Brash, and W. A. Haseltine.** 1981. Distribution of UV light-induced damage in a defined sequence of human DNA: detection of alkaline-sensitive lesions at pyrimidine nucleoside-cytidine sequences. *Proc. Natl. Acad. Sci. USA* **78:**3388–3392.

254. **Ljungman, M., and P. C. Hanawalt.** 1992. Efficient protection against oxidative DNA damage in chromatin. *Mol. Carcinog.* **5:**264–269.

255. **Löber, G., and L. Kittler.** 1977. Selected topics in photochemistry of nucleic acids. Recent results and perspectives. *Photochem. Photobiol.* **25:**215–233.

256. **Loeb, L. A., and T. A. Kunkel.** 1981. Fidelity of DNA synthesis. *Annu. Rev. Biochem.* **52:**429–457.

257. **Loeb, L. A., and B. D. Preston.** 1986. Mutagenesis by apurinic/apyrimidinic sites. *Annu. Rev. Genet.* **20:**201–230.

258. **Loeb, L. A., L. A. Weymouth, T. A. Kunkel, K. P. Gopinathan, R. A. Beckman, and D. K. Dube.** 1978. On the fidelity of DNA replication. *Cold Spring Harbor Symp. Quant. Biol.* **43:**921–927.

259. **Love, J. D., H. T. Nguyen, A. Or, A. K. Attri, and K. W. Minton.** 1986. UV-induced interstrand cross-linking of d(GT)n·d(CA)n is facilitated by a structural transition. *J. Biol. Chem.* **261:**10051–10057.

260. **Loveless, A.** 1966. *Genetic and Allied Effects of Alkylating Agents*. Butterworths, London.

261. **Lutz, W. K.** 1990. Endogenous genotoxic agents and processes as a basis of spontaneous carcinogenesis. *Mutat. Res.* **238:**287–295.

262. **Magaña-Schwencke, N., J. A. P. Henriques, R. Chanet, and E. Moustacchi.** 1982. The fate of 8-methoxypsoralen photoinduced cross-links in nuclear and mitochondrial yeast DNA: comparison of wild type and repair deficient strain. *Proc. Natl. Acad. Sci. USA* **79:**1722–1726.

263. **Makino, F., and N. Munakata.** 1978. Deoxyuridine residues in DNA of thymine-requiring *Bacillus subtilis* strains with defective *N*-glycosidase activity for uracil-containing DNA. *J. Bacteriol.* **134:**24–29.

264. **Male, R., V. M. Fosse, and K. Kleppe.** 1982. Polyamine-induced hydrolysis of apurinic sites in DNA and nucleosomes. *Nucleic Acids Res.* **10:**6305–6318.

265. Marmur, J., and L. Grossman. 1961. Ultraviolet light induced linking of deoxyribonucleic acid strands and its reversal by photoreactivating enzyme. *Proc. Natl. Acad. Sci. USA* **47:**778–787.

266. Martin-Bertram, H. 1981. S1-sensitive sites in DNA after γ-irradiation. *Biochim. Biophys. Acta* **652:**261–265.

267. Maxam, A. M., and W. Gilbert. 1980. Sequencing end-labeled DNA with base-specific chemical cleavages. *Methods Enzymol.* **65:**499–560.

268. McCord, J. M., and I. Fridovich. 1969. Superoxide dismutase. *J. Biol. Chem.* **244:**6049–6055.

269. McCormick, J., R. H. Heflich, and V. M. Maher. 1983. Use of S1 endonuclease to quantify carcinogen-induced lesions in DNA, p. 3–11. *In* E. C. Friedberg and P. C. Hanawalt (ed.), *DNA Repair—A Laboratory Manual of Research Procedures,* vol. 2. Marcel Dekker, Inc., New York.

270. McLaren, A. D., and D. Shugar. 1964. *Photochemistry of Proteins and Nucleic Acids.* Permagon Press, Oxford.

271. MelloFilho, A. C., and R. Menegini. 1984. *In vivo* formation of single-strand breaks in DNA by hydrogen peroxide is mediated by the Haber-Weiss reaction. *Biochim. Biophys. Acta* **781:**56–63.

272. Menichini, P., G. Fronza, S. Tornaletti, S. Galiègue-Zouitina, B. Bailleul, M.-H. Loucheux-Lefebvre, A. Abbondandolo, and A. M. Pedrini. 1989. *In vitro* DNA modification by the ultimate carcinogen of 4-nitroquinoline-1-oxide: influence of superhelicity. *Carcinogenesis* **10:**1589–1593.

273. Miller, E. C., and J. A. Miller. 1947. The presence and significance of bound aminoazo dyes in the livers of rats fed p-dimethylaminoazobenzene. *Cancer Res.* **7:**468–480.

274. Miller, J. A., and Y.-J. Surh. 1994. Historical perspectives on conjugation-dependent bioactivation of foreign compounds. *Adv. Pharmacol.* **27:**1–16.

275. Miller, R. D., L. Prakash, and S. Prakash. 1982. Genetic control of excision of *Saccharomyces cerevisiae* interstrand DNA cross-links induced by psoralen plus near-UV light. *Mol. Cell. Biol.* **2:**939–948.

276. Mitchell, D. L., and J. M. Clarkson. 1984. Use of synthetic polynucleotides to characterise an antiserum made against UV-irradiated DNA. *Photochem. Photobiol.* **40:**743–748.

277. Mitchell, D. L., C. A. Haipek, and J. M. Clarkson. 1985. Further characterisation of a polyclonal antiserum for DNA photoproducts: the use of different labelled antigens to control its specificity. *Mutat. Res.* **146:**129–133.

278. Mitchell, D. L., J. Jen, and J. E. Cleaver. 1992. Sequence specificity of cyclobutane pyrimidine dimers in DNA treated with solar (ultraviolet B) radiation. *Nucleic Acids Res.* **20:**225–229.

279. Mitchell, D. L., and R. S. Nairn. 1989. The biology of the (6-4) photoproduct. *Photochem. Photobiol.* **49:**805–819.

280. Mitchell, D. L., T. D. Nguyen, and J. E. Cleaver. 1990. Nonrandom induction of pyrimidine-pyrimidone (6-4) photoproducts in ultraviolet-irradiated human chromatin. *J. Biol. Chem.* **265:**5353–5356.

281. Modrich, P. 1991. Mechanisms and biological effects of mismatch repair. *Annu. Rev. Genet.* **25:**229–254.

282. Moore, P. D., S. D. Rabkin, A. L. Osborn, C. M. King, and B. S. Strauss. 1982. Effect of acetylated and deacetylated 2-aminofluorene adducts on *in vitro* DNA synthesis. *Proc. Natl. Acad. Sci. USA.* **79:**7166–7170.

283. Moore, P. D., S. D. Rabkin, and B. S. Strauss. 1980. Termination of *in vitro* DNA synthesis at AAF adducts in DNA. *Nucleic Acids Res.* **8:**4473–4484.

284. Mori, T., T. Matsunaga, C.-C. Chang, J. E. Trosko, and O. Nikaido. 1990. In situ (6-4) photoproduct determination by laser cytometry and autoradiography. *Mutat. Res.* **236:**99–105.

285. Mori, T., T. Matsunaga, T. Hirose, and O. Nikaido. 1988. Establishment of a monoclonal antibody recognizing ultraviolet light-induced (6-4) photoproducts. *Mutat. Res.* **194:**263–270.

286. Moyer, R., D. Briley, A. Johnsen, U. Stewart, and B. R. Shaw. 1993. Echinomycin, a bis-intercalating agent, induces C → T mutations via cytosine deamination. *Mutat. Res.* **288:**291–300.

287. Moyer, R., K. Mariën, K. van Holde, and G. Bailey. 1989. Site-specific aflatoxin B1 adduction of sequence-positioned nucleosome core particles. *J. Biol. Chem.* **264:**12226–12231.

288. Muench, K. F., R. P. Misra, and M. Z. Humayun. 1983. Sequence specificity in aflatoxin B1-DNA interactions. *Proc. Natl. Acad. Sci. USA* **80:**6–10.

289. Müller, R., and M. F. Rajewsky. 1980. Antibodies specific for DNA components structurally modified by chemical carcinogens. *J. Cancer Res. Clin. Oncol.* **102:**99–113.

290. Musajo, L., G. Rodighiero, G. Caporale, F. Dall'acqua, S. Marciani, F. Bordin, F. Baccichetti, and R. Bevilacqua. 1974. Photoreactions between skin-photosensitizing furocoumarins and nucleic acid, p. 369–387. *In* M. A. Pathak, L. C. Harber, M. Seiji, and A. Kukita (ed.), *Sunlight and Man.* University of Tokyo Press, Tokyo.

291. Muzyczka, N., R. L. Poland, and M. J. Bessman. 1972. Studies on the biochemical basis of spontaneous mutation. I. A comparison of the deoxyribonucleic acid polymerase of mutator, antimutator, and wild type strains of bacteriophage T4. *J. Biol. Chem.* **247:**7116–7122.

292. Myers, C. E., R. C. Young, and B. A. Chabner. 1975. Biochemical determinants of 5-fluorouracil response *in vivo.* The role of deoxyuridylate pool expansion. *J. Clin. Invest.* **56:**1231–1238.

293. Nackerdien, Z., G. Rao, M. A. Cacciuttolo, E. Gajewski, and M. Dizdaroglu. 1991. Chemical nature of DNA-protein crosslinks produced in mammalian chromatin by hydrogen peroxide in the presence of iron or copper ions. *Biochemistry* **30:**4873–4879.

294. Nebert, D. W., M. Negishi, L. W. Enquist, and D. C. Swan. 1982. Use of recombinant DNA technology in the study of genetic differences in drug metabolism affecting individual risk of malignancy, p. 351–362. *In* C. C. Harris and P. A. Cerutti (ed.), *Mechanism of Chemical Carcinogenesis,* Alan R. Liss, Inc., New York.

295. Nguyen, H. T., and K. W. Minton. 1988. Ultraviolet-induced dimerization of non-adjacent pyrimidines. A potential mechanism for the targeted −1 frameshift mutation. *J. Mol. Biol.* **200:**681–693.

296. Nietert, W. C., L. M. Kellicutt, and H. Kubinski. 1974. DNA-protein complexes produced by a carcinogen, β-propiolactone. *Cancer Res.* **34:**859–864.

297. Niggli, H. J., and P. A. Cerutti. 1982. Nucleosomal distribution of thymine photodimers following far- and near-ultraviolet irradiation. *Biochem. Biophys. Res. Commun.* **105:**1215–1223.

298. Obe, G., C. Johannes, and D. Schulte-Frohlinde. 1992. DNA double-strand breaks induced by sparsely ionizing radiation and endonucleases as critical lesions for cell death, chromosomal aberrations, mutations and oncogenic transformation. *Mutagenesis* **7:**3–12.

299. O'Day, C. L., P. M. J. Burgers, and J.-S. Taylor. 1992. PCNA-induced DNA synthesis past *cis*-syn and *trans*-syn-I thymine dimers by calf thymus DNA polymerase δ in vitro. *Nucleic Acids Res.* **20:**5403–5406.

300. Olivera, B. M. 1978. DNA intermediates at the *Escherichia coli* replication fork. Effect of dUTP. *Proc. Natl. Acad. Sci. USA* **75:**238–242.

301. Orr, W. C., and R. S. Sohal. 1994. Extension of life-span by overexpression of superoxide dismutase and catalase in *Drosophila melanogaster. Science* **263:**1128–1130.

302. Pant, G. S., and N. Kamada. 1977. Chromosome aberrations in normal leukocytes induced by the plasma of exposed individuals. *Hiroshima J. Med. Sci.* **26:**240–244.

303. Pathak, M. A., D. M. Kramer, and T. B. Fitzpatrick. 1974. Photobiology and photochemistry of furocoumarins, p. 335–368. *In* M. A. Pathak, L. C. Harber, M. Seiji, and A. Kukita (ed.), *Sunlight and Man.* University of Tokyo Press, Tokyo.

304. Patrick, M. H., and R. O. Rahn. 1976. Photochemistry of DNA and polynucleotides: photoproducts, p. 35–95. *In* S. Y. Wang (ed.), *Photochemistry and Photobiology of Nucleic Acids,* vol. II. Academic Press, Inc., New York.

305. Peak, M. J., and J. G. Peak. 1986. DNA to protein crosslinks and backbone breaks caused by far- and near-ultraviolet and visible radiations in mammalian cells, p. 193–202. *In* M. G. Simic, L. Grossman and A. C. Upton (ed.), *Mechanisms of DNA Damage and Repair. Implications for Carcinogenesis and Risk Assessment.* Plenum Publishing Corp., New York.

306. Pearlman, D. A., S. R. Holbrook, D. H. Pirkle, and S.-H. Kim. 1985. Molecular models for DNA damaged by photoreaction. *Science* **227:**1304–1308.

307. Pegg, A. E. 1984. Methylation of the O^6 position of guanine in DNA is the most likely initiating event in carcinogenesis by methylating agents. *Cancer Invest.* **2:**223–231.

308. Perrino, F. W., and L. A. Loeb. 1990. Hydrolysis of 3'-terminal mispairs *in vitro* by the 3' → 5' exonuclease of DNA polymerase δ permits subsequent extension by polymerase α. *Biochemistry* **29:**5226–5231.

309. Petruska, J., and M. F. Goodman. 1985. Influence of neighboring bases on DNA polymerase insertion and proofreading fidelity. *J. Biol. Chem.* **260:**7533–7539.

310. Pettijohn, D., and P. C. Hanawalt. 1964. Evidence for repair replication of ultraviolet damaged DNA in bacteria. *J. Mol. Biol.* **9:**395–410.

311. Pfeifer, G. P., R. Drouin, A. D. Riggs, and G. P. Holmquist. 1991. *In vivo* mapping of a DNA adduct at nucleotide resolution: detection of pyrimidine (6-4) pyrimidone photoproducts by ligation-mediated polymerase chain reaction. *Proc. Natl. Acad. Sci. USA* **88:**1374–1378.

312. Pfeifer, G. P., R. Drouin, A. D. Riggs, and G. P. Holmquist. 1992. Binding of transcription factors creates hot spots for UV photoproducts in vivo. *Mol. Cell. Biol.* **12:**1798–1804.

313. Phillips, D. H. 1983. Fifty years of benzo(*a*)pyrene. *Nature* (London) **303:**468–472.

314. Phillips, D. H. 1990. Modern methods of DNA adduct determination, p. 503–546. *In* C. S. Cooper and P. L. Grover (ed.), *Chemical Carcinogenesis and Mutagenesis I.* Springer-Verlag, KG, Berlin.

315. Phillips, D. H., M. Castegnaro, and H. Bartsch (ed.). 1993. *Postlabelling Methods for Detection of DNA Adducts.* IARC Press, Lyon, France.

316. Piette, J., M.-P. Merville-Louis, and J. Decuyper. 1986. Damages induced in nucleic acids by photosensitization. *Photochem. Photobiol.* **44:**793–802.

317. Pinto, A. L., and S. J. Lippard. 1985. Binding of the antitumor drug *cis*-diamminedichloroplatinum(II) (cisplatin) to DNA. *Biochim. Biophys. Acta* **780:**167–180.

318. Poirier, M. C., J. R. Stanley, J. B. Beckwith, I. B. Weinstein, and S. H. Yuspa. 1982. Indirect immunofluorescent localization of benzo[*a*]pyrene adducted to nucleic acids in cultured mouse keratinocyte nuclei. *Carcinogenesis* **3:**354–348.

319. Porschke, D. 1973. A specific photoreaction in polyadenylic acid. *Proc. Natl. Acad. Sci. USA* **70:**2683–2686.

320. Porter, T. D., and M. J. Coon. 1991. Cytochrome P-450. Multiplicity of isoforms, substrates, and catalytic and regulatory mechanisms. *J. Biol. Chem* **266:**13469–13472.

321. Price, A. 1993. The repair of ionising radiation-induced damage to DNA. *Semin. Cancer Biol.* **4:**61–71.

322. Price, A. R. 1976. Bacteriophage-induced inhibitor of a host enzyme, p. 290–294. *In* D. Schlessinger (ed.), *Microbiology—1976.* American Society for Microbiology, Washington, D.C.

323. Radany, E. H., J. D. Love, and E. C. Friedberg. 1981. The use of direct photoreversal of UV-irradiated DNA for the demonstration of pyrimidine dimer-DNA glycosylase activity, p. 91–95. *In* E. Seeberg and K. Kleppe (ed.), *Chromosome Damage and Repair.* Plenum Publishing Corp., New York.

324. Radman, M., C. Dohet, M.-F. Bourgingnon, O. P. Doubleday, and P. Lecomte. 1981. High fidelity devices in the reproduction of DNA, p. 431–445. *In* E. Seeberg and K. Kleppe (ed.), *Chromosome Damage and Repair.* Plenum Publishing Corp., New York.

325. Radman, M., G. Villani, S. Boiteux, A. R. Kinsella, B. W. Glickman, and S. Spadari. 1978. Replicational fidelity: mechanisms of mutation avoidance and mutation fixation. *Cold Spring Harbor Symp. Quant. Biol.* **43:**937–946.

326. Radman, M., and R. Wagner. 1986. Mismatch repair in *Escherichia coli. Annu. Rev. Genet.* **20:**523–538.

327. Raleigh, J. A., A. F. Fuciarelli, and C. R. Kulatunga. 1987. Potential limitation to hydrogen atom donation as a mechanism of repair in chemical models of radiation damage, p. 33–39. *In* P. A. Cerutti, O. F. Nygaard, and M. G. Simic (ed.), *Anticarcinogenesis and Radiation Protection.* Plenum Publishing Corp., New York.

328. Randerath, K., J. G. Liehr, A. Gladek, and E. Randerath. 1990. Age-related DNA modifications (I-compounds): modulation by physiological and pathological processes. *Mutat. Res.* **238:**245–253.

329. Reddy, M. V., and K. Randerath. 1986. Nuclease P1-mediated enhancement of sensitivity of ^{32}P-postlabeling test for structurally diverse DNA adducts. *Carcinogenesis* **7:**1543–1551.

330. Reha-Krantz, L. J. 1988. Amino acid changes coded by bacteriophage T4 polymerase mutator mutants. Relating structure to function. *J. Mol. Biol.* **202:**711–724.

331. Rehn, L. 1895. Blasengeschwulste bei Fuchsin-Arbeitern. *Arch. Klin. Chir.* **50:**588.

332. Rhaese, H.-J., and E. Freese. 1968. Chemical analysis of DNA alterations. I. Base liberation and backbone breakage of DNA and oligodeoxyadenylic acid induced by hydrogen peroxide and hydroxylamine. *Biochim. Biophys. Acta* **155:**476–490.

333. Rich, A., A. Nordheim, and A. H. Wang. 1984. The chemistry and biology of left-handed Z-DNA. *Annu. Rev. Biochem.* **53:**791–846.

334. Richardson, F. C., and K. K. Richardson. 1990. Sequence-dependent formation of alkyl DNA adducts: a review of methods, results, and biological correlates. *Mutat. Res.* **233:**127–138.

335. Richter, C., J.-W. Park, and B. N. Ames. 1988. Normal oxidative damage to mitochondrial and nuclear DNA is extensive. *Proc. Natl. Acad. Sci. USA* **85:**6465–6467.

336. Riley, P. A. 1994. Free radicals in biology: oxidative stress and the effects of ionizing radiation. *Int. J. Radiat. Biol.* **65:**27–33.

337. Ripley, L. S. 1981. Influence of diverse gene 43 DNA polymerases on the incorporation and replication *in vivo* of 2-aminopurine at A·T base-pairs in bacteriophage T4. *J. Mol. Biol.* **150:**197–216.

338. Roberts, J. J. 1978. The repair of DNA modified by cytotoxic, mutagenic, and carcinogenic chemicals. *Adv. Radiat. Biol.* **7:**211–435.

339. Roberts, J. J., and J. M. Pascoe. 1972. Cross-linking of complementary strands of DNA in mammalian cells by anti-tumour platinum compounds. *Nature* (London) **235:**282–284.

340. Rodighiero, G., F. Dall'Acqua, and D. Averbeck. 1988. New psoralen and angelicin derivatives, p. 37–114. *In* F. P. Gasparro (ed.), *Psoralen DNA Photobiology.* CRC Press, Inc., Boca Raton, Fla.

341. Rosenstein, B. S., and J. M. Ducore. 1983. Induction of DNA strand breaks in normal human fibroblasts exposed to monochromatic ultraviolet and visible wavelengths in the 240–546 nm range. *Photochem. Photobiol.* **38:**51–55.

342. Ross, W. C. J. 1962. *Biological Alkylating Agents.* Butterworths, London.

343. Rotilio, G. (ed). 1986. *Superoxide and Superoxide Dismutase in Chemistry, Biology and Medicine.* Elsevier Biomedical Press, Amsterdam.

344. Royer-Pokora, B., L. K. Gordon, and W. A. Haseltine. 1981. Use of exonuclease III to determine the site of stable lesions in defined sequences of DNA: the cyclobutane pyrimidine dimer and cis and trans dichlorodiammine platinum II examples. *Nucleic Acids Res.* **9:**4595–4609.

345. Rycyna, R., and J. Alderfer. 1985. UV irradiation of nucleic acids: formation, purification, and solution conformational analysis of the '6-4 lesion' of dTpdT. *Nucleic Acids Res.* **13:**5949–5963.

346. Sage, E. 1993. Distribution and repair of photolesions in DNA: genetic consequences and the role of sequence context. *Photochem. Photobiol.* **57:**163–174.

347. Sage, E., and M. Leng. 1980. Conformation of poly(dG·dC)· poly(d·G·dC) modified by the carcinogens N-acetoxy-N-2-acetylaminofluorene and N-hydroxy-N-2-aminofluorene. *Proc. Natl. Acad. Sci. USA* **77:**4597–4601.

348. Salts, Y., and A. Ronen. 1971. Neighbor effects in the mutation of ochre triplets in the T4rII gene. *Mutat. Res.* **13:**109–113.

349. Sambrook, J., E. F. Fritsch, and T. Maniatis. 1989. *Molecular Cloning: A Laboratory Manual,* 2nd ed. Cold Spring Harbor Laboratory, Cold Spring Harbor, N.Y.

350. Saran, M., and W. Bors. 1990. Radical reactions in vivo—an overview. *Radiat. Environ. Biophys.* **29:**249–262.

351. Satoh, M. S., C. J. Jones, R. D. Wood, and T. Lindahl. 1993. DNA excision-repair defect of xeroderma pigmentosum prevents removal of a class of oxygen free radical-induced base lesions. *Proc. Natl. Acad. Sci. USA* **90:**6335–6339.

352. Schaaper, R. M. 1993. Base selection, proofreading, and mismatch repair during DNA replication in *Escherichia coli. J. Biol. Chem.* **268:**23762–23765.

353. Schneider, J. E., S. Price, L. Maidt, J. M. C. Gutteridge, and R. A. Floyd. 1990. Methylene blue plus light mediates 8-hydroxy-2'-deoxyguanosine formation in DNA preferentially over strand breakage. *Nucleic Acids Res.* **18:**631–635.

354. Schuster, H. 1960. The reaction of nitrous acid with deoxyribonucleic acid. *Biochem. Biophys. Res. Commun.* **2:**320–323.

355. Scott, D. 1969. The effect of irradiated plasma on normal chromosomes and its relevance to the longlived lymphocyte hypothesis. *Cell Tissue Kinet.* **2:**295–305.

356. Seawell, P. C., and A. K. Ganesan. 1981. Measurement of strand breaks in supercoiled DNA by gel electrophoresis, p. 425–430. *In* E. C. Friedberg and P. C. Hanawalt (ed.), *DNA Repair—A Laboratory Manual of Research Procedures,* vol. 1B. Marcel Dekker, Inc., New York.

357. Seidman, M., H. Slor, and M. Bustin. 1983. The binding of a carcinogen to the nucleosomal and non-nucleosomal regions of the simian virus 40 chromosome *in vivo. J. Biol. Chem.* **258:**5215–5220.

358. Selkirk, J. K., M. C. Macleod, C. J. Moore, B. K. Mansfield, A. Nikbakht, and K. Dearstone. 1982. Species variance in the metabolic activation of polycyclic hydrocarbons, p. 331–349. *In* C. C. Harris and P. A. Cerutti (ed.), *Mechanisms of Chemical Carcinogenesis,* Alan R. Liss, Inc., New York.

359. Setlow, J. K. 1966. The molecular basis of biological effects of ultraviolet radiation and photoreactivation. *Curr. Top. Radiat. Res.* **2:**195–248.

360. Setlow, P. 1992. I will survive: protecting and repairing spore DNA. *J. Bacteriol.* **174:**2737–2741.

361. Setlow, R. B. 1966. Cyclobutane-type pyrimidine dimers in polynucleotides. *Science* **153:**379–386.

362. Setlow, R. B. 1968. The photochemistry, photobiology, and repair of polynucleotides. *Prog. Nucleic Acid Res. Mol. Biol.* **8:**257–295.

363. Setlow, R. B. 1974. The wavelengths in sunlight effective in producing skin cancer: a theoretical analysis. *Proc. Natl. Acad. Sci. USA* **71:**3363–3366.

364. Setlow, R. B., and W. L. Carrier. 1966. Pyrimidine dimers in ultraviolet-irradiated DNA's. *J. Mol. Biol.* **17:**237–254.

365. Shaham, M., Y. Becker, and M. M. Cohen. 1980. A diffusable clastogenic factor in ataxia telangiectasia. *Cytogenet. Cell Genet.* **27:**155–161.

366. Shapiro, R. 1981. Damage to DNA caused by hydrolysis, p. 3–12. *In* E. Seeberg and K. Kleppe (ed.), *Chromosome Damage and Repair.* Plenum Publishing Corp., New York.

367. Shapiro, R., S. Dubelman, A. M. Feinberg, P. F. Crain, and J. A. M. Closkey. 1977. Isolation and identification of cross-linked nucleosides from nitrous acid treated deoxyribonucleic acid. *J. Am. Chem. Soc* **99:**302–303.

368. Shapiro, R., and R. S. Klein. 1966. The deamination of cytidine and cytosine by acidic buffer solutions. Mutagenic implications. *Biochemistry* **5:**2358–2362.

369. Shetlar, M. D., J. Christensen, and K. Hom. 1984. Photochemical addition of amino acids and peptides of DNA. *Photochem. Photobiol.* **39:**125–133.

370. Shetlar, M. D., K. Hom, J. Carbone, D. Moy, E. Steady, and M. Watanabe. 1984. Photochemical addition of amino acids and peptides to homopolyribonucleotides of the major DNA bases. *Photochem. Photobiol.* **39:**135–140.

371. Shibutani, S., M. Takeshita, and A. P. Grollman. 1991. Insertion of specific bases during DNA synthesis past the oxidation-damaged base 8-oxodG. *Nature* (London) **349:**431–434.

372. Shlomai, J., and A. Kornberg. 1978. Unpublished data (quoted in reference 422).

373. Siddiqi, M. A., and E. Bothe. 1987. Single- and double-strand break formation in DNA irradiated in aqueous solution: dependence on dose and radical scavenger concentration. *Radiat. Res.* **112:**449–463.

374. Sies, H., W. Stahl, and A. R. Sundquist. 1992. Antioxidant functions of vitamins. Vitamins E and C, beta-carotene, and other carotenoids. *Ann. N. Y. Acad. Sci.* **669:**7–20.

375. Simic, M. G. 1994. DNA markers of oxidative processes *in vivo*: relevance to carcinogenesis and anticarcinogenesis. *Cancer Res.* **54 (Suppl.):**1918S–1923S.

376. Simic, M. G., and M. Dizdaroglu. 1985. Formation of radiation-induced cross-links between thymine and tyrosine: possible model for cross-linking of DNA and proteins by ionizing radiation. *Biochemistry* **24:**233–236.

377. Simic, M. G., K. A. Taylor, J. F. Ward, and C. von Sonntag (ed). 1988. *Oxygen radicals in biology and medicine,* Basic Life Sciences, vol. 49. Plenum Publishing Corp., New York.

378. Sinden, R. R., and R. S. Cole. 1978. Repair of cross-linked DNA and survival of *Escherichia coli* treated with psoralen plus lights: effects of mutation influencing genetic recombinations and DNA metabolism. *J. Bacteriol.* **136:**538–547.

379. Singer, B. 1975. The chemical effects of nucleic acid alkylation and their relation to mutagenesis and carcinogenesis. *Prog. Nuceic Acid Res. Mol. Biol.* **15:**219–284.

380. Singer, B. 1986. O-Alkyl pyrimidines in mutagenesis and carcinogenesis: occurrence and significance. *Cancer Res.* **46:**4879–4885.

381. Singer, B., and D. Grunberger. 1983. *Molecular Biology of Mutagens and Carcinogens.* Plenum Publishing Corp., New York.

382. Singer, B., and J. T. Kuśmierek. 1982. Chemical mutagenesis. *Annu. Rev. Biochem.* **51:**655–693.

383. Sinha, N. K. 1987. Specificity and efficiency of editing of mismatches involved in the formation of base-substitution mutations by the 3' → 5' exonuclease activity of phage T4 DNA polymerase. *Proc. Natl. Acad. Sci. USA* **84:**915–919.

384. Sirover, M. D., and L. A. Loeb. 1976. Infidelity of DNA synthesis *in vitro*: screening for potential metal mutagens or carcinogens. *Science* **194:**1434–1436.

385. Slater, T. F. 1972. *Free Radical Mechanisms in Tissue Injury.* Pion Press, London.

386. Smith, C. A. 1988. Repair of DNA containing furocoumarin adducts, p. 87–116. *In* F. P. Gasparro (ed.), *Psoralen Photobiology II.* CRC Press, Inc., Boca Raton, Fla.

387. Smith, K. C. 1968. The biological importance of UV-induced DNA-protein cross-linking *in vivo* and its probable chemical mechanism. *Photochem. Photobiol.* **7:**651–660.

388. Smith, K. C. 1974. Molecular changes in the nucleic acids produced by ultraviolet and visible radiation, p. 67–77. *In* M. A. Pathak, L. C. Harber, M. Seiji, and A. Kukita (ed.), *Sunlight and Man.* University of Tokyo Press, Tokyo.

389. Smith, K. C. 1976. The radiation-induced addition of proteins and other molecules to nucleic acids, p. 187–218. *In* S. Y. Wang (ed.), *Photochemistry and Photobiology of Nucleic Acids,* vol. II. Academic Press, Inc., New York.

390. Smith, K. C., and P. C. Hanawalt. 1969. *Molecular Photobiology.* Academic Press, Inc., New York.

391. **Smith, K. C., and M. E. O'Leary.** 1967. Photoinduced DNA-protein cross-links and bacterial killing: a correlation at low temperatures. *Science* **155:**1024–1026.

392. **Snow, E. T., and S. Mitra.** 1987. Do carcinogen-modified deoxynucleotide precursors contribute to cellular mutagenesis? *Cancer Invest.* **5:**119–125.

393. **Sowers, L. C., B. R. Shaw, M. L. Veigl, and W. D. Sedwick.** 1987. DNA base modification: ionized base pairs and mutagenesis. *Mutat. Res.* **177:**201–218.

394. **Speyer, J. F.** 1965. Mutagenic DNA polymerase. *Biochem. Biophys. Res. Commun.* **21:**6–8.

395. **Speyer, J. F., and D. Rosenberg.** 1968. The function of T4 DNA polymerase. *Cold Spring Harbor Symp. Quant. Biol.* **33:**345–350.

396. **Stortz, G., L. A. Tartaglia, and B. N. Ames.** 1990. Transcriptional regulator of oxidative stress-inducible genes: direct activation by oxidation. *Science* **248:**189–194.

397. **Stortz, G., L. A. Tartaglia, S. B. Farr, and B. N. Ames.** 1990. Bacterial defenses against oxidative stress. *Trends Genet.* **6:**363–368.

398. **Strauss, B. S.** 1981. Use of benzoylated naphthoylated DEAE cellulose, p. 319–339. *In* E. C. Friedberg and P. C. Hanawalt (ed.), *DNA Repair—A Laboratory Manual of Research Procedures*, vol. 1B. Marcel Dekker, Inc., New York.

399. **Strickland, P. T.** 1985. Immunoassay of DNA modified by ultraviolet radiation: a review. *Environ. Mutagen.* **7:**599–607.

400. **Strickland, P. T., M. N. Routledge, and A. Dipple.** 1993. Methodologies for measuring carcinogen adducts in humans. *Cancer Epidemiol. Biomed. Prev.* **2:**607–619.

401. **Stryer, L.** 1988. *Biochemistry.* W. H. Freeman & Co., New York.

402. **Swain, C. G., and C. B. Scott.** 1953. Quantitative correlation of relative rates. Comparison of hydroxide ion with other nucleophilic reagents toward alkyl halides, esters, epoxides and acyl halides. *J. Am. Chem. Soc.* **75:**141–147.

403. **Takahashi, I., and J. Marmur.** 1963. Replacement of thymidylic acid by deoxyuridylic acid in the deoxyribonucleic acid of a transducing phage for *Bacillus subtilis. Nature* (London) **197:**794–795.

404. **Tamanoi, F., and T. Okazaki.** 1978. Uracil incorporation into nascent DNA of thymine-requiring mutant of *B. subtilis* 168. *Proc. Natl. Acad. Sci USA* **75:**2195–2199.

405. **Tamm, C., H. S. Shapiro, R. Lipschitz, and E. Chargaff.** 1953. Distribution density of nucleotides within a deoxyribonucleic acid chain. *J. Biol. Chem.* **203:**673–688.

406. **Tartaglia, L. A., G. Storz, S. B. Farr, and B. A. Ames.** 1991. The bacterial adaptation to hydrogen peroxide stress, p. 155–169. *In* H. Sies (ed.), *Oxidative Stress: Oxidants and Antioxidants.* Academic Press, Ltd., London.

407. **Taylor, J.-S., and I. R. Brockie.** 1988. Synthesis of a *trans-syn* thymine dimer building block. Solid phase synthesis of CGTAT[*t,s*]TATGC. *Nucleic Acids Res.* **16:**5123–5136.

408. **Taylor, J.-S., and M. Cohrs.** 1987. DNA, light, and Dewar pyrimidones: the structure and biological significance of TpT3. *J. Am. Chem. Soc.* **109:**2834–2835.

409. **Taylor, J.-S., D. S. Garrett, I. R. Brockie, D. L. Svoboda, and J. Telser.** 1990. ^{1}H NMR assignment and melting temperature study of cis-syn and trans-syn thymine dimer containing duplexes of d(CGTATTATGC)·d(GCATAATACG). *Biochemistry* **29:**8858–8866.

410. **Taylor, J.-S., D. S. Garrett, and M. P. Cohrs.** 1988. Solution-state structure of the Dewar pyrimidinone photoproduct of thymidylyl-(3'-5')-thymidine. *Biochemistry* **27:**7206–7215.

411. **Taylor, J.-S., and C. L. O'Day.** 1990. Cis-syn thymine dimers are not absolute blocks to replication by DNA polymerase I of *Escherichia coli* in vitro. *Biochemistry* **29:**1624–1632.

412. **Teebor, G. W., K. Frenkel, and M. S. Goldstein.** 1984. Ionizing radiation and tritium transmutation both cause formation of 5-hydroxymethyl-2'-deoxyuridine in cellular DNA. *Proc. Natl. Acad. Sci. USA* **81:**318–321.

413. **Téoule, R.** 1987. Radiation-induced DNA damage and its repair. *Int. J. Radiat. Biol.* **51:**573–589.

414. **Tessman, I., and M. A. Kennedy.** 1991. The two-step model of UV mutagenesis reassessed: deamination of cytosine as the likely source of the mutations associated with deamination. *Mol. Gen. Genet.* **227:**144–148.

415. **Thoma, F.** 1992. Nucleosome positioning. *Biochim. Biophys. Acta* **1130:**1–19.

416. **Thomas, D. C., A. G. Morton, V. A. Bohr, and A. Sancar.** 1988. General method for quantifying base adducts in specific mammalian genes. *Proc. Natl. Acad. Sci. USA* **85:**3723–3727.

417. **Thrall, B. D., D. B. Mann, M. J. Smerdon, and D. L. Springer.** 1992. DNA polymerase, RNA polymerase and exonuclease activities on a DNA sequence modfied by benzo[*a*]pyrene diolepoxide. *Carcinogenesis* **13:**1529–1534.

418. **Thrall, B. D., D. B. Mann, M. J. Smerdon, and D. L. Springer.** 1994. Nucleosome structure modulates benzo[*a*]pyrenediol epoxide adduct formation. *Biochemistry* **33:**2210–2216.

419. **Tomita, F., and I. Takahashi.** 1969. A novel enzyme dCTP deaminase, found in *Bacillus subtilis* infected with phage PBS1. *Biochim. Biophys. Acta* **179:**18–27.

420. **Topal, M. D.** 1988. DNA repair, oncogenes and carcinogenesis. *Carcinogenesis* **9:**691–696.

421. **Tornaletti, S., D. Rozek, and G. P. Pfeifer.** 1993. The distribution of UV photoproducts along the human p53 gene and its relation to mutations in skin cancer. *Oncogene* **8:**2051–2057.

422. **Tye, B.-K., J. Chien, I. R. Lehman, B. K. Duncan, and H. R. Warner.** 1978. Uracil incorporation: a source of pulse-labeled DNA fragments in the replication of the *Escherichia coli* chromosome. *Proc. Natl. Acad. Sci. USA* **75:**233–237.

423. **Tye, B.-K., and I. R. Lehman.** 1977. Excision repair of uracil incorporated in DNA as a result of a defect in dUTPase. *J. Mol. Biol.* **117:**293–306.

424. **Tye, B.-K., P.-O. Nyman, I. R. Lehman, S. Hochhauser, and B. Weiss.** 1977. Transient accumulation of Okazaki fragments as a result of uracil incorporation into nascent DNA. *Proc. Natl. Acad. Sci. USA* **74:**154–157.

425. **Tyrrell, R. M.** 1991. UVA (320–380 nm) radiation as an oxidative stress, p. 57–83. *In* H. Sies (ed.), *Oxidative Stress: Oxidants and Antioxidants.* Academic Press, Ltd., London.

426. **Ullman, J. S., and B. J. McCarthy.** 1973. Alkali deamination of cytosine residues in DNA. *Biochim. Biophys. Acta* **294:**396–404.

427. **Ullrich, V., I. Roots, A. Hildebrandt, R. W. Estabrook, and A. H. Conney (ed.).** 1977. *Microsomes and Drug Oxidations.* Pergamon Press, Oxford.

428. **Umlas, M. E., W. A. Franklin, G. L. Chan, and W. A. Haseltine.** 1985. Ultraviolet light irradiation of defined-sequence DNA under conditions of chemical photosensitization. *Photochem. Photobiol.* **42:**265–273.

429. **Vairapandi, M., and N. J. Duker.** 1994. Excision of ultraviolet-induced photoproducts of 5-methylcytosine from DNA. *Mutat. Res.* 315:85–94.

430. **Van Holde, K. E.** 1989. *Chromatin.* Springer Verlag, KG, Berlin.

431. **Varghese, A. J.** 1970. 5-Thyminyl-5,6-dihydrothymine from DNA irradiated with ultraviolet light. *Biochem. Biophys. Res. Commun.* **38:**484–490.

432. **Varghese, A. J.** 1972. Photochemistry of nucleic acids and their constituents. *Photophysiology* **7:**207–274.

433. **Varghese, A. J., and S. Y. Wang.** 1968. Thymine-thymine adduct as a photoproduct of thymine. *Science* **160:**186–187.

434. **Varghese, A. J., and S. Y. Wang.** 1967. Ultraviolet irradiation of DNA *in vitro* and *in vivo* produced a third thymine-derived product. *Science* **156:**955–957.

435. **Vogel, F.** 1992. Risk calculations for hereditary effects of ionizing radiation in humans. *Hum. Genet.* **89:**127–146.

436. von Sonntag, C. 1987. *The Chemical Basis of Radiation Biology.* Taylor & Francis, London.

437. Wacker, A., H. Dellweg, L. Träger, A. Kornhauser, E. Lodemann, G. Türck, R. Selzer, P. Chandra, and M. Ishimoto. 1964. Organic photochemistry of nucleic acids. *Photochem. Photobiol.* **3:**369–394.

438. Wallace, S. S. 1988. AP endonucleases and DNA glycosylases that recognize oxidative DNA damage. *Environ. Mol. Mutagen.* **12:**431–477.

439. Wallace, S. S. 1988. Detection and repair of DNA base damages produced by ionizing radiation. *Environ. Mutagen.* **5:**769–788.

440. Wang, A. H.-J., G. J. Quigley, F. J. Kolpak, G. van der Marel, and J. H. van Boom. 1981. Left-handed double helical DNA: variations in the backbone conformation. *Science* **211:**171–176.

441. Wang, C.-I., and J.-S. Taylor. 1991. Site-specific effect of thymine dimer formation on dA$_n$·dT$_n$ tract bending and its biological implications. *Proc. Natl. Acad. Sci. USA* **88:**9072–9076.

442. Wang, S. Y. 1965. Photochemical reactions of nucleic acid components in frozen solutions. *Fed. Proc.* **24:**71.

443. Wang, S. Y. (ed). 1976. *Photochemistry and Photobiology of Nucleic Acids.* Academic Press, Inc., New York.

444. Ward, J. F. 1985. Biochemistry of DNA lesions. *Radiat. Res.* **103:**103–111.

445. Ward, J. F. 1988. DNA damage produced by ionizing radiation in mammalian cells: identities, mechanisms of formation and reparability. *Prog. Nucleic Acid Res. Mol. Biol.* **35:**95–125.

446. Ward, J. F. 1990. The yield of DNA double-strand breaks produced intracellularly by ionizing radiation: a review. *Int. J. Radiat. Biol.* **57:**1141–1150.

447. Wardman, P. 1993. Radiation chemistry applied to drug design. *Int. J. Radiat. Biol.* **65:**35–41.

448. Warner, H. R., and B. K. Duncan. 1978. *In vivo* synthesis and properties of uracil-containing DNA. *Nature* (London) **272:**32–34.

449. Watson, J. D. 1976. *Molecular Biology of the Gene.* W. A. Benjamin Inc., Menlo Park, Calif.

450. Wei, C.-F., G. A. Alianell, H. B. Gray, Jr., R. J. Legerski, and D. L. Robberson. 1983. Use of *Altermonas* BAL31 nuclease as probe for covalent alterations in duplex DNA, p. 13–40. *In* E. C. Friedberg and P. C. Hanawalt (ed.), *DNA Repair—A Laboratory Manual of Research Procedures,* vol. 2. Marcel Dekker, Inc., New York.

451. Wei, H., and K. Frenkel. 1991. *In vivo* formation of oxidized DNA bases in tumor promoter-treated mouse skin. *Cancer Res.* **51:**4443–4449.

452. Weinblum, D., and H. Johns. 1966. Isolation and properties of isomeric thymine dimers. *Biochim. Biophys. Acta* **114:**450–459.

453. Weinstein, I. B., A. M. Jeffrey, K. W. Jenette, S. H. Blobstein, R. G. Harvey, C. Harris, H. Autrup, H. Kasai, and K. Nakanishi. 1976. Benzo[*a*]pyrene diol epoxides as intermediates in nucleic acid binding in vitro and in vivo. *Science* **193:**592–594.

454. Westra, J. G., E. Kriek, and H. Hittenhausen. 1976. Identification of the persistently bound form of the carcinogen N-acetyl-2-aminofluorene to rat liver DNA in vivo. *Chem-Biol. Interact.* **15:**149–164.

455. Weymouth, L. A., and L. A. Loeb. 1978. Mutagenesis during *in vitro* DNA synthesis. *Proc. Natl. Acad. Sci. USA* **75:**1924–1928.

456. Wild, C. P. 1990. Antibodies to DNA alkylation adducts as analytical tools in chemical carcinogenesis. *Mutat. Res.* **233:**219–233.

457. Wilkins, R. J. 1984. Sequence specificities in the interactions of chemicals and radiations with DNA. *Mol. Cell. Biochem.* **64:**111–126.

458. Williams, J. I., and E. C. Friedberg. 1979. Deoxyribonucleic acid excision repair in chromatin after ultraviolet irradiation of human fibroblasts in culture. *Biochemistry* **18:**3965–3972.

459. Williams, L. D., and B. R. Shaw. 1987. Protonated base pairs explain the ambiguous pairing properties of O^6-methylguanine. *Proc. Natl. Acad. Sci. USA* **84:**1779–1783.

460. Witkin, E. M. 1976. Ultraviolet mutagenesis and inducible DNA repair in *Escherichia coli. Bacteriol. Rev.* **40:**869–907.

461. Yamane, T., B. J. Wyluda, and R. G. Shulman. 1967. Dihydrothymine from UV-irradiated DNA. *Proc. Natl. Acad. Sci. USA* **58:**439–442.

462. Zoltewicz, J. A., F. O. Clark, T. W. Sharpless, and G. Grahe. 1970. Kinetics and mechanism of the acid-catalyzed hydrolysis of some purine nucleosides. *J. Am. Chem. Soc.* **92:**1741–1750.

2

Introduction to Mutagenesis

One of the major reasons that so much effort has been devoted to the analysis of DNA damage and to understanding the biological mechanisms for its repair is that mutations can be introduced into an organism's DNA as a consequence of such damage. Depending on where in the organism's genome a mutation occurs, it can have a variety of direct molecular effects, for example altering the structure of a protein or RNA or causing a change in gene expression. The subsequent physiological consequences of these changes on the organism's capacity for growth, survival, development, and reproduction can range from the undetectable to the profound and have been the object of intensive experimental investigation for many years. As Jan Drake has put it, "Mutation fascinates because of its three faces: the variability it generates that conditions all evolutionary change, the disease it generates that consumes a substantial proportion of our resources, and the means it offers for dissecting all facets of biological phenomena" (46). With particular respect to human health, an interest in understanding the molecular basis of mutagenesis has been strongly stimulated by the discovery that mutations occurring in somatic cells can be important in the development of cancer and by interest in understanding the origin of genetic diseases.

In this chapter, we will define a number of terms that are used when discussing mutations and their biological consequences. We will also describe examples of systems that have been developed to allow convenient analysis of mutations and summarize some of the simpler mechanisms that have been discovered that can lead to the introduction of mutations into DNA. More-extensive discussions of the molecular mechanisms of chemical and radiation mutagenesis in prokaryotes and eukaryotes can be found in chapters 11 and 12, respectively.

Mutations and Mutants: Some Definitions

The terms that are used to discuss the occurrence of mutations and their subsequent biological consequences can be confusing and are frequently misused. We will therefore begin with some definitions. A *mutation* is a heritable change in the sequence of an organism's genome; the full complement of an organism's genetic material is referred to as its *genome*. An organism that carries one or more mutations in its genome is referred to as a *mutant*. Different mutations that have been shown to be located at the same genetic locus (for example, the *lacZ* gene of *Escherichia coli*) are said to be *alleles* of each other. Each mutation at a particular locus is assigned a unique designator, usually referred to as an *allele number*. The particular mutation one is describing can thus be unambiguously specified by stating the name of its genetic locus and its allele number or designator (for example, *lacZ32* refers to a particular mutation located in the *lacZ* gene).

The genetic information that an organism encodes in its genome is referred to as its *genotype*. The ensemble of observable characteristics of an organism is referred to as its *phenotype*. The consequences of a single mutation on the phenotype of a mutant organism are highly dependent on where in the organism's DNA (or RNA in the case of organisms that use RNA as their genetic material) the mutation occurred, on the exact nature of the sequence alteration caused by the mutation, and on various genetic characteristics of the affected organism. A single mutation in the genetic material of an organism may not alter the phenotype at all, it may change one or more aspects of the phenotype, or it may even lead to inviability of the organism. A mutation that has multiple effects on an organism's phenotype is said to be *pleiotropic*. When comparing two organisms that differ by one or more mutations, the "normal" or "parental" organism is often referred to as having a *wild-type* phenotype. This should be regarded as an operational definition, not as a statement regarding some aspect of fundamental biology. Individuals of the same species found in nature do not exhibit identical phenotypes, so that there is no unique "wild-type" phenotype. Furthermore, particularly for organisms that are intensively studied in laboratories, the "wild-type" parent may already have acquired various mutations that distinguish it from nonlaboratory examples of the same species. In this book we have adopted the convention of using a superscript plus sign when we are referring to the wild-type gene of a prokaryote (for example, *lacZ*$^+$ indicates that we are referring to the wild-type form of the *lacZ* gene) or italicized capital letters when referring to the wild type gene of a eukaryote (for example, *RAD3*).

A mutation that changes the phenotype from wild type to a mutant phenotype is said to be a *forward* mutation, whereas a mutation that causes a change of the phenotype from mutant to wild type is said to be a *reversion* mutation. If a condition can be arranged in which all the members of the population die or fail to grow except for the desired class of mutants, the mutants can be obtained by *selection*. If it is necessary to examine all the members of the population to identify the mutants having the desired phenotype, the mutants must be identified by *screening*.

A *mutagen* is an agent that leads to an increase in the frequency of occurrence of mutations. The majority of known mutagens are physical or chemical agents, and these will be the focus of this book. However, it should be remembered that biological agents, for example certain viruses, bacteriophages, or transposable genetic elements, can also increase the frequency of occurrence of mutations and can thus act as mutagens. Most chemical and physical mutagens act by introducing lesions into the DNA. There may be a very direct connection between the initial action of the mutagen and the introduction of mutations into the DNA. However, as will be discussed in chapters 11 and 12 in particular, the processes that occur after the initial action of the mutagen and the ultimate introduction of the mutation can be very complex.

The process by which mutations are produced is referred to as *mutagenesis* (Table 2–1). Mutagenesis that occurs without treatment of the organism with an exogenous mutagen is referred to as *spontaneous mutagenesis*. Spontaneous mutations can occur because of *replication errors*, as discussed in chapter 1, or can arise as a consequence of lesions that are introduced in DNA during normal growth of the cell. A number of examples of such spontaneously arising lesions are discussed

Table 2–1 Key definitions

Term	Definition
Mutation	A heritable change in the sequence of an organism's genome
Mutant	An organism that carries one or more mutations in its genome
Genotype	The genetic information that an organism encodes in its genome
Phenotype	The ensemble of observable characteristics of an organism
Mutagen	An agent that leads to an increase in the frequency of occurrence of mutations
Mutagenesis	The process by which mutations are produced

in chapter 1. In addition, spontaneous mutations can arise as a consequence of other types of mechanisms such as the insertion of a transposable genetic element or by recombination between partially homologous sequences (see chapter 9). Mutagenesis that results from the treatment of an organism with a chemical mutagen or a physical mutagen, such as UV irradiation, is often referred to as *chemical mutagenesis* or *UV mutagenesis*, respectively. It is also sometimes referred to as *induced mutagenesis*, to distinguish it from spontaneous mutagenesis, but this term can be confusing in certain contexts because of the many usages of the word "induction" in modern biology. If the presence of a particular lesion in DNA can result in mutation at the site where it is located, the lesion is termed a *premutagenic* lesion. If a mutation occurs at a site where there was a premutagenic DNA lesion, that mutation is said to be a *targeted* mutation. Mutations occurring at sites where there was not known to be a DNA lesion are referred to as *untargeted* mutations.

The *mutant frequency* refers to the proportion of mutants in a population. The *mutation rate* expresses mutations giving a particular scorable phenotype per DNA replication; in a few cases this can be expressed as mutations per base pair per replication (44, 47, 116). The *genomic mutation rate* expresses mutations per genome per DNA replication (45, 47, 48).

Point Mutations and Other Classes of Mutations

For the purposes of this book, it is most useful to classify mutations initially on the basis of the sequence alterations they cause in an organism's DNA and only then to consider the direct effects they cause on cellular molecules or any subsequent phenotypic manifestations of the mutations. A simple way of classifying mutations on the basis of the DNA sequence changes they cause is by specifying whether they are point mutations or not. *Point mutations* are mutations that result from the substitution of one base pair for another or from the addition or deletion of a small number of base pairs. These are the types of mutations that most commonly result from exposure to physical or chemical mutagens and will be the major focus of this book.

It is important to remember, however, that there are other biologically important classes of mutations that involve more extensive changes in DNA sequence. These include deletions, insertions, duplications, and inversions and can involve large pieces of DNA encoding many genes. Furthermore, in organisms that have multiple chromosomes, mutational events can occur that involve the joining of the DNA of one chromosome to the DNA of another, thereby causing chromosomal rearrangements. In addition, mutations can occur that affect the number of chromosomes present. The molecular basis of at least some mutations involving extensive DNA rearrangements is understood. For example, some arise as the result of recombination between homologous or partially homologous DNA sequences within an organism's genome while others are mediated by the action of transposable genetic elements. DNA damage may influence these events, for example by leading to the production of nicks, gaps, and double strand breaks in DNA, but in general the relationship between DNA damage and the subsequent introduction of these types of mutations is not as well understood as in the case of point mutations. Also, as discussed in chapters 9, 11, and 12, the frequency of occurrence of some classes of mutational events is modulated by DNA repair systems. In addition to the mutations involving extensive DNA rearrangements, complex mutations involving multiple sequence changes have been identified. As will be pointed out below, some of these are thought to have originated as a consequence of a single event involving the slippage of the DNA template during replication.

A mutation that results in the complete loss of function of a genetic locus is termed a *null* mutation. Deletion and insertion mutations are most frequently null mutations. In addition, as described below, certain types of point mutations (nonsense, frameshift, and some missense) can be null mutations. Mutations that result in a partial loss of function of a genetic locus are termed *partial-loss-of-function mutations*, whereas mutations that result in a gain of function of a genetic locus are termed *gain-of-function mutations*.

Base Substitution Mutations

Point mutations that result from the substitution of one base pair for another (or one base for another in the case of single-stranded DNA genomes) are termed *base substitution mutations*. These can be further subclassified into transition mutations and transversion mutations. *Transition mutations* are those that involve a change of one purine for another or one pyrimidine for another (Table 2–2). In contrast, *transversion mutations* involve the interchange of a purine for a pyrimidine or the interchange of a pyrimidine for a purine (Table 2–2).

Many of the terms that are commonly used to refer to base substitution mutations reflect the molecular or phenotypic consequences of the mutation. Consequently, their applicability to a given base substitution mutation is a function of the information content of the DNA sequence that was mutated. A particularly important set of terms apply to base substitution mutations that occur within the protein-coding regions of a genome. *Missense mutations* are base substitution mutations that change the codon for one amino acid to that for another. The possible ultimate effects of the amino acid substitution on the resulting mutant protein include (i) no detectable effect, (ii) a partial loss of function, (iii) a gain of function, (iv) an alteration of function, (v) a change in protein stability with respect to either temperature or proteolytic degradation, and (vi) a complete loss of function. Note that, because of the degeneracy of the genetic code, some base substitution mutations have the effect of changing a codon for an amino acid to a different codon for the same amino acid and thus cause no change in the sequence of the resulting protein. These are sometimes referred to as *silent* or *neutral* mutations.

Nonsense mutations are mutations that change the codon for an amino acid to one of the three stop codons: TAG (*amber*), TAA (*ochre*), or TGA (*opal*). Most frequently, the premature termination of protein synthesis caused by a nonsense mutation leads to a partial or complete loss of protein function, but other outcomes are possible. Nonsense mutations are of particular utility in genetic studies, because their phenotypic consequences can be *suppressed* by a second type of mutation termed a nonsense suppressor. A *nonsense suppressor* is a mutation occurring within a gene encoding a tRNA and alters the anticodon of that tRNA so that it can recognize a nonsense codon. In a cell that contains both a nonsense mutation and a nonsense suppressor, when the nonsense codon is read, at least a fraction of the time the amino acid carried by the nonsense suppressor tRNA is inserted into the growing polypeptide chain. This prevents chain termination from occurring, and the resulting protein is very often functional. In such a case, the phe-

Table 2–2 Types of point mutations

Term	Mutation
Transition	G → A, A → G, C → T, T → C
Transversion	G → T, G → C, A → T, A → C, T → A, T → G, C → A, C → G
Frameshift	$\pm (3n \pm 1)$, where *n* is an integer

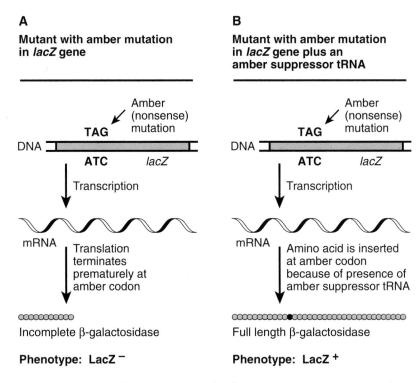

A

**Mutant with amber mutation
in *lacZ* gene**

Amber
(nonsense)
mutation

TAG

DNA

ATC   *lacZ*

Transcription

mRNA   Translation
terminates
prematurely at
amber codon

Incomplete β-galactosidase

Phenotype: LacZ $^-$

B

**Mutant with amber mutation
in *lacZ* gene plus an
amber suppressor tRNA**

Amber
(nonsense)
mutation

TAG

DNA

ATC   *lacZ*

Transcription

mRNA   Amino acid is inserted
at amber codon
because of presence of
amber suppressor tRNA

Full length β-galactosidase

Phenotype: LacZ $^+$

Figure 2–1   Suppression of a nonsense mutation by a nonsense suppressor mutation.

notype of the mutant organism is wild type even though its genotype differs from that of its wild-type parent by the presence of both the nonsense and the nonsense suppressor mutations (Fig. 2–1).

Base substitution mutations can have equally important molecular consequences when they occur in genomic locations other than protein-coding sequences. Table 2–3 gives examples of effects that such base substitution mutations can have on the regulation of gene expression, on transcription termination, on mRNA stability, on RNA splicing, and on the initiation of DNA replication.

Many other terms are used to categorize mutations on the basis of their phenotypic consequences. Table 2–4 provides an illustrative list of such terms. In addition, mutations can be categorized with respect to such things as their relationship to other alleles of the same gene (e.g., recessive, dominant), their relationship to alleles of other genes that affect the same phenotypic trait (e.g., epistatic), or their chromosomal location (e.g., sex-linked, autosomal). Detailed discussions of the terms used to categorize mutations based on these criteria can be found in standard genetics textbooks.

Table 2–3   Examples of molecular effects of base substitution mutations in genetic contexts other than protein-coding regions

Location of mutation	Effect of mutation
Promoter	Reduced or increased gene expression
Regulatory sequence of gene	Alteration of regulation of gene expression
3′ of protein-coding region	Defective transcription termination or alteration of mRNA stability
Certain locations within intron	Defective mRNA splicing
Origin of DNA replication	Defect in initiation of DNA replication

Table 2–4 Examples of terms used to categorize mutations on the basis of their phenotypic consequences

Term	Meaning
Conditional	The organism displays the phenotypic change under one condition (the *restrictive condition*) but does not display it under some other condition (the *permissive condition*)
Temperature sensitive	An example of a conditional mutation in which the restrictive condition is elevated temperature
Auxotroph	A mutation causing a defect in the synthesis of some essential metabolite so that the organism has an increased growth requirement(s)
Lethal	The presence of the mutation renders the organism nonviable

Mutations Resulting from the Addition or Deletion of Small Numbers of Base Pairs

The second category of point mutations contains those that involve the addition or deletion of small numbers of base pairs (Table 2–2). Additions or deletions of $3n \pm 1$ bp (usually 1, 2, 4, or 5 bp) that occur within the protein-coding portion of a gene have the effect of shifting the translational reading frame (Fig. 2–2). In the majority of cases, this results in a failure to synthesize a functional protein, thus allowing the mutation to be identified by its phenotypic consequences. Because these mutations cause a shift in the translational reading frame, they are termed *frameshift* mutations. Historically, frameshift mutations were of particular importance since they were used by Crick et al. (33) in their classic study in which they determined the triplet nature of the genetic code. It is worth noting that additions or deletions of 3 (or 6 or $3n$) bp do not shift the reading frame but nevertheless can, depending on the protein and their location within the protein, have significant effects on the function of the protein. Some particularly important examples of this phenomenon are discussed in chapter 9.

As in the case of base substitution mutations, mutations resulting from the addition or deletion of small numbers of base pairs can occur in regions of the genome other than those that encode proteins and can similarly have a wide variety of phenotypic consequences. Occasionally, these mutations are loosely referred to as being "frameshift" mutations even though they are not occurring in a genetic context in which they can cause a shift of a translational reading frame. The frequency of frameshift mutagenesis increases in the presence of intercalating agents such as proflavine (172), ICR-191 (22), and a number of carcinogens (see chapters 11 and 12).

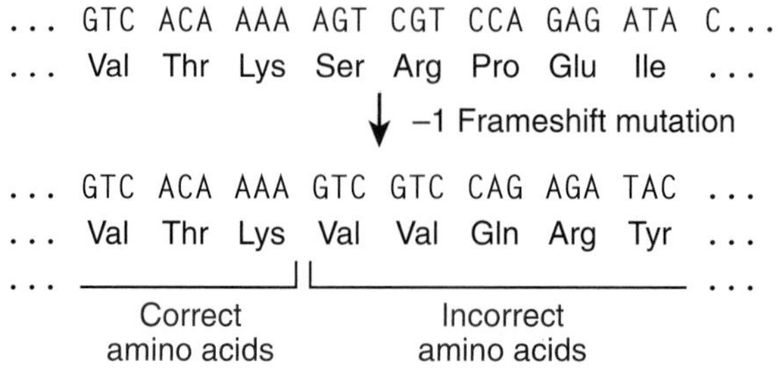

```
... GTC ACA AAA AGT CGT CCA GAG ATA C...
... Val Thr Lys Ser Arg Pro Glu Ile  ...
                      ↓ −1 Frameshift mutation

... GTC ACA AAA GTC GTC CAG AGA TAC ...
... Val Thr Lys Val Val Gln Arg Tyr ...
...  ─────────┘ └──────────────────── ...
      Correct          Incorrect
    amino acids       amino acids
```

Figure 2–2 Illustration of how a frameshift mutation results in a shift in the translational reading frame.

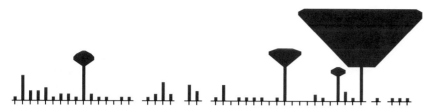

Figure 2–3 A portion of the T4 *rII* gene showing the number of mutations isolated at each site. Each square represents one occurrence at the indicated site. (*Adapted from Benzer [16] with permission.*)

Systems Used To Detect and Analyze Mutations

The First Systems for Analysis of Mutagenesis

To study mutagenesis as a process and gain an understanding of the molecular mechanisms responsible for it, it has been necessary to develop detection systems that permit the convenient detection and analysis of mutations (57, 63, 129). Over the years, enormous strides have been made in development of such detection systems. In their classic study that demonstrated the random nature of spontaneous mutations, Luria and Delbrück (116) utilized resistance to bacteriophage T1 as a genetic marker in their experiments. In subsequent studies in the early 1950s (38), a streptomycin-dependent strain of *Escherichia coli* that could revert to streptomycin independence was used to survey various chemicals for mutagenic activity. These studies were extended (81) to include the use of spot tests to screen for the mutagenicity of certain compounds.

During the 1960s, bacteriophage T4 mutants were used to study the action of mutagens (16, 17, 33, 43, 44, 65, 172). As shown in Fig. 2–3 (16), such studies demonstrated that spontaneous mutations were not randomly distributed. A site at which the number of mutations isolated significantly exceeds the number at other sites is called a *hot spot*. These studies additionally showed that mutations induced by chemical mutagens are also nonrandomly distributed. Furthermore, they showed that the distributions of sites of induced mutations are different from the distribution of sites of spontaneous mutations.

Before DNA sequencing techniques were developed, the sequencing of mutant proteins allowed the identity of certain codons in well-studied systems to be deduced. This allowed the development of reversion systems that permitted more accurate determinations of mutagen specificity. One important system, developed by Charles Yanofsky and his colleagues (18, 186, 187), involved strains carrying specific alleles of the *trpA* locus of *E. coli*. Unlike wild-type *E. coli*, strains carrying a *trpA* allele cannot grow on minimal medium unless tryptophan is added. Such mutant cells that have lost their ability to synthesize essential metabolites are called *auxotrophs*, while wild-type cells that do not require nutritional supplements are called *prototrophs*. Thus, it was possible to *select* for revertants that had regained the wild-type phenotype of synthesizing tryptophan and therefore could grow on minimal medium. Characterizations of the phenotypes of the resulting revertants, together with knowledge of the genetic code, allowed deduction of the mutational changes that occurred with various mutagens. The *trpA* gene was subsequently sequenced (139), and the nature of the original mutations and of the revertants was directly determined. The use of a set of *trpA* alleles has allowed the monitoring of all possible base substitution events, and these have been used in a variety of studies of mutagen specificity (57). In addition, since frameshift mutations revert by undergoing a second frameshift mutation, certain *trpA* alleles have been used to analyze frameshift mutagenesis. For example, the *trpA9777* allele arose by the addition of an A·T base pair to a run of four A·T base pairs. Thus reversion of *trpE9777* monitors a −1 frameshift in a run of five A·T base pairs (57).

Another reversion system that played an important role in early attempts to elucidate the specificity of various mutagens was based on the reversion of a set

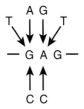

Figure 2–4 Base substitutions required to restore the glutamic acid codon at position 461 of β-galactosidase. Missense or nonsense mutations at coding position 461 result in the Lac⁻ phenotype but they can revert back to the GAG codon by one of six base substitutions. In each case, one specific substitution restores the GAG codon. (*Adapted from Cupples and Miller [34] and Miller [129] with permission.*)

of nine mutants of the yeast *Saccaromyces cerevisiae*, each of which had a single base mutation in the ATG initiation codon of the *CY1* gene (149).

A very convenient reversion system that takes advantage of the specific requirement for a glutamic acid at position 461 in *E. coli* β-galactosidase was developed more recently (34). Six *lacZ* mutations were constructed (Fig. 2–4), each of which can give rise to a LacZ⁺ revertant only by a single known base substitution mutation. Since each strain can revert only by a single known mutational change, unlike most other reversion systems, it is not necessary to further classify or sequence the resulting revertants in order to know the specific mutation that has occurred. Another system that has been exploited recently in the isolation of new mutator strains of *E. coli* is based on the reversion of the *galK2* allele (141).

The Ames *Salmonella* Test: a Widely Used Reversion System

A great many short-term tests have been developed that allow compounds to be screened for potential genotoxicity. The first such test to be optimized for convenience and sensitivity and to come into widespread use was developed in the 1970s by Bruce Ames and his colleagues. The system is based on the reversion of histidine auxotrophs of *Salmonella typhimurium* and is called the *Ames test* (Fig. 2–5) (8, 120, 122). Several different *S. typhimurium* strains, together with the agent to be tested, are plated on minimal medium plates containing a limiting amount of histidine. This amount of histidine allows the bacteria to undergo several generations of growth so that DNA lesions can be processed into mutations and the mutant phenotype can be expressed. After the histidine has been exhausted, His⁺ revertants

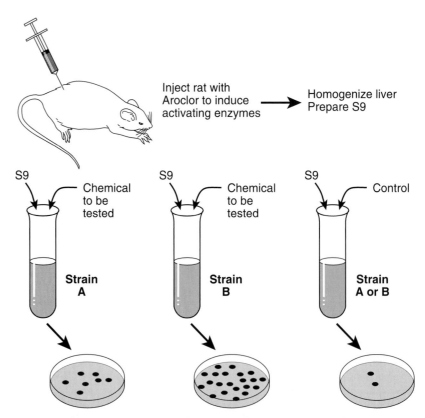

Figure 2–5 The Ames test. A set of *his* auxotrophs of *Salmonella typhimurium* are mixed with the compound to be tested and plated on minimal-glucose plates containing a limiting amount of histidine. After 2 days of incubation, His⁺ revertants on each plate are counted. Mammalian metabolism is simulated by the addition of an extract of rat liver, termed the S9 supernatant. The S9 supernatant is prepared from rats that have been injected with a polychlorinated biphenyl mixture, Aroclor (8).

grow as distinct colonies, each one representing an independent mutational event. As with the *trpA* reversion system discussed above, strains that carry a base substitution *his* allele can revert by a variety of base substitution mutations whereas strains that carry a frameshift *his* allele revert by various frameshift mutations (Fig. 2–6).

The sensitivity of the test was increased by several refinements. First, the *uvrB* gene was deleted, thereby inactivating an accurate excision repair system (see chapter 5). Second, the bacterial lipopolysaccharide was genetically modified to increase the permeability of the cells to the agents to be tested (7). Third, plasmid pKM101 was introduced into some of the strains (123). As discussed in detail in chapter 11, pKM101 carries the *mucA+B+* genes, which are particularly active analogs of the *E. coli umuD+C+* genes, and the presence of these genes greatly increases the sensitivity of the cells to mutagenesis by a wide variety of mutagens. Finally, since many compounds are genotoxic only after enzymatic conversion to a more active derivative (see chapter 1), a method was devised to simulate mammalian metabolism in the bacterial system by adding an extract of rat liver (6, 8). Liver was chosen since much of the important processing of chemicals in mammals occurs in the liver. In some cases, the action of the liver enzymes creates a toxic or mutagenic compound from a substance that was not originally dangerous. The rats are first injected with Aroclor, a polychlorinated biphenyl mixture, to induce the expression of various enzymes involved in metabolic activation of chemicals. The livers of these rats are then homogenized and centrifuged. The supernatant, termed S9 mix, is then added to the bacteria along with the chemical being tested.

Although the Ames test has been extremely widely used to screen chemicals for potential genotoxicity, it was not originally designed to yield information about the precise molecular nature of the revertants that were obtained. However, a series of diagnostic phenotypic tests for classifying revertants were then devised, and these allowed an assessment of the mutagenic specificity of various DNA-damaging agents (112). Subsequently, oligonucleotide probes that permitted the convenient identification of many specific classes of revertants by colony hybridization were developed (28, 107, 130).

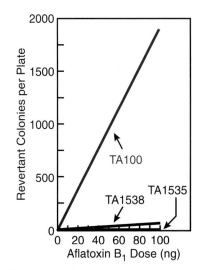

Figure 2–6 The mutagenicity in the Ames test of aflatoxin B$_1$. Strain TA1538 bears a different *his* mutation from that in strains TA100 and TA1535. Strain TA100 is a derivative of TA1535 that carries the plasmid pKM101 (see chapter 11). (*Adapted from McCann and Ames [121] with permission.*)

E. coli lacI: an Example of a Forward Mutational System

To determine the nature of the mutational changes caused by a particular mutagen and also the site specificity of those changes, a large and unbiased collection of mutations is required. In general, the number of mutational events that can cause a given mutant allele to revert is quite limited, so that all reversion systems are biased for the detection of a few specific events at a few specific sites. In principle, a much more unbiased collection of mutants can be obtained by using a *forward* mutational system, which monitors the mutation of a wild-type gene. Although such forward mutational systems are considerably less biased than mutational systems based on reversion, as discussed in chapter 11, these too are subject to various types of bias, some of them subtle.

The *lacI* nonsense system of *E. coli* developed by Jeffrey Miller and his colleagues (32, 127) is an example of a forward system that played an important role in the analyses of the specificity of a number of mutagens (Fig. 2–7). The *lacI+*

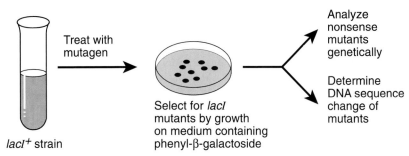

Figure 2–7 The *E. coli lacI* system for the analysis of mutations (127, 128).

gene codes for the repressor of the *lac* operon of *E. coli*; this operon codes for the proteins required to metabolize lactose. In its original implementation, the *lacI* system focused on the analysis of nonsense mutations in the *lacI* gene of *E. coli* (which was present on an F' episome) and involved the use of a set of purely genetic techniques to determine the identity of each particular nonsense mutation. There are more than 80 characterized sites in the *lacI⁺* gene where a nonsense mutation can arise by a single base change. For many mutagens, nonsense mutations constitute 20 to 30% of all the mutations that they induce in the *lacI* gene. Since the *lacI⁺* gene codes for the repressor of the *lac* operon of *E. coli*, cells that are deficient in *lacI* function express the *lac* operon constitutively. Thus, *lacI* mutants that have lost LacI function could be *selected* because of their ability to grow on phenyl-β-galactoside, a lactose analog that can be metabolized but cannot induce the operon. The subclass of mutants that carried nonsense mutations in the *lacI* gene were then recognized by their ability to be suppressed by various tRNA suppressors, and then their individual identities were determined by subsequent genetic analyses. The original *lacI* system had the limitations of being able to detect only base substitution mutations and of not being able to detect A·T → ·GC transitions.

The distribution of the mutations could then be arranged according to the base substitution that causes each nonsense mutation to give what is known as a *mutational spectrum*. Figure 2–8 shows examples of the data that can be obtained in such experiments. Note that, even though the same G·C → A·T transition is involved, there are sites at which the number of mutations isolated significantly exceeds the number at other sites (*hot spots*) and sites at which the number of mutations isolated is significantly smaller than the number at other sites (*cold spots*). Also note that the location of hot spots and cold spots is not the same for different mutagens. In the case of spontaneous mutations, the amber hot spots were found to correlate with the positions of 5-methylcytosine (see chapter 9). Further examples of mutational spectra obtained by using this system are discussed in chapter 11.

Subsequently, as techniques for DNA sequencing became more refined, it became feasible to determine the nature of the *lacI* mutations directly by DNA sequencing (71, 128, 160). This enabled the analysis to include other classes of mutations besides nonsense mutations and also allowed mutagenesis to be monitored at all possible sites within the *lacI* gene. A drawback associated with this type of approach, however, is that it involves a great deal of work, so that the number of mutations that can be sequenced is usually limited by pragmatic considerations. Thus, unlike reversion systems, in which mutation rates are easily computed on the basis of the number of selected revertants, and systems like the *lacI* nonsense system, in which it is easy to obtain multiple occurrences at the same site to study hot spots, forward sequencing systems require an enormous amount of work to collect enough determinations to give a full distribution with significant mutation rates at each site (129).

Other Examples of Forward Mutational Systems

Table 2–5 lists examples of other genes that have been used as targets in forward mutational systems. The nature of the mutation is usually determined directly by DNA sequencing. In many cases, the system has been optimized to make the sequencing as quick and convenient as possible, for example by placing the target gene on a plasmid, bacteriophage, or viral vector. These optimized systems offer some additional experimental flexibility in that they permit the damage to the DNA of the target gene to be manipulated independently of the damage to the host genome. However, they also introduce the formal possibilities that the replication of the extrachromosomal DNA differs in some respect from that of genomic DNA and that this will influence the nature of the mutagenesis that occurs. The use of the *supF* gene as a target for forward mutagenesis is interesting in that it encodes an amber suppressor tRNA rather than a protein and has been used to analyze mutagenesis in a variety of biological contexts. Figure 2–9 shows examples of classes of mutations that have been detected that inactivate *supF* function (98).

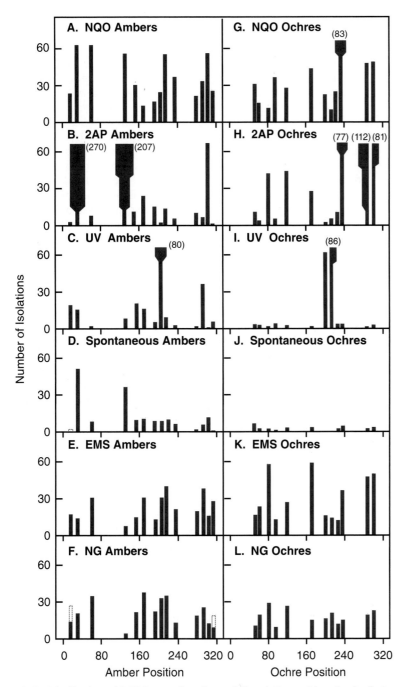

Figure 2–8 Distribution of 3,738 mutations from G·C → A·T transitions in the *lacI* gene of *E. coli*. The number of independent occurrences at each site is indicated by the bar height. One amber and one ochre mutation were analyzed from each mutagenized culture, so that hot spots can be identified by comparing the frequency of mutation at an amber site with that at other amber sites and similarly by comparing the frequency of mutation at an ochre site with that at other ochre sites. See reference 32 for a discussion of amber mutation frequencies relative to ochre mutation frequencies. Large areas, instead of bar height, indicate a number of occurrences greater than 69, with the actual number shown in parentheses. Abbreviations: NQO, 4-nitroquinoline-1-oxide; 2AP, aminopurine; UV, ultraviolet irradiation; EMS, ethyl methanesulfonate; NG, *N*-methyl-*N*'-nitro-*N*-nitrosoguanidine. (*Adapted from Coulondre and Miller [32] with permission.*)

Table 2–5 Examples of genes used as targets in forward mutational systems

Gene	Function encoded	Representative reference
$lacI^+$	Repressor of *lac* operon	127
$lacZ^+$	β-Galactosidase	72
$lacZ'$	α-complementing fragment of β-galactosidase	111
λ cI^+	λ repressor	184
$galK^+$	Galactose kinase	152
$supF$	Tyrosine suppressor tRNA	98
tet^+	Tetracycline resistance	91
HPRT	Hypoxanthine-guanine phosphoribosyltransferase	25
APRT	Adenine phosphoribosyltransferase	51
gpt^+	Guanine-hypoxanthine phosphoribosyltransferase	156
HSV^a tk^+	HSV thymidine kinase	50

^{a} HSV, herpes simplex virus.

A number of these well-studied target genes have been used to analyze mutagenesis in both prokaryotic and eukaryotic systems.

Special Systems To Detect Frameshift or Deletion Mutations

Forward mutagenesis systems allow the detection of frameshift mutations along with base substitution, insertion, and deletion mutations. As mentioned above, it is possible to detect frameshift mutations specifically by inducing reversion of an existing frameshift mutant. However, the reversion of some frameshift mutations (such as *trpA9777*) can occur at only a very few nucleotide positions. Studies of the mechanism of frameshift mutagenesis have been facilitated by the development of more sophisticated frameshift reversion assays that permit reversion of a frameshift mutation by reversions occurring at many positions. For example, for

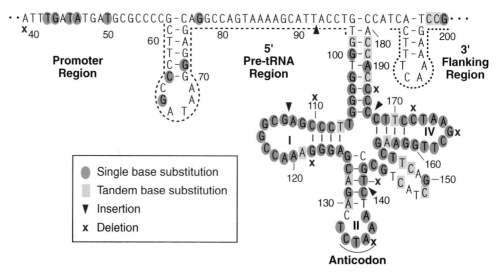

Figure 2–9 Hypothetical secondary structure of a single-stranded DNA containing the *supF* tRNA gene sequence and showing the location of mutations that inactivate *supF* function. Sites of single (red circles) and tandem (grey rectangles) base substitutions, insertions (red triangles), and deletions (red X) are indicated. (*Adapted from Kraemer and Seidman [98] with permission.*)

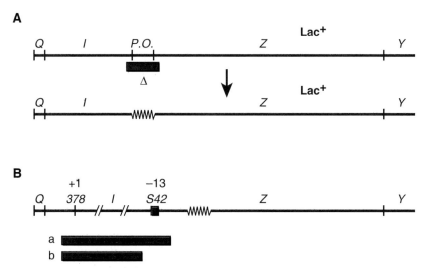

Figure 2–10 Fusions of *lacZ* to *lacI* and the selection system used to detect deletions in *lacI*. (A) The deletion fuses the *lacI* gene to the *lacZ* gene. The resulting hybrid protein is missing the last five (62) residues of the *lac* repressor and the first 23 residues of β-galactosidase but retains β-galactosidase activity (4, 134). *Q* indicates the *lacI*^q promoter. (B) Frameshift mutations *378* and *S42*, which are separated by 697 bp, have been crossed into the *lacI-lacZ* fusion. Only deletions can restore the Lac⁺ character. The principal deletions that were detected were of the a or b type (4). (*Adapted from Albertini et al. [4] with permission.*)

genes that encode proteins with amino-terminal regions that are insensitive to amino acid sequence changes, a frameshift at one site can be made to revert by frameshifts of the opposite sign at many alternative sites (37, 76, 151, 154, 155). This type of selection combines the selectivity of a reversion assay with the power of a forward assay to examine numerous sites (155).

It has also been possible to develop systems that allow the specific detection of large deletion mutations. One such system takes advantage of a specific deletion that fuses the beginning of the *lacZ* gene to the end of the *lacI* gene (4) so that a hybrid protein is produced that lacks the last five (62) residues of the *lac* repressor and the first 23 residues of β-galactosidase. Since the hybrid protein has β-galactosidase activity, the *lacI* portion of the fused genes can be used as a dispensable gene because mutations that do not interrupt transcription or translation should not interfere with β-galactosidase activity. Two widely separated frameshift mutations (*378*, a +1 frameshift, and *S42*, a −13 frameshift) were then introduced into the *lacI* region of the fused genes (Fig. 2–10). Frameshifts or small deletion mutations do not restore the normal reading frame and give Lac⁺ colonies, because there are too many nonsense codons in the +1 and −1 reading frames between the two frameshift mutations. Large deletions that eliminate both frameshifts and restore the original reading frame result in a Lac⁺ phenotype (Fig. 2–10, type a), as can deletions that eliminate only one of the starting mutations but themselves generate a change in phase of the opposite sign to the remaining frameshift mutation (Fig. 2–10, type b) (4).

Analyses of Mutagenesis in Mammalian Cells

Shuttle Vectors

Until this point we have focused on systems that have been, and are still being, used to analyze mutagenesis in prokaryotes and lower eukaryotes. Beginning in the 1980s, it became feasible to develop systems that would allow detailed analyses of mutagenesis in mammalian cells. The first strategy to be used was to analyze mutations that occurred in a target gene of a virus such as simian virus 40 (SV40) (20, 67). However, this approach was soon superseded by the development of

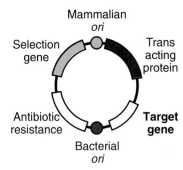

Figure 2–11 Schematic representation of a shuttle vector genetic map. Mammalian *ori* represents DNA sequences from animal virus replication origins allowing replication in mammalian cells by using a viral *trans* acting protein. The selection gene codes for a protein allowing selection and maintenance of this vector in mammalian cells. The bacterial *ori* represents DNA sequences from a bacterial plasmid or bacteriophage. The antibiotic resistance gene codes for a protein allowing the selection and maintenance of the vector in bacteria. The target gene represents the DNA sequences used for detecting mutants (see also Table 2–5). (*Adapted from Sarasin [159] with permission.*)

shuttle vectors to simplify analyses of mutagenesis (52, 159). *Shuttle vectors* consist of a target gene that is used to monitor mutagenesis and additional sequences that permit replication and selection in either mammalian cells or bacteria (Fig. 2–11). The sequences permitting replication in the mammalian cell are usually derived from a virus, while the sequences permitting replication in bacterial cells are usually derived from a plasmid or bacteriophage. As diagrammed in Fig. 2–12, the use of shuttle vectors permits mutagenesis of the target gene to occur in the mammalian cell of interest while allowing the subsequent analyses of the resulting mutations to take advantage of the ease and speed of *E. coli* molecular biology.

Three main classes of shuttle vectors have been developed. One class is termed *transiently replicating* because such shuttle vectors, which are usually based on the SV40 replicon, replicate only transiently in permissive mammalian cells, giving rise to a very large number of vector molecules (around 10^4 to 10^5) per cell (52, 159, 164). DNA is recovered from the mammalian cells about 3 to 5 days after infection and then shuttled back to appropriate bacterial cells for analysis (Fig. 2–12). The first experiments involving such shuttle vectors, which utilized the *E. coli galK*+ gene (152) or the *E. coli lacI*+ gene (21) as a target, revealed a very high spontaneous mutation frequency ($\geq 1\%$) upon passage of the vectors through mammalian cells. These mutations, most of which were deletions and insertions, were thought to be due to DNA lesions that were introduced into the vector DNA during the transfection process and during the migration of the vectors through the cytoplasm to the nucleus. Two working solutions were developed to circumvent the problem. One was to flank the marker gene with sequences that are needed and therefore selected in both mammalian cells and bacteria, such as replication origin sequences or antibiotic resistance genes. For example, the target *supF* gene in the pZ189 shuttle vector was flanked by the plasmid pBR322 origin of replication and by the ampicillin resistance gene (165). The other solution was to identify mammalian cell lines, for example a human embryonic kidney cell line transformed by adenovirus (Ad293), which support vector replication but have a significantly decreased background mutation frequency (79, 110).

One advantage of transiently replicating shuttle vectors is that it is possible to damage the vector DNA independently of the host cellular DNA and therefore to use them as probes to monitor the induction of DNA repair and mutagenesis processes (159). They are also useful for examining mutagenesis in various mammalian cell lines that have DNA repair deficiencies (114, 118, 163). A disadvantage to most shuttle vectors is that, if the same mutation is observed more than once in the same experiment, it is normally not possible to tell whether the mutants are siblings or whether they arose independently. This problem has been circumvented by the development of a new shuttle vector (pS189) that is a population of plasmids, each of which contains a unique 8-bp signature sequence (145). Since each 8-bp sequence provides a unique identification tag for each plasmid, independent mutations can be identified by their unique signature sequences (145).

A second class of shuttle vectors is composed of the *episomal* shuttle vectors. With these vectors, the aim is to establish cell lines in which the shuttle vector replicates permanently as an episome; i.e., the vector replicates autonomously rather than becoming integrated into the cell's genome (Fig. 2–12) (9, 40, 52, 117, 159, 188). These vectors are usually based on Epstein-Barr virus and bovine papilloma virus replication systems that allow the DNA to replicate as a stable episome. These vectors are normally present between 10 and 100 copies per cell nucleus and replicate synchronously with the host cell. The shuttle vectors carry a gene conferring a resistance, for example to G418 or hygromycin, that allows selection for cell lines that contain the vector. At any time, the low-molecular-weight DNA can be recovered and shuttled back to bacterial cells for analysis. The Epstein-Barr virus-based shuttle vectors exhibited a very low spontaneous mutation frequency (10^{-5} to 10^{-6}) when the herpes simplex virus thymidine kinase gene (50) or the *E. coli lacI*+ gene (53) was used as the target. Although episomal shuttle vectors have been useful in analyses of mutagenesis, the low copy number of the plasmid DNA in mammalian cells and the usual necessity to treat both the vector molecules and the host cell at the same time imposes some limitations on these systems (159).

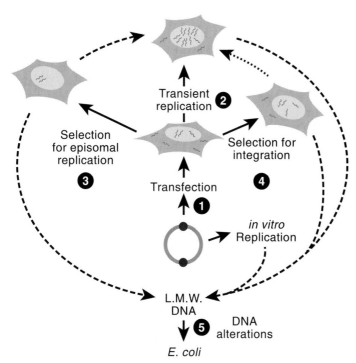

Figure 2–12 General scheme of the various protocols involving shuttle vectors. 1, Vector DNA is first transfected into the mammalian host cell. 2, By using an SV40-based plasmid, transient replication occurs in 2 to 4 days after transfection. 3, By using an Epstein-Barr virus-based plasmid and selecting for antibiotic resistance, established cell lines replicating the vector as an episome are isolated. 4, By using a nonreplicating vector and selecting for a gene on the vector, cell lines containing integrated shuttle vector sequences are isolated. 5, Replicated and mutated vector sequences are recovered as low-molecular-weight (L.M.W.) DNA, which are then transfected into host bacteria cells. DNA alterations in the vector target gene are analyzed from isolated colonies or plaques. (*Adapted from Sarasin [159] with permission.*)

A third class of shuttle vectors is composed of the *integrated* shuttle vectors (11, 52, 159). These result in the integration of the target gene, for example via retrovirus long terminal repeat sequences, into a chromosomal location in the mammalian cell (Fig. 2–12). The DNA must then be recovered prior to analysis in bacterial cells. One strategy that has been used to accomplish this is to include the SV40 *ori* (but not the *trans*-acting SV40 T-antigen gene) in the vector that becomes integrated. The integrated sequences can then be recovered by fusing the cells with permissive COS cells (monkey kidney cells transformed by the SV40 T-antigen gene which can replicate all sequences linked to an SV40 replication origin). This procedure produces full-size vectors after gene amplification and sequence excision. However, the fusion process itself appears to be mutagenic (10). An alternative strategy has been to include bacteriophage λ sequences in the integrated DNA. This allows the integrated sequences to be recovered later as λ phage by exposing the DNA from the mammalian cells to an in vitro λ packaging extract (68). Initially the recovery of DNA by this procedure was relatively inefficient, but recent technical improvements have led to greatly increased efficiencies of recovery (72, 94).

Sequencing of Mutated Target Genes in Mammalian Cells by Use of PCR

For many years it has been possible to *select* for mutants of mammalian cell lines that carry mutations in known genes. For example, cells that have acquired resistance to 6-thioguanine have mutations in the *HPRT* gene. However, despite the ease with which mutants of mammalian cells lines could be obtained, until the development of PCR (158), direct molecular analyses of the mutations in these mutant cell lines involved a formidable amount of work prior to sequencing each mutation. For example, mutations in the endogenous *APRT* gene of Chinese hamster ovary (CHO) cells were analyzed by a procedure that involved cloning the target gene from each mutant (51). The strategy of using shuttle vectors was initially developed in part to circumvent this difficulty.

The development of PCR revolutionized the analysis of mutagenesis in mammalian cell lines. Two strategies are in common use (Fig. 2–13). One involves isolation of mRNA from each mutant to be analyzed and preparation of cDNA. The gene of interest is then amplified either in its entirety or in segments by the use of appropriate PCR primers (27, 119, 181, 185). A second approach involves using appropriate PCR primers to amplify the target gene directly from the genomic DNA, either in its entirety or in segments (23, 24, 27, 156). A variation is to choose the PCR primers so as to amplify only exon DNA. The first method and the variation of the second method have the advantage that less DNA has to be sequenced since only the exons are sequenced. Thus, they are particularly useful in analyses of large genes that contain a great deal of intron DNA. However, additional work is required to identify mutations that affect mRNA splicing (26, 27).

In some situations, the use of an endogenous gene, such as *HPRT*, as a target can be advantageous because it allows comparisons of mutagenesis in naturally occurring variants (e.g., xeroderma pigmentosum [XP] cells) or in cell lines defective in DNA repair and mutagenesis, without the need for prior manipulation of the cells (179, 181). In another situation, to gain the advantage offered by a smaller target gene, a derivative of an *hprt* mutant of a CHO cell line was constructed in which a single copy of the *E. coli gpt⁺* gene was incorporated into the CHO cell genome (156). The *gpt⁺* gene is very small, only 459 bp compared with about 3,300 bp for the endogenous *APRT* gene (74) and over 30,000 bp for the endogenous *HPRT* gene (124). Nevertheless it can replace certain functions of the *HPRT*

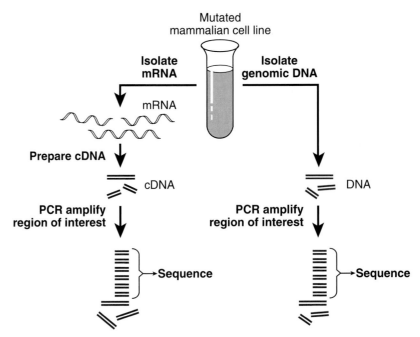

Figure 2–13 Strategies involving PCR to analyze mutations in a target gene in a mammalian cell line. These involve first selecting for mutants (e.g., 6-thioguanine resistance) and then either making cDNA and amplifying the sequence or directly amplifying fragments of the genomic DNA and sequencing.

gene product; in particular, *gpt* mutants of the cell line can be selected on the basis of 6-thioguanine resistance.

A similar PCR-based approach can be used not only to analyze mutations selected in cultured mammalian cells but also to analyze mutations in cancers that are thought to have been caused by some particular carcinogen. For example, basal cell carcinomas are thought to be caused by exposure to sunlight. Mutations in the p53 tumor suppressor gene of a set of clinically isolated basal cell carcinomas have been analyzed by using PCR to amplify exons from the p53 gene of each tumor and then sequencing them (189). Mutations in the p53 tumor suppressor gene of a set of skin tumors from patients with xeroderma pigmentosum have similarly been examined by using PCR to amplify specific segments from p53 cDNA and further analyzing them (54).

Chemical and Physical Methods for Identifying DNA Fragments Carrying a Mutation

Recently, a number of chemical and physical methods that allow the detection of point mutations in defined DNA or RNA molecules have been developed and are being used to help facilitate analyses of mutagenesis (29). For example, RNase A has been shown to cleave preferentially RNA-RNA heteroduplexes that contain a mismatch (182). Furthermore, chemical methods which involve either carbodiimide (140) or a combination of hydroxylamine and osmium tetroxide (30, 31) to specifically modify DNA-DNA heteroduplexes containing a mismatch have been developed. The latter procedure has been used to localize mutations to specific regions of the target gene, thereby simplifying the DNA sequence analysis (177). Briefly, the strategy is first to use PCR to amplify DNA fragments from the target gene of the mutant and then to form heteroduplexes between the PCR fragments derived from the mutant and the corresponding fragments from wild-type DNA, which are ^{32}P end labeled. If the fragment contains a base substitution mutation, a base mismatch will be produced at the site of the mutation. Mismatched cytosine or thymine residues are sensitive to modification by hydroxylamine or osmium tetroxide, respectively. The modified heteroduplex is then sensitive to piperidine cleavage. The cleavage product can be separated on a denaturing polyacrylamide sequencing gel and visualized by autoradiography.

A very interesting development has been the application of denaturing gradient gel electrophoresis (DGGE) to analyses of mutagenesis. DGGE, which was developed by Fischer and Lerman (60, 61), separates DNA fragments according to their melting properties. If a DNA fragment is subjected to electrophoresis through a gel that contains a linearly increasing concentration of denaturants, the fragment will remain double stranded until it reaches a concentration of denaturing agent that causes the lower temperature domain of the fragment to melt. This partial melting greatly decreases the mobility of the fragment in the gel. DNA fragments in which the lower-temperature domains differ by as little as a single base pair substitution will melt at slightly different concentrations because of differences in stacking interactions between adjacent bases in each strand. If they melt at slightly different denaturant concentrations, they will separate in the gel (60, 61, 137). Mutations in higher-temperature domains can be similarly analyzed by a modification in which PCR is used to attach a G + C-rich sequence (termed a GC clamp [1, 135, 136]) to the fragment of interest (166). The technique of DGGE is extremely useful and is being applied to a number of problems (see, e.g., reference 2).

In the method outlined in Fig. 2–14, DGGE is used to facilitate the analysis of mutagenesis of the *HPRT* gene of human cells (23, 24, 89, 142). Human β-lymphoblastoid cells are first treated with the mutagen of interest. Mutants with mutations at the *HPRT* locus are then selected en masse by adding the purine analog 6-thioguanine directly to the bulk culture. After 6-thioguanine-resistant mutants induced by the treatment have overgrown the culture, their DNA is isolated. The DNA segment of interest, in this case exon 3 of *HPRT*, is then amplified by high-fidelity PCR. This yields both fragments that have mutations in exon 3 and fragments that have wild-type exon 3 sequences. The fragments are then

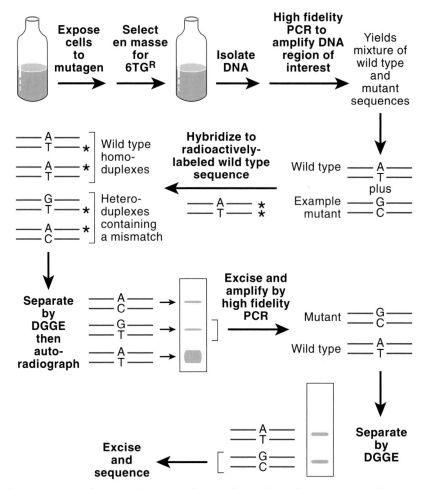

Figure 2–14 Application of DGGE analysis to the analysis of mutagenesis in human cells. 6TG^R, 6-thioguanine resistance. See the text for details.

denatured and mixed with a radioactively labeled wild-type exon 3 fragment. As shown in Fig. 2–14, each mutant exon 3 fragment will give rise to two different heteroduplexes. Upon DGGE, each of these heteroduplexes will melt at a position in the gel that is different from that of the wild-type sequences. The positions of the heteroduplexes on the gel are then identified by autoradiography, and the bands are excised. Further amplification of each such heteroduplex by high-fidelity PCR yields both a wild-type and a mutant exon 3 fragment as illustrated. These are separated by DGGE, and the mutant fragment is then directly sequenced. The procedure illustrated in Fig. 2–14 allows analysis of the low-temperature domain (bp 300 to 403). The addition of GC clamps allows a similar analysis of the high-temperature domain (bp 220 to 299). The method has been used to analyze mutational spectra produced by N-methyl-N'-nitro-N-nitroso-guanidine (MNNG) (23), ICR-191 (23), UV (88), benzo[a]pyrene 7,8-diol-9,10-epoxide (BPDE) (88), oxygen (142), hydrogen peroxide (142), and cis-platinum (25).

Analyses of Mutagenesis in Whole Animals

There has been a great deal of interest in examining the mutagenic processes that occur in whole animals that have been exposed to a mutagen or carcinogen. One approach has been to develop a strategy that is based on the cloning of *hprt* mutant T cells from mice (39, 84, 169), rats (3), monkeys (190), or humans (5, 133). Once the *hprt* mutant cells have been cloned, they can be analyzed by the types of approaches discussed above.

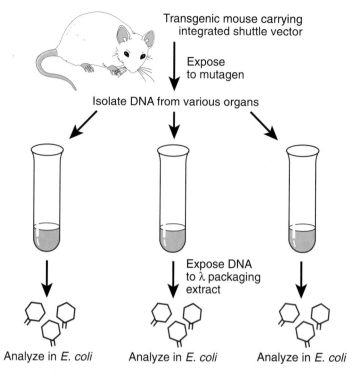

Figure 2–15 Analyses of mutagenesis in transgenic mice. See the text for details.

A recent alternative has been to develop transgenic mice that allow a more convenient analysis of mutagenesis in whole animals. As outlined in Fig. 2–15, transgenic animals have been generated that contain multiple copies of an integrated shuttle vector that can be rescued by exposing the genomic DNA to an in vitro lambda packaging extract (see above) (72, 78, 92–94, 138, 174). The transgenic animals are exposed to the genotoxic agent of interest and, after a time, DNA samples from various organs of the mice are isolated and incubated with a high-efficiency lambda packaging extract. The resulting bacteriophages are then screened in *E. coli* to detect those that carry mutations in the target gene. To date, the genes used as targets have been the *E. coli lacZ*⁺ (72, 78) and *lacI*⁺ (93, 94) genes. It seems likely that other target genes will be used in the future as the transgenic systems for analyzing mutagenesis in whole animals are further refined. Efficient recovery of the shuttle vector from the genomic DNA is required in these experiments, as is an efficient procedure for the identification of mutants; therefore, various technical improvements have played important roles in making this type of analysis feasible (72, 73, 75, 93, 94, 138).

Use of Site-Specific Adducts

Mutational spectra yield information about the sites at which mutations can occur when the DNA is damaged by a specific agent and about the molecular nature of the resulting mutations. As discussed in chapters 11 and 12, when this information is coupled with information concerning the chemical nature of the lesions introduced and the distribution of these lesions in DNA, hypotheses can be formulated as to which lesions are *premutagenic* lesions and as to the type of mutations they cause. However, most mutagens cause a variety of chemically different lesions that range from abundant to extremely rare. Since mutation is in general a fairly rare occurrence, even when cells are exposed to a mutagen, it was difficult for many years to be sure whether a lesion suspected of being a premutagenic lesion was, in fact, responsible for the appearance of the mutation.

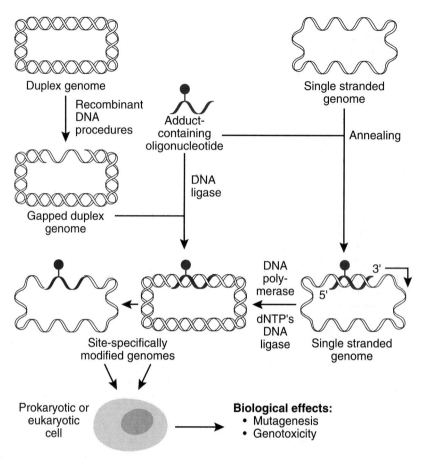

Figure 2–16 Methods for the construction of site-specifically modified genomes. The procedure shown on the left-hand side of the diagram uses DNA ligase to insert an adducted oligonucleotide into a complementary site in a gapped heteroduplex genome. The product can be used directly to study the genetic effect of the lesion in double-stranded DNA, or if the strand opposing the adduct contains a nonligatable nick, the genome can be denatured and the biological effects of an adduct situated in a single-stranded DNA can be determined. One modification of the approach depicted on the right-hand side of the figure has been to use DNA polymerase to join an adducted deoxyribonucleoside triphosphate (dNTP) onto the 3′ end of an unmodified oligonucleotide previously annealed to a single-stranded genome. Synthesis of the site-specifically modified duplex vector is completed upon subsequent addition of unmodified deoxyribonucleoside triphosphates. (*Adapted from Essigmann and Wood [58] with permission.*)

It became possible to directly test hypotheses concerning premutagenic lesions when sophisticated technologies were developed that allow the synthesis of DNA molecules containing a single specific adduct of known chemical structure at a specific place in the DNA molecule (12, 13, 58, 168). These are sometimes referred to as *site-specific adducts*. The two most commonly used procedures for constructing site-specifically modified genomes are summarized in Fig. 2–16. Applications of this approach toward understanding the molecular basis of mutagenesis are discussed in chapters 11 and 12.

Insights into the Molecular Basis of Mutagenesis from the Study of Purified Polymerases

As will be discussed in chapters 11 and 12, the in vivo processing of damaged DNA that gives rise to mutations may be quite complex and in at least some (if not many) cases requires the participation of a number of proteins besides a DNA polymerase.

It is a formidable challenge to establish systems that mimic, in vitro, all of the physiologically relevant events that are responsible for the introduction of mutations in vivo. Consequently, a great deal of effort has been devoted to the study of purified polymerases in order to gain insights into the molecular basis of polymerase fidelity and into the mechanisms by which certain classes of mutations arise.

Error Discrimination by DNA Polymerases: Mechanisms Responsible for Polymerase Fidelity

As pointed out in chapter 1, three different components contribute to the overall fidelity of DNA replication: (i) the polymerization reaction itself, (ii) proofreading by a $3' \rightarrow 5'$ exonuclease activity, and (iii) postreplicational mismatch repair (see chapter 9). Although many DNA polymerases have been purified from a variety of sources (97), with only the simplest of these has it been possible to carry out the detailed investigations necessary to examine fidelity at the mechanistic level. Three polymerases that have been studied especially intensively have been *E. coli* DNA polymerase I, bacteriophage T7 DNA polymerase, and bacteriophage T4 DNA polymerase. Particularly important advances have been the solving of the crystal structures of the Klenow fragment of DNA polymerase I and the human immunodeficiency virus type 1 reverse transcriptase (82, 86, 95, 143) and the development of both pre-steady-state and steady-state kinetic methods of analyzing polymerase action (55, 70, 83, 102).

DNA polymerase I of *E. coli* consists of a single polypeptide chain containing three domains. One domain contains the polymerase active site, a second contains the $3' \rightarrow 5'$ proofreading activity, and the third contains a $5' \rightarrow 3'$ exonuclease activity. The widely studied *Klenow fragment* contains only the polymerase and the $3' \rightarrow 5'$ proofreading exonuclease domains (97). Figure 2–17 shows the crystal structure of the DNA polymerase I Klenow fragment (86). The exonuclease active site is located on a separate domain approximately 30 Å (3 nm) from the polymerase active site. This led to the proposal that the DNA melts and slides to move the 3' terminus from the polymerase site to the exonuclease site (143). Such intramolecular transfers from the polymerase site to the proofreading exonuclease site have been shown to occur for the T7 and T4 DNA polymerases (41, 153). However, in the case of DNA polymerase I, such an intramolecular transfer mechanism is insignificant because the dissociation of DNA from the polymerase is fast compared with the of $3' \rightarrow 5'$ exonuclease action (85, 99).

The application of transient kinetic methods has permitted the identification of various steps involved in DNA polymerase action and quantitation of the contribution of each step to the overall fidelity (55, 83). A general model for DNA polymerase action in shown in Fig. 2–18 (35, 41, 56, 102, 146, 183). Once the DNA polymerase is bound to the DNA substrate, step A occurs, in which a deoxyribonucleoside triphosphate binds to the enzyme. For the T7 DNA polymerase and human immunodeficiency virus type 1 reverse transcriptase, the discrimination against incorrect nucleotides is approximately 200- to 400-fold. This factor is fairly close to what would have been predicted from studies of nucleotide pairing in solution and suggests that mismatches (Fig. 2–19) form much as they do in solution (83). In contrast, the DNA polymerase I Klenow fragment does not discriminate much between correct and incorrect nucleotides at this stage (56). In step B, the polymerase then undergoes a rate-limiting conformational change in which it is thought to clamp down over the primer/template-deoxyribonucleoside triphosphate complex so that phosphodiester bond formation (step C) can take place (83). This conformational change to the catalytically active form of the polymerase evidently demands that the base pair formed by the template base and the paired deoxyribonucleoside triphosphate conform to the geometry of a standard Watson-Crick base pair (with respect to distances between the C-1' carbons of the deoxyribose sugars and with respect to bond angles) (55, 83). As shown in Fig. 2–19, A·T and G·C pairs display a remarkable geometric equivalence, but A·C and G·T wobble pairs or G(*anti*)-A(*syn*) pairs have a very different geometry. This selection for the correct geometry is one of the most important factors contributing to the

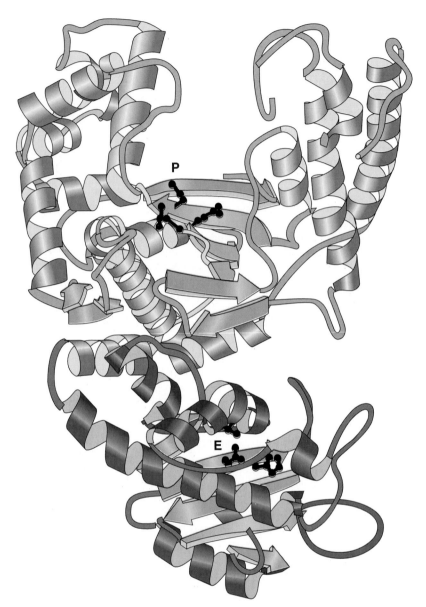

Figure 2–17 Schematic representation of the structure of the Klenow fragment of *E. coli* DNA polymerase I. The α-helices are shown as spiral ribbons, and the β-sheets are shown as arrows. The extent of the separate 3′ → 5′ (proofreading) exonuclease is indicated by the darker shading. The catalytically important carboxylate side chains at the polymerase (P) and exonuclease sites are shown. (*Adapted from Joyce and Steitz [86] with permission.*)

high fidelity of DNA polymerases. In the case of the DNA polymerase I Klenow fragment, if a misincorporation has occurred, there is a slow step (step D) after phosphodiester bond formation but prior to release of PP_i, (step F) (35). Finally, the proofreading exonuclease may remove the 3′ terminal nucleotide (step E_2) if an incorrect nucleotide has been incorporated.

As mentioned above, most of the proofreading by DNA polymerase I involves the dissociation of the DNA from the polymerase followed by rebinding at the exonuclease site. However, T7 and T4 DNA polymerases have been shown to be examples of polymerases for which the DNA shifts from the polymerase site to the proofreading exonuclease site without dissociating from the polymerase. As shown in Fig. 2–20, at each step during processive synthesis a polymerase has four dif-

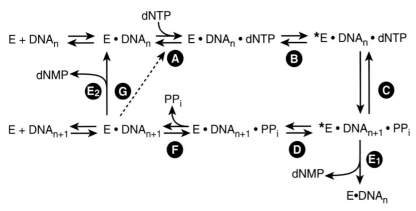

Figure 2–18   Reaction pathway for an exonuclease-proficient DNA polymerase. Asterisks represent the enzyme in a different conformation. The enzyme conformation(s) in the ternary E·DNA$_n$·dNTP and E·DNA$_{n+1}$· PP$_i$ complexes are unknown and are not necessarily the same. Entry into the next cycle of polymerization is indicated by G. Translocation has not been assigned to a particular step in the reaction. (*Adapted from Kunkel [102] with permission.*)

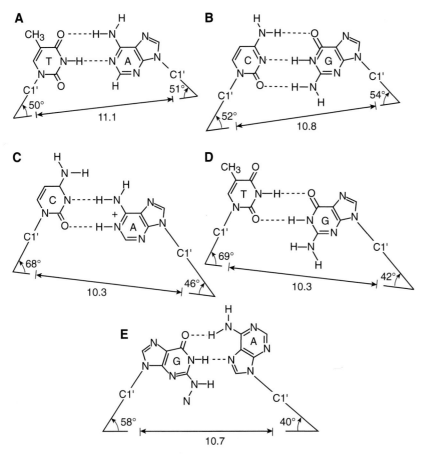

Figure 2–19   Geometric characteristics of Watson-Crick and mismatched base pairs. The figure is based on X-ray crystallography of duplex B-DNA oligonucleotides. The striking geometric identity of the Watson-Crick base pairs (a and b) is not matched by the A·C (c) and G·T (d) wobble pairs or by the G(*anti*)-A(*syn*) pair (e). (*Adapted from Kennard [87] and Echols and Goodman [55] with permission.*)

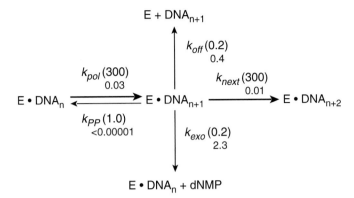

Figure 2–20 Kinetic partitioning in exonuclease error correction. The rates of the reactions that govern the movement of DNA between the polymerase and exonuclease sites on T7 DNA polymerase are shown. The numbers shown are the rate constants for T7 DNA polymerase under appropriate physiological conditions of nucleotides and PP$_i$s. All units are in reciprocal seconds. The numbers in black are those that apply if a *correct* nucleotide has been incorporated, while the numbers in red are those that apply if an *incorrect* nucleotide has been incorporated. (*Adapted from Johnson [83] with permission.*)

ferent options: continued polymerization, pyrophosphorolysis, dissociation of the DNA, or exonucleolytic hydrolysis.

In the case of T7 DNA polymerase, the E-DNA complex normally continues to carry out processive synthesis at a rate of 300 nucleotides s^{-1}, while only occasionally dissociating (0.2 nucleotides s^{-1}) or sliding into the exonuclease site (0.2 nucleotides s^{-1}) (83). However, once an incorrect nucleotide has been misincorporated, striking changes occur. One is that the rate of hydrolysis at the exonuclease site is increased to 2.3 nucleotides s^{-1}. This represents only a 10-fold increase in the rate of hydrolysis and, by itself, would be insufficient to provide efficient proofreading if polymerization continued at a rate of 300 nucleotides s^{-1}. However, the rate of polymerization of a correct deoxyribonucleoside triphosphate just after the misincorporated base is reduced to 0.01 nucleotides s^{-1}, thus greatly increasing the chance that the incorrect base pair will be removed by the proofreading exonuclease. This decrease in the rate of polymerization is due to the mismatch in the primer/template blocking the conformational change of the polymerase that is necessary for it to reach its catalytically active state. Mismatches at the −2 and −3 positions from the 3′ terminus also inhibit subsequent polymerization, thereby further amplifying the contribution of the proofreading exonuclease to fidelity. Also, if a polymerase were to have an additional slow step, corresponding to step D in Fig. 2–18, that would provide another opportunity for exonucleolytic proofreading (step E$_1$).

Another type of approach that has proved to be extremely useful for studying aspects of DNA polymerase fidelity in vitro has been to extend radioactively labeled primers by various DNA polymerases and then to use polyacrylamide gel electrophoresis to analyze the products (42, 77, 131, 132, 157, 162, 178). Such gel assays allow fidelity to be analyzed at many template sites. Two steady-state kinetic methods based on the use of polyacrylamide gel assays have been developed (reviewed in references 55 and 70). One type of assay involves direct competition between the right and wrong deoxyribonucleoside triphosphate substrates for insertion into the DNA by the polymerase being examined (19). Such competition assays have been used to analyze nucleotide insertion frequency (19, 125, 147, 150) and to study the efficiency of extending mispaired 3′ termini relative to the extension of correctly paired 3′ termini (126, 148). The other assay measures the kinetics of insertion for right and wrong deoxyribonucleoside triphosphates in separate reactions (55, 70). The ratio V_{max}/K_m measures the efficiency of nucleotide insertion by polymerase (59). At equimolar concentrations of right (*r*) and wrong (*w*) substrates, the ratio of misinsertion efficiencies for wrong versus right base pairs is equal to the misinsertion efficiency (f_{ins}) (55, 70) so that

$$f_{ins} = (V_{max,w}/K_{m,w})/(V_{max,r}/K_{m,r})$$

Mechanisms Contributing to Spontaneous Mutagenesis

Base Substitution Mutations Resulting from Misincorporation during DNA Synthesis

Given the complexities of DNA polymerase action summarized above, there are obviously a number of ways in which a failure of a DNA polymerase to discriminate between a right and a wrong nucleotide could result in an introduction of a base substitution mutation when the enzyme is copying an undamaged template. Surveys of the fidelities of various DNA polymerases by polyacrylamide gel-based assays (55, 70) and by the use of phage-based mutation assays (100, 115) have made it clear that DNA polymerases that have a proofreading exonuclease are more accurate than those that do not (55, 70, 104). On average, proofreading appears to contribute about 100-fold to polymerase fidelity. However, the greatest contribution to fidelity comes from the discrimination exercised by the polymerase during the polymerization step itself (102). At least some of the mechanisms that DNA polymerases use to achieve this discrimination have been revealed by the analyses of the intensively studied polymerases discussed above. The available data make it clear that different DNA polymerases use different discrimination mechanisms to different extents (55) and that the fidelities of DNA polymerases that lack proofreading can vary widely (102). Studies of the fidelities of various DNA polymerases in vitro have also highlighted the fact that individual DNA polymerases vary in their propensity to make certain types of base substitution errors. They have also shown that, for each polymerase, the sequence context can have a major influence on the frequency with which these errors are made (55, 70, 102).

The presence of various lesions in the DNA being copied can greatly increase the frequency with which base substitution errors are made. Examples of bases, such as 2-aminopurine, that can cause mutations by directly mispairing during DNA synthesis have been described in chapter 1. The more general subject of how lesions can cause mutations in vivo is discussed in detail in chapters 11 and 12.

Although studies of the action of purified polymerases on damaged DNA can yield important insights into fundamental aspects of DNA polymerase behavior, caution must be exerted in extrapolating such results directly to in vivo situations. An illustrative example is provided by abasic sites. Studies of the action of various DNA polymerases on the addition of bases opposite abasic or UV-damaged sites had led to the conclusion that A's are inserted preferentially at "noninstructive" sites as a result of inherent polymerase preference (170, 171). This rule was first suggested by Tessman (175) on the basis of his early studies on the mutagenic specificity of φX174. However, as discussed in more detail in chapters 11 and 12, with one apparent exception (176), the types of mutations caused by abasic sites in vivo in either prokaryotes (105, 109) or eukaryotes (66, 90) do not conform closely to the predictions of the A rule.

Mutations Resulting from Misalignments during DNA Synthesis

The concept that mutations could result from misalignment of the primer and the template was initially proposed by Streisinger et al. (172, 173) on the basis of their studies of frameshift mutagenesis. This concept has since been extended to a general theory that explains how primer template misalignments can account for a variety of types of mutations, including even certain base substitution mutations (46, 55, 101, 106, 155) (Fig. 2–21). For example, if, as originally proposed (172), a misalignment occurs in a repetitive sequence such as a homopolymeric run (Fig.

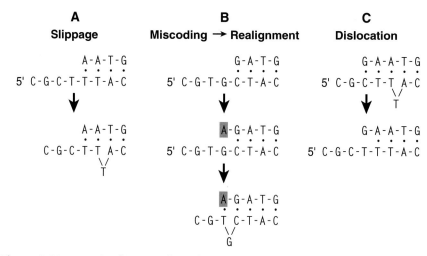

Figure 2–21   Mutational intermediates for substitution and frameshift errors that involve primer-template misalignments. (*Adapted from Kunkel [102] with permission.*)

2–21A), the resulting intermediate can be stabilized by correct base pairs. Continued polymerization from this intermediate will result in a deletion if the unpaired nucleotide is in the template strand (as shown in Fig. 2–21A) or an addition if the unpaired base is in the primer strand. If a misincorporation precedes the strand slippage but the misincorporated base realigns to generate a properly paired terminus, a frameshift error will result (Fig. 2–21B) (15). A third possibility involves an initial slippage, a correct incorporation, and then a realignment (Fig. 2–21C). This mechanism has been termed *dislocation mutagenesis* (64, 100, 101, 103, 106), the analogy being to a dislocated joint which pops out of alignment and then returns (102). Support for the model of dislocation mutagenesis has been provided from in vitro experiments in which alterations of putative templating bases yielded the results predicted by the theory of dislocation mutagenesis (14, 106). In vitro experiments have also indicated that sequence-specific pausing by DNA polymerases is one of the factors increasing the probability that a particular sequence will participate in frameshift mutagenesis (101) or dislocation mutagenesis (144). The central concepts of dislocation mutagenesis have also been extended to DNA containing lesions and used to explain frameshift and deletion errors in vitro (167, 180) and various mutations in vivo (108, 113, 161) (see chapter 11).

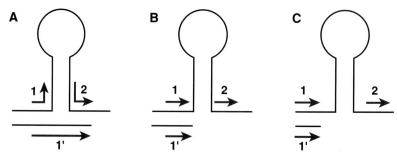

Figure 2–22   Misalignment between tandem repeats improved by a palindrome. In each case the deletion is produced by misaligning arrow 1′ with arrow 2. Arrow 1 is a repeat of arrow 2; arrow 1′ is the normal complement of arrow 1. (A) The palindrome brings arrow 2 to precisely the misaligned position that produces a deletion (80). (B) The palindrome brings arrow 2 adjacent to the misaligned position (96). (C) The palindrome brings arrow 2 closer (4). (*Adapted from Ripley [155] with permission.*)

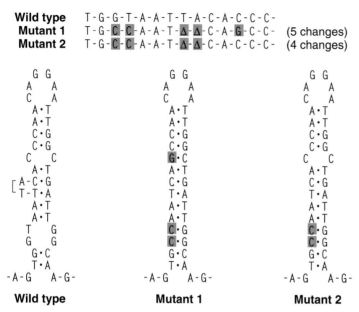

Figure 2–23 Model for the production of templated T4 *rIIB* mutations. The complex mutations termed mutant 1 and mutant 2 are rationalized as being due to the formation of a palindromic intermediate between two quasi-homologous sequences. Changed bases are in red, and Δ indicates a deleted base. (*Adapted from de Boer and Ripley [36] and Drake [46] with permission.*)

Primer-template misalignments have also been invoked to explain more-complex classes of mutations. For example, strand slippage that allows pairing between direct repeats of sequences that are homologous or quasi-homologous can explain large deletion mutations (4, 155). The presence of palindromic sequences in association with repeated sequences could also increase the frequency with which deletions between repeats are seen (69, 155) (Fig. 2–22). More complicated variations on this theme can explain more complex deletion and addition mutations made by DNA polymerases (101, 155).

The formation of hairpins between quasi-homologous inverted repeats has been postulated to account for rare mutations observed in spontaneous mutational spectra that are *complex*, in the sense that they consist of *a cluster of discrete mutations* (36, 46, 49, 155). For example, as diagrammed in Fig. 2–23, it is possible that a hairpin forms between two quasi-homologous sequences during DNA replication, repair, or recombination. If mismatches were eliminated by any of a variety of mechanisms, complex mutations could result.

REFERENCES

1. **Abrams, E. S., S. E. Murdaugh, and L. S. Lerman.** 1990. Comprehensive detection of single base changes in human genomic DNA using denaturing gradient gel electrophoresis and a GC clamp. *Genomics* **7:**463–475.

2. **Abrams, E. S., S. E. Murdaugh, and L. S. Lerman.** 1993. Comprehensive screening of the human *KRAS2* gene for sequence variants. *Genes Chromosomes Cancer* **6:**73–85.

3. **Aidoo, A., L. E. Lyn-Cook, R. A. Mittelstaedt, R. H. Heflich, and D. A. Casciano.** 1991. Induction of 6-thioguanine resistant lymphocytes in Fischer 344 rats following *in vivo* exposure to N-ethyl-N-nitrosourea and cyclophosphamide. *Environ. Mol. Mutagen.* **17:**141–151.

4. **Albertini, A. M., M. Hofer, M. P. Calos, and J. H. Miller.** 1982. On the formation of spontaneous deletions: the importance of short sequence homologies in the generation of large deletions. *Cell* **29:**319–326.

5. **Albertini, R. J., K. L. Castle, and W. R. Borcherding.** 1982. T-cell cloning to detect the mutant 6-thioguanine-resistant lymphocytes present in human peripheral blood. *Proc. Natl. Acad. Sci. USA* **79:**6617–6621.

6. **Ames, B. N., W. E. Durston, E. Yamasaki, and F. D. Lee.** 1973. Carcinogens are mutagens: A simple test system combining liver homogenates for activation and bacteria for detection. *Proc. Natl. Acad. Sci. USA* **70:**2281–2285.

7. **Ames, B. N., F. D. Lee, and W. E. Durston.** 1973. An improved bacterial test system for the detection and classification of mutagens and carcinogens. *Proc. Natl. Acad. Sci. USA* **70:**782–786.

8. **Ames, B. N., J. McCann, and E. Yamasaki.** 1975. Methods for detecting carcinogens and mutagens with the *Salmonella*-microsome mutagenicity test. *Mutat. Res.* **31:**347–364.

9. **Ashman, C. R., and R. L. Davidson.** 1985. High spontaneous mutation frequency of BPV shuttle vector. *Somatic Cell Mol. Genet.* **11:**499–504.

10. **Ashman, C. R., and R. L. Davidson.** 1987. DNA base sequence changes induced by ethyl methanesulfonate in a chromosomally integrated shuttle vector gene in mouse cells. *Somat. Cell Mol. Genet.* **13:**563–568.

11. **Ashman, C. R., P. Jagadeeswaran, and R. L. Davidson.** 1986. Efficient recovery and sequencing of mutant genes from mammalian chromosomal DNA. *Proc. Natl. Acad. Sci. USA* **83:**3356–3360.

12. **Basu, A. K., and J. M. Essigmann.** 1988. Site-specifically modified oligonucleotides as probes for the structural and biological effects of DNA-damaging agents. *Chem. Res. Toxicol.* **1:**1–18.

13. **Basu, A. K., and J. M. Essigmann.** 1990. Site-specific alkylated oligodeoxynucleotides: probes for mutagenesis, DNA repair and the structural effects of DNA damage. *Mutat. Res.* **233:**189–201.

14. **Bebenek, K., J. Abbotts, S. H. Wilson, and T. A. Kunkel.** 1993. Error-prone polymerization by HIV-1 reverse transcriptase. *J. Biol. Chem.* **268:**10324–10334.

15. **Benebek, K., and T. A. Kunkel.** 1990. Frameshift errors initiated by nucleotide misincorporation. *Proc. Natl. Acad. Sci. USA* **87:**4946–4950.

16. **Benzer, S.** 1961. Genetic fine structure. *Harvey Lect.* **56:**1–21.

17. **Benzer, S.** 1961. On the topography of the genetic fine structure. *Proc. Natl. Acad. Sci. USA* **47:**403–415.

18. **Berger, H., W. J. Brammar, and C. Yanofsky.** 1968. Analysis of amino acid replacements resulting from frameshift and missense mutations in the tryptophan synthetase *A* gene of *Escherichia coli. J. Mol. Biol.* **34:**219–238.

19. **Boosalis, M. S., J. Petruska, and M. F. Goodman.** 1987. DNA polymerase insertion fidelity: gel assay for site-specific kinetics. *J. Biol. Chem.* **262:**14689–14696.

20. **Bourre, F., and A. Sarasin.** 1983. Targeted mutagenesis of SV40 DNA induced by UV light. *Nature* (London) **305:**68–70.

21. **Calos, M. P., J. S. Lebkowski, and M. R. Botchan.** 1983. High mutation frequency in DNA transfected into mammalian cells. *Proc. Natl. Acad. Sci. USA* **80:**3015–3019.

22. **Calos, M. P., and J. H. Miller.** 1981. Genetic and sequence analysis of frameshift mutations induced by ICR-191. *J. Mol. Biol.* **153:**39–66.

23. **Cariello, N. F., P. Keohavong, A. G. Kat, and W. G. Thilly.** 1990. Molecular analysis of complex human cell populations: mutational spectra of MNNG and ICR-191. *Mutat. Res.* **231:**165–176.

24. **Cariello, N. F., J. K. Scott, A. G. Kat, W. G. Thilly, and P. Keohavong.** 1988. Resolution of a missense mutant in human genomic DNA by denaturing gradient gel electrophoresis and direct sequencing using *in vitro* DNA amplification: HPRT$_{Munich}$. *Am. J. Hum. Genet.* **42:**726–734.

25. **Cariello, N. F., J. A. Swenberg, and T. R. Skopek.** 1992. *In vitro* mutational specificity of cisplatin in the human hypoxanthine guanine phosphoribosyl transferase gene. *Cancer Res.* **52:**2866–2873.

26. **Carothers, A. M., G. Urlaub, D. Grunberger, and L. A. Chasin.** 1993. Splicing mutants and their second-site suppressors at the dihydrofolate reductase locus in Chinese hamster ovary cells. *Mol. Cell. Biol.* **13:**5085–5098.

27. **Carothers, A. M., G. Urlaub, J. Mucha, D. Grunberger, and L. A. Chasin.** 1989. Point mutation analysis in a mammalian gene: rapid preparation of total RNA, PCR amplification of cDNA, and Taq sequencing by a novel method. *BioTechniques* **7:**494–496.

28. **Cebula, T. A., and W. H. Koch.** 1990. Analysis of spontaneous and psoralen-induced *Salmonella typhimurium hisG46* revertants by oligodeoxynucleotide colony hybridization: use of psoralens to cross-link probes to target sequences. *Mutat. Res.* **229:**79–87.

29. **Cotton, R. G. H.** 1989. Detection of single base changes in nucleic acids. *Biochem. J.* **263:**1–10.

30. **Cotton, R. G. H., and R. D. Campbell.** 1989. Chemical reactivity of matched cytosine and thymine bases near mismatched and unmatched bases in a heteroduplex between DNA strands with multiple differences. *Nucleic Acids Res.* **17:**4223–4233.

31. **Cotton, R. G. H., N. R. Rodrigues, and R. D. Campbell.** 1988. Reactivity of cytosine and thymine in single-base-pair mismatches with hydroxylamine and osmium tetroxide and its application to the study of mutations. *Proc. Natl. Acad. Sci. USA* **85:**4397–4401.

32. **Coulondre, C., and J. H. Miller.** 1977. Genetic studies of the *lac* repressor. IV. Mutagenic specificity in the *lacI* gene of *Escherichia coli. J. Mol. Biol.* **117:**577–606.

33. **Crick, F. H. C., L. Barnett, S. Brenner, and R. J. Watts-Tobin.** 1961. General nature of the genetic code for proteins. *Nature* (London) **192:**1227–1232.

34. **Cupples, C., and J. H. Miller.** 1989. A set of *lacZ* mutations in *Escherichia coli* allows rapid detection of each of the six base substitutions. *Proc. Natl. Acad. Sci. USA* **86:**5345–5349.

35. **Dahlberg, M. E., and S. J. Benkovic.** 1991. Kinetic mechanism of DNA polymerase I (Klenow fragment): identification of a second conformation change and evaluation of the internal equilibrium constant. *Biochemistry* **30:**4835–4843.

36. **de Boer, J. G., and L. S. Ripley.** 1984. Demonstration of the production of frameshift and base-substitution mutations by quasipalindromic DNA sequences. *Proc. Natl. Acad. Sci. USA* **81:**5528–5531.

37. **de Boer, J. G., and L. S. Ripley.** 1988. An *in vitro* assay for frameshift mutations: hotspots for deletions of 1 bp by Klenow-fragment polymerase share a consensus DNA sequence. *Genetics* **118:**181–191.

38. **Demerec, M., G. Bertani, and J. Flint.** 1951. A survey of chemicals for mutagenesis action on *E. coli. Proc. Natl. Acad. Sci. USA* **85:**119–136.

39. **Dempsey, J. L., and A. A. Morley.** 1986. Measurement of *in vivo* mutant frequency in lymphocytes in the mouse. *Environ. Mutagen.* **8:**385–391.

40. **DiMaio, D., R. Treisman, and T. Maniatis.** 1982. Bovine papillomavirus vector that propogates as a plasmid in both mouse and bacterial cells. *Proc. Natl. Acad. Sci. USA* **79:**4030–4034.

41. **Donlin, M. J., S. S. Patel, and K. A. Johnson.** 1991. Kinetic partitioning between the exonuclease and polymerase sites in DNA error correction. *Biochemistry* **30:**538–546.

42. **Dosanijh, M. K., J. M. Essigmann, M. F. Goodman, and B. Singer.** 1990. Comparative efficiency of forming m⁴T·G versus m⁴T·A base pairs at a unique site by use of *Escherichia coli* DNA polymerase I (Klenow fragment) and *Drosophila melanogaster* polymerase α-primase complex. *Biochemistry* **29:**4698–4703.

43. **Drake, J. W.** 1963. Properties of ultraviolet-induced *rII* mutants of bacteriophage T4. *J. Mol. Biol.* **6:**268–283.

44. **Drake, J. W.** 1970. *The Molecular Basis of Mutation.* Holden-Day, San Francisco.

45. **Drake, J. W.** 1991. A constant rate of spontaneous mutation in DNA-based microbes. *Proc. Natl. Acad. Sci. USA* **88:**7160–7164.

46. **Drake, J. W.** 1991. Spontaneous mutation. *Annu. Rev. Genet.* **25:**125–146.

47. **Drake, J. W.** 1992. Mutation rates. *Bioessays* **14:**137–140.

48. **Drake, J. W.** 1993. Rates of spontaneous mutation among RNA viruses. *Proc. Natl. Acad. Sci. USA* **90:**4171–4175.

49. **Drake, J. W., B. W. Glickman, and L. W. Ripley.** 1983. Updating the theory of mutation. *Am. Sci.* **71:**621–630.

50. **Drinkwater, N. R., and D. K. Klinedinst.** 1986. Chemically induced mutagenesis in a shuttle vector with a low background mutant frequency. *Proc. Natl. Acad. Sci. USA* **83:**3402–3406.

51. **Drobetsky, E. A., A. J. Grosovsky, and B. W. Glickman.** 1987. The specificity of UV induced mutation at an endogenous locus in mammalian cells. *Proc. Natl. Acad. Sci. USA* **84:**9103–9107.

52. **DuBridge, R. B., and M. P. Calos.** 1988. Recombinant shuttle vectors for the study of mutagenesis. *Mutagenesis* **3:**1–9.

53. **DuBridge, R. B., P. Tang, H. C. Hsia, P. M. Leong, J. H. Miller, and M. P. Calos.** 1987. Analysis of mutation in human cells using an Epstein-Barr virus shuttle system. *Mol. Cell. Biol.* **7:**379–387.

54. **Dumaz, N., C. Drougard, A. Sarasin, and L. Daya-Grosjean.** 1993. Specific UV-induced mutation spectrum in the p53 gene of skin tumors from DNA-repair-deficient xeroderma pigmentosum patients. *Proc. Natl. Acad. Sci. USA* **90:**10529–10533.

55. **Echols, H., and M. F. Goodman.** 1991. Fidelity mechanisms in DNA replication. *Annu. Rev. Biochem.* **60:**477–511.

56. **Eger, B. T., and S. J. Benkovic.** 1992. Minimal kinetic mechanism for misincorporation by DNA polymerase I (Klenow fragment). *Biochemistry* **31:**9227–9236.

57. **Eisenstadt, E.** 1987. Analysis of mutagenesis, p. 1016–1033. *In* F. Neidhardt, J. L. Ingraham, K. B. Low, B. Magasanik, M. Schaechter, and H. E. Umbarger (ed.), *Escherichia coli and Salmonella typhimurium: Cellular and Molecular Biology.* American Society for Microbiology, Washington, D.C.

58. **Essigmann, J. M., and M. L. Wood.** 1993. The relationship between the chemical structure and mutagenic specificities of the DNA lesions formed by chemical and physical mutagens. *Toxicol. Lett.* **67:**29–39.

59. **Fersht, A. R.** 1985. *Enzyme Kinetics and Mechanism.* W. H. Freeman & Co., New York, NY.

60. **Fischer, S. G., and L. S. Lerman.** 1979. Length-independent separation of DNA restriction fragments in two-dimensional gel electrophoresis. *Cell* **16:**191–200.

61. **Fischer, S. G., and L. S. Lerman.** 1983. DNA fragments differing by single base-pair substitutions are separated in denaturing gradient gels: correspondence with melting theory. *Proc. Natl. Acad. Sci. USA* **80:**1579–1583.

62. **Foster, P. L.** Personal communication.

63. **Foster, P. L.** 1992. *Escherichia coli* and *Salmonella typhimurium,* mutagenesis, p. 107–114. *In* J. Lederberg (ed.), *Encylopedia of Microbiology.* Academic Press, Inc., New York.

64. **Fowler, R. G., G. E. Degnen, and E. C. Cox.** 1974. Mutational specificity of a conditional *Escherichia coli* mutator, *mutD5. Mol. Gen. Genet.* **133:**179–191.

65. **Freese, E.** 1959. On the molecular explanation of spontaneous and induced mutations. *Brookhaven Symp. Biol.* **12:**63–73.

66. **Gentil, A., J. B. Cabral-Neto, R. Mariage-Samson, A. Margot, J. L. Imbach, B. Rayner, and A. Sarasin.** 1992. Mutagenicity of a unique apurinic/apyrimidinic site in mammalian cells. *J. Mol. Biol.* **227:**981–984.

67. **Gentil, A., A. Margot, and A. Sarasin.** 1986. 2-(N-acetoxy N-acetylamino)fluorene mutagenesis in mammalian cells: sequence specific hot spot. *Proc. Natl. Acad. Sci. USA* **83:**9556–9560.

68. **Glazer, P. M., S. N. Sarkar, and W. C. Summers.** 1986. Detection and analysis of UV-induced mutations in mammalian cell DNA using a λ phage shuttle system. *Proc. Natl. Acad. Sci. USA* **83:**1041–1044.

69. **Glickman, B. W., and L. S. Ripley.** 1984. Structural intermediates of deletion mutagenesis: a role for palindromic DNA. *Proc. Natl. Acad. Sci. USA* **81:**512–516.

70. **Goodman, M. F., S. Creighton, L. B. Bloom, and J. Petruska.** 1993. Biochemical basis of DNA replication fidelity. *Crit. Rev. Biochem. Mol. Biol.* **28:**83–126.

71. **Gordon, A. J. E., P. A. Burns, D. F. Fix, F. Yatagi, F. L. Allen, M. J. Horsfall, J. A. Halliday, J. Gray, C. Bernelot-Moens, and B. W. Glickman.** 1988. Missense mutation in the *lacI* gene of *Escherichia coli*: inferences on the structure of the repressor protein. *J. Mol. Biol.* **200:**239–251.

72. **Gossen, J. A., W. J. F. De Leeuw, C. H. Tan, E. C. Zwarthoff, F. Berends, P. H. M. Lohman, D. L. Knook, and J. Vijg.** 1989. Efficient rescue of integrated shuttle vectors from transgenic mice: a model for studying mutations *in vivo. Proc. Natl. Acad. Sci. USA* **86:**7971–7975.

73. **Gossen, J. A., A. C. Molijn, G. R. Douglas, and J. Vijg.** 1992. Application of galactose-sensitive *E. coli* strains as selective hosts for LacZ⁻ plasmids. *Nucleic Acids Res.* **20:**3254.

74. **Grosovsky, A. J., E. A. Drobetsky, P. J. deJong, and B. W. Glickman.** 1986. Southern analysis of genomic alterations in gamma–ray-induced *aprt* hamster cell mutants. *Genetics* **113:**405–415.

75. **Gunther, E. J., N. E. Murray, and P. M. Glazer.** 1993. High-efficiency, restriction-deficient *in vitro* packaging extracts for bacteriophage lambda DNA using a new *E. coli* lysogen. *Nucleic Acids Res.* **21:**3903–3904.

76. **Hampsey, D. M., J. F. Ernst, J. W. Stewart, and F. Sherman.** 1988. Multiple base-pair mutations in yeast. *J. Mol. Biol.* **201:**471–486.

77. **Hillebrand, G. G., and K. L. Beattie.** 1985. Influence of template primary and secondary structure on the rate and fidelity of DNA synthesis. *J. Biol. Chem.* **260:**3116–3125.

78. **Hoorn, A. J. W., L. L. Custer, B. C. Myhr, D. Brusick, J. Gossen, and J. Vijg.** 1993. Detection of chemical mutagens using Muta Mouse: a transgenic mouse model. *Mutagenesis* **8:**7–10.

79. **Hsia, H. C., J. S. Lekowski, P.-M. Leong, M. P. Calos, and J. H. Miller.** 1989. Comparison of ultraviolet irradiation-induced mutagenesis of the *lacI* gene in *Escherichia coli* and in human 293 cells. *J. Mol. Biol.* **205:**103–113.

80. **Ikehata, H., T. Akagi, H. Kimura, S. Akasaka, and T. Kato.** 1989. Spectrum of spontaneous mutations in a cDNA of the human *hprt* gene integrated in chromosomal DNA. *Mol. Gen. Genet.* **219:**349–358.

81. **Iyer, V. N., and W. Szybalski.** 1958. Two simple methods for the detection of chemical mutagens. *Appl. Microbiol.* **6:**23–29.

82. **Jacobo-Molina, D., J. Ding, R. G. Nanni, A. D. Clark Jr., X. Lu, C. Tantillo, R. L. Williams, G. Kamer, A. L. Ferris, P. Clark, A. Hizi, S. H. Hughes, and E. Arnold.** 1993. Crystal structure of human immunodeficiency virus type I reverse transcriptase complexed with double-stranded DNA at 3.0 Angstrom resolution shows bent DNA. *Proc. Natl. Acad. Sci. USA* **90:**6320–6324.

83. **Johnson, K. A.** 1993. Conformational coupling in DNA polymerase fidelity. *Annu. Rev. Biochem.* **62:**685–713.

84. **Jones, I. M., K. Burkhart-Schultz, C. L. Strout, and T. L. Crippen.** 1987. Factors that affect the frequency of thioguanine-resistant lymphocytes in mice following exposure to ethylnitrosourea. *Environ. Mutagen.* **9:**317–329.

85. Joyce, C. M. 1989. How DNA travels between the separate polymerase and 3′-5′-exonuclease sites of DNA polymerase I (Klenow fragment). *J. Biol. Chem.* **264:**10858–10866.

86. Joyce, C. M., and T. A. Steitz. 1994. Function and structural relationships in DNA polymerases. *Annu. Rev. Biochem.* **63:**777–822.

87. Kennard, O. 1987. The molecular structure of base pair mismatches. *Nucleic Acids Mol. Biol.* **1:**25–52.

88. Keohavong, P., V. F. Liu, and W. G. Thilly. 1991. Analysis of point mutations induced by ultraviolet light in human cells. *Mutat. Res.* **249:**147–159.

89. Keohavong, P., and W. G. Thilly. 1992. Mutational spectrometry: a general approach for hot-spot point mutations in selectable genes. *Proc. Natl. Acad. Sci. USA* **89:**4623–4627.

90. Klinedinst, D. K., and N. R. Drinkwater. 1992. Mutagenesis by apurinic sites in normal and ataxia telangiectasia human lymphoblastoid lines. *Mol. Carcinog.* **6:**32–42.

91. Koffel-Schwartz, P., J. M. Verdier, J. F. Lefevre, M. Bichara, A. M. Fruend, M. P. Daune, and R. P. P. Fuchs. 1984. Carcinogen-induced mutation spectrum in wild-type, and *uvrA*, and *umuC* strains of *E. coli*: strain specificity and mutation-prone sequences. *J. Mol. Biol.* **177:**33–51.

92. Kohler, S. W., G. S. Provost, A. Fieck, P. L. Kretz, W. O. Bullock, D. L. Putman, J. A. Sorge, and J. M. Short. 1991. Analysis of spontaneous and induced mutations in transgenic mice using a lambda ZAP/lacI shuttle vector. *Environ. Mol. Mutagen.* **18:**316–321.

93. Kohler, S. W., G. S. Provost, A. Fieck, P. L. Kretz, W. O. Bullock, J. A. Sorge, D. L. Putman, and J. M. Short. 1991. Spectra of spontaneous and mutagen-induced mutations in the *lacI* gene in transgenic mice. *Proc. Natl Acad. Sci. USA* **88:**7958–7962.

94. Kohler, S. W., G. S. Provost, P. L. Kretz, M. J. Dycaico, J. A. Sorge, and S. M. Short. 1990. Development of a short-term, *in vivo* mutagenesis assay: the effects of methylation on the recovery of lambda phage shuttle vector from transgenic mice. *Nucleic Acids Res.* **18:**3007–3013.

95. Kohlstaedt, L. A., J. Wang, J. M. Friedman, P. A. Rice, and T. A. Steitz. 1992. Crystal structure at 3.5 Ångstrom resolution of HIV-1 resverse transcriptase complexed with an inhibitor. *Science* **256:**1783–1790.

96. Korn, D., P. A. Fisher, and T. S.-F. Wang. 1983. Enzymological characterization of human DNA polymerases-α and β, p. 17–55. *In* A. M. de Recondo (ed.), *New Approaches in Eukaryotic DNA Replication*. Plenum Publishing Corp., New York.

97. Kornberg, A., and T. A. Baker. 1992. *DNA Replication*. W. H. Freeman & Co., New York.

98. Kraemer, K. H., and M. M. Seidman. 1989. Use of *supF*, an *Escherichia coli* tyrosine suppressor tRNA gene, as a mutagenic target in shuttle-vector plasmids. *Mutat. Res.* **220:**61–72.

99. Kuchta, R. D., V. Mizrahi, P. A. Benkovic, K. A. Johnson, and S. J. Benkovic. 1987. Kinetic mechanism of DNA polymerase I (Klenow). *Biochemistry* **26:**8410–8417.

100. Kunkel, T. A. 1985. The mutational specificity of DNA polymerase-β during *in vitro* DNA synthesis. Production of frameshift, base substitution, and deletion mutations. *J. Biol. Chem.* **260:**5787–5796.

101. Kunkel, T. A. 1990. Misalignment mediated DNA synthesis errors. *Biochemistry* **29:**8003–8011.

102. Kunkel, T. A. 1992. DNA replication fidelity. *J. Biol. Chem.* **267:**18251–18254.

103. Kunkel, T. A., and P. S. Alexander. 1986. The base substitution fidelity of eucaryotic polymerases. *J. Biol. Chem.* **261:**160–166.

104. Kunkel, T. A., and K. Bebenek. 1988. Recent studies of the fidelity of DNA synthesis. *Biochim. Biophys. Acta* **951:**1–15.

105. Kunkel, T. A., and L. A. Loeb. 1984. Mutational specificity of depurination. *Proc. Natl. Acad. Sci. USA* **81:**1494–1498.

106. Kunkel, T. A., and A. Soni. 1988. Mutagenesis by transient misaligment. *J. Biol. Chem.* **29:**14784–14789.

107. Kupchella, E., and T. A. Cebula. 1991. Analysis of *Salmonella typhimurium hisD3052* revertants: the use of oligodeoxynucleotide colony hybridization, PCR, and direct sequencing in mutational analysis. *Environ. Mol. Mutagen.* **18:**224–230.

108. Lambert, I. B., R. L. Napolitano, and R. P. P. Fuchs. 1992. Carcinogen-induced frameshift mutagenesis in repetitive sequences. *Proc. Natl. Acad. Sci. USA* **89:**1310–1314.

109. Lawrence, C. W., A. Borden, S. K. Banerjee, and J. E. LeClerc. 1990. Mutation frequency and spectrum resulting from a single abasic site in a single-stranded vector. *Nucleic Acids Res.* **18:**2153–2157.

110. Lebkowski, J. S., R. B. DuBridge, E. A. Antell, K. S. Greisen, and M. P. Calos. 1984. Transfected DNA is mutated in monkey, mouse, and human cell lines. *Mol. Cell Biol.* **4:**1951–1960.

111. LeClerc, J. E., N. L. Istock, B. R. Saran, and R. Allen Jr. 1984. Sequence analysis of ultraviolet-induced mutations in M13*lacZ* hybrid phage DNA. *J. Mol. Biol.* **180:**217–237.

112. Levin, D. E., and B. N. Ames. 1986. Classifying mutagens as to their possible specificity in causing the six possible transitions and transversions: a simple analysis using the *Salmonella* mutagenicity assay. *Environ. Mutagen.* **8:**9–28.

113. Levine, J. G., R. M. Schaaper, and D. M. DeMarini. 1994. Complex frameshift mutations mediated by plasmid pKM101: mutational mechanisms deduced from 4-aminobiphenyl-induced mutational spectra in *Salmonella*. *Genetics* **136:**731–746.

114. Levy, D. D., J. D. Groopman, S. E. Lim, M. M. Seidman, and K. H. Kraemer. 1992. Sequence specificity of aflatoxin B₁-induced mutations in a plasmid replicated in Xeroderma pigmentosum and DNA repair proficient human cells. *Cancer Res.* **52:**5668–5673.

115. Loeb, L. A., and T. A. Kunkel. 1982. Fidelity of DNA synthesis. *Annu. Rev. Biochem.* **52:**429–457.

116. Luria, S. E., and M. Delbrück. 1943. Mutations of bacteria from virus sensitivity to virus resistance. *Genetics* **28:**491–511.

117. MacGregor, G. R., and J. F. Burke. 1987. Stability of a bacterial gene in a bovine papilloma virus based shuttle vector maintained extrachromosomally in mammalian cells. *J. Gen. Virol.* **183:**273–278.

118. Madzak, C., J. Armier, A. Stary, L. Daya-Grosjean, and A. Sarasin. 1993. UV-induced mutations in a shuttle vector replicated in repair deficient trichothiodystrophy cells differ with those in genetically-related cancer prone xeroderma pigmentosum. *Carcinogenesis* **14:**1255–1260.

119. Maher, V. M., J. L. Yang, R.-H. Chen, W. G. McGregor, L. Lukash, J. M. Scheid, D. S. Reinhold, and J. J. McCormick. 1991. Use of PCR amplification of cDNA to study mechanism of human cell mutagenesis and malignant transformation. *Environ. Mol. Mutagen.* **18:**239–244.

120. McCann, J., and B. N. Ames. 1976. Detection of carcinogens as mutagens in the *Salmonella*/microsome test: assay of 300 chemicals: discussion. *Proc. Natl. Acad. Sci. USA* **73:**950–954.

121. McCann, J., and B. N. Ames. 1978. The *Salmonella*/microsome mutagenicity test: predictive value for animal carcinogenicity, p. 87–108. *In* W. G. Flamm and M. A. Mehlman (ed.), *Advances in Modern Toxicology*. Hemisphere Publishing Corp., Washington, D.C.

122. McCann, J., E. Choi, E. Yamasaki, and B. N. Ames. 1975. Detection of carcinogens as mutagens in the *Salmonella*/microsome test: assay of 300 chemicals. *Proc. Natl. Acad. Sci. USA* **72:**5135–5139.

123. McCann, J., N. E. Spingarn, J. Kobori, and B. N. Ames. 1975. Detection of carcinogens as mutagens: bacterial tester strains with R factor plasmids. *Proc. Natl. Acad. Sci. USA* **72:**979–983.

124. Melton, D. W., D. S. Konecki, J. Brennand, and C. T. Caskey. 1984. Structure, expression, and mutation of the hypoxanthine phosphoribosyl transferase gene. *Proc. Natl. Acad. Sci. USA* **81:**1484–1488.

125. Mendelman, L. V., M. S. Boosalis, J. Petruska, and M. F. Goodman. 1989. Nearest neighbour influences on DNA polymerases insertion fidelity. *J. Biol. Chem.* **264:**14415–14423.

126. Mendelman, L. V., J. Petruska, and M. F. Goodman. 1990. Base mispair kinetics: comparison of DNA polymerase α and reverse transcriptase. *J. Biol. Chem.* **265:**2338–2346.

127. **Miller, J. H.** 1983. Mutational specificity in bacteria. *Annu. Rev. Genet.* **17:**215–238.

128. **Miller, J. H.** 1985. Mutagenic specificity of ultraviolet light. *J. Mol. Biol.* **182:**45–68.

129. **Miller, J. H.** 1992. *A Short Course in Bacterial Genetics. A Laboratory Manual for Escherichia coli and Related Bacteria.* Cold Spring Harbor Laboratory Press, Cold Spring Harbor, N.Y.

130. **Miller, J. K., and W. M. Barnes.** 1986. Colony probing as an alternative to standard sequencing as a means of direct analysis of chromosomal DNA to determine the spectrum of single-base changes in regions of known sequence. *Proc. Natl. Acad. Sci. USA* **83:**1026–1030.

131. **Moore, P. D., K. K. Bose, S. D. Rabkin, and B. S. Strauss.** 1981. Sites of termination of *in vitro* DNA synthesis on ultraviolet- and N-acetylaminofluorene-treated φX174 templates by prokaryotic and eukaryotic DNA polymerases. *Proc. Natl. Acad. Sci. USA* **78:**110–114.

132. **Moore, P. D., S. D. Rabkin, A. L. Osborn, C. M. King, and B. S. Strauss.** 1982. Effect of acetylated and deacetylated 2-aminofluorene adducts on *in vitro* DNA synthesis. *Proc. Natl. Acad. Sci. USA* **79:**7166–7170.

133. **Morley, A. A., K. J. Trainor, R. Seshadri, and R. G. Ryall.** 1983. Measurement of *in vivo* mutations in human lymphocytes. *Nature* (London) **302:**155–156.

134. **Muller-Hill, B., and J. Kania.** 1974. *Lac* repressor can be fused to β-galactosidase. *Nature* (London) **249:**561–563.

135. **Myers, R. M., S. G. Fisher, L. S. Lerman, and T. Maniatis.** 1985. Nearly all single base substitutions in DNA fragments joined to a GC-clamp can be detected by denaturing gradient gel electrophoresis. *Nucleic Acids Res.* **13:**3131–3145.

136. **Myers, R. M., S. G. Fisher, T. Maniatis, and L. S. Lerman.** 1985. Modification of the melting properties of duplex DNA by attachment of a GC-rich DNA sequence as determined by denaturing gradient gel electrophoresis. *Nucleic Acids Res.* **13:**3111–3129.

137. **Myers, R. M., N. Lumelsky, L. S. Lerman, and T. Maniatis.** 1985. Detection of single base substitutions in total genomic DNA. *Nature* (London) **313:**495–498.

138. **Myhr, B.** 1991. Validation studies with Muta Mouse—a transgenic mouse model for detecting mutations *in vivo*. *Environ. Mol. Mutagen.* **18:**308–315.

139. **Nichols, B. P., and C. Yanofsky.** 1979. Nucleotide sequences of *trpA* of *Salmonella typhimurium* and *Escherichia coli*: an evolutionary comparison. *Proc. Natl. Acad. Sci. USA* **76:**5244–5248.

140. **Novack, D. F., N. J. Casna, S. G. Fischer, and J. P. Ford.** 1986. Detection of single-base pair mismatches in DNA by chemical modification followed by electrophoresis in 15% polyacrylamide gel. *Proc. Natl. Acad. Sci. USA* **83:**586–590.

141. **Oller, A. R., I. J. Fijalkowska, and R. M. Schaaper.** 1993. The *Escherichia coli galK2* papillation assay; its specificity and application to seven newly isolated mutator strains. *Mutat. Res.* **292:**175–185.

142. **Oller, A. R., and W. G. Thilly.** 1992. Mutational spectra in human B-cells: spontaneous, oxygen, and hydrogen peroxide-induced mutations at the *hprt* gene. *J. Mol. Biol.* **228:**813–826.

143. **Ollis, D. L., P. Brick, R. Hamlin, N. G. Xuong, and T. A. Steitz.** 1985. Structure of large fragment of *Escherichia coli* DNA polymerase I complexed with dTMP. *Nature* (London) **313:**762–766.

144. **Papanicolaou, C., and L. S. Ripley.** 1991. An *in vitro* approach to identifying specificity determinants of mutagenesis mediated by DNA misalignments. *J. Mol. Biol.* **221:**805–821.

145. **Parris, C. N., and M. M. Seidman.** 1992. A signature element distinguishes sibling and independent mutations in a shuttle vector plasmid. *Gene* **117:**1–5.

146. **Patel, S. S., I. Wong, and K. A. Johnson.** 1991. Pre-steady state kinetic analysis of processive DNA replication including complete characterization of an exonuclease-deficient mutant. *Biochemistry* **30:**511–525.

147. **Perrino, F. W., and L. A. Loeb.** 1989. Differential extension of 3' mispairs is a major contribution to the high fidelity of calf thymus DNA polymerase-alpha. *J. Biol. Chem.* **264:**2898–2905.

148. **Perrino, F. W., B. D. Preston, L. L. Sandell, and L. A. Loeb.** 1989. Extension of mismatched 3' termini of DNA is major determinant of the infidelity of human immunodeficiency virus type I reverse transcriptase. *Proc. Natl. Acad. Sci. USA* **86:**8343–8347.

149. **Prakash, L., and F. Sherman.** 1973. Mutagenic specificity: reversion of iso-1-cytochrome *c* mutants of yeast. *J. Mol. Biol.* **79:**65–82.

150. **Preston, B. D., B. J. Poiesz, and L. A. Loeb.** 1988. Fidelity of HIV-1 reverse transcriptase. *Science* **242:**1168–1171.

151. **Pribnow, D., D. C. Sigurdson, L. Gold, B. S. Singer, C. Napoli, J. Brosius, T. J. Dull, and H. F. Noller.** 1981. *rII* cistrons of bacteriphage T4: DNA sequence around the intercistronic divide and positions of genetic landmarks. *J. Mol. Biol.* **149:**337–376.

152. **Razzaque, A., H. Mizusawa, and M. M. Seidman.** 1983. Rearrangements and mutagenesis of a shuttle vector plasmid after passage in mammalian cells. *Proc. Natl. Acad. Sci. USA* **80:**3010–3014.

153. **Reddy, M. K., S. E. Weitzel, and P. H. von Hippel.** 1992. Processive proofreading is intrinsic to T4 DNA polymerase. *J. Biol. Chem.* **267:**14157–14166.

154. **Ripley, L., A. Clark, and J. G. de Boer.** 1986. Spectrum of spontaneous frameshift mutations. Sequences of bacteriophage T4 *rII* frameshifts. *J. Mol. Biol.* **191:**601–613.

155. **Ripley, L. S.** 1990. Frameshift mutation: determinants of specificity. *Annu. Rev. Genet.* **24:**189–213.

156. **Romac, S., P. Leong, H. Sockett, and F. Hutchinson.** 1989. DNA base sequence changes induced by ultraviolet light mutagenesis of a gene on a chromosome in Chinese hamster ovary cells. *J. Mol. Biol.* **209:**195–204.

157. **Sagher, D., and B. Strauss.** 1983. Insertion of nucleotides opposite purinic/apyrimidinic sites in deoxyribonucleic acid during *in vitro* synthesis: uniqueness of adenine nucleotides. *Biochemistry* **22:**4518–4526.

158. **Saiki, R. K., D. Gefland, S. Stoffel, S. J. Scharf, R. Higuchi, G. T. Horn, K. B. Mullis, and H. A. Erlich.** 1988. Primer-directed enzymatic amplification of DNA with a thermostable DNA polymerase. *Science* **239:**487–491.

159. **Sarasin, A.** 1989. Shuttle vectors for studying mutagenesis in mammalian cells. *J. Photochem. Photobiol. B* **3:**143–155.

160. **Schaaper, R. M., and R. L. Dunn.** 1991. Spontaneous mutation in the *Escherichia coli lacI* gene. *Genetics* **129:**317–326.

161. **Schaaper, R. M., N. Koeffel-Schwartz, and R. P. P. Fuchs.** 1990. N-acetoxy-N-acetylaminofluorene-induced mutagenesis in the *lacI* gene of *Escherichia coli*. *Carcinogenesis* **11:**1087–1095.

162. **Schaaper, R. M., T. A. Kunkel, and L. A. Loeb.** 1983. Infidelity of DNA synthesis associated with bypass of apurinic sites. *Proc. Natl. Acad. Sci. USA* **80:**487–491.

163. **Seetharam, S., K. H. Kraemer, H. L. Waters, and M. M. Seidman.** 1991. Ultraviolet mutational spectrum in a shuttle vector propagated in xeroderma pigmentosum lymphoblastoid cells and fibroblasts. *Mutat. Res.* **254:**97–105.

164. **Seidman, M.** 1989. The development of transient SV40 based shuttle vectors for mutagenesis studies: problems and solutions. *Mutat. Res.* **220:**55–60.

165. **Seidman, M. M., K. Dixon, A. Razzaque, R. J. Zagursky, and M. L. Berman.** 1985. A shuttle vector plasmid for studying carcinogen-induced point mutations in mammalian cells. *Gene* **38:**233–237.

166. **Sheffield, V. C., D. R. Cox, L. S. Lerman, and R. M. Myers.** 1989. Attachment of a 40-base-pair G + C-rich sequence (GC-clamp) to genomic DNA fragments by the polymerase chain reaction results in improved detection of single-base changes. *Proc. Natl. Acad. Sci. USA* **86:**232–236.

167. **Shibutani, S., and A. P. Grollman.** 1993. On the mechanism of frameshift (deletion) mutagenesis *in vitro*. *J. Biol. Chem.* **268:**11703–11710.

168. **Singer, B., and J. M. Essigman.** 1991. Site-specific mutagenesis: retrospective and prospective. *Carcinogenesis* **12:**949–955.

169. **Skopek, T. R., V. E. Walker, J. E. Cochrane, T. R. Craft, and N. F. Cariello.** 1992. Mutational spectrum at the *Hprt* locus in

splenic T cells of B6C3F$_1$ mice exposed to *N*-ethyl-*N*-nitrosourea. *Proc. Natl. Acad. Sci. USA* **89**:7866–7870.

170. Strauss, B., S. Rabkin, D. Sagher, and P. Moore. 1982. The role of DNA polymerase in base-substitution mutagenesis on non-instructional templates. *Biochimie* **64**:829–838.

171. Strauss, B. S. 1991. The "A rule" of mutagen specificity: a consequence of DNA polymerase bypass of non-instructional lesions. *Bioessays* **13**:79–84.

172. Streisinger, G., Y. Okada, J. Emrich, J. Newton, A. Tsugita, E. Terzaghi, and M. Inouye. 1966. Frameshift mutations and the genetic code. *Cold Spring Harbor Symp. Quant. Biol.* **31**:77–84.

173. Streisinger, G., and J. Owen. 1985. Mechanisms of spontaneous and induced frameshift mutation in bacteriophage T4. *Genetics* **109**:633–659.

174. Summers, W. C., P. M. Glazer, and D. Malkevich. 1989. λ phage shuttle vectors for analysis in mammalian cells in culture and in transgenic mice. *Mutat. Res.* **220**:263–268.

175. Tessman, I. 1976. A mechanism of UV reactivation, p. 87. *In* A. I. Bukhari and E. Ljungquist (ed.), *Abst. Bacteriophage Meet.* Cold Spring Harbor Laboratory, Cold Spring Harbor, N.Y.

176. Tessman, I., and M. A. Kennedy. 1994. DNA polymerase II of *Escherichia coli* in the bypass of apurinic sites *in vivo. Genetics* **136**:439–448.

177. Tindall, K. R., and R. A. Whitaker. 1991. Rapid localization of point mutations in PCR products by chemical (HOT) modification. *Environ. Mol. Mutagen.* **18**:231–238.

178. Toorchen, D., and M. D. Topal. 1983. Mechanisms of chemical mutagenesis and carcinogenesis: effects on DNA replication of methylation at the O^6-guanine position of dGTP. *Carcinogenesis* **4**:1591–1597.

179. Vrieling, H., L.-H. Zhang, A. A. van Zeeland, and M. Z. Zdzienicka. 1992. UV-induced *hprt* mutations in a UV-sensitive hamster cell line from complementation group 3 are biased towards the transcribed strand. *Mutat. Res.* **274**:147–155.

180. Wang, C.-I., and J. S. Taylor. 1992. In vitro evidence that UV-induced frameshift and substitution mutations at T tracts are the result of misalignment-mediated replication past a specific thymine dimer. *Biochemistry* **31**:3671–3681.

181. Wang, Y.-C., V. M. Maher, D. L. Mitchell, and J. J. McCormick. 1993. Evidence from mutation spectra that UV hypermutability of xeroderma pigmentosum variant cells reflects abnormal, error-prone replication on a template containing photoproducts. *Mol. Cell. Biol.* **13**:4276–4283.

182. Winter, F., F. Yamamoto, C. Almoguera, and M. Perucho. 1985. A method to detect and characterize point mutations in transcribed gene: amplification and overexpression of the mutants c-Ki-ras allele in human tumor cells. *Proc. Natl. Acad. Sci. USA* **82**:7575–7579.

183. Wong, I., S. S. Patel, and K. A. Johnson. 1991. An induced-fit kinetic mechanism for DNA replication fidelity: direct measurement by single-turnover kinetics. *Biochemistry* **30**:526–537.

184. Wood, R. D., T. R. Skopek, and F. Hutchinson. 1984. Changes in DNA base sequence induced by targeted mutagenesis of lambda phage by ultraviolet light. *J. Mol. Biol.* **173**:273–291.

185. Yang, J.-L., R.-H. Chen, V. M. Maher, and J. J. McCormick. 1991. Kinds and location of mutations induced by (±)-7β,8α-dihydroxy-9α,10α-epoxy-7,8,9,10-tetrahydrobenzo[a]pyrene in the coding region of the hypoxanthine (guanine) phosphoribosyltransferase gene in diploid human fibroblasts. *Carcinogenesis* **12**:71–75.

186. Yanofsky, C. 1971. Mutagenesis studies with *Escherichia coli* mutants with known amino acid (and base-pair) changes, p. 283–287. *In* A. Hollaender (ed.), *Chemical Mutagens: Principles and Methods for Their Detection.* Plenum Publishing Corp., New York.

187. Yanofsky, C., J. Ito, and V. Horn. 1966. Amino acid replacements and the genetic code. *Cold Spring Harbor Symp. Quant. Biol.* **31**:151–162.

188. Yates, J. L., N. Warren, D. Reisman, and B. Sugden. 1985. Plasmids derived from Epstein-Barr virus replicate stably in in a variety of mammalian cells. *Nature* (London) **313**:812–815.

189. Ziegler, A., D. J. Leffell, S. Kunala, H. W. Sharma, M. Gailani, J. A. Simon, A. J. Halperin, H. P. Baden, P. E. Shapiro, A. E. Bale, and D. E. Brash. 1993. Mutation hotspots due to sunlight in the p53 gene of nonmelanoma skin cancers. *Proc. Natl. Acad. Sci. USA* **90**:4216–4220.

190. Zimmer, D. M., C. S. Aaron, J. P. O'Neill, and R. J. Albertini. 1991. Enumeration of 6-thioguanine-resistant T-lymphocytes in peripheral blood of nonhuman primates (Cynomolgus monkeys). *Environ. Mol. Mutagen.* **18**:161–167.

3

DNA Repair by Reversal of Damage

Now that we have some idea of the spectrum of damage that the cellular genome may sustain, we can begin to consider how living cells respond to such damage. In this regard we would like to clarify an important semantic issue at the very outset. The term *DNA repair* is used throughout this text in a strict biochemical sense, to identify only those cellular responses that are associated with the restoration of the normal base sequence and structure of damaged DNA. In the literature, this phrase frequently explicitly or implicitly connotes other cellular responses to DNA damage which are not necessarily accompanied by the restoration of normal base sequence and chemistry. This relaxed definition was presumably motivated by the recognition that such cellular responses often result in biological end points (e.g., enhanced survival or reduced mutagenesis of cells) which mitigate the potentially harmful effects of DNA damage. Some of these responses may indeed be accompanied by the biochemical repair of DNA strand breaks and/or gaps. However, in this text, cellular responses that do not include the removal of the *primary damage* to the genome are considered *DNA damage tolerance* rather than DNA repair mechanisms. Many DNA damage tolerance mechanisms result in permanent mutations in the genome and hence are discussed in the chapters on mutagenesis. The ability of cells to tolerate DNA damage is biologically as important as their ability to repair such damage. Indeed, it has been suggested that when DNA damage became an evolutionarily prevalent problem for living cells, genetic outcrossing or sexual recombination evolved as an efficient means for exchanging "good" bits of DNA for "bad" (12, 13) without cells having to permanently retain a redundant copy of the entire genome. Several studies have attempted to address the hypothesis that the ability of bacterial cells to exchange genetic information may reflect their need to successfully tolerate DNA damage (143, 144, 180).

Table 3–1 summarizes the spectrum of responses to DNA damage in living cells. Note that DNA repair as defined here can occur by one of two fundamental

Table 3–1 Cellular responses to DNA damage

Response	Mechanism
Reversal of DNA damage	Enzymatic photoreactivation Repair of spore photoproduct Repair of O^6-alkylguanine, O^4-alkylthymine, and alklyphosphotriesters Ligation of DNA strand breaks
Excision of DNA damage	Base excision repair Nucleotide excision repair Mismatch repair
Tolerance of DNA damage	Replicative bypass of template damage with gap formation and recombination Translesion DNA synthesis

cellular responses that involve either the *direct reversal* of DNA damage or its *excision*. To use a simplistic analogy to clarify this distinction at the outset, consider damage to be represented by a knot in a length of twine (DNA). In some cases it is possible to simply undo the knot (reversal of damage); in others it is necessary to cut out a piece of string containing the knot and replace it with a new segment of twine (excision of damage and synthesis of new nucleotides).

In principle, the simplest biochemical mechanism by which damage to DNA might be repaired is one in which a single enzyme composed of a single polypeptide chain catalyzes a single-step reaction which restores the structure of the genome to its normal state. The requirement for only one biochemical event catalyzed by one polypeptide would be expected to provide kinetic and energetic advantages over multistep reactions involving a number of different enzymes or multi-protein complexes. In addition, the potential for introducing errors into the coding elements of damaged genes during the course of repair itself would be expected to be reduced, or, for some mechanisms, eliminated entirely. In this chapter we will consider four examples of the direct reversal of DNA damage, all of which are catalyzed by single polypeptides.

In this chapter, as in later discussions on excision repair, particular emphasis is placed on base damage in DNA. The repair of these particular chemical moieties has attracted most attention in view of their biological importance as coding elements in the genome. However, it should be kept in mind that other forms of DNA damage, such as strand breaks and damage to the sugar moieties of the DNA backbone are of considerable importance, since they can interfere with fundamental DNA functions such as replication, recombination, and transcription. Some aspects of strand break repair are dealt with in this chapter. Others are considered during discussions of DNA damage tolerance, later in the text.

Photoreactivation of DNA

As indicated in the previous chapter, both cyclobutane pyrimidine dimers and 6-4 pyrimidine-pyrimidones [(6–4) lesions] constitute major sources of base damage following the exposure of living cells to UV radiation at wavelengths near the absorption maximum of DNA. Both of these photoproducts, as well as other photoproducts in DNA, can interfere with the function of the constitutive DNA replication and transcription machinery and hence pose serious threats to the viability and functional integrity of cells (9, 22, 146, 234, 252, 267). In view of the central role that UV radiation has played as a source of DNA damage during evolution, it is not surprising that living organisms have evolved an extraordinarily specific mechanism for the repair of the quantitatively major photoproduct, pyrimidine dimers. This repair mechanism is called *photoreactivation (PR) of DNA* (28, 188, 190,

242, 243) and is the first example we will discuss of DNA repair by the direct reversal of base damage.

PR is a light-dependent process involving the enzyme-catalyzed monomerization of *cis-syn*-cyclobutyl pyrimidine dimers (Fig. 3–1). The *trans-syn*-cyclobutyl dimer is also repaired by PR, but the enzyme involved in these repair processes has about a 10^4-fold-lower affinity for the latter type of dimer (106). PR must not be confused with a number of other nonenzymatic light-dependent processes by which pyrimidine dimers can be monomerized. For example, the continued irradiation of DNA containing pyrimidine dimers at wavelengths between 200 and 300 nm results in the direct reversal of some fraction of the dimers until a new equilibrium between monomerization and dimerization is attained (233). This phenomenon, referred to as *direct photoreversal*, is wavelength dependent, with photomonomerization favored at wavelengths of ~235 nm and photodimerization favored at wavelengths of ~280 nm (168). Direct photoreversal provides a means of perturbing the pyrimidine dimer content of DNA for selected experimental purposes but is not a biologically relevant process. Another non-enzyme-dependent reaction which splits pyrimidine dimers in DNA is called *sensitized photoreversal*. This process has been observed in the presence of tryptophan and tryptophan-containing oligopeptides such as Lys-Trp-Lys (27, 70, 71, 249, 262), as well as the protein encoded by gene 32 of phage T4 (72). The mechanism of this reaction is thought to involve an electron transfer from the excited indole ring of tryptophan to the dimer, possibly mediated by base stacking interactions of the tryptophanyl residue with the pyrimidine dimer in DNA. This process has limited biological significance and is not considered a distinct form of DNA repair, although the simple tripeptide-mediated reaction may have served as the starting point for the evolution of PR by more complex polypeptides. Finally, it has been demonstrated that mutants of *Escherichia coli* that are defective in PR can exhibit enhanced survival when illuminated at 334 nm following exposure to UV irradiation at wavelengths that produce pyrimidine dimers. This phenomenon is referred to as *indirect PR*. It does not involve the monomerization of pyrimidine dimers. Rather, it has been suggested that indirect PR inactivates a number of tRNAs by photo-cross-linking, thereby causing a temporary inhibition of growth and division and widening the time window for excision repair of DNA before DNA replication proceeds (77).

PR was the first DNA repair mode to be discovered. In the late 1940s Albert Kelner, then at the Cold Spring Harbor Laboratory, was studying the effects of UV radiation on certain strains of the fungus *Streptomyces griseus* (102). While investigating the influence of postirradiation temperature on colony survival, Kelner was plagued by an experimental variable, the systematic exploration of which ultimately led to the recognition of a specific role of light in the recovery of UV-irradiated cells. Kelner's own description of this discovery merits quotation (102):

> Careful consideration was made of variable factors which might have accounted for such tremendous variation. We were using a glass-fronted water bath placed on a table near a window, in which were suspended transparent bottles containing the irradiated spores. The fact that some of the bottles were more directly exposed to light than others suggested that light might be a factor. Moreover, the greatest and most consistent recovery in our preliminary experiments had taken place in suspensions stored in transparent bottles at room temperature on an open shelf exposed to diffuse light from a window. Experiment showed that exposure of ultra-violet irradiated suspensions to light resulted in an increase in survival rate or a recovery of 100,000- to 400,000-fold. Controls kept in the dark . . . showed no recovery at all. The magnitude of the light effect can hardly be overemphasized. The recovery was so much more complete than any previously observed, that we felt we were dealing here with a key factor in the mechanism causing inactivation and recovery from ultra-violet irradiation.

At about the same time, Renato Dulbecco, then at the Department of Bacteriology, Indiana University, reported the same phenomenon in the T group of coliphages (39). He wrote that:

> the occurrence of photo-reactivation of ultra-violet irradiated phage was noticed accidentally a few weeks after receiving a personal communication from Dr. A.

1. Native DNA

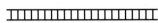

2. Pyrimidine dimer in UV DNA

3. Complex of DNA with photoreactivating enzyme

4. Absorption of light (>300nm)

5. Release of enzyme to restore native DNA

Figure 3–1 Schematic illustration of the enzyme-catalyzed monomerization of pyrimidine dimers by DNA photolyase (photoreactivating enzyme); showing an example of DNA repair by the reversal of base damage. The colored symbols (square and triangle) represent the two chromophores which are required for catalytic activity in all DNA photolyases. (*Adapted from Friedberg [49] with permission.*)

Kelner that he had discovered recovery of ultra-violet treated spores of *Actinomycetes* upon exposure to visible light. I am informed by Dr. Kelner that his results are in course of publication. My observation indicates the correctness of Dr. Kelner's suggestion that the phenomenon discovered by him may be of general occurrence for a number of biological objects.

It was not until the late 1950s, however, that studies with extracts of *E. coli* and the yeast *Saccharomyces cerevisiae* demonstrated that PR of pyrimidine dimers is in fact an enzyme-catalyzed phenomenon (185, 189).

DNA Photolyase (Photoreactivating Enzyme)

Enzymes that catalyze the PR of pyrimidine dimers in DNA are referred to as *DNA photolyases, deoxyribodipyrimidine photolyases*, or *photoreactivating enzymes*. DNA photolyase activity is widely distributed in nature and has been detected in extracts of a large number of microorganisms and multicellular plant and animal cells. In some cases the presence of the enzyme has additionally been inferred from studies on intact cells in which the light-dependent loss of pyrimidine dimers from DNA or the light-dependent biological recovery from the effects of UV radiation has been observed. Its ubiquitous distribution notwithstanding, it is interesting that DNA photolyase activity is lacking in placental mammals (see below), although it has been suggested that a protein similar to DNA photolyase may have been co-opted for other repair reactions in mammalian cells (see chapter 8). Hence, it would appear that PR was lost as a specific DNA repair modality during recent evolution.

MEASUREMENT OF DNA PHOTOLYASE ACTIVITY

DNA photolyase activity can be detected and quantified by a variety of techniques, some of which merit brief consideration.

Restoration of the Transforming Ability of DNA

In the assay involving restoration of the transforming ability of DNA, purified transforming DNA carrying one or more genetic markers is UV irradiated and incubated with DNA photolyase in the presence of light at wavelengths that support PR (see below). The DNA is then used to transform an appropriate recipient cell defective in the same genetic marker(s) (185–187, 189). The extent of PR can be expressed quantitatively as the enhanced acquisition of the transformed phenotype by the recipient cells.

Membrane-Binding Assay

Photoreactivating wavelengths of light are required for the splitting (monomerization) of pyrimidine dimers by all known DNA photolyases. However, binding of the enzyme to dimers in DNA occurs in the absence of photoreactivating light (28, 188, 190, 242, 243). By using this property, it is possible to assay photolyase activity by measuring the binding of complexes of enzyme and radioactively labeled DNA to filters under conditions in which noncomplexed DNA is not retained (133, 134). While this assay is useful for measuring the binding and dissociation constants of purified enzyme, it has limited application for enzyme purification since other proteins that bind to DNA may cause interference.

DNA-Binding Assay

In a related technique, the binding of DNA photolyase to pyrimidine dimers in DNA in the dark can be conveniently measured by a gel retardation assay involving oligonucleotides with and without pyrimidine dimers (76).

Light-Dependent Loss of Thymine-Containing Pyrimidine Dimers

Procedures have been developed for the chromatographic separation of radiolabeled thymine from thymine-containing pyrimidine dimers following the hydrolysis of UV-irradiated DNA (Fig. 3–2) (50, 51). Following the incubation of UV-

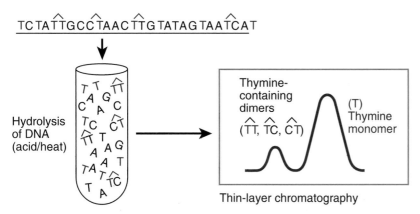

Figure 3–2   Schematic illustration of the measurement of thymine-containing pyrimidine dimers in DNA. DNA radiolabeled in thymine (T) is exposed to UV radiation to produce pyrimidine dimers, some of which contain radiolabeled thymine. The DNA is hydrolyzed in strong acid at high temperature, resulting in the preservation of structurally intact radiolabeled thymine monomer and radiolabeled thymine-containing pyrimidine dimers. These species can be resolved and the amount of thymine-containing pyrimidine dimers can be quantitated by thin layer chromatography. The thymine dimer content of the DNA is then expressed as the fraction of the total radioactivity in thymine present as thymine-containing pyrimidine dimers. (*Adapted from Friedberg [49] with permission.*)

irradiated DNA with photolyase, the light-dependent loss of dimers from DNA can be directly quantitated (Fig. 3–3). Since the most convenient way of specifically radiolabeling DNA is with [^{3}H]thymine or [^{14}C]thymine, such procedures can measure the photoreactivation of T-T and C-T dimers but not of C-C dimers. This limitation can be surmounted by using a modified form of this assay in which the DNA is labeled in the deoxyribose-phosphate backbone with ^{32}P (Fig. 3–4) (44, 244). In UV-irradiated DNA the intradimer phosphodiester bond is uniquely resistant to hydrolysis by a combination of the enzymes pancreatic DNase and snake venom phosphodiesterase, so these phosphates are left in covalent linkage to two pyrimidine nucleoside residues after treatment with the enzymes (44). The remainder of the DNA can be digested to mononucleotides, which can then be digested to nucleosides by further incubation with alkaline phosphatase. Following the complete enzymatic digestion of such DNA, the only labeled phosphate that binds to activated charcoal is that associated with pyrimidine dimers, since P$_i$ does not bind to this matrix. The charcoal-bound radioactivity can then be eluted and quantitated as a measure of all pyrimidine dimer species in DNA before and after PR (44, 244).

Restriction Site Analysis

Not surprisingly, the presence of pyrimidine dimers in a defined oligonucleotide sequence will interfere with the ability of certain restriction enzymes to cut the oligonucleotide at or near dimer sites. The repair of the dimers by PR restores sensitivity to these restriction enzymes, thus altering the size distribution of oligonucleotide fragments during subsequent electrophoresis (122).

Properties and Mechanism of Action of DNA Photolyases

DNA PHOTOLYASE FROM *E. COLI*

DNA photolyase from *E. coli* is the most extensively characterized enzyme of this type. *E. coli* DNA photolyase activity was first identified in vitro in 1958 (189). However, the enzyme was not extensively purified and characterized until relatively recently because it is present in cells at very low levels under normal growth conditions and hence was extremely difficult to isolate. In fact, it has been calculated that there are only about 10 to 20 molecules of DNA photolyase per *E. coli* cell (63). With the advent of recombinant DNA technology, the limitations to

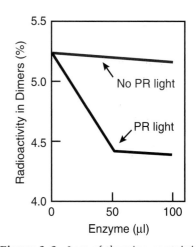

Figure 3–3   Loss of thymine-containing pyrimidine dimers from acid-precipitable DNA after incubation of UV-irradiated DNA with DNA photolyase in the presence of photoreactivating light (PR light). (*Adapted from Friedberg [49] with permission.*)

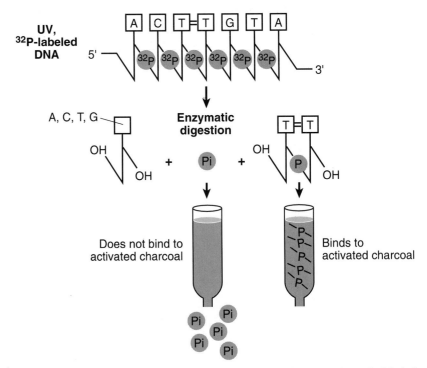

Figure 3–4 Pyrimidine dimers in DNA can be measured by using DNA radiolabeled with ^{32}P. The DNA is enzymatically digested with nucleases and alkaline phosphatase to yield free P_i and nucleoside monophosphates which contain radioactive phosphate associated with the pyrimidine dimers (organic phosphate). The P_i does not bind to activated charcoal whereas the organic phosphate does. Thus, the amount of charcoal-bound ^{32}P is representative of the concentration of all pyrimidine dimers in the DNA. (*Adapted from Friedberg [49] with permission.*)

obtaining sufficient *E. coli* DNA photolyase for biochemical and enzymologic studies were circumvented by cloning the gene which encodes this polypeptide (designated *phr*⁺). The *phr*⁺ gene has been mapped to 15.7 min on the *E. coli* circular chromosome (203, 283).

Transfection of *E. coli* with a multicopy plasmid carrying the cloned gene results in markedly increased levels of DNA photolyase activity commensurate with the plasmid copy number (192, 202, 204). The gene has also been placed under control of the *tac* promoter. Induction of this promoter with the gratuitous *lac* inducer isopropyl-β-D-thiogalactopyranoside (IPTG) results in even higher levels of overexpression, to the extent that DNA photolyase can constitute as much as 15% of the total *E. coli* protein (217).

At this juncture it is relevant to point out that the presence of extremely low levels of enzymes required for a number of DNA repair modes is a theme that is common to many organisms, and the limitations posed to biochemists by this phenomenon will be reiterated several times in this book. In the case of the *E. coli* DNA photolyase, the nucleotide sequence of the cloned gene revealed a pronounced bias for rare codons for which the relative intracellular abundance of corresponding tRNAs is low. In general, strongly expressed *E. coli* genes have a bias for codons for abundant tRNAs, whereas weakly expressed genes frequently show no codon bias (10). It has been suggested that the requirement for large amounts of some cellular proteins is accommodated by rapid translation. This would presumably be favored by the evolution of abundant codons in the genes encoding such proteins, since the concentration of charged cognate tRNAs is rate limiting for the addition of each amino acid to the growing polypeptide chain (10). Hence, the absence of codon bias or the presence of a bias in favor of rare codons in genes is an indication that the genes are weakly expressed.

Many DNA repair genes fall into this category. Perhaps this reflects a metabolic conservation mechanism, since cells are called upon to use DNA repair proteins only when they sustain genomic injury. However, as we shall see, the anticipated corollary to this viewpoint, i.e., that the expression of such genes would be expected to be induced when cells are exposed to genomic injury, is by no means universal. Considerations of the biological relevance of the weak expression of many DNA repair genes notwithstanding, it is a general observation that the genetic basis of many cellular responses to DNA damage in *E. coli* and other organisms was established many years ago. However, productive inroads into the biochemistry of DNA repair constitute relatively recent advances and in almost all cases were facilitated by the overexpression of recombinant genes, as in the case of the *E. coli phr+* gene.

Properties of the *E. coli* Enzyme

Purified *E. coli* DNA photolyase is a protein of 49 kDa under both denaturing and nondenaturing conditions (206). This value is in good agreement with the anticipated size of 54 kDa calculated from the coding region of the cloned gene (215). The apoenzyme consists of a single polypeptide. It has no requirement for divalent cation but has a strict requirement for light at photoreactivating wavelengths (300 to 500 nm). The enzyme has a turnover rate of 50 pyrimidine dimers per DNA photolyase molecule per min in vitro (123), a considerably higher rate than that calculated from in vivo studies (~5 dimers per photolyase molecule per min) (61).

The reaction catalyzed by DNA photolyase can be described by the following reaction scheme.

$$\text{Dark reaction} \qquad\qquad \text{Light reaction}$$

$$\text{E} + \text{DNA[Py-Py]} \underset{k_2}{\overset{k_1}{\rightleftharpoons}} \text{E-DNA[Py-Py]} \overset{\rightarrow}{\underset{k_3}{}} \text{E} + \text{DNA[PyPy]}$$

We shall consider the so-called *dark reaction* first. In the absence of light at wavelengths between 300 and 500 nm (*photoreactivating light*), *E. coli* DNA photolyase (E) binds specifically to DNA containing pyrimidine dimers. At subsaturating concentrations the enzyme binds this substrate at least 100 times more efficiently than unirradiated DNA (218). The enzyme binds with equal efficiency to double-stranded linear, supercoiled, and relaxed circular DNA and single-stranded DNA (218). Furthermore, the efficiency of PR of dimers in single-stranded and duplex DNA is the same (206). These observations suggest that DNA photolyase does not induce a conformational change in the substrate DNA but recognizes *cis-syn*-pyrimidine dimers directly (218). Indeed, no other substrate [including pyrimidine-pyrimidone (6–4) photoproducts (18)] has been identified for this enzyme, and DNA photolyase can therefore be used as an extremely specific and sensitive analytical tool for demonstrating the presence of pyrimidine dimers in DNA.

The action spectrum for PR in vivo exhibits maxima in the near-UV or visible range, at wavelengths that do not directly excite pyrimidine dimers. Thus, it has long been suggested that the light reaction during PR is a photosensitized reaction and that all DNA photolyases contain chromophores which absorb photoreactivating light. The purification of large quantities of enzyme from *E. coli* facilitated the identification of two chromophores, neither of which is covalently bound to the protein. One is a stable neutral radical which has been identified as 1,5-dihydroflavin adenine dinucleotide ($FADH_2$ or $FADH^-$) (Fig. 3–5) (91, 200, 205). The forms FAD_{ox} and FADH are also encountered but are considered to be artifacts of enzyme purification (200). The second chromophore is (5,10-methenyltetrahydrofolyl)polyglutamate (MTHF) (Fig. 3–5), a reduced conjugated *pterin* with appended polyglutamate residues (88). This cofactor has been shown to act catalytically and does not decompose or dissociate during multiple turnovers of the

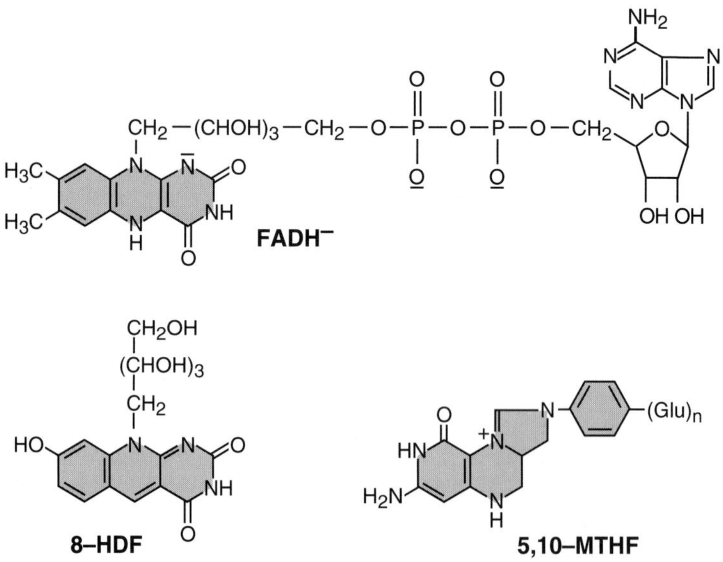

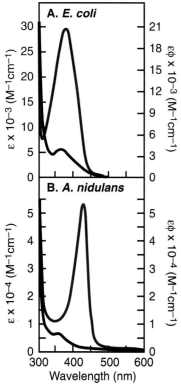

Figure 3–5 Structures of chromophores found in DNA photolyases. The folate class of DNA photolyases contains the anionic form of FADH⁻ and 5,10-MTHF with appended polyglutamate residues. The deazaflavin class of DNA photolyases contains FADH⁻ and 8-HDF. (*Adapted from Sancar [200] with permission.*)

enzyme (60). Each chromophore is present in a 1:1 stoichiometry with the apoenzyme. It is possible to physically resolve photolyase and its chromophores (171). In the absence of the chromophores, the apoenzyme does not bind specifically to UV-irradiated DNA. However, such binding specificity can be restored in the presence of FAD or 5-deaza-FAD. Furthermore, the reduction of enzyme-bound FAD restores catalytic activity (171). The absorption and absolute action spectra of the folate and deazaflavin classes of enzyme are shown in Fig. 3–6.

Under experimental conditions in which FADH₂ and 5,10-MTHF are not acting as photosensitizers, the direct excitation of Trp-277 in the *E. coli* polypeptide at 280 nm leads to the monomerization of pyrimidine dimers with a relatively high quantum yield. However, it is unlikely that this mechanism normally contributes to the repair of pyrimidine dimers in vivo. Possibly it conferred some selective advantage in early life when higher fluxes of short-wavelength radiation reached the Earth's surface (105).

Mechanism of PR Catalyzed by *E. coli* DNA Photolyase

The mechanism of PR catalyzed by the *E. coli* enzyme and the role of the two chromophores in this process have been deciphered in considerable detail. It is instructive to begin this discussion by reviewing the various methods that have been brought to bear on kinetic analyses of DNA photolyases in general. The observation that the light-dependent step of PR (light reaction) is preceded by a light-independent one (dark reaction) during which the enzyme-substrate (ES) complex forms, has facilitated a detailed kinetic analysis of these steps by the use of intense flashes of photoreactivating light of millisecond duration (62, 65, 191).

A single extremely brief flash of light of adequate intensity permits PR by all ES complexes existing at that time. Therefore, the number of complexes can be determined by measuring the disappearance of pyrimidine dimers from DNA after a single light flash. As an alternative to physically measuring a reduction in the pyrimidine dimer content of the substrate DNA, this parameter is often calculated from the extent of a selected biological effect, i.e., increased survival of UV-irradiated bacteria or phage or increased activity of UV-irradiated transforming DNA. This biological effect can be expressed quantitatively by the parameter ΔD, (the difference between the UV dose D actually applied and a smaller UV dose D' that

Figure 3–6 The absorption (ε) and absolute action ($\varepsilon\phi$) spectra of DNA photolyases of the folate class (*E. coli* DNA photolyase) and the deazaflavin class (*A. nidulans* DNA photolyase). The colored lines show the spectra of the holoenzymes. The black lines are the spectra of the enzyme with just FADH⁻. Hence, the shape and maximum wavelengths of the absorption and action spectra in the range from 300 to 500 nm are determined primarily by the second chromophore (5,10-MTHF in the folate class, and 8-HDF in the deazaflavin class). (*Adapted from Sancar [200] with permission.*)

would have led to the same effect without subsequent PR) (Fig. 3–7). Thus, ΔD (*the dose decrement*) is an expression of the amount of UV radiation (or number of pyrimidine dimers) whose effect is mitigated by a single light flash. Under conditions where all enzyme is bound to substrate as ES complex (achieved by establishing a state of substrate excess), the maximal level of ΔD obtained by a single light flash is a measure of the number of DNA photolyase molecules.

The three rate constants k_1, k_2 and k_3 in the reaction scheme shown above can be evaluated in a number of ways. Let us first consider k_1, the rate constant for the binding of DNA photolyase to the dimer substrate. From the reaction scheme, the change in the concentration of substrate [S] is given by the rate law

$$d[\text{S}]/dt = -k_1[\text{E}][\text{S}] + (k_2 - k_3)[\text{ES}] \qquad (1)$$

where S is DNA[Py-Py] and ES is E-DNA[Py-Py] in the reaction scheme. k_1 may be measured under the following reaction conditions: (i) the enzyme and substrate are not preequilibrated in the dark, and (ii) the photoreactivating light flashes are delivered in rapid sequence such that the very short time intervals between flashes permits the formation of only a few ES complexes relative to the total number of enzyme molecules ($\sim$E) and substrate dimers (S). Under these circumstances, the concentration of free enzyme [E] is roughly constant and is approximately equal to the total concentration of enzyme. Consequently, early in the reaction the [ES]-containing term can be neglected and equation 1 reduces to

$$d[\text{S}]/dt = -k_1[\text{E}][\text{S}] \qquad (2)$$

Equation 2 has the integrated form

$$[\text{S}]_t/[\text{S}]_0 = e^{-k_1[\text{E}]t} \qquad (3)$$

where $[\text{S}]_t$ is the substrate concentration at time t, and $[\text{S}]_0$ is the starting substrate concentration. The term $[\text{S}]_t/[\text{S}]_0$ in equation 3 is readily obtained experimentally as the measurement $1 - (\Delta D/\Delta D_{\max})$. ($\Delta D_{\max}$ corresponds to the complete loss of dimers from DNA, or to $[\text{S}]_t = 0$). Hence, the measurement of $1 - (\Delta D/\Delta D_{\max})$ at early times allows for the calculation of k_1, which is the only unknown in equation 1.

To measure the rate constant k_2, a great excess of competing substrate is introduced into the reaction after the dark equilibrium of complex formation has been reached, so that enzyme that dissociates in the dark is likely to bind with the competing DNA, whose repair is not measured by the assay. The resultant decrease in the number of ES complexes is measured as a decrease in PR achieved after the light flash. Typically, high k_1 and low k_2 values are obtained and indicate a high degree of stability of ES complexes under nonphotoreactivating conditions. The equilibrium constant for complete complex formation in the dark is in the order of 10^{10} liters mol^{-1} in vivo (62).

The rate constant k_3 for photolysis of pyrimidine dimers in the ES complex can be expressed by the product $k_p I$, where I is the light intensity and k_p is the *photolytic constant*. When all substrate is complexed with the enzyme prior to illumination (experimentally achieved by using excess enzyme), the application of continuous subsaturating illumination results in the loss of ES complex as a function of the light dose (L). Thus,

$$[\text{ES}]_L/[\text{ES}]_0 = e^{-k_p L}$$

(where $L = I \times t$), from which k_p can be calculated. The parameter k_p is a measure of the efficiency of the use of light in the enzyme reaction and reflects two characteristics (62); (i) the extent to which complexes absorb incident photons (molar extinction coefficient) and (ii) the probability with which an absorbed photon leads to monomerization of a dimer (the quantum yield).

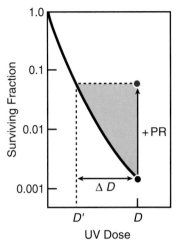

Figure 3–7 Derivation of the parameter ΔD. A survival curve is generated over a range of UV radiation doses. For a selected dose D, the survival of cells exposed to photoreactivating conditions (+PR) after UV irradiation is also measured to yield the datum point shown in color. To establish the dose of UV light that would have resulted in the same level of survival without PR, the extrapolation indicated by the dotted lines is made. This generates the UV dose D'. ΔD is the difference between D' and D. (*Adapted from Friedberg [49] with permission.*)

The interaction of *E. coli* DNA photolyase with DNA containing pyrimidine dimers has been studied by several techniques including nitrocellulose filter binding, conventional, picosecond, and nanosecond (laser) flash photolysis, DNA footprinting; and fluorescence quenching. The enzyme binds to dimer-containing DNA with an association constant (k_1) of 1.4×10^6 to $4.2 \times 10^6 \, \text{mol}^{-1} \, \text{s}^{-1}$. The dissociation of the enzyme displays biphasic kinetics, with k_2 values of 2×10^{-2} to $3 \times 10^{-2} \, \text{s}^{-1}$ for the rapidly dissociating form and 1.3×10^{-3} to $6 \times 10^{-4} \, \text{s}^{-1}$ for the slowly dissociating form. The measured equilibrium association constant K_A (4.7×10^7 to $6 \times 10^7 \, \text{M}^{-1}$) is in good agreement with the values predicted from these rate constants (216), and the discrimination ratio [$K_{A(\text{specific})}/K_{A(\text{nonspecific})}$] is about 10^4 to 10^5 (209). These parameters measured with the purified enzyme concur reasonably with those determined in vivo (61).

The association constant for *E. coli* DNA photolyase in vitro is relatively unaffected by ionic strength, suggesting that electrostatic interactions play a minor role in the binding of the enzyme to pyrimidine dimers in DNA. It is proposed that the protein makes electrostatic contacts with only one or two phosphate groups on the DNA backbone (216). This is consistent with the observation that the enzyme can monomerize pyrimidine dimers in very small oligonucleotides, e.g., oligo(dT$_2$), and that the bound enzyme does not sterically interfere with the incision of UV-irradiated DNA by the UvrABC enzyme of *E. coli*, which is involved in the specific incision of DNA-containing pyrimidine dimers during an alternative repair process called *nucleotide excision repair* (see chapter 5). In fact, DNA photolyase actually stimulates damage-specific incision of UV-irradiated DNA by the UvrABC enzyme, suggesting that binding of *E. coli* DNA photolyase to pyrimidine dimers may alter DNA helix parameters such that the photolyase-DNA complex is a better substrate for the UvrABC enzyme (201).

Evidence that the interaction of these two DNA repair enzymes may indeed be physiologically relevant stems from several observations, some of which go back more than 30 years. *E. coli* cells that are proficient for both nucleotide excision repair and DNA photolyase activity are more UV resistant under non-PR conditions than are cells that are proficient for nucleotide excision repair but defective in DNA photolyase activity (64). Additionally, whereas the introduction of the cloned *E. coli* *phr* gene into a *phr* mutant enhances the UV resistance of such cells under non-PR conditions, the introduction of genes encoding DNA photolyases from other organisms reduces the UV resistance of *E. coli*, suggesting that these photolyases actually inhibit the UvrABC enzyme system (113, 213).

UV-irradiated oligothymidylates [UV-oligo(dT)$_n$] constitute informative model substrates for *E. coli* DNA photolyase (90, 92). Binding studies with very-short-chain oligonucleotides (where $n = 2$ to 4) show that ~80% of the binding energy observed with DNA as the substrate can be attributed to the interaction of the enzyme with a dimer-containing region that spans only 4 nucleotides (89). More-detailed binding studies have used model DNA substrates containing a single thymine dimer at a unique location. Since DNA photolyase can protect this substrate from methylation and from degradation by DNase I or methidium propyl-EDTA-Fe(II), *E. coli* DNA photolyase probably binds to the dimer-containing strand in duplex DNA, making close contacts with the phosphodiester backbone and the cyclobutane ring of the dimer. The enzyme extends along the DNA backbone from the first phosphodiester bond 5' to the dimer to the third phosphodiester bond 3' to the dimer (78) (Color Plate 1).

It is relevant to attempt to distinguish whether *E. coli* DNA photolyase recognizes pyrimidine dimers specifically and exclusively in DNA or whether it recognizes an altered DNA structure that is the unique consequence of the presence of such lesions. Support for the latter model derives from the observation that whereas the enzyme can indeed catalyze the monomerization of dimers in oligonucleotides as short as (dT)$_2$, the K_A for this substrate is reduced by about 4 orders of magnitude (109). A contribution of flanking residues to the specificity of the enzyme-substrate interaction has also been shown with the yeast enzyme (see below) (7).

Measurements of the quantum yield (ϕ) for PR of pyrimidine dimers in *E. coli* in vivo have been estimated to approach 1. Similar results have been obtained in vitro with purified *E. coli* DNA photolyase with a fully reduced flavin chromophore (210). These results indicate that essentially every quantum of light absorbed by the enzyme is used to monomerize dimers. Studies with deoxyribonucleotide homopolymers have shown that the absolute quantum yield (independent of binding affinity) for the reversal of dimers by *E. coli* DNA photolyase is not uniform for all *cis-syn* dimers. The quantum yield in the presence of light at 366 nm (ϕ_{366}) is 0.9 for T-T dimers but only 0.05 for C-C dimers (109). The relative binding affinity of the enzyme for various dimers in these homopolymers also differs, with T-T > U-T > U-U > C-C (109).

Role of the *E. coli* DNA Photolyase Chromophores in PR

What are the roles of the two chromophores in the *E. coli* DNA photolyase? These two moieties are not covalently linked (69, 93, 169). The irreversible reduction of the pterin chromophore (MTHF) is accompanied by complete loss of its absorption and fluorescence at photoreactivating wavelengths, resulting in a lowered extinction coefficient for the enzyme. However, when this decreased absorption is taken into account, the quantum yield for the monomerization of dimers is unchanged, indicating that while the light energy absorbed by the pterin can be used in photolysis, the chromophore is not essential for enzyme catalysis (93). The available evidence suggests that MTHF functions as a sort of photoantenna, since it is the primary light-harvesting cofactor, which contributes 60 to 80% of the A_{385} of the enzyme (211). It absorbs a photon and transfers energy from the excited singlet state to the semiquinone state of FAD (FADH^0) (105). The excited doublet state of FADH^0 is converted by electron transfer to the excited quartet state and yields FADH$_2$ following the abstraction of a hydrogen atom from a tryptophan residue in the apoenzyme. The excited singlet state of FADH$_2$ then monomerizes the pyrimidine dimer in an electron transfer reaction (see below) (68, 105, 166, 271).

Hence, FADH$_2$ acts as a photocatalyst (105). In summary, it has been suggested that evolutionary selection for optimal photoreactivating efficiency has yielded an enzyme which uses one chromophore principally for capturing light and a second chromophore principally to initiate electron transfers that destabilize pyrimidine dimers in DNA (211). Several experimental observations support the notion of energy transfer from MTHF to the flavin (110). In particular time-resolved fluorescence and absorption spectroscopy studies of the chromophores have revealed that the fluorescence lifetime of MTHF which is bound to the enzyme is dramatically reduced in the presence of FADH^0 (104). Measurement of this reduction led to calculation of the rate of energy transfer from MTHF to FADH^0 as 3.0×10^{10} s^{-1} (110). Similar results have been obtained by others using steady-state fluorescence measurements (130). The mechanism of energy transfer is proposed to be the Forster's dipole-dipole mechanism rather than an exchange mechanism (110).

There are indications that FADH$_2$ plays a structural as well as a catalytic role in PR. Whereas DNA photolyase that is depleted of MTHF binds to pyrimidine dimers with the same affinity as native photolyase does, apoenzyme without flavin has no affinity for pyrimidine dimers (170). This lost affinity is regained upon the stoichiometric binding of the enzyme to FAD (170). It has been suggested that FAD in DNA photolyase participates in generating the specific DNA binding site both by its direct interaction with the substrate and by inducing conformational changes in the photolyase polypeptide (110).

Evidence for the involvement of the singlet excited state of FADH$_2$ in the catalytic reaction comes from several experimental sources (110). The photochemical mechanism of photorepair has been proposed to be a light-dependent redox reaction between the singlet excited state of FADH$_2$ and the pyrimidine dimer (110). Direct evidence for photoinduced electron transfer in catalysis stems from the observation of a radical intermediate in a picosecond laser photolysis reaction (166).

These studies, in addition to thermodynamic and other considerations (including isotope effect studies on DNA photolyase and photosensitizer-catalyzed splitting of pyrimidine dimers [5, 271]), are consistent with the concept of a pyrimidine dimer radical anion as a reaction intermediate (Fig. 3–8). The rate of initial electron transfer to the T-T dimer is estimated at $\sim 5 \times 10^9$ s^{-1}, with a quantum efficiency close to 1.0 (110). There is competition between the subsequent monomerization of the pyrimidine dimer radical anion and back electron transfer (Fig. 3–8). However, the latter reaction is considered to be insignificant in reactions catalyzed by true photolyases. In this class of enzymes the active site is highly hydrophobic, thus slowing nonproductive back electron transfer. Indeed, the polarity of the medium is an important determinant for efficient PR (103, 108).

The primary physical, spectroscopic and photochemical properties of the *E. coli* DNA photolyase are summarized in Table 3–2. For comparison, these properties are also shown for another class of DNA photolyases, all of which use a

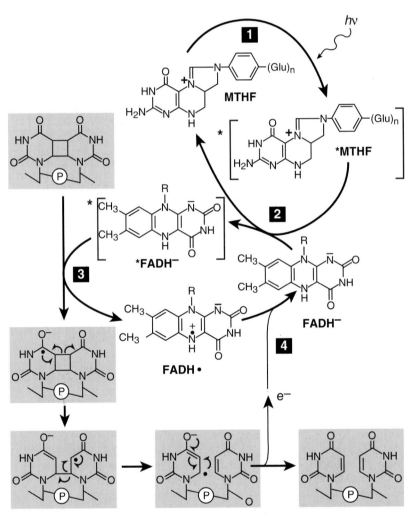

Figure 3–8 Reaction mechanism of DNA photolyase. Absorption of a photon by the chromophore MTFH (step 1) yields the the excited state (*MTFH), which transfers the excitation energy to the catalytic cofactor FADH$^-$ (step 2). (The folate [MTFH] class of enzymes transfer energy less efficiently than the deazaflavin [HDF] class and has an overall lower quantum yield.) The FADH$^-$ excited singlet state (*FADH$^-$) initiates monomerization of the pyrimidine dimer by electron transfer (cycloreversion) (step 3), regenerating the catalytically active flavin (step 4). Back electron transfer from the dimer radical pair to the donor (step 4) is slowed by the apolar environment of the flavin. Hence, the back transfer is very slow relative to dimer monomerization. (*Adapted from Sancar [200] with permission.*)

Table 3–2 Physical, spectroscopic, and photochemical properties of DNA photolyases

Property	Value for enzyme in:	
	Escherichia coli	*Anacystis nidulans*
Class	Folate	Deazaflavin
Protein size (amino acids)	471	484
M_r	53,994	54,475
Subunit	Monomer	Monomer
Cofactors	$FADH^-$ + MTHF	$FADH^-$ + 8-HDF
Absorption maxima (nm)		
Enz-FADH^0 (second chromophore)	384, 480, 580	438, 480, 588
Enz-FADH$^-$ (second chromophore)	384	438
Enz (second chromophore)	384	438
Enz-FADH$^-$	366	355
Fluorescence maxima (nm)[a]		
Enz-FADH^0 (second chromophore (weak))	465, 505	470, 505
Enz-FADH$^-$ (second chromophore)	465	470, 505
Enz (second chromophore)	465	470
Binding constant (M^{-1})	2.4×10^8	3.0×10^8
Quantum yield of repair (ϕ)		
Enz-FADH$^-$ (second chromophore)	0.5–0.6	0.9–1.0
Enz-FADH$^-$	0.8–0.9	0.9–1.0
Fluorescence lifetime (ns)		
Enz (second chromophore)	0.35	2.0
Enz-FADH$^-$ (second chromophore)	0.13	0.05
Enz-FADH^0 (second chromophore)	<0.03	<0.03
Enz-FADH$^-$	1.50	1.80
Enz-FADH$^-$ + T-T	0.16	0.14
Rate of energy transfer (s^{-1})		
Second chromophore → FADH$^-$	4.6×10^9	1.9×10^{10}
Efficiency of energy transfer (%)		
Second chromophore → FADH$^-$	62	98
Rate of electron transfer (s^{-1})		
FADH$^-$ → T-T	5.5×10^9	6.5×10^9
Efficiency of electron transfer (%)		
FADH$^-$ → T-T	89	92

[a] Data from uncorrected emission spectra.
Source: Adapted from Kim and Sancar (110) with permission.

different second chromophore, 8-hydroxy-5-deazaflavin, instead of MTHF (see next section).

DNA PHOTOLYASES FROM OTHER ORGANISMS

DNA photolyases have been purified and characterized from a number of other organisms, including *Salmonella typhimurium*, the cyanobacterium *Anacystis nidulans*, the archaebacteria *Halobacterium halobium* and *Methanobacterium thermoautotrophicum*, the actinomycete *Streptomyces griseus*, the green alga *Scenedesmus acutus*, and the yeast *Saccharomyces cerevisiae* (Table 3–3) (209). In all cases studied, two noncovalently associated chromophores have been identified, one of which is reduced FAD. However, two classes of DNA photolyases are distinguishable by the nature of the second chromophore. The enzymes purified from *E. coli*, *Salmonella typhimurium*, and *Saccharomyces* (see below) fall into the so-called *folate class*, in which the second chromophore is the pterin MTHF. The enzymes that have been purified and characterized from *Streptomyces griseus*, *A. nidulans*, *M. thermoautotro-*

Table 3–3 Properties of multiple DNA photolyases

Source of enzyme	Mol wt	No. of amino acids/no. of Trp residues	PR (λ_{max}) in vivo	in vitro
Escherichia coli	53,994	471/15	365–400	380
Saccharomyces cerevisiae	66,189	565/16	365–385	377
Anacystis nidulans	54,475	484/17	436	436–438
Streptomyces griseus	50,594	455/12	436	443–445
Halobacterium halobium	53,065	481/12	435	NK[a]
Methanobacterium thermoautotrophicum	55,000–60,000	NC[b]	433	434
Scendesmus acutus	56,000	NC	NK	437

[a] NK, not known.
[b] NC, not cloned.
Source: Adapted from Sancar (209) with permission.

phicum, and *Scenedesmus acutus* contain an 8-hydroxy-5-deazaflavin (8-HDF) moiety as the second chromophore (Fig. 3–5) and are referred to as the *deazaflavin class* of photolyases (209). This distinction is reflected primarily in the different action spectra for these two classes of enzymes. The *E. coli* and yeast DNA photolyases have maximal PR activity at a wavelength of ~380 nm, while members of the second group have maximal activity at ~440 nm (Fig. 3–6). These differences notwithstanding, the presence of two chromophores, one of which is reduced FAD, suggests that the fundamental photochemistry of PR is highly conserved. Among the lower eukaryotes mentioned above, the most extensively studied is the yeast *Saccharomyces cerevisiae*.

DNA Photolyase from the Yeast *Saccharomyces cerevisiae*

You will recall that historically the yeast *Saccharomyces cerevisiae* is one of the two organisms in which DNA photolyase was first discovered in cell extracts (189). The early literature contains reports of photoreactivating enzymes with different properties, suggesting the possible existence of multiple enzymes in this yeast (49). However, in recent years a single *PHR* gene has been cloned and its overexpression has facilitated the detailed characterization of what appears to be the exclusive DNA photolyase in *Saccharomyces cerevisiae*.

A yeast mutant defective in PR established the existence of the *PHR1* gene (181, 182). The *PHR1* gene was cloned by phenotypic complementation of a yeast strain carrying the *phr1–1* mutation (224, 278) and was mapped to the right arm of chromosome 15 (224). The *PHR1* gene contains an open reading frame of 1,695 bp (208, 280), which encodes a polypeptide of 66.2 kDa (212). Following overexpression of the yeast gene in *E. coli*, a monomeric protein of 60 kDa was purified (214) and shown to have a turnover number of 0.7 pyrimidine dimer monomerized min^{-1} molecule^{-1}. The *PHR1* gene complements the *E. coli phr* mutant defective in PR (207) and vice versa (118). These observations suggest that the two enzymes share common chromophores, a suggestion that is borne out by the direct isolation and characterization of the chromophoric elements from the purified yeast enzyme (214). It has been estimated that there are ~250 to 300 molecules of DNA photolyase in yeast cells under constitutive conditions (281).

Transcription of the *PHR1* gene is up-regulated when cells are exposed to UV radiation. Curiously, the gene is also regulated in response to treatment with a variety of chemical agents that interact with DNA (225, 226), even though yeast DNA photolyase is not directly involved in the repair of chemical base damage. Conceivably, genes involved in discrete modes of DNA repair have evolved responsiveness to multiple types of DNA damage, which constitute a common signal for their induction. Indeed, as will be seen in chapter 4, a gene that encodes an enzyme activity required specifically for the excision repair of alkylation damage in *Saccharomyces cerevisiae* is induced by agents that do not result in alkylation damage in DNA (274).

As mentioned above, there are indications that DNA photolyase can contribute to the efficiency of nucleotide excision repair in *E. coli*. Similar observations have been made in *Saccharomyces cerevisiae*. Specifically, a functional *PHR1* gene enhances the survival of yeast cells in the dark, i.e., in the absence of photoreactivating light, when these cells are defective in genes which determine sensitivity to UV radiation but which are unrelated to nucleotide excision repair. However, this result is not observed in cells defective in genes which are required for nucleotide excision repair in *Saccharomyces cerevisiae* (213). Hence, regulation of the expression of yeast *PHR1* gene may be relevant to this role. A DNA-binding protein which acts as a repressor of the *PHR1* gene has been identified and is designated PRP (for photolyase regulatory protein).

The predicted amino acid sequences of the cloned yeast and *E. coli* enzymes are very similar (Fig. 3–9) (208, 280). Overall there is 36.2% amino acid sequence identity between them. Two short regions near the amino and carboxyl termini of the two polypeptides show even greater sequence conservation, suggesting that the yeast and *E. coli* enzymes possess common structural and functional domains which are involved in the binding of substrate and/or chromophores. The two genes do not cross-hybridize, and neither has provided a useful probe for the detection of homologous sequences in higher organisms (141).

By using chemical modification (such as protection of specific residues from reductive methylation during its interaction with substrate) and site-directed mutagenesis, it has been shown that the amino acid residues Trp-387, Lys-463, Arg-507, and Lys-517, all of which are located in the highly conserved C-terminal half of the polypeptide, are important for substrate recognition in DNA (7). This region of the polypeptide has also been shown to contain the domain that binds the folate chromophore, while the flavin-binding site is located in the conserved N-terminal domain (135).

Other DNA Photolyase Genes

The structural genes for DNA photolyase from *A. nidulans* (282), *Streptomyces griseus* (112), and the archaebacterium *H. halobium* (255) have also been cloned and sequenced. The deduced amino acid sequences of these genes are remarkably conserved (Fig. 3–9). Additionally, the cloned genes from *H. halobium* (255), *Streptomyces griseus* (112), and *A. nidulans* (256) are also able to complement *phr* mutants of *E. coli*, suggesting that DNA photolyases of eukaryotes, eubacteria, and archaebacteria are derived from a common origin. Of 13 different yeast species surveyed, 5, including the fission yeast *Schizosaccharomyces pombe*, were found to be lacking detectable DNA photolyase activity (279). However, when the cloned *PHR1* gene of *Saccharomyces cerevisiae* was expressed in *Schizosaccharomyces pombe*, cells acquired enzyme activity, indicating that the requisite light-absorbing cofactors are present in this yeast (279).

DNA photolyases are extraordinarily efficient enzymes, since relatively small numbers of enzyme molecules per cell can protect organisms from the lethal effects of doses of UV radiation which produce thousands of pyrimidine dimers per genome (209). The discrimination ratio for binding of purified DNA photolyase from *E. coli* and *Saccharomyces cerevisiae* to DNA with and without pyrimidine dimers is $\sim 10^5$, a value close to that observed for sequence-specific DNA binding proteins such as Lac repressor and the restriction endonuclease *Eco*RI (209).

DNA Photolyases in Higher Organisms

As indicated above, most studies have consistently failed to provide biological or biochemical evidence for PR in organisms evolutionarily more advanced than marsupials. The literature documents biochemical studies suggesting the possible existence of DNA photolyase or photolyase-like proteins in mammalian cells (240, 241, 248). Additionally, light-dependent loss of pyrimidine dimers from DNA has been found in living cells in culture and in intact human skin (245–247, 269). These reports have been sporadic over the past two decades and have not led to definitive conclusions. Hence, at present it must be considered that convincing evidence for PR in any mammalian cells remains to be established. Indeed, studies

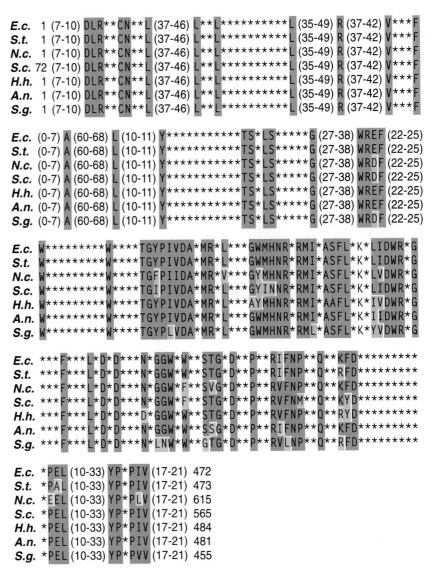

Figure 3–9 Amino acid sequence comparison of DNA photolyases from *Escherichia coli* (*E.c.*), *Salmonella typhimurium* (*S.t.*), *Neurospora crassa* (*N.c.*), *Saccharomyces cerevisiae* (*S.c.*), *Halobacterium halobium* (*H.h.*), *Anacystis nidulans* (*A.n.*), and *Streptomyces griseus* (*S.g.*). Identical amino acids in all seven polpeptides are indicated in color, and conservative substitutions are indicated in grey. The asterisks represent nonidentical amino acids. The numbers in parentheses indicate distances (in amino acids) separating conserved domains. The first 71 amino acids of the *Saccharomyces cerevisiae* polypeptide are not shown. (*Adapted from Ahmad and Cashmore [1] with permission.*)

with a highly sensitive assay which can detect the enzyme-catalyzed monomerization of a single pyrimidine dimer have failed to reveal the presence of DNA photolyase activity in human cells (122).

An example of an apparently true DNA photolyase in metazoans was isolated from the goldfish *Carassius auratus* (277). The cloned cDNA for the gene encoding this enzyme restored PR activity to *E. coli phr* mutant strains. The gene is novel in two notable respects. First, it is substantially induced by exposure of cells to visible light. Second, there is very limited if any amino acid sequence homology between the goldfish gene and that isolated from the other sources discussed above (277). It will certainly be intriguing to determine whether the product of this gene represents a hitherto unknown new family of DNA photolyases that may be more widely represented in vertebrates.

Aside from the evolutionary conservation of DNA photolyases in prokaryotes and eukaryotes, this family is also conserved in other proteins that interact with light. For example, the *HY4* locus of the plant *Arabidopsis thaliana* encodes a protein with the characteristics of a blue-light receptor which is required for normal hypocotyl elongation. The amino acid sequence of the HY4 protein shows striking sequence homology with that of the DNA photolyases. Interestingly, amino acid sequence identity and similarity are greatest in the regions of the photolyases known to be involved in the binding of chromophores (1).

Light-Dependent "Repair" of (6-4) Photoproducts in DNA

The term *enzymatic PR* has been exclusively associated with the monomerization of cyclobutane pyrimidine dimers in DNA (and as will be seen presently, possibly in RNA). Recently an enzyme activity from *Drosophila melanogaster* embryos has been described that carries out a light-dependent reaction during which pyrimidine-pyrimidone (6-4) lesions in DNA disappear, at least to the extent that they are no longer recognized by a specific antibody and are no longer labile to hot-alkali treatment (261). The products of this reaction have not yet been determined, and so it is not at all clear that this reaction represents an example of the reversal of base damage, as is the case with PR of dimers. Nor for that matter is it clear that this represents a true DNA reaction, although UV-irradiated DNA treated with the *Drosophila* fraction did regain enhanced transforming activity (261). As indicated in chapter 1, pyrimidine-pyrimidone (6-4) lesions are not dimers in the strict chemical sense, since the amino group of the 3' pyrimidine is transferred to the C-5 of the 5' pyrimidine. The enzyme-catalyzed reversal of such lesions has been confirmed with a (6-4) photoproduct DNA photolyase purified from *Drosophila melanogaster* nuclear extracts and a defined substrate containing a single T[6,4]T lesion contained in a 49 bp duplex (107). The Dewar isomer of the (6-4) photoproduct is not recognized by this enzyme, which, surprisingly, restores the lesion to native monomeric products through a proposed oxetane intermediate (107).

Photoreactivation of RNA

While the essential focus of this book is on the repair of DNA damage, it should be borne in mind that genetic information in some viruses is encoded in an RNA genome. Damage and repair of RNA has not received anything like the attention that of DNA has. Nonetheless, PR mechanisms for the repair of pyrimidine dimers in RNA have been documented. For example, when a number of UV-irradiated RNA plant viruses and/or free viral RNAs are assayed on appropriate hosts, an increase in specific infectivity occurs when the assay plant is illuminated immediately after the application of the infectious material (53, 75, 263). This result is not observed when heat-inactivated infectious material is used; additionally, preillumination of the assay host has no effect (263). An activity in cell extracts of tobacco plants catalyzes the in vitro PR of tobacco mosaic virus RNA but has not been purified or characterized because of its extreme lability (75). DNA photolyases from *Saccharomyces cerevisiae* or from pinto bean seedlings are inactive on UV-irradiated tobacco mosaic virus RNA (75).

It has been observed that the E-FADH$_2$ form of purified *E. coli* DNA photolyase can catalyze the monomerization of uracil dimers in poly(U) in vitro. The affinity of the enzyme for uracil dimers in RNA is about 10^4-fold lower that for U-U in DNA. However, once the enzyme is bound to its substrate, these dimers are repaired with the same quantum yield as that observed with U-U in DNA (109).

Repair of Spore Photoproduct

In *Bacillus subtilis* the process of sporulation is typically initiated by the depletion of one or more nutrients. Spores have no detectable metabolism, but they can survive for extended periods. During this period of dormancy the spores are exposed to the environment and may sustain damage to their genome, which must be repaired when they germinate. The repair events which operate during spore

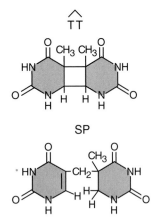

Figure 3–10 Structure of the spore photo-product (SP) (bottom). The structure of a cyclobutane thymine dimer (top) is shown for comparison. (*Adapted from Setlow [232].*)

germination include a process which is able to repair a unique photoproduct generated exclusively in spore DNA exposed to UV radiation (232). The details of this process are still incompletely understood, but the process resembles PR in many respects and represents a second example of the repair of DNA damage by its direct reversal.

Formation of Spore Photoproduct

The exposure of *B. subtilis* spores to UV radiation at ~254 nm does not result in the formation of conventional cyclobutyl-type pyrimidine dimers and produces only low levels of (6-4) photoproducts (232). Instead, a unique type of thyminyl-thymine adduct is formed, which is termed the *spore photoproduct* (SP) (Fig. 3–10). This altered photochemistry occurs because the DNA in spores is in an A-like conformation, a conformation caused in part by the relatively dehydrated state of spores relative to growing cells. The switch to an A-type conformation can also be effected by the interaction of spore DNA with a particular class of polypeptides termed *small acid-soluble proteins*. There are two major types of these proteins, designated a and b, which are encoded by the $sspA^+$ and $sspB^+$ genes, respectively (232). These proteins bind to double-stranded DNA, which is capable of adopting an A-like conformation.

When *B. subtilis* spores germinate, some of the SP and probably also (6-4) photoproducts are repaired by the nucleotide excision repair system (see chapter 5). However, most of the SP is repaired by a poorly defined SP-specific repair process. This process takes place during early germination and appears to involve the monomerization of SP back to native thymine in DNA in a reaction that requires energy but not light. This process was discovered almost 20 years ago (264) but remains biochemically undefined. This photoreversal process is determined by a single gene called spl^+ (for SP lyase) (formerly called ssp^+). The spl^+ gene has been cloned by phenotypic correction of the UV radiation sensitivity of a spl mutant and screening for transformants whose spores exhibited UV radiation resistance (43). The spl^+ gene contains an open reading frame which can encode a polypeptide of ~40 kDa. The translated amino acid sequence shows regional homology with DNA photolyases from a number of bacteria and fungi (43).

Repair of O^6-Guanine and O^4-Thymine Alkylations and Phosphotriesters in DNA

We will now consider two particular forms of chemical damage to the DNA bases and to the phosphodiester linkages in the sugar-phosphate backbone, which provide a third example of repair by the direct reversal of base damage. The discussions of DNA damage in chapter 1 mentioned that certain mutagenic monofunctional alkylating agents, such as *N*-methyl-*N'*-nitro-*N*-nitrosoguanidine (MNNG), *N*-methyl-*N*-nitrosourea (MNU), and to a lesser extent methyl methanesulfonate (MMS), react with DNA to produce both O-alkylated and N-alkylated products. Among the former class, O^6-alkylguanine and O^4-alkylthymine are potentially mutagenic lesions because they can mispair during semi-conservative DNA synthesis (29, 131, 237). It has been generally assumed that O^6-alkylguanine stably mispairs with thymine, and that O^4-alkylthymine stably mispairs with guanine. Indeed, in a self-complementary dodecanucleotide containing two O^6-methyl-G·T mispairs, the mispairs were found by X-ray diffraction to closely resemble a Watson-Crick base pair (121). However, other studies have shown that alkyl-G·T and alkyl-T·G pairs are in fact less stable than their nonmispaired counterparts (250). Hence, it has been suggested that the mutagenic potential of these alkylated bases might be realized by the tendency for DNA polymerases to read O^6-alkylguanine as adenine in DNA and O^4-alkylthymine as cytosine (250).

The repair of these potentially miscoding lesions in *E. coli* is understood in considerable detail. In addition to further illustrating the phenomenon of DNA repair by the reversal of base damage, the molecular biology of the repair of these particular types of alkylation damage provides the first example we will encounter of the regulated expression of genes required for DNA damage processing. Other examples of regulated responses to DNA damage by induction of specific gene functions are considered in later chapters.

Adaptation to Alkylation Damage in *E. coli*

For many years it was known that a curious relationship exists between the induction of mutations in *E. coli* and the time of exposure to mutagens such as MNNG. Specifically, the mutation frequency increases with time of exposure and then dramatically reaches a plateau as if some sort of *antimutagenic* mechanism were at work (24, 87, 162, 163). Trivial explanations such as decay of the mutagen in the medium in which the *E. coli* cells were grown were ruled out. Additionally, it was shown that this result was not due to saturation of all possible mutable sites in DNA, since the use of small doses of mutagen that are unlikely to be saturating yields the same qualitative result with a lower plateau level (87). A definitive explanation for this phenomenon was provided by Leona Samson and John Cairns and led to the discovery of a novel and unexpectedly complex DNA repair pathway. Like many significant scientific discoveries, this one has a serendipitous element worth recounting as an anecdotal diversion. In a lecture to the Royal Society of London in 1978, Cairns was quick to point out that

> scientific reports are hardly ever written so that the reader can get much idea of the real origins of an investigation. In this respect, at least, the scientist is like any politician who wishes by appropriate omissions to conceal from the electorate his many past ineptitudes.

In this lecture Cairns recounted that the experiments that Samson and he carried out originally stemmed from their interest in using *E. coli* as a model for studying mutagenesis and its possible relationship to carcinogenesis in self-renewing basal epithelial cells in higher organisms. They had developed the hypothesis that the rate of accumulation of mutations in such cells might be normally controlled by the strict operation of rules governing the segregation of sister chromatids at cell division (23). To examine this hypothesis, they devised an experimental system for studying the segregation of mutant and nonmutant DNA strands in bacteria briefly exposed to mutagens. Unfortunately, it soon became evident that the segregation of sister chromosomes was random. This result was of course disappointing for the carcinogenesis model they wished to test. Nonetheless, he recognized that the experimental system was potentially useful for exploring the relationship between mutation rates and mutagen concentrations, even at very low levels of mutagen.

Continuous exposure to low doses of MNNG produced mutations for the first 60 min but not thereafter. Further exploration of this phenomenon led to the intriguing discovery that when *E. coli* cells were exposed to very low levels of MNNG and subsequently challenged with a much higher dose of the alkylating agent, there was a marked resistance to both the lethal and mutagenic effects of the chemical in the "adapted" cells relative to unadapted controls (84, 86, 197) (Fig. 3–11). This resistance was dependent on active protein synthesis by the cells prior to the challenge dose, suggesting that it involved the induction of one or more genes in response to low levels of the alkylating agent (197). The phenomenon was designated as the *adaptive response to alkylation damage*.

Adaptation to killing and to mutagenesis result from biochemically and mechanistically distinct repair processes involving different genetic loci. Strains of *E. coli* which are defective in the *polA*$^+$ gene (which codes for DNA polymerase I) and those defective in a gene called *alkA*$^+$ (which encodes a member of a novel class of DNA repair enzymes called DNA glycosylases [discussed in chapter 4]) are deficient in the adaptive response to cell killing (42, 83, 85, 99). Other mutants designated as *ada* (for adaptive minus) are also deficient in the adaptation to killing (82, 83). Additionally, they are defective in adaptation to mutagenesis. This chapter is concerned primarily with the latter phenomenon, since it provides illustrative examples of DNA repair by the reversal of alkylation base damage. Adaptation

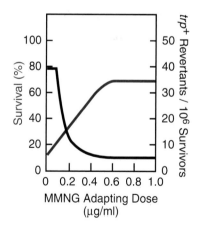

Figure 3–11 Adaptation to cell killing and to mutagenesis in *E. coli*. A culture of *trp* cells was grown for 90 mins in the presence of various small amounts of MNNG (adapting doses). At the end of this time, samples were exposed to a much larger dose (100 μg/ml) of MNNG for 5 min (challenging dose) and the surviving fraction and *trp*$^+$ reversion frequency were determined. In adapted cells, survival increased and mutation frequency (normalized to survival) decreased. (*Adapted from Friedberg [49] with permission.*)

to cell killing involves a different DNA repair pathway by which bases carrying alkylations at positions other than O^6-guanine and O^4-thymine are excised from the genome during a process called *base excision repair*, which is considered in chapter 4.

Repair of O^6-Alkylguanine and O^4-Alkylthymine in DNA

Quantitative comparisons of the major alkylation products in high-molecular-weight DNA in adapted and non-adapted cells subsequently challenged with radioactively-labeled MNNG, showed that the levels of 7-methylguanine and 3-methyladenine were indistinguishable. However, a significant reduction in the amount of O^6-methylguanine (labeled in the methyl group) was observed (223) (Fig. 3–12). At the time that these experiments were carried out, the selective excision of damaged bases from DNA (*excision repair*) was a well-established biochemical phenomenon. Hence, it was reasonably anticipated that the loss of O^6-methylguanine from high-molecular-weight DNA would be accompanied by its stoichiometric recovery in the fraction containing excised low-molecular-weight products (183, 223). Surprisingly, this was not the case and provided the first hint of an entirely distinct DNA repair mechanism.

The selective loss of O^6-methylguanine from DNA was reproduced with cell-free preparations of previously adapted strains of *E. coli* (46, 97). When such extracts were incubated with DNA containing radiolabeled O^6-methylguanine, labeled methyl groups were recovered bound to a protein (97). The protein was shown to be an unusual methyltransferase activity that both removes the methyl group from O^6-methylguanine in DNA, hence resulting in the direct reversal of damage to this base, and transfers it to one of its own cysteine residues, generating *S*-methylcysteine in the protein (167) (Fig. 3–13).

Confirmatory evidence for the repair of O^6-methylguanine by direct enzymatic demethylation came from studies with the synthetic DNA polymer poly (dC·dG-8[^{3}H]me^{6}dG) containing O^6-methyl-8-[^{3}H]guanine, in which the radiolabel was present in the purine ring rather than in the O^6-methyl group. Following incubation with extracts of *E. coli* containing methyltransferase activity, this polymer substrate was shown to contain unsubstituted radiolabeled guanine, hence verifying the direct reversal of guanine modification by removal of the methyl group (46).

O^6-Methylguanine-DNA Methyltransferase I of *E. coli*

The *E. coli* enzyme which selectively removes methyl groups from the O^6 position of guanine was originally designated as O^6*-methylguanine-DNA methyltransferase*. It

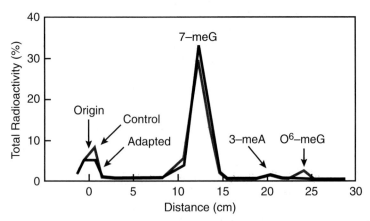

Figure 3–12 Preferential loss of O^6-methylguanine from the DNA of adapted cells. *E. coli* cells were exposed to either adapting or nonadapting conditions and then challenged with a larger dose of radiolabeled alkylating agent. Following a period of incubation, DNA from both groups of cells was isolated and hydrolyzed. The alkylated bases 3-methyladenine (3-meA), 7-methylguanine (7-meG) and O^6-methylguanine (O^6-meG) in the hydrolysate were identified by chromatography.(*Adapted from Friedberg [49] with permission.*)

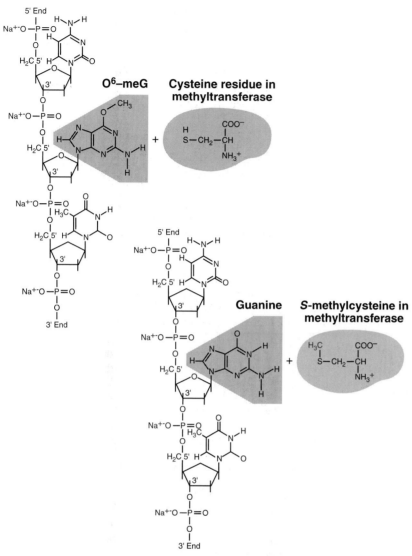

Figure 3–13 The enzyme O^6-MGT transfers a methyl group from the O^6 position of guanine in DNA to a cysteine residue in the protein, thereby restoring the native chemistry of guanine and repairing the base damage by direct reversal. (*Adapted from Friedberg [49] with permission.*)

is also sometimes called *Ada protein* because of its central regulatory role in the adaptive response to alkylation damage (196) (see below). Subsequent studies showed that the methyltransferase activity is endowed with a broader substrate specificity which includes O^4-methylthymine as well as methylphosphotriesters (2, 138–140, 222). Furthermore, alkyl groups larger than methyl groups are recognized as substrates by this enzyme (147). Finally, as mentioned below, *E. coli* cells contain a second, alkyltransferase activity which also catalyzes the removal of methyl groups from O^6-methylguanine and O^4-methylthymine (175, 177, 235). These caveats notwithstanding, in the interests of historical accuracy we will retain the original name of O^6-methylguanine-DNA methyltransferase, abbreviated to O^6-MGT I. The designation "I" is in deference to the existence of the second enzyme in *E. coli* (see below).

O^6-MGT I has been purified to physical homogeneity from an *E. coli* mutant which constitutively expresses the adaptive response (35, 36). The enzyme has also been purified following overexpression of the cloned ada^+ gene (14, 140, 154).

The purified protein has a molecular mass of ~39 kDa but is readily and specifically degraded to an 18 kDa form (35, 36, 124, 126, 128, 167) which retains the ability to transfer methyl groups from O^6-methylguanine and O^4-methylthymine (2, 259). The active form of the enzyme is a monomer with an isoelectric point of 7.1. The circular dichroism spectrum suggests a low α-helical content (14). It has a radius of gyration of 23 Å (2.3 nm), suggesting a compact, globular three-dimensional conformation (14).

Adaptation in *E. coli* confers resistance to mutation by ethylating, propylating, and butylating agents as well as by methylating agents (221). Consistent with these observations, O^6-MGT I repairs O^6-ethylguanine in DNA in vitro, with the concomitant formation of *S*-ethylcysteine residues in the protein (230). However, in vivo, the rate of disappearance of this lesion from DNA is only about 1/10 that of methyl groups in guanine (230), and substrate analogs such as O^6-hydroxyethylguanine are dealkylated 100 to 500 times more slowly in vitro (184). The enzyme apparently can also catalyze the removal of O^6-chloroethyl groups from guanine in DNA, since its presence prevents the appearance of interstrand cross-links in DNA treated with the chloroethylating agent 1,3-bis(2-chloroethyl)-1-nitrosourea (BCNU) or *N*-(2-chloroethyl)-*N*'-cyclohexyl-*N*-nitrosourea (CCNU) (128). Presumably the inhibition of cross-linking results from repair of O^6-chloroethylguanine monoadducts before the second step of the cross-linking reaction can occur (128).

During the course of enzyme catalysis, the enzyme must overcome a significant energy barrier, since O^6-methylguanine is a relatively stable chemical entity at neutral pH (129). However, the enzyme has no requirement for divalent cations or other known cofactors. There is no indication of peptide bond cleavage associated with the transfer of methyl groups, since the size of the methylated protein is indistinguishable from that of the unmethylated form. The direct reversal of an O^6-methylguanine residue by O^6-MGT I is a rapid and error-free process, which occurs in less than 1s at 37°C (126). DNA containing O^4-methylthymine (which is present in alkylated DNA at about 1/10 the concentration of O^6-methylguanine), is rapidly repaired by O^6-MGT I (129). However, at limiting enzyme concentrations the repair of O^6-methylguanine is more efficient than that of O^4-methylthymine (139). This might reflect a preference of the enzyme for one of these two substrates. Interestingly, in a model in vitro system in which mutagenesis at O^6-methylguanine and O^4-methylthymine residues in oligonucleotides of otherwise identical composition in *E. coli* were compared, the mutation frequency at the former lesion was extremely low (0 to 1.7%), whereas that at the latter site was as high as 12%, suggesting that O^6-methylguanine is repaired much more efficiently than O^4-methylthymine (38).

Alkyl groups in the *O*-6 position of G and the *O*-4 position of T project into the major groove of DNA. However, O^6-MGT I does not catalyze the removal of alkyl groups in other positions on bases which also project into the major groove. Hence, in addition to the major groove of the double helix, enzyme specificity is apparently conferred by oxygen atoms of the nitrogenous bases carrying bound alkyl groups (139).

Native O^6-MGT I also catalyzes the removal of methyl groups from phosphotriesters in DNA. DNA substrates containing mainly methylphosphotriesters can be prepared by heating alkylated DNA under conditions that promote the preferential depurination of 7-methylguanine, 7-methyladenine, and 3-methyladenine (140). By using this substrate, it was established that the 18-kDa proteolytic product of O^6-MGT I is inactive against methylphosphotriesters (140). This observation afforded the opportunity to use the truncated polypeptide as a reagent to prepare an alkylated DNA substrate devoid of O^6-methylguanine and of O^4-methylthymine and containing almost exclusively methylphosphotriesters. When this substrate was incubated with the 39-kDa native form of the enzyme, approximately half of the methyl groups were transferred from methylphosphotriester to cysteine residues in the protein in a stoichiometric reaction (140).

Methylphosphotriesters in DNA exist in one of two isomeric forms designated *R* and *S* (140) (Fig. 3–14). The observation that only half the methyl groups in phosphotriesters were removed from the polymer substrate suggested that only one of these isomers was recognized as the substrate. Quantitation of each confirmed that only the *S* isomer is repaired (140, 270). The solution structure of an N-terminal 10-kDa fragment of Ada protein which retains the zinc-binding and phosphotriester activities provides some interesting insights concerning this isomeric preference (153). These studies indicate that Cys-69 is embedded in a solvent-exposed surface of the protein and is readily accessible from the outside. For B-form DNA the *S* isomer of methylphosphotriester is expected to project into solution from the edge of the phosphodiester backbone, whereas the *R* isomer is expected to project inward to the major groove. Hence the *S* isomer is expected to be more accessible to the active-site thiolate of the Ada protein (153).

Incubation of the purified native (39-kDa) form of O^6-MGT I with alkylated DNA substrates containing mainly either O^6-methylguanine or methylphosphotriesters showed that the enzyme can accept one methyl group per molecule from each type of substrate. In contrast, the proteolytic (19-kDa) form of the enzyme containing the C-terminal domain exclusively accepts methyl groups from O^6-methylguanine. These results suggested that the active sites for the repair of the two substrates are distinct and that the site for repair of O^6-methylguanine (and presumably O^4-methylthymine) is located in the C-terminal half of the polypeptide. In support of this model, when the purified enzyme was incubated with excess substrate containing either O^6-methylguanine or methylphosphotriester, both substrates were repaired and the protein contained approximately two methyl groups per molecule (140).

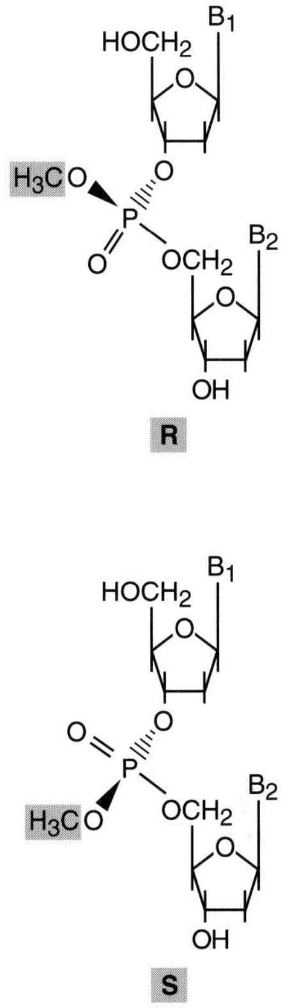

O^6-MGT I IS A SUICIDE ENZYME

O^6-MGT I activity is consumed in the reactions it catalyzes (124, 126, 140). Hence, when a limiting amount of enzyme is present, efficient transfer of some of the available methyl groups occurs very rapidly and then essentially ceases (124, 126, 140). These results strongly suggest that the methylated protein acceptor cannot be regenerated and that the enzyme is expended in the reaction, i.e., that the enzyme is not catalytic in the usual sense of the term. This is a curious phenomenon, since by conventional biochemical wisdom an enzyme is not consumed during reaction with its substrate. The reaction catalyzed by O^6-MGT I is in fact analogous to those between suicide enzyme inactivators and their target enzymes, other examples of which are known (125, 128). It is also not unprecedented in biochemistry for a protein-modifying enzyme to use itself as the main target for modification (125). Notable examples are protein kinases that catalyze autophosphorylation and the major acceptor for poly(ADP-ribose) in mammalian cell nuclei, which is the poly(ADP-ribose) synthetase itself (95, 165). Although these reactions do not involve inactivation of the mediating enzyme molecule, several enzymes are known to be irreversibly inactivated by formation of dead-end complexes as a result of enzyme-catalyzed covalent binding of substrate analogs at the active site (128).

It has been pointed out that the unusual feature of the suicide inactivation exhibited by O^6-MGT I is that in this case the reaction occurs between the enzyme and its ''natural'' substrate, rather than with a substrate analog (128). It is certainly curious that evolution has selected for such a metabolically expensive mechanism for repairing mutagenic forms of alkylation damage to DNA. Nor at first glance is this repair reaction a particularly efficient means of protecting against mutagenesis, since the repair capacity of the cell is saturated if exposure to alkylating agents results in a larger number of O^6- or O^4-alkyl residues in DNA than the number of available transferase molecules (34, 129). However, from the point of view of population fitness in a rapidly changing environment, one must avoid the misconception that antimutagenic mechanisms are necessarily advantageous. As will be pointed out in later chapters, environmental stress responses are often characterized by the potential for *genetic diversification* in cell populations by mutation. Additionally, as noted below, *E. coli* and at least some other prokaryotes

Figure 3–14 Schematic representation of the *R* and *S* stereoisomers of methylphosphotriesters produced by the alkylation of phosphate residues in DNA. B_1 and B_2 connote bases in DNA. (*Adapted from McCarthy and Lindahl [140] with permission.*)

have evolved mechanisms for rapidly increasing the amount of O^6-MGT I activity in cells which are chronically exposed to alkylation.

Unadapted wild-type bacteria contain about 100 to 200-fold-lower levels of DNA alkyltransferase than adapted cells do (127, 142). It is not clear whether the unadapted level reflects constitutive levels of both O^6-MGT I and O^6-MGT II or the response of cells to persistent low-grade inducing stimuli which result in continuous overexpression of O^6-MGT I protein. It is unlikely that the presence of low levels of O^6-MGT activity in unadapted cells is due to the presence of small amounts of alkylating agents in the growth medium, since these should be destroyed at the high temperatures associated with sterilization of media (127). Persistent adaptation might conceivably result from metabolically determined (i.e., nonenvironmental) alkylation events in DNA effected by S-adenosylmethionine and/or other intracellular methyl donors. Indeed, when highly purified radiolabeled S-adenosylmethionine is incubated with DNA in vitro, small amounts of alkylation products can be detected in DNA (193).

There are indications that in addition to S-adenosylmethionine, other products of normal cellular metabolism in *E. coli* (and presumably other organisms) can result in O alkylations in DNA. In particular, it has been observed that *E. coli* strains deleted for both the *ada*$^+$ gene and a gene called *ogt*$^+$, which encodes O^6-MGT II activity (see below) have an elevated spontaneous mutation rate (132). This effect was not observed with mutants defective in one or the other of these genes and could be corrected by introduction of either one of the cloned genes into *E. coli ada ogt* mutants (132). As discussed below, the adaptive response to alkylation damage is conserved in many bacteria. The widespread occurrence of this phenomenon suggests that environmental sources of alkylation might also be both prevalent and biologically relevant (265). Several possible examples of such environmental sources have been documented. For example, chemical nitrosation of some natural compounds under mildly acidic conditions can give rise to MNU and MNNG. Methyl chloride, methylhydrazines, and the naturally occurring antibiotic streptozotocin can all induce the adaptive response, as can methyl chloride and methyl iodide (265). Methyl chloride is the most abundant halocarbon in the atmosphere, with an estimated annual global emission of 5×10^6 tons, most of which is from natural sources. Although methyl chloride is both mutagenic and carcinogenic, it is very volatile, and its ability to directly alkylate DNA has not been demonstrated (265).

O^6-MGT I acts poorly on single-stranded DNA containing O^6-alkylguanine moieties (126). This property of the enzyme provides a reasonable explanation for the well-known propensity of agents such as MNNG and MNU to produce a greater concentration of mutations near replication forks in *E. coli* than in nonreplicating regions of the genome (25). Thus, O^6-alkylguanine and O^4-alkylthymine present in parental DNA strands at replication forks may be relatively refractory to repair until replication restores the duplex structure by misincorporation in the daughter strand opposite the alkylated nucleotides (128).

Role of Ada Protein in the Adaptive Response to Mutagenesis

The gene which complements *E. coli* mutants defective in the adaptive response to alkylation-induced mutagenesis and cell killing is termed *ada*$^+$ (for adaptive response). As discussed above, overexpression of the cloned *ada* gene results in the accumulation of a 37-kDa protein with O^6-MGT I activity (227). It was initially believed that the *ada*$^+$ gene encoded a 37-kDa regulatory protein and that a different gene encoded an 18-kDa protein with alkyltransferase activity. However, the failure to isolate mutants defective exclusively in O^6-MGT I activity led to a reexamination of this model. It is now firmly established that the *ada*$^+$ gene product is in fact both a DNA alkyltransferase and a regulatory protein, and that the 18-kDa protein is a proteolytic degradation product of the larger 37-kDa gene product.

An elegant series of experiments by Tomas Lindahl and Mutsuo Sekiguchi and their respective colleagues (34, 129) reconciled the various substrate specificities of O^6-MGT I (or Ada protein, as we shall refer to it during this discussion of its regulatory function) with its repair and regulatory functions and established the following detailed picture of the molecular mechanism of regulation of the adaptive response to mutagenesis following alkylation damage in *E. coli* (Fig. 3–15).

Ada protein transfers alkyl groups to two different cysteine residues following the repair of alkylphosphotriesters and of O^6- and O^4-alkylguanine (Fig. 3–16). Cys-321, the most carboxyl-terminal of the 12 cysteines in the Ada polypeptide, is the acceptor site for O alkylations in the bases of DNA (37, 196, 253). Replacement of Cys-321 with His or inversion of the adjacent amino acids Cys-321 and His-322 inactivates O^6-methylguanine-transferase activity in the O^6-MGT I enzyme

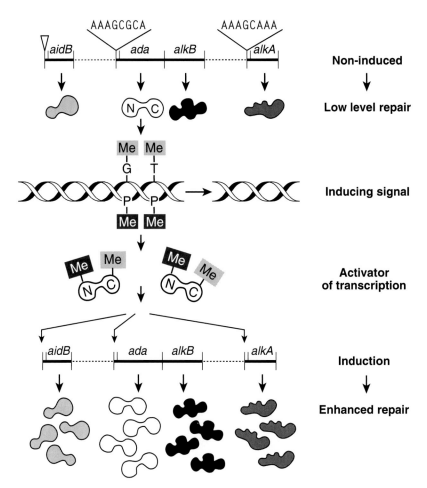

Figure 3–15 Regulation of the adaptive response to alkylating agents. The Ada regulon, consisting of the *ada*$^+$ gene as well as the *alkB*$^+$, *alkA*$^+$, and *aidB*$^+$ genes (see chapter 4), is shown. The sequence of the "Ada boxes" in the promoters of the *ada*,$^+$ *alkB*$^+$ operon and the *alkA*$^+$ gene are also indicated, and the polypeptides encoded by these genes are shown schematically. The polypeptide encoded by the *ada*$^+$ gene is represented to show the N-terminal and C-terminal domains containing receptor cysteine residues for alkyl groups (see the text for details). Following the exposure of *E. coli* cells to methylating agents the cellular DNA is alkylated at several sites, including the O^6 position of guanine, the O^4 position of thymine, and phosphate residues in the sugar-phosphate backbone, to form phosphotriesters. Ada protein catalyzes the transfer of methyl groups from phosphotriesters to the N-terminal cysteine (Cys-69) and from O^6-alkylguanine or O^4-alkylthymine to the C-terminal cysteine (Cys-321) (see Fig. 3–17). The Cys-69 alkylation results in a conformational change in the protein that converts it to a strong transcriptional activator which binds to the promoters of genes in the Ada regulon, resulting in enhanced transcription and translation. The increased levels of Ada protein and of the products of the other genes (see chapter 4) result in enhanced repair of multiple forms of alkylation damage in DNA. (*Adapted from Lindahl et al. [129] with permission.*)

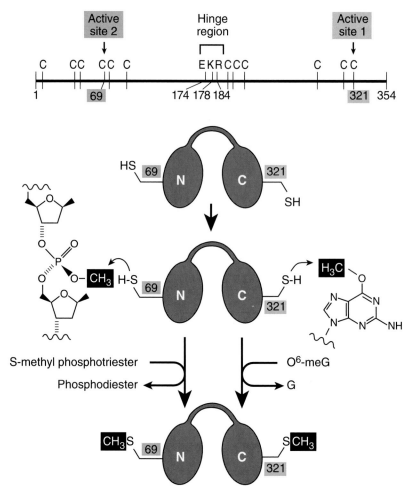

Figure 3–16 Acceptor cysteine residues in the Ada protein involved in removal of simple alkyl groups from DNA. The 354-amino-acid Ada polypeptide (top) is represented to show the 12 cysteine residues in the polypeptide and the relative positions of Cys-69 (in the N-terminal half) and Cys-321 (in the C-terminal half), separated by a central hinge region. Amino acids (E, K, R) that constitute sites which are particularly sensitive to endogenous cleavage of the polypeptide are also indicated. The N-terminal and C-terminal domains are represented diagrammatically below to show the specific abstraction of methyl groups from phosphotriesters (left) and O^6-alkylguanine (right). (*Adapted from Lindahl et al. [129] and Myers et al. [152] with permission.*)

(257). The second alkylated cysteine residue derives from abstraction of the *S* diastereoisomer of alkylphosphotriesters in DNA. This event occurs at Cys-69, near the amino terminus of the Ada polypeptide (196, 231, 253) (Fig. 3–16).

There is considerable amino acid sequence similarity in the regions flanking these two cysteine residues (37, 231) (Fig. 3–17). It has been suggested that the amino acid context in which these residues reside might generate a particular conformation of Ada protein which facilitates activation of cysteine for the acceptance of alkyl groups. The basic residues (Lys or His) near the cysteines could serve as proton acceptors in a charge transfer reaction to generate a reactive thiolate anion (37). Thymidylate synthase represents another example of an enzyme which accepts methyl groups at a specific cysteine residue. This enzyme, purified from a variety of sources, also contains the Pro-Cys-His motif which is conserved in Ada protein, suggesting that this motif has general significance for the generation of reactive cysteines in proteins (34, 129) (Fig. 3–18).

The N-terminal domain of Ada protein contains a high-affinity binding site for a single Zn^{2+} ion (152). The metal possibly coordinates the motif Cys-X_3-Cys-

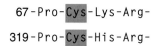

Figure 3–17 Amino acid similarities in the regions of the *E. coli* Ada polypeptide bearing the alkyl acceptor cysteine residues Cys-69 and Cys-321.

O^6-MGT	*E. coli*	-Ala-Ile-Val-Ile- Pro-Cys-His -Arg-Val-Val-Arg-
Thymidylate synthase	*E. coli*	-Met-Ala-Leu-Ala- Pro-Cys-His -Ala-Phe-Phe-Gln-
	T4 phage	-Met-Ala-Leu-Pro- Pro-Cys-His -Met-Phe-Tyr-Gln-
	L. casei	-Met-Ala-Leu-Pro- Pro-Cys-His -Thr-Leu-Tyr-Gln-
	Yeast	-Met-Ala-Leu-Pro- Pro-Cys-His -Ile-Phe-Ser-Gln-

Figure 3–18 The Cys-69 active site region of the *E. coli* Ada protein compared with the active sites of thymidylate synthases from various sources. (*Adapted from Demple et al. [37].*)

X_{26}-Cys-X_2-Cys (which includes Cys-69) (Fig. 3–19), reminiscent of (Cys)$_4$ zinc-binding elements in other DNA-binding proteins, in particular the eukaryotic nuclear hormone receptor family of transcription factors. Indeed, thus far Ada protein represents the sole example of such a factor in prokaryotes and has prompted a consideration that *E. coli* Ada protein might be structurally and evolutionarily related to regulatory proteins in higher organisms (11). It has been suggested that a conformational switch that converts Ada from a non-sequence-specific to a sequence-specific DNA-binding protein may be effected by the methylation-dependent reorganization of the polypeptide ligand about the metal. Studies involving site-directed mutagenesis have indeed shown that zinc participates in the autocatalytic activation of the active-site cysteine (151).

The alkylation of *E. coli* Ada protein at Cys-69 converts it to a strong transcriptional activator of several genes, including the *ada* gene itself (Fig. 3–15). (The other genes which are activated by Ada protein are discussed in chapter 4.) This is the essential molecular basis for inducing *ada* expression, thereby dramatically increasing the amount of O^6-MGT I activity in adapted cells. Enhanced expression of the *ada*$^+$ gene has been demonstrated both at the transcriptional level and by DNA-dependent protein synthesis in vitro (156, 260). Additionally, it has been shown that Ada protein alkylated at Cys-69 binds specifically to the *ada* promoter immediately upstream of the putative RNA polymerase-binding site (228, 260). Hence, bound alkylated Ada protein functions as a positive regulator of the *ada*$^+$ gene by facilitating the binding of RNA polymerase (34, 129, 196).

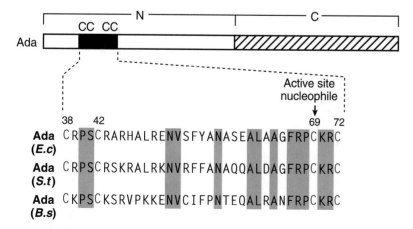

Figure 3–19 The zinc finger domain (Cys-X_3-Cys-X_{26}-Cys-X_2-Cys) of Ada protein is located in the N-terminal domain of the protein. Amino acid sequences in this region are conserved in the proteins from *E. coli* (*E.c*), *S. typhimurium* (*S.t*), and *B. subtilis* (*B.s*). Identical amino acids in all three polypeptides are highlighted. (*Adapted from Myers et al. [152].*)

Figure 3–20 The *E. coli ada⁺* promoter contains the Ada box (highlighted) in a region of dyad symmmetry (arrows).

The transcriptional regulatory element ("Ada box") (Fig. 3–20) in the *ada* gene to which Ada protein binds is in a region of dyad symmetry, consistent with the role of this region in regulating gene expression (34, 129). This regulatory element has been precisely defined by deletion analysis and by site-specific mutagenesis (157). It consists of the octanucleotide sequence AAAGCGCA. The extended sequence AAANNAAAGCGCA is shared by at least one other gene known to be regulated by activated Ada protein, called *alkA⁺*, which encodes a particular DNA glycosylase involved in the repair of other types of alkylation damage (see chapter 4). This sequence is not present in other known *E. coli* promoters, however.

The Ada regulon consists of four known genes (129). In addition to the *ada⁺* gene itself, Ada protein is known to regulate the *alkA⁺* gene just mentioned and the *aidB⁺* and the *alkB⁺* genes (see chapter 4). Not all genes under *ada* control are regulated identically. For example, in living cells, lower levels of alkylated Ada protein are required to activate the *ada⁺* gene than the *alkA⁺* gene (236). Additionally, just the N-terminal half of alkylated Ada protein is sufficient to activate efficient transcription of the *alkA⁺* gene but not of *ada⁺*. Finally, some truncated derivatives of the protein are powerful constitutive inducers of *ada* but activate the *alkA* gene only under inducible conditions (236).

Studies with truncated Ada proteins and with Ada proteins derived by site-specific mutagenesis suggest that the protein has distinct recognition sites for binding to the promoter and for interaction with RNA polymerase (34, 129). Polypeptides deleted of the C-terminal region and hence missing Cys-321 remain strong activators of *ada⁺* gene expression, suggesting that activation of only Cys-69 is important (34, 120, 129). However, conversion of Cys-321 to Ala-321 by site-specific mutagenesis results in activation of Ada protein in the absence of adaptive treatment with alkylating agents (253). Mutations in other regions of the polypeptide can also apparently activate Ada protein. For example, four independently isolated mutants which are constitutive for the adaptive response have mutations in the coding region of the *ada⁺* gene (34, 74, 129). A truncated Ada protein deleted of 12% of the C-terminal domain also conveys this phenotype (235).

It is not obvious why abstraction of a methyl group from a particular stereoisomer of methylphosphotriester in DNA activates Ada protein for transcriptional regulation of *ada⁺* (and other genes; see chapter 4) whereas abstraction of a methyl group from O^6-methylguanine or O^4-methylthymine apparently does not. It has been suggested that this mechanism may have evolved because phosphotriesters are rarer products of simple alkylation of DNA than are O^6-alkylguanine and O^4-alkylthymine. Hence, exposure to low levels of alkylation with the attendant higher probability of guanine and thymine damage could be handled by constitutive levels of O^6-MGT I and O^6-MGT II. Induction of the *ada⁺* gene would be required only for more extensive levels of DNA alkylation with the increased probability of phosphotriester formation (34, 129).

If indeed the activated form of Ada protein is required for the induction of *ada⁺*, at most only a few (~one to four) molecules are constitutively available for *ada* gene activation in cells exposed to alkylating agents. This might explain the observation that adaptation is a relatively slow process, taking at least 60 min of chronic exposure to alkylating agents before peak expression of O^6-MGT I activity and maximal adaptation are achieved (33, 155, 197). Alternatively, one must consider the possibility that constitutively there is no Ada protein in *E. coli* cells at all, in which case there must be some other mechanism for initiating induction of the *ada* gene. This would seem unlikely, since no induction of β-galactosidase can be detected in *ada* mutants transformed with plasmids carrying *ada-lacZ* fusion genes (33, 120).

STRUCTURE OF THE O^6-MGT I (Ada) PROTEIN

The X-ray crystal structure of the 19-kDa, 178-amino-acid C-terminal domain of the *E. coli* Ada protein which retains the O^6-MGT activity (referred to as AdaC protein) has been solved (145). The N-terminal region (88 amino acid residues) of this polypeptide has a fold that resembles part of RNase H. The remaining C-terminal region (90 amino acids) consists predominantly of α-helix and random coiled domains (Color Plate 2). The three helices in the C-terminal region of the polypeptide (red in Color Plate 2) may reflect a variant of the helix-turn-helix motif. However, the observation that this region is some distance from the Pro-Cys-His-Arg active site suggests that it may be involved in generalized binding to DNA rather than specifically to the O^6-alkyl substrate (145). The active-site thiol of the Cys residue in the Pro-Cys-His-Arg active site (corresponding to Cys-321 in the native polypeptide) is buried in the structure revealed (Color Plate 2), suggesting that a conformational change must take place in the protein to allow this cysteine residue to carry out nucleophilic attack of the substrate alkyl group (145). One model that has been advanced is that the most C-terminal helix (green in Color Plate 2) might swivel to expose a DNA-binding surface of the protein and render Cys-321 in the active site accessible to the substrate (145).

Termination of the Adaptive Response

Alkylation of Ada protein is apparently irreversible. How, then, is the adaptive response switched off? One possibility is that activated Ada protein is simply diluted out during growth in the absence of alkylating agents (34, 129). A second possibility is that activated Ada protein is proteolytically degraded. The latter would appear to be a viable mechanism for down-regulating *ada* gene expression, since it has been shown that the Ada protein is indeed sensitive to cleavage by a distinct cellular protease in cell extracts (259). The 20-kDa N-terminal proteolysis product is able to bind to the *ada* promoter but is ineffective in activating gene expression (34, 129). Truncated Ada fragments encoding 42 to 66% of the intact protein compete favorably with intact protein for DNA binding. Hence, down-regulation of the *ada* gene by proteolysis of activated Ada protein might be facilitated by the dominant action of Ada degradation products (235). Consistent with this model, independent studies have shown that a methylated 20-kDa peptide encoded by a truncated *ada* gene binds to the *ada* promoter but strongly inhibits transcription (3).

Studies with *E. coli* strains which are defective in specific proteases suggest that the degradation of Ada protein to yield 20- and 19-kDa polypeptide products in vitro is effected by a protease encoded by the *ompT*+ gene, a membrane-bound protease known to degrade the *E. coli* ferric enterobactin receptor, and T7 DNA polymerase after cell lysis (229). However, an *ompT* mutant expresses normal levels of O^6-MGT I activity, and the kinetics of induction of the enzyme are indistinguishable from those in an *ompT*+ strain. Hence, evidence for specific degradation of Ada protein in vivo is still lacking.

In summary, the *E. coli* Ada protein apparently plays at least four roles in the adaptive response to alkylation damage.

1. It prevents transition mutations by repairing O-alkylation damage in guanine and thymine.

2. It acts as a sensor of alkylation damage in DNA by monitoring alkylations at phosphotriesters.

3. It activates the transcription of genes (including itself) whose products are required for the adaptive response.

4. It terminates the adaptive response.

DNA Alkyltransferase II

E. coli cells normally contain only one or two molecules of O^6-MGT I protein. It takes about 1 h of growth in the presence of nontoxic levels of alkylating agents to induce expression of O^6-MGT I to the optimal level of ~3,000 molecules per cell. During this time, cells replicate their DNA at least once and are potentially at risk for unrepaired alkylation damage. Therefore, it is perhaps not surprising that this organism has evolved a second methyltransferase activity which can provide immediate protection against such damage. This enzyme is called DNA alkyltransferase II (O^6-MGT II). The presence of a second enzyme was suggested by the observation that extracts of *E. coli* deleted of the entire ada^+ gene or containing insertional mutations in the gene have detectable DNA alkyltransferase activity (177, 235). This enzyme differs from O^6-MGT I in a number of respects.

1. O^6-MGT II is not induced by treating cells with low levels of alkylating agents.

2. O^6-MGT II is a smaller protein with a molecular mass of ~19 kDa.

3. O^6-MGT II does not catalyze the removal of methyl groups from methylphosphotriesters.

The biochemical properties of O^6-MGT II are very similar to those of the 19-kDa proteolytic form of O^6-MGT I (179). However, O^6-MGT II is much more heat labile, and heat lability studies suggest that most, if not all, DNA alkyltransferase activity in constitutive *E. coli* (~30 molecules per cell) is in the O^6-MGT II form. Definitive evidence for two O^6-MGT enzymes came from the molecular cloning of a gene which is distinct from *ada* and which encodes O^6-MGT activity (174, 175, 196). This gene (designated ogt^+ [for O^6-alkylguanine-DNA alkyltransferase]), contains an open reading frame of 171 codons which can encode a polypeptide of 19.1 kDa (175). The cloned gene maps at 29.4 min on the *E. coli* chromosome (254). Hence, it is not linked to the ada^+ gene, which maps at ~47 min.

The nucleotide sequence of the ogt^+ gene is distinct from that of *ada*. However, the translated amino acid sequences reveal extensive regions of homology, including the region containing Cys-321 in O^6-MGT I (Fig. 3–21 and 3–22). Direct evidence that the *ogt* gene encodes O^6-MGT II was provided by showing identity of the N-terminal sequence of O^6-MGT II protein with that predicted from the cloned *ogt* gene (175). The 5′ untranslated region of the ogt^+ gene does not contain an Ada box, consistent with the observation that O^6-MGT II is not induced in cells exposed to chronic alkylation damage. However, since induction of O^6-MGT II was examined in an *ada* mutant, one cannot eliminate the possibility that induction of O^6-MGT II requires the ada^+ gene.

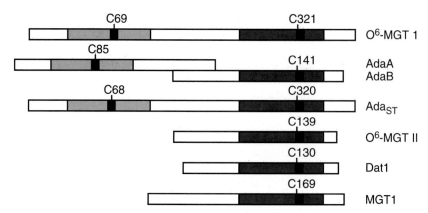

Figure 3–21 Diagrammatic representation of regions of amino acid sequence homology in the N- and C-terminal regions of seven microbial O^6-MGT proteins. These are O^6-MGT I and O^6-MGTII of *E. coli*, AdaA, AdaB, and Dat1 proteins of *B. subtilis*; Ada$_{ST}$ protein of *Salmonella typhimurium*; and MGT1 protein of the yeast *Saccharomyces cerevisiae*. The alkyl-acceptor cysteine residues are indicated in each protein. (*Adapted from Samson [196] with permission.*)

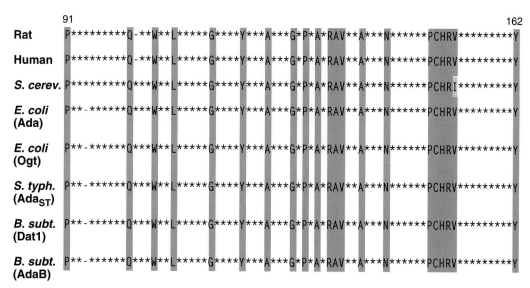

Figure 3–22 Conservation of the amino acid sequence in the C-terminal regions of multiple O^6-MGT proteins from prokaryotes and eukaryotes. The amino acids are numbered according to the rat polypeptide.

Support for the notion that the *ogt*⁺ gene protects cells from the mutagenic consequences of alkylation damage during the period prior to the optimal induction of the *ada*⁺ gene stems from the observation that *ada ogt*⁺ cells manifest a threshold in resistance to mutation induction by low doses of MNNG, whereas mutation induction in *ada ogt* cells is linear with very low doses of MNNG (178) (Fig. 3–23). At higher doses this effect is not observed, because O^6-MGT II is depleted and cells become effectively *ogt*. Additionally, *ada*⁺ *ogt* cells are more resistant than *ada ogt* cells to mutations induced by MNNG (178, 196, 254). Mutants defective in the *ogt* gene are also more sensitive to killing by MNNG in an *ada* background (178). Cells that are defective in both the *ada*⁺ and *ogt*⁺ genes have a higher spontaneous mutation rate than wild-type, *ada*⁺ *ogt*, or *ada ogt*⁺ cells do (Fig. 3–23). This provides indirect evidence that alkylation damage of DNA can occur spontaneously (178, 196).

The relative affinity of O^6-MGT I and O^6-MGT II for O^6-alkylguanine and O^4-alkylthymine was evaluated by measuring the amount of an oligonucleotide carrying one or the other lesion required to inactivate 50% of the enzyme activity in cell extracts. Interestingly, these experiments demonstrated a distinct preference of O^6-MGT I for O^6-methylguanine, while O^6-MGT II has a preference for O^4-methylthymine (219). Increased levels of O^6-MGT I activity have been detected in tissues of mice carrying an *E. coli ada*⁺ transgene (137, 160, 176). The gene expresses enzyme activity in a variety of tissues and hence potentially affords the opportunity for examining the role of O^6-alkylguanine and O^4-alkylthymine in mutagenesis and carcinogenesis in an animal model system. Such animals can be expected to yield interesting information on the role of this enzyme in protection from the mutagenic and carcinogenic effects of alkylating agents. Indeed, it has already been demonstrated that transgenic mice that express recombinant human O^6-MGT I activity are significantly protected from developing thymic lymphomas and liver tumors following exposure to known carcinogenic alkylating agents (40, 159).

DNA Alkyltransferases in Other Organisms

PROKARYOTES

A number of other prokaryotes have been examined for the adaptive response to both alkylation-induced killing and mutagenesis and for DNA alkyltransferase activities in cell extracts. Extracts from *Micrococcus luteus* contain an inducible O^6-MGT activity which removes methyl groups from the O^6 position of guanine in

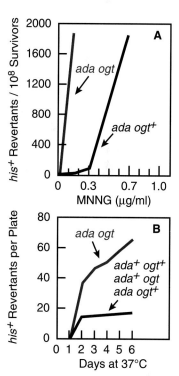

Figure 3–23 (A) MNNG-induced mutagenesis in *E. coli ada ogt* and *ada ogt*⁺ mutant strains. *E. coli* cells (*his*) were treated with MNNG and plated on minimal plates lacking histidine (to quantitate *his*⁺ revertants) and on minimal plates supplemented with histidine (to quantitate survivors). (B) Spontaneous mutagenesis in various *ada* and *ogt* mutants. *E. coli* cells (*his*) were plated on minimal plates containing very small amounts of histidine and incubated at 37°C. Revertants (*his*⁺) were scored daily. (*Adapted from Rebeck and Samson [178]*.)

alkylated DNA (4). Two overlapping open reading frames designated *adaA*[+] and *adaB*[+], which could encode polypeptides of 24.3 and 20.1 kDa, respectively, have been identified in the *B. subtilis* genome (149) (Fig. 3–21). Both genes show regions of amino acid sequence conservation with the *E. coli ada*[+] and *ogt*[+] genes and with a human gene (Fig. 3–22) which encodes *O*[6]-MGT activity. *B. subtilis* mutants defective in the *adaB*[+] gene produce *ada* transcripts following treatment of cells with low doses of alkylating agents. However, mutants defective in the *adaA*[+] gene do not. These results suggest that in *B. subtilis* the product of the *adaA*[+] gene functions as the transcriptional activator of the *ada* operon, while the product of the *adaB*[+] gene functions as the active alkyltransferase (150).

A gene (designated *dat1*[+]) which encodes an *O*[6]-MGT activity was cloned from *B. subtilis* by screening an *E. coli ada* mutant (114) and a *B. subtilis dat1* mutant (148) for enhanced resistance to alkylating agents. This may be the functional homolog of the *ogt*[+] gene in *E. coli*. The gene can encode a polypeptide of 18.8 kDa (Fig. 3–21). The penta-amino acid sequence Pro-Cys-His-Arg-Val, present in the *E. coli O*[6]-MGT I and *O*[6]-MGT II proteins, is conserved in the *B. subtilis* Dat1 protein (Fig. 3–22).

A survey of 33 species of gram-negative bacteria belonging to 19 genera has shown the adaptive response to be quite general (45, 196) (Table 3–4). The re-

Table 3–4 Properties of microbial *O*[6]-methylguanine DNA methyltransferases

Organism	Enzyme	*O*[6]-Methyl-guanine	*O*[4]-Methyl-thymine	Methylphos-photriesters	Alkylation-inducible methyltransferase	Alkylation-inducible resistance
Escherichia coli	Ada	+	+	+	+	+
	Ogt	+	+	−	−	NT
Bacillus subtilis	AdaA	−	NT[a]	+	+	+
	AdaB	+	NT	−	+	NT
	Dat1	+	NT	−	−	NT
Salmonella typhimurium	Ada$_{ST}$	+	NT	+	−/+	−
	Ogt$_{ST}$	+	+	−	−	NT
Saccharomyces cerevisiae	MGT1	+	+[b]	−	−	−
Micrococcus luteus	TI	+	−	−	+	+
	TII	−	+	−	+	NT
	TIII	−	−	+	+	NT
Anacystis nidulans	ANAT1	+	NT	−	−	+
	ANAT2	+	NT	−	+	NT
	ANAT3	+	NT	−	+	NT
	ANAT4	−	NT	+	+	NT
Escherichia alkalescens	TBN[c]	+	NT	+	+	NT
Escherichia hermanii	TBN	+	NT	+	+	NT
Escherichia fergusonii	TBN	+	NT	+	+	NT
Klebsiella aerogenes	TBN	+	NT	+	+	NT
Shigella sonnei	TBN	+	NT	+	+	NT
Shigella boydii	TBN	+	NT	+	+	NT
Citrobacter intermediens	TBN	+	NT	+	+	NT
Aerobacter aerogenes	TBN	+	NT	+	+	NT
Pseudomonas aeruginosa	TBN	+	NT	NT	+	+
Xanthomonas maltophilia	TBN	+	NT	NT	+	NT
Streptomyces fradiae	TBN	+	NT	NT	+	NT
Bacillus thuringiensis	TBN	+	NT	NT	+	NT
Staphylococcus aureus	TBN	NT	NT	NT	−	NT
Rhizobium meliloti	TBN	+	NT	NT	−	NT
Proteus mirabilis	TBN	NT	NT	NT	+	NT

[a] NT, not tested.
[b] Very low affinity for *O*[4]-methylthymine in vitro.
[c] TBN, to be named (may represent more than one enzyme).

Source: Adapted from Samson (196) with permission.

sponse is apparently not conserved in *H. influenzae* (111) (Table 3–4) or *Salmonella typhimurium* (57). However, the latter organism contains an activity very similar in size and properties to O^6-MGT II of *E. coli*. In particular, the enzyme from *Salmonella typhimurium* can abstract methyl groups with the appropriate base specificity but not from methylphosphotriesters (179).

A gene called ada^+_{ST} has been cloned from *Salmonella typhimurium* by functional complementation of an *E. coli ada* mutant (59). The protein encoded by this gene repairs alkylation damage efficiently. However, the protein encoded by the ada^+_{ST} gene is not a strong positive regulator of the *ada* regulon (59, 266).

EUKARYOTES

Although difficult to detect in extracts of some organisms, DNA-alkyltransferases are widespread in nature. The enzyme activity has been detected in extracts of fish (161), *Drosophila melanogaster* (56), the yeast *Saccharomyces cerevisiae* (220), the filamentous fungus *Aspergillus nidulans* (8), and mammalian cells. The activity from *Saccharomyces cerevisiae* is expressed at a level of only ~150 molecules per cell in exponentially growing cells and is undetectable in stationary-phase cultures. In contrast, when the release of O^6-methylguanine from DNA was measured in yeast cells in vivo, rapid release of this alkylated base was detected in stationary-phase cells (55). There is no evidence that the yeast enzyme is regulated by exposure of cells to alkylating agents (220). However, in contrast to the *E. coli* O^6-MGT II enzyme, the yeast enzyme has a much lower affinity for O^4-methylthymine than for O^6-methylguanine (219).

A gene for O^6-alkylguanine DNA alkyltransferase has been cloned from *Saccharomyces cerevisiae* by functional complementation of an *E. coli ada ogt* double mutant; it has been designated *MGT1* (272). Mutants carrying a disruption in the *MGT1* gene lack O^6-MGT activity and are sensitive to killing and mutagenesis following treatment with alkylating agents. Additionally, *mgt1* mutants have an increased rate of spontaneous mutagenesis, suggesting that an endogenous source of alkylation damage operates in *Saccharomyces cerevisiae* (273). The *MGT1* gene maps to chromosome IV. Consistent with the absence of an adaptive response to alkylation damage in *Saccharomyces cerevisiae*, *MGT1* transcript levels are not increased in response to alkylation treatment (272). However, deletion analysis has defined an upstream repression sequence, whose removal results in increased levels of basal expression (273). The observation that yeast mutants defective in the *MGT1* gene contain no additional O^6-MGT activity suggests that the product of the *MGT1* gene also participates in the repair of O^4-methylthymine (272).

The predicted amino acid sequences of the yeast, human (see below), and bacterial O^6-alkylguanine DNA alkyltransferase genes show extensive conservation. An 88-amino-acid stretch in the C-terminal half shows >40% amino acid identity between the yeast *MGT1* gene, the *E. coli ada+* and *ogt+* genes, and the *B. subtilis dat1+* and *adaB+* genes and 34.1% identity between the yeast and human genes (272). A comparison of the amino acid sequences of eight alkyltransferases from bacterial, yeast, and mammalian species reveals the presence of 21 conserved amino acids in the 87-amino-acid C-terminal domain (Fig. 3–22) (195). In particular, the penta-amino acid sequence Pro-Cys-His-Arg-Val/Ile, which includes the known methyl-accepting Cys-321 residue in the *E. coli* Ada protein, is conserved in all eight polypeptides (Fig. 3–22). Immediately upstream of this penta-amino acid sequence, a second highly conserved domain consisting of 10 amino acids with the consensus sequence PXA(A/V)RAV(G/A)XA has been identified (Fig. 3–22) (195).

Treatment of *Aspergillus nidulans* with low, nonlethal doses of agents such as MNNG results in a substantial increase in DNA alkyltransferase activity (8). Four polypeptides with O^6-MGT activity have been detected in this organism. Two inducible species of 18.5 and 21 kDa are active primarily against O^6-methylguanine. A third, 19.5-kDa species is also inducible but acts primarily against methylphosphotriesters. This represents the only known example of a eukaryotic O^6-MGT activity which recognizes methylphosphotriesters (8). A fourth species, of 16 kDa, is weakly induced.

DNA ALKYLTRANSFERASES AND THE ADAPTIVE RESPONSE TO MUTAGENESIS IN MAMMALIAN CELLS

A single O^6-MGT activity has been isolated from various mammalian cells. As is true in prokaryotes, the mammalian enzyme uses one of its cysteine residues as a methyl group acceptor and is inactivated in the course of the reaction (15, 16, 66, 73, 117, 172, 173). Like the constitutively expressed O^6-MGT II form of *E. coli*, the human enzyme (21.7 kDa) is smaller than *E. coli* O^6-MGT I and is unable to catalyze the removal of alkyl groups from alkylphotriesters in DNA (66, 101, 275). In contrast to the *E. coli* enzyme, however, the substrate specificity of the mammalian enzyme is confined mainly to alkylation at O^6-guanine in DNA (21). O^4-alkylthymine is a very weak substrate for this enzyme (115). Purified human and rat O^6-MGT both repaired O^6-methylguanine in alkylated poly[d(G-C)] or poly(dG-dC) with a rate constant of 1×10^9 M^{-1} min^{-1}, whereas the repair of O^4-methylthymine from alkylated poly[d(A-T)] or poly(dA-dT) was only 1.8×10^5 M^{-1} min^{-1} (284). There is no evidence for a distinct second enzyme that repairs the latter substrate better.

Screening of human cDNA libraries in *E. coli* mutant *ada* hosts yielded a cDNA which encodes a ~22-kDa O^6-MGT activity (158, 258). Similarly, screening Mer$^-$ cells (which are deficient in O^6-MGT activity [see below]) with a cDNA library prepared from Mer$^+$ cells (which are proficient in O^6-MGT activity [see below]) yielded a cDNA which expresses the same enzyme activity (67). A human cDNA was also isolated by screening a library with oligonucleotide probes derived from the active-site amino acid sequence of the purified bovine enzyme (194).

The human O^6-MGT gene is located on human chromosome 10 (194). It contains five exons and spans more than 170 kb (158). The amino acid product of the cloned human gene shows extensive regions of sequence similarity with both the *E. coli* O^6-MGT I and O^6-MGT II polypeptides and with one of the *B. subtilis* O^6-MGT proteins (67, 258). However, there is little similarity at the nucleotide sequence level, and the human cDNA does not hybridize to *E. coli* genomic DNA (258). Overexpression of the human cDNA results in a 50-fold increase in the level of O^6-MGT activity and facilitated purification of the human enzyme to apparent physical homogeneity (115). The recombinant protein and that purified from untransformed cultured human lymphoblasts are identical (268).

O^6-Alkylguanine is a lesion of considerable biological importance in mammalian cells and tissues. This lesion is implicated in mutagenesis by alkylating agents in cultured mammalian cells (164), and there is considerable evidence that O^6-alkylguanine may be involved in the production of tumors by alkylating carcinogens in experimental animals. In general, alkylating agents that produce little O^6-alkylguanine in DNA are weak carcinogens (251). In addition, different rates of repair of this lesion in target and nontarget tissues can have a profound effect on tumor production. For example, the production of brain tumors in young rats treated with *N*-ethyl-*N*-nitrosourea is correlated with the persistence of O^6-alkylguanine in the target organ (54). Similarly, chronic treatment with MNU specifically results in neural tumors in experimental animals and is accompanied by a progressive accumulation of O^6-methylguanine in the brain without any concomitant accumulation in other tissues (136).

The adaptive response in mammalian cells and tissues has been addressed in numerous studies in efforts to draw direct parallels with the results in prokaryotes. Both in Chinese hamster ovary (CHO) cells and in human skin fibroblasts transformed by simian virus 40, exposure to very low doses of MNNG renders the cells resistant to the induction of sister chromatid exchanges (believed to be an indicator of persisting DNA damage) by further alkylation damage (199). CHO cells also become more resistant to killing, however, no adaptation to mutation has been observed (94). An enhanced resistance to cell killing also occurs in Chinese hamster V79 cells exposed to nontoxic doses of MNU before challenge with a toxic dose of this alkylating agent (41). However, neither the frequency of mutation to 6-thioguanine resistance nor the loss of O^6-methylguanine from DNA is altered in these cells (41). Several studies have shown increased levels of O^6-MGT activity in rat hepatoma cells exposed to ionizing radiation (26). Additionally, cells pretreated with ionizing radiation are more resistant to BCNU (58).

The regulation of the mammalian O^6-MGT gene is further complicated by the observation that cell lines in culture differ in their capacity to repair O^6-alkylguanine in DNA. This phenomenon was first observed with respect to the competence of cells to reactivate adenovirus treated with MNNG (30, 276). Cells able to carry out such host cell reactivation are designated as Mer$^+$ (30–32, 276) or Mex$^+$ (6, 238, 239). The majority of human fibroblast cell lines fall into this category, but many cell lines transformed with DNA viruses such as simian virus 40 are Mer$^-$ (or Mex$^-$) (129, 227). In addition, a number of human tumor cell lines are Mer$^-$ (30–32, 276). Extracts of Mer$^+$ HeLa cells contain ~100,000 molecules of O^6-MGT activity per cell (47), assuming that each molecule is catalytically active only once.

The availability of the cloned O^6-MGT gene facilitated investigation of the Mer phenotype and of the adaptive response in mammalian cells. Mer$^-$ cells either are deleted of the gene or contain unstable or unexpressed transcripts (194, 258). No major differences have been found in the coding or promoter regions of this gene in Mer$^+$ and Mer$^-$ cells, however (158). Lower levels of methylation are present in the promoter region of Mer$^-$ cells (158). Hence, regulation of the Mer phenotype is apparently determined epigenetically, a fact that provides a satisfactory explanation for its high frequency in populations of cultured cells. The steady-state levels of mRNA are not increased in various Mer$^+$ and Mer$^-$ cell lines exposed to a variety of "adaptive" treatments (48). Increased transcription of the gene, accompanied by enhanced enzyme activity and reduced MNNG-induced mutation frequency, has been found in rat liver cells exposed to a variety of DNA-damaging agents, including alkylating agents. This response was not detected in several human cell lines examined, and it remains to be determined whether these results reflect a specificity for rat liver cells. Nevertheless, the pattern of induction and its responsiveness to nonalkylation damage to DNA distinguish this inducible response from the adaptive response observed in *E. coli* and other bacteria.

In summary, mammalian cells contain a O^6-MGT enzyme activity which appears to be similar to the constitutive form of the *E. coli* enzyme (O^6-MGT II) in terms of its noninducibility but resembles O^6-MGT I in its substrate specificity. However, there is no good evidence that the expression of this activity is regulated in mammalian cells or that its regulation is analogous to the inducible adaptive response to mutation in *E. coli* and other prokaryotes.

Mammalian cells can acquire resistance to killing and to chromosomal damage, but not to mutagenesis, following exposure to alkylating agents. This phenotype of *methylation tolerance* is less well defined (96). However, insights into the phenomenon of tolerance are beginning to emerge (96). One interesting model invokes the loss of a DNA mismatch correction mechanism (52, 96, 98) (see chapter 9). During the replication of DNA containing O^6-methylguanine, there is a high probability that T will be inserted opposite the alkylated base. It has been suggested that both O^6-methyl-G·C and O^6-methyl-G·T are recognized as mispaired bases by the mismatch repair system, which carries out excision and degradation of the strand opposite the alkylated base. (Mismatch repair is a specialized excision repair system that operates with specificity for the newly synthesized strand [see chapter 9.]) In view of this strand specificity, such repair would indeed be futile, because the mismatch repair system would continue to excise the newly synthesized DNA opposite the alkylated base. This continuing futile degradation of DNA with its attendant persistence of strand breaks in DNA is presumed to increase the potential of cell killing. On the basis of this model, one might anticipate that cells which are incapable of mismatch repair would become tolerant to alkylating agents that introduce O^6-alkylguanine into DNA. Consistent with this idea, cell lines that are deficient in the correction of mispaired bases have indeed been shown to be more resistant to killing by alkylating agents than normal cells (17, 100). Similarly, in *E. coli*, mutational inactivation of the mechanism for strand discrimination during mismatch correction results in hypersensitivity to killing by MNNG. However, this increased sensitivity can be abrogated by a second mutation, which inactivates the process of mismatch correction itself (98) (see chapter 9).

The stable integration and expression of the cloned *E. coli ada* gene into Mer⁻ mammalian cells result in enhanced resistance to killing and mutagenesis following treatment with a variety of alkylating agents (20, 79, 80, 101, 198). Phenotypic complementation is also effected by truncated proteins containing just the C-terminal domain which includes the critical Cys-321 residue, whereas proteins deleted of this region of the Ada protein are defective for correction (19, 79, 101).

Repair of Single-Strand Breaks by Direct Rejoining

Agents that promote the hydrolysis of phosphodiester bonds in duplex DNA are discussed in chapter 1. Primary among these are ionizing radiation such as X rays and γ rays. The repair of DNA strand breaks typically requires DNA synthetic and/or recombinative events that are discussed in later chapters. As discussed in chapter 1, most radiation-induced single-strand breaks in DNA are accompanied by end group damage of the adjacent sugars and/or bases. Hence, the repair of DNA strand breaks typically requires various processing reactions that remove such damage before the ends can be rejoined. However, at least in *E. coli* and possibly in other organisms as well, some of the single-strand breaks in DNA produced by ionizing radiation under anoxic conditions are repaired by simple rejoining of the ends (81), and such repair may be considered an example of the direct reversal of DNA damage. In these in vitro reactions, the incubation of irradiated DNA with the enzyme *DNA ligase* results in the loss of a fraction of the total DNA strand breaks as measured by sedimentation velocity in alkaline sucrose gradients (81). DNA ligase is a highly specific enzyme which is ubiquitous in its distribution and plays a role in most known biochemical pathways which require the rejoining of strand breaks in DNA (116, 119). A detailed discussion of this enzyme is deferred to chapter 4, when we will consider the rejoining of newly synthesized DNA to extant DNA during the process of excision repair. The enzyme from *E. coli* has an absolute requirement for NAD and for Mg²⁺ as cofactors (116, 119). All DNA ligases require free ends in duplex DNA with no missing nucleotides at the site of the break and the presence of adjacent 3′ OH and 5′ P termini (116) (Fig. 3–24). Thus, only strand breaks with these particular characteristics produced by DNA damage are subject to repair by direct reversal.

Summary and Conclusions

The reversal of damage in DNA is obviously the most direct mode of DNA repair and suggests a number of distinct possible advantages to a living cell.

1. In each of the four examples discussed above, only a single gene product is required—a highly economical use of genetic information. However, the alkyltransferase mode of repair is clearly energetically expensive, since an entire protein molecule is expended for each reaction.
2. As a corollary of the previous statement, the reversal mode of DNA repair is kinetically advantageous since it presumably occurs more rapidly than multistep biochemical pathways such as excision repair (see chapters 4, 6 and 8).
3. In general, these processes would tend to be relatively error free because of their high degree of specificity.

In the light of these considerations, one might anticipate that reversal of DNA damage in single-step reactions would be a highly selected mode of DNA repair and that further examples are yet to be discovered in both prokaryotes and eukaryotes.

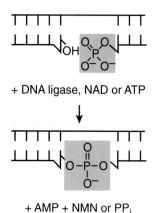

+ DNA ligase, NAD or ATP

+ AMP + NMN or PPᵢ

Figure 3–24 DNA ligase catalyzes the joining of strand breaks that contain juxtaposed 3′OH or 5′P termini in DNA. The enzyme from *E. coli* requires NAD as a cofactor; that encoded by phage T4 requires ATP. (*From Friedberg [49] with permission.*)

REFERENCES

1. **Ahmad, M., and A. R. Cashmore.** 1993. *HY4* gene of *A. thaliana* encodes a protein with characteristics of a blue-light photoreceptor. *Nature* (London) **366:**162–166.

2. **Ahmmed, Z., and J. Laval.** 1984. Enzymatic repair of O-alkylated thymidine residues in DNA: involvement of a O^4-methylthymine-DNA methyltransferase and a O^2-methylthymine DNA glycosylase. *Biochem. Biophys. Res. Commun.* **120:**1–8.

3. **Akimaru, H., K. Sakumi, T. Yoshikai, M. Anai, and M. Sekiguchi.** 1990. Positive and negative regulation of transcription by a cleavage product of Ada protein. *J. Mol. Biol.* **216:**261–273.

4. **Ather, A., Z. Ahmed, and S. Riazuddin.** 1984. Adaptive response of *Micrococcus luteus* to alkylating chemicals. *Nucleic Acids Res* **12:**2111–2126.

5. **Austin, R., M. S., and T. Begley.** 1992. Mechanistic studies on DNA photolyase. 5. Secondary deuterium isotope effects on the cleavage of the uracil photodimer radical cation and anion. *J. Am. Chem. Soc.* **114:**1886–1887.

6. **Ayres, K., R. Sklar, K. Larson, V. Lindgren, and B. Strauss.** 1982. Regulation of the capacity for O^6-methylguanine removal from DNA in human lymphoblastoid cells studied by cell hybridization. *Mol. Cell. Biol.* **2:**904–913.

7. **Baer, M. E., and G. B. Sancar.** 1993. The role of conserved amino acids in substrate binding and discrimination by photolyase. *J. Biol. Chem.* **268:**16717–16724.

8. **Baker, S. M., G. P. Margison, and P. Strike.** 1992. Inducible alkyltransferase DNA repair proteins in the filamentous fungus *Aspergillus nidulans. Nucleic Acids Res.* **20:**645–651.

9. **Benbow, R. M., A. J. Zuccarelli, and R. L. Sinsheimer.** 1974. A role for single-strand breaks in bacteriophage phi-X174 genetic recombination. *J. Mol. Biol.* **88:**629–651.

10. **Bennetzen, J. L., and B. D. Hall.** 1982. Codon selection in yeast. *J. Biol. Chem.* **257:**3026–3031.

11. **Berg, J. M.** 1990. Zinc fingers and other metal-binding domains. Elements for interactions between macromolecules. *J. Biol. Chem.* **265:**6513–6516.

12. **Bernstein, H., H. C. Byerly, F. A. Hopf, and R. E. Michod.** 1984. Origin of sex. *J. Theor. Biol.* **110:**323–351.

13. **Bernstein, H., H. C. Byerly, F. A. Hopf, and R. E. Michod.** 1985. The evolutionary role of recombinational repair and sex. *Int. Rev. Cytol.* **96:**1–24.

14. **Bhattacharyya, D., K. Tano, G. J. Bunick, E. C. Uberbacher, and W. D. Behnke.** 1988. Rapid, large-scale purification and characterization of 'Ada protein' (O^6-methylguanine-DNA methyltransferase) of *E. coli. Nucleic Acids Res.* **16:**6397–6410.

15. **Bogden, J. M., A. Eastman, and E. Bresnick.** 1981. A system in mouse liver for the repair of O^6-methylguanine lesions in methylated DNA. *Nucleic Acids Res.* **9:**3089–3103.

16. **Boulden, A. M., R. S. Foote, G. S. Fleming, and S. Mitra.** 1987. Purification and some properties of human DNA-O^6-methylguanine methyltransferase. *J. Biosci.* **11:**215–224.

17. **Branch, P., G. Aquilina, M. Bignami, and P. Karran.** 1993. Defective mismatch binding and a mutator phenotype in cells tolerant to DNA damage. *Nature* (London) **362:**652–654.

18. **Brash, D. E., W. A. Franklin, G. B. Sancar, A. Sancar, and W. A. Haseltine.** 1985. *Escherichia coli* DNA photolyase reverses cyclobutane pyrimidine dimers but not pyrimidine-pyrimidone (6–4) photoproducts. *J. Biol. Chem.* **260:**11438–11441.

19. **Brennand, J., and G. P. Margison.** 1986. Expression in mammalian cells of a truncated *Escherichia coli* gene coding for O^6-alkylguanine alkyltransferase reduces the toxic effects of alkylating agents. *Carcinogenesis* **7:**2081–2084.

20. **Brennand, J., and G. P. Margison.** 1986. Reduction of the toxicity and mutagenicity of alkylating agents in mammalian cells harboring the *Escherichia coli* alkyltransferase gene. *Proc. Natl. Acad. Sci.* USA **83:**6292–6296.

21. **Brent, T. P., M. E. Dolan, H. Fraenkel-Conrat, J. Hall, P. Karran, L. Laval, G. P. Margison, R. Montesano, A. E. Pegg, and P. M. Potter.** 1988. Repair of O-alkylpyrimidines in mammalian cells: a present consensus. *Proc. Natl. Acad. Sci.* USA **85:**1759–1762.

22. **Caillet-Fauquet, P., M. Defais, and M. Radman.** 1977. Molecular mechanisms of induced mutagenesis. Replication in vivo of bacteriophage phi-X174 single-stranded, ultraviolet light-irradiated DNA in intact and irradiated host cells. *J. Mol. Biol.* **117:**95–110.

23. **Cairns, J.** 1980. The Leeuwenhoek lecture, 1978. Bacteria as proper subjects for cancer research. *Proc. R. Soc. London Ser. B* **208:**121–133.

24. **Cerda-Olmedo, E., and P. C. Hanawalt.** 1968. Diazomethane as the active agent in nitrosoguanidine mutagenesis and lethality. *Mol. Gen. Genet.* **101:**191–202.

25. **Cerda-Olmedo, E., P. C. Hanawalt, and N. Guerola.** 1968. Mutagenesis of the replication point by nitrosoguanidine: map and pattern of replication of the *Escherichia coli* chromosome. *J. Mol. Biol.* **33:**705–719.

26. **Chan, C. L., Z. Wu, A. Eastman, and E. Bresnick.** 1992. Irradiation-induced expression of O^6-methylguanine-DNA methyltransferase in mammalian cells. *Cancer Res.* **52:**1804–1809.

27. **Chen, J., C. W. Huang, L. Hinman, M. P. Gordon, and D. A. Deranleau.** 1976. Photomonomerization of pyrimidine dimers by indoles and proteins. *J. Theor. Biol.* **62:**53–67.

28. **Cook, J. S.** 1970. Photoreactivation in animal cells. *Photophysiology* **5:**191–233.

29. **Coulondre, C., and J. H. Miller.** 1977. Genetic studies of the lac repressor. III. Additional correlation of mutational sites with specific amino acid residues. *J. Mol. Biol.* **117:**525–567.

30. **Day, R., III, and C. Ziolkowski.** 1979. Human brain tumor cell strains with deficient host-cell reactivation of N-methyl-N'-nitro-N-nitrosoguanidine-damaged adenovirus 5. *Nature* (London) **279:**797–799.

31. **Day, R., III, and C. Ziolkowski.** 1981. MNNG-pretreatment of a human kidney carcinoma cell strain decreases its ability to repair MNNG-treated adenovirus 5. *Carcinogenesis* **2:**213–214.

32. **Day, R., III, and C. H. Ziolkowski.** 1980. Defective repair of alkylated DNA by human tumour and SV40-transformed human cell strains. *Nature* (London) **288:**724–727.

33. **Demple, B.** 1986. Mutant *Escherichia coli* Ada proteins simultaneously defective in the repair of O^6-methylguanine and in gene activation. *Nucleic Acids Res.* **14:**5575–5589.

34. **Demple, B.** 1990. Self-methylation by suicide DNA repair enzymes, p. 285–304. *In* W. K. Paik and S. Kim (ed.), *Protein Methylation*, vol. 2. CRC Press, Inc., Boca Raton, Fla.

35. **Demple, B., A. Jacobsson, A. Olsson, M. Karran, and T. Lindahl.** 1983. Isolation of O^6-methylguanine-DNA methyltransferase from *E. coli*, p. 41–52. *In* E. C. Friedberg and P. C. Hanawalt (ed.), *DNA Repair—Laboratory Manual of Research Procedures*, vol. 2. Marcel Dekker, Inc., New York.

36. **Demple, B., A. Jacobsson, M. Olsson, P. Robins, and T. Lindahl.** 1982. Repair of alkylated DNA in *Escherichia coli*. Physical properties of O^6-methylguanine-DNA methyltransferase. *J. Biol. Chem.* **257:**13776–13780.

37. **Demple, B., B. Sedgwick, P. Robins, N. Totty, M. D. Waterfield, and T. Lindahl.** 1985. Active site and complete sequence of the suicidal methyltransferase that counters alkylation mutagenesis. *Proc. Natl. Acad. Sci.* USA **82:**2688–2692.

38. **Dosanjh, M. K., B. Singer, and J. M. Essigmann.** 1991. Comparative mutagenesis of O^6-methylguanine and O^4-methylthymine in *Escherichia coli. Biochemistry* **30:**7027–7033.

39. **Dulbecco, R.** 1949. Reactivation of ultraviolet inactivated bacteriophage by visible light. *Nature* (London) **163:**949–950.

40. **Dumenco, L. L., E. Allay, K. Norton, and S. L. Gerson.** 1993. The prevention of thymic lymphomas in transgenic mice by human O^6-alkylguanine-DNA alkyltransferase. *Science* **259:**219–222.

41. Durrant, L. G., G. P. Margison, and J. M. Boyle. 1981. Pretreatment of Chinese hamster v79 cells with MNU increases survival without affecting DNA repair or mutagenicity. *Carcinogenesis* **2:**55–60.

42. Evensen, G., and E. Seeberg. 1982. Adaption to alkylation resistance involves the induction of a DNA glycosylase. *Nature* (London) **296:**773–775.

43. Fajardo-Cavazos, P., C. Salazar, and W. L. Nicholson. 1993. Molecular cloning and characterization of the *Bacillus subtilis* spore photoproduct lyase (*spl*) gene, which is involved in repair of UV radiation-induced DNA damage during spore germination. *J. Bacteriol.* **175:**1735–1744.

44. Farland, W. H., and B. M. Sutherland. 1981. Analysis of pyrimidine dimer content of isolated DNA by nuclease digestion, p. 45–56. *In* E. C. Friedberg and P. C. Hanawalt (ed.), *DNA repair—a Laboratory Manual of Research Procedures*, vol. 1, part A. Marcel Dekker, Inc., New York.

45. Fernandez de Henestrosa, A. R., and J. Barbe. 1991. Induction of the *alkA* gene of *Escherichia coli* in gram-negative bacteria. *J. Bacteriol.* **173:**7736–7740.

46. Foote, R. S., S. Mitra, and B. C. Pal. 1980. Demethylation of O^6-methylguanine in a synthetic DNA polymer by an inducible activity in *Escherichia coli*. *Biochem. Biophys. Res. Commun.* **97:**654–659.

47. Foote, R. S., B. C. Pal, and S. Mitra. 1983. Quantitation of O^6-methylguanine-DNA methyltransferase in HeLa cells. *Mutat. Res.* **119:**221–228.

48. Fornace, A. J., Jr., M. A. Papathanasiou, M. C. Hollander, and D. B. Yarosh. 1990. Expression of the O^6-methylguanine-DNA methyltransferase gene MGMT in MER + and MER − human tumor cells. *Cancer Res.* **50:**7908–7911.

49. Friedberg, E. C. 1985. *DNA Repair*. W. H. Freeman & Co., New York.

50. Friedberg, E. C., and P. C. Hanawalt. 1981. *DNA Rrepair—a Laboratory Manual of Research Procedures*, vol. 1, part A. Marcel Dekker, Inc., New York.

51. Friedberg, E. C., and P. C. Hanawalt. 1983. *DNA Repair—a Laboratory Manual of Research Procedures*, vol. 2. Marcel Dekker, Inc., New York.

52. Goldmacher, V. S., R. A. Cuzick, Jr., and W. G. Thilly. 1986. Isolation and partial characterization of human cell mutants differing in sensitivity to killing and mutation by methylnitrosourea and N-methyl-N′-nitro-N-nitrosoguanidine. *J. Biol. Chem.* **261:**12462–12471.

53. Gordon, M. P. 1975. Photorepair of RNA, p. 115–121. *In* P. C. Hanawalt and R. B. Setlow (ed.), *Molecular Mechanisms for Repair of DNA*. Plenum Publishing Corp., New York.

54. Goth, R., and M. Rajewsky. 1974. Persistence of O^6-ethylguanine in rat-brain DNA: correlation with nervous system-specific carcinogenesis by ethylnitrosourea. *Proc. Natl. Acad. Sci USA* **71:**639–643.

55. Goth-Goldstein, R., and P. L. Johnson. 1990. Repair of alkylation damage in *Saccharomyces cerevisiae*. *Mol. Gen. Genet.* **221:**353–357.

56. Green, D. A., and W. A. Deutsch. 1983. Repair of alkylated DNA: *Drosophila* have DNA methyltransferases but not DNA glycosylases. *Mol. Gen. Genet.* **192:**322–325.

57. Guttenplan, J. B., and S. Milstein. 1982. Resistance of *Salmonella typhimurium* TA 1535 to O^6-guanine methylation and mutagenesis induced by low doses of N-methyl-N′-nitro-N-nitrosoguanidine: an apparent constitutive repair activity. *Carcinogenesis* **3:**327–331.

58. Habrakan, Y., and F. Laval. 1991. Enhancement of 1,3-bis(2-chloroethyl)-1-nitrosourea resistance by gamma-irradiation or drug pretreatment in rat hepatoma cells. *Cancer Res* **51:**1217–1220.

59. Hakura, A., K. Morimoto, T. Sofuni, and T. Nohmi. 1991. Cloning and characterization of the *Salmonella typhimurium ada* gene, which encodes O^6-methylguanine-DNA methyltransferase. *J. Bacteriol.* **173:**3663–3672.

60. Hamm-Alvarez, S., A. Sancar, and K. V. Rajagopalan. 1990. The folate cofactor of *Escherichia coli* DNA photolyase acts catalytically. *J. Biol. Chem.* **265:**18656–18662.

61. Harm, W. 1970. Analysis of photoenzymatic repair of UV lesions in DNA by single light flashes. V. Determination of the reaction-rate constants in *E. coli* cells. *Mutat. Res.* **10:**277–290.

62. Harm, W. 1975. Kinetics of photoreactivation, p. 89–101. *In* P. C. Hanawalt and R. B. Setlow (ed.), *Molecular Mechanisms for Repair of DNA*. Plenum Publishing Corp., New York.

63. Harm, W., H. Harm, and C. S. Rupert. 1968. Analysis of photoenzymatic repair of UV lesions in DNA by single light flashes. II. In vivo studies with *Escherichia coli* cells and bacteriophage. *Mutat. Res.* **6:**371–385.

64. Harm, W., and B. Hillebrandt. 1962. A non-photoreactivable mutant of *E. coli* B. *Photochem. Photobiol.* **1:**271–272.

65. Harm, W., C. S. Rupert, and H. Harm. 1972. Photoenzymatic repair of DNA. I. Investigation of the reaction by flash illumination, p. 53–78. *In* J. R. F. Beers, R. M. Herriott, and R. C. Tilghman (ed.), *Molecular and Cellular Repair Processes*. Johns Hopkins University Press, Baltimore.

66. Harris, A. L., P. Karran, and T. Lindahl. 1983. O^6-methylguanine-DNA methyltransferase of human lymphoid cells: structural and kinetic properties and absence in repair-deficient cells. *Cancer Res* **43:**3247–3252.

67. Hayakawa, H., G. Koike, and M. Sekiguchi. 1990. Expression and cloning of complementary DNA for a human enzyme that repairs O^6-methylguanine in DNA. *J. Mol. Biol.* **213:**739–747.

68. Heelis, P. F., T. Okamura, and A. Sancar. 1990. Excited-state properties of *Escherichia coli* DNA photolyase in the picosecond to millisecond time scale. *Biochemistry* **29:**5694–5698.

69. Heelis, P. F., G. Payne, and A. Sancar. 1987. Photochemical properties of *Escherichia coli* DNA photolyase: selective photodecomposition of the second chromophore. *Biochemistry* **26:**4634–4640.

70. Helene, C., and H. Charlier. 1977. Photosensitized splitting of pyrimidine dimers by indole derivatives and by tryptophan-containing oligopeptides and proteins. *Photochem. Photobiol.* **25:**429–434.

71. Helene, C., M. Charlier, J. J. Toulme, and F. Toulme. 1978. Photosensitized splitting of thymine dimers in DNA by peptides and protein containing tryptophanyl residues, p. 141–146. *In* P. C. Hanawalt, E. C. Friedberg, and C. F. Fox (ed.), *DNA Repair Mechanisms*. Academic Press, Inc., New York.

72. Helene, C., F. Toulme, M. Charlier, and M. Yaniv. 1976. Photosensitized splitting of thymine dimers in DNA by gene 32 protein from phage T4. *Biochem. Biophys. Res. Commun.* **71:**91–98.

73. Hora, J. F., A. Eastman, and E. Bresnick. 1983. O^6-methylguanine methyltransferase in rat liver. *Biochemistry* **22:**3759–3763.

74. Hughes, S. J., and B. Sedgwick. 1989. The adaptive response to alkylation damage. Constitutive expression through a mutation in the coding region of the *ada* gene. *J. Biol. Chem.* **264:**21369–21375.

75. Hurter, J., M. P. Gordon, J. P. Kirwan, and A. D. McLaren. 1974. In vitro photoreactivation of ultraviolet-inactivated ribonucleic acid from tobacco mosaic virus. *Photochem. Photobiol.* **19:**185–190.

76. Husain, I., and A. Sancar. 1987. Binding of *E. coli* DNA photolyase to a defined substrate containing a single T<>T dimer. *Nucleic Acids Res.* **15:**1109–1120.

77. Husain, I., and A. Sancar. 1987. Photoreactivation in *phr* mutants of *Escherichia coli* K-12. *J. Bacteriol.* **169:**2367–2372.

78. Husain, I., G. B. Sancar, S. R. Holbrook, and A. Sancar. 1987. Mechanism of damage recognition by *Escherichia coli* DNA photolyase. *J. Biol. Chem.* **262:**13188–13197.

79. Ishizaki, K., T. Tsujimura, C. Fujio, Y. P. Zhang, H. Yawata, and Y. Nakabeppu. 1987. Expression of the truncated *E. coli* O^6-methylguanine methyltransferase gene in repair-deficient human cells and restoration of cellular resistance to alkylating agents. *Mutat. Res.* **184:**121–128.

80. Ishizaki, K., T. Tsujimura, H. Yawata, C. Fujio, Y. Nakabeppu, M. Sediguchi, and M. Ikenaga. 1986. Transfer of the *E. coli* O^6-methylguanine methyltransferase gene into repair-deficient human cells and restoration of cellular resistance to N-methyl-N′-nitro-N-nitrosoguanidine. *Mutat. Res.* **166:**135–141.

81. Jacobs, A., A. Bopp, and U. Hagen. 1972. In vitro repair of single-strand breaks in γ-irradiated DNA by polynucleotide ligase. *Int. J. Radiat. Biol.* **22:**431–435.

82. Jeggo, P. 1979. Isolation and characterization of *Escherichia coli* K-12 mutants unable to induce the adaptive response to simple alkylating agents. *J. Bacteriol.* **139:**783–791.

83. Jeggo, P. 1980. The adaptive response of *E. coli*: a comparison of its two components, killing and mutagenic adaptation, p. 153–160. *In* M. Alecevic (ed.), *Progress in Environmental Mutagenesis.* Elsevier Biomedical Press, Amsterdam.

84. Jeggo, P., M. Defais, L. Samson, and P. Schendel. 1978. An adaptive response of *E. coli* to low levels of alkylating agent, p. 1011–1024. *In* I. Molineux and M. Kohiyama (ed.), *DNA Synthesis, Present and Future.* Plenum Publishing Corp., New York.

85. Jeggo, P., M. Defais, L. Samson, and P. Schendel. 1978. The adaptive response of *E. coli* to low levels of alkylating agent: the role of *polA* in killing adaptation. *Mol. Gen. Genet.* **162:**299–305.

86. Jeggo, P., T. M. Defais, L. Samson, and P. Schendel. 1977. An adaptive response of *E. coli* to low levels of alkylating agent: comparison with previously characterised DNA repair pathways. *Mol. Gen. Genet.* **157:**1–9.

87. Jimenez-Sanchez, A., and E. Cerda-Olmedo. 1975. Mutation and DNA replication in *Escherichia coli* treated with low concentrations of N-methyl-N'-nitro-N-nitrosoguanidine. *Mutat. Res.* **28:**337–345.

88. Johnson, J. L., S. Hamm-Alvarez, G. Payne, G. B. Sancar, K. V. Rajagopalan, and A. Sancar. 1988. Identification of the second chromophore of *Escherichia coli* and yeast DNA photolyases as 5,10-methnyltetrahydrofolate. *Proc. Natl. Acad. Sci. USA* **85:**2046–2050.

89. Jordan, S. P., J. L. Alderfer, L. P. Chanderkar, and M. S. Jorns. 1989. Reaction of *Escherichia coli* and yeast photolyases with homogeneous short-chain oligonucleotide substrates. *Biochemistry* **28:**8149–8153.

90. Jordan, S. P., and M. S. Jorns. 1988. Evidence for a singlet intermediate in catalysis by *Escherichia coli* DNA photolyase and evaluation of substrate binding determinats. *Biochemistry* **27:**8915–8923.

91. Jorns, M. S., G. B. Sancar, and A. Sancar. 1984. Identification of a neutral flavin radical and characterization of a second chromophore in *Escherichia coli* DNA photolyase. *Biochemistry* **23:**2673–2679.

92. Jorns, M. S., G. B. Sancar, and A. Sancar. 1985. Identification of oligothymidylates as new simple substrates for *Escherichia coli* DNA photolyase and their use in a rapid spectrophotometric enzyme assay. *Biochemistry* **24:**1856–1861.

93. Jorns, M. S., B. Wang, and S. P. Jordan. 1987. DNA repair catalyzed by *Escherichia coli* DNA photolyase containing only reduced flavin: elimination of the enzyme's second chromophore by reduction with sodium borohydride. *Biochemistry* **26:**6810–6816.

94. Jostes, R., L. Samson, and J. L. Schwartz. 1981. Kinetics of mutation and sister-chromatid exchange induction by ethyl methanesulfonate in Chinese hamster ovary cells. *Mutat. Res.* **91:**255–258.

95. Jump, D. B., and M. Smulson. 1980. Purification and characterization of the major nonhistone protein acceptor for poly(adenosine diphosphate ribose) in HeLa cell nuclei. *Biochemistry* **19:**1024–1030.

96. Karran, P., and M. Bignami. 1992. Self-destruction and tolerance in resistance of mammalian cells to alkylation damage. *Nucleic Acids Res.* **20:**2933–2940.

97. Karran, P., T. Lindahl, and B. Griffin. 1979. Adaptive response to alkylating agents involves alteration *in situ* of O^6-methylguanine residues in DNA. *Nature* (London) **280:**76–77.

98. Karran, P., and M. G. Marinus. 1982. Mismatch correction at O^6-methylguanine residues in *E. coli* DNA. *Nature* (London) **296:**868–869.

99. Karran, P., S. Stevens, and B. Sedgwick. 1982. The adaptive response to alkylating agents: the removal of O^6-methylguanine from DNA is not dependent on DNA polymerase1. *Mutat. Res.* **104:**67–73.

100. Kat, A., W. G. Thilly, W. H. Fang, M. J. Longley, and G. M. Li. 1993. An alkylation-tolerant, mutator human cell line is deficient in strand-specific mismatch repair. *Proc. Natl. Acad. Sci. USA* **90:**6424–6428.

101. Kataoka, H., J. Hall, and P. Karran. 1986. Complementation of sensitivity to alkylating agents in *Escherichia coli* and Chinese hamster ovary cells by expression of a cloned bacterial DNA repair gene. *EMBO J.* **5:**3195–3200.

102. Kelner, A. 1949. Effect of visible light on the recovery of *Streptomyces griseus* conidia from ultra-violet irradiation injury. *Proc. Natl. Acad. Sci. USA* **35:**73–79.

103. Kim, S. T., T. F. Hartman, and S. D. Rose. 1990. Solvent dependence of pyrimidine dimer splitting in a covalently linked dimer-indole system. *Photochem. Photobiol.* **52:**789–794.

104. Kim, S. T., P. F. Heelis, T. Okamura, Y. Hirata, N. Mataga, and A. Sancar. 1991. Determination of rates and yields of interchromophore (folate----flavin) energy transfer and intermolecular (flavin----DNA) electron transfer in *Escherichia coli* photolyase by time-resolved fluorescence and absorption spectroscopy. *Biochemistry* **30:**11262–11270.

105. Kim, S. T., Y. F. Li, and A. Sancar. 1992. The third chromophore of DNA photolyase: Trp-277 of *Escherichia coli* DNA photolyase repairs thymine dimers by direct electron transfer. *Proc. Natl. Acad. Sci. USA* **89:**900–904.

106. Kim, S. T., K. Malhotra, C. A. Smith, J. S. Taylor, and A. Sancar. 1993. DNA photolyase repairs the trans-syn cyclobutane thymine dimer. *Biochemistry* **32:**7065–7068.

107. Kim, S. T., K. Malhotra, C. A. Smith, J. S. Taylor, and A. Sancar. 1994. Characterization of (6–4) photoproduct DNA photolyase. *J. Biol. Chem.* **269:**8535–8540.

108. Kim, S. T., and S. D. Rose. 1992. Pyrimidine dimer splitting in covalently linked dimer-arylamine systems. *Photochem. Photobiol. B. Biol.* **12:**179–191.

109. Kim, S. T., and A. Sancar. 1991. Effect of base, pentose, and phosphodiester backbone structures on binding and repair of pyrimidine dimers by *Escherichia coli* DNA photolyase. *Biochemistry* **30:**8623–8630.

110. Kim, S. T., and A. Sancar. 1993. Photochemistry, photophysics, and mechanism of pyrimidine dimer repair by DNA photolyase. *Photochem. Photobiol.* **57:**895–904.

111. Kimball, R. F. 1980. Further studies on the induction of mutation in *Haemophilus influenzae* by N-methyl-N'-nitro-N-nitrosoguanidine: lack of an inducible error-free repair system and the effect of exposure medium. *Mutat. Res.* **72:**361–372.

112. Kobayashi, T., M. Takao, A. Oikawa, and A. Yasui. 1989. Molecular characterization of a gene encoding a photolyase from *Streptomyces griseus. Nucleic Acids Res.* **17:**4731–4744.

113. Kobayashi, T., M. Takao, A. Oikawa, and A. Yasui. 1990. Increased UV sensitivity of *Escherichia coli* cells after introduction of foreign photolyase genes. *Mutat. Res.* **236:**27–34.

114. Kodama, K. I., Y. Nakabeppu, and M. Sekiguchi. 1989. Cloning and expression of the *Bacillus subtilis* methyltransferase gene in *Escherichia coli ada⁻* cells. *Mutat. Res.* **218:**153–163.

115. Koike, G., H. Maki, H. Takeya, H. Hayakawa, and M. Sekiguchi. 1990. Purification, structure, and biochemical properties of human O^6-methylguanine-DNA methyltransferase. *J. Biol. Chem.* **265:** 14754–14762.

116. Kornberg, A., and T. Baker. 1992. *DNA Synthesis,* 2nd ed. W. H. Freeman & Co., New York.

117. Krokan, H., A. Haugen, B. Myrnes, and P. H. Guddal. 1983. Repair of premutagenic DNA lesions in human fetal tissues: evidence for low levels of O^6-methylguanine-DNA methyltransferase and uracil-DNA glycosylase activity in some tissues. *Carcinogenesis* **4:**1559–1564.

118. Langeveld, S. A., A. Yasui, and A. P. Eker. 1985. Expression of an *Escherichia coli phr* gene in the yeast *Saccharomyces cerevisiae. Mol. Gen. Genet.* **199:**396–400.

119. Lehman, I. R. 1974. DNA ligase: structure, mechanism, and function. *Science* **186:**790–797.

120. LeMotte, P. K., and G. C. Walker. 1985. Induction and autoregulation of *ada,* a positively acting element regulating the response of *Escherichia coli* K-12 to methylating agents. *J. Bacteriol.* **161:**888–895.

121. Leonard, G. A., J. Thomson, W. O. Watson, and T. Brown. 1990. High-resolution structure of a mutagenic lesion in DNA. *Proc. Natl. Acad. Sci. USA* **87:**9573–9576.

122. Li, Y. F., S. T. Kim, and A. Sancar. 1993. Evidence for lack of DNA photoreactivating enzyme in humans. *Proc. Natl. Acad. Sci. USA* **90:**4389–4393.

123. Li, Y. F., and A. Sancar. 1991. Cloning, sequencing, expression and characterization of DNA photolyase from *Salmonella typhimurium. Nucleic Acids Res.* **19:**4885–4890.

124. Lindahl, T. 1981. DNA methyl transferase acting on O^6-methylguanine residues in adapted *E. coli,* p. 207–217. *In* E. Seeberg and K. Kleppe (ed.), *Chromosome Damage and Repair.* Plenum Publishing Corp., New York.

125. Lindahl, T. 1982. DNA repair enzymes. *Annu. Rev. Biochem.* **51:**61–87.

126. Lindahl, T., B. Demple, and P. Robins. 1982. Suicide inactivation of the *E. coli* O^6-methylguanine-DNA methyltransferase. *EMBO J.* **1:**1359–1363.

127. Lindahl, T., B. Rydberg, T. Hjelmgren, M. Olsson, and A. Jacobsson. 1982. Molecular and cellular mechanisms of mutagenesis, p. 89–102. *In* J. F. Lemontt and W. M. Generoso (ed.), *Cellular Defense Mechanisms against Alkylation of DNA.* Plenum Publishing Corp., New York.

128. Lindahl, T., B. Sedgwick, B. Demple, and P. Karran. 1983. Enzymology and regulation of the adaptive response to alkylating agents, p. 241–253. *In* E. C. Friedberg and B. A. Bridges (ed.), *Cellular Responses to DNA Damage.* Alan R. Liss, Inc., New York.

129. Lindahl, T., B. Sedgwick, M. Sekiguchi, and Y. Nakabeppu. 1988. Regulation and expression of the adaptive response to alkylating agents. *Annu. Rev. Biochem.* **57:**133–157.

130. Lipman, R. S., and M. S. Jorns. 1992. Direct evidence for singlet-singlet energy transfer in *Escherichia coli* DNA photolyase. *Biochemistry* **31:**786–791.

131. Loveless, A. 1969. Possible relevance of O-6 alkylation of deoxyguanosine to the mutagenicity and carcinogenicity of nitrosamines and nitrosamides. *Nature* (London) **223:**206–207.

132. Mackay, W. J., S. Han, and L. D. Samson. 1994. Personal communication.

133. Madden, J. J., and H. Werbin. 1974. Use of membrane binding technique to study the kinetics of yeast deoxyribonucleic acid photolyase reactions. Formation of enzyme-substrate complexes in the dark and their photolysis. *Biochemistry* **13:**2149–2154.

134. Madden, J. J., H. Werbin, and J. Denson. 1973. A rapid assay for DNA photolyase using a membrane-binding technique. *Photochem. Photobiol.* **18:**441–445.

135. Malhotra, K., M. Baer, Y. F. Li, G. B. Sancar, and A. Sancar. 1992. Identification of chromophore binding domains of yeast DNA photolyase. *J. Biol. Chem.* **267:**2909–2914.

136. Margison, G. P., and P. Kleihues. 1975. Chemical carcinogenesis in the nervous system. Preferential accumulation of O^6-methylguanine in rat brain deoxyribonucleic acid during repetitive administration of N-methyl-N-nitrosourea. *Biochem. J.* **148:**521–525.

137. Matsukuma, S., Y. Nakatsuru, K. Nakagawa, T. Utakoji, H. Sugano, H. Kataoka, M. Sekiguchi, and T. Ishikawa. 1989. Enhanced O^6-methylguanine-DNA methyltransferase activity in transgenic mice containing an integrated *E. coli ada* repair gene. *Mutat. Res.* **218:**197–206.

138. McCarthy, J. G., B. V. Edington, and P. F. Schendel. 1983. Inducible repair of phosphotriesters in *Escherichia coli. Proc. Natl. Acad. Sci. USA* **80:**7380–7384.

139. McCarthy, T. V., P. Karran, and T. LIndahl. 1984. Inducible repair of O-alkylated DNA pyrimidines in *Escherichia coli. EMBO J.* **3:**545–550.

140. McCarthy, T. V., and T. Lindahl. 1985. Methyl phosphotriesters in alkylated DNA are repaired by the Ada regulatory protein of *E. coli. Nucleic Acids* Res. **13:**2683–2698.

141. Meechan, P. J., K. M. Milam, and J. E. Cleaver. 1986. Evaluation of homology between cloned *Escherichia coli* and yeast DNA photolyase genes and higher eukaryotic genomes. *Mutat. Res.* **166:**143–147.

142. Mitra, S., B. C. Pal, and R. S. Foote. 1982. O^6-Methylguanine-DNA methyltransferase in wild-type and *ada* mutants of *Escherichia coli. J. Bacteriol.* **152:**534–537.

143. Mongold, J. A. 1992. DNA repair and the evolution of transformation in *Haemophilus influenzae. Genetics* **132:**893–898.

144. Mongold, J. A. 1993. DNA repair and the evolution of sex in bacteria: the abilities of bacteria to exchange genes may reflect their need to repair DNA damage. *ASM News* **59:**397–400.

145. Moore, M. H., J. M. Gulbis, E. J. Dodson, B. Demple, and P. C. E. Moody. 1994. Crystal structure of a suicidal DNA repair protein: the Ada O^6-methylguanine-DNA methyltransferase from *E. coli. EMBO J.* **13:**1495–1501.

146. Moore, P., and B. S. Strauss. 1979. Sites of inhibition of in vitro DNA synthesis in carcinogen- and UV-treated phi-X174 DNA. *Nature* (London) **278:**664–666.

147. Morimoto, K., M. E. Dolan, D. Scicchitano, and A. E. Pegg. 1985. Repair of O^6-propylguanine and O^6-butylguanine in DNA by O^6-alkylguanine-DNA alkyltransferases from rat liver and *E. coli. Carcinogenesis* **6:**1027–1031.

148. Morohoshi, F., K. Hayashi, and N. Munakata. 1989. *Bacillus subtilis* gene coding for constitutive O^6-methylguanine-DNA alkyltransferase. *Nucleic Acids Res.* **17:**6531–6543.

149. Morohoshi, F., K. Hayashi, and N. Munakata. 1990. *Bacillus subtilis ada* operon encodes two DNA alkyltransferases. *Nucleic Acids Res.* **18:**5473–5480.

150. Morohoshi, F., K. Hayashi, and N. Munakata. 1991. Molecular analysis of *Bacillus subtilis ada* mutants deficient in the adaptive response to simple alkylating agents. *J. Bacteriol.* **173:**7834–7840.

151. Myers, L. C., M. P. Terranova, A. E. Ferentz, G. Wagner, and G. L. Verdine. 1993. Repair of DNA methylphosphotriesters through a metalloactivated cysteine nucleophile. *Science* **261:**1164–1167.

152. Myers, L. C., M. P. Terranova, H. M. Nash, M. A. Markus, and G. L. Verdine. 1992. Zinc binding by the methylation signaling domain of the *Escherichia coli* Ada protein. *Biochemistry* **31:**4541–4547.

153. Myers, L. C., G. L. Verdine, and G. Wagner. 1993. Solution structure of the DNA methyl phosphotriester repair domain of *Escherichia coli* Ada. *Biochemistry* **32:**14089–14094.

154. Nakabeppu, Y., H. Kondo, S. Kawabata, S. Iwanaga, and M. Sekiguchi. 1985. Purification and structure of the intact Ada regulatory protein of *Escherichia coli* K12, O^6-methylguanine-DNA methyltransferase. *J. Biol. Chem.* **260:**7281–7288.

155. Nakabeppu, Y., Y. Mine, and M. Sekiguchi. 1985. Regulation of expression of the cloned *ada* gene in *Escherichia coli. Mutat. Res.* **146:**155–167.

156. Nakabeppu, Y., and M. Sekiguchi. 1986. Regulatory mechanisms for induction of synthesis of repair enzymes in response to alkylating agents: ada protein acts as a transcriptional regulator. *Proc. Natl. Acad. Sci. USA* **83:**6297–6301.

157. Nakamura, T., Y. Tokumoto, K. Sakumi, G. Koike, and Y. Nakabeppu. 1988. Expression of the *ada* gene of *Escherichia coli* in response to alkylating agents. Identification of transcriptional regulatory elements. *J. Mol. Biol.* **202:**483–494.

158. Nakatsu, Y., K. Hattori, H. Hayakawa, K. Shimizu, and M. Sekiguchi. 1993. Organization and expression of the human gene for O^6-methylguanine-DNA methyltransferase. *Mutat. Res.* **293:**119–132.

159. Nakatsuru, Y., S. Matsukuma, N. Nemoto, H. Sugano, M. Sekiguchi, and T. Ishikawa. 1993. O^6-methylguanine-DNA methyltransferase protects against nitrosamine-induced hepatocarcinogenesis. *Proc. Natl. Acad. Sci USA* **90:**6468–6472.

160. Nakatsuru, Y., S. Matsukuma, M. Sekiguchi, and T. Ishikawa. 1991. Characterization of O^6-methylguanine-DNA methyltransferase in transgenic mice introduced with the *E. coli ada* gene. *Mutat. Res.* **254:**225–230.

161. Nakatsuru, Y., N. Nemoto, K. Nakagawa, P. Masahito, and T. Ishikawa. 1987. O^6-methylguanine DNA methyltransferase activity in liver from various fish species. *Carcinogenesis* **8:**1123–1127.

162. Neale, S. 1972. Effect of pH and temperature on nitrosamide-induced mutation in *Escherichia coli. Mutat. Res.* **14:**155–164.

163. Neale, S. 1976. Mutagenicity of nitrosamides and nitrosamidines in micro-organisms and plants. *Mutat. Res.* **32:**229–266.

164. Newbold, R. F., W. Warren, A. S. Medcalf, and J. Amos. 1980. Mutagenicity of carcinogenic methylating agents is associated with a specific DNA modification. *Nature* (London) **283:**596–599.

165. Ogata, N., K. Ueda, M. Kawaichi, and O. Hayaishi. 1981. Poly (ADP-ribose) synthetase, a main acceptor of poly (ADP-ribose) in isolated nuclei. *J. Biol. Chem.* **256:**4135–4137.

166. Okumura, T., A. Sancar, P. F. Heelis, T. P. Begley, Y. Hirata, and N. Mataga. 1991. Picosecond laser photolysis studies on the photorepair of pyrimidine dimers by DNA photolyase. 1. Laser photolysis of photolyase-2-deoxyuridine dinucleotide photodimer complex. *J. Am. Chem. Soc.* **113:**3143–3145.

167. Olsson, M., and T. Lindahl. 1980. Repair of alkylated DNA in *Escherichia coli.* Methyl group transfer from O^6-methylguanine to a protein cysteine residue. *J. Biol. Chem.* **255:**10569–10571.

168. Patrick, M., and R. Rahn. 1976. Photochemistry of DNA and polynucleotides: photoproducts, p. 35–95. *In* S. Y. Wang (ed.), *Photochemistry and Photobiology of the Nucleic Acids.* Academic Press, Inc., New York.

169. Payne, G., P. F. Heelis, B. R. Rohrs, and A. Sancar. 1987. The active form of *Escherichia coli* DNA photolyase contains a fully reduced flavin and not a flavin radical, both in vivo and in vitro. *Biochemistry* **26:**7121–7127.

170. Payne, G., and A. Sancar. 1990. Absolute action spectrum of E-FADH2 and E-FADH2-MTHF forms of *Escherichia coli* DNA photolyase. *Biochemistry* **29:**7715–7727.

171. Payne, G., M. Wills, C. Walsh, and A. Sancar. 1990. Reconstitution of *Escherichia coli* photolyase with flavins and flavin analogues. *Biochemistry* **29:**5706–5711.

172. Pegg, A. E. 1990. Mammalian O^6-alkylguanine-DNA alkyltransferase: regulation and importance in response to alkylating carcinogenic and therapeutic agents. *Cancer Res.* **50:**6119–6129.

173. Pegg, A. E., L. Wiest, R. S. Foote, S. Mitra, and W. Perry. 1983. Purification and properties of O^6-methylguanine-DNA transmethylase from rat liver. *J. Biol. Chem.* **258:**2327–2333.

174. Potter, P. M., J. Brennand, and G. P. Margison. 1986. Lack of sequence homology between a fragment of *E. coli* DNA encoding an O^6-methylguanine methyltransferase and the *ada* gene. *Br. J. Cancer* **54:**366–367.

175. Potter, P. M., M. C. Wilkinson, J. Fitton, F. J. Carr, J. Brennand, and D. P. Cooper. 1987. Characterization and nucleotide sequence of *ogt*, the O^6-alkylguanine-DNA-alkyltransferase gene of *E. coli. Nucleic Acids Res.* **15:**9177–9193.

176. Rafferty, J. A., C. Y. Fan, P. M. Potter, A. J. Watson, L. Cawkwell, P. J. O'Connor, and G. P. Margison. 1992. Tissue-specific expression and induction of human O^6-alkylguanine-DNA alkyltransferase in transgenic mice. *Mol. Carcinog.* **6:**26–31.

177. Rebeck, G. W., S. Coons, P. Carroll, and L. Samson. 1988. A second DNA methyltransferase repair enzyme in *Escherichia coli. Proc. Natl. Acad. Sci. USA* **85:**3039–3043.

178. Rebeck, G. W., and L. Samson. 1991. Increased spontaneous mutation and alkylation sensitivity of *Escherichia coli* strains lacking the *ogt* O^6-methylguanine DNA repair. *J. Bacteriol.* **173:**2068–2076.

179. Rebeck, G. W., C. M. Smith, D. L. Goad, and L. Samson. 1989. Characterization of the major DNA repair methyltransferase activity in unadapted *Escherichia coli* and identification of a similar activity in *Salmonella typhimurium. J. Bacteriol.* **171:**4563–4568.

180. Redfield, R. J. 1993. Evolution of natural transformation: testing the DNA repair hypothesis in *Bacillus subtilis* and *Haemophilus influenzae. Genetics* **133:**755–761.

181. Resnick, M. A. 1969. A photoreactivationless mutant of *Saccharomyces cerevisiae. Photochem. Photobiol.* **9:**307–312.

182. Resnick, M. A., and J. Setlow. 1972. Photoreactivation and gene dosage in yeast. *J. Bacteriol.* **109:**1307–1309.

183. Robins, P., and J. Cairns. 1979. Quantitation of the adaptive response to alkylating agents. *Nature* (London) **280:**74–76.

184. Robins, P., A. L. Harris, I. Goldsmith, and T. Lindahl. 1983. Cross-linking of DNA induced by chloroethylnitrosourea is presented by O^6-methylguanine-DNA methyltransferase. *Nucleic Acids Res.* **11:**7743–7758.

185. Rupert, C. S. 1960. Photoreactivation of transforming DNA by an enzyme from baker's yeast. *J. Gen. Physiol.* **43:**573–595.

186. Rupert, C. S. 1962. Photoenzymatic repair of ultraviolet damage in DNA. I. Kinetics of the reaction. *J. Gen. Physiol.* **45:**703–724.

187. Rupert, C. S. 1962. Photoenzymatic repair of ultraviolet damage in DNA. II. Formation of an enzyme-substrate complex. *J. Gen. Physiol.* **45:**725–741.

188. Rupert, C. S. 1975. Enzymatic photoreactivation: overview, p. 73–87. *In* P. C. Hanawalt and R. B. Setlow (ed.), *Molecular Mechanisms for Repair of DNA.* Plenum Publishing Corp., New York.

189. Rupert, C. S., S. Goodgal, and R. M. Herriott. 1958. Photoreactivation *in vitro* of ultraviolet inactivated *Hemophilus influenzae* transforming factor. *J. Gen. Physiol.* **41:**451–471.

190. Rupert, C. S., and W. Harm. 1966. Reactivation after photobiological damage. *Adv. Radiat. Biol.* **2:**1–81.

191. Rupert, C. S., W. Harm, and H. Harm. 1972. Photoenzymatic repair of DNA. II. Physical-chemical characterization of the process, p. 64–78. *In* R. F. Beers, Jr., R. M. Herriott, and R. C. Tilghman (ed.), *Molecular and Cellular Repair Processes.* Johns Hopkins University Press, Baltimore.

192. Rupert, C. S., and A. Sancar. 1978. Cloning the *phr* gene of *Escherichia coli,* p. 159–162. *In* P. C. Hanawalt, E. C. Friedberg, and C. F. Fox (ed.), *DNA Repair Mechanisms.* Academic Press, Inc., New York.

193. Rydberg, B., and T. Lindahl. 1982. Nonenzymatic methylation of DNA by the intracellular methyl group donor S-adenosyl-L-methionine is a potentially mutagenic reaction. *EMBO J.* **1:**211–216.

194. Rydberg, B., N. Spurr, and P. Karran. 1990. cDNA cloning and chromosomal assignment of the human O^6-methylguanine-DNA methyltransferase. cDNA expression in *Escherichia coli* and gene expression in human cells. *J. Biol. Chem.* **265:**9563–9569.

195. Sakumi, K., A. Shiraishi, H. Hayakawa, and M. Sekiguchi. 1991. Cloning and expression of cDNA for rat O^6-methylguanine-DNA methyltransferase. *Nucleic Acids Res.* **19:**5597–5601.

196. Samson, L. 1992. The suicidal DNA repair methyltransferases of microbes. *Mol. Microbiol.* **6:**825–831.

197. Samson, L., and J. Cairns. 1977. A new pathway for DNA repair in *Escherichia coli. Nature* (London) **267:**281–283.

198. Samson, L., B. Derfler, and E. A. Waldstein. 1986. Suppression of human DNA alkylation-repair defects by *Escherichia coli* DNA-repair genes. *Proc. Natl. Acad. Sci USA* **83:**5607–5610.

199. Samson, L., and J. L. Schwartz. 1980. Evidence for an adaptive DNA repair pathway in CHO and human skin fibroblast cell lines. *Nature* (London) **287:**861–863.

200. Sancar, A. 1994. Structure and function of DNA photolyase. *Biochemistry* **33:**2–9.

201. Sancar, A., K. A. Franklin, and G. B. Sancar. 1984. *Escherichia coli* DNA photolyase stimulates uvrABC excision nuclease in vitro. *Proc. Natl. Acad. Sci. USA* **81:**7397–7401.

202. Sancar, A., and C. S. Rupert. 1978. Cloning of the *phr* gene and amplification of photolyase in *Escherichia coli. Gene* **4:**295–308.

203. Sancar, A., and C. S. Rupert. 1978. Correction of the map location of the *phr* gene in *Escherichia coli* K-12. *Mutat. Res.* **51:**139–143.

204. Sancar, A., and C. S. Rupert. 1983. The *phr* gene of *Escherichia coli*, p. 241–252. *In* E. C. Friedberg and P. C. Hanawalt (ed.), *DNA Repair—a Laboratory Manual of Research Procedures*, vol. 2. Marcel Dekker, Inc., New York.

205. Sancar, A., and G. B. Sancar. 1984. *Escherichia coli* DNA photolyase is a flavoprotein. *J. Mol. Biol.* **172:**223–227.

206. Sancar, A., F. W. Smith, and G. B. Sancar. 1984. Purification of *Escherichia coli* DNA photolyase. *J. Biol. Chem.* **259:**6028–6032.

207. Sancar, G. B. 1985. Expression of a *Saccharomyces cerevisiae* photolyase gene in *Escherichia coli. J. Bacteriol.* **161:**769–771.

208. Sancar, G. B. 1985. Sequence of the *Saccharomyces cerevisiae* PHR1 gene and homology of the PHR1 photolyase to *E. coli* photolyase. *Nucleic Acids Res.* **13:**8231–8246.

209. Sancar, G. B. 1990. DNA photolyases: physical properties, action mechanism, and roles in dark repair. *Mutat. Res.* **236:**147–160.

210. Sancar, G. B., M. S. Jorns, G. Payne, D. J. Fluke, C. S. Rupert, and A. Sancar. 1987. Action mechanism of *Escherichia coli* DNA photolyase. III. Photolysis of the enzyme-substrate complex and the absolute action spectrum. *J. Biol. Chem.* **262:**492–498.

211. Sancar, G. B., and A. Sancar. 1987. Structure and function of DNA photolyases. *Trends Biochem. Sci.* **12:**259–261.

212. Sancar, G. B., and F. W. Smith. 1988. Construction of plasmids which lead to overproduction of yeast PHR1 photolyase in *Saccharomyces cerevisiae* and *Escherichia coli. Gene* **64:**87–96.

213. Sancar, G. B., and F. W. Smith. 1989. Interactions between yeast photolyase and nucleotide excision repair proteins in *Saccharomyces cerevisiae* and *Escherichia coli. Mol. Cell. Biol.* **9:**4767–4776.

214. Sancar, G. B., F. W. Smith, and P. F. Heelis. 1987. Purification of the yeast PHR1 photolyase from an *Escherichia coli* overproducing strain and characterization of the intrinsic chromophores of the enzyme. *J. Biol. Chem.* **262:**15457–15465.

215. Sancar, G. B., F. W. Smith, M. C. Lorence, C. S. Rupert, and A. Sancar. 1984. Sequences of the *Escherichia coli* photolyase gene and protein. *J. Biol. Chem.* **259:**6033–6038.

216. Sancar, G. B., F. W. Smith, R. Reid, G. Payne, M. Levy, and A. Sancar. 1987. Action mechanism of *Escherichia coli* DNA photolyase. I. Formation of the enzyme-substrate complex. *J. Biol. Chem.* **262:**478–485.

217. Sancar, G. B., F. W. Smith, and A. Sancar. 1983. Identification and amplification of the *E. coli phr* gene product. *Nucleic Acids Res.* **11:**6667–6678.

218. Sancar, G. B., F. W. Smith, and A. Sancar. 1985. Binding of *Escherichia coli* DNA photolyase to UV-irradiated DNA. *Biochemistry* **24:**1849–1855.

219. Sassanfar, M., M. Dosanjh, J. Essigmann, and L. Samson. 1991. Relative efficiencies of the bacterial, yeast, and human DNA methyltransferases for the repair of O^6-methylguanine and O^4-methylthymine. *J. Biol. Chem.* **266:**2767–2771.

220. Sassanfar, M., and L. Samson. 1990. Identification and preliminary characterization of an O^6-methylguanine DNA repair methyltransferase in the yeast *Saccharomyces cerevisiae. J. Biol. Chem.* **265:**20–25.

221. Schendel, P. 1981. Inducible repair systems and their implications for toxicology. *Crit. Rev. Toxicol.* **8:**311–362.

222. Schendel, P. F., B. V. Edington, J. G. McCarthy, and M. L. Todd. 1983. Repair of alkylation damage in *E. coli*, p. 227–240. *In* E. C. Friedberg and B. A. Bridges (ed.), *Cellular Responses to DNA Damage.* Alan R. Liss, Inc., New York.

223. Schendel, P. F., and P. E. Robins. 1978. Repair of O^6-methylguanine in adapted *Escherichia coli. Proc. Natl. Acad. Sci. USA* **75:**6017–6020.

224. Schild, D., J. Johnston, C. Chang, and R. K. Mortimer. 1984. Cloning and mapping of *Saccharomyces cerevisiae* photoreactivation gene PHR1. *Mol. Cell. Biol.* **4:**1864–1870.

225. Sebastian, J., B. Kraus, and G. B. Sancar. 1990. Expression of the yeast PHR1 gene is induced by DNA-damaging agents. *Mol. Cell. Biol.* **10:**4630–4637.

226. Sebastian, J., and G. B. Sancar. 1991. A damage-responsive DNA binding protein regulates transcription of the yeast DNA repair gene PHR1. *Proc. Natl. Acad. Sci. USA* **88:**11251–11255.

227. Sedgwick, B. 1983. Molecular cloning of a gene which regulates the adaptive response to alkylating agents in *Escherichia coli. Mol. Gen. Genet.* **191:**466–472.

228. Sedgwick, B. 1987. Molecular signal for induction of the adaptive response to alkylation damage in *Escherichia coli. J. Cell Sci. Suppl.* **6:**215–223.

229. Sedgwick, B. 1989. In vitro proteolytic cleavage of the *Escherichia coli* Ada protein by the *ompT* gene product. *J. Bacteriol.* **171:**2249–2251.

230. Sedgwick, B., and T. Lindahl. 1982. A common mechanism for repair of O^6-methylguanine and O^6-ethylguanine in DNA. *J. Mol. Biol.* **154:**169–175.

231. Sedgwick, B., P. Robins, N. Totty, and T. Lindahl. 1988. Functional domains and methyl acceptor sites of the *Escherichia coli* ada protein. *J. Biol. Chem.* **263:**4430–4433.

232. Setlow, P. 1992. I will survive: protecting and repairing spore DNA. *J. Bacteriol.* **174:**2737–2741.

233. Setlow, R. B. 1968. The photochemistry, photobiology, and repair of polynucleotides. *Prog. Nucleic. Acid Res. Mol. Biol.* **8:**257–295.

234. Setlow, R. B., P. A. Swenson, and W. L. Carrier. 1963. Thymine dimers and inhibition of DNA synthesis by ultraviolet irradiation of cells. *Science* **142:**1464–1466.

235. Shevell, D. E., A. M. Abou-Zamzam, B. Demple, and G. C. Walker. 1988. Construction of an *Escherichia coli* K-12 *ada* deletion by gene replacement in a *recD* strain reveals a second methyltransferase that repairs alkylated DNA. *J. Bacteriol.* **170:**3294–3296.

236. Shevell, D. E., and G. C. Walker. 1991. A region of the Ada DNA-repair protein required for the activation of ada transcription is not necessary for activation of *alkA. Proc. Natl. Acad. Sci. USA* **88:**9001–9005.

237. Singer, B., and M. K. Dosanjh. 1990. Site-directed mutagenesis for quantitation of base-base interactions at defined sites. *Mutat. Res.* **233:**45–51.

238. Sklar, R., K. Brady, and B. Strauss. 1981. Limited capacity for the removal of O^6-methylguanine and its regeneration in a human lymphoma line. *Carcinogenesis* **2:**1293–1298.

239. Sklar, R., and B. Strauss. 1981. Removal of O^6-methylguanine from DNA of normal and xeroderma pigmentosum-derived lymphoblastoid lines. *Nature* (London) **289:**417–420.

240. Sutherland, B. M. 1974. Photoreactivating enzyme from human leukocytes. *Nature* (London) **248:**109–112.

241. Sutherland, B. M. 1975. The human leukocyte photoreactivating enzyme, p. 107–113. *In* P. C. Hanawalt and R. B. Setlow (ed.), *Molecular Mechanisms for Repair of DNA.* Plenum Publishing Corp., New York.

242. Sutherland, B. M. 1977. Symposium on molecular mechanisms in photoreactivation. Introduction: fundamentals of photoreactivation. *Photochem. Photobiol.* **25:**413–414.

243. Sutherland, B. M. 1978. Enzymatic photoreactivation of DNA, p. 113–122. *In* P. C. Hanawalt, E. C. Friedberg, and C. F. Fox (ed.), *DNA Repair Mechanisms.* Academic Press, Inc., New York.

244. Sutherland, B. M., and M. J. Chamberlin. 1973. A rapid and sensitive assay for pyrimidine dimers in DNA. *Anal. Biochem.* **53:**168–176.

245. Sutherland, B. M., L. C. Harber, and I. E. Kochevar. 1980. Pyrimidine dimer formation and repair in human skin. *Cancer Res.* **40:**3181–3185.

246. Sutherland, B. M., and R. Oliver. 1976. Culture conditions affect photoreactivating enzyme levels in human fibroblasts. *Biochim. Biophys. Acta* **442:**358–367.

247. Sutherland, B. M., R. Oliver, C. O. Fuselier, and J. C. Sutherland. 1976. Photoreactivation of pyrimidine dimers in the DNA of normal and xeroderma pigmentosum cells. *Biochemistry* **15:**402–406.

248. **Sutherland, B. M., P. Runge, and J. C. Sutherland.** 1974. DNA photoreactivation enzyme from placental mammals. Origin and characteristics. *Biochemistry* **13:**4710–4715.

249. **Sutherland, J. C., and K. P. Griffin.** 1980. Monomerization of pyrimidine dimers in DNA by tryptophan-containing peptides: wavelength dependence. *Radiat. Res.* **83:**529–536.

250. **O. Swann, P. F.** 1990. Why do O^6-alkylguanine and O^4-alkylthymine miscode? The relationship between the structure of DNA containing O^6-alkylguanine and O^4-alkylthymine and the mutagenic properties of these bases. *Mutat. Res.* **233:**81–94.

251. **Swann, P. F., and P. N. Magee.** 1969. Induction of rat kidney tumours by ethyl methylsulfonate and nervous tissue tumours by methylmethane sulfonate. *Nature* (London) **223:**947–949.

252. **Swenson, P. A., and R. B. Setlow.** 1966. Effects of ultraviolet radiation on macromolecular synthesis in *Escherichia coli*. *J. Mol. Biol.* **15:**201–219.

253. **Takano, K., Y. Nakabeppu, and M. Sekiguchi.** 1988. Functional sites of the Ada regulatory protein of *Escherichia coli*. Analysis by amino acid substitutions. *J. Mol. Biol.* **201:**261–271.

254. **Takano, K., T. Nakamura, and M. Sekiguchi.** 1991. Roles of two types of O^6-methylguanine-DNA methyltransferases in DNA repair. *Mutat. Res.* **254:**37–44.

255. **Takao, M., T. Kobayashi, A. Ooikawa, and A. Yasui.** 1989. Tandem arrangement of photolyase and superoxide dismutase genes in *Halobacterium halobium*. *J. Bacteriol.* **171:**6323–6329.

256. **Takao, M., A. Oikawa, A. P. Eker, and A. Yasui.** 1989. Expression of an *Anacystis nidulans* photolyase gene in *Escherichia coli*: functional complementation and modified action spectrum of photoreactivation. *Photochem. Photobiol.* **50:**633–637.

257. **Tano, K., D. Bhattacharyya, R. S. Foote, R. J. Mural, and S. Mitra.** 1989. Site-directed mutation of the *Escherichia coli ada* gene: effects of substitution of methyl acceptor cysteine-321 by histidine in Ada protein. *J. Bacteriol.* **171:**1535–1543.

258. **Tano, K., S. Shiota, J. Collier, R. S. Foote, and S. Mitra.** 1990. Isolation and structural characterization of a cDNA clone encoding the human DNA repair protein for O^6-alkylguanine. *Proc. Natl. Acad. Sci. USA* **87:**686–690.

259. **Teo, I., B. Sedgwick, B. Demple, B. Li, and T. Lindahl.** 1984. Induction of resistance to alkylating agents in *E. coli*: the *ada$^+$* gene product serves both as a regulatory protein and as an enzyme for repair of mutagenic damage. *EMBO J.* **3:**2151–2157.

260. **Teo, I., B. Sedgwick, M. W. Kilpatrick, T. V. McCarthy, and T. Lindahl.** 1986. The intracellular signal for induction of resistance to alkylating agents in *E. coli*. *Cell* **45:**315–324.

261. **Todo, T., H. Takemori, H. Ryo, M. Ihara, T. Matsunaga, O. Nikaido, K. Sato, and T. Nomura.** 1993. A new photoreactivating enzyme that specifically repairs ultraviolet light-induced (6-4) photoproducts. *Nature* (London) **361:**371–374.

262. **Toulme, J. J., and C. Helene.** 1977. Specific recognition of single-stranded nucleic acids. Interaction of tryptophan-containing peptides with native, denatured, and ultraviolet-irradiated DNA. *J. Biol. Chem.* **252:**244–249.

263. **Towill, L., C. W. Huang, and M. P. Gordon.** 1977. Photoreactivation of DNA-containing cauliflower mosaic virus and tobacco mosaic virus RNA on Datura. *Photochem. Photobiol.* **25:**249–257.

264. **Van Wang, T. C., and C. S. Rupert.** 1977. Evidence for the monomerization of spore photoproduct to two thymines by the light-independent "spore repair" process in *Bacillus subtilis*. *Photochem. Photobiol.* **25:**123–127.

265. **Vaughan, P., T. Lindahl, and B. Sedgwick.** 1993. Induction of the adaptive response of *Escherichia coli* to alkylation damage by the environmental mutagen, methyl chloride. *Mutat. Res.* **293:**249–257.

266. **Vaughan, P., and B. Sedgwick.** 1991. A weak adaptive response to alkylation damage in *Salmonella typhimurium*. *J. Bacteriol.* **173:**3656–3662.

267. **Villani, G., S. Boiteux, and M. Radman.** 1978. Mechanism of ultraviolet-induced mutagenesis: extent and fidelity of in vitro DNA synthesis on irradiated templates. *Proc. Natl. Acad. Sci USA* **75:**3037–3041.

268. **von Wronski, M. A., S. Shiota, K. Tano, S. Mitra, D. D. Bigner, and T. P. Brent.** 1991. Structural and immunological comparison of indigenous human O^6-methylguanine-DNA methyltransferase with that encoded by a cloned cDNA. *J. Biol. Chem.* **266:**1064–1070.

269. **Wagner, E. K., M. Rice, and B. M. Sutherland.** 1975. Photoreactivation of herpes simplex virus in human fibroblasts. *Nature* (London) **254:**627–628.

270. **Weinfeld, M., A. F. Drake, J. K. Saunders, and M. C. Paterson.** 1985. Stereospecific removal of methyl phosphotriesters from DNA by an *Escherichia coli ada$^+$* extract. *Nucleic Acids Res.* **13:**7067–7077.

271. **Whitmer, M. R., E. Altmann, H. Young, T. P. Begley, and A. Sancar.** 1989. Mechanistic studies on DNA photolyase. 1. Secondary deuterium isotope effects on the cleavage of 2'-deoxyuridine dinucleotide photodimers. *J. Am. Chem. Soc.* **111:**9264–9265.

272. **Xiao, W., B. Derfler, J. Chen, and L. Samson.** 1991. Primary sequence and biological functions of a *Saccharomyces cerevisiae* O^6-methylguanine/O^4-methylthymine DNA repair methyltransferase gene. *EMBO J.* **10:**2179–2186.

273. **Xiao, W., and L. Samson.** 1992. The *Saccharomyces cerevisiae MGT1* DNA repair methyltransferase gene: its promoter and entire coding sequence, regulation and in vivo biological functions. *Nucleic Acids Res.* **20:**3599–3606.

274. **Xiao, W., K. K. Singh, B. Chen, and L. Samson.** 1993. A common element involved in transcriptional regulation of two DNA alkylation repair genes (*MAG* and *MGT1*) of *Saccharomyces cerevisiae*. *Mol. Cell. Biol.* **13:**7213–7221.

275. **Yarosh, D. B., M. Rice, and R. S. Day III.** 1984. O^6-methylguanine-DNA methyltransferase in human cells. *Mutat. Res.* **131:**27–36.

276. **Yarosh, D. B., M. Rice, C. H. J. Ziolkowski, R. S. Day III, and D. A. Scudiero.** 1983. O^6-methylguanine-DNA methyltransferase in human tumor cells, p. 261–270. *In* E. C. Friedberg and B. A. Bridges (ed.), *Cellular Responses to DNA Damage*. Alan R. Liss, Inc., New York.

277. **Yasuhira, S., and A. Yasui.** 1992. Visible light-inducible photolyase gene from the goldfish *Carassius auratus*. *J. Biol. Chem.* **267:**25644–25647.

278. **Yasui, A., and M. R. Chevallier.** 1983. Cloning of photoreactivation repair gene and excision repair gene of the yeast *Saccharomyces cerevisiae*. *Curr. Genet.* **7:**191–194.

279. **Yasui, A., A. P. Eker, and M. Koken.** 1989. Existence and expression of photoreactivation repair genes in various yeast species. *Mutat. Res.* **217:**3–10.

280. **Yasui, A., and S. A. Langeveld.** 1985. Homology between the photoreactivation genes of *Saccharomyces cerevisiae* and *Escherichia coli*. *Gene* **36:**349–355.

281. **Yasui, A., and W. Laskowski.** 1975. Determination of the number of photoreactivating enzyme molecules per haploid *Saccharomyces* cells. *J. Radiat. Biol.* **28:**511–518.

282. **Yasui, A., M. Takao, A. Oikawa, A. Kiener, C. T. Walsh, and A. P. Eker.** 1988. Cloning and characterization of a photolyase gene from the cyanobacterium *Anacystis nidulans*. *Nucleic Acids Res.* **16:**4447–4463.

283. **Youngs, D. A., and K. C. Smith.** 1978. Genetic location of the *phr* gene of *Escherichia coli* K-12. *Mutat. Res.* **51:**133–137.

284. **Zak, P., K. Kleibl, and F. Laval.** 1994. Repair of O^6-methylguanine and O^4-methylthymine by the human and rat O^6-methylguanine-DNA methyltransferases. *J. Biol. Chem.* **269:**730–733.

Base Excision Repair

The types of DNA damage that are repaired by direct reversal are limited. The most general DNA repair mode observed in nature is one in which damaged or inappropriate bases are excised from the genome and replaced by the normal nucleotide sequence and chemistry. This type of cellular response to DNA damage is referred to as *excision repair* and is the subject of this and the succeeding four chapters.

Excision Repair of DNA—General Considerations

We know of a number of distinct enzymatic mechanisms by which the excision of damaged bases occurs in living cells. The excision of some forms of base damage is initiated by the action of a specific class of DNA repair enzymes called *DNA glycosylases* (78, 164, 165), which catalyze the hydrolysis of the N-glycosylic bonds linking particular types of chemically altered (or inappropriate) bases to the deoxyribose-phosphate backbone (Fig. 4–1). Excision repair that is initiated by DNA glycosylases is called *base excision repair* (80), because the chemically modified moieties are excised as *free bases*. This initial enzymatic event during base excision repair actually generates another type of DNA damage, because it results in the formation of sites of base loss called *apurinic or apyrimidinic (AP) sites*. AP sites can also result from the depurination or depyrimidination of DNA owing to spontaneous hydrolysis of N-glycosylic bonds (see chapter 1). The repair of AP sites requires further biochemical events to complete base excision repair, which is a multistep process. The removal of AP sites is initiated by a second class of base excision repair enzymes called *apurinic/apyrimidinic* (AP) endonucleases (165), which specifically recognize these sites in duplex DNA. AP endonucleases produce *incisions* or *nicks* in duplex DNA by hydrolysis of one or another of the phosphodiester bonds immediately 5' or 3' to each AP site (Fig. 4–1). As discussed below, most AP endonucleases have specificity for the 5' phosphodiester bond.

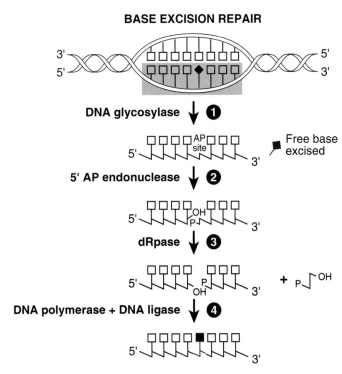

Figure 4–1 Schematic representation of base excision repair. For simplicity, only the relevant DNA strand is shown in most of the figure. Certain forms of base damage are recognized by DNA glycosylases (step 1) that catalyze excision of the free base by hydrolysis of the N-glycosyl bond linking the base to the sugar-phosphate backbone. This reaction leaves an AP site in the DNA. Attack at such sites by a 5′ AP endonuclease (step 2) results in a strand break with a 5′ terminal deoxyribose-phosphate moiety. These are excised by the action of a DNA deoxyribophosphodiesterase (dRpase) (step 3). The resulting single nucleotide gap is filled by repair synthesis and DNA repair is completed by DNA ligase (step 4). (*Adapted from Friedberg [87] with permission.*)

Hydrolysis of the phosphodiester bond immediately 5′ to an AP site generates a 5′ terminal deoxyribose-phosphate residue (Fig. 4–1). The removal of these residues during base excision repair requires yet another class of enzymes: nucleases that can specifically initiate the degradation of DNA at free ends. Such enzymes are called *exonucleases*. Unlike DNA glycosylases and AP endonucleases, exonucleases are not repair-specific enzymes. They can degrade DNA with free ends in a variety of DNA transactions, including repair, replication, and recombination (148). Exonucleases which are restricted to the degradation of single-stranded DNA and are additionally not inhibited by the presence of deoxyribose-phosphate moieties in place of complete nucleotides, are sometimes referred to as *DNA-deoxyribophosphodiesterases*; they can participate in the process of base excision repair. Hence, as shown in Fig. 4–1, the sequential action of a DNA glycosylase, a 5′ AP endonuclease, and a DNA deoxyribophosphodiesterase can generate a single nucleotide gap in the DNA duplex during base excision repair.

A second excision repair mode with which many cells are endowed is called *mismatch excision repair* (simply *mismatch repair*) or *mismatch correction*. Whereas the term "base excision repair" has a specific mechanistic connotation, the terms "mismatch repair" and "mismatch correction" refer to the repair of a particular type of DNA damage, i.e., mispaired bases in DNA. This can occur by several biochemical pathways, including base excision repair. DNA glycosylases that participate in mismatch repair are considered in this chapter. The general phenomenon of mismatch correction and the details of other biochemical pathways for the excision of mispaired bases are considered in chapter 9.

Many of the known forms of base damage, particularly those caused by the interaction of DNA with environmental agents, are not recognized by specific DNA glycosylases. There is yet another class of repair-specific enzymes which generate incisions in DNA at or near sites of base damage (112, 166) and define a third mode of excision repair. Unlike the DNA glycosylases, AP endonucleases, or mismatch repair enzymes, which recognize specific types of DNA damage, this class of enzymes can identify substrate determinants in DNA that are common to a wide variety of chemically altered bases. These enzymes attack the DNA in an *endonucleolytic* fashion, i.e., they do not require free ends to initiate the hydrolysis of phosphodiester bonds. We refer to this class of enzymes as *damage-specific endonucleases*. Most, and possibly all, damage-specific endonucleases incise the affected DNA strand on both sides of the base damage, thereby generating an oligonucleotide fragment which includes the damaged base (Fig. 4–2). Indeed, the ultimate release (*excision*) of these oligonucleotide fragments effects the removal of the damage from the genome. Hence, this general mode of excision repair is called *nucleotide excision repair*.

In summary, excision repair of DNA can be initiated by three distinct biochemical mechanisms. If the genome contains damaged, mispaired or, inappropriate bases (such as uracil), which are specifically recognized by DNA glycosylases, incision of DNA occurs by a two-step reaction involving the sequential action of a DNA glycosylase and an AP endonuclease (Fig. 4–1). Incision of DNA containing certain mispaired bases can occur by the *direct* action of a mismatch repair-specific endonuclease activity (see later chapters), and incision of DNA containing many different types of base damage can be effected by a more generic damage-specific endonuclease (Fig. 4–2).

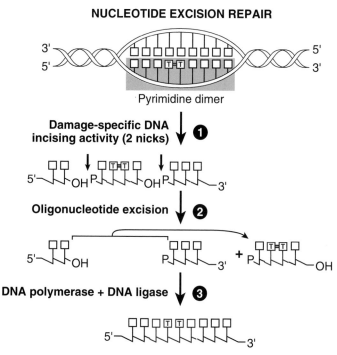

Figure 4–2 Schematic representation of nucleotide excision repair. For simplicity, only the relevant DNA strand is shown in most of the figure. Base damage such as pyrimidine dimers is recognized by a damage-specific endonuclease that nicks the DNA on each side of the lesion, generating a potential oligonucleotide fragment (step 1). (The size of the oligonucleotide shown here is purely schematic.) Subsequent enzyme-catalyzed events result in the release (excision) of this fragment (step 2). The resulting gap (which is always larger than 1 nucleotide) is filled by repair synthesis, and DNA repair is completed by DNA ligase (step 3) as in base excision repair. (*Adapted from Friedberg [87] with permission.*)

Regardless of the particular mechanism by which the excision of damaged nucleotides or deoxyribose-phosphate moieties is effected, the repair of double-stranded DNA is not complete until the missing nucleotides are replaced by DNA synthesis and the covalent integrity of the genome is restored by joining of the last newly synthesized nucleotide to the extant (parental) DNA. These events are referred to as *repair synthesis of DNA* and *DNA ligation*, respectively (Fig. 4–1 and 4–2) (112, 166). Repair synthesis is sometimes called *non-semiconservative DNA synthesis* or *unscheduled DNA synthesis* (UDS) to distinguish it from the DNA synthesis mode which operates during DNA replication. In the remainder of this chapter we will consider the phenomenon of base excision repair in detail. Nucleotide excision repair and the general topic of mismatch repair are discussed in subsequent chapters.

Despite recent rapid strides in deciphering the molecular mechanisms of DNA repair and mutagenesis in eukaryotes, including mammalian cells, much fundamental information on how living cells respond to genomic insult and injury is historically derived from studies on the prokaryote *Escherichia coli*. Hence, throughout the book this organism serves as a basic paradigm for explaining the essential features and, in many cases, the only known details of most known cellular responses to DNA damage. *E. coli* also serves as a useful model for contrasting the complexity of the biochemistry of DNA repair and mutagenesis in bacteria and in more highly evolved forms. Table 4–1 lists the genes which are known to be involved in the excision repair modes that have been defined in *E. coli* to date. Even though this list is certain to be expanded as new genes are discovered, the total number of genes tabulated constitutes a significant fraction of the total genome of *E. coli* and attests to the importance of just this limited aspect of cellular metabolism in the maintenance and control of genetic fidelity.

Much of our knowledge about excision repair has come from the identification and purification of relevant enzymes and the characterization of their prop-

Table 4–1 Known *E. coli* genes involved in excision repair of DNA

Gene	Gene product
ada[+]	Regulator of adaptive response to alkylation damage
tag-1[+]	3-mA-DNA glycosylase I
alkA[+]	3-mA-DNA glycosylase II (Hx-DNA glycosylase)
ung[+]	ura-DNA glycosylase
fpy[+]	FaPy DNA glycosylase
mutY[+]	Adenine-DNA glycosylase
nth[+]	Endonuclease III, involved in BER[a]
xthA[+]	Exonuclease III, involved in BER
nfo[+]	Endonuclease IV, involved in BER
recJ[+]	DNA deoxyribophosphodiesterase, involved in BER
uvrA[+]	Required for NER[b]
uvrB[+]	Required for NER
uvrC[+]	Required for NER
uvrD[+]	DNA helicase II
mfd[+]	Transcription repair coupling factor
polA[+]	DNA polymerase I
polB[+]	DNA polymerase II
polC[+]	DNA polymerase III
lig[+]	DNA ligase
dam[+]	DNA methylase required for mismatch repair
mutH[+]	Required for mismatch repair
mutL[+]	Required for mismatch repair
mutS[+]	Required for mismatch repair
recA[+]	RecA protein, required for regulation of NER
lexA[+]	LexA protein, required for regulation of NER

[a] BER, base excision repair.
[b] NER, nucleotide excision repair.

erties in vitro. Such studies, particularly when accompanied by physiological and genetic correlations in organisms which are defective or deficient in these enzymes, have considerably advanced our understanding of the molecular biology and biochemistry of excision repair. However, it should be borne in mind that in most of these studies purified DNA has been used as the substrate, whereas in the living cell the substrate is the genome. This distinction is more than just semantic, because even in prokaryotes the genome is a metabolically and structurally more complex entity than naked DNA in the test tube, and the genome of eukaryotes is possessed of profound structural complexity associated with the packaging of three billion base pairs of DNA into the tiny volume of the nucleus. Studies on the role of DNA topoisomerases, DNA helicases, and other DNA-binding proteins, which may facilitate the access of excision repair enzymes to sites of base damage in the highly compact structure of chromatin, are still in their infancy. However, in recent years considerable attention has been devoted to the consideration of the role of chromatin conformation in excision repair in mammalian cells (see chapter 7).

Aside from these structural considerations, the functional dynamics of DNA must be constantly borne in mind. The information garnered from studies of cell-free systems may be valid for excision repair in nonreplicating regions of the genome which are located well ahead of or behind the replication fork or in which replication and postreplicational modifications of DNA (such as methylation) are completed. However, there are indications that excision repair of damaged bases may occur at or near replication forks (112), and we know very little about how such repair may differ from that of bases in nonreplicating regions of the genome. In addition to replication, the genome of living cells undergoes frequent *transcription* and *recombination*. As discussed in later chapters, there are compelling indications that at least some forms of excision repair are dramatically influenced by these processes.

DNA Glycosylases

As indicated above, the enzymes that initiate and particularly characterize the multistep process of base excision repair are called *DNA glycosylases*. A list of the known DNA glycosylases and their substrates is provided in Table 4–2. Like the enzymes involved in the reversal of base damage, many DNA glycosylases recognize only a particular form of base damage, a particular inappropriate base (such as uracil incorporated during semiconservative DNA synthesis [Fig. 4–3]), or a particular mispairing. Some of these enzymes have a more relaxed substrate specificity, however, suggesting the recognition of more general determinants in DNA. Most DNA glycosylases are highly specific for a particular form of *monoadduct* base damage in DNA, although the DNA glycosylases that recognize pyrimidine dimers are prototypic examples of enzymes which recognize more-complex base damage.

Most of the currently known DNA glycosylases were first identified in *E. coli*, although the majority of the enzymes we will discuss are ubiquitously distributed. In general, DNA glycosylases are small proteins which are <30 kDa in size, with no subunit structure and no requirement for cofactors (78, 112, 165, 166). The latter property has greatly facilitated their detection in crude extracts, since in the absence of added divalent cation and in the presence of EDTA, nonspecific degradation of the substrate DNA by nucleases (most of which require metal cofactors) is prevented. However, one should bear in mind the possibility that some DNA glycosylases require metal cofactors and therefore would not be detected under such restrictive assay conditions.

Considering for the moment just monoadduct base damage to DNA, the essential feature of all known DNA glycosylases is the catalytic release of free bases as products of their reaction with DNA. The direct demonstration of the release of the free base is thus the most definitive assay of DNA glycosylase activity. This is usually achieved by chromatographic analysis of either the entire incubation mixture containing radiolabeled DNA or the fraction containing just the acid- or

Table 4–2 DNA glycosylases

Enzyme	Substrate	Products
Ura-DNA glycosylase	DNA containing uracil	Uracil + AP sites
Hmu-DNA glycosylase	DNA containing hydroxymethyluracil	Hydroxymethyluracil + AP sites
5-mC-DNA glycosylase	DNA containing 5-methylcytosine	5-methylcytosine + AP sites
Hx-DNA glycosylase	DNA containing hypoxanthine	Hypoxanthine + AP sites
Thymine mismatch-DNA glycosylase	DNA containing G·T mispairs	Thymine + AP sites
MutY-DNA glycosylase	DNA containing G·A mispairs	Adenine + AP sites
3-mA-DNA glycosylase I	DNA containing 3-methyladenine	3-Methyladenine + AP sites
3-mA-DNA glycosylase II	DNA containing 3-methyladenine,7-methylguanine, or 3-methylguanine	3-Methyladenine, 7-methylguanine, or 3-methylguanine + AP sites
FaPy-DNA glycosylase	DNA containing formamidopyrimidine moieties, or 8-hydroxyguanine	2,6-Diamino-4-hydroxy-5-*N*-methylformamido-pyrimidine and 8-hydroxyguanine + AP sites
5,6-HT-DNA glycosylase (endonuclease III)	DNA containing 5,6-hydrated thymine moieties	5,6-Dihydroxydi-hydrothymine or 5,6-dihydrothymine + AP sites
PD-DNA glycosylase	DNA containing pyrimidine dimers	Pyrimidine dimers in DNA with hydrolyzed 5' glycosyl bonds + AP sites

ethanol-soluble oligonucleotides and mononucleotides (78). For many years the release of acid- or ethanol-soluble radioactivity from radiolabeled DNA during its incubation with cell extracts was interpreted as an indication of DNA degradation exclusively by endonucleases and/or exonucleases. The discovery of DNA glycosylases led to an appreciation that the enzymatic degradation of DNA can result by mechanisms other than the hydrolysis of phosphodiester bonds.

3-Methyladenine-DNA Glycosylase

It is appropriate to begin our discussion of DNA glycosylases in the context of the adaptive response to alkylation damage that was featured in the previous chapter as an example of DNA repair by the reversal of base damage. The *3-methyladenine-DNA glycosylases (3-mA-DNA glycosylases)* represent the second example of repair enzymes which are specific for alkylation damage in DNA. It is not obvious what selective pressures have operated for the evolution in cells (particularly in bacteria, which lack the activating enzymes necessary to convert many weakly reactive alkylating agents into strong electrophiles) of enzymes specific for alkylation damage. As discussed in chapter 3, a provocative hypothesis is that reactive methylating agents are continuously generated intracellularly from methyl donors such as *S*-adenosylmethionine, which serve functions in normal cellular metabolism (234). Additionally, the observation that both *E. coli* and yeast cells that are totally defective in the repair of O^6-methylguanine suffer elevated rates of spontaneous mutation provides further indications that products of normal metabolism can result in alkylation damage to DNA.

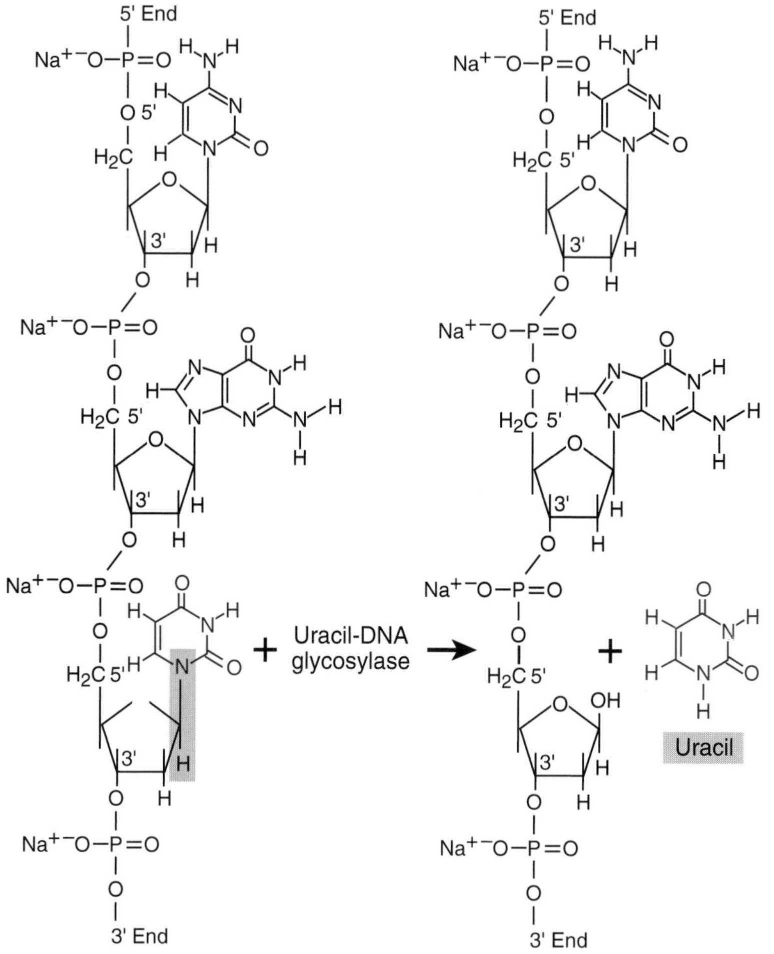

Figure 4–3 Structure of a single polynucleotide chain showing the action of a DNA glycosylase, in this case uracil-DNA glycosylase, which hydrolyzes the N-glycosylic bond linking the base to the sugar-phosphate backbone, thereby releasing free uracil. The deoxyribose moiety at the site of base loss (AP site) is shown in the closed (furanose) form but actually exists in equilibrium with the open (aldehyde) form. (*Adapted from Friedberg [87] with permission.*)

The existence of enzymes capable of excising methylated purines from DNA was inferred from early studies which showed that following the exposure of cells to alkylating agents, 3-methyladenine and 3-methylguanine were lost from DNA much more rapidly than could be accounted for from measurements of the spontaneous release of these bases (154, 155, 175) (Fig. 4–4). In fact, the initial rate of the loss of 3-methyladenine from *E. coli* DNA in vivo cannot be precisely determined, because after exposure of this organism to alkylating agents for only a very brief period at 37°C, most of the 3-methyladenine residues initially formed during the period of alkylation are already excised.

E. COLI HAS TWO 3-mA-DNA GLYCOSYLASES

A 3-mA-DNA glycosylase has been purified from extracts of wild-type *E. coli* (15, 227, 239). This enzyme is called *3-mA-DNA glycosylase I*. It is a small protein with a molecular mass of ~21 kDa, in agreement with the predicted size of the polypeptide encoded by the cloned gene for this enzyme, designated *tag*+ (for three methyl adenine-DNA glycosylase) (266) located at 79.3 min on the *E. coli* genetic map. The enzyme has a broad pH optimum between 6 and 8.5 and is stimulated by Mg^{2+}, Mn^{2+}, and Ca^{2+}. Here, then, is an example of a DNA glycosylase whose

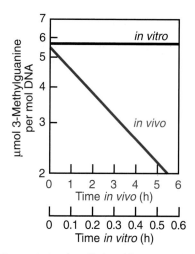

Figure 4–4 The alkylated base 3-methylguanine is lost more rapidly from the DNA of living cells exposed to alkylating agents than from alkylated DNA incubated in vitro. This suggests that loss of this base in vivo is effected by enzyme catalysis rather than by spontaneous depurination. (*Adapted from Friedberg [87] with permission.*)

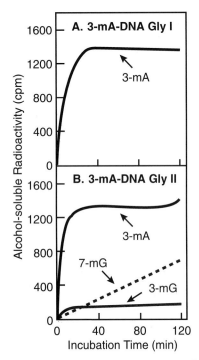

Figure 4–5 The enzyme 3-mA-DNA glycosylase I catalyzes the selective excision of free 3-methyladenine (3-mA) from DNA (top). The enzyme 3-mA-DNA glycosylase II catalyzes the additional excision of 7-methylguanine (7-mG) and 3-methylguanine (3-mG) (bottom). (*Adapted from Friedberg [87] with permission.*)

catalytic activity is influenced by divalent cations. The enzyme is sensitive to inhibition by agents which inactivate SH groups and has a strong preference for double-stranded DNA. The enzyme has a stringent substrate specificity. When alkylated DNA containing 7-methylguanine, 7-methyladenine, 3-methyladenine, or O^6-methylguanine is incubated with the purified enzyme, only free 3-methyladenine is released (Fig. 4–5) (227). The enzyme also catalyzes the excision of 3-ethyladenine and 3-methylguanine (14) from alkylated DNA. However, no other modified (or normal) bases are known to be excised by this activity. The enzyme is product inhibited by free 3-methyladenine with an apparent K_i of ~1.0 mM (227).

Mutants of *E. coli* which are defective in the *tag*$^+$ gene are deficient but not totally defective in 3-mA-DNA glycosylase activity (138). The residual enzyme activity is not the result of leakiness of the mutant gene but represents the existence of a *second form of 3-mA-DNA glycosylase*, encoded by a different gene. The latter enzyme (called *3-mA-DNA glycosylase II*) is present at similar levels in both wild-type cells and *tag* mutants and differs from the *tag*$^+$ gene product in being considerably more heat stable and insensitive to product inhibition by free 3-methyladenine (135, 274). This enzyme has also been purified (193, 274). It is a protein of ~30 kDa and has the typical general characteristics of a DNA glycosylase. However, unlike 3-mA-DNA glycosylase I, it has a broader substrate specificity. In addition to 3-methyladenine, the enzyme catalyzes the excision of 3-methylguanine, 7-methylguanine (Fig. 4–5) and 7-methyladenine from alkylated DNA. Additionally, it catalyzes the release of both N^1-carboxyethyladenine and N^7-carboxyethylguanine from DNA (274).

3-mA-DNA GLYCOSYLASE II OF *E. COLI* IS INVOLVED IN THE ADAPTIVE RESPONSE TO ALKYLATION DAMAGE

During the discussion of the adaptive response to alkylation damage in *E. coli* in the previous chapter, it was pointed out that adaptation to survival and adaptation to mutation are governed by distinct biochemical pathways, as evidenced by the observation that adaptation to survival requires a functional *polA* gene whereas adaptation to mutagenesis does not. The *polA*$^+$ gene encodes DNA polymerase I, which is involved in repair synthesis during base excision repair in *E. coli*. Therefore, the requirement for a functional *polA* gene was an early clue that the enzymatic mechanism of adaptation to survival might involve excision repair of DNA, in contrast to the adaptation to mutation, which occurs by the direct reversal of base damage.

Adaptation to survival is associated with the *induction* of increased synthesis of 3-mA DNA glycosylase II. This enzyme accounts for only 5 to 10% of the total 3-mA-DNA glycosylase enzyme activity in extracts of unadapted cells (135). However, in adapted cells the enzyme accounts for 50 to 70% of the total activity (135) (Fig. 4–6). The identification of the induced enzyme as 3-mA-DNA glycosylase II was confirmed by experiments with a mutant defective in 3-mA-DNA glycosylase I, in which 3-mA-DNA glycosylase activity was induced by exposure to adapting levels of *N*-methyl-*N*′-nitro-*N*-nitrosoguanidine (MNNG). The induced activity had identical properties to 3-mA-DNA glycosylase II (135).

The biochemical demonstration of two 3-mA-DNA glycosylases in *E. coli*, one of which is constitutive and the other inducible by alkylation adaptation, has been confirmed by extensive genetic studies (84). A mutant of *E. coli* which is extremely sensitive to the alkylating agent methyl methanesulfonate (MMS) has been isolated. This mutant, which is unable to repair phage λ previously treated with MMS and is deficient in total 3-mA-DNA glycosylase activity (138), is defective in 3-mA-DNA glycosylase I and in both the mutational and survival adaptive responses to alkylation damage; i.e., it is phenotypically Tag$^-$Ada$^-$. Restoration of the adaptive function by transduction of the *ada*$^+$ gene rendered the strain resistant to the lethal effect of MMS, but only if the cells were grown under conditions in which protein synthesis (and hence expression of induced genes) could occur, for example, growth in complete medium (84). These observations are consistent with the presence in *E. coli* of two pathways for the repair of 3-methyladenine: a constitutive pathway mediated by the *tag*$^+$ gene (involving 3-mA-DNA glycosylase l) and an inducible pathway (involving 3-mA-DNA glycosylase II), which is a component of the adaptive response controlled by the *ada⁺* locus.

The observation that mutations in the *ada⁺* gene prevent induction of 3-mA-DNA glycosylase II is consistent with its regulatory function in the adaptive response (see chapter 3) and suggests that a different structural gene codes for the glycosylase. As already indicated, the structural gene for 3-mA-DNA glycosylase I is *tag⁺*. A gene called *alkA⁺* (312, 313) is the structural gene for 3-mA-DNA glycosylase II. This gene maps at approximately 43 min on the *E. coli* genetic map. The cloned *alkA⁺* gene contains an open reading frame of 282 amino acids and putatively encodes a polypeptide of 31.4-kDa (42, 194). Mutants defective in the *alkA⁺* gene are very sensitive to MMS, despite the presence of normal levels of 3-mA-DNA glycosylase I. This suggests that 3-methyladenine and 3-methylguanine are not the only lethal lesions in DNA caused by this alkylating agent and that one or more of the other alkylation products that are recognized as substrates by the second DNA glycosylase are lethal (135).

Which lesions are responsible for the killing of *E. coli* following exposure to alkylating agents? Other than 3-methyladenine, 3-methylguanine and 7-methylguanine are the only known alkylation products whose loss is significantly enhanced in adapted cells (135). Although 3-methylguanine is a minor alkylation product in DNA, it may have profound biological consequences if it is not removed from living cells. In DNA containing 3-methylguanine and 3-methyladenine the methyl groups occupy the minor groove of DNA, which, unlike the major groove, is normally free of methyl groups (Fig. 4–7). 3-Methylguanine accounts for only about 1% of the total alkylation products in DNA bases after treatment with methylating agents, whereas 3-methyladenine constitutes about 15 times as much product (135). Thus, it seems reasonable to speculate that after the rapid removal of most of the 3-methyladenine by the constitutive DNA glycosylase I of *E. coli*, approximately equal (though small) amounts of 3-methyladenine and 3-methylguanine would be present in the minor groove of the genome, and that these lesions are lethal unless removed by 3-mA-DNA glycosylase II (135). Indeed, by using a synthetic DNA substrate with a high G + C content, it has been shown that 3-methylguanine is excised from duplex DNA by both 3-mA-DNA glycosylases but that the efficiency of enzyme II is considerably greater than that of enzyme I (14).

At first glance the broad substrate specificity of 3-mA-DNA glycosylase II is surprising, since, as discussed below, most other known *E. coli* DNA glycosylases are quite specific for limited types of base damage in DNA. It has been suggested that the enzyme might recognize a positively charged purine residue rather than alkylation at a particular site (166). 3-mA-DNA glycosylase II also removes O^2-methylthymine and O^2-methylcytosine but not O^4-methylthymine from alkylated DNA (178).

There is no homology between the amino acid sequences deduced from the cloned *tag⁺* and *alkA⁺* genes (238), suggesting that the precise mechanisms of action of their products are distinct. This is consistent with the observation that only 3-mA-DNA glycosylase I is product inhibited and with their different substrate specificities. Overexpression of the *tag⁺* gene almost completely suppresses the alkylation sensitivity of *alkA* mutants (132). This observation is surprising, given the narrow substrate specificity of 3-mA-DNA glycosylase I (Tag protein) and the broader substrate specificity of 3-mA-DNA glycosylase II (AlkA protein). However, this suppression can be rationalized in light of the considerations (i) that 3-methylguanine and 3-methyladenine are the primary lethal adducts following alkylation of DNA (as suggested above) and (ii) that although 3-mA-DNA glycosylase I removes 3-methylguanine with substantially lower efficiency than it removes 3-methyladenine, this limitation can be overcome by overexpression of the *tag⁺* gene. Overexpresssion of the *tag⁺* gene in wild-type cells sensitizes them to alkylation damage (132).

ALKYL-DNA GLYCOSYLASES IN OTHER ORGANISMS

3-mA-DNA glycosylase activity has been identified in other prokaryotes and in various eukaryotes, but it is not clear whether two distinct forms of this enzyme

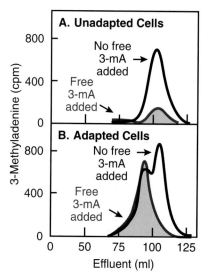

Figure 4–6 Activity of 3-mA-DNA glycosylases I and II following gel filtration of extracts of unadapted (A) and adapted (B) *E. coli* cells. Enzyme I is sensitive to product inhibition by free 3-methyladenine. Hence, it can be shown by the addition of this base that most of the activity present in unadapted cells represents 3-mA-DNA glycosylase I. In adapted cells, 3-mA-DNA glycosylases II activity (which elutes earlier from the gel filtration column) can be readily detected because it is resistant to inhibition by free 3-methyladenine. (*Adapted from Friedberg [87] with permission.*)

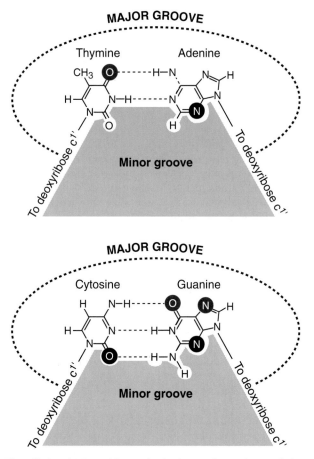

Figure 4–7   The alkylated N³ positions of adenine and guanine and the O² position of cytosine occupy the minor groove of the DNA helix. Other sites of alkylation in bases such as O⁶ in guanine, O⁴ in thymine, and N⁷ in guanine (shown in red) occupy the major groove. (*Adapted from Friedberg [87] with permission.*)

exist in all of these. By analogy with *E. coli*, in which one enzyme form has a much broader substrate specificity, a number of observations suggest that a type II enzyme is present in other cells. A DNA glycosylase from *Micrococcus luteus*, which catalyzes the release of free 7-methylguanine from DNA, is apparently distinct from a 3-mA-DNA glycosylase I activity in that organism (153). Like 3-mA DNA glycosylase II of *E. coli*, the activity from *M. luteus* that excises 7-methylguanine is quite heat stable. Five chromatographically distinct forms of DNA glycosylase activity against alkylated DNA have been identified in extracts of adapted *M. luteus* cells (224), but these have not yet been characterized as distinct enzymes.

The question whether there is more than one 3-mA-DNA glycosylase activity in mammalian cells with the same range of substrate specificities as the two *E. coli* enzymes was controversial for some time (27, 32, 99). There are now clear indications of a 3-mA-DNA glycosylase activity in mammalian cells, with the broad range of substrate specificity characteristic of the *E. coli* 3-mA-DNA glycosylase II activity (see below), but there are no compelling indications that expression of this enzyme is transcriptionally regulated. A 3-mA-DNA glycosylase activity is "induced" in rat liver during regeneration after partial hepatectomy (101); however, there is no evidence that this enzyme is directly comparable to 3-mA DNA glycosylase II of *E. coli*, and the increased enzyme activity may simply reflect the increased proliferation of cells associated with liver regeneration rather than a regulated transcriptional response.

A cDNA that encodes 3-mA-DNA glycosylase activity in rats was isolated by phenotypic complementation of an *E. coli tag alkA* mutant (202). The gene encodes a presumed polypeptide of 27.9 kDa. Human and mouse cDNAs that encode a similar activity have also been isolated (33, 83, 241), and the translated amino acid sequences of these three genes show extensive amino acid identity (Fig. 4–8), and limited similarities to the *E. coli* AlkA and yeast Mag (see below) 3-mA-DNA glycosylase proteins (33, 241). The human gene (variously designated as *AAG* [for 3-alkyladenine DNA-glycosylase] or *MPG* [for *N*-methylpurine-DNA glycosylase]) maps to human chromosome 16 (241) and comprises five exons (282).

The full-length human *AAG/MPG* cDNA can encode a polypeptide of 293 amino acids, and when expressed in *E. coli* it yields a polypeptide with an apparent molecular mass of 32 kDa (201). A different cDNA has been identified (282), which results from variant splicing of the primary transcript between the first and second exons and yields an expected polypeptide of 298 amino acids (201). The amino acid sequence predicted from the full-length human cDNA contains a 12-amino-acid repeat near the N terminus (Fig. 4–8) which (surprisingly) is not present in the predicted rat or mouse polypeptides (201). The human enzyme has been extensively purified following expression of the cloned full-length cDNA in *E. coli* (201). The purified protein catalyzes the excision of 3-methyladenine, 7-methylguanine, and 3-methylguanine from alkylated DNA. However, the kinetic constants for these activities suggest that the enzyme is more specific for 3-methyladenine than for 7-methylguanine (201). It is estimated that there are 1,000 to 2,000 molecules of the enzyme per cell.

A partially purified extract of human placenta was shown to catalyze the release of free $1,N^6$-ethenoadenine from DNA (233, 235). This modified base is present in the DNA of cells from animals exposed to urethane carbamate, vinyl carbamate, or vinyl halides, and its enzyme-catalyzed excision initially suggested the presence of a new DNA glycosylase which specifically and uniquely recognizes this lesion (233, 235). However, it was subsequently demonstrated that the putative $1,N^6$-ethenoadenine-DNA glycosylase is also able to excise 3-methyladenine from alkylated DNA and that *E. coli* DNA glycosylase II is able to excise $1,N^6$-ethenoadenine. Hence, this observation reflects the extended substrate versatility of 3-mA-DNA glycosylase II (233, 235). Indeed, in addition to $1,N^6$-ethenoadenine, extracts of cells in which the human 3-mA-DNA glycosylase activity was over-expressed from the cloned *AAG/MPG* cDNA were also able to excise $3,N^4$-ethenocytosine, $3,N^2$-ethenoguanine, and $1,N^2$-ethenoguanine (72). Furthermore, there are indications that partially purified human 3-mA-DNA glycosylase releases $1,N^6$-ethenoadenine with considerably faster kinetics than it releases 3-methyladenine (71). This raises interesting questions about the "natural" or primary substrate for this enzyme.

A 3-mA-DNA glycosylase activity from *Saccharomyces cerevisiae* was also identified by screening *E. coli tag alkA* mutants for recombinant plasmids which confer protection against cell killing following exposure to alkylation damage. The gene for this enzyme is designated *MAG* (for 3-mA-DNA glycosylase) (36). As indicated above, the predicted amino acid sequence of the translated *MAG* gene shares similarity with that of the *E. coli alkA*+ gene, and both genes protect cells against killing following treatment with alkylating agents (36). Like the *alkA*+ gene of *E. coli*, the

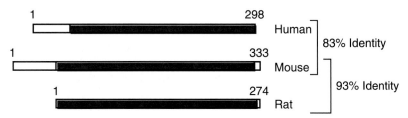

Figure 4–8   Schematic illustration of the extensive amino acid homology between the 3-mA-DNA glycosylases encoded by the human, mouse, and rat *AAG/MPG* genes. (*Adapted from Engelward et al. [83] with permission.*)

yeast *MAG* gene is transcriptionally induced when cells are exposed to alkylating agents (37). However, in contrast to the adaptive response in *E. coli*, regulation of the yeast *MAG* gene is also responsive to treatment with a diverse array of DNA-damaging agents, including UV radiation and UV mimetic chemicals, that do not produce alkylation damage in DNA (38). Additionally, the yeast *MGT1* gene, which encodes O^6-alkylguanine DNA methyltransferase activity (the yeast equivalent of Ada protein, see chapter 3), is not involved in the regulation of the *MAG* gene.

Analysis of the promoter region of *MAG* has revealed the presence of several *cis*-acting regulatory elements (309). One of these appears to be involved in positive regulation, but at least one and possibly two other sequences are apparently negative regulators and have been named upstream repressing sequences (309). One of these upstream repressing sequences matches a decamer sequence identified in multiple other regulatable yeast genes, including *MGT1* (see chapters 3 and 13). Yeast proteins of 26 and 39 kDa, which specifically bind to this upstream repressing sequence in *MAG* and *MGT1*, respectively, have been identified (309).

The availability of the cloned yeast *MGT1*, the *MAG* gene, and a gene designated *APN1*, which encodes an AP endonuclease that participates in base excision repair in *S. cerevisiae* (see below), has facilitated sensitive manipulations of the capacity of yeast cells to repair spontaneous alkylation damage to DNA (308). In the absence of the *MGT1* gene the spontaneous mutation rate increases, suggesting that O^6-alkylguanine arises spontaneously in *S. cerevisiae* cells. Similarly, in the absence of the *APN1* gene, overexpression of the *MAG1* gene increases the frequency of spontaneous mutations, suggesting that the *MAG1*-encoded DNA glycosylase acts on other types of alkylated bases that arise spontaneously. The resultant apurinic sites are mutagenic because they are not repaired (as a result of the *apn1* mutation). These studies provide persuasive evidence that eukaryotic cells harbor endogenous metabolites which can alkylate DNA (308).

OTHER ADAPTIVE RESPONSE GENES IN *E. COLI*

The adaptive response to alkylation damage in *E. coli* has thus far been discussed primarily from the perspective of two enzymes which catalyze the repair of different types of alkylation damage in DNA. These enzymes (O^6-alkylguanine-DNA alkyltransferase and 3-mA-DNA glycosylase II) are both encoded by members of the *ada* regulon. Several other genes have been identified as members of this regulon in *E. coli*. Unfortunately, the proteins encoded by these genes have not yet been isolated, and at present very little is known about the specific function(s) of these genes in cellular responses to alkylation damage. This remains a challenging area for further studies.

The *alkB⁺* Gene

The *alkB⁺* gene is located downstream of the *ada⁺* gene in a single operon. It encodes a 24-kDa protein of unknown function (139, 146, 158). Partial deletion or disruption of the *alkB⁺* gene has no discernable effect on the regulation of other members of the *ada* regulon (139, 285). An *alkA alkB* double mutant is more sensitive to killing by MMS than is either mutant alone, suggesting that the products of these genes function in different biochemical pathways (255). Interestingly, *alkB* mutants are more sensitive to killing by MMS than by MNNG (140).

The *aidB⁺* and *aidC⁺* Genes

Two genes designated *aid* (for alkylation inducible) were identified by screening for alkylation-dependent induction of β-galactosidase activity after random fusions of *E. coli* genomic DNA to the *lac* operon. A gene designated *aidB⁺* is regulated by the *ada⁺* gene (285). The *aidB⁺* gene is located at 95 min on the *E. coli* genetic map. Its function and the nature of its product are unknown. The *aidC⁺* gene maps at 92 min on the *E. coli* chromosome, very close to the *uvrA⁺* and *ssb⁺* genes, in the order *uvrA ssb aidC* (283). Induction of *aidC⁺* by alkylating agents is not blocked by mutations in either the *ada⁺* or *recA⁺* genes, indicating that it is not regulated by either the adaptive or SOS responses in *E. coli*. Additionally, in contrast to the members of the *ada* regulon, *aidC⁺* is strongly induced by ethylating and propylating agents, and induction is blocked by extensive aeration of cultures during alkylation treatment (283). Some of the principal features of the known members of the ada regulon are summarized in Table 4–3.

Possible Regulatory Links between the Adaptive and SOS Systems

The so-called SOS system in *E. coli* includes a large number of genes which are under *negative* regulatory control by a gene called *lexA*⁺. The induction of SOS genes occurs in response to a variety of DNA-damaging agents and involves the degradation of the LexA repressor protein facilitated by the activation of a protein encoded by the *recA*⁺ gene. This complex regulon is discussed in detail in chapter 10. For the purposes of the present discussion, the positively regulated adaptive response genes and the negatively regulated SOS genes are generally distinct. However, mutations in a gene called *recF*⁺, which render cells hypersensitive to UV radiation and to nalidixic acid, also confer a phenotype of hypermutability after exposure to MMS and MNNG (255). Additionally, such mutants are defective in induction of selected SOS genes, including *recA*⁺, and result in reduced induction of the *ada/alkB* operon, *alkA* and *aidB*, when cells are exposed to alkylating agents (284).

Further evidence in support of a relationship between these two regulatory systems stems from the observations that expression of the adaptive response inhibits the subsequent induction of some SOS genes and that expression of the SOS response can reduce the adaptive response. It has been suggested that the processing of DNA damage by gene products which are expressed during one of these regulatory responses may somehow interfere with the generation of a signal required for the response of the other (255). Alternatively, both regulatory responses may require interactions with the product of the *recF*⁺ gene, resulting in competition for RecF protein when both are activated simultaneously (284).

Uracil-DNA Glycosylase

The various mechanisms whereby uracil can arise in DNA were discussed in chapter 1. Of these, the deamination of cytosine is particularly important since it results in the formation of a U·G mispair in DNA. Consequently, G·C → A·T transitions are increased in *E. coli* mutants (called *ung*), which are defective in ura-DNA glycosylase (79). Therefore, it is likely that organisms, particularly those existing at high temperatures, which favor the hydrolytic deamination of cytosine, have experienced selective pressure to evolve mechanisms for the repair of such potentially mutagenic lesions.

The evolution of a specific DNA glycosylase which removes uracil from DNA may have precluded the general presence of this base in DNA and provides a provocative explanation for the presence of thymine in most genomes. As noted

Table 4–3 Inducible genes of the adaptive response to alkylation damage

Inducible gene	Chromosomal location	Mutant phenotype	Gene product	Lesions repaired	Other properties
ada⁺	47 min	Sensitive to mutagenesis	39-kDa DNA methyl-transferase	O^6-Methylguanine O^4-Methylthymine methylphospho-triester	Positive regulator of adaptive and killing response
alkB⁺	47 min	Sensitive to killing	24-kDa protein	NK[a]	
alkA⁺	45 min	Sensitive to killing	31-kDa DNA glycosylase	3-Methyladenine 3-Methylguanine O^2-Methylcytosine O^2-Methylthymine	
aidB⁺	95 min	Resistant to killing	NK		

[a] NK, not known.
Source: Adapted from Lindahl and Sedgwick (167a).

in chapter 1, some life forms, such as the *Bacillus subtilis* bacteriophages PBS1 and PBS2, do contain uracil instead of thymine in their DNA (270). Surprisingly, the *B. subtilis* host for these phages contains a ura-DNA glycosylase activity (46). How, then, is the DNA of these phages able to survive and replicate? We shall return to this question a little later. These limited exceptions notwithstanding, it is possible to genetically manipulate an organism such as *E. coli* to tolerate uracil in its genome to the extent that as much as 93 to 96% of the thymine in DNA is replaced by uracil. However, such a mutant strain is only capable of limited growth (82), indicating that even though the uracil is ''appropriately'' base paired with adenine, this base cannot fulfill the requirement for metabolic transactions of DNA such as transcription.

Ura-DNA glycosylase was the first DNA glycosylase to be discovered (46, 90, 163). The enzyme has since been shown to be present in many prokaryotes and eukaryotes, including a variety of mammalian and human cells and tissues (78, 238). Interestingly, this enzyme is not detectable in extracts of the fruit fly *Drosophila melanogaster* or in extracts of other insects which undergo pupation during development (76). This observation has led to the suggestion that DNA containing relatively large amounts of uracil might be required for normal developmental stages in pupating insects (76).

PROPERTIES OF Ura-DNA GLYCOSYLASES

Ura-DNA glycosylase has been extensively purified and characterized from many prokaryotic and eukaryotic sources. Table 4–4 summarizes the principal properties of the enzyme from some of these. All ura-DNA glycosylases are monomeric proteins with molecular masses ranging from about 19 to 35 kDa. None has a requirement for any known cofactor, and all are fully active in the presence of EDTA. All ura-DNA glycosylases utilize as substrate either double- or single-stranded DNA or deoxyribopolymers containing deoxyuridine. RNA containing uracil is not recognized as a substrate, and this enzyme cannot excise uracil from a ribopolymer annealed to a deoxyribopolymer. Neither 5-bromouracil nor 5-hydroxymethyluracil is recognized by the *B. subtilis* enzyme; however, the *E. coli* enzyme catalyzes the excision of 5-fluorouracil from DNA (296). Additionally, 5-hydroxy-2'-deoxyuridine, an oxidation product of uracil is recognized by the enzyme (116). It has been shown that 5-hydroxy-2'-deoxyuridine can be efficiently incorporated into DNA from 5-hydroxy-2'-deoxyuridine triphosphate (116). The size distribution of oligonucleotide fragments generated from polynucleotide substrates carrying uracil in known positions was determined after incubation with ura-DNA glycosylase. A comparison of this size distribution with that theoretically predicted for distributive and processive modes of enzyme action was consistent with the former mode of action (213).

The strict substrate specificity of ura-DNA glycosylase notwithstanding, *ung* mutants of *E. coli* are more resistant than wild-type strains to the lethal effects of UV light when cultures are grown in medium containing 5-bromouracil or 5-bromodeoxyunidine (311). Additionally, *ung* cells with 5-bromouracil in their DNA have a higher mutation frequency following UV irradiation. These results suggest that ura-DNA glycosylase recognizes a lesion in UV-irradiated 5-bromouracil-substituted DNA, possibly debrominated residues.

The enzyme from *E. coli* is encoded by a gene designated *ung*$^+$, which maps at 55 min on the *E. coli* circular map (78). This enzyme has a turnover number of more than 800 uracil residues released per min, and it is estimated that about 300 enzyme molecules are present per cell (167). The enzyme has a K_m of 4×10^{-8} for dUMP residues in PBS2 DNA (167); a similar value for dUMP residues in poly(dU) has been measured for the enzyme from *B. subtilis* (46). The *E. coli* enzyme is inhibited in the presence of free uracil, which acts as a noncompetitive inhibitor with a K_i of 10^{-4} M. Deoxyuridine, dUMP, thymine, 5-bromouracil, 5-aminouracil, 2-thiouracil, and orotic acid do not inhibit the enzyme (167). Sites of base loss in DNA lower the V_{max} of ura-DNA glycosylase for uracil in DNA (23, 77), suggesting that the enzyme has an affinity for such sites even though no associated AP endonuclease activity has been detected in vitro. We shall return to

Table 4–4 Properties of uracil-DNA glycosylases purified from different sources

Source	M_r (10³)	pI	K_m (µM)	ss- or ds-DNA[a]	Optimum temp (°C)	Optimum pH	Required cofactors	Inhibitors/activators
Herpes simplex virus type 1	36.3							
Herpes simplex virus type 2	32.5							
Equine herpes virus								
Varicella-zoster virus	34.4							
Epstein-Barr virus	28.6							
Human cytomegalovirus	28.3							
Mycoplasma lactucae	41.0							
	28.5	6.4	1.05	ss/ds			None	
E. coli	25.6		0.04			8.0	None	Inhibited by NaCl
B. subtilis	24.0		0.0011			7.3–7.8	None	Inhibited by heavy metals, stimulated by NaCl
M. luteus	19.4		0.07			5.0–7.0		Activated by spermine and spermidine. Inhibited by NaCl and spermidine at high concentrations
B. stearothermophilus	30.0		0.4		~44			Inhibited by NaCl
Thermothrix thiopara	26.0					7.5–9.0		Inhibited by *N*-methylmaleimide
S. pneumoniae	24.0			ss > ds			None	
Dictyostelium discoideum	55.0				30–37	6.5–8.5	None	Inhibited by 1 mM *p*-(hydroxymercury)benzoate
Artemia salina						8.5–9.0		Inhibited by low and stimulated by high concentrations of NaCl
S. cerevisiae	40.5							
	33.0							
	27.8							
Hordeum vulgare (barley)				ss > ds	26, 37	7.2		
Zea mays (maize)								
Mitochondria	18.0							
Nuclei	18.0							
Triticum vulgare (wheat)	27.0							
Rat liver								
Mitochondria	24.0	10.3	1.1	ss > ds			None	Inhibited by uracil and AP sites, stimulated by <25 mM NaCl
	29.0							
Nuclei	33.0	9.3		ss > ds	>37	~7.5	None	Inhibited by NaCl
Calf thymus	28.3		~1					Inhibited by AP sites
Human placenta	25.8	2						Stimulated by <70 mM NaCl
	33.8							
	37.0							
HeLa S3 cells								
Mitochondria	29.0	~10.3	2.2	ss > ds		7.5	None	Stimulated by NaCl, Sodium-Acetate, NH$_4$Cl, and <60 mM KCl
Nuclei	30.0		1.4	ss > ds		7.5	None	Stimulated by NaCl, sodium-acetate, NH$_4$Cl, and <60 mM KCl
Acute leukemia blast cells	30.0		0.8			7.2–8.0	Possible	Inhibited by NaCl, KCl, K$_2$SO$_4$, CaCl$_2$, MgCl$_2$, and MnCl$_2$, stimulated by spermine and spermidine

[a] ss, single-stranded; ds, double-stranded.
Source: Adapted from Slupphaug [262] with permission.

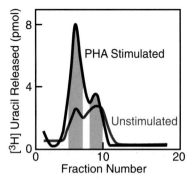

Figure 4–9 Extracts of quiescent lymphocytes show two peaks of ura-DNA glycosylase activity identified by phosphocellulose chromatography. One of these is increased in cells stimulated to divide with phytohemagglutinin (PHA) and represents increased expression of the nuclear enzyme. The other peak probably represents the mitochondrial form of the enzyme. (*Adapted from Friedberg [87] with permission.*)

this observation below when considering the biological relevance of some DNA glycosylases that do have associated AP endonuclease activity in vitro.

The *E. coli* ura-DNA glycosylase appears to function in a partially processive mode. Thus, the addition of enzyme to form I plasmid DNA containing uracil results in the formation of linear (form III) molecules before all the form I DNA is used up. However, in the presence of larger amounts of enzyme, there is a further accumulation of form III DNA after all form I substrate is used up (119). This mode of action is also characteristic of the pyrimidine dimer-specific DNA glycosylase encoded by bacteriophage T4 (see below).

The enzyme from *E. coli* or *B. subtilis* does not digest single-stranded oligonucleotides smaller than d(U)$_4$ and apparently requires a region at least 4 nucleotides long for stable binding (46, 53, 163). Release of uracil by the *E. coli* enzyme has been detected with a substrate as small as pd(UTT) (280). The *E. coli* enzyme can remove 5'-terminal uracil residues from short synthetic oligodeoxyribonucleotides provided that the 5' end is phosphorylated. However, it does not remove uracil residues from the 3' terminus of such substrates (280).

Studies on human cells in culture have demonstrated the presence of both nucleus- and mitochondria-associated activities (139). Two chromatographically distinguishable forms of the enzyme have been observed in phytohemagglutinin-stimulated human lymphocytes (Fig. 4–9) (260). In quiescent lymphocytes, low levels of both forms of the enzyme are detected. However, following stimulation of cell division by phytohemagglutinin a significant increase in the level of one of these occurs (Fig. 4–9). This observation suggests that one form, present in the nucleus, is regulated in the cell cycle and is induced during the S phase, whereas the other form, present in mitochondria, exists at a uniform level throughout the cell cycle. The two activities isolated from stimulated lymphocytes have molecular masses of ~42.5 and 40 kDa, respectively (260). Nuclear and mitochondrial forms of ura-DNA glycosylase have been extensively purified from rat liver (69, 70). Bovine nuclear ura-DNA glycosylase has also been extensively purified (81) and has a 1.7-fold preference for uracil in single-stranded as opposed to double-stranded DNA (81).

GENES FOR Ura-DNA GLYCOSYLASES

The gene that encodes ura-DNA glycosylase has been cloned from a number of organisms (238). The *ung*$^+$ gene from *E. coli* contains an ORF of 229 amino acids and could encode a polypeptide of 25.6 kDa, very close to the size of the purified protein. A yeast mutant defective in ura-DNA glycosylase activity was isolated by selecting strains which could support the propagation of yeast plasmids containing uracil in their DNA (29). This mutant (*ung1-1*) facilitated the cloning of the yeast *UNG1* gene, which maps to the left arm of chromosome XIII (210). The sequence of the C-terminal two-thirds of the predicted yeast polypeptide shows blocks of amino acid identity with the entire length of the predicted *E. coli* polypeptide (Fig. 4–10). The yeast *UNG1* gene is cell cycle regulated in exactly the same way as genes required for DNA replication in yeast. Steady-state levels of *UNG1* mRNA increase dramatically during the late G$_1$ and S phases of the cell cycle and decrease in the late S phase (122).

Consistent with the postulated role of ura-DNA glycosylase in vivo, yeast *ung1* deletion mutants showed a 5- to 100-fold increase in the frequency of spontaneous mutations at various loci (29). A detailed analysis of 69 independently derived spontaneous mutations generated in an *E. coli ung* mutant showed that all but 5 were due to G·C → A·T transitions, as expected from the failure to repair uracil which arises from cytosine deamination (122).

UNG genes have also been cloned from human cells and various mammalian cell viruses (52, 179, 207, 306). The human *UNG1* gene maps to chromosome 12 (1) and has been shown to complement the weak mutator phenotype of *E. coli ung* mutants, as well as the ability of these mutants to support the growth of phages containing uracil in their DNA (206). A comparison of extended regions of the predicted polypeptides encoded by the human, herpes simplex virus type 1 and 2, Epstein-Barr virus, varicella-zoster virus, yeast, *E. coli*, and several other genes

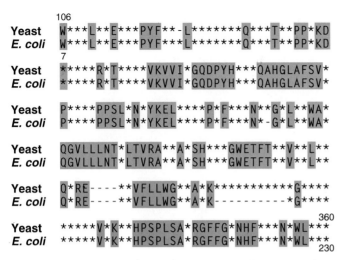

```
       106
Yeast   W***L**E***PYF**-L*******Q***T**PP*KD
E. coli W***L**E***PYF***L*******Q***T**PP*KD
        7
Yeast   *****R*T****VKVVI*GQDPYH***QAHGLAFSV*
E. coli *****R*T****VKVVI*GQDPYH***QAHGLAFSV*

Yeast   P****PPSL*N*YKEL****P*F***N**G*L**WA*
E. coli P****PPSL*N*YKEL****P*F***N*-G*L**WA*

Yeast   QGVLLLNT*LTVRA**A*SH***GWETFT**V**L**
E. coli QGVLLLNT*LTVRA**A*SH***GWETFT**V**L**

Yeast   Q*RE----**VFLLWG**A*K**********G****
E. coli Q*RE----**VFLLWG**A*K---------*G****
                                          360
Yeast   *****V*K**HPSPLSA*RGFFG*NHF***N*WL***
E. coli *****V*K**HPSPLSA*RGFFG*NHF***N*WL***
                                          230
```

Figure 4–10 Comparison of the amino acid sequences of the *E. coli* and yeast ura-DNA glycosylases. Only identical amino acids are shown. (*Adapted from Percival et al. [210] with permission.*)

shows significant conservation of amino acid sequence (Fig. 4–11). Remarkably, amino acid sequence conservation is most extensive between the human and *E. coli* genes, with about 55% amino acid identity over a 230-amino-acid region (206). Allowing for conservative amino acid substitutions, there is about 73% similarity between the translated human and *E. coli ung* genes.

Cell cycle analysis has demonstrated an eight- to twelve-fold increase in the steady-state levels of mRNA encoded by the human *UNG1* gene late in the G_1 phase. This increase in transcriptional activity precedes a two- to three-fold increase in the level of total ura-DNA glycosylase activity in cells in the succeeding S phase (263). Both RNA synthesis and protein synthesis are required for this induction. Inhibition of DNA replication by the DNA synthesis inhibitor aphidicolin does not prevent this response. Hence, DNA synthesis during the S phase is apparently not required for the induction of the *UNG* gene during the cell cycle.

The use of inhibiting antibodies specific for ura-DNA glycosylase activity has facilitated the isolation of several cDNAs which are different from that described above and from each other and which are also apparently regulated in the cell cycle (190, 286). One of these cDNAs has been shown to encode the 37-kDa subunit of human glyceraldehyde-3-phosphate dehydrogenase (183). This activity is extremely weak (see below) and is probably not biologically relevant to the metabolism of uracil in DNA. A second cDNA encodes a 36-kDa polypeptide which shows a striking amino acid sequence homology with several eukaryotic cyclins over a 150-amino-acid region referred to as the cyclin box (191). Additionally, the promoter region of this gene contains two cell cycle box regulatory elements found in yeast cyclin genes. Unlike the *UNG* gene, which maps to chromosome 12, the gene which encodes the 36-kDa polypeptide shows limited amino acid similarity to ura-DNA glycosylases from prokaryotes and viruses. It has been noted that mRNA levels for this gene increase three- to four-fold during the G_1 phase of the cell cycle in synchronized human fibroblasts (191). The ura-DNA glycosylase activity associated with the 37-kDa subunit of human glyceraldehyde-3-phosphate dehydrogenase also appears to be cell cycle regulated (174).

How many biologically relevant ura-DNA glycosylases exist in mammalian cells? The demonstration of a particular enzymatic activity in a purified protein does not necessarily imply that it operates in living cells. For example, a direct comparison of the specific activity of the ura-DNA glycosylase encoded by the *UNG* gene, which maps to chromosome 12, and that of the 37-kDa subunit of commercial glyceraldehyde-3-phosphate dehydrogenase indicates that the latter is at least 6 orders of magnitude lower (262). Furthermore, antisense mRNA tran-

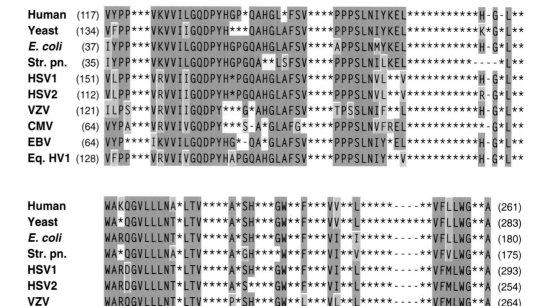

Human	(117)	VYPP***VKVVILGQDPYHGP*QAHGL*FSV****PPPSLNIYKEL**********H-G-L**	
Yeast	(134)	VFPP***VKVVIIGQDPYH***QAHGLAFSV****PPPSLNIYKEL**********K*G*L**	
E. coli	(37)	IYPP***VKVVILGQDPYHGPGQAHGLAFSV****APPSLNMYKEL**********H-G*L**	
Str. pn.	(35)	IYPP***VKVVILGQDPYHGPGQA**LSFSV****PPPSLNILKEL**********----*L**	
HSV1	(151)	VLPP***VRVVIIGQDPYH*PGQAHGLAFSV****PPPSLNVL**V**********H-G*L**	
HSV2	(112)	VLPP***VRVVIIGQDPYH*PGQAHGLAFSV****PPPSLNVL**V**********R-G*L**	
VZV	(121)	ILPS***VRVVIIGQDPY***G*AHGLAFSV****TPSSLNIF**L**********H-G*L**	
CMV	(64)	VYPA***VRVVIVGQDPY***S-A*GLAFG****PPPSLNVFREL**********-G*L**	
EBV	(64)	VYP****IKVVILGQDPYHG*-QA*GLAFSV****PPPSLNIY*EL**********H-G*L**	
Eq. HV1	(128)	VFPP***VRVVIVGQDPYHAPGQAHGLAFSV****PPPSLNIY**V**********H-G*L**	

Human	WAKQGVLLLNA*LTV****A*SH***GW**F***VV**L****----**VFLLWG**A	(261)
Yeast	WA*QGVLLLNT*LTV****A*SH***GW**F***VV**L**********VFLLWG**A	(283)
E. coli	WARQGVLLLNT*LTV****A*SH***GW**F***VI**I*****----**VFLLWG**A	(180)
Str. pn.	WA*QGVLLLNA*LTV****A*GH***W***F***VI**V*****----**VFVLWG**A	(175)
HSV1	WARDGVLLLNT*LTV****A*SH***GW**F***VI**L*****----**VFMLWG**A	(293)
HSV2	WARDGVLLLNT*LTV****A*S****GW**F***VI**L*****----**VFMLWG**A	(254)
VZV	WARQGVLLLNT*LTV****P*SH***GW**L***VL**L*****----**VFMLWG**A	(264)
CMV	WAR*GVLLLNT*FTV****P*SH***GW**L***VI**L*****----**VFMLWG**A	(207)
EBV	WA*QGVLLLNT*LTV****P*SH***GW**F***VI**L*****----**VFMLWG**A	(207)
Eq. HV1	WA*QGVLLLNT*LTV****P*SH***GW**L***VI**L*****----**VFMLWG**A	(270)

Figure 4–11 Comparison of the amino acid sequences in the most highly conserved regions of ura-DNA glycosylases from various prokaryotes and eukaryotes. Abbreviations: Str. pn., *Streptococcus pneumoniae*; HSV1, herpes simplex virus type 1; HSV2, herpes simplex virus type 2; VZV, varicella-zoster virus; CMV, human cytomegalovirus; EBV, Epstein-Barr virus; Eq. HV1, equine herpesvirus type 1. Numbers in parentheses refer to amino acid positions (*Adapted from Slupphaug [262] with permission.*)

scribed from the human *UNG* gene results in almost complete loss of ura-DNA glycosylase activity generated by in vitro translation of total mRNA from human fibroblasts (263). Hence, *UNG1* appears to be the major and probably the exclusive physiologically relevant ura-DNA glycosylase activity in mammals.

PROTEIN INHIBITORS OF Ura-DNA GLYCOSYLASE

Since bacteriophages PBS1 and PBS2 normally contain uracil rather than thymine in their DNA, it is not surprising that very early after infection of their natural host *B. subtilis*, a potent inhibitor of the host ura-DNA glycosylase is expressed (45, 90) (Fig. 4–12). This protein also inhibits the *E. coli* enzyme, as well as ura-DNA glycosylases from a variety of other sources (134). However, it does not inhibit other DNA glycosylases from *E. coli*. These results suggest that the mechanisms of action of all ura-DNA glycosylases are similar, if not identical.

The gene for this inhibitor (designated *ugi+*, for ura-DNA glycosylase inhibitor) was cloned by screening a PBS2 genomic library for recombinant plasmids which could support the growth of M13 phage containing uracil (291). The gene contains an open reading frame which can encode an acidic polypeptide with a calculated M_r of ~9.5-kDa (292). The polypeptide purified from the cloned gene expressed in *E. coli* has an aberrant electrophoretic mobility, yielding an apparent M_r of ~18 kDa (292). The inhibitor protein inactivates *E. coli* ura-DNA glycosylase by binding to the protein, yielding a complex of 36-kDa with a 1:1 stoichiometry. The inhibitor protein prevents binding of the enzyme to DNA and also dissociates ura-DNA glycosylase from preformed enzyme-DNA complexes (10). Kinetic studies have revealed a K_i of 0.14 μM, a value approximately 12-fold lower than the K_m for the glycosylase (293). Addition of as much as a 30-fold molar excess of [^{35}S]-Ugi protein to preformed [^{3}H]glycosylase-Ugi complex did not result in the detectable incorporation of ^{35}S into the complex (11). Hence, the enzyme-inhibitor

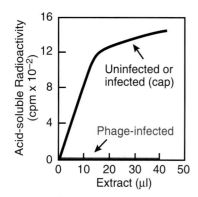

Figure 4–12 Infection of *B. subtilis* with bacteriophage PBS2 results in the loss of ura-DNA glycosylase activity. However, in the presence of chloramphenicol (cap) (which inhibits synthesis of phage proteins), the level of ura-DNA glycosylase activity is the same as that in uninfected cells. (*Adapted from Friedberg [87] with permission.*)

complex is extremely stable and the process is essentially irreversible under physiological conditions (11).

An inhibitor of the *E. coli* ura-DNA glycosylase has been reported in extracts of bacteriophage T5-infected cells (295), even though this phage contains thymine in its DNA as a principal pyrimidine. Interestingly, the T5 protein inhibits ura-DNA glycosylase from this organism only and has no effect on the enzyme from a number of other sources. The biological role of the T5 inhibitor and its distinction from the PBS2-induced protein are not known.

Hydroxymethyluracil-DNA Glycosylase

Ura-DNA glycosylase does not catalyze the excision of free hydroxymethyluracil. Indeed, no such activity has been found in extracts of prokaryotic cells. However, a low level of 5-hydroxymethyluracil-DNA glycosylase activity has been detected in extracts of several eukaryotic cells and has been partially purified (22, 120). Hydroxymethyluracil could conceivably arise in DNA containing 5-methylcytosine, by oxidation of this base to 5-hydroxymethylcytosine followed by deamination (162). The enzyme has been reported to be reduced in activity in senescent human fibroblasts (100).

A mutant mammalian cell line lacking 5-hydroxymethyluracil-DNA glycosylase activity was isolated by selection for resistance to the thymidine analog 5-hydroxymethyl-2'-deoxyuridine. The mutant is able to incorporate 5-hydroxymethyl-2'-deoxyuridine into its DNA but, lacking the glycosylase, is unable to excise it (20). Hence, the toxicity of 5-hydroxymethyl-2'-deoxyuridine in wild-type cells is apparently not the result of the substitution of this thymine analog for thymine as much as the result of extensive degradation of DNA by base excision repair initiated by 5-hydroxymethyluracil-DNA glycosylase (20).

Hypoxanthine-DNA Glycosylase

As indicated in chapter 1, hypoxanthine can arise from the deamination of adenine in DNA. In theory, this base could also be generated in DNA following the incorporation of dIMP instead of dGMP during semiconservative DNA synthesis. However, this phenomenon appears unlikely in *E. coli*, since in contrast to the potential for dUMP incorporation from the dUTP pool, this organism is devoid of detectable dIMP and does not have demonstrable deoxyinosine kinase activity (165). Nonetheless, dITP is readily incorporated into *E. coli* DNA in vitro and is associated with significant fragmentation of nascent DNA (273).

A hypoxanthine-DNA glycosylase (Hx-DNA glycosylase) has been found in extracts of *E. coli* (136), HeLa cells (192), and calf thymus (137). The activity in extracts of *E. coli* is much lower than that of ura-DNA glycosylase. It is a small protein, ~30-kDa in size. The enzyme catalyzes the release of free hypoxanthine but not xanthine, adenine, guanine, or uracil from nitrous acid-treated DNA. Free hypoxanthine is also released from the single-stranded polymer poly(dA-[³H]dI). DNA purine residues with other alterations in the C-6 position are substrates for the enzyme. In contrast to ura-DNA glycosylase, Hx-DNA glycosylase is not product inhibited. Hypoxanthine analogs such as caffeine, xanthine, and deoxyinosine also do not inhibit enzyme activity (136).

Hx-DNA glycosylase has also been extensively purified from calf thymus. Consistent with the proposed role of this enzyme in the repair of hypoxanthine which arises from the deamination of adenine rather than from the misincorporation of dIMP, the purified calf thymus enzyme is about 20 times more efficient with an oligonucleotide substrate containing I·T base pairs than one containing I·C base pairs (61).

Recent studies suggest that Hx-DNA glycosylase is not a distinct enzymatic entity, but, rather reflects a previously unrecognized property of 3-mA-DNA glycosylase II, the product of the *alkA⁺* gene discussed above (245). Specifically, it has been demonstrated that proteins that catalyze these two DNA glycosylase activities have the same chromatographic behavior and the identical molecular weight and that both are apparently induced during the adaptive response to alkylation damage. Furthermore, transformation of *E. coli* cells with a plasmid bearing the cloned *alkA*⁺ gene results in the overexpression of both 3-mA-DNA glycosylase and Hx-DNA glycosylase activities. Finally, an *alkA* mutant is defective in Hx-DNA glycosylase activity (245). 3-mA-DNA glycosylase activities from yeast, rat, and human cells also catalzye the release of dIMP residues from appropriate substrates (245).

5-Methylcytosine-DNA Glycosylase

Incubation of human HeLa cell extracts with the alternating copolymer poly(dG–5-methyl-dC):poly (dG–5-methyl-dC) radiolabeled in the methyl group of 5-methylcytosine, results in the release of free 5-methylcytosine and the formation of AP sites (277). Most of this product is recovered as thymine, suggesting extensive deamination of the free 5-methylcytosine. This putative DNA glycosylase has not yet been purified, but the existence of an enzyme that possibly specifically removes 5-methylcytosine from DNA provides a potential mechanism for perturbing the 5-methylcytosine content of DNA (277). Such regulation is possibly augmented by the degradation of the free base to thymine, thereby preventing recycling of this base into DNA through a triphosphate intermediate.

Thymine Mismatch-DNA Glycosylase

A DNA glycosylase activity, which specifically excises free thymine from G·T mispairs in DNA that arise from the deamination of 5-methylcytosine, has been detected in extracts of HeLa cells, (28, 302) (see chapter 10). This most interesting enzyme is a 55-kDa protein (197). In addition to G·T mismatches, the purified protein selectively removes thymine from C·T and T·T mispairs with an order of preference that is G·T > C·T > T·T. The protein does not catalyze the excision of T from single-stranded DNA and does not recognize T·A base pairs. It also has no detectable endonuclease activity on AP sites (197). It remains to be determined how this enzyme is able to discriminate between normally paired T and T that is mispaired with G, C, or T in duplex DNA.

DNA Glycosylases with Associated AP Lyase Activity

The DNA glycosylases that we have discussed thus far have no other known catalytic activities. Following the release of damaged bases, incision of DNA at the resulting AP sites is effected by a physically distinct 5′ AP endonuclease activity (see below). However, the incubation of some DNA glycosylases with substrate DNA is associated with the breakage of phosphodiester bonds in vitro. It has been shown that in these instances DNA incision occurs by β-elimination 3′ to the AP site. β-Elimination at AP sites can also be facilitated by certain basic proteins. However, it is not clear that these basic proteins act catalytically as true enzymes do, in the sense that true enzymatic catalysis enormously accelerates the rate of reactions (7, 173, 176). The same phenomenon has been demonstrated with oligopeptides containing aromatic amino acids, such as Lys-Trp-Lys (45, 174, 262). These observations have generated controversy concerning the definition of "true" AP endonucleases and their distinction from proteins which result in the spurious cleavage of phosphodiester bonds at sites of base loss. It is also controversial whether DNA glycosylases that are endowed with these properties effect the incision at AP sites during base excision repair in vivo.

CHEMISTRY AND STRUCTURE OF AP SITES

AP sites in DNA exist as an equilibrium mixture of the open-chain α,β unsaturated aldehyde, the α- and β-hemiacetals (which are 2-deoxy-D-*erythro*-pentofuranoses), and the open chain α,β unsaturated hydrate (66) (Fig. 4–13). In duplex DNA the hemiacetal species constitute the vast majority of these four species (173). Open-chain aldehydes constitute only ~1% of the total AP sites, however, they are the most highly reactive chemically (303). AP sites can undergo several chemical reactions which lead to cleavage of phosphodiester bonds in the absence of added proteins (66).

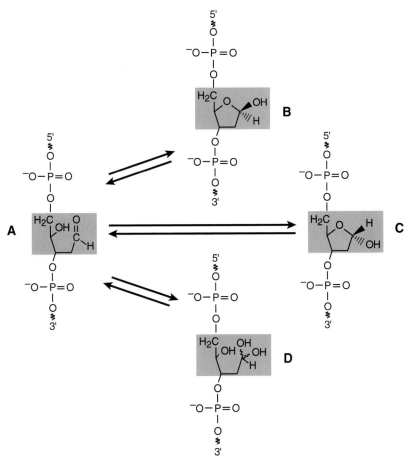

Figure 4–13 The chemical structure of an AP site; (A) Open aldehyde form; (B and C) α-and β-hemiacetals, respectively; (D) open-chain hydrate. (*Adapted from Doetsch and Cunningham [66] with permission.*)

β-Elimination

β-Elimination reactions are effected by nucleophiles and can occur by two distinct chemical pathways. In one pathway (Fig. 4–14), a proton is transferred from the CH_2 group of the deoxyribose α to the carbonyl group at C-1. In the other pathway a Schiff base is formed between an amine and the C-1 carbonyl group of the ring-open aldehyde. Both reactions are followed by β-elimination which leaves 3′ α,β unsaturated aldehyde, 4-hydroxy-2-pentenal, and 5′ phosphate termini (303). β-Elimination reactions are a relevant mechanism for cleavage of the phosphodiester backbone at AP sites.

δ-Elimination

Under some conditions the 3′ α,β unsaturated aldehyde can undergo an additional δ elimination reaction, resulting in the release of a 4-hydroxy-pent-2,4-dienal and yielding a single nucleotide gap flanked by 3′ and 5′ phosphate termini (Fig. 4–14).

Rearrangement

Under alkaline conditions the 3′ α,β unsaturated aldehyde can rearrange to form a 3′-2-oxocyclopent-1-enyl terminus (Fig. 4–14).

The structure of AP sites has been examined in several model oligonucleotides. It is likely that AP sites constitute a family of structures depending on the opposing base and on the sequence context (66). Nuclear magnetic resonance

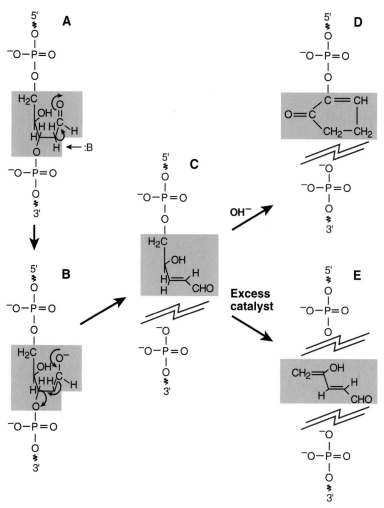

Figure 4–14 Chemical reactions that can occur at AP sites. The ring-open form of an AP site (A) can undergo a β-elimination reaction through an intermediate (B), resulting in cleavage of the 3′ phosphodiester bond (C). The 3′ terminus generated by this reaction is an α,β-unsaturated aldehyde, 4-hydroxy-2-pentenal. In the presence of alkali, this unsaturated aldehyde can rearrange to generate a 3′-2-oxocyclopent-1-enyl terminus (D). Under conditions of excess catalyst, the α,β-unsaturated aldehyde can also undergo δ-elimination, resulting in cleavage of the 5′ phosphodiester bond, yielding free 4-hydroxy-pent-2,4-dienal (E) and the formation of a single nucleotide gap in the DNA duplex flanked by 5′- and 3′-phosphoryl termini. (*Adapted from Doetsch and Cunningham [66] with permission.*)

studies indicate that if the base opposite an AP site is a purine, interstrand stacking interactions predominate and the structure is relatively extended. On the other hand, if the opposite base is a pyrimidine, stacking of the flanking bases predominates and the helix collapses. The propensity for purines to stack into the helix and hence extend the collapsed structure around AP sites may facilitate their incorporation during DNA synthesis and may explain the preferential incorporation of purines (especially adenine) relative to pyrimidines by many DNA polymerases (47).

All DNA glycosylases that can effect the hydrolysis of phosphodiester bonds at sites of base loss do so by a β-elimination mechanism. These reactions result in strand breaks 3′ to AP sites, leaving 3′-terminal unsaturated sugar derivatives, which cannot be directly utilized as primers by DNA polymerases (66). It has been suggested that these and other proteins which facilitate β-elimination reactions at AP sites should be designated as *AP lyases* rather than AP endonucleases. Aside from the niceties of the precise mechanistic differences involved in the cleavage of phosphodiester bonds by AP lyases and AP endonucleases, the important biological question is whether AP lyases play a functional role in base excision repair in living cells. We shall return to this issue presently. For the moment, let us consider the DNA glycosylases with associated AP lyase activity.

MutY-DNA Glycosylase

As indicated in chapter 1, mispaired bases that arise during semiconservative DNA synthesis constitute a prevalent form of spontaneous DNA damage. The postreplicative repair of such lesions can occur by a variety of mismatch repair mechanisms (see chapter 10). Some of these are dependent on the state of methylation of the DNA. We have already encountered the recognition and specific excision of T from G·T mispairs by a DNA glycosylase (see above). In *E. coli*, G·A mispairs can also be corrected by the action of a specific DNA glycosylase called *mutY-DNA glycosylase*, which catalyzes the removal of free adenine, independent of the state of methylation of the DNA. This 36-kDa protein is encoded by a gene called *mutY+* (or *micA+*) (5). *mutY* mutants confer hypermutability, reflecting G·C → T·A transitions.

The *mutY+* gene has been cloned and contains an open reading frame which encodes a polypeptide of 39.1-kDa (185). The predicted amino acid sequence of the polypeptide shows homology with that of endonuclease III, the product of the *nth+* gene of *E. coli* (Fig. 4–15) (185). A gene from *Salmonella typhimurium* with 91% amino acid identity to *mutY+* (Fig. 4–15) has also been isolated and is designated *mutB+* (60). The Nth protein is also involved in the repair of ring saturation and ring fragmentation derivatives of thymine (see later discussion).

MutY protein has been purified to near homogeneity (276). The purified protein removes adenine from DNA containing either A·G or A·C mispairs and has an associated 3′ AP lyase activity (276). The glycosylase activity against A·C mispairs is ~20-fold weaker than that against A·G mispairs. Both the DNA gly-

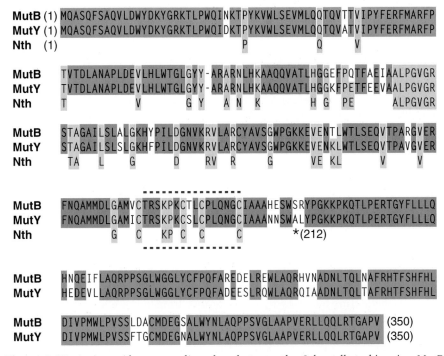

Figure 4–15 Amino acid sequence homology between the *Salmonella typhimurium* MutB and the *E. coli* MutY and Nth (endonuclease III) DNA glycosylase proteins. The Nth sequence terminates at amino acid 212. The region between the dotted lines shows the location of four conserved cysteine residues which may be involved in binding a [4Fe-4S]$^{2+}$ cluster. (*Adapted from Michaels et al. [185] with permission.*)

cosylase and AP lyase activities can be recovered by the renaturation of MutY protein recovered from sodium dodecyl sulfate-polyacrylamide gels. The renaturation process requires the presence of iron and sulfide, suggesting that MutY protein is an iron-sulfur protein, as has been shown for *E. coli* endonuclease III (see below).

Formamidopyrimidine-DNA Glycosylase

Treatment of DNA with dimethyl sulfate followed by prolonged incubation at pH 11.4 results in the formation of 7-methylguanine with opened imidazole rings, generating 2,6-diamino-4-hydroxy-5-*N*-methylformamidopyrimidine (39, 40) (see chapter 1). Similar substituted formamidopyrimidines can occur in DNA as a consequence of the exposure of cells to ionizing radiation under both anoxic and oxic conditions (98, 117, 278). A DNA glycosylase which catalyzes the excision of 2,6-diamino-4-hydroxy-5-*N*-methylformamidopyrimidine as well as other imidazole ring-opened forms from DNA has been identified in extracts of *E. coli* and mammalian cells (Fig. 4–16). This enzyme was originally designated *formamidopyrimidine-DNA glycosylase (FaPy-DNA glycosylase)* (16, 39).

Subsequent studies showed that the properties of FaPy-DNA glycosylase and an enzyme called 8-hydroxyguanine endonuclease (41), which recognizes 8-hydroxyguanine residues in DNA, are identical. Indeed, purified FaPy-DNA glycosylase efficiently recognizes the oxidized purine as a substrate (271). The 8-hydroxyguanine lesion is ubiquitous and can arise from both endogenously and exogenously induced oxy-radical damage to DNA (271). Nonetheless, it has been shown that oxidation damage to DNA and exposure to γ radiation results in the formation of substantial amounts of both 2,6-diamino-4-hydroxy-5-*N*-methylformamidopyrimidine (a lethal lesion) and 8-hydroxyguanine (a mutagenic lesion) (3). Hence, both of these lesions may reasonably be identified as "natural" substrates for the enzyme. Since the protein was originally named formamidopyrimidine-DNA glycosylase, we shall retain this nomenclature here.

The enzyme from *E. coli* has been extensively purified following overexpression of a cloned gene called *fpg*⁺, which maps at 81.7 min on the *E. coli* chromosome (18, 44). Like other known DNA glycosylases, the enzyme from *E. coli* is a relatively small protein, with a molecular mass of ~31 kDa. The enzyme has a strong and possibly absolute specificity for duplex DNA as a substrate.

The cloned *fpg*⁺ gene contains an open reading frame of 269 amino acids and encodes a polypeptide of 30.2 kDa. The deduced amino acid sequence of the *fpg*⁺ gene shows no significant similarities to those of other DNA glycosylase genes, except for two limited regions of similarity with the phage T4 pyrimidine dimer-DNA glycosylase (PD-DNA glycosylase) (see below) (18). The predicted amino acid sequence reveals the presence of a putative zinc finger domain, and the wild-type protein contains one atom of Zn per protein molecule (272). Amino acid substitutions in this domain result in a loss of both DNA glycosylase and DNA-binding activities, suggesting that the zinc finger is required for normal function (272).

There is no indication that the *fpg*⁺ gene is regulated (17). Mutational inactivation of the gene does not render *E. coli* cells sensitive to ionizing radiation or alkylating agents (17). However, mutants that are defective in both the *fpg*⁺ and *uvrA*⁺ genes are highly sensitive to treatment with methylene blue plus visible light, which is known to produce singlet oxygen damage to guanine residues in DNA (51).

Purified FaPy-DNA glycosylase effects the incision of DNA at apurinic sites (203), leaving both 3'- and 5'-phosphoryl groups. Detailed examination of the incision mechanism indicates that this results from a β,δ-elimination reaction which leaves a single-nucleotide gap in the DNA (9). Double-stranded DNA containing chemically reduced AP sites (which are not subject to β-elimination) are not degraded by the FaPy-DNA glycosylase. However, the protein binds very stably to such sites (31).

The *fpg*⁺ gene is identical to a gene designated *mutM*⁺, which was isolated as a mutator gene. Inactivation of *mutM*⁺ (*fpg*⁺) results in an increased frequency of G → T transversion mutations (30). This mutator effect is consistent with the

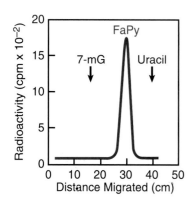

Figure 4–16 Release of free FaPy from radiolabeled DNA by a specific DNA glycosylase (FaPy-DNA glycosylase) in *E. coli* can be detected by chromatography of ethanol-soluble material after incubation of the enzyme with suitably prepared substrate DNA. The chromatographic positions of authentic free 7-methylguanine and of uracil are shown. (*Adapted from Friedberg [87] with permission.*)

demonstration that 8-oxyguanine lesions in viral DNA lead primarily to G → T transversions (304) and that various DNA polymerases preferentially insert A opposite such lesions in vitro, leading to A·OG (8-oxyguanine) mispairs (256). Hence, the mutator genes *mutY* and *mutM* (*fpg*), which lead to G·C → T·A transversion mutations, both encode DNA glycosylases. However, whereas the former gene product directly excises A from G·A mispairs, the latter removes DNA damage that has the potential for generating G·C → T·A transversion mutations.

The observation that both *mutM* (*fpg*) and *mutY* mutant strains manifest an increased frequency of G·C → T·A transversions led to the notion that they participate in a common repair pathway. Indeed, it has been shown that MutY protein is able to catalyze the excision of A from A·OG mispairs, a substrate that could arise from the error-prone replicative bypass of oxidized G residues (184). Furthermore, MutY protein remains bound to DNA after removal of A residues in A·OG mispairs, thereby protecting 8-OG sites from attack by the FaPy-DNA glycosylase and averting double-strand breaks (184). The biochemical indications that the MutM (FaPy-DNA glycosylase) and MutY proteins participate in a common biochemical pathway are supported by genetic evidence. Overexpression of the *mutM*[+] (*fpg*[+]) gene completely corrects the phenotype of a *mutY* mutant (184). Additionally, an extragenic suppressor of *mutY* turned out to be an allele of *mutM* (*fpg*). Finally, a *mutY mutM* double mutant has a G·C → T·A transversion rate that is 20-fold greater than the sum of the rates for each single mutant strain.

Collectively, these observations have led to the following scenario for the repair of OG damage in DNA (Fig. 4–17). If 8-OG lesions are removed by the FaPy-DNA glycosylase prior to DNA replication at OG·C sites, repair is effected by the conventional base excision pathway operating on the damaged strand. If the 8-OG lesion is not removed prior to replication, accurate translesion synthesis by a replicative DNA polymerase(s) affords another opportunity for FaPy-DNA glycosylase-catalyzed repair. Translesion synthesis by replicative polymerases is highly error prone, however, and can lead to the formation of OG·A mispairs. Excision of A from these mispairs by the MutY-DNA glycosylase would initiate base excision repair on the undamaged strand, during which the excised adenine nucleotide can be replaced by accurate repair synthesis, resulting in the reformation of a OG·C base pair that is once again available for repair by the FaPy-DNA glycosylase. Following the removal of A by MutY protein, attack at the 8-OG site in the damaged strand by FaPy-DNA glycosylase (MutM protein) is prevented by binding of MutY to the site. It remains to be determined how accurate base excision repair of the undamaged strand proceeds in the presence of bound MutY protein.

THE *mutT*[+] GENE ALSO PROTECTS CELLS FROM THE MUTAGENIC EFFECTS OF 8-HYDROXYGUANINE DAMAGE

Regardless of whether the details of the proposed interaction between the *mutY* and *mutM* (*fpg*) gene products described above are correct, there is compelling evidence that both genes protect *E. coli* cells from the potentially mutagenic effects of 8-hydroxyguanine. Hence, it has been suggested that *mutY*[+] and *mutM*[+] (*fpg*[+]) participate in an error avoidance system called the "GO (8-oxyguanine) system" (184). A third participant in the GO system is a gene called *mutT*[+], which is not a DNA glycosylase but which contributes to error avoidance by influencing the frequency of A·T → C·G transversions through another mechanism (Fig. 4–17).

The *mutT*[+] gene maps to about 2.5 min on the *E. coli* chromosome. Inactivation of the gene causes a very strong mutator phenotype and can result in a 10,000-fold increase in the level of A·T → C·G transversion mutations. The protein encoded by *mutT*[+] is small (M_r = 15,000). The protein is a weak GTPase triphosphatase but is about 1,000-fold more active on 8-oxo-7,8-dihydro-2'-dGTP (8-oxo-dGTP), which can result from the spontaneous oxidation of dGTP (172). The evidence suggests that the incorporation of the corresponding mononucleotide leads to the mutator phenotype of *mutT* mutants (184). Incorporation of this monophosphate opposite template A during DNA replication would lead to OG·A mispairs and ultimately to A·T → G·C transversions (Fig. 4–17). An ~18-kDa protein with properties very similar to those of the *E. coli* MutT protein has been partially

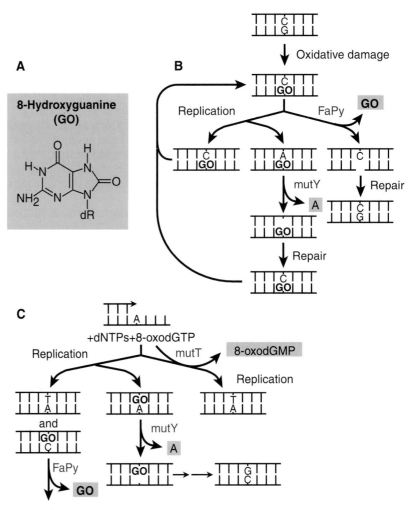

Figure 4–17 Schematic representation of enzymatic systems for the protection of cells against the presence of 7,8-dihydro-8-oxoguanine (8-hydroxyguanine) represented here as GO. (A) GO is shown as the predominant tautomeric form. (B) This lesion can be excised from DNA by FaPy-DNA glycosylase (MutM protein), and subsequent completion of base excision repair can restore the original G·C base pair (right). If the GO lesion is not removed prior to DNA replication, DNA synthesis can retain the "normal" GO·C base pair (left), with subsequent opportunity for repair as before. Alternatively, DNA replication may result in the formation of an GO·A mispair (center). The misincorporated A can be removed by the MutY-DNA glycosylase and replaced by C, yielding further opportunity for repair of the GO lesion by the FaPy-DNA glycosylase (MutM protein). (C) Oxidative damage can additionally lead to the formation of 8-oxo-dGTP. This triphosphate is a substrate for MutT protein, which can remove it from the triphosphate pool (top right). If replication occurred with some 8-oxo-dGTP in the triphosphate pool, replication would be largely accurate because T is preferentially inserted opposite A (top left). However, in the event that 8-oxo-dGMP is incorporated into DNA during replication, it may be either correctly base paired opposite C (bottom left) or mispaired opposite A (center). In the latter event, excision of the mispaired A by the MutY-DNA glycosylase can result in an A·T → G·C transition mutation (bottom right). dNTP, deoxynucleoside triphosphate. (*Adapted from Michaels and Miller [184].*)

partially purified from extracts of human cells (187). Purification of an 8-oxo-dGTPase from human cells facilitated the cloning of a cDNA which encodes this protein (237). The cDNA can encode a polypeptide of 156 amino acids with limited homology to the *E. coli* MutT protein. Introduction of the human gene into *E. coli mutT* cells results in the expression of substantial levels of 8-oxo-dGTPase activity and reduces the frequency of spontaneous mutations in such cells (237).

FaPy-DNA GLYCOSYLASE ACTIVITY IN MAMMALIAN CELLS

During the discussion of DNA glycosylases which can remove simple alkylations from DNA, we alluded to several such activities in mammalian cells. One such enzyme isolated from mouse cells has been shown to also catalyze the removal of 8-hydroxyguanine from DNA (12). Indeed, the mouse so-called *N*-methylpurine-DNA glycosylase can rescue the mutator phenotype of *E. coli* cells lacking FaPy-DNA glycosylase (MutM protein) from 8-hydroxyguanine damage (12). A similar activity has been shown for human *N*-methylpurine-DNA glycosylases, and a human DNA glycosylase has been isolated from HeLa cells primarily on the basis of its ability to excise 8-hydroxyguanine from DNA (13). Whether this enzyme will turn out to have a broader substrate specificity that includes N-methylpurines, remains to be determined.

Thymine Glycol-DNA Glycosylase

In the late 1970's an enzyme activity from *E. coli* that endonucleolytically degrades UV-irradiated DNA was purified and designated as *endonuclease III* (219). Subsequent studies showed that the purified enzyme is in fact a DNA glycosylase/AP lyase which attacks pyrimidine adducts other than pyrimidine dimers and (6-4) photoproducts, generated by the exposure of DNA to UV radiation or ionizing radiation. This DNA glycosylase activity was independently discovered in a number of laboratories, and the literature carries numerous designations for the enzyme, including *thymine glycol-DNA glycosylase, urea-DNA glycosylase, X-ray (or λ ray) endonuclease, redoxyendonuclease*, and simply *UV endonuclease* (66, 87). We will stick with the designation *thymine glycol-DNA glycosylase (TG-DNA glycosylase)*.

It is now known that TG-DNA glycosylase recognizes pyrimidine residues damaged by ring saturation, ring fragmentation, or ring contraction (Fig. 4–18) (21, 26). These include thymine glycol, 5,6-dihydrothymine, urea, β-ureidoisobutyric acid, 5-hydroxy-6-hydrothymine, uracil glycol, 5,6-dihydrouracil, and 5-hydroxy-6-hydrouracil. Additionally, it has been demonstrated that 5-hydroxy-2′-deoxycytidine and 5-hydroxy-2′-deoxyuridine in DNA are recognized as substrates by this enzyme (116). The structural features common to these perturbations of pyrimidines are not known precisely. They may include saturation of the 5,6 double bond or loss of the planar structure in pyrimidine derivatives lacking this bond (26). It has also been reported that TG-DNA glycosylase attacks guanine residues in DNA damaged by oxidizing agents (105).

TG-DNA glycosylase is encoded by the *nth*⁺ (for endonuclease three) gene of *E. coli* and maps at ~36 min on the *E. coli* chromosome. Overexpression of the cloned (4, 50) *nth*⁺ gene greatly facilitated purification of the protein to physical

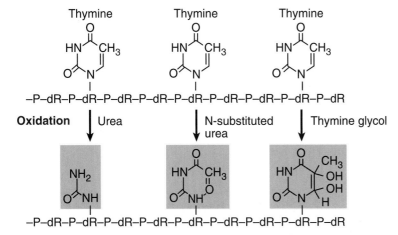

Figure 4–18 Schematic representation of a single polynucleotide chain showing various substrates for the TG-DNA glycosylase, all of which are ring-saturated or ring fragmentation products of thymine. (*Adapted from Friedberg [87] with permission.*)

homogeneity (4). The amino acid sequence of the cloned gene predicts a poly-peptide of 23.4 kDa. The purified protein absorbs light at 410 nm. Further physical characterization showed that TG-DNA glycosylase is in fact an iron-sulfur protein containing a [4Fe-4S]$^{2+}$ cluster (48).

The crystal structure of the protein has been solved to 2-Å (0.2-nm) resolution (Color Plate 3) (152). The protein is elongated and bilobal with a deep cleft separating two similarly sized "lobes," one of which includes a six-helix domain and the other a four-helix domain as well as the iron-sulfur cluster (152). The [4Fe-4S]$^{2+}$ cluster is believed to have a structural role rather than a role that determines binding of the protein to DNA, as is the case with other Zn-S enzymes (97). The surface of the long axis of the enzyme has charge characteristics suitable for binding a 46-Å (4.6-nm) length of duplex B-DNA. A comparison between the free TG-DNA glycosylase structure and that soaked with the enzyme inhibitor thymine glycol (151) localized the thymine glycol-binding site to a 7-amino acid β-hairpin formed by amino acid residues 113 to 119 (152). MutY, an *E. coli* protein homologous to TG-DNA glycosylase (see above), shows complete conservation of the hairpin, suggesting an important functional role for this region. TG-DNA glycosylase has both N-glycosylase and AP lyase activities. However, unlike the phage T4 PD-DNA glycosylase, in TG-DNA glycosylase these activities appear to function in a concerted manner (151), since no unnicked AP sites are found and inhibitor studies also suggest a sequential concerted reaction.

Kinetic analyses (151) indicate that the first step in the TG-DNA glycosylase-catalyzed reaction is the protonation of the ring oxygen of the deoxyribose residue followed by the formation of a Schiff base between the deoxyribose and the base. This is followed by a transminimization reaction in which a Schiff base is formed between the enzyme and the C-1 aldehyde of the sugar, resulting in the release of the damaged base and β-elimination of the 3′ phosphate. In support of this reaction scheme, the sugar product produced by TG-DNA glycosylase is identical to that produced by alkali-mediated β-elimination (7) and has been identified as 4-hydroxy-2-pentenal (144, 177). This proposed mechanism of action has been confirmed by showing that TG-DNA glycosylase catalyzes strand cleavage by *syn*-β-elimination through abstraction of the 2′-pro-S proton to form a 3′-α,β-unsaturated aldehyde (177). Interestingly, the crystal structure of TG-DNA glycosylase identifies two charged amino acid side chains (Glu-112 and Lys-120) at either end of the β-hairpin which binds thymine glycol. These could serve in the proposed catalytic role. Glu-112 is a likely nucleophile for the DNA glycosylase reaction, while Lys-120 is a likely candidate for the formation of the Schiff base in the AP lyase reaction. This reaction mechanism, which is similar to that proposed for the T4 PD-DNA glycosylase (see below), awaits direct confirmation by site-directed mutagenesis.

Mutants defective in the *nth* gene are not abnormally sensitive to killing by hydrogen peroxide or γ radiation (50). However, such strains are mutators. Mutations presumably result from mispairings during DNA replication at sites of spontaneous base damage that cannot be excised in *nth* mutants. Surprisingly, however, the mutator effect is not enhanced after exposure to agents which produce oxidative base damage (299). As is the case with the FaPy-DNA glycosylase, incubation of the appropriate substrate DNA with TG-DNA glycosylase is associated with DNA incision. This incision occurs 3′ to sites of base loss by a β-elimination mechanism (7).

Enzymes with catalytic properties very similar to TG-DNA glycosylase have been identified in a variety of other organisms, including *M. luteus*, *Drosophila*, and bovine and human cells (24, 25, 67, 118, 130, 131, 156). Three chromatographically distinct TG-DNA glycosylase activities have been detected in mitochondrial extracts from mammalian cells (275). Similar activities have been observed in nuclear fractions (145, 275). However, the relative levels of these activities are different from those observed with the mitochondrial activities.

Endonucleases VIII and IX

As was the case with the TG-DNA glycosylase, DNA glycosylases with associated AP lyase activity are not infrequently discovered by their latter function and hence are first named as endonucleases rather than DNA glycosylases. The availability of *nth* mutants defective in TG-DNA glycosylase activity facilitated the identification of two apparently distinct enzymes with very similar properties, designated *endonucleases VIII and IX* (290). Like the TG-DNA glycosylase, the DNA glycosylase function of endonuclease VIII releases modified T and C residues. Additionally, the enzyme attacks DNA containing β-ureidoisobutyric acid, an alkali cleavage product of dihydrothymine and a radiolysis product which is a poor substrate for TG-DNA glycosylase. Endonuclease VIII is present in extracts of *E. coli* at 5 to 10% of the level of TG-DNA glycosylase (289, 290).

A more-detailed examination of the properties of the purified enzyme (28 to 30 kDa in size) indicates many similarities to those of endonuclease III (Nth protein) (182). Analysis of the reaction products of endonuclease VIII with damaged DNA has shown the presence of free thymine glycol and dihydrothymine, indicating that the enzyme is a true DNA glycosylase (182). The additional observation that DNA containing AP sites is degraded by the enzyme but DNA containing reduced AP sites is not, further suggests that the DNA glycosylase has an associated AP lyase function. Consistent with this notion, the products of endonuclease VIII-mediated degradation of DNA containing AP sites are poor substrates for DNA synthesis by DNA polymerase I of *E. coli*, suggesting that phosphodiester bond cleavage occurs 3' to the AP sites (182).

Endonuclease IX also recognizes urea residues and β-ureidoisobutyric acid in DNA, but DNA containing thymine glycol or dihydrothymine is not a substrate for this enzyme. The multiplicity of DNA glycosylases that recognize and attack sites of oxidative damage in DNA attests to the importance of this form of base damage (see below).

Pyrimidine Dimer-DNA Glycosylase

Enzyme activities that catalyze the selective incision of DNA at sites of pyrimidine dimers were detected in extracts of *M. luteus* (133, 257, 267) and phage T4-infected *E. coli* (93, 315) before the discovery of any DNA glycosylases. For a number of years it was believed that these enzymes represented prototypic examples of damage-specific DNA endonucleases of the class that are involved in nucleotide excision repair, since early experimental evidence was consistent with the idea that these enzymes directly catalyzed the cleavage of phosphodiester bonds 5' to sites of dimers in DNA (186, 225, 226, 316).

In chapter 1 we discussed the use of DNA-sequencing gels for studying the distribution of pyrimidine dimers in a small fragment of DNA of known nucleotide sequence. In these studies a preparation of the *M. luteus* "endonuclease" was used as a specific probe to detect pyrimidine dimers in DNA (110, 115). On the basis of the assumption that the *M. luteus* enzyme hydrolyzes phosphodiester bonds directly, i.e., without the formation of an AP intermediate, a particular size distribution of DNA fragments after gel electrophoresis was expected. Instead, the fragments electrophoresed as if they were approximately 1 nucleotide larger. At the time of these experiments, DNA glycosylases were a well-described enzymological entity. Hence, it was suggested that the two-step DNA glycosylase-AP endonuclease/lyase reaction shown in Fig. 4–19 could explain the unexpected electrophoretic observation (110, 115). This hypothesis turned out to be correct, and this mode of action satisfactorily explains the mechanism of attack of UV-irradiated DNA at pyrimidine dimers by both the *M. luteus* and the phage T4 "endonucleases" (110, 115, 215, 250, 294). Thus, both enzymes have been renamed as *pyrimidine dimer-DNA glycosylases (PD-DNA glycosylases)*. Both the *M. luteus* and the T4 enzymes have associated AP lyase activity, which explains their ability to incise DNA.

The phage T4 and *M. luteus* PD-DNA glycosylases are the only two examples of this enzyme known, although there is evidence of a similar if not identical enzyme in yeast cells (see below). These enzymes catalyze the hydrolysis of N-glycosyl bonds in DNA, thereby creating AP sites as products, just as all other DNA glycosylases do. However, they effect the hydrolysis of only one of the two glycosyl bonds in pyrimidine dimers (108, 217) (Fig. 4–19). Hence, unlike other DNA glycosylases, they do not excise free bases, because the "liberated" bases are still

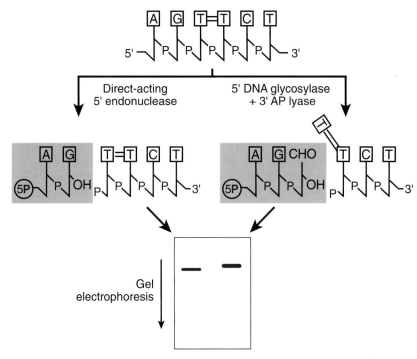

Figure 4–19 Schematic illustration of the way in which the *M. luteus* PD-DNA glycosylase activity was discovered. For simplicity, only one DNA strand is shown. If DNA radiolabeled at its 5' end (color) and containing a pyrimidine dimer is cleaved 5' to the dimer by a conventional (direct-acting) endonuclease (left), DNA fragments of a particular size distribution are expected in the sequencing gel. However, the fragments observed following incubation with the *M. luteus* enzyme migrate in sequencing gels as if they are approximately one nucleotide larger, suggesting the cleavage mechanism shown on the right. (*Adapted from Friedberg [87] with permission.*)

covalently bound to the DNA through the cyclobutane ring which characterizes the structure of the dimers. Excision of dimers that is initiated by PD-DNA glycosylases therefore occurs during postincisional degradation of the DNA, and the dimerized pyrimidines are eventually excised as part of a nucleotide structure, not as free bases. Since the mechanism of DNA incision is typical of base excision repair, however, this particular case is still considered to be an example of this mode of excision repair.

Bacteriophage T4 PD-DNA Glycosylase

Bacteriophage T4 is the only bacterial virus known to encode an enzyme specifically for the excision of pyrimidine dimers in DNA. All other phages rely on host enzymes for the repair of their genome, a phenomenon called *host cell reactivation*. The discovery of the *denV+* gene of phage T4 goes back nearly 50 years, and its history represents an interesting development in the field of DNA repair.

In 1947, Salvador Luria made the observation that phage T4 is about twice as resistant to UV light at 254 nm as phages T2 and T6 are (171) (Fig. 4–20). At that time, he speculated that this might be due to an absence of one or more genetic loci in T4, the presence of which in phages T2 and T6 was associated with increased UV radiation sensitivity. Genetic crosses between T2 and T4 phages revealed that the UV radiation sensitivity phenotype behaved genetically as a unit character; i.e., both the T2-like and the T4-like offspring fell into two distinct classes, whose levels of UV radiation sensitivity corresponded to those of the parental types. This locus was mapped and designated *u* (for ultraviolet). In view of the original interpretation of Luria, the T2 state (i.e., the UV radiation-sensitive state) was considered the wild-type or *u+* allele and the T4 state was considered the mutant or *u* allele (268).

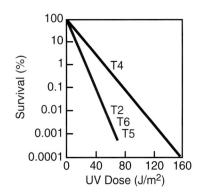

Figure 4–20 The relative sensitivity of bacteriophages T4, T2, T6, and T5 to UV radiation. Phage T4 is significantly more resistant than the other phages. (*Adapted from Friedberg [87] with permission.*)

The survival of UV-irradiated u^+ phages in the presence of the u state was investigated by coinfecting *E. coli* with lightly UV-irradiated T2 phages (u^+) and heavily UV-irradiated T4 phages (u) (113). At the dose used for the heavy irradiation, the T4 phage sustained about 70 lethal hits per phage and its survival was negligible. However, the target size of the u gene itself was sufficiently small to ensure that many of the infected cells contained at least one undamaged copy. These experiments revealed a significant increase in the survival of the lightly irradiated phage T2, approximating the survival curve for T4 (Fig. 4–21). This effect (termed u-gene reactivation) was not observed during infection with UV-inactivated phage T2, nor was the survival of lightly irradiated phage T4 significantly affected by infection with inactivated phage T4. Single plaques originating from the surviving u-gene-reactivated T2 were isolated, and the phages were UV irradiated again. The phage survival was now typical of the u^+ (T2) state, thereby precluding the possibility of marker rescue during the mixed infection as an explanation of u-gene reactivation. Therefore, it was concluded that the u gene (T4 allelic state) actually encodes a function that actively reduces the lethality of UV-irradiated T2 phage. Since this interpretation was clearly inconsistent with the original allelic designation of u, the name of the gene was changed to v, with the allelic state v^+ being that of T4 and v being the mutant or T2 state (113).

In the late 1960s, it was shown that the v gene is required for the excision of thymine-containing pyrimidine dimers from UV-irradiated T4 DNA in vivo (254). Subsequently, a dependency on the gene for thymine dimer excision was demonstrated to occur in a cell-free system (269), and it was also shown that the v^+ gene codes for an enzyme activity that incises UV-irradiated DNA but not unirradiated DNA in vitro (93, 253). This activity was variously referred to as T4 "UV endonuclease" (94) or endonuclease V of T4 (315). Studies with temperature-sensitive mutants defective in the v gene confirmed that it is the structural gene that encodes the UV endonuclease activity (249). When a revised nomenclature for a number of T4 genes was introduced (305), the designation v was changed to $denV^+$ (for DNA endonuclease V). We will continue to refer to this enzyme as *T4 PD-DNA glycosylase*.

T4 PD-DNA glycosylase has been purified to physical homogeneity (195). Like other DNA glycosylases, the enzyme is a small monomeric protein of ~18 kDa, with no requirement for divalent cation or other cofactors (186, 195). It is absolutely specific for cyclobutane dipyrimidines in DNA (86, 124, 186, 200, 316). In fact, dimers are attacked when they are present in single-strand substrates, in which the secondary structure of the DNA is lost (102, 186, 316). Additionally, in duplex DNA containing dimers on only one of the two strands, the dimer-containing strand is attacked uniquely (259). Consistent with this evidence, T4 *denV* mutants are abnormally sensitive to UV radiation (114, 205, 279) but not to chemicals or to ionizing radiation (86, 124, 189, 200).

A highly specific method for detecting PD-DNA glycosylase activity is to subject the enzyme-reacted DNA to conditions that monomerize dimers (e.g., direct photoreversal or incubation with photoreactivating enzyme) (Fig. 4–22). When UV-irradiated DNA radiolabeled in thymine is incubated with the T4 enzyme, subsequent monomerization of the dimers results in the liberation of free thymine (59, 215) (Fig. 4–22). The amount of free thymine liberated is exactly half of that associated with thymine-containing pyrimidine dimers lost from the DNA during the photoreversal (217) (Fig. 4–23). This stoichiometric relation demonstrates that only one of the two N-glycosylic bonds in the dimerized nucleotides is enzymatically hydrolyzed. Additionally, excision of thymine-containing pyrimidine dimers from DNA pretreated with T4 enzyme requires the action of a $5' \rightarrow 3'$ exonuclease activity (95), indicating that the glycosyl bond cleaved is exclusively the 5' one.

The PD-DNA glycosylase activity of the $denV^+$ gene product has been demonstrated in vivo by infecting UV-irradiated strains of *E. coli* defective in the *uvrA* or *uvrB* functions (and hence unable to catalyze incision of DNA by nucleotide excision repair [see chapter 5]) with phage T4 (216). Under these conditions, incision of *E. coli* DNA is mediated by the phage PD-DNA glycosylase. The dimer-

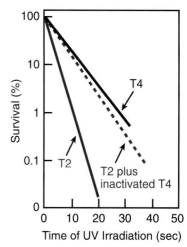

Figure 4–21 When phage T2 is exposed to modest doses of UV radiation and introduced into *E. coli* in the presence of heavily UV irradiated (and hence functionally inactivated) phage T4, the T2 phage shows enhanced survival relative to that observed in the absence of the coinfection. This result suggests that a polypeptide encoded by a phage T4 gene contributes to the survival of the irradiated T2 phage. This effect is called *v* gene reactivation. (*Adapted from Friedberg [87] with permission.*)

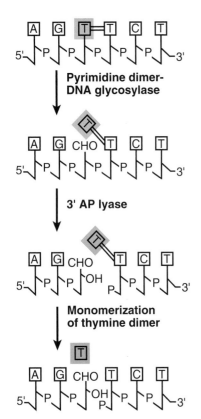

Figure 4–22 Enzyme-mediated cleavage of just one of the N-glycosyl bonds in a pyrimidine dimer by a PD-DNA glycosylase results in the formation of free thymine following the monomerization of the dimer by photoreversal or by the use of DNA photolyase. Measurement of the free thymine provides a sensitive and convenient assay for the PD-DNA glycosylase activity. (*Adapted from Friedberg [87] with permission.*)

containing oligonucleotides subsequently excised from the *E. coli* chromosome contain thymine that is released following photoreversal. The amount of free thymine so formed accounts quantitatively for all thymine-containing pyrimidine dimers excised (216).

It has been proposed that proteins that interact with a specific binding site in DNA do so in a two-step reaction. In the first step, the protein binds weakly to DNA in a nonspecific fashion. The binding energy of this reaction is derived from the entropy changes associated with displacement of small cations from phosphate residues in the DNA backbone (64, 287). This initial binding reaction provides an increased residence time of the protein on DNA and reduces the diffusion complexity from a three-dimensional to a one-dimensional problem, thus leading to productive binding to the target site. Experimental studies support the notion that the T4 PD-DNA glycosylase is processive and scans UV-irradiated DNA by a linear diffusion mechanism until a pyrimidine dimer is encountered (64). Site-directed mutagenesis suggests that a highly charged α-helical region of the protein contributes electrostatically to the ability of the enzyme to interact processively on nontarget DNA (6, 73–75, 199). Additionally, it has been suggested that the enzyme dimerizes in solution and that the dimerized state is more efficient in the processive searching of nontarget DNA (198). Electron microscopic studies indicate that scanning occurs by a looping mechanism rather than by the enzyme sliding along the double helix (64).

The X-ray structure of the T4 PD-DNA glycosylase has been determined at 1.6Å (0.16-nm) resolution (188). The protein consists of a single compact domain classified as an all-α type of structure because it is composed of three α-helices (Fig. 4–24). Site-directed mutagenesis of the *denV*[+] gene has implicated the region Trp-128 to Tyr-131 of the polypeptide (Fig. 4–25) in binding to DNA at a pyrimidine dimer (64). Independent studies have demonstrated that the substitution of Glu-23 (Fig. 4–25) with Gln or Asp or the substitution of Arg-3 (Fig. 4–25) with Gln results in a complete loss of DNA glycosylase activity (68). Additionally Arg-22 and Arg-26 (Fig. 4–25), together with Arg-117 and Lys-121, constitute a cluster of basic amino acids on the surface of the protein, which participate in substrate binding. Glu-23 is the only acidic residue in this cluster (Fig. 4–25). A protein with Gln substituted for Glu-23 retains substantial ability to bind substrate. The loss of DNA glycosylase activity is attributed to the loss of the negative charge at Glu-23 which results in changes in ionic interactions with surrounding water molecules (68). Site-directed mutagenesis has also implicated a sequence composed largely of aromatic residues (WYKYY) located in a loop structure near the C terminus of the protein (Fig. 4–24) in pyrimidine dimer-specific binding and possibly nicking of DNA (123, 223).

The N-terminal α-amino group of the T4 PD-DNA glycosylase has been implicated in the catalytic function of this enzyme (65). It has been proposed that after the glycosylase step, an (imino) covalent enzyme-substrate intermediate is formed between the N-terminal α-amino group of the polypeptide and the C-1' of the 5'-deoxyribose of the dimer substrate (65). In this regard it is interesting that, from the X-ray structure of the protein, the 7 N-terminal residues TRINLTL actually penetrate between helices H1 and H3 (Fig. 4–24).

Hydrolysis of the 5' glycosyl bond in UV-irradiated DNA by the T4 PD-DNA glycosylase is accompanied by DNA incision (186, 315, 316). It has been demonstrated that this occurs by a β-elimination reaction (173, 176). The replacement of Glu-23 with Gln (but not Glu-23 with Asp) abolishes the DNA lyase activity, suggesting that the carboxylate anion at amino acid position 23 acts as a general base in the β-elimination reaction (121).

M. luteus PD-DNA Glycosylase

Like the T4 enzyme, the *M. luteus* PD-DNA glycosylase is a small monomeric protein with a molecular mass of about 18 kDa and has no known cofactor requirement. It prefers duplex DNA-containing dimers to single-stranded DNA with such lesions. This PD-DNA glycosylase also catalyzes the hydrolysis exclusively of the 5' glycosyl bond in dimerized pyrimidines (108). Interestingly, under conditions of substrate excess, the enzyme apparently prefers dimers with 5′ thymines to those containing 5′ cytosines (108). The PD-DNA glycosylase from *M. luteus* has associated AP lyase activity (102, 103, 108, 110, 115).

The *M. luteus* PD-DNA glycosylase participates in the repair of pyrimidine dimers in vivo, as evidenced by the observation that mutants which are defective in both the DNA glycosylase gene and a gene that removes dimers by nucleotide excision repair are more UV radiation sensitive than either mutant alone (196). The gene for the *M. luteus* PD-DNA glycosylase has not yet been cloned, and the extent of the homology between the *M. luteus* and phage T4 enzymes is unknown. However, antibodies raised against the purified T4 PD-DNA glycosylase cross-react with the *M. luteus* protein. Despite these apparent similarities, the *denV*+ gene does not hybridize to *M. luteus* total genomic DNA (314).

A PD-DNA Glycosylase Activity in *S. cerevisiae*

Numerous investigations with many other prokaryotes and eukaryotes have failed to detect PD-DNA glycosylase activity in cell-free extracts. However, studies of the yeast *S. cerevisiae* have identified a purified fraction which generates the expected products of the concerted action of a PD-DNA glycosylase/DNA lyase on UV irradiated DNA (111). The observations that this activity is detected only after extensive fractionation of yeast extracts and only when a particularly sensitive assay was used suggest that this protein is present in very small amounts in yeast cells. Its biological relevance remains to be determined.

Biological Significance of the AP Lyase Activity Associated with DNA Glycosylases

As indicated above, regardless of the precise biochemical mechanism by which incision of DNA at AP sites occurs in vitro, an important biological question is whether the AP lyase function of the DNA glycosylases discussed above is functional in living cells. Concerted hydrolysis of N-glycosyl bonds and cleavage of phosphodiester bonds (the latter by β-elimination) would be expected to generate 5′-phosphoryl and 3′ unsaturated aldehyde termini (Fig. 4–26), or, in the case of β,δ-elimination, 3′-phosphoryl termini (Fig. 4–26). An alternative mechanism for the incision of DNA during base excision repair in vivo would be the hydrolysis of N-glycosyl bonds followed by hydrolysis of phosphodiester bonds catalyzed by an unassociated 5′ AP endonuclease, leaving 3′ OH and 5′ deoxyribose-phosphate termini (Fig. 4–26). Several experimental observations suggest that the latter pathway probably operates during base excision repair in vivo.

(i) When DNA containing pyrimidine dimers is incubated with the T4 PD-DNA glycosylase/AP lyase in the presence of native DNA, this substrate does not compete for DNA glycosylase activity but does compete for AP lyase activity against the UV irradiated DNA (181). The ability of native DNA to compete differentially for these two activities suggests that the hydrolysis of the 5′ glycosyl bond and cleavage of the 3′ phosphodiester bond at a pyrimidine dimer in DNA are not concerted in vitro; i.e., the enzyme dissociates from the DNA between these two events.

(ii) Mutations in both the *xthA*+ and *nfo*+ genes, which code for two well-characterized 5′ AP endonucleases in *E. coli* called exonuclease III and endonuclease IV, respectively (see below), result in a reduced survival of UV-irradiated phage T4. However, mutations in these genes have no effect on the survival of UV irradiated phage T4 which is defective in the *denV*+ gene (247). These observations suggest that these 5′ AP endonucleases function in the same pathway as the T4 PD-DNA glycosylase (247).

(iii) The T4 PD-DNA glycosylase/AP lyase does not complement the alkylation sensitivity of *E. coli xthA* mutants (218), suggesting that the 3′ AP lyase function of the T4 enzyme cannot substitute for the defective 5′ AP endonuclease activity of exonuclease III.

(iv) Cleavage at AP sites by β-elimination requires subsequent removal of 3′ unsaturated aldehyde termini to complete the excision process. Conceivably such

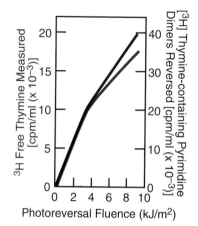

Figure 4–23 The amount of tritium released as free thymine from [³H] thymine-labeled UV irradiated DNA incubated with the phage T4 PD-DNA glycosylase is half that lost from the DNA as thymine-containing pyrimidine dimers (note the different scales on the *y* axes). This result indicates that thymine-containing pyrimidine dimers are the probable source of the free thymine and that only one of the two N-glycosyl bonds in the dimer is cleaved. (*Adapted from Friedberg [87] with permission.*)

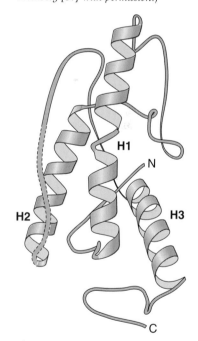

Figure 4–24 Proposed structure of the phage T4 PD-DNA glycosylase (endonuclease V or DenV protein). The protein consists of a single compact domain classified as an all-α type of structure, because it is composed of three α-helices (H1, H2, and H3). Note the N-terminal tail that penetrates between helices H1 and H3 and the C-terminal loop. This loop is thought to constitute a caplike structure that covers the ends of the three helices which are arrayed side by side. (*Reproduced from Morikawa et al. [188] with permission.*)

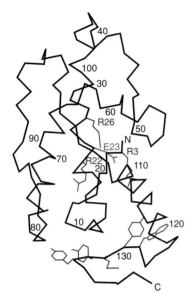

Figure 4–25 Stereo drawing of the Cα backbone of the T4 PD-DNA glycosylase in the same orientation as the ribbon structure shown in Fig. 4–24. The side chains of the three basic residues (Arg-3, Arg-22 and Arg-26) and the single acidic residue (Glu-23) that are believed to be involved in the glycosylase activity are indicated in color. The side chains of the WYKYY sequence in the C-terminal loop, which are also thought to be involved in enzyme catalysis, are shown as well. (*Reproduced from Morikawa et al. [188] with permission.*)

termini can be gratuitously removed by 3′ → 5′ exonucleases or by the action of a 5′ AP endonuclease. However, no enzyme activity has been specifically associated with the removal of such residues. In contrast, as indicated above, the 5′ deoxyribose-phosphate termini which result from cleavage at AP sites by 5′ AP endonucleases can be excised by the DNA deoxyribophosphodiesterase referred to in the introduction to this chapter and discussed in greater detail below.

Collectively, these considerations suggest that the general rule for the incision of DNA at sites of base loss (whether these are generated by spontaneous loss of bases or by their removal by DNA glycosylases) is by a 5′ AP endonuclease and that the 3′ AP lyase function associated with some DNA glycosylases may not be biologically important or, at most, plays a secondary role in the incision of DNA at AP sites.

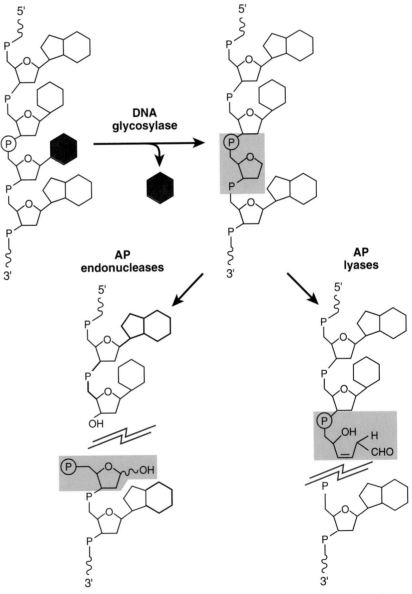

Figure 4–26 Sites of phosphodiester bond cleavage at an AP site by AP endonucleases and AP lyases. AP endonucleases cleave hydrolytically 5′ to AP sites to yield 5′-deoxyribose-5-phosphate (colored) and 3′-OH nucleotide residues. AP lyases cleave by β-elimination to yield 5′-terminal deoxynucleoside-5′-phosphate residues and 3′-terminal α,β-unsaturated aldehydes (colored). (*Adapted from Demple and Harrison [56] with permission.*)

Concluding Comments on DNA Glycosylases

DNA glycosylases are biochemically efficient enzymes for DNA repair. Each enzyme is endowed with limited substrate specificity, and in each case the capacity for specific DNA damage recognition and the capacity for catalytic activity are both embodied in a relatively small polypeptide with no cofactor requirement. However, when considered from the viewpoint of overall genetic economy, this mode of DNA repair is potentially inefficient, since a large number of genes would be required to accommodate the diverse range of base damages encountered in nature. It is likely that additional DNA glycosylases remain to be discovered and that their identification will probably depend on the characterization of new types of base damage which arise spontaneously in DNA. However, it is unrealistic to expect a unique repair enzyme to exist for each of the myriad forms of base damage to DNA which are known to be repaired in organisms such as *E. coli*. Evidence discussed in chapter 5 suggests that many types of base damage are in fact processed by a different mode of DNA incision, involving a single enzyme: the damage-specific DNA-incising endonuclease.

AP Endonucleases

As already indicated several times, base excision repair is a multistep pathway. Following the release of free damaged or inappropriate bases by DNA glycosylases, AP sites are produced. These can also arise spontaneously in native and particularly in alkylated DNA (see chapter 1). The repair of such lesions is initiated by *AP endonucleases* (88, 165, 166), which catalyze the incision of DNA exclusively at AP sites, thereby preparing the DNA for subsequent excision, repair synthesis, and DNA ligation.

Proteins that hydrolyze the C(3')-O-P bond 5' to AP sites, leaving 3'-hydroxyl-nucleotide and 5'-deoxyribose-phosphate termini, are true nucleotidyl hydrolases. To date, no 3' AP endonucleases have been discovered. AP endonucleases and AP lyases can be readily distinguished by relatively simple assays. For example if [^{32}P]DNA is used as a substrate, cleavage by a 5' AP endonuclease leaves radiolabel attached to the DNA in a form which can be released by subsequent alkali-catalyzed β-elimination (Fig. 4–27). On the other hand, cleavage 3' to the AP site effected by the action of an AP lyase leaves the radiolabel attached to the DNA in a form that is insensitive to release by subsequent alkali treatment (Fig. 4–27) (43, 159). However, in the latter case the radiolabel can be released by treatment with a 5' AP endonuclease (Fig. 4–27).

AP Endonucleases in *E. coli*

AP ENDONUCLEASE FUNCTION OF EXONUCLEASE III

E. coli exonuclease III was originally characterized as an exonuclease with an associated phosphatase activity (66, 88, 298, 299). Exonuclease III specifically requires 3'-OH ends in duplex DNA; i.e., it is a 3' → 5' exonuclease. In light of the associated phosphatase activity, DNA containing either 3'-OH or 3'-phosphate termini can be attacked exonucleolytically, since in the latter case the removal of the phosphate group creates a substrate for the exonuclease. In addition to these two catalytic functions, exonuclease III degrades a mixed copolymer of ribo- and deoxyribonucleotides, reflecting an RNase H-like activity (an enzyme that degrades RNA in RNA-DNA hybrids) (141). Exonuclease III also has a 3'-phosphodiesterase activity, which removes 3'-phosphoglycolate residues from DNA (58, 261). This activity may also remove the 3' α,β unsaturated aldehyde residues generated after β-elimination at AP sites, since exonuclease III can activate DNA incised by the AP lyase function of TG-DNA glycosylase to serve as a primer-template for *E. coli* DNA polymerase I (66). (DNA polymerase I requires 3'-terminal nucleotides for priming DNA synthesis.) A final catalytic function of exonuclease III, which appears to be the major physiological role of the enzyme and the one that primarily concerns us in the present discussion, is a *5' AP endonuclease activity.*

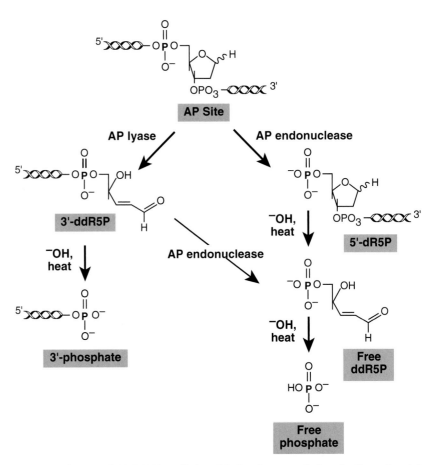

Figure 4–27 The use of [^{32}P] DNA to distinguish the cleavage of phosphodiester bonds by AP endonucleases and AP lyases. An AP site is shown attached to DNA with a ^{32}P-labeled phosphate (colored) 5′ to the AP site. Cleavage by a 5′ AP endonuclease generates a labeled 5′-terminal deoxyribose-5-phosphate residue (5′-dR5P). The radiolabel can be released by alkali-catalyzed β-elimination, yielding free 4-hydroxy-5-[^{32}P]phospho-2-pentenal (ddR5P), which on further δ-elimination yields free radiolabeled phosphate. Cleavage by a 3′ AP lyase (left) leaves the radiolabel as part of the 3′-terminal structure, where it is resistant to release from the DNA by alkali treatment. However, further treatment of this substrate with a 5′ AP endonuclease will release the radiolabel as free ddR5P. (*Adapted from Levin and Demple [159] with permission.*)

Before it was recognized that this AP endonuclease activity is a property of exonuclease III, the endonuclease was variously referred to as *endonuclease II* (91, 92) or *endonuclease VI* (281) of *E. coli*. Mutants of *E. coli* defective in both the 3′-exonuclease and the associated 3′-phosphatase functions of exonuclease III have been identified (310) and mapped to the *xthA*$^+$ gene (301) at approximately 38.5 min on the *E. coli* genetic map. All *xthA* mutants have been shown to also be defective in AP endonuclease activity (310). In addition, exonuclease III purified to greater than 98% homogeneity contains AP endonuclease activity, which cannot be separated from exonuclease and phosphatase activities (297).

The discovery of a new activity in a previously described enzyme is not without precedent. The 3′ → 5′ and 5′ → 3′ exonuclease activities of *E. coli* DNA polymerase I were discovered independently and were originally designated as exonucleases II and VI of *E. coli*, respectively. However, once it was clear that these were in fact catalytic functions associated with DNA polymerase I, these designations were dropped. In keeping with this nomenclatural precedent, the terms endonuclease II and VI are no longer used and we shall refer instead to the *AP endonuclease function of exonuclease III* or simply *exonuclease III*.

Purified exonuclease III is a protein ~28-kDa in size (104, 298). The AP endonuclease activity has an absolute requirement for Mg^{2+} and is inhibited in the presence of EDTA (104, 298). The enzyme catalyzes the hydrolysis of double-stranded AP DNA on the 5' side of sites of base loss, leaving 3'-OH and 5'-phosphate termini (104, 298). It is generally assumed that the AP endonuclease function of exonuclease III attacks sites of pyrimidine or purine loss in duplex DNA with equal facility; however, quantitative comparisons of this parameter have not been specifically documented. The enzyme is at least as active on DNA that contains AP sites reduced with sodium borohydride as on the unreduced substrate (104), suggesting that the aldehyde group of the C-1 in deoxyribose is not required for enzyme action. The mechanism of action of the enzyme is distinct from the β-elimination reaction catalyzed by alkali or AP lyases.

The $xthA^+$ gene has been cloned (232) and sequenced (248). The calculated molecular weight and N-terminal amino acid sequence of the predicted polypeptide are in agreement with those of the purified protein. A homologous gene ($exoA^+$) that encodes a quantitatively major exonuclease with many of the properties of exonuclease III has been isolated from *Streptococcus pneumoniae* (214).

The presence of multiple catalytic functions associated with a relatively small monomeric protein suggests that a single active site catalyzes all the enzymatic reactions of exonuclease III. In this regard, it has been proposed that the enzyme has three important domains in its structure (297) (Fig. 4–28). One is the active site that catalyzes the cleavage of phosphodiester bonds in one strand of duplex nucleic acid. The second is a site for recognizing duplex structure in the nucleic acid and depends on the presence of deoxyribose in the strand opposite to that in which the active site works. (This idea is consistent with the absolute requirement for duplex DNA or for a DNA-RNA hybrid as substrate.) A third domain is postulated to recognize a "space" in the DNA duplex, which may be constituted by a missing base (thereby facilitating its AP endonuclease function), by the region beyond the end of a DNA strand, or by the space created by the partial denatur-

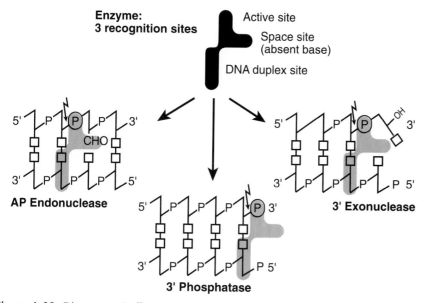

Figure 4–28 Diagrammatic illustration of how exonuclease III (Xth protein) of *E. coli* may function as a 3' phosphatase, a 5' AP endonuclease, or a 3' exonuclease. It is proposed that the enzyme has a domain that functions as the active site for nuclease and/or phosphatase activity. In addition, a second domain recognizes a space in the DNA duplex that can be constituted by either a missing base (AP site) (left), an "absent" base at the end of a DNA molecule (middle), or an "absent" base at the end of a DNA molecule caused by fraying of the end of a DNA duplex (right). A third domain in the enzyme is proposed to recognize duplex DNA. (*Adapted from Friedberg [87] with permission.*)

ation of the DNA duplex at the site of an internal nick (thereby providing substrate for the phosphatase-exonuclease function).

It has been suggested that the structure recognized at the 3′ end of duplex DNA can also include a ring-opened nucleotide (149). Ring-opened sugars allow rotation of a 3′-terminal base or an interior base with a secondary amine at the N-glycosyl bond, thereby generating a ''space'' 5′ to the target nucleoside (66). This model accommodates the observation that exonuclease III can incise DNA 5′ to O-alkylhydroxylamine-N-glycosides, as well as 5′ to urea N-glycosides in DNA (149, 150).

The exact biological role of the AP endonuclease activity of exonuclease III is not clear. xthA mutants are sensitive to killing by agents such as near-UV radiation and H_2O_2, which produce oxidative damage in DNA (55). The nature of the lethal lesion(s) responsible for this phenotype is unknown, but a number of possibilities have been suggested (55). For example, the free radicals arising from the decomposition of H_2O_2 might generate DNA strand breaks with 3′-phosphate termini. Such termini may be refractory to further processing unless the phosphate group is removed by the phosphatase function of exonuclease III. Exonuclease III accounts for >99% of the 3′ phosphatase activity present in E. coli (55). The demonstration that exonuclease III also attacks DNA containing thymine fragmentation products such as urea residues suggests that this enzyme plays a role in the repair of oxidative damage.

Mutations in the katF⁺ gene of E. coli also result in increased sensitivity to near-UV radiation and to H_2O_2 (240). This gene is one of several kat genes which encode catalases involved in protection against oxidative damage in E. coli. Insertional mutagenesis of the katF⁺ gene eliminates exonuclease III activity (236), suggesting that the xthA⁺ gene may be regulated by the katF⁺ gene. Consistent with this suggestion, katF xthA double mutants are no more sensitive to near-UV radiation than is either single mutant. Additionally, transformation of katF mutants with plasmids carrying the katF⁺ gene restores exonuclease III activity (236).

xthA mutants are defective in the induction of heat shock proteins following severe heat shock (shifting from 30 to 50°C) but not following mild heat shock (shifting from 30 to 42°C). The biological significance of this observation is not clear, but it has been suggested that the heat shock response may require a particular conformation of DNA, which can be generated by exonuclease III, as an inducing signal (208). xthA mutants of E. coli (including deletion mutants) are only partially sensitive to treatment with alkylating agents such as MMS (170, 310). This may reflect the ability of other AP endonucleases to assume the essential function(s) of this enzyme during the repair of sites of base loss. In contrast, deficiency of both the xthA⁺ and dut⁺ genes (the latter codes for deoxyuridine triphosphatase) in E. coli is conditionally lethal (300). The increased levels of dUTP associated with the dut mutation (see chapter 1) and the subsequent excision of uracil by ura-DNA glycosylase presumably lead to the formation of apyrimidinic sites. These are apparently lethal lesions in the absence of the AP endonuclease function of exonuclease III.

ENDONUCLEASE IV

Endonuclease IV is a second example of a 5′ AP endonuclease in E. coli. This activity is not readily detectable in wild-type cells, because it constitutes only about 10% of the total AP endonuclease activity in crude extracts. Its identification was facilitated by the isolation of xthA mutants defective in exonuclease III. Endonuclease IV has been extensively purified and characterized (169). It catalyzes the formation of single-strand breaks at sites of base loss in duplex DNA. Like the AP endonuclease function of exonuclease III, endonuclease IV attacks phosphodiester bonds 5′ to the sites of base loss in DNA, leaving 3′-OH groups (8). Additionally, like the AP endonuclease function of exonuclease III, endonuclease IV can remove phosphoglycoaldehyde, phosphate, deoxyribose-5-phosphate, and 4-hydroxy-2-pentenal residues from the 3′ terminus of duplex DNA (66). It also has an endonuclease function active against DNA containing urea residues (289). Indeed, the only catalytic activity that distinguishes endonuclease IV from exonuclease III is the presence of a $3' \rightarrow$ exonuclease activity in the latter (66).

Endonuclease IV has a sedimentation coefficient of 3.4 and an M_r of ~33 kDa. Enzyme activity is not stimulated by $MgCl_2$ or $CaCl_2$. However, endonuclease IV can be inactivated by EDTA in the presence of substrate, suggesting that it contains a tightly bound metal ion (160). The AP endonuclease activity is unusually resistant to high ionic strength (50% activity is retained in the presence of 0.56 M NaCl) and to heat (full activity is retained after heating to 60°C for 5 min). It is inhibited by SH-group inhibitors.

The gene that encodes endonuclease IV (nfo^+) (for endonuclease four) has been cloned (49) and sequenced (246). The predicted gene product has a calculated M_r of 31,562. The cloned gene was used to construct insertion mutants in *E. coli*. These mutants have increased sensitivity to monofunctional alkylating agents such as MMS and to mitomycin and the oxidizing agents *tert*-butyl hydroperoxide and bleomycin. However, sensitivity to these agents and to ionizing radiation is markedly increased in the presence of the *xthA* mutation (which encodes exonuclease III) (49). The double mutants are also more mutable by MMS than the single mutants. These results are consistent with the observation that endonuclease IV and exonuclease III have similar substrate specificities. However, *nfo* mutants are more sensitive to *tert*-butyl hydroperoxide and bleomycin than are *xthA* mutants, suggesting that endonuclease IV might recognize some lesions that exonuclease III does not (49).

A particular mutant allele of the *nfo* gene designated *nfo-186* (127) retains normal AP endonuclease activity, but is defective in the 3'-phosphatase activity. This allele is able to correct the hypersensitivity of an *xthA nfo* double mutant to MMS but not to H_2O_2. This observation suggests that the lethal damage produced by H_2O_2 requires a 3'-phosphatase for its processing, and hence is a 3' blocking type of damage rather than an AP site (127).

Endonuclease IV Is Inducible

Chemical agents such as methyl viologen (paraquat), plumbagin, menadione, and phenazine methosulfate, which are enzymatically reduced in vivo via one-electron transfer reactions and then auto-oxidized to generate superoxide radicals, induce a 10- to 20-fold increase in the level of endonuclease IV (34, 299) (Fig. 4–29). Consistent with the interpretation that induction is effected by superoxide, endonuclease IV is also induced when cells defective in superoxide dismutase activity are grown in the presence of pure oxygen (299). Induction is unaffected by mutations in the $oxyR^+$ gene which controls the peroxide-induced oxidative stress regulon.

The redundancy of AP endonuclease activities in *E. coli* attests to the fundamental importance of base excision repair in living cells. Aside from the high frequency of spontaneous base loss, it has already been pointed out that many of the types of base damage repaired by specific DNA glycosylases are generated by oxidative damage to DNA. This must be recognized as a major form of spontaneous DNA damage in all aerobic life forms which generate free radicals during the course of normal metabolism. Since the repair of these types of base damage involves AP intermediates, AP endonucleases are extremely important for the maintenance of genetic stability. Later discussions will elaborate on the mutagenic and hence carcinogenic potential of sites of base loss in DNA.

AP Endonucleases in Eukaryotes

AP endonucleases are ubiquitous, and the literature contains numerous reports of such enzymes in both lower and higher organisms. In many instances these enzymes have not been extensively purified or characterized, and it is not clear whether these represent AP lyases or AP endonucleases. Hence, this discussion will focus on selected examples of well-characterized AP endonucleases which resemble the 5' AP endonucleases exonuclease III and endonuclease IV of *E. coli*.

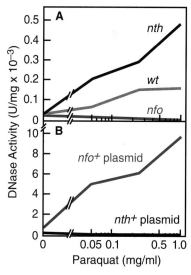

Figure 4–29 Paraquat induction of endonuclease IV activity. (A) AP endonuclease activity is induced in extracts of wild-type and *nth* (endonuclease III) mutant strains but not in *nfo* (endonuclease IV) mutant strains. (B) Cells transfected with a plasmid carrying the cloned *nfo⁺* gene show induction of AP endonuclease activity, while strains that overexpress the *nth* gene do not. (*Adapted from Chan and Weiss [34].*)

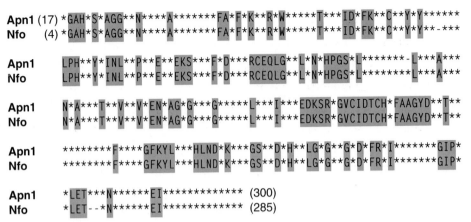

Figure 4–30 The predicted amino acid sequences of the cloned yeast *APN1* (AP endonuclease I) and *E. coli nfo⁺* (endonuclease VI) genes are 41% identical. (*Adapted from Popoff et al. [211].*)

SACCHAROMYCES CEREVISIAE

A number of AP endonuclease activities have been found in extracts of the yeast *S. cerevisiae* (66). The enzyme that has been most extensively characterized and that appears to be the quantitatively major AP endonuclease activity in *S. cerevisiae* has been variously designated as *yeast DNA diesterase* (128, 129), *yeast 3'-phosphoglycoaldehyde diesterase* (128, 129), or *yeast Apn1 apurinic endonuclease/3'-diesterase* (222). For the sake of simplicity and uniformity, we suggest that this protein be referred to as *yeast AP endonuclease I*.

Yeast AP endonuclease I bears a close resemblance to endonuclease IV of *E. coli* (128, 129). The enzyme can excise a variety of 3' esters in DNA, including 3'-phosphoglycoaldehyde, 3'-phosphoryl groups, and 3'-α,β unsaturated aldehydes. The enzyme hydrolyzes DNA 5' to AP sites, generating 3'-OH and 5'-deoxyribose-phosphate termini. The protein has a molecular mass of ~40.5 kDa and is optimally stimulated by Co^{2+}. Both *E. coli* endonuclease IV and yeast AP endonuclease I are metalloenzymes. Atomic absorption spectroscopy has revealed the presence of 2.4 Zn^{2+} and 0.7 Mg^{2+} atoms in the former protein and 3.3 Zn^{2+} atoms in the latter (161).

The structural gene that encodes this AP endonuclease (designated *APN1* [for AP endonuclease]) has been cloned and sequenced (211). Gene disruption confirms that this is the quantitatively major AP endonuclease in yeast. Strains carrying disruptions or deletions of the *APN1* gene are hypersensitive to killing by a variety of oxidative and alkylating agents which damage DNA. Additionally, such mutants have a spontaneous mutation rate which is about an order of magnitude greater than that observed in wild-type yeast cells (222). Indirect-immunofluorescence studies have localized Apn1 protein to the nucleus of yeast cells (220).

The predicted amino acid sequence of the *APN1* gene is homologous with that of the cloned *E. coli nfo⁺* gene (Fig. 4–30). The *APN1* gene can complement sensitivity to killing by oxidant and alkylating agents in *E. coli nfo* mutants (221, 222), but it has no effect on the sensitivity of *E. coli xth* mutants to hydrogen peroxide, MMS, or mitomycin. Hence, yeast Apn1 protein is apparently able to replace the functions of its bacterial homolog endonuclease IV, but not the functions of the catalytically closely related enzyme exonuclease III (221, 222).

DROSOPHILA MELANOGASTER

Two AP endonucleases designated I and II, with distinctive chromatographic properties, have been isolated from *D. melanogaster* (66). Both cleave apurinic DNA 3' to sites of base loss and hence would appear to be examples of AP lyases rather than AP endonucleases. However, there are indications that AP endonuclease I leaves 3'-deoxyribose-phosphate and 5'-OH groups at AP sites (265), in which case it would be the first example of a true 3' AP endonuclease.

Screening a *D. melanogaster* expression library with an antibody raised against a human AP endonuclease resulted in the isolation of a gene designated *AP3* (for AP endonuclease 3) (142). This gene was used to screen a human cDNA library from HeLa cells, yielding a related gene with 66% amino acid sequence identity, which turned out to be a homolog of a previously cloned *D. melanogaster* gene called *PO*. The *PO* gene is believed to encode a ribosome-associated protein (107, 228). It remains to be determined whether this protein functions as both a ribosomal protein and a repair protein in vivo. It is interesting that a similar association between a ribosomal location and an enzyme with apparent AP endonuclease activity has been observed with a human protein (see below).

The preliminary characterization of a possibly fourth distinct AP endonuclease from *D. melanogaster* suggests that at least some enzymes of this class may be endowed with even greater versatility than several of the multifunctional AP endonucleases already discussed. This protein was originally purified from *D. melanogaster* embryo extracts by an assay which measures DNA strand transfer and hence was designated as Rrp1 protein (for recombination repair protein 1). Cloning and sequencing of the *RRP1* cDNA revealed that the 252-amino-acid C-terminal region of the predicted polypeptide shares significant similarity with the amino acid sequence of *E. coli* exonuclease III and the related protein exonuclease A from *Streptococcus pneumoniae* (242, 243). Purified Rrp1 protein has AP endonuclease and 3' diesterase activities (243).

MAMMALIAN CELLS

A bewildering multiplicity of AP endonucleases and AP lyase activities have been reported from a variety of mammalian sources, including human cells and tissues. These enzymes are presumably involved in base excision repair in living cells. However, mutants defective in specific mammalian AP endonucleases have not been identified, and our current knowledge of the function(s) of these enzymes is restricted to biochemical information. The activities all require or are stimulated by divalent cations, and they hydrolyze DNA 5' to AP sites, leaving 5'-deoxyribose-5-phosphate and 3'-OH termini (66, 88, 89).

BAP1, APEX, and HAP1 AP Endonucleases

Following the purification of a 3'-diesterase/AP endonuclease activity from calf thymus (designated *bovine AP endonuclease 1 [BAP1]*), the corresponding cDNA was cloned and sequenced (231). The gene is called *BAP1* (for bovine AP endonuclease 1) and can encode a polypeptide of 35.5 kDa. The predicted amino acid sequence shows 36 and 25% identity with the *Streptococcus pneumoniae* ExoA and *E. coli* ExoIII proteins, respectively (Fig. 4–31) (231).

An enzyme with 3'-phosphatase and 3' → 5' exonuclease activities in addition to AP endonuclease and DNA 3'-diesterase activities has been purified from mouse cells (252) and is designated *APEX nuclease* (for AP endonuclease/exonuclease). A cDNA clone for this protein was isolated from a mouse library by using degenerate oligonucleotide probes deduced from the N-terminal amino acid sequence. Sequencing of the cDNA identified an open reading frame which can encode a protein of 35.4 kDa. The amino acid sequence of this open reading frame is 93.1% identical to that of the protein encoded by the bovine *BAP1* gene described above and shares extensive homology with the *Streptococcus pneumoniae ExoA and E. coli* ExoIII (Xth) proteins but not with *E. coli* endonuclease IV, yeast Apn1 protein, or *Drosophila* AP3 protein (251). When the APEX gene was introduced into *E. coli xth nfo* double mutants, the transformed cells expressed a 36.4-kDa polypeptide and acquired increased resistance to MMS. The cells also demonstrated the ability to prime bleomycin-treated or acid-depurinated DNA for DNA polymerase (229).

A human cDNA with a translated amino acid sequence of 93% identity to that of the bovine BAP1 protein has also been isolated (Fig. 4–31) (57, 229). This

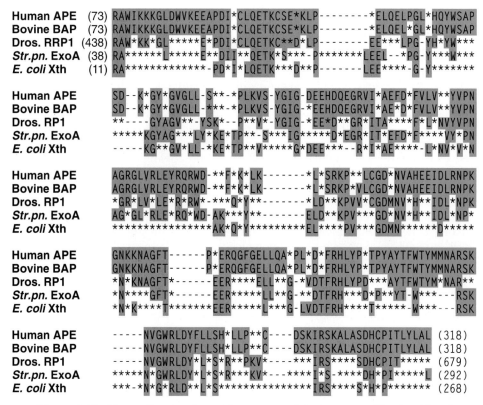

Figure 4–31 The predicted amino acid sequences of the human and bovine AP endonuclease genes *APE* and *BAP* show significant identity with those of the *Drosophila RRP1*, the *Streptococcus pneumoniae exoA*+ (exonuclease A), and the *E. coli xth*+ (exonuclease III) genes. (*Adapted from Demple et al. [57] and Robson et al. [231] with permission.*)

gene represents the human equivalent of the mouse *APEX* and bovine *BAP1* genes and is designated *HAP1* (for human AP endonuclease) (229) or *APE* (for AP endonuclease) (57). The *HAP1* gene consists of five exons (230). The gene maps to human chromosome 14q11.2-12 (230). The predicted amino acid sequences of the *E. coli xth*+, *S. pneumoniae exoA*+, *Drosophila RRP1*, bovine *BAP1*, human *HAP1* (*APE*) (Fig. 4–31), and mouse *APEX* genes share significant similarity (57).

The *HAP1* gene can substitute for some but not all of the mutant phenotypes of *E. coli xth* mutants (229). Additionally, a *dut xth*(Ts) double mutant, which is inviable at 42°C because it is unable to repair AP sites produced by the excision of uracil, can be rescued by the human *HAP1* gene. However, the *HAP1* gene does not complement any of the phenotypes of *nfo* mutants (229).

The polypeptide encoded by the *HAP1* gene turns out to be the same as a nuclear regulatory protein called Ref-1, a redox factor believed to regulate the Fos-Jun heterodimer transcription factor AP1 through reduction of a cysteine residue located in the DNA-binding domain (307). A 61-amino acid N-terminal domain in the HAP1 protein, which is not conserved in prokaryotic and eukaryotic AP endonucleases and which is also dispensable for the AP endonuclease function of HAP1, is essential for the DNA-binding activity of oxidized Jun protein (288). The redox activity of HAP1 protein has in fact been localized to Cys-65. These observations suggest the intriguing possibility that the redox protein plays an important role in transducing cellular oxidative stress signals to a complex cellular gene regulatory pathway. It remains to be determined how such a signal transduction pathway might relate to AP endonuclease activity in particular and base excision repair in general in mammalian cells.

Relevance of Base Excision Repair to Oxidative Damage—General Considerations

Many of the base excision repair enzymes that have just been discussed are critical for the repair of various types of oxidative damage to DNA. This aspect of DNA damage and repair has received relatively modest attention in the literature, and space constraints preclude a more detailed exploration of this topic here. Nonetheless, the reader should understand that since animal life depends on oxygen, oxidant by-products of normal metabolism such as superoxide radicals, hydrogen peroxide, and hydroxyl radicals can cause extensive damage to DNA, to proteins, and to cell membranes (2, 56) (see chapter 1). Indeed, it has been argued that oxidative damage lies at the heart of the pathophysiology of multiple diseases such as cancer, cardiovascular disease, neurological degeneration, immune dysfunction, and even aging (2). Animals have evolved elaborate antioxidant defense mechanisms, but since these are less than perfect, some oxidants find their way to target molecules such as DNA. It has been estimated that the human genome sustains as many as 10,000 oxidative "hits" per day, most of which are removed by DNA glycosylases. Rats, which have an increased rate of oxidative metabolism, may accumulate 10 times as many hits, and by the time a rat is 2 years of age (old for a rat), it may have accumulated several million oxidative lesions in its genome (2).

Since base excision repair is well understood in detailed biochemical terms, it occupies but a single chapter in this book. On the other hand, nucleotide excision repair is the subject of four chapters because of its greater biochemical complexity. However, the relative mechanistic complexity of a repair pathway bears no relationship to its biological significance. Indeed, given the increasing appreciation from epidemiological studies that with the single exception of cigarette smoking, environmental agents make a relatively small contribution to cancer incidence, the importance of spontaneous DNA damage in general, and oxidative damage in particular, cannot be overemphasized.

Postincision Events during Base Excision Repair

DNA Deoxyribophosphodiesterase of *E. coli*

As discussed above, the incision of DNA during base excision repair in living cells is presumed to take place mainly, if not exclusively, 5′ to AP sites by AP endonucleases, which leave 3′-OH and 5′-deoxyribose-phosphate termini. The 3′-OH ends constitute appropriate primer-termini for DNA polymerases which participate in repair synthesis. However, completion of the excision process requires removal of the 5′-deoxyribose-phosphate moieties. For many years it was believed that 5′ → 3′ exonucleases could subserve this requirement. An obvious early candidate for this role in *E. coli* was DNA polymerase I, since this enzyme is endowed with 5′ → 3′ exonuclease activity (see below), and it seemed intuitively obvious to link this exonuclease with the polymerase function in repair synthesis. In vitro studies with purified DNA polymerase have shown that the 5′ → 3′ exonuclease is unable to excise deoxyribose-phosphate residues in free form from the 5′ termini of incised AP sites, however, and can only slowly release an oligonucleotide with a 5′-deoxyribose terminus (85). This observation prompted the search for and eventual discovery of an exonuclease in *E. coli* that has been designated *DNA deoxyribophosphodiesterase (dRpase)*. This exonuclease can excise 2-deoxyribose-5-phosphate residues from the 5′ side of single-strand interruptions in DNA containing sites of base loss (85). The protein is 50–55 kDa in size and requires Mg^{2+} for activity. Unlike AP lyases, which effect 3′ cleavage at AP sites via a β-elimination reaction, the excision of 2-deoxyribo-5-phosphate by dRpase occurs by hydrolytic cleavage, since the enzyme is active even when the aldehyde group at position 1 of the base-free sugar is reduced to the corresponding alcohol and is therefore not susceptible to β-elimination (85).

Highly purified preparations of dRpase appear to be identical to the product of the *E. coli recJ*⁺ gene (63), a gene that has been implicated in recombination repair of DNA. This gene encodes a protein with 5′ → 3′ exonuclease activity for single-stranded DNA. Purified dRpase also has such activity and homogeneous preparations of RecJ protein have high levels of dRpase activity (63). Furthermore, both enzymatic activities are defective in *recJ* mutant strains.

It has been reported that the *E. coli* dRpase activity represents a newly discovered function of a previously characterized enzyme called exonuclease I, an exonuclease that specifically degrades single-stranded DNA in the 3′ → 5′ direction (244). However, these observations have not been substantiated in independent studies (63). The *E. coli* FaPy-DNA glycosylase discussed above is also apparently able to act as a dRpase (109). However, in contrast to the RecJ protein, which requires Mg²⁺, the FaPy-DNA glycosylase does not. In fact, the dRpase activity of this DNA glycosylase is inhibited in the presence of Mg²⁺ (63). The physiological roles of the dRpase activities of RecJ protein and of FaPy DNA glycosylase in base excision repair are not clear. No defect in the repair of AP sites has been detected in vivo in *recJ* or *recJ fpg* mutants (63). Hence, there may be alternative backup modes for the obligatory excision of dRp residues. These may include their excision as part of oligonucleotide fragments by the 5′ → 3′ exonuclease function of DNA polymerase I or even the slow nonenzymatic release of 5′-terminal dRp residues by cellular polyamines and basic proteins (63). A dRpase has also been partially purified from mammalian cells (212). It has a molecular mass of ~47 kDa and is located in the cell nucleus.

The combined actions of a DNA glycosylase, AP endonuclease, and dRpase would leave a single nucleotide gap in duplex DNA, which could presumably be filled by any of the several DNA polymerases in *E. coli*, since there is clearly no special requirement for the 5′ → 3′ exonuclease function of DNA polymerase I. Indeed, when duplex oligonucleotides containing single dUMP residues are incubated with carefully prepared extracts of *E. coli* or mammalian cells, the patch size at the majority of the repair events is a single nucleotide (62).

Completion of Base Excision Repair Initiated by the Phage T4 and *M. luteus* PD-DNA Glycosylases

The completion of base excision repair initiated by PD-DNA glycosylases presents a special problem if we assume that in living cells the AP sites generated by hydrolysis of the 5′ N-glycosyl bonds at pyrimidine dimers are attacked by 5′ AP endonucleases and not by the 3′ AP lyase function intrinsic to the enzyme. Even if the 5′-deoxyribose-phosphate termini were removed by one or more dRpases, the adjacent 3′ thymine nucleotides would still be covalently linked to thymine residues. Excision of these thymine-thymidylate residues would presumably require the action of a 5′ → 3′ exonuclease.

DNA polymerase I is a likely candidate for such a role. Support for this stems from studies on *E. coli* infected with UV-irradiated phage T4. Not surprisingly, loss of thymine-containing pyrimidine dimers from phage DNA has an absolute dependence on a functional *denV*⁺ gene (209). However, no detectable dimer excision is observed in either *polA1* or *polAex* mutants of *E. coli* (which are defective in the 5′ → 3′ exonuclease) (209), although phage T4 expresses an independent 5′ → 3′ exonuclease activity which catalyzes the excision of dimers from specifically preincised UV-irradiated DNA in vitro (96, 204, 258). Apparently this phage T4-encoded exonuclease is not involved in pyrimidine dimer excision in vivo; rather, the repair of the phage genome is dependent on functional *E. coli* DNA polymerase I. Consistent with this interpretation, UV-irradiated phage T4 has a lower survival in *E. coli polA1* mutants than in *polA* cells (189, 264).

Repair Synthesis during Base Excision Repair

Repair synthesis during base excision repair might involve one or more DNA polymerases. *E. coli* contains three such enzymes, referred to as DNA polymerases I, II, and III, encoded by the *polA*⁺, *polB*⁺, and *polC*⁺ genes, respectively. The principal properties of and distinctions between these three enzymes are summarized in Table 4–5. The following sections present some highlights about these polymerases. The reader interested in more detailed information is referred to the comprehensive text by Kornberg and Baker (148).

E. COLI DNA POLYMERASE I

DNA polymerase I (so designated because it was the first DNA polymerase identified in extracts of *E. coli*) was discovered by Arthur Kornberg and his associates (148). The enzyme is the product of the *polA*$^+$ gene, which maps at 85 min on the *E. coli* genetic map. The enzyme has been purified to physical homogeneity and has been extensively characterized. It is a monomeric protein with a molecular mass of 109 kDa and has discrete catalytic functions for DNA polymerization, pyrophosphorolysis, PP$_i$ exchange, 3′ → 5′ exonucleolytic degradation, and 5′ → 3′ exonucleolytic degradation (148). Polymerization, pyrophosphorolysis, and PP$_i$ exchange are all features of the polymerization activity, the last two simply being the reverse of the polymerization reaction. Like all known DNA polymerases, DNA

Table 4–5 Properties of DNA polymerases I, II, and III of *E. coli*

Property	Value for:		
	Pol I	Pol II	Pol III (core)
Functions			
Polymerization: 5′ → 3′	+	+	+
Exonuclease: 3′ → 5′	+	+	+
Exonuclease: 5′ → 3′	+	−	−
Pyrophosphorolysis and PP$_i$ exchange	+		+
Template primer			
Intact duplex	−	−	−
Primed single strands[a]	+	−	−
Stimulation by single strand binding protein	−	+	−
Nicked duplex [poly d(AT)]	+	−	−
Duplex with gaps or protruding single strand 5′ ends of			
<100 nucleotides	+	+	+
>100 nucleotides	+	−	−
Polymer synthesis de novo	+	−	−
Activity			
Effect of KC1 (% of optimal activity)			
20 mM	60	60	100
50 mM	80	100	50
100 mM	100	70	10
150 mM	80	50	0
K_m for dNTPs[b]	Low	Low	High
Stimulation by β subunit	−	+	+
Inhibition by 2′-deoxy analogs	−	+	+
Inhibition by arabinosyl CTP	−	+	−
Inhibition by sulfhydryl-blocking agents	−	+	+
Inhibition by pol I antiserum	+	−	−
General			
Size (kDa)	103	90	130, 27.5, 10[c]
Affinity to phosphocellulose			
molarity of phosphate required for elution	0.15	0.25	0.10
Estimated no. of molecules/cell	400		10–20
Turnover number, estimated	600	30	9,000
Structural genes	*polA*	*polB*	*polC*
Conditional-lethal mutant	Yes	No	Yes

[a] A long single strand with a short length of complementary strand annealed.
[b] dNTPs, deoxynucleoside triphosphates.
[c] Sizes of the α, ε, and φ subunits, respectively.
Source: Adapted from Kornberg and Baker (148) with permission.

polymerase I of *E. coli* catalyzes the growth of DNA chains only in the $5' \rightarrow 3'$ direction. This polarity is dictated by its specificity for 3'-OH primer termini and for deoxynucleoside 5'-triphosphates as substrates. A striking feature of DNA polymerase I, observed in no other *E. coli* polymerase, is its ability to promote replication of DNA at a nick, unaided by other proteins. This requires unwinding of the duplex beyond the nick and progressive strand displacement of the 5' chain. The $5' \rightarrow 3'$ exonuclease function can catalyze exonucleolytic degradation of DNA in the absence of any base damage, from nicks containing 3'-OH termini. DNA synthesis coupled with degradation by the $5' \rightarrow 3'$ exonuclease leads to transfer of the nick along the template (*nick translation*). In so doing, the enzyme cleaves a phosphodiester bond only at base-paired regions (i.e., it requires duplex DNA for activity). In addition, the $5' \rightarrow 3'$ exonuclease can excise terminal DNA fragments containing as many as 10 nucleotides. It is precisely these properties of the exonuclease that endow DNA polymerase I with the ability to remove oligonucleotides containing pyrimidine dimers or other forms of bulky base damage from DNA with nicks located 5' to the sites of damage (95, 143). However, *E. coli* DNA polymerase I is unable to support nick translation at a nick with a 5'-deoxyribosephosphate residue.

The $3' \rightarrow 5'$ exonuclease of DNA polymerase I is a component of the enzyme necessary for recognition of an incorrectly base-paired primer terminus (148). It supplements the capacity of DNA polymerase I to match the nucleotide substrate to the template by base pairing and hence plays a critical role in the fidelity of repair synthesis during excision repair. In addition, this exonuclease function can potentially serve as an excision exonuclease for the removal of 3'-terminal damaged nucleotides. In vitro, *E. coli* DNA polymerase I can be cleaved by gentle proteolysis into two fragments of 76 and 36 kDa (148). The former (*large fragment*) (often called Klenow fragment) retains the polymerizing and $3' \rightarrow 5'$ exonuclease activities, while the latter (*small fragment*) retains the $5' \rightarrow 3'$ exonuclease activity (148).

DNA polymerase I is roughly spherical, with a diameter of nearly 65 Å (6.5 nm), and it can contact a DNA helix across its width for a length of nearly two helical turns. Thus, if the enzyme is processive in its action, i.e., if it does not dissociate from the DNA after each nucleotide incorporation, a single productive binding event could facilitate complete repair synthesis of excision tracts less than 20 nucleotides long. Various experimental approaches have been used to establish whether DNA polymerase I is processive or distributive in its action (148). The general conclusion that has emerged from in vitro studies is that the enzyme binds very rapidly to suitable primer-templates. The binding is followed by a relatively long delay before initiation of polymerization, during which time there is a shift in equilibrium from inactive to active enzyme. A cycle of about 20 polymerization steps takes place before the enzyme slowly dissociates (148). These observations would suggest that during the synthesis of very short patches during base excision repair, the enzyme is essentially processive.

Processivity is influenced by many factors that affect the secondary structure of the DNA or the conformation of the enzyme. These include temperature, ionic conditions, and interactions with other proteins (148). For example, a mutant form of DNA polymerase I (PolA5) incorporates only about one-fifth the number of nucleotides added by the wild-type enzyme before dissociating from the template. These factors are relevant for DNA repair, since the processivity of the enzyme during repair synthesis of DNA may contribute to the relative heterogeneity of the patches synthesized and could affect the kinetics and possibly even the fidelity of repair synthesis in vivo.

E. COLI DNA POLYMERASE II

The isolation by John Cairns and his coworkers of a viable mutant of *E. coli* defective in DNA polymerase I activity (54) established that while this enzyme may be involved in DNA replication, it is clearly not the enzyme that catalyzes the majority of semiconservative DNA synthesis in vivo. A search for other DNA polymerases in *E. coli* uncovered two new enzymes called DNA polymerases II and

III (148). DNA polymerase II, which is encoded by the $polB^+$ gene, is distinguished from DNA polymerase I by a number of criteria (Table 4–5). Most significantly from the point of view of its potential role in DNA repair, it is devoid of $5' \rightarrow 3'$ exonuclease activity and it cannot use a template-primer that is simply nicked or is extensively single-stranded (148). The optimal substrate for DNA polymerase II is a duplex with short gaps of less than 100 nucleotides. This substrate requirement suggests that DNA polymerase II would be ideally suited to a primary role in repair synthesis of the excision gaps generated during base and/or nucleotide excision repair.

The $polB^+$ gene has been isolated by molecular cloning on the basis of its mapped position in the *E. coli* genome (35). The cloned gene has been shown to be a previously identified gene designated $dinA^+$ (19, 126), one of several DNA damage-inducible genes in the *recA/lexA* regulon (see chapter 10). The gene could encode a polypeptide of ~90 kDa (125). Sequence analysis of the cloned gene indicates that it shares remarkable homology with a group of DNA polymerases from both prokaryotic and eukaryotic organisms (125).

E. COLI DNA POLYMERASE III

DNA polymerase III of *E. coli* is a complex multicomponent enzyme that resembles RNA polymerase in its complexity (148). The holoenzyme core has been purified to virtual homogeneity and appears to contain 10 different subunits (180). Polymerase III of *E. coli* has both $3' \rightarrow 5'$ and $5' \rightarrow 3'$ exonuclease activities in vitro. However, some of the properties of these exonucleases differ from those of DNA polymerase I. For example, the $3' \rightarrow 5'$ exonuclease of DNA polymerase III fails to degrade dinucleotides. Additionally, while the $5' \rightarrow 3'$ exonuclease activity can excise oligonucleotides from DNA, it differs from DNA polymerase I in its requirement for a single-stranded substrate (148). However, once hydrolysis is initiated, the exonuclease proceeds into a duplex region. Hence, the $5' \rightarrow 3'$ exonuclease function can catalyze the selective excision of thymine-containing pyrimidine dimers from UV-irradiated DNA previously incised with the *M. luteus* (and presumably the phage T4) PD-DNA glycosylase (168).

Multiple DNA polymerases also exist in eukaryotic cells. These, and their possible roles in both base and nucleotide excision repair, are discussed in chapter 8.

DNA Ligation

The final postincision biochemical event common to all forms of excision repair is the joining of the last newly incorporated nucleotide to the extant polynucleotide chain, i.e., the sealing of the nick left following the completion of repair synthesis. In *E. coli* this event is catalyzed by an extensively characterized enzyme called *DNA ligase*. As is true of nucleases and the DNA polymerases involved in the earlier postincision events in excision repair, DNA ligase is an enzyme that plays a role in aspects of DNA metabolism other than DNA repair, including genetic recombination and DNA replication (148).

DNA ligase of *E. coli* is a product of the $ligA^+$ gene, which maps at ~51 min on the chromosome (106). A conditional-lethal mutant designated *lig*(Ts)7 is abnormally sensitive to killing by alkylating agents (such as MMS) or by UV radiation, suggesting an involvement in DNA repair (147). *E. coli* DNA ligase is a monomeric protein of ~77 kDa that can catalyze ~25 joining events per min. There are about 200 to 400 DNA ligase molecules per cell under optimal growth conditions (157).

E. coli DNA ligase catalyzes the joining of nicks in duplex DNA with nucleotide termini containing 3'-OH and 5'-phosphate groups, in a reaction that has a requirement for Mg^{2+} and for NAD (157). During the reaction, the NAD is hydrolyzed to yield nicotinamide mononucleotide and AMP. The cleavage of a PP_i bond leads to the synthesis of a phosphodiester bond in DNA by a sequence of three steps involving two covalently linked intermediates (157). The first step (Fig. 4–32) consists of a reaction between the enzyme and NAD that generates a DNA ligase-adenylate and nicotinamide mononucleotide. Subsequently (Fig. 4–32), the

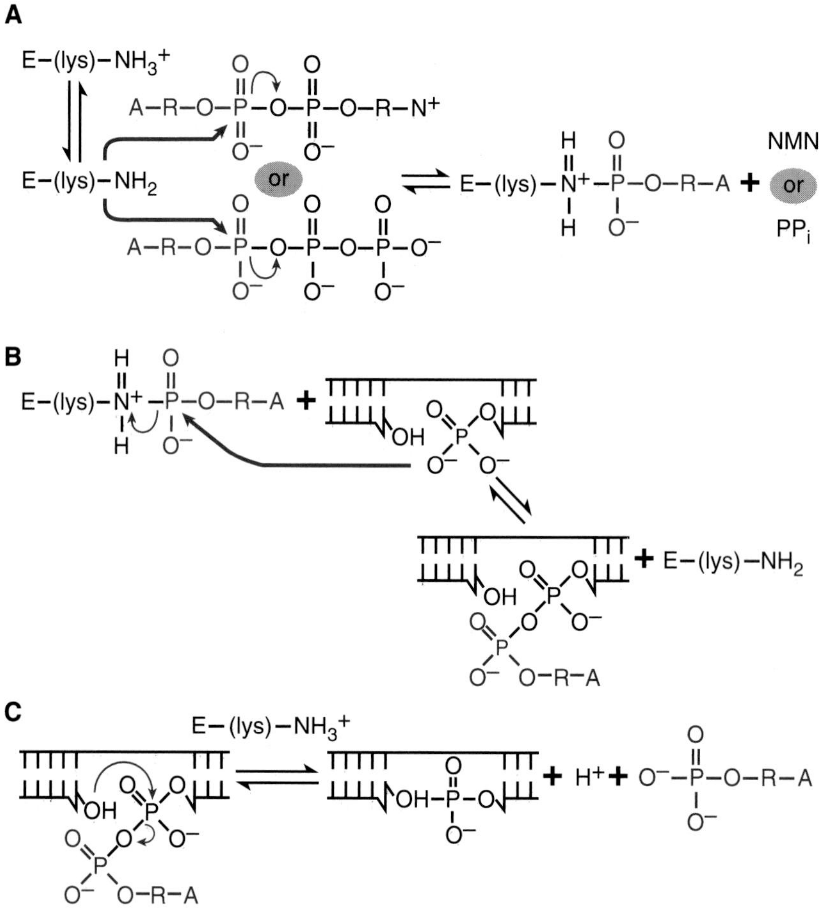

Figure 4–32 Steps in the overall DNA ligase reaction. See the text for details. (*Adapted from Kornberg and Baker [148] with permission.*)

adenylyl group is transferred to DNA to generate a new PP_i linkage between AMP and the 5′-phosphoryl terminus at the nick. The 5′-phosphate is then attacked by the opposing 3′-OH group to generate a phosphodiester bond, and AMP is liberated (Fig. 4–32).

DNA ligase is an ubiquitous enzyme. In higher eukaryotes, multiple forms of the enzyme have been detected, and these are discussed in chapter 8. Another prokaryote form of the enzyme that has been studied in detail is encoded by bacteriophage T4. It, too, is a single polypeptide chain, with a molecular mass of 63 kDa (157). In contrast to the *E. coli* enzyme, the phage T4 enzyme uses ATP rather than NAD as an energy source (157) and yields AMP and PP_i as reaction products. The T4 enzyme can catalyze the joining of oligodeoxyribonucleotides or of oligoribonucleotides in RNA-DNA hybrid duplex molecules and can also promote the joining of DNA duplexes with flush ends. Although the rate of the latter reaction is lower than that of the joining of nicks on duplex DNA, it is an extremely useful reaction that is widely used in recombinant DNA technology.

REFERENCES

1. **Aasland, R., L. C. Olsen, N. K. Spurr, H. E. Krokan, and D. E. Helland.** 1990. Chromosomal assignment of human uracil-DNA glycosylase to chromosome 12. *Genomics* **7:**139–141.

2. **Ames, B. N., M. K. Shigenaga, and T. M. Hagen.** 1993. Oxidants, antioxidants, and the degenerative diseases of aging. *Proc. Natl. Acad. Sci. USA* **90:**7915–7922.

3. **Aruoma, O. I., B. Halliwell, and M. Dizdaroglu.** 1989. Iron ion-dependent modification of bases in DNA by the superoxide radical-generating system hypoxanthine/xanthine oxidase. *J. Biol. Chem.* **264:**13024–13028.

4. **Asahara, H., P. M. Wistort, J. F. Bank, R. H. Bakerian, and R. P. Cunningham.** 1988. Purification and characterization of *Escherichia coli* endonuclease III from the cloned *nth* gene. *Biochemistry* **28:**4444–4449.

5. **Au, K. G., S. Clark, J. H. Miller, and P. Modrich.** 1988. *Escherichia coli mutY* gene encodes an adenine glycosylase active on G-A mispairs. *Proc. Natl. Acad. Sci. USA* **86:**8877–8881.

6. **Augustine, M. L., R. W. Hamilton, M. L. Dodson, and R. S. Lloyd.** 1991. Oligonucleotide site directed mutagenesis of all histidine residues within the T4 endonuclease V gene: role in enzyme-nontarget DNA binding. *Biochemistry* **30:**8052–8059.

7. **Bailly, V., and W. G. Verly.** 1987. *Escherichia coli* endonuclease III is not an endonuclease but a β-elimination catalyst. *Biochem. J.* **242:**565–572.

8. **Bailly, V., and W. G. Verly.** 1989. The multiple activities of *Escherichia coli* endonuclease IV and the extreme lability of 5′-terminal base-free deoxyribose 5-phosphates. *Biochem. J.* **259:**761–768.

9. **Bailly, V., W. G. Verly, T. R. O'Connor, and J. Laval.** 1989. Mechanism of DNA strand nicking at apurinic/apyrimidinic sites by *Escherichia coli* (formamidopyrimidine) DNA glycosylase. *Biochem. J.* **262:**581–589.

10. **Bennett, S. E., and D. W. Mosbaugh.** 1992. Characterization of the *Escherichia coli* uracil-DNA glycosylase inhibitor protein complex. *J. Biol. Chem.* **267:**22512–22521.

11. **Bennett, S. E., M. I. Schimerlik, and D. W. Mosbaugh.** 1993. Kinetics of the uracil-DNA glycosylase/inhibitor protein association. *J. Biol. Chem.* **268:**26879–26885.

12. **Bessho, T., R. Roy, K. Yamamoto, H. Kasai, S. Nishimura, K. Tano, and S. Mitra.** 1993. Repair of 8-hydroxyguanine in DNA by mammalian N-methylpurine-DNA glycosylase. *Proc. Natl. Acad. Sci. USA* **90:**8901–8904.

13. **Bessho, T., K. Tano, H. Kasai, E. Ohtsuka, and S. Nishimura.** 1993. Evidence for two DNA repair enzymes for 8-hydroxyguanine (7,8-dihydro-8-oxoguanine) in human cells. *J. Biol. Chem.* **268:**19416–19421.

14. **Bjelland, S., M. Bjoras, and E. Seeberg.** 1993. Excision of 3-methylguanine from alkylated DNA by 3-methyladenine DNA glycosylase I of *Escherichia coli*. *Nucleic Acids Res.* **21:**2045–2049.

15. **Bjelland, S., and E. Seeberg.** 1987. Purification and characterization of 3-methyladenine DNA glycosylase I from *Escherichia coli*. *Nucleic Acids Res.* **15:**2787–2801.

16. **Boiteux, S., M. Bichara, R. P. Fuchs, and J. Laval.** 1989. Excision of the imidazole ring-opened form of N-2-aminofluorene-C(8)-guanine adduct in poly (dG-dC) by *Escherichia coli* formamidopyrimidine-DNA glycosylase. *Carcinogenesis* **10:**1905–1909.

17. **Boiteux, S., and O. Huisman.** 1989. Isolation of a formamidopyrimidine-DNA glycosylase (*fpg*) mutant of *Escherichia coli* K12. *Mol. Gen. Genet.* **215:**300–305.

18. **Boiteux, S., T. R. O'Connor, and J. Laval.** 1987. Formamidopyrimidine-DNA glycosylase of *Escherichia coli*: cloning and sequencing of the *fpg* structural gene and overproduction of the protein. *EMBO J.* **6:**3177–3183.

19. **Bonner, C. A., S. Hays, K. McEntee, and M. F. Goodman.** 1990. DNA polymerase II is encoded by the DNA damage-inducible *dinA* gene of *Escherichia coli*. *Proc. Natl. Acad. Sci. USA* **87:**7663–7667.

20. **Boorstein, R. J., L. N. Chiu, and G. W. Teebor.** 1992. A mammalian cell line deficient in activity of the DNA repair enzyme 5-hydroxymethyluracil-DNA glycosylase is resistant to the toxic effects of the thymidine analog 5-hydroxymethyl-2′-deoxyuridine. *Mol. Cell. Biol.* **12:**5536–5540.

21. **Boorstein, R. J., T. P. Hilbert, J. Cadet, R. P. Cunningham, and G. W. Teebor.** 1989. UV-induced pyrimidine hydrates in DNA are repaired by bacterial and mammalian DNA glycosylase activities. *Biochemistry* **28:**6164–6170.

22. **Boorstein, R. J., D. D. Levy, and G. W. Teebor.** 1987. 5-Hydroxymethyluracil-DNA glycosylase activity may be a differentiated mammalian function. *Mutat. Res.* **183:**257–263.

23. **Borle, M., F. Campagnari, and D. Creissen.** 1979. Properties of purified uracil-DNA glycosylase from calf thymus. *J. Biol. Chem.* **257:**1208–1214.

24. **Breimer, L. H.** 1983. Urea–DNA glycosylase in mammalian cells. *Biochemistry* **22:**4192–4197.

25. **Breimer, L. H.** 1986. A DNA glycosylase for oxidized thymine residues in *Drosophila melanogaster*. *Biochem. Biophys. Res. Commun.* **134:**201–204.

26. **Breimer, L. H., and T. Lindahl.** 1984. DNA glycosylase activities for thymine residues damaged by ring saturation, fragmentation, or ring contraction are functions of endonuclease III in *Escherichia coli*. *J. Biol. Chem.* **259:**5543–5548.

27. **Brent, T.** 1979. Partial purification and characterization of human 3-methyladenine-DNA glycosylase. *Biochemistry* **18:**911–916.

28. **Brown, T. C., I. Zbinden, P. A. Cerutti, and J. Jiricny.** 1989. Modified SV40 for analysis of mismatch repair in simian and human cells. *Mutat. Res.* **220:**115–123.

29. **Burgers, P. M., and M. B. Klein.** 1986. Selection by genetic transformation of a *Saccharomyces cerevisiae* mutant defective for the nuclear uracil-DNA glycosylase. *J. Bacteriol.* **166:**905–913.

30. **Cabrera, M., Y. Nghiem, and J. Miller.** 1988. *mutM*, a second mutator locus in *Escherichia coli* that generates G·C → T·A transversions. *J. Bacteriol.* **170:**5405–5407.

31. **Castaing, B., S. Boiteux, and C. Zelwer.** 1992. DNA containing a chemically reduced apurinic site is a high affinity ligand for the *Escherichia coli* formamidopyrimidine-DNA glycosylase. *Nucleic Acids Res.* **20:**389–394.

32. **Cathcart, R., and D. A. Goldthwait.** 1981. Enzymatic excision of 3-methyladenine and 7-methylguanine by a rat liver nuclear fraction. *Biochemistry* **20:**273–280.

33. **Chakravarti, D., G. C. Ibeanu, K. Tano, and S. Mitra.** 1991. Cloning and expression in *Escherichia coli* of a human cDNA encoding the DNA repair protein N-methylpurine-DNA glycosylase. *J. Biol. Chem.* **266:**15710–15715.

34. **Chan, E., and B. Weiss.** 1987. Endonuclease IV of *Escherichia coli* is induced by paraquat. *Proc. Natl. Acad. Sci. USA* **84:**3189–3193.

35. **Chen, H., S. K. Bryan, and R. E. Moses.** 1989. Cloning the *polB* gene of *Escherichia coli* and identification of its product. *J. Biol. Chem.* **264:**20591–20595.

36. **Chen, J., B. Derfler, A. Maskati, and L. Samson.** 1989. Cloning a eukaryotic DNA glycosylase repair gene by the suppression of a DNA repair defect in *Escherichia coli*. *Proc. Natl. Acad. Sci. USA* **86:**7961–7965.

37. **Chen, J., B. Derfler, and L. Samson.** 1990. *Saccharomyces cerevisiae* 3-methyladenine DNA glycosylase has homology to the AlkA glycosylase of *E. coli* and is induced in response to DNA alkylation damage. *EMBO J.* **9:**4569–4575.

38. **Chen, J., and L. Samson.** 1991. Induction of *S. cerevisiae MAG* 3-methyladenine DNA glycosylase transcript levels in response to DNA damage. *Nucleic Acids Res.* **19:**6427–6432.

39. **Chetsanga, C. J., and T. Lindahl.** 1979. Release of 7-methylguanine residues whose imidazole rings have been opened from damaged DNA by a DNA glycosylase from *Escherichia coli*. *Nucleic Acids Res.* **6:**3673–3683.

40. Chetsanga, C. J., M. Lozon, C. Makaroff, and L. Savage. 1981. Purification and characterization of *Escherichia coli* formamidopyrimidine-DNA glycosylase that excises damaged 7-methylguanine from deoxyribonucleic acid. *Biochemistry* **20:**5201–5207.

41. Chung, M. H., H. Kasai, D. S. Jones, H. Inoue, H. Ishikawa, E. Ohtsuka, and S. Nishimura. 1991. An endonuclease activity of *Escherichia coli* that specifically removes 8-hydroxyguanine residues from DNA. *Mutat. Res.* **254:**1–12.

42. Clarke, N. D., M. Kvaal, and E. Seeberg. 1984. Cloning of *Escherichia coli* genes encoding 3-methyladenine DNA glycosylases I and II. *Mol. Gen. Genet.* **197:**368–372.

43. Clements, J. E., S. G. Rogers, and B. Weiss. 1978. A DNase for apurinic/apyrimidinic sites associated with exonuclease III of *Haemophilus influenzae. J. Biol. Chem.* **253:**2990–2999.

44. Coleman, S. H., and D. G. Wild. 1991. Location of the *fpg* gene on the *Escherichia coli* chromosome. *Nucleic Acids Res.* **19:**3999.

45. Cone, R., T. Bonura, and E. C. Friedberg. 1980. Inhibitor of uracil-DNA glycosylase induced by bacteriophage PBS2. Purification and preliminary characterization. *J. Biol. Chem.* **255:**10354–10358.

46. Cone, R., J. Duncan, L. Hamilton, and E. C. Friedberg. 1977. Partial purification and characterization of a uracil DNA N-glycosidase from *Bacillus subtilis. Biochemistry* **16:**3194–3201.

47. Cuniasse, P., G. V. Fazakerley, W. Guschlbauer, B. E. Kaplan, and L. C. Sowers. 1990. The abasic site as a challenge to DNA polymerase. A nuclear magnetic resonance study of G, C and T opposite a model abasic site. *J. Mol. Biol.* **213:**303–314.

48. Cunningham, R. P., H. Asahara, J. F. Bank, C. P. Scholes, J. C. Salerno, K. Surerus, E. Munck, J. McCracken, J. Peisach, and M. H. Emptage. 1988. Endonuclease III is an iron-sulfur protein. *Biochemistry* **28:**4450–4455.

49. Cunningham, R. P., S. M. Saporito, S. G. Spitzer, and B. Weiss. 1986. Endonuclease IV (*nfo*) mutant of *Escherichia coli. J. Bacteriol.* **168:**1120–1127.

50. Cunningham, R. P., and B. Weiss. 1985. Endonuclease III (*nth*) mutants of *Escherichia coli. Proc. Natl. Acad. Sci. USA* **82:**474–478.

51. Czeczot, H., B. Tudek, B. Lambert, J. Laval, and S. Boiteux. 1991. *Escherichia coli* Fpg protein and UvrACB endonuclease repair DNA damage induced by methylene blue plus visible light in vivo and in vitro. *J. Bacteriol.* **173:**3419–3424.

52. Davison, A. J., and J. E. Scott. 1986. The complete DNA sequence of varicella-zoster virus. *J. Gen. Virol.* **67:**1759–1816.

53. Delort, A. M., A. M. Duplaa, D. Molko, R. Teoule, J. P. Leblanc, and J. Laval. 1985. Excision of uracil residues in DNA: mechanism of action of *Escherichia coli* and *Micrococcus luteus* uracil-DNA glycosylases. *Nucleic Acids Res.* **13:**319–335.

54. DeLucia, P., and J. Cairns. 1969. Isolation of an *Escherichia coli* strain with a mutation affecting DNA polymerase. *Nature* (London) **224:**1164–1166.

55. Demple, B., J. Halbrook, and S. Linn. 1983. *Escherichia coli xth* mutants are hypersensitive to hydrogen peroxide. *J. Bacteriol.* **153:**1079–1082.

56. Demple, B., and L. Harrison. 1994. Repair of oxidative damage to DNA: enzymology and biology. *Annu. Rev. Biochem.* **63:**915–948.

57. Demple, B., T. Herman, and D. S. Chem. 1991. Cloning and expression of *APE*, the cDNA encoding the major human apurinic endonuclease: definition of a family of DNA repair enzymes. *Proc. Natl. Acad. Sci. USA* **88:**11450–11454.

58. Demple, B., A. Johnson, and D. Fung. 1986. Exonuclease III and endonuclease IV remove 3′ blocks from DNA synthesis primers in H$_2$O$_2$-damaged *Escherichia coli. Proc. Natl. Acad. Sci. USA* **83:**7731–7735.

59. Demple, B., and S. Linn. 1980. DNA N-glycosylases and UV repair. *Nature* (London) **287:**203–208.

60. Desiraju, V., W. G. Shanabruch, and A. L. Lu. 1993. Nucleotide sequence of the *Salmonella typhimurium mutB* gene, the homolog of *Escherichia coli mutY. J. Bacteriol.* **175:**541–543.

61. Dianov, G., and T. Lindahl. 1991. Preferential recognition of I-T base-pairs in the initiation of excision-repair by hypoxanthine-DNA glycosylase. *Nucleic Acids Res.* **19:**3829–3833.

62. Dianov, G., A. Price, and T. Lindahl. 1992. Generation of single-nucleotide repair patches following excision of uracil residues from DNA. *Mol. Cell. Biol.* **12:**1605–1612.

63. Dianov, G., B. Sedgwick, D. Graham, M. Olsson, and S. Lovett. 1994. Release of 5′-terminal deoxyribose-phosphate residues from incised abasic sites in DNA by the *Escherichia coli* RecJ protein. *Nucleic Acids Res.* **22:**993–998.

64. Dodson, M. L., and R. S. Lloyd. 1989. Structure-function studies of the T4 endonuclease V repair enzyme. *Mutat. Res.* **218:**49–65.

65. Dodson, M. L., R. D. Schrock, and R. S. Lloyd. 1993. Evidence for an imino intermediate in the T4 endonuclease V reaction. *Biochemistry* **32:**8284–8290.

66. Doetsch, P. W., and R. P. Cunningham. 1990. The enzymology of apurinic/apyrimidinic endonucleases. *Mutat. Res.* **236:**173–201.

67. Doetsch, P. W., W. D. Henner, R. P. Cunningham, J. H. Toney, and D. E. Helland. 1987. A highly conserved endonuclease activity present in *Escherichia coli*, bovine, and human cells recognizes oxidative DNA damage at sites of pyrimidines. *Mol. Cell. Biol.* **7:**26–32.

68. Doi, T., A. Recktenwald, Y. Karaki, M. Kikuchi, K. Morikawa, M. Ikehara, T. Inaoka, N. Hori, and E. Ohtsuka. 1992. Role of the basic amino acid cluster and Glu-23 in pyrimidine dimer glycosylase activity of T4 endonuclease V. *Proc. Natl. Acad. Sci. USA* **89:**9420–9424.

69. Domena, J. D., and D. W. Mosbaugh. 1985. Purification of nuclear and mitochondrial uracil-DNA glycosylase from rat liver. Identification of two distinct subcellular forms. *Biochemistry* **24:**7320–7328.

70. Domena, J. D., R. T. Timmer, S. A. Dicharry, and D. W. Mosbaugh. 1988. Purification and properties of mitochondrial uracil-DNA glycosylase from rat liver. *Biochemistry* **27:**6742–6751.

71. Dosanjh, M., R. Roy, S. Mitra, and B. Singer. 1994. 1,N6-ethenoadenine is preferred over 3-methyladenine as substrate by a cloned human N-methylpurine-DNA glycosylase (3-methyladenine-DNA glycosylase). *Biochemistry* **33:**1624–1628.

72. Dosanjh, M. K., A. Chenna, E. Kim, H. Fraenkel-Conrat, L. Samson, and B. Singer. 1994. All four known cyclic adducts formed in DNA by the vinyl chloride metabolite chloroacetaldehyde are released by a human DNA glycosylase. *Proc. Natl. Acad. Sci. USA* **91:**1024–1028.

73. Dowd, D. R., and R. S. Lloyd. 1989. Biological consequences of a reduction in the non-target DNA scanning capacity of a DNA repair enzyme. *J. Mol. Biol.* **208:**701–707.

74. Dowd, D. R., and R. S. Lloyd. 1989. Site-directed mutagenesis of the T4 endonuclease V gene: the role of arginine-3 in the target search. *Biochemistry* **28:**8699–8705.

75. Dowd, D. R., and R. S. Lloyd. 1990. Biological significance of facilitated diffusion in protein-DNA interactions. Applications to T4 endonuclease V-initiated DNA repair. *J. Biol. Chem.* **265:**3424–3431.

76. Dudley, B., A. Hammond, and W. A. Deutsch. 1992. The presence of uracil-DNA glycosylase in insects is dependent upon developmental complexity. *J. Biol. Chem.* **267:**11964–11967.

77. Duker, N. J., D. E. Jensen, D. M. Hart, and D. E. Fishbein. 1982. Perturbations of enzymic uracil excision due to purine damage in DNA. *Proc. Natl. Acad. Sci. USA* **79:**4878–4882.

78. Duncan, B. K. 1981. DNA glycosylases, p. 565–586. *In* P. D. Boyer (ed.), *The Enzymes*, 3rd ed. Academic Press, Inc., New York.

79. Duncan, B. K., and B. Weiss. 1982. Specific mutator effects of *ung* (uracil-DNA glycosylase) mutations in *Escherichia coli. J. Bacteriol.* **151:**750–755.

80. Duncan, J., L. Hamilton, and E. C. Friedberg. 1976. Enzymatic degradation of uracil-containing DNA. *J. Virol.* **19:**338–345.

81. Efdedal, I., P. H. Guddal, G. Slupphaug, G. Volden, and H. E. Krokan. 1993. Consensus sequences for good and poor removal of uracil from double stranded DNA by uracil-DNA glycosylase. *Nucleic Acids Res.* **21:**2095–2101.

82. **el-Hajj, H. H., L. Wang, and B. Weiss.** 1992. Multiple mutant of *Escherichia coli* synthesizing virtually thymineless DNA during limited growth. *J. Bacteriol.* **174:**4450–4456.

83. **Engelward, B. P., M. S. Boosalis, B. J. Chen, Z. Deng, J. J. Siciliano, and L. D. Samson.** 1993. Cloning and characterization of a mouse 3-methyladenine/7-methylguanine/3-methylguanine DNA glycosylase cDNA whose gene maps to chromosome 11. *Carcinogenesis* **14:**175–181.

84. **Evensen, G., and E. Seeberg.** 1982. Adaptation to alkylation resistance involves the induction of a DNA glycosylase. *Nature* (London) **296:**773–775.

85. **Franklin, W. A., and T. Lindahl.** 1988. DNA deoxyribophosphodiesterase. *EMBO J.* **7:**3617–3622.

86. **Friedberg, E. C.** 1972. Studies on the substrate specificity of the T4 excision repair endonuclease. *Mutat. Res.* **15:**113–123.

87. **Friedberg, E. C.** 1985. *DNA Repair.* W. H. Freeman & Co., New York.

88. **Friedberg, E. C., T. Bonura, E. H. Radany, and J. D. Love.** 1981. Enzymes that incise damaged DNA, p. 251–279. *In* P. D. Boyer (ed.), *The Enzymes,* 3rd ed. Academic Press, Inc., New York.

89. **Friedberg, E. C., K. H. Cook, J. Duncan, and K. Mortelmans.** 1977. DNA repair enzymes in mammalian cells. *Photochem. Photobiol. Rev.* **2:**263–322.

90. **Friedberg, E. C., A. K. Ganesan, and K. Minton.** 1975. N-Glycosidase activity in extracts of *Bacillus subtilis* and its inhibition after infection with bacteriophage PBS2. *J. Virol.* **16:**315–321.

91. **Friedberg, E. C., and D. A. Goldthwait.** 1968. Endonuclease II of *E. coli. Cold Spring Harbor Symp. Quant. Biol.* **33:**271–275.

92. **Friedberg, E. C., and D. A. Goldthwait.** 1969. Endonuclease II of *E. coli,* I. Isolation and purification. *Proc. Natl. Acad. Sci. USA* **62:**934–940.

93. **Friedberg, E. C., and J. J. King.** 1969. Endonucleolytic cleavage of UV-irradiated DNA controlled by the *V*+ gene in phage T4. *Biochem. Biophys. Res. Commun.* **37:**646–651.

94. **Friedberg, E. C., and J. J. King.** 1971. Dark repair of ultraviolet-irradiated deoxyribonucleic acid by bacteriophage T4: purification and characterization of a dimer-specific phage-induced endonuclease. *J. Bacteriol.* **106:**500–507.

95. **Friedberg, E. C., and I. R. Lehman.** 1974. Excision of thymine dimers by proteolytic and amber fragments of *E. coli* DNA polymerase I. *Biochem. Biophys. Res. Commun.* **58:**132–139.

96. **Friedberg, E. C., K. Minton, G. Pawl, and P. Verzola.** 1974. Excision of thymine dimers in vitro by extracts of bacteriophage-infected *Escherichia coli. J. Virol.* **13:**953–959.

97. **Fu, W., S. O'Handley, R. P. Cunningham, and M. K. Johnson.** 1992. The role of the iron-sulfur cluster in *Escherichia coli* endonuclease III. A resonance Raman study. *J. Biol. Chem.* **267:**16135–16137.

98. **Furiacelli, A. F., B. J. Wegher, W. F. Blakely, and M. Dizdaroglu.** 1990. Yields of radiation-induced base products in DNA: effects of DNA conformation and gassing conditions. *Int. J. Radiat. Biol.* **58:**397–415.

99. **Gallagher, P. E., and T. P. Brent.** 1984. Further purification and characterization of human 3-methyladenine-DNA glycosylase. Evidence for broad specificity. *Biochim. Biophys. Acta* **782:**394–401.

100. **Ganguly, T., and N. J. Duker.** 1990. Glycosylases that excise modified DNA pyrimidines in young and senescent human WI-38 fibroblasts. *Mutat. Res.* **237:**107–115.

101. **Gombar, C. T., E. J. Katz, P. N. Magee, and M. A. Sirover.** 1981. Induction of the DNA repair enzymes uracil DNA glycosylase and 3-methyladenine DNA glycosylase in regenerating rat liver. *Carcinogenesis* **2:**595–599.

102. **Gordon, L. K., and W. A. Haseltine.** 1980. Comparison of the cleavage of pyrimidine dimers by the bacteriophage T4 and *Micrococcus luteus* UV-specific endonucleases. *J. Biol. Chem.* **255:**12047–12050.

103. **Gordon, L. K., and W. A. Haseltine.** 1981. Early steps of excision repair of cyclobutane pyrimidine dimers by the *Micrococcus luteus* endonuclease. A three-step incision model. *J. Biol. Chem.* **256:**8608–8616.

104. **Gossard, F., and W. G. Verly.** 1978. Properties of the main endonuclease specific for apurinic sites of *Escherichia coli* (endonuclease VI). Mechanism of apurinic site excision from DNA. *Eur. J. Biochem.* **82:**321–332.

105. **Gossett, J., K. Lee, R. P. Cunningham, and P. W. Doetsch.** 1988. Yeast redoxyendonuclease, a DNA repair enzyme similar to *Escherichia coli* endonuclease III. *Biochemistry* **27:**2629–2634.

106. **Gottesman, M. M., M. L. Hicks, and M. Gellert.** 1973. Genetics and function of DNA ligase in *Escherichia coli. J. Mol. Biol.* **77:**531–547.

107. **Grabowski, D. T., W. A. Deutsch, D. Derda, and M. R. Kelley.** 1991. *Drosophila* AP3, a presumptive DNA repair protein, is homologous to human ribosomal associated protein P0. *Nucleic Acids Res.* **19:**4297.

108. **Grafstrom, R. H., L. Park, and L. Grossman.** 1982. Enzymatic repair of pyrimidine dimer-containing DNA. A 5′ dimer DNA glycosylase:3′-apyrimidinic endonuclease mechanism from *Micrococcus luteus. J. Biol. Chem.* **257:**13465–13474.

109. **Graves, R. J., I. Felzenszwalk, J. Laval, and T. R. O'Connor.** 1992. Excision of 5′-terminal deoxyribose phosphate from damaged DNA is catalyzed by the Fpg protein of *Escherichia coli. J. Biol. Chem.* **267:**14429–14435.

110. **Grossman, L., S. Riazuddin, W. A. Haseltine, and C. Lindan.** 1979. Nucleotide excision repair of damaged DNA. *Cold Spring Harbor Symp. Quant. Biol.* **43:**947–955.

111. **Hamilton, K. K., P. M. Kim, and P. W. Doetsch.** 1992. A eukaryotic DNA glycosylase/lyase recognizing ultraviolet light-induced pyrimidine dimers. *Nature* (London) **356:**725–728.

112. **Hanawalt, P. C., P. K. Cooper, A. K. Ganesan, and C. A. Smith.** 1979. DNA repair in bacteria and mammalian cells. *Annu. Rev. Biochem.* **48:**783–836.

113. **Harm, W.** 1961. Gene-controlled reactivation of ultraviolet-inactivated bacteriophage. *J. Cell. Comp. Physiol.* **58** (Suppl. 1):69–77.

114. **Harm, W.** 1963. Mutants of phage T4 with increased sensitivity to ultraviolet. *Virology* **19:**66–71.

115. **Haseltine, W. A., L. K. Gordon, C. P. Lindan, R. H. Grafstrom, N. L. Shaper, and L. Grossman.** 1980. Cleavage of pyrimidine dimers in specific DNA sequences by a pyrimidine dimer DNA-glycosylase of *M. luteus. Nature* (London) **285:**634–641.

116. **Hatahet, Z., Y. W. Kow, A. A. Permal, R. P. Cunningham, and S. S. Wallace.** 1994. New substrates for old enzymes: 5-hydroxy-2′-deoxycytidine and 5-hydroxy-2′-deoxyuridine are substrates for *Escherichia coli* endonuclease III and formamidopyrimidine DNA N-glycosylase while 5-hydroxy-2′-deoxyuridine is a substrate for uracil-DNA N-glycosidase. *J. Biol. Chem.* **269:**18814–18820.

117. **Hems, G.** 1960. Effects of ionizing radiation of aqueous solutions of inosine and adenosine. *Radiat. Res.* **13:**777–787.

118. **Hentosh, P., W. D. Henner, and R. J. Reynolds.** 1985. Sequence specificity of DNA cleavage by *Micrococcus luteus* gamma endonuclease. *Radiat. Res.* **102:**119–129.

119. **Higley, M., and R. S. Lloyd.** 1993. Processivity of uracil DNA glycosylase. *Mutat. Res.* **294:**109–116.

120. **Hollstein, M. C., P. Brooks, S. Linn, and B. N. Ames.** 1984. Hydroxymethyluracil DNA glycosylase in mammalian cells. *Proc. Natl. Acad. Sci. USA* **81:**4003–4007.

121. **Hori, N., T. Doi, Y. Karaki, M. Kikuchi, M. Ikehara, and E. Ohtsuka.** 1993. Participation of glutamic acid 23 of T4 endonuclease V in the beta-elimination reaction of an abasic site in a synthetic duplex DNA. *Nucleic Acids Res.* **20:**4761–4764.

122. **Impellizzeri, K. J., B. Anderson, and P. M. Burgers.** 1991. The spectrum of spontaneous mutations in a *Saccharomyces cerevisiae* uracil-DNA-glycosylase mutant limits the function of this enzyme to cytosine deamination repair. *J. Bacteriol.* **173:**6807–6810.

123. **Ishida, M., Y. Kanamori, N. Hori, T. Inaoka, and E. Ohtsuka.** 1990. In vitro and in vivo activities of T4 endonuclease V mutants altered in the C-terminal aromatic region. *Biochemistry* **29:**3817–3821.

124. Ito, M., and M. Sekiguchi. 1976. Repair of DNA damaged by 4-nitroquinoline 1-oxide: a comparison of *Escherichia coli* and bacteriophage T4 repair systems. *Jpn. J. Genet.* **51:**129–133.

125. Iwasaki, H., Y. Ishino, H. Toh, A. Nakata, and H. Shinagawa. 1991. *Escherichia coli* DNA polymerase II is homologous to alpha-like DNA polymerases. *Mol. Gen. Genet.* **226:**24–33.

126. Iwasaki, H., A. Nakata, G. C. Walker, and H. Shinagawa. 1990. The *Escherichia coli* polB gene, which encodes DNA polymerase II, is regulated by the SOS system. *J. Bacteriol.* **172:**6268–6273.

127. Izumu, T., K. Ishizaki, M. Ikenaga, and S. Yonei. 1992. A mutant endonuclease IV of *Escherichia coli* loses the ability to repair lethal DNA damage induced by hydrogen peroxide but not that induced by methyl methanesulfonate. *J. Bacteriol.* **174:**7711–7716.

128. Johnson, A. W., and B. Demple. 1988. Yeast DNA 3′-repair diesterase is the major cellular apurinic/apyrimidinic endonuclease: substrate specificity and kinetics. *J. Biol. Chem.* **263:**18017–18022.

129. Johnson, A. W., and B. Demple. 1988. Yeast DNA diesterase for 3′-fragments of deoxyribose: purification and physical properties of a repair enzyme for oxidative DNA damage. *J. Biol. Chem.* **263:**18009–18016.

130. Jorgensen, T. J., E. A. Furlong, and W. D. Henner. 1988. Gamma endonuclease of *Micrococcus luteus*: action on irradiated DNA. *Radiat. Res.* **114:**556–566.

131. Jorgensen, T. J., Y. W. Kow, S. S. Wallace, and S. S. Henner. 1987. Mechanism of action of *Micrococcus luteus* gamma-endonuclease. *Biochemistry* **26:**6436–6443.

132. Kaasen, I., G. Evensen, and E. Seeberg. 1986. Amplified expression of the tag⁺ and alkA⁺ genes in *Escherichia coli*: identification of gene products and effects on alkylation resistance. *J. Bacteriol.* **168:**642–647.

133. Kaplan, J. C., S. F. Kushner, and L. Grossman. 1969. Enzymatic repair of DNA. 1. Purification of two enzymes involved in the excision of thymine dimers from ultraviolet-irradiated DNA. *Proc. Natl. Acad. Sci. USA* **63:**144–151.

134. Karran, P., R. Cone, and E. C. Friedberg. 1981. Specificity of the bacteriophage PBS2 induced inhibitor of uracil-DNA glycosylase. *Biochemistry* **21:**6092–6096.

135. Karran, P., T. Hjelmgren, and T. Lindahl. 1982. Induction of a DNA glycosylase for N-methylated purines is part of the adaptive response to alkylating agents. *Nature* (London) **296:**770–773.

136. Karran, P., and T. Lindahl. 1978. Enzymatic excision of free hypoxanthine from polydeoxynucleotides and DNA containing deoxyinosine monophosphate residues. *J. Biol. Chem.* **253:**5877–5879.

137. Karran, P., and T. Lindahl. 1980. Hypoxanthine in deoxyribonucleic acid: generation by heat-induced hydrolysis of adenine residues and release in free form by a deoxyribonucleic acid glycosylase from calf thymus. *Biochemistry* **19:**6005–6011.

138. Karran, P., T. Lindahl, I. Ofsteng, G. B. Evensen, and E. Seeberg. 1980. *Escherichia coli* mutants deficient in 3-methyladenine-DNA glycosylase. *J. Mol. Biol.* **140:**101–127.

139. Kataoka, H., and M. Sekiguchi. 1985. Molecular cloning and characterization of the alkB gene of *Escherichia coli*. *Mol. Gen. Genet.* **198:**263–269.

140. Kataoka, H., Y. Yamamoto, and M. Sekiguchi. 1983. A new gene (alkB) of *Escherichia coli* that controls sensitivity to methyl methanesulfonate. *J. Bacteriol.* **153:**1301–1307.

141. Keller, W., and R. Crouch. 1972. Degradation of DNA RNA hybrids by ribonuclease H and DNA polymerases of cellular and viral origin. *Proc. Natl. Acad. Sci. USA* **69:**3360–3364.

142. Kelley, M. R., S. Venugopal, J. Harless, and W. A. Deutsch. 1989. Antibody to a human DNA repair protein allows for cloning of a *Drosophila* cDNA that encodes an apurinic endonuclease. *Mol. Cell. Biol.* **9:**965–973.

143. Kelly, R. B., M. R. Atkinson, J. A. Huberman, and A. Kornberg. 1969. Excision of thymine dimers and other mismatched sequences by DNA polymerase of *Escherichia coli*. *Nature* (London) **224:**495–501.

144. Kim, J., and S. Linn. 1988. The mechanism of action of *E. coli* endonuclease III and T4 endonuclease (endonuclease V) at AP sites. *Nucleic Acids Res.* **16:**1135–1141.

145. Kim, J., and S. Linn. 1989. Purification and characterization of UV endonucleases I and II from murine plasmacytoma cells. *J. Biol. Chem.* **264:**2739–2745.

146. Kondo, H., Y. Nakabeppu, H. Kataoka, S. Kuhara, S. Kawabata, and M. Sekiguchi. 1986. Structure and expression of the alkB gene of *Escherichia coli* related to the repair of alkylated DNA. *J. Biol. Chem.* **261:**15772–15777.

147. Konrad, E. B., P. Modrich, and I. R. Lehman. 1973. Genetic and enzymatic characterization of a conditional lethal mutant of *Escherichia coli* K-12 with a temperature-sensitive DNA ligase. *J. Mol. Biol.* **77:**519–529.

148. Kornberg, A., and T. Baker. 1992. *DNA Replication*. W. H. Freeman & Co., New York.

149. Kow, Y. W. 1989. Mechanism of action of *Escherichia coli* exonuclease III. *Biochemistry* **28:**3280–3287.

150. Kow, Y. W., and S. S. Wallace. 1985. Exonuclease III recognizes urea residues in oxidized DNA. *Proc. Natl. Acad. Sci. USA* **82:**8354–8358.

151. Kow, Y. W., and S. S. Wallace. 1987. Mechanism of action of *Escherichia coli* endonuclease III. *Biochemistry* **26:**8200–8206.

152. Kuo, C. F., D. E. McRee, C. L. Fisher, S. F. O'Handley, R. P. Cunningham, and J. A. Tainer. 1992. Atomic structure of the DNA repair (4Fe-4S) enzyme endonuclease III. *Science* **258:**434–440.

153. Laval, J., J. Pierre, and F. Laval. 1981. Release of 7-methylguanine residues from alkylated DNA by extracts of *Micrococcus luteus* and *Escherichia coli*. *Proc. Natl. Acad. Sci. USA* **78:**852–855.

154. Lawley, P. D., and D. J. Orr. 1970. Specific excision of methylation products from DNA of *Escherichia coli* treated with N-methyl-N′-nitro-N-nitrosoguanidine. *Chem.-Biol. Interact.* **2:**154–157.

155. Lawley, P. D., and W. Warren. 1976. Removal of minor methylation products 7-methyladenine and 3-methylguanine from DNA of *Escherichia coli* treated with dimethyl sulfate. *Chem.-Biol. Interact.* **12:**211–220.

156. Lee, K., W. H. McCray, and P. W. Doetsch. 1987. Thymine glycol-DNA glycosylase/AP endonuclease of CEM-C1 lymphoblasts: a human analog of *Escherichia coli* endonuclease III. *Biochem. Biophys. Res. Commun.* **149:**93–101.

157. Lehman, I. R. 1974. DNA ligase: structure, mechanism, and function. *Science* **186:**790–797.

158. LeMotte, P. K., and G. C. Walker. 1985. Induction and autoregulation of ada, a positively acting element regulating the response of *Escherichia coli* K-12 to methylating agents. *J. Bacteriol.* **161:**888–895.

159. Levin, J. D., and B. Demple. 1990. Analysis of class II (hydrolytic) and class I (beta-lyase) apurinic/apyrimidinic endonucleases with a synthetic DNA substrate. *Nucleic Acids Res.* **18:**5069–5075.

160. Levin, J. D., A. W. Johnson, and B. Demple. 1988. Homogeneous *Escherichia coli* endonuclease IV. *J. Biol. Chem.* **263:**8066–8071.

161. Levin, J. D., R. Shapiro, and B. Demple. 1991. Metalloenzymes in DNA repair. *Escherichia coli* endonuclease IV and *Saccharomyces cerevisiae* Apn1. *J. Biol. Chem.* **266:**22893–22898.

162. Levy, D., and G. Teebor. 1990. Site directed mutagenesis by 5-hydroxymethyluracil following in vivo replication of ox-174am3DNA. *J. Cell Biol.* **14A**(Suppl.):56.

163. Lindahl, T. 1974. An N-glycosidase from *Escherichia coli* that releases free uracil from DNA containing deaminated cytosine residues. *Proc. Natl. Acad. Sci. USA* **71:**3649–3653.

164. Lindahl, T. 1976. New class of enzymes acting on damaged DNA. *Nature* (London) **259:**64–66.

165. Lindahl, T. 1979. DNA glycosylases, endonucleases for apurinic/apyrimidinic sites, and base excision-repair. *Prog. Nucleic Acid Res. Mol. Biol.* **22:**135–192.

166. Lindahl, T. 1982. DNA repair enzymes. *Annu. Rev. Biochem.* **51:**61–87.

167. **Lindahl, T., S. Ljungquist, W. Siegert, B. Nyberg, and B. Sperens.** 1977. DNA N-glycosidases: properties of uracil-DNA glycosidase from *Escherichia coli. J. Biol. Chem.* **252:**3286–3294.

167a. **Lindahl, T., and B. Sedgwick.** 1988. Regulation and expression of the adaptive response to alkylating agents. *Annu. Rev. Biochem.* **57:**133–157.

168. **Livingston, D. M., and C. C. Richardson.** 1975. Deoxyribonucleic acid polymerase III of *Escherichia coli.* Characterization of associated exonuclease activities. *J. Biol. Chem.* **250:**470–478.

169. **Ljungquist, S.** 1977. A new endonuclease from *Escherichia coli* acting at apurinic sites in DNA. *J. Biol. Chem.* **252:**2808–2814.

170. **Ljungquist, S., T. Lindahl, and P. Howard-Flanders.** 1976. Methyl methanesulfonate-sensitive mutant of *Escherichia coli* deficient in an endonuclease specific for apurinic sites in deoxyribonucleic acid. *J. Bacteriol.* **126:**646–653.

171. **Luria, S. E.** 1947. Reactivation of irradiated bacteriophage by transfer of self-reproducing units. *Proc. Natl. Acad. Sci. USA* **33:**253–264.

172. **Maki, H., and M. Sekiguchi.** 1992. MutT protein specifically hydrolyses a potent mutagenic substrate for DNA synthesis. *Nature* (London) **355:**273–275.

173. **Manoharan, M., S. C. Ransom, A. Mazumder, and J. A. Gerlt.** 1988. The characterization of abasic sites in DNA heteroduplexes by site specific labeling with ^{13}C. *J. Am. Chem. Soc.* **110:**1620–1622.

174. **Mansur, N. R., K. Meyer-Siegler, J. C. Wurzer, and M. A. Sirover.** 1993. Cell cycle regulation of the glyceraldehyde-3-phosphate dehydrogenase/uracil DNA glycosylase gene in normal human cells. *Nucleic Acids Res.* **21:**993–998.

175. **Margison, G. P., and P. J. O'Connor.** 1973. Biological implications of the instability of the N-glycosidic bone of 3-methyldeoxyadenosine in DNA. *Biochim. Biophys. Acta.* **331:**349–356.

176. **Mazumder, A., J. Gerlt, L. Rabow, M. Absalon, J. Stubbe, and P. Bolton.** 1989. UV endonuclease V from Bacteriophage T4 catalyzes DNA strand cleavage at aldehydic abasic sites by a syn β-elimination reaction. *J. Am. Chem. Soc.* **111:**8029–8030.

177. **Mazumder, A., J. A. Gerlt, M. A. Absalon, J. A. Stubbe, R. P. Cunningham, J. Withka, and P. H. Bolton.** 1991. Stereochemical studies of the β-elimination reactions at aldehyde abasic sites in DNA: endonuclease III from *Escherichia coli,* sodium hydroxide, and Lys-Trp-Lys. *Biochemistry* **30:**1119–1126.

178. **McCarthy, T., P. Karran, and T. Lindahl.** 1984. Inducible repair of O-alkylated DNA pyrimidines in *Escherichia coli. EMBO J.* **3:**545–550.

179. **McGeoch, D. J., M. A. Dalrymple, A. J. Davison, A. Dolan, M. C. Frame, and D. McNab.** 1988. The complete DNA sequence of the long unique region in the genome of herpes simplex virus type 1. *J. Gen Virol.* **69:**1531–1574.

180. **McHenry, C. S.** 1991. DNA polymerase III holoenzyme. Components, structure, and mechanism of a true replicative complex. *J. Biol. Chem.* **266:**19127–19130.

181. **McMillan, S., H. J. Edenberg, E. H. Radany, R. C. Friedberg, and E. C. Friedberg.** 1981. *denV* gene of bacteriophage T4 codes for both pyrimidine dimer-DNA glycosylase. *J. Virol.* **40:**211–223.

182. **Melamede, R. J., Z. Hatahet, Y. W. Kow, H. Ide, and S. S. Wallace.** 1994. Isolation and characterization of endonuclease VIII from *Escherichia coli. Biochemistry* **33:**1255–1264.

183. **Meyer-Siegler, K., D. Mauro, G. Seal, J. Wurzer, J. DeRiel, and M. Sirover.** 1991. A human nuclear uracil DNA glycosylase is the 37-kDa subunit of glyceraldehyde-3-phosphate dehydrogenase. *Proc. Natl. Acad. Sci. USA* **88:**8460–8464.

184. **Michaels, M. L., and J. H. Miller.** 1992. The GO system protects organisms from the mutagenic effect of the spontaneous lesion 8-hydroxyguanine (7,8-dihydro-8-oxoguanine). *J. Bacteriol.* **174:**6321–6325.

185. **Michaels, M. L., L. Pham, Y. Nghiem, C. Cruz, and J. H. Miller.** 1990. MutY, an adenine glycosylase active on G-A mispairs, has homology to endonuclease III. *Nucleic Acids Res.* **18:**3841–3845.

186. **Minton, K., M. Durphy, R. Taylor, and E. C. Friedberg.** 1975. The ultraviolet endonuclease of bacteriophage T4. Further characterization. *J. Biol. Chem.* **250:**2823–2829.

187. **Mo, J. Y., H. Maki, and M. Sekiguchi.** 1992. Hydrolytic elimination of a mutagenic nucleotide, 8-oxodGTP, by human 18-kilodalton protein: sanitization of nucleotide pool. *Proc. Natl. Acad. Sci. USA* **89:**11021–11025.

188. **Morikawa, K., O. Matsumoto, M. Tsujimoto, K. Katayanagi, M. Ariyoshi, T. Doi, M. Ikehara, T. Inaoka, and E. Ohtsuka.** 1992. X-ray strucuture of T4 endonuclease V: an excision repair enzyme specific for a pyrimidine dimer. *Science* **256:**523–526.

189. **Mortelmans, K., and E. C. Friedberg.** 1972. Deoxyribonucleic acid repair in bacteriophage T4: observations on the roles of the *x* and *v* genes and of host factors. *J. Virol.* **10:**730–736.

190. **Muller, S. J., and S. Caradonna.** 1991. Isolation and characterization of a human cDNA encoding uracil-DNA glycosylase. *Biochim. Biophys. Acta* **1088:**197–207.

191. **Muller, S. J., and S. Caradonna.** 1993. Cell cycle regulation of a human cyclin-like gene encoding uracil-DNA glycosylase. *J. Biol. Chem.* **268:**1310–1319.

192. **Myrnes, B., P. H. Guddal, and H. Krokan.** 1982. Metabolism of dITP in HeLa cell extracts, incorporation into DNA by isolated nuclei and release of hypoxanthine from DNA by a hypoxanthine-DNA glycosylase activity. *Nucleic Acids Res.* **10:**3693–3701.

193. **Nakabeppu, Y., H. Kondo, and M. Sekiguchi.** 1984. Cloning and characterization of the *alkA* gene of *Escherichia coli* that encodes 3-methyladenine DNA glycosylase II. *J. Biol. Chem.* **259:**13723–13729.

194. **Nakabeppu, Y., T. Miyata, H. Kondo, S. Iwanaga, and M. Sekiguchi.** 1984. Structure and expression of the *alkA* gene of *Escherichia coli* involved in adaptive response to alkylating agents. *J. Biol. Chem.* **259:**13730–13736.

195. **Nakabeppu, Y., K. Yamashita, and M. Sekiguchi.** 1982. Purification and chacterization of normal and mutant forms of T4 endonuclease V. *J. Biol. Chem.* **257:**2556–2562.

196. **Nakayama, H., S. Shiota, and K. Umezu.** 1992. UV endonuclease-mediated enhancement of UV survival in *Micrococcus luteus*: evidence revealed by deficiency in the Uvr homolog. *Mutat. Res.* **273:**43–48.

197. **Neddermann, P., and J. Jiricny.** 1993. The purification of a mismatch-specific thymine-DNA glycosylase from HeLa cells. *J. Biol. Chem.* **268:**21218–21224.

198. **Nickell, C., and R. S. Lloyd.** 1991. Mutations in endonuclease V that affect both protein-protein association and target site location. *Biochemistry* **30:**8638–8648.

199. **Nickell, C., M. A. Prince, and R. S. Lloyd.** 1992. Consequences of molecular engineering enhanced DNA binding in a DNA repair enzyme. *Biochemistry* **31:**4189–4198.

200. **Nishida, Y., S. Yasuda, and M. Sekiguchi.** 1976. Repair of DNA damaged by methyl methanesulfonate in bacteriophage T4. *Biochim. Biophys. Acta* **442:**208–215.

201. **O'Connor, T. R.** 1993. Purification and characterization of human 3-methyladenine-DNA glycosylase. *Nucleic Acids Res.* **21:**5561–5569.

202. **O'Connor, T. R., and F. Laval.** 1990. Isolation and structure of a cDNA expressing a mammalian 3-methyladenine-DNA glycosylase. *EMBO J.* **9:**3337–3342.

203. **O'Connor, T. R., and J. Laval.** 1989. Physical association of the 2,6-diamino-4-hydroxy-5N-formamidopyrimidine-DNAglycosylase of *Escherichia coli* and an activity nicking DNA at apurinic/apyrimidinic sites. *Proc. Natl. Acad. Sci. USA* **86:**5222–5226.

204. **Ohshima, S., and M. Sekiguchi.** 1972. Induction of a new enzyme activity to excise pyrimidine dimers in *Escherichia coli* infected with bacteriophage T4. *Biochem. Biophys. Res. Commun.* **47:**1126–1132.

205. **Ohshima, S., and M. Sekiguchi.** 1975. Biochemical studies on radiation-sensitive mutations in bacteriophage T4-1. *J. Biochem.* **77:**303–311.

206. Olsen, L., R. Aasland, H. Krokan, and D. Helland. 1991. Human uracil-DNA glycosylase complements *Escherichia coli ung* mutants. *Nucleic Acids Res.* **19**:4473–4478.

207. Olsen, L. C., R. Aasland, C. U. Wittwer, H. E. Krokan, and D. E. Helland. 1989. Molecular cloning of human uracil-DNA glycosylase, a highly conserved DNA repair enzyme. *EMBO J.* **8**:3121–3125.

208. Paek, K. H., and G. C. Walker. 1986. Defect in expression of heat shock proteins at high temperature in *xthA* mutants. *J. Bacteriol.* **165**:763–770.

209. Pawl, G., R. Taylor, K. Minton, and E. C. Friedberg. 1976. Enzymes involved in thymine dimer excision in bacteriophage T4-infected *Escherichia coli. J. Mol. Biol.* **108**:99–109.

210. Percival, K. J., M. B. Klein, and P. M. Burgers. 1989. Molecular cloning and primary structure of the uracil-DNA-glycosylase gene from *Saccharomyces cerevisiae. J. Biol. Chem.* **264**:2593–2598.

211. Popoff, S. C., A. I. Spira, A. W. Johnson, and B. Demple. 1990. Yeast structural gene (*APN1*) for the major apurinic endonuclease: homology to *Escherichia coli* endonuclease IV. *Proc. Natl. Acad. Sci. USA* **87**:4193–4197.

212. Price, A., and T. Lindahl. 1991. Enzymatic release of 5′-terminal deoxyribose phosphate residues from damaged DNA in human cells. *Biochemistry* **30**:8631–8637.

213. Purmal, A. A., G. W. Lampman, E. I. Pourmal, R. J. Melamede, S. S. Wallace, and Y. W. Kow. 1994. Direct proof that uracil DNA N-glycosidase distributively interacts with uracil-containing DNA. *J. Biol. Chem.* **269**:22046–22053.

214. Puyet, A., B. Greenberg, and S. A. Lacks. 1989. The *exoA* gene of *Streptococcus pneumoniae* and its product, a DNA exonuclease with apurinic endonuclease activity. *J. Bacteriol.* **171**:2278–2286.

215. Radany, E. H., and E. C. Friedberg. 1980. A pyrimidine dimer-DNA glycosylase activity associated with the *v* gene of bacteriophage T4. *Nature* (London) **286**:182–185.

216. Radany, E. H., and E. C. Friedberg. 1982. Demonstration of pyrimidine dimer-DNA glycosylase activity in vivo: bacteriophage T4-infected *Escherichia coli* as a model system. *J. Virol.* **41**:88–96.

217. Radany, E. H., J. D. Love, and E. C. Friedberg. 1981. The use of direct photoreversal of UV-irradiated DNA for the demonstration of pyrimidine dimer-DNA glycosylase activity, p. 91–95. *In* E. Seeberg and K. Kleppe (ed.), *Chromosome Damage and Repair.* Plenum Publishing Corp., New York.

218. Radany, E. H., H. T. Nguyen, and K. W. Minton. 1987. Activitites involved in base excision repair of bacteriophage T4 and lambda DNA in vivo. *Mol. Gen. Genet.* **209**:83–89.

219. Radman, M. 1976. An endonuclease from *Escherichia coli* that introduces single polynucleotide chain scissions in ultraviolet-irradiated DNA. *J. Biol. Chem.* **251**:1438–1445.

220. Ramotar, D., C. Kim, R. Lillis, and B. Demple. 1993. Intracellular localization of the *Apn1* DNA repair enzyme of *Saccharomyces cerevisiae.* Nuclear transport signals and biological role. *J. Biol. Chem.* **268**:20533–20539.

221. Ramotar, D., S. C. Popoff, and B. Demple. 1991. Complementation of DNA repair-deficient *Escherichia coli* by the yeast *Apn1* apurinic/apyrimidinic endonuclease gene. *Mol. Microbiol.* **5**:149–155.

222. Ramotar, D., S. C. Popoff, E. B. Gralla, and B. Demple. 1991. Cellular role of yeast Apn1 apurinic endonuclease/3′-diesterase: repair of oxidative and alkylation DNA damage and control of spontaneous mutation. *Mol. Cell. Biol.* **11**:4537–4544.

223. Recinos, A., III, and R. S. Lloyd. 1988. Site-directed mutagenesis of the T4 endonuclease *V* gene: role of lysine-130. *Biochemistry* **27**:1832–1838.

224. Riazuddin, S., A. Athar, Z. Ahmed, S. M. Lali, and A. Sohail. 1987. DNA glycosylase enzymes induced during chemical adaptation of *M. luteus. Nucleic Acids Res.* **15**:6607–6624.

225. Riazuddin, S., and L. Grossman. 1977. *Micrococcus luteus* correndonucleases. II. Mechanism of action of two endonucleases specific for DNA containing pyrimidine dimers. *J. Biol. Chem.* **252**:6287–6293.

226. Riazuddin, S., and L. Grossman. 1977. *Micrococcus luteus* correndonucleases. I. Resolution and purification of two endonucleases specific for DNA containing pyrimidine dimers. *J. Biol. Chem.* **252**:6280–6286.

227. Riazuddin, S., and T. Lindahl. 1978. Properties of 3-methyladenine-DNA glycosylase from *Escherichia coli. Biochemistry* **17**:2110–2118.

228. Rich, B. E., and J. A. Steitz. 1987. Human acidic ribosomal phosphoproteins P0, P1, and P2: analysis of cDNA clones, in vitro synthesis and assembly. *Mol. Cell. Biol.* **7**:4065–4074.

229. Robson, C. N., and I. D. Hickson. 1991. Isolation of cDNA clones encoding a human apurinic/apyrimidinic endonuclease that corrects DNA repair and mutagenesis defects in *Escherichia coli xth* (exonuclease III) mutants. *Nucleic Acids Res.* **19**:5519–5523.

230. Robson, C. N., D. Hochhauser, R. Craig, K. Rack, V. J. Buckle, and I. D. Hickson. 1992. Structure of the human DNA repair gene *HAP1* and its localization to chromosome 14q 11.2-12. *Nucleic Acids Res.* **20**:4417–4421.

231. Robson, C. N., A. M. Milne, D. J. Pappin, and I. D. Hickson. 1991. Isolation of cDNA clones encoding an enzyme from bovine cells that repairs oxidative DNA damage in vitro: homology with bacterial repair enzymes. *Nucleic Acids Res.* **19**:1087–1092.

232. Rogers, S. G., and B. Weiss. 1980. Cloning of the exonuclease III gene of *Escherichia coli. Gene* **11**:187–195.

233. Rydberg, B., M. K. Dosanjh, and B. Singer. 1992. Human cells contain protein specifically binding to a single 1,N6-ethenoadenine in DNA fragment. *Proc. Natl. Acad. Sci. USA* **88**:6839–6842.

234. Rydberg, B., and T. Lindahl. 1982. Nonenzymatic methylation of DNA by the intracellular methyl group donor S-adenosyl-L-methionine is a potentially mutagenic reaction. *EMBO J.* **1**:211–216.

235. Rydberg, B., Z. H. Qiu, M. K. Dosanjh, and B. Singer. 1992. Partial purification of a human DNA glycosylase acting on the cyclic carcinogen adduct 1, N6-ethenodeoxyadenosine. *Cancer Res.* **52**:1377–1379.

236. Sak, B. D., A. Eisenstark, and D. Touati. 1989. Exonuclease III and the catalase hydroperoxidase II in *Escherichia coli* are both regulated by the *katF* gene product. *Proc. Natl. Acad. Sci. USA* **86**:3271–3275.

237. Sakumi, K., M. Fuuichi, T. Tsuzuki, T. Kakuma, S. Kawabata, H. Maki, and M. Sekiguchi. 1993. Cloning and expression of cDNA of a human enzyme that hydrolyzes 8-oxo-dGTP, a mutagenic substrate for DNA synthesis. *J. Biol. Chem.* **268**:23524–23530.

238. Sakumi, K., and M. Sekiguchi. 1990. Structures and functions of DNA glycosylases. *Mutat. Res.* **236**:161–172.

239. Sakumi, K., Y. Nakabeppu, Y. Yamamoto, S. Kawabata, S. Iwanaga, and M. Sekiguchi. 1986. Purification and structure of 3-methyladenine-DNA glycosylase I of *Escherichia coli. J. Biol. Chem.* **261**:15761–15766.

240. Sammartano, L. J., R. W. Tuveson, and R. Davenport. 1986. Control of sensitivity to inactivation by H_2O_2 and broad-spectrum near-UV radiation by the *Escherichia coli katF* locus. *J. Bacteriol.* **168**:13–21.

241. Samson, L., B. Derfler, M. Boosalis, and K. Call. 1991. Cloning and characterization of a 3-methyladenine DNA glycosylase cDNA from human cells whose gene maps to chromosome 16. *Proc. Natl. Acad. Sci. USA* **88**:9127–9131.

242. Sander, M., K. Lowenhaupt, W. S. Lane, and A. Rich. 1991. Cloning and characterization of *Rrp1*, the gene encoding *Drosophila* strand transferase: carboxy-terminal homology to DNA repair endo/exonucleases. *Nucleic Acids Res.* **19**:4523–4529.

243. Sander, M., K. Lowenhaupt, and A. Rich. 1991. *Drosophila* Rrp1 protein: an apurinic endonuclease with homologous recombination activities. *Proc. Natl. Acad. Sci. USA* **88**:6780–6784.

244. Sandigursky, M., and W. A. Franklin. 1992. DNA deoxyribophosphodiesterase of *Escherichia coli* is associated with exonuclease I. *Nucleic Acids Res.* **20**:4699–4703.

245. Saparbaev, M., and J. Laval. 1994. Excision of hypoxanthine from DNA containing dIMP residues by the *Escherichia coli*, yeast, rat, and human alkylpurine DNA glycosylases. *Proc. Natl. Acad. Sci. USA* **91:**5873–5877.

246. Saporito, S. M., and R. P. Cunningham. 1988. Nucleotide sequence of *nfo* gene of *Escherichia coli* K-12. *J. Bacteriol.* **170:**5141–5145.

247. Saporito, S. M., M. Gedenk, and R. P. Cunningham. 1989. Role of exonuclease III and endonuclease IV in repair of pyrimidine dimers initiated by bacteriophage T4 pyrimidine dimer-DNA glycosylase. *J. Bacteriol.* **171:**2542–2546.

248. Saporito, S. M., B. J. Smith-White, and R. P. Cunningham. 1988. Nucleotide sequence of the *xth* gene of *Escherichia coli* K-12. *J. Bacteriol.* **170:**4542–4547.

249. Sato, K., and M. Sekiguchi. 1976. Studies on temperature-dependent ultraviolet light-sensitive mutants of bacteriophage T4: the structural gene for T4 endonuclease V. *J. Mol. Biol.* **102:**15–26.

250. Seawell, P. C., C. A. Smith, and A. K. Ganesan. 1980. *denV* gene of bacteriophage T4 determines a DNA glycosylase specific for pyrimidine dimers in DNA. *J. Virol.* **35:**790–796.

251. Seki, S., K. Akiyama, S. Watanabe, M. Hatsushika, S. Ikeda, and K. Tsutsui. 1991. cDNA and deduced amino acid sequence of a mouse DNA repair enzyme (APEX nuclease) with significant homology to *Escherichia coli* exonuclease III. *J. Biol. Chem.* **266:**20797–20802.

252. Seki, S., S. Ikeda, S. Watanabe, M. Hatsushika, K. Tsutsui, K. Akiyama, and B. Zhang. 1991. A mouse DNA repair enzyme (APEX nuclease) having exonuclease and apurinic/apyrimidinic endonuclease activities: purification and characterization. *Biochim. Biophys. Acta* **1079:**57–64.

253. Sekiguchi, M., S. Yasuda, S. Okubo, H. Nakayama, K. Shimada, and Y. Takagi. 1970. Mechanism of repair of DNA in bacteriophage. I. Excision of pyrimidine dimers from ultraviolet-irradiated DNA by an extract of T4-infected cells. *J. Mol. Biol.* **47:**231–242.

254. Setlow, R. B., and W. L. Carrier. 1968. The excision of pyrimidine dimers in vivo and in vitro, p. 134–141. *In* W. J. Peacock and R. D. Brock (ed.), *Replication and Recombination of Genetic Material.* Australian Academy of Sciences, Canberra.

255. Shevell, D. E., B. M. Friedman, and G. C. Walker. 1990. Resistance to alkylation damage in *Escherichia coli*: role of the Ada protein in induction of the adaptive response. *Mutat. Res.* **233:**53–57.

256. Shibutani, S., M. Takeshita, and A. P. Grollman. 1991. Insertion of specific bases during DNA synthesis past the oxidation-damaged base 8-oxodG. *Nature* (London) **349:**431–434.

257. Shimada, K., H. Nakayama, S. Okubo, M. Sekiguchi, and Y. Takagi. 1967. An endonucleolytic activity specific for ultraviolet-irradiated DNA in wild type and mutant strains of *Micrococcus lysodeikticus. Biochem. Biophys. Res. Commun.* **27:**539–545.

258. Shimizu, K., and M. Sekiguchi. 1976. 5' leads to 3'-exonucleases of bacteriophage T4. *J. Biol. Chem.* **251:**2613–2619.

259. Simon, T. J., C. A. Smith, and E. C. Friedberg. 1975. Action of bacteriophage T4 ultraviolet endonuclease on duplex DNA containing one ultraviolet-irradiated strand. *J. Biol. Chem.* **250:**8748–8752.

260. Sirover, M. A. 1979. Induction of the DNA repair enzyme uracil-DNA glycosylase in stimulated human lymphocytes. *Cancer Res.* **39:**2090–2095.

261. Siwek, B., S. Bricteux-Gregoire, V. Bailly, and W. G. Verly. 1988. The relative importance of *Escherichia coli* exonuclease III and endonuclease IV for the hydrolysis of 3'-phosphoglycolate ends in polydeoxynucleotides. *Nucleic Acids Res.* **16:**5031–5038.

262. Slupphaug, G. 1992. Regulation and expression of uracil-DNA glycosylase and O6-methylguanine-DNA methyltransferase in mammalian cells. Ph.D. thesis. University of Trondheim, Trondheim, Norway.

263. Slupphaug, G., L. C. Olsen, D. Helland, R. Aasland, and H. E. Krokan. 1991. Cell cycle regulation and in vitro hybrid arrest analysis of the major human uracil-DNA glycosylase. *Nucleic Acids Res.* **19:**5131–5137.

264. Smith, S. M., N. Symonds, and P. White. 1970. The Kornberg polymerase and the repair of irradiated T4 bacteriophage. *J. Mol. Biol.* **54:**391–393.

265. Spiering, A. L., and W. A. Deutsch. 1986. *Drosophila* apurinic/apyrimidinic DNA endonucleases. Characterization of mechanism of action and demonstration of a novel type of enzyme activity. *J. Biol. Chem.* **261:**3222–3228.

266. Steinum, A. L., and E. Seeberg. 1986. Nucleotide sequence of the *tag* gene from *Escherichia coli. Nucleic Acids Res.* **14:**3763–3772.

267. Strauss, B., T. Searashi, and M. Robbins. 1966. Repair of DNA studied with nuclease specific for UV-induced lesions. *Proc. Natl. Acad. Sci. USA* **56:**932–939.

268. Streisinger, G. 1956. The genetic control of ultraviolet sensitivity levels in bacteriophages T2 and T4. *Virology* **2:**1–12.

269. Tagaki, Y., M. Sekiguchi, S. Okubo, H. Nakayama, K. Shimada, S. Yasuda, T. Nishimoto, and H. Yoshihara. 1968. Nucleases specific for ultraviolet light-irradiated DNA and their possible role in dark repair. *Cold Spring Harbor Symp. Quant. Biol.* **33:**219–227.

270. Takahashi, I., and J. Marmur. 1963. Replacement of thymidylic acid by deoxyuridylic acid in the deoxyribonucleic acid of a transducing phage for *Bacillus subtilis. Nature* (London) **197:**794–795.

271. Tchou, J., H. Kasai, S. Shibutani, M. H. Chung, J. Laval, A. P. Grollman, and S. Nishimura. 1991. 8-Oxoguanine (8-hydroxyguanine) DNA glycosylase and its substrate specificity. *Proc. Natl. Acad. Sci. USA* **88:**4690–4694.

272. Tchou, J., M. L. Michaels, J. H. Miller, and A. P. Grollman. 1993. Function of the zinc finger in *Escherichia coli* Fpg protein. *J. Biol. Chem.* **268:**26738–26744.

273. Thomas, K. R., P. Manalapaz-Ramos, and B. M. Olivera. 1978. Formation of Okazaki pieces at the *Escherichia coli* replication fork in vitro. *Cold Spring Harbor Symp. Quant. Biol.* **43:**231–237.

274. Thomas, L., C. H. Yang, and D. A. Goldthwait. 1982. Two DNA glycosylases in *Escherichia coli* which release primarily 3-methyladenine. *Biochemistry* **21:**1162–1169.

275. Tomkinson, A. E., R. T. Bonk, J. Kim, N. Bartfeld, and S. Linn. 1990. Mammalian mitochondrial endonuclease activities specific for ultraviolet-irradiated DNA. *Nucleic Acids Res.* **18:**929–935.

276. Tsai-Wu, J. J., H. F. Liu, and A. L. Lu. 1992. *Escherichia coli* MutY protein has both N-glycosylase and apurinic/apyrimidinic endonuclease activities on A·C and A·G mispairs. *Proc. Natl. Acad. Sci. USA* **89:**8779–8783.

277. Vairapi, M., and N. J. Duker. 1993. Enzymic removal of 5-methylcytosine from DNA by a human DNA-glycosylase. *Nucleic Acids Res.* **21:**5323–5327.

278. van Hemmen, J. J., and J. F. Bleichrodt. 1973. The decomposition of adenine by ionizing radiation. *Radiat. Res.* **46:**444–456.

279. van Minderhout, L., J. Grimbergen, and B. de Groot. 1974. Nonsense mutants in the bacteriophage T4D *v* gene. *Mutat. Res.* **29:**333–348.

280. Varshney, U., and J. H. van de Sande. 1991. Specificities and kinetics of uracil excision from uracil-containing DNA oligomers by *Escherichia coli* uracil DNA glycosylase. *Biochemistry* **30:**4055–4061.

281. Verly, W. G. 1978. Endonucleases specific for apurinic sites in DNA, p. 187–190. *In* P. C. Hanawalt, E. C. Friedberg, and C. F. Fox (ed.), *DNA Repair Mechanisms.* Academic Press, Inc., New York.

282. Vickers, M. A., P. Vyas, P. C. Harris, D. L. Simmons, and D. R. Higgs. 1993. Structure of the human 3-methyladenine DNA glycosylase gene and localization close to the 16p telomere. *Proc. Natl. Acad. Sci. USA* **90:**3437–3441.

283. Volkert, M. R. 1988. Adaptive response of *Escherichia coli* to alkylation damage. *Environ. Mol. Mutagen.* **11:**241–255.

284. Volkert, M. R. 1989. Altered induction of the adaptive response to alkylation damage in *Escherichia coli recF* mutants. *J. Bacteriol.* **171:**99–103.

285. Volkert, M. R., D. C. Nguyen, and K. C. Beard. 1986. *Escherichia coli* gene induction by alkylation treatment. *Genetics* **112:**11–26.

286. Vollberg, T. M., K. M. Siegler, B. L. Cool, and M. A. Sirover. 1989. Isolation and characterization of the human uracil DNA glycosylase gene. *Proc. Natl. Acad. Sci. USA* **86:**8693–8697.

287. Von Hippel, P. H., and O. G. Berg. 1989. Facilitated target location in biological systems. *J. Biol. Chem.* **264:**675–678.

288. Walker, L. J., C. N. Robson, E. Black, D. Gillespie, and I. D. Hickson. 1993. Identification of residues in the human DNA repair enzyme HAP1 (Ref-1) that are essential for redox regulation of Jun DNA binding. *Mol. Cell. Biol.* **13:**5370–5306.

289. Wallace, S. S. 1988. AP endonucleases and DNA glycosylases that recognize oxidative DNA damage. *Environ. Mol. Mutagen.* **12:**431–477.

290. Wallace, S. S., H. Ide, Y. W. Kow, M. F. Laspia, R. J. Melamede, L. A. Petrullo, and E. LeClerc. 1988. Processing of oxidative DNA base damage in *Escherichia coli*, p. 151–157. *In* E. C. Friedberg and P. C. Hanawalt (ed.), *Mechanisms and Consequences of DNA Damage Processing.* Alan R. Liss, Inc., New York.

291. Wang, Z., and D. W. Mosbaugh. 1988. Uracil-DNA glycosylase inhibitor of bacteriophage PBS2: cloning and effects of expression of the inhibitor gene in *Escherichia coli. J. Bacteriol.* **170:**1082–1091.

292. Wang, Z., and D. W. Mosbaugh. 1989. Uracil-DNA glycosylase inhibitor gene of a bacteriophage PBS2 encodes a binding protein specific for uracil-DNA glycosylase. *J. Biol. Chem.* **264:**1163–1171.

293. Wang, Z. G., D. G. Smith, and D. W. Mosbaugh. 1991. Overproduction and characterization of the uracil-DNA glycosylase inhibitor of bacteriophage PBS2. *Gene* **99:**31–37.

294. Warner, H. R., B. F. Demple, W. A. Deutsch, C. M. Kane, and S. Linn. 1980. Apurinic/apyrimidinic endonucleases in repair of pyrimidine dimers and other lesions in DNA. *Proc. Natl. Acad. Sci. USA* **77:**4602–4606.

295. Warner, H. R., L. K. Johnson, and D. P. Snustad. 1980. Early events after infection of *Escherichia coli* by bacteriophage T5. III. Inhibition of uracil-DNA glycosylase activity. *J. Virol.* **33:**535–538.

296. Warner, H. R., and P. A. Rockstroh. 1980. Incorporation and excision of 5-fluorouracil from deoxyribonucleic acid in *Escherichia coli. J. Bacteriol.* **141:**680–686.

297. Weiss, B. 1976. Endonuclease II of *Escherichia coli* is exonuclease III. *J. Biol. Chem.* **251:**1896–1901.

298. Weiss, B. 1981. Exodeoxyribonucleases of *Escherichia coli*, p. 203–231. *In* P. D. Boyer (ed.), *The Enzymes*, 3rd ed. Academic Press, INc., New York.

299. Weiss, B., R. P. Cunningham, E. Chan, and I. R. Tsaneva. 1988. AP endonucleases of *Escherichia coli*, p. 133–142. *In* E. C. Friedberg and P. C. Hanawalt (ed.), *Mechanisms and Consequences of DNA Damage Processing.* Alan R. Liss, Inc., New York.

300. Weiss, B., S. G. Rogers, and A. F. Taylor. 1978. The endonuclease activity of exonuclease III and the repair of uracil-containing DNA in *Escherichia coli*, p. 191–194. *In* P. C. Hanawalt, E. C. Friedberg, and C. F. Fox (ed.), *DNA Repair Mechanisms.* Academic Press, Inc., New York.

301. White, B. J., S. J. Hochhauser, N. M. Cintron, and B. Weiss. 1976. Genetic mapping of *xthA*, the structural gene for exonuclease III in *Escherichia coli* K-12. *J. Bacteriol.* **126:**1082–1088.

302. Wiebauer, K., and J. Jiricny. 1990. Mismatch-specific thymine DNA glycosylase and DNA polymerase β mediate the correction of G·T mispairs in nuclear extracts from human cells. *Proc. Natl. Acad. Sci. USA* **87:**5842–5845.

303. Wilde, J., P. Bolton, A. Mazumder, M. Manoharan, and J. Gerilt. 1989. Characterization of the equilibrating forms of the aldehydic abasic site in duplex DNA by ^{17}O NMR. *J. Am. Chem. Soc.* **111:**1894–1896.

304. Wood, M. L., M. Dizdaroglu, E. Gajewski, and J. M. Essigmann. 1990. Mechanistic studies of ionizing radiation and oxidative mutagenesis: genetic effects of a single 8-hydroxyguanine (7-hydro-8-oxoguanine) residue inserted at a unique site in a viral genome. *Biochemistry* **29:**7024–7032.

305. Wood, W. B., and H. R. Revel. 1976. The genome of bacteriophage T4. *Bacteriol. Rev.* **40:**847–868.

306. Worrad, D. M., and S. Caradonna. 1988. Identification of the coding sequence for herpes simplex virus uracil-DNA glycosylase. *J. Virol.* **62:**4774–4777.

307. Xanthoudakis, S., G. Miao, F. Wang, Y. C. Pan, and T. Curran. 1992. Redox activation of Fos-Jun DNA binding activity is mediated by a DNA repair enzyme. *EMBO J.* **11:**3323–3335.

308. Xiao, W., and L. Samson. 1993. In vivo evidence for endogenous DNA alkylation damage as a source of spontaneous mutation in eukaryotic cells. *Proc. Natl. Acad. Sci. USA* **90:**2117–2121.

309. Xiao, W., K. K. Sing, B. Chen, and L. Samson. 1993. A common element involved in transcriptional regulation of two DNA alkylation repair genes (*MAG* and *MGT1*) of *Saccharomyces cerevisiae. Mol. Cell. Biol.* **13:**7213–7221.

310. Yajko, D. M., and B. Weiss. 1975. Mutations simultaneously affecting endonuclease II and exonuclease III in *Escherichia coli. Proc. Natl. Acad. Sci. USA* **72:**688–692.

311. Yamamoto, Y., and Y. Fujiwara. 1990. Uracil-DNA glycosylase causes 5-bromodeoxyuridine photosensitization in *Escherichia coli* K-12. *J. Bacteriol.* **172:**5278–5285.

312. Yamamoto, Y., M. Katsuki, M. Sekiguchi, and N. Otsuji. 1978. *Escherichia coli* gene that controls sensitivity to alkylating agents. *J. Bacteriol.* **135:**144–152.

313. Yamamoto, Y., and M. Sekiguchi. 1979. Pathways for repair of DNA damaged by alkylating agent in *Escherichia coli. Mol. Gen. Genet.* **171:**251–256.

314. Yarosh, D. B., and J. Ceccoli. 1989. Comparison of phage T4 *denV* endonuclease V and *Micrococcus luteus* UV-DNA endonuclease by serology and DNA hybridization. *Photochem. Photobiol.* **49:**53–57.

315. Yasuda, S., and M. Sekiguchi. 1970. T4 endonuclease involved in repair of DNA. *Proc. Natl. Acad. Sci. USA* **67:**1839–1845.

316. Yasuda, S., and M. Sekiguchi. 1976. Further purification and characterization of T4 endonuclease V. *Biochim. Biophys. Acta* **442:**197–207.

5

Nucleotide Excision Repair in Prokaryotes

Years before the discovery of DNA glycosylases, apurinic/apyrimidinic (AP) endonucleases, and the general phenomenon of base excision repair, the most extensively studied model lesion for excision repair was the pyrimidine dimer, a major photoproduct produced in DNA by UV radiation at ~254 nm (see chapter 1). This form of DNA damage is very convenient to study, since it is easy to generate in the laboratory (requiring only a germicidal lamp), it is chemically stable in DNA, and reliable and sensitive methods exist for its detection and quantitation (48, 49). Additionally, it is a biologically relevant form of DNA damage because the same type of base damage is produced in vivo when cells are exposed to sunlight (see chapter 1).

It has long been known that strains of *Escherichia coli* exposed to UV radiation undergo recovery processes which involve the repair of DNA containing pyrimidine dimers (47, 69, 95). Such recovery can be demonstrated as an increased survival of the bacteria themselves or as their ability to promote the survival of UV-irradiated bacteriophages for which the bacteria serve as hosts. The latter phenomenon is called *host cell reactivation* and is a sensitive indicator of DNA repair, both in prokaryotes and in mammalian cells infected with viruses (47, 69, 95). More recently it has become appreciated that the repair of a second major photoproduct in DNA [the so-called (6-4) lesion (see chapter 1)] is also important for biological recovery from UV radiation.

Following the discovery of enzymatic photoreactivation (see chapter 3), DNA repair modes associated with recovery from the effects of UV radiation were distinguished as *light repair* (to indicate the dependence of photoreactivation on visible light) and *dark repair* (to indicate DNA repair processes which are independent of light). During the early 1960s a series of experiments reported by Setlow and Carrier (176) and by Boyce and Howard-Flanders (17) led to important insights into the nature of dark repair of UV radiation-induced damage. Following the irradiation of UV-resistant (wild-type) and UV-sensitive (mutant) *E. coli* strains, it was noted that in the wild-type strains pyrimidine dimers were lost from the high-molecular-weight (acid-insoluble) fraction of DNA and appeared in the low-molecular-weight (acid-soluble) phase during post-UV incubation. The kinetics of this process correlated well with the resumption of DNA synthesis (178) and led to the suggestion that the loss of dimers from DNA (i.e., their *excision*) was directly related to the recovery of replicative DNA synthesis and hence to survival. This hypothesis was supported by the observation that in mutant UV radiation-sensitive and host-cell reactivation-defective (*hcr*) strains, thymine-containing dimers were not lost from high molecular weight DNA during post-UV incubation in the absence of photoreactivating light.

These seminal observations, together with the demonstration by Pettijohn and Hanawalt (136, 137) of repair synthesis of DNA in cells exposed to UV radiation, defined the process that is called *nucleotide excision repair,* a process by which damaged bases such as pyrimidine dimers and (6-4) photoproducts are enzymatically excised from DNA as intact nucleotides (actually as part of *oligonucleotide fragments*) rather than as free bases. Like base excision repair, nucleotide excision repair is a multistep process which eventually leads to the formation of a gap in the DNA duplex. This gap is filled by repair synthesis and covalently sealed by DNA ligase, as discussed in chapter 4. However, nucleotide excision repair is a biochemically more complex process than base excision repair and involves many more gene products, particularly in eukaryotic cells. This chapter is devoted to a consideration of nucleotide excision repair in prokaryotes, in which *E. coli* will once again constitute the primary paradigm. The next three chapters are devoted to a consideration of nucleotide excision repair in lower eukaryotes and in mammalian cells.

UvrABC Damage-Specific Endonuclease of *E. coli*

Early genetic analyses identified three loci, designated *uvrA+*, *uvrB+* and *uvrC+*, that are required for the excision of pyrimidine dimers in *E. coli* (73, 74, 105, 209). Subsequently, two other genes, *uvrD+* and *polA+*, were implicated in this process as well (152, 154, 210). We have already encountered the *polA+* gene, which encodes DNA polymerase I (PolI), in the context of the discussion of repair synthesis during base excision repair (see chapter 4). The use of techniques (some of which are discussed below) that reveal the presence of single-strand breaks (nicks) in DNA established that the *uvrA+*, *uvrB+*, and *uvrC+* genes are specifically required for *endonucleolytic incision of DNA* containing pyrimidine dimers (145, 182). These observations reinforced the notion that the dark repair of pyrimidine dimers does indeed occur by an excision-resynthesis mode and that this must somehow involve breakage of the polynucleotide chain specifically at or near sites of pyrimidine dimers; i.e., the DNA must somehow be subject to *damage-specific incision.*

One mechanism for the specific incision of DNA containing pyrimidine dimers that was discussed in chapter 4 is by the sequential action of a pyrimidine dimer-DNA glycosylase (PD-DNA glycosylase) and an AP endonuclease. However, DNA glycosylase/AP endonuclease-mediated incision of DNA containing pyrimidine dimers is a highly specialized form of base excision repair which is confined to *Micrococcus luteus* and to *E. coli* infected with bacteriophage T4 and possibly to the yeast *Saccharomyces cerevisiae.* The stringent substrate specificity of the T4 PD-DNA glycosylase is consistent with the observation that phage T4 mutants which are defective in the *denV+* gene (which encodes this DNA glycosylase) are not abnormally sensitive to any other type of base damage. On the other hand, it is well established that *E. coli* mutants defective in the *uvrA+*, *uvrB+*, or *uvrC+* gene(s) are sensitive not only to UV radiation but also to a diverse array of chemical agents, including mitomycin, nitrogen mustard, photoactivated psoralen, and *N*-methyl-*N*'-nitro-*N*-nitrosoguanidine (MNNG) (73, 74, 105, 115), to name just a few (Fig. 5–1).

Hence, a reasonable answer to the question why *E. coli* and many other organisms do not possess a PD-DNA glycosylase to deal with the potentially lethal

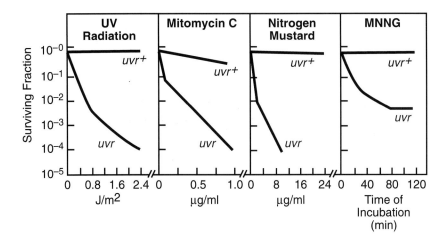

Figure 5–1 Nucleotide excision repair-defective (*uvr*) strains of *E. coli* are abnormally sensitive to killing by a wide variety of DNA-damaging agents, including UV radiation, mitomycin, nitrogen mustard, and *N*-methyl-*N'*-nitro-*N*-nitrosoguanidine (MNNG). (*Adapted from Friedberg [45a] with permission.*)

and mutagenic effects of pyrimidine dimers is that the presence of many other types of base damage that challenged the survival of cells during biological evolution provided the selection for a more general excision repair mode which could serve to identify these multiple types of damage and effect their removal from DNA. As discussed here and in subsequent chapters, we still do not understand precisely how the nucleotide excision repair machinery is able to identify diverse types of base damage which apparently have few chemical and even structural features in common. Nor is it fully understood how this machinery is able to specifically recognize such damage in the general background of other perturbations in DNA structure which naturally arise from its dynamic state and its participation as a substrate in other aspects of DNA metabolism.

Although biological studies on nucleotide excision repair in *E. coli* were initiated over 30 years ago, it is only in the last decade that major insights have been obtained concerning the detailed biochemistry of this process. Several factors impeded progress in this area of biochemistry. For one thing, all cells, including *E. coli*, contain multiple nuclease activities which rapidly degrade DNA in cell extracts. The preparation of cell extracts by Seeberg et al. (162, 165) which could support the preferential incision (nicking) of UV-irradiated DNA relative to unirradiated DNA was an important milestone that was fundamental to this research endeavor. This process of specific DNA incision was shown to be dependent on the presence of divalent cations such as Mg^{2+} or Mn^{2+} and to have an absolute requirement for ATP. Consistent with the cellular biology and genetics of nucleotide excision repair, the specific nicking of UV-irradiated DNA was not observed with extracts of *uvrA*, *uvrB*, or *uvrC* mutant cells. However, extracts of these mutant cells could be complemented for nicking activity by addition of extract from a different *uvr* mutant strain (162, 163, 165) (Fig. 5–2).

Aside from the importance of appropriate preparation of cell extracts, it is now well recognized that the *uvrA⁺*, *uvrB⁺* and *uvrC⁺* gene products are constitutively expressed in extremely small amounts in *E. coli*, and for many years attempts to purify these proteins were additionally frustrated by these quantitative limitations. The purification of the UvrA, UvrB, and UvrC proteins in high yields was enormously facilitated by the molecular cloning of the genes that encode them and by their extensive overexpression (63, 102, 142, 147, 152, 154, 170, 210, 226). The amplification of *uvr* gene expression can result in yields of Uvr proteins approaching 25 to 30% of the total soluble protein of *E. coli* cells, allowing the purification of milligram quantities of these proteins to physical homogeneity (205, 226, 232). As discussed in later chapters, the overexpression of recombinant nucleotide excision repair proteins from eukaryotic cells has been far less successful.

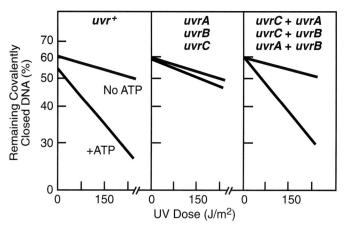

Figure 5–2 UV-irradiated DNA is incised by extracts of *E. coli* cells which are proficient for nucleotide excision repair (*uvr⁺*) in the presence of ATP and Mg²⁺ (left panel). Extracts of *uvr* mutant cells fail to demonstrate the preferential incision of UV-irradiated DNA (center panel); however, mixtures of extracts from different *uvr* cells complement each other for nicking activity (right panel). Incision (nicking) of DNA was measured by the conversion of covalently closed circular (form I) DNA to the relaxed-circular or linear configurations. (*Adapted from Friedberg [45a] with permission.*)

The enzyme-catalyzed incision of damaged DNA during nucleotide excision repair in *E. coli* absolutely requires the UvrA, UvrB, and UvrC proteins. Regrettably, a standard nomenclature for this enzyme has not been developed. The enzyme has been variously referred to as the *UvrABC endonuclease*, the *UvrABC excinuclease*, the *Uvr(A)BC endonuclease*, and the *Uvr(A)BC excinuclease*. The term "excinuclease" was introduced in deference to the fundamental observation, discussed in detail below, that the enzyme catalyzes damage-specific incision of DNA and thereby facilitates the *excision* of an oligonucleotide generated during the incision reactions. The reason for denoting the UvrA protein in parentheses will become apparent later. In keeping with standard biochemical nomenclature, we prefer the term *endonuclease*, since the endonucleolytic incision of DNA is a primary biochemical event catalyzed by this enzyme. Hence, in this book we will use the name *UvrABC endonuclease*. Before discussing how these proteins effect damage-specific incision of DNA, let us consider the proteins themselves and the genes that encode them.

The *uvrA⁺* Gene and UvrA Protein

The *uvrA⁺* gene is located at 92 min on the *E. coli* genetic map. *uvrA* is one of a series of genes collectively referred to as *SOS genes*, which are induced to increased levels of transcription by agents that cause DNA damage. Details of the SOS response to DNA damage are discussed in chapter 10. For the purposes of the present discussion, it is relevant to note that the expression of SOS genes is regulated by a repressor which is the product of a gene called *lexA⁺*. A specific binding site for the LexA repressor protein at the operator-promoter region of the *uvrA⁺* gene (the so-called *LexA box* or *SOS box*) has been identified (153). Not unexpectedly, similar sequences have been identified in the regulatory regions of other genes involved in the SOS response, including the *uvrB* gene (see below).

The *uvrA⁺* gene constitutes a single operon and is not cotranscribed with any other gene (170). The gene is separated by 253 bp from a gene called *ssb* (which encodes a single-stranded DNA-binding protein), but the two genes are transcribed divergently (170). The *uvrA* LexA binding site is located in this intergenic region and is required for inducible regulation of *uvrA* (13), but it is not involved in the expression of the *ssb* gene (170). There are only ~25 molecules of UvrA protein constitutively in *E. coli* cells, but SOS-mediated induction of the *uvrA* gene increases the intracellular content of UvrA protein to ~250 molecules per cell (170, 210).

Table 5–1 Properties of UvrA, UvrB, and UvrC proteins of *E. coli*

Property	UvrA	UvrB	UvrC
Mol mass (Da)	103,874	76,118	66,038
No. of amino acids	940	673	610
No. of Trp residues/molecule	3	0	2
Molar extinction coefficient	46,680	27,699	36,200
pI	6.5	5.0	7.3
Stokes radius (Å)	59	41	41
Sedimentation coefficient	7.4	4.2	4.9
Intrinsic metal	2 Zn	None	None
DNA binding	Yes	No	Yes
Nucleotide-binding motifs	2	1	None
ATPase activity	Yes	No	No
SOS regulation	Yes	Yes	No
No of molecules/cell	20 (250)[a]	250 (1,000)	10

[a] Numbers in parentheses indicate values after SOS induction.
Source: Adapted from Van Houten (210).

The cloned *uvrA*[+] gene contains an open reading frame of 940 amino acids, which can encode a polypeptide of about 103.9 kDa (78, 170, 210), in reasonable agreement with the molecular mass of the purified protein (~114 kDa) (Table 5–1) (205, 232). The protein has a Stokes radius of 59 Å (5.9 nm) (Table 5–1) and an S value of 7.4 (Table 5–1). It is a slightly acidic protein, with a measured pI of 6.5 (Table 5–1) (170).

UvrA protein is a DNA-independent ATPase and a DNA-binding protein. Both of these functional attributes correlate with specific structural motifs in the translated nucleotide sequence. The amino acid sequence reveals the presence of two consensus so-called Walker type A and type B purine nucleotide recognition motifs (223), found in many proteins which bind and hydrolyze ATP and/or GTP (152, 210). The type A sequences are the most highly conserved (Fig. 5–3). More detailed analysis of the predicted UvrA polypeptide and comparisons with the sequences of a number of other ATPases revealed regions of homology extending over approximately 250 amino acids, suggesting that UvrA protein is a member of a superfamily of prokaryotic ATPases, many of which are plasma membrane-bound proteins involved in active transport, multidrug resistance, cell division, nodulation during nitrogen fixation, protein export, recombination, and replication (40, 55, 71). The amino acid sequence of the UvrA protein also contains two zinc finger motifs (Fig. 5–3) (see below) and it has been suggested that a consensus

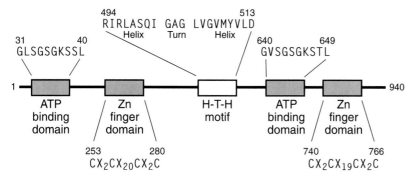

Figure 5–3 Diagrammatic representation of functional motifs identified in the amino acid sequence of the *E. coli* UvrA protein (940 amino acids). The location of the two Walker type A consensus sequences for nucleotide (ATP) binding are shown, as well as the two zinc finger domains and the helix-turn-helix (H-T-H) motif. (*Adapted from Grossman and Thiagalingam [62] with permission.*)

helix-turn-helix motif is present as well (Fig. 5–3) (224). Both the zinc finger and helix-turn-helix motifs are thought to be important determinants for the ability of the protein to bind to DNA. The C-terminal 44 amino acids of UvrA protein are rich in glycine residues (31) and are thought to be important for DNA damage recognition (see below).

Site-directed mutagenesis of UvrA protein suggests that it has two functional ATPase motifs located near the N-terminal and C-terminal regions of the polypeptide and corresponding to those identified by inspection of the amino acid sequence (203). The K_m for ATP of the wild-type protein is 149 μM. This value is intermediate between those independently calculated for the N-terminal site (60 μM) and for the C-terminal domain (312 μM). The apparent second order rate constant for the association of ATP with the C-terminal ATP-binding site is only slightly less than that of the full-length wild-type protein. However, the rate constant for association with the N-terminal site is about an order of magnitude lower. Hence, it would appear that the former site has a higher affinity for ATP than the latter has (203). The ATPase activity of UvrA protein has a k_{cat} of ~1 s^{-1} (205). However, there is a substantial decrease in the k_{cat} of the ATPase activity for each individual site compared with the native protein, suggesting that the two sites act cooperatively in ATP hydrolysis (203). The ATPase activity is stimulated ca. twofold by the simultaneous presence of unirradiated DNA and UvrB protein (but not by either alone) (205). The roles of ATP hydrolysis during the damage-specific incision of DNA by the UvrABC endonuclease are considered more specifically below.

When radioiodination of UvrA protein is carried out in the presence of ATP, significantly higher levels and initial rates of tyrosine iodination are observed than in the absence of ATP, suggesting that the binding of ATP to UvrA protein is associated with a conformational change in the protein (63, 226). In support of this suggestion, UvrA protein can exist as a monomer in the absence of ATP but dimerizes in its presence (152, 226). High concentrations of UvrA protein and the binding (but not the hydrolysis) of ATP favor dimerization of the protein (107), and the presence of the nonhydrolyzable ATP analog ATP[γS] also favors dimerization (107). At monomer-dimer equilibrium, the apparent association constant is $K_a = $ ~10^8 M^{-1} (152). In the presence of ATP[γS] UvrA protein becomes trapped in its dimeric state, suggesting that ATP hydrolysis is linked to the reversal of UvrA dimerization (226). The exact stoichiometry of ATP binding to UvrA protein is 1ATP/(UvrA)$_2$ (117). In summary, UvrA protein dimerizes in solution and the dimerization constant is increased by the binding of ATP. ATP binding (and possibly dimerization) results in a conformational change of the (UvrA)$_2$ protein complex.

The amino acid sequence domains that are involved in the binding and hydrolysis of ATP by UvrA protein are separated by two zinc finger motifs with the prototypic sequence CXXCX$_{18-20}$CXXC (12) (Fig. 5–3). These are believed to be involved in DNA-binding (152). Extended X-ray absorption fine-structure analysis of purified UvrA protein has shown that it does indeed contain two bound Zn atoms per molecule, each coordinated with four sulfur atoms (in cysteine residues) at a distance of 2.32 Å (0.23 nm) (121). The functional significance of these motifs is not certain. Substitution of Cys-253, the first cysteine residue in the N-terminal zinc finger (Fig. 5–3) by histidine, serine, or alanine results in a modest loss of the ability of the mutant gene to fully complement the UV radiation sensitivity of *uvrA* strains (121). On the other hand, independent studies suggest that the substitution of Cys-253 or Cys-256 (Fig. 5–3) with serine has little or no significant effect on the resistance of the mutant *uvrA* strains to UV radiation or on the activity of purified UvrA protein (219). However, substitution of Cys-763 in the C-terminal zinc finger motif (Fig. 5–3) confers extreme sensitivity to UV radiation and inactivates UvrA protein (219). Hence, it is possible that the second (C-terminal) zinc finger motif is required for protein binding to DNA, while the first (N-terminal motif) is required for the proper dimerization of UvrA protein.

Incomplete polypeptides comprising the 70-kDa N-terminal and 35-kDa C-terminal regions of UvrA protein and carrying one or the other of these two zinc finger domains have been purified and characterized (117). When mixed, the two fragments fail to reconstitute functional UvrABC endonuclease activity in the presence of UvrB and UvrC proteins. Hence, it is likely that the specificity for binding to damaged DNA is provided by the proper orientation of the two zinc finger motifs relative to one another and is not an intrinsic property of the individual domains (117). A third zinc finger motif has also been identified in the UvrA protein, but it has diverged extensively, and its functional significance is unknown (152). Further evidence of a requirement for zinc for the functional integrity of UvrA protein derives from studies showing that when recombinant UvrA enzyme is expressed in an insoluble form, it can be solubilized by procedures that denature the protein and allow it to refold. Refolding requires zinc to reconstitute functional UvrA protein (30). As indicated above, in addition to the zinc finger motifs, UvrA protein has a putative consensus helix-turn-helix motif. Base substitution experiments with this motif suggest that it is required for the protein to recognize base damage in DNA (224).

The domains of the UvrA protein that are required for its various functional activities have been mapped by comparing the properties of various mutant polypeptides generated by systematic deletion mutagenesis (31). A region located in the N-terminal 230 amino acids is believed to be required for interaction with UvrB protein, whereas the region required for self-dimerization is within the first 680 amino acids. Almost the entire polypeptide (940 amino acids) is required for binding to DNA (31).

The binding of $(UvrA)_2$ protein to DNA has been demonstrated by a variety of experimental techniques (82, 107, 152, 205, 213, 219, 232). The protein binds nonspecifically to duplex DNA, with relatively low affinity. This nonspecific binding affinity is apparently higher for ends in duplex DNA and for single-stranded DNA (107, 152). In the absence of ATP, the protein binds more specifically to damaged DNA, binding to a psoralen adduct in duplex DNA with a binding constant $K_S = \sim 5 \times 10^7\ M^{-1}$ (152). The specificity of this binding increases another twofold in the presence of ATP hydrolysis. By using substrates containing the major products of UV radiation, it has been observed that the *off* (dissociation) rate for UvrA protein is exceedingly low in the absence of ATP hydrolysis but is stimulated in its presence (139). Thus, it appears that ATP hydrolysis increases the specificity of binding to damaged DNA but lowers the equilibrium binding constant by stimulating dissociation. DNA-binding affinity is also increased in the presence of ATP[γS] and is inhibited in the presence of ADP, suggesting that ATP hydrolysis is not required for this event (126, 166).

In its dimeric (and possibly ATP-bound) form, UvrA protein binds DNA containing various forms of base damage, ranging from AP sites to cross-link-initiated triple helixes (149, 170, 193, 215). However, these $(UvrA)_2$-DNA complexes are short lived and dissociate rapidly. Further selectivity for the binding of UvrA protein to damaged DNA is achieved by its specific interaction with UvrB protein. Before discussing this interaction, let us consider the second player in the triad of molecular components required for the damage-specific incision of DNA in *E. coli*, the *uvrB*+ gene and its polypeptide product.

The *uvrB*+ Gene and UvrB Protein

The *uvrB*+ gene maps at 17 min on the *E. coli* chromosome and hence is unlinked to *uvrA*+. Like the *uvrA*+ gene, *uvrB*+ is monocistronic. Regulation of the *uvrB*+ gene (which, as indicated above is also a member of the SOS regulon and is inducible by DNA damage) is complex (44, 158). Transcription of *uvrB*+ appears to be controlled by both SOS-dependent and SOS-independent promoters (210). In vitro the gene is transcribed from two overlapping promoters called P_1 and P_2 (Fig. 5–4). A LexA protein binding site is present in the promoter region. In vitro this site apparently functions as the operator for P_2, since transcription from P_2 is inhibited in the presence of LexA repressor protein while that from P_1 is unaffected. On the other hand, there are indications that in vivo P_1 (the promoter closest to the coding region of *uvrB*) determines both constitutive and induced levels of

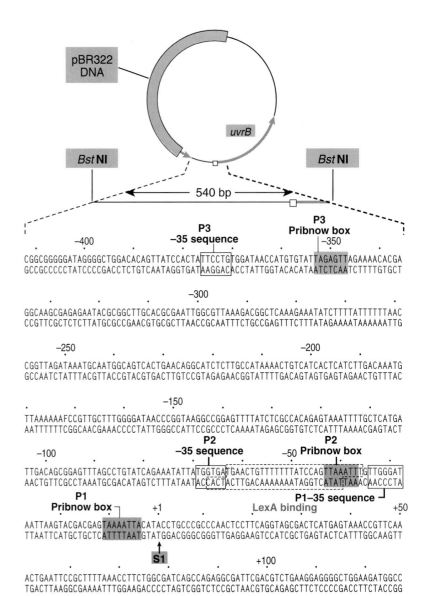

Figure 5–4 Nucleotide sequence of a 540-bp *Bst*NI fragment containing the *E. coli uvrB*$^+$ gene regulatory regions. The numbering of nucleotides is relative to a transcriptional start site designated S1. Pribnow boxes (-10 sequences) and -35 sequences for three promoters (P1, P2, and P3) are identified, as is a region protected from DNase I attack by the binding of LexA protein. (*Adapted from Friedberg [45a] with permission.*)

uvrB$^+$ transcription and that P$_2$ serves mainly to regulate transcription from P$_1$. In vitro the binding of LexA protein to P$_2$ may interfere with the binding of RNA polymerase to P$_1$ or, alternatively, may interfere with the local unwinding of DNA that precedes initiation of transcription (9, 208).

A third promoter, designated P$_3$, has been identified 320 bp upstream of P$_2$ (Fig. 5–4). In vitro, transcription from P$_3$ is directed toward the *uvrB*$^+$ structural gene but terminates in the region of the LexA-binding site even in the absence of LexA protein (155). The physiological role (if any) of this promoter and the nature of its transcript are unknown. Plasmid vectors carrying the P$_3$ region can be propagated only in certain media or in strains deleted of *uvrB* (148, 155, 207). Mutations that modify a potential stem-loop structure in the P$_3$ region stabilize the propagation of plasmids (207). It has therefore been suggested that transcription from P$_3$ leads to plasmid loss and that the stem-loop structure acts as a regulatory

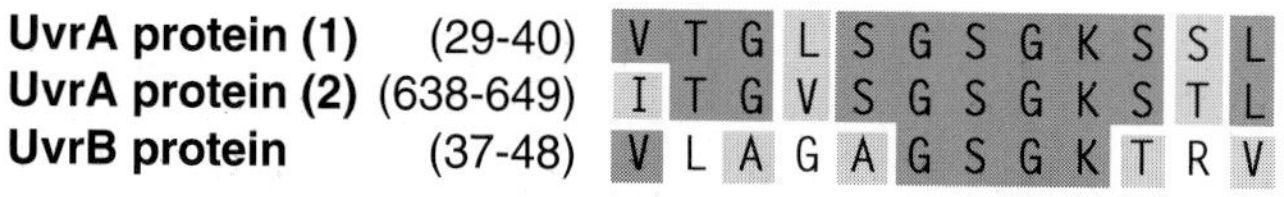

UvrA protein (1) (29-40) V T G L S G S G K S S L
UvrA protein (2) (638-649) I T G V S G S G K S T L
UvrB protein (37-48) V L A G A G S G K T R V

Figure 5–5 The Walker type A nucleotide binding motifs in the *E. coli* UvrA and UvrB proteins share extensive amino acid identity and similarity.

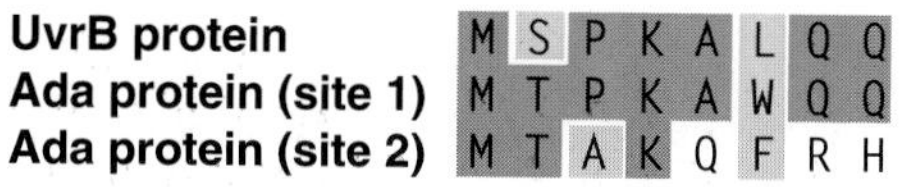

UvrB protein M S P K A L Q Q
Ada protein (site 1) M T P K A W Q Q
Ada protein (site 2) M T A K Q F R H

Figure 5–6 The *E. coli* UvrB protein has an amino acid sequence motif that resembles those observed in the *E. coli* Ada protein (see chapter 3). Those in the Ada protein are sites of preferred proteolytic cleavage of the protein.

site for this promoter (207). Interestingly, the putative stem-loop structure in P_3 shares sequence homology with the binding site for DnaA protein in the *E. coli* replication origin *oriC* (50), suggesting the interesting possibility that expression of the *uvrB*$^+$ gene is somehow coupled to DNA replication (207). In this regard, it may be relevant that *polA uvrB* mutants are inviable (111).

The 3′ untranslated region of the *uvrB*$^+$ gene contains so-called repetitive extragenic palindromic sequences (6, 8). Such sequences in other genes have been shown to be binding sites for DNA gyrase (229), suggesting that they may be involved in determining higher-order structure in the *E. coli* chromosome (210).

The *uvrB*$^+$ gene can encode a polypeptide of 673 amino acids with a calculated molecular mass of 76.6 kDa (6, 8). The amino acid sequence of UvrB shows several interesting features. First, like the UvrA amino acid sequence, the protein has a consensus Walker type A nucleotide-binding motif (Fig. 5–5). The purified protein has no detectable ATPase activity. However, as discussed below, there are indications that UvrB may be a cryptic ATPase which becomes activated only when bound to UvrA protein or when degraded by specific proteolysis (23).

A second notable feature of UvrB protein is the presence of an amino acid sequence motif which is very similar to that known to function as a specific proteolytic cleavage site in the *E. coli* Ada protein (DNA alkyltransferase I) (Fig. 5–6) (see chapter 3) (6, 202). Indeed, a proteolysis product of UvrB protein designated UvrB* has been identified in extracts of *E. coli*. The suggested physiological significance of this degradation reaction is discussed below. Finally, the amino acid sequence of the UvrB protein shows homology with two limited stretches of the sequence of the UvrC protein (210). The more extensive of these is a stretch of 14 amino acids located close to the C terminus of the UvrB polypeptide, 13 of which are identical to a region in the middle of the UvrC polypeptide (Table 5–2). As discussed below, a UvrC polypeptide that contains just the C-terminal 314 amino acids and excludes this region of homology with UvrB protein is sufficient to reconstitute functional UvrABC activity in vitro (100). However, the resistance to UV radiation of mutant strains carrying this deleted UvrC protein is considerably reduced, although it is not clear that this reduced survival specifically reflects a requirement for the UvrB-like motif in UvrC protein.

Purified UvrB protein has a relative molecular mass of 84 kDa (Table 5–1), slightly larger than that predicted from the open reading frame of the *uvrB* gene

Table 5–2 Homology between the UvrB and UvrC polypeptides

Polypeptide (nt)[a]	Sequence
UvrB (649–664)	A Q N L G F E E A A Q I N D Q L
UvrC (198–213)	S Q N L G F E E A A N I N D Q I

[a] nt, nucleotides.

(148, 205, 230). It is a globular acidic protein with a Stokes radius of 41 Å (4.1 nm), an S value of 5.2, and a pI of 5.0 (Table 5–1) (170). The protein is monomeric in solution and does not bind ATP or DNA in isolation (81, 213). However, UvrB protein interacts specifically with UvrA protein to form protein-protein and protein-DNA complexes which are important intermediates in the biochemistry of the damage-specific incision of DNA. UvrB protein is constitutively expressed at a level of ~250 molecules per cell (significantly higher than that of UvrA protein [see above]). Following SOS induction, the level of UvrB protein increases to as much as ~1,000 molecules per cell (170).

INTERACTIONS BETWEEN UvrA AND UvrB PROTEINS—THE PHENOMENON OF DAMAGE-SPECIFIC RECOGNITION

As indicated above, a fundamental aspect of the specificity of nucleotide excision repair is the ability of the repair machinery to locate and identify base damage in the genome, a process referred to as *damage-specific recognition*. In *E. coli*, this recognition problem is apparently solved by interactions between a complex of the UvrA and UvrB proteins and their interactions with damaged DNA.

Purified UvrA protein associates with UvrB protein to form a $(UvrA)_2(UvrB)_1$ complex (Fig. 5–7). At physiological concentrations (~10 nM UvrA protein and ~200 nM UvrB protein), all UvrA protein is in the form of $(UvrA)_2(UvrB)_1$ complexes (130). This interaction is strictly dependent on the presence of ATP (130). ATP[γS] cannot be substituted for ATP in this binding reaction, suggesting that ATP hydrolysis rather than ATP binding is critical for $(UvrA)_2(UvrB)_1$ complex

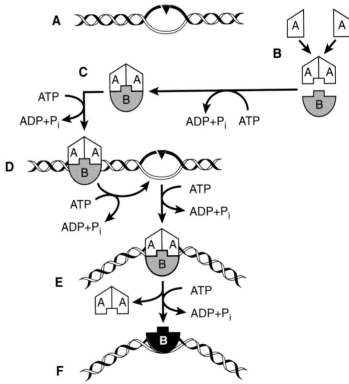

Figure 5–7 Model for the formation of a stable $(UvrB)_1$–damaged-DNA complex during nucleotide excision repair in *E. coli*. A site of base damage on one DNA strand is represented diagrammatically (A). A $(UvrA)_2(UvrB)_1$ protein complex forms in solution (B and C) and initially binds to DNA at a site removed from the damage (D). Promoters in genes might constitute preferential sites for the initial ''docking'' of such complexes. The $(UvrA)_2(UvrB)_1$ complex tracks along the DNA using a DNA helicase activity (D). When the site of base damage is encountered (E), UvrA protein dissociates from the complex, leaving a stable $(UvrB)_1$-DNA complex (F). This is associated with bending and kinking of the DNA.

formation (Fig. 5–7) (152). $(UvrA)_2(UvrB)_1$ complexes have a higher binding affinity for damaged DNA than does UvrA alone. In contrast, there are indications that the reverse is true for undamaged DNA (211). As indicated previously, the association constant for the specific binding of UvrA protein to a psoralen adduct in duplex DNA is $\sim 5 \times 10^7$ M^{-1} in the absence of ATP and UvrB protein (152). In the presence of these factors it increases to $\sim 1 \times 10^9$ M^{-1} (152), resulting in a significantly increased affinity for damaged relative to undamaged DNA.

It has been suggested that the fundamental role of UvrA protein in nucleotide excision repair is to function as a "molecular matchmaker" which delivers UvrB protein to sites of distortive damage in DNA by the formation of a transient $(UvrA)_2(UvrB)_1$-DNA complex from which it rapidly dissociates, leaving a highly stable $UvrB_1$-DNA complex (Fig. 5–7) (102, 130, 150, 154). The concept of molecular matchmaking embodies the notion that UvrA protein serves as a transient component of a multiprotein complex and does not itself directly participate in biochemical events after the stable binding of UvrB protein at sites of base damage. It is for this reason that the UvrA protein is designated parenthetically in the term Uvr(A)BC endonuclease. This concept is not without precedent. Molecular matchmakers have been defined as a class of proteins which use the energy of ATP hydrolysis to effect conformational changes in one or both components of a *binding protein–DNA target pair*. They are believed to participate in key protein-DNA interactions in several distinct aspects of DNA metabolism, including replication, transcription, mismatch repair, and possibly recombination and transposition; it has been hypothesized that they provide a fundamental mechanism for high-avidity binding to DNA by proteins which cannot rely on sequence specificity for such interactions (130, 150).

This model derives from several observations with respect to the UvrA-UvrB interaction. First, in the presence of UvrB protein, the transient 33-bp $(UvrA)_2$ footprint observed at a psoralen monoadduct is reduced to a stable smaller (19-bp) footprint thought to represent the binding of UvrB alone to DNA (14, 152, 213). Second, when UvrA and UvrB protein are incubated in different proportions in the presence of UV-irradiated DNA and DNA-protein complexes are isolated and analyzed for protein content, the ratio of UvrA to UvrB (as bound protein) decreases as the amount of UvrA protein in the reaction mixtures is reduced, and it eventually becomes vanishingly small, suggesting that only UvrB protein is bound to the irradiated DNA. Under conditions of saturating UvrB protein, one UvrB molecule is bound per damaged site in a reaction that requires only catalytic amounts of UvrA protein (130).

While there is a consensus among several laboratories that the $(UvrA)_2$ $(UvrB)_1$ complex participates in the process of damage-specific recognition, there is less consensus about exactly how this process transpires and the precise events by which a stable $(UvrB)_1$-DNA preincision complex ultimately assembles at sites of base damage. Using gel shift analysis with a 49-bp duplex oligonucleotide containing a benzoxyamine-modified AP site, it has been observed that the interaction of the $(UvrA)_2(UvrB_1)$ protein complex with a benzoxyamine site results in three detectable intermediates, consistent with the presence of a $Uvr(A)_2$-DNA complex, a $(UvrA)_2(UvrB)_1$-DNA complex, and a $(UvrB)_1$-DNA complex (215). Similar results have been obtained with a 96-bp oligonucleotide substrate containing a cisplatin adduct in a defined position (218). The $(UvrA)_2$ complex bound to DNA containing a benzoxyamine site is short-lived, with a half life of ~ 15 s. However, the other two complexes, which include UvrB protein, have half lives of ~ 2 h and appear to be in equilibrium (215). The relative amount of each is suggested to be dependent on the concentrations of UvrA and UvrB proteins, the nature and the amount of the damaged DNA, and the DNA concentration (215). These variables may contribute to the different efficiencies of incision by the UvrABC endonuclease observed for different types of DNA damage (see below).

The delivery of UvrB protein to sites of damage in UV-irradiated DNA has an absolute requirement for both UvrA protein and ATP hydrolysis (131). Effective loading of UvrB protein occurs in a limited range of UvrA protein concentration

(5 to 50 nM) and is inhibited at high concentrations. The rate of loading of UvrB protein is low ($k_{on} = 6 \times 10^{-4}$ M/s), suggesting that this step may be limiting for damage-specific incision of DNA. The K_m of ATP hydrolysis during the loading of UvrB protein is very similar to that for ATP hydrolysis by UvrA protein alone, suggesting that ATP hydrolysis during this loading reaction is effected by UvrA protein (131). Consistent with this interpretation, substitution of the Lys residue with Ala in the invariant tripeptide GKS/T of the most N-terminal and C-terminal Walker type A boxes inactivates the ability of UvrA protein to hydrolyze ATP. These amino acid substitutions have a marginal effect on the ability of UvrA protein to bind to damaged DNA, but they drastically reduce the ability of the protein to deliver UvrB protein to sites of base damage (117).

Transient (UvrA)$_2$(UvrB)$_1$-DNA complexes have not been unambiguously detected, nor has the binding constant for (UvrB)$_1$-DNA been measured. However, isolated (UvrB)$_1$-damaged DNA complexes are very stable, with a k_{off} of 1.1×10^{-4} s^{-1}, which extrapolates to a half-life of 2 to 3 h, depending on the particular type of base damage (130). Once a relatively stable (UvrA)$_2$(UvrB)$_1$-DNA complex has formed, ATP hydrolysis by the activated UvrB ATPase function is thought to be accompanied by kinking and local denaturation of the DNA and a concomitant change in the conformation of UvrB protein. These changes are believed to result in the tight binding of UvrB protein to sites of base damage and to the release of UvrA protein (Fig. 5–7) (154). Indeed, experiments that have exploited the ability of dinuclear platinum compounds to both damage DNA and cross-link proteins have resulted in the identification of UvrB-platinum-DNA ternary complexes (214). Hence, while DNA damage recognition by UvrA protein is a process of relatively low specificity, extremely high specificity is attained by the UvrA-dependent binding of UvrB protein at sites of base damage. (UvrB)$_1$-DNA complexes have not been detected with undamaged DNA (170).

Experimental evidence in support of conformational alterations associated with the binding of UvrB protein to damaged DNA comes from various sources (97, 226). Measurements of DNA unwinding in the presence of UvrA and UvrB proteins and ATP suggest the unwinding of as much as one full helical turn per Uvr protein-DNA complex (126, 226). This unwinding might be a direct effect of the binding of the Uvr proteins to DNA. There are certainly ample precedents in the literature for this. For example, during the formation of a stable protein-DNA complex, *E. coli* RNA polymerase has been shown to unwind the DNA helix by ~17 bp. Similarly, catabolite gene activator protein and *Eco*RI endonuclease have been shown to unwind the helix during binding to specific cognate sequences (210). Alternatively, the Uvr proteins could unwind DNA via a DNA helicase activity. Evidence for such an activity as an intrinsic property of a UvrA-UvrB complex is discussed below. In addition to evidence of local denaturation of DNA, electron microscopy indicates that after formation of the (UvrA)$_2$(UvrB)$_1$-DNA ternary complex the DNA is kinked by ~130° (150, 180) (Color Plate 4). Further evidence in support of alterations in the structure of DNA upon its interaction with UvrB protein derives from flow linear dichroism studies (198).

The UvrB footprint at sites of psoralen monoadducts in DNA is most pronounced on the undamaged strand, suggesting that UvrB protein makes more intimate contact with that strand. In support of this conclusion, it has been observed that *E. coli* DNA photolyase (which binds to the damaged strand at pyrimidine dimer sites [see chapter 3]) does not inhibit damage-specific incision of UV-irradiated DNA in the presence of the *E. coli* Uvr proteins; in fact, it stimulates the incision reaction (76, 149). Thus, the endonuclease apparently binds at the DNA face opposite to the one containing the adduct. The only distinctive effect on the damaged strand is the appearance of a DNase-hypersensitive phosphodiester bond (170).

What specific determinants guide the precise interactions of the (UvrA)$_2$(UvrB)$_1$ protein complex with damaged DNA? One possibility is that (UvrA)$_2$(UvrB)$_1$ protein complexes randomly patrol the genome and that specific binding is guided by the affinity of UvrA for particular types of DNA distortions.

The more "correct" (i.e., distinctive from native DNA) the distortion, the greater is the probability of productive $(UvrA)_2(UvrB)_1$-DNA complex formation, i.e., one that leads to the dissociation of UvrA protein and the formation of a stable $(UvrB)_1$-DNA complex. The formation and stability of productive intermediates may also be influenced by the extent to which the DNA distortion initially recognized is progressively modified during the course of these reactions. Stated differently, it has been suggested (62, 226) that the DNA substrate may be progressively determined by conformations imposed on it as the result of its interactions with the UvrA-UvrB protein complex. Hence, the initial conformational distortion in damaged DNA may simply serve to lower the K_m for the binding of $(UvrA)_2(UvrB)_1$ complexes (62, 226). This concept predicts a distinctive hierarchy for the kinetics of damage-specific incision of DNA substrates with different types of helix-distortive damage. Experiments with different types of base damage suggest that this is indeed the case (226).

As indicated above, the difference in the affinity of UvrA protein for damaged and undamaged double-stranded DNA is low (a factor of only 10^3 to 10^4). Statistical considerations indicate that under conditions where only a single nucleotide in 10^7 (or fewer) is damaged, initial Uvr protein-DNA interactions will more probably occur at undamaged than at damaged sites in DNA (64). These considerations have prompted consideration of a mechanism by which the initial formation of a nonspecific $(UvrA)_2(UvrB)_1$-DNA complex is followed by translocation of the Uvr proteins to sites of distortive damage, yielding specific $(UvrA)_2(UvrB)_1$-DNA complexes (58, 60, 61, 63, 64).

It has in fact been shown that a UvrA-UvrB complex [presumably the $(UvrA)_2(UvrB)_1$ complex discussed above] is capable of limited DNA helicase activity in the presence of ATP (127, 128). This helicase can unidirectionally unwind very short stretches of DNA duplexes and DNA D-loop structures with strict $5' \rightarrow 3'$ polarity, and it is inhibited by the presence of bulky base damage in UV irradiated DNA. While a DNA helicase could potentially function in several roles during nucleotide excision repair, the properties described above are consistent with a role in the translocation of a $(UvrA)_2(UvrB)_1$ complex from a site of relatively weak-affinity binding in DNA to the actual site of base damage, where the inhibitory effect of a bulky adduct might arrest the complex. Consistent with this model, mutational alteration of the Lys-45 residue located in the Walker type A nucleotide binding motif of the UvrB polypeptide leads to defective ATPase and DNA helicase activity in UvrA-UvrB complexes and a failure to support the incision of UV-irradiated DNA (167).

The ATPase that presumably drives the DNA helicase function of the UvrA-UvrB complex is believed to be cryptic in UvrB protein (6, 205) but becomes activated when it undergoes a conformational change associated with binding to UvrA protein and DNA (23). In keeping with this notion of a cryptic ATPase activity, UvrB* protein generated by proteolysis of UvrB protein in vivo possesses a DNA-dependent ATPase activity in the absence of UvrA protein (23).

An interesting question is whether the $(UvrA)_2(UvrB)_1$ complex has preferred sites of docking with DNA. By footprinting analysis it has been observed that the complex binds preferentially to the nontranscribed strand of promoter sites downstream from the RNA polymerase-promoter complex. This strand preference is consistent with the $5' \rightarrow 3'$ directionality of the UvrAB helicase activity (3) and suggests that the binding of RNA polymerase to promoter sites in DNA may provide a signal for selective landing sites for the $(UvrA)_2(UvrB)_1$ complex on DNA (3).

In summary, a reasonable working model (Fig. 5–7) which is consistent with the consensus experimental evidence is that the binding of dimeric UvrA protein to UvrB protein in solution leads to the assembly of a $(UvrA)_2(UvrB)_1$ complex. This complex binds to DNA nonspecifically but is able to unwind DNA by using the energy of ATP hydrolysis (possibly by the bound UvrB protein), and it translocates unidirectionally (62). This may represent a fundamental DNA patrolling mechanism whereby the integrity of the genome is constantly monitored for base damage in living cells. Such a patrolling mechanism is not limited to a three-dimensional diffusion process in solution but utilizes a mechanism which effectively decreases the volume of solution that must be "searched" by proteins while in a nonspecifically bound state (221). It has been suggested that when a site of base damage is encountered the $(UvrA)_2(UvrB)_1$ complex unwinds and kinks the DNA in a particular manner, UvrA protein dissociates, and a stable $(UvrB)_1$-DNA complex is formed in which the DNA is now conformationally primed for incision (102, 150, 210).

It remains to be determined precisely how sites of base damage facilitate the formation of stable $(UvrB)_1$-DNA complexes once such sites have been recognized and how these sites are distinguished from temporary changes in the local dynamics of undamaged DNA. This question is particularly pertinent when we consider the extremely broad range of substrates that are recognized by the UvrABC system in vitro (see below). An intriguing possibility is that the $(UvrA)_2(UvrB)_1$ complex recognizes lesion-imposed restrictions of the normal dynamic range of DNA (138). In other words, the complex may systematically scan the DNA for normal helical parameters by constantly inducing limited conformational changes in structure and testing these. Registration ("sensing") of normal binding interactions is followed by dissociation of the complex. However, any lesion which constrains some subset of conformations available to native DNA results in "jamming" of the protein-DNA complex in a high-affinity conformation (138). Any structural modification of DNA that causes the $(UvrA)_2(UvrB)_1$ complex to unload UvrB protein onto DNA is recognized as a substrate for subsequent endonucleolytic incision (98). The attractive feature of this type of model is the suggestion that it requires alterations in the *dynamics* of the DNA helix imposed by base damage, rather than alterations in its shape or *static* structure to determine substrate recognition (98, 138). It has been suggested that a particular aspect of the dynamic state of DNA that may determine the formation of stable $(UvrB)_1$-DNA complexes in the presence of multiple diverse types of base damage is an alteration in the energy of base stacking interactions in DNA (215). UvrB protein (which is quite hydrophobic) may "read" destabilized base stacking interactions in the damaged strand. In this regard it may be relevant that UvrB protein and single-stranded DNA-binding protein share limited regions of amino acid sequence homology (215).

The *uvrC*+ Gene and UvrC Protein

The final step in the assembly of the catalytically active damage-specific endonuclease in *E. coli* is the binding of UvrC protein to specific $(UvrB)_1$–damaged-DNA complexes (Fig. 5–8). UvrC protein is the product of the *uvrC*+ gene, which maps at 42 min on the *E. coli* genetic map. *uvrC*+ is one of two genes in a single operon. The more proximal gene encodes a protein of 23 kDa. It has been pointed out that the amino acid sequence of this 23-kDa protein is homologous with several known prokaryotic positive regulators of gene expression (122, 170). However, the protein is apparently not involved in the regulation of *uvrC*+, since its inactivation has no effect on the sensitivity of *E. coli* cells to UV radiation (110).

In addition to its cotranscription with the gene for the 23-kDa protein, *uvrC*+ is monocistronically transcribed from an internal promoter (54, 179). Like the *uvrA*+ and *uvrB*+ genes, *uvrC*+ is weakly expressed constitutively and there are only 10 to 20 molecules of the protein per cell (170). Unlike the other two *uvr* genes, however, *uvrC*+ is not inducible by DNA damage and is not a member of the SOS regulon (45, 56, 156).

The *uvrC*+ gene is expected to encode a polypeptide of 610 amino acids with a calculated size of ~68.5 kDa (Table 5–1) (99). The amino acid sequence shows no remarkable features; nucleotide binding domains and zinc fingers are not present in the UvrC polypeptide (170). Purified UvrC protein is essentially neutral, with a pI of 7.3 (Table 5–1). It is a globular protein with a Stokes radius of 41 Å (4.1 nm) and an S value of 4.9 (Table 5–1) (170). The purified protein has a strong tendency to aggregate and becomes inactivated on prolonged storage (204).

UvrC protein does not associate with either UvrA or UvrB protein in solution, but it has a high affinity for the $(UvrB)_1$-DNA complex. The stoichiometry of UvrC

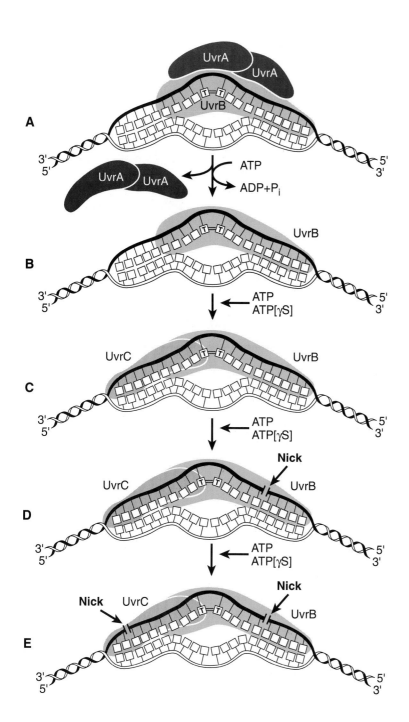

Figure 5–8 Diagrammatic representation of bimodal damage-specific nicking of DNA by the *E. coli* UvrABC endonuclease. Following the formation of a stable (UvrB)$_1$–damaged-DNA complex (A and B) (see Fig. 5–7), UvrC protein binds at the site (C) and induces a conformational change which enables bound UvrB protein to nick the DNA 4 nucleotides 3' to the site of damage (D) (shown as a pyrimidine dimer). This reaction requires the binding of ATP (or ATP[γS]) by UvrB protein, but no ATP hydrolysis occurs at this step. Following the 3' incision, UvrC protein catalyzes nicking of the DNA 7 nucleotides 5' to the dimer (E).

binding in this reaction is uncertain. When the UvrA, UvrB, and UvrC proteins are incubated with damaged DNA in the presence of ATP, specific incision of the DNA results. Incision can also be effected by the addition of purified UvrC protein to UV-irradiated DNA carrying a single molecule of UvrB protein specifically bound at each site of base damage (130). This incision reaction is relatively slow; it takes about 2 min to complete in the presence of a vast molar excess of UvrC protein, a curious observation since in vivo UvrC protein is present in very low abundance (Table 5–1). ATP is required for the incision of damaged DNA when UvrC protein is added to the (UvrB)$_1$-DNA complex (Fig. 5–8), but there is no ATP hydrolysis. Indeed, the requirement for a nucleotide cofactor in the incision reaction can be supplanted by nonhydrolyzable analogs of ATP, suggesting that nucleotide binding causes a conformational change which is required for damage-specific incision (131).

DNA INCISION IS BIMODAL

For a disquieting number of years (in retrospect), molecular models of nucleotide excision repair in *E. coli* invoked one strand break exclusively on the 5′ side of the sites of base damage. The chief imperative for this model was the known propensity for *E. coli* DNA polymerase I to degrade DNA at nicked sites in the 5′ → 3′ direction by a process called *nick translation*. Hence, it was intuitively persuasive that the combined actions of a 5′ DNA damage-specific endonuclease, the 5′ → 3′ exonuclease of DNA polymerase I, and DNA polymerase I-catalyzed repair synthesis, could account fully for the mechanism of nucleotide excision repair of damaged DNA, and this biochemical issue was considered essentially solved. However, when DNA sequencing gels were used to define the precise location of UvrABC protein-dependent incisions relative to pyrimidine dimers in DNA, the surprising and unexpected observation was first made by Sancar, Rupp, and coworkers (146, 151) that the damaged strand contained nicks on both sides of each dimer (Fig. 5–8), an observation confirmed by others (230, 231). Historically it is now amusing to recall that these investigators were as surprised by this result as the rest of the scientific community. Wedded to the 5′ incision model in vogue at the time, they sequenced the damaged DNA strand labeled at one end. Sequencing of the same strand carrying a radiolabel at the other end was carried out to confirm the location of the presumed single incision but, instead, led to the seminal observation that there were in fact two nicks flanking each dimer site. As discussed in later chapters, bimodal incision is apparently universal in nature during nucleotide excision repair.

The complete UvrC polypeptide is not required for this incision reaction. The presence of the C-terminal 314 amino acids is sufficient to confer resistance to killing by UV radiation in *uvrC* mutants (100) (Fig. 5–9). Additionally, when the portion of the *uvrC* gene that encodes this region is fused to a gene that encodes maltose-binding protein, the fusion protein results in efficient nicking of UV-irradiated DNA in the presence of the UvrA and UvrB proteins (100).

The UvrABC damage-specific endonuclease typically hydrolyzes the eighth phosphodiester bond 5′ to pyrimidine dimers or (6-4) photoproducts in UV-irradiated DNA (6, 130) (Fig. 5–8 and Color Plate 5). The sites of incision 3′ to these lesions are more variable. Incision occurs at the fifth phosphodiester bond 3′ to all pyrimidine dimers (Fig. 5–8) and at the fourth phosphodiester bond 3′ to (6-4) photoproducts (151, 195, 230). In DNA containing cisplatin diadducts the 3′ incision site is almost exclusively at the fourth phosphodiester bond (10). DNA containing either psoralen monoadducts (either pyrone or furone), *N*-acetoxyacetylaminofluorene, cyclohexylcarbodiimide, 4-nitroquinoline-1-oxide, or mitomycin monoadducts is cut at the fifth 3′ phosphodiester bond.

The precise locations of both 5′ and 3′ incisions can also be affected by DNA sequence context. For example, with both (6-4) photoproducts and pyrimidine dimers, incisions have been observed as close as the sixth bond on the 5′ side of lesions and as removed as the sixth phosphodiester bond on the 3′ side depending on the particular DNA sequence context in which these lesions are located (118). Additionally, when DNA fragments were constructed in which acetylaminofluorene adducts were specifically introduced at each of three different guanine residues in the sequence -GGCGCC-, major differences were detected in the efficiency of DNA incision following incubation with the UvrABC endonuclease (164). In contrast, all three adducts were recognized with equal efficiency as judged by DNase I protection experiments (164).

Site-specific mutagenesis of the UvrC and UvrB proteins has led to the important conclusion that the incision process is not a concerted one. The 3′ incisions are apparently effected by UvrB protein, and this incision precedes the 5′ cut catalyzed by UvrC protein (Fig. 5–8) (97, 101). Hence, in effect there are two physically distinct endonuclease activities in the *E. coli* UvrABC enzyme. This, too, might turn out to be universal. As will be seen in the next chapter, two endonu-

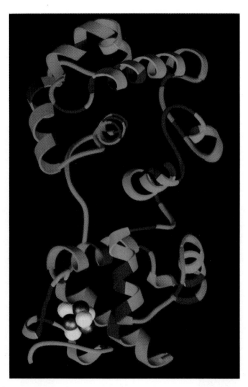

Color Plate 3 Ribbon representation of the proposed structure of the *E. coli* Nth protein (endonuclease III) showing the bilobal organization of the elongated polypeptide separated by a deep cleft. The amino acid residues shown in red are identical in the Nth and homologous MutY proteins (see Fig. 4–15). Green residues are conserved in the two proteins, while blue residues are not homologous. Endonuclease III contains an iron (shown in brown)-sulfur (shown in yellow) cluster, which is bound entirely within a C-terminal loop. (Photograph courtesy of R. Cunningham.)

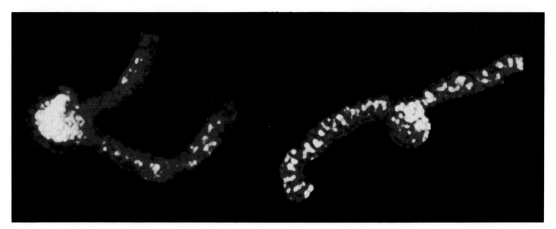

Color Plate 4 Electron micrograph showing kinking of DNA molecules at the point of UvrB protein binding. DNA bound to wild-type protein is severely kinked (left). A mutant form of UvrB protein in which the amino acid residue Asp-478 was replaced by Ala eliminates the bending of DNA while still allowing loading of the protein by UvrA protein at a site of base damage (right). (Reproduced with permission from D. S. Hsu, M. Takahashi, E. Delagoutte, E. Bertrand-Burggraf, Y. H. Wang, B. Norden, R. P. P. Fuchs, J. Griffith, and A. Sancar. 1994. Flow linear dichroism and electron microscopic analysis of protein-DNA complexes of a mutant UvrB protein which binds to but cannot kink DNA. *J. Mol. Biol.* **241:**645–650, and Q. Shi, R. Thresher, A. Sancar, and J. Griffith. 1992. Electron microscopic study of (A)BC excinuclease. DNA is sharply bent in the UvrB-DNA complex. *J. Mol. Biol.* **226:**425–432. Photograph courtesy of A. Sancar.).

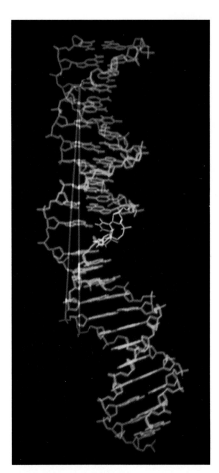

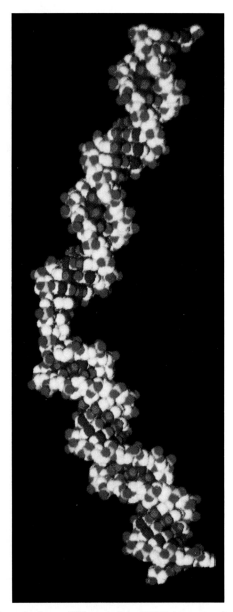

Color Plate 5 Energy-minimized structure for duplex DNA containing a thymine-thymine cyclobutane dimer. The structure has been oriented to display a helical kink of 27° and unwinding of 19.7° associated with the presence of the dimer (magenta). Sites of anticipated UvrABC-mediated endonucleolytic cleavage are shown at the eighth phosphodiester bond (red) 5′ to the dimer and the fourth or fifth phosphodiester bonds (red) 3′ to the dimer. (Reproduced with permission from B. Van Houten, H. Gamper, S. R. Holbrook, J. E. Hearst, and A. Sancar. 1986. Action mechanism of ABC excision nuclease on a DNA substrate containing a psoralen crosslink at a defined position. *Proc. Natl. Acad. Sci. USA* **83:**8077–8081. Photograph courtesy of A. Sancar.)

Color Plate 6 Space-filling model of DNA incorporating a psoralen cross-link, based on energy minimization structure. The cross-link unwinds the helix by 87.7° and produces a 46.5° helical kink. (Reproduced with permission from D. A. Pearlman, and S. R. Holbrook. 1985. Molecular models for DNA damaged by photoreaction. *Science* **227:**1304–1308. Photograph courtesy of A. Sancar.)

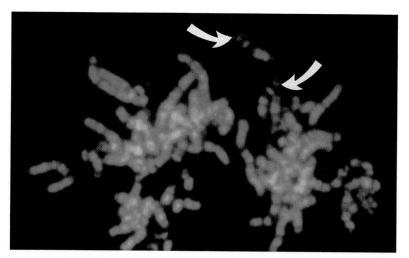

Color Plate 7 Mapping of the *XPG* gene to human chromosome 13q33 by fluorescence in situ hybridization. Both of the chromosomes showing the positive hybridization signal (red dots) are chromosome 13. (Reproduced with permission from S. Samec, T. A. Jones, J. Corlet, D. Scherly, R. D. Wood, and S. G. Clarkson. 1994. The human gene for xeroderma pigmentosum complementation group G (*XPG*) maps to 13q33 by fluorescence in situ hybridization. *Genomics* **21**:283–285. Photograph courtesy of R. Wood.)

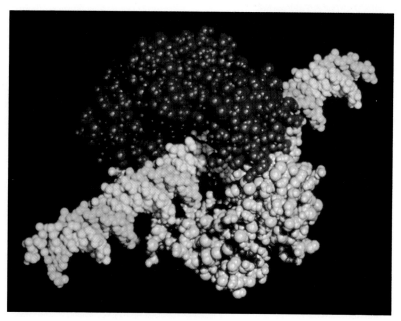

Color Plate 8 Space-filling model of the *E. coli* DNA polymerase III β-subunit dimer with B-form DNA. One monomer is colored red, and the other is colored yellow. The radii of the spheres correspond to the van der Waals radii of the corresponding atoms. Hydrogen atoms are not explicitly displayed but manifest as increased radii for atoms to which they are bonded. The B-form DNA is shown passing through the hole in the β subunit dimer with no steric repulsions. (Reproduced with permission from X.-P. Kong, R. Onrust, M. O'Donnell and J. Kuriyan. 1992. Three-dimensional structure of the β subunit of *E. coli* DNA polymerase III holoenzyme: A sliding DNA clamp. *Cell* **69**:425–437. Photograph courtesy of John Kuriyan.)

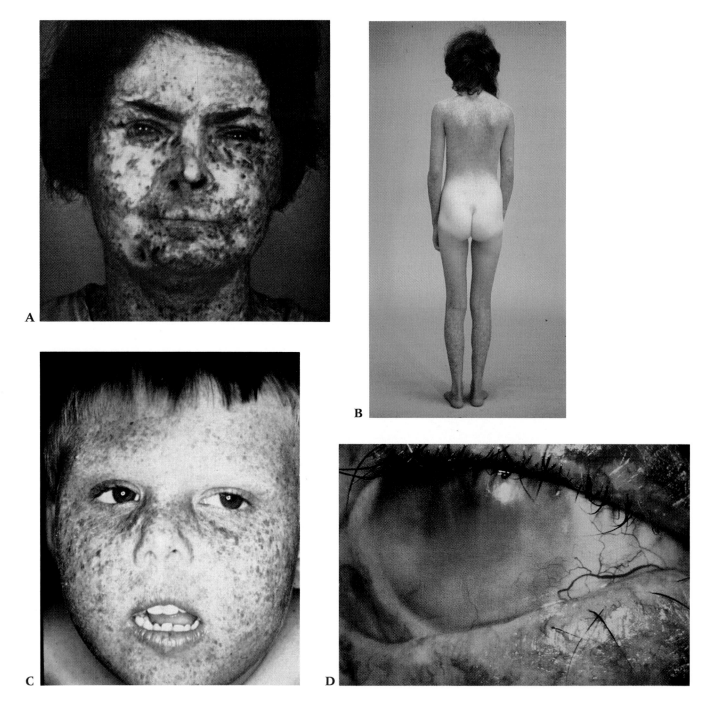

Color Plate 9 Clinical features of xeroderma pigmentosum (XP). (A) The disease typically presents with severe pigmentary disturbances on the sun-exposed areas of the skin. (B) Note that areas not normally exposed to sunlight are not affected. (C) These clinical features can present at a very early age. (D) The eyes are also frequently affected. The individual shown here has clouding of the cornea and atrophy and loss of lashes from the lower eye lid. (Reproduced with permission from K. H. Kraemer. Progressive degenerative diseases associated with defective DNA repair: xeroderma pigmentosum and ataxia telangiectasia. p. 37–71. *In* W. W. Nichols and D. G. Murphy (ed.), *DNA repair processes, Symposia Specialists, Miami*, 1977).

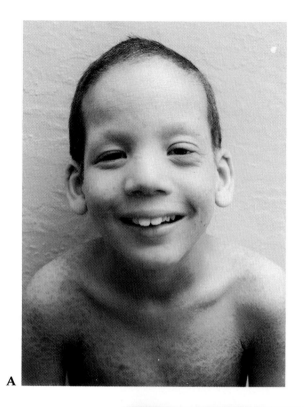

A

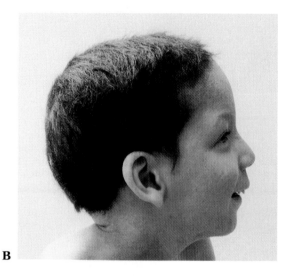

B

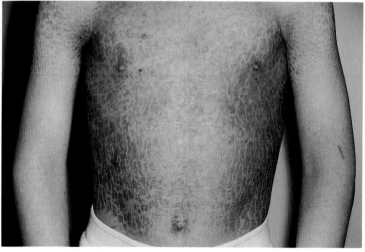

C

Color Plate 10 (A and B) Frontal and side views of the face of a young boy with trichothiodystrophy (TTD). Note the sparse broken scalp hair, eyebrows and eyelashes, the thickened epicanthal folds, the protruding ears and the receding chin. (C) View of the trunk of this same patient. Note the large ichthyotic (fish-like) scales on the anterior part of the trunk and the atopic eczema in the armpits. (Reproduced with permission from V. H. Price, R. B. Odom, W. H. Ward, and F. T. Jones. 1980. Trichothiodystrophy: sulfur-deficient brittle hair as a marker for a neuroectodermal symptom complex. *Arch. Dermatol.* **166:**1375–1384. Photograph courtesy of V. Price.)

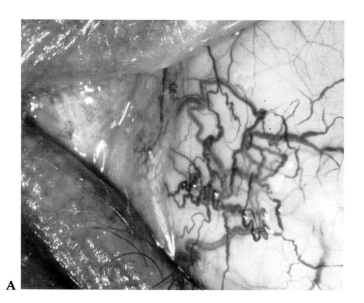

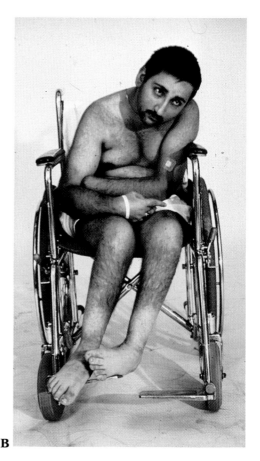

Color Plate 11 Some of the clinical features of ataxia telangiectasia (AT). (A) The conjunctiva of the eye of an individual showing telangiectasia (abnormal dilatation of blood vessels). (B) This 22-year-old patient has severe ataxia and is wheelchair bound. (Reproduced with permission from K. H. Kraemer. Progressive degenerative diseases associated with defective DNA repair: xeroderma pigmentosum and ataxia telangiectasia, p. 37–71. *In* W. W. Nichols and D. G. Murphy (ed.) *DNA repair processes. Symposia Specialists, Miami.* 1977.)

A

B

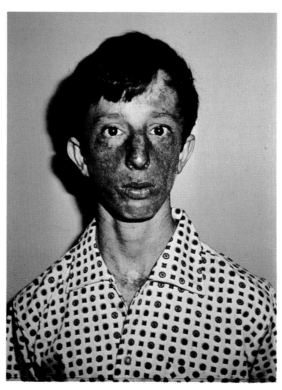

Color Plate 12 Patients with Bloom's syndrome (BS) typically have a butterfly distribution of light-induced capillary dilatation of the skin of the face. (Photograph courtesy of J. German.)

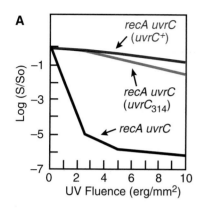

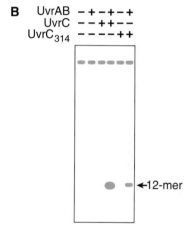

Figure 5–9 The C-terminal 314-amino-acid UvrC polypeptide is able to participate in nucleotide excision repair. (A) Transformation of a *recA uvrC* mutant with a plasmid that encodes the truncated UvrC protein restores resistance to UV radiation to levels comparable to that observed following transformation with a plasmid that encodes the wild-type UvrC protein (610 amino acids). (B) A 137-mer duplex DNA fragment containing ^{32}P radiolabel at the sixth phosphodiester bond 5' to a psoralen monoadduct was incubated with the indicated proteins, and the reaction was analyzed on a DNA sequencing gel. Following autoradiography of the gel, an excised radiolabeled 12-mer oligonucleotide was observed as a product of the reactions with both the normal and the truncated UvrC protein. (*Adapted from Lin and Sancar [100].*)

cleases, in this case encoded by three distinct genes, have been implicated in nucleotide excision repair in the yeast *S. cerevisiae*.

These observations in *E. coli* provide a reasonable explanation for the observation of 3'-uncoupled incision of DNA with aged UvrC protein and with certain DNA substrates (102). Furthermore, the notion of uncoupled incisions provides a tenable explanation for the long-standing observation that *uvrC* mutant strains can still support some nicking of UV irradiated DNA, whereas *uvrA* and *uvrB* mutants cannot. The stoichiometry of incision by UvrB and UvrC proteins is one molecule of each protein per damaged site.

Incision is a relatively slow reaction which proceeds at a rate of 10^4 M^{-1}s^{-1} (132). Like many of the biochemical events that precede incision, the incision step itself is dependent on the presence of ATP. However, in contrast to the formation of the (UvrA)$_2$(UvrB)$_1$ and (UvrB)$_1$-DNA complexes, the hydrolysis of ATP is not essential for DNA incision, and ATP[γS] can be substituted for ATP during this step (154).

Substrate Specificity of the UvrABC Endonuclease

In contrast to the largely nondistortive modifications of bases that are recognized by DNA glycosylases, the majority of the numerous chemicals to which *uvr* mutants are sensitive share with UV radiation the capacity for generating bulky base adducts which can cause significant distortion of the DNA helix (Table 5–3). One study has shown that a pyrimidine dimer unwinds the helix by 19.7° and introduces a kink of 27° which protrudes into the major groove of DNA (134). Additionally, one- and two-dimensional gel electrophoresis and quantitative electron microscopy studies of DNA fragments of known sequence containing thymine dimers in defined positions indicate that pyrimidine dimers cause a bend of ~30°

Table 5–3 Substrates for the UvrABC endonuclease of *E. coli*

DNA-damaging agent	Adduct(s)
N-acetoxy-2-acetylaminofluorene	*C*-8-Guanine
Anthramycin	*N*-2-Guanine
AP sites	Base loss
AP sites (reduced)	Ring opened AP site
Alkoxamine-modified AP sites	AP site analog
Benzo[*a*]pyrenediolepoxide	*N*-2-Guanine
CC-1065	*N*-3-Adenine
Cisplatin and transplatin	*N*-7-Guanine
Cyclohexylcarbodiimide	Unpaired G and T residues
Ditercalanium	Noncovalent bisintercalator
Doxorubicin and AD32	Intercalated compounds
N-hydroxyaminofluorene	*C*-8-Guanine
N,*N*′-bis(2-chloroethyl)-*N*-nitrosourea	Bifunctional alkylation
N-methyl-*N*′-nitro-*N*-nitrosoguanidine	O^6-Methylguanine
Mitomycin	*N*-7-Guanine
Nitrogen mustard	Bifunctional alkylation
4-Nitroquinoline-1-oxide	*C*-8, *N*-2-Guanine
	N-6-Adenine
6-4 Photoproduct	*C*-6, *C*-4-PyC
Psoralen	*C*-5, *C*-6-Thymine
Pyrimidine dimer	*C*-5, *C*-6-Pyrimidine
Thymine glycol	*C*-5, *C*-6-Thymine

Source: Adapted from Van Houten (210).

in DNA (75). There is also evidence for structural deformation of DNA containing the bulky monoadduct acetylaminofluorene (159).

DNA cross-links are particularly distortive. Space-filling models suggest that a psoralen cross-link unwinds the DNA duplex by 87.7°, causes a 46.5° helical kink, and displaces the helical axis by 3.49 Å (0.35 nm) (134) (Color Plate 6). This model has been largely confirmed by using two-dimensional nuclear magnetic resonance spectroscopy (NMR) (206) and electron microscopy (181). Finally, studies on DNA treated with a different cross-linking agent (cisplatin) indicate that the adduct *cis*-[Pt(NH$_3$)$_2$-(dpGpG)] kinks the DNA into the major groove by 50 to 60° and unwinds it by 40 to 60° (85).

There has been considerable emphasis on conformational distortion of DNA structure as a crucial element for damage-specific recognition leading to the productive incision of DNA. However, in vitro studies have demonstrated that the *E. coli* damage-specific endonuclease has a range of substrate specificity that includes base damage which is not considered distortive (138, 170, 210, 215). Thus, for example, lesions such as AP sites, thymine glycols, and O^6-methylguanine are recognized as substrates by the enzyme (84, 93, 138, 168, 170, 193, 199, 210, 215, 220). Single-nucleotide mismatches and even larger loops caused by mispairing, as well as naturally bent DNA, are not productive substrates for the UvrABC endonuclease, however (170). Intercalating agents such as ethidium bromide and chloroquine result in binding sites in DNA for UvrA protein, but the addition of UvrB and UvrC proteins does not lead to incision in these instances. In fact, trapping of UvrA protein in nonproductive protein-DNA complexes containing intercalated chemicals can result in the inhibition of nucleotide excision repair. This phenomenon possibly explains the well-known observation that caffeine is an inhibitor of DNA repair (169).

In general, these observations are consistent with the suggestion that many types of base damage are recognized with differing affinities by the UvrABC endonuclease and that the K_m for binding varies as a function of the type and extent of helix distortion produced. Alternatively, as discussed above, perhaps productive binding of UvrB protein to DNA results only when base damage precludes the adoption of specific conformational constraints imposed by the protein on native

DNA, thereby freezing the protein-damaged DNA complex in place, or when UvrB protein specifically recognizes alterations in base-stacking interactions in DNA. Earlier in the chapter, we referred to the presence of a glycine-rich region at the C terminus of the UvrA protein. Glycine and alanine residues allow for more accessible conformations of proteins (106). Hence, it has been suggested that this feature of UvrA protein might facilitate a flexible induced-fit mechanism for the specific recognition of many different types of base damage (62).

Roles of ATP in Damage-Specific Incision

It is evident from the preceding discussion that ATP is essential for damage-specific incision during nucleotide excision repair in *E. coli* (62, 64, 152, 210). This high-energy compound appears to play several roles in this process, reflecting the fact that ATP is not only a source of energy but is also a ligand whose binding causes conformational changes in the Uvr proteins. The key steps in which ATP participates can be summarized as follows.

(i) ATP increases the dimerization constant of the UvrA subunit and is apparently bound rather than hydrolyzed in this process, since nonhydrolyzable ATP analogs can substitute for ATP.

(ii) ATP is essential for the formation of the $(UvrA)_2(UvrB)_1$ complex. Nonhydrolyzable ATP analogs cannot substitute for ATP in this event. Hence, more than simple binding of ATP is apparently involved here.

(iii) ATP is necessary for the formation of a specific $(UvrB)_1$-DNA complex.

(iv) Some form of a UvrAUvrB complex [possibly $(UvrA)_2(UvrB)_1$] hydrolyzes ATP in order to drive a DNA helicase activity.

(v) ATP is necessary for the bimodal incision of DNA.

Role of Proteolysis in the Regulation of DNA Incision by the UvrABC Endonuclease

As already indicated, a proteolysis product of UvrB protein has been detected in cell extracts of *E. coli*. This degradation product, UvrB* protein (24), is capable of interacting with UvrA protein and exhibits a DNA-dependent ATPase activity. However, the UvrAUvrB* complex is devoid of DNA helicase-ATPase activity and does not facilitate incision of DNA in the presence of UvrC protein (24). A protease that cleaves UvrB protein in vitro has been identified as OmpT protease, which is also involved in the cleavage of Ada protein (see chapter 3) (6, 161). It has been suggested that specific proteolysis of UvrB protein might regulate incision of DNA in vivo, thereby protecting the genome from spurious cutting of undamaged DNA if the intracellular levels of Uvr proteins are abnormally elevated for any reason. Additionally, or alternatively, proteolysis of UvrB protein in vivo might conceivably facilitate turnover of the UvrB-UvrC-DNA complex following damage-specific incision and of unproductive complexes that are improperly constituted or bind to sites that cannot be cleaved (24). However, many proteins have been observed to be cleaved by the OmpT protease in vitro during cell lysis (70), and at present there is no convincing biological evidence in support of proteolysis as a mechanism for regulating nucleotide excision repair in *E. coli*.

Summary of Damage-Specific Incision during Nucleotide Excision Repair in *E. coli*

Damage-specific incision of DNA is a biochemically complex process, even in a simple prokaryote like *E. coli*. This complexity derives from the stochastic nature of base damage to DNA. As a consequence, living cells cannot rely on DNA sequence for the specificity of protein-DNA interactions and have evolved an elaborate multistep process during which the ordered assembly of DNA-protein complexes provides an architectural specificity which discriminates damaged from undamaged sites. Much is still to be learned about precisely how this specificity is achieved, and the UvrABC endonuclease of *E. coli* will certainly continue to serve as a primary instructional model for understanding the detailed biochemical mechanisms of nucleotide excision repair.

Postincisional Events—Excision of Damaged Nucleotides, Repair Synthesis, and DNA Ligation

The demonstration of a bimodal incision mechanism led to the realization that the actual excision of base damage during nucleotide excision repair in *E. coli* requires the release of the oligonucleotide fragment defined by these incisions and of the bound UvrB and UvrC proteins. The UvrBUvrC complex does not turn over at a detectable rate in vitro (25, 152, 170). However, such turnover clearly takes place in living cells. For example, when UV-irradiated *E. coli* cells are held in liquid after exposure to UV radiation (*liquid holding recovery*), thousands of pyrimidine dimers are excised over several hours (200).

The phenomenon of liquid holding recovery is not observed in *E. coli* mutants which are defective in the *polA*⁺or *uvrD*⁺ genes (200). This observation prompted an examination of Pol I and UvrD proteins in the presence of the UvrABC proteins. Purified Pol I and UvrD proteins increase the extent of DNA incision, consistent with increased turnover of the UvrBUvrC incision complex (25, 77, 89). It is not clear whether both proteins are absolutely required for this effect, since stimulation can be achieved with just one or the other under certain experimental conditions (170). However, maximal stimulation requires both Pol I and UvrD proteins.

These observations are supported by biological studies. Whereas *uvrD* mutants of *E. coli* are not totally defective in DNA incision, these mutants display abnormally low rates of incision in semi-in vivo experimental systems, e.g., in cells permeabilized with detergents (11). In addition, intact UV-irradiated *uvrD* mutant cells manifest a reduced rate of excision of pyrimidine dimers (87, 140, 141, 216) (Fig. 5–10). Consistent with an auxiliary role of UvrD protein in nucleotide excision repair, mutant strains that are entirely defective in UvrD protein are not as sensitive to UV radiation as are *uvrA*, *uvrB*, or *uvrC* mutants (225) (Fig. 5–10).

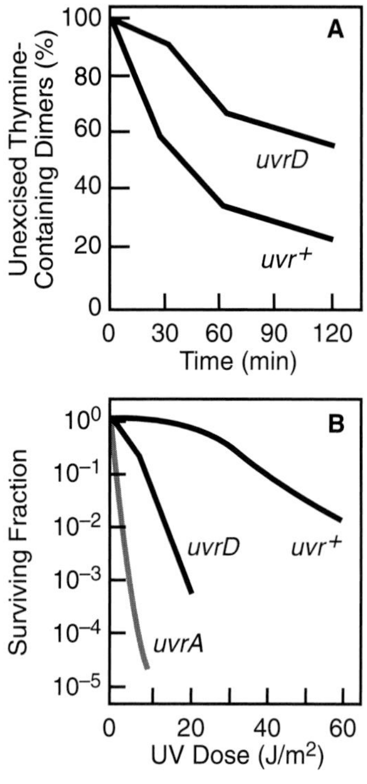

Role of UvrD Protein (DNA Helicase II) and Pol I in Excision and Repair Synthesis

THE *uvrD*⁺ GENE

The gene that encodes UvrD protein was independently identified in a number of different studies and hence was initially given redundant genetic designations. A UV-sensitive mutant of *E. coli* was shown to be distinct from mutants defective at the *uvrA*, *uvrB*, and *uvrC* loci, and the relevant gene was designated *uvrD*⁺ (125). Subsequently a strain containing a mutant allele called *mutU* which conferred increased UV radiation sensitivity and an increased spontaneous mutation frequency, was isolated (186). At about the same time, other mutant alleles called *uvrE*, which conferred very similar phenotypes, were identified (190), and yet another locus that affects genetic recombination was designated *recL* (72). This genetic complexity was resolved when it was shown that *uvrE*, *recL*, *mutU*, and *uvrD* are all alleles of the same gene, which is located at about 84 min on the *E. coli* genome (92).

The *uvrD*⁺ gene has been cloned (7, 103, 123), and plasmid or phage vectors carrying the gene complement the UV radiation sensitivity of recessive *uvrD*, *uvrE*, and *recL* mutations (7, 103, 123). Analysis of the polypeptides expressed from these vectors has identified the product of the *uvrD*⁺ gene as a single polypeptide of ~75-kDa, with a DNA-dependent ATPase activity (7, 88, 91, 103, 123, 124). A known DNA-dependent ATPase of *E. coli* with the same molecular mass was previously designated as DNA helicase II (91). It is now firmly established that UvrD protein and DNA helicase II are one and the same, and henceforth we shall refer to this protein as *DNA helicase II*.

Like the *uvrA*⁺ and *uvrB*⁺ genes, the *uvrD*⁺ gene is inducible by DNA damage. When *E. coli* cells are treated with mitomycin or nalidixic acid (well-established inducers of SOS-regulated genes [see chapter 10]), the level of DNA-dependent ATPase activity increases four- to sixfold (88). This increase is not observed in *recA* mutants defective in the SOS response (103). Additionally, the promoter region

Figure 5–10 (A) Mutants defective in the *uvrD* gene show a reduced rate and extent of loss of thymine-containing pyrimidine dimers from DNA. (B) Such mutants are also abnormally sensitive to UV radiation, but they are not as sensitive as mutants defective in the *uvrA* gene. (*Adapted from Friedberg [45a] with permission.*)

of the *uvrD⁺* gene has been shown to contain a canonical LexA protein-binding
site (91).

The addition of Pol I and DNA helicase II to the UvrABC endonuclease-
mediated reaction in vitro does not alter the initial rate of DNA incision. This rate
(~0.3 incision per min) is apparently determined by the kinetics of assembly of
the UvrBUvrC complex, after which a lower secondary rate (~0.1 incision per
min) is established. This lower rate is the product of multiple reactions, some of
which involve Pol I and DNA helicase II proteins and include displacement of UvrB
and UvrC proteins, dissociation of the oligonucleotide, repair synthesis of DNA,
and the reloading of Uvr proteins on the DNA at new sites of damage (Fig. 5–11).
This second steady-state rate results in the excision of ~0.08 dimers per min, a
rate that reasonably approximates the rate measured in vivo (~0.25 dimer excised
per min) (77, 152).

The precise mechanism by which Pol I and DNA helicase II effect the turnover of UvrB and UvrC proteins and displacement of the damage-containing oligonucleotide is not clear. Footprinting studies with the Uvr proteins plus Pol I and
DNA helicase II do not reveal the presence of a large multiprotein complex. Therefore, there is no experimental evidence in support of a stable large protein machine
("repairosome") comprising the UvrB and UvrC proteins, Pol I, DNA helicase II,
and possibly DNA ligase (152). However, DNA footprinting would not detect short-
lived complexes, and such a large multiprotein complex may exist transiently.

DNA helicase II is required for the release of the oligonucleotide fragment
and of UvrC protein from the postincision complex. Other *E. coli* DNA helicases

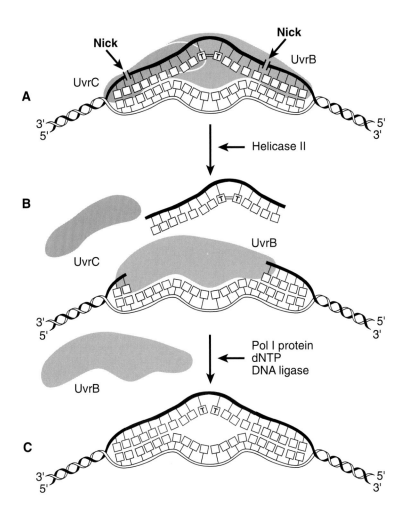

Figure 5–11 Model for postincisional
events during nucleotide excision repair in
E. coli. DNA helicase II (UvrD protein) is required for the release of an oligonucleotide
fragment (excision) following bimodal incisions generated by the UvrABC endonuclease and for the displacement of UvrC
protein (A and B). UvrB protein remains
bound to the gapped DNA during the excision reaction and is released during the
repair synthesis reaction catalyzed by Pol I
(C). DNA ligation completes the nucleotide
excision repair reaction (C).

cannot substitute for this function, suggesting that this is a specific function of DNA helicase II (132). It is not known whether these displacement reactions require that DNA helicase II specifically interact with the Uvr proteins in the postincision complex or with the nicks that are generated during the incision process. The latter possibility is certainly feasible, since studies with purified DNA helicase II have demonstrated its ability to initiate the unwinding of duplex regions at nicks when it is present in sufficiently high concentrations (143, 144). UvrB protein is not displaced during the excision reaction but apparently remains bound to the gapped DNA (132) and is released only when Pol I and substrate deoxynucleoside triphosphates (dNTPs) for repair synthesis are present (Fig. 5–11). It appears, then, that following incision, DNA helicase II releases UvrC protein and the damage-bearing oligonucleotide, leaving a UvrB–gapped-DNA complex in which UvrB protein presumably protects the single-stranded DNA from nonspecific degradation. Pol I binds to the 3′ OH terminus generated at the 5′ incision and displaces bound UvrB protein during the course of repair synthesis (132) (Fig. 5–11).

When Pol I and DNA helicase II are present in an excision repair reaction together with DNA ligase, a repair patch of 12 nucleotides is observed ~90% of the time. The repair patch size has been accurately determined in vitro by a novel

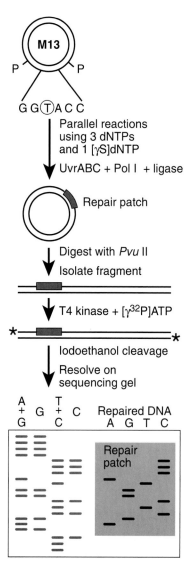

Figure 5–12 Assay to measure the size of the repair patch generated during in vitro nucleotide excision repair in *E. coli*. The plasmid DNA substrate contains a single psoralen adduct (circled T) in a defined position contained within a *Pvu*II (P) restriction fragment. This substrate is incubated in four separate reactions, each containing the UvrABC endonuclease, *E. coli* Pol I, DNA ligase, three unlabeled dNTPs, and one [γS]dNTP. The dNTPs and [γS]dNTP are varied between reactions. Following the reaction the DNA is digested with *Pvu*II. The resulting 322-bp DNA fragment is end labeled with polynucleotide kinase and heated in the presence of iodoethanol. This compound ethylates the phosphorothioate groups incorporated during repair synthesis and makes them sensitive to hydrolysis at high temperature. The DNA fragment is then analyzed on a sequencing gel, generating a sequence ladder of the repair patch. The sequence of the entire repair patch can be unambiguously read, and its size can be accurately measured. (*Adapted from Sibghat-Ullah, et al. [185]*).

and elegant technique for measuring repair synthesis (185) (Fig. 5–12). A plasmid substrate containing a single psoralen monoadduct in a unique location was used as the substrate for nucleotide excision repair in the presence of UvrABC endonuclease, Pol I, and DNA ligase (without added DNA helicase II). Repair synthesis is carried out in the presence of a dNTP carrying sulfur instead of phosphorus in the α position, resulting in a DNA product with phosphorothioate bonds. These bonds are preferentially cleaved by heating the DNA in the presence of iodoethanol. Hence, if a restriction fragment of the DNA bearing the entire repair patch is end labeled with ^{32}P, sites of phosphorothioate cleavage can be identified on a DNA sequencing gel. When repair synthesis is carried out in four sequential reactions, each in the presence of a single thiotriphosphate, the composite sequence ladder will be confined to the region of repair synthesis.

From these results it has been concluded that nick translation catalyzed by the $5' \rightarrow 3'$ exonuclease of *E. coli* Pol I is not required during nucleotide excision repairs in vitro. Several factors have been implicated in the generally strict avoidance of nick translation during repair synthesis. For one thing, it turns out that the excision gap size of ~12 nucleotides is about the length of the optimal processivity of Pol I (222). It has also been suggested that DNA helicase II may move with Pol I in the direction of repair synthesis (but on the opposite strand), and may somehow facilitate displacement of Pol I once the gap is filled (154).

The failure to observe nick translation during repair synthesis associated with nucleotide excision repair is at first glance difficult to reconcile with the observation that mutants (called *polAex*) which are defective in the $5' \rightarrow 3'$ exonuclease function of Pol I are abnormally sensitive to UV radiation (34). However, it would appear that this phenotype is not the result of defective repair synthesis associated with defective nick translation but, rather, stems from qualitatively abnormal repair synthesis by the Pol I encoded by the *polAex* allele, resulting in a frequent failure to insert the last nucleotide during the gap-filling process and, at a lower frequency, in strand displacement which allows for prolonged synthesis by the *polAex* polymerase (222) (see the next section). Another possible explanation for the UV radiation sensitive phenotype in *polAex* mutants is that the $5' \rightarrow 3'$ exonuclease function of *E. coli* Pol I plays a role in some other aspect of excision repair. One such possible role is suggested by the phenomenon of long patch excision repair.

LONG-PATCH EXCISION REPAIR OF DNA

A small fraction of the repair synthesis patches in *E. coli* are much longer than 12 nucleotides and can in fact be longer than 1,500 nucleotides (35, 86). The process whereby these tracts are generated is referred to as *long patch excision repair* to distinguish it from the more general *short patch repair mode* (36) associated with the conventional nucleotide excision repair pathway which is the central theme of this chapter. The observation of long repair patches implies that under some circumstances repair synthesis is preceded or accompanied by extensive degradation of DNA.

Long-patch repair differs from short patch repair in its absolute requirement for the *SOS inducible response* (35) already alluded to on several occasions and presented fully in chapter 10. Independent evidence for an inducible pathway of nucleotide excision repair comes from the observation that the completion of at least some fraction of the repair events initiated by the *uvrA*$^+$, *uvrB*$^+$, and *uvrC*$^+$ genes in vivo requires the *recA*$^+$ and *lexA*$^+$ genes, as well as the capacity for protein synthesis (233).

The frequency of long patches during nucleotide excision repair is increased in certain *uvrD* mutants (140) and in mutants defective in the $5' \rightarrow 3'$ exonuclease function of Pol I (34). The molecular mechanism and physiological significance of normally occurring long patch excision repair are not known. In *E. coli* cells in which the SOS system is induced, this excision repair mode correlates with a more rapid recovery of semiconservative DNA synthesis and with cell survival after UV irradiation, compared with uninduced cells (68). Thus, this mode of repair syn-

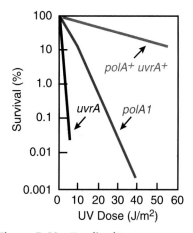

Figure 5–13 *E. coli polA* mutants are abnormally sensitive to killing by UV radiation. However, they are not as sensitive as *uvrA* mutants. (*Adapted from Friedberg [45a] with permission.*)

thesis may reflect a distinct form of nucleotide excision repair associated with the ability of *E. coli* to bypass pyrimidine dimers at or near replication forks (see chapter 11).

It is of historical interest that long-patch repair was discovered by the paradoxical observation of increased total repair synthesis of DNA in UV-irradiated cells which are defective in Pol I, despite the repair of fewer pyrimidine dimers in these mutants (37). This led to the speculation that the primary biochemical distinction between short- and long-patch repair is that the former is mediated by Pol I and the latter is mediated by other DNA polymerases (37). However, strains defective in Pol II or III carry out essentially normal amounts of long-patch repair under conditions where this repair synthesis mode is optimally observed (see below) (35). It appears, then, that Pol I is required for both short- and long-patch repair.

Repair Synthesis

A description of the *E. coli* DNA polymerases was presented in the previous chapter. The complete absence of repair synthesis of DNA in UV-irradiated *E. coli* is observed only in mutants defective in all three DNA polymerases, suggesting that all of these enzymes are potentially able to perform this function (104). This is consistent with the observation that in vitro all three enzymes can utilize gapped DNA substrates of the type generated during nucleotide excision repair. It is therefore difficult to assess the relative contribution of each to repair synthesis in wild-type cells. Nonetheless, several observations suggest that *E. coli* Pol I occupies a primary role in repair synthesis under normal conditions. First, *polA* mutants are abnormally sensitive to UV radiation (57), although, not surprisingly in view of the redundancy just described, these mutants are not as sensitive as *uvrA*, *uvrB*, or *uvrC* mutants (Fig. 5–13). Second, *polB* mutants, which are defective in Pol II, are not abnormally UV sensitive (22), and Pol II cannot substitute for Pol I in the turnover of UvrB protein. Since *polC* mutants (defective in DNA Pol III) exist only as conditional-lethal mutants, it is not possible to assess their UV sensitivity under conditions in which the enzyme is not functional (69). However, the participation of Pol III in dimer excision in *E. coli* is suggested by the observation that a mutant (*polA polC*) deficient in both Pol I and III has much less excision capability than does the *polA* strain alone, even at low UV doses (38).

DNA Ligation

The final postincisional biochemical event in all forms of excision repair is the joining of the last newly incorporated nucleotide to the polynucleotide chain, i.e., the sealing of the nick left following the completion of repair synthesis. In *E. coli*, this event is catalyzed by an extensively characterized enzyme called DNA ligase (see chapter 4).

Nucleotide Excision Repair of Cross-Links in DNA—a Special Situation

The excision of cross-links in DNA presents a special problem, since adducts are covalently attached to both DNA strands. Early genetic studies revealed a requirement for the *recA⁺* gene of *E. coli*. In addition to its regulatory role in the SOS response, the product of this gene, RecA protein, is involved in recombination (see chapters 10 and 11). This suggested that the dilemma of avoiding cell death while having to cut both strands of the genome might be solved by recombinational events (187, 188, 210). An incisional-recombinational model proposed that the UvrABC complex effects the incision of one DNA strand, generating an oligonucleotide that remains covalently attached to DNA through the cross-link (Fig. 5–14). Displacement of the oligonucleotide would then generate a gapped structure which could be repaired by recombinational repair (see chapter 11). Subsequent incision on the other strand would result in release of a cross-linked oligonucleo-

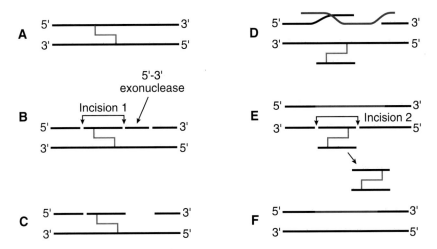

Figure 5–14 Model for the repair of DNA interstrand cross-links. (A) Cross-linked DNA. (B) The UvrABC endonuclease cuts just the furan strand of the DNA, generating incisons which flank the cross-link (incision 1). (C) It is proposed that the 5′ → 3′ exonuclease of Pol I then generates a gap 3′ to the site of the cross-link, which is a substrate for RecA-mediated recombination. (D) During recombinational repair of the gap, an invading homologous DNA strand displaces the cross-linked oligonucleotide. (E) Once recombination is completed, the pyrone DNA strand is incised by the UvrABC endonuclease (incision 2), leading to the excision of a complex 11-mer or 12-mer oligonucleotide structure. (F) DNA repair is completed by repair synthesis and ligation. (*Adapted from Van Houten [210].*)

tide structure, leaving a second gap to be repaired by conventional repair synthesis and DNA ligation (33) (Fig. 5–14).

Biochemical experiments have provided considerable support for this model. As indicated previously, studies with purified Uvr proteins have shown that incision of psoralen cross-links does indeed occur exclusively on the furan strand (212), although other studies have shown that the incised strand is sometimes the pyrone strand (79). The UvrABC endonuclease is also able to incise a model three-stranded structure in which one of the strands is an oligonucleotide covalently attached to double-stranded DNA (28, 29). Additionally, in the presence of purified RecA protein and a "transiently" positioned third strand at the cross-linked site (provided by an oligonucleotide homologous to the furan side strand), the UvrABC endonuclease catalyzes incision of the pyrone side strand (27). Further support for the model described above derives from in vitro studies (189) showing that RecA protein can mediate repair of a cross-link if there is a gapped structure adjacent to the cross-linked DNA. The cross-link is first incised bimodally by the UvrABC endonuclease on one strand, and the gap flanking the incised region is generated in vitro by exonucleolytic degradation with the 5′ → 3′ exonuclease of Pol I. The gap is required for the initiation of RecA polymerization, and the final RecA recombinant product includes the short triple-stranded region due to the presence of the cross-link (189).

The incisions that accompany excision repair of interstrand cross-links have been analyzed with DNA containing psoralen. Thymine-psoralen-thymine cross-links are asymmetrical because one strand is linked via the furan ring of the psoralen while the other strand is attached to the pyrone ring (152) (Fig. 5–15). Incision of this lesion occurs at the ninth phosphodiester bond 5′ to the furan adducted strand and at the third bond 3′ to the adduct on the same strand, while the pyrone strand remains intact (212) (Fig. 5–15). With psoralen monoadducts there is no apparent preference for the furan ring. Hence, the unique incision reaction observed with psoralen cross-links apparently reflects an affinity imposed by the conformation of the cross-link (152). The mechanism of incision of DNA cross-linked with psoralen may be influenced by sequence context. Enrichment of the G + C content of the region immediately 5′ to the modified thymine residue

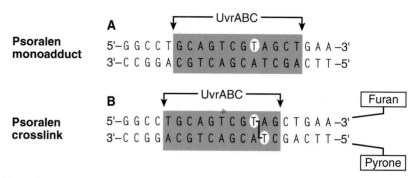

Figure 5–15 The incisions that accompany nucleotide excision repair of psoralen monoadducts and interstrand cross-links. (A) Incision of monoadduct base damage by the UvrABC endonuclease occurs at the eighth phosphodiester bond 5' and the fifth phosphodiester bond 3' to the lesion, on the strand containing the base damage. (B) A psoralen cross-link is cut at the ninth phosphodiester bond 5' and the third phosphopdiester bond 3' to the cross-link only on the furan-adducted strand, while the pyrone-adducted strand remains intact. (*Adapted from Van Houten, et al. [212].*)

on the furan strand results in preferential incision of that strand. However, when the 3' side of this lesion is enriched for G + C, incision occurs on either strand (80).

The incision of cross-linked DNA by the UvrABC endonuclease is also influenced by the superhelical density of substrate DNA. The rate of incision of supercoiled DNA containing the furan monoadduct of psoralen is relatively insensitive to superhelix density. However, efficient incision of cross-linked (by psoralen) DNA requires underwound DNA (114). The $(UvrA)_2(UvrB)_1$ complex binds with equal efficiency at the site of both monoadducts and cross-links. Hence, the requirement for negative supercoiling for efficient nicking at the latter site must be at a step subsequent to lesion recognition (114). It has been pointed out that the formation of an underwound structure could not occur if either of the DNA strands was already nicked in the immediate vicinity. Hence, such underwinding provides a potential mechanism for avoiding the formation of double-strand breaks (114).

A Uvr Protein-Independent Pathway for the Repair of Cross-Links in DNA?

Repair of psoralen cross-links in DNA has been demonstrated to occur in *uvr* mutants of *E. coli* (19–21). Additionally, an *E. coli* mutant with enhanced resistance to photoactivated psoralen has been isolated and shown to overproduce a 55-kDa protein (2), and a novel enzymatic activity which exclusively incises DNA containing psoralen monoadducts has been identified in *E. coli* (189, 210). Preliminary characterization of this activity suggests that it might be a DNA glycosylase (210).

Coupling of Transcription and Nucleotide Excision Repair in *E. coli*

Studies first carried out with mammalian cells demonstrated that nucleotide excision repair of DNA occurs preferentially in actively transcribed genes (see chapters 7 and 8). A similar phenomenon has been demonstrated in *E. coli*, for which it has been shown that in actively transcribed DNA nucleotide excision repair has a pronounced bias for the transcribed strand (108). These observations led to the initial suggestion that transcriptional arrest caused by the presence of bulky adducts in DNA may serve as a signal for the binding of Uvr proteins, leading to the preferential repair of the template strand in genes which are presumably important for cell survival, since they are transcriptionally active (108). This could not only explain the bias for the transcribed strand but also provide an additional structural element for evaluating the complex issue of substrate recognition discussed above.

The phenomenon of transcriptional arrest by base damage in template strands of DNA has been extensively described (109, 157). When the "stalled RNA polymerase" hypothesis was directly tested by adding purified *E. coli* RNA polymerase to nucleotide excision repair reactions in the presence of UvrA, UvrB, and UvrC proteins, it was paradoxically observed that the presence of the polymerase actually inhibited nucleotide excision repair (171). A search for an activity that could both overcome this inhibition and direct strand-specific repair in a cell-free system led to the discovery of a protein designated *transcription-repair coupling factor* (TRCF) in extracts of *E. coli* (172).

TRCF is the product of a gene called *mfd*$^+$ (for mutation frequency decline), which maps at 25.3 min on the *E. coli* chromosome (173,174). For improved understanding of mutation frequency decline and how it relates to strand-specific repair, we must briefly recount some key observations which date back to the mid-1950s. Almost exactly 35 years prior to the discovery of TRCF, Witkin (227) reported the curious observation that wild-type strains of *E. coli* manifest a decrease in the frequency of DNA damage-induced mutations when protein synthesis is transiently inhibited immediately after exposure of the cells to DNA damage. She subsequently noted that this phenomenon is dependent on the *uvr*$^+$ genes and on the *mfd*$^+$ gene (defined by the isolation of an *mfd* mutant strain) (228). The *mfd* mutant was found to have a distinctly lower rate of excision of pyrimidine dimers from DNA (53), suggesting that mutation frequency decline might involve a specialized form of nucleotide excision repair.

The phenomenon of mutation frequency decline and the function of the *mfd*$^+$ gene were largely ignored, except for a series of provocative experiments reported by Bockrath and his colleagues during the 1970s, the details of which are well reviewed elsewhere (174). These experiments led Bockrath and his collaborators to conclude that "MFD is a unique process involving excision repair of premutational lesions located only in the transcribed strand of DNA" (15). Remarkably, this conclusion was proffered 12 years before strand-specific repair of DNA was formally described in *E. coli*.

Direct examination of extracts of *mfd* mutant cells led to the observation that they are defective in strand-specific repair and that this defect can be corrected by the addition of partially purified TRCF (175) (Table 5–4). Further support for the conclusion that the *mfd*$^+$ gene encodes TRCF came from several studies demonstrating that whereas UV radiation-induced mutations at sites of adjacent pyrimidines in the *lacI*$^+$ gene arise largely in the nontranscribed strand in *mfd*$^+$ strains, they arise largely in the transcribed strand in *mfd* mutant cells (90, 129).

The *mfd*$^+$ gene was enriched from an *E. coli* genomic library by PCR with degenerate oligonucleotide primers deduced from the N-terminal amino acid sequence of partially purified TRCF protein, and was cloned by functional complementation of a UV-sensitive *mfd* mutant (173) (Fig. 5–16). The cloned gene can encode a protein of ~130 kDa. The translated amino acid sequence of the *mfd*$^+$ gene reveals the presence of multiple consensus domains observed in many nu-

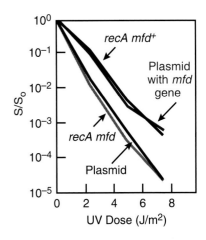

Figure 5–16 Phenotypic complementation of the UV radiation sensitivity of an *E. coli mfd* mutant by introduction of the cloned *mfd*$^+$ gene. (*Adapted from Selby and Sancar [173] with permission.*)

Table 5–4 Repair synthesis in the transcribed and coding strands of DNA

Strain	Strand[a]	Repair synthesis[b]		
		With Rif	Without Rif	Without Rif, with TRCF
WU3620 (*mfd*$^+$)	t	111	470	575
	c	100	103	147
	t/c ratio	1.1	4.5	3.9
WU3610-45 (*mfd-1*)	t	111	116	351
	c	100	117	145
	t/c ratio	1.1	1.0	2.4

[a] t, transcribed; c, coding.
[b] Repair synthesis is shown in *mfd*$^+$ and *mfd* mutant strains of *E. coli* in the presence or absence of rifampicin (Rif) and TRCF. The t/c ratio reflects the strand preference for repair.
Source: Adapted from Selby, et al. (175).

cleotide binding proteins. Purified TRCF is indeed a weak ATPase (k_{cat} = ~3 min^{-1}), but the ATPase activity is not DNA dependent and the purified protein does not have detectable DNA helicase activity in isolation (173). Mfd protein also has a 140-amino-acid stretch near the N terminus which is homologous with *E. coli* UvrB protein and related proteins from several other prokaryotes, and it also has a leucine zipper motif near the C terminus (173). It has been estimated that there are about 500 copies of TRCF protein per cell (174). The protein is a monomer and binds weakly to DNA.

Experiments with purified TRCF, Uvr proteins, and RNA polymerase suggest that TRCF is able to recognize and interact with a stalled RNA polymerase–damaged DNA–mRNA ternary complex, resulting in displacement of the stalled polymerase and the truncated transcript and in binding of TRCF to DNA at or near the site of base damage. TRCF has a demonstrated binding affinity for UvrA protein, suggesting that when bound to damaged DNA it might be especially efficient in recruiting (UvrA)$_2$(UvrB)$_1$ protein complexes to DNA and in facilitating the formation of productive UvrB-DNA complexes. This results in strand-selective nucleotide excision repair (Fig. 5–17) (173).

In summary, in *E. coli* (and as we shall see in later chapters, in higher organisms) there are at least two modes of nucleotide excision repair. One mode does not involve the coupling of repair to transcription and is used for nucleotide excision repair of transcriptionally silent genes and presumably for the repair of the nontranscribed strand of transcriptionally active genes. A second mode appears to be mechanistically very similar, if not identical, but has a specific requirement for a protein which can target the nucleotide excision repair machinery to sites of base damage in the transcribed strand with a high degree of preference. Mutants defective in the *mfd* gene have a significant but limited sensitivity to UV radiation damage (129, 173) which is less than that exhibited by *uvr* mutants. This may explain the fact that mutants selected on the basis of reduced survival after exposure to DNA damage were readily related to nucleotide excision repair, independent of a strand bias for such repair. Had nucleotide excision repair-defective mutants been originally selected on the basis of a phenotype of DNA damage-dependent mutation frequency, the history of our understanding of this DNA repair mode might be quite different.

Miscellaneous Functions Possibly Associated with Nucleotide Excision Repair

The involvement of other functions in nucleotide excision repair (either prior to, concomitant with, or following DNA incision) is suggested by a number of interesting observations which have not yet been fully explored. For example, mutants of *E. coli* with mutations in a gene designated *top*$^+$, which encodes DNA topoisomerase I, are abnormally sensitive to killing by UV radiation (194). In addition, the presence of pyrimidine dimers in DNA is associated with a reduction in the rate of the relaxation of supercoiled plasmid DNA by *E. coli* DNA topoisomerase I in vitro (135). This has led to the suggestion that altered relaxation of damaged superhelical DNA in vivo resulting from reduced topoisomerase I activity might inhibit recognition and binding of the UvrABC endonuclease to damaged DNA if this enzyme is sensitive to the superhelical density of substrate DNA (135). Other DNA topoisomerases such as DNA gyrase might also participate in nucleotide excision repair of DNA in some kinetically relevant fashion. Little is known about the role of DNA topology in DNA repair, and it is hoped that future years will provide enlightening information in this area, although it is unlikely that DNA topology plays a crucial mechanistic role.

Are Uvr Proteins Involved in Other Cellular Processes in *E. coli*?

As indicated above, *uvrB* mutants are inviable in the presence of mutations in the *polA* gene (111). Similar observations have been made with *uvrD polA* (111) and *uvrA polA* double mutants (94). These observations await a definitive explanation.

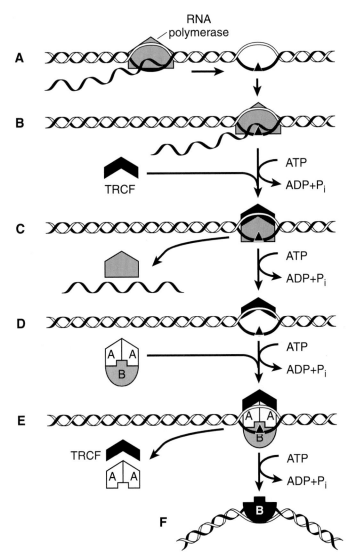

Figure 5–17 Model for strand-specific nucleotide excision repair in *E. coli*. (A) RNA polymerase is shown transcribing a template DNA strand which contains base damage ahead of the transcription complex. (B to D) Stalling of RNA polymerase at the site of base damage in the transcribed strand (B) results in the binding of TRCF (C) and displacement of the polymerase and the truncated transcript, leaving TRCF bound at the site of damage (D). (E and F) TRCF binds to UvrA protein, resulting in the recruitment of the $(UvrA)_2(UvrB)_1$ complex to the site of damage (E), where nucleotide excision repair takes place (F). (*Adapted from Selby and Sancar[173] with permission.*)

However, superficially they suggest that in the absence of functional Pol I, one or more Uvr proteins are required for some essential aspect of cellular metabolism. Since the $uvrB^+$ gene contains a binding site for DnaA protein, it has been suggested that UvrB protein may be involved in DNA replication under some circumstances (46).

Detection and Measurement of Nucleotide Excision Repair

It is useful to know something about the commonly used experimental techniques for detecting and measuring biochemical events associated with nucleotide excision repair, particularly since many of these techniques are frequently referred to in the literature. The following section presents the essential principles of several of these techniques, with a special emphasis on those used for studies with intact bacterial cells. As will be seen in later chapters, other techniques have lent themselves well to the qualitative and quantitative evaluation of nucleotide excision repair in eukaryotic cells.

EXCISION OF DAMAGED BASES

Loss of Radiolabeled Pyrimidine Dimers from DNA

In DNA radiolabeled in thymine (usually with ^{14}C or ^{3}H), excision of thymine-containing pyrimidine dimers can be monitored by demonstrating the transfer of radiolabeled dimers to an ethanol- or acid-soluble fraction, since high-molecular-weight DNA is readily precipitated by ethanol or acid. It is of course necessary to resolve the radiolabel specifically associated with dimers from that associated with nondimer nucleotides that are also transferred to the soluble phase after precipitation of DNA. These techniques have a number of important limitations that must be kept in mind. For one thing, it is very difficult to reliably detect the excision of thymine-containing pyrimidine dimers from DNA when these lesions represent a very small fraction of the total radioactivity in thymine ($<0.05\%$), unless highly sensitive separation techniques (such as high-pressure liquid chromatography) are used. Second, the loss of dimers from high molecular weight DNA does not distinguish dimer excision catalyzed by a PD-DNA glycosylase/AP endonuclease from that catalyzed by a UvrABC-type of endonuclease. Hence, the simple demonstration of the transfer of thymine dimers from the insoluble to the soluble fraction has no specific mechanistic implications.

Loss of DNA Sites Sensitive to Specific Enzyme Probes

The *M. luteus* and phage T4 PD-DNA glycosylases are absolutely specific for pyrimidine dimers in DNA (see chapter 4). If these lesions are removed during nucleotide excision repair and the covalent integrity of the DNA is restored by repair synthesis and DNA ligation, DNA isolated from such cells will obviously no longer be sensitive to incision by these enzymes. Thus, when such DNA is incubated with one of these DNA glycosylases and then denatured and sedimented in alkaline sucrose gradients (or treated in some other way that allows the measurement of the molecular weight of single-stranded DNA), it will have a higher single-stranded molecular weight than will DNA in which dimers are still present (Fig. 5–18) (51, 133). This assay of nucleotide excision repair is sometimes called the *loss of enzyme (or endonuclease-) sensitive site assay*.

This general technique is quantitatively more sensitive than the direct measurement of the loss of thymine-containing pyrimidine dimers from DNA and is particularly well suited to studies in which cells are exposed to very low levels of UV radiation. Furthermore, the loss of enzyme-sensitive sites measures the excision of all pyrimidine dimers and hence is not limited to those containing thymine, which of course is necessarily the case when pyrimidine dimers are tracked through radioactivity in thymine. In principle, this general technology is applicable to any lesion in DNA for which specific enzyme probes exist. However, like the measurement of the transfer of damaged bases to the acid- or ethanol-soluble fraction of DNA, it does not yield information about the precise mechanism of the excision repair.

The emergence of sensitive techniques for detecting specific DNA sequences by nucleic acid hybridization combined with the use of enzyme probes specific for particular types or classes of DNA damage has facilitated an extremely sensitive technique for measuring nucleotide excision repair of pyrimidine dimers and other forms of base damage in single-copy DNA sequences. This technique is discussed in detail in chapter 7.

Appearance and Disappearance of Strand Breaks in DNA

Damage-specific incision of DNA can be monitored by a variety of techniques which directly detect the strand breaks (or nicks) that are enzymatically produced in vivo, rather than detecting excision of the substrate lesions themselves (4, 18,

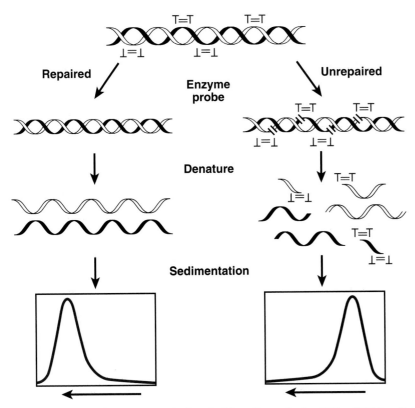

Figure 5–18 The presence of unrepaired pyrimidine dimers in the DNA of UV-irradiated cells can be detected with the use of dimer-specific enzyme probes such as the *M. luteus* or phage T4 PD-DNA glycosylase/AP lyases (see chapter 4). Radiolabeled DNA is extracted from cells and incubated with the enzyme. The enzyme catalyzes the formation of strand breaks at pyrimidine dimer (enzyme-sensitive) sites. DNA containing dimers (enzyme-sensitive sites) will sediment more slowly in alkaline sucrose gradients than will DNA containing no (or fewer) dimers. (*Adapted from Friedberg [45a] with permission.*)

32, 67, 83, 96, 160). When cells have undergone complete nucleotide excision repair and the covalent integrity of the DNA has been restored, these strand breaks obviously disappear. The disappearance of damage-specific strand breaks is therefore a useful indicator of nucleotide excision repair. Since this method does not require identification of the particular lesion being repaired, strand breaks that are spuriously introduced into DNA (and are thus unrelated to the repair of base damage) may not be readily distinguished from those that are related to repair. The extensively studied model of DNA containing pyrimidine dimers serves as an example of how this problem can be overcome in specific instances. Pyrimidine dimers are subject to repair by enzymatic photoreactivation (see chapter 2). One can therefore ask whether the strand breaks that appear during nucleotide excision repair fail to appear if the cells are subjected to photoreactivation prior to excision repair. If so, the breaks are presumably incisions at pyrimidine dimers. Alternatively, the specificity of DNA strand breaks observed in wild-type cells can be evaluated by direct comparison with mutants defective in incision of DNA at the lesions in question.

Quantitative PCR

Many alterations in DNA bases result in inhibition of the PCR catalyzed by *Taq* DNA polymerase. Hence, the presence of base damage can be detected by reduced PCR amplification when this is measured quantitatively. Since only nondamaged templates participate in PCR, this assay can accurately measure the fraction of template DNA molecules that contain no damage (26). Assuming a random distribution of base damage, the Poisson equation allows the calculation of the average frequency of base damage per DNA strand (26).

Measurement of Repair Synthesis

The most general method used for measuring repair synthesis is by density labeling of the DNA (191) (Fig. 5–19). In this procedure, DNA synthesis following damage is carried out in the presence of bromo-dUTP. The incorporation of sufficient amounts of 5-bromouracil (5-BU) instead of thymine imparts an increased buoyant density to the DNA. Such increases are generally too small to be detectable in regions of nonsemiconservative (repair) synthesis because the incorporated 5-BU constitutes only a tiny fraction of the mass of the DNA fragments isolated by usual procedures. However, a significant increase in density can usually be detected in regions of semiconservative replication. Thus, with the use of radiolabeled 5-BU, DNA undergoing semiconservative synthesis can be separated from DNA exclusively undergoing repair synthesis by sedimentation in isopycnic gradients, because in the former case the 5-BU is typically incorporated into an entire

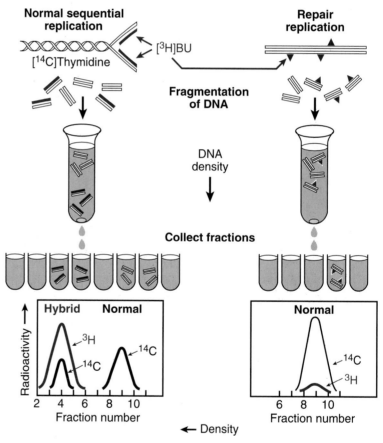

Figure 5–19 Schematic illustration of the detection of repair synthesis by buoyant density centrifugation of DNA containing 5-BU. DNA is prelabeled with [^{14}C]thymidine to provide a uniform label. Following exposure to UV radiation (or some other form of DNA damage), repair synthesis during nucleotide excision repair takes place in the presence of [^{3}H]BU. DNA synthesized both by semiconservative and by non-semiconservative modes will thus be density labeled (red). To distinguish these, the DNA is fragmented (by shearing) and sedimented to equilibrium density. Fragments of DNA-containing strands that were synthesized semiconservatively will have a hybrid density detected by the position of the ^{3}H radiolabel (left). Repair synthesis patches are too small to alter the density of the DNA, and hence the ^{3}H radiolabel appears at the position of normal density DNA (right). (*Adapted from Friedberg [45a] with permission.*)

strand of a given DNA fragment and hence constitutes an appreciable fraction of its mass. The radioactivity incorporated into unreplicated (parental density) DNA is then a measure of the total amount of repair synthesis (Fig. 5–19).

The accuracy of this measurement is influenced chiefly by the amount of background semiconservative DNA synthesis. Thus, wherever feasible, selective inhibition of the latter is attempted. In prokaryotes such as *E. coli*, such selective inhibition is difficult to achieve, although semiconservative replication is reduced considerably by DNA damage itself (196). Refinements of the density-labeling technique allow for estimates of the size of the regions (patches) of repair synthesis in DNA. Most simply, the amount of radiolabel incorporated during repair, together with an independent determination of the number of repair events, can be used to calculate the average repair patch size (36). An alternative procedure involves shearing the DNA to known small size by sonication of the isolated parental density fraction containing repair patches. The repair patches now constitute an appreciable fraction of the length of the DNA fragments. The DNA is then analyzed in alkaline isopycnic gradients so that the density shift of only the affected strands is measured. The observed increase in the density of the DNA fragments, together with the measured average size of the fragments, yields an estimate of the average size of the repair patches (192).

Another procedure for measuring the size of repair synthesis patches in DNA exploits the photolytic sensitivity of DNA containing 5-BU (177). When DNA containing this thymine analog is exposed to radiation at 313 nm, debromination followed by free radical attack of the deoxyribose or deoxyribose-phosphate backbone occurs, resulting in DNA strand breaks and/or formation of alkali-labile sites (177). Alkaline sucrose gradient sedimentation can then be used to measure the extent of DNA fragmentation (Fig. 5–20). If enough 313-nm UV radiation can be delivered to achieve a plateau level of fragmentation (i.e., produce at least one break or alkali-labile site per repair patch), the average patch size can be derived directly from the known efficiency of the 5-BU photolysis (177).

Nucleotide Excision Repair in Other Prokaryotes

M. LUTEUS

As indicated in chapter 4, *M. luteus* is endowed with a DNA-glycosylase/AP endonuclease (UV endonuclease) which specifically repairs pyrimidine dimers by a base excision repair mode. However, mutants that are defective in this enzymatic activity are not abnormally sensitive to UV radiation under nonphotoreactivating conditions (201), suggesting that this organism may possess alternative mechanisms for excising pyrimidine dimers from DNA. Consistent with this suggestion, highly UV-sensitive mutants were identified with normal levels of PD-DNA glycosylase activity (197, 201). Like the *uvr* mutants of *E. coli*, these mutants are abnormally sensitive to killing to other agents such as mitomycin and 4-nitroquinoline 1-oxide (59, 197, 201).

M. luteus genes homologous to the *uvrA*[+] and *uvrB*[+] genes of *E. coli* have been cloned (119, 183, 184). The *uvrA*[+] gene complements the multiple sensitivities of a *M. luteus* strain designated DB7, which is defective in *uvrA*. The cloned gene has an open reading frame of 992 codons [slightly longer than that of the *E. coli uvrA*[+] gene (940 codons)]. The *M. luteus uvrA*[+] gene shares extensive amino acid sequence homology with the translated sequence of its *E. coli* homolog. Additionally, all of the presumed functional elements revealed by the sequence of the *E. coli uvrA*[+] gene (nucleotide binding domains and zinc fingers) are conserved (119, 184). Similar homology exists between the *E. coli* and *M. luteus uvrB*[+] genes. The latter complements the phenotypes of a *uvrB* mutant strain designated UV[s]N[1] and has an open reading frame of 709 amino acids; again somewhat longer than the 672 codons in the *E. coli uvrB* open reading frame (119, 183). One would predict that a homolog of the *E. coli uvrC*[+] gene also exists in *M. luteus*. However, to date no mutants distinct from those defective in the *uvrA* and *uvrB* genes have been identified.

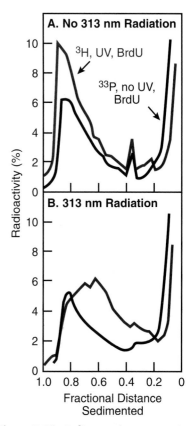

Figure 5–20 Sedimentation patterns in alkaline sucrose of labeled DNA from normal human fibroblasts treated or not treated with UV radiation and allowed to repair in the presence of 5-BU before exposure to 313-nm photolysis. Photolysis of incorporated 5-BU results in strand breakage of the DNA when sedimented in alkali (top). No degradation of DNA is observed in the absence of photolytic irradiation (bottom). The amount of 313-nm radiation required to cause strand breakage at all sites of 5-BU incorporation provides a means for estimating the size of the DNA synthesis (repair) patches. (*Adapted from Friedberg [45a] with permission.*)

DEINOCOCCUS RADIODURANS

Deinococcus (formerly *Micrococcus*) *radiodurans* and other members of the same genus are extremely resistant to the lethal and mutagenic effects of most agents that damage DNA, including UV radiation. Hence, this organism presents interesting opportunities for exploring the biochemistry and molecular biology of nucleotide excision repair in fully viable populations of cells that have sustained large numbers of "hits" in their genome. *D. radiodurans* lacks DNA photolyase activity (112). Nonetheless, the D_{37} for UV radiation exposure (the dose of UV radiation that yields 37% survival) is much higher than that observed with *E. coli* (~880 and <50 J/m², respectively) (65). There is nothing unusual about the efficiency of the formation of photoproducts in the DNA of *D. radiodurans* cells exposed to UV radiation, except that there is a higher relative ratio of (6-4) photoproducts to pyrimidine dimers (217), probably reflecting the high G + C content of the DNA of this organism, which results in many more TC dinucleotide sequences (116).

Within 90 min following exposure of *D. radiodurans* cells to UV radiation, the content of thymine-containing pyrimidine dimers is reduced from 1.7 to 0.3% (217). This extraordinarily efficient (presumably excision) repair of pyrimidine dimers was first demonstrated by Bolling and Setlow (16), who showed that the many pyrimidine dimers excised from the genome of *D. radiodurans* appear in the medium rather than in the acid-soluble fraction of DNA. It is now established that *D. radiodurans* is endowed with two independent excision repair pathways, either of which alone is sufficient to endow the organism with wild-type resistance to UV radiation at 254 nm. One of these pathways appears to operate by nucleotide excision and is believed to require at least two genes, originally designated *mtcA*⁺ and *mtcB*⁺ (5, 41, 113). In addition to pyrimidine dimers, this pathway can process mitomycin-induced DNA cross-links and bulky chemical adducts. The proteins encoded by the *mtcA*⁺ and *mtcB*⁺ genes have not yet been physically identified. However, it was proposed that these proteins are required for the activity of an endonuclease called UV endonuclease-α. It has since been determined that *mtcA* and *mtcB* are in fact portions of a single gene and that this gene has extensive amino acid sequence homology with *uvrA*⁺ of *E. coli* and *M. luteus* (1). Hence, *mtcA* and *mtcB* have been renamed *uvrA1* and *uvrA2*, respectively, and the full-length wild-type allele is designated *uvrA*⁺, to conform with the *E. coli* nomenclature for nucleotide excision repair genes. The UvrA protein of *D. radiodurans* presumably participates in a pathway formally analogous to that of the UvrABC-mediated pathway of *E. coli*. However, to date *uvrB*⁺ and *uvrC*⁺ homologs have not been identified in *D. radiodurans*.

In contrast to the situation in *E. coli*, *uvrA* mutants of *D. radiodurans* retain essentially wild-type levels of resistance to UV radiation and are able to remove pyrimidine dimers from their DNA by the action of a second endonuclease called UV endonuclease-β, which requires the gene(s) *uvsCDE*⁺. To date, no other substrates have been demonstrated for this endonuclease activity. Mutants defective in both endonucleases (e.g., *uvrA1 uvsE* double mutants) are highly sensitive to killing by UV radiation and are defective in the incision of UV-irradiated DNA (41, 113).

The limited substrate specificity of UV endonuclease-β has led to the suggestion that it is a PD-DNA glycosylase, analogous to the phage T4 and *M. luteus* PD-DNA glycosylases discussed in chapter 4. UV endonuclease-β has been partially purified and found to have a molecular mass of ~36 kDa. The endonuclease activity has a novel requirement for Mn^{2+} ions (42), and, in contrast to the T4 and *M. luteus* PD-DNA glycosylases, it is inactive in the presence of EDTA. Furthermore, its mode of action appears to be directly endonucleolytic rather than glycosylic, since incisions in DNA catalyzed by the enzyme do not generate free thymine following photoreversal (43) (see chapter 4). In addition to these distinguishing features, the *D. radiodurans* UV endonuclease-β activity confers wild-type levels of resistance to UV radiation in *uvrA* mutants. In contrast, in *M. luteus* the UV sensitivity of *uvrA* mutants is only marginally corrected by introduction of the gene which encodes the endogenous PD-DNA glycosylase (120). These and other similar observations (66) have led to the speculation that UV endonuclease-β might also recognize (6-4) photoproducts in DNA, which account for as much as 17% of the thymine-containing bimolecular photoproducts in the DNA of *D. radiodurans* cells exposed to UV radiation. The definitive conclusion that UV endonuclease-β is a true direct-acting repair endonuclease, as distinct from a PD-DNA glycosylase, must await more-intensive study of this interesting enzyme.

OTHER ORGANISMS

In several other prokaryotes, including *Salmonella typhimurium, Pseudomonas fluorescens, Streptococcus pneumoniae, Bacillus subtilis,* and *Mycoplasma genitalium,* pyrimidine dimers are excised as oligonucleotide fragments of 13 nucleotides by hydrolysis of the eighth phosphodiester bond 5' to the dimer and the fifth phosphodiester bond 3' to the lesion (28). Hence, it seems reasonable to conclude that the mechanism of nucleotide excision repair so elegantly deciphered in *E. coli* is conserved in prokaryotes.

Other Damage-Specific DNA-Incising Activities?

ENDONUCLEASE V OF *E. COLI*

A small protein that degrades a variety of substrates which share no obvious structural similarity has been purified from *E. coli* and is termed *endonuclease V* (39, 52). This enzyme has an absolute requirement for $MgCl_2$ for activity. Among the substrates for the purified enzyme are duplex UV-irradiated DNA (at sites other than pyrimidine dimers), DNA treated with osmium tetroxide, heat- or acid-depurinated DNA, and DNA containing adducts of 7-bromomethylbenz[*a*]anthracene (39, 52). In addition, native DNA from phage PBS2 (a DNA that contains uracil instead of thymine [see chapter 1]) and phage T5 DNA containing thymine substituted by uracil are degraded by the enzyme (39, 52). The enzyme also attacks the single-stranded DNA from phage fd and duplex fd RFI DNA (39, 52).

It is difficult to identify a common determinant in all of these substrates, and the basis for the apparent broad substrate specificity of endonuclease V remains unclear. An interesting possibility is that the enzyme is single strand specific and therefore recognizes single-stranded areas in duplex DNA created by relatively nonspecific base damage. Endonuclease V is optimally active at pH 9.5 and has little if any detectable activity at neutral pH. The alkaline pH optimum might facilitate helix destabilization in areas of localized conformational distortion in duplex DNA. This could explain the observed degradation of PBS2 DNA, since at pH 9.5 the disruption of A·U base pairs may occur more readily than that of A·T base pairs.

When endonuclease V reactions with various DNAs are sedimented in alkali to monitor single-strand breaks (nicks) and in neutral sucrose to measure double strand breaks, the ratio of single-strand nicks to double-strand breaks is only 8:1 over a wide range of nicks and a sixfold range of enzyme concentration (39). Thus, the double-strand breaks apparently do not arise from the random juxtaposition of two nicks or from an enzyme concentration-dependent alteration in the mechanism of endonucleolytic cleavage of DNA (39).

Endonuclease V is present at normal levels in all *E. coli* mutants examined thus far, including a number of strains defective in functions required for replication, recombination, and repair. Although the multiplicity of substrates recognized by this activity might be interpreted in terms of its role in DNA repair, the determination of the true cellular function of endonuclease V must await the isolation of mutants defective in the activity (39).

Summary and Conclusions

Damage-specific endonucleases such as the UvrABC enzyme of *E. coli* and the products of certain *RAD* gene products in *S. cerevisiae* and of the genes in human cells that are defective in human patients with the disease xeroderma pigmentosum (see later chapters) are singularly important enzymes for excision repair. Cellular and biological studies with mutants defective in the genes encoding these products suggest that these endonucleases are required for the incision of DNA containing a large spectrum of base damage. Thus, it is probable that the nucleotide excision repair pathway initiated by these damage-specific enzymes is a much more general mode for excision repair of base damage than is the base excision repair mechanism in which limited types of base damage are excised by specific DNA glycosylases.

It is noteworthy that some types of base damage, including sites of base loss generated by DNA glycosylases, are recognized by both the base and nucleotide excision repair pathways, allowing for possible convergence of these two general pathways of excision repair (193). Such convergence may exclude repair by one pathway in the presence of the other, leading to unexpected phenotypes for certain mutants. For example, the binding of the $(UvrA)_2(UvrB)_1$ complex to AP sites can inhibit base excision repair of such lesions (193). If the completion of the repair of AP sites by nucleotide excision repair was inhibited by a mutant *uvrC* gene, it is possible that such mutant strains would confer abnormal sensitivity to types of damage typically repaired by the base excision repair pathway.

The availability of highly purified gene products encoded by genes cloned into recombinant DNA molecules has unquestionably overcome the quantitative problems that hampered progress in this area of nucleic acid enzymology for so long. Despite the impressive progress made with *E. coli*, much remains to be learned about the details of the biochemistry of nucleotide excision repair in this organism and particularly in other prokaryotes.

REFERENCES

1. **Agostini, H., and K. W. Minton.** 1994. Personal communication.

2. **Ahmed, S. I., and I. B. Holland.** 1985. Isolation and analysis of a mutant of *Escherichia coli* hyper-resistant to near-ultraviolet light plus 8-methoxypsoralen. *Mutat. Res.* **151**:43–47.

3. **Ahn, B. C., and L. Grossman.** 1994. Personal communication.

4. **Ahnström, G., and K. Erixon.** 1981. Measurement of strand breaks by alkaline denaturation and hydroxyapatite chromatography, p. 403–418. *In* E. C. Friedberg and P. C. Hanawalt (ed.), *DNA Repair—A Laboratory Manual of Research Procedures*, vol. 1, part A. Marcel Dekker, Inc., New York.

5. **Al-Bakri, G. H., M. W. Mackay, P. A. Whittaker, and B. E. B. Moseley.** 1985. Cloning of the DNA repair genes *mtcA*, *mtcB*, *uvsC*, *uvsE* and the *leuB* gene from *Deinococcus radiodurans*. *Gene* **33**:305–311.

6. **Arikan, E., M. S. Kulkarni, D. C. Thomas, and A. Sancar.** 1986. Sequences of the *E. coli uvrB* gene and protein. *Nucleic Acids Res.* **14**:2637–2650.

7. **Arthur, H. M., D. Bramhill, P. B. Eastlake, and P. T. Emmerson.** 1982. Cloning of the *uvrD* gene of *E. coli* and identification of the product. *Gene* **19**:285–295.

8. **Backendorf, C., H. Spaink, A. P. Barbeiro, and P. Van de Putte.** 1986. Structure of the *uvrB* gene of *Escherichia coli*. Homology with other DNA repair enzymes and characterization of the *uvrB5* mutation. *Nucleic Acids Res.* **14**:2877–2890.

9. **Backendorf, C. M., E. A. Van den Berg, J. A. Brandsma, T. Kartasova, C. A. Van Sluis, and P. Van de Putte.** 1983. In vivo regulation of the *uvr* and *ssb* genes in *Escherichia coli*, p. 161–171. *In* E. C. Friedberg and B. A. Bridges (ed.), *Cellular Responses to DNA Damage*. Alan R. Liss, Inc., New York.

10. **Beck, D. J., S. Popoff, A. Sancar, and W. D. Rupp.** 1985. Reactions of the UVRABC excision nuclease with DNA damaged by diamminedichloroplatinum(II). *Nucleic Acids Res.* **13**:7395–7412.

11. **Ben-Ishai, R., and R. Sharon.** 1981. On the nature of the repair deficiency in *E. coli uvrE*, p. 147–151. *In* E. Seeberg and K. Kleppe (ed.), *Chromosome Damage and Repair*. Plenum Publishing Corp, New York.

12. **Berg, J. M.** 1986. Potential metal-binding domains in nucleic acid binding proteins. *Science* **232**:485–487.

13. **Bertrand-Burggraf, E., S. Hurstel, M. Daune, and M. Schnarr.** 1987. Promoter properties and negative regulation of the *uvrA* gene by the LexA repressor and its amino-terminal DNA binding domain. *J. Mol. Biol.* **193**:293–302.

14. **Bertrand-Burggraf, E., C. P. Selby, J. E. Hearst, and A. Sancar.** 1991. Identification of the different intermediates in the interaction of (A)BC excinuclease with its substrates by DNase I footprinting on two uniquely modified oligonucleotides. *J. Mol. Biol.* **218**:27–36.

15. **Bockrath, R. C., and J. E. Palmer.** 1977. Differential repair of premutational UV-lesions at tRNA genes in *E. coli*. *Mol. Gen. Genet.* **156**:133–140.

16. **Bolling, M. E., and J. K. Setlow.** 1966. The resistance of *Micrococcus radiodurans* to ultraviolet radiation a repair mechanism. *Biochim. Biopys. Acta.* **123**:26–33.

17. **Boyce, R. P., and P. Howard-Flanders.** 1964. Release of ultraviolet light-induced thymine dimers from DNA in *E. coli* K-12. *Proc. Natl. Acad. Sci. USA* **51**:293–300.

18. **Braun, A. G.** 1981. Measurement of strand breaks by nitrocellulose membrane filtration, p. 447–455. *In* E. C. Friedberg and P. C. Hanawalt (ed.), *DNA Repair—A Laboratory Manual of Research Procedures*, vol. 1, part B. Marcel Dekker, Inc., New York.

19. **Bridges, B. A.** 1983. Psoralens and serendipity: aspects of the genetic toxicology of 8-methoxypsoralen. *Environ. Mutagen.* **5**:329–339.

20. **Bridges, B. A.** 1984. Further characterization of repair of 8-methoxypsoralen crosslinks in UV-excision-defective *Escherichia coli*. *Mutat. Res.* **132**:153–160.

21. **Bridges, B. A., and M. Stannard.** 1982. A new pathway for repair of cross-linkable 8-methoxypsoralen mono-adducts in *uvr* strains of *Escherichia coli*. *Mutat. Res* **92**:9–14.

22. **Campbell, J. L., L. Soll, and C. C. Richardson.** 1972. Isolation and partial characterization of a mutant of *Escherichia coli* deficient in DNA polymerase II. *Proc. Natl. Acad. Sci. USA* **69**:2090–2094.

23. Caron, P. R., and L. Grossman. 1988. Involvement of a cryptic ATPase activity of UvrB and its proteolysis product, UvrB*, in DNA repair. *Nucleic Acids Res.* **16:**9651–9662.

24. Caron, P. R., and L. Grossman. 1988. Potential role of proteolysis in the control of UvrABC incision. *Nucleic Acids Res.* **16:**9641–9650.

25. Caron, P. R., S. R. Kushner, and L. Grossman. 1985. Involvement of helicase II (*uvrD* gene product) and DNA polymerase I in excision mediated by the uvrABC protein complex. *Proc. Natl. Acad. Sci. USA* **82:**4925–4929.

26. Chandrasekhar, D., and B. Van Houten. 1994. High resolution mapping of UV-induced photoproducts in the *Escherichia coli lacI* gene. Inefficient repair of the non-transcribed strand correlates with high mutation frequency. *J. Mol. Biol.* **238:**319–332.

27. Cheng, S., A. Sancar, and J. E. Hearst. 1991. RecA-dependent incision of psoralen-crosslinked DNA by (A)BC excinuclease. *Nucleic Acids Res.* **19:**657–663.

28. Cheng, S., B. Van Houten, H. B. Gamper, A. Sancar, and J. E. Hearst. 1988. Use of psoralen-modified oligonucleotides to trap three-stranded RecA-DNA complexes and repair of these cross-linked complexes by ABC excinuclease. *J. Biol. Chem.* **263:**15110.

29. Cheng, S., B. Van Houten, H. B. Gamper, A. Sancar, and J. E. Hearst. 1988. Use of triple–stranded DNA complexes to study crosslink repair, p. 105–113. *In* E. C. Friedberg and P. C. Hanawalt (ed.), *Mechanisms and Consequences of DNA Damage Processing.* Alan R. Liss, Inc., New York.

30. Claassen, L. A., B. Ahn, H. S. Koo, and L. Grossman. 1991. Construction of deletion mutants of the *Escherichia coli* UvrA protein and their purification from inclusion bodies. *J. Biol. Chem.* **266:**11380–11387.

31. Claassen, L. A., and L. Grossman. 1991. Deletion mutagenesis of the *Escherichia coli* UvrA protein localizes domains for DNA binding, damage recognition, and protein-protein interactions. *J. Biol. Chem.* **266:**11388–11394.

32. Clayton, D. A. 1981. Measurement of strand breaks in supercoiled DNA by electron microscopy, p. 419–424. *In* E. C. Friedberg and P. C. Hanawalt (ed.), *DNA repair—A Laboratory Manual of Research Procedures,* vol. 1, part B. Marcel Dekker, Inc., New York.

33. Cole, R. S. 1973. Repair of DNA containing interstrand crosslinks in *Escherichia coli:* sequential excision and recombination. *Proc. Natl. Acad. Sci. USA* **70:**1064–1068.

34. Cooper, P. K. 1977. Excision-repair in mutants of *Escherichia coli* deficient in DNA polymerase I and/or its associated 5′ → 3′ exonuclease. *Mol. Gen. Genet.* **150:**1–12.

35. Cooper, P. K. 1982. Characterization of long patch excision repair of DNA in ultraviolet-irradiated *Escherichia coli:* an inducible function under *rec-lex* control. *Mol. Gen. Genet.* **185:**189–197.

36. Cooper, P. K., and P. C. Hanawalt. 1972. Heterogeneity of patch size in repair replicated DNA in *Escherichia coli. J. Mol. Biol.* **67:**1–10.

37. Cooper, P. K., and P. C. Hanawalt. 1972. Role of DNA polymerase I and the *rec* system in excision-repair in *Escherichia coli. Proc. Natl. Acad. Sci. USA* **69:**1156–1160.

38. Cooper, P. K., and J. G. Hunt. 1978. Alternative pathways for excision and resynthesis in *Escherichia coli:* DNA polymerase III role? p. 255–260. *In* P. C. Hanawalt, E. C. Friedberg, and C. F. Fox (ed.), *DNA Repair Mechanisms.* Academic Press, Inc., New York.

39. Demple, B., and S. Linn. 1982. On the recognition and cleavage mechanism of *Escherichia coli* endodeoxyribonuclease V, a possible DNA repair enzyme. *J. Biol. Chem.* **257:**2848–2855.

40. Doolittle, R. F., M. S. Johnson, I. Husain, B. Van Houten, D. C. Thomas, and A. Sancar. 1986. Domainal evolution of a prokaryotic DNA repair protein and its relationship to active-transport proteins. *Nature* (London) **323:**451–453.

41. Evans, D. M., and B. E. B. Moseley. 1983. Roles of *uvsC, uvsD, uvsE,* and *mtcA* genes in the two pyrimidine dimer excision repair pathways of *Deinococcus radiodurans. J. Bacteriol.* **156:**576–583.

42. Evans, D. M., and B. E. B. Moseley. 1985. Identification and initial characterization of a pyrimidine dimer endonuclease (UV endonuclease b) from *Deinococcus radiodurans;* a DNA-repair enzyme that requires manganese ions. *Mutat. Res.* **145:**119–128.

43. Evans, D. M., and B. E. B. Moseley. 1988. *Deinococcus radiodurans* UV endonuclease β DNA incisions do not generate photoreversible thymine residues. *Mutat. Res.* **207:**117–119.

44. Fogliano, M., and P. F. Schendel. 1981. Evidence for the inducibility of the *uvrB* operon. *Nature* (London) **289:**196–198.

45. Forster, J. W., and P. Strike. 1985. Organization and control of the *Escherichia coli uvrC* gene. *Gene* **35:**71–82.

45a. Friedberg, E. C. 1985. *DNA Repair.* W. H. Freeman Co., New York.

46. Friedberg, E. C. 1987. The molecular biology of nucleotide excision repair of DNA: recent progress. *J. Cell Sci. Suppl.* **6:**1–23.

47. Friedberg, E. C., and B. A. Bridges. 1983. *Cellular Responses to DNA Damage.* Alan R. Liss, Inc., New York.

48. Friedberg, E. C., and P. C. Hanawalt. 1981. *DNA Repair—A Laboratory Manual of Research Procedures,* vol. 1, Part A. Marcel Dekker, Inc., New York.

49. Friedberg, E. C., and P. C. Hanawalt. 1981. *DNA Repair—A Laboratory Manual of Research Procedures,* vol. 2. Marcel Dekker, Inc., New York.

50. Fuller, R. S., B. E. Funnell, and A. Kornberg. 1984. The dnaA protein complex with the *E. coli* chromosomal replication origin (oriC) and other DNA sites. *Cell* **38:**889–900.

51. Ganesan, A. K., C. A. Smith, and A. A. Van Zeeland. 1981. Measurement of the pyrimidine dimer content of DNA in permeabilized bacterial or mammalian cells with endonuclease V of bacteriophage T4, p. 89–97. *In* E. C. Friedberg and P. C. Hanawalt (ed.), *DNA Repair—A Laboratory Manual of Research Procedures,* vol. 1, part A. Marcel Dekker, Inc., New York.

52. Gates, F. T., III, and S. Linn. 1977. Endonuclease V of *Escherichia coli. J. Biol. Chem.* **252:**1647–1653.

53. George, D. L., and E. M. Witkin. 1975. Ultraviolet light-induced responses of an *mfd* mutant of *Escherichia coli* B/r having a slow rate of dimer excision. *Mutat. Res.* **28:**347–354.

54. Gopalakrishnan, A. S., Y. C. Chen, M. Temkin, and W. Dowhan. 1986. Structure and expression of the gene locus encoding the phosphatidylglycerophosphate synthase of *Escherichia coli. J. Biol. Chem.* **261:**1329–1338.

55. Gorbalenya, A. E., and E. V. Koonin. 1990. Superfamily of UvrA-related NTP-binding proteins. Implications for rational classification of recombination/repair systems. *J. Mol. Biol.* **213:**583–591.

56. Granger-Schnarr, M., M. Schnarr, and C. A. Van Sluis. 1986. In vitro study of the interaction of the LexA repressor and the UvrC protein with a *uvrC* regulatory region. *FEBS Lett.* **198:**61–65.

57. Gross, J., and M. Gross. 1969. Genetic analysis of an *E. coli* strain with a mutation affecting DNA polymerase. *Nature* (London) **224:**1166–1168.

58. Grossman, L., P. R. Caron, S. J. Mazur, and E. Y. Oh. 1988. Repair of DNA-containing pyrimidine dimers. *FASEB J.* **2:**2696–2701.

59. Grossman, L., J. C. Kaplan, S. R. Kushner, and I. Mahler. 1967. Enzymes involved in the early stages of repair of ultraviolet-irradiated DNA. *Cold Spring Harbor Symp. Quant. Biol.* **33:**229–234.

60. Grossman, L., S. J. Mazur, P. R. Caron, and E. Y. Oh. 1988. Dynamics of the *E. coli uvr* DNA repair system, p. 73–77. *In* E. C. Friedberg and P. C. Hanawalt (ed.), *Mechanisms and Consequences of DNA Damage Processing.* Alan R. Liss, Inc., New York.

61. Grossman, L., E. Y. Oh, S. Mazur, and P. Caron. 1989. Macromolecular physiology of the *Escherichia coli* Uvr proteins, p. 11–15. *In* A. Castellani (ed.), *DNA Damage and Repair.* Plenum Publishing Corp., New York.

62. Grossman, L., and S. Thiagalingam. 1993. Nucleotide excision repair, a tracking mechanism in search of damage. *J. Biol. Chem.* **268:**16871–16874.

63. Grossman, L., and A. T. Yeung. 1990. The UvrABC endonuclease of *Escherichia coli*. *Photochem. Photobiol.* **51:**749–755.

64. Grossman, L., and A. T. Yeung. 1990. The UvrABC endonuclease system of *Escherichia coli*—a view from Baltimore. *Mutation Res.* **236:**213–221.

65. Gutman, P. D., P. Fuchs, L. Ouyang, and K. W. Minton. 1993. Identification, sequencing and targeted mutagenesis of a DNA polymerase gene required for the extreme radioresistance of *Deinococcus radiodurans*. *J. Bacteriol.* **175:**3581–3590.

66. Gutman, P. D., H. L. Yao, and K. W. Minton. 1991. Partial complementation of the UV sensitivity of *Deinococcus radiodurans* excision repair mutants by the cloned *denV* gene of bacteriophage T4. *Mutat. Res.* **254:**207–215.

67. Hagen, U. F. W. 1981. Measurement of strand breaks by end labeling, p. 431–445. *In* E. C. Friedberg and P. C. Hanawalt (ed.), *DNA Repair—A Laboratory Manual of Research Procedures*, vol. 1, part B. Marcel Dekker, Inc., New York.

68. Hanawalt, P. C. 1982. Perspectives on DNA repair and inducible recovery phenomena. *Biochimie* **64:**847–851.

69. Hanawalt, P. C., P. K. Cooper, A. K. Ganesan, and C. A. Smith. 1979. DNA repair in bacteria and mammalian cells. *Annu. Rev. Biochem.* **48:**783–836.

70. Henderson, T. A., P. M. Dombrosky, and K. D. Young. 1994. Artifactual processing of penicillin-binding proteins 7 and 1b by the OmpT protease of *Escherichia coli*. *J. Bacteriol.* **176:**256–259.

71. Higgins, C. F., I. D. Hiles, G. P. Salmond, D. R. Gill, J. A. Downie, I. J. Evans, I. B. Holland, L. Gray, S. D. Buckel, and A. W. Bell. 1986. A family of related ATP-binding subunits coupled to many distinct biological processes in bacteria. *Nature* (London) **323:**448–450.

72. Horii, Z. I., and A. J. Clark. 1973. Genetic analysis of the *recF* pathway to genetic recombination in *Escherichia coli* K12: isolation and characterization of mutants. *J. Mol. Biol.* **80:**327–344.

73. Howard-Flanders, P., and R. P. Boyce. 1966. DNA repair and genetic recombination: studies on mutants of *Escherichia coli* defective in these processes. *Radiat. Res.* **6**(Suppl.):156–184.

74. Howard-Flanders, P., R. P. Boyce, and L. Theriot. 1966. Three loci in *Escherichia coli* K-12 that control the excision of pyrimidine dimers and certain other mutagen products from DNA. *Genetics* **53:**1119–1136.

75. Husain, I., J. Griffith, and A. Sancar. 1988. Thymine dimers bend DNA. *Proc. Natl. Acad. Sci. USA* **85:**2558–2562.

76. Husain, I., G. B. Sancar, S. R. Holbrook, and A. Sancar. 1987. Mechanism of damage recognition by *Escherichia coli* DNA photolyase. *J. Biol. Chem.* **262:**13188–13197.

77. Husain, I., B. Van Houten, D. C. Thomas, M. Abdel-Monem, and A. Sancar. 1985. Effect of DNA polymerase I and DNA helicase II on the turnover rate of UvrABC excision nuclease. *Proc. Natl. Acad. Sci. USA* **82:**6774–6778.

78. Husain, I., B. Van Houten, D. C. Thomas, and A. Sancar. 1986. Sequences of *Escherichia coli uvrA* gene and protein reveal two potential ATP binding sites. *J. Biol. Chem.* **261:**4895–4901.

79. Jones, B. K., and A. T. Yeung. 1988. Repair of 4,5',8-trimethylpsoralen monoadducts and cross-links by the *Escherichia coli* UvrABC endonuclease. *Proc. Natl. Acad. Sci. USA* **85:**8410–8414.

80. Jones, B. K., and A. T. Yeung. 1990. DNA base composition determines the specificity of UvrABC endonuclease incision of a psoralen cross-link. *J. Biol. Chem.* **265:**3489–3496.

81. Kacinski, B. M., and W. D. Rupp. 1982. *E. coli* uvrB protein binds to DNA in the presence of uvrA protein. *Nature* (London) **294:**480–481.

82. Kacinski, B. M., A. Sancar, and W. D. Rupp. 1981. A general approach for purifying proteins encoded by cloned genes without using a functional assay: isolation of the *uvrA* gene product from radiolabeled maxicells. *Nucleic Acids Res.* **9:**4495–4508.

83. Kohn, K. W., R. A. G. Ewig, L. C. Erickson, and L. A. Zwelling. 1981. Measurement of strand breaks and cross-links by alkaline elution, p. 379–401. *In* E. C. Friedberg and P. C. Hanawalt (ed.), *DNA Repair—A Laboratory Manual of Research Procedures*, vol. 1, part B. Marcel Dekker, Inc., New York.

84. Kow, Y. W., S. S. Wallace, and B. Van Houten. 1990. UvrABC nuclease complex repairs thymine glycol, an oxidative DNA base damage. *Mutat. Res.* **235:**147–156.

85. Kozelka, J., G. A. Petsko, G. J. Quigley, and S. J. Lippard. 1986. High-salt and low-salt models for kinked adducts of cis-diamminedichloroplatinum (II) with oligonucleotide duplexes. *Inorg. Chem.* **25:**1075–1077.

86. Kuemmerle, N., R. Ley, and W. Masker. 1981. Analysis of resynthesis tracts in repaired *Escherichia coli* deoxyribonucleic acid. *J. Bacteriol.* **147:**333–339.

87. Kuemmerle, N. B., and W. E. Masker. 1980. Effect of the *uvrD* mutation on excision repair. *J. Bacteriol.* **142:**535–546.

88. Kumura, K., K. Oeda, M. Akiyama, T. Horiuchi, and M. Sekiguchi. 1983. The *uvrD* gene of *E. coli*: molecular cloning and expression, p. 51–62. *In* E. C. Friedberg and B. A. Bridges (ed.), *Cellular Responses to DNA Damage*. Alan R. Liss, Inc., New York.

89. Kumura, K., M. Sekiguchi, A. L. Steinum, and E. Seeberg. 1985. Stimulation of the UvrABC enzyme-catalyzed repair reactions by the UvrD protein (DNA helicase II). *Nucleic Acids Res.* **13:**1483–1492.

90. Kunala, S., and D. E. Brash. 1992. Excision repair at individual bases of the *Escherichia coli lacI* gene: relation to mutation hot spots and transcription coupling activity. *Proc. Natl. Acad. Sci. USA* **89:**11031–11035.

91. Kushner, S. R., V. F. Maples, A. Easton, I. Farrance, and P. Peramachi. 1983. Physical, biochemical, and genetic characterization of the *uvrD* gene, p. 153–159. *In* E. C. Friedberg and B. A. Bridges (ed.), *Cellular Responses to DNA Damage*. Alan R. Liss, Inc., New York.

92. Kushner, S. R., J. Shepherd, G. Edwards, and V. F. Maples. 1978. *uvrD, uvrE*, and *recL* represent a single gene, p. 251–254. *In* P. C. Hanawalt, E. C. Friedberg, and C. F. Fox (ed.), *DNA Repair Mechanisms*. Academic Press, Inc., New York.

93. Lambert, B., B. K. Jones, B. P. Roques, J. B. Le Pecq, and A. T. Yeung. 1989. The noncovalent complex between DNA and the bifunctional intercalator ditercalinium is a substrate for the UvrABC endonuclease of *Escherichia coli*. *Proc. Natl. Acad. Sci. USA* **86:**6557–6561.

94. Lambert, B., B. P. Roques, and J. B. Le Pecq. 1988. Induction of an abortive and futile DNA repair process in *E. coli* by the antitumor DNA bifunctional intercalator, ditercalinium: role of *polA* in death induction. *Nucleic Acids Res.* **16:**1063–1078.

95. Lehmann, A. R., and P. Karran. 1981. DNA repair. *Int. Rev. Cytol.* **72:**101–146.

96. Lett, J. T. 1981. Measurement of single-strand breaks by sedimentation in alkaline sucrose gradients, p. 363–378. *In* E. C. Friedberg and P. C. Hanawalt (ed.), *DNA Repair—A Laboratory Manual of Research Procedures*, vol. 1, part B. Marcel Dekker, Inc., New York.

97. Lin, J. J., A. M. Phillips, J. E. Hearst, and A. Sancar. 1992. Active site of (A)BC excinuclease. II. Binding, bending, and catalysis mutants of UvrB reveal a direct role in 3' and an indirect role in 5' incision. *J. Biol. Chem.* **267:**17693–17700.

98. Lin, J. J., and A. Sancar. 1989. A new mechanism for repairing oxidative damage to DNA: (A)BC excinuclease removes AP sites and thymine glycols from DNA. *Biochemistry* **28:**7979–7984.

99. Lin, J. J., and A. Sancar. 1990. Reconstitution of nucleotide excision nuclease with UvrA and UvrB proteins from *Escherichia coli* and UvrC protein from *Bacillus subtilis*. *J. Biol. Chem.* **265:**21337–21341.

100. Lin, J. J., and A. Sancar. 1991. The C-terminal half of UvrC protein is sufficient to reconstitute (A)BC excinuclease. *Proc. Natl. Acad. Sci. USA* **88:**6824–6828.

101. Lin, J. J., and A. Sancar. 1992. Active site of (A)BC excinuclease. I. Evidence for 5' incision by UvrC through a catalytic site involving Asp399, Asp438, Asp466, and His538 residues. *J. Biol. Chem.* **267:**17688–17692.

102. Lin, J. J., and A. Sancar. 1992. (A)BC excinuclease: the *Escherichia coli* nucleotide excision repair enzyme. *Mol. Microbiol.* **6:**2219–2224.

103. **Maples, V. F., and S. R. Kushner.** 1982. DNA repair in *Escherichia coli*: identification of the *uvrD* gene product. *Proc. Natl. Acad. Sci. USA* **79:**5616–5620.

104. **Masker, W., P. C. Hanawalt, and H. Shizuya.** 1973. Role of DNA polymerase II in repair replication in *Escherichia coli. Nature* (London) *New Biol.* **244:**242–243.

105. **Mattern, I. E., M. P. Van Winden, and A. Rörsch.** 1965. The range of action of genes controlling radiation sensitivity in *Escherichia coli. Mutat. Res.* **2:**111–131.

106. **Matthews, B. W., H. Nicholson, and W. J. Becktel.** 1987. Enhanced protein thermostability from site-directed mutations that decrease the entropy of unfolding. *Proc. Natl. Acad. Sci. USA* **84:**6663–6667.

107. **Mazur, S., and L. Grossman.** 1991. Dimerization of *Escherichia coli* UvrA and its binding to undamaged and ultraviolet light damaged DNA. *Biochemistry* **30:**4432–4443.

108. **Mellon, I., and P. C. Hanawalt.** 1989. Induction of the *Escherichia coli* lactose operon selectively increases repair of its transcribed DNA strand. *Nature* (London) **342:**95–98.

109. **Michalke, H., and H. Bremer.** 1969. RNA synthesis in *Escherichia coli* after irradiation with ultraviolet light. *J. Mol. Biol.* **41:**1–23.

110. **Moolenaar, G. F., C. A. Van Sluis, C. Backendorf, and P. Van de Putte.** 1987. Regulation of the *Escherichia coli* excision repair gene *uvrC*. Overlap between the *uvrC* structural gene and the region coding for a 24 kD protein. *Nucleic Acids Res.* **15:**4273–4289.

111. **Morimyo, M., and Y. Shimazu.** 1976. Evidence that the gene *uvrB* is indispensable for a polymerase I deficient strain of *Escherichia coli* K-12. *Mol. Gen. Genet.* **147:**243–250.

112. **Moseley, B. E. B.** 1983. Photobiology and radiobiology of *Micrococcus* (*Deinococcus*) *radiodurans*. *Photochem. Photobiol. Rev.* **7:**223–274.

113. **Moseley, B. E. B., and D. M. Evans.** 1983. Isolation and properties of strains of *Micrococcus* (*Deinococcus*) *radiodurans* unable to excise ultraviolet light-induced pyrimidine dimers from DNA: evidence for two excision pathways. *J. Gen. Microbiol.* **129:**2437–2445.

114. **Munn, M. M., and W. D. Rupp.** 1991. Interaction of the UvrABC endonuclease with DNA containing a psoralen monoadduct or cross-link. Differential effects of superhelical density and comparison of preincision complexes. *J. Biol. Chem.* **36:**24748–24756.

115. **Murray, M. L.** 1979. Substrate-specificity of *uvr* excision repair. *Environ. Mutagen.* **1:**347–352.

116. **Murray, R. G. E.** 1992. The family *Deinococcaceae*, p. 3732–3744. *In* A. Balows, H. G. Trüper, M. Dworkin, W. Harder, and K. H. Schleifer (ed.), *The Prokaryotes.* 2nd ed. Springer-Verlag, New York.

117. **Myles, G. M., and A. Sancar.** 1991. Isolation and characterization of functional domains of UvrA. *Biochemistry* **30:**3834–3840.

118. **Myles, G. M., B. Van Houten, and A. Sancar.** 1987. Utilization of DNA photolyase, pyrimidine dimer endonucleases, and alkali hydrolysis in the analysis of aberrant ABC excinuclease incisions adjacent to UV-induced DNA photoproducts. *Nucleic Acids Res.* **15:**1227–1243.

119. **Nakayama, H., and S. Shiota.** 1988. Excision repair in *Micrococcus luteus*: evidence for a UvrABC homolog, p. 115–120. *In* E. C. Friedberg and P. C. Hanawalt (ed.), *Mechanisms and Consequences of DNA Damage Processing.* Alan R. Liss, Inc., New York.

120. **Nakayama, H., S. Shiota, and K. Umezu.** 1992. UV endonuclease-mediated enhancement of UV survival in *Micrococcus luteus. Mutat. Res.* **273:**43–48.

121. **Navaratnam, S., G. M. Myles, R. W. Strange, and A. Sancar.** 1989. Evidence from extended X-ray absorption fine structure and site-specific mutagenesis for zinc fingers in UvrA protein of *Escherichia coli. J. Biol. Chem.* **264:**16067–16071.

122. **Nonho, T., S. Noji, S. Taniguchi, and T. Saito.** 1989. The *narX* and *narL* genes encoding the nitrate-sensing regulators of *Escherichia coli* are homologous to a family of prokaryotic two-component regulatory genes. *Nucleic Acids Res.* **17:**2947–2957.

123. **Oeda, K., T. Horiuchi, and M. Sekiguchi.** 1981. Molecular cloning of the *uvrD* gene of *Escherichia coli* that controls ultraviolet sensitivity and spontaneous mutation frequency. *Mol. Gen. Genet.* **184:**191–199.

124. **Oeda, K., T. Horiuchi, and M. Sekiguchi.** 1982. The *uvrD* gene of *E. coli* encodes a DNA-dependent ATPase. *Nature* (London) **298:**98–100.

125. **Ogawa, H., K. Shimada, and J. Tomizawa.** 1968. Studies on radiation-sensitive mutants of *E. coli*. I. Mutants defective in the repair synthesis. *Mol. Gen. Genet.* **101:**227–244.

126. **Oh, E. Y., and L. Grossman.** 1986. The effect of *Escherichia coli* Uvr protein binding on the topology of supercoiled DNA. *Nucleic Acids Res.* **14:**8557–8571.

127. **Oh, E. Y., and L. Grossman.** 1987. Helicase properties of the *Escherichia coli* UvrAB protein complex. *Proc. Natl. Acad. Sci. USA* **84:**3638–3642.

128. **Oh, E. Y., and L. Grossman.** 1989. Characterization of the helicase activity of the *Escherichia coli* UvrAB protein complex. *J. Biol. Chem.* **264:**1336–1343.

129. **Oller, A. R., I. J. Fijalkowska, R. L. Dunn, and R. M. Schaaper.** 1992. Transcription-repair coupling determines the strandedness of ultraviolet mutagenesis in *Escherichia coli. Proc. Natl. Acad. Sci. USA* **89:**11036–11040.

130. **Orren, D. K., and A. Sancar.** 1989. The (A)BC excinuclease of *Escherichia coli* has only the UvrB and UvrC subunits in the incision complex. *Proc. Natl. Acad. Sci. USA* **86:**5237–5241.

131. **Orren, D. K., and A. Sancar.** 1990. Formation and enzymatic properties of the UvrB DNA complex. *J. Biol. Chem.* **265:**15796–15803.

132. **Orren, D. K., C. P. Selby, J. E. Hearst, and A. Sancar.** 1992. Post-incision steps of nucleotide excision repair in *Escherichia coli*. Diassembly of the UvrBC-DNA complex by helicase II and DNA polymerase I. *J. Biol. Chem.* **267:**780–788.

133. **Paterson, M. C., B. P. Smith, and P. J. Smith.** 1981. Measurement of enzyme-sensitive sites in UV- or γ-irradiated human cells using *Micrococcus luteus* extracts, p. 99–111. *In* E. C. Friedberg and P. C. Hanawalt (ed.), *DNA Repair—A Laboratory Manual of Research Procedures.* vol. 1, part A. Marcel Dekker, Inc., New York.

134. **Pearlman, D. A., and S. R. Holbrook.** 1985. Molecular models for DNA damaged by photoreaction. *Science* **227:**1304–1308.

135. **Pedrini, A. M., and G. Ciarrochi.** 1983. Inhibition of *Micrococcus luteus* DNA topoisomerase I by UV photoproducts. *Proc. Natl. Acad. Sci. USA* **80:**1787–1791.

136. **Pettijohn, D. E., and P. C. Hanawalt.** 1963. Deoxyribonucleic acid replication in bacteria following ultraviolet irradiation. *Biochim. Biophys. Acta* **72:**127–129.

137. **Pettijohn, D. E., and P. C. Hanawalt.** 1964. Evidence for repair-replication of ultraviolet damaged DNA in bacteria. *J. Mol. Biol.* **9:**395–410.

138. **Pu, W. T., R. Kahn, M. M. Munn, and W. D. Rupp.** 1989. UvrABC incision of N-methylmitomycin A-DNA monoadducts and cross-links. *J. Biol. Chem.* **264:**20697–20704.

139. **Reardon, J. T., A. F. Nichols, S. Keeney, C. A. Smith, J. S. Taylor, and S. Linn.** 1993. Comparative analysis of binding of human damaged DNA-binding protein (XPE) and *Escherichia coli* damage recognition protein (UvrA) to the major ultraviolet photoproducts: T[c,s]T, T[6-4]T, and T[Dewar]T. *J. Biol. Chem.* **268:**21301–21308.

140. **Rothman, R. H.** 1978. Dimer excision and repair replication patch size in *recL152* mutant of *Escherichia coli* K-12. *J. Bacteriol.* **136:**444–448.

141. **Rothman, R. H., and A. J. Clark.** 1977. Defective excision and postreplication repair of UV-damaged DNA in a *recL* mutant strain of *E. coli* K-12. *Mol. Gen. Genet.* **155:**267–277.

142. **Rubin, J. S.** 1988. The molecular genetics of the incision step in the DNA excision repair process. *Int. J. Radiat. Biol.* **54:**309–365.

143. **Runyon, G. T., D. G. Bear, and T. M. Lohman.** 1990. *Escherichia coli* helicase II (UvrD) protein initiates DNA unwinding at nicks and blunt ends. *Proc. Natl. Acad. Sci. USA* **87:**6383–6387.

144. Runyon, G. T., and T. M. Lohman. 1989. *Escherichia coli* helicase II (UvrD) protein can completely unwind fully duplex linear and nicked circular DNA. *J. Biol. Chem.* **264:**17502–17512.

145. Rupp, W. D., and P. Howard-Flanders. 1968. Discontinuities in the DNA synthesized in an excision-defective strain of *Escherichia coli* following ultraviolet irradiation. *J. Mol. Biol.* **31:**291–304.

146. Rupp, W. D., A. Sancar, and G. B. Sancar. 1982. Properties and regulation of the UVRABC endonuclease. *Biochimie* **64:**595–598.

147. Sancar, A. 1987. DNA repair in vitro. *Photobiochem. Photobiophys.* **11:**301–315.

148. Sancar, A., N. D. Clarke, J. Griswold, W. J. Kennedy, and W. D. Rupp. 1981. Identification of the *uvrB* gene product. *J. Mol. Biol.* **148:**63–76.

149. Sancar, A., K. A. Franklin, and G. B. Sancar. 1984. *Escherichia coli* DNA photolyase stimulates uvrABC excision nuclease in vitro. *Proc. Natl. Acad. Sci. USA* **81:**7397–7401.

150. Sancar, A., and J. E. Hearst. 1993. Molecular matchmakers. *Science* **259:**1415–1420.

151. Sancar, A., and W. D. Rupp. 1983. A novel repair enzyme: UVRABC excision nuclease of *Escherichia coli* cuts a DNA strand on both sides of the damaged region. *Cell* **33:**249–260.

152. Sancar, A., and G. B. Sancar. 1988. DNA repair enzymes. *Annu. Rev. Biochem.* **57:**29–67.

153. Sancar, A., G. B. Sancar, W. D. Rupp, J. W. Little, and D. W. Mount. 1982. LexA protein inhibits transcription of the *E. coli uvrA* gene in vitro. *Nature* (London) **298:**96–98.

154. Sancar, A., and M. S. Tang. 1993. Nucleotide excision repair. *Photochem. Photobiol.* **57:**905–921.

155. Sancar, G. B., A. Sancar, J. W. Little, and W. D. Rupp. 1982. The *uvrB* gene of *Escherichia coli* has both lexA-repressed and lexA-independent promoters. *Cell* **28:**523–530.

156. Sancar, G. B., A. Sancar, and W. D. Rupp. 1984. Sequences of the *E. coli uvrC* gene and protein. *Nucleic Acids Res.* **12:**4593–4608.

157. Sauerbier, W., R. L. Millette, and P. B. Hackett, Jr. 1970. The effects of ultraviolet irradiation on the transcription of T4 DNA. *Biochim. Biophys. Acta* **209:** 368–386.

158. Schendel, P. F., M. Fogliano, and L. D. Strausbaugh. 1982. Regulation of the *Escherichia coli* K-12 *uvrB* operon. *J. Bacteriol.* **150:**676–685.

159. Schwartz, A. L., L. Marrot, and M. Leng. 1989. The DNA bending by acetylaminofluorene residues and by apurinc sites. *J. Mol. Biol.* **207:**445–450.

160. Seawell, P. C., and A. K. Ganesan. 1981. Measurement of strand breaks in supercoiled DNA by gel electrophoresis, p. 425–430. *In* E. C. Friedberg and P. C. Hanawalt (ed.), *DNA Repair—A Laboratory Manual of Research Procedures.* vol. 1, part B. Marcel Dekker, Inc., New York.

161. Sedgwick, B. 1989. In vitro proteolytic cleavage of the *Escherichia coli* Ada protein by the *ompT* gene product. *J. Bacteriol.* **171:**2249–2251.

162. Seeberg, E. 1978. Reconstitution of an *Escherichia coli* repair endonuclease activity from the separated *uvrA*⁺ and *uvrB*⁺/*uvrC*⁺ gene products. *Proc. Natl. Acad. Sci. USA* **75:**2569–2573.

163. Seeberg, E. 1981. Multiprotein interactions in the strand cleavage of DNA damaged by UV and chemicals. *Prog. Nucleic Acid Res. Mol. Biol.* **26:**217–226.

164. Seeberg, E., and R. P. Fuchs. 1990. Acetylaminofluorene bound to different guanines of the sequence -GGCGCC- is excised with different efficiencies by the UvrABC excision nuclease in a pattern not correlated to the potency of mutation induction. *Proc. Natl. Acad. Sci. USA* **87:**191–194.

165. Seeberg, E., J. Nissen-Meyer, and P. Strike. 1976. Incision of ultraviolet-irradiated DNA by extracts of *E. coli* requires three different gene products. *Nature* (London) **263:**524–526.

166. Seeberg, E., and A. L. Steinum. 1982. Purification and properties of the uvrA protein from *Escherichia coli*. *Proc. Natl. Acad. Sci. USA* **79:**988–992.

167. Seeley, T. W., and L. Grossman. 1990. The role of *Escherichia coli* UvrB in nucleotide excision repair. *J. Biol. Chem.* **265:**7158–7165.

168. Selby, C. P., and A. Sancar. 1988. ABC excinuclease incises both 5′ and 3′ to the CC-1065-DNA adduct and its incision activity is stimulated by DNA helicase II and DNA polymerase I. *Biochemistry* **27:**7184–7188.

169. Selby, C. P., and A. Sancar. 1990. Molecular mechanisms of DNA repair inhibition by caffeine. *Proc. Natl. Acad. Sci. USA* **87:**3522–3525.

170. Selby, C. P., and A. Sancar. 1990. Structure and function of the (A)BC excinuclease of *Escherichia coli*. *Mutat. Res.* **236:**203–211.

171. Selby, C. P., and A. Sancar. 1990. Transcription preferentially inhibits nucleotide excision repair of the template DNA strand in vitro. *J. Biol. Chem.* **265:**21330–21336.

172. Selby, C. P., and A. Sancar. 1991. Gene- and strand-specific repair in vitro: partial purification of a transcription-repair coupling factor. *Proc. Natl. Acad. Sci. USA* **88:**8232–8236.

173. Selby, C. P., and A. Sancar. 1993. Molecular mechanism of transcription-repair coupling. *Science* **260:**53–58.

174. Selby, C. P., and A. Sancar. 1993. Transcription-repair coupling and mutation frequency decline. *J. Bacteriol.* **175:** 7509–7514.

175. Selby, C. P., E. M. Witkin, and A. Sancar. 1991. *Escherichia coli mfd* mutant deficient in "mutation frequency decline" lacks strand-specific repair: in vitro complementation with purified coupling factor. *Proc. Natl. Acad. Sci. USA* **88:**11574–11578.

176. Setlow, R. B., and W. L. Carrier. 1963. The disappearance of thymine dimers from DNA: an error-correcting mechanism. *Proc. Natl. Acad. Sci. USA* **51:**226–231.

177. Setlow, R. B., and J. D. Regan. 1981. Measurement of repair synthesis by photolysis of bromouracil, p. 307–318. *In* E. C. Friedberg and P. C. Hanawalt (ed.), *DNA Repair—A Laboratory Manual of Research Procedures.* Vol. 1, part B. Marcel Dekker, Inc., New York.

178. Setlow, R. B., P. A. Swenson, and W. L. Carrier. 1963. Thymine dimers and inhibition of DNA synthesis by ultraviolet irradiation of cells. *Science* **142:**1464–1466.

179. Sharma, S., T. F. Stark, W. G. Beattie, and R. E. Moses. 1986. Multiple control elements for the *uvrC* gene unit of *Escherichia coli*. *Nucleic Acids Res.* **14:**2301–2318.

180. Shi, Q., R. Thresher, A. Sancar, and J. Griffith. 1992. Electron microscopic study of (A)BC excinuclease. DNA is sharply bent in the UvrB-DNA complex. *J. Mol. Biol.* **226:**425–432.

181. Shi, Y. B., J. Griffith, H. Gamper, and J. E. Hearst. 1988. Evidence for structural deformation of the DNA helix by a psoralen diadduct but not by a monoadduct. *Nucleic Acids Res.* **16:**8945–8952.

182. Shimada, K., H. Ogawa, and J. Tomizawa. 1968. Studies on radiation-sensitive mutants of *E. coli*. II. Breakage and repair of ultraviolet irradiated intracellular DNA of phage lambda. *Mol. Gen. Genet.* **101:**245–256.

183. Shiota, S., and H. Nakayama. 1988. Evidence for a *Micrococcus luteus* gene homologous to *uvrB* of *Escherichia coli*. *Mol. Gen. Genet.* **213:**21–29.

184. Shiota, S., and H. Nakayama. 1989. *Micrococcus luteus* homolog of the *Escherichia coli uvrA* gene: identification of a mutation in the UV-sensitive mutant DB7. *Mol. Gen. Genet.* **217:**332–340.

185. Sibghat-Ullah, A. Sancar, and J. E. Hearst. 1990. The repair patch of *E. coli* (A)BC excinuclease. *Nucleic Acids Res.* **18:**5051–5053.

186. Siegel, E. C. 1973. Ultraviolet-sensitive mutator strain of *Escherichia coli* K-12. *J. Bacteriol.* **113:**145–160.

187. Sinden, R. R., and R. S. Cole. 1978. Repair of cross-linked DNA and survival of *Escherichia coli* treated with psoralen and light: effects of mutations influencing genetic recombination and DNA metabolism. *J. Bacteriol* **136:**538–547.

188. Sinden, R. R., and R. S. Cole. 1978. Topography and kinetics of genetic recombination in *Escherichia coli* treated with psoralen and light. *Proc. Natl. Acad. Sci. USA* **75:**2373–2377.

189. **Sladek, F. M., B. Dynlacht, and P. Howard-Flanders.** 1988. Characterization and partial purification of a novel activity from *E. coli* that acts on psoralen monoadducts. *J. Cell Biochem.* **12a:**276.

190. **Smirnov, G. B., and A. G. Skavronskaya.** 1971. Location of *uvr502* mutation on the chromosome of *Escherichia coli* K-12. *Mol. Gen. Genet.* **113:**217–221.

191. **Smith, C. A., P. K. Cooper, and P. C. Hanawalt.** 1981. Measurement of repair replication by equilibrium sedimentation, p. 289–305. *In* E. C. Friedberg and P. C. Hanawalt (ed.), *DNA Repair—A Laboratory Manual of Research Procedures*, vol. 1, part B. Marcel Dekker, Inc., New York.

192. **Smith, C. A., and P. C. Hanawalt.** 1978. Phage T4 endonuclease V stimulates DNA repair replication in isolated nuclei from ultraviolet-irradiated human cells, including xeroderma pigmentosum fibroblasts. *Proc. Natl. Acad. Sci. USA* **75:**2598–2602.

193. **Snowden, A., Y. W. Kow, and B. Van Houten.** 1990. Damage repertoire of the *Escherichia coli* UvrABC nuclease complex includes abasic sites, base-damage analogues, and lesions containing adjacent 5' or 3' nicks. *Biochemistry* **29:**7251–7259.

194. **Sternglanz, R., S. DiNardo, K. A. Voelkel, Y. Nishimura, Y. Hirota, K. Becherer, L. Zumstein, and J. C. Wang.** 1981. Mutations in the gene coding for *Escherichia coli* DNA topoisomerase I affect transcription and transposition. *Proc. Natl. Acad. Sci. USA* **78:**2747–2751.

195. **Svoboda, D. L., C. A. Smith, J. S. Taylor, and A. Sancar.** 1993. Effect of sequence, adduct type, and opposing lesions on the binding and repair of ultraviolet photodamage by DNA photolyase and (A)BC excinuclease. *J. Biol. Chem.* **268:**10694–10700.

196. **Swenson, P. A., and R. B. Setlow.** 1966. Effects of ultraviolet radiation on macromolecular synthesis in *Escherichia coli*. *J. Mol. Biol.* **15:**201–219.

197. **Takagi, Y., M. Sekiguchi, S. Okubo, H. Nakayama, K. Shimada, S. Yasuda, T. Nishimoto, and H. Yoshihara.** 1967. Nucleases specific for ultraviolet light-irradiated DNA and their possible role in dark repair. *Cold Spring Harbor Symp. Quant. Biol.* **33:**219–227.

198. **Takahashi, M., E. Bertrand-Burggraf, R. P. Fuchs, and B. Norden.** 1992. Structure of UvrABC excinuclease-UV-damaged DNA complexes studied by flow linear dichroism. DNA curved by UvrB and UvrC. *FEBS Lett.* **314:**10–12.

199. **Tang, M. S., C. S. Lee, R. Doisy, L. Ross, D. R. Needham-VanDevanter, and L. H. Hurley.** 1988. Recognition and repair of the CC-1065-(N3-adenine)-DNA adduct by the UVRABC nucleases. *Biochemistry* **27:**893–901.

200. **Tang, M. S., and K. C. Smith.** 1981. The effects of *lexA*, *recB21*, *recF143*, and *uvrD3* mutations on liquid-holding recovery in ultraviolet-irradiated *Escherichia coli* K12 *recA56*. *Mutat. Res.* **80:**15–25.

201. **Tao, K., A. Noda, and S. Yonei.** 1987. The roles of different excision-repair mechanisms in the resistance of *Micrococcus luteus* to UV and chemical mutagens. *Mutat. Res.* **183:**231–239.

202. **Teo, I. A.** 1987. Proteolytic processing of the Ada protein that repairs DNA O⁶-methylguanine residues in *E. coli*. *Mutat. Res.* **183:**123–127.

203. **Thiagalingam, S., and L. Grossman.** 1991. Both ATPase sites of *Escherichia coli* UvrA have functional roles in nucleotide excision repair. *J. Biol. Chem.* **266:**11395–11403.

204. **Thomas, D. C., T. A. Kunkel, N. J. Casna, J. P. Ford, and A. Sancar.** 1986. Activities and incision patterns of ABC excinuclease on modified DNA containing single-base mismatches and extrahelical bases. *J. Biol. Chem.* **261:**14496–14505.

205. **Thomas, D. C., M. Levy, and A. Sancar.** 1985. Amplification and purification of UvrA, UvrB, and UvrC proteins of *Escherichia coli*. *J. Biol. Chem.* **260:**9875–9883.

206. **Tomic, M. T., D. E. Wemmer, and S. H. Kim.** 1987. Structure of a psoralen cross-linked DNA in solution by nuclear magnetic resonance. *Science* **238:**1722–1725.

207. **Van den Berg, E. A., R. H. Geerse, J. Memelink, R. A. Bovenberg, F. A. Magnee, and P. Van de Putte.** 1985. Analysis of regulatory sequences upstream of the *E. coli uvrB* gene: involvement of the DnaA protein. *Nucleic Acids Res.* **13:**1829–1840.

208. **Van den Berg, E. A., R. H. Geerse, H. Pannekoek, and P. Van de Putte.** 1983. In vivo transcription of the *E. coli uvrB* gene: both promoters are inducible by UV. *Nucleic Acids Res.* **11:**4355–4363.

209. **Van de Putte, P., C. A. Van Sluis, J. Van Dillewijn, and A. Rörsch.** 1965. The location of genes controlling radiation sensitivity in *Escherichia coli*. *Mutat. Res.* **2:**97–110.

210. **Van Houten, B.** 1990. Nucleotide excision repair in *Escherichia coli*. *Microbiol. Rev.* **54:**18–51.

211. **Van Houten, B.** 1994. Personal communication.

212. **Van Houten, B., H. Gamper, S. R. Holbrook, J. E. Hearst, and A. Sancar.** 1986. Action mechanism of ABC excision nuclease on a DNA substrate containing a psoralen crosslink at a defined position. *Proc. Natl. Acad. Sci. USA* **83:**8077–8081.

213. **Van Houten, B., H. Gamper, A. Sancar, and J. E. Hearst.** 1987. DNase I footprint of ABC excinuclease. *J. Biol. Chem.* **262:**13180–13187.

214. **Van Houten, B., S. Illenye, Y. Qu, and N. Farrell.** 1993. Homodinuclear (Pt, Pt) and heterodinuclear (Ru, Pt) metal compounds as DNA-protein cross-linking agents: potential suicide DNA lesions. *Biochemistry* **32:**11794–11801.

215. **Van Houten, B., and A. Snowden.** 1993. Mechanism of action of the *Escherichia coli* UvrABC nuclease: clues to the damage recognition problem. *Bioessays* **15:**51–59.

216. **Van Sluis, C. A., I. E. Mattern, and M. C. Paterson.** 1974. Properties of *uvrE* mutants of *Escherichia coli* K12. I. Effects of UV irradiation on DNA metabolism. *Mutat. Res.* **25:**273–279.

217. **Varghese, A. J., and R. S. Day.** 1970. Excision of cytosine-thymine adducts from the DNA of ultraviolet-irradiated *Micrococcus radiodurans*. *Photochem. Photobiol.* **11:**511–517.

218. **Visse, R. M., M. De Ruijter, G. F. Moolenaar, and P. Van de Putte.** 1992. Analysis of UvrABC endonuclease reaction intermediates on cisplatin-damaged DNA using mobility shift gel electrophoresis. *J. Biol. Chem.* **267:**6736–6742.

219. **Visse, R. M., M. De Ruijter, M. Ubbink, J. A. Brandsma, and P. Van de Putte.** 1993. The first zinc-binding domain of UvrA is not essential for UvrABC-mediated DNA excision repair. *Mutat. Res.* **294:**263–274.

220. **Voight, J. M., B. Van Houten, A. Sancar, and M. D. Topal.** 1989. Repair of O⁶-methylguanine by ABC excinuclease of *Escherichia coli*. *J. Biol. Chem.* **264:**5172–5176.

221. **von Hippel, P. H., and O. G. Berg.** 1989. Facilitated target location in biological systems. *J. Biol. Chem.* **264:**675–678.

222. **Wahl, A. F., J. W. Hockensmith, S. Kowalski, and R. A. Bambara.** 1983. Alternative explanation for excision repair deficiency caused by the *polAex1* mutation. *J. Bacteriol.* **155:**922–925.

223. **Walker, J. E., M. Saraste, M. J. Runswick, and N. J. Gay.** 1982. Distantly related sequences in the α- and β-subunits of ATP synthase, myosin, kinases and other ATP-requiring enzymes and a common nucleotide binding fold. *EMBO J.* **1:**945–951.

224. **Wang, J., and L. Grossman.** 1993. Mutations in the helix-turn-helix motif of the *Escherichia coli* UvrA protein eliminate its specificity for UV-damaged DNA. *J. Biol. Chem.* **268:**5323–5331.

225. **Washburn, B. K., and S. R. Kushner.** 1991. Construction and analysis of deletions in the structural gene (*uvrD*) for DNA helicase II of *Escherichia coli*. *J. Bacteriol.* **178:**2569–2575.

226. **Weiss, B., and L. Grossman.** 1987. Phosphodiesterases involved in DNA repair. *Adv. Enzymol. Relat. Areas Mol. Biol.* **60:**1–34.

227. **Witkin, E. M.** 1956. Time, temperature, and protein synthesis: a study of ultraviolet-induced mutation in bacteria. *Cold Spring Harbor Symp. Quant. Biol.* **21:**123–140.

228. **Witkin, E. M.** 1966. Radiation-induced mutations and their repair. *Science* **152:**1345–1353.

229. **Yang, Y., and G. F. Ames.** 1988. DNA gyrase binds to the family of prokaryotic repetitive extragenic palindromic sequences. *Proc. Natl. Acad. Sci. USA* **85:**8850–8854.

230. Yeung, A. T., W. B. Mattes, E. Y. Oh, and L. Grossman. 1983. Enzymatic properties of purified *Escherichia coli* uvrABC proteins. *Proc. Natl. Acad. Sci. USA* **80:**6157–6161.

231. Yeung, A. T., W. B. Mattes, E. Y. Oh, and L. Grossman. 1983. Enzymatic properties of the purified *Escherichia coli* uvrABC complex, p. 77–86. *In* E. C. Friedberg and B. A. Bridges (ed.), *Cellular Responses to DNA Damage.* Alan R. Liss, Inc., New York.

232. Yeung, A. T., W. B. Mattes, E. Y. Oh, G. H. Yoakum, and L. Grossman. 1986. The purification of the *Escherichia coli* UvrABC incision system. *Nucleic Acids Res.* **14:**8535–8556.

233. Youngs, D. A., E. Van der Schueren, and K. C. Smith. 1974. Separate branches of the uvr gene-dependent excision repair process in ultraviolet-irradiated *Escherichia coli* K-12 cells: their dependence upon growth medium and the *polA, recA, recB,* and *exrA* genes. *J. Bacteriol.* **117:**717–725.

6

Nucleotide Excision Repair in Lower Eukaryotes

As we ascend the evolutionary ladder and encounter increasingly more complex organisms, our information about the molecular biology and biochemistry of nucleotide excision repair becomes less precise. Several aspects of the biology of higher cells contribute to this paucity of information. For instance, the structure of the eukaryotic genome is more complex than that of the prokaryotes considered in the previous chapter. Eukaryotic genomes are considerably larger, and the requirements for packaging such large amounts of DNA into the limited confines of the cell nucleus have been solved by a complex structural organization. In chapter 1 we noted that a consideration of the distribution of genomic injury in eukaryotes must take into account that nuclear DNA exists in intimate association with both histones and nonhistone chromosomal proteins and that the structure of chromosomes reflects various levels of folding and coiling of the basic chromatin structural unit, the *nucleosome*. The same considerations apply to cellular responses to genomic injury, including nucleotide excision repair.

Our current understanding of how chromosome structure and nucleosome conformation influence the enzymology of DNA incision and of postincision events during nucleotide excision repair in eukaryotes is still scanty. Nonetheless, it is likely that structural elements limit the access of repair enzymes to sites of damage and that specific perturbations of chromatin structure are necessary to facilitate nucleotide excision repair in eukaryotic cells. The precise mechanism(s) by which access of repair enzymes is provided to sites of base damage is uncertain. Nor is it clear whether the phenomenon of accessibility is specifically linked to DNA repair or whether nucleotide excision repair and other repair processes gratuitously exploit open chromatin conformations that evolve during other aspects of DNA metabolism, such as replication and transcription. As discussed below, relatively recent observations that directly link nucleotide excision repair to transcription in *Saccharomyces cerevisiae* provide a compelling basis for considering such gratuitous exploitation as a major mechanism for solving the "accessibility problem" during repair.

Eukaryotic Models for Nucleotide Excision Repair

A number of eukaryotic systems have proven to be extremely informative for the study of nucleotide excision repair. Understandably, the choice of particular organisms has been dictated largely by experimental considerations. Among the lower eukaryotes, the genetic and molecular versatility of the yeast *S. cerevisiae* (and more recently *Schizosaccharomyces pombe*), and the fruit fly *Drosophila melanogaster* have identified them as particularly useful model systems, and these organisms are considered in detail in this chapter. Mammalian cells have been less tractable to genetic and molecular analysis. Nonetheless, recent years have witnessed impressive inroads into the genetics and molecular biology of nucleotide excision repair in rodent cells in culture. Additionally, the human diseases xeroderma pigmentosum (XP) and Cockayne's syndrome (see chapters 8 and 14) have provided highly informative model systems for investigating nucleotide excision repair in human cells, as well as the complex relationships between these two diseases. Nucleotide excision repair in mammalian cells, including those from humans, is considered in detail in the following two chapters.

Nucleotide Excision Repair in the Yeast
Saccharomyces cerevisiae

Overview of the Genetics of Nucleotide Excision Repair

The yeast *S. cerevisiae* (which we will frequently refer to simply as the yeast, with no intended slight to its distantly related cousin *S. pombe*) is genetically well characterized, especially with respect to its response to agents which cause DNA damage. Over 30 *RAD* loci (for radiation sensitivity) that confer resistance to killing by UV and/or ionizing radiation have been identified. These loci have been classified into three (largely) nonoverlapping *epistasis groups*, which are thought to reflect three fundamentally distinct cellular responses to DNA damage (48–51, 52, 67, 71, 132). Epistatic interactions are operationally defined by the use of mutant strains. If the presence of mutations at two different genetic loci confers a phe-

Table 6–1 Epistasis groups for yeast genes involved in cellular responses to DNA damage[a]

RAD3 group	RAD52 group	RAD6 group
RAD1	*RAD50*	*RAD5 (REV2) (SNM2)*
RAD2	*RAD51*	*RAD6*
RAD3	*RAD52*	*RAD8**
RAD4	*RAD53*	*RAD9**
RAD7	*RAD54*	*RAD15**
RAD10	*RAD55*	*RAD18*
RAD14	*RAD56*	*RADH**
SSL1	*RAD57*	*REV1*
SSL2 (RAD25)	*RAD24*	*REV3 (PSO1)*
TFB1	*XRS2*	*CDC9*
RAD16		*REV5**
RAD23		*REV6**
CDC8		*REV7*
CDC9		*CDC7**
MMS19		*CDC8*
PSO2 (SNM1)		*MMS3**
PSO3		*PSO4**
UVS12		*UMR1–7*

[a] The genes are shown as wild-type alleles for simplicity. The assignment of genes to a particular epistasis group is now based on comprehensive analysis of UV or ionizing radiation sensitivity in all cases. In some instances (indicated by asterisks), genes have been assigned strictly on the basis of limited phenotypic characterization. Genes shown in parentheses are allelic to those primarily listed.
Source: From Friedberg et al. (52) with permission.

notype (such as sensitivity to UV radiation) that is quantitatively the same as that conferred by each single mutation alone, the two genes are said to be *epistatic* with respect to one another. The simplest (but by no means the exclusive) interpretation of epistatic interactions between different mutations is that the genes they define are involved in sequential steps of a single multistep biochemical pathway or that they encode components of a multimeric or multiprotein complex. More-complex interpretations of epistasis are tenable, and some of them are alluded to later in the chapter, but it is not important to consider these in any detail.

The epistasis groups that have been defined with respect to radiation sensitivity in *S. cerevisiae* are shown in Table 6–1. The *RAD3* epistasis group is named after one of its member loci, the *RAD3* gene. The mutational inactivation of any of these genes results in increased sensitivity to UV radiation and/or to many chemicals which produce bulky base adducts (48–51, 52, 67, 71, 132). The majority of mutants in the *RAD3* epistasis group have been shown to be defective in damage-specific incision of UV-irradiated DNA and consequently in the excision of pyrimidine dimers. These principal phenotypes provide the most persuasive evidence that many, if not all, the loci in the *RAD3* epistasis group participate in nucleotide excision repair in *S. cerevisiae*.

In wild-type yeast cells the incision of DNA at pyrimidine dimers appears to be followed rapidly by excision/resynthesis and DNA ligation. Hence, the number of damage-specific nicks that can be detected in the genome at any given time is small relative to the total number of repairable lesions (140). This makes it difficult to be certain whether or not a given *rad* mutant is totally defective in DNA incision. To circumvent this problem, a temperature-sensitive mutation in the gene for DNA ligase (*CDC9*) was introduced into a number of *rad* mutant strains (194). Because of the defect in DNA ligase activity, incisions associated with nucleotide excision repair cannot be sealed and many more strand breaks can be detected than in wild-type strains (Fig. 6–1). It was therefore possible to demonstrate that mutational inactivation of five *RAD* genes that were initially characterized (*RAD1*, *RAD2*, *RAD3*, *RAD4*, and *RAD10*) results in a severe, presumably a total, defect in the incision of UV-irradiated DNA in vivo (140, 194) (Table 6–2).

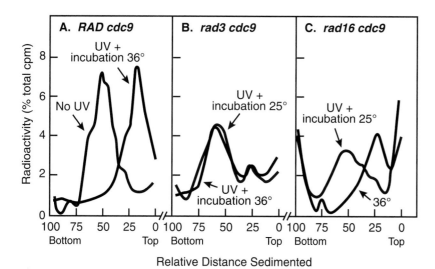

Figure 6–1 Defective incision of UV-irradiated DNA in a *rad3* mutant in vivo. After UV irradiation, the wild-type (*RAD*) strain (A) and a *rad16* mutant strain (C) generate DNA incisions after UV irradiation that shift the sedimentation of the DNA toward the top of the sucrose gradients. These incisions persist at 36°C because of the presence of the temperature-sensitive *cdc9* mutation that prevents DNA ligation and hence the completion of nucleotide excision repair. At 25°C DNA ligase is active and excision repair can be completed. Thus, the higher-molecular-weight DNA of wild-type and *rad16* cells sediments faster in the alkaline gradients. No DNA strand breaks are detected in the DNA of the *rad3* mutant (B), even at 36°C, indicating that this mutant is defective in the incision of UV-irradiated DNA. (*Adapted from Friedberg [47a] with permission.*)

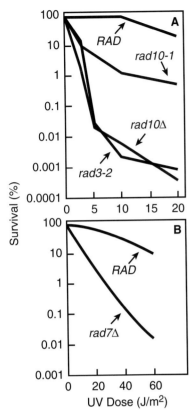

Figure 6–2 (A) Deletion of the *RAD10* gene of *S. cerevisiae* results in a significant increase in sensitivity to killing by UV radiation over that observed in the *rad10-1* point mutant, to a level comparable to that sustained by a *rad3* mutant. (B) Deletion of the *RAD7* gene results in less sensitivity to killing by UV radiation than deletion of the *RAD10* gene does. (Note the different UV dose scales in (A) and (B)). (*Adapted from Weiss and Friedberg [191a] and from Perozzi and Prakash with permission [127].*)

Consistent with these observations, comparisons of the UV radiation sensitivity of *rad1*, *rad2*, *rad3*, *rad4*, and *rad10* point mutants (which may be leaky) and deletion or disruption mutants in which gene function is totally obliterated have sometimes revealed a marked increase, under the latter conditions, in the sensitivity of cells to killing by many DNA-damaging agents known to be processed by nucleotide excision repair (Fig. 6–2). This phenotype is also produced by deletion of the *RAD14* gene (10), an allele of which carrying a point mutation confers a significant residual capacity for the formation of DNA damage-specific strand breaks in vivo (Table 6–2), suggesting that the *RAD14* gene is also required for early events during nucleotide excision repair. Disruption or deletion of *RAD3* has a lethal phenotype in haploid yeast cells, demonstrating that in addition to its involvement in nucleotide excision repair, this gene is *essential for viability* in yeast cells in the absence of DNA damage. As discussed below, this phenotypic theme has been repeated following the disruption of other more recently discovered genes whose products participate in nucleotide excision repair. The disruption or deletion of several other members of the *RAD3* epistasis group, designated *RAD7*, *RAD16*, and *RAD23*, results in a more modest sensitivity to killing by DNA damaging agents (Fig. 6–2). Such mutant strains appear to be deficient but not totally defective in nucleotide excision repair (140, 194) (Fig. 6–1).

Further loci that determine the sensitivity of yeast cells to UV radiation, designated *RAD19*, *RAD20*, *RAD21* and *RAD22*, have also been identified (40) but have not been characterized with respect to their ability to carry out nucleotide excision repair, primarily because mutant strains are weakly UV sensitive and are not inviting for detailed molecular and genetic studies. Hence, they are not included in the *RAD3* epistasis group at this time. If these genes are indeed involved in nucleotide excision repair, and if the (limited) mutant alleles thus far isolated do not simply reflect leaky mutations, they probably constitute further members of the *RAD7*, *RAD16*, and *RAD23* subclass of genes in the *RAD3* epistasis group.

Some Nucleotide Excision Repair Genes Were Discovered Incidentally

There is evidence that the polypeptide products of at least five other genes are absolutely required for nucleotide excision repair in *S. cerevisiae*. Unlike the genes mentioned above, these were not identified by a deliberate search for UV radiation-sensitive mutants but turned up unexpectedly in the course of studies on other aspects of nucleic acid metabolism. The discoveries of the *SSL2* and *SSL1* genes constitute excellent examples of such studies. A team of investigators interested primarily in the molecular mechanism of the initiation of translation in yeast cells undertook a search for mutant alleles which could suppress a stable stem-loop structure engineered into the leader sequence of mRNA transcribed from the *HIS4* gene (Fig. 6–3). *HIS4* is required for histidine biosynthesis and hence for the

Table 6–2 Strand breaks in DNA of UV-irradiated cells defective in *RAD* and *CDC9* genes

Strain	No. of breaks observed (%)[a]
RAD CDC	0
RAD cdc9	100
rad1 cdc9	0
rad2 cdc9	0
rad3 cdc9	0
rad4 cdc9	0
rad10 cdc9	0
rad14 cdc9	29
rad16 cdc9	63

[a] Cells were irradiated and held at 36°C for 1 h before strand breaks were measured by sedimentation velocity.

Source: From Friedberg (47a) with permission.

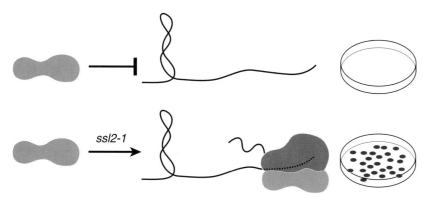

Figure 6–3 Schematic representation of the selection scheme used for identifying genes involved in translation initiation. The 43S preinitiation complex (shown on the left as a dumbbell structure) is not capable of initiating translation from *HIS4* mRNA (top) because the stable stem-loop structure in the leader sequence of the mRNA blocks ribosomal scanning and binding. This results in a His⁻ phenotype when cells are plated on synthetic medium lacking histidine. Suppressors of this phenotype allow the ribosome to initiate protein synthesis on the defective mRNA (bottom), resulting in colonies which can grow on medium unsupplemented with histidine. This selection scheme identified the *ssl2-1* suppressor mutant. (*Adapted from Gulyas and Donahue [56] with permission.*)

survival of cells plated on medium lacking this essential amino acid. The rationale of this experiment was that the stem-loop structure would interfere with translation of *HIS4* mRNA and that suppression by mutant so-called *SSL* genes (for suppressor of stem loop) encoding the lethal phenotype imposed by the stem-loop, might identify genes whose products participate in translation initiation. Among others, this search yielded two genes that are immediately relevant to our story, the wild-type alleles of which are designated *SSL2* and *SSL1* (56, 195).

The translated nucleotide sequence of the cloned *SSL2* gene revealed that it is the yeast homolog of a human gene originally designated *ERCC3* (now called *XPB*) (Fig. 6–4), which had previously been implicated in nucleotide excision repair in human cells and in the nucleotide excision repair-defective disease XP (see chapters 8 and 14). Indeed, use of the human *XPB* (*ERCC3*) gene as a hybridization probe led to the independent cloning of *SSL2* as the *RAD25* gene (126). Like *RAD3*, *SSL2* (*RAD25*) is essential for the growth and viability of haploid yeast cells in the absence of DNA damage (56, 126).

Further evidence that *SSL2* is involved in nucleotide excision repair derives from several observations. Some viable mutant *ssl2* alleles confer a UV radiation-sensitive phenotype, and such mutants fall into the *RAD3* (nucleotide excision repair) epistasis group and not in the *RAD50* or *RAD6* epistasis groups involved in other aspects of DNA repair (see chapter 12) (126) (Fig. 6–5). Additionally, these mutants have been directly shown to be defective in nucleotide excision repair both in vivo (176) and in an in vitro cell-free system (188) (Fig. 6–6). In the latter system the repair-defective phenotype can be specifically corrected by supplementing the mutant extracts with Ssl2 protein (188). This result implicates a direct role for Ssl2 protein in the biochemistry of nucleotide excision repair in yeast cells.

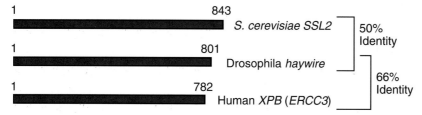

Figure 6–4 The translated nucleotide sequence of the *SSL2* gene of *S. cerevisiae* reveals extensive amino acid identity with the *Drosophila haywire* and human *XPB* (*ERCC3*) genes.

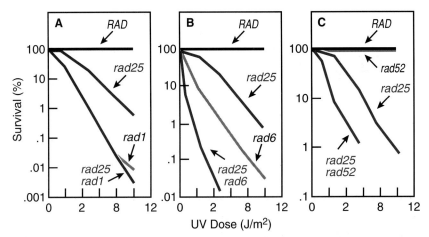

Figure 6–5 A viable mutant strain defective in the *SSL2* (*RAD25*) gene falls into the *RAD3* epistasis group of nucleotide excision repair genes as evidenced by the observation that a *rad1 ssl2* double mutant is no more sensitive to UV radiation than is the *rad1* strain alone (A). Hence, the *SSL2* and *RAD1* genes are presumed to operate in the same biochemical pathway. When the *ssl2* mutant strain additionally carries mutations in the *RAD6* or *RAD52* genes (B and C), which are not involved in nucleotide excision repair, sensitivity to UV radiation is increased. (*Adapted from Park, et al. [126].*)

Hence, the product of the *SSL2* gene appears to have (at least) two distinct functions. One function is required for the growth and viability of yeast cells; the other is required for nucleotide excision repair.

The screen for stem loop suppressor genes just described also yielded a second *SSL* gene called *SSL1*. This gene, as well as another essential gene called *TFB1* has also been shown to be involved in nucleotide excision repair in yeast cells (187). In addition to the *SSL1*, *SSL2*, and *TFB1* genes, two other genes, designated *TFB2* and *TFB3* which encode polypetides of 55 and 38 kDa, respectively, are believed to be essential genes which are indispensable for nucleotide excision repair. However, at the time of writing the *TFB2* and *TFB3* genes have not been formally characterized. All five of these genes will be discussed in greater detail when we consider the specific nature of the essential function of nucleotide excision repair gene products in yeast cells.

In summary, there are two major categories of genes involved in nucleotide excision repair in yeast cells. Some of these (*RAD1*, *RAD2*, *RAD3*, *RAD4*, *RAD10*, *RAD14*, *SSL2* [*RAD25*], *SSL1*, *TFB1*, and probably *TFB2* and *TFB3*) are indispensable for this process. Other genes (*RAD7*, *RAD16*, *RAD23*, and possibly others) are not absolutely required for nucleotide excision repair but apparently play some role in this process in vivo. Mutants with mutations in the latter group are less sensitive to UV radiation and other DNA-damaging agents and show a residual capacity for both DNA incision and the excision of damaged nucleotides.

Potential Roles of *RAD7*, *RAD16*, and *RAD23* in Perturbing Chromatin Structure

As mentioned above, deletion of the *RAD7*, *RAD16*, and *RAD23* genes has confirmed that the resulting mutants are not as sensitive as are strains with mutations in genes in the first group (127, 153). There are indications that *RAD7*, *RAD16*, and *RAD23* may be involved in perturbations of chromatin structure that provide greater accessibility of the catalytically active repair complex encoded by the indispensable group of genes to sites of base damage. These indications derive from studies on the role of these genes in the repair of two homologous genes which exist in different conformational states. The *MATα* locus, a gene involved in mating-type specification in yeast cells, is transcriptionally active and exists in an ''open'' chromatin configuration. This gene is repaired faster than the transcrip-

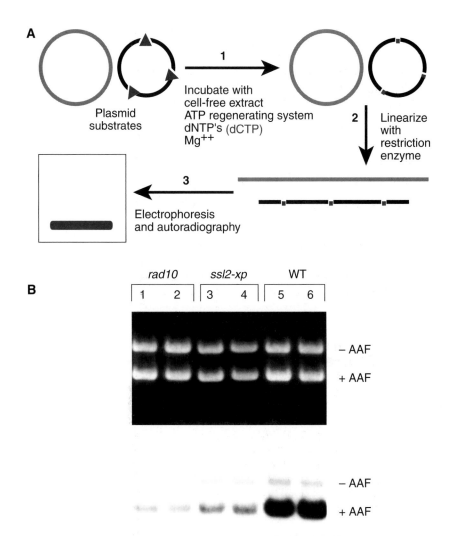

Figure 6–6 Extracts of cells from an *ssl2* mutant strain are defective in nucleotide excision repair. (A) The assay used for measuring nucleotide excision repair. This assay (see chapter 8 for details) monitors repair synthesis of plasmid DNA previously treated with the chemical acetylaminofluorene (AAF) and incubated in the presence of a ^{32}P-labeled deoxynucleoside triphosphate (dNTP) (step 1). Radioactivity incorporated into the damaged plasmid during nucleotide excision repair is observed by autoradiography following linearization and agarose gel electrophoresis of the plasmid (steps 2 and 3). A second (smaller) plasmid which was not treated with AAF and which migrates more slowly in the gel during electrophoresis is included in the assay to monitor background levels of nonspecific DNA synthesis. (B) Extracts of wild-type (WT) yeast strains (*RAD10 SSL2*) are proficient in nucleotide excision repair, as evidenced by the autoradiographic signal in the AAF-treated plasmid DNA (+AAF) relative to the control plasmid (−AAF) (lanes 5 and 6, bottom panel). Extracts of either *ssl2* mutants (lanes 3 and 4, bottom panel) or *rad10* mutants (lanes 1 and 2, bottom panel) are defective in nucleotide excision repair. The top panel shows the gel stained with ethidium bromide prior to autoradiography, in order to demonstrate that equal amounts of plasmid DNA were loaded into each of the lanes. (*Adapted from Hoeijmakers et al. [71a] and Wang et al. [188] with permission.*)

tionally silent *HMLα* locus, which has a closed heterochromatin-like configuration (28) (Fig. 6–7). Several experiments indicate that the differential rate of nucleotide excision repair of these two genes reflects their different conformation rather than the difference in their transcriptional activity (28). For example, in cells carrying a deletion of the regulatory upstream activator sequence (UAS) in the *MATα* locus transcription from this gene is abolished. Nonetheless, the silenced *HMLα* locus is still repaired more slowly than the transcriptionally inactive *MATα* locus (Fig. 6–

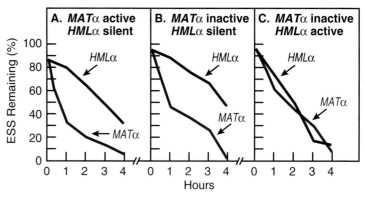

Figure 6–7 Nucleotide excision repair is more rapid and complete in the *MATα* gene of *S. cerevisiae* than in the homologous *HMLα* gene, regardless of whether *MATα* is transcriptionally active (A) or inactive (B). When the transcriptionally silent *HMLα* gene is derepressed, it undergoes rapid repair (C). Nucleotide excision repair is monitored by the loss of pyrimidine dimers from DNA, evidenced by the observation that such lesions are no longer detected by probing the DNA with pyrimidine dimer-DNA glycosylase. This is quantitated as the loss of enzyme-sensitive sites (ESS). (*Adapted from Brouwer et al. [28] with permission.*)

7). Activation of transcription in *HMLα* by disrupting a gene (*SIR3*) that controls silencing of this locus restores repair in this gene to the high rate observed at *MATα* (Fig. 6–7).

The relatively slow repair of the *HMLα* locus is almost totally abolished in *rad16* and *rad7* mutants, whereas the rapid repair of the *MATα* locus is unaffected in these strains (28, 177, 178). Hence, it would appear that the *RAD7* and *RAD16* genes are required for the processing of the heterochromatin-like structure of genes such as *HMLα*. Recent studies involving the two hybrid genetic system described later in the chapter have demonstrated a specific interaction between the Rad7 and Sir3 proteins (125). The *SIR3* gene is known to be involved in the transcriptional silencing of the cryptic mating-type cassettes in yeast cells and of telomeric sequences (89, 147). Hence, this interaction suggests that the interaction between Rad7 and Sir3 may be directly relevant to alterations in chromatin structure which facilitate nucleotide excision repair in transcriptionally repressed regions. Indeed, deletion of the *SIR3* gene results in reduced UV radiation sensitivity of a *rad7* deletion mutant (125).

The *RAD7* and *RAD16* genes are apparently also required for the removal of pyrimidine dimers from the *nontranscribed* strand of transcriptionally active genes (185a). Hence, the repair of heterochromatin-like regions of the yeast genome such as *HMLα* may be mechanistically related to the phenomenon of strand-specific repair observed in transcriptionally active genes that was discussed in chapter 5 in relation to nucleotide excision repair in *Escherichia coli*. Strand-specific repair in yeast cells will be discussed more fully below.

Possible Presence of More Genes for Nucleotide Excision Repair

There are indications of further potential genetic complexity for nucleotide excision repair in yeast cells. As already mentioned, the roles of several genes identified by existing mutants (such as *RAD19*, *RAD20*, *RAD21*, and *RAD22*) have not yet been determined, although of itself the phenotype of UV radiation sensitivity by no means implicates these genes in nucleotide excision repair as opposed to some other cellular response mode to DNA damage. This caveat aside, during the analysis of the many UV radiation-sensitive mutants isolated during an intensive screening process that identified most of the *RAD* genes thus far discussed, it was observed that of the 22 different genetic complementation groups that emerged from this screen, only 12 were represented by more than a single allele (40).

Statistical considerations therefore suggest that it is unlikely that this collection of UV radiation-sensitive mutants reflects all the genes in *S. cerevisiae* which determine resistance to DNA damage. Indeed, the identification of the essential *SSL2* and *SSL1* genes by strategies that are independent of the sensitivity of mutants to DNA-damaging agents provides convincing support for this contention.

Genes That Are Indispensable for Nucleotide Excision Repair

The *RAD1*, *RAD2*, *RAD3*, *RAD4*, *RAD10*, *RAD14*, *SSL2*, *SSL1*, *TFB1* (and probably the *TFB2* and *TFB3*) genes are believed to be involved in biochemical events associated with damage-specific recognition and damage-specific incision of DNA during nucleotide excision repair. All nine of these genes have been cloned and sequenced. The *RAD1*, *RAD2*, *RAD3*, *RAD4*, *RAD10*, and *RAD14* genes were isolated by functional complementation of the UV radiation sensitivity of mutant strains (48–51, 52, 132). As already indicated, the *SSL2*, *SSL1*, and *TFB1* genes were identified primarily in terms of their essential function and secondarily in terms of their role in nucleotide excision repair. The sizes of the open reading frames (ORFs) of these genes and of the polypeptides they are expected to encode are summarized in Table 6–3. None of the genes are related in terms of amino acid sequence, and, with the exception of *RAD3* and *SSL2* (see below), computer-assisted analysis of the sequenced genes has not been particularly informative about the possible biochemical function(s) of the polypeptides they encode.

The translated sequences of the *RAD1*, *RAD2*, *RAD3*, *RAD4*, and *RAD10* genes reveal a low codon bias index. It has been noted that many strongly expressed yeast genes have a bias for codons which are homologous to the anticodons of the major isoacceptor tRNA species and are biased against codons for rare tRNAs (20, 72, 156). Calculation of the fraction of the 22 most frequently used codons in highly expressed genes yields codon bias indices of ~1.0 (20). In contrast, weakly expressed genes show no particular bias against codons for rare tRNAs and yield correspondingly lower bias indices.

Table 6–3   Genes absolutely required for DNA incision during nucleotide excision repair

Gene	Coding region (no. of codons)	Predicted polypeptide size (kDa)	Biochemical function of purified protein
RAD1	1,100	126.2	Endonuclease (with Rad10 protein)
RAD10	210	24.3	Endonuclease (with Rad1 protein)
RAD2	1,031	117.1	Endonuclease
RAD3[a]	778	89.7	DNA-dependent ATPase; DNA-DNA and DNA-RNA helicase with 5′ → 3′ polarity; Pol II basal transcription protein
SSL2 (RAD25)[a]	843	95.3	DNA-DNA helicase with 3′ → 5′ polarity; Pol II basal transcription protein
RAD4	754	87.1	Protein not purified
RAD14	247	29.3	DNA-binding protein with preferential affinity for damaged DNA
SSL1[a]	461	52.2	Pol II basal transcription protein
TFB1[a]	642	73.0	Pol II basal transcription protein
TFB2		55.0	Pol II basal transcription protein
TFB3		38.0	Pol II basal transcription protein

[a] Essential gene.

A possible explanation for these correlations is that the requirement for large amounts of particular cellular proteins is met by rapid translation with abundant tRNAs (20). Rapid translation is expected to be favored by the availability of codons for abundant tRNAs, since the addition of amino acids to the growing polypeptide chain is rate limited by the concentration of charged cognate tRNAs. It has been suggested that in highly expressed genes, i.e., those which encode large amounts of polypeptides, there has been selection for codons that utilize abundant tRNAs (20). The observation that all the *RAD* genes under consideration here have little or no codon bias suggests that they are weakly expressed. The presence of small amounts of intracellular proteins for nucleotide excision repair was discussed in chapter 5 with respect to the Uvr proteins of *E. coli*. Apparently, this phenomenon is quite general.

THE *RAD1* AND *RAD10* GENES AND THE Rad1 AND Rad10 PROTEINS

For reasons that will be soon become obvious, the *RAD1* and *RAD10* genes are conveniently discussed together. Translation of the *RAD1* coding region indicates that the putative polypeptide is acidic, with a calculated net charge of -36. Additionally, the Rad1 amino acid sequence shows the presence of a bipartite motif believed to represent a nuclear localization signal, indicative of the expected nuclear location of this protein (43, 51a) (Table 6–4). As discussed above, *rad1* mutants are highly UV sensitive and are defective in damage-specific incision of DNA during nucleotide excision repair. The *RAD1* gene is also required for certain types of mitotic recombination involving repeated sequences. For example, recombination in the rDNA repeat unit is stimulated by the rDNA Pol1 promoter *HOT1*. However, this recombination is reduced in a *rad1* mutant (186, 196). These transcriptionally stimulated recombination events are reduced to a greater extent in *rad52* mutants (which are defective in another mode of recombinational repair [see chapter 12]), but a synergistic decrease in recombination is observed in both transcriptionally active and inactive DNA in *rad1 rad52* double mutants, suggesting that *RAD1* and *RAD52* operate in separate recombinational pathways.

The *RAD1* gene is also required for normal levels of intrachromosomal recombination. This has been demonstrated by using two mutant *his3* alleles placed in the same chromosome, one of which carries a deletion at the 3′ end and the other carries a deletion at the 5′ end. Intrachromosomal recombination is thus required to restore a functional *HIS3* gene. Mutations in the *RAD2*, *RAD3* or *RAD4* genes elicit no effect on the frequency of His⁺ prototrophs, indicating that this type of recombination is not dependent on the nucleotide excision repair pathway per se but on the *RAD1* and *RAD10* genes in particular (151). *RAD1* is also required for mitotic ectopic recombination between the duplicated *SAM1* and *SAM2* genes located on chromosomes XII and IV respectively (6). These homeologous (similar but not identical) *SAM* genes are 83% identical at the nucleotide sequence level. In related studies it was observed that the rate of recombination between the *SAM1*

Table 6–4   Bipartite nuclear transport sequences in yeast genes involved in nucleotide excision repair

Gene	Amino acid	Bipartite motif
RAD1	517	KR ********** K**RK
RAD2	343	RH ********** KR**R
RAD4	31	RR ********** *KKK*
	377	RR ********** KRR**
	388	RK ********** K**RK
RAD14	169	RR ********** K**KK
SSL2	680	RR ********** R*KRR
SSL1	47	HR ********** KKR**
TFB1	71	KR ********** K**RH
RAD7	137	KK ********** RRRKR

Source: Adapted from Friedberg (51a) and Dingwall and Laskey (43) with permission.

and *SAM2* genes was sixfold higher in a mutant defective in the *TOP3* gene which encodes a DNA topoisomerase, and this effect was also dependent on *RAD1* (5). A study of recombination between specific constructs has also demonstrated a requirement for *RAD1* in a class of intrachromosomal recombinational events thought to result from mismatch repair of short heteroduplex regions (83). *RAD1* is apparently not required for intrachromosomal recombination between all types of heteroalleles or for all forms of interchromosomal mitotic recombination or meiotic recombination (83, 151).

The *RAD10* gene is also involved in mitotic recombination involving repeated sequences (152). The role of this gene has not been as thoroughly studied as that of *RAD1*. However, in the one system in which both *rad1* and *rad10* mutants were

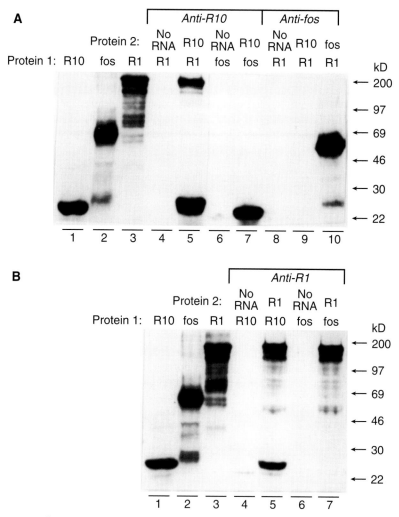

Figure 6–8 The Rad1 and Rad10 proteins of *S. cerevisiae* form a stable and specific complex in vitro. *RAD1* and *RAD10* mRNA was translated in vitro in the presence of [^{35}S] methionine. The translation reaction mixtures were mixed and were precipitated with polyclonal antisera to Rad10 protein (A) or Rad1 protein (B). The immunoprecipitates were loaded onto SDS-gels, and autoradiograms were prepared after electrophoresis. Lanes 5 in panels A and B show coimmunoprecipitation of Rad1 and Rad10 proteins. Lanes 1 to 3 of both autoradiograms show the electrophoretic positions of in vitro translated Rad10, Fos, and Rad1 proteins respectively. The remaining lanes are controls that show the specificity of the coimmunoprecipitation of Rad1 and Rad10. For example, lanes 4 show that Rad1 alone is not precipitated by antibodies to Rad10 protein and vice versa, and lane 10 in panel A shows that antibodies to an irrelevant protein such as Fos, do not result in nonspecific coprecipitation of Rad1 or Rad10 protein. (*Adapted from Bardwell et al. [16].*)

The two hybrid system

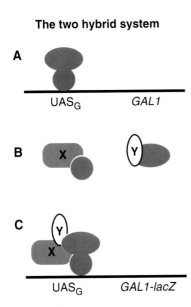

Figure 6–9 Schematic representation of the two-hybrid system for detecting protein-protein interactions in vivo. (A) Yeast Gal4 protein activates transcription by binding specifically to the upstream activator sequences (UAS$_G$) of genes involved in galactose metabolism such as *GAL1*. (B) Yeast cells are transformed with two plasmids carrying the DNA-binding domain and the transcriptional activation of the *GAL4* gene fused to genes encoding two potentially interacting proteins X and Y, respectively. (C) If X and Y interact in vivo, functional Gal4 protein can be reconstituted, thereby activating expression of a *GAL1-lacZ* reporter gene and resulting in the expression of β-galactosidase. (*Adapted from Fields and Song [46] with permission.*)

examined either singly or in combination, their effects were indistinguishable. These observations suggest that the Rad1 and Rad10 proteins act at the same or consecutive biochemical steps during recombination in yeast cells and provided early indications that the products of at least some *RAD* genes required for nucleotide excision repair may also be involved in other aspects of DNA metabolism. Hence, it is likely that the Rad1 and Rad10 proteins can be assembled into different multiprotein complexes in which they execute the same biochemical functions in distinct aspects of DNA metabolism.

The Rad1 and Rad10 Proteins Form a Stable Complex

The Rad1 and Rad10 proteins form a complex in vitro in the absence of damaged DNA. This was demonstrated by in vitro transcription and translation of the cloned genes followed by mixing of the translation reactions and coimmunoprecipitation of the Rad1 and Rad10 polypeptides (Fig. 6–8) and also by coimmunoprecipitation of these proteins from extracts of lysed cells (7, 16). The Rad1-Rad10 complex is very stable in vitro with a dissociation constant (K_d) of ~3 × 10^{-9} M and a half-life of ~15 h (12). This interaction is also highly specific, since neither Rad1 nor Rad10 protein complexes with functionally unrelated proteins (Fig. 6–8). The complex formed in vitro consists of one molecule of Rad1 and one molecule of Rad10 (180). It is estimated that there are ~50 molecules of Rad10 protein per cell. If one assumes that Rad1 protein is present at about the same level, the measured dissociation constant suggests that in vivo most if not all Rad1 and Rad10 molecules exist as a heterodimeric complex.

The N-terminal 35% of Rad10 protein is dispensable for its interaction with Rad1 protein. This result can be reconciled with the observation that this region of Rad10 is poorly conserved between the yeast and the homologous human (183) and mouse (184) polypeptides and is dispensable for the functional activity of the human protein (183). The C-terminal 65% of the Rad10 protein is sufficient and the C-terminal 20% of the polypeptide is necessary for its interaction with Rad1 protein (16), consistent with the observation that the C-terminal two-thirds of the Rad10 polypeptide is evolutionarily conserved (183, 184). The C-terminal 29% of Rad1 protein is necessary and largely sufficient for interaction with Rad10 in vitro. This region is also evolutionarily conserved.

The interaction between the Rad1 and Rad10 proteins in vitro has been confirmed by using an experimental system called the two-hybrid system (Fig. 6–9), which was designed for detecting protein-protein interactions in vivo (12, 46). It exploits the separability of the yeast Gal4 transcriptional activator protein into two functional domains: a DNA-binding domain that specifically recognizes the upstream activation sequence UAS$_G$ (a promoter element required for transcription of *GAL* genes in yeast cells), and a transcriptional activation domain. To determine whether two proteins of interest interact in vivo, the gene encoding one of them is constructed as a *GAL4* DNA-binding fusion gene carried on one plasmid and the other is constructed as a *GAL4* transcriptional activation fusion gene carried on a second plasmid. Following the introduction of these plasmids into yeast cells, interaction between the two test proteins can reconstitute functional Gal4 activity, which can be assayed by monitoring transcription from a reporter gene such as the *E. coli lacZ*$^+$ gene which encodes β-galactosidase activity. By using this system it can be shown that the Rad1 and Rad10 proteins interact in vivo (Fig. 6–10).

More-precise localization of the Rad1-Rad10 interacting domains has shown that the Rad10-binding domain of Rad1 protein is situated between amino acids 809 and 997 and that the Rad1-binding domain of Rad10 protein is located between amino acids 90 and 210 (Fig. 6–11) (12). Both of these regions are evolutionarily conserved. With respect to *RAD1*, a homologous gene from the fission yeast *S. pombe* designated *rad16*$^+$ (35) shares 59% amino acid identity with the Rad10-binding domain of the *RAD1* gene over a 90-amino-acid stretch and 43% identity over a 53-amino-acid stretch (Fig. 6–11) (12, 35). The homology between these two genes overall is ~31% amino acid identity (35). With respect to Rad10, the region between amino acids 92 and 210 shares ~30% identity and ~50% similarity with the human ERCC1 protein (Fig. 6–11). This region of the Rad10 polypeptide is also 37% identical to the predicted amino acid sequence of the *swi10+* gene of *S. pombe* (Fig. 6–11). There is no significant homology between the N-terminal 91 amino acids of the Rad10 polypeptide and the corresponding regions of the human ERCC1 and *S. pombe* Swi10 proteins (Fig. 6–11).

The interaction domains of both Rad1 and Rad10 proteins identified in these studies are distinctly hydrophobic. Consistent with the notion of hydrophobic interactions between these polypeptides, Rad1-Rad10 complex formation in vitro is stimulated by increasing ionic strength (12). In this regard it may be of significance that a potential leucine-isoleucine zipper motif with the sequence (L-X_6-L-X_6-L-X_6-I-X_6-L-X_6-L) has been identified between amino acids 931 and 966 in the Rad10-binding domain of Rad1 protein (132). Similarly, a potential helix-turn-helix motif and specifically spaced pairs of aromatic residues, both of which may

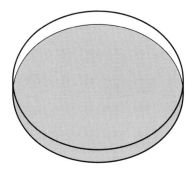

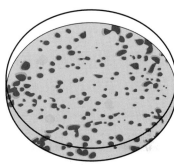

Figure 6–10 Rad1 and Rad10 proteins interact in vivo. A yeast strain carrying a *GAL1-lacZ* fusion gene was transformed with plasmids in which proteins X and Y in Figure 6–9 were encoded by the *RAD1* and *RAD10* genes, respectively. The cells were grown as colonies on agar plates, and colony lifts imprinted on nitrocellulose filters were frozen in liquid nitrogen and then transferred to filter paper soaked in a buffer containing a substrate for β-galactosidase. Following incubation, colony lifts that express β-galactosidase turn a deep blue color (bottom), whereas those transformed with plasmids that encode non-interacting proteins do not (top).

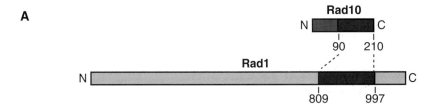

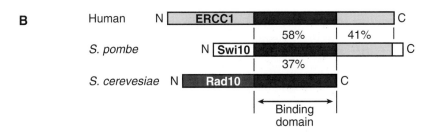

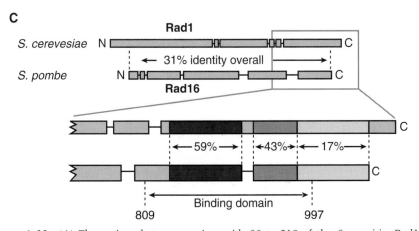

Figure 6–11 (A) The regions between amino acids 90 to 210 of the *S. cerevisiae* Rad10 polypeptide and amino acids 809 to 997 of the Rad1 polypeptide are required for specific interaction between the two polypeptides. (B and C) The Rad1-binding domain of Rad10 protein of *S. cerevisiae* (B) and the Rad10-binding domain of Rad1 protein (C) are evolutionarily conserved in homologous genes in the fission yeast *S. pombe* and in human cells. The percentages refer to amino acid identity in the indicated regions. (*Panel C adapted from Bardwell et al. [12] with permission.*)

constitute protein interaction regions, have been identified in the Rad1-binding domain of the Rad10 polypeptide (138, 183).

The Rad1 and Rad10 Proteins Are Subunits of an Endonuclease Activity

Both Rad1 (171, 180) and Rad10 (15, 169) have been purified to apparent physical homogeneity. Purified Rad10 protein binds to single-stranded DNA and promotes the annealing of homologous single-stranded DNA (169). It is not clear what biological role this latter biochemical function subserves. When present together, the two proteins catalyze the endonucleolytic degradation of M13 bacteriophage DNA (Fig. 6–12) (171, 179). Neither protein alone is endowed with endonuclease activity. The endonuclease has an absolute requirement for divalent cations such as Mg^{2+} and a pH optimum of ~8.5. It is inhibited in the presence of >20 mM NaCl. Efficient labeling (by polynucleotide kinase) of the 5' ends of the products of degradation of M13 bacteriophage DNA requires prior treatment of the DNA with a phosphatase, indicating that the Rad1-Rad10 endonuclease generates 5' phosphate and (presumably) 3' OH termini (179). In addition to M13 bacteriophage DNA, the Rad1-Rad10 endonuclease degrades single-stranded regions present in some supercoiled plasmids (171, 180).

More-refined analyses of the Rad1-Rad10 endonuclease activity by using polymer substrates of defined length and sequence have demonstrated that the endonuclease specifically recognizes the junction between double-stranded DNA and 3'-single-strand tails (13). Single-stranded 49-mer polymers, duplex polymers, and partially duplex polymers with duplex/5'-single-stranded tails are not recognized by the enzyme. Hence, it is likely that the degradation of M13 bacteriophage DNA reflects the recognition of duplex/3'-single-strand junctions in regions of DNA with secondary structure rather than of single-stranded DNA per se. The Rad1-Rad10 complex is a weak endonuclease in vitro, and does not recognize sites of base damage such as pyrimidine dimers in duplex DNA (179). These observations suggest that in living cells the endonuclease activity is modulated and

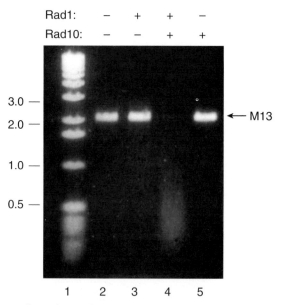

Figure 6–12 The Rad1-Rad complex is an endonuclease that degrades M13 circular DNA, resulting in small oligonucleotide products that form a smear during agarose gel electrophoresis (lane 4). These incisions are presumed to occur at single-strand/duplex junctions generated by transient secondary structure in the M13 molecules. Electrophoresis was performed in the presence of ethidium bromide so that the DNA could be visualized by UV transillumination. The numbers on the left are size markers in kilobases.

presumably stimulated by other protein-protein and/or specific protein-DNA interactions.

The specificity of the Rad1-Rad10 endonuclease for duplex/3'-single-strand junctions suggests a plausible model for its participation in DNA damage-specific incision during nucleotide excision repair (13). The model requires the elaboration of Y-shaped duplex/single-strand junctions flanking sites of base damage. Such junctions would mark the limits of a region of localized unwinding of the DNA duplex conceptually similar to the denaturation "bubbles" postulated to occur during transcription and replication (Fig. 6–13). The specific recognition of the duplex/3'-single-strand junction 5' to a site of base damage is expected to constitute a substrate for the Rad1-Rad10 endonuclease. The polypeptide product of the *RAD2* gene, which is also known to be a single-stranded endonuclease (see below), is an attractive candidate for a second junction-specific endonuclease endowed with duplex/5'-single-strand polarity. As such, this enzyme would be expected to recognize the duplex/5'-single-strand junction 3' to a site of base damage (Fig. 6–13). At the time of writing, this junctional specificity has not been reported for the yeast Rad2 protein. However, purified homologous human XPG protein has been shown to have such substrate specificity in vitro (see chapter 8). To the extent that *rad1*, *rad10*, and *rad2* mutants have been studied, there is no evidence that they manifest a residual capacity for incision of DNA. Hence, the mechanism of nucleotide excision repair in vivo presumably provides for coordinated incisions at both Y junctions in the model substrate shown in Fig. 6–13, such that Rad1-Rad10-mediated incisions do not occur in the absence of functional Rad2 protein and vice versa.

There is no formal proof of a bimodal mechanism during damage-specific incision in yeast cells. However, such a mechanism is likely, since it has been shown to operate in human cells (174) (see chapter 8) in addition to *E. coli*. If indeed the yeast Rad1-Rad10 endonuclease turns out to be the functional homolog of the *E. coli* UvrC protein, it is remarkable that there is no hint of amino acid sequence conservation between the Rad1, Rad10, and UvrC proteins. However, as will be noted in chapter 8, there is limited amino acid sequence homology between UvrC and the human ERCC1 protein, the homolog of Rad10.

The junction-specific endonuclease activity of the Rad1-Rad10 protein complex also accommodates its known role in recombination between repeated sequences (13). A model plasmid substrate has been generated containing repeated copies of the *E. coli lacZ*⁺ gene, one of which contains a 117-bp cutting site for the mating-type endonuclease designated HO (47). HO endonuclease-induced double-strand breakage therefore introduces about 60 bp of 3'-terminal DNA into one of the *lacZ* sequences which is not present in the other sequence. Mutants defective in the *RAD1* gene are not able to effect recombination between these repeated *lacZ* sequences (47). However, when both *lacZ* sequences contain HO endonuclease cutting sites (i.e., when complete homology between the repeated sequences is restored), recombination is effected in both the wild type and *rad1* mutants. A recombination model of HO endonuclease-mediated cleavage followed by single-strand assimilation accounts for the generation of a duplex/3'-single-strand junction generated by the ~60 bp of nonhomologous sequence. This junction is presumably recognized by the Rad1-Rad10 endonuclease, resulting in the removal of nonhomologous DNA (13, 47).

THE *RAD2* GENE AND Rad2 PROTEIN

Translation of the *RAD2* coding region shows that it, too, is expected to encode an acidic protein with a calculated net charge of − 35. The C-terminal tail of the Rad2 polypeptide is rich in basic amino acids; 18 of the last 41 amino acids (~44%) are either Arg or Lys (48). This region of the polypeptide is important for its function, since deletion of the terminal 78 codons of the gene inactivates its ability to complement *rad2* mutants (70). As discussed below, Rad2 protein is known to interact with both Ssl2 and Tfb1 proteins, encoded respectively by the essential *SSL2* and *TFB1* genes referred to previously (11). The Rad2 amino acid sequence also shows

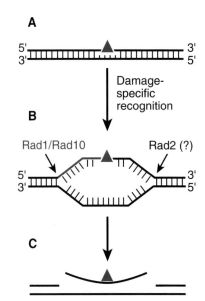

Figure 6–13 Model of bimodal incision of damaged DNA during nucleotide excision repair in *S. cerevisiae*. The model postulates that following the specific recognition of base damage (A) by the nucleotide excision repair machinery, a denaturation "bubble" incorporating the site of base damage is generated (B). The Rad1-Rad10 endonuclease recognizes the duplex/3'-single-strand junction in the bubble structure 5' to the site of base damage, while the Rad2 protein is presumed to recognize the duplex/5'-single-strand junction located 3' to the site of base damage. The oligonucleotide generated by this bimodal incision reaction is subsequently excised (C).

the presence of a bipartite motif believed to represent a nuclear localization signal (43) (Table 6–4).

The *RAD2* gene is represented in human cells by a homologous gene designated *XPG*, which has been implicated in genetic complementation group G of the disease XP (123, 150) (see chapters 8 and 14). The regions of homology are confined largely to two domains, one located near the N termini and one located near the C termini of the yeast and human translated sequences (Fig. 6–14). The N-terminal domain is about 100 amino acids long, and the C-terminal domain is about 140 amino acids long (Fig. 6–14). These two domains are also conserved in homologous genes from the fission yeast *S. pombe* (see below) and from the frog *Xenopus laevis*. Thus, it is likely that these domains are important for the function of Rad2 protein.

These two domains are also conserved in a much smaller gene present in the genome of both *S. cerevisiae* and *S. pombe* (Fig. 6–14) (36, 150). The gene from *S. cerevisiae* was originally identified as an unassigned ORF during the systematic sequencing of chromosome XI from this organism (75). This ORF (designated *YKL510*) can potentially encode a polypeptide of 382 amino acids. Deletion of the *S. cerevisiae* gene (independently designated as *RAD27* [136] and *RTH1* [for RAD two homolog] [132]) confers modest sensitivity to UV radiation (136). In combination with a mutation in the sequence-related *RAD2* gene, the *rad27* deletion mutant yields further sensitivity to UV radiation, suggesting that the *RAD27* gene product participates in a cellular response(s) to DNA damage other than nucleotide excision repair (136). Epistasis analysis has placed this gene in the *RAD6* epistasis group (136). Consistent with this view, the *rad27* mutant manifests marked sensitivity to killing by the alkylating agent methyl methanesulfonate (MMS) (136). However, the mutant is not unusually sensitive to ionizing radiation (132, 136), suggesting that it is not sensitive to DNA strand breaks per se. At present the precise biological role of *RAD27* in DNA repair is unclear. However, there are indications that the product of this gene is a nuclease that participates in DNA replication (136). Consistent with this view, a *rad27* deletion mutant is temperature sensitive for growth and undergoes cell cycle arrest in the G_2 phase of the cell cycle (136). Additionally, the *RAD27* gene is cell cycle regulated and undergoes optimal expression in early S phase (136). Hence, *RAD27* apparently has no direct functional role in nucleotide excision repair in *S. cerevisiae*, and the regions of amino acid homology between the Rad2 and Rad27 proteins presumably reflect the presence of domains that are important for related but distinct nuclease activities in these two polypeptides (136).

The *S. pombe* homolog of *RAD27* is called *rad2*+ (not to be confused with the

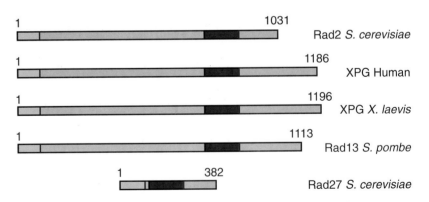

Figure 6–14 Rad2 family of proteins. The members of this family include the Rad2 protein of *S. cerevisiae*, the human and *X. laevis* XPG proteins, the Rad13 protein of *S. pombe*, and the Rad27 protein of *S. cerevisiae*. All five proteins have two motifs in common. The N-terminal motif is ~100 amino acids long and represents overall 21.9% amino acid identity and 56.2% amino acid similarity. The more C-terminal motif is ~140 amino acids in length and represents 21.0% amino acid identity and 52.2% amino acid similarity overall. (*Adapted from Scherly et al. [150] with permission.*)

RAD2 gene of *S. cerevisiae*). Like *RAD27* of *S. cerevisiae*, *rad2+* of *S. pombe* does not appear to participate exclusively (if at all) in nucleotide excision repair. However, *rad2* mutants are defective in the excision of photoproducts from DNA exposed to UV radiation (103), suggesting the possible existence of a second nucleotide excision repair pathway in this organism. The *S. pombe rad2* mutant shows the additional interesting phenotype of chromosomal instability (114), a phenotype that is shared with the *S. cerevisiae rad27* deletion mutant (136).

Human (114) and murine (63) homologs of the *S. cerevisiae/S. pombe RAD27/ rad2+* gene have been isolated. The human gene is able to complement the UV radiation sensitivity of the *S. pombe rad2* mutant (114). Additionally, the polypeptide encoded by the murine gene has a structure-specific nuclease activity that is identical to that previously described for a purified murine protein called FEN-1 (63). Indeed, it is now evident that the murine *FEN-1* gene, the *S. cerevisiae RAD27/ RTH1* gene, the *S. pombe rad2+* gene, and the human equivalent of *rad2+* are homologous. The FEN-1 protein cleaves duplex DNA molecules containing 5′-single-stranded overlapping flaps at sites of nicks (63). Such intermediates may be generated during particular modes of recombination and/or repair in cells. Purified FEN-1 protein also has a $5′ \rightarrow 3′$ exonuclease activity (63). Both the structure-specific endonuclease activity and the exonuclease activity are evident in extracts of *E. coli* cells transformed with a plasmid expressing the cloned *RAD27* gene (63). As is the case with the *S. cerevisiae RAD27* gene product, there are indications that murine FEN-1 protein is related (if not identical to) a protein from human cells that is strongly implicated in semiconservative DNA replication (63a, 141a).

Rad2 Protein Is Also a DNA Endonuclease

Like the Rad1 and Rad10 proteins, Rad2 protein of *S. cerevisiae* has been purified to apparent physical homogeneity (60). The properties of this endonuclease in isolation are very similar to those of the Rad1-Rad10 endonuclease. The Rad2 endonuclease specifically cleaves single-stranded bacteriophage DNA in the presence of Mg^{2+}. The enzyme has a pH optimum of 8.0 and appears to leave 3′ OH and 5′ phosphate groups at sites of endonucleolytic cleavage (60) (Fig. 6–15). Rad2 protein is an attractive candidate for the second endonuclease alluded to above with respect to the postulated bimodal damage-specific incision reaction in yeast cells (Fig. 6–13). Indeed as indicated above, purified XPG protein, the human homolog of Rad2 protein, has been shown to be a structure-specific endonuclease that cleaves DNA at duplex/5′-single-strand junctions (122) (see chapter 8). This substrate specificity would provide for a role in mediating incision 3′ to sites of base damage during nucleotide excision repair (Fig. 6–13).

If the Rad2 endonuclease indeed catalyzes damage-specific incision on one side of base damage, as suggested above (Fig. 6–13) and the Rad1-Rad10 endonuclease catalyzes the 5′ incision, one might expect that strains carrying mutations in just the *RAD1*, *RAD10*, or *RAD2* gene would show residual leakiness for incision during the postirradiation incubation of cells. As argued above with respect to the Rad1-Rad10 endonuclease, the fact that such residual incisions have not been observed in single gene mutants suggests that a defect in either one of the endonuclease activities precludes the action of the second. Hence, in contrast to the situation in *E. coli*, bimodal incision during nucleotide excision repair in yeast cells may be a concerted reaction, even though two different proteins are involved.

The *RAD2* gene has been tailored by genetic engineering to mimic (but not precisely duplicate) the *RAD27* homolog discussed above (63). The protein encoded by this tailored Rad2 polypeptide acquires a new endonuclease activity that resembles the structure-specific activity of the murine FEN-1 endonuclease (63).

The *RAD2* Gene Is DNA Damage Inducible

The limited constitutive expression of the many *RAD* genes involved in nucleotide excision repair prompted an examination of their inducible expression analogous to that of some of the *E. coli uvr* genes discussed in chapter 5. *RAD2-lacZ* (Fig. 6–16), *RAD7-lacZ*, and *RAD23-lacZ* fusion genes express increased levels of β-galactosidase following the exposure of cells to UV radiation (98, 142). The

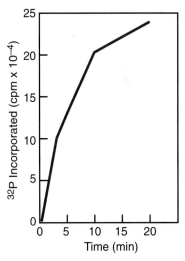

Figure 6–15 The Rad2 endonuclease generates 5′ phosphate and 3′ OH termini at sites of cleavage in single stranded bacteriophage DNA. The number of these increase as a function of the time of incubation with Rad2 protein. The formation of 5′ phosphate terminal residues is inferred from the fact that T4 polynucleotide kinase-mediated incorporation of ^{32}P at the 5′ termini of the oligonucleotide products of Rad2 degradation requires prior incubation of these products with alkaline phosphatase. (*Adapted from Habraken et al. [60] with permission.*)

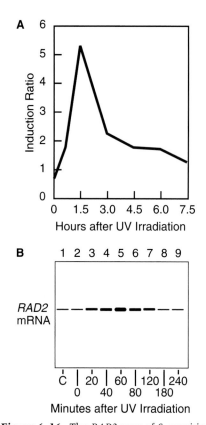

Figure 6–16 The *RAD2* gene of *S. cerevisiae* is inducible by exposure of cells to DNA-damaging agents. Induction can be demonstrated by measuring β-galactosidase activity expressed from a *RAD2-lacZ* fusion gene integrated into the genome of yeast cells (A) or by measuring the level of *RAD2* mRNA in cells by Northern (RNA) analysis (B). (*Adapted from Robinson et al. [142] and Madura and Prakash [98].*)

RAD1, RAD3, RAD4, and *RAD10* genes are not DNA damage inducible. *RAD14, SSL1, SSL2,* and *TFB1* have not been specifically examined in this regard. These results have been confirmed by examining the steady-state levels of *RAD2* mRNA in uninduced and induced cells (161). The steady-state levels of *RAD2* transcripts increase as much as sixfold after treatment with a variety of DNA-damaging agents (Fig. 6–16), including some that do not confer sensitivity to *rad2* mutants, (e.g., γ rays). This suggests that the inducing signal for enhanced transcription does not stem from a particular type of DNA damage.

Deletion analysis of the promoter regions of the *RAD2* gene has demonstrated that induction by DNA damage is positively regulated (161). Four A + T-rich tracts are required for constitutive expression of this weakly expressed gene. Deletion of these elements results in loss of UV inducibility and reduced constitutive expression. Two other sequences, which share the consensus motif AGGNATTPuAAA, located ~70 and ~140 bp upstream of the translational start site in *RAD2* have been identified as *cis*-acting regions that are required for DNA damage-specific regulation of *RAD2* and are designated damage-responsive elements 1 and 2 (DRE1 and DRE2) (161). These are presumably analogous to yeast UASs identified in many positively regulated yeast promoters. Deletion of either of the DRE sequences results in defective or reduced inducibility following UV radiation, and deletion of DRE1 also leads to reduced constitutive expression of *RAD2* (161).

Gel retardation studies have demonstrated the specific binding of proteins to sequences in DRE1 and DRE2 and to a consensus A + T-rich sequence representative of the four A + T blocks described above (160). DNA-protein complex formation is observed under constitutive conditions and is increased in amount in cells exposed to DNA damage. Increased complex formation is inhibited in the presence of cycloheximide, suggesting that at least one step in the activation of enhanced transcription requires de novo protein synthesis (160).

The basal transcription of the *RAD2* gene is not cell cycle regulated (99, 161). However, induction of *RAD2* is dramatically reduced if cells are held in the stationary phase or are arrested in the G₁ phase of the cell cycle following the exposure of cells to UV radiation. Hence, induction apparently requires a stage of the cell cycle outside the G₁ phase (161). Increased levels of *RAD2* transcripts are observed during meiosis (99, 161). However, no *cis*-acting sequences which are required for this increase have been identified (161), suggesting that this reflects a general increase in transcriptional activity rather than a specifically regulated phenomenon. Additionally, there is no known meiotic phenotype of *rad2* disruption mutants.

Failure to induce the *RAD2* gene following deletion of DRE1 results in increased sensitivity to UV radiation exclusively in the G₁/S phase of the cell cycle (160). This mutant phenotype provides further support for the role of DRE1 defined by deletion analysis. Additionally, the observation of increased UV sensitivity in the G₁/S phase of the cycle, coupled with the observation that induction of *RAD2* is ablated in cells arrested in the G₁ phase, suggests that increased expression of *RAD2* may be necessary for the repair of DNA damage prior to the onset of semiconservative DNA synthesis and that an event(s) in early S phase may trigger increased expression of the gene.

In addition to *RAD2*, a number of other yeast genes are responsive to a signal(s) associated with DNA damage. Two of these DNA damage-inducible genes (*RAD7* and *RAD23*) are known to be involved in nucleotide excision repair, and their regulation is discussed below. Several others are known to be involved in other DNA repair modes, and still others have no obvious relationship to DNA repair at all. The induction of these genes after exposure of cells to DNA damage is discussed in chapter 13.

THE *RAD3, SSL2, SSL1,* AND *TFB1* GENES

Like the *RAD1* and *RAD10* genes, which are functionally related, the *RAD3, SSL2, SSL1,* and *TFB1* genes are considered collectively, since, as indicated on several occasions, all of them are essential genes that have additional roles in nucleotide excision repair. As discussed below, their essential function involves their participation in RNA polymerase II-dependent basal transcription. Hence, the polypeptides encoded by these four genes are bifunctional proteins.

The *RAD3* Gene Encodes a DNA Helicase Activity

The *RAD3* ORF comprises 778 codons and is expected to encode a polypeptide of ~89.7 kDa (Table 6–3) (119, 137). The amino acid sequence of the translated *RAD3* gene reveals the presence of a consensus nucleotide binding motif near the N-terminus (Fig. 6–17), characteristic of many proteins that bind adenine nucleotides and catalyze the hydrolysis of nucleoside triphosphates, particularly ATP (48, 49–51, 52, 132). The Rad3 sequence additionally shows the presence of other conserved regions identified as helicase consensus motifs (165). These observations provided early clues that Rad3 protein may be a DNA helicase.

The predicted DNA helicase activity of Rad3 protein has been borne out by direct studies. The purified protein (62, 170) catalyzes the hydrolysis of ATP (or dATP) to ADP (or dADP) and P_i in the presence of Mg^{2+} (or Mn^{2+}) and single-stranded DNA. Double-stranded DNA does not support ATP hydrolysis. The ATPase activity has an extremely narrow pH dependence, with a sharp optimum at pH 5.6 (Fig. 6–18). Very little ATPase activity is detectable at pHs below 5.0 or above 6.0, thereby distinguishing it from many other yeast ATPases. The K_m for ATP hydrolysis has been measured at ~50 μM (62, 170).

The hydrolysis of ATP or dATP by purified Rad3 protein drives a DNA helicase activity with the same narrow pH dependence (62, 168a) (Fig. 6–18). The Rad3 DNA helicase requires single-stranded regions to initiate unwinding of duplex DNA. Both circular and linear partially duplex molecules are used as substrates. The enzyme unwinds duplex DNA unidirectionally, with $5' \rightarrow 3'$ polarity relative to the single strand to which it is bound (Fig. 6–18), the same polarity as that of the DNA helicase function of the *E. coli* $(UvrA)_2(UvrB)_1$ complex (see chapter 5). However, in contrast to the *E. coli* UvrA/B helicase, which can displace only short oligonucleotides from partial duplex substrates, Rad3 protein can unwind duplex regions as long as 850 bp (168a). The enzyme can unwind oligonucleotides from gapped regions as small as 21 nucleotides. However, single stranded gaps that are only 4 nucleotides long are not used as substrates. Therefore, the minimum

Rad3	*(S. cerev.)*	G	T	G	K	T	V	S	L	L	S	=	V	I	F	D	E	A	H	N	I	D
XPD	*(Human)*	G	T	G	K	T	V	S	L	L	A	=	V	V	F	D	E	A	H	N	I	D
Rad15	*(S. pombe)*	G	T	G	K	T	I	S	L	L	A	=	V	V	F	D	E	A	H	N	I	D
Ssl2	*(S. cerev.)*	G	A	G	K	T	L	V	G	I	T	=	I	I	L	D	E	V	H	V	V	P
XPB	*(Human)*	G	A	G	K	S	L	V	G	V	T	=	M	I	L	D	E	V	H	T	I	P
haywire	*(Drosoph.)*	G	A	G	K	S	L	V	G	V	T	=	M	V	L	D	E	V	H	T	I	P
Rad16	*(S. cerev.)*	G	M	G	K	T	I	Q	T	I	A	=	V	I	L	D	E	A	H	N	I	K
CSB	*(Human)*	G	L	G	K	T	I	Q	I	I	A	=	V	I	L	D	E	G	H	K	I	R
RadH	*(S. cerev.)*	G	T	G	K	T	K	V	L	T	S	=	V	L	V	D	E	F	Q	D	T	N
Rad54	*(S. cerev.)*	G	L	G	K	T	L	Q	T	I	S	=	M	L	A	D	E	G	H	R	L	K
Rad5	*(S. cerev.)*	G	L	G	K	T	V	A	A	Y	S	=	I	I	I	D	E	G	H	N	I	R

Figure 6–17 The predicted amino acid sequence of the Rad3 polypeptide reveals a conserved Walker type A (GKT/S) amino acid sequence motif which is shared with other known (XPD, XPB, RadH) or suspected (Rad16, Ssl2, haywire, CSB, Rad54, Rad5) DNA helicases, all of which are involved in cellular responses to DNA damage. The XPD and Rad15 proteins are the human and *S. pombe* homologs, respectively, of Rad3. The XPB and haywire polypeptides are the human and *Drosophila* homologs, respectively, of Ssl2. All of the proteins shown also share a consensus DE–H/Q sequence motif (so-called DEAD box), which is also characteristic of many DNA helicases.

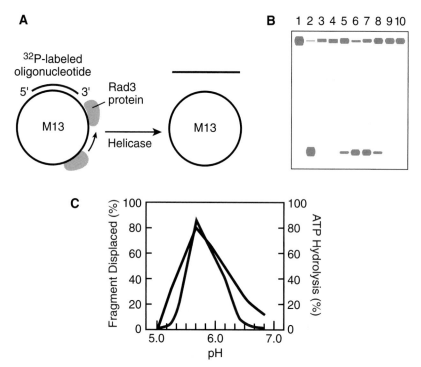

Figure 6–18 (A) The Rad3 DNA helicase activity can be demonstrated by measuring the displacement of a radiolabeled oligonucleotide annealed to complementary single-stranded circular DNA. (B and C) Rad3 protein displaces such oligonucleotides with a strict 5′ → 3′ polarity with respect to the strand to which it is bound (the single-stranded circle) and with a pronounced pH dependence which is optimum at ~5.6. Panel B is a diagram of an autoradiogram following gel electrophoesis, showing the displacement of the radiolabeled oligonucleotide fragment. In lane 1 no Rad3 protein was added. Lanes 5 to 8 represent reactions at pH 5.3, 5.6, 5.9, and 6.2 respectively. Lane 2 is a control in which the substrate was heat denatured. This pH profile is graphically shown in panel C. The ATPase activity of Rad3 protein has the same pH requirements (C). (*Adapted from Sung et al. [168a] with permission.*)

single-stranded region apparently required for binding of Rad3 protein to DNA is somewhere between 4 and 21 nucleotides. The enzyme does not catalyze unwinding of duplex nicked DNA in vitro (62).

Rad3, Ssl1, Ssl2, and Tfb1 Proteins Are Required for Transcription by RNA Polymerase II

As discussed more fully below, there is persuasive evidence that the DNA helicase function of Rad3 protein is required for its role in nucleotide excision repair. However, the essential function of the *RAD3* gene in yeast cells is explained by the additional involvement of Rad3 protein in the initiation of basal transcription by RNA polymerase II (45). The helicase activity is apparently not involved in subserving the role of Rad3 protein in transcription initiation. At this juncture, a few general comments on transcription in eukaryotes are warranted to provide an intelligible context for discussing the specific involvement of nucleotide excision repair proteins in this process and the particular relationship between transcription and nucleotide excision repair.

For RNA polymerase II to transcribe genes, a basal transcription complex comprising a large number of proteins must be assembled at the promoters of these genes. More than 20 proteins are believed to be involved in the assembly of this complex in yeast cells (31). It remains to be firmly established precisely how this complex is assembled and precisely which polypeptides are involved (31, 37, 172). Regardless, once this complex is correctly assembled transcription by RNA polymerase II can be initiated (31, 37). The process of transcription initiation itself can be dissected into various phases (31, 37, 54), which are not especially relevant to the present discussion. However, it is important to recognize that once transcription by RNA polymerase II is properly initiated, the ensuing process of *transcript elongation* is believed to involve the association of RNA polymerase II with a different set of proteins, the *transcription elongation complex*. Some members of the elongation complex may also be involved in transcriptional initiation (37, 54).

Various named components of the multiprotein basal transcription initiation complex have been purified from yeast cells as individual polypeptides or as stable protein complexes. Additionally, other components have been purified as recombinant proteins following the molecular cloning of genes known to be involved in the process of transcription initiation (Fig. 6–19). Among these is a subcomplex originally called *factor b*, now designated as *TFIIH* (for transcription factor IIH) (Fig. 6–19) (37, 45, 85). Factor TFIIH is the yeast homolog of a human RNA polymerase II transcription initiation complex, also called TFIIH or basal transcription factor 2 (BTF2) (37, 85, 149). Human TFIIH is discussed further in chapter 8, which discusses the association between nucleotide excision repair and transcription in mammalian cells. Of the multiple protein factors that make up the basal transcription machinery in eukaryotes, TFIIH is the focus of particular interest in the present discussion because its component polypeptides (plus at least one other polypeptide that associates with it) are known nucleotide excision repair proteins.

Highly purified TFIIH is composed of five polypeptides, of 85, 75, 55, 50, and 38 kDa (Fig. 6–19). However, there are indications that other polypeptides are intimately associated with these five proteins during transcription initiation in vivo (173). Hence, TFIIH should not be thought of as a functionally discrete entity but, rather, as a particularly stable core complex of polypeptides which remain associated in vitro during multiple fractionation steps. The 85-kDa polypeptide is Rad3 protein (45), the 75-kDa polypeptide is the product of the *TFB1* gene (53), and the 50-kDa polypeptide is the product of the essential *SSL1* gene (45, 195) (Fig. 6–19). As discussed earlier in the chapter, the 55- and 38-kDa subunits are encoded by the *TFB2* and *TFB3* genes, respectively. The circumstantial evidence that these proteins are also involved in nucleotide excision repair is heavily influenced by the knowledge that all three of the other components of core TFIIH are involved. Hence, the essential function of all of these genes is presumably their requirement for transcription in yeast cells. Consistent with this conclusion, temperature-sensitive *rad3* (57) and *ssl2* (*rad25*) (134) mutants have been shown to be specifically defective in RNA polymerase II-dependent transcription.

Like *SSL1*, the essential gene *SSL2* was originally identified through a presumed role in translation initiation. It should be noted parenthetically that the

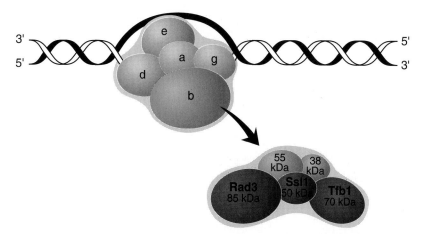

Figure 6–19 RNA polymerase II basal transcription in *S. cerevisiae* requires multiple proteins designated factors a, b, d, e, and g. These correspond to the human transcription factors TFIIE, TFIIH, TFIID, TFIIB, and TFIIF, respectively. Yeast factor b (core TFIIH) consists of five subunits of 85, 70, 55, 50, and 38 kDa. The 85-, 70-, and 50-kDa subunits are encoded by the *RAD3*, *TFB1* and *SSL1* genes, respectively.

translational role of the *SSL2* and *SSL1* gene products is still undefined. While not an integral component of highly purified core TFIIH, Ssl2 protein binds to purified TFIIH in vitro (Fig. 6–20) (45) and also specifically interacts with Rad3 protein (14). These observations, coupled with the essential function of the *SSL2* gene and the fact that conditional-lethal *ssl2* mutants are defective in transcription (134), indicate that Ssl2 protein is required for transcription initiation. Indeed, this has been independently shown by using a defined in vitro transcription system (172, 173). Therefore, it is probable that, like Rad3, Ssl1, and Tfb1 proteins, Ssl2 protein participates in transcription initiation as a component of TFIIH but is not as tightly associated as the other subunits of this multiprotein complex and dissociates during purification of the complex.

Earlier in the chapter, we indicated that Rad3, Ssl2, Ssl1, and Tfb1 proteins have all been shown to participate directly in the process of nucleotide excision repair. This evidence stems both from studies of the repair of specific genes (see below) in viable *ssl2* mutants (176) and from in vitro studies with a cell-free yeast system, (187, 188). In the latter system defective repair synthesis of damaged plasmid DNA can be complemented in extracts of viable *rad3* (Fig. 6–21), *ssl1*, *ssl2*, and *tfb1* mutants by the addition of purified core TFIIH (factor b) or of TFIIH complexed with Ssl2 protein (188). It has additionally been shown that in *rad3* mutant extracts, defective nucleotide excision repair cannot be corrected by the addition of a large excess of purified Rad3 protein (188) (Fig. 6–21). Thus, it appears that Rad3, Ssl2, Ssl1, and Tfb1 proteins (as well as the proteins encoded by the *TFB2* and *TFB3* genes) operate in the biochemistry of nucleotide excision repair as part of a stable multiprotein complex which is common both to repair and transcription. There is evidence that this complex forms the core of an even larger collection of proteins, which we refer to as the *nucleotide excision repairosome* (see below).

SSL2 (RAD25) Also Encodes DNA Helicase Activity

The *SSL2* gene can encode a polypeptide with an expected size of ~95 kDa (Table 6–3). The translated amino acid sequence of the *SSL2* gene suggests that it also encodes a DNA helicase, and in fact both purified Ssl2 protein (57a) and the

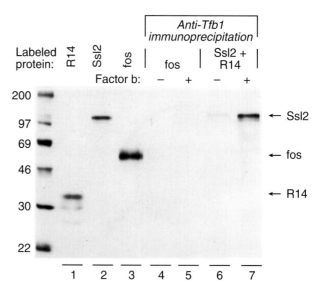

Figure 6–20 Ssl2 protein interacts with purified factor b (core TFIIH) in vitro. In vitro translated ^{35}S-labeled Rad14, Ssl2, and Fos proteins were immunoprecipitated with antibodies against Tfb1 protein (a component of factor b) (lanes 4 to 7) in the presence (+) or absence (−) of purified factor b. The immunoprecipitates were harvested and analyzed by SDS-PAGE, and an autoradiogram of the gel was prepared. Note that Ssl2 protein coimmunoprecipitates with factor b but Rad14 and Fos proteins do not. Lanes 1 to 3 show the electrophoretic positions of in vitro translated Rad14, Ssl2, and Fos proteins, respectively. (*Adapted from Feaver et al. [45] with permission.*)

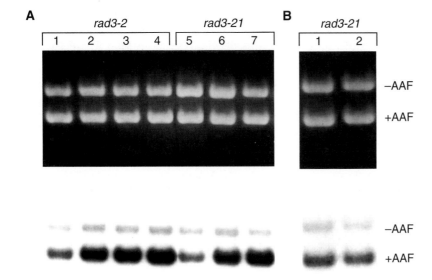

Figure 6–21 Purified core TFIIH (factor b) (lanes 2 to 4 and 6 and 7 in panel A), but not purified Rad3 protein itself (lane 2 in panel B), corrects defective nucleotide excision repair in extracts from two different *rad3* mutants. See Fig. 6–6 for details of the experimental system used for measuring nucleotide excision repair in cell extracts. (*Adapted from Wang et al. [188] with permission.*)

homologous human XPB protein (146, 149) (see chapter 8) have been shown to be DNA helicases in vitro. The polarity of both helicases is 3' → 5' with respect to the strand to which the helicase is bound, the opposite of that of Rad3 protein.

Roles of the Rad3 and Ssl2 Helicases in Transcription and in Nucleotide Excision Repair

Substitution of the Lys codon with Arg in the highly conserved Gly-Lys-Thr(Ser) codon sequence in the nucleotide-binding motif of the *SSL2* gene is lethal (126). Hence, the DNA helicase activity of Ssl2 protein is apparently essential and is probably required for transcription initiation in yeast cells. The fact that the 3' → 5' polarity of the Ssl2 helicase is the same as that of the template strand during transcription suggests that this helicase may be involved in DNA unwinding during transcription initiation. However, the specific role of the Ssl2 helicase during transcription initiation remains to be experimentally determined. It also remains to be established whether the helicase function of Ssl2 protein is required for its function in nucleotide excision repair. In this regard it has been demonstrated that certain plasmid-borne genes that encode Ssl2-GAL4 fusion proteins can complement the nucleotide excision repair defect but not the death of *ssl2* mutants (14). Hence, the repair and essential functions of the *SSL2* gene appear to be separable. On the other hand, studies on the human XPB homolog of Ssl2 have shown that mutations engineered into various helicase motifs of the *XPB* gene inactivate the ability of the gene to support nucleotide excision repair in rodent cell lines that are defective in the rodent equivalent of *XPB* (*ERCC3*) (97).

The role of Rad3 protein in transcription is unknown. Conceivably, the protein is simply an indispensable structural component of TFIIH. While discussing Rad3 protein, it should be noted that in addition to the role of the Rad3 helicase activity in nucleotide excision repair (which is discussed in greater detail below), and its essential function in transcription, certain mutations in the *RAD3* gene result in an increased frequency of *spontaneous mutagenesis, increased mitotic recombination,* and cell death in the presence of mutations in other *RAD* genes which are required for the repair of double-strand breaks in yeast DNA (73, 102, 106). This combination of properties is referred to as the Rem (for recombination/mutation) phenotype of *rad3* mutants (52). Mutations in the well-characterized *rem1-1* and *rem1-2* alleles have been mapped to codons 237 and 661, respectively, in the Rad3 polypeptide, and these alleles have been renamed *rad3-101* and *rad3-102*, respectively (107). The remaining discussion in this chapter will focus heavily on the roles of Rad3 protein in nucleotide excision repair and in basal transcription by RNA polymerase. However, the reader would do well to remember that the as yet unexplained Rem phenotype hints strongly at the participation of Rad3 protein in some other function(s) associated with recombination and possibly DNA replication. In this regard, it may be relevant that a 63-kDa DNA-dependent ATPase activity has been purified from yeast cells and shown to stimulate the activity of yeast DNA polymerase α (see below). This ATPase (which has DNA helicase activity) is not detected in extracts of *rad3* mutant cells (166), suggesting that the ATPase protein may normally interact with Rad3 protein and become labile or inactive in its absence (166). Alternatively, Rad3 protein may be required for expression of the 63-kDa ATPase (166).

In an effort to dissect the multiple functions of the *RAD3* gene, a series of alleles carrying mutations that yield one or more of the phenotypes described above has been isolated and characterized (120, 165, 168). The protein encoded by one of these mutant alleles carrying a substitution of Lys to Arg at codon 48 has been purified (168). As indicated above, this Lys residue is conserved in a consensus sequence that defines the nucleotide binding motif of many DNA helicases (Fig. 6–17). Mutations in this codon of the *RAD3* gene render cells sensitive to killing by UV radiation (165, 168) (Fig. 6–22). The Rad3 protein containing Arg-48 is totally inactivated for both ATPase and DNA helicase activities (168) (Fig. 6–22), although it retains the ability to weakly bind ATP (168). Consistent with these results, *rad3* mutant alleles that carry mutations in other regions identified as consensus motifs for DNA helicases are also highly sensitive to UV radiation (165) (Fig. 6–22). Collectively, these observations suggest that the DNA helicase function of Rad3 protein is indeed required for nucleotide excision repair. Since the *rad3* mutant containing Arg-48 is viable, the absence of detectable DNA helicase activity in extracts of this mutant further indicates that, in contrast to the Ssl2 helicase, the Rad3 helicase activity is not required for the transcription function of the *RAD3* gene. In fact, TFIIH purified from a *rad3* mutant strain carrying a mutation in the nucleotide binding domain (165) (and hence presumably defective in Rad3 helicase activity) is able to support normal transcription in vitro (45).

How might a DNA helicase be involved in nucleotide excision repair of DNA? In light of the evidence that Rad3 protein alone cannot correct defective nucleotide excision repair in *rad3* mutant extracts, whereas TFIIH can (188), any consideration of the role of Rad3 protein in this process should appropriately be considered in the context of the TFIIH complex. This is supported by the observation that at pH 5.6, purified TFIIH retains both the single-stranded DNA-dependent ATPase and DNA helicase activities, which are characteristic of purified Rad3 protein (190). The association of Rad3 protein with TFIIH leads to the obvious assumption that Rad3 protein is obligatorily loaded onto DNA at promoter sites during transcription initiation. We shall consider the implications of this phenomenon for the preferential repair of actively transcribed genes below. However, the observation that viable *rad3* mutant cells are highly UV sensitive and show a total defect in both damage-specific incision and the excision of pyrimidine dimers leads to the suggestion that TFIIH also probably participates in nucleotide excision repair in transcriptionally silent regions of the genome (bulk DNA). This has indeed been shown to be the case for *ssl2* mutants (176).

One possible role for the Rad3 helicase is that it facilitates the release (excision) of damage-containing oligonucleotides following damage-specific incision of DNA. A second possible role of Rad3 protein in nucleotide excision repair stems from observations on the biochemistry of this process in *E. coli*. In this organism, the UvrD protein is involved in nucleotide excision repair (see chapter 5). This protein is also a DNA helicase and plays a role in the displacement and turnover of the UvrABC protein complex, which catalyzes damage-specific incision. A similar function might be entertained for Rad3 protein during nucleotide excision

A

B

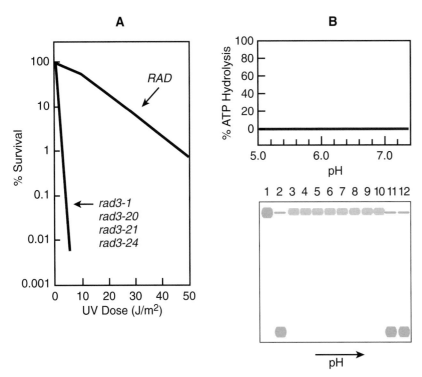

Figure 6–22 (A) Yeast stains with mutations in the *RAD3* gene that encode polypeptides carrying amino acid substitutions in conserved helicase motifs are defective in nucleotide excision repair, as evidenced by a marked sensitivity to killing by UV radiation. The *rad3-20* mutant allele encodes a polypeptide in which Gly-47 in the highly conserved GKT helicase motif (see Fig. 6–17) is replaced with Asp; the *rad3-21* allele encodes a polypeptide in which Lys-48 in the GKT helicase motif is replaced with Glu; the *rad3-1* allele encodes a polypeptide in which Glu-236 in the DEAH helicase motif II (see Fig. 6–17) is replaced with Lys, and the *rad3-24* allele encodes a polypeptide in which Gly-604 (helicase motif VI) is replaced with Arg. (B) Rad3 protein was purified from a mutant strain in which Lys-48 was replaced with Arg. This protein has no detectable ATPase (top) or DNA helicase (bottom, lanes 3 to 10) activity measured over a broad pH range. DNA helicase activity was assayed as shown in Fig. 6–18. Lanes 1, 11, and 12 are control reactions with no Rad3 protein or wild-type Rad3 protein assayed at pH 5.6 and 5.9, respectively. Lane 2 is a heat-denatured control with no Rad3 protein. (*A is adapted from Song et al. [165] and B is from Sung et al. [188] with permission.*)

repair in yeast cells. However, in contrast to the weakly UV radiation-sensitive phenotype of *uvrD* mutants (including deletion mutants), most *rad3* mutants are highly sensitive to UV radiation and appear to be totally defective in damage-specific incision of DNA.

A third possible role for a DNA helicase in nucleotide excision repair, especially in transcriptionally silent regions of the genome, is suggested by other observations on the *E. coli* Uvr proteins. The purified (UvrA)$_2$(UvrB)$_1$ protein complex has also been demonstrated to function as a DNA helicase in vitro. This observation led to the suggestion that the complex may translocate from sites of initial binding (docking) on DNA to sites of base damage by unwinding DNA and that the presence of base damage might lead to the arrest of the helicase, thereby facilitating specific recognition of such damage (55, 124, 185) (see chapter 5). Some of the properties of the Rad3 DNA helicase are consistent with a related damage-specific recognition role. The helicase activity is strongly inhibited by base damage in partially duplex DNA (Fig. 6–23) (62, 115). This inhibition is quite general. In addition to photoproducts generated by UV radiation, various types of chemical damage and even sites of base loss inhibit the helicase (116, 118). Furthermore, the inhibition phenomenon is strand specific (Fig. 6–23). The helicase activity is inhibited when base damage is present on the strand to which Rad3 protein binds and on

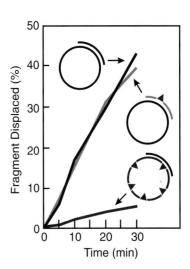

Figure 6–23 The DNA helicase activity of Rad3 protein is inhibited by the presence of base damage in the DNA strand to which it binds and translocates in the 5′ → 3′ direction (red). When the base damage is placed on the opposite strand (grey), no inhibition is observed. (*Adapted from Naegeli et al. [115] with permission.*)

which it translocates in the $5' \rightarrow 3'$ direction. When base damage is confined to the opposite strand, no inhibition is observed. In this regard it is interesting that Rad3 protein can also displace ribonucleotide fragments from partially duplex RNA-DNA hybrids (117). This property is also strand specific and is realized only when the protein is bound to the DNA strand. No helicase or ATPase activity is detected when Rad3 protein is incubated with single-stranded RNA to which a complementary deoxyribonucleotide is annealed (115). Hence, Rad3 protein is not an RNA helicase in vitro, and the ability to unwind RNA-DNA hybrids may simply reflect the fact that the protein is "responsive" to the presence of base damage (or, for RNA, altered chemistry) only on the strand to which it is bound and on which it translocates and not on the opposite strand. The inhibition of DNA helicase activity in the presence of damaged DNA is not unique to Rad3 protein (118). However, although several other DNA helicases have been examined, Rad3 alone manifests the property of strand selectivity. It is therefore tempting to speculate that during the evolution of nucleotide excision repair in yeast cells the Rad3 DNA helicase was usurped to specifically serve as a *DNA damage recognition protein*. In this sense it may be incorrect to view Rad3 protein as a helicase. Perhaps it would be conceptually more accurate to view its function as a DNA "translocase" or scanning protein which searches for base damage by transiently unwinding DNA and "reading" its chemistry.

A final possible role of the Rad3 helicase in nucleotide excision repair is to generate a region of localized denaturation surrounding sites of base damage, perhaps in concert with the Ssl2 DNA helicase, thereby resulting in the formation of a particular DNA conformation that is specifically recognized by other components of the nucleotide excision repair machinery, i.e., the Rad1-Rad10 endonuclease and the Rad2 endonuclease, as suggested above (Fig. 6–13).

The obligatory binding of TFIIH at the promoters of transcriptionally active genes provides a tenable explanation of how Rad3 (TFIIH) protein is loaded onto DNA in transcriptionally active regions (see below). How, though, is this achieved in silent regions of the genome? One possibility is that such loading transpires during DNA replication (particularly during lagging-strand DNA synthesis). Indeed, the coupling of Rad3 protein (TFIIH) to replicational initiation seems intuitively as compelling as its coupling to transcription, although there is as yet no experimental evidence in support of this. Such a scenario is unlikely to be the sole means for the access of TFIIH to DNA, however, since stationary phase cells which are not replicating DNA can carry out extensive nucleotide excision repair.

In summary, on the basis of the known biochemical properties of Rad3 protein, it seems reasonable to suggest that this protein is important for the process of damage-specific recognition during nucleotide excision repair in bulk, i.e., in transcriptionally silent DNA, and may play an additional or different role in nucleotide excision repair in transcriptionally active DNA. As discussed below, other Rad proteins can preferentially recognize at least some forms of base damage in DNA and may contribute to the damage recognition process.

Role of TFIIH during Nucleotide Excision Repair in Transcriptionally Active DNA: Transcriptionally Coupled Repair

The previous discussion focused specifically on accommodating the known helicase function of Rad3 protein in nucleotide excision repair. This section shifts the focus to the accommodation of the larger core transcription initiation complex TFIIH, of which Rad3 is an integral component. There is good evidence that in yeast cells, as in mammalian cells (see chapter 7), transcriptionally active regions of the genome are repaired faster than transcriptionally silent regions and that, as is the case in *E. coli* (see chapter 5), during nucleotide excision repair in transcriptionally active genes the template strand is (in general) repaired faster than the coding strand. The phenomenon of the *preferential repair of transcriptionally active DNA* in eukaryotes may primarily if not exclusively reflect events that are specific to the processing of chromatin structure that is intrinsic to the process of transcription. Hence, one might consider that the coupling of transcription to nucleotide excision repair through the elaboration of a class of proteins required for both processes is a mechanism by which the otherwise relative inaccessibility of the multiprotein repair machinery (repairosome) to sites of base damage in DNA has been solved in genes that are crucial for the normal function of eukaryotic cells. Indeed, in some experimental systems, increased transcription of certain regions of the genome has been predicted on the basis of the observation of increased nucleotide excision repair in these regions (163).

The preferential repair of the template strand over the nontranscribed strand in actively transcribed genes may be thought of as an extension of this phenomenon during which the obligatory loading of nucleotide excision repair proteins (such as TFIIH) onto the transcribed strand provides a kinetic advantage to that strand. Such strand-specific repair has been observed in at least two yeast genes, *GAL7* and *RPB2*. The repair of both pyrimidine dimers and thymine glycols in the former gene has been shown to occur two to three times faster in the template strand than in the nontranscribed strand or in the genome overall (90). Similarly, the template strand of the *RPB2* gene, which encodes the second largest subunit of RNA polymerase II, is repaired faster than the coding strand of this gene (175). This result has been observed both in the chromosomal *RPB2* gene and when the gene is carried in cells on a centromeric yeast plasmid (Fig. 6–24). The specific relationship of this result to transcription was provided by the observation that this kinetic difference is abolished at the restrictive temperature in a mutant that is thermoconditional for RNA polymerase II-mediated transcription (175) (Fig. 6–24). The strand distribution of sunlight-induced mutations in the yeast tRNA gene *SUP4-o* is also consistent with the preferential repair of lesions on the transcribed strand (4).

There are fundamental differences in the precise relationships between transcription and nucleotide excision repair in prokaryotes such as *E. coli* and in eukaryotic cells; these are consistent with the notion that the bifunctional roles (in transcription and repair) of proteins in eukaryotes serve a unique purpose that is not required in prokaryotes. In *E. coli* a single RNA polymerase complex is responsible for catalyzing all transcription, and only one additional protein (sigma factor) is required for initiating basal transcription at promoter sites. As far as is known, neither the RNA polymerase complex nor the sigma factor participates directly in the biochemistry of nucleotide excision repair. Therefore, in prokaryotes transcription initiation (and presumably transcription elongation) does not require the explicit participation of nucleotide excision repair proteins. The linkage between transcription and repair in such organisms does, however, emerge once the process of transcription is arrested by the presence of base damage in template strands, because excision repair of such damage cannot transpire until the nucleotide excision repair machinery is recruited to sites of such damage (see Fig. 5–17). In *E. coli*, and presumably in many other prokaryotes, this is effected through the action of a specific *transcription-repair coupling factor* (see chapter 5). Transcription-repair coupling factor also apparently dissociates the aborted transcript.

On the basis of these considerations, the phrase *transcriptionally coupled DNA repair* has two distinct mechanistic connotations, and one must distinguish between the true coupling of nucleotide excision repair to the process of transcription by the positioning of nucleotide excision repair proteins in the transcription machinery, as apparently occurs in yeast and in mammalian cells, and the recruitment of a transcription-independent nucleotide excision repair apparatus to sites of base damage in the transcribed DNA strand. In eukaryotes both processes may take place, whereas in prokaryotes such as *E. coli* only the latter occurs.

What Happens following the Arrest of Transcription by Base Damage?

The issue of exactly how and at which stage of the entire process transcription becomes coupled to nucleotide excision repair in eukaryotes is more than just a semantic one. As discussed above, there is good experimental evidence for the involvement of the yeast nucleotide excision repair proteins Rad3, Ssl1, Ssl2, Tfb1

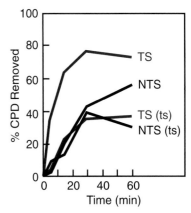

Figure 6–24 The transcribed (template) strand (TS) of the *RBP2* (RNA polymerase II B subunit) gene of *S. cerevisiae* is repaired faster than the non-transcribed (coding) strand (NTS) during incubation of cells for the times indicated after their exposure to UV radiation. (Details of the experimental technique for measuring gene- and strand-specific repair are provided in chapter 7 and Fig. 7-13.) When the same experiment was carried out with a strain with a temperature-sensitive (ts) *rbp2* allele incubated at the restrictive temperature (36°C), the preferential repair of the transcribed strand was abolished. (CPD = cyclobutane pyrimidine dimers) (*Adapted from Sweder and Hanawalt [175].*)

(and probably Tfb2 and Tfb3) in the process of transcription initiation. However, these proteins presumably do not exercise their required roles in nucleotide excision repair until the transcription elongation complex actually encounters sites of base damage in the template DNA strand. The biochemical events that ensue once transcription is initiated and elongation proceeds are poorly understood (37). It is likely that the initiation complex is partially or fully replaced with new proteins (*elongation factors*) that now become active players in the transcription scenario. The discovery that TFIIH and other associating proteins in yeast cells comprise a series of known nucleotide excision repair proteins leads to the (intuitive) assumption that in addition to its role in transcription initiation, TFIIH is coopted to the elongation complex. However, this remains to be experimentally proved. In fact, the results of experiments in the literature at the time of this writing indicate that in yeast cells once transcription is fully initiated and the transcription machinery has successfully cleared the promoter, TFIIH is not required for elongation of the transcript (54). This leads to the (initially counterintuitive) suggestion that TFIIH may in fact dissociate from the transcription complex once elongation proceeds and that the elongation complex actually may contain no nucleotide excision repair proteins. How, then, is nucleotide excision repair effected in the transcribed strand? Two possible scenarios merit consideration. One is that when the elongation complex stalls or arrests at sites of base damage, TFIIH reassociates with the complex and that this is followed by the progressive assembly of the remaining proteins required to constitute a complete, functional repairosome. Alternatively, the entire preassembled repairosome may load onto the template strand at sites of transcriptional arrest. The latter model is supported by evidence for the existence of a large repairosome in yeast extracts, comprising at least factor TFIIH together with bound Ssl2, Rad1, Rad10, Rad2, Rad14, and possibly Rad4 proteins (173) (see below). Transcription-repair coupling factor may play a specific role in guiding the reassociation of TFIIH (or the complete repairosome) with the transcription machinery at this stage.

The counterintuitive nature of the possibility that having achieved a formal coupling to transcription during the initiation phase of this process, nucleotide excision repair proteins may actually "desert" the transcriptional machinery during the phase that they are most needed, is of course a consequence of teleogical reasoning. More considered reasoning suggests that there may in fact be mechanistic parallels between the requirements for initiating transcription at promoters on undamaged templates and the restarting of transcription after polymerase stalling at sites of base damage, both of which involve TFIIH. Hence, the issue of exactly when during transcription nucleotide excision repair is coupled to transcription may indeed be a semantic one.

Aside from the accessibility function alluded to above, the coupling of transcription and nucleotide excision repair in eukaryotic cells may serve other functions. TFIIH may serve as the specific "molecular antenna" which recognizes the presence of base damage in template strands once transcription is arrested and, as suggested above, may constitute the biochemical nucleation site for the assembly of a functional repairosome to effect nucleotide excision repair. This contrasts with the situation in prokaryotes such as *E. coli*, in which the molecular antenna for the recognition of base damage is apparently the stalled RNA polymerase complex itself or a conformation induced by the stalled polymerase-RNA-DNA ternary complex (see chapter 5). Yet another purpose for the coupling of nucleotide excision repair to transcription in eukaryotes derives from indications that TFIIH also participates in nucleotide excision repair in transcriptionally silent regions of the genome. The participation of TFIIH in transcription and in nucleotide excision repair of nontranscribed DNA suggests that one process could limit the rate of the other. If TFIIH had a higher binding affinity for sites of damage than for the transcription initiation complex, the recruitment of TFIIH for nucleotide excision repair in bulk DNA could limit transcription initiation when cells are exposed to DNA damage. This seems a useful way of optimizing the potential for error avoidance during transcription in eukaryotes.

We have suggested that the participation of TFIIH in the resolution of arrested transcription provides a mechanism for the biochemical coupling of nucleotide excision repair to transcription in yeast cells. At present nothing is known about the specific role(s) of individual repair proteins in TFIIH during the preferential repair of the template strand. The ultimate clarification of this role(s) must accommodate the fact that the Rad3 subunit is a DNA helicase with strict $5' \rightarrow 3'$ polarity with respect to the strand to which it is bound. Assuming that TFIIH joins the arrested elongation complex on the template strand, the polarity of its movement is necessarily $3' \rightarrow 5'$ with respect to that strand (Fig. 6–25). Conceivably, once the repairosome is assembled at sites of base damage, the combined action of the $5' \rightarrow 3'$ Rad3 helicase and the $3' \rightarrow 5'$ Ssl2 helicase generates a localized site of dentauration (denuration bubble) that defines precise sites of nicking by the Rad1-Rad10 and Rad2 endonucleases.

SSL1 and *TFB1* Genes and Other Genes Involved in Both Transcription and Nucleotide Excision Repair

At the time of writing, the specific biochemical functions of Ssl1 and Tfb1 proteins in nucleotide excision repair (and for that matter in transcription) are unknown. The *SSL1* gene has an ORF of 461 amino acids (195) that can encode a polypeptide consistent with the 50-kDa size of the Ssl1 subunit of factor TFIIH. Searches of DNA and protein databases have not been informative with respect to a specific biochemical function of Ssl1 protein (195). However, the translated amino acid sequence of the cloned *SSL1* gene reveals the presence of six Cys-X$_2$-Cys, one Cys-X$_3$-Cys, and one Cys-X$_2$-His motifs, all of which are localized to the C-terminal region of the polypeptide (195) (Fig. 6–26). Computer-assisted analysis of the cloned *TFB1* gene, which encodes the 77-kDa subunit of factor TFIIH, has been equally uninformative (53).

As indicated above, highly purified TFIIH also comprises stoichiometric amounts of two other polypeptides, 55 and 38 kDa in size (Table 6–3). The cloned *TFB2* and *TFB3* genes for these polypeptides have not yet been characterized. However, the observation that all the other subunits of TFIIH, as well as the associated protein Ssl2, are encoded by essential genes and participate directly in the process of nucleotide excision repair suggests that these proteins will probably be shown to have similar properties.

Conversion of TFIIH to a Repairosome

Regardless of the precise mechanism by which sites of bulky base damage are "flagged" in the template strand during transcription, the reasonable assump-

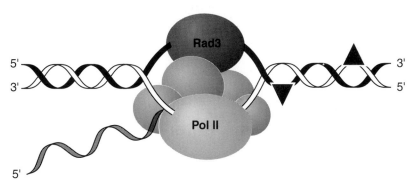

Figure 6–25 The known $5' \rightarrow 3'$ polarity of the Rad3 DNA helicase suggests that if this helicase activity is used during transcriptional elongation, the protein would have to translocate on the nontranscribed rather than the transcribed strand. The figure also suggests that in addition to Rad3 protein, all the other subunits of factor b (TFIIH) are intrinsic components of the transcription elongation complex. However, this has not been experimentally demonstrated. In fact, experiments with human transcription factors (see chapter 8) suggest that TFIIH may not be involved in transcription elongation at all. The red triangles represent sites of base damage that are expected to cause transcriptional arrest.

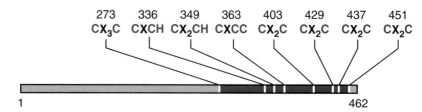

Figure 6–26 The C-terminal region between amino acids 273 and 454 of the Ssl1 polypeptide of *S. cerevisiae* is rich in cysteine residues and Cys-His motifs encountered in zinc finger proteins. The numbers refer to the starting amino acid positions of these potential zinc fingers.

tion is that a functional repairosome is assembled at or near such sites and that nucleotide excision repair takes place by the bimodal paradigm deciphered in *E. coli*. The assembly of this repair complex presumably requires the participation of the other Rad proteins known to be involved in nucleotide excision repair in yeast cells, including at least the Rad1, Rad2, Rad4, Rad10, Rad14, and Ssl2 proteins. Three of these proteins, Rad2, Rad4, and Ssl2, have been shown to interact specifically with TFIIH in vitro (Fig. 6–27) (11). Indeed Rad2 protein binds specifically to Tfb1 and Ssl2 proteins (Fig. 6–27). The particular components of TFIIH that interact with Rad4 protein have not yet been identified. In addition to these elements of repairosome assembly, it is of course known that Rad10 complexes with Rad1 protein in the nucleus of living yeast cells (12). Therefore, the mechanistic framework for complete repairosome assembly is emerging. Studies have actually identified a multiprotein complex in vitro that includes TFIIH, Ssl2 protein, Rad14 protein, Rad2 protein, Rad4 protein, Rad1 protein, and Rad10 protein (173). Such complexes were identified in the absence of DNA damage, leading to the intriguing possibility that in yeast cells the repairosome exists constitutively as an assembled complex.

Clearly much remains to be learned about the biochemical relationships between the proteins that participate in nucleotide excision repair and transcription and those that participate uniquely in the repair or transcription processes. Indeed, recent evidence indicates that core TFIIH comprising the six polypeptides already discussed can be assembled in cells to yield either a TFIIH holocomplex which is the form active in RNA polymerase II transcription initiation, or a repairosome which is active in nucleotide excision repair (173). We have already alluded to the mysteries that currently surround the regulation of the proposed interconversion of core TFIIH.

THE *RAD4* GENE

Little is known about the function of Rad4 protein. From the size of the *RAD4* ORF, the protein has a calculated molecular mass of ~87 kDa (Table 6–3), with

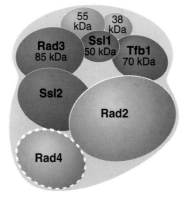

Figure 6–27 The Rad2 and Rad4 proteins of *S. cerevisiae* are known to interact with yeast factor b (TFIIH). Rad2 protein interacts specifically with Tfb1 and Ssl2, proteins. No specific protein-protein interactions have been identified for Rad4. Hence, the polypeptide is circled with a dotted line. The resulting complex comprising Rad3, Ssl1, Tfb1, Rad2, Ssl2 and Rad4 proteins, together with the genetically unidentified 55- and 38-kDa subunits, may form an eight-member repairosome intermediate for nucleotide excision repair in yeast cells.

no distinctive features that emerge from the amino acid sequence. A region of the Rad4 polypeptide which spans ~370 amino acids near the C terminus shares about 25% amino acid sequence identity and about 42% amino acid sequence similarity with a region located near the C terminus of the translated human and *Drosophila* *XPC* genes (Fig. 6–28) (69, 91). The human *XPC* gene corrects the phenotype of cells from genetic complementation group C of the human hereditary excision repair-defective disease XP (91) (see chapters 8 and 14). A comparison of the translated amino acid sequences of the *XPC*^DM, *RAD4*, and *XPC* genes reveals ~19% amino acid identity and ~42% amino acid sequence similarity among the three proteins in this region (69). Several UV radiation-sensitive *rad4* mutant alleles have been shown to carry frameshift mutations which are expected to result in truncated polypeptides missing the conserved C-terminal region (39). Hence, this conserved region is probably functionally important.

As indicated previously, Rad4 protein binds to TFIIH in vitro (11). Since the *RAD4* gene is not essential, it is unlikely that this protein is involved in transcription. A more plausible explanation for this interaction is that it represents an intermediate in the conversion of TFIIH to a repairosome. It is not known which particular subunit of TFIIH interacts with Rad4. Conceivably it requires the complete TFIIH complex in order to bind.

THE *RAD14* GENE

Initially it was not recognized that the *RAD14* gene is absolutely required for damage-specific incision of DNA during nucleotide excision repair, because the first *rad14* mutant characterized was only moderately sensitive to UV radiation. However, the availability of the cloned gene facilitated the construction of a *rad14* deletion mutant which turned out to be considerably more UV sensitive, suggesting that *RAD14* is a member of the group of genes which is indispensable for damage-specific incision of DNA (10). In support of this conclusion, the translated amino acid sequence of the *RAD14* gene is homologous with that of a human gene designated *XPA*, which was isolated by functional complementation of UV radiation-sensitive cells from XP genetic complementation group A (see chapter 8) (10). The *XPA* amino acid sequence predicts a polypeptide of ~29 kDa (10) and reveals a zinc finger domain that is conserved in the putative yeast Rad14 polypeptide (Fig. 6–29), suggesting that XPA protein and, by inference, Rad14 protein are DNA binding proteins. This has been confirmed biochemically with purified XPA protein (see chapter 8).

Rad14 protein has been purified to homogeneity (58). The purified protein migrates in denaturing polyacrylamide gels as a doublet with apparent molecular masses of 35 and 36 kDa. This feature is also observed following electrophoresis of the homologous human XPA protein (see chapter 8). The Rad14 protein contains a single zinc atom and binds to UV-irradiated DNA but not to undamaged DNA. The binding to UV-irradiated DNA is unaffected by the removal of pyrimidine dimers by prior enzymatic photoreactivation, suggesting that Rad14 protein specifically recognizes (6-4) photoproducts in UV-irradiated DNA (58). These observations imply a role for Rad14 protein in the process of damage-specific

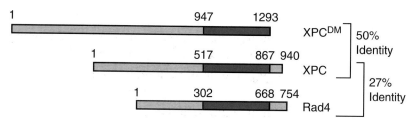

Figure 6–28 The predicted amino acid sequence of the Rad4 polypeptide of *S. cerevisiae* reveals a region of ~340 amino acids near the C terminus which shares homology with similarly located regions in the predicted human XPC and *Drosophila* XPC^DM proteins. (*Adapted from Henning et al. [61] with permission.*)

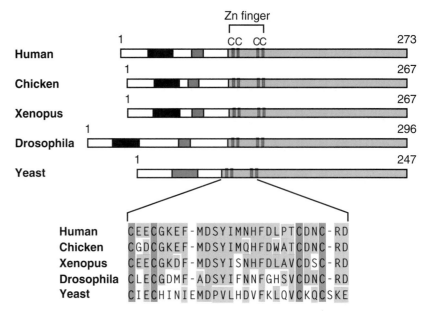

Figure 6–29 The translated *RAD14* gene of *S. cerevisiae* shares an extended region of homology (pink) with the *XPA* gene from human, chicken, *Xenopus*, and *Drosophila* cells. The N-terminal portion of this region of homology is characterized by a zinc finger motif, whose amino acid sequence is shown below. (*Adapted from Shimamoto et al. [157] with permission.*)

recognition. However, it remains to be determined whether the ability of the protein to bind damaged DNA is more general or is restricted to DNA containing (6-4) photoproducts.

Other Genes Involved in Nucleotide Excision Repair

THE *RAD7* AND *RAD23* GENES

Cloning the *RAD7*, *RAD16*, *RAD23*, and *RAD24* genes by functional complementation proved considerably more difficult than for the group of genes discussed above, because their limited sensitivity to UV radiation provides a much smaller window for phenotypic correction. The *RAD7* gene is one of at least seven tightly linked genes clustered on the right arm of chromosome X (105). This ~38-kb region of DNA was isolated by chromosome walking from a cloned region and divided into separate fragments (127). One of these was shown to contain *RAD7* by phenotypic correction of a *rad7* deletion mutant (127). The *RAD7* ORF consists of 565 codons and is expected to encode a polypeptide of ~63.7 kDa. Like many of the other Rad proteins involved in nucleotide excision repair, the putative Rad7 polypeptide has a bipartite nuclear transport sequence motif (Table 6–4).

A mutant *RAD7* allele deleted of the amino-terminal 99 amino acids confers only partial correction of *rad7* mutants when present on a low-copy-number plasmid (127). However, if these mutants are additionally defective in the *RAD23* gene, no complementation is observed. Hence, a function(s) of *RAD7* can apparently be partially subserved by *RAD23*. This functional relationship is consistent with evidence suggesting an evolutionary relationship between these two genes. Both are closely linked to the conserved cytochrome oxidase genes *CYC1* (*RAD7*) and *CYC7* (*RAD23*) (105). The regions of the yeast genome containing these genes might represent duplications during evolution. However, *RAD7* and *RAD23* are not homologous genes.

RAD23 is located on chromosome V between the *ANP1* and *CYC7* genes (100). Nucleotide sequence analysis of a region of DNA containing *RAD23* revealed an ORF of 398 codons which could encode an acidic polypeptide (pI, 4.07) of 42.3 kDa (191). The N-terminal 76 amino acids of the translated *RAD23* sequence shows homology with yeast ubiquitin (Fig. 6–30) (191). Several eukaryotic proteins have been found to have ubiquitin-like sequences. It has been suggested that these might function as signal sequences for the degradation of these proteins (77). Examples include the human BAT3 protein, a member of the human major histocompatability complex (8), and a protein called GDX, which is encoded by an X-linked gene in human cells (182). Unlike ubiquitin and other ubiquitinated proteins, in which the ubiquitin moiety is cleaved from the rest of the polypeptide, Rad23 protein and the GDX and BAT3 polypeptides lack a Gly residue at the C terminus of the ubiquitin moiety, which is necessary for this processing. Furthermore, Rad23 protein is very stable in vivo, so there are no indications that the ubiquitin moiety is related to the degradation of the protein. This region of the polypeptide in Rad23 is certainly required for its function, since deletion of an N-terminal region that includes amino acids 2 to 60 results in loss of the ability to complement the UV radiation sensitivity of *rad23* mutants (191).

Interestingly, the expression of both *RAD7* and *RAD23* is up-regulated when cells are exposed to DNA-damaging agents (see below), as is the case with the *RAD2* gene discussed above.

THE *RAD16* GENE

The *RAD16* gene was cloned by functional complementation of a *rad16* mutant (9). It too is DNA damage inducible. The gene was independently cloned during a search for ORFs which could encode polypeptides with homology to the yeast *RAD54* and *SNF2* genes (153). *RAD54* is required for recombinational repair in yeast cells (see chapter 12). Its translated amino acid sequence shares extensive similarity with that of a gene called *SNF2*, which is required for the transcriptional activation of a number of diversely regulated yeast genes (153), and with an ORF located very near the *LYS2* locus. The homology between the translated amino acid sequence of this ORF and those mentioned above prompted an examination of the phenotype of mutants disrupted in this region. The resulting strains, including some which were completely deleted of the ORF, were found to be moderately UV sensitive. The further observation that the *RAD16* gene maps near *LYS2* prompted a determination of whether the ORF near *LYS2* was in fact *RAD16*. This turned out to be the case (153). As noted above, mutants completely deleted of the chromosomal *RAD16* gene retain the moderate UV sensitivity observed in the original *rad16* point mutant, characteristic of genes in this secondary group. The homology between the Rad16 and Snf2 polypeptides is intriguing. This homology extends to a family of proteins (Fig. 6–31) that includes the human homolog of Snf2, called Hsnf2↓, the yeast Mot1 protein, the *Drosophila* Brm protein, and the mammalian Chd1 protein, all of which are implicated in transcriptional regulation. This family also includes the *Drosophila* Lds protein implicated in chromosome stability, the human ERCC6 and yeast Rad5 and Rad54 proteins implicated in various DNA repair modes, and the yeast Sth1 and Fun30 proteins, which have no known function. The predicted Rad16 polypeptide also contains a cysteine-rich motif that may bind zinc; this motif is located outside the region of homology with the putative Rad54 and Snf2 polypeptides (153).

As noted above, there are indications that the products of the *RAD16* and *RAD7* genes are specifically required for nucleotide excision repair in genes with a "closed" heterochromatin-like structure and for the repair of the nontranscribed

Figure 6–30 The N-terminal 76 amino acids of the *S. cerevisiae* translated *RAD23* sequence is homologous with the N-terminal region of ubiquitin. Identical amino acids are in red, and related amino acids are in grey.

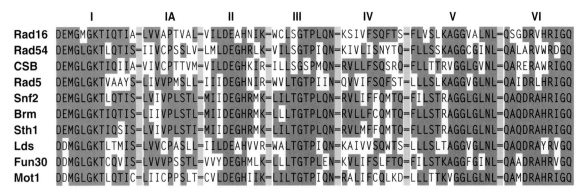

	I	IA	II	III	IV	V	VI
Rad16	DEMGMGKTIQTIA=	LVVAPTVAL=	VILDEAHNIK=	WCLSGTPLQN=	KSIVFSQFTS=	FLVSLKAGGVALNL=	QSGDRVHRIGQ
Rad54	DEMGLGKTLQTIS=	IIVCPSSLV=	LMLDEGHRLK=	VILSGTPIQN=	KIVLISNYTQ=	FLLSSKAGGCGINL=	QALARVWRDGQ
CSB	DEMGLGKTIQIIA=	VIVCPTTVM=	VILDEGHKIR=	ILLSGSPMQN=	RVLLFSQSRQ=	FLLTTRVGGLGVNL=	QARERAWRIGQ
Rad5	DEMGLGKTVAAYS=	LIVVPMSLL=	IIIDEGHNIR=	WVLTGTPIIN=	QVVIFSQFST=	LLLSLKAGGVGLNL=	QAIDRLHRIGQ
Snf2	DEMGLGKTLQTIS=	LVIVPLSTL=	MIIDEGHRMK=	LILTGTPLQN=	RVLIFFQMTQ=	FILSTRAGGLGLNL=	QAQDRAHRIGQ
Brm	DEMGLGKTIQTIS=	LIIVPLSTL=	MIIDEGHRMK=	LLLTGTPLQN=	RVLLFCQMTQ=	FLLSTRAGGLGLNL=	QAQDRAHRIGQ
Sth1	DEMGLGKTIQSIS=	LVIVPLSTI=	MIIDEGHRMK=	LILTGTPLQN=	RVLMFFQMTQ=	FLLSTRAGGLGLNL=	QAQDRAHRIGQ
Lds	DDMGLGKTLTMIS=	LVVCPASLL=	IILDEAHVVR=	WALTGTPIQN=	KAIVVSQWTS=	LLLSLTAGGVGLNL=	QAQDRAYRVGQ
Fun30	DDMGLGKTCQVIS=	LVVVPSSTL=	VVYDEGHMLK=	LLLTGTPLEN=	KVLIFSLFTQ=	FILSTKAGGFGINL=	QAADRAHRVGQ
Mot1	DDMGLGKTLQTIC=	LIICPPSLT=	CVLDEGHIIK=	LILTGTPIQN=	RALIFCQLKD=	LLLTTKVGGLGLNL=	QAMDRAHRIGQ

Figure 6–31 The predicted Rad16 polypeptide of *S. cerevisiae* has multiple consensus helicase motifs (I to VI). These show extensive amino acid sequence identity with the other members of the family of proteins shown here (see the text for details). Amino acids shown in red are identical in all or nearly all 10 polypeptides. Among the remaining amino acids, most of the differences reflect conservative substitutions. The grey bars represent discontinuities in the amino acid sequences.

strand of transcriptionally active genes. In this regard it is interesting that the homologous (with *RAD16*) *SNF2* gene also appears to be involved in the organization of chromatin structure. Suppressors of a *snf2* deletion mutant have been isolated, and several of them appear to be chromatin-associated proteins. Furthermore, the transcription defects in strains that lack the *SNF2* gene can be suppressed by deletion of one of the two sets of genes that encode histones H2A and H2B (70a). These observations suggest that *SNF2* (and possibly other genes) can effect changes in the organization of chromatin that facilitate transcription. Conceivably the *RAD16* gene product also perturbs chromatin structure through an as yet unidentified role in transcription.

Summary of Nucleotide Excision Repair in Yeast Cells

The previous edition of this book, entitled *DNA Repair* and published in 1985, contained a mere five pages devoted to nucleotide excision repair in *S. cerevisiae*. The claim to most of an entire chapter in this edition affords testimony to the rapid inroads that have been made in elucidating the genetic and biochemical complexity of nucleotide excision repair in this organism. The indispensable (for nucleotide excision repair) *RAD1*, etc., group of genes is now known to have at least 10 and as many as 12 or more members. These genes encode polypeptides that are apparently required for early and biochemically specific events, including the location and recognition of base damage in DNA, the (presumably) bimodal incision of the damaged strand, and possibly some of the postincisional events. On the basis of existing knowledge, it appears that these biochemical events may be mediated by a large multiprotein complex or *repairosome*, which includes the RNA polymerase II transcription initiation factor TFIIH, which may serve as a fundamental macromolecular nucleation site for repairosome assembly.

However, even the correct assembly of a complex of core TFIIH-Ssl2 with Rad1-Rad10, Rad2, Rad4, and Rad14 proteins may not effect damage-specific recognition and incision of DNA in vitro. As will be seen when we consider nucleotide excision repair in mammalian cells (see chapter 8), yet other proteins may be required for damage-specific incision during nucleotide excision repair. Some of these that have already been identified include single-stranded DNA-binding protein and proliferating-cell nuclear antigen. These proteins are also required for repair synthesis of DNA. Homologs of these proteins exist in yeast cells, but their role in nucleotide excision repair has not yet been explored. Therefore, conceivably the true functional repairosome in eukaryotes is an enormous complex that includes proteins involved in damage recognition, DNA incision, oligonucleotide excision, DNA resynthesis, and possibly even DNA ligation. The challenge for the future is not only to test these notions in a defined in vitro system but also to elucidate in more specific terms the role of the Rad7, Rad16, and Rad23 proteins during nucleotide excision repair in living cells.

Previous discussions have mentioned that the majority of the yeast *RAD* genes that are required for the damage-specific incision of DNA during nucleotide excision repair are conserved in other eukaryotes, including humans. This homology between yeast and human nucleotide excision repair genes provides persuasive evidence of the utility of *S. cerevisiae* as a model system for dissecting the molecular biology and biochemistry of nucleotide excision repair in eukaryotes. Such conservation also suggests that yeast and human nucleotide excision repair proteins may be functionally related to some extent. Evidence in support of this idea comes from studies with the yeast *RAD10* and human *ERCC1* genes. Expression of Rad10 protein in Chinese hamster mutant cells defective in the rodent *ERCC1* homolog results in partial complementation of both UV sensitivity and defective nucleotide excision repair (87). Similarly, the human *XPD* cDNA can rescue the lethality of a *rad3* disruption mutation in yeast cells (167). However, the human *RAD3* homolog does not rescue defective nucleotide excision repair in viable *rad3* mutants (167).

Other Members of the *RAD3* Epistasis Group

THE *CDC8* and *CDC9* GENES

The *CDC8* and *CDC9* genes fall into the *RAD3* epistasis group by conventional analysis. *CDC8* was originally defined as a gene which is required for normal cell cycle progression in yeast cells (65). A conditional-lethal *cdc8* mutant was shown to cease DNA replication under restrictive conditions (66). The mutant is moderately sensitive to UV radiation at restrictive temperatures (131). The cloned gene (21, 86) contains an ORF which could encode a polypeptide of ~24.8 kDa. The product of the *CDC8* gene has been shown to be *thymidylate kinase* (81, 155). The involvement of this gene in the biosynthesis of precursors for DNA synthesis (including repair synthesis) provides an obvious explanation for its peripheral relationship to DNA repair.

A similar explanation holds for the *CDC9* gene, which encodes DNA ligase (17, 79), a readily predicted participant in nucleotide excision repair in yeast cells. The *CDC9* gene was cloned by phenotypic complementation of its temperature-sensitive growth (18). The translated nucleotide sequence of the gene shows limited overall similarity to that of the genes for DNA ligase from phage T4 or T7. The cloned gene is, however, able to complement a DNA ligase-deficient mutant of *S. pombe* (18). Further evidence of the general conservation of DNA ligase I in eukaryotes stems from the observations that both a human cDNA which contains the ORF for human DNA ligase I (19) and the vaccinia virus DNA ligase I gene (82) correct the phenotypes of yeast *cdc9* mutants. Like *RAD2*, *RAD7*, *RAD16*, and *RAD23*, the *CDC8* and *CDC9* genes are DNA damage inducible (see below).

NUCLEOTIDE EXCISION REPAIR OF CHEMICAL DAMAGE

The genes in the *RAD3* epistasis group have been characterized principally in terms of their processing of UV radiation damage. They are also required for the repair of certain types of chemical damage. However, their involvement in the repair of alkylation damage is genetically complicated and in fact poses a serious challenge to the consistency of the epistasis analysis presented above. Mutant alleles of *RAD1*, *RAD2*, *RAD4*, and *RAD14* (but, surprisingly, not *RAD3*, *RAD10*, and *RAD16*) are hypersensitive to ethylating agents (38). The situation is further complicated by the observation that different mutant alleles are sensitive to different ethylating agents. For example, the mutants *rad1-1* and *rad1-2* are sensitive to diethyl sulfate, ethylnitrosoguanidine, and ethylnitrosourea but not to ethyl methanesulfonate. On the other hand the *rad4-4* and *rad4-12* mutants are sensitive to ethylnitrosoguanidine and ethylnitrosourea but are resistant to diethyl sulfate (38).

Among the many chemically induced bulky base adducts that can result in DNA, those produced by monofunctional and bifunctional psoralen and by the cross-linking agent nitrogen mustard have been extensively investigated (48, 52). In addition to the *RAD* genes required for damage-specific incision of UV radiation damage, genes in other epistasis groups are involved in the repair of cross-links in DNA. These include genes in the *RAD52* group, which are required for recombinational repair of DNA damage and the repair of double-strand breaks (see chapter 12), as well as genes which appear to be uniquely involved in the repair of this type of damage. The latter category includes the *PSO2* and *PSO3* genes (Table 6–1), which are required for the repair of bifunctional psoralen damage but not for the repair of UV radiation or ionizing-radiation damage (48, 52). For example, *pso2* mutants are unable to incise DNA containing psoralen cross-links and are defective in the repair of double-strand breaks (101). However, unlike genes in the *RAD52* epistasis group, *PSO2* is not required for generalized recombination in yeast cells (112).

The requirement of the *PSO2* gene for the repair of DNA cross-links has been confirmed by showing that a gene independently isolated and originally designated *SNM1* (for sensitivity to nitrogen mustard) (144, 145) is in fact *PSO2* (26). Mutants defective in *PSO2* are abnormally sensitive to other cross-linking agents such as cis-platin (26). The *SNM1* gene has been cloned and sequenced (59, 141). The predicted Snm1 polypeptide can encode a polypeptide of ~76 kDa which contains a possible zinc finger domain.

Little more can be said about the repair mechanisms of cross-links and other chemical damage in yeast cells. They bear some resemblance to nucleotide excision repair of UV radiation damage, at least in terms of some of their gene requirements. However, there are apparent differences between nucleotide excision repair of pyrimidine dimers and of some chemicals, but their nature and mechanistic implications are not understood at present.

REGULATION OF OTHER NUCLEOTIDE EXCISION REPAIR GENES IN YEAST CELLS

As mentioned previously, as is the case with *RAD2*, the *RAD7*, *RAD16*, *RAD23*, *CDC8*, and *CDC9* genes in the *RAD3* epistasis group are DNA damage inducible (44, 80, 100, 128, 193). The last two genes are also cell cycle regulated (128, 193). Despite a significant increase in the steady-state levels of *CDC9* mRNA in the late G_1 phase, however, the level of DNA ligase activity increases only ~50% (193).

The evaluation of the apparent inducibility of genes following exposure to DNA-damaging agents requires special attention if they are also regulated during the cell cycle, because DNA damage can result in the partial synchronization of cells. If in such cells a gene is cell cycle regulated, this synchronization can lead to the erroneous conclusion of specific induction following exposure to DNA damage. However, increased expression of the *CDC9* gene after exposure to UV radiation has been demonstrated in stationary-phase cultures when cells are clearly not cycling (76). Despite a ~10 fold increase in mRNA levels, DNA ligase activity was again increased by only about 50%. Hence, it is not at all clear why this gene is inducible. The *RAD7*, *RAD16*, and *RAD23* genes are not cell cycle regulated.

Enzymes Involved in Postincision Events during Nucleotide Excision Repair in Yeast Cells

Unlike the situation in *E. coli* (see chapter 5), little is known about the specific biochemical events which facilitate postincisional excision of a (putative) oligonucleotide fragment, the possible turnover of stably bound incision proteins, and the initiation and completion of repair synthesis in yeast cells. However, a lot is known about the DNA polymerases that participate in DNA synthesis in yeast cells, and these merit some discussion with respect to their possible roles in repair synthesis, for both base and nucleotide excision repair.

YEAST DNA POLYMERASES

At present there is evidence that three distinct DNA polymerases are involved in the replication, recombination, and/or repair of nuclear DNA in yeast cells (34)

(Table 6–5). In keeping with the general use of Greek letter designations for eukaryotic DNA polymerases, these are referred to as DNA polymerases yPol α, yPol δ and yPol ε, (where y stands for yeast) (34). A yeast homolog of DNA polymerase β has also been identified (158), but mutants deleted of the gene that encodes this enzyme (referred to as *POLIV* or *POLX*) are not abnormally sensitive to UV light, ionizing radiation, or alkylating agents. Additionally, such mutants are proficient in sporulation and are perfectly viable (133). A fifth nuclear DNA polymerase is inferred from the amino acid sequence of the cloned *REV3* gene but is not yet biochemically defined (109). A comprehensive discussion of yeast DNA polymerases is beyond the province of this text. The biochemical properties of these enzymes and of the genes encoding them have been extensively reviewed elsewhere (32–34). The following paragraphs highlight some of the principal characteristics of these enzymes, with a particular emphasis on their possible role in repair synthesis.

Yeast DNA Polymerase α

On the basis of several criteria, including subunit composition, biochemical properties, and its tight association with primase activity, yPol α appears to be the yeast equivalent of DNA polymerase α in mammalian cells (32–34) (see chapter 8). The gene encoding the 180-kDa catalytic subunit of yPol α is designated *POL1* or *CDC17* (78, 95) and has been cloned by screening a λgt11 genomic library with polyclonal antibodies raised against the purified protein (78, 95). The gene maps to chromosome IV (30, 96). A 74-kDa B subunit has no identified activity. Subunits of 58 and 48 kDa are required for DNA primase activity. As expected for a DNA polymerase involved in semi-conservative DNA synthesis, the *POL1* gene is an essential gene (78, 95).

A temperature-sensitive conditional lethal mutation in the *POL1* gene designated *pol1-17* confers a "quick stop" DNA replication phenotype (34). Both the rate and extent of the repair of single- or double-strand breaks produced by ionizing radiation are unaffected in this mutant (30). Additionally, the *pol1-17* mutant is not abnormally sensitive to killing by UV radiation (30). Hence, it would appear that yPol α is not involved in the repair of DNA damage or that other DNA polymerases can compensate for its deficiency (30). However, a temperature-sensitive allele of *POL1* designated *hpr3* confers increased sensitivity to killing by MMS at semipermissive temperatures (1). The UV radiation sensitivity of this mutant is comparable to that in wild-type strains (1).

Yeast DNA Polymerase δ

DNA polymerase δ is the designation given to a class of eukaryotic DNA polymerases which, unlike DNA polymerase α, contain an associated $3' \rightarrow 5'$

Table 6–5 DNA polymerases from the yeast *S. cerevisiae*

Characteristic	Value for:		
	yPol α	yPol δ	yPol ε
Catalytic subunit (kDa)	167	124	170
Encoding gene	*POL1* (*CDC17*)	*POL3* (*CDC2*)	*POL2*
Size of accessory subunit(s) (kDa)	70	50	?
	58 (PRI2) primase		
	48 (PRI1) primase		
Exonuclease	No	Intrinsic	Intrinsic
I_{50}[a] aphidicolin (μg/ml)	1	0.6	1
I_{50} BuPdGTP (μg/ml)	1	>100	>100
Processivity	100	<100	>1,000
PCNA	Independent	Dependent	Stimulated
Major function	Replication	Replication	Repair (?)

[a] I_{50}, the amount of aphidicolin that results in 50% inhibition of the polymerase activity.
Source: Adapted from Campbell and Newlon (34) with permission.

exonuclease activity (see chapter 8). Despite numerous similarities, it appears that this enzyme actually represents two distinct DNA polymerases, in both yeast and mammalian cells, called Pol δ and Pol ε (34). For yeast cells this distinction is supported by the demonstration that the polymerases are the products of two different genes. One of these enzymes is strictly dependent on the presence of proliferating-cell nuclear antigen in mammalian cells. Its yeast equivalent was originally designated as DNA polymerase III (34). The identification of the *CDC2* gene which encodes yPol δ established that this enzyme is distinct from yPol α and that it is an essential gene required for DNA replication in yeast cells. The distinction between yPol δ and yPol ε is reflected in the observation that yPol ε is present in normal amounts in *cdc2 pol1-17* double mutants (23, 162). Its role in DNA repair is not clearly defined. However, it is noteworthy that, like the *hpr3* mutant (see above), a mutant allele of *CDC2* designated *hpr6* is sensitive to MMS but not to UV radiation (1).

Purified yPol δ has a molecular mass of ~124 kDa. Its electrophoretic mobility suggests that unlike yPol α, y-Pol δ is not posttranslationally modified (29). The purified enzyme is endowed with a $3' \rightarrow 5'$ exonuclease activity which is believed to be intrinsic to the core catalytic subunit of this polymerase and not a contaminating polypeptide (29). A mutant that is defective in this exonuclease activity has a spontaneous mutator phenotype (110).

Yeast DNA Polymerase ε

yPol ε was originally designated yeast DNA polymerase II. A form of the enzyme designated DNA polymerase II* consists of a complex of polypeptides of 200, 80, 34, 39, and 29 kDa which coelute through several different purification steps (61). As indicated above, yPol ε is found at normal levels in mutants defective in both the *POL1* and *CDC2* genes and is not inhibited by antibodies to yPol α or yPol δ (34). Its physical and biochemical properties suggest that it is the yeast homolog of the proliferating-cell nuclear antigen-independent form of DNA polymerase δ in mammalian cells, which is variously referred to as DNA polymerase δ2 or DNA polymerase ε (see chapter 8). The yeast yPol ε gene (*POLII*) has been cloned by immunoscreening of a yeast expression library and independently by oligonucleotide probing of a yeast genomic library (108). The gene contains an ORF which could encode a polypeptide of about 255 kDa. Like *POLI* and *POLIII*, the yeast *POLII* gene is essential (108).

The 80-kDa subunit of yPol ε is encoded by a gene designated *DPB2* (for DNA polymerase B subunit 2) (3). The DNA sequence of this gene reveals an ORF encoding a potential polypeptide of 79.4 kDa. Disruption of the yeast chromosomal allele is lethal to haploid cells, indicating that, like *POLII* itself, *DPB2* is essential (3). The gene which encodes both the 34- and 30-kDa subunits has also been cloned and is designated *DPB3* (for DNA polymerase B subunit 3) (3). The 30-kDa subunit is apparently derived by proteolysis of the larger 34-kDa subunit. The translated amino acid sequence suggests that the product of the *DPB3* gene may be a nucleotide-binding protein.

Results of studies with the equivalent enzyme in mammalian cells suggest that yPol ε may play a role in DNA repair in yeast cells. The mammalian enzyme is able to complement the phenotype of permeabilized human cells defective in DNA repair and, in fact, was partially purified by using this complementation assay (121) (see chapter 8).

Like yPol δ discussed above, yPol ε is endowed with an intrinsic $3' \rightarrow 5'$ exonuclease activity, and like the yPol δ exonuclease mutant, a mutant defective in the $3' \rightarrow 5'$ exonuclease function of yPol ε suffers an increased spontaneous mutation rate (110). Indeed, double mutants that are defective in the $3' \rightarrow 5'$ exonuclease function of both yPol δ and yPol ε are inviable, presumably as a result of the large spontaneous DNA damage load (110).

A DNA Polymerase Encoded by the Yeast *REV3* Gene?

Cloning and sequencing of the yeast *REV3* gene has revealed the features of yet another DNA polymerase in *S. cerevisiae* (109). This gene maps to chromosome

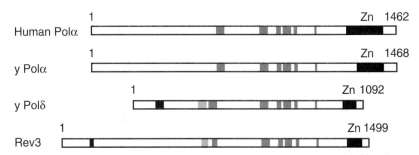

Figure 6–32 The translated sequence of the *REV3* gene of *S. cerevisiae* suggests that the Rev3 polypeptide may be a DNA polymerase, on the basis of regions of amino acid sequence homology with the human DNA polymerase α and yeast Pol α and Pol δ proteins. The C-terminal region of each of these proteins contains a zinc finger motif. (*Adapted from Campbell and Newlon [34] with permission.*)

XVI and contains an ORF which could encode a polypeptide of ~173 kDa. The amino acid sequence suggests that this enzyme belongs to the δ class of DNA polymerases (Fig. 6–32). The *REV3* gene is one of a number of genes identified by the isolation of mutants with a decreased frequency of UV radiation-induced mutations (93, 94, 135), implicating them in the tolerance of unexcised UV radiation damage to DNA (see chapter 12). It is of interest that the amino acid sequence of the *REV1* gene encodes a polypeptide which resembles that of the *umuCD* gene which is required for UV-induced mutagenesis in *E. coli* (see chapter 11).

Complete deletion of the *REV3* gene is not lethal to haploid yeast cells, suggesting that the putative DNA polymerase encoded by this gene is not involved in DNA replication and supporting its role in some type of repair-related event such as translesion DNA synthesis (see chapter 12). Mutants with mutations in *REV1* or *REV3* are slightly more sensitive to killing by DNA-damaging agents but are not as sensitive to UV radiation as are mutants in the *RAD3* epistasis group. The failure to detect a DNA polymerase associated with this gene has led to the suggestion that the product of the *REV3* gene may have a different catalytic activity, despite the structural retention of an active site for DNA polymerase. One possibility is that the polymerase function is lost, leaving an intact nuclease function (34).

Repair Synthesis during Base Excision Repair Is Catalyzed by Polymerase ε

The availability of mutants that are temperature sensitive for the three known yeast DNA polymerases has facilitated the systematic evaluation of which enzyme catalyzes repair synthesis during base excision repair in a defined cell-free system derived from *S. cerevisiae* (189). Repair synthesis at restrictive temperatures is defective exclusively in extracts from *pol2* (yPol ε) mutants (Fig. 6–33), and the defective repair synthesis can be complemented by the addition of purified yPol ε. Hence, it appears that this DNA polymerase is principally if not exclusively involved in repair synthesis during base excision repair in yeast cells. Studies on repair synthesis during nucleotide excision repair in yeast cells are not completed but are expected to yield similar results because it has been shown that DNA polymerase ε is the principal polymerase involved in nucleotide excision repair in mammalian cells (see chapter 8).

DNA LIGASE

As discussed above, the yeast *CDC9* gene encodes a DNA ligase. This enzyme has been extensively purified, and its detailed biochemical characterization indicates that is very similar to the mammalian form of DNA ligase I, which is discussed in detail in chapter 8 (181). At present there are suggestions that eukaryotic cells contain multiple DNA ligases (88). These are best characterized in mammalian

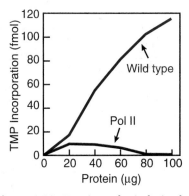

Figure 6–33 Repair synthesis during base excision repair (and hence possibly in nucleotide excision repair) in cell extracts of the yeast *S. cerevisiae* is defective in *pol2* mutants which encode a temperature-sensitive DNA polymerase ε. Various amounts of wild-type or *pol2* mutant extract were preincubated at a restrictive temperature (37°C) for 30 min. Substrate plasmid DNA and deoxynucleoside triphosphates were then added, and repair reactions were performed at 37°C. Repair synthesis was measured by using the system described in the legend to Fig. 6–6, except that prior to autoradiography radiolabeled DNA bands were cut out of the gel and the radioactivity was quantitated. (*Adapted from Wang et al. [189].*)

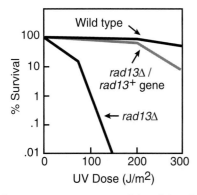

Figure 6–34 A mutant deleted for the *rad13+* gene (the *S. pombe* homologue of the *S. cerevisiae RAD2* gene) is only marginally sensitive to UV radiation compared with *S. cerevisiae rad2* mutants (compare the UV radiation dose to that in Fig. 6–2 and 6–22). A plasmid carrying the cloned *rad13+* gene corrects the UV radiation sensitivity of the *S. pombe rad13* mutant. (*Adapted from Carr et al. [36] with permission.*)

cells and will be discussed later (see chapter 8). Additional DNA ligases from yeast cells have not yet been purified or characterized.

Nucleotide Excision Repair in Other Lower Eukaryotes

Schizosaccharomyces pombe

The molecular biology of DNA repair and mutagenesis in the fission yeast *S. pombe* has not been as extensively characterized as in the budding yeast *S. cerevisiae*. Nonetheless, recent years have witnessed significant progress in the cloning of genes apparently required for nucleotide excision repair in this yeast. *S. pombe* is significantly more resistant to UV radiation and γ rays than is *S. cerevisiae*, especially in the G₂ phase of the cell cycle. DNA repair mutant genes have been mapped to ~35 genetic complementation groups in *S. pombe* (2, 92, 129). As in *S. cerevisiae*, the *S. pombe* mutants can be grouped phenotypically as being sensitive mainly to UV radiation, to γ rays, or to both (2, 129). The existence of multiple repair pathways is suggested by limited epistatic analysis (Table 6–6). The group of highly UV radiation-sensitive strains comprises seven mutants with mutations in the *rad2+*, *rad5+*, *rad7+*, *rad11+*, *rad13+*, *rad16+*, and *rad17+* genes and a mutant with a mutation in a gene which is homologous to the human *XPB* gene (see chapter 8), designated *XPB^sp* (92). A mutant gene originally designated as *rad5* has been shown to be allelic to *rad15*, and *rad10*, *rad20*, and *swi9* are alleles of *rad16*. The *rad17+* gene (as well as *rad1+*, *rad3+*, and *rad9+*) is a checkpoint gene that participates in cellular processes whereby *S. pombe* cells undergo cell cycle arrest in response to DNA damage exposure (see chapter 13). The remaining six mutants presumably reflect defects in a repair pathway formally analogous to the nucleotide excision repair pathway of *S. cerevisiae*. However, the *S. pombe* mutants are less sensitive to UV radiation than are those from *S. cerevisiae* (Fig. 6–34) Furthermore, none of the *S. pombe* mutants thus far characterized is totally defective in pyrimidine dimer excision (103). This is not the result of leakiness of mutants, since the same phenotype is demonstrated by strains bearing deletions in the *rad13+* or *rad16+* gene. Hence, it has long been speculated that there may be several partially redundant excision repair pathways for bulky base damage in *S. pombe* (22).

These differences notwithstanding, phenotypic complementation of *S. pombe* mutants presumably defective in nucleotide excision repair has been attempted with the *S. cerevisiae RAD1* and *RAD2* genes (104). Complementation of UV radiation resistance to near wild-type levels was observed following transformation of the *S. pombe rad13* mutant with the *S. cerevisiae RAD2* gene (Fig. 6–34). As is the case with *rad2* mutations in *S. cerevisiae*, the *rad13* mutation in *S. pombe* enhances UV mutability, suggesting that nucleotide excision repair is an error-avoiding mechanism in both organisms.

Table 6–6 Phenotypic groups of DNA repair genes in *S. pombe*

Excision repair	Recombination repair	γ-Ray repair
rad2+	*rad1+*	*rad21+*
rad5+ (*rad15+*)	*rad3+*	*rad22+*
rad7+	*rad4+*	
swi10+	*rad6+*	
rad11+	*rad8+*	
rad13+	*rad9+*	
rad15+	*rad12+*	
rad16+ (*rad10+*)(*swi9+*)(*rad20+*)	*rad18+*	
rad17+	*rad19+*	
XPB^sp	*rad20+*	
	rad23+	
	rhp6+	

Source: Adapted from Lehmann et al. (92) with permission.

Phenotypic correction of *S. pombe* mutants with genomic DNA from this yeast has facilitated the cloning of several *rad*+ genes. Note that the genetic nomenclature for wild-type genes in *S. pombe* uses lowercase italics with a plus superscript. Unfortunately, DNA repair genes from both organisms have been designated as *RAD* and *rad*+ for *S. cerevisiae* and *S. pombe*, respectively, with an arbitrary numbering system for both, so that inevitably genes with the same name and number are not homologs (Tables 6–1 and 6–6).

A number of the *S. pombe* genes are clearly homologs of genes involved in nucleotide excision repair in *S. cerevisiae*. The *S. pombe rad15*+ gene (also called *rhp3*+ [139]) was cloned by using the homologous *RAD3* gene from *S. cerevisiae* as a hybridization probe (113). Indeed, the *S. pombe rad15*+ gene fully corrects the UV sensitivity of *S. cerevisiae rad3* mutants (139). The *S. pombe rad13*+ and *rad16*+ genes are homologs of the *S. cerevisiae RAD2* and *RAD1* genes, respectively (113, 143, 154). Mutant genes designated *rad10* and *swi9* are allelic with *rad16*. Hence, the *rad10*+ (*rad16*+) and *swi9*+ genes are one and the same. The *swi10*+ gene was cloned by phenotypic correction of the mating-type switching defect of a *swi10* mutant (143). The translated amino acid sequence of the *swi10*+ ORF reveals significant homology with Rad10 protein of *S. cerevisiae* and with the human ERCC1 protein (see chapter 8). The ORF is interrupted by three introns. The positions of introns I and III of *swi10*+ are evolutionarily conserved with introns III and IV of the human *ERCC1* gene (143).

The designation *swi* for some of the genes of *S. pombe* reflects their involvement in the process of mating-type switching, a process by which a haploid cell converts from one mating type to another. The molecular biology of mating type switching in both *S. pombe* and *S. cerevisiae* has been extensively studied and involves recombinational events during which certain mating type genes are transposed between specific locations. A detailed discussion of the switching process is beyond the province of this book. As previously indicated, two *S. pombe swi* genes, *swi9*+ and *swi10*+, are also involved in nucleotide excision repair (154). This is consistent with the fact that the homologs of these genes in *S. cerevisiae*, *RAD1* and *RAD10*, respectively, are known to be involved in recombinational events (see above). However, the mechanisms of mating-type switching in *S. pombe* and *S. cerevisiae* are sufficiently distinct that *S. cerevisiae rad1* and *rad10* mutants do not manifest a defect in mating-type switching.

The *S. pombe* homolog of the *S. cerevisiae POL3* gene was cloned by DNA sequence homology. The translated amino acid sequences indicate 52% identity, with as much as 60% identity in a central region (130). An *S. pombe* cell-free system that apparently supports the repair of both UV radiation and γ radiation damage has been developed (159).

Drosophila melanogaster

Another organism that offers the attractions of sophisticated genetics and molecular versatility is the fruit fly, *Drosohila melanogaster*, now widely recognized for its contributions to developmental biology. Genetic analysis has identified ~30 genes that are required for resistance to various mutagens (25). Mutations at 24 of these loci have been introduced into cells in tissue culture, considerably increasing their utility for biochemical and molecular studies, and alterations in cellular responses to DNA damage have been identified in at least 13 of these mutants (25).

Genes involved in protection against mutagen sensitivity have been designated *MUS* (25). Additionally, meiosis-defective mutants designated *mei* have been found to be required for DNA damage processing. Several assay systems have demonstrated that the mutants *mei-9* and *mus201* are severely defective in the excision of pyrimidine dimers and presumably represent two of the *Drosophila* genes involved in nucleotide excision repair. Other *mus* mutants have reduced dimer excision capacity relative to wild-type controls, although they are not as defective as the *mei-9* and *mus201* mutants. These include mutants designated *mus205*, *mus302*, *mus304*, *mus306*, and *mus308* (25).

Another series of repair-defective mutants was isolated independently (164). In this study 16 mutagen-sensitive *Drosophila* strains were organized into three nonoverlapping groups (Table 6–7). Genes in strains in group 2 are believed to be involved in nucleotide excision repair of UV radiation and alkylation damage to DNA. In addition to the *MEI9, MUS201, MUS205, MUS306,* and *MUS308* genes mentioned above, this group includes genes designated *MUS204, MUS206, MUS207,* and *MUS310* (164).

These studies provide evidence of a genetic complexity that certainly approaches that described for the yeast *S. cerevisiae.* The potential for cloning some of these genes has received major impetus from the technology of transposon tagging and has facilitated the cloning of the *mei-41* gene (24). The spectrum of mutagen sensitivity exhibited by mutants with mutations in this gene is unusually broad, suggesting that the *mei-41* gene is involved in some fundamental aspect of DNA metabolism rather than in any single specific repair pathway (24). In general, however, the isolation of nucleotide excision repair genes from *D. melanogaster* and the biochemical characterization of the products they encode is still very much in its infancy.

D. melanogaster homologs of several known mammalian and yeast nucleotide excision repair genes have been cloned by direct DNA hybridization. These include the *Drosophila* homolog of the human *XPC* gene, designated *XPC^DM* (69), which is homologous with the yeast *RAD4* gene. The *Drosophila* homolog of the *XPD* gene has also been cloned and is designated *XPD^DM* (84). *Drosophila* homologs of the human *XPB* (yeast *SSL2*) and human *XPA* (157) (yeast *RAD14*) genes have also been isolated. The former *Drosophila* gene is called *haywire* (111). Unfortunately, there is no information on whether any of these genes correspond to the known *Drosophila* mutants mentioned above. Table 6–8 provides a comprehensive comparison of the nucleotide excision repair genes in various eukaryotes and demonstrates the extraordinary conservation of these genes in the yeasts *S. cerevisiae* and *S. pombe,* in *D. melanogaster,* and in humans. This table is reproduced in chapter 8.

In contrast to the results obtained in yeast and mammalian cells, transcriptionally coupled strand-specific repair of UV radiation damage has not been observed in a *Drosophila* cell line that has been well characterized with respect to the transcriptional activity of several genes (41, 42). This curious exception to the apparently general rule of the preferential repair of the template strand of transcriptionally active genes is puzzling and merits more study to determine its possible generality. If indeed this is generally true in *D. melanogaster,* it has profound implications for the composition of the nucleotide excision repairosome in *D. melanogaster* compared with that in other eukaryotes such as yeast and mammalian cells, on the basis of what we currently know about the dual function of basal transcription factors. On the other hand, it is conceivable that the genes thus far examined for strand-specific repair all undergo divergent transcription.

Other Lower Eukaryotes

In addition to *D. melanogaster,* other lower eukaryotes have been used to various extents as model systems for studies of nucleotide excision repair. These include the nematode *Caenorhabditis elegans,* the slime mold *Dictyostelium discoideum,* and

Table 6–7 Groups of *Drosophila* genes involved in DNA repair

Group 1 (alkylation repair [?])	Group 2 (alkylation and NER^a [?])	Group 3 (?)
MUS101	*MEI9*	*MEI41*
MUS103	*MUS201*	*MUS102*
MUS106	*MUS204*	*MUS105*
	MUS205	*MUS304*
	MUS207	
	MUS306	
	MUS308	
	MUS310	

a NER, nucleotide excision repair.

Table 6–8 Known genetic correlations of nucleotide excision repair in eukaryotes

S. cerevisiae gene	Human gene	Human complementation group	Human chromosome	Rodent complementation group	S. pombe gene	Drosophila gene
RAD3	XPD	XP-D (CS-XPD) (TTD-XPD)	19q13	RCG-2	rad15+	XPD^dm
SSL2 (RAD25)	XPB	XP-B (CS-XPB) (TTD-XPB)	2q21	RCG-3	ERCC3^sp	XPB^dm (hwr)
RAD2	XPG	XP-G (CS-XPG)	13q32-33	RCG-5	rad13+	
RAD14	XPA	XP-A	9q34			XPA^dm
RAD4	XPC	XP-C	3p25			XPC^dm
RAD1	ERCC4/?XPF	XP-F	16p13	RCG-4	rad16+	
RAD26	CSB	CS-B	10q11	RCG-6		
RAD10	ERCC1		19q13	RCG-1	swi10+	
RAD23	HHR23A		3p25			
	HHR23B		19p31·2			
TFB1	p62		11p14-15·1			
SSL1	p44					
RAD7						
RAD16						
		XP-E				
		XP-V				
		CS-A				
		TTD-A				
			7(?)	RCG-9		
				RCG-7		
				RCG-8		
				RCG-10		
				RCG-11		

the fungus *Neurospora crassa*. Mutants defective in cellular responses to DNA damage, including the excision of pyrimidine dimers, have been identified in several of these organisms (27, 64, 74, 192). However, at present little detailed information is available about these systems.

Nucleotide excision repair of UV-irradiated plasmid DNA has also been demonstrated following microinjection of the DNA into *Xenopus laevis* oocytes (68, 148). The mechanism of nucleotide excision repair in these extracts appears to be processive on individual plasmid molecules and is highly efficient. As many as 10^{10} dimers are excised per oocyte (148) This system has the potential for realizing the purification of large quantities of proteins for nucleotide excision repair, since frog oocytes are known to store a vast excess of nuclear proteins.

Conclusions

It is singularly evident that nucleotide excision repair is much more complex in eukaryotes than in prokaryotes such as *E. coli*. The various eukaryotic models explored in this chapter have become powerful experimental tools for exploring the molecular biology and biochemistry of this DNA repair mode at this level of biological organization and have set the stage for considering nucleotide excision repair in mammalian, including human, cells; this is the topic of the next two chapters.

REFERENCES

1. **Aguilera, A., and H. L. Klein.** 1988. Genetic control of intrachromosomal recombination in *Saccharomyces cerevisiae*. I. Isolation and genetic characterization of hyper-recombination mutations. *Genetics* **119:**779–790.

2. **Al-Khodairy, F., E. Fotou, K. S. Sheldrick, D. J. F. Griffiths, A. R. Lehmann, and A. M. Carr.** 1994. Identification and characterization of new elements involved in checkpoint and feedback controls in fission yeast. *Mol. Biol. Cell,* **5:**147–160.

3. **Araki, H., R. K. Hamatake, L. H. Johnston, and A. Sugino.** 1991. *DPB2,* the gene encoding DNA polymerase II subunit B, is required for chromosome replication in *Saccharomyces cerevisiae. Proc. Natl. Acad. Sci. USA* **88:**4601–4605.

4. **Armstrong, J. D., and B. A. Kunz.** 1992. Excision repair influences the site and strand specificity of sunlight mutagenesis in yeast. *Mutat. Res.* **274:**123–133.

5. **Bailis, A. M., L. Arthur, and R. Rothstein.** 1992. Genome rearrangement in *top3* mutants of *Saccharomyces cerevisiae* requires a functional *RAD1* excision repair gene. *Mol. Cell. Biol.* **12:**4988–4993.

6. **Bailis, A. M., and R. Rothstein.** 1990. A defect in mismatch repair in *Saccharomyces cerevisiae* stimulates ectopic recombination between homeologous genes by an excision repair dependent process. *Genetics* **126:**535–547.

7. **Bailly, V., C. H. Sommers, P. Sung, L. Prakash, and S. Prakash.** 1992. Specific complex formation between proteins encoded by the yeast DNA repair and recombination genes *RAD1* and *RAD10. Proc. Natl. Acad. Sci. USA* **89:**8273–8277.

8. **Banerji, J., J. Sands, J. L. Strominger, and T. Spies.** 1990. A gene pair from the human major histocompatibility complex encodes large proline-rich proteins with multiple repeated motifs and a single ubiquitin-like domain. *Proc. Natl. Acad. Sci. USA* **87:**2374–2378.

9. **Bang, D. D., R. Verhage, N. Goosen, J. Brouwer, and P. van de Putte.** 1992. Molecular cloning of *RAD16,* a gene involved in differential repair in *Saccharomyces cerevisiae. Nucleic Acids Res.* **20:**3925–3931.

10. **Bankmann, M., L. Prakash, and S. Prakash.** 1992. Yeast *RAD14* and human xeroderma pigmentosum group A DNA-repair genes encode homologous proteins. *Nature* (London) **355:**555–558.

11. **Bardwell, A. J., L. Bardwell, N. Iyer, J. Q. Svefstrup, W. J. Feaver, R. D. Kornberg, and E. C. Friedberg.** 1994. Yeast nucleotide excision repair proteins Rad2 and Rad4 interact with RNA polymerase II basal transcription factor b (TFIIH). *Mol. Cell. Biol.* **14:**3569–3576.

12. **Bardwell, A. J., L. Bardwell, D. K. Johnson, and E. C. Friedberg.** 1993. Yeast DNA recombination and repair proteins Rad1 and Rad10 constitute a complex in vivo mediated by localized hydrophobic domains. *Mol. Microbiol.* **8:**1177–1188.

13. **Bardwell, A. J., L. Bardwell, A. E. Tomkinson, and E. C. Friedberg.** 1994. Specific cleavage of model recombination and repair intermediates by the yeast Rad1-Rad10 DNA endonuclease. *Science,* **265:**2082–2085.

14. **Bardwell, L., A. J. Bardwell, W. J. Feaver, J. Q. Svejstrup, R. D. Kornberg, and E. C. Friedberg.** 1994. Yeast RAD3 protein binds directly to both SSL2 and SSL1 proteins: implications for the structure and function of transcription/repair factor b. *Proc. Natl. Acad. Sci. USA* **91:**3926–3930.

15. **Bardwell, L., H. Burtscher, W. A. Weiss, C. M. Nicolet, and E. C. Friedberg.** 1990. Characterization of the *RAD10* gene of *Saccharomyces cerevisiae* and purification of Rad10 protein. *Biochemistry* **29:**3119–3126.

16. **Bardwell, L., A. J. Cooper, and E. C. Friedberg.** 1992. Stable and specific association between the yeast recombination and DNA repair proteins RAD1 and RAD10 in vitro. *Mol. Cell. Biol.* **12:**3041–3049.

17. **Barker, D. G., A. L. Johnson, and L. H. Johnston.** 1985. An improved assay for DNA ligase reveals temperature-sensitive activity in *cdc9* mutants of *Saccharomyces cerevisiae. Mol. Gen. Genet.* **200:**458–462.

18. **Barker, D. G., and L. H. Johnston.** 1983. *Saccharomyces cerevisiae CDC9,* a structural gene for yeast DNA ligase which complements *Schizosaccharomyces pombe cdc17. Eur. J. Biochem.* **134:**315–319.

19. **Barnes, D. E., L. H. Johnston, K.-I. Kodama, A. E. Tomkinson, D. D. Lasko, and T. Lindahl.** 1990. Human DNA ligase I cDNA: cloning and functional expression in *Saccharomyces cerevisiae. Proc. Natl. Acad. Sci. USA* **87:**6679–6683.

20. **Bennetzen, J. L., and B. D. Hall.** 1982. Codon selection in yeast. *J. Biol. Chem.* **257:**3026–3031.

21. **Birkenmeyer, L. G., J. C. Hill, and L. B. Dumas.** 1984. *Saccharomyces cerevisiae CDC8* gene and its product. *Mol. Cell. Biol.* **4:**583–590.

22. **Birnboim, H. C., and A. Nasim.** 1975. Excision of pyrimidine dimers by several UV-sensitive mutants of *S. pombe. Mol. Gen. Genet.* **136:**1–8.

23. **Boulet, A., M. Simon, G. Faye, G. A. Bauer, and P. M. Burgers.** 1989. Structure and function of the *Saccharomyces cerevisiae CDC2* gene encoding the large subunit of DNA polymerase III. *EMBO J.* **8:**1849–1854.

24. **Boyd, J. B., S. S. Banga, A. H. Yamamoto, J. M. Mason, D. R. Oliveri, D. S. Henderson, A. Velazquez, E. Leonhardt, and A. W. Rosenstein.** 1990. Cloning *Drosophila* repair genes by transposon tagging, p. 205–211. *In* M. L. Mendelsohn, and R. J. Albertini (ed.), *Mutation and the Environment,* part A. Wiley-Liss, New York.

25. **Boyd, J. B., P. V. Harris, J. M. Presley, and M. Narachi.** 1983. *Drosophila melanogaster*: a model eukaryote for the study of DNA repair, p. 107–123. *In* E. C. Friedberg and B. A. Bridges (ed.), *Cellular Responses to DNA Damage.* Alan R. Liss, Inc., New York.

26. **Brendel, M., and A. Ruhland.** 1984. Relationships between functionality and genetic toxicology of selected DNA–damaging agents. *Mutat. Res.* **133:**51–85.

27. **Bronner, C. E., D. L. Welker, and R. A. Deering.** 1992. Mutations affecting sensitivity of the cellular slime mold *Dictyostelium discoideum* to DNA-damaging agents. *Mutat. Res.* **274:**187–200.

28. **Brouwer, J., D. D. Bang, R. Verhage, and P. van de Putte.** 1992. Preferential repair in *Saccharomyces cerevisiae,* p. 274–283. *In* V. A. Bohr, K. Wassermann, and K. H. Kraemer (ed.), *DNA Repair Mechanisms.* Munksgaard, Copenhagen.

29. **Brown, W. C., J. A. Duncan, and J. L. Campbell.** 1993. Purification and characterization of the *Saccharomyces cerevisiae* DNA polymerase δ overproduced in *Escherichia coli. J. Biol. Chem.* **268:**982–990.

30. **Budd, M. E., K. D. Wittrup, J. E. Bailey, and J. L. Campbell.** 1989. DNA polymerase I is required for premeiotic DNA replication and sporulation but not for X-ray repair in *Saccharomyces cerevisiae. Mol. Cell. Biol.* **9:**365–376.

31. **Buratowski, S.** 1994. The basics of basal transcription by RNA polymerase II. *Cell* **77:**1–3.

32. **Burgers, P. M.** 1989. Eukaryotic DNA polymerases α and δ: conserved properties and interactions, from yeast to mammalian cells. *Prog. Nucleic Acid Res. Mol. Biol.* **37:**235–280.

33. **Campbell, J. L.** 1986. Eukaryotic DNA replication. *Annu. Rev. Biochem.* **55:**733–771.

34. **Campbell, J. L., and C. S. Newlon.** 1991. Chromosomal DNA replication, p. 41–146. *In* J. R. Broach, J. R. Pringle and E. W. Jones (ed.), *The Molecular and Cellular Biology of the Yeast Saccharomyces. Genome Dynamics, Protein Synthesis, and Energetics.* Cold Spring Harbor Laboratory, Cold Spring Harbor, N.Y.

35. **Carr, A. M., H. Schmidt, S. Kirchhoff, W. J. Muriel, K. S. Sheldrick, D. J. Griffiths, C. N. Basmacioglu, M. Clegg, A. Nasim, and A. R. Lehmann.** 1994. The *rad16* gene of *Schizosaccharomyces pombe*: a homolog of the *RAD1* gene of *Saccharomyces cerevisiae. Mol. Cell. Biol.* **14:**2029–2040.

36. **Carr, A. M., K. S. Sheldrick, J. M. Murray, R. al-Harithy, F. Z. Watts, and A. R. Lehmann.** 1993. Evolutionary conservation of excision repair in *Schizosaccharomyces pombe*: evidence for a family of sequences related to the *Saccharomyces cerevisiae RAD2* gene. *Nucleic Acids Res.* **21:**1345–1349.

37. **Conaway, R. C., and J. W. Conaway.** 1993. General initiation factors for RNA polymerase II. *Annu. Rev. Biochem.* **62:**161–190.

38. Cooper, A. J., and R. Waters. 1987. A complex pattern of sensitivity to simple monofunctional alkylating agents exists amongst the *rad* mutants of *Saccharomyces cerevisiae*. *Mol. Gen. Genet.* **209:**142–148.

39. Couto, L. B., and E. C. Friedberg. 1989. Nucleotide sequence of the wild-type *RAD4* gene of *Saccharomyces cerevisiae* and characterization of mutant *rad4* alleles. *J. Bacteriol.* **171:**1862–1869.

40. Cox, B. S., and J. M. Parry. 1968. The isolation, genetics and survival characteristics of ultraviolet light-sensitive mutants in yeast. *Mutat. Res.* **6:**37–55.

41. de Cock, J. G., E. C. Klink, W. Ferro, P. H. M. Lohman, and J. C. Eeken. 1992. Neither enhanced removal of cyclobutane pyrimidine dimers nor strand-specific repair is found after transcription induction of the β 3-tubulin gene in a *Drosophila* embryonic cell line Kc. *Mutat. Res.* **293:**11–20.

42. de Cock, J. G., A. van Hoffen, J. Wijnands, G. Molenaar, P. H. M. Lohman, and J. C. Eeken. 1992. Repair of UV-induced (6-4)photoproducts measured in individual genes in the *Drosophila* embryonic Kc cell line. *Nucleic Acids Res.* **20:**4789–4793.

43. Dingwall, C., and R. A. Laskey. 1991. Nuclear targeting sequences—a consensus? *Trends Biochem. Sci.* **16:**478–481.

44. Elledge, S. J., and R. W. Davis. 1987. Identification and isolation of the gene encoding the small subunit of ribonucleotide reductase from *Saccharomyces cerevisiae*: DNA damage-inducible gene required for mitotic viability. *Mol. Cell. Biol.* **7:**2783–2793.

45. Feaver, W. J., J. Q. Svejstrup, L. Bardwell, A. J. Bardwell, S. Buratowski, K. D. Gulyas, T. F. Donahue, E. C. Friedberg, and R. D. Kornberg. 1993. Dual roles of a multiprotein complex from *S. cerevisiae* in transcription and DNA repair. *Cell* **75:**1379–1387.

46. Fields, S., and O. Song. 1989. A novel genetic system to detect protein-protein interactions. *Nature* (London) **340:**245–246.

47. Fishman-Lobell, J., and J. E. Haber. 1992. Removal of non-homologous DNA ends in double-strand break recombination: the role of the yeast ultraviolet repair gene *RAD1*. *Science* **258:**480–484.

47a. Friedberg, E. C. 1985. *DNA Repair*, W. H. Freeman & Co., New York.

48. Friedberg, E. C. 1988. Deoxyribonucleic acid repair in the yeast *Saccharomyces cerevisiae*. *Microbiol. Rev.* **52:**70–102.

49. Friedberg, E. C. 1990. The genetic and biochemical complexity of DNA repair in eukaryotes: the yeast *Saccharomyces cerevisiae* as a paradigm, p. 255–273. *In* P. R. Strauss and S. H. Wilson (ed.), *The Eukaryotic Nucleus. Molecular Biochemistry and Macromolecular Assemblies*. Telford, New York.

50. Friedberg, E. C. 1991. Eukaryotic DNA repair: glimpses through the yeast *Saccharomyces cerevisiae*. *Bioessays* **13:**295–302.

51. Friedberg, E. C. 1991. Yeast genes involved in DNA-repair processes: new looks on old faces. *Mol. Microbiol.* **5:**2303–2310.

51a. Friedberg, E. C. 1992. Nuclear targeting sequences. *Trends Biochem. Sci.* **17:**347.

52. Friedberg, E. C., W. Siede, and A. J. Cooper. 1991. Cellular responses to DNA damage in yeast, p. 147–192. *In* J. R. Broach, J. R. Pringle and E. W. Jones (ed.), *The Molecular and Cellular Biology of the Yeast Saccharomyces. Genome Dynamics, Protein Synthesis, and Energetics*. Cold Spring Harbor Laboratory, Cold Spring Harbor, N.Y.

53. Gileadi, O., W. J. Feaver, and R. D. Kornberg. 1992. Cloning of a subunit of yeast RNA polymerase II transcription factor b and CTD kinase. *Science* **257:**1389–1392.

54. Goodrich, J. A., and R. Tjian. 1994. Transcription factors IIE and IIH and ATP hydrolysis direct promoter clearance by RNA polymerase II. *Cell* **77:**145–156.

55. Grossman, L., and A. T. Yeung. 1990. The UvrABC endonuclease system of *Escherichia coli*—a view from Baltimore. *Mutat. Res.* **236:**2–3.

56. Gulyas, K. D., and T. F. Donahue. 1992. *SSL2*, a suppressor of a stem-loop mutation in the *HIS4* leader encodes the yeast homolog of human *ERCC-3*. *Cell* **69:**1031–1042.

57. Guzder, S. N., H. Qiu, C. H. Sommers, P. Sung, L. Prakash, and S. Prakash. 1994. DNA repair gene *RAD3* of *S. cerevisiae* is essential for transcription by RNA polymerase II. *Nature* (London) **367:**91–94.

57a. Guzder, S. N., P. Sung, V. Bailly, L. Prakash, and S. Prakash. 1994. RAD25 is a DNA helicase required for DNA repair and RNA polymerase II transcription. *Nature* (London) **369:**578–581.

58. Guzder, S. N., P. Sung, L. Prakash, and S. Prakash. 1993. Yeast DNA-repair gene *RAD14* encodes a zinc metalloprotein with affinity for ultraviolet-damaged DNA. *Proc. Natl. Acad. Sci. USA* **90:**5433–5437.

59. Haase, E., D. Riehl, M. Mack, and M. Brendel. 1989. Molecular cloning of *SNM1*, a yeast gene responsible for a specific step in the repair of cross-linked DNA. *Mol. Gen. Genet.* **218:**64–71.

60. Habraken, Y., P. Sung, L. Prakash, and S. Prakash. 1993. Yeast excision repair gene *RAD2* encodes a single-stranded DNA endonuclease. *Nature* (London) **366:**365–368.

61. Hamatake, R. K., H. Hasegawa, A. B. Clark, K. Bebenek, T. A. Kunkel, and A. Sugino. 1990. Purification and characterization of DNA polymerase II from the yeast *Saccharomyces cerevisiae*. Identification of the catalytic core and a possible holoenzyme form of the enzyme. *J. Biol. Chem.* **265:**4072–4083.

62. Harosh, I., L. Naumovski, and E. C. Friedberg. 1989. Purification and characterization of Rad3 ATPase/DNA helicase from *Saccharomyces cerevisiae*. *J. Biol. Chem.* **264:**20532–20539.

63. Harrington, J. J., and M. R. Lieber. 1994. The characterization of a mammalian DNA structure-specific endonuclease. *EMBO J.* **13:**1235–1246.

63a. Harrington, J. J., and M. R. Lieber. 1994. Functional domains within *FEN-1* and *RAD2* define a family of structure-specific endonucleases: implications for nucleotide excision repair. *Genes Dev.* **8:**1344–1355.

64. Hartman, P. S., J. Hevelone, V. Dwarakanath, and D. L. Mitchell. 1989. Excision repair of UV radiation-induced DNA damage in *Caenorhabditis elegans*. *Genetics* **122:**379–385.

65. Hartwell, L. H. 1971. Genetic control of the cell division cycle in yeast. II. Genes controlling DNA replication and its initiation. *J. Mol. Biol.* **59:**183–194.

66. Hartwell, L. H. 1973. Three additional genes required for deoxyribonucleic acid synthesis in *Saccharomyces cerevisiae*. *J. Bacteriol.* **115:**966–974.

67. Haynes, R. H., and B. A. Kunz. 1981. DNA repair and mutagenesis in yeast, p. 371–414. *In* J. N. Strathern, E. W. Jones, and J. R. Broach (ed.), *The Molecular Biology of the Yeast Saccharomyces. Life Cycle and Inheritance*. Cold Spring Harbor Laboratory, Cold Spring Harbor, N.Y.

68. Hays, J. B., E. J. Ackerman, and Q. S. Pang. 1990. Rapid and apparently error-prone excision repair of nonreplicating UV-irradiated plasmids in *Xenopus laevis* oocytes. *Mol. Cell. Biol.* **10:**3505–3511.

69. Henning, K. A., C. Peterson, R. Legerski, and E. C. Friedberg. 1994. Cloning the *Drosophila* homolog of the xeroderma pigmentosum complementation group C gene reveals homology between the predicted human and *Drosophila* polypeptides and that encoded by the yeast *RAD4* gene. *Nucleic Acids Res.* **22:**257–261.

70. Higgins, D. R., L. Prakash, P. Reynolds, and S. Prakash. 1984. Isolation and characterization of the *RAD2* gene of *Saccharomyces cerevisiae*. *Gene* **30:**121–128.

70a. Hirschhorn, J. N., S. A. Brown, C. D. Clark, and F. Winston. 1992. Evidence that *SNF2/SWI2* and *SNF5* activate transcription in yeast by altering chromatin structure. *Genes Dev.* **6:**2288–2298.

71. Hoeijmakers, J. H. J. 1993. Nucleotide excision repair I: from *E. coli* to yeast. *Trends Genet.* **9:**173–177.

71a. Hoeijmakers, J. H. J., A. P. M. Eker, R. D. Wood, and P. Robins. 1990. Use of in vivo and in vitro assays for the characterization of mammalian excision repair and isolation of repair proteins. *Mutat. Res.* **236:**223–238.

72. Hoekema, A., R. A. Kastelein, M. Vasser, and H. A. de Boer. 1987. Codon replacement in the *PGK1* gene of *Saccharomyces cer-*

evisiae: experimental approach to study the role of biased codon usage in gene expression. *Mol. Cell. Biol.* **7**:2914–2924.

73. Hoekstra, M. F., and R. E. Malone. 1987. Hyper-mutation caused by the *rem1* mutation in yeast is not dependent on error-prone or excision repair. *Mutat. Res.* **178**:201–210.

74. Ishii, C., and H. Inoue. 1989. Epistasis, photoreactivation and mutagen sensitivity of DNA repair mutants *upr-1* and *mus-26* in *Neurospora crassa*. *Mutat. Res.* **218**:95–103.

75. Jacquier, A., P. Legrain, and B. Dujon. 1992. Sequence of a 10.7 kb segment of yeast chromosome XI identifies the *APN1* and the *BAF1* loci and reveals one tRNA gene and several new open reading frames including homologs to *RAD2* and kinases. *Yeast* **8**:121–132.

76. Johnson, A. L., D. G. Barker, and L. H. Johnston. 1986. Induction of yeast DNA ligase genes in exponential and stationary phase cultures in response to DNA damaging agents. *Curr. Genet.* **11**:107–112.

77. Johnson, E. S., B. Bartel, W. Seufert, and A. Varshavsky. 1992. Ubiquitin as a degradation signal. *EMBO J.* **11**:497–505.

78. Johnson, L. M., M. Snyder, L. M. Chang, R. W. Davis, and J. L. Campbell. 1985. Isolation of the gene encoding yeast DNA polymerase I. *Cell* **43**:369–377.

79. Johnston, L. H., and K. A. Nasmyth. 1978. *Saccharomyces cerevisiae* cell cycle mutant *cdc9* is defective in DNA ligase. *Nature* (London) **274**:891–893.

80. Jones, J. S., L. Prakash, and S. Prakash. 1990. Regulated expression of the *Saccharomyces cerevisiae* DNA repair gene *RAD7* in response to DNA damage and during sporulation. *Nucleic Acids Res.* **18**:3281–3285.

81. Jong, A. Y., C. L. Kuo, and J. L. Campbell. 1984. The *CDC8* gene of yeast encodes thymidylate kinase. *J. Biol. Chem.* **259**:11052–11059.

82. Kerr, S. M., L. H. Johnston, M. Odell, S. A. Duncan, K. M. Law, and G. L. Smith. 1991. Vaccinia DNA ligase complements *Saccharomyces cerevisiae cdc9*, localizes in cytoplasmic factories and affects virulence and virus sensitivity to DNA damaging agents. *EMBO J.* **10**:4343–4350.

83. Klein, H. L. 1988. Different types of recombination events are controlled by the *RAD1* and *RAD52* genes of *Saccharomyces cerevisiae*. *Genetics* **120**:367–377.

84. Koken, M. H., C. Vreeken, S. A. Bol, N. C. Cheng, I. Jaspers-Dekker, J. H. J. Hoeijmakers, J. C. Eeken, G. Weeda, and A. Pastink. 1992. Cloning and characterization of the *Drosophila* homolog of the xeroderma pigmentosum complementation-group B correcting gene, *ERCC3*. *Nucleic Acids Res.* **20**:5541–5548.

85. Kornberg, R. D. 1994. Personal communication.

86. Kuo, C. L., and J. L. Campbell. 1983. Cloning of *Saccharomyces cerevisiae* DNA replication genes: isolation of the *CDC8* gene and two genes that compensate for the *cdc8-1* mutation. *Mol. Cell. Biol.* **3**:1730–1737.

87. Lambert, C., L. B. Couto, W. A. Weiss, R. A. Schultz, L. H. Thompson, and E. C. Friedberg. 1988. A yeast DNA repair gene partially complements defective excision repair in mammalian cells. *EMBO J.* **7**:3245–3253.

88. Lasko, D. D., A. E. Tomkinson, and T. Lindahl. 1990. Eukaryotic DNA ligases. *Mutat. Res.* **236**:2–3.

89. Laurenson, P., and J. Rine. 1992. Silencers, silencing, and heritable transcriptional states. *Microbiol. Rev.* **56**:543–560.

90. Leadon, S. A., and D. A. Lawrence. 1992. Strand-selective repair of DNA damage in the yeast *GAL7* gene requires RNA polymerase II. *J. Biol. Chem.* **267**:23175–23182.

91. Legerski, R., and C. Peterson. 1992. Expression cloning of a human DNA repair gene involved in xeroderma pigmentosum group C. *Nature* (London) **359**:70–73.

92. Lehmann, A. R., A. M. Carr, F. Z. Watts, and J. M. Murray. 1991. DNA repair in the fission yeast, *Schizosaccharomyces pombe*. *Mutat. Res.* **250**:205–210.

93. Lemontt, J. F. 1971. Mutants of yeast defective in mutation induced by ultraviolet light. *Genetics* **68**:21–33.

94. Lemontt, J. F. 1971. Pathways of ultraviolet mutability in *Saccharomyces cerevisiae*. I. Some properties of double mutants involving *uvs9* and *rev*. *Mutat. Res.* **13**:311–317.

95. Lucchini, G., A. Brandazza, G. Badaracco, M. Bianchi, and P. Plevani. 1985. Identification of the yeast DNA polymerase I gene with antibody probes. *Curr. Genet.* **10**:245–252.

96. Lucchini, G., C. Mazza, E. Scacheri, and P. Plevani. 1988. Genetic mapping of the *Saccharomyces cerevisiae* DNA polymerase I gene and characterization of a *pol1* temperature-sensitive mutant altered in DNA primase-polymerase complex stability. *Mol. Gen. Genet.* **212**:459–465.

97. Ma, L., A. Westbroek, A. G. Jochemsen, G. Weeda, A. Bosch, D. Bootsma, J. H. J. Hoeijmakers, and A. J. van der Eb. 1994. Mutational analysis of *ERCC3*, which is involved in DNA repair and transcription initiation: identification of domains essential for the DNA repair function. *Mol. Cell. Biol.* **14**:4126–4134.

98. Madura, K., and S. Prakash. 1986. Nucleotide sequence, transcript mapping, and regulation of the *RAD2* gene of *Saccharomyces cerevisiae*. *J. Bacteriol.* **166**:914–923.

99. Madura, K., and S. Prakash. 1990. The *Saccharomyces cerevisiae* DNA repair gene *RAD2* is regulated in meiosis but not during the mitotic cell cycle. *Mol. Cell. Biol.* **10**:3256–3257.

100. Madura, K., and S. Prakash. 1990. Transcript levels of the *Saccharomyes cerevisiae* DNA repair gene *RAD23* increase in response to UV light and in meiosis but remain constant in the mitotic cell cycle. *Nucleic Acids Res.* **18**:4737–4742.

101. Magana-Schwencke, N., J. A. Henriques, R. Chanet, and E. Moustacchi. 1982. The fate of 8-methoxypsoralen photoinduced crosslinks in nuclear and mitochondrial yeast DNA: comparison of wild-type and repair-deficient strains. *Proc. Natl. Acad. Sci. USA* **79**:1722–1726.

102. Malone, R. E., and M. F. Hoekstra. 1984. Relationships between a hyper-rec mutation (*rem1*) and other recombination and repair genes in yeast. *Genetics* **107**:33–48.

103. McCready, S., A. M. Carr, and A. R. Lehmann. 1993. Repair of cyclobutane pyrimidine dimers and 6-4 photoproducts in the fission yeast *Schizosaccharomyces pombe*. *Mol. Microbiol.* **10**:885–890.

104. McCready, S. J., H. Burkill, S. Evans, and B. S. Cox. 1989. The *Saccharomyces cerevisiae RAD2* gene complements a *Schizosaccharomyces pombe* repair mutation. *Curr. Genet.* **15**:27–30.

105. McKnight, G. L., T. S. Cardillo, and F. Sherman. 1981. An extensive deletion causing overproduction of yeast iso-2-cytochrome c. *Cell* **25**:409–419.

106. Montelone, B. A., M. F. Hoekstra, and R. E. Malone. 1988. Spontaneous mitotic recombination in yeast: the hyper-recombinational rem1 mutations are alleles of the *RAD3* gene. *Genetics* **119**:289–301.

107. Montelone, B. A., and R. E. Malone. 1994. Analysis of the *rad3-101* and *rad3-102* mutations of *Saccharomyces cerevisiae*: implications for structure/function of Rad3 protein. *Yeast* **10**:13–27.

108. Morrison, A., H. Araki, A. B. Clark, R. K. Hamatake, and A. Sugino. 1990. A third essential DNA polymerase in *S. cerevisiae*. *Cell* **62**:1143–1151.

109. Morrison, A., R. B. Christensen, J. Alley, A. K. Beck, E. G. Bernstine, J. F. Lemontt, and C. W. Lawrence. 1989. REV3, a *Saccharomyces cerevisiae* gene whose function is required for induced mutagenesis, is predicted to encode a nonessential DNA polymerase. *J. Bacteriol.* **171**:5659–5667.

110. Morrison, A., and A. Sugino. 1994. The $3' \rightarrow 5'$ exonucleases of both DNA polymerases δ and ε participate in correcting errors of DNA replication in *Saccharomyces cerevisiae*. *Mol. Gen. Genet.* **242**:289–296.

111. Mounkes, L. C., R. S. Jones, B. C. Liang, W. Gelbart, and M. T. Fuller. 1992. A *Drosophila* model for xeroderma pigmentosum and Cockayne's syndrome: *haywire* encodes the fly homolog of *ERCC3*, a human excision repair gene. *Cell* **71**:925–937.

112. Moustacchi, E., C. Cassier, R. Chanet, N. Magaña-Schwencke, T. Saeki, and J. A. P. Henriques. 1983. Biological role of photo-induced crosslinks and monoadducts in yeast DNA: genetic control and steps involved in their repair, p. 87–106. *In* E. C. Friedberg and B. A. Bridges (ed.), *Cellular Responses to DNA Damage*. Alan R. Liss, Inc., New York.

113. Murray, J. M., C. L. Doe, P. Schenk, A. M. Carr, A. R. Lehmann, and F. Z. Watts. 1992. Cloning and characterisation of the *S. pombe rad15* gene, a homologue to the *S. cerevisiae RAD3* and human *ERCC2* genes. *Nucleic Acids Res.* **20**:2673–2678.

114. Murray, J. M., M. Tavassoli, R. Al-Harithy, K. S. Sheldrick, A. R. Lehmann, A. M. Carr, and F. Z. Watts. 1994. Structural and functional conservation of the human homolog of the *Schizosaccharomyces pombe rad2* gene, which is required for DNA repair and chromosome segregation. *Mol. Cell. Biol.* **14**:4878–4888.

115. Naegeli, H., L. Bardwell, and E. C. Friedberg. 1992. The DNA helicase and adenosine triphosphatase activities of yeast Rad3 protein are inhibited by DNA damage. A potential mechanism for damage-specific recognition. *J. Biol. Chem.* **267**:392–398.

116. Naegeli, H., L. Bardwell, and E. C. Friedberg. 1993. Inhibition of Rad3 DNA helicase activity by DNA adducts and abasic sites: implications for the role of a DNA helicase in damage-specific incision of DNA. *Biochemistry* **32**:613–621.

117. Naegeli, H., L. Bardwell, I. Harosh, and E. C. Friedberg. 1992. Substrate specificity of the Rad3 ATPase/DNA helicase of *Saccharomyces cerevisiae* and binding of Rad3 protein to nucleic acids. *J. Biol. Chem.* **267**:7839–7844.

118. Naegeli, H., P. Modrich, and E. C. Friedberg. 1993. The DNA helicase activities of Rad3 protein of *Saccharomyces cerevisiae* and helicase II of *Escherichia coli* are differentially inhibited by covalent and noncovalent DNA modifications. *J. Biol. Chem.* **268**:10386–10392.

119. Naumovski, L., G. Chu, P. Berg, and E. C. Friedberg. 1985. *RAD3* gene of *Saccharomyces cerevisiae*: nucleotide sequence of wild-type and mutant alleles, transcript mapping, and aspects of gene regulation. *Mol. Cell. Biol.* **5**:17–26.

120. Naumovski, L., and E. C. Friedberg. 1986. Analysis of the essential and excision repair functions of the *RAD3* gene of *Saccharomyces cerevisiae* by mutagenesis. *Mol. Cell. Biol.* **6**:1218–1227.

121. Nishida, C., P. Reinhard, and S. Linn. 1988. DNA repair synthesis in human fibroblasts requires DNA polymerase δ. *J. Biol. Chem.* **263**:501–510.

122. O'Donovan, A., A. A. Davies, J. G. Moggs, S. C. West, and R. D. Wood. 1994. XPG endonuclease makes the 3′ incision in human DNA nucleotide excision repair. *Nature* (London). **371**:432–435.

123. O'Donovan, A., and R. D. Wood. 1993. Identical defects in DNA repair in xeroderma pigmentosum group G and rodent ERCC group 5. *Nature* (London) **363**:185–188.

124. Oh, E. Y., and L. Grossman. 1987. Helicase properties of the *Escherichia coli* UvrAB protein complex. *Proc. Natl. Acad. Sci. USA* **84**:3638–3642.

125. Paetkau, D. W., J. A. Riese, and R. D. Geitz. 1994. Interaction of the yeast RAD7 and SIR3 proteins; implications for DNA repair and chromatin structure. *Genes Dev.* **8**:2035–2045

126. Park, E., S. N. Guzder, M. H. Koken, I. Jaspers-Dekker, G. Weeda, J. H. J. Hoeijmakers, S. Prakash, and L. Prakash. 1992. *RAD25* (*SSL2*), the yeast homolog of the human xeroderma pigmentosum group B DNA repair gene, is essential for viability. *Proc. Natl. Acad. Sci. USA* **89**:11416–11420.

127. Perozzi, G., and S. Prakash. 1986. *RAD7* gene of *Saccharomyces cerevisiae*: transcripts, nucleotide sequence analysis, and functional relationship between the *RAD7* and *RAD23* gene products. *Mol. Cell. Biol.* **6**:1497–1507.

128. Peterson, T. A., L. Prakash, S. Prakash, M. A. Osley, and S. I. Reed. 1985. Regulation of *CDC9*, the *Saccharomyces cerevisiae* gene that encodes DNA ligase. *Mol. Cell. Biol.* **5**:226–235.

129. Phipps, J., A. Nasim, and D. R. Miller. 1985. Recovery, repair, and mutagenesis in *Schizosaccharomyces pombe*. *Adv. Genet.* **23**:1–72.

130. Pignede, G., D. Bouvier, A. M. de Recondo, and G. Baldacci. 1991. Characterization of the *POL3* gene product from *Schizosaccharomyces pombe* indicates inter-species conservation of the catalytic subunit of DNA polymerase δ. *J. Mol. Biol.* **222**:209–218.

131. Prakash, L., D. Hinkle, and S. Prakash. 1979. Decreased UV mutagenesis in *cdc8*, a DNA replication mutant of *Saccharomyces cerevisiae*. *Mol. Gen. Genet.* **172**:249–258.

132. Prakash, S., P. Sung, and L. Prakash. 1993. DNA repair genes and proteins of *Saccharomyces cerevisiae*. *Annu. Rev. Genet.* **27**:33–70.

133. Prasad, R., S. G. Widen, R. K. Singhal, J. Watkins, L. Prakash, and S. H. Wilson. 1993. Yeast open reading frame YCR14C encodes a DNA β-polymerase-like enzyme. *Nucleic Acids Res.* **21**:5301–5307.

134. Qiu, H., E. Park, L. Prakash, and S. Prakash. 1993. The *Saccharomyces cerevisiae* DNA repair gene *RAD25* is required for transcription by RNA polymerase II. *Genes Dev.* **7**:2161–2171.

135. Quah, S. K., R. C. von Borstel, and P. J. Hastings. 1980. The origin of spontaneous mutation in *Saccharomyces cerevisiae*. *Genetics* **96**:819–839.

136. Reagan, M., C. Pittenger, W. Siede, and E. C. Friedberg. 1995. Characterization of a yeast strain delected for the *RAD27* gene: a structural homolog of the *RAD2* nucleotide excision repair gene. *J. Bacteriol.* in press.

137. Reynolds, P., D. R. Higgins, L. Prakash, and S. Prakash. 1985. The nucleotide sequence of the *RAD3* gene of *Saccharomyces cerevisiae*: a potential adenine nucleotide binding amino acid sequence and a nonessential acidic carboxyl terminal region. *Nucleic Acids Res.* **13**:2357–2372.

138. Reynolds, P., L. Prakash, D. Dumais, G. Perozzi, and S. Prakash. 1985. Nucleotide sequence of the *RAD10* gene of *Saccharomyces cerevisiae*. *EMBO J.* **4**:3549–3552.

139. Reynolds, P. R., S. Biggar, L. Prakash, and S. Prakash. 1992. The *Schizosaccharomyces pombe rhp3*+ gene required for DNA repair and cell viability is functionally interchangeable with the *RAD3* gene of *Saccharomyces cerevisiae*. *Nucleic Acids Res.* **20**:2327–2334.

140. Reynolds, R. J., and E. C. Friedberg. 1981. Molecular mechanisms of pyrimidine dimer excision in *Saccharomyces cerevisiae*: incision of ultraviolet-irradiated deoxyribonucleic acid in vivo. *J. Bacteriol.* **146**:692–704.

141. Richter, D., E. Niegemann, and M. Brendel. 1992. Molecular structure of the DNA cross-link repair gene *SMN1* (*PSO2*) of the yeast *Saccharomyces cerevisiae*. *Mol. Gen. Genet.* **231**:194–200.

141a. Robins, P., D. J. C. Pappin, R. D. Wood, and T. Lindahl. 1994. Structural and functional homology between mammalian DNaseIV and the 5′ nuclease of *Escherichia coli* DNA polymerase I. *J. Biol. Chem.* **269**:28535–28538.

142. Robinson, G. W., C. M. Nicolet, D. Kalainov, and E. C. Friedberg. 1986. A yeast excision-repair gene is inducible by DNA damaging agents. *Proc. Natl. Acad. Sci. USA* **83**:1842–1846.

143. Rödel, C., S. Kirchhoff, and H. Schmidt. 1992. The protein sequence and some intron positions are conserved between the switching gene *swi10* of *Schizosaccharomyces pombe* and the human excision repair gene *ERCC1*. *Nucleic Acids Res.* **20**:6347–6353.

144. Ruhland, A., E. Haase, W. Siede, and M. Brendel. 1981. Isolation of yeast mutants sensitive to the bifunctional alkylating agent nitrogen mustard. *Mol. Gen. Genet.* **181**:346–351.

145. Ruhland, A., M. Kircher, F. Wilborn, and M. Brendel. 1981. A yeast mutant specifically sensitive to bifunctional alkylation. *Mutat. Res.* **91**:457–462.

146. Sancar, A. 1994. Personal communication.

147. Sandell, L. E., and V. A. Zakian. 1992. Telomere position effect in yeast. *Trends Cell Biol.* **2**:10–14.

148. Saxena, J. K., J. B. Hays, and E. J. Ackerman. 1990. Excision repair of UV-damaged plasmid DNA in *Xenopus oocytes* is mediated by DNA polymerase α (and/or δ). *Nucleic Acids Res.* **18**:7425–7432.

149. Schaeffer, L., R. Roy, S. Humbert, V. Moncollin, W. Vermeulen, J. H. J. Hoeijmakers, P. Chambon, and J.-M. Egly. 1993. DNA repair helicase: a component of BTF2 (TFIIH) basic transcription factor. *Science* **260**:58–63.

150. Scherly, D., T. Nouspikel, J. Corlet, C. Ucla, A. Bairoch, and S. G. Clarkson. 1993. Complementation of the DNA repair defect in xeroderma pigmentosum group G cells by a human cDNA related to yeast *RAD2. Nature* (London) **363:**182–185.

151. Schiestl, R. H., and S. Prakash. 1988. *RAD1*, an excision repair gene of *Saccharomyces cerevisiae*, is also involved in recombination. *Mol. Cell. Biol.* **8:**3619–3626.

152. Schiestl, R. H., and S. Prakash. 1990. *RAD10*, an excision repair gene of *Saccharomyces cerevisiae*, is involved in the *RAD1* pathway of mitotic recombination. *Mol. Cell. Biol.* **10:**2485–2491.

153. Schild, D., B. J. Glassner, R. K. Mortimer, M. Carlson, and B. C. Laurent. 1992. Identification of *RAD16*, a yeast excision repair gene homologous to the recombinational repair gene *RAD54* and to the *SNF2* gene involved in transcriptional activation. *Yeast* **8:**385–395.

154. Schlake, C., K. Ostermann, H. Schmidt, and H. Gutz. 1993. Analysis of DNA repair pathways of *Schizosaccharomyces pombe* by means of *swi-rad* double mutants. *Mutat. Res.* **294:**59–67.

155. Sclafani, R. A., and W. L. Fangman. 1984. Yeast gene *CDC8* encodes thymidylate kinase and is complemented by herpes thymidine kinase gene *TK. Proc. Natl. Acad. Sci. USA* **81:**5821–5825.

156. Sharp, P. M., T. M. Tuohy, and K. R. Mosurski. 1986. Codon usage in yeast: cluster analysis clearly differentiates highly and lowly expressed genes. *Nucleic Acids Res.* **14:**5125–5143.

157. Shimamoto, T., K. Kohno, K. Tanaka, and Y. Okada. 1991. Molecular cloning of human *XPAC* gene homologs from chicken, *Xenopus laevis* and *Drosophila melanogaster. Biochem. Biophys. Res. Commun.* **181:**1231–1237.

158. Shimizu, K., C. Santocanale, P. A. Ropp, M. P. Longhese, P. Plevani, G. Lucchini, and A. Sugino. 1993. Purification and characterization of a new DNA polymerase from budding yeast *Saccharomyces cerevisiae*. A probable homolog of mammalian DNA polymerase β. *J. Biol. Chem.* **268:**27148–27153.

159. Sidik, K., H. B. Lieberman, and G. A. Freyer. 1992. Repair of DNA damaged by UV light and ionizing radiation by cell-free extracts prepared from *Schizosaccharomyces pombe. Proc. Natl. Acad. Sci. USA* **89:**12112–12116.

160. Siede, W., and E. C. Friedberg. 1992. Regulation of the yeast *RAD2* gene: DNA damage-dependent induction correlates with protein binding to regulatory sequences and their deletion influences survival. *Mol. Gen. Genet.* **232:**247–256.

161. Siede, W., G. W. Robinson, D. Kalainov, T. Malley, and E. C. Friedberg. 1989. Regulation of the *RAD2* gene of *Saccharomyces cerevisiae. Mol. Microbiol.* **3:**1697–1707.

162. Sitney, K. C., M. E. Budd, and J. L. Campbell. 1989. DNA polymerase III, a second essential DNA polymerase, is encoded by the *S. cerevisiae CDC2 gene. Cell* **56:**599–605.

163. Smerdon, M. J., R. Gupta, and A. O. Murad. 1992. DNA repair in transcriptionally active chromatin, p. 258–270. *In* V. A. Bohr, K. Wassermann, and K. H. Kraemer (ed.), *DNA Repair Mechanisms.* Munksgaard, Copenhagen.

164. Smith, P. D., and R. L. Dusenbery. 1988. Twelve genes of the alkylation excision repair pathway of *Drosophila melanogaster*, p. 251–255. *In* E. C. Friedberg and P. C. Hanawalt (ed.), *Mechanisms and Consequences of DNA Damage Processing.* Alan R. Liss, Inc., New York.

165. Song, J. M., B. A. Montelone, W. Siede, and E. C. Friedberg. 1990. Effects of multiple yeast *rad3* mutant alleles on UV sensitivity, mutability, and mitotic recombination. *J. Bacteriol.* **172:**6620–6630.

166. Sugino, A., B. H. Ryu, T. Sugino, L. Naumovski, and E. C. Friedberg. 1986. A new DNA-dependent ATPase which stimulates yeast DNA polymerase I and has DNA-unwinding activity. *J. Biol. Chem.* **261:**11744–11750.

167. Sung, P., V. Bailly, C. Weber, L. H. Thompson, L. Prakash, and S. Prakash. 1993. Human xeroderma pigmentosum group D gene encodes a DNA helicase. *Nature* (London) **365:**852–855.

168. Sung, P., D. Higgins, L. Prakash, and S. Prakash. 1988. Mutation of lysine-48 to arginine in the yeast RAD3 protein abolishes its ATPase and DNA helicase activities but not the ability to bind ATP. *EMBO J.* **7:**3263–3269.

168a. Sung, P., L. Prakash, S. W. Matson, and S. Prakash. 1987. RAD3 protein of *Saccharomyces cerevisiae* is a DNA helicase. *Proc. Natl. Acad. Sci. USA* **84:**8951–8955.

169. Sung, P., L. Prakash, and S. Prakash. 1992. Renaturation of DNA catalysed by yeast DNA repair and recombination protein RAD10. *Nature* (London) **355:**743–745.

170. Sung, P., L. Prakash, S. Weber, and S. Prakash. 1987. The *RAD3* gene of *Saccharomyces cerevisiae* encodes a DNA-dependent ATPase. *Proc. Natl. Acad. Sci. USA* **84:**6045–6049.

171. Sung, P., P. Reynolds, L. Prakash, and S. Prakash. 1993. Purification and characterization of the *Saccharomyces cerevisiae* RAD1/RAD10 endonuclease. *J. Biol. Chem.* **268:**26391–26399.

172. Svejstrup, J. Q., W. J. Feaver, J. LaPointe, and R. D. Kornberg. 1994. RNA polymerase II transcription factor IIH holoenzyme from yeast. *J. Biol. Chem.*, in press.

173. Svejstrup, J. Q., Z. Wang, W. J. Feaver, X. Wu, F. Donahue, E. C. Friedberg, and R. D. Kornberg. 1995. Different forms of yeast RNA polymerase transcription factor IIH (TFIIH) for transcription and DNA repair: holoTFIIH and a nucleotide excision repairosome. *Cell,* in press.

174. Svoboda, D. L., J.-S. Taylor, J. E. Hearst, and A. Sancar. 1993. DNA repair by eukaryotic nucleotide excision nuclease. Removal of thymine dimer and psoralen monoadduct by HeLa cell-free extract and of thymine dimer by *Xenopus laevis* oocytes. *J. Biol. Chem.* **268:**1931–1936.

175. Sweder, K. S., and P. C. Hanawalt. 1992. Preferential repair of cyclobutane pyrimidine dimers in the transcribed strand of a gene in yeast chromosomes and plasmids is dependent on transcription. *Proc. Natl. Acad. Sci. USA* **89:**10696–10700.

176. Sweder, K. S., and P. C. Hanawalt. 1994. The COOH terminus of suppressor of stem loop (*SSL2/RAD25*) in yeast is essential for overall genomic excision repair and transcription-coupled repair. *J. Biol. Chem.* **269:**1852–1857.

177. Terleth, C., P. Schenk, R. Poot, J. Brouwer, and P. van de Putte. 1990. Differential repair of UV damage in *rad* mutants of *Saccharomyces cerevisiae*: a possible function of G_2 arrest upon UV irradiation. *Mol. Cell. Biol.* **10:**4678–4684.

178. Terleth, C., R. Waters, J. Brouwer, and P. van de Putte. 1990. Differential repair of UV damage in *Saccharomyces cerevisiae* is cell cycle dependent. *EMBO J.* **9:**2899–2904.

179. Tomkinson, A. E., A. J. Bardwell, L. Bardwell, N. J. Tappe, and E. C. Friedberg. 1993. Yeast DNA repair and recombination proteins Rad1 and Rad10 constitute a single-stranded-DNA endonuclease. *Nature* (London) **362:**860–862.

180. Tomkinson, A. E., A. J. Bardwell, N. Tappe, W. Ramos, and E. C. Friedberg. 1994. Purification of Rad1 protein from *Saccharomyces cerevisiae* and further characterization of the Rad1/Rad endonuclease complex. *Biochemistry* **33:**5305–5311.

181. Tomkinson, A. E., N. J. Tappe, and E. C. Friedberg. 1992. DNA ligase I from *Saccharomyces cerevisiae*: physical and biochemical characterization of the *CDC9* gene product. *Biochemistry* **31:**11762–11771.

182. Toniolo, D., M. Persico, and M. Alcalay. 1988. A "housekeeping" gene on the X chromosome encodes a protein similar to ubiquitin. *Proc. Natl. Acad. Sci. USA* **85:**851–855.

183. van Duin, M., J. de Wit, H. Odijk, A. Westerveld, A. Yasui, H. M. Koken, J. H. J. Hoeijmakers, and D. Bootsma. 1986. Molecular characterization of the human excision repair gene *ERCC-1*: cDNA cloning and amino acid homology with the yeast DNA repair gene *RAD10. Cell* **44:**913–923.

184. van Duin, M., J. van den Tol, P. Warmerdam, H. Odijk, D. Meijer, A. Westerveld, D. Bootsma, and J. H. J. Hoeijmakers. 1988. Evolution and mutagenesis of the mammalian excision repair gene *ERCC-1. Nucleic Acids Res.* **16:**5305–5322.

185. Van Houten, B. 1990. Nucleotide excision repair in *Escherichia coli. Microbiol. Rev.* **54:**18–51.

185a. Verhage, R., A.-M. Zeeman, N. de Groot, F. Gleig, D. D. Bang, P. van de Putte, and J. Brouwer. 1994. The *RAD7* and *RAD16* genes, which are essential for pyrimidine removal from the silent mating type loci, are also required for repair of the nontranscribed strand of an active gene in *Saccharomyces cerevisiae. Mol. Cell. Biol.* **14:**6135–6142.

186. Voelkel-Meiman, K., R. L. Keil, and G. S. Roeder. 1987. Recombination-stimulating sequences in yeast ribosomal DNA correspond to sequences regulating transcription by RNA polymerase I. *Cell* **48:**1071–1079.

187. Wang, Z., S. Buratowski, J. Q. Svejstrup, W. J. Feaver, X. Wu, R. D. Kornberg, T. F. Donahue, and E. C. Friedberg. 1995. Yeast *TFB1* and *SSL1* genes encoding subunits of transcription factor IIH (TFIIH) are required for nucleotide excision repair and transcription. *Mol. Cell. Biol.* in press.

188. Wang, Z., J. Q. Svejstrup, W. J. Feaver, X. Wu, R. D. Kornberg, and E. C. Friedberg. 1994. Transcription factor b (TFIIH) is required during nucleotide-excision repair in yeast. *Nature* (London) **368:**74–76.

189. Wang, Z., X. Wu, and E. C. Friedberg. 1993. DNA repair synthesis during base excision repair in vitro is catalyzed by DNA polymerase ε and is influenced by DNA polymerases α and δ in *Saccharomyces cerevisiae. Mol. Cell. Biol.* **13:**1051–1058.

190. Wang, Z., X. Wu, and E. C. Friedberg. Unpublished observations.

191. Watkins, J. F., P. Sung, L. Prakash, and S. Prakash. 1993. The *Saccharomyces cerevisiae* DNA repair gene *RAD23* encodes a nuclear protein containing a ubiquitin-like domain required for biological function. *Mol. Cell. Biol.* **13:**7757–7765.

191a. Weiss, W. A., and E. C. Friedberg. 1985. Molecular cloning and characterization of the yeast *RAD10* gene and expression of Rad10 protein in *E. coli. EMBO J.* **4:**1575–1582.

192. Welker, D. L., and R. A. Deering. 1978. Genetics of radiation sensitivity in the slime mould *Dictyostelium discoideum. J. Gen. Microbiol.* **109:**11–23.

193. White, J. H., D. G. Barker, P. Nurse, and L. H. Johnston. 1986. Periodic transcription as a means of regulating gene expression during the cell cycle: contrasting modes of expression of DNA ligase genes in budding and fission yeast. *EMBO J.* **5:**1705–1709.

194. Wilcox, D. R., and L. Prakash. 1981. Incision and postincision steps of pyrimidine dimer removal in excision-defective mutants of *Saccharomyces cerevisiae. J. Bacteriol.* **148:**618–623.

195. Yoon, H., S. P. Miller, E. K. Pabich, and T. F. Donahue. 1992. *SSL1*, a suppressor of a *HIS4* 5'-UTR stem-loop mutation, is essential for translation initiation and affects UV resistance in yeast. *Genes Dev.* **6:**2463–2477.

196. Zehfus, B. R., A. D. McWilliams, Y. H. Lin, M. F. Hoekstra, and R. L. Keil. 1990. Genetic control of RNA polymerase I-stimulated recombination in yeast. *Genetics* **126:**41–52.

7

Nucleotide Excision Repair in Mammalian Cells: General Considerations and Chromatin Dynamics

This chapter discusses nucleotide excision repair in mammalian cells. Unlike the situation in some prokaryotic cells, there is no evidence for the existence of pyrimidine dimer-DNA glycosylases (PD-DNA glycosylases) in mammalian cells. Hence, it is likely that all excision of pyrimidine dimers and of other types of bulky base damage in higher organisms is mediated by the action of a general damage recognition-specific DNA incision activity formally equivalent to the UvrABC endonuclease of *Escherichia coli* (see chapter 5) and the analogous enzyme activity predicted in *S. cerevisiae* (see chapter 6). As indicated in the previous chapter, at the time of writing such an activity has not yet been fully characterized in any eukaryotic system, and the extent to which the basic biochemistry of nucleotide excision repair defined in prokaryotes is conserved in mammalian cells remains an open question.

Biology of Nucleotide Excision Repair in Mammalian Cells

The basic phenomenon of nucleotide excision repair in mammalian cells is documented by a wealth of radiobiology and cell biology studies dating back almost 30 years. These studies have been aided by the availability of a large number of mutant cell lines whose phenotypes have in general corroborated the results of experiments in normal cells. These mutants are discussed more fully in chapter 8 in the context of their utility for cloning mammalian genes involved in nucleotide excision repair. Studies on nucleotide excision repair in mammalian cells have involved a variety of experimental strategies and techniques, many of which have been adapted from methods developed for studies with bacteria. Some of these procedures measure damage-specific incision of DNA, either directly or indirectly. Others measure the physical excision of damaged nucleotides, repair synthesis of DNA, or the rejoining of strand breaks (DNA ligation). A description of most of

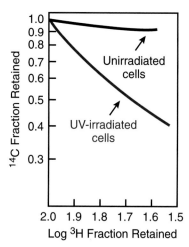

Figure 7–1 Measurement of nucleotide excision repair by the alkaline elution technique. The DNA of CHO cells was prelabeled with [³H]thymidine, and the cells were incubated in the presence of hydroxyurea and 1-β-ᴅ-arabinofuranosylcytosine prior to exposure to UV radiation (or not) to inhibit repair synthesis, thereby accumulating strand breaks introduced by damage-specific incision of DNA. Cells were allowed to carry out repair for 6 min prior to alkaline elution. The alkaline elution profiles of the cells were normalized to that of an internal standard [¹⁴C]DNA sample treated with a known amount of γ-irradiation. The initial slopes of the curves are a measure of the relative number of strand breaks introduced during nucleotide excision repair. (*Adapted from Thompson et al. [176] with permission.*)

these techniques was presented in chapter 5 with a particular emphasis on their utility for bacterial cells. The following section stresses their utility for studies in mammalian cells, particularly with respect to UV radiation-induced DNA damage. However, many of these techniques are readily applicable to the study of other forms of DNA damage. A detailed and comprehensive review of various techniques used for measuring nucleotide excision repair and other types of DNA repair in mammalian cells is presented in a published multivolume series (50–53).

Damage-Specific Incision of DNA

The presence of repair-specific single-strand breaks in the DNA of mammalian cells exposed to agents such as UV radiation can be demonstrated by the use of techniques which essentially involve the denaturation of DNA and the resolution of fragments based on their relative size. The technique of alkaline elution is a particularly sensitive and useful procedure. In this procedure, cells in alkali are placed on a filter that is subject to a controlled vacuum (81). The alkali lyses the cells and also denatures the DNA. The rate at which the DNA is eluted from the filters is a direct function of its single-strand molecular weight; i.e., the smaller the DNA (the more nicks), the faster it elutes (81) (Fig. 7–1).

Loss of Radiolabeled Thymine-Containing Pyrimidine Dimers from DNA

As described in chapter 5, the excision of damage-containing oligonucleotides from UV irradiated DNA radiolabeled in thymine residues can be conveniently demonstrated by the transfer of labeled thymine to an ethanol- or acid-soluble fraction. It is of course necessary to resolve the radiolabel specifically associated with pyrimidine dimers from that associated with nondimer nucleotides simultaneously transferred to the soluble phase. A number of techniques for achieving such resolution are discussed in chapter 5.

Loss of Sites in DNA Sensitive to Pyrimidine Dimer-Specific Enzymes

In chapter 4 we stated that the *Micrococcus luteus* and phage T4 pyrimidine dimer-DNA glycosylases are specific for pyrimidine dimers in DNA. If these lesions are removed during nucleotide excision repair and the covalent integrity of the DNA is restored by repair synthesis and DNA ligation, such DNA will no longer be sensitive to attack by these enzyme probes. Thus, when such DNA is incubated with one of these enzymes and then denatured and sedimented in alkaline sucrose gradients, it will have a higher single stranded-molecular weight than will DNA in which dimers are still present (see Fig. 5–18). This so-called loss of enzyme-sensitive sites (ESS) assay is discussed in detail in chapter 5. As will be seen later in this chapter, this enzyme specificity coupled with the technology of Southern hybridization facilitated the development of a novel and elegant technique for measuring the loss of ESS in single genes, thereby permitting research into several aspects of the intra- and intergenic heterogeneity of nucleotide excision and other types of repair in mammalian cells.

Ligation-Mediated PCR

Ligation-mediated PCR is a technique that enables the precise detection of incisions introduced into genomic DNA by any endonuclease that generates ligatable (by DNA ligase) nicks (113, 131). Three oligonucleotide primers that are complementary to three overlapping sites in a sequenced region of the genome are synthesized. After genomic DNA is cut with an endonuclease that generates 5' phosphate termini, the fragments are denatured and annealed to the first primer. Primer extension with a polymerase used for PCR terminates at the nicks, generating ligatable blunt ends. The ligation of oligonucleotide linkers to these ends provides a common sequence at the 5' end of each DNA fragment. The second oligonucleotide primer (whose sequence partially overlaps that of the first primer) is then used in a PCR in conjunction with a primer to the common linker sequence.

The amplified molecules are resolved on a DNA sequencing gel and hybridized with a radiolabeled probe for the region in question generated by PCR with the third primer. Following autoradiography, each band in the gel represents a nucleotide position where a ligatable nick was originally introduced.

This technique has been adapted to measurement of the repair of pyrimidine dimers by a refinement of the ESS assay. If pyrimidine dimers persist in DNA after an arbitrary repair period, they are of course sensitive to nicking by PD-DNA glycosylase. The incisions so generated are ligatable after releasing the 5′ pyrimidine in the dimer by monomerizing the dimer with DNA photolyase (see Fig. 4–22). Hence, following ligation-mediated PCR, bands detected by autoradiography of a DNA sequencing gel reflect the positions of persisting pyrimidine dimers in the region of DNA in question. This technique allows for not only the overall measurement of repair in a defined region of the genome but also the relative efficiency of the repair of lesions at different nucleotide positions in that region (56, 178).

Measurement of Repair Synthesis in DNA

The most general method used for measuring repair synthesis involves a combination of tritium and density labeling of the DNA (see chapter 5 and Fig. 5–19). The accuracy of this measurement is influenced chiefly by the amount of background semiconservative DNA synthesis. Thus, whenever feasible, selective inhibition of such synthesis is attempted. In mammalian cells, *hydroxyurea* is widely used for this purpose (34). This compound is an inhibitor of ribonucleotide diphosphate reductase (ribonucleotide reductase), an enzyme that facilitates the biosynthesis of deoxynucleoside triphosphates from the corresponding ribonucleoside diphosphates (171). Hydroxyurea has little or no effect on repair synthesis of DNA (34), probably because in mammalian cells there is a relatively small requirement of nucleotide excision repair for DNA precursors. Thus, the preexisting precursor pool is sufficient to allow repair synthesis without the need for further precursor production (34). Refinements of the density labeling technique allow for estimates of the size of the regions (patches) of repair synthesis in DNA (see chapter 5 and Fig. 5–20). As discussed in chapter 8, repair synthesis of DNA provides a sensitive and specific means of monitoring nucleotide excision repair in cell extracts of mammalian cells.

About the time that the technique for measuring DNA repair synthesis in bacteria was developed, an autoradiographic procedure provided the first demonstration of repair synthesis (and, incidentally, of nucleotide excision repair) in mammalian cells in culture (135). This procedure takes advantage of the fact that eukaryotic cells carry out semiconservative DNA synthesis only during a limited period of the cell cycle, a period designated as the S phase. If mammalian cells in culture are allowed to grow for a short period in the presence of [³H]thymidine and are then examined autoradiographically, S-phase cells show an intense labeling of their nuclei with silver grains, while cells not in the S phase show essentially no detectable grains (32) (Fig. 7–2). However, if the cells in culture undergo repair synthesis of DNA, non-S-phase cells also manifest the presence of silver grains (Fig. 7–2). By appropriate adjustment of the labeling and autoradiographic development times, the number of silver grains in cells carrying out repair (unscheduled) DNA synthesis, can be quantitated (32). This value reflects the amount of repair synthesis in the cells. Despite the rapid progress in the technology of molecular biology, this simple, albeit tedious, procedure is still widely used for demonstrating and quantitating nucleotide excision repair in vivo in mammalian cells.

Kinetics of Nucleotide Excision Repair in Mammalian Cells

While all the above procedures fully support the existence of a nucleotide excision repair mode(s) in higher organisms, the relative kinetics of DNA incision, the loss of thymine-containing dimers from high-molecular-weight DNA, and repair syn-

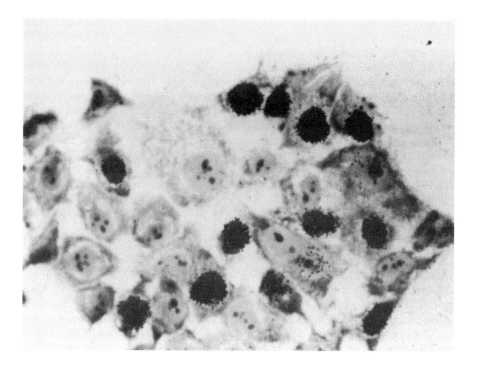

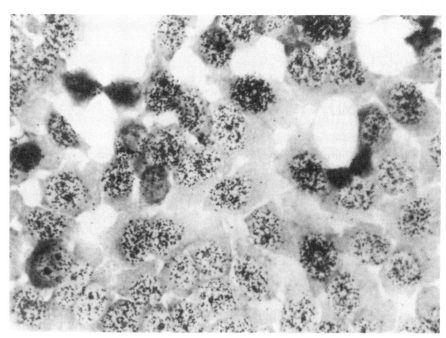

Figure 7–2   Repair synthesis of DNA during nucleotide excision repair can be visualized by autoradiography. Cultured human HeLa cells were either unirradiated (top) or exposed to UV radiation (bottom) and were then pulse-labeled with [³H]thymidine. Following autoradiography, cells in the S phase of the cell cycle engaged in active semiconservative DNA synthesis show intense autoradiographic labeling of their nuclei. The remaining unirradiated cells show virtually no background labeling except in the occasional cell just entering or leaving the S phase. On the other hand, all non-S-phase cells exposed to UV radiation show autoradiographic stippling, reflecting repair synthesis (so-called unscheduled DNA synthesis). (*Reproduced from Cleaver and Thomas [32] with permission.*)

thesis in mammalian cells are still controversial and have led to conflicting hypotheses about the precise mechanism of nucleotide excision repair in mammalian cells (49). In part these controversies have arisen from the use of different experimental procedures, which, though designed to measure the same end points, have optimal reliabilities at different extents of DNA damage and hence may not be strictly comparable. Nonetheless, some of these apparent "experimental inconsistencies" may actually reflect genuine distinctions in nucleotide excision repair associated with different cell types, different functional states of cells and/or particular regions of the genome, and different types of damage. Hence, they merit some consideration.

Following low levels of UV radiation, i.e., <20 J/m², it is difficult to accurately monitor nucleotide excision repair by simply measuring the loss of thymine-containing pyrimidine dimers from the acid-precipitable fraction of cells, because the absolute amount of radioactivity associated with the dimers at these doses becomes limiting (140). Hence, nucleotide excision repair at low doses of UV radiation is generally monitored by measuring the loss of ESS or by measuring repair synthesis. In general, such measurements are in reasonable agreement; a significant fraction of the total sites present are lost early after irradiation, and by 8 to 10 h at least half of the ESS are no longer detectable (45, 77, 106, 127, 194, 195) (Fig. 7–3). However, the time required for the complete removal of these sites may be much longer and can be as long as 10 to 12 days in cells irradiated with 1 J/m² and 15 to 20 days in cells irradiated at 10 J/m² (77). In the few studies in which direct comparisons have been made under similar experimental conditions, the kinetics of the loss of ESS correlates well with that of repair synthesis (82) (Fig. 7–3), suggesting that the two observations measure related biochemical events during nucleotide excision repair.

Some of the discrepancies in the literature on the kinetics of nucleotide excision repair have emerged from the phenotypic characterization of nucleotide excision repair-defective rodent and human cell lines. Several mutant Chinese hamster ovary (CHO) cell lines are very sensitive to UV radiation but retain high residual levels of nucleotide excision repair (200). Examination of one such mutant showed that it is defective in the repair of pyrimidine dimers but retains substantial capacity for repairing (6-4) photoproducts (107, 108). A second nucleotide excision repair-defective mutant from a different genetic complementation group (see chapter 8) has intermediate levels of UV radiation sensitivity and a normal capacity for DNA incision and is fully proficient in the removal of (6-4) photoproducts (107, 108). However, measurement of the loss of pyrimidine dimers showed a profound defect in the repair of this photoproduct (176). A final example of the same phenomenon is provided by studies on yet a third CHO mutant line, called UV20, which is defective in the rodent homolog of a human nucleotide excision repair gene called *ERCC1* (see chapter 8). The *ERCC1* gene maps to human chromosome 19. UV20 cells were fused to normal human lymphocytes, and a hybrid line carrying human chromosome 19 was isolated. As anticipated, this hybrid line was substantially complemented for UV radiation sensitivity, for excision of pyrimidine dimers, and for UV radiation-induced hypermutability at three different loci examined. However, nucleotide excision repair of (6-4) photoproducts was only partially restored (119).

It is now generally acknowledged that (6-4) photoproducts are removed from the genome of human and rodent cells (Fig. 7–4) more rapidly than are cyclobutane pyrimidine dimers (105). It is unlikely that these results reflect distinct mechanisms for the repair of these two types of photoproducts in mammalian cells. A more probable explanation is that, as is the case with the UvrABC endonuclease of *E. coli*, the nucleotide excision repair apparatus in mammalian cells recognizes different types of DNA damage with different binding affinities which determine the relative kinetics of the repair of different lesions. Indeed, on the basis of the results of studies with a mammalian cell-free system (see chapter 8), it has been suggested that the more rapid repair of (6-4) photoproducts relative to pyrimidine dimers (Fig. 7–4) is due in part to a significant difference in the

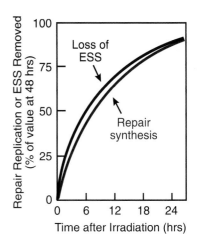

Figure 7–3 Comparison of the kinetics of repair synthesis (after UV irradiation at 10 J/m²) and the loss of sites in DNA sensitive to the T4 pyrimidine dimer-specific enzyme (ESS) (after irradiation at 5 J/m²) in human diploid fibroblasts. (*Adapted from Friedberg [49] with permission.*)

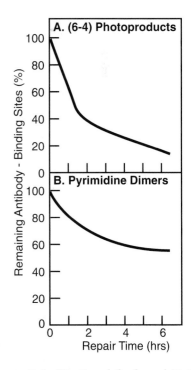

Figure 7–4 Kinetics of the loss of (6-4) photoproducts (A) and cyclobutane pyrimidine dimers (B) from the DNA of UV-irradiated CHO cells. (*Adapted from Nairn et al. [119] with permission.*)

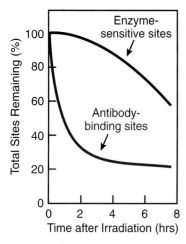

Figure 7–5 The rates of loss of sites sensitive to the T4 pyrimidine dimer-specific DNA glycosylase (enzyme-sensitive sites [ESS]) and to a pyrimidine dimer-specific antibody probe in human HeLa cells exposed to UV radiation are different. Sites sensitive to the antibody disappear faster than do those sensitive to the enzyme. (*Adapted from Friedberg [49] with permission.*)

ability of the nucleotide excision repair complex to locate and incise these lesions in chromatin (168).

As just discussed, the kinetics of the loss of pyrimidine dimers from DNA uses well characterized techniques which directly measure these photoproducts. Most studies of the repair of (6-4) photoproducts rely on the use of a sensitive immunological assay involving antibodies to this photoproduct. To eliminate the possibility that loss of antibody-sensitive sites does not reflect loss of antigen epitopes rather than excision repair, a procedure was developed to photochemically convert (6-4) photoproducts to strand breaks in DNA (103). The kinetics of the repair of these breaks was similar to that of the loss of antibody-sensitive sites. The kinetics of very early events during nucleotide excision repair measured by a variety of techniques correlates well with that of (6-4) photoproduct removal (26, 104, 141, 175), suggesting that these events do in fact reflect the repair of these lesions.

Another example of the complexity that emerges from measurements of the kinetics of nucleotide excision repair in mammalian cells concerns the use of pyrimidine dimer-specific antibodies as a probe for these lesions in DNA. Sites of binding for a dimer-specific antibody disappear more rapidly than do sites sensitive to the T4 pyrimidine dimer-DNA glycosylase probe (106). Within 1 h of irradiation 50% of the antibody sites are lost from the DNA of UV-irradiated CHO or HeLa cells, and within 4 h 80% of such sites are lost (106) (Fig. 7–5). This contrasts with the loss after 8 h of only about 40% of the sites sensitive to the T4 enzyme probe (106) (Fig. 7–5). The loss of antibody binding is not due to prior cleavage of the 5' glycosyl bonds of the dimers or to hydrolysis of the phosphodiester bonds 3' to the AP sites created by the T4 PD-DNA glycosylase, since when dimer-containing DNA is treated with T4 enzyme to effect cleavage at these bonds, dimers are still recognized by the antibody probe (106). Conceivably, pyrimidine dimers and/or the regions of DNA in which they reside in mammalian cell chromatin must undergo some sort of structural modification before they become accessible to the enzyme probe but are recognized by the antibody in the absence of such modification (106).

In general it would appear that most, if not all, differences in the kinetics of specific repair events associated with nucleotide excision repair in mammalian cells can be reconciled with the relative sensitivity of the techniques used and with the fact that certain lesions are indeed apparently repaired faster than others. Nonetheless, there are persistent but as yet incompletely substantiated indications that there are real differences in the physiology, if not the mechanism, of nucleotide excision repair in different cells, for example rodent cells and human cells.

Repair Patch Sizes in Mammalian Cells

Repair synthesis in mammalian cells is sometimes described in terms of short- and long-patch repair modes (63, 64). These do not have the mechanistic equivalence of the short and long patches in *E. coli* (see chapter 5). Short patches in mammalian cells are estimated to be 1 or 2 nucleotides long. They are thought to be associated with the *base excision repair* of damage caused by ionizing radiation and by certain alkylating agents known to produce small alkyl adducts that can be repaired by the action of specific DNA glycosylases during base excision repair (126, 139, 148) (see chapter 4). Long-patch repair in mammalian cells is equivalent to conventional nucleotide excision repair in *E. coli* (63, 64). These patches (in the order of 30 nucleotides) reflect nucleotide excision repair that presumably operates following the exposure of cells to UV radiation and to "UV radiation-like" agents such as *N*-2-acetyl-2-aminofluorene (AAF), aflatoxin, or psoralen plus UV light. No repair patches corresponding to the very long tracts (>1,000 nucleotides) associated with the putative inducible nucleotide excision repair system in *E. coli* (see chapter 5) have been observed over a wide range of UV doses and postirradiation incubation times in mammalian cells (63, 64).

Influence of Chromatin and Higher-Order Structure on Nucleotide Excision Repair in Mammalian Cells

In chapter 1 we briefly discussed some aspects of chromatin structure in the context of the influence of such structure on the accessibility of sites in DNA, particularly the bases, to damage by exogenous agents. This theme is reiterated here with specific reference to the influence of chromatin structure on DNA repair. A typical mammalian cell contains almost 50 cm of genomic DNA. Fitting this amount of DNA in to the confines of the nucleus requires a >50,000-fold reduction in length. This feat is accomplished by various levels of folding (179, 199). About 160 bp of DNA is wrapped around an octomer of histones comprising the histones H2A, H2B, H3 and H4, to form *nucleosome cores* separated by linker regions of DNA of different lengths (Fig. 7–6). Histone H1 binds regions where DNA enters or exits the core particles, and the highly basic C-terminal tail of histone H1 interacts with DNA in the linker region (47, 179, 199). The nucleosome core particle, together with the linker DNA and linker histone H1, is called a *nucleosome* (Fig. 7–6); the nucleosome can be identified electrophoretically following gentle digestion of chromatin with micrococcal nuclease (MNase). More-extensive digestion with MNase eliminates the linker DNA and identifies nucleosome core particles. Chemical cross-linking studies have revealed that the core histones bind to nucleosomal DNA in a symmetrically linear array in which histones H2A and H2B bind at the periphery and H3 and H4 bind at the center of the core particle (179) (Fig. 7–6). Histone H3 has weak interactions with the DNA at the points where the DNA enters and leaves the core particle (179) (Fig. 7–6).

The winding of two turns of helical DNA around the nucleosome core results in the formation of a 55-Å (5.5-nm) coil, achieving a sevenfold compaction (179). Further coiling results in the formation of helical structures with six to nine nucleosomes per turn, yielding so-called *thick fibers* with a diameter of 25 to 30 nm (Fig. 7–7) (173, 179). The organization of DNA into thick fibers achieves a further sevenfold compaction factor (179). Only a small fraction of the eukaryotic genome is actively engaged in transcription at any one time. Hence, it is believed that most of the nonexpressed genome is packaged into 30-nm chromatin fibers. This accomplishes a packing ratio of ~50. Further structural alterations increase this ratio by a factor of a further 250, to yield the highly condensed state typical of the metaphase chromosome. In interphase nuclei and in metaphase chromosomes, the thick fiber is apparently folded into loops (5, 36) or domains (128), comprising 30 to 100 kb of DNA (Fig. 7–7). These loops are believed to be anchored to a nuclear support structure variously termed the *nuclear matrix, nuclear cage,* or *nuclear scaffold* (1, 128, 132). There are indications that the nuclear matrix is not static but, rather, is a dynamic scaffold system intimately associated with metabolic transactions of DNA such as replication, transcription, and repair. These loops are possibly further structurally organized in some special way to yield the final chromosomal state (Fig. 7–7).

As indicated in chapter 6, one might reasonably expect that the organization of eukaryotic DNA into nucleosomes and the folding of these to yield higher order levels of chromosome structure would limit the accessibility of DNA-processing enzymes, particularly those required for nucleotide excision repair, to sites of base damage in DNA. The problem of accessibility to DNA in higher-order structures has been extensively addressed with respect to transcription (47, 83, 112, 172, 180). The solution to this problem may be directly relevant to nucleotide excision repair in mammalian cells. As discussed more extensively in chapter 6 and reiterated below, it is likely that the direct biochemical coupling of nucleotide excision repair to transcription in eukaryotes affords cells the ability to exploit the accessibility to individual nucleotides in DNA that is provided during the transcription of chromatin.

The question of precisely how higher-order chromosome structures are "processed" to allow the faithful passage of the transcription machinery along the DNA template is not well understood (47, 83, 112, 172, 180). Acetylation of core histones may play a role in the decondensation of the 30-nm thick fibers (65, 179,

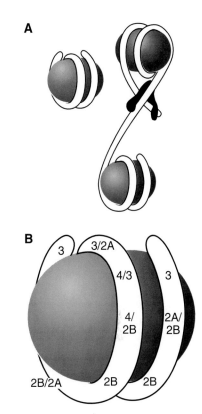

Figure 7–6 (A) The nucleosome core particle (left) consists of 160 bp of DNA wrapped around a histone octamer comprising two molecules each of the core histones H2A, H2B, H3, and H4. Also shown (right) is a representation of the nucleosome, consisting of the core particle plus the linker DNA between two core particles and bound histone H1 (shown in black). (B) Diagrammatic representation of the positions of the contacts made between the core histones (indicated by their numbers) and DNA as it wraps around the histone octamer. (*Adapted from Wolffe [199] with permission.*)

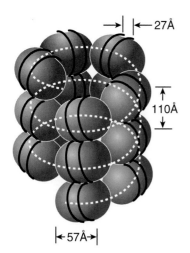

Figure 7–7 The higher-order organization of nucleosomes showing thick (30-nm) chromatin fibers which have a helical coil of six to nine nucleosomes per turn. These thick fibers are organized into loops, each of which contains 30 to 100 kb of DNA and which are anchored to a nuclear support matrix. The loops are further organized into microscopically visible chromosome bands. (*Adapted from Lewin [91a] with permission.*)

199). Additionally, phosphorylation of linker histones is believed to effect changes in the stability of condensed chromatin (65, 179, 199). At the level of nucleosomes, histones can impede the access of transcription factors to DNA and, in so doing, act as transcriptional regulators (83, 199). Indeed, activator proteins have been implicated in the solution to the nucleosome accessibility problem for transcription based on their proposed ability to bring about the disruption of nucleosome structure, either directly or possibly by recruiting other factors (83). A general model based on the ability of *trans*-acting transcriptional activators to displace histone H1 by one means or another has been called the *dynamic competition model* (47). An alternative general mechanism, referred to as the *preemptive competition model*, suggests that sites for the binding of transcription factors are blocked in the chromatin of resting cells and become accessible only during DNA replication. Once established, such sites of accessibility can be maintained in resting cells because of the limited size of the pools of free histones (47).

Once transcription is initiated, the question of what happens to nucleosomes in the path of the advancing RNA polymerase and associated transcription elongation factors is also controversial. A particularly favored hypothesis has been referred to as the *twin supercoil domain model* (94). This model essentially posits that since rotation of the RNA polymerase around the DNA is restricted by the viscous drag of the large transcription complex and the nascent RNA chain, and since rotation of the template DNA duplex is restricted by its attachment to (presumed) nuclear scaffolds, the passage of the transcription machinery must induce local domains of positive supercoiling ahead of the polymerase and of negative supercoiling behind it (Fig. 7–8). The wrapping of DNA around the histone octomers

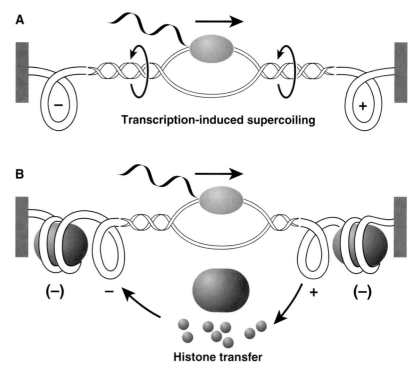

Figure 7–8 Schematic illustration of the twin supercoil domain model. (A) A transcription complex together with a nascent mRNA molecule is shown moving ∼10 bp along the DNA duplex. This movement creates a single positive supercoil in front of the polymerase and a negative supercoil behind it. (B) Histones are shown transferring (possibly as intact octamers or as subunits). This transfer is facilitated by the fact that negatively supercoiled DNA preferentially binds histone octamers compared with positively supercoiled DNA. An additional two nucleosome cores are shown more remote from the polymerase, each storing one negative supercoil associated with the organization of two turns of DNA. (*Adapted from Thoma [172] with permission.*)

results in each nucleosome storing one negative supercoil of DNA. Hence, the presence of positive supercoils is expected to be destabilizing to nucleosome structure, while the presence of negative supercoils is expected to favor stable nucleosomes (Fig. 7–8). It has been experimentally demonstrated that nucleosome formation is indeed favored on negatively supercoiled DNA (25, 130). Exactly what transpires once nucleosomes become destabilized by the presence of positively supercoiled DNA is not clear. There is evidence for the entire spectrum of possible answers, ranging from no displacement of histones from DNA to partial displacement to complete displacement (47, 65, 83, 112, 172, 180).

Nucleotide Excision Repair in Chromatin

With respect to nucleotide excision repair specifically, evidence in support of the concept of relative inaccessibility associated with chromatin structure stems from several studies. Attack at sites of pyrimidine dimer formation in chromatin by the phage T4 PD-DNA glycosylase/AP lyase (see chapter 4) was compared in simian virus 40 minichromosomes and naked DNA (46). Although this enzyme is a relatively small protein (~18 kDa), only about half the dimers in the minichromosomes were recognized as substrates by the T4 enzyme immediately after irradiation. Furthermore, when UV-irradiated mammalian cells were permeabilized to the *M. luteus* PD-DNA glycosylase/AP lyase activity, fewer sites were incised than in purified UV-irradiated DNA (193). However, additional enzyme-catalyzed nicks were detected by exposing the cells to salt concentrations known to promote nucleosome unfolding and histone displacement (193). Similar results were obtained with the phage T4 enzyme in UV-irradiated mammalian cells permeabilized by freezing and thawing (182).

In chapter 8 we will discuss a mammalian cell-free system which supports nucleotide excision repair of exogenously added plasmid DNA. In this system, most, if not all, of the added plasmid DNA is not folded into nucleosomes and excision repair can therefore be monitored by the demonstration and measurement of repair synthesis in naked DNA. However, if histones are added to the cell-free system or if the plasmid DNA is preincubated with extracts of *Xenopus laevis* under conditions which promote nucleosome formation, nucleotide excision repair is inhibited (189), once again suggesting that chromatin structure impedes access of repair enzymes to sites of base damage. This "repressed" nucleotide excision repair substrate provides a potential opportunity for exploring the specific roles of transcriptional derepression and transcriptional activation in the restoration of repair.

Regardless of the specific mechanism(s) by which the accessibility problem is solved during nucleotide (and possibly also base) excision repair in transcriptionally active and transcriptionally silent regions of the genome, it is instructive to recount the experimental observations on the distribution of selected types of DNA damage in nucleosomes and how this relates to the distribution of repair events observed in chromatin.

Our current understanding of the organization of histones and DNA into nucleosomes stems in part from studies on the susceptibility of DNA in chromatin to digestion with specific nucleases (57, 74, 97, 179, 192, 199). The linker regions of DNA in nucleosomes are most sensitive to MNase. Indeed, this sensitivity is what led to the operational definition of linker DNA. As indicated above, the core nucleosomal DNA is more resistant to digestion with this enzyme. The sensitivity to MNase digestion of DNA in chromatin in mammalian cells exposed to DNA damage has been examined to establish whether the distribution of repair synthesis is the same or different in MNase-sensitive and MNase-resistant regions of the genome (10, 11, 27, 28, 92, 125, 157, 159, 160, 177, 196, 203). One way of carrying out such experiments is as follows. First, DNA is uniformly labeled with [^{14}C] thymidine so that the radiolabel is present both in core and in linker DNA. Then the cells are exposed to a DNA-damaging agent and incubated in the presence of [^{3}H]thymidine to differentially label regions of repair synthesis under conditions that are restrictive for semiconservative DNA synthesis. Finally, isolated

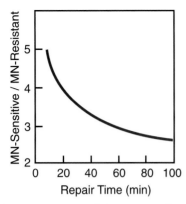

Figure 7–9 Ratio of the amount of repair synthesis per unit DNA in micrococcal nuclease (MN)-sensitive and -resistant regions as a function of repair time after UV irradiation of cells. Note that very soon after UV irradiation, repair synthesis label is distributed largely in MN-sensitive regions of the genome. However, at later times the label is distributed more uniformly between MN-sensitive and -resistant regions. (*Adapted from Friedberg [49] with permission.*)

nuclei are incubated with MNase, and the relative sensitivities of the ^{3}H and ^{14}C radiolabels to enzymatic digestion are compared.

When mammalian cells are exposed to UV radiation, pyrimidine dimers are thought to be uniformly distributed in chromatin (196), although not within nucleosome subdomains (54). However, the interpretation of the results of MNase digestion studies is complicated (158). The radioactivity incorporated immediately after irradiation is significantly more sensitive to MNase digestion than is the radioactivity associated with bulk DNA (Fig. 7–9). However, with increasing times after irradiation, this difference becomes less apparent, until at ~24 h almost as much repair synthesis is present in nuclease-resistant as in nuclease-sensitive regions of DNA (157, 158, 196) (Fig. 7–9).

A number of interpretations of these results are tenable. For example, DNA repair synthesis may occur in both linker and core nucleosomal DNA but initially may be faster in the former as a result of specific structural influences. Alternatively, at early times after UV radiation exposure, repair patches may be significantly smaller in core than in linker regions. A third possible explanation is that the position of nucleosomes may shift during DNA excision repair. Thus, for instance, repair label could be inserted preferentially into core particles and then shift rapidly into nuclease-sensitive linker regions. Perhaps one approach to reconciling some of these apparent contradictions is to recognize at the outset that the assumption that MNase sensitivity of nucleosomes exclusively reflects the degradation of linker DNA may not be valid during nucleotide excision repair (92, 156). It is possible that lesions and/or repair events produce local perturbations in nucleosome structure which result in increased sensitivity of core nucleosomal DNA to the nuclease probe (92, 156). This model of transient alteration in nucleosome conformation associated with excision repair events resulting in an obligatory local sensitivity to MNase is supported by some experiments. A nuance of the repair of UV-irradiated DNA in chromatin that has received relatively little attention relates to the fact (mentioned above) that pyrimidine dimers and (6-4) photoproducts are apparently repaired with different kinetics. Additionally, the distribution of these photoproducts in chromatin may be different.

It has been suggested that in cells exposed to UV radiation, DNA that is synthesized during the course of nucleotide excision repair can exist in several states associated with transient rearrangements in nucleosome structure (156). These rearrangements in chromatin structure that eventually restore the native sensitivity of repaired DNA to MNase may involve multiple steps (72). Inhibition of repair synthesis results in inhibition of ligation of repair patches and chromatin rearrangement, consistent with the view that completed excision repair is a prerequisite for nucleosomal rearrangement (154). Independent studies that examined the relationship between DNA ligation and the formation of native nucleosomes during nucleotide excision repair led to similar conclusions (154). There are also hints that the mechanism of reassembly of nucleosomes following nucleotide excision repair of DNA is different from that following semiconservative DNA synthesis. Inhibition of protein synthesis by cycloheximide had no effect on the kinetics of the appearance or disappearance of MNase sensitivity of repair-labeled DNA in human fibroblasts (92). However, protein synthesis was required for the synthesis of new histones for assembly of nucleosomes following replicative synthesis (191).

More detailed studies have helped clarify the types and mechanisms of nucleosome rearrangements associated with nucleotide excision repair in mammalian cells (156). Examination of repair synthesis in different regions of the nucleosomal core shows that it is not random. During the early, rapid phase of nucleotide excision repair in human chromatin, there is preferential repair of regions near the 5' end of the nucleosome core. Lesser amounts of repair synthesis also occur near the 3' end, leaving a region essentially devoid of repair synthesis in the middle of the core (84). With time, this distribution of repair synthesis becomes random (122), suggesting a slow repositioning of core histones along the DNA after repair. Analysis of the extent of this repositioning indicates that core histones would have to slide an average distance of ~50 bp (4). Two models have been advanced to explain the nonrandom distribution of repair patches in nucleosome cores during early nucleotide excision repair. One model assumes a nonrandom formation of bulky base damage, and the other assumes a nonrandom repair in nucleosomes (156). Experimental support has been provided for the nonrandom distribution of base damage caused by UV radiation (75).

A reasonable working model for the changes that occur in nucleosome conformation and in higher-order structure during nucleotide excision repair in mammalian cells can be summarized as follows (155, 156) (Fig. 7–10). Absent other aspects of DNA metabolism, such as transcription or replication, that facilitate the unfolding of chromatin structure, it has been suggested (156) that the specific recognition of base damage by a nucleotide excision repair protein or protein complex can result in the relaxation of the chromatin fiber. This relaxation may involve postsynthetic modifications of histones and/or nonhistone proteins (e.g., acetylation and or/ADP ribosylation) over a relatively large area of the chromatin fiber (156). In this regard it is relevant that access of the repair machinery to sites of base damage may be greatly facilitated by the specific distribution of lesions in DNA folded into nucleosomes. For example, it has been reported that the formation of pyrimidine dimers in DNA packaged into chromatin prior to exposure to UV radiation results in a distribution of these photoproducts such that they are

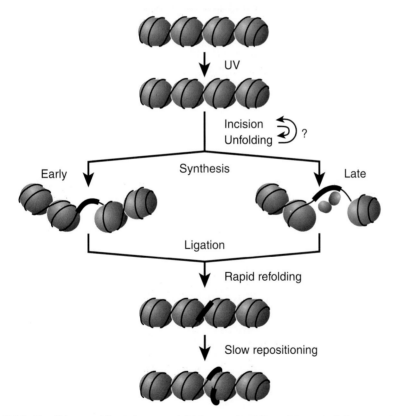

Figure 7–10 Possible transitions that occur during nucleotide excision repair in transcriptionally silent chromatin. Base damage resulting from UV radiation is recognized by the nucleotide excision repair machinery, resulting in the relaxation of the thick (30-nm) chromatin fiber. This relaxation process may be aided by specific perturbations of histone and nonhistone proteins. Early in the process of excision repair, local nucleosome structure is unfolded. This may involve different mechanisms during different phases of repair. Once ligation is completed, nucleosome structure is restored such that repair patches generated during the early phases of repair are weighted toward the 5′ ends of the nucleosome cores. The positioning of histones then slowly changes during the refolding of the thick fibers. Black bars represent sites of repair synthesis. (*Adapted from Smerdon [156] with permission.*)

preferentially oriented away from the histone face in nucleosomes (54). A similar distribution of these photoproducts has been noted following the assembly of nucleosomes on previously UV-irradiated DNA (167). During the incision process itself, or at the time of initiation of repair synthesis, local nucleosome structure is believed to be disrupted by an unfolding mechanism (Fig. 7–10). This unfolding may involve displacement of histone H1, an induced sliding mechanism during which core histones migrate down the DNA away from the site of the lesion, or possibly the complete dissociation of core histones. This disruption phenomenon may occur by different mechanisms during the rapid early and slower later phases of repair. Following DNA ligation, the DNA is believed to be rapidly folded into a nucleosome structure. At this time, repair patches synthesized during the early, rapid phase of nucleotide excision repair in mammalian cells are heavily weighted to the 5′ end of the nucleosome cores. However, the positioning of core histones slowly changes, perhaps during the repair that takes place at other neighboring sites or during refolding of the chromatin fiber into higher-order structures (156). During the refolding process, a reversal of earlier protein modifications may take place (156).

Transcriptionally active regions of chromatin contain hyperacetylated histone moieties (138, 179). The increased accessibility of such regions appears to reflect the neutralization of positive lysine charges on histones by acetylation, resulting in an "opening up" of the nucleosomal solenoid structure (138, 179). It has been suggested (129) that histone acetylation may be a general mechanism for relaxing chromatin domains, thereby facilitating the access of DNA-processing enzymes. Hyperacetylation of histones can be effected by treatment of cells with *n*-butyrate, a short-chain fatty acid which inhibits the enzymatic deacetylation of histones (138, 179). In cells treated with *n*-butyrate, the accessibility of chromatin to the *M. luteus* PD-DNA glycosylase/AP lyase is increased more than twofold and there is a corresponding increase in repair in intact cells exposed to UV radiation (162). Additionally, pretreatment of normal human fibroblasts with sodium butyrate increases the levels of repair synthesis after UV radiation (197). Examination of DNA repair in nucleosome subpopulations with different degrees of histone acetylation showed significantly enhanced repair synthesis in hyperacetylated mononucleosomes during the first 30 min after irradiation (134). This effect is transient, lasting ~12 h, as evidenced by pulse-chase experiments. The enhanced repair synthesis is associated primarily with core regions and does not appear to be the result of increased lesions in the DNA of hyperacetylated chromatin (134). These results must be interpreted with caution, since *n*-butyrate has other effects on cells, including inhibition of DNA synthesis and cell growth. Nonetheless, they are consistent with the operation of a histone acetylation-based chromatin surveillance system in mammalian cells which increases the accessibility to repair (and other) DNA-processing enzymes (156, 162).

In short, the important question of how the large multiprotein nucleotide excision repair machinery gains access to all sites of base damage in the chromatin of mammalian cells is still unresolved. The direct biochemical coupling of nucleotide excision repair to transcription initiation and (presumably) transcription elongation (see below), provides a useful conceptual framework for understanding how repair transpires in transcriptionally active regions of the genome. However, we have no complete explanation of how the accessibility problem is solved in transcriptionally silent regions of DNA.

Heterogeneity of Nucleotide Excision Repair in the Mammalian Genome

Intragenomic Heterogeneity

Complexities associated with (but not restricted to) nucleotide excision repair in mammalian cells can be viewed in the general context of genomic heterogeneity beyond the level of nucleosome conformation (for reviews, see references 11a–12, 14, 15a, 42, 60a–62, 161, 170). One example involves the repair of repetitive

DNA. The mammalian genome contains regions of highly repetitive DNA, which, in some cases, can be readily isolated from bulk DNA by digestion with selected restriction enzymes (14). A repetitive 172-bp unit of DNA from African green monkey cells, designated α-DNA (152), has been extensively studied in terms of its susceptibility to both DNA damage and repair relative to bulk DNA (14, 161). No differences were found in the repair of UV radiation damage in α-DNA and bulk genomic DNA. However, significant differences have been observed with other agents that produce bulky adducts in DNA. For example, various psoralens have been observed to be removed with a markedly reduced efficiency in α-DNA (202, 204) (Fig. 7–11). This is apparently not due to differences in the relative frequency of adduct formation in the two types of DNA or to the preferential formation of a class of psoralen adducts which are inherently refractory to nucleotide excision repair. Similar results have been obtained after treatment of cells with N-acetoxy-AAF (202) or activated aflatoxin B_1 (89).

The observation that UV radiation damage was proficiently repaired in α-DNA stimulated the hypothesis that UV radiation itself might somehow improve the efficiency of nucleotide excision repair in such DNA, either by altering its conformation or by stimulating factors which increase the accessibility of repair enzymes or both (161). To test this model, nucleotide excision repair in α-DNA was examined in cells exposed to both UV radiation and aflatoxin B_1 (86, 89). At doses as low as 1 J/m^2 or greater, an enhanced efficiency of the repair of aflatoxin adducts was observed (Fig. 7–12). Increasing the time between UV irradiation and chemical treatment of cells decreased the enhancement in a manner suggesting that the phenomenon was dependent on the absolute number of photoproducts remaining in the DNA at the time of treatment with aflatoxin. Increased efficiency of repair of aflatoxin adducts in α-DNA was also observed in actively growing cells relative to G_0 cells (86).

In summary, it would appear that a repair system(s) for several bulky chemical adducts is somehow excluded from interacting with α-DNA, even though the damaging agents themselves are not. This exclusion seems to be confined to bulky adducts, since the repair of AP sites, thymine glycol damage, and simple alkylation damage repaired by the base excision repair pathway is indistinguishable in α-DNA and bulk DNA (161). It remains to be determined whether this represents a special example of restricted accessibility for nucleotide excision repair or reflects differences in the biochemistry of nucleotide excision repair in different regions of the genome.

Another interesting example of *intragenomic heterogeneity* during nucleotide excision repair stems from the demonstration that repair synthesis in normal human cells in culture appears to be localized preferentially to the nuclear matrix compartment, the presumed site of DNA replication and transcription (117). Deviations from this distribution have been noted in two hereditary human diseases, xeroderma pigmentosum (XP) and Cockayne's syndrome (CS). The utility of XP cells for studies on the genetics, molecular biology, and biochemistry of nucleotide excision repair in humans is discussed more fully in chapters 8 and 14. For the purposes of the present discussion, it is relevant that XP cells from genetic complementation group C are characterized by low levels of nucleotide excision repair relative to normal cells but are not as defective as cells from other genetic complementation groups. Several studies have demonstrated that most, if not all, of the residual repair in these cells is confined to a limited region or domain of the genome (76, 78, 96). These domains have not yet been characterized, but there are indications that they represent transcriptionally active DNA associated with the nuclear matrix (116, 117, 183).

CS is a second human hereditary disease characterized by defective nucleotide excision repair (see chapters 8 and 14). In this syndrome the repair defect appears to be more pronounced in nuclear matrix-associated DNA (117). As noted below, the association of nucleotide excision repair and transcription with the nuclear matrix compartment is consistent with the preferential repair of actively transcribed genes.

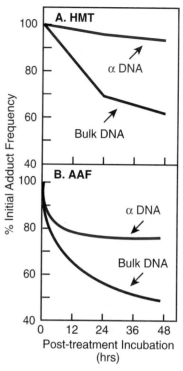

Figure 7–11 Both the monofunctional psoralen 4'-(hydroxymethyl)-4,5',8-trimethylpsoralen (HMT) (A) and *N*-2-acetyl-2-aminofluorene (AAF) (B) are removed more slowly from αDNA sequences of African green monkey cells than from bulk DNA. (*Adapted from Zolan et al. [204] and Leadon and Hanawalt [86] with permission.*)

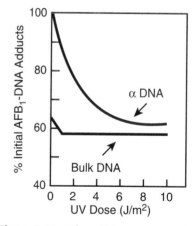

Figure 7–12 When African green monkey cells are treated with aflatoxin B_1 (AFB$_1$), the aflatoxin adducts are removed from α-DNA more slowly than from bulk DNA (zero UV radiation dose). However, if the cells are additionally exposed to UV radiation, the rate of removal of the aflatoxin adducts from α-DNA increases in a UV dose-dependent manner. (*Adapted from Leadon and Hanawalt [86] with permission.*)

Intragenic Heterogeneity of Repair: Preferential Repair of Actively Transcribed DNA

During the past decade our understanding of nucleotide excision repair in transcriptionally active regions of the genome has increased enormously. This topic has already been discussed with respect to the prokaryote *E. coli* (see chapter 5) and the yeast *S. cerevisiae* (see chapter 6). Historically the phenomenon of *preferential repair of transcriptionally active genes* was first discovered and characterized in mammalian cells. The following discussion will recount the evidence that in transcriptionally active regions of the mammalian genome the rate of nucleotide excision repair is increased and possibly the mechanism of repair is altered. However, we should not forget the importance of repair in transcriptionally silent (bulk) DNA. The rapid and preferential repair of transcriptionally active regions may preclude the formation of aborted or truncated transcripts and hence alleviate a threat to the immediate viability of injured cells. However, repair of transcriptionally silent DNA may be critical for reducing the mutational load of cells and hence for protecting against longer-term problems such as neoplastic transformation in subsequent generations of cells. Indeed, as will be discussed more fully in chapter 14, it is noteworthy that cells from individuals with CS, which are defective in the repair of transcriptionally active genes but not bulk DNA, are not unusually cancer prone.

The availability of single-copy DNA probes for mammalian genes facilitated the development of an elegant technique for quantitating the repair of various types of base damage in individual genes and within different regions of individual genes. As already indicated, the technique exploits the specificity of selected DNA repair enzymes for particular forms of base damage (for reviews, see references 11a–12, 14–15a, 42, 60a–62, 161, 170). The essential features of this experimental procedure are illustrated in Fig. 7–13 with respect to assessing the repair of pyrimidine dimers specifically. Cells exposed to UV radiation are incubated for different periods. Cellular DNA is density labeled to facilitate the separation of replicated from unreplicated DNA by equilibrium sedimentation, thereby avoiding dilution of regions containing pyrimidine dimers by DNA replication. Unreplicated DNA isolated at various times after irradiation is treated with an appropriate restriction endonuclease to generate fragments of suitable length. The restricted DNA is then treated (or not) with the phage T4 (or *M. luteus*) PD-DNA glycosylase (endonuclease V of T4), which, as discussed in chapter 4, is specific for pyrimidine dimers in DNA, and fragments are resolved by gel electrophoresis. The DNA is blotted and hybridized with a radiolabeled probe for a selected gene or a particular region(s) of the gene by using conventional Southern analysis. Restriction fragments that do not contain pyrimidine dimers, i.e., undamaged or fully repaired DNA (so-called *zero class fragments*), retain maximal length and yield the strongest hybridization signal. Fragments in which pyrimidine dimers were retained because of a lack of repair are degraded by endonuclease V, and the extent of the degradation (i.e., the number of dimers) is reflected by correspondingly weaker hybridization signals at the position of the zero class fragments. The fraction of zero class molecules can be quantitated to estimate the frequency of pyrimidine dimers and hence the extent of their repair as a function of postirradiation incubation (11a, 11c, 12, 15).

Various refinements of this basic technique have facilitated the selective examination of nucleotide excision repair in different strands (transcribed versus nontranscribed) of the same gene by hybridization with strand-specific probes and have expanded the repertoire of base damages and repair modes that can be examined. For example, the repair of cross-links in DNA is reflected by a loss of the rapid renaturability in defined DNA sequences, a property of DNA fragments containing cross-links (184). The use of the *E. coli* UvrABC endonuclease, a general damage-specific endonuclease (see chapter 5) instead of T4 endonuclease V has increased the range of bulky base adducts that can be examined (169, 174). When UV-irradiated DNA containing (6-4) photoproducts is exposed to UV radiation at

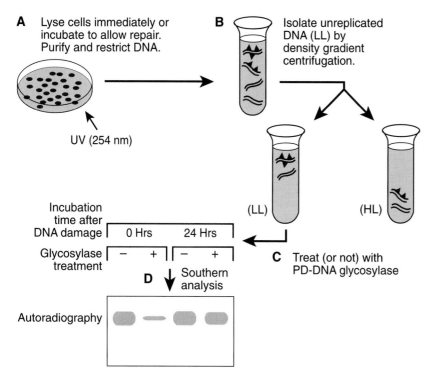

Figure 7–13 Detection of nucleotide excision repair in specific gene sequences. (A) Cells are exposed to a DNA-damaging agent which produces base adducts for which a specific detection method is available that can convert the damage to DNA strand breaks. In the example shown, DNA is damaged with UV radiation, thereby generating pyrimidine dimers that can be probed with a PD-DNA glycosylase/AP lyase. Some of the cells are lysed immediately while others are allowed to carry out nucleotide excision repair in the presence of the heavy thymine analog bromodeoxyuridine. DNA is extracted, purified, and treated with a restriction enzyme to generate fragments. (B) The restricted DNA is subjected to equilibrium density centrifugation in CsCl gradients to separate replicated (HL) DNA from unreplicated (LL) DNA. (C) The unreplicated (LL) DNA is then treated (or not) with PD-DNA glycosylase/AP lyase, resulting in the formation of strand breaks at pyrimidine dimer sites. (D) The DNA fragments are resolved by gel electrophoresis, transferred to a membrane, and probed with a [^{32}P]DNA probe of interest. Following autoradiography, DNA fragments not exposed to the enzyme (−glycosylase) yield an autoradiographic signal. In contrast, DNA isolated from cells immediately after exposure to UV radiation is degraded by the PD-DNA glycosylase/AP lyase (+glycosylase) and will yield a faint or no autoradiographic signal. The intensity of the autoradiographic signal after incubation (24 h) reflects the extent to which dimers were removed by nucleotide excision repair. The ratio of the intensity of the signal in the treated and untreated samples for a given time point indicates the fraction of DNA molecules that were free of damage (P_0). Using the Poisson expression ($S = -\log_n P_0$), the number of damaged sites per fragment (S) can be derived. (*Adapted from Bohr [11c] with permission.*)

wavelengths above 320 nm, these photoproducts are converted to the so-called Dewar photoisomer (103). This photoisomer is highly labile to alkali treatment, generating single-strand breaks which can be quantified by Southern analysis (93).

The use of monoclonal antibodies against bromouracil-containing DNA has facilitated the detection of repair by immunological methods when repair synthesis occurs in the presence of bromodeoxyuridine, rather than by the more restrictive use of repair enzymes with limited substrate specificities (85). Following the separation of fully hybrid-density DNA generated by semiconservative replication from unreplicated DNA containing repair patches, the latter fraction is reacted against the monoclonal antibody and antibody-bound DNA is isolated by ammonium sulfate precipitation and subsequent centrifugation. This DNA can then be electrophoresed and subjected to Southern analysis with a desired DNA probe(s) (85). The immunodetection of repair in individual genes has also been facilitated by the availability of antibodies against specific types of base damage (190). A variation of this basic theme takes advantage of the potential of bulky lesions in DNA to inhibit PCR. Hence, the extent of PCR in a defined region of the genome reflects the amount of base damage initially present in that region and the extent of its repair (58). Finally, the observation that the cutting of DNA by restriction enzymes is inhibited by the presence of base damage at substrate sites has facilitated the monitoring of the presence and loss of base damage in selected regions of the genome (7).

Transcription-Dependent Nucleotide Excision Repair

It has long been known that rodent cells in culture are deficient in the excision of pyrimidine dimers (and presumably other bulky base adducts) relative to human cells (63). Typically, such cells lose only 10 to 20% of the dimers in their genomic DNA during the first 24 h after irradiation, compared with the loss of ~80% of these lesions in human cells (15, 55, 182, 201) (Table 7–1). However, rodent cells in culture are not particularly sensitive to UV radiation (Table 7–1). This has been referred to as the *rodent repair paradox* (42). Resistance to killing by DNA-damaging agents obviously reflects the sum of all cellular responses to such damage in living cells. Hence, one possible explanation for the relative UV resistance of rodent cells in the face of apparently restricted nucleotide excision repair capacity is that these cells rely on other mechanisms to subvert the potentially lethal effects of DNA damage. However, another explanation explored by Philip Hanawalt and Vilhelm Bohr and their colleagues is that rodent cells might selectively repair genes that are essential for their survival.

A comparison of nucleotide excision repair in expressed and unexpressed genes in rodent cells supports this latter hypothesis. When the loss of T4 endonuclease V-sensitive sites in the coding region of the essential *DHFR* gene in CHO cells was compared with the loss of such sites in an unexpressed region of the genome near the coding region, the former region was found to be preferentially repaired (Tables 7–1 and 7–2) (15). This result was subsequently shown to be independent of whether the *DHFR* gene was amplified. Similar results were obtained by comparing the transcriptionally active proto-oncogene c-*abl* and the silent c-*mos* proto-oncogene in Swiss mouse 3T3 cells (Table 7–2). Approximately 85% of the pyrimidine dimers in a fragment of the former gene were lost during the first 24 h after UV irradiation. However, under comparable conditions only 10 to 20% of the dimers were removed from the DNA of the latter gene (95). Furthermore, transcriptional activation of the CHO metallothionein (*MT*) gene by treatment of cells with heavy metals resulted in an increased efficiency of repair relative to that in uninduced cells (124) (Table 7–2).

Refined studies on the *DHFR* gene of CHO cells revealed interesting differences in repair efficiency in different regions of the gene (11a, 13). Loss of T4 endonuclease V-sensitive sites was most efficient in the 5′ region of the gene and in 5′-flanking sequences. A region upstream of the gene and sequences downstream of the gene were repaired with an efficiency comparable to that observed in bulk DNA (Fig. 7–14) (11a, 13). The *DHFR* gene in CHO cells has DNase I-hypersensitive sites at its 5′ end (109), and in both CHO (109) and mouse (38) cells, a divergent transcript originates from the *DHFR* promoter. Thus, the 5′ region

Table 7–1 Comparison between survival and repair in the genome overall and in an essential gene in human and rodent cells

Survival or repair	Value (%) in:	
	Human cells	CHO cells
Survival after UV irradiation (3 J/m²)	85	85
DNA repair in genome overall	85	15
DNA repair in *DHFR* gene	85	70

Source: Adapted from Bohr et al. (14) with permission.

Table 7–2 Preferential repair in mammalian cells after exposure to UV radiation

Cell	Repair (%)	
	After 8 h	After 24 h
CHO cells		
Bulk DNA		15
Noncoding region		15
Transcriptionally active gene	50	75
MT^a gene (no induction)	17	31
MT gene (after induction with $ZnCl_2$)	35	52
Mouse cells		
c-*abl* proto-oncogene (active)		80
c-*mos* proto-oncogene (inactive)		20
Human cells		
Bulk genomic DNA	35	80
DHFR gene	75	80

[a] MT = metallothionein
Source: Adapted from Bohr (11b) with permission.

of this gene might be particularly accessible to nucleotide excision repair enzymes (61). Alternatively, it has been suggested that these results may reflect a polarity of repair associated with decreasing processivity as the nucleotide excision repair complex moves down the gene in a $5' \rightarrow 3'$ direction (61).

Despite the general efficiency of nucleotide excision repair in the bulk DNA of human cells relative to rodent cells, human cells also have a repair bias for expressed genes, although this bias is less pronounced than in rodent cells. At early times after UV irradiation, the loss of endonuclease V-sensitive sites occurs more rapidly in the *DHFR* gene of human cells than in bulk DNA. However, at later times these differences are less pronounced (100) (Table 7–2). Kinetic differences in repair were also observed in the *MT* family of human genes (Fig. 7–15) (88). Following exposure to either UV radiation or aflatoxin B_1, repair was initially faster in fragments containing the actively transcribed $MT\text{-}I_A$, $MT\text{-}I_E$ and $MT\text{-}II_A$ genes than in the genome overall. The kinetics of repair of UV radiation damage in the $MT\text{-}I_B$ gene (which manifests cell-specific expression) and in the $MT\text{-}II_B$ gene

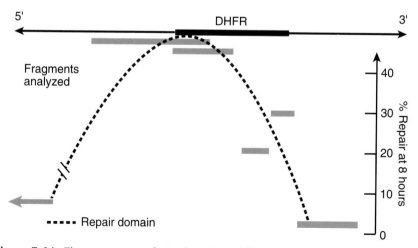

Figure 7–14 Fine-structure analysis of repair in different fragments of the *DHFR* gene of CHO cells. The use of hybridization probes (colored bars) for different regions of the *DHFR* gene and flanking 5′ and 3′ sequences 8 h after exposure to UV radiation reveals a repair domain that is most efficient in the 5′ region of the gene and less efficient on either side of this region. (*Adapted from Bohr [11a] with permission.*)

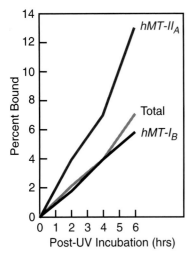

Figure 7–15 Nucleotide excision repair in the human metallothionein (*hMT*) genes. DNA with bromouracil-containing repair patches was isolated using a monoclonal antibody against bromodeoxyuridine. This DNA was then analyzed for gene-specific repair by Southern analysis as described for Fig. 7–13. Repair in the actively transcribed *MT-II$_A$* gene is faster than in the transcriptionally silent *MT-I$_B$* gene. The latter gene shows the same rate of repair as total human DNA. (*Adapted from Leadon and Snowden [88].*)

(a nontranscribed pseudogene) were approximately the same as in bulk DNA (88) (Fig. 7–15). Increasing the levels of transcription of the three expressed *MT* genes by induction with cadmium chloride further increased the initial rate of repair in these genes (88).

The general phenomenon of the increased rate of repair of actively transcribing genes compared with their transcriptionally silent equivalents or other silent regions of the genome has been extended to a wide range of bulky DNA damage, including cisplatin intra- and inter-strand cross-links, AAF, benzo-[*a*]pyrene diol-epoxide and 4-nitroquinoline 1-oxide (12, 62). The development and refinement of techniques for exploring the heterogeneity of repair in the genome of cells has also been extended to the removal of nondistortive DNA damage, typically handled by base excision repair. Several studies have demonstrated no differences in the repair of damage produced by the alkylating agent dimethyl sulfate in transcriptionally active and silent genes (123, 147). On the other hand, the extent of the loss of alkali-labile sites (i.e., AP sites presumably generated by DNA glycosylase-mediated excision of alkylated bases) in the transcriptionally active insulin gene was increased compared with that in the gene in a transcriptionally inactive state (90).

The preferential repair of actively transcribing genes is not observed in all cell types. Thus, for example, nucleotide excision repair of pyrimidine dimers was not detected in the creatine kinase gene in either myoblasts (in which the gene is silent) or during differentiation to myotubes (when the gene becomes transcriptionally active) (80). Furthermore, in proliferating mouse proadipocyte stem (3T3-T) cells, the rate and extent of the removal of pyrimidine dimers in several actively transcribed genes were found to be higher than in transcriptionally silent genes. However, these differences disappeared when these cells differentiated into mature adipocytes (8). Finally, as indicated in chapter 6, studies with *Drosophila melanogaster* have thus far failed to identify any differences in the rate or extent of nucleotide excision repair in actively transcribed and silent genes.

Aside from the cell-specific examples quoted above, other exceptions are noteworthy. The repair of AAF adducts (169) and of 4-nitroquinoline 1-oxide (164), both of which are expected to constitute substrates for conventional nucleotide excision repair, is not enhanced by transcription in the *DHFR* gene of CHO cells. As pointed out above, consistent with the notion that base excision repair is not impeded by the organization of DNA in chromatin and is not coupled to transcription, the repair of alkylpurines is not biased by the transcriptional activity of genes (91, 123, 147).

Collectively, these findings suggest a general correlation between the transcriptional state of a gene and the efficiency of nucleotide excision repair. Such a relationship might be provided by the more "open" configuration of chromatin during transcription of genes. Hence, the phenomenon described in the preceding paragraphs is usefully referred to as *transcription-dependent repair*. In the previous chapter we mentioned that the available data on the differential repair of transcriptionally active and silent mating genes in yeast cells are also most consistent with this interpretation.

PREFERENTIAL REPAIR OF THE TRANSCRIBED STRAND

In addition to DNA repair that is dependent on active transcription, there is compelling evidence that, as is the case in *E. coli* (see chapter 5) and in the yeast *S. cerevisiae* (see chapter 6), nucleotide excision repair in mammalian cells is biochemically linked to the process of transcription (so-called *transcriptionally coupled repair*). Nucleotide excision repair of pyrimidine dimers has been compared in the transcribed and nontranscribed strands of the same gene. In initial experiments, genomic fragments of the CHO and human *DHFR* genes were cloned into vectors containing bacteriophage promoters oriented in opposite directions, so that RNA probes could be generated to quantify repair in the transcribed and nontranscribed strands. In CHO cells ~80% of the dimers (enzyme-sensitive sites) were lost from the transcribed strand of the *DHFR* gene within 4 h, while essentially no loss of dimers was observed in the nontranscribed strand (Table 7–3) (101). Similar, although quantitatively less dramatic results were obtained with the human *DHFR* gene (101). This basic observation has also been extended to a number of different genes and a number of different DNA-damaging agents (12, 16, 18, 62, 98, 114).

A prediction of the preferential repair of different strands is that the frequency of mutations would be increased in the unrepaired relative to the repaired strand in repair-proficient cells and would show no bias in repair-defective cells. In general, this expectation has been borne out experimentally (42). For example, in repair-proficient rodent cells exposed to physiologically relevant doses of UV radiation (2 J/m²), >85% of the mutations can be attributed to dipyrimidine lesions [pyrimidine dimers or (6-4) photoproducts] in the nontranscribed strand of the *HPRT* gene (Fig. 7–16) (187). Under these experimental conditions, >90% of the cyclobutane pyrimidine dimers were lost from the transcribed strand within 8 h after irradiation of the cells, whereas essentially no dimer loss was detected from the nontranscribed strand (187). Similar conclusions have emerged from the distribution of mutations in the *APRT* gene of CHO cells treated with psoralen (142) and from studies on human cells (19, 99). One would also predict that the strand bias for mutations would be lost if cells were exposed to DNA damage just prior to the S phase, when DNA is replicated, because there would be too little time for repair before mutations were fixed. This is indeed the case (99).

An unexpected and unexplained anomaly of strand-specific repair is the observation in both human and hamster cells which are defective in nucleotide excision repair that the bias toward mutations in the nontranscribed strand is not only lost but actually reversed, resulting in a bias toward the transcribed strand (42, 181). In CHO cells which were defective in nucleotide excision repair and which had a sevenfold enhanced mutation frequency after exposure to UV radiation, 90% of the base pair changes were caused by photoproducts in the transcribed strand (102, 186) (Fig. 7–16). This surprising bias in favor of one strand suggests that other factors may determine the relative distribution of DNA damage, repair, and mutation fixation in different strands. One possibility is that different mechanisms of DNA replication may operate on the two DNA strands, resulting in differing replication fidelities of different DNA polymerases in the presence of DNA damage (42, 118, 186, 188). For example, if the transcribed strand is also the template strand for leading-strand replication, an error-prone translesion DNA synthesis mode operating in the absence of nucleotide excision repair might generate a high error rate in that strand. Additionally or alternatively, discontinuous DNA replication in the nontranscribed lagging strand might allow replicative arrest

Table 7–3 Nucleotide excision repair in the *DHFR* gene of CHO cells

Strand probed	Time (h)	Pyrimidine dimers removed (%)
Both	0	0
	2	13
	4	51
	8	57
	24	62
Transcribed	0	0
	2	30
	4	82
	8	85
	24	89
Nontranscribed	0	0
	2	5
	4	12
	8	4
	24	10

Source: Adapted from Mellon (101) with permission.

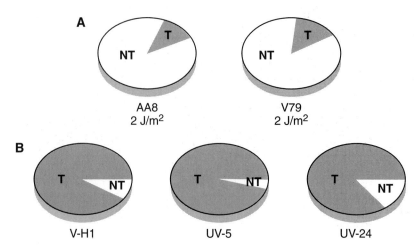

Figure 7–16 Strand specificity of UV radiation-induced mutations in CHO cells. (A) After exposure of CHO cells (strains AA8 and V79 [proficient for nucleotide excision repair]) to UV radiation at the modest dose of 2 J/m², ~90% of the mutations at dipyrimidine sites are in the nontranscribed (NT) strand of the *HPRT* gene. (B) In contrast, in CHO cells that are defective in nucleotide excision repair (strains V-H1, UV-5, and UV-24) mutations at dipyrimidine sites are not equally distributed between the two strands but, rather, are strongly biased in favor of the transcribed (T) strand. (*Adapted from Van Zeeland et al. [181] with permission.*)

at sites of base damage with error-free gap filling in the absence of nucleotide excision repair (42, 118, 186, 188) (see chapter 12). This model is compatible with the notion that the leading and lagging strands are replicated by different DNA polymerases. A second but somewhat less likely explanation for this reversed strand bias is that there is a bias for the formation of mutagenic lesions particularly in the transcribed strand of the *HPRT* gene, the gene that has been most extensively explored with respect to mutational bias (41).

It has been pointed out that there is a conceptual problem in directly relating the strand selectivity for mutagenesis and that for nucleotide excision repair in all cases, because mutations necessarily require the fixation of base damage by replication, whereas the conventional assay for measuring strand-specific repair necessarily requires the isolation of unreplicated DNA to avoid dilution of the damage due to replication (42). However, the selectivity of nucleotide excision repair in the *DHFR* gene of CHO cells has been shown to be the same in replicated and in unreplicated DNA (165).

MECHANISM OF TRANSCRIPTIONALLY COUPLED REPAIR

A number of experimental approaches have been used to associate strand-specific repair with transcription by RNA polymerase II (62). For example, strand-specific repair of DNA damage in the human *MT* gene is inhibited in the presence of α-amanitin, an inhibitor of RNA polymerase II (87). In related experiments, both in vivo and in vitro results obtained with α-amanitin have confirmed the anticipated inhibition of transcriptional elongation in the *DHFR* gene of CHO cells (21). Switching back to *S. cerevisiae* for a moment, in a mutant carrying a temperature-sensitive mutation in the largest subunit of RNA polymerase II, preferential repair of the transcribed strand of the *RPB2* gene occurs only at the permissive temperature (see chapter 6). Other observations are consistent with the conclusion that the phenomenon of transcriptionally coupled repair is RNA polymerase II dependent. Strand-selective repair has not been observed in rRNA genes (22). Furthermore, the residual nucleotide excision repair in XP cells from genetic complementation group C is abolished in the presence of α-amanitin, consistent with the view that these cells are defective in the repair of nontranscribed (bulk) DNA and not in transcriptionally coupled repair (17).

The results described above suggest that features of chromatin structure in mammalian cells are not absolute determinants of the selective repair of the template strand in actively transcribing genes. Rather, it would appear that nucleotide excision repair is directly associated with transcription, as has been demonstrated with yeast factor TFIIH (see chapter 6). Much of what is known about this connection can probably be directly extrapolated from the studies with yeast cells. The precise subunit composition of mammalian basal transcription factor TFIIH is not clearly defined at the time of writing. However, in at least one study (146), this factor was shown to include the known nucleotide excision repair protein XPB (ERCC3), the human homolog of Ssl2 protein. Human TFIIH also includes the human homologs of the yeast Rad3 (XPD) and Ssl1 proteins (see chapters 8 and 14).

This complex in mammalian cells presumably serves as the nidus for the assembly of a repairosome to specifically clear expressed genes of bulky adducts that block transcription. The mechanism of the recognition of base damage during this process and the precise coupling of repair to arrested transcription is not known, but, as discussed in chapter 6, the existence of a specific transcription-repair coupling factor in *E. coli* (see chapter 5) suggests a plausible general biochemical paradigm through its human homolog, the ERCC6 protein. The presence of arrested transcriptional complexes at or near sites of base damage may impose conformational alterations in the genome which facilitate preferential recognition by, and binding of, components of the nucleotide excision repair machinery. A prediction of this model is that nucleotide excision repair-defective mutants that are defective in one or another of these repair modes should exist. The human excision repair-defective hereditary diseases XP and CS have been referred to in several contexts and are discussed more fully in chapter 14. As indicated in chapter 6, CS represents a compelling paradigm for a defect in the preferential repair of transcribed DNA. There is indeed evidence that CS cells exhibit a lack of preferential repair (182a) (Fig. 7–17). Furthermore, both CS complementation group A and B cells additionally do not show a difference in the rate of repair of the transcribed and nontranscribed strands of the active *ADA* and *DHFR* genes (62, 115).

As indicated above, the preferential repair of transcriptionally active genes appears to be confined to genes whose transcription is dependent on RNA polymerase II. Genes that encode rRNA, which are transcribed by RNA polymerase I, are not preferentially repaired (22, 166, 185), although the situation with respect to tRNA genes transcribed by RNA polymerase III remains to be fully clarified. In fact, in general, the repair of pyrimidine dimers and interstrand cross-links induced by nitrogen mustard has been observed to be considerably less efficient in rRNA genes than in the *DHFR* gene of mouse cells (166). On the other hand, monofunctional alkylation damage induced by methyl methanesulfonate (MMS) and even interstrand cross-links induced by treatment of cells with nitrogen mustard were found to be efficiently repaired in rRNA genes (166). The lack of preferential repair in rRNA genes is consistent with the fact that distinct factors are used for the initiation of transcription by different RNA polymerases. Evidently a dual function in transcription and nucleotide excision repair is confined to transcription factors associated with RNA polymerase II. Both nondividing and proliferating XP-C cells have been shown to be defective in the repair of rDNA, consistent with the notion that XP-C cells are defective in the repair of all regions of the genome that are not transcribed by RNA polymerase II (23).

In summary, there are compelling indications of a kinetic hierarchy of nucleotide excision repair in mammalian genes. Transcriptionally silent (repressed) genes are repaired the most slowly, transcriptionally active genes are repaired more rapidly, and in such genes the transcribed strand is repaired fastest (118). However, the molecular determinants for this hierarchy are unknown. It is interesting that transcriptionally repressed genes, such as the *754* and factor *IX* genes on the X chromosome of humans, are localized to the cytogenetically defined R-bands of human chromosomes, while transcriptionally active housekeeping genes are localized in the G-bands. DNA in the R-bands is typically replicated early, while G-band DNA is replicated later. Hence, the rate of nucleotide excision repair of genes may be related to the timing of their replication in the S phase (118).

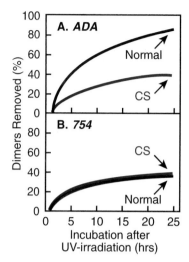

Figure 7–17 Nucleotide excision repair of pyrimidine dimers in normal human and CS cells. (A) Nucleotide excision repair in the transcriptionally active adenosine deaminase (*ADA*) gene is considerably slower in CS cells, suggesting a defect in the preferential repair of this gene. (B) Consistent with this notion, the rate of repair in a nontranscribed locus (designated *754*) located on the X chromosome is lower in both normal and CS cells and is comparable to the low rate observed in the *ADA* gene of CS cells. (*Adapted from Venema et al. [182a].*)

As is the case with the yeast *S. cerevisiae*, much remains to be learned about the detailed biochemistry of nucleotide excision repair in template strands of actively transcribed genes in mammalian cells and how this differs from that in transcriptionally silent DNA. As discussed in chapter 8, genes and their purified polypeptide products that are involved in nucleotide excision repair in mammalian cells are being detected at an astonishing rate, and this aspect of cellular responses to DNA damage is likely to be deciphered in the not too distant future, perhaps before this writing is translated into a published book. For the moment, it seems reasonably safe to predict that the basic paradigm (extensively discussed in chapter 6) of a core repairosome provided by the transcription initiation factor TFIIH operates in mammalian cells. It is possible that TFIIH is obligatorily loaded onto promoter sites during basal transcription initiation and is coopted into the transcription elongation complex. Arrested transcription at sites of base damage during transcriptional elongation would then provide the immediate availability of this core repairosome for assembly to a complete, functionally active repair complex. What is considerably less certain is the role of CSB (ERCC6) protein and the putative CSA protein (the product of the as yet uncloned *CSA* gene) in this process. We will reconsider some of these conundrums in the next chapter and also in chapter 14, when we consider the molecular pathology of CS and its relationship to XP.

As discussed in chapter 6, in our view the primary evolutionary significance of this biochemical coupling is the advantage afforded to nucleotide excision repair by the "open" conformation of transcribed chromatin. The fact that the transcribed strand itself derives an added kinetic advantage over the nontranscribed strand for repair by virtue of this coupling may indeed contribute to the viability of cells but can be viewed as a consequence of rather a primary selective force for the coupling of transcription to repair.

Other Aspects of Genomic Organization during Excision Repair

Methylation of DNA during Excision Repair

The level and/or distribution of 5-methylcytosine in DNA is important for the regulation of gene expression in higher organisms (24, 35, 59, 120, 136, 137). Normally, methylation patterns in parental DNA are preserved during DNA replication by rapid methylation of newly synthesized daughter strand DNA at sites of methylation in the parental strands (9). This mechanism stabilizes the methylation pattern from one cell generation to the next. A question of considerable interest is whether in the course of repair synthesis, newly incorporated cytosine becomes normally methylated and, if so, by what means and with what kinetics? If we assume that modified (by methylation) gene-controlling sequences exist in DNA, DNA damage in or adjacent to such sequences could have profound effects on gene expression. It has been proposed (70) that higher organisms contain enzymes that recognize DNA with one methylated strand and add a methyl group to the other strand. However, nonmethylated DNA is not a substrate for the modification enzyme. Under normal conditions, replication of DNA results in half-methylated DNA and the daughter strands become methylated by the action of a maintenance methylase soon after DNA synthesis. If a DNA sequence has no methyl groups, it remains unmethylated in the presence of this enzyme. It has been proposed that in one state genes are transcribed and in the other they are repressed (70). This difference is, in effect, stably inherited, since a change in gene expression would occur only if the methylating enzyme was lost by mutation.

This regulatory mechanism could be upset during excision repair. If base damage located just ahead of the replication fork is removed by nucleotide excision repair, the new patch of DNA generated during repair synthesis will not be immediately methylated. Hence, if replication occurs before this can be corrected, alterations in gene expression may result. Damage to a parental strand immediately after replication could produce the same effect (69) (Fig. 7–18). Consistent

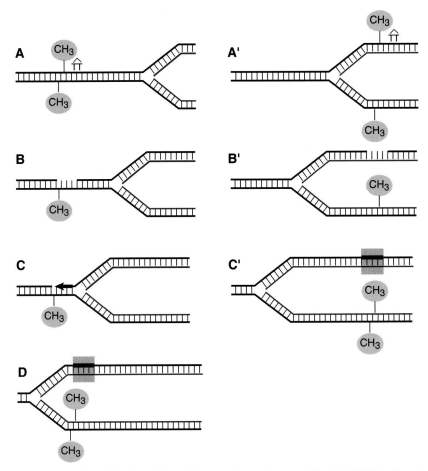

Figure 7–18 Effect of nucleotide excision repair on the methylation of DNA. Repair of damage immediately in front of a replication fork is shown on the left. If excision is initiated close to a methylated controlling sequence (A), a methylated base may be removed (B) and replaced (by repair synthesis) with a nonmethylated base (C). Replication of this region before methylation can occur gives rise to a nonmethylated DNA duplex which is not a substrate for the maintenance methylase. The other half-methylated DNA duplex will be normally methylated (D). Repair of damage to DNA close to a controlling sequence immediately after replication is shown on the right. Excision of damage before the daughter strands have been methylated (A′ and B′), gives rise to a nonmethylated DNA molecule (C′), which is not a substrate for the maintenance methylase. (*Adapted from Friedberg [49] with permission.*)

with this expectation, in confluent (nondividing) human diploid fibroblasts exposed to UV radiation, *N*-methyl-*N*-nitrosourea (MNU), or *N*-acetoxy-AAF-acetylaminofluorene, methylation of deoxycytidine incorporated by repair synthesis has been shown to be slow and incomplete (79) (Fig. 7–19). In cells from cultures in logarithmic growth, 5-methylcytosine formation in repair synthesis patches associated with the excision of pyrimidine dimers is faster and more extensive but still does not attain the level observed in replicating nondamaged DNA. The hypomethylated repair patches in confluent cells are further methylated when the cells are stimulated to divide; however, such regions may still not be fully methylated before cell division occurs (79). Thus, DNA damage and repair apparently can lead to changes in methylation patterns in daughter cells.

There are also indications that methylation of cytosine in the DNA of mammalian cells is directly affected by the presence of certain forms of DNA damage. Thus, a diverse range of chemical carcinogens inhibit the transfer of methyl groups from *S*-adenosylmethionine to hemimethylated DNA, in a reaction catalyzed by

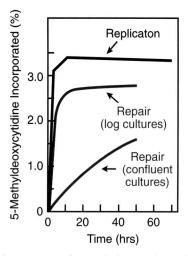

Figure 7–19 The methylation of DNA following nucleotide excision repair of UV radiation damage in confluent and in logarithmic phase human fibroblasts is slower and less complete than in undamaged cells undergoing normal semiconservative DNA synthesis. (*Adapted from Friedberg [49] with permission.*)

mouse spleen methyltransferase in vitro (198). Furthermore, some carcinogens have been shown to directly modify and inactivate the methyltransferase enzyme. In addition, DNA containing sites of base loss has a reduced ability to accept methyl groups. Carcinogenic agents may therefore cause heritable changes in 5-methylcytosine patterns by a variety of mechanisms (198).

The effect of DNA methylation on repair has also been studied at the level of a single gene. A hypomethylated derivative of a CHO cell line was derived by passaging cells in the presence of increasing concentrations of 5-azacytidine for several months (68). These cells showed >60% demethylation of their DNA. Following exposure to UV radiation, the amount of repair replication during 24 h was increased ca. twofold overall in these cells. However, this was not accompanied by any change in resistance to UV radiation. Loss of T4 endonuclease V-sensitive sites was unaffected in the 5' half of the *DHFR* gene, in a nontranscribed region immediately downstream of the gene, and in an extragenic region even farther downstream. However, increased repair was detected in the 3' half of the gene.

ADP-Ribosylation and DNA Repair in Mammalian Cells

NAD is the most abundant of the respiratory coenzymes (66). The classic biochemical studies of Harden, Young, Warburg, and others established that the major biochemical function of NAD is as an electron carrier in various biological oxidation-reduction systems (66). Its structure was determined in 1936 by Von Euler and his collaborators and was shown to consist of two mononucleotides (5' AMP and nicotinamide mononucleotide) linked together by a PP_i bond. The structure of NAD can also be viewed as an ADP-ribosyl moiety (ADP-ribose) attached covalently to the vitamin nicotinamide through an *N*-glycosylic linkage (Fig. 7–20). This linkage constitutes a high-energy bond, since its free energy of hydrolysis is ca −8.2 kcal/mol (ca. −34.3 kJ/ml) at pH 7 and 25°C (66).

The energy of this bond is used to enzymatically transfer the ADP-ribosyl moiety of NAD to macromolecules. Thus, ADP-ribosyl transferase catalyzes the transfer of the ADP-ribosyl moiety of NAD to chromosomal proteins, with the concomitant release of nicotinamide and a proton. This reaction is termed *mono-ADP-ribosylation* (66). Additionally, a polymer of ADP-ribose can be formed as a moiety attached to a macromolecular acceptor such as a chromosomal protein, in a reaction termed *poly-ADP-ribosylation* (66). The catalytic activity responsible for these reactions is termed *poly(ADP-ribose) synthetase* or *poly(ADP-ribose) polymerase*. In normal (undamaged) cells, about 90% of the ADP-ribose is present in the monomeric form; the small percentage present as polymers may contain up to 70 ADP-ribose units in a single chain. However, the total amount of constitutive ADP-ribose in the chromatin of undamaged cells is relatively small (1 ADP-ribose residue per 5,000 nucleotides) (66).

In eukaryotic cells, all NAD synthesis occurs in the nucleus, and the vast majority of NAD formed is used for the biosynthesis of ADP-ribose and poly(ADP-ribose) (149). Poly(ADP-ribose) synthetase is tightly associated with chromatin and catalyzes the synthesis of acceptor-bound poly-(ADP-ribose) in a reaction that has an absolute requirement for DNA (149). Furthermore, this DNA must have

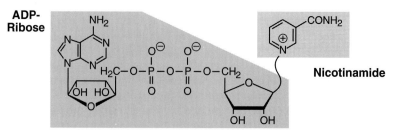

Figure 7–20 The structure of NAD can also be viewed as ADP-ribosyl nicotinamide. (*Adapted from Friedberg [49] with permission.*)

free ends. Covalently closed circular DNA does not activate the enzyme, but fragmentation of DNA increases its ability to activate the enzyme, and the activation is proportional to the number of nicks in the DNA (60, 133, 149). In light of these considerations, the possibility has been entertained that ADP-ribosylation is involved in DNA repair by the modification of chromosomal proteins and that such modification is effected by the specific relationship of the activation of poly(ADP-ribose) synthetase to the presence of strand breaks in DNA (hence the inclusion of this topic in the present chapter). The following paragraphs recount some of the history of these investigations. In the final analysis, the current evidence would suggest that poly(ADP-ribose) polymerase influences the efficiency of DNA ligation during base excision repair of DNA but does not have any effect on nucleotide excision repair (39, 110, 144).

A variety of compounds inhibit poly(ADP-ribose) biosynthesis. These include 5-methylnicotinamide, methylxanthines, thymine, thymidine, and benzamides such as 3-aminobenzamide (149). With the use of these inhibitors it can be shown that the biosynthesis of poly(ADP-ribose) is not needed for cell growth or for DNA, RNA, or protein synthesis in cultures of exponentially growing cells not subject to DNA damage (149). However, a variety of DNA-damaging agents, in particular those that cause damage which is repaired by the base excision repair mode, cause a marked lowering of cellular NAD levels (44, 153). In addition, the specific activity of the synthetase increases in these cells in a dose-dependent fashion. This effect can be prevented by inhibitors of poly(ADP-ribose) synthesis (44) (Fig. 7–21). These observations suggested that DNA-damaging agents somehow promoted the utilization of cellular NAD for the biosynthesis of poly(ADP-ribose) and led to the hypothesis that this biosynthesis is a cellular response to DNA damage required for efficient DNA repair (149, 150).

This hypothesis has been tested in a number of ways. When DNA incision and rejoining were monitored as parameters of base excision repair, there was a marked inhibition of the reconstitution of high-molecular-weight DNA if cells exposed to MNU were incubated in the presence of inhibitors of poly(ADP-ribose) synthetase (Fig. 7–21) (44, 150). Additionally, the ability of cells to carry out base excision repair was depressed when cellular NAD levels were severely reduced by nutritional deprivation (149). When the cellular NAD level was reduced below K_m levels for the synthetase and the cells were exposed to γ radiation or to alkylation damage, they still incorporated precursors for DNA, RNA, and protein synthesis

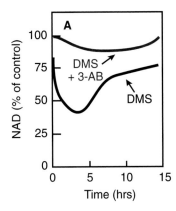
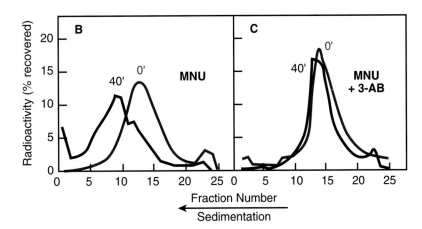

Figure 7–21 (A) Dimethyl sulfate (DMS) causes a decrease in the NAD content of mouse L cells in culture. This effect can be largely mitigated by inhibitors of poly(ADP-ribose) polymerase such as 3-aminobenzamide (3-AB). (B) When cells treated with N-methyl-N-nitrosourea (MNU) are incubated for about 40 min and then lysed and sedimented in alkaline sucrose gradients, the DNA (40') sediments faster than DNA from unincubated cells (0'), indicating that repair of strand breaks associated with MNU treatment took place. (C) However, in the presence of 3-aminobenzamide, repair of DNA strand breaks is inhibited. (*Adapted from Friedberg [49] with permission.*)

at normal rates but had a marked defect in the reconstitution of DNA strand breaks (149). Finally, in the presence of the inhibitor 5-methylnicotinamide, cell killing induced by treatment of cells with the alkylating agent MNU was markedly potentiated (43, 44, 121, 150). Similar results were obtained in the presence of other inhibitors of poly(ADP-ribose) synthetase, such as methylxanthines, theobromine, and theophylline. These inhibitors also potentiated a synergistic increase in cell toxicity after treatment of cells with radiation or neocarzinostatin (150). Other biological end points possibly related to DNA repair have been examined. In CHO cells grown in nicotinamide-free medium (which results in a marked decrease in cellular NAD levels), the spontaneous frequency of certain chromosomal abnormalities termed *sister chromatid exchanges* was enhanced (71). Furthermore, treatment of cells with inhibitors of poly(ADP-ribose) synthetase in the absence of any exogenous DNA-damaging agents resulted in a marked increase in the frequency of sister chromatid exchanges and other chromosomal aberrations (71).

Poly(ADP-ribose) is apparently not required for the incision of damaged DNA or for repair synthesis. However, it was originally suggested that this compound may activate a DNA ligase required for DNA repair in mammalian cells (37). Mammalian cells contain three forms of this enzyme, called DNA ligases I, II, and III (see chapter 8). When cells are exposed to the alkylating agent dimethylsulfate, the total DNA ligase activity increases about twofold; most, if not all, of this reflects an increase in DNA ligase II activity (37). This increase can be prevented by all known inhibitors of poly(ADP-ribose) synthetase. This led to the hypothesis that DNA ligase may be an acceptor protein for ADP-ribose, in which state it is activated to carry out more-efficient rejoining of nicks in DNA (37). However, this model fails to explain why the removal of pyrimidine dimers during nucleotide excision repair (which results in approximately the same frequency of DNA strand breaks over the same time intervals as do low doses of alkylating agents) does not involve a ligation step that is sensitive to perturbations of the level of poly(ADP-ribose) by 3-aminobenzamide (29, 111). Either the precise chemical nature of the incisions produced by the nucleotide excision repair of pyrimidine dimers makes them considerably less effective as a stimulus for synthesis of poly(ADP-ribose), or single-strand breaks by themselves are not the effective trigger for such synthesis (111). Thus, some other change associated with DNA damage by alkylating agents and/or its repair may be important (43).

Results of direct measurements on the role of ADP ribosylation in DNA ligation during base excision repair have been contradictory. An examination of the rate of ligation of intracellular repair patches and their size in MMS-treated cells showed no alterations when poly(ADP-ribose) polymerization was inhibited, leading to the conclusion that poly(ADP-ribose) does not regulate DNA ligation during the repair of damage caused by this alkylating agent (30). A similar conclusion was reached from studies in which cells were exposed to UV radiation (33). However, the use of different parameters for measuring DNA ligation in MMS-treated cells led to the conclusion that inhibition of poly(ADP-ribose) synthesis by 3-aminobenzamide actually facilitated DNA ligation (31).

The latter notion has received support from more recent studies in which the relationship of poly(ADP-ribose) to the repair of strand breaks has been examined by using a human cell-free system. In this cell-free system (whose general principles are discussed more extensively in chapter 8), plasmid DNA damaged by selected physical or chemical agents was used as a substrate for base excision repair. When γ-irradiated plasmid DNA containing strand breaks was incubated with human cell extracts in the presence of NAD^+, a marked stimulation in the rate of conversion of nicked circular molecules to covalently closed circular molecules was noted (143). This effect was accompanied by a rapid increase in the level of poly(ADP-ribose) and of ribosylated poly(ADP-ribose) synthetase. The automodification of this enzyme is known to result in a decreased binding affinity of the enzyme for DNA. Hence, it has been suggested that the stimulation of strand break rejoining is not the result of a direct effect on DNA ligase but, rather, the result of the dissociation of bound poly(ADP-ribose) polymerase from sites of such breaks,

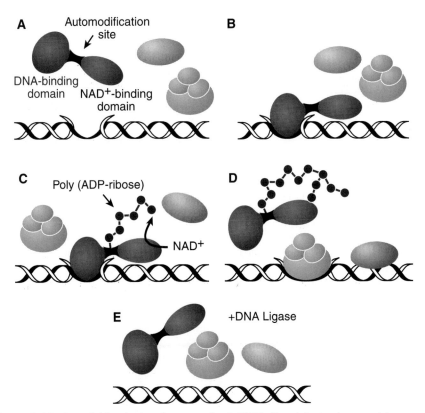

Figure 7–22 A model for the involvement of poly(ADP-ribose) during base excision repair of DNA. (A) Following the introduction of strand breaks by agents such as γ-radiation, poly(ADP-ribose) polymerase, base excision repair enzymes (pink), and DNA polymerase(s) (grey) compete for binding to strand breaks in the DNA. (B) The binding of poly(ADP-ribose) polymerase interferes with access to the strand breaks by enzymes required for base excision repair. (C) The binding of poly(ADP-ribose) polymerase activates the synthesis of poly(ADP-ribose), resulting in automodification of the molecule and resultant reduced binding affinity for the damaged DNA. (D) Automodified poly(ADP-ribose) polymerase is released, allowing base excision repair to occur to completion. (E) The degradation of poly(ADP-ribose) by a specific glycohydrolase results in the regeneration of unmodified poly(ADP-ribose) polymerase.

thereby allowing strand rejoining to take place (143) (Fig. 7–22). Consistent with this view, cell extracts depleted of poly(ADP-ribose) synthetase efficiently catalyze the rejoining of strand breaks in γ-irradiated plasmid DNA in the presence or absence of NAD$^+$. No stimulatory effect was observed by supplementing extracts with either purified mammalian DNA ligase I or II (145). NAD$^+$-promoted DNA repair of strand breaks was also observed with alkylated and UV-irradiated plasmid DNA (145). In the latter situation, the effect is ascribed to the presence of pyrimidine hydrates rather than to the repair of bulky adducts such as pyrimidine dimers or (6-4) photoproducts. Indeed, the available evidence indicates that nucleotide excision repair is not influenced by the levels of NAD$^+$ in cells and does not involve the activation of poly(ADP-ribose) polymerase. The presence of a large multiprotein complex for nucleotide excision repair may preclude the binding of the polymerase to ligatable nicks during this excision repair mode (144). Similar conclusions emerged from the results of independent studies in which it was shown that microinjection of a recombinant polypeptide containing the DNA-binding domain of poly(ADP-ribose) polymerase into cells in culture blocked repair synthesis in cells treated with the alkylating agent *N*-methyl-*N*′-nitro-*N*-nitrosoguanidine but not in cells treated with UV radiation (110). Consistent with this model, the addition of recombinant poly(ADP-ribose) polymerase deleted of the region of the protein required for the binding of NAD but retaining the N-terminal DNA-binding domain inhibited the repair of ionizing radiation damage in extracts of human cells depleted of endogenous poly(ADP-ribose) polymerase (163). However, when a protein defective in the DNA-binding domain but not in the NAD-binding domain was added to the extracts, the inhibition of repair was not alleviated by the addition of NAD (163).

These experimental observations provide a reasonable explanation of how the levels of poly(ADP-ribose) polymerase and hence ADP-ribose formation are influenced by the repair status of cells exposed to certain types of DNA damage. However, there is no indication that poly(ADP-ribose) polymerase is obligatory for base excision repair. Therefore, the central function of this interesting enzyme in cells remains unclear. Conceivably, it is required for the protection of DNA strand breaks that are introduced during the course of a specific biological process, such as DNA replication or the association of DNA with the nuclear scaffold (143). Alternatively, and closer to the topic of repair, it has been suggested that the binding of the enzyme to strand breaks introduced by DNA damage might effect a specific signal that shuts down DNA replication until base excision repair is completed (143). Indeed, poly(ADP-ribose) polymerase has been reported to copurify with replication forks (6), topoisomerase I (48), and polymerase α (151).

The notion that poly(ADP-ribose) polymerase influences DNA repair by modifying chromatin structure remains an interesting one. Electron-microscopic examination of chromatin indicates that the superstructure relaxes when histone H1 is poly-ADP-ribosylated (40). This and the reverse reaction effected by poly(ADP-ribose) glycohydrolase during reassembly of the histone have been termed the "histone shuttle" (3) and may help provide access for nucleotide excision repair enzymes to sites of base damage (110).

The Poly(ADP-Ribose) Polymerase Gene

The human poly(ADP-ribose) polymerase gene was cloned from a phage λ expression library by immunological screening (2). The nucleotide sequence revealed a single open reading frame which can encode a protein of ~113 kDa. Computer-derived analysis of the predicted polypeptide indicates that it belongs to a subfamily of DNA/NAD-binding proteins (20). The gene maps to human chromosome 1q42 (67) and is highly conserved in vertebrates (39). Comparisons of the amino acid sequences of the cloned human and mouse genes reveal extensive similarities, particularly in domains believed to be required for DNA and NAD$^+$ binding (73).

Expression of the gene at the transcriptional level is not affected by the presence of strand breaks introduced by X-irradiation of cells or by treatment with a variety of alkylating agents, suggesting that, in contrast to the catalytic requirement of the polymerase, induction of the gene which encodes the enzyme is not a strand break-dependent process (6). The availability of the cloned gene affords the opportunity for studies in which its regulated expression can be effected in cells under a variety of experimental conditions. Such studies might help further elucidate the role of poly(ADP-ribose) in DNA repair.

Conclusions

This chapter has focused on selected aspects of the biology of nucleotide excision repair in mammalian cells, with a particular emphasis on how structural attributes of the mammalian genome may influence its repairability. The following chapter considers the genes and proteins which participate in the biochemistry of nucleotide excision repair in mammals and what we currently know about this process.

REFERENCES

1. **Adolph, K. W., S. M. Cheng, J. R. Paulson, and U. K. Laemmli.** 1977. Isolation of a protein scaffold from mitotic HeLa cell chromosomes. *Proc. Natl. Acad. Sci. USA* **74:**4937–4941.

2. **Alkhatib, H. M., D. Chen, B. Cherney, K. Bhatia, V. Notario, C. Giri, G. Stein, E. Slattery, and R. G. Roeder.** 1987. Cloning and expression of cDNA for human poly(ADP-ribose) polymerase. *Proc. Natl. Acad. Sci. USA* **84:**1224–1228.

3. **Althaus, F. R.** 1992. Poly ADP-ribosylation: a histone shuttle mechanism in DNA excision repair. *J. Cell Sci.* **102:**663–670.

4. **Arnold, G. E., A. K. Dunker, and M. J. Smerdon.** 1987. Limited nucleosome migration can completely randomize DNA repair patches in intact human cells. *J. Mol. Biol.* **196:**433–436.

5. **Benyajati, C., and A. Worcel.** 1976. Isolation, characterization, and structure of the folded interphase genome of *Drosophila melanogaster.* *Cell* **9:**393–407.

6. **Bhatia, K., Y. Pommier, C. Giri, A. J. Fornace, Jr., M. Imaizumi, T. R. Breitman, B. W. Cherney, and M. E. Smulson.** 1990. Expression of the poly(ADP-ribose) polymerase gene following natural and induced DNA strand breakage and effect of hyperexpression on DNA repair. *Carcinogenesis* **11:**123–128.

7. **Bianchi, N. O., M. S. Bianchi, K. Alitalo, and A. de la Chapelle.** 1991. UV damage and repair in the domain of the human *c-myc* oncogene. *DNA Cell Biol.* **10:**125–132.

8. **Bill, C. A., B. M. Grochan, R. E. Meyn, V. A. Bohr, and P. J. Tofilon.** 1991. Loss of intragenomic DNA repair heterogeneity with cellular differentiation. *J. Biol. Chem.* **266:**21821–21826.

9. **Bird, A. P.** 1978. Use of restriction enzymes to study eukaryotic DNA methylation. II. The symmetry of methylated sites supports semiconservative copying of the methylation pattern. *J. Mol. Biol.* **118:**49–60.

10. **Bodell, W. J.** 1977. Nonuniform distribution of DNA repair in chromatin after treatment with methyl methanesulfonate. *Nucleic Acids Res.* **4:**2619–2628.

11. **Bodell, W. J., and M. R. Banerjee.** 1979. The influence of chromatin structure on the distribution of DNA repair synthesis studied by nuclease digestion. *Nucleic Acids Res.* **6:**359–370.

11a. **Bohr, V. A.** 1987. *Preferential DNA Repair in Active Genes,* p. 1–16. Laegeforeningens, Copenhagen.

11b. **Bohr, V. A.** 1988. DNA repair and transcriptional activity in genes. *J. Cell Sci.* **90:**175–178.

11c. **Bohr, V. A.** 1991. Gene specific DNA repair. *Carcinogenesis* **12:**1983–1991.

12. **Bohr, V. A.** 1992. Gene-specific DNA repair: characteristics and relations to genomic instability, p. 217–227. *In* V. A. Bohr, K. Wassermann, and K. H. Kraemer (ed.), *DNA Repair Mechanisms.* Munksgaard, Copenhagen.

13. **Bohr, V. A., D. S. Okumoto, L. Ho, and P. C. Hanawalt.** 1986. Characterization of a DNA repair domain containing the dihydrofolate reductase gene in Chinese hamster ovary cells. *J. Biol. Chem.* **261:**16666–16672.

14. **Bohr, V. A., D. H. Phillips, and P. C. Hanawalt.** 1987. Heterogeneous DNA damage and repair in the mammalian genome. *Cancer Res.* **47:**6426–6436.

15. **Bohr, V. A., C. A. Smith, D. S. Okumoto, and P. C. Hanawalt.** 1985. DNA repair in an active gene: removal of pyrimidine dimers from the *DHFR* gene of CHO cells is much more efficient than in the genome overall. *Cell* **40:**359–369.

15a. **Bohr, V. A., and K. Wassermann.** 1988. DNA repair at the level of the gene. *Trends Biochem. Sci.* **13:**429–432.

16. **Carothers, A. M., W. Zhen, J. Mucha, Y.-J. Zhang, R. M. Santella, D. Grunberger, and V. A. Bohr.** 1992. DNA strand-specific repair of (+ −)-3α,4β-dihydroxy-1α,2α-epoxy-1,2,3,4-tetrahydrobenzo[c]phenanthrene adducts in the hamster dihydrofolate reductase gene. *Proc. Natl. Acad. Sci. USA* **89:**11925–11929.

17. **Carreau, M., and D. Hunting.** 1992. Transcription-dependent and independent DNA excision repair pathways in human cells. *Mutat. Res.* **274:**57–64.

18. **Chen, R. H., V. M. Maher, J. Brouwer, P. van de Putte, and J. J. McCormick.** 1992. Preferential repair and strand-specific repair of benzo[a]pyrene diolepoxide adducts in the *HPRT* gene of diploid human fibroblasts. *Proc. Natl. Acad. Sci. USA* **89:**5413–5417.

19. **Chen, R.-H., V. M. Maher, and J. J. McCormick.** 1990. Effect of excision repair by diploid human fibroblasts on the kinds and locations of mutants induced by (+ −)-7β,8α-dihydroxy-9α,10α-epoxy-7,8,9,10-tetrahydrobenzo[a]pyrene in the coding region of the *HPRT* gene. *Proc. Natl. Acad. Sci. USA* **87:**8680–8684.

20. **Cherney, B. W., O. W. McBride, D. Chen, H. Alkhatib, K. Bhatia, P. Hensley, and M. E. Smulson.** 1987. cDNA sequence, protein structure, and chromosomal location of the human gene for poly(ADP-ribose) polymerase. *Proc. Natl. Acad. Sci. USA* **84:**8370–8374.

21. **Christians, F. C., and P. C. Hanawalt.** 1992. Inhibition of transcription and strand-specific DNA repair by α-amanitin in Chinese hamster ovary cells. *Mutat. Res.* **274:**93–101.

22. **Christians, F. C., and P. C. Hanawalt.** 1993. Lack of transcription-coupled repair in mammalian ribosomal RNA genes. *Biochemistry* **32:**10512–10518.

23. **Christians, F. C., and P. C. Hanawalt.** 1994. Repair in ribosomal RNA is deficient in xeroderma pigmentosum group C and in Cockayne's syndrome cells. *Mutat. Res.* **323:**179–187.

24. **Christman, J. K., P. Price, L. Pedrinan, and G. Acs.** 1977. Correlation between hypomethylation of DNA and expression of globin genes in Friend erythroleukemia cells. *Eur. J. Biochem.* **81:**53–61.

25. **Clark, D. J., and G. Felsenfeld.** 1991. Formation of nucleosomes on positively supercoiled DNA. *EMBO J.* **10:**387–395.

26. **Clarkson, J. M., D. L. Mitchell, and G. M. Adair.** 1983. The use of an immunological probe to measure the kinetics of DNA repair in normal and UV-sensitive mammalian cell lines. *Mutat. Res.* **112:**287–299.

27. **Cleaver, J. E.** 1977. DNA repair processes and their impairment in some human diseases, p. 29–42. *In* D. Scott, B. A. Bridges, and F. Sobels (ed.), *Progress in Genetic Toxicology.* Elsevier Biomedical Press, Amsterdam.

28. **Cleaver, J. E.** 1977. Nucleosome structure controls rates of excision repair in DNA of human cells. *Nature* (London) **270:**451–453.

29. **Cleaver, J. E., W. J. Bodell, W. F. Morgan, and B. Zelle.** 1983. Differences in the regulation by poly(ADP-ribose) of repair of DNA damage from alkylating agents and ultraviolet light according to cell type. *J. Biol. Chem.* **258:**9059–9068.

30. **Cleaver, J. E., and W. F. Morgan.** 1985. Poly(ADP-ribose) synthesis is involved in the toxic effects of alkylating agents but does not regulate DNA repair. *Mutat. Res.* **150:**69–76.

31. **Cleaver, J. E., and S. D. Park.** 1986. Enhanced ligation of repair sites under conditions of inhibition of poly(ADP-ribose) synthesis by 3-aminobenzamide. *Mutat. Res.* **173:**287–290.

32. **Cleaver, J. E., and G. H. Thomas.** 1981. Measurement of unscheduled synthesis by autoradiography, p. 277–287. *In* E. C. Friedberg and P. C. Hanawalt (ed.), *DNA Repair—a Laboratory Manual of Research Procedures,* vol. 1, part B. Marcel Dekker, Inc., New York.

33. **Collins, A.** 1985. Poly(ADP-ribose) is not involved in the rejoining of DNA breaks accumulated to high levels in u.v.-irradiated HeLa cells. *Carcinogenesis* **6:**1033–1036.

34. **Collins, A. R. S., and R. T. Johnson.** 1981. Use of metabolic inhibitors in repair studies, p. 341–360. *In* E. C. Friedberg and P. C. Hanawalt (ed.), *DNA Repair—a Laboratory Manual of Research Procedures,* vol. 1, part B. Marcel Dekker, Inc., New York.

35. **Compere, S. J., and R. D. Palmiter.** 1981. DNA methylation controls the inducibility of the mouse metallothionein-I gene in lymphoid cells. *Cell* **25:**233–240.

36. **Cook, P. R., and I. A. Brazell.** 1975. Supercoils in human DNA. *J. Cell Sci.* **19:**261–279.

37. **Creissen, D., and S. Shall.** 1982. Regulation of DNA ligase activity by poly(ADP-ribose). *Nature* (London) **296:**271–272.

38. **Crouse, G. F., E. J. Leys, R. N. McEwan, E. G. Frayne, and R. E. Kellems.** 1985. Analysis of the mouse *dhfr* promoter region: existence of a divergently transcribed gene. *Mol. Cell. Biol.* **5:**1847–1858.

39. **de Murcia, G., and J. M. de Murcia.** 1994. Poly(ADP ribose) polymerase: a molecular nick sensor. *Trends Biochem. Sci.* **19:**172–176.

40. **de Murcia, G., A. Huletsky, D. Lamarre, A. Gaudreau, J. Pouyet, M. Daune, and G. G. Poirier.** 1986. Modulation of chromatin superstructure induced by poly(ADP-ribose) synthesis and degradation. *J. Biol. Chem.* **261:**7011–7017.

41. **Dorado, G., H. Steingrimsdottir, C. F. Arlett, and A. R. Lehmann.** 1991. Molecular analysis of ultraviolet-induced mutations in a xeroderma pigmentosum cell line. *J. Mol. Biol.* **217:**217–222.

42. **Downes, C. S., A. J. Ryan, and R. T. Johnson.** 1993. Fine tuning of DNA repair in transcribed genes: mechanisms, prevalence and consequences. *Bioessays* **15:**209–216.

43. **Durkacz, B. W., N. Nduka, O. Omidiji, S. Shall, and A.-A. Zia'ee.** 1980. ADP-ribose and DNA repair, p. 207–224. In M. Smulson and T. Sugimura (ed.), *Novel ADP-Ribosylations of Regulatory Enzymes and Proteins.* Elsevier/North Holland, Amsterdam.

44. **Durkacz, B. W., O. Omidiji, D. A. Gray, and S. Shall.** 1980. (ADP-ribose)$_n$ participates in DNA excision repair. *Nature* (London) **283:**593–596.

45. **Ehmann, U. K., and E. C. Friedberg.** 1980. An investigation of the effect of radioactive labeling of DNA on excision repair in UV-irradiated human fibroblasts. *Biophys. J.* **31:**285–291.

46. **Evans, D. H., and S. Linn.** 1984. Excision repair of pyrimidine dimers from simian virus 40 minichromosomes in vitro. *J. Biol. Chem.* **259:**10252–10259.

47. **Felsenfeld, G.** 1992. Chromatin as an essential part of the transcriptional mechanism. *Nature* (London) **355:**219–224.

48. **Ferro, A. M., N. P. Higgins, and B. M. Olivera.** 1983. Poly(ADP-ribosylation) of a DNA topoisomerase. *J. Biol. Chem.* **258:**6000–6003.

49. **Friedberg, E. C.** 1985. *DNA Repair.* W. H. Freeman & Co., New York.

50. **Friedberg, E. C., and P. C. Hanawalt (ed.).** 1981. *DNA Repair—a Laboratory Manual of Research Procedures*, vol. 1, part A. Marcel Dekker, Inc., New York.

51. **Friedberg, E. C., and P. C. Hanawalt (ed.).** 1981. *DNA Repair—a Laboratory Manual of Research Procedures*, Vol 1, part B. Marcel Dekker, Inc., New York.

52. **Friedberg, E. C., and P. C. Hanawalt (ed.).** 1983. *DNA Repair—a Laboratory Manual of Research Procedures*, vol. 2. Marcel Dekker, Inc., New York.

53. **Friedberg, E. C., and P. C. Hanawalt (ed.).** 1988. *DNA Repair—a Laboratory Manual of Research Procedures*, vol. 3. Marcel Dekker, Inc., New York.

54. **Gale, J. M., K. A. Nissen, and M. J. Smerdon.** 1987. UV-induced formation of pyrimidine dimers in nucleosome core DNA is strongly modulated with a period of 10.3 bases. *Proc. Natl. Acad. Sci. USA* **84:**6644–6648.

55. **Ganesan, A., G. Spivak, and P. C. Hanawalt.** 1983. Expression of DNA repair genes in mammalian cells, p. 45–54. In P. Nagley, A. W. Linnar, W. J. Peacock, and J. A. Pateman (ed.), *Manipulation and Expression of Genes in Eukaryotics.* Academic Press, Sydney, Australia.

56. **Gao, S., R. Drouin, and G. P. Holmquist.** 1994. DNA repair rates mapped along the human *PGK1* gene at nucleotide resolution. *Science* **263:**1438–1440.

57. **Gottesfeld, J. M., W. T. Garrard, G. Bagi, R. F. Wilson, and J. Bonner.** 1974. Partial purification of the template-active fraction of chromatin: a preliminary report. *Proc. Natl. Acad. Sci. USA* **71:**2193–2197.

58. **Govan, H. L., III, Y. Valles-Ayoub, and J. Braun.** 1990. Fine-mapping of DNA damage and repair in specific genomic segments. *Nucleic Acids Res.* **18:**3823–3830.

59. **Groudine, M., R. Eisenman, and H. Weintraub.** 1981. Chromatin structure of endogenous retroviral genes and activation by an inhibitor of DNA methylation. *Nature* (London) **292:**311–317.

60. **Halldorsson, H., D. A. Gray, and S. Shall.** 1978. Poly(ADP-ribose) polymerase activity in nucleotide permeable cells. *FEBS Lett.* **85:**349–352.

60a. **Hanawalt, P. C.** 1986. Intragenomic heterogeneity in DNA damage processing: potential implications for risk assessment. p 489–498. In M. G. Simic, L. Grossman, and A. C. Upton (ed.), *Mechanisms of DNA Damage and Repair.* Plenum, New York.

61. **Hanawalt, P. C.** 1987. Preferential DNA repair in expressed genes. *Environ. Health Perspect.* **76:**9–14.

61a. **Hanawalt, P. C.** 1991. Heterogeneity of DNA repair at the gene level. *Mutat. Res.* **247:**203–211.

62. **Hanawalt, P. C.** 1992. Transcription-dependent and transcription-coupled DNA repair responses, p. 231–242. In V. A. Bohr, K. Wassermann, and K. H. Kraemer (ed.), *DNA Repair Mechanisms.* Munksgaard, Copenhagen.

63. **Hanawalt, P. C., P. K. Cooper, A. K. Ganesan, and C. A. Smith.** 1979. DNA repair in bacteria and mammalian cells. *Annu. Rev. Biochem.* **48:**783–836.

64. **Hanawalt, P. C., P. K. Cooper, and C. A. Smith.** 1981. Repair replication schemes in bacteria and human cells. *Prog. Nucleic Acid Res. Mol. Biol.* **26:**181–196.

65. **Hansen, J. C., and J. Ausio.** 1992. Chromatin dynamics and the modulation of genetic activity. *Trends Biochem. Sci.* **17:**187–191.

66. **Hayaishi, O., and K. Ueda.** 1977. Poly(ADP-ribose) and ADP-ribosylation of proteins. *Annu. Rev. Biochem.* **46:**95–116.

67. **Herzog, H., B. U. Zabel, R. Schneider, B. Auer, M. Hirsch-Kauffmann, and M. Schweiger.** 1989. Human nuclear NAD$^+$ ADP-ribosyltransferase: localization of the gene on chromosome 1q41-q42 and expression of an active human enzyme in *Escherichia coli*. *Proc. Natl. Acad. Sci. USA* **86:**3514–3518.

68. **Ho, L., V. A. Bohr, and P. C. Hanawalt.** 1989. Demethylation enhances removal of pyrimidine dimers from the overall genome and from specific DNA sequences in Chinese hamster ovary cells. *Mol. Cell. Biol.* **9:**1594–1603.

69. **Holliday, R.** 1980. Possible relationships between DNA damage, DNA modification and carcinogenesis, p. 93–100. In M. Alecevic (ed.), *Progress in Environmental Mutagenesis.* Elsevier Biomedical Press, Amsterdam.

70. **Holliday, R., and J. E. Pugh.** 1975. DNA modification mechanisms and gene activity during development. *Science* **187:**226–232.

71. **Hori, T.** 1981. High incidence of sister chromatid exchanges and chromatid interchanges in the conditions of lowered activity of poly(ADP-ribose) polymerase. *Biochem. Biophys. Res. Commun.* **102:**38–45.

72. **Hunting, D. J., S. L. Dresler, and M. W. Lieberman.** 1985. Multiple conformational states of repair patches in chromatin during DNA excision repair. *Biochemistry* **24:**3219–3226.

73. **Huppi, K., K. Bhatia, D. Siwarski, D. Klinman, B. Cherney, and M. Smulson.** 1989. Sequence and organization of the mouse poly(ADP-ribose) polymerase gene. *Nucleic Acids Res.* **17:**3387–3401.

74. **Igo-Kemenes, T., W. Hörz, and H. G. Zachau.** 1982. Chromatin. *Annu. Rev. Biochem.* **51:**89–121.

75. **Jensen, K. A., and M. J. Smerdon.** 1990. DNA repair within nucleosome cores of UV-irradiated human cells. *Biochemistry* **29:**4773–4782.

76. **Kantor, G. J., and A. N. Player.** 1986. A further definition of characteristics of DNA-excision repair in xeroderma pigmentosum complementation group A strains. *Mutat. Res.* **166:**79–88.

77. **Kantor, G. J., and R. B. Setlow.** 1981. Rate and extent of DNA repair in nondividing human diploid fibroblasts. *Cancer Res.* **41:**819–825.

78. **Karentz, D., and J. E. Cleaver.** 1986. Excision repair in xeroderma pigmentosum group C but not group D is clustered in a small fraction of the total genome. *Mutat. Res.* **165:**165–174.

79. **Kastan, M. B., B. J. Gowans, and M. W. Lieberman.** 1982. Methylation of deoxycytidine incorporated by excision-repair synthesis of DNA. *Cell* **30:**509–516.

80. **Kessler, O., and R. Ben-Ishai.** 1988. Lack of preferential DNA repair of a muscle specific gene during myogenesis, p. 267–272. *In* E. C. Friedberg and P. C. Hanawalt (ed.), *Mechanisms and Consequences of DNA Damage Processing.* Alan R. Liss, Inc., New York.

81. **Kohn, K. W., R. A. G. Ewig, L. C. Erickson, and L. A. Zwelling.** 1981. Measurement of strand breaks and cross-links by alkaline elution, p. 379–401. *In* E. C. Friedberg and P. C. Hanawalt (ed.), *DNA Repair—a Laboratory Manual of Research Procedures,* vol. 1, part B. Marcel Dekker, Inc., New York.

82. **Konze-Thomas, B., J. W. Levinson, V. M. Maher, and J. J. McCormick.** 1979. Correlation among the rates of dimer excision, DNA repair replication, and recovery of human cells from potentially lethal damage induced by ultraviolet radiation. *Biophys. J.* **28:**315–325.

83. **Kornberg, R. D., and Y. Lorch.** 1991. Irresistible force meets immovable object: transcription and the nucleosome. *Cell* **67:**833–836.

84. **Lan, S. Y., and M. J. Smerdon.** 1985. A nonuniform distribution of excision repair synthesis in nucleosome core DNA. *Biochemistry* **24:**7771–7783.

85. **Leadon, S. A.** 1986. Differential repair of DNA damage in specific nucleotide sequences in monkey cells. *Nucleic Acids Res.* **14:**8979–8995.

86. **Leadon, S. A., and P. C. Hanawalt.** 1984. Ultraviolet irradiation of monkey cells enhances the repair of DNA adducts in alpha DNA. *Carcinogenesis* **5:**1505–1510.

87. **Leadon, S. A., and D. A. Lawrence.** 1991. Preferential repair of DNA damage on the transcribed strand of the human metallothionein genes requires RNA polymerase II. *Mutat. Res.* **255:**67–78.

88. **Leadon, S. A., and M. M. Snowden.** 1988. Differential repair of DNA damage in the human metallothionein gene family. *Mol. Cell. Biol.* **8:**5331–5338.

89. **Leadon, S. A., M. E. Zolan, and P. C. Hanawalt.** 1983. Restricted repair of aflatoxin B1 induced damage in alpha DNA of monkey cells. *Nucleic Acids Res.* **11:**5675–5689.

90. **LeDoux, S. P., N. J. Patton, J. W. Nelson, V. A. Bohr, and G. L. Wilson.** 1990. Preferential DNA repair of alkali-labile sites within the active insulin gene. *J. Biol. Chem.* **265:**14875–14880.

91. **LeDoux, S. P., M. Thangada, V. A. Bohr, and G. L. Wilson.** 1991. Heterogeneous repair of methylnitrosourea-induced alkali-labile sites in different DNA sequences. *Cancer Res.* **51:**775–779.

91a. **Lewin, B.** 1987. *Genes III.* John Wiley & Sons, Inc., New York.

92. **Lieberman, M. W., M. J. Smerdon, T. D. Tlsty, and F. B. Oleson.** 1979. The role of chromatin structure in DNA repair in human cells damaged with chemical carcinogens and ultraviolet radiation, p. 345–363. *In* P. Emmelot and E. Kriek (ed.), *Environmental Carcinogenesis.* Elsevier Biomedical Press, Amsterdam.

93. **Link, C. J., Jr., D. L. Mitchell, R. S. Nairn, and V. A. Bohr.** 1992. Preferential and strand-specific DNA repair of (6-4) photoproducts detected by a photochemical method in the hamster *DHFR* gene. *Carcinogenesis* **13:**1975–1980.

94. **Liu, L. F., and J. C. Wang.** 1987. Supercoiling of the DNA template during transcription. *Proc. Natl. Acad. Sci. USA* **84:**7024–7027.

95. **Madhani, H. D., V. A. Bohr, and P. C. Hanawalt.** 1986. Differential DNA repair in transcriptionally active and inactive proto-oncogenes: *c-abl* and *c-mos. Cell* **45:**417–423.

96. **Mansbridge, J. N., and P. C. Hanawalt.** 1983. Domain-limited repair of DNA in ultraviolet irradiated fibroblasts from xeroderma pigmentosum complementation group C, p. 195–207. *In* E. C. Friedberg and B. A. Bridges (ed.), *Cellular Responses to DNA Damage.* Alan R. Liss, Inc., New York.

97. **Mathis, D., P. Oudet, and P. Chambon.** 1980. Structure of transcribing chromatin. *Prog. Nucleic Acid Res. Mol. Biol.* **24:**1–55.

98. **May, A., R. S. Nairn, D. S. Okumoto, K. Wasserman, T. Stevsner, J. C. Jones, and V. A. Bohr.** 1993. Repair of individual DNA strands in the hamster dihydrofolate reductase gene after treatment with ultraviolet light, alkylating agents, and cisplatin. *J. Biol. Chem.* **268:**1650–1657.

99. **McGregor, W. G., R.-H. Chen, L. Lukash, V. M. Maher, and J. J. McCormick.** 1991. Cell cycle-dependent strand bias for UV-induced mutations in the transcribed strand of excision repair-proficient human fibroblasts but not in repair-deficient cells. *Mol. Cell. Biol.* **11:**1927–1934.

100. **Mellon, I., V. A. Bohr, C. A. Smith, and P. C. Hanawalt.** 1986. Preferential DNA repair of an active gene in human cells. *Proc. Natl. Acad. Sci. USA* **83:**8878–8882.

101. **Mellon, I., G. Spivak, and P. C. Hanawalt.** 1987. Selective removal of transcription-blocking DNA damage from the transcribed strand of the mammalian *DHFR* gene. *Cell* **51:**241–249.

102. **Menichini, P., H. Vrieling, and A. A. van Zeeland.** 1991. Strand-specific mutation spectra in repair-proficient and repair-deficient hamster cells. *Mutat. Res.* **251:**143–155.

103. **Mitchell, D. L., D. E. Brash, and R. S. Nairn.** 1990. Rapid repair kinetics of pyrimidine(6-4)pyrimidone photoproducts in human cells are due to excision rather than conformational change. *Nucleic Acids Res.* **18:**963–971.

104. **Mitchell, D. L., R. M. Humphrey, G. M. Adair, L. H. Thompson, and J. M. Clarkson.** 1988. Repair of (6-4)photoproducts correlates with split-dose recovery in UV-irradiated normal and hypersensitive rodent cells. *Mutat. Res.* **193:**53–63.

105. **Mitchell, D. L., and R. S. Nairn.** 1989. The biology of the (6-4) photoproduct. *Photochem. Photobiol.* **49:**805–819.

106. **Mitchell, D. L., R. S. Nairn, J. A. Alvillar, and J. M. Clarkson.** 1982. Loss of thymine dimers from mammalian cell DNA. The kinetics for antibody-binding sites are not the same as that for T4 endonuclease V sites. *Biochim. Biophys. Acta* **697:**270–277.

107. **Mitchell, D. L., J. E. Vaughan, and R. S. Nairn.** 1989. Inhibition of transient gene expression in Chinese hamster ovary cells by cyclobutane dimers and (6-4) photoproducts in transfected ultraviolet-irradiated plasmid DNA. *Plasmid* **21:**21–30.

108. **Mitchell, D. L., M. Z. Zdzienicka, A. A. van Zeeland, and R. S. Nairn.** 1989. Intermediate (6-4) photoproduct repair in Chinese hamster V79 mutant V-H1 correlates with intermediate levels of DNA incision and repair replication. *Mutat. Res.* **226:**43–47.

109. **Mitchell, P. J., A. M. Carothers, J. H. Han, J. D. Harding, E. Kas, L. Venolia, and L. A. Chasin.** 1986. Multiple transcription start sites, DNase I-hypersensitive sites, and an opposite-strand exon in the 5′ region of the CHO *dhfr* gene. *Mol. Cell. Biol.* **6:**425–440.

110. **Molinete, M., W. Vermeulen, A. Burkle, J. Menissier-de Murcia, J. H. Kupper, J. H. Hoeijmakers, and G. de Murcia.** 1993. Overproduction of the poly(ADP-ribose) polymerase DNA-binding domain blocks alkylation-induced DNA repair synthesis in mammalian cells. *EMBO J.* **12:**2109–2117.

111. **Morgan, W. F., and J. E. Cleaver.** 1983. Effect of 3-aminobenzamide on the rate of ligation during repair of alkylated DNA in human fibroblasts. *Cancer Res.* **43:**3104–3107.

112. **Morse, R. H.** 1992. Transcribed chromatin. *Trends Biochem. Sci.* **17:**23–26.

113. **Mueller, P. R., and B. Wold.** 1989. In vivo footprinting of a muscle specific enhancer by ligation mediated PCR. *Science* **246:**780–786.

114. **Mullenders, L. H. F., A. M. Hazekamp-van Dokkum, W. H. Kalle, H. Vrieling, M. Z. Zdzienicka, and A. A. van Zeeland.** 1993. UV-induced photolesions, their repair and mutations. *Mutat. Res.* **299:**271–276.

115. **Mullenders, L. H. F., R. J. Sakker, A. van Hoffen, J. Venema, A. T. Natarajan, and A. A. van Zeeland.** 1992. Genomic heterogeneity of UV-induced repair: relationship to chromatin structure and transcriptional activity, p. 247–254. *In* V. A. Bohr, K. Wassermann, and K. H. Kraemer (ed.), *DNA Repair Mechanisms.* Munksgaard, Copenhagen.

116. **Mullenders, L. H. F., A. C. van Kesteren, C. J. Bussmann, A. A. van Zeeland, and A. T. Natarajan.** 1984. Preferential repair of nuclear matrix associated DNA in xeroderma pigmentosum complementation group C. *Mutat. Res.* **141:**75–82.

117. **Mullenders, L. H. F., A. C. van Kesteren van Leeuwen, A. A. van Zeeland, and A. T. Natarajan.** 1988. Nuclear matrix associated DNA is preferentially repaired in normal human fibroblasts,

exposed to a low dose of ultraviolet light, but not in Cockayne's syndrome fibroblasts. *Nucleic Acids Res.* **16**:10607–10622.

118. Mullenders, L. H. F., H. Vrieling, J. Venema, and A. A. van Zeeland. 1991. Hierarchies of DNA repair in mammalian cells: biological consequences. *Mutat. Res.* **250**:223–228.

119. Nairn, R. S., D. L. Mitchell, G. M. Adair, L. H. Thompson, M. J. Siciliano, and R. M. Humphrey. 1989. UV mutagenesis, cytotoxicity and split-dose recovery in a human-CHO cell hybrid having intermediate (6-4) photoproduct repair. *Mutat. Res.* **217**:193–201.

120. Naveh-Many, T., and H. Cedar. 1981. Active gene sequences are undermethylated. *Proc. Natl. Acad. Sci. USA* **78**:4246–4250.

121. Nduka, N., C. J. Skidmore, and S. Shall. 1980. The enhancement of cytotoxicity of N-methyl-N-nitrosourea and of gamma-radiation by inhibitors of poly(ADP-ribose) polymerase. *Eur. J. Biochem.* **105**:525–530.

122. Nissen, K. A., S. Y. Lan, and M. J. Smerdon. 1986. Stability of nucleosome placement in newly repaired regions of DNA. *J. Biol. Chem.* **261**:8585–8588.

123. Nose, K., and O. Nikaido. 1984. Transcriptionally active and inactive genes are similarly modified by chemical carcinogens or x-ray in normal human fibroblasts. *Biochim. Biophys. Acta* **781**:273–278.

124. Okumoto, D. S., and V. A. Bohr. 1987. DNA repair in the metallothionein gene increases with transcriptional activation. *Nucleic Acids Res.* **15**:10021–10030.

125. Oleson, F. B., B. L. Mitchell, A. Dipple, and M. W. Lieberman. 1979. Distribution of DNA damage in chromatin and its relation to repair in human cells treated with 7-bromomethylbenz(a)anthracene. *Nucleic Acids Res.* **7**:1343–1361.

126. Painter, R. B., and B. R. Young. 1972. Repair replication in mammalian cells after x-irradiation. *Mutat. Res.* **14**:225–235.

127. Paterson, M. C., P. H. M. Lohman, and M. L. Sluyter. 1973. Use of UV endonuclease from *Micrococcus luteus* to monitor the progress of DNA repair in UV-irradiated human cells. *Mutat. Res.* **19**:245–256.

128. Paulson, J. R., and U. K. Laemmli. 1977. The structure of histone-depleted metaphase chromosomes. *Cell* **12**:817–828.

129. Perry, M., and R. Chalkley. 1982. Histone acetylation increases the solubility of chromatin and occurs sequentially over most of the chromatin. *J. Biol. Chem.* **257**:7336–7347.

130. Pfaffle, P., and V. Jackson. 1990. Studies on rates of nucleosome formation with DNA under stress. *J. Biol. Chem.* **265**:16821–16829.

131. Pfeifer, G. P., S. D. Steigerwald, P. R. Mueller, B. Wold, and A. D. Riggs. 1989. Genomic sequencing and methylation analysis by ligation mediated PCR. *Science* **246**:810–813.

132. Pienta, K. J., and D. S. Coffey. 1984. A structural analysis of the role of the nuclear matrix and DNA loops in the organization of the nucleus and chromosome. *J. Cell Sci. Suppl.* **1**:123–135.

133. Purnell, M. R., P. R. Stone, and W. J. D. Whish. 1980. ADP-ribosylation of nuclear proteins. *Biochem. Soc. Trans.* **8**:215–227.

134. Ramanathan, B., and M. J. Smerdon. 1989. Enhanced DNA repair synthesis in hyperacetylated nucleosomes. *J. Biol. Chem.* **264**:11026–11034.

135. Rasmussen, R. E., and R. B. Painter. 1964. Evidence for repair of ultra-violet damaged deoxyribonucleic acid in cultured mammalian cells. *Nature* (London) **203**:1360–1362.

136. Razin, A., and J. Friedman. 1981. DNA methylation and its possible biological roles. *Prog. Nucleic Acid Res. Mol. Biol.* **25**:33–52.

137. Razin, A., and A. D. Riggs. 1980. DNA methylation and gene function. *Science* **210**:604–610.

138. Reeves, R. 1984. Transcriptionally active chromatin. *Biochim. Biophys. Acta* **782**:343–393.

139. Regan, J. D., and R. B. Setlow. 1974. Two forms of repair in the DNA of human cells damaged by chemical carcinogens and mutagens. *Cancer Res.* **34**:3318–3325.

140. Reynolds, R. J., K. H. Cook, and E. C. Friedberg. 1981. Measurement of thymine-containing pyrimidine dimers by one-dimensional thin-layer chromatography, p. 11–21. *In* E. C. Friedberg and P. C. Hanawalt (ed.), *DNA Repair—a Laboratory Manual of Research Procedures*, vol. 1, part A. Marcel Dekker, Inc., New York.

141. Roza, L., W. Vermeulen, J. B. Bergen Henegouwen, A. P. M. Eker, N. G. J. Jaspers, P. H. M. Lohman, and J. H. J. Hoeijmakers. 1990. Effects of microinjected photoreactivating enzyme on thymine dimer removal and DNA repair synthesis in normal human and xeroderma pigmentosum fibroblasts. *Cancer Res.* **50**:1905–1910.

142. Sage, E., E. A. Drobetsky, and E. Moustacchi. 1993. 8-Methoxypsoralen induced mutations are highly targeted at crosslinkable sites of photoaddition on the non-transcribed strand of a mammalian chromosomal gene. *EMBO J.* **12**:397–402.

143. Satoh, M. S., and T. Lindahl. 1992. Role of poly(ADP-ribose) formation in DNA repair. *Nature* (London) **356**:356–358.

144. Satoh, M. S., and T. Lindahl. 1994. Enzymatic repair of oxidative DNA damage. *Cancer Res.* **54**:1899s–1901s.

145. Satoh, M. S., G. G. Poirier, and T. Lindahl. 1993. NAD(+)-dependent repair of damaged DNA by human cell extracts. *J. Biol. Chem.* **268**:5480–5487.

146. Schaeffer, L., R. Roy, S. Humbert, V. Moncollin, W. Vermeulen, J. H. J. Hoeijmakers, P. Chambon, and J.-M. Egly. 1993. DNA repair helicase: a component of BTF2 (TFIIH) basic transcription factor. *Science* **260**:58–63.

147. Scicchitano, D. A., and P. C. Hanawalt. 1989. Repair of N-methylpurines in specific DNA sequences in Chinese hamster ovary cells: absence of strand specificity in the dihydrofolate reductase gene. *Proc. Natl. Acad. Sci. USA* **86**:3050–3054.

148. Setlow, R. B., F. M. Faulcon, and J. D. Regan. 1976. Defective repair of gamma-ray induced DNA damage in xeroderma pigmentosum cells. *Int. J. Radiat. Biol. Relat. Stud. Phys. Chem. Med.* **29**:125–136.

149. Shall, S., B. Durkacz, D. Ellis, J. Irwin, P. Lewis, and M. Perera. 1981. (ADP-ribose)n, a new component in DNA repair, p. 477–487. *In* E. Seeberg and K. Kleppe (ed.), *Chromosome Damage and Repair.* Plenum Publishing Corp, New York.

150. Shall, S., B. W. Durkacz, D. A. Gray, J. Irwin, P. J. Lewis, M. Perera, and M. Tavassoli. 1982. (ADP-ribose)n participates in DNA excision repair, p. 389–407. *In* C. C. Harris and P. A. Cerutti (ed.), *Mechanisms of Chemical Carcinogenesis.* Alan R. Liss, Inc., New York.

151. Simbulan, C. M., M. Suzuki, S. Izuta, T. Sakurai, E. Savoysky, K. Kojima, K. Miyahara, Y. Shizuta, and S. Yoshida. 1993. Poly(ADP-ribose) polymerase stimulates DNA polymerase α by physical association. *J. Biol. Chem.* **268**:93–99.

152. Singer, M. F. 1982. Highly repeated sequences in mammalian genomes. *Int. Rev. Cytol.* **76**:67–112.

153. Skidmore, C. J., M. I. Davies, P. M. Goodwin, H. Halldorsson, P. J. Lewis, S. Shall, and A. A. Zia'ee. 1979. The involvement of poly(ADP-ribose) polymerase in the degradation of NAD caused by gamma-radiation and N-methyl-N-nitrosourea. *Eur. J. Biochem.* **101**:135–142.

154. Smerdon, M. J. 1986. Completion of excision repair in human cells. Relationship between ligation and nucleosome formation. *J. Biol. Chem.* **261**:244–252.

155. Smerdon, M. J. 1991. DNA repair and the role of chromatin structure. *Curr. Opin. Cell Biol.* **3**:422–428.

156. Smerdon, M. J. 1992. DNA excision repair at the nucleosome level of chromatin, p. 271–294. *In* M. W. Lambert and J. Laval (ed.), *DNA Repair Mechanisms and Their Biological Implications in Mammalian Cells.* Plenum Publishing Corp., New York.

157. Smerdon, M. J., M. B. Kastan, and M. W. Lieberman. 1979. Distribution of repair-incorporated nucleotides and nucleosome rearrangement in the chromatin of normal and xeroderma pigmentosum human fibroblasts. *Biochemistry* **18**:3732–3739.

158. Smerdon, M. J., and M. W. Lieberman. 1978. Nucleosome rearrangement in human chromatin during UV-induced DNA-repair synthesis. *Proc. Natl. Acad. Sci. USA* **75**:4238–4241.

159. **Smerdon, M. J., T. D. Tlsty, and M. W. Lieberman.** 1978. Distribution of ultraviolet-induced DNA repair synthesis in nuclease sensitive and resistant regions of human chromatin. *Biochemistry* 17:2377–2386.

160. **Smerdon, M. J., J. F. Watkins, and M. W. Lieberman.** 1982. Effect of histone H1 removal on the distribution of ultraviolet-induced deoxyribonucleic acid repair synthesis within chromatin. *Biochemistry* 21:3879–3885.

161. **Smith, C. A.** 1987. DNA repair in specific sequences in mammalian cells. *J. Cell Sci. Suppl.* 6:225–241.

162. **Smith, P. J.** 1986. n-Butyrate alters chromatin accessibility to DNA repair enzymes. *Carcinogenesis* 7:423–429.

163. **Smulson, M., N. Instock, R. Ding, and B. Cherney.** 1994. Deletion mutants of poly(ADP-ribose) polymerase support a model of cyclic association and dissociation of enzyme from DNA ends during DNA repair. *Biochemistry* 33:6186–6191.

164. **Snyderwine, E. G., and V. A. Bohr.** 1992. Gene- and strand-specific damage and repair in Chinese hamster ovary cells treated with 4-nitroquinoline 1-oxide. *Cancer Res.* 52:4183–4189.

165. **Spivak, G., and P. C. Hanawalt.** 1992. Translesion DNA synthesis in the dihydrofolate reductase domain of UV-irradiated CHO cells. *Biochemistry* 31:6794–6800.

166. **Stevnsner, T., A. May, L. N. Petersen, F. Larminat, M. Pirsel, and V. A. Bohr.** 1993. Repair of ribosomal RNA genes in hamster cells after UV irradiation, or treatment with cisplatin or alkylating agents. *Carcinogenesis* 14:1591–1596.

167. **Suquet, C., and M. J. Smerdon.** 1993. UV damage to DNA strongly influences its rotational setting on the histone surface of reconstituted nucleosomes. *J. Biol. Chem.* 268:23755–23757.

168. **Szymkowsli, D. E., C. W. Lawrence, and R. D. Wood.** 1993. Repair by human cell extracts of single (6-4) and cyclobutyl thymine-thymine photoproducts in DNA. *Proc. Natl. Acad. Sci. USA* 90:9823–9827.

169. **Tang, M.-S., V. A. Bohr, X.-S. Zhang, J. Pierce, and P. C. Hanawalt.** 1989. Quantification of aminofluorene adduct formation and repair in defined DNA sequences in mammalian cells using the UVRABC nuclease. *J. Biol. Chem.* 264:14455–14462.

170. **Terleth, C., P. van de Putte, and J. Brouwer.** 1991. New insights in DNA repair: preferential repair of transcriptionally active DNA. *Mutagenesis* 6:103–111.

171. **Thelander, L.** 1973. Physicochemical characterization of ribonucleoside diphosphate reductase from *Escherichia coli*. *J. Biol. Chem.* 248:4591–4601.

172. **Thoma, F.** 1991. Structural changes in nucleosomes during transcription: strip, split or flip? *Trends Genet.* 7:175–177.

173. **Thoma, F., T. Koller, and A. Klug.** 1979. Involvement of histone H1 in the organization of the nucleosome and of the salt-dependent superstructures of chromatin. *J. Cell Biol.* 83:403–427.

174. **Thomas, D. C., D. S. Okumoto, A. Sancar, and V. A. Bohr.** 1989. Preferential DNA repair of (6-4) photoproducts in the dihydrofolate reductase gene of Chinese hamster ovary cells. *J. Biol. Chem.* 264:18005–18010.

175. **Thompson, L. H., K. W. Brookman, L. E. Dillehay, C. L. Mooney, and A. V. Carrano.** 1982. Hypersensitivity to mutation and sister-chromatid-exchange induction in CHO cell mutants defective in incising DNA containing UV lesions. *Somatic Cell Genet.* 8:759–773.

176. **Thompson, L. H., D. L. Mitchell, J. D. Regan, S. D. Bouffler, S. A. Stewart, W. L. Carrier, R. S. Nairn, and R. T. Johnson.** 1989. CHO mutant UV61 removes (6-4) photoproducts but not cyclobutane dimers. *Mutagenesis* 4:140–146.

177. **Tlsty, T. D., and M. W. Lieberman.** 1978. The distribution of DNA repair synthesis in chromatin and its rearrangement following damage with N-acetoxy-2-acetylaminofluorene. *Nucleic Acids Res.* 5:3261–3273.

178. **Tornaletti, S., and G. P. Pfeifer.** 1994. Slow repair of pyrimidine dimers at p53 mutation hotspots in skin cancer. *Science* 263:1436–1438.

179. **van Holde, K. E. (ed.).** 1989. *Chromatin.* Springer-Verlag, New York.

180. **van Holde, K. E., D. E. Lohr, and C. Robert.** 1992. What happens to nucleosomes during transcription? *J. Biol. Chem.* 267:2837–2840.

181. **van Zeeland, A. A., L. H. F. Mullenders, M. Z. Zdzienicka, and H. Vrieling.** 1992. Influence of strand-specific repair of DNA damage on molecular mutation spectra, p. 117–122. *In* V. A. Bohr, K. Wassermann, and K. H. Kraemer (ed.), *DNA Repair Mechanisms.* Munksgaard, Copenhagen.

182. **van Zeeland, A. A., C. A. Smith, and P. C. Hanawalt.** 1981. Sensitive determination of pyrimidine dimers in DNA of UV-irradiated mammalian cells. Introduction of T4 endonuclease V into frozen and thawed cells. *Mutat. Res.* 82:173–189.

182a. **Venema, J., L. H. F. Mullenders, A. T. Natarajan, A. A. van Zeeland, and L. V. Mayne.** 1990. The genetic defect in Cockayne syndrome is associated with a defect in repair of UV-induced DNA damage in transcriptionally active DNA. *Proc. Natl. Acad. Sci. USA* 87:4707–4711.

183. **Venema, J., A. van Hoffen, A. T. Natarajan, A. A. van Zeeland, and L. H. F. Mullenders.** 1990. The residual repair capacity of xeroderma pigmentosum complementation group C fibroblasts is highly specific for transcriptionally active DNA. *Nucleic Acids Res.* 18:443–448.

184. **Vos, J.-M., and P. C. Hanawalt.** 1987. Processing of psoralen adducts in an active human gene: repair and replication of DNA containing monoadducts and interstrand cross-links. *Cell* 50:789–799.

185. **Vos, J.-M., and E. L. Wauthier.** 1991. Differential introduction of DNA damage and repair in mammalian genes transcribed by RNA polymerases I and II. *Mol. Cell. Biol.* 11:2245–2252.

186. **Vrieling, H., M. L. Van Rooijen, N. A. Groen, M. Z. Zdzienicka, J. W. Simons, P. H. M. Lohman, and A. A. van Zeeland.** 1989. DNA strand specificity for UV-induced mutations in mammalian cells. *Mol. Cell. Biol.* 9:1277–1283.

187. **Vrieling, H., J. Venema, M. L. van Rooyen, A. van Hoffen, P. Menichini, M. Z. Zdzienicka, J. W. Simons, L. H. F. Mullenders, and A. A. van Zeeland.** 1991. Strand specificity for UV-induced DNA repair and mutations in the Chinese hamster HPRT gene. *Nucleic Acids Res.* 19:2411–2415.

188. **Vrieling, H., L. H. Zhang, A. A. van Zeeland, and M. Z. Zdzienicka.** 1992. UV-induced *hprt* mutations in a UV-sensitive hamster cell line from complementation group 3 are biased towards the transcribed strand. *Mutat. Res.* 274:147–155.

189. **Wang, Z., X. Wu, and E. C. Friedberg.** 1991. Nucleotide excision repair of DNA by human cell extracts is suppressed in reconstituted nucleosomes. *J. Biol. Chem.* 266:22472–22478.

190. **Wani, A. A., and J. Arezina.** 1991. Immunoanalysis of ultraviolet radiation induced DNA damage and repair within specific gene segments of plasmid DNA. *Biochim. Biophys. Acta* 1090:195–203.

191. **Weintraub, H.** 1976. Cooperative alignment of nu bodies during chromosome replication in the presence of cycloheximide. *Cell* 9:419–422.

192. **Weintraub, H., and M. Groudine.** 1976. Chromosomal subunits in active genes have an altered conformation. *Science* 193:848–856.

193. **Wilkins, R. J., and R. W. Hart.** 1974. Preferential DNA repair in human cells. *Nature* (London) 247:35–36.

194. **Williams, J. I., and J. E. Cleaver.** 1978. Excision repair of ultraviolet damage in mammalian cells. Evidence for two steps in the excision of pyrimidine dimers. *Biophys. J.* 22:265–279.

195. **Williams, J. I., and J. E. Cleaver.** 1979. Removal of T4 endonuclease V-sensitive sites from SV40 DNA after exposure to ultraviolet light. *Biochim. Biophys. Acta* 562:429–437.

196. **Williams, J. I., and E. C. Friedberg.** 1979. Deoxyribonucleic acid excision repair in chromatin after ultraviolet irradiation of human fibroblasts in culture. *Biochemistry* 18:3965–3972.

197. Williams, J. I., and E. C. Friedberg. 1982. Increased levels of unscheduled DNA synthesis in UV-irradiated human fibroblasts pretreated with sodium butyrate. *Photochem. Photobiol.* **36:**423–427.

198. Wilson, V. L., and P. A. Jones. 1983. Inhibition of DNA methylation by chemical carcinogens in vitro. *Cell* **32:**239–246.

199. Wolffe, A. P. (ed.) 1992. *Chromatin: Structure and Function.* Academic Press, Inc., New York.

200. Zdzienicka, M. Z., G. P. van der Schans, A. Westerveld, A. A. van Zeeland, and J. W. Simons. 1988. Phenotypic heterogeneity within the first complementation group of UV-sensitive mutants of Chinese hamster cell lines. *Mutat. Res.* **193:**31–41.

201. Zelle, B., R. J. Reynolds, M. J. Kottenhagen, A. Schuite, and P. H. M. Lohman. 1980. The influence of the wavelength of ultraviolet radiation on survival, mutation induction and DNA repair in irradiated Chinese hamster cells. *Mutat. Res.* **72:**491–509.

202. Zolan, M. E., G. A. Cortopassi, C. A. Smith, and P. C. Hanawalt. 1982. Deficient repair of chemical adducts in α DNA of monkey cells. *Cell* **28:**613–619.

203. Zolan, M. E., C. A. Smith, N. M. Calvin, and P. C. Hanawalt. 1982. Rearrangement of mammalian chromatin structure following excision repair. *Nature* (London) **299:**462–464.

204. Zolan, M., C. A. Smith, and P. C. Hanwalt. 1984. Formation and repair of furocoumarin adducts in α deoxyribonucleic acid and bulk deoxyribonucleic acid of monkey kidney cells. *Biochemistry* **23:**63–69.

8

Nucleotide Excision Repair: Mammalian Genes and Proteins

The isolation of genes involved in nucleotide excision repair in bacteria and in yeasts by phenotypic complementation of DNA-damage-sensitive repair-defective mutants prompted similar studies with mammalian cells. Two mammalian cell systems have been extensively investigated in this regard. One consists of a series of rodent cell lines selected in the laboratory for defective nucleotide excision repair following the mutagenesis of established cell lines. The second consists of cell lines derived from three repair-defective human hereditary diseases, xeroderma pigmentosum (XP), Cockayne's syndrome (CS), and a photosensitive form of trichothiodystrophy (TTD).

Genes Involved in Nucleotide Excision Repair

Excision Repair Mutants from Rodent Cells

Among the mutant rodent cell lines used for studies on nucleotide excision repair, the majority originate from Chinese hamsters, although a number of murine lines have also been isolated (14, 15, 27, 28, 268). Many of the known mutant strains and their properties are listed in recent comprehensive reviews (28, 268), to which the interested reader is referred for detailed information. Most of these mutants were generated by exposure of exponentially growing cell populations to agents such as ethyl methanesulfonate or UV radiation and allowing suitable periods for fixation of damage-induced mutations. Selection of mutants has been carried out in several ways, in most cases predicated on the assumption that defective nucleotide excision repair is reflected by an increased sensitivity of the cells to DNA-damaging agents (14–16, 27, 28, 227, 268).

The largest and most completely characterized collection of the rodent mutant cell lines has been derived from the Chinese hamster ovary (CHO) cell line AA8. One massive study used a semiautomated procedure designed for the large-scale screening of mutagenized cells in culture (14–16). Large agar trays were inoculated with mutagenized cells, which grew to form visible colonies. The colonies were irradiated with low levels of UV light controlled by an integrating radiation meter. The dishes were photographed at various times after the UV challenge, and by observation of the colony images from sequential photographs on the screen of a microfilm reader, colonies exhibiting little or no visible growth between photographies were identified as possible UV radiation-sensitive mutants. These colonies were isolated and further screened to determine colony-forming ability after exposure to UV radiation (224). Since then, additional CHO mutant lines have been isolated in independent studies (28, 227, 228, 276).

Rodent Genetic Complementation Groups

Collectively, the above studies yielded ~220 UV-sensitive mutants. A further 35 mutants were derived from the mitomycin-sensitive line MC5 (227), and eight UV-sensitive mutant lines were derived from an ethyl methanesulfonate-sensitive line designated EM9 (16). Approximately 56% of these mutants were sufficiently sensitive to UV radiation to permit rapid and reliable assignment to *genetic complementation groups*. This exercise yielded six genetic complementation groups (14). Subsequent analysis of 74 less-sensitive mutants did not identify any new complementation groups, but several novel phenotypes of existing groups emerged (15). Only eight mutant lines with less than twice the UV sensitivity of wild-type cells remain unassigned. A mutant line originally derived from Chinese hamster V79 cells (287) was assigned to a seventh genetic complementation group. A mouse mutant line derived from the mouse lymphoma line L5178Y (186) was shown to complement mutants from all seven groups, hence establishing an eighth genetic complementation group (78, 229). Mutants constituting the 9th, 10th, and 11th genetic complementation groups were also isolated from CHO cell lines (28, 169, 198, 268). In the interests of nomenclatural simplicity, we refer to these genetic complementation groups as *rodent complementation groups 1 to 11* (RCG1 to RCG11). These complementation groups are identified in Table 8–1 and Table 8–2. Table 8–2 (reproduced in chapter 6 in Table 6–8) also summarizes the known evolutionary relationships between nucleotide excision repair genes in the yeasts *Saccharomyces cerevisiae* and *Schizosaccharomyces pombe*, the fruit fly *Drosophila melanogaster*, and rodent and human cells.

Accurate assignment of rodent cell lines to different genetic complementa-

Table 8–1 Rodent genetic complementation groups for nucleotide excision repair

RCG	Representative mutants	Parental cell type	Sensitivity to[a]: UV	Mitomycin
1	UV20 43-3B	CHO	+ +	+ + +
2	UV5 VH-1	CHO V79	+ +	+
3	UV24 27-1	CHO	+ +	+
4	UV41	CHO	+ +	+ + +
5	UV135	CHO	+ (+)	+ / −
6	UV61	CHO	+	+
7	VB11	V79	+	+ / −
8	US31	Mouse lymphoma	+	+
9	CHO4PV	CHO	+	+
10	CHO7PV	CHO	+	+
11	UVS1	CHO	+ / + +	+

[a] The number of + signs corresponds to the extent of the UV radiation or mitomycin sensitivity as follows: +, 2 to 5 times more sensitive than wild-type cells; + +, 5 to 10 times more sensitive; + + +, >10 times more sensitive.

Source: Adapted from Hoeijmakers (76a) with permission.

Table 8–2 Genetic complexity of nucleotide excision repair in eukaryotes

S. cerevisiae gene	Human gene	Human complementation group	Human chromosome	Rodent complementation group	S. pombe gene	Drosophila gene
RAD3	XPD	XP-D (CS-XPD) (TTD-XPD)	19q13	RCG-2	rad103$^+$	XPD^{dm}
SSL2 (RAD25)	XPB	XP-B (CS-XPB) (TTD-XPB)	2q21	RCG-3	ERCC3^{sp}	XPB^{dm} (hwr)
RAD2	XPG	XP-G (CS-XPG)	13q32-33	RCG-5	rad13$^+$	
RAD14	XPA	XP-A	9q34			XPA^{dm}
RAD4	XPC	XP-C	3p25			XPC^{dm}
RAD1	ERCC4/?XPF	XP-F	16p13	RCG-4	rad16$^+$	
RAD26	CSB	CS-B	10q11	RCG-6		
RAD10	ERCC1		19q13	RCG-1	swi10	
RAD23	HHR23A		3p25			
	HHR23B		19p31·2			
TFB1	p62		11p14-15·1			
SSL1	p44					
RAD7						
RAD16						
		XP-E				
		XP-V				
		CS-A				
		TTD-A				
			7(?)	RCG-9		
				RCG-7		
				RCG-8		
				RCG-10		
				RCG-11		

tion groups is crucial. These groups are presumed to define the genetic complexity of nucleotide excision repair in normal rodent (and possibly all mammalian) cells, and, as will be seen below, they have provided the essential starting material for gene cloning by phenotypic correction. Generally, genetic complementation is tested by fusing different cell lines and examining the hybrids for the correction of mutant phenotypes. However, this exercise is not without pitfalls and complications, and in the absence of independent confirmation, the extrapolation of equivalence between different genetic complementation groups and different genes must be tempered with reservation. The following complications, among others, have been encountered:

1. Depending on the nature of the mutation, widely differing phenotypes can appear within a single genetic complementation group. In some cases this reflects allele-specific variations.

2. Phenotypic complementation in hybrid cell lines is rarely complete and can be difficult to interpret.

3. A given cell line *A* may complement a second line *B*, but not a third line *B'* which is apparently from the same genetic complementation group as *B*, leading to the erroneous assumption that *B* and *B'* are in different genetic complementation groups.

4. Intragenic complementation can result in the erroneous assignment of a larger number of genetic loci than actually exist for nucleotide excision repair.

Finally, the frequency with which mammalian cell mutants can be isolated by phenotypic selection is influenced by the ploidy of the genome. The isolation of rodent mutants defective in different nucleotide excision repair genes has been greatly facilitated by the high degree of *hemizygosity* of the cell lines used. The identification of new genes may obviously be limited by the extent of such hemizygosity, and the 11 genetic complementation groups currently identified may reflect the minimum genetic complexity of nucleotide excision repair in mammalian cells.

Cloning Human *ERCC* Genes

Representative mutant clones from the 11 known rodent genetic complementation groups have been examined for their ability to carry out nucleotide excision repair. Many of the UV-sensitive mutants are indeed defective in parameters that measure nucleotide excision repair in mammalian cells and therefore suggest themselves as candidate cell lines for gene cloning by phenotypic complementation. Transfection of many of these CHO mutants with human genomic DNA has resulted in the identification and molecular cloning of a series of complementing genes designated *ERCC* (for excision repair cross complementing), denoting the fact that rodent cells were corrected for mutant phenotypes by human DNA (28, 76a, 266, 268). In all cases studied thus far, the human genes so isolated have turned out to be involved in nucleotide excision repair in human cells.

The cloning strategy typically used in these studies (Fig. 8–1) involved cotransfection of cells with human genomic DNA together with a plasmid carrying a dominant selectable marker (e.g., the *Escherichia coli gpt*$^+$ gene, which encodes resistance to mycophenolic acid). Mycophenolic acid-resistant transfectants were screened for enhanced resistance to UV radiation conferred by integrated sequences of interest in primary *transfectants* (or in some cases secondary or tertiary transfectants, carrying much smaller amounts of irrelevant human DNA). UV radiation-resistant transfectants were used to construct small cosmid or phage λ genomic libraries from which human sequences were isolated by selecting for the cointegrated dominant selectable marker, by direct functional complementation, or by strategies which specifically select for human sequences. In many cases human probes so derived were used to isolate appropriate cDNAs from cDNA libraries.

This general experimental strategy resulted in the molecular cloning of genes called *ERCC1*, *ERCC2*, *ERCC3*, *ERCC4*, *ERCC5*, and *ERCC6*, named for the corresponding RCG1 through RCG6. At this point it should be noted that the nomenclature for these genes has been changed according to the following recommendations (118).

1. The designation *ERCC* is retained as long as a human gene is not yet proven to be involved in a known human repair-defective disease such as XP, CS, or TTD.

2. If an *ERCC* gene is proven to be an XP, CS, or TTD (or other) gene (by appropriate experiments discussed below), it is given the designation *XP-*, *CS-*, or *TTD-*, depending on the genetic complementation group. For example, it is now known that human genes originally called *ERCC2*, *ERCC3*, *ERCC5*, and *ERCC6* are in fact the *XPD*, *XPB*, *XPG*, and *CSB* genes, respectively. From now on in this text, these genes will be referred to primarily or exclusively by this genetic nomenclature.

3. A designation for proven XP genes originally used in the literature appended a terminal letter C to the gene name to stand for "correcting/complementing." For instance, the gene now called *XPA* was originally called *XPAC*. This designation is no longer recommended.

At first glance, the use of rodent cells to clone human genes would seem unnecessarily indirect, especially in light of the availability of human repair-defective cell lines. However, the enormous success of this experimental strategy derives from the fact that foreign DNA sequences are stably integrated into the genome of rodent cells with far greater efficiency than is the case with human cells (see later discussion).

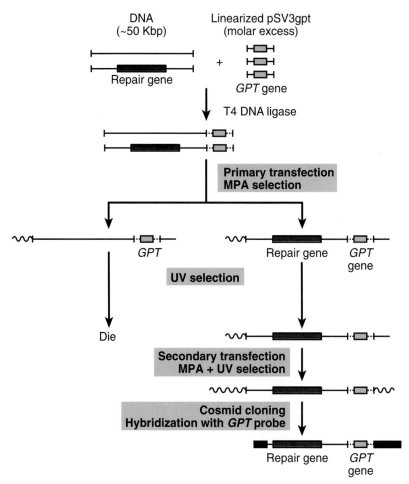

Figure 8–1 Schematic representation of a cloning strategy used for the isolation of human *ERCC* genes. High-molecular-weight DNA from human HeLa cells is cleaved with a restriction enzyme to an average size of ~50 kbp. This DNA is then ligated to a molar excess of restriction-digested plasmid DNA containing the *E. coli gpt* gene, which confers resistance to mycophenolic acid (MPA). The cloning strategy is essentially predicated on the (random) close physical linkage of the *gpt* gene to the *ERCC* gene of interest. The hybrid molecules are transfected (primary transfection) into rodent cells which are defective in a particular *ERCC* gene. These mutant cell lines are hemizygous for multiple chromosomes. Hence, mutational inactivation of just one *ERCC* allele renders them homozygous recessive for the gene of interest. The cells are placed under selection for resistance to MPA, thereby eliminating cells that did not stably integrate DNA sequences that include the *E. coli gpt* gene. These primary transfectants are screened for resistance to killing by UV radiation (or some other appropriate form of DNA damage), thereby selecting for the *ERCC* gene of interest. Individual UV radiation-resistant colonies are expanded and are phenotypically characterized in detail to confirm UV radiation resistance and competence for nucleotide excision repair. DNA is extracted from these cells and used to transfect mutant cells again, generating secondary transfectants which are selected simultaneously for both MPA and UV radiation resistance. Following detailed phenotypic characterization secondary transfectants are used to generate a cosmid library. The library is screened by DNA hybridization for the presence of the *gpt* gene. Positive cosmids are then tested for correction of UV radiation sensitivity by the cloned *ERCC* gene in mutant rodent cells. (*Adapted from Westerveld et al. [270a] with permission.*)

THE *ERCC1* GENE

The human *ERCC1* gene is ~15 kb in size and is located on human chromosome 19 (191, 243) (Tables 8–2 and 8–3), where it is closely linked (within ~250 kb) to another human nucleotide excision repair gene designated *XPD* (formerly called *ERCC2*) (144, 191). This assignment is consistent with the results of earlier somatic cell fusion studies, which demonstrated that a mutant CHO cell line from RCG1

Table 8–3 Cloned human nucleotide excision repair genes

Gene	Chromosomal location	Approx. gene size (kb)	No. of codons	Size of protein[a] (kDa)
ERCC1	19q13.2	15–17	297	32.5
ERCC4/XPF	16p13-13	?	?	?
ERCC2/XPD	19q13.2	20	760	86.9
ERCC3/XPB	2q21	45	782	89.2
ERCC5/XPG	13q32-33	32	1,186	133.3
ERCC6/CSB	10q11-21	85	1,493	168
XPA	9q34	25	273	31
XPC	3p25	?	823	106

[a] Predicted from the cloned gene.

Source: Adapted from Hoeijmakers (76a) with permission.

was complemented uniquely by human chromosome 19 (226). *ERCC1* precursor RNA is subject to alternative splicing of an internal 72-bp coding exon. Only the cDNA of the larger transcript confers resistance to UV radiation in cells from RCG1 (243).

The *ERCC1* cDNA has an open reading frame (ORF) that can encode a polypeptide of 297 amino acids, with a calculated molecular mass of ~32.5 kDa (Table 8–3). The cloned gene also complements the sensitivity of RCG1 cells to mitomycin, 4-nitroquinoline 1-oxide, *N*-acetoxy-*N*-2-acetyl-2-aminofluorene, and ethylnitrosourea (ENU) and restores normal levels of UV-induced mutagenesis at the ouabain locus (243, 286). Hence, it would appear that this gene represents the human equivalent of the gene mutated in CHO cells from RCG1 and is involved in nucleotide excision repair in human cells. The specificity of the gene was demonstrated by showing that *ERCC1* has no effect on the phenotype of mutants from several other rodent genetic complementation groups for nucleotide excision repair (244). Tests of protein-protein interactions in vitro and in vivo (121, 157) have demonstrated specific interaction between the human ERCC1 protein and those encoded by the *XPA* and *ERCC4* genes (see below).

The human *ERCC1* gene was used as a hybridization probe to isolate the corresponding mouse gene (245). The mouse and human genes are highly conserved (Fig. 8–2), although the mouse gene encodes one more amino acid. Overall, there is 85% amino acid identity between these two genes. Most amino acid divergence is concentrated in the N-terminal one-third of the two proteins. There is 70% identity in the first 100 amino acids, but this increases to 97 and 89% in amino acids 100 to 200 and 200 to 298, respectively (245). Mouse *ERCC1* transcripts are present in low abundance in all organs and stages of development examined.

The translated sequences of the mouse and human *ERCC1* genes are homologous with those of the *S. cerevisiae RAD10* gene (243, 245) and the *S. pombe swi10⁺* gene (Fig. 8–2). As discussed in chapter 6, Rad10 protein is known to specifically associate with Rad1 protein to constitute an endonuclease that is an attractive candidate for the enzyme that catalyzes specific incision at 5' sites of base damage in DNA during nucleotide excision repair. The homology between the human and mouse ERCC1 proteins and the yeast Rad10 and Swi10 proteins is most pronounced in the middle region of the polypeptides; the genes are not conserved in the N-terminal regions. These results are consistent with the observation that removal of the N-terminal region of both the human and mouse cDNAs has little effect on phenotypic complementation of appropriate rodent mutants (243, 245). The amino acid sequences of the C-terminal region of the human and mouse ERCC1 polypeptides are similar to regions of the *E. coli* UvrA and UvrC polypeptides (see chapter 5).

Overexpression of a mutant allele of *ERCC1* carrying an insertion of two codons in the conserved C-terminal region has a dominant negative effect and results in a marked increase in the sensitivity of normal human cells to the bifunctional alkylating agent mitomycin, presumably because of competition with endogenous wild-type ERCC1 protein (6). Interestingly, overexpression of the

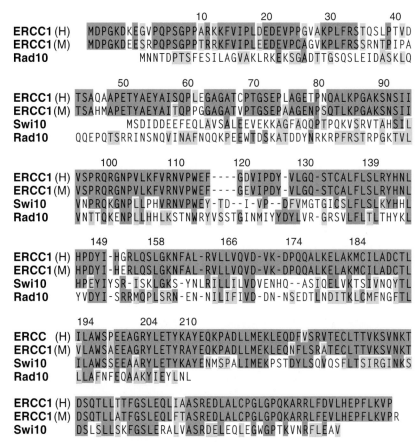

Figure 8–2   The amino acid sequences of the predicted human (H) and mouse (M) ERCC1 polypeptides and the *S. pombe* Swi10 and *S. cerevisiae* Rad10 polypeptides share extensive homology.

wild-type *ERCC1* gene results in an increased sensitivity of CHO cells to bifunctional nitrogen mustard and to cisplatin, both of which are expected to result in cross-links in DNA, but not to UV radiation. Hence, overexpression of ERCC1 protein appears to specifically interfere with a pathway for the repair of cross-links in DNA (10). Since *RAD10*, the *S. cerevisiae* homolog of *ERCC1*, is known to be required for specialized forms of mitotic recombination (see chapter 6), this result may reflect interference with an *ERCC1*-dependent recombinational repair mechanism.

A question of obvious importance is whether the human *ERCC1* gene can complement human cell lines from patients with the hereditary diseases XP, CS, or TTD that are defective in nucleotide excision repair. A systematic exploration of this question has thus far failed to reveal complementation of any of the XP or CS cell lines examined (246). Hybridization studies also failed to demonstrate gross deletions or rearrangement of the gene or its transcript in XP cells. There are persuasive indications that the *ERCC1* gene is involved in nucleotide excision repair in human cells. Hence, it would appear that the genetic complexity of this repair mode in human cells is not limited to the XP cell lines currently available and that further XP genetic complementation groups await discovery.

On the other hand, the phenotype of mouse strains inactivated of both copies of the mouse *ERCC1* gene suggests that such human individuals might be impossible or extremely difficult to detect. Such mice were isolated by the so-called embryonic stem (ES) cell knock-out strategy (137, 182, 267). In this technique a gene of interest is isolated from a genomic library of an ES cell derived from a selected strain of mice and is cloned into a specifically designed targeting vector.

This vector is used to disrupt or delete (by homologous recombination) one of the two equivalent genes in ES cells in culture, under conditions that allow for selection of the mutated ES cells. These cells, carrying a heterozygous mutation in the gene of interest, are injected into a blastocyst from a selected different mouse strain, and the blastocyst is brought to developmental completion in a pseudopregnant foster mother (Fig. 8–3). If the ES cells participated in the process of embryogenesis, the resulting pups will manifest phenotypes that reflect genetic chimerism between the two strains (such as mixed coat color) (Fig. 8–3). Males that are chimeric in the germ line are then mated to the strain from which the ES

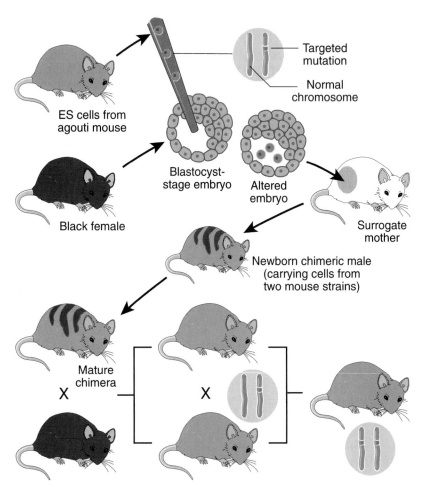

Figure 8–3   Targeted gene replacement in mice. ES cells (shown in pink) from an inbred mouse strain with an agouti coat color (shown here as grey) are propagated in culture and mutated in a selected gene of interest by homologous recombination between the wild-type resident gene and a cloned mutated copy of that gene. These ES cells (heterozygous for the mutation of interest) are microinjected into the blastocyst of a pure-bred strain with a black coat, where they may (or may not) participate in the maturation of the blastocyst to embryos. This maturation process takes place in the uterus of a pseudopregnant foster (surrogate) mother (shown here in white), whose genetic constitution is irrelevant. If the ES cells do indeed participate in embryogenesis, the resultant pups will be *chimeric,* since they have cells and tissues of both genetic backgrounds (agouti and black). This chimerism can be phenotypically recognized by the presence of a mixed coat color. Chimeric males are crossed with females of the black strain from which the blastocyst was originally derived. Agouti coat color is dominant in such crosses. Hence, if the chimeric males carry the agouti (i.e., the ES) strain genetic profile in their germ line, their offspring will have an agouti coat color. Such offspring are screened by Southern analysis to verify the presence of the mutated gene of interest. Following further breeding to outcross the unwanted black genetic profile, pure-bred agouti heterozygous mutant males and females are mated to yield the homozygous mutant strain. (*Adapted from Capecchi [18a] with permission.*)

cells were derived. Some of the progeny of such matings are expected to be the pure ES strain carrying the heterozygous mutation of interest. Heterozygous mutant males and females are then bred to yield homozygous mutant mouse strains (Fig. 8–3).

Homozygous *ercc1* mutant mice generated in this fashion were observed to be runted at birth and died of liver failure before weaning. Histological and histochemical examination of the livers of these mice revealed subtle but distinctive abnormalities, and several tests of liver function were found to be abnormal. The mice also showed increased levels of p53 protein in the kidneys, liver, and brain. These phenotypes are surprising for a defect in a nucleotide excision repair gene, especially since the mice were not deliberately exposed to exogenous DNA-damaging agents. These results suggest that the *ERCC1* gene may have an as yet unexplored critical role in mammalian development. Since *RAD10*, the *S. cerevisiae* homolog of *ERCC1*, is not an essential gene (see chapter 6), an alternative possibility that has been suggested is that inactivation of nucleotide excision repair renders the development and growth of these mice extremely sensitive to endogenous oxidative damage (137). This hypothesis is difficult to reconcile with the observation that knockout of the *XPA* gene (216, 267) (see below), which is also indispensable for nucleotide excision repair, does not apparently result in developmental disturbances or premature death of the mice. Regardless, the striking failure of development and growth of *ERCC1* mutant mice suggests that if the same phenotype holds in human homozygous mutants, such individuals may not come to term during embryogenesis or, if born, may die before XP is clinically recognizable. In independent studies a set of *ERCC1* mouse strains with different mutations have been generated (267). Mice carrying more subtle point mutations in the *ERCC1* gene have an increased life span compared with mice with a complete knockout of the gene. The former animals also display mutant phenotypes at a later stage in life (267), suggesting that subtle mutations in this gene may be tolerated in humans.

THE *ERCC4* GENE

As already alluded to above and discussed in detail in chapter 6, the *S. cerevisiae* ERCC1 homolog, Rad10 protein, forms a stable and specific complex with Rad1 protein, another component of the nucleotide excision repair family of genes. Since human cells possess a *RAD10* homolog, it is reasonable to expect that they also have a homolog of the *S. cerevisiae RAD1* gene. This is indeed the case, and the gene is designated *ERCC4* (223). The *ERCC4* gene has been mapped to human chromosome 16p13.13-p13.2 (Table 8–3) by somatic cell hybridization (124). This gene was also cloned by functional correction of a rodent cell line, this time from RCG4 (223). At the time of writing, this gene is not completely sequenced. However, there are clear indications that it shares amino acid sequence homology with the homologous *RAD1* and *rad16+* genes of *S. cerevisiae* and *S. pombe*, respectively (223).

THE ERCC1 AND ERCC4 POLYPEPTIDES APPEAR TO FORM A STABLE PROTEIN COMPLEX

There are also indications that the polypeptides encoded by the human *ERCC1* and *ERCC4* genes interact to form a stable complex in living cells. Cell extracts from RCG1 and RCG4 fail to correct each other for defective repair synthesis that is specific to nucleotide excision repair (see below), although they readily complement defective repair synthesis in cell extracts from most of the other genetic complementation groups tested (8, 168, 247). Additionally, the amount of ERCC1 protein is reduced in RCG4 extracts, consistent with an increased lability of this protein in the absence of the interacting ERCC4 protein (8). Protein fractions from human HeLa cells that correct defective repair synthesis in extracts of mutant *ercc1* cells also contain correcting activity for *ercc4* mutant extracts, and an antibody against ERCC1 protein can simultaneously deplete correcting activities for both *ercc1* and *ercc4* mutant cells.

Results of independent experiments using a different experimental approach are consistent with an interaction between ERCC1 and ERCC4 proteins (157). In these studies, human (HeLa) cell extracts were passed through a column containing immobilized recombinant XPA protein. The protein fraction that bound to the XPA protein was recovered and shown to complement defective repair synthesis in extracts of human *xpa* and rodent *ercc1* and *ercc4* mutants. As indicated previously, ERCC1 and XPA proteins interact specifically (121, 157). Hence, it is likely that the ternary XPA-ERCC1-ERCC4 complex is formed by interaction of the ERCC1-ERCC4 complex with XPA protein (157). Deletion mapping indicates that the N-terminal 40% of the ERCC1 protein is sufficient for binding to XPA protein (121).

In several of the studies just summarized, it was observed that extracts of human cells from XP complementation group F (XP-F) behaved as if they were allelic to the rodent RCG4 cells. Hence, it is likely that *ERCC4* is indeed the *XPF* gene (Table 8–2). However, formal proof of this supposition will require the demonstration that the cloned *ERCC4* gene specifically complements XP-F cells and that such cells carry mutations in the *ERCC4* gene. Like many of the rapidly breaking fronts in the field, this proof may well be in hand by the time this writing is translated to a published book.

THE *XPD* (*ERCC2*) GENE

The basic experimental strategy already considered for the isolation of the *ERCC1* and *ERCC4* genes resulted in the molecular cloning of a cosmid carrying a second human nucleotide excision repair gene which uniquely complements the cellular phenotypes of mutants from RCG2. This gene was designated *ERCC2* (230, 263) (Tables 8–2 and 8–3) but is now known to be the *XPD* gene (see below). The gene is ~15 to 20 kb and, like *ERCC1* is located on chromosome 19, where it is separated from *ERCC1* by a physical distance of ~250 kb (144). The gene is organized into 22 exons separated by 23 relatively small introns (54a).

Sequencing of an incomplete *XPD* cDNA, as well as genomic sequence corresponding to the missing portion of the gene, revealed an ORF that could encode a polypeptide of 760 amino acids with a calculated size of 86.9 kDa (264). Like the *ERCC1* translated sequence, that of the *XPD* gene is homologous with a second yeast nucleotide excision repair gene, in this case the yeast *RAD3* gene, which encodes an 87.9-kDa 5' → 3' helicase with DNA-dependent ATPase activity and which is also an essential gene in *S. cerevisiae* (see chapter 6). Overall, 51% of the amino acids are identical in the *XPD* and *RAD3* translated sequences (Fig. 8–4), and conservative substitutions extend the homology to >75% similarity (264). The gene is most remarkably conserved in vertebrates. There is 98% amino acid sequence identity between the human and hamster *XPD* homologs (262) (Fig. 8–4) and 83% identity between the human gene and its homolog from fish (255) (Fig. 8–4). Recombinant XPD protein has been purified ~20,000-fold after its over-

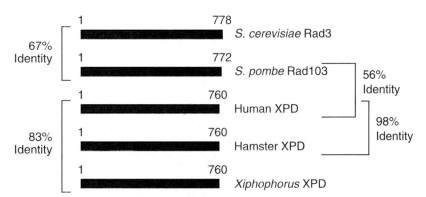

Figure 8–4 The amino acid sequence of the predicted human XPD polypeptide reveals extensive homology over the entire sequence with that from other vertebrates and the *S. pombe* and *S. cerevisiae* Rad103 and Rad3 polypeptides. The predicted polypeptides from all of these organisms are also very similar in size.

expression in yeast cells. It has a molecular mass of ~86.9 kDa, consistent with the size of the *XPD* ORF (204). The purified protein has been shown to be a DNA-dependent ATPase with DNA helicase activity (204), whose properties are very similar to those of the yeast Rad3 protein (see chapter 6).

If the sole function of the yeast Rad3 protein was nucleotide excision repair, one would not expect *RAD3* to be an essential gene. However, in addition to its indispensable role in nucleotide excision repair in yeast cells, *RAD3* is involved in RNA polymerase II basal transcription (see chapter 6). Hence, while direct experimental proof that the *XPD* gene is essential in human cells is not yet available, such is almost certainly the case. In this regard it is interesting that a retrospective review of the results of the initial hunt for rodent cell lines defective in nucleotide excision repair revealed that whereas RCG1 mutants were isolated in approximately equal numbers when using mutagens expected to yield frameshift or single missense mutations in these cell lines, agents expected to cause frameshift mutations failed to yield RCG2 mutants (14). Since chain-terminating mutations in an essential gene are expected to be lethal, this observation is consistent with the notion that the *XPD* gene is essential in human cells.

Both a cosmid and a cDNA carrying the *XPD* gene correct the UV radiation sensitivity of cells from the XP genetic complementation group D (53; quoted in reference 119). This result is consistent with the observation that a single rearranged human chromosome carrying a small region of chromosome 19, which includes the repair gene cluster containing *ERCC1* and *XPD*, also corrects the cellular phenotypes of XP-D cells (52, 53, 119). Mutations have been detected in the *XPD* gene in several XP-D cell lines, suggesting that this gene alone accounts for the human disease XP group D (54a, 265). Additionally, it has been demonstrated that mutant *XPD* alleles isolated from two different XP-D cell lines are unable to correct the UV radiation-sensitive phenotype of XP-D cells when introduced in autonomously replicating extrachromosomal plasmids (54a) (Fig. 8–5), whereas the wild-type allele can. Surprisingly, one of the mutant alleles is the conservative amino acid substitution Leu-461 → Val. The other is a chain-terminating mutation at codon 726 in the XPD polypeptide (54a). These results provide definitive evidence that the *ERCC2* and *XPD* genes are indeed the same.

THE *XPB* (*ERCC3*) GENE

The human *XPB* gene was cloned by phenotypic complementation of cells from RCG3 (270). The gene has been localized to human chromosome 2 (Tables 8–2 and 8–3) (224, 225). The cloned gene is 35 to 45 kb and contains at least 14 exons (269). *XPB* corrects the UV radiation sensitivity of the rodent cell line CHO27-1 from RCG3. Consistent with its role in the repair of bulky base damage, the gene does not complement the sensitivity of some RCG3 cells to alkylating agents which produce nondistortive adducts in DNA. Hence, this second phenotype presumably reflects the presence of at least one other mutation in these cells.

The *XPB* cDNA has an ORF of 782 codons, which could encode a polypeptide of 89.2 kDa (Table 8–3) (270). The translational start codon is believed to be the second ATG in this ORF. Examination of the amino acid sequence suggests that, like the polypeptide encoded by the *XPD* gene, the putative XPB polypeptide is a DNA helicase but is endowed with $3' \rightarrow 5'$ polarity (41) as opposed to the $5' \rightarrow 3'$ polarity of the XPD helicase. This has been confirmed by biochemical characterization of the purified protein (41, 172, 178). The *S. cerevisiae* and *Drosophila* homologs of this gene designated *SSL2* (*RAD25*) and *haywire*, respectively, have been isolated and characterized (see chapter 6). Both of these genes are essential for viability, suggesting that the human homolog is also an essential gene. Support for this conclusion stems from the demonstration that the human basal transcriptional initiation factor BTF2 (TFIIH) not only contains XPD protein (the human homolog of yeast Rad3) as a subunit but also contains XPB protein (see below).

The *XPB* cDNA complements the UV sensitivity and nucleotide excision repair defect in cells from the XP-B genetic complementation group (see chapter 14) (78). Additionally, mutational analysis of the *ERCC3* gene in XP-B cells confirms the contention that *ERCC3* and *XPB* are synonymous (251, 270). The patient from

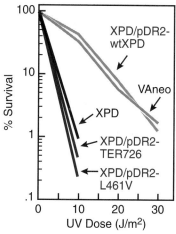

Figure 8–5 A plasmid carrying the cloned wild-type *ERCC2* gene (pDR2-wtXPD) fully complements the UV radiation sensitivity of cells from an individual with XP complementation group D to the same level as cells that carry wild-type *XPD* genes (VAneo). On the other hand, plasmids carrying mutant alleles (pDR2–TER726 and pDR2-L461V) of the *ERCC2* gene are unable to correct the UV radiation sensitivity of XP-D cells. Hence, the *ERCC2* and *XPD* genes are identical. (*Adapted from Frederick et al. [54a] with permission.*)

whom this cell line was derived also manifested the clinical features of CS (see chapter 14), suggesting that a single gene defect is associated with two clinically distinct hereditary diseases. However, the possibility cannot be eliminated that mutational inactivation of the *XPB* gene alone is necessary but not sufficient for XP/CS and that a second as yet unidentified gene is also involved in the association of XP with CS. The specificity of the *XPB* gene for XP-B cells has facilitated the identification of new XP-B individuals in whom the clinical features of CS were not as obvious (119) (see chapter 14).

THE NUCLEOTIDE EXCISION REPAIR/TRANSCRIPTION CONNECTION IN MAMMALIAN CELLS

In chapter 6 we discussed evidence that polypeptides encoded by the *RAD3*, *SSL2*, *TFB1*, and *SSL1* genes of *S. cerevisiae* are components of the multiprotein TFIIH complex required for the initiation of basal transcription by RNA polymerase II in this yeast and that these proteins (and presumably at least two other subunits of TFIIH) are also required for nucleotide excision repair. Consistent with the general conservation of many aspects of nucleic acid metabolism in eukaryotes, the human homologs of all these proteins have been identified as components of human TFIIH and there are strong indications that most (and presumably all) are involved in both RNA polymerase II-dependent basal transcription and nucleotide excision repair (41, 87, 91, 168, 178, 179, 248).

Several studies have now shown that XPD protein is an integral component of human factor TFIIH and is required for transcription and nucleotide excision repair (41, 178, 252). The protein has been identified by amino acid sequence analysis of the 80-kDa subunit of purified TFIIH (178). Additionally, the use of antibodies to XPD protein has demonstrated copurification and coimmunodepletion of the protein with other components of the TFIIH complex (178, 252). Purified preparations of TFIIH also contain the nucleotide excision repair protein XPB (ERCC3) (41, 179, 248) as well as a polypeptide of 62 kDa (41, 179) (the human homolog of Tfb1 protein in yeast cells [see chapter 6]). Additionally, a protein in TFIIH designated p44 appears to be the human homolog of the yeast Ssl1 protein (87), and yet another subunit of TFIIH called p34 has features suggestive of a structural relationship to p44 (87). There are indications that both of these proteins are required for nucleotide excision repair and for transcription (252). Consistent with these observations, in vitro translated recombinant XPD, XPB, and p62 proteins interact with one another when tested in all pairwise combinations (91). Parenthetically, it is interesting that, as is the case in yeast cells, in which Rad2 protein is not a component of purified TFIIH but specifically interacts with subunits of this complex (Ssl2 and Tfb1 proteins) (see chapter 6), XPD, XPB, and p62 proteins have each been shown to specifically coimmunoprecipitate with in vitro translated XPG protein (the human homolog of Rad2 protein [see below]) (91).

There are indications that in human cells, as in yeast cells, XPD, XPB, p62, p44, and p34 proteins are tightly associated and function as a stable complex (41, 87, 91, 178, 252). Interestingly, none of the genes that encode these and other basal transcription factors are linked in the human genome (73). Cell extracts of XP-D cells can be corrected for defective nucleotide excision repair by the addition of purified TFIIH (41, 178, 252) but not by purified XPD protein alone (41). By using a cell-free system for nucleotide excision repair, it has been shown that extracts of XP-D and XP-B cells marginally complement each other, yet both extracts are able to efficiently complement defective repair synthesis in several other XP extracts (41, 168). These results are consistent with the notion that the basal transcription factor TFIIH contains both XPD and XPB proteins in a complex (41, 248). Hence, mixing extracts in which the TFIIH complex is inactivated in both cases does not allow for substitution of the defective XPD and XPB polypeptides in either complex. These results are also consistent with the observation that the yeast homologs of XPD and XPB proteins, Rad3 and Ssl2, respectively, physically interact in vitro (see chapter 6). Extracts from some RCG2 and RCG3 cells are able to complement each other (8, 247), suggesting that the interaction between XPB and XPD proteins may not be very tight in living cells.

As is the case with the homologous yeast protein Ssl2, XPB protein has been shown directly to be involved in both transcription and nucleotide excision repair (41, 248). In particular, microinjection of purified TFIIH into XP-B cells corrects defective unscheduled DNA synthesis (248) (Fig. 8–6). In contrast to the inability to observe correction of defective nucleotide excision repair in a cell-free system complemented with purified XPD protein (41), microinjection of purified XPB protein effectively corrects defective unscheduled DNA synthesis (248). These results suggest that in vivo free XPB protein is able to exchange with that bound in the TFIIH complex.

As indicated above, the properties of purified recombinant XPD protein are very similar to those of yeast Rad3 protein. In chapter 6 we discussed the notion that the possible specific roles of Rad3 protein include the following: (i) facilitating the release of oligonucleotides containing sites of bulky base damage; (ii) facilitating release of the incision complex from sites of damage after specific endonuclease-mediated cleavage of the region of DNA carrying base damage; (iii) using its helicase activity to translocate along transcriptionally silent regions of the genome and hence locate sites of base damage by arresting at such sites; and (iv) generating a "denaturation bubble" or localized region of denaturation at sites of base damage in DNA, thereby providing duplex/single-strand junctions that are specifically recognized by the Rad1/Rad10 and Rad2 endonucleases.

There is intriguing evidence that a protein designated p56, which is a member of the TFIIE transcription complex, interacts with TFIIH and in so doing inhibits all DNA helicase activity of TFIIH (41). Apparently, p56 has no effect on the helicase activity of purified XPD protein, suggesting that this helicase is dormant in

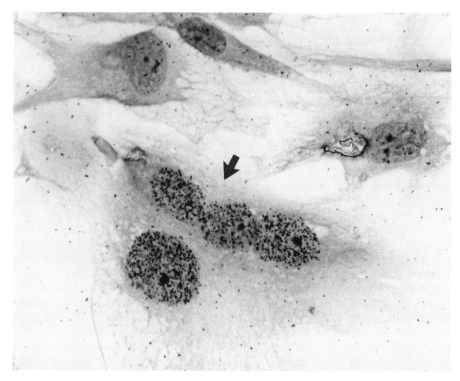

Figure 8–6 Correction of defective repair synthesis in XP-B cells by microinjection of purified transcription factor TFIIH. Fibroblasts from an XP group B individual were grown in monolayer culture. Selected cells (arrow) were microinjected with TFIIH, exposed to UV radiation, and cultured in the presence of [³H]thymidine. Following autoradiography, silver grains in the nuclei represent repair synthesis (unscheduled DNA synthesis) in the injected cells. Other cells nearby that were not injected show defective repair typical of the *XPB* gene defect. (*Photo courtesy of Wim Vermeulen and Jan Hoeijmakers.*).

the TFIIH complex and that the helicase activity of the TFIIH complex reflects that of XPB protein exclusively (41). A particular mutant form of XPB protein in which the $3' \to 5'$ DNA helicase activity of XPB protein is believed to be inactivated (248) has a dominant negative phenotype when expressed in normal cells and completely abrogates both transcription (leading to an apparent collapse of the normal chromatin configuration) and nucleotide excision repair. Other mutations in the *XPB* gene have been found to inactivate its ability to support normal nucleotide excision repair in rodent cells that are defective in the homologous gene (128). In particular, the substitution of Arg for the invariant Lys residue in the ATPase motif of the first helicase domain of the XPB polypeptide, mutations in other helicase domains, and a deletion in a putative helix-turn-helix motif believed to be involved in DNA binding of the protein all failed to complement the UV radiation sensitivity of the rodent mutant cell line (128).

In summary, there is a general parallel between yeast and human cells with respect to the bifunctionality of proteins for transcription and nucleotide excision repair. However, it remains to be determined whether the basal transcription complex that operates in yeast cells and in human cells is made up of the identical set of homologous polypeptides. In this regard it is provocative that the ternary complex between TFIIH, DNA, and nascent RNA formed during the process of transcription elongation has been reported to be devoid of XPD and XPB proteins (41). If this is indeed the case, the mechanism of assembly of a functional repairosome (which presumably includes XPD and XPB proteins) at sites of base damage in the transcribed strand remains an unsolved conundrum (see chapter 6).

The bifunctional association of human proteins with RNA polymerase II-dependent transcription and nucleotide excision repair provides the intriguing possibility that human diseases characterized by complex clinical phenotypes which include, but are not confined to, defective nucleotide excision repair may be explained by subtle defects in transcription. This discussion will be more fully elaborated in chapter 14, when we discuss the clinical features and aspects of the molecular pathology of XP, CS, TTD, and syndromes that appear to be related to TTD.

THE *CSB* (*ERCC6*) GENE

The *CSB* gene has also been cloned by phenotypic complementation and may be as large as 100 kb, with at least 21 exons (78, 81, 237, 238). A cDNA clone corrects the UV sensitivity of cells from RCG6 and has an ORF which could encode a protein of 168-kDa (Table 8–3) (239). Cells from a patient with CS from genetic complementation group B carry a deletion in one chromosome in the region 10q11-21.1, which includes the *CSB* gene. The cloned gene corrects the UV sensitivity of CS-B cells, suggesting that CS group B represents a defect in this gene (239). The *CSB* gene has been mapped to chromosome 10q11 (Table 8–2) and encodes two weakly expressed mRNAs of 6.5 and 8.5 kb (78).

Examination of the amino acid sequence of the translated *CSB* ORF suggests that it too might be a DNA helicase (see Fig. 6–31) (239). The translated sequence of *CSB* shares striking amino acid homology with a distinctive family of proteins that includes the ones encoded by the yeast *RAD16*, *RAD54*, and *RAD5* genes, the last two being involved in recombinational and error-prone repair modes (see Fig. 6–31). In the case of the Rad16 polypeptide, there are strong experimental indications that along with the product of the *RAD7* gene, the protein somehow perturbs chromatin structure in closed heterochromatic regions on the genome that are otherwise inaccessible to the nucleotide excision repair machinery (see chapter 6). All the polypeptides in this family have consensus helicase domains. Proteins in this family have diverse functions, including transcriptional regulation and chromosome stability in lower eukaryotes (239) (see chapter 6).

THE *XPG* (*ERCC5*) GENE

The human *XPG* gene has an interesting history, which provides an example of the unpredictability of scientific research and the enormous value of related work in different biological systems. Somatic cell analyses revealed that human chromosome 13q contains a gene capable of correcting the UV radiation sensitivity of cells from RCG5 (78, 224, 225). Transfection of human genomic DNA has long been known to correct this phenotype of RCG5 mutants (129, 202). However, the correcting activity was first isolated as two overlapping cosmids which yielded a functional ~32-kb *ERCC5* gene only by intercosmid recombination (145). In totally independent studies an incomplete cDNA was isolated by screening a *Xenopus laevis* cDNA expression library with antiserum from a human patient with the autoimmune disease systemic lupus erythematosis (180). When the full-length frog and human cDNAs were isolated, their translated sequences were found to be homologous to two regions of the *S. cerevisiae RAD2* and *S. pombe rad13+* genes (see chapter 6). Furthermore, the human cDNA was shown to correct the UV radiation sensitivity of cells from an individual with XP-G. In parallel studies in vitro, complementation was found to be absent when extracts of human XP-G and rodent RCG5 cells were combined (154). A purified fraction from extracts of normal human cells was also shown to correct defective nucleotide excision repair in extracts of both cell types (154). This correction was shown to be inhibitable by an antibody to XPG protein (154). This biochemical evidence strongly suggested that the *ERCC5* and *XPG* genes are in fact identical, a conclusion confirmed by the isolation of the *ERCC5* cDNA (130, 185) and by the in situ mapping of the *XPG* gene to chromosome 13q33 (174)(Color Plate 7; Tables 8–2 and 8–3), the same site to which the *ERCC5* gene was mapped (211). Definitive evidence that the *XPG* (*ERCC5*) gene is indeed causally implicated in XP-G was provided by studies showing that mutant alleles of the gene are not able to correct the UV radiation sensitive phenotype of XP-G cells (152a). The cloned gene has an ORF of 1,186 codons and is expected to encode a polypeptide of 133 kDa (Table 8–3).

XPG PROTEIN

Recombinant XPG protein has been purified to apparent physical homogeneity (153). The purified protein corrects defective nucleotide excision repair in extracts of human XP-G cells and rodent ERCC5 cells but has no effect when added to extracts of other repair-defective cell lines. Like the yeast Rad1-Rad10 and Rad2 endonucleases, XPG protein cleaves bacteriophage M13 DNA (see chapter 6). In contrast to the case with the homologous yeast Rad2 protein, which specifically requires Mg^{2+} as a metal cofactor, degradation of bacteriophage M13 DNA by XPG protein can utilize either Mg^{2+} or Mn^{2+} (153).

When discussing the substrate specificity of the Rad1-Rad10 endonuclease of *S. cerevisiae*, we pointed out that this enzyme specifically recognizes duplex/3′-single-strand DNA junctions. This property makes the Rad1-Rad10 endonuclease an attractive candidate to cut duplex DNA with such a junction at a site of base damage, specifically 5′ to the site of damage (see Fig. 6–13). We also suggested that the yeast Rad2 endonuclease might function as the mirror enzyme and be constrained to cut on the 3′ side of such sites of base damage because it specifically recognizes duplex/5′-single-strand junctions in DNA (see Fig. 6–13). While such (anticipated) substrate specificity has not been reported for the purified yeast Rad2 protein at the time of writing, it has indeed been shown to be a property of the homologous human XPG protein (152b).

Use of Rodent Cell Lines for Cloning Human Repair Genes: Concluding Comments

At the time of writing, the strategy of screening the human genome for genes which correct cellular phenotypes of mutant rodent cell lines presumed to be defective in nucleotide excision repair has paid handsome dividends. This strategy has resulted in the molecular cloning and detailed characterization of the *ERCC1*, *ERCC4*, *XPD*, *XPB*, *XPG*, and *CSB* genes. All of these genes are clearly implicated in nucleotide excision repair by the direct demonstration that they can complement repair-defective human cell lines and/or that they have yeast homologs known to be involved in nucleotide excision repair. What about the remaining six RCGs? As discussed below, two other human genes for nucleotide excision repair

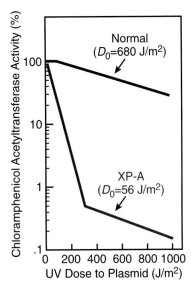

Figure 8–7 XP complementation group A cells are defective in the repair of UV radiation damage in the *cat* gene carried on a plasmid. The D_0 values for XP-A and normal cells are derived from the initial slopes of the gene inactivation curves. The value obtained for XP-A cells extrapolates to a dose of UV radiation that is expected to yield one pyrimidine dimer per *cat* gene on the average. (*Adapted from Cleaver and Kraemer [24] with permission.*)

have been isolated by the direct phenotypic correction of XP cells. They may account for two of these six rodent cell types. The genes that account for the remaining four are as yet unidentified.

Mutant Cell Lines from Patients with XP

The human disease XP has already been mentioned in several contexts and is fully discussed in chapter 14, which addresses human hereditary diseases associated with abnormal cellular responses to DNA damage. It is relevant to consider selected aspects of this disease in the present chapter, since XP cells in culture represent the human equivalents of some of the yeast and rodent nucleotide excision repair-defective mutants already described, and in the context of the present discussion some XP cells have served investigators well as tools for the functional cloning of several other human nucleotide excision repair genes.

Fusions in various pairwise combinations between cells from different patients with this disease have established the existence of seven distinct genetic complementation groups designated A to G, for the *"classical form"* of the disease (Table 8–2). Indeed, it is interesting historically that these experiments initiated by Dirk Bootsma and his colleagues in 1972 (36) were the first indication of the utility of cell fusion for analysis of genetic complementation in higher eukaryotes. Early literature addressed nine genetic complementation groups in the classical form of XP (106). However, genetic complementation groups H and I were subsequently shown to be examples of the mistaken identity of XP-D and XP-C respectively (9). An XP variant (XP-V) form has also been described (24, 60, 67, 106, 107).

Members of each of the genetic complementation groups from the classical form of the disease are defective to various extents in the excision repair of a large spectrum of DNA damage (24, 60, 67, 106, 107). Indeed, by using host cell reactivation of a plasmid-borne gene encoding the chloramphenicol acetyltransferase gene (*cat*) as an assay of the nucleotide excision repair capacity of human cells, it has been shown that a single pyrimidine dimer inactivates expression of the *cat* gene in XP cells (165) (Fig. 8–7). Most studies on XP variant cells have failed to reveal a clear defect in nucleotide excision repair. However, experiments with cell extracts do suggest that defective nucleotide excision repair occurs in these cells under some circumstances (see below).

As is the case with many of the yeast *RAD* genes discussed in chapter 6, it appears that much, if not all, of the genetic complexity of XP is concentrated at early steps in nucleotide excision repair, specifically those required for the recognition of base damage and the incision of damaged DNA (57). The presence of strand breaks in the DNA of UV-irradiated normal human cells generated during post-UV incubation has been demonstrated by the technique of alkaline elution discussed previously (104) (see chapter 7). Such cells show a rapid accumulation of strand breaks, followed by a gradual restoration to normal molecular weight, indicative of the completion of repair (54) (Fig. 8–8). Cells from XP groups A, B, C, and D fail to accumulate significant numbers of breaks in their DNA during postirradiation incubation, suggesting a defect in the incision of DNA in these cells (54) (Fig. 8–8). Similarly, when purified PD-DNA glycosylase is introduced into permeabilized UV-irradiated cells or microinjected into cells, normal levels of repair synthesis of DNA are restored to cells from all complementation groups examined (34, 219) (Table 8–4). Use of T4 endonuclease V or the equivalent *Micrococcus luteus* enzyme to measure the loss of endonuclease-sensitive sites from the DNA of XP cells has also shown a defect in incision in cells from genetic complementation groups A, B, C, D, and G (288). Cells from XP complementation groups E and F are leaky by this parameter (288) and may indeed represent leaky mutants.

Phenotypic correction of the UV radiation sensitivity of XP-A (72) and XP-D (2) cells has been demonstrated following stable transfection with the cloned phage T4 *denV* gene. Presumably the T4 PD-DNA glycosylase catalyzes incisions in DNA at pyrimidine dimers, effectively circumventing the endogenous incision defect. On the other hand, the microinjection of purified *E. coli* UvrA, UvrB, UvrC, and UvrD proteins into XP cells did not restore normal levels of repair synthesis

Table 8–4 Repair synthesis in homopolykaryons after microinjection of *M. luteus* PD-DNA glycosylase

Cell Type	Repair synthesis (% of normal) for:	
	Noninjected	Injected
Normal	103	100
XP-A	1	79
XP-B	9	54
XP-C	20	37
XP-D	17	55
XP-E	50	65
XP-F	19	50

Source: Adapted from de Jonge et al. (34) with permission.

(291). Negative results were also obtained following the transfection of XP-A cells with the cloned *E. coli uvrA⁺* gene (37). These failures presumably reflect the inability of these *E. coli* proteins to reconstitute a functional endonuclease in mammalian cells.

MOLECULAR CLONING OF XP GENES

As is true of the nucleotide excision repair-defective bacterial, yeast, and rodent mutants discussed above and in chapters 5 and 6, the UV radiation-sensitive phenotype of XP cells in culture invites a molecular approach to the study of this disease through the cloning of XP genes by phenotypic complementation of human cells. Attempts to achieve this goal by transfection of XP cells with total human genomic DNA have been largely unsuccessful (56, 173, 181, 212, 222). Similar difficulties have been encountered with other human cell lines with potentially selectable phenotypes. In an attempt to clone the ataxia telangiectasia (AT) group D gene (see chapter 14), AT cells from genetic complementation group D (which are abnormally sensitive to killing by ionizing radiation) have been transfected with total genomic DNA in numerous laboratories. The transfection frequency of the cotransfected selectable marker *neo* was found to be reasonable and within the range typically observed in human cell lines. But, the screening of >65,000 independent Neoʳ transfectants (which represents the statistical distribution of an amount of DNA equivalent to more than three times the haploid human genome [127]) failed to yield a single X-ray-resistant clone, even though the screening protocol was able to rescue one wild-type cell from 10⁶ AT cells in a reconstruction experiment (127).

The precise reasons for the disappointing differences between the efficiency of the stable transfection of rodent and human cells are not known. Several detailed studies suggest that the overall efficiency with which DNA sequences are stably integrated and expressed in human cells is considerably reduced relative to that in rodent cells. A direct comparison between several CHO and human cell lines showed significant differences within both groups (80). However, in general, the average amount of DNA stably incorporated into the human lines was 20- to 100-fold less than in the CHO lines. A related study showed that murine cells integrated 9 to 186 kb of DNA on average, whereas a primate kidney line integrated only 3 to 6 kb of DNA on average (25).

The integration efficiency of human cell lines specifically defective in cellular responses to DNA damage was examined in a study with immortalized XP, AT, and CS cells. The cells were transfected with DNA fragments containing the dominant selectable markers *hyg* (which confers resistance to the antibiotic hygromycin) and *neo* (which confers resistance to the antibiotic geneticin [G418]), separated by distances ranging between 0 and 40 kb. The transfection frequency for the initially selected marker (*neo*) was comparable to that observed in many mammalian, including rodent, cells. Approximately 50% of these cells were also ini-

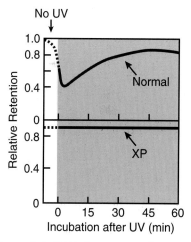

Figure 8–8 DNA damage-specific incision in UV-irradiated normal and XP fibroblasts. The figure shows the relative retention of DNA on filters as measured by the alkaline elution technique (see chapter 7). Normal human fibroblasts generate strand breaks in their DNA during postirradiation incubation. This results in a reduced retention of the DNA on the filters following denaturation of the DNA in alkali, because of the smaller size of the DNA fragments. The gradual return to higher-molecular-weight DNA associated with the completion of nucleotide excision repair is reflected by increased filter retention. DNA from XP cells does not show evidence of DNA incision by this technique. (*Adapted from Friedberg [55] with permission.*)

tially resistant to hygromycin. However, when individual clones were expanded over an extended period under continual selection for G418^R, about one to three copies of the *neo* marker were retained, whereas only 0.1 copy of the unselected marker was retained (135).

Expression of stably integrated genes can also vary from one human cell line to another. Plasmids containing the bacterial *gpt* gene under control of the simian virus 40 (SV40) promoter were transfected into a human fibroblast line, and two stable Gpt$^+$ transfectants were selected for further study. Gpt$^-$ segregants were detected at a frequency of ~10^{-4}. A detailed study of these showed that 14 of 19 had undergone rearrangements of the *gpt*-containing sequences. In others the gene was expressed at drastically reduced levels, possibly resulting from altered patterns of DNA methylation (61).

It would clearly be of interest to determine the particular attributes that confer an advantage to rodent cell lines with respect to their stable transfection. A pragmatic attempt to confer such attributes to human XP lines was based on the fusion of XP-D cells to a HeLa cell line selected for high transfection efficiency. A hybrid line that retained both the repair-defective phenotypes of XP-D cells and the improved transfection efficiency of the parental HeLa line was successfully isolated (2, 93). It has also been observed that stable transfection of DNA into human cells is improved by previous exposure of the transfecting DNA to UV or ionizing radiation (111, 160, 196, 197). Interestingly, CHO cells did not show increased transfection with UV-irradiated plasmids (146).

These difficulties notwithstanding, the direct transfection of human XP cells with total genomic DNA did yield one of the anticipated XP genes, *XPA*. A second XP gene, *XPC*, was cloned by using a human expression cDNA library rather than total genomic DNA. This important wrinkle in the history of the isolation of human nucleotide excision repair genes is taken up later in the chapter.

The *XPA* Gene

The logistics of cloning the mouse *XPA* gene by direct phenotypic complementation of XP-A cells transfected with total genomic mouse DNA were formidable and underscore the magnitude of the problems discussed above. Two primary UV-resistant XP transfectants were isolated after screening 160,000 Gpt$^+$ XP-A cells. Cotransfection of DNA from one of these primary transfectants with a plasmid carrying the *gpt* gene yielded a single secondary transfectant among 480,000 Gpt$^+$ colonies screened. A phage λ library was constructed from this secondary transformant (which was now considerably enriched for mouse sequences), and mouse sequences were isolated and tested for phenotypic complementation of XP-A cells. Two phage clones carrying overlapping regions of a complementing mouse gene were so identified. Cotransfection of these clones did not alter the UV radiation sensitivity of XP group C, D, F, or G cells. Northern (RNA) analysis revealed transcripts of ~1 to 1.3 kb in mouse and human cells. These transcripts were not detected in XP-A cells (218).

The genomic *XPA* gene is ~25 kb in size and is split into six exons (175, 176). Sequencing the human and mouse *XPA* cDNAs revealed a single ORF predicted to encode a polypeptide of 273 amino acids with a calculated size of ~31 kDa (217). The two predicted polypeptides share ~95% identity. The *XPA* gene is highly conserved in eukaryotes. Homologs have been isolated from *Xenopus*, chicken, and *Drosophila* DNA (see Fig. 6–29) (184). The homologous gene from the yeast *S. cerevisiae* is *RAD14*. As discussed in chapter 6, Rad14 protein has been purified to homogeneity and has been shown to be a DNA-binding protein that specifically recognizes (6-4) photoproducts but not pyrimidine dimers in DNA. The amino acid sequence of the putative XPA protein reveals a zinc finger motif (see Fig. 6–29), suggesting that it is a DNA-binding protein (217). The *XPA* gene is located on human chromosome 9q34.1 and on mouse chromosome 4C2 (217). XP group A homozygotes are defective in expression of *XPA* mRNA (176, 217). In the case of several Japanese patients, this reduction in mRNA levels probably results from defective processing of mRNA owing to a G → C transition mutation that alters a consensus AG dinucleotide sequence in the 3' splice acceptor site of intron 3 of the *XPA* gene (Fig. 8–9) (217). More-extensive studies on XP patients from genetic complementation group A showed this mutation to be present in 16 of 21 unrelated Japanese homozygous individuals. However, 11 Caucasians and 2 blacks with XP group A did not have this mutation (176). Further mutational analyses have demonstrated that the majority of cases of XP group A in Japanese patients are caused by one or more of three mutations: a splicing mutation in intron 3 or nonsense mutations at either codon 116 (Fig. 8–9) or 228 (177). In contrast, an analysis of mutations among XP-A individuals in Tunisia revealed a prevalence of point mutations in exon 6 of the gene. In these patients, 86% of the mutations in exon 6 affected codon 228 (152).

The XPA Protein

The first indication that exogenous proteins can enter the nucleus of XP cells and correct their phenotypes came from the fusion of XP-A and XP-C cells with cytoplasts from different repair-proficient human cells. Such "cybrids" showed transient correction of defective nucleotide excision repair, with a half-life of ~12 h in the case of XP-A cells (103). Subsequent studies have shown that proteins in

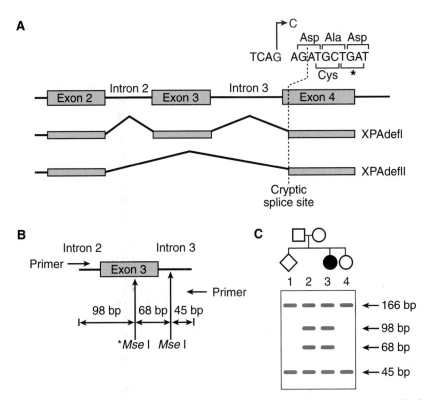

Figure 8–9 Mutations in the *XPA* gene. (A) A large percentage of Japanese individuals with XP complementation group A harbor a G → C substitution (top) at the 3′ splice acceptor site (TCAG) of intron 3 of the *XPA* gene, which alters the obligatory AG acceptor dinucleotide to AC. As a consequence, a cryptic 3′ splice acceptor site (dashed vertical line) is activated in exon 4, resulting in a shift in the reading frame. This shift generates a codon for the amino acid cysteine followed by a TGA stop codon (*). The 3′ splice acceptor site of intron 2 and the 5′ splice acceptor site of intron 3 are often not used. As a result, two *XPA* mRNA species called *XPA*defI and *XPA*defII are observed in these individuals. (B) Individuals with the *XPA* allele shown in panel A frequently have compound heterozygosity in the *XPA* gene. The second mutation sometimes involves a T → A transversion in exon 3 that changes Tyr-116 to a nonsense codon. This mutation generates a new *Mse*I restriction site (asterisk) which can be detected by Southern analysis of the *XPA* gene. (C) Such analysis shows that the proband (lane 3) and the proband's mother (lane 2) are heterozygous for this allele. The mutant allele is not present in the genome of the proband's father (lane 1) or sister (lane 4). (*Adapted from Satokata et al. [176 and 177] with permission.*)

Figure 8–10   Schematic representation of the human ERCC1 and XPA polypeptides. The thick dark lines in each structure indicate regions of amino acid (aa) conservation between ERCC1 and the homologous yeast Rad10 polypeptides and between XPA and the homologous yeast Rad14 polypeptides. The colored boxes represent amino acid motifs (XPA IR [for XPA-interacting region] and ERCC1 IR [for ERCC1-interacting region]) which are required for interaction between these two proteins. (*Adapted from Li et al. [121]*).

cell extracts of normal human cells can complement XP cells following microinjection (35, 79, 82, 250, 283).

The human and bovine proteins that correct the phenotype of XP-A cells have been extensively purified in several laboratories by using either the assay just described or the complementation of defective nucleotide excision repair in cell extracts (see below) (79, 171). Additionally, recombinant XPA fusion proteins (fused with either *E. coli* maltose-binding protein [156] or an N-terminal polyhistidine tag [94]) have been purified. XPA protein specifically complements XP-A cells, although weak correction of XP-C cells (presumably reflecting the presence of a contaminating protein) is observed in partially purified preparations (79). As indicated previously a specific interaction has been observed between XPA protein and the protein encoded by the *ERCC1* gene both in vitro, and in vivo by using the two-hybrid genetic screen (see chapter 6) (121, 157). Deletion mapping indicates that the N-terminal 42% of the XPA polypeptide is required for this interaction and that amino acid residues 75 to 114 are necessary and possibly also sufficient (121) (Fig. 8–10). However, the presence of a second domain required for interaction with ERCC1 protein has not been excluded (121).

These and the other protein-protein interactions that have been discussed in this chapter presumably reflect the composition and organization of the (putative) human multisubunit repairosome. In this regard it is interesting that there are differences in the interactions that have been identified between human and yeast nucleotide excision repair proteins. For example, interactions between the yeast XPA homolog (Rad14) and the yeast ERCC1 homolog (Rad10) have not been observed. Negative results are always difficult to evaluate. Nonetheless, it is provocative that the ERCC1-binding domain in XPA protein between amino acids 75 and 114 is not conserved in the yeast Rad14 protein. Furthermore, yeast Tfb1 protein, a subunit of TFIIH, has not been observed to interact with yeast Ssl2 protein, whereas, as mentioned above, coimmunoprecipitation experiments with in vitro translated recombinant proteins have demonstrated a specific interaction between the human homologs of these two proteins, p62 and XPB, respectively (91). Similarly, in vitro translated XPD protein coimmunoprecipitates with XPG protein (91), whereas no interaction has been observed between the homologous yeast Rad3 and Rad2 proteins. Hence, despite the overall general conservation of nucleotide excision repair proteins in eukaryotes, the precise architecture of the repairosome and the detailed mechanism of its assembly apparently differ between yeast and human cells.

XPA protein migrates in denaturing polyacrylamide gels with a mobility of a protein of 40 to 45 kDa, slightly larger than the size predicted from the cloned *XPA* gene (217). The protein binds to both unirradiated single-stranded DNA-agarose and UV-irradiated DNA-cellulose (79, 94, 171). Binding to UV-irradiated DNA is about 1,000 times stronger than to native DNA on a per nucleotide basis (171) (Fig. 8–11), suggesting that, like its yeast homolog (Rad14 protein), XPA protein may participate in some aspect of DNA damage recognition during nucleotide excision repair. A further remarkable resemblance to the yeast Rad14 protein derives from the observation that XPA protein binds mainly to (6-4) photoproducts in UV-irradiated DNA rather than to pyrimidine dimers (94) (Fig. 8–11). The dissociation constant (K_d) for binding to UV-irradiated DNA is ~3 × 10⁻⁶ M, whereas that for binding to unirradiated DNA is ~6 × 10⁻⁵ M (94). The protein also binds to DNA treated with cisplatin (94). Collectively, these observations suggest that XPA protein functions in the process of DNA damage-specific recognition during nucleotide excision repair.

XPA protein is destroyed by proteinase K but can be heated to 95°C for 5 min in the presence of sodium dodecyl sulfate (SDS) and β-mercaptoethanol without loss of complementing activity after microinjection (79). However, this treatment destroys the complementing activity of XPA protein in cell extracts (171). XPA protein purified from calf thymus migrates in denaturing gels as a doublet of 42 and 40 kDa (49, 171). Similar results have been independently observed with protein expressed from the endogenous *XPA* gene in human cells, detected by immunoblotting, and in vitro translation of the cloned gene (143) and with XPA protein expressed in *E. coli* (94). The two bands apparently do not reflect post-translational modification of XPA protein. The two polypeptides were also detected in XP-A cells transfected with the *XPA* gene but were both absent in untransfected XP-A cells (217). A definitive explanation for this duplication of the XPA polypeptide in SDS-gels is lacking. However, it is strongly suspected that they represent alternatively unfolded conformers (171). Interestingly, deletion of 70 to 86 amino acids from the C terminus eliminates one of the bands, suggesting that this region of the polypeptide is responsible for the possible dual conformation (278). The recombinant protein yields a Stokes radius of 33 Å (3.3 nm) and an S value of 2.2. On the basis of these two hydrodynamic parameters, a molecular mass of 30 kDa was calculated, a value close to that of 31 kDa predicted from the size of the translated *XPA* ORF (94). Hence, XPA protein exists in solution as a monomer.

The *XPC* Gene

An alternative to phenotypic complementation of XP cells with genomic DNA is to transfect cells with cDNA. The major problem with this strategy concerns the uncertainties regarding the completeness of such libraries, especially for single-copy genes which are weakly expressed. Nonetheless, several such libraries have been constructed, one of which yielded an *XPC* cDNA (115). This particular library was constructed in a plasmid vector carrying the Epstein-Barr virus replication origin *oriP* and the gene for the Epstein-Barr virus nuclear antigen, *EBNA1*. The presence of these genes confers the ability of plasmids to replicate as relatively stable extrachromosomal episomes in mammalian cells. Hence, plasmids carrying cDNAs of interest can be conveniently rescued from complemented cells. This property enormously simplifies the logistics of gene cloning compared with the use of genomic DNA described previously, since it is no longer necessary to go through the laborious exercise of rescuing integrated sequences from the genome of the recipient cell line. Regrettably, the Epstein-Barr virus-based episome technology cannot be efficiently used with genomic DNA because the available generation of autonomously replicating plasmids can accept only relatively small (by mammalian genome dimensions) recombinant inserts. Hence, it is most useful with cDNA inserts.

The *XPC* cDNA originally isolated has an ORF that can encode a polypeptide of ~93 kDa (115). Subsequently a more complete *XPC* cDNA was independently isolated which contains coding sequence for an additional 117 amino acids at the N terminus, which are apparently not essential for the DNA repair function of the protein. Hence, the complete *XPC* ORF is expected to encode a hydrophilic polypeptide of ~106 kDa (Table 8–3) with a basic pI (133). The translated amino acid sequence shows a limited region of homology with the *S. cerevisiae RAD4* gene, extending over a region of ~230 amino acids (see Fig. 6–28). This region of homology was confirmed, and in fact extended, by including a comparison with the cloned *Drosophila* homolog of *XPC* (74) (see Fig. 6–28). Northern analysis of normal human cells has revealed a single transcript of ~3.8 kb. The level of this transcript

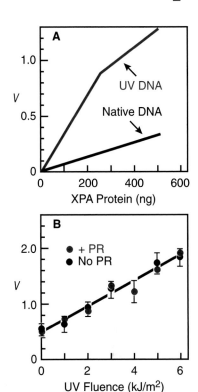

Figure 8–11 Binding of XPA protein to UV-irradiated and native DNA. (A) Following gel mobility shift analysis of protein-DNA binding, the amount of unbound radiolabeled DNA can be quantitated. The parameter *V* (the negative logarithm of the fraction of unbound DNA) is proportional to the average number of XPA protein molecules bound per DNA fragment. The reciprocal of the protein concentration at which 50% of the DNA is bound gives an estimate of the equilibrium association constant. (B) When the binding of XPA protein to UV-irradiated DNA is carried out with (red circles) or without (black circles) prior treatment of the DNA with DNA photolyase to remove pyrimidine dimers, the extents of the binding are identical. This suggests that the protein binds mainly to (6-4) photoproducts in UV-irradiated DNA. (*Adapted from Jones and Wood [94] with permission.*)

was extensively reduced or was undetectable in several XP-C cell lines (115). More direct confirmation that the *XPC* cDNA is relevant to the molecular pathology of XP group C stems from the identification of mutations, many of which are chain terminating in nature, in the *XPC* gene from several XP-C cell lines (120) (Fig. 8–12). Interestingly, one mutant allele was found to have an insertion of the sequence GTG. Since the two codons preceding this also have this sequence (which normally encodes valine), this may represent a limited example of trinucleotide repeat expansion that has been associated with other human hereditary diseases (120). The *XPC* gene maps to human chromosome 3p25 (116).

The XPC Protein

The molecular cloning of the complete *XPC* cDNA described above was facilitated by the initial purification of XPC protein and the derivation of partial amino acid sequence (133). The protein was purified from repair-proficient human extracts on the basis of its ability to correct defective repair synthesis in extracts of XP group C cells. Following extensive purification, the XP-C-correcting activity was found to be composed of two polypeptides that could not be further resolved. The XPC polypeptide has an apparent molecular mass of 125 kDa based on its electrophoretic mobility in denaturing gels (133). The second, tightly associated protein is 58 kDa in size based on this criterion and proved to be the human homolog of a yeast gene called *RAD23*, which is involved in but not absolutely required for nucleotide excision repair (see chapter 6). In fact, further analysis revealed the presence of two highly conserved but nonidentical copies of this gene (designated *HHR23A* and *HHR23B* [for human homolog of *RAD23*]) in the human genome (133).

We mentioned in chapter 7, in a discussion about the heterogeneity of DNA repair in the human genome and of transcriptionally coupled nucleotide excision repair, that XP-C cells are apparently unique in that they manifest only a slight defect in nucleotide excision repair in transcriptionally active genes but appear to

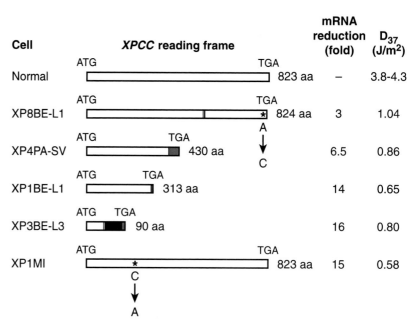

Figure 8–12 Correlation between various mutations in the *XPC* gene and phenotypic consequences. Regions of the *XPC* ORF shown in grey represent altered codons downstream of frameshifting mutations. Regions in red represent insertional mutations. (Note the small trinucleotide insertion in XP8BE-L1, which results in an extra codon in addition to the point mutation in this allele.) Missense mutations are indicated by asterisks. The D_{37} values shown represent the dose of UV radiation required to reduce the survival of cells to 37%. aa, amino acids. (*Adapted from Li et al. [120] with permission.*)

be more severely defective in the repair of transcriptionally silent regions of DNA (95). Similarly, near-normal levels of repair of UV radiation damage have been observed in the transcribed strand of the transcriptionally active *ADA* gene (Fig. 8–13), whereas repair in the nontranscribed strand of the gene is severely deficient (249) (Fig. 8–13). One implication of this intriguing observation is that XPC protein is not required for, and hence does not participate in, the preferential repair of template strand damage in transcriptionally active genes. On the other hand, while the homologous Rad4 protein has yet to be purified to homogeneity, there is evidence to suggest that this protein interacts with the yeast RNA polymerase II basal transcription factor TFIIH (see chapter 6). Hence, Rad4 protein may be a participant in a repairosome that includes transcription factors but exclusively participates in nucleotide excision repair that is not directly coupled to transcription. Alternatively, in XP-C cells the low levels of XPC protein may be preferentially recruited to a repairosome complex that participates in the repair of transcriptionally active DNA, and this may leave insufficient amounts of the protein for complexes that are active on bulk DNA (5a).

The XPE Protein

XP-E cells manifest a very modest sensitivity to UV radiation and hence have not lent themselves to functional cloning of the *XPE* gene. However, a protein that may be encoded by this gene has been identified and partially characterized. The binding of cellular proteins to damaged DNA was examined by gel retardation assays. These studies identified a defect in a DNA-binding protein in XP-E cells, designated XP-E binding factor (Fig. 8–14) (21, 158). However, defective protein binding is not a universal biochemical phenotype of XP-E cells. At least four fibroblast strains from unrelated Japanese patients with this form of XP do not show this defect (96), and in total only 3 of 12 XP-E cell lines examined show this defect (102). These cells are designated as being phenotypically Ddb+ (for DNA damage binding), whereas XP-E cells that are defective in the binding activity are called Ddb− (102). Several explanations are consistent with the notion that this DNA damage-binding protein is functionally defective in all XP-E cell lines. Of these, the most compelling is the suggestion that the protein in the majority of cell lines is altered at some domain other than that required for binding to DNA (102).

Related studies have identified XP-E factor as an inducible DNA-binding protein with specific affinity for (6-4) photoproducts in DNA and not for pyrimidine dimers (76, 236). XP-E binding factor presumably plays an auxiliary role in nu-

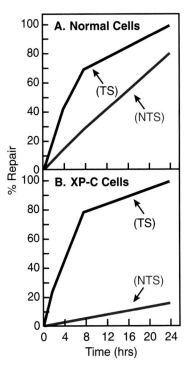

Figure 8–13 XP-C cells (B) are defective in the repair of the nontranscribed strand (NTS) of a transcriptionally active gene (in this case the human *ADA* gene) relative to normal cells (A) but repair the transcribed strand (TS) at near normal levels. (*Adapted from Venema et al. [249].*)

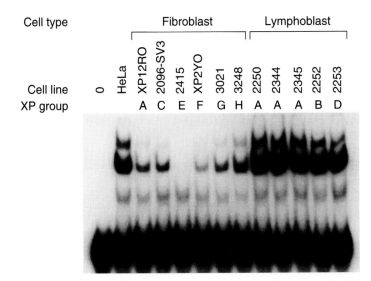

Figure 8–14 Gel shift analysis shows defective binding of protein to UV-irradiated DNA in extracts of XP-E cells. Extracts of normal cells and of cells from other XP genetic complementation groups show the binding of protein to UV-irradiated DNA. (*From Chu and Chang [21] with permission. Photo courtesy of Gil Chu.*)

cleotide excision repair since, as indicated above, cells from XP-E individuals are only slightly UV sensitive and only moderately defective in nucleotide excision repair. A protein that appears to be identical to that just discussed has been independently isolated in several studies on the basis of its high binding affinity for UV irradiated DNA. The protein has been variously termed UV-DDB (for UV-damaged DNA-binding) protein (76) and UVBP1 (for UV damage binding protein 1) (242). It recognizes (6-4) photoproduct dimers in DNA but not pyrimidine dimers or bulky chemical adducts. A monkey cDNA that encodes this protein and the cognate human cDNA have been isolated (213). This gene is expressed normally in all cells examined, including XP-E cells in which the binding of the protein to UV-irradiated DNA is known to be defective. The human cDNA maps to chromosome 11. Computer-assisted searches of the databases revealed homology of the protein encoded by the UV-DDB cDNA with as yet uncharacterized genes from *Dictyostelium discoideum* and *Oryza sativa* (213).

The so-called XPE protein has been extensively purified (88, 99, 242). The denatured protein has an apparent molecular mass of 125 kDa based on analysis by SDS-polyacrylamide gel electrophoresis (PAGE) (88). However, gel filtration and glycerol gradient sedimentation suggest a molecular mass of 154 to 166 kDa because of the presence of a copurifying 41-kDa subunit (99). Binding to UV-irradiated DNA occurs at a level significantly greater than to native DNA (166). The protein has been shown to bind to *trans-syn*-thymine-thymine and Dewar thymine-thymine dimers and to (6-4) photoproducts but not to *cis-syn*-thymine-thymine dimers (166).

Microinjection of the purified protein into two unrelated XP-E cell lines results in the restoration of repair synthesis to wild-type levels (100). This correction is specific for XP-E cells; injection of the protein into cells from other XP complementation groups had no effect. Hence, these results provide persuasive evidence that the deficiency in nucleotide excision repair in at least two XP-E cells is due to a defect in XPE protein (100).

Nucleotide Excision Repair in Cell-Free Systems

Over the years, numerous laboratories have attempted to define a mammalian cell-free system that could specifically support nucleotide excision repair by using either XP or nucleotide-excision repair-defective rodent cell extracts as negative controls (97). Such a system would have obvious utility for exploring the biochemistry of nucleotide excision repair in mammalian cells. First and foremost, perhaps, is the potential of such a system as an assay to facilitate the purification of relevant proteins from crude extracts. Under any circumstances, such a biological assay is essential to evaluate the biological integrity of proteins purified by the use of nonfunctional assays. Finally, as indicated in chapter 6, the availability of a cell-free system provides an indispensable means of sorting out structural proteins (which are directly involved in the biochemistry of nucleotide excision repair) from regulatory proteins (which may be implicated in this process indirectly through their effect on the transcription and/or translation of *bone fide* nucleotide excision repair genes).

The first successful development of such a cell-free system by Richard Wood and his colleagues in the late 1980s (280) was a major contribution to this field. A cell-free system that measures the excision of damage-containing oligonucleotide fragments (84) as well as repair synthesis (167, 188) was independently developed by other investigators. These two cell-free systems have provided convincing evidence for the conservation of the essential biochemical parameters of nucleotide excision repair defined in *E. coli* (see chapter 5). Indeed, defective repair synthesis of DNA in extracts of XP cells can be complemented by the addition of purified *E. coli* UvrABC endonuclease (68), indicating that the XP cells in question are indeed specifically defective in nucleotide excision repair.

The cell-free system typically used monitors repair synthesis of plasmid DNA following damage-specific incision and excision of oligonucleotides from DNA (68,

70, 71, 84, 167, 188, 277, 279, 280). Plasmids damaged with one of a variety of agents can be used. Extracts are prepared essentially by following a protocol originally established to monitor mammalian cell transcription in vitro (131). As discussed in chapter 6, during incubation of the plasmid DNA with extracts in the presence of deoxyribonucleoside triphosphates (one of which is radiolabeled), Mg^{2+}, ATP, and an ATP-regenerating system, repair synthesis patches are generated in the plasmid DNA (see Fig. 6–6). To specifically distinguish this DNA synthesis mode from nonspecific background synthesis, a second undamaged plasmid of different size is included in the incubations as an internal control (see Fig. 6–6). Following incubation, plasmid DNA is recovered and linearized by cutting with a suitable restriction enzyme(s). This converts all possible plasmid topoisomers to two single unit lengths distinguished only by the different sizes of the two plasmids. These can be readily resolved by gel electrophoresis. The intensity of bands detected by autoradiography of the gels reflects the extent of incorporation of radiolabel into each plasmid during DNA synthesis (see Fig. 6–6).

Extracts of repair-proficient cells catalyze repair synthesis in plasmids damaged by UV radiation, *cis*- or *trans*-diaminedichloroplatinum(II), psoralens (Fig. 8–15) or *N*-acetoxy-2-acetyl-2-aminofluorene. In UV-irradiated DNA both photoreactivable lesions (pyrimidine dimers) and (6-4) photoproducts are recognized as substrates. Interestingly, when plasmids carrying one or another of these lesions were examined, pyrimidine dimers were found to generate a much weaker repair synthesis signal than (6-4) photoproducts did (210), suggesting that the human nucleotide excision endonuclease recognizes the latter lesions more efficiently. This caveat was alluded to in chapter 7 when considering apparent discrepancies in the kinetics of various parameters which measure nucleotide excision repair in mammalian cells in vivo. The observation that different photoproducts are recognized with different efficiencies by the human nucleotide excision repair system

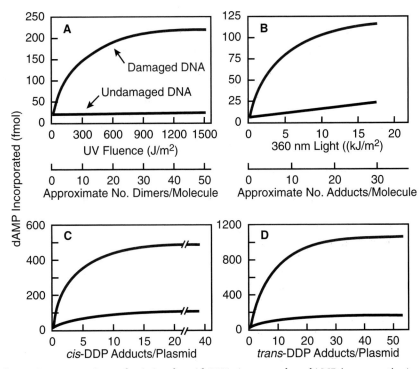

Figure 8–15 Repair synthesis in plasmid DNA (measured as dAMP incorporation) containing various types of base damage (measured by the amount of radiolabeled dAMP incorporated) is supported by human cell extracts. The plasmid substrate DNA was treated with UV radiation (A), 8-methoxypsoralen plus 360 nm UV light (B), *cis*-diamminedichloroplatinum(II) (*cis*-DDP) (C), or *trans*-diamminedichloroplatinum(II) (*trans*-DDP) (D). (*Adapted from Hoeijmakers et al. [79] with permission.*)

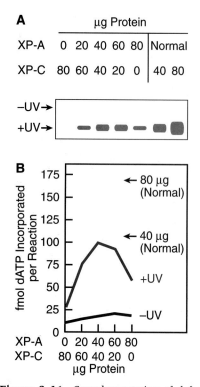

Figure 8–16 Complementation of defective nucleotide excision repair (repair synthesis) in vitro by mixing extracts from XP-A and XP-C cells. (A) Diagrammatic representation of an autoradiogram showing that the defective (or deficient) repair synthesis in extracts from either cell type alone is corrected by mixing (see Fig. 6–6 for details of the autoradiographic assay). (B) Optimal correction is obtained with 40 μg of protein from each mutant extract. This yields ~60% of the level of repair synthesis observed with an equivalent amount (80 μg) of extract from a normal human cell. (*Adapted from Wood et al. [280] with permission.*)

in vitro also provides a simple and compelling explanation for the finding in several studies that mutant and revertant (from mutant) cell lines sometimes retain or regain the ability to repair one type of major photoproduct and not the other. Hence, it is unnecessary to invoke the existence of alternative or multiple different nucleotide excision repair pathways to deal with different types of base damage.

Minor photoproducts such as thymine glycols are also recognized as substrates in the repair synthesis assay. However, because of their relatively low yield and the smaller size of the repair synthesis patches associated with base excision repair of these photoproducts in mammalian cells (see chapter 4), these products usually constitute an acceptably low background at doses of UV radiation below 450 J/m² (79).

Extracts of a variety of XP cells from different genetic complementation groups show reduced and in many cases totally absent repair synthesis or oligonucleotide excision. Similar results have been obtained with XP-V cells (280), even though in vivo assays of nucleotide excision repair in such cells indicate no defect in nucleotide excision repair (see above). However, the results with extracts of XP-V cells are variable, suggesting that the putative XPV protein is unusually labile rather than absolutely defective in vitro (69). Conceivably, defective XPV protein may render some other nucleotide excision repair protein particularly labile, thus yielding a phenotype of defective repair synthesis in vitro.

Mixing extracts from different XP complementation groups results in restoration of functional nucleotide excision repair in both cell-free systems used (Fig. 8–16). Hence, these systems do indeed have the potential for use as a specific assay for proteins which are defective in different XP cells. As discussed above, XPA has been purified by such an assay. Additionally, a protein that specifically corrects defective repair synthesis in human XP-G cells and CHO cells from RCG5 has been considerably enriched, hence providing direct support for the contention that the *ERCC5* and *XPG* gene products are one and the same thing (154). The availability of purified proteins will set the stage for the synthesis of oligonucleotides of defined sequence derived from amino acid sequence information. Such oligonucleotides can be used as hybridization probes for screening human DNA libraries to clone the remaining genes that have been intractable to functional complementation, i.e., *XPE, XPV, ERCC7, ERCC8, ERCC9, ERCC10,* and *ERCC11*. Additionally, the avail-

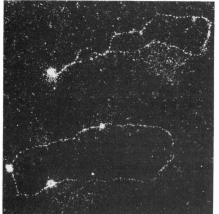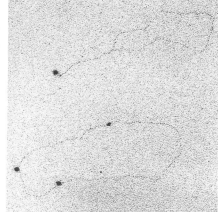

Figure 8–17 Electron micrographs showing repair synthesis patches in plasmid DNA. The plasmid DNA was damaged with UV radiation and then repaired in the standard cell-free system (see text for details), which included biotinylated dUTP. The plasmid was extensively purified and prepared for electron microscopy. Grids were soaked in the presence of streptavidin conjugated to colloidal gold (diameter 10 nm). The plasmids are 1 μm in circumference. The figure on the left is a dark-field view, and that on the right is a bright-field view. (*From Szymkowski et al. [209] with permission. Original electron micrographs provided by R. D. Wood.*)

Human cells

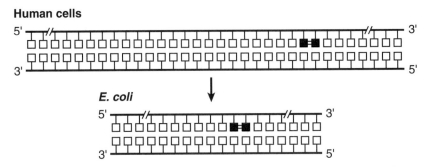

Figure 8–18 Extracts of human cells support the excision of oligonucleotide fragments from damage-containing plasmid DNA. The sizes of the recovered oligonucleotides predict that the mechanism of nucleotide excision repair in human cells involves the generation of nicks at the 22nd to 24th phosphodiester bonds 5' to sites of damage, and the 5th phosphodiester bond 3' to such sites. The exact locations of these nicks with respect to the site of damage are different from those generated by the *E. coli* UvrABC endonuclease (see chapter 5), which is shown for comparison.

ability of specific antibodies raised against purified XP proteins affords opportunities for screening expression libraries for the relevant genes.

The repair synthesis patches generated in the mammalian cell-free system can be visualized in individual plasmid molecules by electron microscopy, based on the binding of streptavadin-colloidal gold to biotinylated deoxyuridine incorporated from biotinylated dUTP during repair synthesis (Fig. 8–17) (209). This elegant technique has yielded estimates of ~30 nucleotides as the average repair patch size in UV-irradiated DNA, a value that is in excellent agreement with those independently derived by other repair synthesis measurements (187) and with independent measurements of the size of the oligonucleotide fragment excised (27 to 32 nucleotides) during nucleotide excision repair of either UV-irradiated DNA (84) or psoralen-containing DNA (206). Additionally it has been shown with the latter system that bimodal incisions are generated in both substrates just mentioned, such that the 5' incision is located at the 22nd to 24th phosphodiester from a pyrimidine dimer and the 3' incision is generated at the 5th phosphodiester bond from the lesion (84, 206) (Fig. 8–18). Similar results were obtained with extracts from *Xenopus laevis* oocytes, suggesting that an incision pattern which generates an oligonucleotide fragment of ca. 27 to 30 nucleotides is common to all vertebrates, if not all eukaryotes (206).

Repair synthesis in repair-proficient human cell extracts (but not in extracts from various XP genetic complementation groups) has also been observed in UV-irradiated SV40 minichromosomes. However, the extent of the repair synthesis signal is reduced compared with that observed with naked plasmid DNA (132, 203). Consistent with the apparent rate limitation imposed by the presence of chromosomal proteins, the addition of histones to human cell extracts containing naked plasmid substrate DNA drastically reduces the amount of repair synthesis (see chapter 7) (261).

Requirements for Nucleotide Excision Repair in a Cell-Free System

The availability of this cell-free system has facilitated an analysis of the requirement for particular proteins during nucleotide excision repair in mammalian cells. Repair synthesis of UV-irradiated plasmid DNA is inhibited in the presence of monoclonal antibodies against human single-strand-binding protein (RPA) (31). This inhibition can be reversed by addition of excess purified RPA. Additionally, the addition of RPA to extracts of normal human cells stimulates the levels of repair synthesis but has no effect on defective repair synthesis in extracts of XP cells (31). The precise role(s) of RPA in nucleotide excision repair is not clear. When damaged DNA is incised with the *E. coli* UvrABC endonuclease, the completion of repair that is mediated by factors in the human cell-free system is insensitive to inhibition by antibodies to RPA. This suggests that RPA also functions in an early step of repair. Conceivably it aids in the recognition of DNA damage, perhaps by interacting with short single-stranded regions that are generated by bulky adducts, thereby placing damaged DNA in a conformation required for incision and/or excision (31).

In addition to RPA, there is an apparent requirement for proliferating-cell nuclear antigen (PCNA) in human cell-free systems that support nucleotide excision repair (Fig. 8–19) (141, 148, 187). This requirement is at the level of the repair synthesis event itself and suggests that repair synthesis in human extracts is mediated by DNA polymerases δ and/or ε, since PCNA is known to stimulate these two polymerases. The ability to examine nucleotide excision repair in fractions that are depleted of PCNA provides the potential for dissecting this multistep process in vitro. The direct measurement of oligonucleotide excision in vitro has led to the suggestion that PCNA may additionally play a role in the actual excision event by increasing the turnover rate of the nucleotide excision repair complex (148).

In addition to its requirement in vitro, there are indications that PCNA plays a role in nucleotide excision repair in living cells. For example anti-PCNA antibodies reveal the presence of the protein exclusively in the nuclei of cells in the S phase. However, after exposure to UV radiation, cells in culture show evidence of PCNA in other phases of the cycle (19). Similarly, quiescent cells are positive for PCNA after UV irradiation (235). However, quiescent XP-A cells are not, consistent with their excision-defective phenotype (142). Finally, it has been shown that repair synthesis in cells is inhibited by the microinjection of PCNA antibodies (215).

It is not known exactly how PCNA participates in repair synthesis during nucleotide excision repair. It has been suggested (281) that PCNA might be a component of the DNA polymerase holoenzyme and that specific protein-protein interactions between PCNA, the particular DNA polymerase involved in repair synthesis (see below) and other accessory proteins such as RPA and RF-C proteins may facilitate protection of the gapped region generated by excision of a ~30-mer oligonucleotide and may also facilitate controlled gap filling during repair synthesis.

Permeabilized Cell Systems for Identifying Factors Required for Nucleotide Excision Repair

Several techniques have been developed to permeabilize mammalian cells in culture, including the use of detergents, hypotonic lysis, and mechanical disruption (97, 101). Many of these techniques were originally designed to investigate replicative DNA synthesis and were subsequently adapted for the study of nucleotide excision repair. These systems have been useful for elucidating the requirements for small molecules such as ATP, divalent cations, and deoxyribonucleoside triphosphates during repair and for larger macromolecules in cells depleted of soluble components (97, 101).

DNA polymerase ε (see below) complements defective repair synthesis of DNA in permeabilized XP-A cells when added in combination with fractionated extracts from repair-proficient cells (150, 151). Addition of purified DNA polymerase ε has no effect on XP-A cells (101, 149). This result provided an assay for identifying and purifying proteins defective in XP-A cells. Consistent with the expectation that XP-A cells are defective in damage-specific incision of DNA, addition of both DNA polymerase ε and purified T4 UV endonuclease (which circumvents the incision defect) also complements defective repair synthesis following UV radiation damage in XP-A cells (150).

Extracts from human HeLa cells correct repair synthesis in XP-A, XP-C, and XP-E cells, and extracts from normal human fibroblasts or from a murine cell line correct repair synthesis in XP-C and XP-D cells (101, 149). None of the correcting factors complements repair synthesis in more than a single XP complementation group (101).

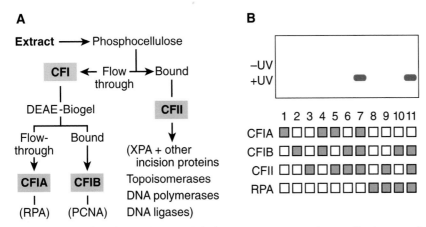

Figure 8–19 Nucleotide excision repair in human extracts requires replication protein A (RPA) and PCNA. (A) Fractionation of human cell extracts as shown yields fractions designated CFIA (containing RPA), CFIB (containing PCNA), and CFII (containing various XP proteins and other proteins required for nucleotide excision repair). (B) Repair synthesis is observed by autoradiography only when all three fractions (CFIA, CFIB, and CFII) are mixed or when fraction CFIA is replaced with purified RPA. (*Adapted from Shivji et al. [187] with permission.*)

Mammalian DNA Polymerases

The preceding discussions indicate that although considerable progress has been made toward unraveling the biochemical complexity of nucleotide excision repair in mammalian cells, there is still quite a way to go, and at present the precise biochemistry of damage-specific recognition and incision of DNA in higher organisms is unknown. Repair synthesis and DNA ligation are obviously fundamental components of nucleotide excision repair in mammalian cells. There are also indications that enzymes such as topoisomerases may be involved in this process. The following sections examine the properties of the DNA polymerases, DNA ligases, and topoisomerases that are believed to play a role in nucleotide excision repair.

The topic of mammalian cell DNA polymerases has been comprehensively addressed in several excellent reviews (12, 59, 105, 117, 123, 126, 195, 256, 273). The interested reader is referred to these accounts for detailed information. It is also not the intention of this text to review the mechanism of DNA replication in eukaryotes. Considerable progress has been achieved in this particular area of biochemistry, and once again the interested reader to referred to specialized reviews and to the authoritative text by Kornberg and Baker (105).

Mammalian cells contain three biochemically distinct nuclear DNA polymerases designated α, β, and δ (Table 8–5). DNA polymerase also exists in a form called DNA polymerase ε. A fifth polymerase designated DNA polymerase γ is localized in mitochondria and is believed to be involved in mitochondrial DNA replication (Table 8–5).

DNA POLYMERASE α

DNA polymerase α is involved in semiconservative DNA synthesis in mammalian cells. It accounts for more than 85% of the total DNA polymerase activity in growing cells but for only about 5% of the activity in quiescent cells. Thus, the activity of this enzyme undergoes a marked increase in actively proliferating cells (105). It is now firmly established by the use of monoclonal antibodies that the enzyme from human cells has a nuclear location (7, 220), although extraction procedures during enzyme purification frequently result in leakage into the cytoplasm (48).

DNA polymerase α is a complex enzyme composed of multiple subunits. Despite extensive studies for over 25 years, there is still uncertainty concerning its precise subunit composition. It is present in cells (even actively proliferating cells)

Table 8–5 Eukaryotic DNA polymerases

Property	Value for DNA polymerase				
	α	δ	ε	β	γ
Previous mammalian designation	α	δ, δ1	δ, δ2, δII, δ*	β	γ
Yeast designation	Pol I	Pol III	Pol II		Mitochondrial
Yeast gene	POL1 (CDC17)	POL3 (CDC2)	POL2		MIP1
Mass (kDa)					
Native	>250	170	256	36–38	160–300
Catalytic core	165–180	125	215	36–38	125
Other subunits	70, 50, 60	48	55	None	35, 47
Location	Nucleus	Nucleus	Nucleus	Nucleus	Mitochondria
Associated functions					
3′ → 5′ exonuclease	No[a]	Yes	Yes	No	Yes
Primase	Yes	No	No	No	No
Properties					
Responses to auxiliary factors	Yes	No	No	No	No
Response to PCNA	No	Yes	No	No	No
Preferred template	Gap	Poly(dA)·oligo(dT)	Poly(dA)·oligo(dT)	Gap	Poly(rA)·oligo(dT)
Divalent cation	Mg^{2+}	Mg^{2+}	Mg^{2+}	Mg^{2+}/Mn^{2+}	Mg^{2+}/Mn^{2+}
Processivity	Low	High[b]	High	Low	High
Fidelity	High	High	High	Low	High
Inhibitors					
NaCl (0.15 M)	Strong	Strong	Strong	None	None
Aphidicolin	Strong	Strong	Strong	None	None
N-Ethylmaleimide	Strong	Strong	Strong	None	Strong
Butylphenyl dGTP (I_{50})	1 μM	100 μM	100 μM	100 μM	None
Dideoxy-NTPs	None	Weak	Weak	Strong	Strong
Replication	Yes	Yes	Yes	No	Yes

[a] Cryptic in *Drosophila* cells.
[b] In the presence of PCNA.
Source: Adapted from Kornberg and Baker (105) with permission.

at relatively low concentration, and, despite heroic efforts in a number of laboratories, it has been extremely difficult to purify without the complications of proteolysis and possible loss of subunits (117). The development of rapid purification techniques, particularly the use of affinity chromatography, has greatly reduced proteolysis of polypeptides and facilitated the isolation of what is believed to be largely intact enzyme from a variety of sources.

DNA polymerase α from a variety of mammalian cells consists of four subunits: a large subunit of ~180 kDa, a subunit of ~70 kDa, and two small subunits of ~60 and ~50 kDa (117). The largest subunit is frequently isolated as a family of polypeptides ranging in size from 140 to 185 kDa. However, immunological studies with a battery of monoclonal antibodies suggest that these are related and probably reflect persistent proteolysis (257). DNA polymerase activity is associated with the 180-kDa subunit, and the 60- and 50-kDa subunits are associated with DNA primase activity, an integral component of DNA polymerase α (63, 117).

The DNA primase is an oligoribonucleotide polymerase which synthesizes short ribonucleotide primers which can be extended by DNA polymerase α (85, 240, 257, 282). The ability to synthesize a primer of defined length is inherent in the primase subunits and is not dependent on their physical association with the polymerase subunit (117). However, in all cases studied, coupling of the two subunits suppresses multimeric primer synthesis and attenuates primer length (117). The mechanism whereby DNA polymerase α-primase complex switches from a primer synthesis mode to DNA chain elongation is not known. The preferred primer-template for DNA polymerase α is duplex DNA with gaps of 20 to 70 nucleotides (258). Nicked DNA is ineffective as a substrate (258). The polymerization mechanism is apparently the same as that for prokaryotic enzymes, and both pyro-phosphorolysis and PP_i exchange have been demonstrated (105).

Mammalian DNA polymerase α does not exhibit any exonuclease activities (51). However, when the four subunits of the *Drosophila* DNA polymerase α complex are dissociated, a cryptic $3' \rightarrow 5'$ proofreading exonuclease activity is revealed (29). The unmasking of this exonuclease appears to depend specifically on dissociation of the 182- and 73-kDa *Drosophila* subunits. Attempts to isolate the ~180-kDa mammalian subunit free of the ~70-kDa subunit have not been successful. Hence, it is not clear whether a cryptic $3' \rightarrow 5'$ exonuclease is a general feature of DNA polymerase α. Interestingly, addition of the ε subunit of *E. coli* DNA polymerase III (which contains the $3' \rightarrow 5'$ proofreading activity) to calf thymus DNA polymerase α results in improved replicational fidelity, even though there is no physical association between the two proteins (161). Hence, it has been suggested that in mammalian cells chain elongation by DNA polymerase α, excision of mismatched nucleotides by a separate proofreading exonuclease, and continued chain elongation could constitute sequential events in vivo (161).

Although purified DNA polymerase α-primase complex is devoid of other catalytic activities, DNA-dependent ATPase, RNase H, $3' \rightarrow 5'$ exonuclease, and other enzymatic activities have been observed in higher-molecular-weight DNA polymerase complexes (155, 190). The physiological significance of these higher-molecular-weight forms is unknown.

The probable existence of accessory factors for the four-subunit form of DNA polymerase α was predicted from the poor efficiency with which the enzyme replicates template-primers with long stretches of single-stranded DNA (117). Factors that interact with the enzyme have been detected both in association with and separate from it. Two factors designated C_1 and C_2, which interact with each other and with DNA polymerase α, have been purified from human cells (108, 164). These factors stimulate the enzyme activity on single-stranded DNA. A mammalian single-stranded-DNA-binding protein has also been identified as an essential component of a multienzyme system that promotes SV40 DNA replication in vitro (117). A factor designated alpha accessory factor has been isolated from mouse cells (64, 65). It contains polypeptides of ~132 and 44 kDa and stimulates DNA polymerase α-primase complex on unprimed poly(dT) and in the self-primed reaction with unprimed single-stranded DNA. It is highly specific for DNA polymerase α-primase. No effect was observed with a variety of other eukaryotic and prokaryotic DNA polymerases.

Human DNA polymerase α is phosphorylated in a cell cycle-dependent manner. The α catalytic polypeptide is phosphorylated throughout the cycle and is hyperphosphorylated during the M phase. There are indications that this phosphorylation is effected by the p34^{cdc2} kinase, a key mitotic regulatory kinase (256).

By using a panel of rodent-human hybrids in which human-specific DNA polymerase α activity could be detected immunologically, the human gene was mapped to the X chromosome at the region p21.3-22.1 (260). A cDNA clone of the polymerase subunit of the human enzyme has been isolated (275). Comparisons of the primary amino acid sequence of the cloned gene with sequences from other eukaryotic and prokaryotic DNA polymerases have revealed multiple regions of similarity (Fig. 8–20). Three of these possibly define the active site required for interaction with deoxynucleotides. Two putative DNA-binding domains have also been identified (275).

Studies on the steady-state levels of DNA polymerase α mRNA show a striking correlation with cell proliferation and transformation (275). Cell cycle studies have shown that during the activation of quiescent (G_0 phase) cells the levels of polymerase α mRNA increase just prior to the peak of DNA synthesis and that this increase correlates with increased enzymatic activity and synthesis of DNA polymerase (254). However, other studies have demonstrated that in human fibroblasts stimulated to reenter the cell cycle the large increase in DNA polymerase α activity is the result predominantly of its increased phosphorylation rather than the synthesis of new enzyme (32).

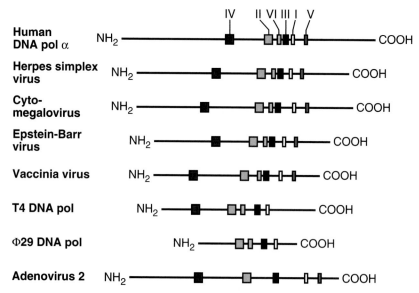

Figure 8–20 Conserved amino acid sequence motifs in human DNA polymerase α and various other DNA polymerases. (*Adapted from Wong et al. [275] with permission.*)

In actively growing cells separated into discrete phases of the cell cycle, levels of transcripts and enzymatic activity are constitutively expressed in all stages of the cycle, with only a slight increase prior to the S phase and a moderate decline in the G_2 phase (254). These findings suggest that regulation of human DNA polymerase α is at the translational or posttranslational level. Furthermore, the regulatory mechanisms which operate during entrance into the mitotic cycle are apparently distinct from those that operate during the meiotic cycle (254).

DNA POLYMERASES δ AND ε

DNA polymerase δ was originally defined as a mammalian DNA polymerase distinguishable from DNA polymerase α by the presence of a $3' \rightarrow 5'$ exonuclease activity (17). Interest in the enzyme was heightened by the finding that it is an essential component of the complex of enzymes involved in SV40 DNA replication in vitro (50, 90, 274). A protein of 36 kDa purified from calf thymus stimulates the activity of homologous DNA polymerase δ and has been shown to be identical to PCNA, which, as discussed above, is apparently required in human cell-free systems to support nucleotide excision repair (11, 214). Both DNA polymerases α and δ are sensitive to inhibition by aphidicolin and by *N*-ethylmaleimide (117), and both enzymes are resistant to dideoxynucleoside triphosphates. In contrast, the nucleotide analog butylphenyl-dGTP strongly inhibits DNA polymerase α but has only a marginal effect on the activity of DNA polymerase δ (117).

DNA polymerase δ has been isolated in two forms distinguished primarily by their response to PCNA. One form has a low degree of processivity on templates containing long stretches of single-stranded DNA, but its activity and processivity are markedly stimulated in the presence of PCNA (117, 214). This form, isolated from calf thymus, consists of two subunits of 125 and 48 kDa (113). The second form of the enzyme is moderately processive in the absence of PCNA and is not significantly affected by its presence. The form isolated from calf thymus has associated DNA primase activity (33). These two forms of DNA polymerase δ are sometimes referred to as δ1 and δ2 or as DNA polymerases δ and ε, respectively (12). At salt concentrations closer to those encountered physiologically, DNA polymerase ε is also PCNA dependent (13, 114). Hence, the PCNA dependence of nucleotide excision repair referred to above is consistent with an involvement of either of these polymerases.

Several studies suggest that DNA polymerases δ and ε are distinct enzyme forms (5, 208). DNA polymerase δ isolated from human HeLa cells is thought to consist of two subunits of 134 and 47 kDa, whereas DNA polymerase ε consists of subunits of 215 and 55 kDa (101). DNA polymerases δ isolated from a variety of other organisms has vastly different sizes and subunit compositions, and there is still uncertainty about the precise structure of this enzyme and its distinction from DNA polymerase ε (117). Fractions containing purified DNA polymerase ε from calf thymus have been reported to contain 5' → 3' exonuclease activity (189). DNA polymerases δ and ε represent the mammalian equivalents of the yeast enzymes Pol II (yPol δ) and Pol III (yPol ε) discussed in chapter 6.

DNA primase activity has not been reproducibly associated with DNA polymerase δ. This observation, together with that of its relatively high processivity on single-stranded DNA, has led to the suggestion that the enzyme functions in replication of the leading DNA strand during semiconservative DNA synthesis, while DNA polymerase α, with its associated DNA primase, functions in lagging-strand synthesis (40). Indeed, during DNA synthesis in vitro from the SV40 replication origin, DNA polymerase α-primase complex was shown to initiate DNA synthesis at the replication origin and to continue as the lagging-strand polymerase. Subsequently, DNA polymerase δ initiates replication on the leading-strand template (241). Despite the presence of DNA polymerase ε in crude cell extracts that support the semiconservative replication of SV40 DNA from a true replication origin, this polymerase apparently does not participate in DNA replication in vitro (253). This is consistent with the postulated role of DNA polymerase ε in repair synthesis in yeast cell extracts (see chapter 6) and human cell extracts (see below).

Human and bovine cDNAs for the catalytic subunit of DNA polymerase δ have been isolated and shown to share 94% amino acid identity. The human cDNA can encode a polypeptide of ~124 kDa (22). The translated amino acid sequence reveals the presence of two putative zinc finger domains in the C-terminal region. The human gene has been mapped to chromosome 19 (22).

DNA POLYMERASE β

DNA polymerase β is also located in the nucleus. The purified enzyme from a variety of sources has a molecular mass of ~40 kDa. The enzyme is less sensitive than DNA polymerase α to SH-blocking agents and is stimulated by salt. Like DNA polymerase α, its primer-template preference is for gapped DNA (259). It, too, has no associated exonuclease activity. Early studies showed that the primary structure of DNA polymerase β from various vertebrates is highly conserved. In mammalian cells in culture, the level of enzymatic activity is low and is independent of the cell cycle and state of growth. Hence, this enzyme is not believed to be essential for semiconservative DNA synthesis but has consistently been identified as a prime candidate for repair synthesis of DNA (273).

The genes for DNA polymerase β from both humans and rats have been cloned (1, 183, 289). In humans, DNA polymerase β is specified by a single-copy gene located on chromosome 8 (136). The amino acid sequence of the translated cloned gene shares no obvious similarity to other DNA polymerases. However, there is a striking similarity to that of the terminal transferase gene (273). Levels of DNA polymerase β mRNA are maintained at constitutive levels during the cell cycle and during various stages of cell growth in culture (290). Down-regulation of the DNA polymerase β gene by transfection of cells with antisense mRNA resulted in a threefold increase in the doubling time of cells (290).

ROLE OF DNA POLYMERASES α, δ, ε, AND β IN DNA REPAIR

The question of the relative roles of DNA polymerases α, δ, ε, and β in repair synthesis of DNA is still controversial despite many years of study (101). A major limitation is the lack of mutants which are totally defective in one or other of these enzymes, necessitating the use of less direct experimental approaches to this issue. Perhaps the most physiological of these are experiments on terminally differentiated cells such as neurons or skeletal muscle cells, which do not carry out semiconservative DNA synthesis and hence presumably contain little or no DNA polymerases α and δ (86, 200, 201, 271). Such cells carry out repair synthesis following UV radiation and other forms of DNA damage, suggesting that an enzyme(s) other than DNA polymerase α or δ is, or at least can be, involved in excision repair.

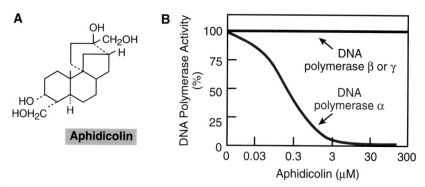

Figure 8–21 The DNA polymerase inhibitor aphidicolin (A) inhibits DNA polymerase α (as well as polymerases δ and ε) activity in vitro. (B) DNA polymerases β and γ are not affected. (*Adapted from Friedberg [55] with permission.*)

However, it is not clear that the residual polymerase α or δ that may be present in these cells is not adequate to carry out this synthesis. In addition, the absence of DNA polymerase α or δ in terminally differentiated cells does not necessarily imply that these enzymes are not involved in repair synthesis of DNA in cells that do possess adequate levels of activity.

Aphidicolin is a tetracyclic diterpinoid, obtained from *Cephalosporium aphidicola*, that selectively inhibits DNA polymerases α, δ, and ε (5, 89, 256) (Fig. 8–21). This selectivity offers the possibility that the inhibitor will be useful for establishing whether these DNA polymerases play an exclusive role in repair synthesis during excision repair in mammalian cells. However, the results of these and related studies have led to equivocal and sometimes apparently contradictory conclusions. For example, UV-irradiated HeLa cells incubated in the presence of aphidicolin carry out repair synthesis of DNA, suggesting that this synthesis is a function of DNA polymerase β activity (159). Furthermore, in mitotic HeLa cells that are not carrying out replicative DNA synthesis, aphidicolin does not reduce the level of repair synthesis detected in UV-irradiated cells, leading to the conclusion that DNA polymerases α and δ are not essential for repair synthesis in mitosis (62). On the other hand, other studies show that this compound inhibits both semiconservative DNA synthesis in unirradiated cells and repair synthesis in UV-irradiated cells, implying that DNA polymerase α or δ is responsible for the latter phenomenon (66).

In an attempt to measure the effect of aphidicolin specifically on DNA repair synthesis, a semi-in vitro system consisting of isolated nuclei from G₁ or G₂ (i.e., non-S-phase) cells has been used (66). Nuclei were incubated in the presence of hydroxyurea and cytosine arabanoside to inhibit residual semiconservative DNA synthesis, and ATP (an essential component for semiconservative DNA synthesis in isolated nuclei) was omitted. Aphidicolin inhibited the incorporation of [³H]dTMP into the nuclei of UV-irradiated cells by more than 90%. In view of the apparent specificity of aphidicolin for DNA polymerases α, δ, and ε, these experiments support the contention that one or all of these enzymes are required for repair synthesis. A possible (though not necessarily exclusive) role for DNA polymerase α or δ in DNA repair is also suggested by the isolation of mutant CHO cells resistant to aphidicolin (20, 125). One such mutant contains aphidicolin-resistant DNA polymerase α activity in vitro and is UV sensitive (Fig. 8–22) (125).

Some experiments have led to the conclusion that DNA polymerases α, δ, and β function selectively in the repair of different kinds of DNA damage (139, 140). For example, aphidicolin and cytosine arabinoside were used as specific inhibitors of DNA polymerase α (or δ), and dideoxythymidine triphosphate was used as a selective inhibitor of DNA polymerase β. On the basis of the presumed selectivity of these inhibitors of DNA synthesis, it was concluded that the latter enzyme is primarily responsible for repair synthesis induced by agents such as bleomycin or neocarzinostatin, whereas polymerase α or (δ) plays a more prominent role in

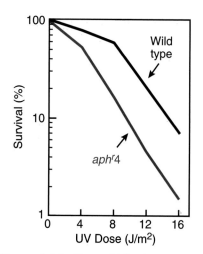

Figure 8–22 The aphidicolin-resistant mutant *aph^r4* is abnormally sensitive to killing by UV radiation, suggesting an involvement of DNA polymerase α in nucleotide excision repair. (*Adapted from Friedberg [55] with permission.*)

repair synthesis associated with the repair of UV radiation and alkylation damage (139, 140).

There are indications that at low doses of DNA damage to human fibroblasts in culture, repair synthesis of DNA is mediated largely by an enzyme other than DNA polymerase α or δ, presumably DNA polymerase β (45). However, with increasing amounts of DNA damage, there is increased repair synthesis and a greater participation in this process by DNA polymerase α and/or δ. At high levels of damage, the fraction of repair synthesis mediated by DNA polymerase α (or δ) reaches a maximum, which varies as a function of the particular DNA-damaging agent (45). With agents such as UV radiation and N-acetoxy-2-acetyl-2-amino-fluorene, which produce bulky base damage, DNA polymerase α (or δ) is estimated to contribute to about 80% of the total repair synthesis.

One of the problems associated with the interpretation of these and other experiments on aphidicolin-treated cells is that the amount of DNA synthesis (and perhaps the amount of DNA polymerase) required for DNA repair is very small relative to that observed during semiconservative replication of DNA. Thus, even though the compound inhibits ~98% of the total DNA synthesis in undamaged cells, residual synthesis may still be catalyzed by DNA polymerase α or δ and may be sufficient for DNA repair. Similarly, one can never be certain that the DNA synthesis that is inhibitable by aphidicolin in damaged cells is exclusively or even wholly repair synthesis, since very low levels of residual replicative synthesis are difficult to distinguish from repair synthesis. It has also been argued that much of the controversy surrounding the effects of cytosine arabinoside and aphidicolin can be reconciled by considering the particular cells or state of the cell used experimentally (192). Thus, little or no inhibition of repair synthesis is observed in exponentially growing cells; however, in contact-inhibited cells, repair synthesis is reduced to 20 to 30% of normal over a wide concentration range of both agents. Inhibition of repair synthesis in actively growing cells is observed in the presence of hydroxyurea, suggesting that in cells actively involved in semiconservative DNA synthesis, the levels of DNA polymerase α and δ and of dCTP are too high for the inhibitors to have any noticeable effect on repair synthesis.

Repair synthesis associated with nucleotide excision repair in vitro is sensitive to inhibition by aphidicolin (30), whereas such synthesis is not sensitive to inhibition to neutralizing antibodies to DNA polymerase α (30). All repair synthesis in vitro is also PCNA dependent. Collectively, these observations argue that at least in vitro, repair synthesis during nucleotide excision repair is catalyzed by either DNA polymerase δ or ε. As discussed in chapter 6, repair synthesis during base excision repair in yeast cell extracts is catalyzed primarily by yPol ε. However, mutations in the genes encoding DNA polymerase δ and α can somehow influence the levels of repair synthesis in this system.

Use of Permeabilized Cell Systems To Study Repair Synthesis

As indicated above, permeabilized systems have been useful for identifying the requirements for various soluble factors in nucleotide excision repair. All such systems in which repair of UV radiation damage occurs (presumably by nucleotide excision repair) have a requirement for, or are stimulated by, ATP. This is consistent with the observation that nucleotide excision repair in vitro is ATP dependent and indeed requires an ATP-regenerating system (280). By measuring the accumulation of strand breaks in DNA in the presence or absence of ATP, it has been shown that at least one ATP-dependent step occurs prior to or during DNA incision (45, 98). ATP also appears to play a role in postincision events. Repair synthesis itself does not have an absolute requirement for this compound in permeabilized human fibroblasts. However, ATP is clearly required for DNA ligation and possibly for nucleosome rearrangements (101).

Permeable cell systems have provided useful insights into the roles of various DNA polymerases in DNA repair, particularly in the years since the discovery of DNA polymerase δ and its identification as being distinct from DNA polymerase α. Butylphenyl-dGTP, a potent inhibitor of DNA polymerase α but not of DNA polymerase δ, inhibits repair synthesis in UV-irradiated cells much less strongly than it inhibits DNA polymerase α itself, suggesting that DNA polymerase δ plays a key role in repair synthesis during nucleotide excision repair (42, 43). Additionally, it has been shown that repair synthesis in UV-irradiated cells is more sensitive to inhibition by ddTTP than is inhibition of DNA polymerase α by this compound. DNA polymerase δ has a sensitivity to ddTTP which is intermediate between those of DNA polymerases α and β, again suggesting a major role of DNA polymerase δ in repair synthesis during nucleotide excision repair (44, 101). A comparison of the effective concentration of inhibitors for repair synthesis in a semi-in vitro system with the K_i values of various DNA polymerases in vitro suggests a best fit with DNA polymerases δ and ε (162).

Many of the studies just alluded to were carried out before it was recognized that what has been referred to as DNA polymerase δ actually represents two distinct forms of DNA polymerase now called δ and ε. As indicated above, the use of permeabilized cells depleted of soluble components has facilitated in vitro complementation by the addition of soluble extracts from other cells. These studies have reinforced the notion that DNA polymerase ε plays a major role in repair synthesis during nucleotide excision repair. A factor from human HeLa cells that restores the ability to carry out repair synthesis after UV irradiation was extensively purified and shown to be associated with a 220-kDa polypeptide with DNA polymerase and 3′ → 5′ exonuclease activities (151). This DNA polymerase is not sensitive to inhibition by monoclonal antibodies against DNA polymerase α or to inhibition by butylphenyl-dGTP. Furthermore, neither purified DNA polymerase α or β can substitute for this DNA polymerase in the permeabilized system (101). This DNA polymerase is not stimulated by PCNA and is highly processive (207), properties characteristic of DNA polymerase ε (101).

In summary, it is likely that different DNA polymerases have different relative affinities for substrates produced by endonucleases and/or exonucleases during excision repair of DNA damage. It is also probable that under different physiological conditions, different DNA polymerases can substitute for each other during nucleotide excision repair and other repair modes in mammalian cells. Thus, mammalian DNA polymerases may respond to signals to replicate DNA with suitable primer-templates, regardless of whether the polymerization is concerned ultimately with maintenance or with propagation of the genome.

Mammalian DNA Ligases

Mammalian cells contain multiple DNA ligases, which are almost certainly encoded by distinct genes (109, 122) (Table 8–6).

DNA LIGASE I

DNA ligase I from calf thymus is the most extensively studied mammalian DNA ligase. It has been purified to homogeneity and has an apparent molecular mass of ~130 kDa (110, 122, 221, 233). This size is considerably larger than that predicted from the size of the ORF identified in the cloned DNA ligase I cDNA (~102,000 kDa) (see below). This size discrepancy has been attributed to multiple factors. The protein has a very high proline content, which is believed to lead to anomalous migration in denaturing gels. The polypeptide is also posttranslationally modified. The literature on the size of DNA ligase I has also been historically plagued by several other factors that are now quite well understood (122). The enzyme is a monomer with a markedly asymmetric shape (the frictional ratio is 1.9), which causes anomalous estimates of molecular weight in gel filtration experiments (122). In addition, DNA ligase is highly sensitive to proteolysis. The calf thymus enzyme is frequently degraded by an endogenous protease which removes the N-terminal 216 amino acids, leaving a catalytically active C-terminal fragment of 703 amino acids. Several early studies probably focused on this C-terminal fragment. Reports in the literature on a form of DNA ligase I which is considerably larger than the native enzyme probably stem from the fact that several antisera against purified DNA ligase I cross-react with a 200-kDa protein in crude extracts. This led to the suggestion that DNA ligase I might be synthesized as a protein with a molecular mass of ~200 kDa and then proteolytically processed to a smaller form. Isolation and sequencing of the DNA ligase I cDNA have demonstrated un-

Table 8–6   Mammalian DNA ligases

Property	Value in:		
	Ligase I	Ligase II	Ligase III
Molecular mass estimated by SDS-PAGE	125 kDa	72 kDa	100 kDa (associated with 46-kDa polypeptide)
cDNA sequence	102 kDa		
Chromosomal localization	19q13.2-13.3		
Ligation			
Oligo(dT)·poly(dA)	Yes	Yes	Yes
Oligo(dT)·poly(rA)	No	Yes	Yes
Oligo(rA)·poly(dT)	Yes	No	Yes
K_m for ATP	Low	High	Low
Adsorption to hydroxylapatite	Weak	Strong	Weak
Recognition by DNA ligase I-specific antisera	Yes	No	No
Subcellular localization	Nucleus	Nucleus	Nucleus/cytoplasm
Induction upon cell proliferation	Yes	No	No
Proportion of DNA ligase activity in calf thymus extract	~85%	5–10%	5–10%

Source: Adapted from Lindahl and Barnes (122) with permission.

equivocally that this is not the case, since an in-frame termination codon is present 99 bp upstream of the ORF (122).

Like most known DNA ligases, DNA ligase I of mammalian cells participates in the formation of a covalent enzyme-adenylate complex, during which ATP is cleaved to AMP and PP$_i$. In this reaction, the adenylyl residue is linked via a phosphoramidate bond to the ε-amino group of a specific lysine residue at the active site of the protein (234) (see Fig. 4–32). Several enzymes that interact with nucleotide substrates, including DNA ligases, contain an unusually reactive lysine residue in their active site (Fig. 8–23). This residue can form a Schiff base with pyridoxal phosphate. Hence, mammalian DNA ligase I is inhibited by this compound in vitro (122). The ligase-adenylate complex formed in the presence of radioactive ATP can be detected as a radiolabeled species by electrophoresis and autoradiography (Fig. 8–24). This provides a convenient and sensitive assay of DNA ligases. The activated AMP residue of the DNA ligase-adenylate intermediate is then transferred to the 5'-phosphate terminus of a nick in duplex DNA, to generate a covalent DNA-AMP intermediate with a 5'-5' phosphoanydride bond (see Fig. 4–32). Finally, unadenylated DNA ligase I catalyzes displacement of the AMP residue through attack on the adenylylated site by the adjacent 3'-OH group, leading to DNA ligation (see Fig. 4–32) (122).

The putative active site for enzyme-adenylate formation of mammalian DNA ligase I has been identified by isolating and sequencing a radiolabeled tryptic peptide-AMP complex from the bovine enzyme and identifying the same peptide in the predicted human cDNA that encodes this enzyme (233). This region is highly conserved in multiple other DNA and RNA ligases (Fig. 8–23). The distance between this putative adenylylation site and the C termini of all of these enzymes is also highly invariant, except for the NAD-dependent *E. coli* enzyme.

DNA ligase I effectively joins nicks in DNA and joins staggered ends produced by restriction enzymes. The enzyme is also able to catalyze the joining of blunt ends in duplex DNA (3). It can join oligo(dT) molecules that are hydrogen bonded to poly(dA) and oligo(rA) molecules base paired with poly(dT). However, it cannot join oligo(dT) molecules hydrogen bonded to poly(rA), a property that has been very useful in distinguishing this enzyme from other mammalian DNA ligases.

DNA ligase I is a phosphoprotein. Dephosphorylation with various phosphatases leads to a significant reduction in enzyme activity, suggesting that phos-

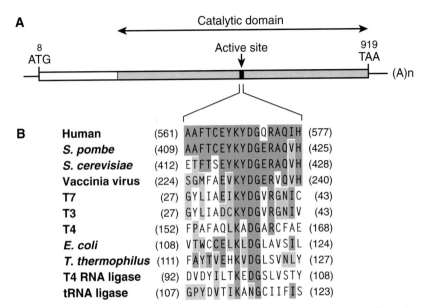

Figure 8–23 Schematic representation of the human DNA ligase I ORF. The active site comprises a 17-amino-acid peptide motif that is conserved in multiple DNA (and RNA) ligases. The invariant Lys (K) residue is believed to participate in the adenylation reaction. (*Adapted from Lindahl and Barnes [122] with permission.*)

phorylation of the native protein is biologically important. Indeed, the formation and accumulation of the enzyme-AMP complex occur only with the phosphorylated form of the protein (122). Expression of full-length DNA ligase I in *E. coli* yielded largely inactive protein. However, subsequent incubation of this protein with casein kinase II resulted in phosphorylation of the N-terminal region and activation of ligase activity (163). Immunofluorescence studies have localized DNA ligase I to the nucleus. The enzyme has a half-life of ~7 h and is present at considerably higher levels in proliferating than in nongrowing cells.

A cDNA for human DNA ligase I has been isolated by functional complementation of a yeast mutant defective in DNA ligase and by screening of human cDNA libraries with degenerate oligonucleotides predicted from peptide sequences derived from cleavage of purified DNA ligase I. Southern hybridization with the cDNA as a probe shows the presence of a single genomic gene estimated to be ~50 kb (122). A single mRNA of 3.2 kb is transcribed from this gene. The human gene maps to chromosome 19, in the region 19q13.2-13.3. As discussed above, this

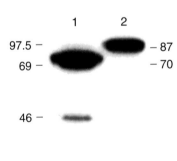

Figure 8–24 DNA ligase activity can be assayed by the formation of an enzyme-adenylate complex in which radiolabeled AMP is transferred to the protein from ATP. The figure shows an autoradiogram of an SDS-polyacrylamide gel after adenylylation and electrophoresis of 87-kDa yeast DNA ligase I (Cdc9 protein) (lane 2) and a slightly smaller (70-kDa) proteolytic fragment (lane 1). The positions of molecular weight markers (in thousands) are shown on the left. (*Adapted from Tomkinson et al. [233a] with permission.*)

region contains a DNA repair gene cluster that includes the *ERCC1* and *XPD* genes, as well as another repair gene, *XRCC1*, involved in the repair of ionizing radiation damage (see chapter 12). These genes have been ordered in the centromere-to-telomere direction as *XRCC1-XPD-ERCC1-LIG1*.

As already indicated, the C-terminal region of mammalian DNA ligase I is believed to contain the catalytic domain. This region of the predicted amino acid sequence of the *LIG1* gene is homologous with the DNA ligases from *S. pombe* and *S. cerevisiae* (Fig. 8–25). The role of the nonconserved N-terminal region is unclear. It contains sequences which may function for nuclear localization and for interaction with chromatin or the nuclear scaffold. It has been suggested that this region of the polypeptide might also be required for protein-protein interactions during DNA replication (122).

A defect in DNA ligase I has been implicated in a cell line designated 46BR from a patient with an as yet undefined clinical syndrome. Additionally, partially purified DNA ligase I from cells isolated from a patient with Bloom's syndrome exhibits reduced activity. These topics are discussed in greater detail in chapter 14.

A protein inhibitor of DNA ligase I has been isolated from human cells (284, 285). The protein is between 55 and 75 kDa in size and forms a reversible complex with DNA ligase I, but it has no effect on DNA ligase II. The inhibitor appears to block the transfer of the AMP moiety from the ligase-AMP intermediate to DNA. The physiological significance of this inhibitor is not known.

DNA LIGASE II

A second form of DNA ligase designated DNA ligase II has been purified from calf thymus extracts as a protein of ~70 kDa (122). Several features suggest that this enzyme is distinct from DNA ligase I. First, the two enzymes show no evidence of serological cross-reactivity. Second, the second enzyme does not catalyze the formation of phosphodiester bonds with an oligo(rA)·poly(dT) substrate (122) (Table 8–6). Third, the active sites of DNA ligases I and II can be distinguished by peptide mapping (170). Finally, several peptides derived from bovine DNA ligase II exhibit weak homology with the catalytic domain of DNA ligase I, suggesting that these two enzymes are encoded by distinct genes which evolved from a common ancestral gene (232). Consistent with this speculation, DNA ligase II, which apparently functions by the same basic mechanism as other DNA ligases (122), has been found to contain the evolutionarily conserved C-terminal peptide motif (232).

DNA LIGASE III

Yet a third protein with DNA ligase activity and a molecular mass of ~100 kDa has been designated DNA ligase III (122). Unlike DNA ligase II, DNA ligase III

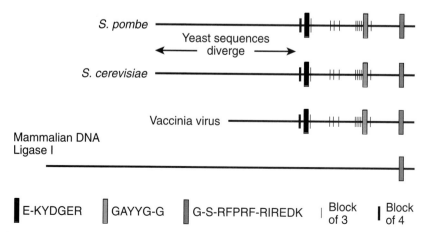

Figure 8–25 Certain amino acid sequence motifs in the C-terminal half of the mammalian DNA ligase I polypeptide are highly conserved. (*Adapted from Lasko et al. [109] with permission.*)

shares with DNA ligase I the property of utilizing oligo(rA)·poly(dT) as a substrate for joining (Table 8–6). Although not recognized by a polyclonal antiserum against DNA ligase I, DNA ligase III is also recognized by an antiserum raised against the conserved C-terminal peptide sequence present in all eukaryotic DNA ligases examined to date (232). A study of DNA ligase active sites by proteolytic mapping has led to the conclusion that DNA ligases II and III are related to each other and are distinct from DNA ligase I (231). A comparison of peptide sequences from DNA ligases II and III confirmed that these polypeptides are related but are probably encoded by distinct genes (232).

During the purification of DNA ligase III from calf thymus extracts, a 46-kDa polypeptide has been observed to copurify with the 100-kDa protein that participates in the formation of an enzyme-adenylate complex (122). DNA ligase III has in fact been identified as a component of several multiprotein complexes. A high-molecular-weight recombination complex designated RC1, that contains DNA ligase III and DNA polymerase ε has been partially purified from calf thymus extracts (92). DNA ligase has also been detected in association with a protein encoded by the *XRCC1* gene (18), which is believed to be involved in the repair of DNA strand breaks in mammalian cells (see chapter 12). In cells that are mutated in *XRCC1*, the level of DNA ligase III activity is reduced; it is restored following the transfection of the mutant cells with a cosmid carrying the wild-type *XRCC1* gene (18). The observations are consistent with an increased lability of DNA ligase III in the absence of XRCC1 protein. This association of DNA ligase III with XRCC1 protein suggests that the ligase is also involved in the repair of DNA strand breaks. It has been speculated that certain DNA repair proteins involved in such DNA repair have been recruited to function in immunoglobulin gene rearrangements. This may account for the presence of DNA ligase III in a calf thymus complex (92).

Mammalian DNA Topoisomerases and DNA Repair

Type I and type II DNA topoisomerases are present in eukaryotic cells, including mammalian cells (4, 26, 83, 105, 138). Type I topoisomerases show distinct differences from those present in *E. coli* and *M. luteus*, however (105). The eukaryotic enzymes do not have an absolute requirement for Mg^{2+}. In addition, the bacterial enzymes act only on negative superhelical turns in DNA and lose catalytic efficiency as the degree of superhelicity decreases, whereas the eukaryotic enzymes can work on DNA with either negative or positive superhelicity (105). Finally, whereas the prokaryotic nicking and closing enzymes form transient nicks by covalent linkage at the 5′ phosphoryl group, the eukaryotic enzymes have been found linked to the 3′ phosphoryl group.

A role for topoisomerase II in excision repair of DNA in eukaryotic cells is suggested by observations on the sensitivity of repair to inhibitors of this enzyme. Thus, the accumulation of DNA strand breaks in UV-irradiated HeLa cells undergoing excision repair is reduced by treatment of cells with novobiocin, suggesting that topoisomerase II may be involved in incision or preincision steps of excision repair (26). In addition, novobiocin results in a time-dependent reversible inhibition of excision of thymine-containing pyrimidine dimers and of repair synthesis of DNA (134). The specificity of these effects by novobiocin is questionable, however (193). This antibiotic has also been shown to be an effective inhibitor of partially purified DNA polymerase α from monkey cells (47) and rat brain (205) and of DNA polymerases α and δ from *S. cerevisiae* (147). There are also indications that the inhibition of semiconservative DNA synthesis observed in novobiocin-treated cells may not be related to the relaxing of supercoils in DNA, and in *E. coli* novobiocin appears to inhibit a number of other ATPase-dependent reactions (23). At concentrations similar to those required to inhibit DNA gyrase, it has been shown to inhibit nick translation by *E. coli* DNA polymerase I and DNA ligation by phage T4 DNA ligase (23). All of these observations suggest that the inhibitory effects of novobiocin in complex cellular systems may not be useful indicators of the specific involvement of topoisomerase II in DNA repair.

The antitumor agent etoposide has been found to be a powerful direct inhibitor of topoisomerase II (38). In intact human cells, etoposide does not inhibit nucleotide excision repair (39). Similar results have been obtained with a second more specific inhibitor of topoisomerase II, 4'-(9-acridylamino)methanesulfon-*m*-aniside (*m*-AMSA) (272). The use of the cell-free nucleotide excision repair system which appears to faithfully model the process in living cells and which has been alluded to frequently in this chapter, has shown no evidence of inhibitory effects in the presence of well-characterized inhibitors of topoisomerases (58). On the other hand, studies with permeabilized systems have reached the opposite conclusion. The accumulation of strand breaks was prevented in the DNA of UV-irradiated cells treated with novobiocin, nalidixic acid, or etoposide (46). A detailed analysis of these and related studies suggests that while a topoisomerase II-dependent step may be involved in cellular responses to UV radiation in mammalian cells, it is required only after very high levels of DNA damage (38). At present, there is no good evidence that topoisomerase II is involved in the normal nucleotide excision repair pathway which operates within the physiological UV radiation dose range and which is presumably defective in XP cells (38). The role of topoisomerase I in nucleotide excision repair remains similarly unproved.

Gene-specific repair is not affected by inhibitors of either topoisomerase I or II following the UV radiation of CHO cells. However, when both enzymes were simultaneously inhibited, inhibition of both gene-specific and strand-specific repair was observed in transcriptionally active genes but not in bulk DNA (199). These observations suggest that at least one topoisomerase must be present for normal nucleotide excision repair in transcriptionally active mammalian genes. In support of this conclusion, inhibition of just topoisomerase II in a cell line that is deficient in topoisomerase I results in defective gene-specific repair (199).

In summary, while it is obvious that many elements of DNA metabolism that encompass the cutting, resynthesis, and ligation of DNA probably require the action of one or more DNA topoisomerases, there is no evidence that these enzymes participate directly in nucleotide excision repair (75).

Nucleotide Excision Repair of Mitochondrial DNA

Ever since it was discovered that the mitochondria of eukaryotic cells contain a genome that is independent of the nuclear genome, there has been conjecture about whether all of the known forms of DNA repair that operate on the nuclear genome also operate in mitochondrial DNA. There is no unequivocal evidence for nucleotide excision repair in mitochondrial DNA of various forms of bulky base damage that have thus far been investigated, including pyrimidine dimers and cisplatin intrastrand and interstrand cross-links (112). One study suggests that damage introduced by the UV radiation-mimetic chemical 4-NQO may be repaired in mitochondrial DNA (194). It is likely that the redundancy of mtDNA in mammalian cells precludes a requirement for such repair. Alternatively, recombination modes of DNA damage tolerance may be the preferred mode of handling base damage in the mitochondrial genome.

Conclusions

It is evident that nucleotide excision repair is a ubiquitous process that is required for the excision of multiple types of base damage in mammalian cells. This process is genetically and biochemically complex. However, rapid strides have been made toward elucidating the mechanism of this important cellular defense against DNA damage and hence against the phenotypic consequences of mutations that can arise as a consequence of unexcised base damage. The cancer-prone hereditary disease XP represents a prototypic example of such consequences, and we will return to a consideration of the relationship between defective nucleotide excision repair and human disease in a later chapter. Meanwhile, our focus now shifts to a consideration of yet another form of excision repair known to exist in cells: the repair of mismatched bases.

REFERENCES

1. Abbotts, J., D. N. SenGupta, B. Zmudzka, S. G. Widen, V. Noratio, and S. H. Wilson. 1988. Expression of human DNA polymerase beta in *Escherichia coli* and characterization of the recombinant enzyme. *Biochemistry* **27**:901–909.

2. Arrand, J. E., S. Squires, N. M. Bone, and R. T. Johnson. 1987. Restoration of u.v.-induced excision repair in xeroderma D cells transfected with the *denV* gene of bacteriophage T4. *EMBO J.* **6**:3125–3131.

3. Arrand, J. E., A. E. Willis, I. Goldsmith, and T. Lindahl. 1986. Different substrate specificities of the two DNA ligases of mammalian cells. *J. Biol. Chem.* **261**:9079–9082.

4. Baldi, M. I., P. Benedetti, E. Mattoccia, and G. P. Tocchini-Valentini. 1980. In vitro catenation and decatenation of DNA and a novel eucaryotic ATP-dependent topoisomerase. *Cell* **20**:461–467.

5. Bambara, R. A., and C. B. Jessee. 1991. Properties of DNA polymerases δ and ε, and their roles in eukaryotic DNA replication. *Biochim. Biophys. Acta* **1088**:11–24.

5a. Bardwell, A. J., L. Bardwell, N. Iyer, J. Q. Svejstrup, W. J. Feaver, R. D. Kornberg, and E. C. Friedberg. 1994. Yeast nucleotide excision repair proteins Rad2 and Rad4 interact with RNA polymerase II basal transcription factor b. *Mol. Cell. Biol.* **14**:3569–3576.

6. Belt, P. B. G. M., M. F. van Oosterwijk, H. Odijk, J. H. J. Hoeijmakers, and C. Backendorf. 1991. Induction of a mutant phenotype in human repair proficient cells after overexpression of a mutated human DNA repair gene. *Nucleic Acids Res.* **19**:5633–5637.

7. Bensch, K. G., S. Tanaka, S. Z. Hu, T. S. Wang, and D. Korn. 1982. Intracellular localization of human DNA polymerase α with monoclonal antibodies. *J. Biol. Chem.* **257**:8391–8396.

8. Biggerstaff, M., D. E. Szymkowski, and R. D. Wood. 1993. Co-correction of the *ERCC1, ERCC4* and xeroderma pigmentosum group F DNA repair defects in vitro. *EMBO J.* **12**:3685–3692.

9. Bootsma, D., W. Keijzer, E. G. Jung, and E. Bohnert. 1989. Xeroderma pigmentosum complementation group XP-I withdrawn. *Mutat. Res.* **218**:149–151.

10. Bramson, J., and L. C. Panasci. 1993. Effect of ERCC-1 overexpression on sensitivity of Chinese hamster ovary cells to DNA damaging agents. *Cancer Res.* **53**:3237–3240.

11. Bravo, R., R. Frank, P. A. Blundell, and H. Macdonald-Bravo. 1987. Cyclin/PCNA is the auxiliary protein of DNA polymerase-δ. *Nature* (London) **326**:515–517.

12. Burgers, P. M. 1989. Eukaryotic DNA polymerases α and δ: conserved properties and interactions, from yeast to mammalian cells. *Prog. Nucleic Acid Res. Mol. Biol.* **37**:235–280.

13. Burgers, P. M. 1991. *Saccharomyces cerevisiae* replication factor C. II. Formation and activity of complexes with the proliferating cell nuclear antigen and with DNA polymerases δ and ε. *J. Biol. Chem.* **266**:22698–22706.

14. Busch, D., C. Greiner, K. Lewis, R. Ford, G. Adair, and L. Thomson. 1989. Summary of complementation groups of UV-sensitive CHO cell mutants isolated by large-scale screening. *Mutagenesis* **4**:349–354.

15. Busch, D., C. Greiner, K. L. Rosenfeld, R. Ford, J. de Wit, J. H. J. Hoeijmakers, and L. H. Thompson. 1994. Complementation group assignments of moderately UV-sensitive CHO mutants isolated by large-scale screening. *Mutagenesis* **9**:301–306.

16. Busch, D. B., J. E. Cleaver, and D. A. Glaser. 1980. Large-scale isolation of UV-sensitive clones of CHO cells. *Somatic Cell Genet.* **6**:407–418.

17. Byrnes, J. J., K. M. Downey, V. L. Black, and A. G. So. 1976. A new mammalian DNA polymerase with 3′ to 5′ exonuclease activity: DNA polymerase δ. *Biochemistry* **15**:2817–2823.

18. Caldecott, K. W., C. K. McKeown, J. D. Tucker, S. Ljungquist, and L. H. Thompson. 1994. An interaction between the mammalian DNA repair protein XRCC1 and DNA ligase III. *Mol. Cell. Biol.* **14**:68–76.

18a. Capecchi, M. R. 1994. Targeted gene replacement. *Sci. Am.* **270**:52–59.

19. Celis, J. E., and P. Madsen. 1986. Increased nuclear cyclin/PCNA antigen staining of non S-phase transformed human amnion cells engaged in nucleotide excision DNA repair. *FEBS Lett.* **209**:277–83.

20. Chang, C. C., J. A. Boezi, S. T. Warren, C. L. Sabourin, P. K. Liu, L. Glatzer, and J. E. Trosko. 1981. Isolation of a UV-sensitive hypermutable aphidicolin-resistant Chinese hamster cell line. *Somatic Cell Genet.* **7**:235–253.

21. Chu, G., and E. Chang. 1988. Xeroderma pigmentosum group E cells lack a nuclear factor that binds to damaged DNA. *Science* **242**:564–567.

22. Chung, D. W., J. Zhang, C.-K. Tan, E. W. Davie, A. G. So, and K. M. Downey. 1991. Primary structure of the catalytic subunit of human DNA polymerase δ and chromosomal location of the gene. *Proc. Natl. Acad. Sci. USA* **88**:11197–11201.

23. Cleaver, J. E. 1982. Specificity and completeness of inhibition of DNA repair by novobiocin and aphidicolin. *Carcinogenesis* **3**:1171–1174.

24. Cleaver, J. E., and K. H. Kraemer. 1989. Xeroderma pigmentosum, p. 2949–2971. *In* C. R. Scriver, A. L. Beaudet, W. S. Sly, and D. Valle (ed.), *The Metabolic Basis of Inherited Disease* 6th ed. McGraw-Hill Book Co., New York.

25. Colbère-Garapin, F., and M.-L. Ryhiner. 1986. Patterns of integration of exogenous DNA sequences transfected into mammalian cells of primate and rodent origin. *Gene* **50**:279–288.

26. Collins, A., and R. T. Johnson. 1979. Novobiocin; an inhibitor of the repair of UV-induced but not X-ray induced damage in mammalian cells. *Nucleic Acids Res.* **7**:1311–1320.

27. Collins, A., and R. T. Johnson. 1987. The molecular biology of nucleotide excision repair of DNA: recent progress. *J. Cell Sci. Suppl.* **6**:61–82.

28. Collins, A. R. S. 1993. Mutant rodent cell lines sensitive to ultraviolet light, ionizing radiation and cross-linking agents: a comprehensive survey of genetic and biochemical characteristics. *Mutat. Res.* **293**:99–118.

29. Cotterill, S. M., M. E. Reyland, L. A. Loeb, and I. R. Lehman. 1987. A cryptic proofreading 3′ → 5′ exonuclease associated with the polymerase subunit of the DNA polymerase-primase from *Drosophila melanogaster*. *Proc. Natl. Acad. Sci. USA* **84**:5635–5639.

30. Coverley, D., M. K. Kenny, D. P. Lane, and R. D. Wood. 1992. A role for the human single-stranded DNA binding protein HSSB/RPA in an early stage of nucleotide excision repair. *Nucleic Acids Res.* **20**:3873–3880.

31. Coverley, D., M. K. Kenny, M. Munn, W. D. Rupp, D. P. Lane, and R. D. Wood. 1991. Requirement for the replication protein SSB in human DNA excision repair. *Nature* (London) **349**:538–541.

32. Cripps-Wolfman, J., E. C. Henshaw, and R. A. Bambara. 1989. Alterations in the phosphorylation and activity of DNA polymerase α correlate with the change in replicative DNA synthesis as quiescent cells reenter the cell cycle. *J. Biol. Chem.* **264**:19478–19486.

33. Crute, J. J., A. F. Wahl, and R. A. Bambara. 1986. Purification and characterization of two new high molecular weight forms of DNA polymerase δ. *Biochemistry* **25**:26–36.

34. de Jonge, A. J. R., W. Vermeulen, W. Keijzer, J. H. J. Hoeijmakers, and D. Bootsma. 1985. Microinjection of *Micrococcus luteus* UV-endonuclease restores UV-induced unscheduled DNA synthesis in cells of 9 xeroderma pigmentosum complementation groups. *Mutat. Res.* **150**:99–105.

35. de Jonge, A. J. R., W. Vermeulen, B. Klein, and J. H. J. Hoeijmakers. 1983. Microinjection of human cell extracts corrects xeroderma pigmentosum defect. *EMBO J.* **2**:637–641.

36. De Weerd-Kastelein, E. A., W. Keijzer, and D. Bootsma. 1972. Genetic heterogeneity of xeroderma pigmentosum demonstrated by somatic cell hybridization. *Nature* (London) *New Biol.* **238**:80–83.

37. **Dickstein, R., N. D. Huh, I. Sandlie, and L. Grossman.** 1988. The expression of the *Escherichia coli uvrA* gene in human cells. *Mutat. Res.* **193:**75–86.

38. **Downes, C. S., and R. T. Johnson.** 1988. DNA topoisomerases and DNA repair. *Bioessays* **8:**179–184.

39. **Downes, C. S., A. M. Mullinger, and R. T. Johnson.** 1987. Action of etoposide (VP-16-123) on human cells: no evidence for topoisomerase II involvement in excision repair of u.v.-induced DNA damage, nor for mitochondrial hypersensitivity in ataxia telangiectasia. *Carcinogenesis* **8:**1613–1618.

40. **Downey, K. M., C.-K. Tan, D. M. Andrews, X. Li, and A. G. So.** 1988. Proposed roles for DNA polymerases α and δ at the replication fork. *Cancer Cells* **6:**403–410.

41. **Drapkin, R., J. T. Reardon, A. Ansari, J.-C. Huang, L. Zawel, K. Ahn, A. Sancar, and D. Reinberg.** 1994. Dual role of TFIIH in DNA excision repair and in transcription by RNA polymerase II. *Nature* (London) **368:**769–772.

42. **Dresler, S. L., and M. G. Frattini.** 1986. DNA replication and UV-induced DNA repair synthesis in human fibroblasts are much less sensitive than DNA polymerase α to inhibition by butylphenyl-deoxyguanosine triphosphate. *Nucleic Acids Res.* **14:**7093–7102.

43. **Dresler, S. L., and M. G. Frattini.** 1988. Analysis of butyl-phenyl-guanine, butylphenyl-deoxyguanosine, and butylphenyl-deoxyguanosine triphosphate inhibition of DNA replication and ultraviolet-induced DNA repair synthesis using permeable human fibroblasts. *Biochem. Pharmacol.* **37:**1033–1037.

44. **Dresler, S. L., and K. S. Kimbro.** 1987. 2′,3′-Dideoxythymidine 5′-triphosphate inhibition of DNA replication and ultraviolet-induced DNA repair synthesis in human cells: evidence for involvement of DNA polymerase δ. *Biochemistry* **26:**2664–2668.

45. **Dresler, S. L., and M. W. Lieberman.** 1983. Identification of DNA polymerases involved in DNA excision repair in diploid human fibroblasts. *J. Biol. Chem.* **258:**9990–9994.

46. **Dresler, S. L., and R. M. Robinson-Hill.** 1987. Direct inhibition of u.v.-induced DNA excision repair in human cells by novobiocin, coumermycin and nalidixic acid. *Carcinogenesis* **8:**813–817.

47. **Edenberg, H. J.** 1980. Novobiocin inhibition of simian virus 40 DNA replication. *Nature* (London) **286:**529–531.

48. **Eichler, D. C., P. A. Fisher, and D. Korn.** 1977. Effect of calcium on the recovery and distribution of DNA polymerase α from cultured human cells. *J. Biol. Chem.* **252:**4011–4014.

49. **Eker, A. P. M., W. Vermeulen, N. Miura, K. Tanaka, N. G. J. Jaspers, J. H. J. Hoeijmakers, and D. Bootsma.** 1992. Xeroderma pigmentosum group A correcting protein from calf thymus. *Mutat. Res.* **274:**211–224.

50. **Fairman, M. P., and B. Stillman.** 1988. Cellular factors required for multiple stages of SV40 DNA replication in vitro. *EMBO J.* **7:**1211–1218.

51. **Fisher, P. A., T. S. Wang, and D. Korn.** 1979. Enzymological characterization of DNA polymerase α. Basic catalytic properties processivity, and gap utilization of the homogeneous enzyme from human KB cells. *J. Biol. Chem.* **254:**6128–6137.

52. **Flejter, W. L., L. D. McDaniel, M. Askari, E. C. Friedberg, and R. A. Schultz.** 1992. Characterization of a complex chromosomal rearrangement maps the locus for in vitro complementation of xeroderma pigmentosum group D to human chromosome band 19q13. *Genes Chromosomes Cancer* **5:**335–342.

53. **Flejter, W. L., L. D. McDaniel, D. Johns, E. C. Friedberg, and R. A. Schultz.** 1992. Correction of xeroderma pigmentosum complementation group D mutant cell phenotypes by chromosome and gene transfer: involvement of the human *ERCC2* DNA repair gene. *Proc. Natl. Acad. Sci. USA* **89:**261–265.

54. **Fornace, A. J., Jr., K. W. Kohn, and H. E. Kann, Jr.** 1976. DNA single-strand breaks during repair of UV damage in human fibroblasts and abnormalities of repair in xeroderma pigmentosum. *Proc. Natl. Acad. Sci. USA* **73:**39–43.

54a. **Frederick, G. D., R. H. Amirkhan, R. A. Schultz, and E. C. Friedberg.** 1994. Structural and mutational analysis of the xeroderma pigmentosum group D (*XPD*) gene. *Hum. Mol. Genet.,* **3:**1783–1788.

55. **Friedberg, E. C.** 1985. *DNA Repair.* W.H. Freeman & Co., New York.

56. **Friedberg, E. C., C. Backendorf, J. Burke, A. Collins, L. Grossman, J. H. J. Hoeijmakers, A. R. Lehmann, E. Seeberg, G. P. van der Schans, and A. A. van Zeeland.** 1987. Molecular aspects of DNA repair. *Mutat. Res.* **184:**67–86.

57. **Friedberg, E. C., U. K. Ehmann, and J. I. Williams.** 1979. Human diseases associated with defective DNA repair. *Adv. Radiat. Biol.* **8:**85–174.

58. **Frosina, G., and O. Rossi.** 1992. Effect of topoisomerase poisoning by antitumor drugs VM 26, fostriecin and camptothecin on DNA repair replication by mammalian cell extracts. *Carcinogenesis* **13:**1371–1377.

59. **Fry, M., and L. A. Loeb.** 1986. *Animal Cell DNA Polymerases.* CRC Press, Inc., Boca Raton, Fla.

60. **Fujiwara, Y., A. Matsumoto, M. Ichihashi, and Y. Satoh.** 1987. Heritable disorders of DNA repair: xeroderma pigmentosum and Fanconi's anemia. *Curr. Probl. Dermatol.* **17:**182–198.

61. **Gebara, M. M., C. Drevon, S. A. Harcourt, H. Steingrims-dottir, M. R. James, J. F. Burke, C. F. Arlett, and A. R. Lehmann.** 1987. Inactivation of a transfected gene in human fibroblasts can occur by deletion, amplification, phenotypic switching, or methylation. *Mol. Cell. Biol.* **7:**1459–1464.

62. **Giulotto, E., and C. Mondello.** 1981. Aphidicolin does not inhibit the repair synthesis of mitotic chromosomes. *Biochem. Biophys. Res. Commun.* **99:**1287–1294.

63. **Goulian, M., and C. J. Heard.** 1989. Intact DNA polymerase α/primase from mouse cells: purification and structure. *J. Biol. Chem.* **264:**19407–19415.

64. **Goulian, M., and C. J. Heard.** 1990. The mechanism of action of an accessory protein for DNA polymerase α/primase. *J. Biol. Chem.* **265:**13231–13239.

65. **Goulian, M., C. J. Heard, and S. L. Grimm.** 1990. Purification and properties of an accessory protein for DNA polymerase α/primase. *J. Biol. Chem.* **265:**13221–13230.

66. **Hanaoka, F., H. Kato, S. Ikegami, M. Ohashi, and M.-A. Yamada.** 1979. Aphidicolin does inhibit repair replication in HeLa cells. *Biochem. Biophys. Res. Commun.* **87:**575–580.

67. **Hanawalt, P. C., and A. Sarasin.** 1986. Cancer-prone hereditary diseases with DNA processing abnormalities. *Trends Genet.* **2:**124–129.

68. **Hansson, J., L. Grossman, T. Lindahl, and R. D. Wood.** 1990. Complementation of the xeroderma pigmentosum DNA repair synthesis defect with *Escherichia coli* UvrABC proteins in a cell-free system. *Nucleic Acids Res.* **18:**35–40.

69. **Hansson, J., S. M. Keyse, T. Lindahl, and R. D. Wood.** 1991. DNA excision repair in cell extracts from human cell lines exhibiting hypersensitivity to DNA-damaging agents. *Cancer Res.* **51:**3384–90.

70. **Hansson, J., M. Munn, W. D. Rupp, R. Kahn, and R. D. Wood.** 1989. Localization of DNA repair synthesis by human cell extracts to a short region at the site of a lesion. *J. Biol. Chem.* **264:**21788–21792.

71. **Hansson, J., and R. D. Wood.** 1989. Repair synthesis by human cell extracts in DNA damaged by cis- and trans-diamminedichloroplatinum(II). *Nucleic Acids Res.* **17:**8073–8091.

72. **Henderson, E. E., K. Valerie, A. P. Green, and J. K. de Riel.** 1989. Host cell reactivation of CAT-expression vectors as a method to assay for cloned DNA-repair genes. *Mutat. Res.* **220:**151–160.

73. **Heng, H. H., H. Xiao, X. M. Shi, J. Greenblatt, and L. C. Tsui.** 1994. Genes encoding general initiation factors for RNA polymerase II transcription are dispersed in the human genome. *Hum. Mol. Genet.* **3:**61–64.

74. Henning, K. A., C. Peterson, R. Legerski, and E. C. Friedberg. 1994. Cloning the *Drosophila* homolog of the xeroderma pigmentosum complementation group C gene reveals homology between the predicted human and *Drosophila* polypeptides and that encoded by the yeast *RAD4* gene. *Nucleic Acids Res.* **22:**257–261.

75. Hickson, I. D., S. L. Davies, S. M. Davies, and C. N. Robson. 1990. DNA repair in radiation sensitive mutants of mammalian cells: possible involvement of DNA topoisomerases. *Int. J. Radiat. Biol.* **58:**561–568.

76. Hirschfeld, S., A. S. Levine, K. Ozato, and M. Protic. 1990. A constitutive damage-specific DNA-binding protein is synthesized at higher levels in UV-irradiated primate cells. *Mol. Cell. Biol.* **10:**2041–2048.

76a. Hoeijmakers, J. H. J. 1993. Nucleotide excision repair II: from yeast to mammals. *Trends Genet.* **9:**211–217.

77. Hoeijmakers, J. H. J. 1993. Nucleotide excision repair I: from *E. coli* to yeast. *Trends Genet.* **9:**173–177.

78. Hoeijmakers, J. H. J., and D. Bootsma. 1990. Molecular genetics of eukaryotic DNA excision repair. *Cancer Cells* **2:**311–320.

79. Hoeijmakers, J. H. J., A. P. M. Eker, R. D. Wood, and P. Robins. 1990. Use of in vivo and in vitro assays for the characterization of mammalian excision repair and isolation of repair proteins. *Mutat. Res.* **236:**223–238.

80. Hoeijmakers, J. H. J., H. Odijk, and A. Westerveld. 1987. Differences between rodent and human cell lines in the amount of integrated DNA after transfection. *Exp. Cell Res.* **169:**111–119.

81. Hoeijmakers, J. H. J., M. van Duin, G. Weeda, A. J. van der Eb, C. Troelstra, A. P. Eker, N. G. Jaspers, A. Westerveld, and D. Bootsma. 1988. Analysis of mammalian excision repair: from mutants to genes and gene products, p. 281–287. *In* E. C. Friedberg and P. C. Hanawalt (ed.), *Mechanisms and Consequences of DNA Damage Processing.* Alan R. Liss, Inc., New York.

82. Hoeijmakers, J. H. J., J. C. M. Zwetsloot, W. Vermeulen, A. J. R. de Jonge, C. Backendorf, B. Klein, and D. Bootsma. 1983. Phenotypic correction of xeroderma pigmentosum cells by microinjection of crude extracts and purified proteins, p. 173–181. *In* E. C. Friedberg and B. A. Bridges (ed.), *Cellular Responses to DNA Damage.* Alan R. Liss, Inc., New York.

83. Hsieh, T.-S., and D. Brutlag. 1980. ATP-dependent DNA topoisomerase from *D. melanogaster* reversibly catenates duplex DNA rings. *Cell* **21:**115–125.

84. Huang, J.-C., D. L. Svoboda, J. T. Reardon, and A. Sancar. 1992. Human nucleotide excision nuclease removes thymine dimers from DNA by incising the 22nd phosphodiester bond 5′ and the 6th phosphodiester bond 3′ to the photodimer. *Proc. Natl. Acad. Sci. USA* **89:**3664–3668.

85. Hübscher, U. 1983. The mammalian primase is part of a high molecular weight DNA polymerase α polypeptide. *EMBO J.* **2:**133–136.

86. Hübscher, U., C. C. Kuenzle, W. Limacher, P. Scherrer, and S. Spadari. 1979. Functions of DNA polymerases α, β, and γ in neurons during development. *Cold Spring Harbor Symp. Quant. Biol.* **43:**625–629.

87. Humbert, S., H. van Vuuren, Y. Lutz, J. H. J. Hoeijmakers, J.-M. Egly, and V. Moncollin. 1994. Characterization of p44/SSL1 and p34 subunits of the BTF2/TFIIH transcription/repair factor. *EMBO J.* **13:**2393–2398.

88. Hwang, B. J., and G. Chu. 1993. Purification and characterization of a human protein that binds to damaged DNA. *Biochemistry* **32:**1657–1666.

89. Ikegami, S., T. Taguchi, M. Ohashi, M. Oguro, H. Nagano, and Y. Mano. 1978. Aphidicolin prevents mitotic cell division by interfering with the activity of DNA polymerase-α. *Nature* (London) **275:**458–460.

90. Ishimi, Y., A. Claude, P. Bullock, and J. Hurwitz. 1988. Complete enzymatic synthesis of DNA containing the SV40 origin of replication. *J. Biol. Chem.* **263:**19723–19733.

91. Iyer, N., and E. C. Friedberg. 1994. Unpublished observations.

92. Jessberger, R., V. Podust, U. Hübscher, and P. Berg. 1993. A mammalian protein complex that repairs double-strand breaks and deletions by recombination. *J. Biol. Chem.* **268:**15070–15079.

93. Johnson, R. T., S. Squires, G. C. Elliott, G. L. Koch, and A. J. Rainbow. 1985. Xeroderma pigmentosum D-HeLa hybrids with low and high ultraviolet sensitivity associated with normal and diminished DNA repair ability, respectively. *J. Cell Sci.* **76:**115–133.

94. Jones, C. J., and R. D. Wood. 1993. Preferential binding of the xeroderma pigmentosum group A complementing protein to damaged DNA. *Biochemistry* **32:**12096–12104.

95. Kantor, G. J., L. S. Barsalou, and P. C. Hanawalt. 1990. Selective repair of specific chromatin domains in UV-irradiated cells from xeroderma pigmentosum complementation group C. *Mutat. Res.* **235:**171–180.

96. Kataoka, H., and Y. Fujiwara. 1991. UV damage-specific DNA-binding protein in xeroderma pigmentosum complementation group E. *Biochem. Biophys. Res. Commun.* **175:**1139–1143.

97. Kaufmann, W. K. 1988. In vitro complementation of xeroderma pigmentosum. *Mutagenesis* **3:**373–380.

98. Kaufmann, W. K., and L. P. Briley. 1987. Reparative strand incision in saponin-permeabilized human fibroblasts. *Mutat. Res.* **184:**237–243.

99. Keeney, S., G. J. Chang, and S. Linn. 1993. Characterization of a human DNA damage binding protein implicated in xeroderma pigmentosum E. *J. Biol. Chem.* **268:**21293–21300.

100. Keeney, S., A. P. Eker, T. Brody, W. Vermeulen, D. Bootsma, J. H. J. Hoeijmakers, and S. Linn. 1994. Correction of the DNA repair defect in xeroderma pigmentosum group E by injection of a DNA damage binding protein. *Proc. Natl. Acad. Sci. USA* **91:**4053–4056.

101. Keeney, S., and S. Linn. 1990. A critical review of permeabilized cell systems for studying mammalian DNA repair. *Mutat. Res.* **236:**239–252.

102. Keeney, S., H. Wein, and S. Linn. 1992. Biochemical heterogeneity in xeroderma pigmentosum complementation group E. *Mutat. Res.* **273:**49–56.

103. Keijzer, W., A. Verkerk, and D. Bootsma. 1982. Phenotypic correction of the defect in xeroderma pigmentosum after fusion with isolated cytoplasts. *Exp. Cell Res.* **140:**119–125.

104. Kohn, K. W., R. A. Ewig, L. C. Erickson, and L. A. Zwelling. 1981. Measurement of strand breaks and cross-links by alkaline elution, p. 379–401. *In* E. C. Friedberg and P. C. Hanawalt (ed.), *DNA Repair—A Laboratory Manual of Research Procedures*, vol. 1, part B. Marcel Dekker, Inc., New York.

105. Kornberg, A., and T. Baker. 1991. *DNA Replication.* W. H. Freeman & Co., New York.

106. Kraemer, K. H., M. M. Lee, and J. Scotto. 1987. Xeroderma pigmentosum. *Arch. Dermatol.* **123:**241–250.

107. Kraemer, K. H., and H. Slor. 1985. Xeroderma pigmentosum. *Clin. Dermatol.***3:**33–69.

108. Lamothe, P., B. Baril, A. Chi, L. Lee, and E. Baril. 1981. Accessory proteins for DNA polymerase α activity with single-strand DNA templates. *Proc. Natl. Acad. Sci. USA* **78:**4723–4727.

109. Lasko, D. D., A. E. Tomkinson, and T. Lindahl. 1990. Eukaryotic DNA ligases. *Mutat. Res.* **236:**277–287.

110. Lasko, D. D., A. E. Tomkinson, and T. Lindahl. 1990. Mammalian DNA ligases. Biosynthesis and intracellular localization of DNA ligase I. *J. Biol. Chem.* **265:**12618–12622.

111. Leadon, S. A., A. K. Ganesan, and P. C. Hanawalt. 1987. Enhanced transforming activity of ultraviolet-irradiated pSV2-gpt is due to damage outside the *gpt* transcription unit. *Plasmid* **18:**135–141.

112. LeDoux, S. P., G. L. Wilson, E. J. Beecham, T. Stevsner, K. Wassermann, and V. A. Bohr. 1992. Repair of mitochondrial DNA after various types of DNA damage in Chinese hamster ovary cells. *Carcinogenesis* **13:**1967–1973.

113. Lee, M. Y. W. T., C.-K. Tan, K. M. Downey, and A. G. So. 1984. Further studies on calf thymus DNA polymerase δ purified to homogeneity by a new procedure. *Biochemistry* **23:**1906–1913.

114. **Lee, S. H., Z. Q. Pan, A. D. Kwong, P. M. Burgers, and J. Hurwitz.** 1991. Synthesis of DNA by DNA polymerase ε in vitro. *J. Biol. Chem.* **266:**22707–22717.

115. **Legerski, R., and C. Peterson.** 1992. Expression cloning of a human DNA repair gene involved in xeroderma pigmentosum group C. *Nature* (London) **359:**70–73.

116. **Legerski, R. J., P. Liu, L. Li, C. A. Peterson, Y. Zhao, R. J. Leach, S. L. Naylor, and M. J. Siciliano.** 1994. Assignment of xeroderma pigmentosum group C (*XPC*) gene to chromosome 3p25. *Genomics* **21:**266–269.

117. **Lehman, I. R., and L. S. Kaguni.** 1989. DNA polymerase α. *J. Biol. Chem.* **264:**4265–4268.

118. **Lehmann, A. R., D. Bootsma, S. G. Clarkson, J. E. Cleaver, P. J. McAlpine, K. Tanaka, L. H. Thompson, and R. D. Wood.** 1994. Nomenclature of human DNA repair genes. *Mutat. Res.* **315:**41–42.

119. **Lehmann, A. R., J. H. J. Hoeijmakers, A. A. van Zeeland, C. M. Backendorf, B. A. Bridges, A. Collins, R. P. Fuchs, G. P. Margison, R. Montesano, E. Moustacchi, et al.** 1992. Workshop on DNA repair. *Mutat. Res.* **273:**1–28.

120. **Li, L., E. S. Bales, C. A. Peterson, and R. J. Legerski.** 1993. Characterization of molecular defects in xeroderma pigmentosum group C. *Nat. Genet.* **5:**413–417.

121. **Li, L., S. J. Elledge, C. A. Peterson, E. S. Bales, and R. J. Legerski.** 1994. Specific association between the human DNA repair proteins XPA and ERCC1. *Proc. Natl. Acad. Sci. USA* **91:**5012–5016.

122. **Lindahl, T., and D. E. Barnes.** 1992. Mammalian DNA ligases. *Annu. Rev. Biochem.* **61:**251–281.

123. **Linn, S.** 1991. How many pols does it take to replicate nuclear DNA? *Cell* **66:**185–187.

124. **Liu, P., J. Siciliano, B. White, R. Legerski, D. Callen, S. Reeders, M. J. Siciliano, and L. H. Thompson.** 1993. Regional mapping of human DNA excision repair gene *ERCC4* to chromosome 16p13.13-p13.2. *Mutagenesis* **8:**199–205.

125. **Liu, P. K., C. C. Chang, J. E. Trosko, D. K. Dube, G. M. Martin, and L. A. Loeb.** 1983. Mammalian mutator mutant with an aphidicolin-resistant DNA polymerase α. *Proc. Natl. Acad. Sci. USA* **80:**797–801.

126. **Loeb, L. A., P. K. Liu, and M. Fry.** 1986. DNA polymerase-alpha: enzymology, function, fidelity, and mutagenesis. *Prog. Nucleic Acid Res. Mol. Biol.* **33:**57–110.

127. **Lohrer, H., M. Blum, and P. Herrlich.** 1988. Ataxia telangiectasia resists gene cloning: an account of parameters determining gene transfer into human recipient cells. *Mol. Gen. Genet.* **212:**474–480.

128. **Ma, L., A. Westbroek, A. G. Jochemsen, G. Weeda, A. Bosch, D. Bootsma, J. H. J. Hoeijmakers, and A. van der Eb.** 1994. Mutational analysis of *ERCC3*, which is involved in DNA repair and transcription: identification of domains essential for the DNA repair function. *Mol. Cell. Biol.* **14:**4126–4134.

129. **MacInnes, M. A., J. M. Bingham, L. H. Thompson, and G. F. Strniste.** 1984. DNA-mediated cotransfer of excision repair capacity and drug resistance into Chinese hamster ovary mutant cell line UV-135. *Mol. Cell. Biol.* **4:**1152–1158.

130. **MacInnes, M. A., J. A. Dickson, R. R. Hernandez, D. Learmonth, G. Y. Lin, J. S. Mudgett, M. S. Park, S. Schauer, R. J. Reynolds, G. F. Strniste, and J. Y. Yu.** 1993. Human *ERCC5* cDNA-cosmid complementation for excision repair and bipartite amino acid domains conserved with RAD proteins of *Saccharomyces cerevisiae* and *Schizosaccharomyces pombe*. *Mol. Cell. Biol.* **13:**6393–6402.

131. **Manley, J. L., A. Fire, A. Cano, P. A. Sharp, and M. L. Gefter.** 1980. DNA-dependent transcription of adenovirus genes in a soluble whole-cell extract. *Proc. Natl. Acad. Sci. USA* **77:**3855–3859.

132. **Masutani, C., K. Sugasawa, H. Asahina, K. Tanaka, and F. Hanaoka.** 1993. Cell-free repair of UV-damaged simian virus 40 chromosomes in human cell extracts. II. Defective DNA repair synthesis by xeroderma pigmentosum cell extracts. *J. Biol. Chem.* **268:**9105–9109.

133. **Masutani, C., K. Sugasawa, J. Yanigasawa, T. Soniyama, M. Ui, T. Enomoto, K. Takio, K. Tanaka, P. J. van der Spek, D. Bootsma, J. H. J. Hoeijmakers, and F. Hanaoka.** 1994. Purification and cloning of a nucleotide excision repair complex involving the xeroderma pigmentosum group C protein and a human homologue of yeast *RAD23*. *EMBO J.* **13:**1831–1843.

134. **Mattern, M. R., and D. A. Scudiero.** 1981. Dependence of mammalian DNA synthesis on DNA supercoiling. III. Characterization of the inhibition of replicative and repair-type DNA synthesis by novobiocin and nalidixic acid. *Biochim. Biophys. Acta* **653:**248–258.

135. **Mayne, L. V., T. Jones, S. W. Dean, S. A. Harcourt, J. E. Lowe, A. Priestly, H. Steingrimsdottir, H. Sykes, M. H. L. Green, and A. R. Lehmann.** 1988. SV40-transformed normal and DNA-repair-deficient human fibroblasts can be transfected with high frequency but retain only limited amounts of integrated DNA. *Gene* **66:**65–76.

136. **McBride, O. W., C. A. Kozak, and S. H. Wilson.** 1990. Mapping of the gene for DNA polymerase β to mouse chromosome 8. *Cytogenet. Cell Genet.* **53:**108–111.

137. **McWhir, J., J. Selfridge, D. J. Harrison, S. Squires, and D. W. Melton.** 1993. Mice with DNA repair gene (*ERCC-1*) deficiency have elevated levels of p53, liver nuclear abnormalities and die before weaning. *Nat. Genet.* **5:**217–224.

138. **Miller, K. G., L. F. Liu, and P. T. Englund.** 1981. A homogeneous type II DNA topoisomerase from HeLa cell nuclei. *J. Biol. Chem.* **256:**9334–9339.

139. **Miller, M. R., and D. N. Chinault.** 1982. Evidence that DNA polymerases α and β participate differentially in DNA repair synthesis induced by different agents. *J. Biol. Chem.* **257:**46–49.

140. **Miller, M. R., and D. N. Chinault.** 1982. The roles of DNA polymerases α, β, and γ in DNA repair synthesis induced in hamster and human cells by different DNA damaging agents. *J. Biol. Chem.* **257:**10204–10209.

141. **Miura, M., M. Domon, T. Sasaki, S. Kondo, and Y. Takasaki.** 1992. Restoration of proliferating cell nuclear antigen (PCNA) complex formation in xeroderma pigmentosum group A cells following cis-diamminedichloroplatinum(II)-treatment by cell fusion with normal cells. *J. Cell. Physiol.* **152:**639–645.

142. **Miura, M., M. Domon, T. Sasaki, and Y. Takasaki.** 1992. Induction of proliferating cell nuclear antigen (PCNA) complex formation in quiescent fibroblasts from a xeroderma pigmentosum patient. *J. Cell. Physiol.* **150:**370–376.

143. **Miura, N., I. Miyamoto, H. Asahina, I. Satokata, K. Tanaka, and Y. Okada.** 1991. Identification and characterization of xpac protein, the gene product of the human *XPAC* (xeroderma pigmentosum group A complementing) gene. *J. Biol. Chem.* **266:**19786–19789.

144. **Mohrenweiser, H. W., A. V. Carrano, A. Fertitta, B. Perry, L. H. Thompson, J. D. Tucker, and C. A. Weber.** 1989. Refined mapping of the three DNA repair genes, *ERCC1*, *ERCC2*, and *XRCC1*, on human chromosome 19. *Cytogenet. Cell Genet.* **52:**11–14.

145. **Mudgett, J. S., and M. A. MacInnes.** 1990. Isolation of the functional human excision repair gene *ERCC5* by intercosmid recombination. *Genomics* **8:**623–633.

146. **Nairn, R. S., R. M. Humphrey, and G. M. Adair.** 1988. Transformation of UV-hypersensitive Chinese hamster ovary cell mutants with UV-irradiated plasmids. *Int. J. Radiat. Biol.* **53:**249–260.

147. **Nakayama, K., and A. Sugino.** 1980. Novobiocin and nalidixic acid target proteins in yeast. *Biochem. Biophys. Res. Commun.* **96:**306–312.

148. **Nichols, A. F., and A. Sancar.** 1992. Purification of PCNA as a nucleotide excision repair protein. *Nucleic Acids Res.* **20:**2441–2446.

149. **Nishida, C., S. Y. Choi, J. Kim, S. Keeney, and S. Linn.** 1988. DNA polymerase δ plus HeLa- or human fibroblast cell-free extracts complement permeabilized xeroderma pigmentosum (XP) fibroblasts: application for purification of XP correcting factors, p. 337–341. *In* E. C. Friedberg and P. C. Hanawalt (ed.), *Mechanisms and Consequences of DNA Damage Processing*. Alan R. Liss, Inc., New York.

150. **Nishida, C., and S. Linn.** 1988. DNA repair synthesis in permeabilized human fibroblasts mediated by DNA polymerase δ and ap-

plication for purification of xeroderma pigmentosum factors. *Cancer Cells* **6**:411–415.

151. Nishida, C., P. Reinhard, and S. Linn. 1988. DNA repair synthesis in human fibroblasts requires DNA polymerase delta. *J. Biol. Chem.* **263**:501–510.

152. Nishigori, C., M. Zghal, T. Yagi, S. Imamura, M. R. Komoun, and H. Takebe. 1993. High prevalence of the point mutation in exon 6 of the xeroderma pigmentosum group A-complementating (*XPAC*) gene in xeroderma pigmentosum group A patients in Tunisia. *Am. J. Hum. Genet.* **53**:1001–1006.

152a. Nouspikel, T., and S. Clarkson. 1994. Mutations that disable the DNA repair gene *XPG* in a xeroderma pigmentosum group G patient. *Hum. Mol. Genet.* **3**:963–967.

152b. O'Donovan, A., A. A. Davies, J. G. Moggs, S. C. West, and R. D. Wood. 1994. XPG endonuclease makes the 3′ incision in human DNA nucleotide excision repair. *Nature* (London), **371**:432–435.

153. O'Donovan, A., D. Scherly, S. G. Clarkson, and R. D. Wood. 1994. Isolation of active recombinant XPG protein, a human repair endonuclease. *J. Biol. Chem.* **269**:15965–15968.

154. O'Donovan, A., and R. D. Wood. 1993. Identical defects in DNA repair in xeroderma pigmentosum group G and rodent ERCC group 5. *Nature* (London) **363**:185–188.

155. Ottiger, H.-P., and U. Hübscher. 1984. Mammalian DNA polymerase α holoenzymes with possible functions at the leading and lagging strand of the replication fork. *Proc. Natl. Acad. Sci. USA* **81**:3993–3997.

156. Park, C. H., and A. Sancar. 1993. Reconstitution of mammalian excision repair activity with mutant cell-free extracts and XPAC and ERCC1 proteins expressed in *Escherichia coli*. *Nucleic Acids Res.* **21**:5110–5116.

157. Park, C. H., and A. Sancar. 1994. Formation of a ternary complex between human XPA, ERCC1, and ERCC4 (XPF) excision repair proteins. *Proc. Natl. Acad. Sci. USA* **91**:5017–5021.

158. Patterson, M., and G. Chu. 1989. Evidence that xeroderma pigmentosum cells from complementation group E are deficient in a homolog of yeast photolyase. *Mol. Cell. Biol.* **9**:5105–5112.

159. Pedrali-Noy, G., and S. Spadari. 1979. Effect of aphidicolin on viral and human DNA polymerases. *Biochem. Biophys. Res. Commun.* **88**:1194–1202.

160. Perez, C. F., M. R. Botchan, and C. A. Tobias. 1985. DNA-mediated gene transfer efficiency is enhanced by ionizing and ultraviolet irradiation of rodent cells in vitro. I. Kinetics of enhancement. *Radiat. Res.* **104**:200–213.

161. Perrino, F. W., and L. A. Loeb. 1989. Proofreading by the ε subunit of *Escherichia coli* DNA polymerase III increases the fidelity of calf thymus DNA polymerase α. *Proc. Natl. Acad. Sci. USA* **86**:3085–3088.

162. Popanda, O., and H. W. Thielmann. 1992. The function of DNA polymerases in DNA repair synthesis of ultraviolet-irradiated human fibroblasts. *Biochim. Biophys. Acta* **1129**:155–160.

163. Prigent, C., D. D. Lasko, K.-I. Kodama, J. R. Woodgett, and T. Lindahl. 1992. Activation of mammalian DNA ligase I through phosphorylation by casein kinase II. *EMBO J.* **11**:2925–2933.

164. Pritchard, C. G., D. T. Weaver, E. F. Baril, and M. L. DePamphilis. 1983. DNA polymerase α cofactors C_1C_2 function as primer recognition proteins. *J. Biol. Chem.* **258**:9810–9819.

165. Protic-Sabljic, M., and K. H. Kraemer. 1985. One pyrimidine dimer inactivates expression of a transfected gene in xeroderma pigmentosum cells. *Proc. Natl. Acad. Sci. USA* **82**:6622–6626.

166. Reardon, J. T., A. F. Nichols, S. Keeney, C. A. Smith, J.-S. Taylor, S. Linn, and A. Sancar. 1993. Comparative analysis of binding of human damaged DNA-binding protein (XPE) and *Escherichia coli* damage recognition protein (UvrA) to the major ultraviolet photoproducts: T[c,s]T, T[t,s]T, T[6-4]T, and T[Dewar]T. *J. Biol. Chem.* **268**:21301–21308.

167. Reardon, J. T., P. Spielmann, J. C. Huang, S. Sastry, A. Sancar, and J. E. Hearst. 1991. Removal of psoralen monoadducts and crosslinks by human cell free extracts. *Nucleic Acids Res.* **19**:4623–4629.

168. Reardon, J. T., L. H. Thompson, and A. Sancar. 1993. Excision repair in man and the molecular basis of xeroderma pigmentosum syndrome. *Cold Spring Harbor Symp. Quant. Biol.* **58**:605–617.

169. Riboni, R., E. Botta, M. Stefanini, M. Numata, and A. Yasui. 1992. Identification of the eleventh complementation group of UV-sensitive excision repair-defective rodent mutants. *Cancer Res.* **52**:6690–6691.

170. Roberts, E., R. A. Nash, P. Robins, and T. Lindahl. 1994. Different active sites of mammalian DNA ligases I and II. *J. Biol. Chem.* **269**:3789–3792.

171. Robins, P., C. J. Jones, M. Biggerstaff, T. Lindahl, and R. D. Wood. 1991. Complementation of DNA repair in xeroderma pigmentosum group A cell extracts by a protein with affinity for damaged DNA. *EMBO J.* **10**:3913–3921.

172. Roy, R., L. Schaeffer, S. Humbert, W. Vermeulen, G. Weeda, and J.-M. Egly. 1994. The DNA-dependent ATPase activity associated with the class II basic transcription factor BTF2/TFIIH. *J. Biol. Chem.* **269**:9826–9832.

173. Royer-Pokora, B., and W. A. Haseltine. 1984. Isolation of UV-resistant revertants from a xeroderma pigmentosum complementation group A cell line. *Nature* (London) **311**:390–392.

174. Samec, S., T. A. Jones, J. Corlet, D. Scherly, R. D. Wood, and S. G. Clarkson. 1994. The human gene for xeroderma pigmentosum complementation group G (*XPG*) maps to 13q33 by fluorescence in situ hybridization. *Genomics* **21**:283–285.

175. Satokata, I., K. Iwai, T. Matsuda, Y. Okada, and K. Tanaka. 1993. Genomic characterization of the human DNA excision repair-controlling gene XPAC. *Gene* **136**:345–348.

176. Satokata, I., K. Tanaka, N. Miura, I. Miyamoto, Y. Satoh, S. Kondo, and Y. Okada. 1990. Characterization of a splicing mutation in group A xeroderma pigmentosum. *Proc. Natl. Acad. Sci. USA* **87**:9908–9912.

177. Satokata, I., K. Tanaka, N. Miura, M. Narita, T. Mimaki, Y. Satoh, S. Kondo, and Y. Okada. 1992. Three nonsense mutations responsible for group A xeroderma pigmentosum. *Mutat. Res.* **273**:193–202.

178. Schaeffer, L., V. Moncollin, R. Roy, A. Staub, M. Mezzina, A. Sarasin, G. Weeda, J. H. Hoeijmakers, and J. M. Egly. 1994. The ERCC2/DNA repair protein is associated with the class II BTF2/TfIIH transcription factor. *EMBO J.* **13**:2388–2392.

179. Schaeffer, L., R. Roy, S. Humbert, V. Moncollin, W. Vermeulen, J. H. Hoeijmakers, P. Chambon, and J. M. Egly. 1993. DNA repair helicase: a component of BTF2 (TFIIH) basic transcription factor. *Science* **260**:58–63.

180. Scherly, D., T. Nouspikel, J. Corlet, C. Ucla, A. Bairoch, and S. G. Clarkson. 1993. Complementation of the DNA repair defect in xeroderma pigmentosum group G cells by a human cDNA related to yeast RAD2. *Nature* (London) **363**:182–185.

181. Schultz, R. A., D. P. Barbis, and E. C. Friedberg. 1985. Studies on gene transfer and reversion to UV resistance in xeroderma pigmentosum cells. *Somatic Cell Mol. Genet.* **11**:617–624.

182. Selfridge, J., A. M. Pow, J. McWhir, T. M. Magin, and D. W. Melton. 1992. Gene targeting using a mouse *HPRT* minigene/HPRT-deficient embryonic stem cell system: inactivation of the mouse *ERCC-1* gene. *Somatic Cell Mol. Genet.* **18**:325–336.

183. SenGupta, D. N., B. Z. Zmudzka, P. Kumar, F. Cobianchi, J. Skowronski, and S. H. Wilson. 1986. Sequence of human DNA polymerase β mRNA obtained through cDNA cloning. *Biochem. Biophys. Res. Commun.* **136**:341–347.

184. Shimamoto, T., K. Kohno, K. Tanaka, and Y. Okada. 1991. Molecular cloning of human *XPAC* gene homologs from chicken, *Xenopus laevis* and *Drosophila melanogaster*. *Biochim. Biophys. Acta* **181**:1231–1237.

185. Shiomi, T., Y. Harada, T. Saito, N. Shiomi, Y. Okuno, and M. Yamaizumi. 1994. An *ERCC5* gene with homology to yeast *RAD2* is involved in group G xeroderma pigmentosum. *Mutat. Res.* **314**:167–175.

186. Shiomi, T., N. Hieda-Shiomi, and K. Sato. 1982. Isolation of UV-sensitive mutants of mouse L5178Y cells by a cell suspension spotting method. *Somatic Cell Genet.* **8:**329–345.

187. Shivji, M. K. K., M. K. Kenny, and R. D. Wood. 1992. Proliferating cell nuclear antigen is required for DNA excision repair. *Cell* **69:**367–374.

188. Sibghat-Ullah, I. Husain, W. Carlton, and A. Sancar. 1989. Human nucleotide excision repair in vitro: repair of pyrimidine dimers, psoralen and cisplatin adducts by HeLa cell-free extract. *Nucleic Acids Res.* **17:**4471–4484.

189. Siegal, G., J. J. Turchi, T. W. Myers, and R. A. Bambara. 1992. A 5′ to 3′ exonuclease functionally interacts with calf DNA polymerase ε. *Proc. Natl. Acad. Sci. USA* **89:**9377–9381.

190. Skarnes, W., P. Bonin, and E. Baril. 1986. Exonuclease activity associated with a multiprotein form of HeLa cell DNA polymerase α. Purification and properties of the exonuclease. *J. Biol. Chem.* **261:** 6629–6636.

191. Smeets, H., L. Bachinski, M. Coerwinkel, J. Schepens, J. H. J. Hoeijmakers, M. van Duin, K. H. Grzeschik, C. A. Weber, P. de Jong, M. J. Siciliano, et al. 1990. A long-range restriction map of the human chromosome 19q13 region: close physical linkage between *CKMM* and the *ERCC1* and *ERCC2* genes. *Am. J. Hum. Genet.* **46:**492–501.

192. Smith, C. A. 1984. Analysis of repair synthesis in the presence of inhibitors, p. 51–71. *In* A. R. S. Collins, C. S. Downes, and R. T. Johnson (ed.), *DNA Repair and Its Inhibition.* IRL Press, Oxford.

193. Snyder, R. D., B. Van Houten, and J. D. Regan. 1982. Studies on the inhibition of repair of ultraviolet- and methyl methanesulfonate-induced damage in the DNA of human fibroblasts by novobiocin. *Nucleic Acids Res.* **10:**6207–6219.

194. Snyderwine, E. G., and V. A. Bohr. 1992. Gene- and strand-specific damage and repair in Chinese hamster ovary cells treated with 4-nitroquinoline 1-oxide. *Cancer Res.* **52:**4183–4189.

195. So, A. G., and K. M. Downey. 1988. Mammalian DNA polymerases α and δ: current status in DNA replication. *Biochemistry* **27:** 4591–4595.

196. Spivak, G., A. K. Ganesan, and P. C. Hanawalt. 1984. Enhanced transformation of human cells by UV-irradiated pSV2 plasmids. *Mol. Cell. Biol.* **4:**1169–1171.

197. Spivak, G., S. A. Leadon, J.-M. Vos, S. Meade, P. C. Hanawalt, and A. K. Ganesan. 1988. Enhanced transforming activity of pSV2 plasmids in human cells depends upon the type of damage introduced into the plasmid. *Mutat. Res.* **193:**97–108.

198. Stefanini, M., A. R. S. Collins, R. Riboni, M. Klaude, E. Botta, D. L. Mitchell, and F. Nuzzo. 1991. Novel Chinese hamster ultraviolet-sensitive mutants for excision repair form complementation groups 9 and 10. *Cancer Res.* **51:**3965–3971.

199. Stevsner, T., and V. A. Bohr. 1993. Studies on the role of topoisomerases in general, gene- and strand-specific DNA repair. *Carcinogenesis* **14:**1841–1850.

200. Stockdale, F. E. 1971. DNA synthesis in differentiating skeletal muscle cells: initiation by ultraviolet light. *Science* **171:**1145–1147.

201. Stockdale, F. E., and M. C. O'Neill. 1972. Repair DNA synthesis in differentiated embryonic muscle cells. *J. Cell Biol.* **52:**589–597.

202. Striniste, G. F., D. J. Chen, D. de Bruin, J. A. Luke, L. S. McCoy, J. S. Mudgett, J. W. Nickols, R. T. Okinaka, J. G. Tesmer, and M. A. MacInnes. 1988. Restoration of the chinese hamster cell radiation resistance by the human repair gene ERCC-5 and progress in molecular cloning of this gene, p. 301–306. *In* E. C. Friedberg and P. C. Hanawalt (ed.), *Mechanisms and Consequences of DNA Damage Processing.* Alan R. Liss, Inc., New York.

203. Sugasawa, K., C. Masutani, and F. Hanaoka. 1993. Cell-free repair of UV-damaged simian virus 40 chromosomes in human cell extracts. I. Development of a cell-free system detecting excision repair of UV-irradiated SV40 chromosomes. *J. Biol. Chem.* **268:**9098–9104.

204. Sung, P., V. Bailly, C. Weber, L. H. Thompson, L. Prakash, and S. Prakash. 1993. Human xeroderma pigmentosum group D gene encodes a DNA helicase. *Nature* (London) **365:**852–855.

205. Sung, S. C. 1974. Effect of novobiocin on DNA-dependent DNA polymerases from developing rat brain. *Biochim. Biophys. Acta* **361:**115–117.

206. Svoboda, D. L., J.-S. Taylor, J. E. Hearst, and A. Sancar. 1993. DNA repair by eukaryotic nucleotide excision nuclease. Removal of thymine dimer and proralen monoadduct by HeLa cell-free extract and of thymine dimer by *Xenopus laevis* oocytes. *J. Biol. Chem.* **268:** 1931–1936.

207. Syväoja, J., and S. Linn. 1989. Characterization of a large form of DNA polymerase delta from HeLa cells that is insensitive to proliferating cell nuclear antigen. *J. Biol. Chem.* **264:**2489–2497.

208. Syväoja, J., S. Suomensaari, C. Nishida, J. S. Goldsmith, G. S. J. Chui, S. Jain, and S. Linn. 1990. DNA polymerases α, δ, and ε: three distinct enzymes from HeLa cells. *Proc. Natl. Acad. Sci. USA* **87:**6664–6668.

209. Szymkowski, D. E., M. A. N. Hajibagheri, and R. D. Wood. 1993. Electron microscopy of DNA excision repair patches produced by human cell extracts. *J. Mol. Biol.* **231:**251–260.

210. Szymkowski, D. E., C. W. Lawrence, and R. D. Wood. 1993. Repair by human cell extracts of single (6-4) and cyclobutane thymine-thymine photoproducts in DNA. *Proc. Natl. Acad. Sci. USA* **90:**9823–9827.

211. Takahashi, E., N. Shiomi, and T. Shiomi. 1992. Precise localization of the excision repair gene, *ERCC5*, to human chromosome 13q32.3-q33.1 by direct R-banding fluorescence in situ hybridization. *Jpn. J. Cancer Res.* **83:**1117–1119.

212. Takano, T., M. Noda, and T.-A. Tamura. 1982. Transfection of cells from a xeroderma pigmentosum patient with normal human DNA confers UV resistance. *Nature* (London) **296:**269–270.

213. Takao, M., M. Abramic, M. Moos Jr., V. R. Otrin, J. C. Wootton, M. McLenigan, A. S. Levine, and M. Protic. 1993. A 127 kDa component of a UV-damaged DNA-binding complex, which is defective in some xeroderma pigmentosum group E patients, is homologous to a slime mold protien. *Nucleic Acids Res.* **21:**4111–4118.

214. Tan, C.-K., C. Castillo, A. G. So, and K. M. Downey. 1986. An auxiliary protein for DNA polymerase-δ from fetal calf thymus. *J. Biol. Chem.* **261:**12310–12316.

215. Tanaka, K. 1992. Joint workshop on DNA repair mechanisms and embryo manipulation. Report of the third annual workshop of the Institute for Molecular and Cellular Biology, Osaka University, held in Osaka (Japan), 21–23 January 1991. *Mutat. Res.* **273:**237–241.

216. Tanaka, K. 1994. Personal communication.

217. Tanaka, K., N. Miura, I. Satokata, I. Miyamoto, M. C. Yoshida, Y. Satoh, S. Kondo, A. Yasui, H. Okayama, and Y. Okada. 1990. Analysis of a human DNA excision repair gene involved in group A xeroderma pigmentosum and containing a zinc-finger domain. *Nature* (London) **348:**73–76.

218. Tanaka, K., I. Satokata, Z. Ogita, T. Uchida, and Y. Okada. 1989. Molecular cloning of a mouse DNA repair gene that complements the defect of group-A xeroderma pigmentosum. *Proc. Natl. Acad. Sci. USA* **86:**5512–5516.

219. Tanaka, K., M. Sekiguchi, and Y. Okada. 1975. Restoration of ultraviolet-induced unscheduled DNA synthesis of xeroderma pigmentosum cells by the concomitant treatment with bacteriophage T4 endonuclease V and HVJ (Sendai virus). *Proc. Natl. Acad. Sci. USA* **72:**4071–4075.

220. Tanaka, S., S. Z. Hu, T. S.-F. Wang, and D. Korn. 1982. Preparation and preliminary characterization of monoclonal antibodies against human DNA polymerase α. *J. Biol. Chem.* **257:**8386–8390.

221. Teraoka, H., and K. Tsukada. 1982. Eukaryotic DNA ligase. Purification and properties of the enzyme from bovine thymus, and immunochemical studies of the enzyme from animal tissues. *J. Biol. Chem.* **257:**4758–4763.

222. Thacker, J. 1986. The use of recombinant DNA techniques to study radiation-induced damage, repair and genetic change in mammalian cells. *Int. J. Radiat. Biol.* **50:**1–30.

223. Thompson, L. H., K. W. Brookman, C. A. Weber, E. P. Salazar, J. T. Reardon, A. Sancar, Z. Deng, and M. J. Siciliano. 1994.

Molecular cloning of the human nucleotide excision-repair gene *ERCC4*. *Proc. Natl. Acad. Sci. USA* **91**:6855–6859.

224. **Thompson, L. H., D. B. Busch, K. Brookman, C. L. Mooney, and D. A. Glaser.** 1981. Genetic diversity of UV-sensitive DNA repair mutants of Chinese hamster ovary cells. *Proc. Natl. Acad. Sci. USA* **78**:3734–3737.

225. **Thompson, L. H., A. V. Carrano, K. Sato, E. P. Salazar, B. F. White, S. A. Stewart, J. L. Minkler, and M. J. Siciliano.** 1987. Identification of nucleotide-excision-repair genes on human chromosomes 2 and 13 by functional complementation in hamster-human hybrids. *Somatic Cell Mol. Genet.* **13**:539–551.

226. **Thompson, L. H., C. L. Mooney, K. Burkhart-Schultz, A. V. Carrano, and M. J. Siciliano.** 1985. Correction of a nucleotide-excision-repair mutation by human chromosome 19 in hamster-human hybrid cells. *Somatic Cell Mol. Genet.* **11**:87–92.

227. **Thompson, L. H., J. S. Rubin, J. E. Cleaver, G. F. Whitmore, and K. Brookman.** 1980. A screening method for isolating DNA repair-deficient mutants of CHO cells. *Somatic Cell Genet.* **6**:391–405.

228. **Thompson, L. H., E. P. Salazar, K. W. Brookman, C. C. Collins, S. A. Stewart, D. B. Busch, and C. A. Weber.** 1987. Recent progress with the DNA repair mutants of Chinese hamster ovary cells. *J. Cell Sci. Suppl.* **6**:97–110.

229. **Thompson, L. H., T. Shiomi, E. P. Salazar, and S. A. Stewart.** 1988. An eighth complementation group of rodent cells hypersensitive to ultraviolet radiation. *Somatic Cell Mol. Genet.* **14**:605–612.

230. **Thompson, L. H., C. A. Weber, and A. V. Carrano.** 1988. Human DNA repair genes, p. 289–293. *In* E. C. Friedberg and P. C. Hanawalt (ed.), *Mechanisms and Consequences of DNA Damage Processing*. Alan R. Liss, Inc., New York.

231. **Tomkinson, A. E.** 1994. Personal communication.

232. **Tomkinson, A. E., I. Husain, J.-W. Chen, and J. M. Besterman.** 1994. Personal communication.

233. **Tomkinson, A. E., D. D. Lasko, G. Daly, and T. Lindahl.** 1990. Mammalian DNA ligases. Catalytic domain and size of DNA ligase I. *J. Biol. Chem.* **265**:12611–12617.

233a. **Tomkinson, A. E., N. Tappe, and E. C. Friedberg.** 1992. DNA ligase I from *Saccharomyces cerevisiae*: physical and biochemical characterization of the *CDC9* gene product. *Biochemistry* **31**:11762–11771.

234. **Tomkinson, A. E., N. F. Totty, M. Ginsburg, and T. Lindahl.** 1991. Location of the active site for enzyme-adenylate formation in DNA ligases. *Proc. Natl. Acad. Sci. USA* **88**:400–404.

235. **Toschi, L., and R. Bravo.** 1988. Changes in cyclin/proliferating cell nuclear antigen distribution during DNA repair synthesis. *J. Cell Biol.* **107**:1623–1628.

236. **Treiber, D. K., Z. Chen, and J. M. Essigmann.** 1992. An ultraviolet light-damaged DNA recognition protein absent in xeroderma pigmentosum group E cells binds selectively to pyrimidine (6-4) pyrimidone photoproducts. *Nucleic Acids Res.* **20**:5805–5810.

237. **Troelstra, C., W. Hesen, D. Bootsma, and J. H. J. Hoeijmakers.** 1993. Structure and expression of the excision repair gene *ERCC6*, involved in the human disorder Cockayne's syndrome group B. *Nucleic Acids Res.* **21**:419–426.

238. **Troelstra, C., H. Odijk, J. de Wit, A. Westerveld, L. H. Thompson, D. Bootsma, and J. H. J. Hoeijmakers.** 1990. Molecular cloning of the human DNA excision repair gene *ERCC-6*. *Mol. Cell. Biol.* **10**:5806–5813.

239. **Troelstra, C., A. van Gool, J. de Wit, W. Vermeulen, D. Bootsma, and J. H. J. Hoeijmakers.** 1992. *ERCC6*, a member of a subfamily of putative helicases, is involved in Cockayne's syndrome and preferential repair of active genes. *Cell* **71**:939–953.

240. **Tseng, B. Y., and C. N. Ahlem.** 1982. DNA primase activity from human lymphocytes. Synthesis of oligoribonucleotides that prime DNA synthesis. *J. Biol. Chem.* **257**:7280–7283.

241. **Tsurimoto, T., T. Melendy, and B. Stillman.** 1990. Sequential initiation of lagging and leading strand synthesis by two different polymerase complexes at the SV40 DNA replication origin. *Nature* (London) **346**:534–539.

242. **van Assendelft, G. B., E. M. Rigney, and I. D. Hickson.** 1993. Purification of a HeLa cell nuclear protein that binds selectively to DNA irradiated with ultra-violet light. *Nucleic Acids Res.* **21**:3399–3404.

243. **van Duin, M., J. de Wit, H. Odijk, A. Westerveld, A. Yasui, H. M. Koken, J. H. J. Hoeijmakers, and D. Bootsma.** 1986. Molecular characterization of the human excision repair gene *ERCC-1*: cDNA cloning and amino acid homology with the yeast DNA repair gene *RAD10*. *Cell* **44**:913–923.

244. **van Duin, M., J. H. Janssen, J. de Wit, J. H. J. Hoeijmakers, L. H. Thompson, D. Bootsma, and A. Westerveld.** 1988. Transfection of the cloned human excision repair gene *ERCC-1* to UV-sensitive CHO mutants only corrects the repair defect in complementation group-2 mutants. *Mutat. Res.* **193**:123–130.

245. **van Duin, M., J. van den Tol, P. Warmerdam, H. Odijk, D. Meijer, A. Westerveld, D. Bootsma, and J. H. J. Hoeijmakers.** 1988. Evolution and mutagenesis of the mammalian excision repair gene *ERCC-1*. *Nucleic Acids Res.* **16**:5305–5322.

246. **van Duin, M., G. Vredeveldt, L. V. Mayne, H. Odijk, W. Vermeulen, B. Klein, G. Weeda, J. H. J. Hoeijmakers, D. Bootsma, and A. Westerveld.** 1989. The cloned human DNA excision repair gene *ERCC-1* fails to correct xeroderma pigmentosum complementation groups A through I. *Mutat. Res.* **217**:83–92.

247. **van Vuuren, A. J., E. Appeldoorn, H. Odijk, A. Yasui, N. G. J. Jaspers, D. Bootsma, and J. H. J. Hoeijmakers.** 1993. Evidence for a repair enzyme complex involving ERCC1 and complementing activities of ERCC4, ERCC11 and xeroderma pigmentosum group F. *EMBO J.* **12**:3693–3701.

248. **van Vuuren, A. J., W. Vermeulen, L. Ma, G. Weeda, E. Appeldoorn, N. G. Jaspers, A. J. van der Eb, D. Bootsma, J. H. J. Hoeijmakers, S. Humbert, L. Schaeffer, and J. M. Egly.** 1994. Correction of xeroderma pigmentosum repair defect by basal transcription factor BTF2 (TFIIH). *EMBO J.* **13**:1645–1653.

249. **Venema, J., A. van Hoffen, V. Karcagi, A. T. Natarajan, A. A. van Zeeland, and L. H. Mullenders.** 1991. Xeroderma pigmentosum complementation group C cells remove pyrimidine dimers selectively from the transcribed strand of active genes. *Mol. Cell. Biol.* **11**:4128–4134.

250. **Vermeulen, W., P. Osseweijer, A. J. de Jonge, and J. H. J. Hoeijmakers.** 1986. Transient correction of excision repair defects in fibroblasts of 9 xeroderma pigmentosum complementation groups by microinjection of crude human cell extracts. *Mutat. Res.* **165**:199–206.

251. **Vermeulen, W., R. J. Scott, S. Rodgers, H. J. Muller, J. Cole, C. F. Arlett, W. J. Kleijer, D. Bootsma, J. H. J. Hoeijmakers, and G. Weeda.** 1994. Clinical heterogeneity within xeroderma pigmentosum associated with mutations in the DNA repair and transcription gene *ERCC3*. *Am. J. Hum. Genet.* **54**:191–200.

252. **Vermeulen, W., A. J. van Vuuren, E. Appeldoorn, G. Weeda, N. G. J. Jaspers, A. Priestly, C. F. Arlett, A. R. Lehmann, M. Stefanini, M. Mezzina, A. Sarasin, D. Bootsma, J.-M. Egly, and J. H. J. Hoeijmakers.** 1994. Three unusual repair deficiencies associated with transcription factor BTF2(TFIIH). Evidence for the existence of transcription syndromes. Cold Spring Harbor Symp. Quant. Biol. **59**: In press.

253. **Waga, S., and B. Stillman.** 1994. Anatomy of a DNA replication fork revealed by reconstitution of SV40 DNA replication in vitro. *Nature* (London) **369**:207–212.

254. **Wahl, A. F., A. M. Geis, B. H. Spain, S. W. Wong, D. Korn, and T. S.-F. Wang.** 1988. Gene expression of human DNA polymerase α during cell proliferation and the cell cycle. *Mol. Cell. Biol.* **8**:5016–5025.

255. **Walter, R.** 1994. Personal communication.

256. **Wang, T. S.-F.** 1991. Eukaryotic DNA polymerases. *Annu. Rev. Biochem.* **60**:513–552.

257. **Wang, T. S.-F., S. Z. Hu, and D. Korn.** 1984. DNA primase from KB cells. Characterization of a primase activity tightly associated with immunoaffinity-purified DNA polymerase-α. *J. Biol. Chem.* **259**:1854–1865.

258. **Wang, T. S.-F., and D. Korn.** 1980. Reactivity of KB cell deoxyribonucleic acid polymerases α and β with nicked and gapped deoxyribonucleic acid. *Biochemistry* **19**:1782–1790.

259. Wang, T. S.-F., and D. Korn. 1982. Specificity of the catalytic interaction of human DNA polymerase β with nucleic acid substrates. *Biochemistry* **21:**1597–1608.

260. Wang, T. S.-F., B. E. Pearson, H. A. Suomalainen, T. Mohandas, L. J. Shapiro, J. Schroder, and D. Korn. 1985. Assignment of the gene for human DNA polymerase α to the X chromosome. *Proc. Natl. Acad. Sci. USA* **82:**5270–5274.

261. Wang, Z., X. Wu, and E. C. Friedberg. 1991. Nucleotide excision repair of DNA by human cell extracts is suppressed in reconstituted nucleosomes. *J. Biol. Chem.* **266:**22472–22478.

262. Weber, C. 1994. Personal communication.

263. Weber, C. A., E. P. Salazar, S. A. Stewart, and L. H. Thompson. 1988. Molecular cloning and biological characterization of a human gene, *ERCC2*, that corrects the nucleotide excision repair defect in CHO UV5 cells. *Mol. Cell. Biol.* **8:**1137–1146.

264. Weber, C. A., E. P. Salazar, S. A. Stewart, and L. H. Thompson. 1990. *ERCC2:* cDNA cloning and molecular characterization of a human nucleotide excision repair gene with high homology to yeast *RAD3. EMBO J.* **9:**1437–1447.

265. Weber, C. A., and K. Takayama. 1994. Personal communication.

266. Weeda, G., and J. H. J. Hoeijmakers. 1993. Genetic analysis of nucleotide excision repair in mammalian cells. *Semin. Cancer Biol.* **4:**105–117.

267. Weeda, G., and J. H. J. Hoeijmakers. 1994. Personal communication.

268. Weeda, G., J. H. J. Hoeijmakers, and D. Bootsma. 1993. Genes controlling nucleotide excision repair in eukaryotic cells. *Bioessays* **15:**249–258.

269. Weeda, G., L. B. Ma, R. C. van Ham, A. J. van der Eb, and J. H. J. Hoeijmakers. 1991. Structure and expression of the human *XPBC/ERCC-3* gene involved in DNA repair disorders xeroderma pigmentosum and Cockayne's syndrome. *Nucleic Acids Res.* **19:**6301–6308.

270. Weeda, G., R. C. van Ham, W. Vermeulen, D. Bootsma, A. J. van der Eb, and J. H. J. Hoeijmakers. 1990. A presumed DNA helicase encoded by *ERCC-3* is involved in the human repair disorders xeroderma pigmentosum and Cockayne's syndrome. *Cell* **62:**777–791.

270a. Westerveld, A., J. H. J. Hoeijmakers, M. van Duin, J. de Wit, H. Odijk, A. Pastink, R. D. Wood, and D. Bootsma. 1984. Molecular cloning of a human DNA repair gene. *Nature* (London) **310:**425–429.

271. Wicha, M., and F. E. Stockdale. 1972. DNA-dependent DNA polymerases in differentiating embryonic muscle cells. *Biochem. Biophys. Res. Commun.* **48:**1079–1087.

272. Wilkins, R. J. 1983. Failure of the intercalating agent 4′-(9-acridylamino)-methanesulphon-m-aniside to induce DNA-repair replication in cultured human cells. *Mutat. Res.* **122:**211–216.

273. Wilson, S., J. Abbotts, and S. Widen. 1988. Progress toward molecular biology of DNA polymerase β. *Biochim. Biophys. Acta* **949:**149–157.

274. Wold, M. S., and T. Kelly. 1988. Purification and characterization of replication protein A, a cellular protein required for in vitro replication of simian virus 40 DNA. *Proc. Natl. Acad. Sci. USA* **85:**2523–2527.

275. Wong, S. W., A. F. Wahl, P.-M. Yuan, N. Arai, B. E. Pearson, K.-I. Arai, D. Korn, M. W. Hunkapiller, and T. S.-F. Wang. 1988. Human DNA polymerase α gene expression is cell proliferation dependent and its primary structure is similar to both prokaryotic and eukaryotic replicative DNA polymerases. *EMBO J.* **7:**37–47.

276. Wood, R., and H. J. Burki. 1982. Repair capability and the cellular age response for killing and mutation induction after UV. *Mutat. Res.* **95:**505–514.

277. Wood, R. D. 1989. Repair of pyrimidine dimer ultraviolet light photoproducts by human cell extracts. *Biochemistry* **28:**8287–8292.

278. Wood, R. D. 1994. Personal communication.

279. Wood, R. D., and D. Coverley. 1991. DNA excision repair in mammalian cell extracts. *Bioessays* **13:**447–453.

280. Wood, R. D., P. Robins, and T. Lindahl. 1988. Complementation of the xeroderma pigmentosum DNA repair defect in cell-free extracts. *Cell* **53:**97–106.

281. Wood, R. D., and D. E. Szymkowski. 1992. A system to dissect nucleotide excision repair in vitro and its use to study repair of DNA modified by cisplatin, p. 314–324. *In* V. A. Bohr, K. Wassermann, and K. H. Kraemer (ed.), *DNA Repair Mechanisms.* Munksgaard, Copenhagen.

282. Yamaguchi, M., E. A. Hendrickson, and M. L. DePamphilis. 1985. DNA primase-DNA polymerase α from simian cells. Modulation of RNA primer synthesis by ribonucleotide triphosphates. *J. Biol. Chem.* **260:**6254–6263.

283. Yamaizumi, M., T. Sugano, H. Asahina, Y. Okada, and T. Uchida. 1986. Microinjection of partially purified protein factor restores DNA damage specificallly in group A of xeroderma pigmentosum cells. *Proc. Natl. Acad. Sci. USA* **83:**1476–1479.

284. Yang, S., F. F. Becker, and J. Y.-H. Chan. 1993. Biochemical characterization of a protein inhibitor for DNA ligase I from human cells. *Biochem. Biophys. Res. Commun.* **191:**1004–1013.

285. Yang, S.-W., F. F. Becker, and J. Y.-H. Chan. 1992. Identification of a specific inhibitor for DNA ligase I in human cells. *Proc. Natl. Acad. Sci. USA* **89:**2227–2231.

286. Zdzienicka, M. Z., L. Roza, A. Westerveld, D. Bootsma, and J. W. Simons. 1987. Biological and biochemical consequences of the human *ERCC-1* repair gene after transfection into a repair-deficient CHO cell line. *Mutat. Res.* **183:**69–74.

287. Zdzienicka, M. Z., and J. W. Simons. 1987. Mutagen-sensitive cell lines are obtained with a high frequency in V79 Chinese hamster cells. *Mutat. Res.* **178:**235–244.

288. Zelle, B., and P. H. Lohman. 1979. Repair of UV-endonuclease-susceptible sites in the 7 complementation groups of xeroderma pigmentosum A through G. *Mutat. Res.* **62:**363–368.

289. Zmudzka, B. Z., A. Fornace, Jr., J. Collins, and S. H. Wilson. 1988. Characterization of DNA polymerase β mRNA: cell-cycle and growth response in cultured human cells. *Nucleic Acids Res.* **16:**9587–9596.

290. Zmudzka, B. Z., and S. H. Wilson. 1990. Deregulation of DNA polymerase β by sense and antisense RNA expression in mouse 3T3 cells alters cell growth. *Somatic Cell Mol. Genet.* **16:**311–320.

291. Zwetsloot, J. C., A. P. Barbeiro, W. Vermeulen, H. M. Arthur, J. H. J. Hoeijmakers, and C. Backendorf. 1986. Microinjection of *Escherichia coli* UvrA, B, C and D proteins into fibroblasts of xeroderma pigmentosum complementation groups A and C does not result in restoration of UV-induced unscheduled DNA synthesis. *Mutat. Res.* **166:**89–98.

9

Mismatch Repair

B oth prokaryotic and eukaryotic cells are capable of repairing mismatched base pairs in their DNA. Mismatched base pairs in DNA can arise by several processes. One of the most important is by *replication errors*. In this case, the correct base of the mispair is located in the parental strand of the newly replicated DNA and proper correction of the mismatch contributes to the maintenance of the fidelity of the genetic information. Another mechanism is by the *formation of a heteroduplex* between two homologous DNA molecules as part of a recombinational process. If the two DNAs differ slightly in their sequence, as a consequence either of a mutation used as a genetic marker or of sequence changes acquired during evolutionary divergence, mismatches can be formed. In these cases, repair of mismatched base pairs influences the nature of the products that are obtained from the recombinational event. Mismatches can also be formed if hairpins form between imperfect palindromes. An additional specialized but important way that mismatched base pairs can arise is by the *deamination of 5-methylcytosine*. This modified base is present in the DNA of many organisms, both prokaryotic and eukaryotic. Deamination of 5-methylcytosine converts a G·5-mC base pair to a G·T base pair. Correct conversion of this class of G·T base pairs to G·C base pairs is important for the maintenance of genetic fidelity. However, unlike the deamination of cytidine, in this case the deamination product, T, is a normal DNA base and consequently cannot be repaired by ura-DNA glycosylase (see chapter 4). In addition, mismatched base pairs in which one of the mismatched bases is a base analog or a chemically modified derivative of a normal base can be formed. Such cases represent a separate class of mismatched base pairs because one of the participating bases is not a normal constituent of DNA.

Recent work has made it clear that the molecular mechanisms of mismatch repair processes are related to the molecular mechanisms of base excision repair and nucleotide excision repair (see chapters 4 to 8). Furthermore, the basic enzymology of the major mismatch repair process appears to be very similar between prokaryotic and eukaryotic organisms. However, a consideration of mechanisms for repair of mismatched base pairs introduces a new concept that did not arise during our earlier discussions of base and nucleotide excision repair. For a mismatch repair event to contribute to the maintenance of genetic fidelity, the correct base in the mispair must be distinguished from the incorrect base. Since both bases are normal constituents of DNA, this cannot be achieved by an enzyme scanning the DNA for a lesion or structure that is not a normal constituent of DNA. In this chapter, we will summarize the evidence that led to the recognition that cells have the capability of repairing mismatched base pairs and the insights that have been gained into the mechanisms of these mismatch repair processes. We will also discuss the insights these basic studies have offered into a hereditary predisposition to cancer.

Early Biological Evidence for the Existence of Mismatch Repair

Genetic Phenomena Suggesting the Existence of Mismatch Repair

Historically, an interest in the fate of mismatched sequences in DNA arose from studies of genetic recombination. Thus, for example, it had been noted since the 1950s that apparent genetic exchanges within very short intervals of certain fungi and bacteriophage genomes take place very frequently, suggesting the occurrence of multiple crossovers at much greater than random frequencies (8, 26, 164). This phenomenon is termed *localized* or *high-negative interference*, because in classical genetic studies it frequently resulted in an overestimate of the linkage distance between two closely linked markers. During the 1960s, the concept emerged that this as well as other genetic "peculiarities," such as the phenomenon of *gene conversion* (nonreciprocal transfer of genetic information from one DNA molecule to another) observed in certain fungi, might be accounted for by the correction of mismatched sequences present in the heteroduplex regions of DNA generated during exchanges between two genomes (218). The notion that such correction might take place by excision repair stemmed directly from the basic model of excision repair of pyrimidine dimers that evolved in the early 1960s. An instructional example of this extrapolation is contained in a paper on the topic of gene conversion published by Robin Holliday in 1964. In exploring various mechanisms by which gene conversion might occur, Holliday referred to the formation of "hybrid" DNA molecules by homologous pairing between portions of single strands from two different DNA duplexes. He stated (79) that if

> this part of the genetic material is homozygous then normal base pairing will occur in the hybrid region, but if the annealed region spans a point of heterozygosity—a mutant site—then mispairing of bases will occur at this site. It is further postulated that this condition of mispaired bases is unstable. . . . One or both of such bases may get involved in exchange reactions. . . . [and] it is most reasonable to suppose that such exchange reactions would be enzyme mediated. There is a rather obvious connection between this suggestion and the growing evidence for mechanisms in the cell which can repair DNA damaged by mutagens. . . . If there are enzymes which can repair points of damage in DNA, it would seem possible that the same enzymes could recognize the abnormality of base pairing, and by exchange reactions rectify this.

Interestingly, in the same year that Robin Holliday's paper appeared, Evelyn Witkin independently suggested the existence of mismatch repair on the basis of experiments on bacterial mutagenesis (223). She had found that lactose-negative mutants induced by 5-bromouracil in *Escherichia coli* and detected without selection were often found as pure clones in sectors of otherwise lactose-positive colonies. The existence of the pure clones indicated that both strands of the DNA carried the same genetic error after the incorporation of 5-bromouracil. She suggested that this could come about, for example, if the incorporation of 5-bromouracil in place of cytosine carries with it a high probability of replacement by adenine of the guanine paired improperly with the 5-bromouracil erroneously incorporated. Such replacement might reflect the operation of intracellular repair mechanisms functioning to restore normal hydrogen bonding in damaged DNA.

Long-Patch Mismatch Repair in Prokaryotes

The most complex, but also the best understood, bacterial mismatch repair systems are the *methyl-directed mismatch repair system* of *E. coli* (69, 139, 145–147, 171, 173) and the related *Hex-dependent mismatch repair system* of *Streptococcus pneumoniae* (28, 103). These two systems have a similar specificity for different mismatches, which they process in a strand-specific manner. Furthermore, for both systems the strand specificity of the repair is determined by secondary signals that can be located a considerable distance from the actual mismatch. The excision tracts associated with these pathways can be large, 10^3 bp or more, so that these systems are usually referred to as *long-patch mismatch DNA repair systems*. However, the use of this terminology does not imply that all the patches are necessarily long, and the lower limit for the excision-repair tract size has not been established for either organism (146). Given the similarities in their biochemical mechanism, it is not too surprising that certain of the key proteins for mismatch repair in each organism appear to be evolutionarily related.

As will be discussed below, evidence is accumulating rapidly to suggest that eukaryotic cells have mismatch repair systems that are closely related to the long-patch mismatch repair systems of *E. coli* and *S. pneumoniae*, in terms of both mechanism and evolutionary origin. In this section, we will summarize the key observations and experiments that have led to the current model for long-patch mismatch repair.

Mismatch Repair after Transformation of *S. pneumoniae*

HIGH-EFFICIENCY AND LOW-EFFICIENCY MARKERS IN TRANSFORMATION OF *S. PNEUMONIAE*

The first evidence for the repair of mismatched bases in prokaryotes came from studies of transformation in *S. pneumoniae* (28). During the transformation of this organism, as in other gram-positive bacteria, the DNA from the donor cell is converted to single-stranded segments upon entry into a recipient cell (103). By a recombinational process, these donor segments then replace homologous segments in the recipient chromosome to generate a heteroduplex region (57, 101). If there is a genetic difference between the donor and recipient, this results in heteroduplex DNA containing a mismatch. A particularly striking feature of this phenomenon is the *variation in integration efficiencies* of different genetic markers (48, 102) (Table 9–1). High-efficiency markers yield transformants with an efficiency approaching one transformant per genome equivalent of donor DNA entering the cell. In contrast, the transformation efficiencies of other markers are typically in the range of 0.05 to 0.5. Interestingly, reciprocal crosses give generally the same efficiency regardless of whether it is high, low, or intermediate (102). That is to say, either base of a particular mispair can be eliminated, but the one that is eliminated is the one that is on the donor strand. Mismatch repair was postulated to account for these differences, making the assumptions that the repair occurs *preferentially* on the donor strand and that its *frequency* depends on the *identity* of the mismatch (47). In other words, the higher the repair efficiency, the lower the transformation efficiency.

Mismatch repair was also invoked to account for another feature of pneumococcal transformation (48). When the donor DNA carries two closely linked markers, one of low integration efficiency and one of high integration efficiency, the integration efficiency of the high-efficiency markers is lower than it would otherwise be. This effect of low-efficiency markers was explained by postulating

Table 9–1 Base changes at mutated sites and integration efficiency

malM mutation	Sequence location	DNA change	Integration efficiency[a]	Base mismatches[b]	
				l	r
malM564[c]	2710–2802	del93 bp	0.83		
malM594	3042	GC → TA	0.98	AGA	ATA
				TAT	TCT
malM567	2722	GC → AT	0.04	GGC	GAC
				CTG	CCG
malM582	2722	GC → CG	0.50	GGC	GCC
				CGG	CCG

[a] Ratio of Mal+ to Sul^r transformants with DNA from a mal+ sul^r strain. The sul^r reference marker transforms *S. pneumoniae* with an efficiency as high as any known marker.
[b] l (upper) and r (lower) donor strand: mismatched base pairs are shown in boldface type.
[c] malM564 is a deletion of 93 bp.
Source: Data from Lacks (104) and Lacks et al. (105).

that an excision repair process provoked by the low-efficiency marker removed some or all of the donor strand including the high-efficiency marker. The distance dependence of the effect indicated that a donor strand segment 1 to 2 kb long is eliminated together with the mismatched base (102). This length corresponds to the size of the average segment that is normally integrated.

hex MUTANTS ARE DEFECTIVE IN MISMATCH REPAIR AND ARE SPONTANEOUS MUTATORS

S. pneumoniae mutants that no longer discriminate between high-efficiency and low-efficiency markers were isolated after mutagenic treatment of wild-type cells (100). The mutations responsible for this phenotype were designated *hex* (originally for high-efficiency, unknown [x]; now for heteroduplex repair deficiency) (28). In these mutants, low-efficiency markers transformed with the same high efficiency characteristic of high-efficiency markers, thus providing support for the concept of a positively acting mismatch repair system in *S. pneumoniae*. An extremely interesting property of the *hex* mutants was that they had an elevated spontaneous mutation rate (206). This finding raised the possibility that the *hex*-dependent mismatch-repair system might play a role in *mutation avoidance*.

TARGETING OF MISMATCH REPAIR TO ONE STRAND

For mismatch repair to be targeted to the donor strand, the donor strand must be discriminated from the host strand in the heteroduplex. It was postulated that this discrimination was provided by the presence of the strand breaks at the end of the donor strand (71). This model was consistent with the observation that low-efficiency markers were much more sensitive to UV irradiation of the donor DNA than were high-efficiency markers and that this sensitivity required the action of the *hex* genes (28, 122). A possible explanation for this phenomenon is that after DNA ligase has sealed the initial ends of the donor strand, the introduction of breaks into the donor DNA during nucleotide excision repair of the UV radiation damage makes the low-efficiency markers susceptible once again to strand-specific mismatch repair. The hypothesis that strand discrimination involves the recognition of breaks was extended to account for the mutator phenotype of the *hex* genes (30, 56, 105). Since the lagging strand of DNA is synthesized discontinuously, the breaks at the ends of Okazaki fragments (94) could serve to target *hex*-dependent mismatch repair to the daughter strand. This would permit the correction of mistakes introduced as replication errors. The possibility that the breaks in the nascent strands arise solely from uracil incorporation and removal has been ruled out by the demonstration that an *S. pneumoniae ung* mutation, which eliminates ura-DNA glycosylase (see chapter 4), has no effect on mutation avoidance by *hex*-dependent mismatch repair system (27).

IN VIVO ANALYSES OF METHYL-DIRECTED REPAIR IN *E. COLI*

The best-understood of any long-patch mismatch repair system is that of *E. coli*. Although this system is similar to that of *S. pneumoniae*, it has the additional interesting feature of using the state of N^6-methylation of adenine in GATC sequences as the major mechanism for determining the strand to be repaired. This feature of the *E. coli* long-patch mismatch repair system has facilitated both genetic and biochemical analyses, which have resulted in the formulation of a detailed model for the mechanism of long-patch mismatch repair.

REPAIR OF HETERODUPLEX DNA

The possibility that *E. coli* possesses a mismatch repair system was suggested by the observation of high-negative interference in bacteriophage λ crosses (216) (see below). Direct evidence that *E. coli* possesses a mismatch repair system came from experiments with heteroduplex molecules, usually of phage origin. These molecules are usually constructed by techniques in which the two strands of a phage duplex are separated on the basis of differential density and then reannealed with complementary strands of a mutant phage. Following transfection of appropriate hosts, if no mismatch correction occurs, then each DNA strand will serve as a template for the synthesis of two different homoduplex molecules that can be distinguished genetically. In such a case, two genetically distinct populations are obtained, ideally as a 50-50 mixture if there is no mismatch repair. However, if mismatched (heteroduplex) regions are corrected before DNA replication, affected molecules will yield only one type of homoduplex and the phenotype will reflect a corresponding bias in its representation in the progeny phage (Fig. 9–1).

The best-controlled (173) experiments of this type involved multiply marked λ heteroduplexes under conditions where both replication and recombination were blocked (212, 217, 221). The fate of the heterozygotic markers and hence of the corresponding mismatches was determined by analysis of the genotypes of

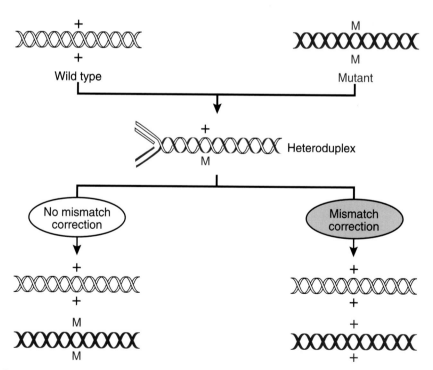

Figure 9–1 Use of heteroduplex molecules for measuring mismatch corrections. If the mismatch is corrected, all molecules generated by semiconservative DNA synthesis will be wild-type homoduplexes (right). However, in the absence of mismatch correction 50% of the progeny molecules will be mutant homoduplexes (left). (*Adapted from Friedberg [59] with permission.*)

phage particles emerging from single infective centers. These experiments demonstrated that heteroduplex correction occurs before the onset of DNA replication. They also showed that the efficiency of correction of various mismatches can vary by as much as an order a magnitude, a finding reminiscent of observations in *S. pneumoniae* transformation. Furthermore, if the heteroduplexes contained two or more closely spaced mismatches, analyses of the resulting progeny suggested that mismatch repair was restricted to one strand and that closely spaced mismatches tended to be cocorrected (212, 221). These observations were interpreted as being due to a type of excision repair mechanism with the excision repair tract extending over an average distance of 3 kb (212). On the basis of these observations, the explicit suggestion was made that mismatch correction could serve to eliminate DNA biosynthetic errors from newly synthesized DNA, pointing out that such a mechanism required that the repair be directed to the newly synthesized strand.

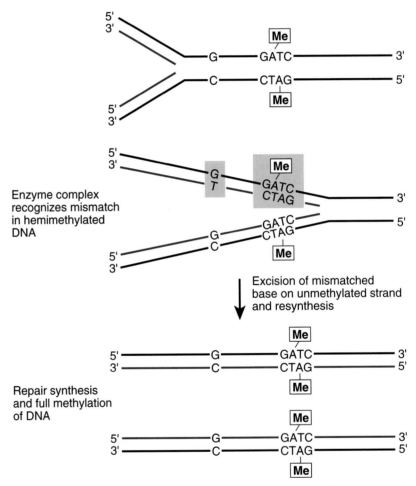

Figure 9–2 Wagner-Meselson model for postreplicative mismatch correction of DNA. GATC sequences in DNA are normally methylated (Me) at the 6 position of adenine. During semiconservative DNA synthesis, a G·T mismatch arises in one of the sister DNA duplexes. The enzymatic mechanism for repairing this lesion depends on discrimination between the newly synthesized (red) and parental (black) strands. This is achieved by recognition of the transient lack of methylation of the newly synthesized strand before postreplicative DNA methylation takes place. The nonmethylated daughter strand containing the incorrect base is enzymatically attacked by mismatch correction enzymes, and the misincorporated base is excised. Repair synthesis and daughter-strand methylation at GATC sequences restore the sister DNA duplexes to their native state. An alternative that Wagner and Meselson considered was that the strand discrimination was based on a special relationship between the mismatch repair system and the replication fork (212). (*Adapted from Friedberg [59] with permission.*)

STRAND DISCRIMINATION IS DETERMINED BY THE STATE OF METHYLATION OF GATC SITES

The possibility that the state of methylation plays a role in strand discrimination during postreplication mismatch correction was first proposed by Miroslav Radman and was discussed during the EMBO Recombination Workshop in 1975 (174). The following year, as part of their model for postreplication mismatch correction, Bob Wagner and Matthew Meselson (212) suggested that undermethylation of the newly synthesized strand might be one means of strand discrimination during mismatch correction and that another might be a special relationship between the repair system and the replication fork (Fig. 9–2). This hypothesis is based on the fact that methylation of DNA by sequence-specific methylases lags somewhat behind DNA replication (129, 131). Thus, immediately after synthesis, the newly-synthesized daughter-strand DNA is *undermethylated* relative to the parental strand. This difference in methylation state between parental and daughter strands just behind the replication fork could permit discrimination between the two strands. This idea was consistent with the observation that *E. coli dam* mutants (which are deficient in the methylation of GATC sites [132, 134]) display a spontaneous mutator phenotype (Table 9–2).

Direct support for the hypothesis that the state of methylation at GATC sequences plays a role in strand discrimination was provided by experiments involving *heteroduplexes* in which the two strands *differed* with respect to their state of *dam*⁺-dependent methylation (95, 126, 139, 168, 174) (Table 9–3). With hemimethylated heteroduplexes, which are methylated at GATC sequences only on one DNA strand, repair is highly biased to the unmethylated strand, with the methylated strand serving as the template for correction. If neither strand is methylated, mismatch correction occurs but shows little strand preference. Heteroduplexes in which both strands are highly methylated at GATC sites undergo mismatch repair at substantially reduced frequency. The bias toward repair of the unmethylated strand was found to occur efficiently even in DNA molecules in which the number of GATC sequences is smaller than that expected on a statistical basis. For example, methyl-directed mismatch repair occurred efficiently in a phage f1 heteroduplex in which there were only four GATC sites, the nearest being 1,000 bp from the mismatch (126). The results of experiments with heteroduplexes of φX174 with and without GATC sequences (wild-type φX174 contains no GATC sequences) indicated that mismatch repair requires the presence of an unmethylated GATC site (106).

An independent indication that the state of GATC methylation directs mismatch repair came from the finding that overproduction of the *dam*-encoded methylase increased the spontaneous mutation rate (78, 135). Since the rate of methylation of GATC sites is limited by the intracellular level of the *dam* methylase (203), an increased level of *dam* methylase activity would have the effect of shortening the temporal window in which daughter strands could be distinguished from parental strands and would thus reduce postreplicational mismatch correction.

THE *mutH, mutL, mutS*, AND *uvrD* GENE PRODUCTS ARE REQUIRED FOR METHYL-DIRECTED MISMATCH REPAIR

The *uvrD* gene product (helicase II) was first recognized as being important for mismatch repair during an experiment in which a phage λ heteroduplex was trans-

Table 9–2 Reversion frequencies of identical markers in *dam-3* and wild-type strains of *E. coli*

Strain	Reversion frequency ($\times 10^{-9}$) of:				
	leu-6	*proA2*	*lacY1*	*strA*	*rif*
GM44 (wild type)	3,67	0.16	4.5	0.16	4.0
GM45 (*dam-3*)	26.3	2.67	161.0	5.0	182.4

Source: Adapted from Marinus and Morris (133) and Friedberg (59) with permission.

Table 9–3 Transfections of wild-type *E. coli* with doubly heterozygous DNA heteroduplexes

Heteroduplex[a]	su+		su−		No. of plaques		Total no. of plaques analyzed
					Genotypes		
	cI+	cI	cI+	cI	cI+ Pam3	cIP+	
1 + Pam3 r ———— cI +	254	258	16	248	238 (46.5%)	248 (48.4%)	512
1 + Pam3 r ———— cI +	584	21	20	13	564 (96.5%)	13 (2%)	605
1 + Pam3 r ———— cI +	40	760	21	758	19 (2.4%)	758 (94.75%)	800
1 + Pam3 r ———— cI +	339	289	58	271	281 (44.7%)	271 (43%)	628

a ————, methylated strands; ———, unmethylated strands.

Source: Adapted from Radman et al. (174) and Friedberg (59) with permission.

fected into various recombination-defective and UV radiation-sensitive mutants (151). The *mutH*, *mutL*, and *mutS* mutants were originally recognized as being spontaneous mutators that exhibited a 100- to 1,000-fold increase in their spontaneous mutability (33). The isolation of *mutH*, *mutL*, *mutS*, and *uvrD* mutants during a screen for mutants with high mutation rates in the presence of the base analog bromouracil suggested that they might play role in mismatch repair. This suspicion was confirmed by transfecting them with a λ heteroduplex (183). Thus, as with *S. pneumoniae*, mutants defective in mismatch repair were found to have a mutator phenotype. This finding led to the subsequent demonstration that *mutH*, *mutL*, *mutS*, and *uvrD* mutants were deficient in their ability to carry out methyl-directed mismatch repair (Table 9–4). As discussed below, the biochemical roles of these gene products are now known in considerable detail.

The involvement of the *mutH*, *mutL*, and *mutS* gene products in methyl-directed mismatch repair was consistent with the other findings that associated these mutations with *dam*-dependent phenotypes. For example, combining *dam* mutations with *recA*, *recB*, *recC*, *recJ*, *lexA*, or *polA* mutations was found to result in

Table 9–4 Mismatch repair of λ heteroduplexes containing highly methylated chains and failure to observe mismatch correction in a *mutL* strain

Heteroduplex[a]	me state of l/ me state of h	% of each type of plaque in *mutL*+ host			No of plaques	% of each type of plaque in *mutL* host			No. of plaques
		c	Mixed	+		c	Mixed	+	
I	me+/me+	33	32	35	210	42	38	20	314
————— c l	me−/me−	64	3	33	235	52	21	27	289
————— + h	me+/me−	95	3	2	190	66	18	16	229
	me−/me+	4	2	94	203	22	28	50	404
II	me+/me+	29	26	45	215	34	35	30	204
————— + l	me−/me−	40	16	44	119	33	27	40	458
————— c h	me+/me−	38	29	32	226	34	37	29	337
	me−/me+	33	30	37	223	30	30	40	276

a Heteroduplexes are drawn with the light chain (*l*) on top and the heavy chain (*h*) below. With respect to the conventional genetic map of λ, the 5′ → 3′ direction is rightward on the light strand and leftward on the heavy strand. c is the *cI27* mutation that causes a "clear-plaque" phenotype. + represents the wild-type allele. Heteroduplex II is refractory to repair.

Source: Adapted from Pukkila et al. (168) with permission.

inviability (135). Most of the suppressors of the inviability of *dam recA*, *dam recB*, and *dam recC* double mutants turned out to be alleles of *mutH*, *mutL*, and *mutS* (138). The inviability appears to result from the accumulation of DNA double-strand breaks since *dam recA*(Ts) and *dam recB*(Ts) mutants were found to accumulate double-strand breaks at 42°C, a temperature at which they are inviable. Both the double-strand break formation and inviability are suppressed by the introduction of *mutL* or *mutS* mutations (213). In addition, *dam* mutants are killed by growth in the presence of certain base analogs such as 2-aminopurine (64). The majority of suppressors that allow the growth of *dam* mutants in the presence of 2-aminopurine turned out to be alleles of *mutH, mutL,* and *mutS*. From these results, it was suggested (64) that mismatch repair could initiate on either strand in the absence of methylation and consequently lead to the formation of double-strand breaks. This explanation was criticized (66) on the grounds that mismatches occur too infrequently to result in the overlap of excision tracts. As discussed below, more-recent work suggests that the cell death results from MutH making a second incision at an unmethylated GATC site if a mismatch is present and thereby producing a double-strand break (13).

Not only do strains defective in mismatch repair exhibit a spontaneous mutator phenotype but they are also hypersensitive to base substitution mutagenesis by 2-aminopurine (63) and bromouracil (183) and to frameshift mutagenesis by 9-aminoacridine (195) and oxazolopyridocarbazole (180). Since mutagenesis by these agents results from effects at the level of DNA synthesis (45), it appears that the elevated mutation rate results from a failure to repair mismatches or small deletions and insertions caused by these agents.

SPECIFICITY OF METHYL-DIRECTED MISMATCH REPAIR IN VIVO

The *specificity* of methyl-directed mismatch repair was investigated in vivo by the use of heteroduplex substrates containing different mismatches (39, 89, 95). Although the efficiency of repair was influenced to some extent by the base composition of the sequence surrounding the mismatch, some mismatches were found to be repaired more efficiently than others. In particular, the G·T and A·C mismatches, which give rise to transition mutations, are very efficiently repaired. Of mismatches that give rise to transversion mutations, G·G and A·A mismatches are usually corrected efficiently, some T·T, C·T, and G·A mismatches are corrected less efficiently, while the C·C mismatch appears to be subject to very little, if any, methyl-directed mismatch repair. Mismatches corresponding to the addition or deletion of a few nucleotides are also subject to efficient methyl-directed mismatch repair (40, 53), but heteroduplexes containing large nonhomologies are not processed by this system (38). The specificity for *hex*-dependent mismatch repair in *S. pneumoniae* was found to be extremely similar to that of the methyl-directed system of *E. coli*, reinforcing the view that the two systems are closely related (28, 29, 104, 105).

The specificity of mismatch repair in *E. coli* deduced from these in vivo experiments with artificially constructed heteroduplexes is consistent with the mutational spectra that have been determined for mutants that are defective in their ability to carry out methyl-directed mismatch repair. The spectra of spontaneous mutations in *mutH*, *mutL*, and *mutS* mutants were found to be remarkably similar (186) with respect to (i) the frequencies with which mutations occurred; (ii) the ratio of base substitution to frameshift mutations, with base substitutions predominating (Table 9–5); and (iii) the predominance of transition mutations over transversion mutations (Table 9–5). These similarities suggested either that the products of the *mutH*⁺, *mutL*⁺, and *mutS*⁺ genes act as a complex or that their action is very tightly coupled so that the loss of any of the three functions results in essentially the same phenotype. The high frequency of transition mutations observed in these spectra indicate that $mutH^+L^+S^+$-dependent mismatch repair makes a particularly important contribution to the maintenance of genetic fidelity by repairing G·T and A·C mismatches that are introduced during DNA replication. The spectrum of mutations seen in *mutH*, *mutL*, and *mutS* strains correlates well with the misincorporation propensities of DNA polymerase III holoenzyme (45), with one excep-

Table 9–5 Analysis of 487 sequenced spontaneous i⁻ᵈ mutations in *mutH*, *mutL*, and *mutS* strains

Mutation	No. of mutations in:			Total no.
	mutH	*mutL*	*mutS*	
Base substitutions				
G·C → A·T	40	41	38	119
A·T → G·C	85	90	56	231
G·C → T·A	3	1	5	9
A·T → T·A	0	0	0	0
G·C → C·G	0	0	2	2
A·T → C·G	2	0	2	4
Total	130	132	103	365
Frameshifts	47	25	50	122
Total	177	157	153	487

Source: Adapted from Schaaper and Dunn (186).

tion: G·A mispairs are common polymerase errors, but levels of mutations resulting from G·A mispairs are not highly elevated in *mutH*, *mutL*, and *mutS* mutants (147). It is possible that this discrepancy is because *E. coli* possesses auxiliary systems for repairing G·A mispairs (see below).

The capacity of the methyl-directed mismatch repair system to correct replication errors is limited and can be *saturated* if the cells carry another mutation that results in a high frequency of replication errors. The best-studied example is that of *mutD5*, which has the strongest spontaneous mutator phenotype known in *E. coli* (184, 185, 187). The *mutD5* mutation is an allele of *dnaQ*⁺, the gene encoding the proofreading exonuclease subunit of DNA polymerase III (94), and strains carrying this allele exhibit an extremely high mutation rate when grown on rich medium. Such cells were found to be unable to carry out *mutH*⁺*L*⁺*S*⁺-dependent mismatch repair of DNA heteroduplexes (184, 185, 187). However, the inability of *mutD5* mutants to carry out mismatch repair was shown to be different from that of *mutH*, *mutL*, and *mutS* mutants in several respects. The deficiency was not observed unless active DNA replication was occurring in the *mutD5* strain, and it could be suppressed by overexpression of either the MutH or the MutL protein (187). Taken together, these observations indicate that the mismatch-repair deficiency of *mutD5* strains results from a saturation of the methyl-directed mismatch repair system by the excess of primary DNA replication errors that arise because of the defect in proofreading. They further suggest that the extremely high mutation rate of *mutD5* strains under conditions of rapid DNA replication (i.e., growing on rich medium) is due to both the elevated frequency of replication errors and the effective loss of methyl-directed mismatch repair (187).

Biochemistry of Methyl-Directed Mismatch Repair in *E. coli*

The in vivo characterization of methyl-directed mismatch repair described above allowed the development of an in vitro assay for methyl-directed mismatch repair. This in turn led directly to the identification and purification of the various proteins required for mismatch repair and to the delineation of central aspects of the biochemical mechanism of mismatch repair. Together with the cloning of the genes encoding the mismatch repair proteins, this has facilitated the reconstitution of an in vitro system for methyl-directed mismatch repair that consists entirely of purified components and has resulted in a detailed model for the molecular mechanism of this process.

AN IN VITRO ASSAY FOR METHYL-DIRECTED MISMATCH REPAIR

A key event in the elucidation of the mechanism of methyl-directed mismatch repair was the development of an assay (126, 127) that allowed mismatch repair

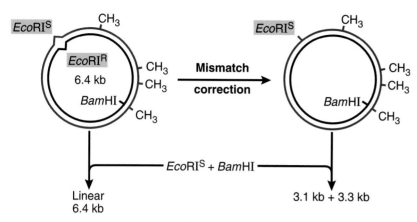

Figure 9–3 In vitro assay for mismatch correction. The substrate for mismatch repair is a covalently closed heteroduplex of f1R229 containing a mismatch within the *Eco*RI site (positions 5616 to 5621). Methyl groups indicate the locations of the four d(GATC) sites within the DNA (positions 216, 1382, 1714, and 2221; the last of these is also a *Bam*HI site). Cleavage of mismatch heteroduplexes with *Eco*RI and *Bam*HI yields the full-length linear *Bam*HI product, since the hybrid *Eco*RI site is resistant. Mismatch correction on the strand containing the mutant *Eco*RI sequence renders the site sensitive. Molecules repaired in this configuration thus yield two products upon hydrolysis with the pair of endonucleases. (*Adapted from Lu et al. [127] with permission.*)

to be detected in crude extracts of *E. coli*. The substrate used in the initial experiments was a *heteroduplex of bacteriophage f1 DNA* that contained a single mismatch (Fig. 9–3). The mismatch was located within a single *Eco*RI restriction enzyme site in this molecule, so that the consequences of the mismatch repair could be monitored by testing whether the products were susceptible to cleavage by the *Eco*RI enzyme. The molecule contains four GATC sites, so that heteroduplexes could be constructed in which one strand or the other was methylated. As summarized in Table 9–6, methyl-directed mismatch repair of these heteroduplexes was indeed observed in extracts of wild-type *E. coli*, but extracts of *mutH*, *mutL*, *mutS*, and *uvrD* mutants did not carry out this repair. The demonstration of both the methyl dependence of the repair and the dependence on the functions of the *mutH*[+], *mutL*[+], *mutS*[+] and *uvrD*[+] gene products indicated that the process observed in this in vitro assay was the same as the one characterized in vivo. As also shown in Table 9-6,

Table 9–6 Extracts of *E. coli* mutator strains are defective in mismatch repair[a]

Source of fraction	No. of *Eco*RI sites repaired (fmol/h/mg of protein)
mut[+]	42 ±9
mutH	4
mutS	<4
uvrD	<4
mutL	<4
mutH + *mutS*	45
mutH + *uvrD*	42
mutS + *uvrD*	36
mutH + *mutL*	36
mutS + *mutL*	50
uvrD + *mutL*	42

[a] Mismatch repair assays utilized hemimethylated f1R229 heteroduplexes containing a G·T mismatch at position 5621 within the *Eco*RI site of the viral DNA. GATC sites were methylated on the strand that contained the wild-type *Eco*RI sequence. Reaction mixtures contained concentrated crude extracts at 12 to 27 mg/ml when single fractions were assayed. In in vitro complementation assays, each extract was present at 12 to 14 mg/ml.

Source: Adapted from Lu et al. (127) with permission.

it was possible to restore mismatch repair in vitro by mixing extracts of two different mutants (e.g., *mutS* and *mutL*) that were deficient in mismatch repair. This latter observation was of great importance since it meant that the MutH, MutL, and MutS proteins could be purified by monitoring their activities in *in vitro complementation assays*. For example, the MutS protein could be purified from a MutS-overproducing strain by assaying its ability to restore mismatch repair to an extract of a *mutS* mutant.

These initial experiments also provided the first direct evidence that the mechanism of methyl-directed mismatch repair involves the *excision and resynthesis of DNA*. Substantial incorporation of dTMP into the heteroduplex DNA required both a mismatch and hemimethylation (Table 9–7). The polymerase required for the resynthesis appeared to be DNA polymerase III, since extracts of a *polA* strain (which is deficient in DNA polymerase I) were proficient in mismatch repair (109) but extracts of a temperature-sensitive *dnaZ* mutant strain (*dnaZ*⁺ encodes the τ and γ subunits of DNA polymerase III holoenzyme) were temperature sensitive for mismatch repair. In addition, antibodies to single-stranded-DNA-binding protein (SSB) inhibited mismatch repair in cell extracts (127), indicating that SSB was required. DNA ligase was also expected to be required to seal the nicks left at the end of the resynthesis step.

The efficiencies of repair of different mismatches by cell extracts were found to be similar to those determined in vivo (127, 201). In addition, methyl-directed mismatch repair was shown to be capable of repairing 1-, 2-, and 3-base insertion and deletion mismatches (112). This finding agrees well with in vivo experiments showing that heteroduplexes with 1-, 2-, or 3-base deletions are repaired as efficiently as G·T mismatches whereas 4-base deletions are marginally repaired and 5-base or larger deletions are not repaired at all (23, 158). It is also consistent with the later finding that MutS protein binds to heteroduplexes with 1, 2, 3, or 4 unpaired bases but not to a heteroduplex with 5 unpaired bases (158). Subsequent work also established that a single hemimethylated site is sufficient to determine the strand specificity of repair but that its activity in directing repair is dependent on both its sequence environment and its proximity to the mispair (22, 108, 123, 214).

PURIFICATION OF THE MutS, MutL, AND MutH PROTEINS AND CLONING OF THEIR GENES

Since the *mutH*⁺, *mutL*⁺, and *mutS*⁺ genes clearly played critical roles in methyl-directed mismatch repair, considerable effort was expended in identifying and purifying their gene products and in analyzing their biochemical activities. The *mutS*⁺ genes of *E. coli* and *Salmonella typhimurium* were cloned and found to encode 97-kDa proteins (155, 156, 202). The MutS protein was then purified from a strain which overproduced the *E. coli* MutS protein; its activity was monitored by in vitro complementation of *mutS* extracts (202). The MutS protein was found to bind to DNA substrates that contained a mismatch. The affinity of this binding was found to correlate roughly with the efficiency with which the various mispairs are repaired by methyl-directed mismatch repair (201). The binding of MutS protein to a mismatch protects about 22 bp of DNA from digestion with DNase I. No other

Table 9–7 Methyl dependence and mismatch dependence of repair DNA synthesis[a]

Mismatch	Methylation	Relative dTMP incorporation
G·T	+ / −	1.0
G·T	+ / +	0.27 ± 0.06
Non (G·C)	+ / −	0.33 ± 0.11

[a] Repair synthesis was measured by monitoring the incorporation of [α-³²P]dTTP into DNA.

Source: Adapted from Lu et al. (126).

protein required for mismatch repair has been shown to bind preferentially to mismatches, so MutS is a critical protein for the recognition of a mismatch in DNA. In addition, MutS was found to possess a weak ATPase activity (70, 74). Sequencing of the *mutS*[+] gene of *S. typhimurium* revealed a putative ATP-binding site in the predicated protein (73). Genetic alteration of this ATP-binding site was found to result in a mutator phenotype in vivo and in a large increase in the K_m for the ATPase activity (74). Interestingly, the presence of ATP influences the behavior of MutS protein on DNA that contains a mismatch. In the absence of ATP, MutS binds to mismatches as described above, but in the presence of ATP, it forms α-shaped loop structures which are stabilized at the DNA junction by the bound protein (147). The formation of the loops might play a role in allowing the mismatch and the GATC sites to interact during the repair process.

The *S. typhimurium mutS*[+] gene (73) and the *S. pneumoniae hexA*[+] gene (163) were sequenced at the same time, and a comparison of their sequences revealed that they were clearly *homologs*, with the region around the ATP-binding site being particularly strongly conserved. This finding provided compelling evidence that the mechanism of *hex*-dependent mismatch repair is related to that of methyl-directed mismatch repair in *E. coli* and also suggested that HexA protein plays a role in the recognition of mismatched base pairs in *S. pneumoniae*. Although the *hexA*[+] gene will not complement an *E. coli mutS* mutation, its expression in a wild-type background causes a mutator phenotype (167). This finding was rationalized by suggesting that either HexA protein bound to mismatches but was unable to interact with the other Mut proteins or that it formed nonfunctional repair complexes. Since the discovery of the homology between MutS and HexA, homologs of the *mutS*[+] gene have been discovered in yeast and mammalian cells; their significance and biological roles are discussed below.

The *mutL*[+] gene was cloned from *E. coli* (127) and *S. typhimurium* (155, 156) and was found to encode a 70-kDa protein. As with MutS protein, MutL protein was purified from a strain that overproduced *E. coli* MutL protein by in vitro complementation of a *mutL* extract (70). A specific biochemical function has not been detected for isolated MutL protein, but it has been shown to bind to MutS-heteroduplex complexes; another property is described below. The *S. typhimurium mutL*[+] gene (130) was sequenced at the same time as the *S. pneumoniae hexB*[+] gene (166) and the *Saccharomyces cerevisiae PMS1* gene (97) (see below), and in this case as well, the three proteins were found to be homologs.

The *E. coli* MutH protein was purified from an overproducing strain and was found to possess an extremely weak Mg^{2+}-dependent *endonuclease activity* that could nick hemimethylated or unmethylated GATC sequences in DNA (214). Cleavage occurred 5' to the G in the GATC sequence. Hemimethylated sequences were found to be cut on the unmethylated strand, unmethylated sequences were found to be cut usually only on one strand, and symmetrically methylated sequences were resistant to cleavage. These observations suggested that MutH protein plays a key role in the strand discrimination stage of mismatch repair and that this stage of the process involves the introduction of a strand break into the unmethylated strand. The function of MutH protein as a GATC-specific endonuclease was confirmed with the demonstration, both in vivo (110) and in vitro (107), that a *persistent strand break* not only circumvents the requirement for a GATC signal in mismatch repair but also bypasses the requirement for MutH protein. In the in vivo experiments, a persistent nick was achieved by transfecting a nicked heteroduplex DNA into a temperature-sensitive DNA ligase mutant at 40°C.

ESTABLISHMENT OF A DEFINED SYSTEM FOR MISMATCH CORRECTION

The availability of the purified MutH, MutL, and MutS proteins made it possible to reconstitute methyl-directed repair in a system composed of purified components (31, 107, 147). In addition to these three proteins, the complete system has been shown to require DNA helicase II (the *uvrD*[+] gene product), SSB, DNA polymerase III holoenzyme, exonuclease I, exonuclease VII (the *xse*[+] gene product), the RecJ exonuclease, DNA ligase, ATP, the four deoxynucleoside triphosphates, and NAD^+, the cosubstrate for *E. coli* DNA ligase. This elegant body of biochemical work, which was carried out primarily by Paul Modrich and his colleagues, has led to a detailed understanding of the mechanism of mismatch repair.

MECHANISM OF THE INITIATION OF MISMATCH REPAIR

The endonuclease activity of purified MutH protein is so low (less than one cleavage per h per MutH monomer equivalent) that it was suggested (214) that the activity might be stimulated during the initial stage of mismatch repair. This hypothesis was strengthened by the observation that the specific activity of MutH protein with respect to its activity in the reconstituted methyl-directed system is 20- to 70-fold higher than its activity as an isolated endonuclease. Activation of the ability of MutH to cleave hemimethylated GATC sites was subsequently demonstrated (13) and was found to require the MutS and MutL proteins, ATP, and a divalent cation (Table 9–8). Furthermore, the degree of activation correlated with the efficiency with which a particular mismatch is subject to mismatch repair (G·T > G·G > A·C > C·C). The cleavage was found to occur on the unmethylated strand immediately 5' to a GATC sequence, leaving a 3'-hydroxyl terminus and a 5' phosphate (pN-3'-OH   pGpApTpC). ATP hydrolysis is apparently required, although it has not yet been established whether the MutS ATPase activity is responsible for this. An important observation was that the incised GATC sequence may lie either 3' or 5' to the mismatch on the unmethylated DNA strand. This finding implied that the mismatch repair system lacks obligate directionality; the implications of this result for the mechanism of excision are discussed below. The rate of cleavage of hemimethylated sites by MutH protein when MutS and MutL are present compares favorably with the rate of mismatch repair, providing strong support for the idea that MutHLS-dependent cleavage of hemimethylated GATC sites represents the initial step in methyl-directed mismatch repair.

An unexpected observation in the course of these studies (13) was that the activated form of MutH protein can cleave both strands at a GATC site in a heteroduplex if both strands are unmethylated. This finding is of particular interest because it provides a compelling alternative to the earlier rationalization (64) (see above) for the sensitivity of *dam* mutants to the base pair analog 2-aminopurine. In this explanation, only a single mismatch is necessary to provoke the formation of a double-strand break in a *dam* mutant. It also provides an explanation for an interesting class of *mutH* mutations that were found to suppress the lethal effect of 2-aminopurine in a *dam* strain but not to cause a defect in methyl-directed mismatch repair (66). According to this model, these mutations would simply reduce the ability of activated MutH protein to carry out cleavage of the second strand. The observation that MutH can cleave both strands of an unmethylated heteroduplex also provides a molecular explanation for the observation (41) that

Table 9–8 Requirements for the activation of MutH endonuclease[a]

Reaction conditions	Endonuclease activity (fmol/20 min)
Complete	17
−H	<0.5
−L	<0.5
−S	<0.5
−H −L	<0.5
−H −S	<0.5
−L −S	<0.5
−MgCl$_2$	<0.5
−ATP	<0.5
−ATP + ATPγS	0.8
+ATP + ATPγS (1 mM each)	1.3

[a] The hemimethylated f1MR1/f1MR3 heteroduplex contained a G·T mismatch and one GATC site. Mapping relative to unique restriction sites within the heteroduplex placed the site of viral strand cleavage at or near the GATC sequence.

Source: Adapted from Au et al. (13) with permission.

a single repairable mismatch in a nonmethylated λ heteroduplex causes mismatch-stimulated killing. It has also been speculated that this phenomenon might account for the conspicuous absence of GATC sequences from bacteriophage that grow in enterobacteria (35).

The fact that activation of the ability of MutH protein to cleave a hemimethylated GATC site requires a mismatch implies that the combination of MutH, MutL, and MutS proteins must be capable of sensing the presence of both the mismatch and the hemimethylated GATC site. A possible explanation of how this is achieved is suggested by the formation of α-like structures when MutS protein is incubated with heteroduplex DNA in the presence of ATP (see above). The formation of such loops could serve to bring a mismatch and a hemimethylated GATC site into close proximity, thereby enabling the activation of the MutH endonuclease (Fig. 9–4). Furthermore, since excision appears to initiate at the site of cleavage by MutH protein (68), the mismatch repair system must be able to assess the absolute orientation of the heteroduplex so that excision from the incised GATC site proceeds with appropriate directionality. This requirement can be met only if the interaction between the mismatch and the GATC site is mediated along the helix contour. One possibility for how this could occur is by directional protein translocation (147), a hypothesis that is compatible with the formation of the α-like structures. According to this model, the interaction of a mismatch and a GATC site is provoked by MutS protein binding to a mismatch followed by subsequent ATP-dependent translocation of one or more of the Mut proteins along the helix leading to the cleavage of a GATC sequence on either side of the mismatch (Fig. 9–4). A second formal possibility for how the absolute orientation is assessed is by polymerization of an asymmetric protein along the helix (147).

EXCISION REACTION IN METHYL-DIRECTED MISMATCH REPAIR

The observation that activated MutH protein can cleave a hemimethylated GATC site located either 3' or 5' to the mismatch suggested that mismatch repair can be initiated by a single-strand break in the unmethylated strand that is located either 3' or 5' of the lesion. This inference was tested directly by examining the excision tracts that were generated by the methyl-directed mismatch repair system on a circular G·T heteroduplex that contained a single hemimethylated GATC site (68) (Fig. 9–5). Dideoxynucleotides were added to the reaction to terminate repair synthesis and allow visualization of the gaps by electron microscopy. Regardless of which strand was methylated, the gap was found to span the shortest path between the GATC site and the mismatch. Analyses of the endpoints of the single-strand gaps indicated that each excision tract initiated at the GATC and terminated within a 100-nucleotide region just beyond the mismatch. As illustrated in Fig. 9–5, this means that excision must have occurred in a 3' → 5' direction in one case and a 5' → 3' direction in the other.

The enzymology of the two classes of excision reactions was examined by using a cleverly designed set of *linear substrates* for methyl-directed mismatch repair (31). Linear duplex DNA is rapidly degraded in cell extracts and is also sensitive to some hydrolysis in the reconstituted system. To circumvent this problem, end-blocked linear heteroduplexes were constructed by ligating a hairpin oligonucleotide onto the termini of linear G·T substrates (31) (Fig. 9–6). Despite the difference in polarity of the unmethylated strand, both substrates were found to be subject to methyl-directed correction in cell extracts (Table 9–9). When the unmethylated GATC site resides 3' to the mismatch, a 3' → 5' exonucleolytic activity is required. Exonuclease I (the *sbcB*⁺ gene product) is sufficient, although at least one as yet unidentified nuclease may be involved (31). In contrast, when the unmethylated GATC sequence is located 5' to the mismatch, a 5' → 3' exonuclease is required. A survey of known proteins with this polarity revealed that either exonuclease VII (the *xse*⁺ gene product) or the RecJ exonuclease could carry out this degradation (31). Furthermore, a cell extract of an *xse recJ* double mutant was found to be deficient in the repair of heteroduplex substrates requiring 5' → 3' excision. The requirement for 5' → 3' exonuclease activity was not observed in the first attempts to reconstitute the methyl-directed mismatch repair system with purified

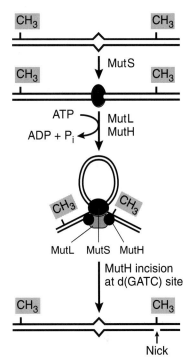

Figure 9–4 Model for initiation of methyl-directed mismatch repair. As discussed in the text, this scheme is based on in vitro properties of the MutH, MutL, and MutS proteins. Since details of stoichiometry and protein-protein interaction are not understood, those shown are for purposes of illustration only. (*Adapted from Modrich [147] with permission.*)

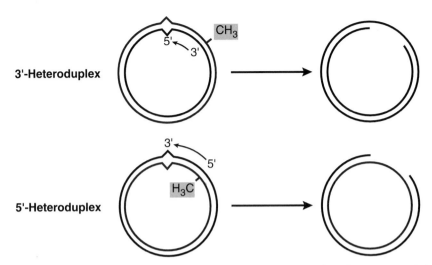

Figure 9–5 Excision tracts generated in cell extracts by using circular substrates containing a mismatch and a single GATC site. The 6,440-bp covalently closed circular heteroduplex used in this work by Grilley et al. (68) contained a G·T mismatch at position 5632 and a single d(GATC) sequence at position 216, which was modified on the viral or complementary DNA strand. Since hemimethylation imposes an asymmetry on the helix, heteroduplexes are designated according to the orientation of the unmodified strand. Molecules bearing complementary-strand methylation are referred to as 3′-heteroduplexes, since the unmodified d(GATC) sequence that directs repair is located 3′ to the mismatch along the shorter path (1,024 bp) separating the two sites in the circular molecule. The substrate with viral strand modification is designated a 5′-heteroduplex for a similar reason. During the reactions in cell extracts, DNA resynthesis was inhibited by dideoxynucleoside triphosphates. Electron microscopic visualization of DNA products produced under these conditions revealed the presence of single-strand gaps that spanned the shorter path between the mismatch and the d(GATC) site in the circular substrate. (*Adapted from Grilley et al. [68] with permission.*)

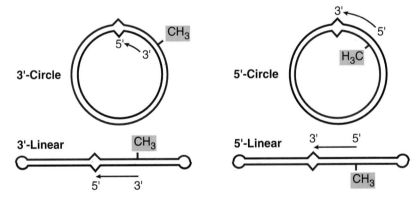

Figure 9–6 Circular and linear heteroduplex DNAs. Circular 6,440-bp and end-blocked linear 6,464-bp heteroduplexes were prepared. A hairpin oligonucleotide (10-bp stem, five thymidylate residues in loop) present at each end renders the linear DNAs resistant to exonuclease attack in *E. coli* extracts. With the exception of topology and terminal sequences of linear molecules, circular and linear substrates are identical, each containing a G·T mismatch and a single hemimethylated d(GATC) sequence. These two DNA sites are separated by 1,024 bp (a shorter distance in circular molecules). Since hemimethylation imposes an asymmetry on the helix, heteroduplexes are designated according to d(GATC)-mismatch orientation on the unmodified strand, with circular molecules specified according to orientation along the shorter path. In 3′-heteroduplexes (modification on the complementary strand), the unmethylated d(GATC) sequence lies 3′ to the mismatch, while in 5′-heteroduplexes (methylation on the viral strand), this sequence is located 5′ to the mismatch. (*Adapted from Cooper et al. [31] with permission.*)

Table 9–9   Requirements for repair of linear heteroduplexes in cell extracts

Repair system[a]	3' linear repair (fmol/mg)[b]	5' linear repair (fmol/mg)[b]
Wild type	180 (12)	160 (17)
mutS	8.8	3.6
mutS + MutS	190	110
mutL	12	5.1
mutL + MutL	100	130
mutH	17	4.6
mutH + MutH	120	100

[a] Extract reaction mixtures were incubated at 37°C for 60 min. Strain AB1157 was the source of wild-type extract, and repair-deficient extracts were prepared from isogenic *mut* derivatives. Reaction mixtures were supplemented with purified MutS, MutL, or MutH protein as indicated.

[b] Parenthetical values indicate repair occurring on the modified DNA strand.

Source: Adapted from Cooper et al. (31) with permission.

components (107), because the DNA polymerase III preparations used were contaminated by a 5' → 3' exonuclease activity (31).

Since exonuclease I, exonuclease VII, and the RecJ exonuclease are highly specific for single-stranded DNA, it seems extremely likely that the role of helicase II (the *uvrD*[+] gene product) is to displace the incised strand, thereby making it sensitive to attack by these single-stranded exonucleases (68, 147). As discussed in chapter 5, helicase II translocates 3' → 5' along a DNA strand in an ATP-driven reaction, unwinding DNA when it encounters regions of secondary structure. One possibility would be for helicase II to load on both strands at the mismatch and proceed bidirectionally (110). However, this would lead to displacement of the unmethylated strand from the site of GATC incision to a point well beyond the mismatch (147), a prediction that appears to be at odds with the analyses of excision tracts discussed above. The alternative, i.e., loading of helicase II at the nicked GATC site, would require that the helicase be loaded onto the unmethylated strand when the GATC site is located 3' to the mismatch and onto the methylated strand when the GATC site is located 5' to the mismatch. Such a discrimination would require prior interaction of the mismatch and GATC site mediated along the helix contour; an example is diagrammed in Fig. 9–4.

MODEL FOR BIDIRECTIONAL METHYL-DIRECTED MISMATCH REPAIR

Fig. 9–7 presents a model for methyl-directed mismatch repair that incorporates all of the observations discussed above. As discussed below, a bidirectional mechanism for strand-specific mismatch repair appears to be highly conserved in nature. The *E. coli* system discriminates between parental and daughter strands in newly synthesized DNA by monitoring the state of methylation of GATC sites, and this particular aspect of the system is not highly conserved in nature. However, even in the *E. coli* system, mismatch repair can be directed by a persistent strand break. The presence of such a break obviates the need for a hemimethylated GATC site and for MutH function, making it clear that strand breaks in the daughter strand could be used as an alternative means of directing the strand specificity of mismatch repair.

EFFECTS OF LONG-PATCH MISMATCH REPAIR ON GENETIC RECOMBINATION

The analysis of long-patch mismatch repair in *S. pneumoniae* discussed above illustrates how the action of a long-patch mismatch repair system operating on a DNA heteroduplex recombination intermediate can influence the number and nature of the recombinants obtained. Studies with *E. coli* have offered additional insights into the relationship between mismatch repair and homologous recombination.

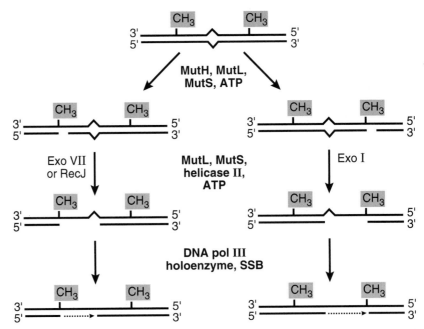

Figure 9–7 Mechanism of bidirectional methyl-directed mismatch repair. Repair is initiated by activation of a latent MutH endonuclease in a reaction that is dependent on a mismatch, MutS, MutL, and ATP hydrolysis. The activated form of MutH cleaves the unmodified strand at a GATC site that can be located on either side of the mismatch. Excision subsequently removes DNA spanning the two sites, with ensuing repair synthesis initiating near the GATC site or the mismatch, depending on the polarity of the unmodified strand. MutS and MutL are shown as required for the excision stage of mismatch correction, since they are necessary for repair of preincised heteroduplex, perhaps for loading the appropriate excision activities. Involvement of SSB in excision has not been tested. (*Adapted from Grilley et al. [68] with permission.*)

Interest in the role of mismatch repair on genetic recombination has focused attention on two different classes of recombinational events: (i) those that involve recombination between highly homologous sequences, such as in a cross between a base substitution mutant and its isogenic parent, and (ii) those that involve recombination between sequences that, although showing overall homology, exhibit substantial sequence divergence.

EFFECT OF LONG-PATCH MISMATCH REPAIR ON RECOMBINATION BETWEEN HIGHLY HOMOLOGOUS SEQUENCES

A particularly clear example of the effect of mismatch repair on genetic recombination between highly homologous sequences in *E. coli* came from experiments which examined the influence of mismatch repair on the recombination of λ bacteriophage (82). These experiments demonstrated that 20% of the recombinant phages emerging from a *rec⁺ red⁺* cross were heterozygous in an interval adjacent to a selected exchange event, provided that the recombinants were selected on a *mutL* indicator strain. The frequency of heterozygous phage was reduced by an order of magnitude on a *mutL⁺* indicator, thus implicating MutL protein in the rectification of mispairs within a multiply marked region of hybrid DNA. In these experiments, the fact that the yield of heterozygotes was independent of the *mutL* genotype of the strain in which the cross was performed was attributed to rapid packaging of the recombinant phage protecting them from mismatch repair. The average length of the heteroduplex formed in these crosses was shown to be about 4 kb, with some as large as 10 kb. The effect of long-patch mismatch repair in these crosses is to reduce the overall recombination frequency because of corepair (53) of two mismatches on the same heteroduplex strand (87, 171). However, it does not appear to contribute significantly to the formation of recombinants involving very closely linked markers (87). These effects appear to be due to the short-patch mismatch repair systems (discussed below).

It was demonstrated that intragenic recombination in an *E. coli* Hfr × F⁻ cross increased 4- to 14-fold when the recipient is a *mutH*, *mutL*, *mutS*, or *uvrD* mutant, with the largest effect noted with *uvrD* mutants (51). The molecular mechanism by which this occurs has not been established, but it does suggest that processes involving key genes for long-patch mismatch repair can reduce the number of recombinants obtained in a cross. One plausible mechanism is that since the donor DNA integrates as a single strand (194), methylated donor DNA might become covalently linked to the undermethylated newly synthesized strands, so that methyl-directed mismatch repair of the incoming donor strand marker might occur (171).

The above observations suggest that mismatch repair would not influence the frequency of recombination that occurs between perfectly homologous DNA sequences. This conclusion is supported by the observation that the integration of a plasmid into phage λ DNA via homology provided by a short cloned sequence was not affected by an *E. coli mutS* mutation when the homology was 100% (190).

EFFECTS OF MISMATCH REPAIR ON RECOMBINATION BETWEEN SUBSTANTIALLY DIVERGED SEQUENCES

The idea that recombination between substantially diverged sequences (sometimes referred to as *quasi-homologous* or *homeologous* sequences) might be differentially inhibited by mismatch repair (owing to overlapping excision tracts [41, 64, 211]) was first proposed by Karimova et al. in 1985 (90) and later independently proposed by others (87, 171, 172, 191). Several striking examples of effects of long-patch mismatch repair on recombination between highly diverged sequences have been reported, and it appears that they may be of great biological significance.

The hypothesis that mismatch repair would *reduce recombination* if the sequences were substantially diverged was tested (191) by using of a model system in which recombination between a plasmid and bacteriophage was examined under conditions where recombination was restricted to a 405-bp region of homology. When the homology within this region was reduced from 100 to 89%, the frequency of recombinants dropped about 240-fold in a wild-type host but only 9-fold in a *mutS* mutant. In contrast, as mentioned above, a *mutS* mutation had no effect on the frequency of recombination if the homology was 100%.

Radman and his colleagues (177) tested the hypothesis that action of the mismatch repair system might be responsible for the very low frequency of conjugal recombination between *E. coli* and *S. typhimurium*. The genetic maps of these two organisms are very similar, and their DNA sequences are approximately 80% homologous. However, a cross between an *E. coli* Hfr donor and a wild-type *S. typhimurium* recipient occurs with an efficiency of 10^{-6} compared with the corresponding cross with a wild-type *E. coli* recipient (Table 9–10). It seemed unlikely that this resulted from an inability to form heteroduplexes, since heteroduplex formation by the RecA protein tolerates up to 30% mismatches (34) (see chapter 10). However, if the *S. typhimurium* recipient carried a *mutS*, *mutL*, *mutH*, or *uvrD* mutation (155, 156), the frequency of recombinants increased by several orders of magnitude. In contrast to the effect on recombination between highly homologous sequences described above (51), the greatest increase was seen in *mutS* and *mutL* mutants, with *uvrD* (*mutU*) having the smallest effect. Subsequent work has revealed that the recombinants correspond to replacements of one continuous block of the recipient genome by the corresponding region of the donor genome (136). The mechanism responsible for this phenomenon (Fig. 9–8) appears to involve an inhibition of the branch migration stage of RecA-catalyzed strand transfer by the MutS and MutL proteins when mismatched base pairs are formed during the course of heteroduplex formation (224). These striking experiments strongly suggest that long-patch mismatch repair is a molecular mechanism for maintaining a genetic barrier between species that have diverged (but not drastically) at the

Table 9–10   Conjugational crosses between *E. coli* and *S. typhimurium*[a]

E. coli Hfr donor	*S. typhimurium* F⁻ recipient	*E. coli* F⁻ recipient	Recombination frequency	
			xyl⁺ per *Hfr*	*metE*⁺ *metA*⁺ per *Hfr*
mut⁺	*mut*⁺		9×10^{-7}	2×10^{-6}
mut⁺	*mutS*		2×10^{-3}	2×10^{-3}
mut⁺	*mutL*		3×10^{-3}	2×10^{-3}
mut⁺	*mutH*		6×10^{-5}	2×10^{-4}
mut⁺	*uvrD*		5×10^{-5}	9×10^{-6}
mut⁺		*mut*⁺	3×10^{-1}	

[a] Heterospecific crosses involved an *E. coli* Hfr strain and *S. typhimurium* (F⁻) *mut*⁺, *mutH, mutL, mutS,* and *uvrD* mutants. Homospecific crosses involved an *E. coli* Hfr strain or PK3 *mutU* and an *E. coli* F⁻ strain. Log-phase bacteria were mixed in a 1:1 (Hfr:F⁻) ratio and immediately deposited on a sterile 0.45-mm-pore-size filter, which was incubated on prewarmed, rich-medium agar. After 40 min at 37°C, the conjugants were resuspended in 10⁻² M MgSO₄ and viciously separated by vortexing. The exconjugants were then plated on synthetic medium either lacking methionine to select for *met*⁺ recombinants or containing xylose as the sole carbons source to select for *xyl*⁺ recombinants. The Hfr donor cells were doubly counterselected with 25 mg of streptomycin ml⁻¹ and, for intergeneric conjugation, by withholding threonine and leucine. Amino acids were added to a final concentration of 100 μg ml⁻¹, thiamine was added to 30 μg ml⁻¹, and xylose was added to 0.4%. Because of the late appearance of intergeneric recombinants from *mut*⁺ and *uvrD* crosses, all recombinants were scored after 88 h of incubation.
[b] Recombination frequencies (*xyl*⁺ or *metE*⁺ *metA*⁺) are expressed per *Hfr* donor after subtraction of unmated revertants.
Source: Adapted from Rayssiguier et al. (177) with permission.

DNA sequence level and thus provides a form of reproductive isolation in the absence of geographical isolation (177).

Evidence that the MutS and MutL functions of mismatch repair can also prevent *intrachromosomal* exchanges between highly diverged sequences has also been provided (162). This was achieved by examining the genetic control of large chromosomal duplications in *E. coli*. *E. coli* and *S. typhimurium* carry large duplications of genomic segments at frequencies that can be as high as 10^{-4} of a bacterial population (9, 161). These duplications can arise from unequal recombination events between homologous sequences within the genome (Fig. 9–9). Two examples of such repeated sequences in the *E. coli* chromosome are the *rrn* operons for RNA and the five *rhs* loci, which have no known genetic function. A characteristic of the *rrn* and *rhs* gene families is a slight divergence in DNA sequences, and it seemed possible that, in a wild-type strain, mismatch repair would reduce the frequency of recombination occurring between these diverged sequences. This supposition received strong support from the finding that *mutS* and *mutL* mutants had a 10-fold-higher frequency of large duplications. Excessive recombination between repeated, interspersed, and diverged DNA sequences is a potential source of genetic instability, and this observation suggests that the mismatch repair system plays a role in suppressing genetic exchange between such partially diverged sequences (162).

EFFECT OF MISMATCH REPAIR ON THE STABILITY OF SHORT REPEATED SEQUENCES

In light of recent findings in eukaryotic cells (see below), a particularly interesting observation was that mismatch repair affects the stability of short repeated sequences in *E. coli*. In the model system used (113), the stability of a tract of poly $(GT)_n$ [paired with $(AC)_n$ on the opposite strand] tandem repeats cloned into bacteriophage M13 was examined. The frequency of detectable frameshift mutations within a 40-bp tract of tandem repeats was found to be greater than 1% and to increase more than linearly with length. The observed insertions and deletions were attributed to slippage during replication rather than to recombination, since the experiments were carried out with a *recA* host. Interestingly, *mutS* and *mutL* mutations were found to increase the frequency of frameshifts by 13-fold, indicating that the mismatch repair system corrects over 90% of the frameshifts that occur.

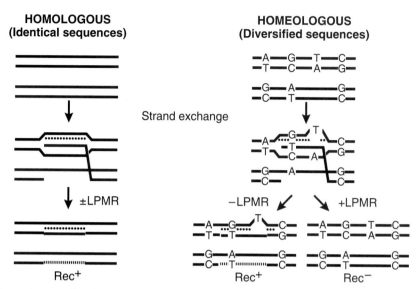

Figure 9–8 Molecular model for selective antirecombination by long-patch mismatch repair proteins. Homologous (identical) sequences recombine with the same efficiency in the presence or absence of active long-patch mismatch repair (LPMR) proteins. Homeologous (diversified) sequences start to recombine by forming a heteroduplex containing mismatches caused by sequence differences in the two parental DNAs. If long-patch mismatch repair proteins are not functional, homeologous recombination proceeds to form recombinant products (Rec+ phenotype). If long-patch mismatch repair proteins are functional, mismatches in the heteroduplex regions are recognized, the reaction is reversed by the removal or destruction of the heteroduplex intermediate, and no viable recombinants arise (Rec− phenotype). Although irrelevant to the mechanism of antirecombination, the initiation of strand exchange is arbitrarily shown in a single-strand invasion mode. Only base pair differences between the parental molecules are indicated: the hybrid region in this figure (shown with dots indicating hydrogen bonding) contains one G·T mismatch plus one unpaired T residue (as in a frameshift mutation). (*Adapted from Rayssiguier et al. [177] with permission.*)

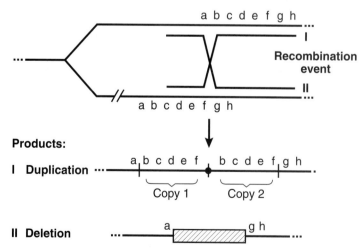

Figure 9–9 A likely mechanism for duplication formation. Genetic exchanges between different sites on sister chromosomes can generate duplications and deletions. Many of the deletions are probably lethal, but large duplications cause no loss of gene function. (*Adapted from Anderson and Roth [10] with permission.*)

Long-Patch Mismatch Repair in Eukaryotes

Evidence is rapidly accumulating that eukaryotic organisms are also able to carry out long-patch mismatch repair and that the overall mechanism of this process has been highly conserved in evolution. It further appears that some of the genes that play important roles in mismatch repair in prokaryotes, particularly *mutS* and *mutL*, have also been conserved during evolution, so that homologs of these genes are present in eukaryotes. As in prokaryotes, mismatch repair systems in eukaryotes appear to play important roles in the maintenance of genetic fidelity during DNA replication, in the outcome of genetic recombinations, and in genome stability. In addition, they appear to be important for preventing the appearance of certain types of cancers.

In Vivo Evidence for Mismatch Repair in Yeasts and Fungi

As indicated above, mismatch repair of a heteroduplex recombination intermediate has been invoked to explain the phenomenon of nonreciprocal genetic transfer during meiotic recombination in fungi, including the yeast *S. cerevisiae*, (79, 140) (Fig. 9–10). In these models, heteroduplex DNA correction can generate 6:2 or 2:6 aberrant segregations or gene conversions. A deficiency in mismatch correction of heteroduplex DNA results in 5:3 or 3:5 segregation, or *postmeiotic segregations*. In addition, mismatch repair can be the cause of gradients of gene conversion that have been observed during meiotic recombination. For example, in the *HIS4* gene of *S. cerevisiae*, most mutant alleles at the 5′ end of the gene have a higher rate of meiotic recombination (gene conversion) than do mutant alleles at

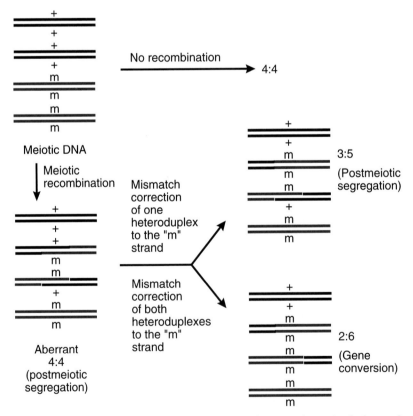

Figure 9–10   Illustration of how reduced proficiency of mismatch repair of a heteroduplex generated by meiotic recombination can lead to a 3:5 or 5:3 segregation (postmeiotic segregation). If no mismatch repair were to occur in the case illustrated, the result would be two postmeiotic segregations from the same meiosis (so-called aberrant 4:4 segregation). 3:5 or 5:3 postmeiotic segregations can also arise from asymmetric strand exchange and no mismatch repair (140).

the 3' end of the gene. These gradients have usually been interpreted as being due to a higher frequency of heteroduplex formation at the high-conversion end of the gene. However, recent studies have suggested that the gradient reflects primarily the direction of mismatch repair rather than the frequency of heteroduplex formation (7, 37, 178).

Wild-type *S. cerevisiae* has also been shown to be capable of carrying out mismatch repair of artificial heteroduplexes, but a general strand-specific pathway has not been described. As judged by transformation frequencies with such heteroduplexes (16, 96) or by postmeiotic segregation frequencies (36, 215), the efficiencies with which *S. cerevisiae* is able to process various base pair mismatches and small frameshift mispairs are essentially identical to those described above for long-patch repair in bacteria and to those described below for higher eukaryotic cells.

In addition, events termed *postswitching segregation* have been described during the switching of mating-type genes in *S. cerevisiae*, a highly efficient gene conversion process initiated by a double-strand break generated by the HO endonuclease (137, 176). When *MATa* switches to *MATα*, a 3'-single-stranded end of HO-cleaved *MAT* DNA invades the homologous donor *HMLα* to form a heteroduplex. If mismatches are present in this heteroduplex, they are subject to a highly preferential form of mismatch repair in which the invading strand from MAT is corrected to the genotype of the resident *HML* donor strand (72). It has been suggested that the mismatch repair may be directed by the end of the invading strand, just as it can be directed by a nick in bacterial mismatch repair (137, 176).

Homologs of *mutL⁺* and *mutS⁺* in Eukaryotic Cells

Mutations causing increased postmeiotic segregation (PMS) frequencies in unselected tetrads were isolated by the use of a selection for hyperrecombination between closely linked markers (222). Three loci were identified: *PMS1*, *PMS2*, and *PMS3*. Mutants defective in any of these loci were found to have a spontaneous mitotic mutator phenotype, suggesting that they might be defective in postreplicational mismatch correction. Furthermore, they were found to be deficient in the repair of heteroduplex DNA. The *pms1* and *pms2* mutants were found to be deficient in the repair of base pair mismatches and a −1 frameshift mismatch (96), whereas the *pms3* mutants were deficient in the repair of the base pair mismatches but proficient in the repair of the −1 frameshift mismatch.

When the *PMS1* locus was cloned and sequenced (97), it was found to be a homolog of the *E. coli mutL⁺* gene (130) and the *S. pneumoniae hexB⁺* gene (166). This finding raised the possibility that the role of the *PMS1* gene product in yeast mismatch repair is related to its role in *E. coli* mismatch repair. An additional *mutL⁺* homolog in *S. cerevisiae*, termed *MLH1* (for *mutL* homolog) has been reported (165). Mutants defective in *MLH1* function have the same phenotype as *pms1* mutants. Furthermore, the mutator and meiotic phenotypes of *pms1 mlh1* double mutants are the same as those of the *pms1* and *mlh1* single mutants. These results indicate that the *MLH1* gene product also plays a role in mismatch repair that is related to the role of MutL protein (165). It is intriguing that two types of MutL-like proteins are required for mismatch repair in *S. cerevisiae* whereas only one type of MutL protein is required in the *E. coli* mismatch repair system. The requirement for both MutL-like functions is explained by the recent findings that the Mlh1 and Pms1 proteins physically associate, possibly forming a heterodimer, and that Mlh1 and Pms1 act in concert to bind a Msh2-heteroduplex complex containing a G·T mismatch (165a). Thus it appears that mismatch repair in *S. cerevisiae* involves a tertiary complex between Msh2, Mlh1, and Pms1.

Human homologs of MutL have been identified (18, 157) by two different experimental strategies (188). In one approach, degenerate PCR primers were designed based on peptides conserved within the bacterial and yeast families of MutL proteins (18). A PCR procedure carried out with these primers and cDNA prepared from poly(A)⁺-enriched primary human fibroblasts allowed the isolation of DNA probes which, in turn, allowed the isolation of two cDNAs encoding human MutL homologs. One of these was termed *hMLH1* because of its similarity to the yeast

MLH1 gene. The other appeared to be more similar to the yeast *PMS1* gene (18). In a different approach (157), three human genes encoding homologs of MutL were identified as follows. First a library of sequences of fragments of randomly chosen cDNA clones was searched (4–6). Three genes that had significant homology to *mutL* at their 5' ends were identified. One of these was the *hMLH1* gene discussed above. The other two had slightly greater similarity to yeast *PMS1* and have been termed *hPMS1* and *hPMS2* (157). Evidence that the *hMLH1* gene is required for mismatch repair in human cells has been provided by analyses of H6, a human colorectal tumor cell line manifesting microsatellite instability (157, 159) (see below). Furthermore, defects in *hMLH1*, *hPMS1*, and *hPMS2* have been found to be associated with hereditary nonpolyposis colorectal cancer (HNPCC) (see discussion below).

Homologs of MutS, the protein that recognizes the mismatch in *E. coli* long-patch mismatch repair, have now been found in yeast, mouse, and human cells. The first eukaryotic homologs of MutS to be reported were encountered during analyses of the *DHFR* locus in human and mouse cells. In both cases a divergently transcribed gene, named *Duc-1* and *REP3* in human and mouse cells, respectively, whose deduced amino acid sequence has a high degree of homology to the MutS protein was identified (60, 121). However, at the time of writing there is no direct evidence that these proteins play a role in mismatch repair.

S. cerevisiae is a particularly interesting case because four homologs of the *mutS*⁺ gene have been identified. Analyses of their physiological functions have begun to offer insights into their specific biological roles and suggest that multiple *mutS*⁺ homologs will be found in other eukaryotes. By using degenerate PCR primers based on protein sequence comparisons of MutS, HexA, and Duc-1, probes were obtained that allowed the isolation of two genes that encode homologs of *mutS*, *MSH1*, and *MSH2* (*mutS* homolog) (179). A third *mutS* homolog (*MSH3*), which is closer in sequence to the mammalian homologs *REP3* and *Duc-1*, has been similarly isolated (152) and was also identified during the *S. cerevisiae* chromosome III gene-mapping project (209). A fourth *mutS*⁺ homolog, *MSH4*, has been identified (182). The relationship, if any, of these loci to *PMS2* and *PMS3* is not yet known.

The phenotype of yeast *msh2* mutants was found to be very similar to that of *pms1* mutants—a strongly elevated spontaneous mitotic mutation rate and an increase in postmeiotic segregation (178, 179). Furthermore, the *msh2* mutant resembled the *pms1* mutant in being defective in the repair of all base pair mismatches as well as 1, 2, and 4 insertion/deletion mispairs (7). These properties suggest that the *MSH2* gene product also participates in a mismatch repair pathway that is similar to bacterial long-patch mismatch repair pathways. This inference is strongly supported by the discovery that *msh2* mutants lack a ~110-kDa protein that binds to at least six different mismatch base pairs in a fashion that is independent of the sequence context (144). The basis for strand discrimination in this mismatch repair system has not been established, but it could result from the recognition of breaks in the daughter strands or an interaction of mismatch repair proteins with the proteins involved in DNA replication. The role of the *MSH3* gene is less clear, since *msh3* mutations cause a relatively weak mitotic mutator phenotype and yield weak postmeiotic segregation (152). To date, no loss of mismatch-binding activity has been detected in *msh3* mutants, but it is possible that future experiments will reveal such an activity (144). These observations have led to the suggestion that the primary function of Msh3 is not in postreplicative mismatch repair but, rather, may be related to one of the other known functions of MutS protein such as preventing recombination between diverged sequences. There is no evidence for an involvement of the *MSH4* gene product in mismatch repair, but *msh4* mutants missegregate their chromosomes during meiosis (182). In contrast to the other three MutS homologs, the *MSH1* gene product appears to play a role in the repair or stability of mitochondrial DNA, since disruption of the *MSH1* gene results in a petite phenotype and also causes point mutagenesis and large-scale rearrangements of mitochondrial DNA (178, 179).

A particularly interesting human homolog of *mutS*, termed *hMSH2*, has been identified (52, 111). A probe to this gene was isolated by the use of the same degenerate PCR primers that were used to identify the *S. cerevisiae MSH1* and *MSH2* genes (52, 111). Once obtained, the probe was used to identify a full-length cDNA that encodes a 934-amino-acid protein. Since this protein is about 40% identical with the 966-amino-acid *S. cerevisiae* Msh2 protein and has some regions that are extremely highly conserved, this gene has been termed *hMSH2*. Expression of this protein in *E. coli* was shown to cause a dominant mutator phenotype (52), an observation reminiscent of the finding that expression of the HexA protein in *E. coli* causes a dominant mutator phenotype (167). The observation is consistent with the idea that the hMsh2 protein can interact with mismatched nucleotides but lacks the ability to interact with other proteins in the *E. coli* MutHLS-dependent mismatch repair pathway. Very strong evidence that the hMsh2 protein can bind mismatches was provided by the sequencing of tryptic peptides derived from a ca. 100-kDa HeLa cell protein that had been shown to bind to oligonucleotide duplexes containing G·T mismatches and other mismatches (81, 84, 85). The sequence of the tryptic peptides derived from this protein indicates that it is the hMsh2 protein (154). Like the *E. coli* MutS protein, the hMsh2 protein has an ATPase activity and a weak helicase activity (81, 84, 154). A second protein from HeLa cells that also binds G-T base pairs has a molecular mass of 160 kDa. However, it is unrelated to hMsh2 and is encoded by a separate gene (81, 84, 86); its physiological role has not yet been established. A ~100-kDa protein has been found in several human cell lines that has been reported to bind to A·C, T·C, and T·T base pairs in a largely sequence-independent fashion (198), but its relationship to hMsh2 has not yet been established. As discussed below, defects in *hMSH2* have been found to be associated with HNPCC.

The dendrogram shown in Fig. 9–11 illustrates the sequence relationships among various MutS homologs (52, 152) and indicates that distinct groupings of MutS homologs can be discerned on the basis of sequence relatedness. The fact that the *S. cerevisiae* Msh1, Msh2, Msh3, and Msh4 proteins fall into different groups suggests that these groups may reflect *functional specialization* rather than phylogeny (52, 152). Two homologs of *mutS* have been identified in the fission yeast *Schizosaccharomyces pombe*, one of which (*swi4*) is involved in the termination of copy synthesis during mating-type switching (54, 209). As indicated in Fig. 9–11, the Swi4 protein appears to be a member of the group that includes Msh3 (*S. cerevisiae*), Dug (human; same as Duc-1), and Rep3 (mouse).

In Vitro Analyses of Mismatch Repair in Eukaryotic Cells: Human Cells Have a Bidirectional Mismatch Repair System

Although the genetic analyses of mismatch repair are most advanced in *S. cerevisiae* and mismatch correction has been detected in cell extracts (148), it has not yet been shown to be strand specific, and there has not yet been a systematic dissection of its biochemical mechanism. However, as mentioned above, it seems likely that the Msh2 protein plays a role in yeast mismatch repair analogous to that of MutS in *E. coli* (144, 178, 179) and that the Pms1 and Mlh1 proteins play roles related to that of MutL protein (96, 97). It is interesting in this regard that the *RADH* (*HPR5, SRS2*) gene of *S. cerevisiae* is homologous to the *E. coli uvrD*+ gene (3, 181) and certain *radH* (*hpr5, srs2*) alleles not only cause increased sensitivity to killing by UV radiation but also cause a hyper-gene conversion phenotype (181). A direct role for the *RADH* (*HPR5, SRS2*) gene product in mismatch repair has not been demonstrated but remains a possibility. A mismatch-dependent endonuclease has been found in yeast cell extracts, but it shows a strong dependence on sequence context, and its relationship to mismatch repair is not yet clear (25).

Transformation of mammalian cells with a variety of heteroduplex-containing viral or plasmid vectors has been used to provide evidence for the existence of a system for general mismatch repair in mammals (2, 21, 55); mammalian systems that repair specific mismatches are discussed below. Extracts of human cells (62) or *Xenopus laevis* eggs (19, 210) were also used to demonstrate the ex-

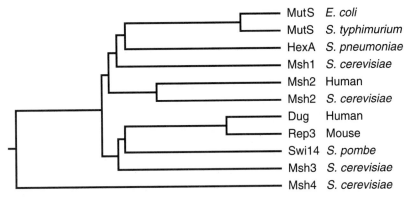

MutS	*E. coli*
MutS	*S. typhimurium*
HexA	*S. pneumoniae*
Msh1	*S. cerevisiae*
Msh2	Human
Msh2	*S. cerevisiae*
Dug	Human
Rep3	Mouse
Swi14	*S. pombe*
Msh3	*S. cerevisiae*
Msh4	*S. cerevisiae*

Figure 9–11 Phylogenetic tree of MutS-related proteins. The first 21 amino acids of *S. cerevisiae* MSH1 that constitute the mitochondrial targeting sequence were removed from the MSH1 sequence prior to construction of the phylogenetic tree. The MSH4 sequence was provided by P. Ross-MacDonald and G.S. Roeder (182). (*Adapted from Fishel et al. [52] with permission.*)

istence of a general process for mismatch repair in their respective systems. In the *Xenopus* system, mismatch-specific DNA synthesis was shown to occur in the region of the mismatch. The first demonstration of the strand specificity of mismatch repair in higher eukaryotes was obtained by using circular heteroduplex substrates containing a mismatch and a strand break (80). The design of the substrate was suggested by the finding discussed above that in the *E. coli* system, a persistent strand break not only is sufficient to direct mismatch correction but also bypasses the requirements for MutH protein and a hemimethylated GATC sequence (107, 110). These substrates were repaired in extracts from both human and *Drosphila* K$_c$ cells, with repair in each case being highly biased to the incised strand (80).

Further analyses of the molecular mechanism of mismatch repair in extracts of human cells have shown that human cells possess a mismatch repair system similar to the *E. coli* methyl-directed pathway with respect to both *mispair specificity* (49, 50, 80, 205), and *bidirectional excision capability* (49, 50). The initial indications that repair is bidirectional were obtained by examining the repair of various circular heteroduplex substrates that contained a nick. The rate of repair of substrates containing a single G·T mismatch was found to decrease monotonically with increasing separation between the mismatch and the strand break that targets repair, as viewed along the shorter path joining the two sites in the circular substrate. The discovery that this decrease was independent of the polarity of the excised strand (50) suggested that the human cells might encode a mismatch repair system with a bidirectional excision capability similar to that of *E. coli* methyl-directed mismatch repair (see above). This possibility was confirmed (50) by analyzing the excision tracts generated when aphidicolon (an inhibitor of DNA polymerases α, δ, and ε [205]) was added to the reactions or when exogenous deoxynucleoside triphosphates were omitted from the reaction. As shown in Fig. 9–12, the gaps formed were found to span the shorter path between the strand break and the mismatch, irrespective of the polarity of the incised strand. Regardless of the orientation of the two sites, these gaps were found to extend from the site of the strand break and to terminate at a number of discrete sites in the region 90 to 170 nucleotides beyond the mismatch. Although these data strongly support the hypothesis that human mismatch repair is bidirectional, the relevant nucleases needed for the specific excisions (see above) have not yet been identified. It should be noted that the use of circular heteroduplex substrates containing a nick may not allow the detection of all of the components of human mismatch repair. In particular, the presence of the preexisting nick would circumvent the requirement for any function conceptually analogous to that served by the MutH protein in the *E. coli* system.

It is worth noting that, as in *E. coli*, G·T mismatches in mammalian cells may

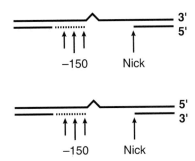

Figure 9–12 Schematic view of excision products observed with 5'- and 3'-circular heteroduplexes in nuclear extracts of HeLa cells. Dashed lines indicate variability in observed excision tract end points. For the heteroduplex in which the incision is 3' to the mismatch, products shown correspond to those derived from aphidicolin-inhibited reactions. (*Adapted from Fang et al. [50] with permission.*)

be repaired by two different mechanisms depending on their sequence context. The more general case, under discussion here, is a bidirectional long-patch mismatch repair mechanism that appears to involve MutS and MutL homologs . The other system (see below) involves a short-patch mismatch repair system with a preference for the repair of G·T base pairs in the context of CpG sites. An activity has been purified from HeLa cells that nicks heteroduplexes containing any of the eight possible base pair mismatches (227). However, since its specificity differs from that of long-patch mismatch repair and it exhibits sequence specificity, it seems unlikely that it participates in long-patch mismatch repair.

An Alkylation-Tolerant Mutator Human Cell Line Is Deficient in Strand-Specific Repair

The first human cell line shown to be deficient in strand-specific mismatch repair (93) was MT1, which was known to be alkylation resistant and a spontaneous mutator (65); a second example is discussed in the following section. The MT1 cell line was derived from TK6 lymphoblastoid cells after frameshift mutagenesis and screening for mutants able to survive the cytotoxic effects of N-methyl-N'-nitro-N-nitrosoguanidine (MNNG) (65). Although the MT1 line is several hundred times more resistant to killing by MNNG than its parent is, it is alkylated normally by MNNG and remains sensitive to mutagenesis by this agent. The MT1 cell line also has a spontaneous mutator phenotype, with spontaneous mutability at the *HPRT* locus increased about 60-fold (65). The spontaneous mutations occurring at the locus include single nucleotide insertions, transversions, and A·T → G·C transitions (93). In addition, the MT1 line displays poor clone-forming ability at low cell densities and differs from the parental line with respect to cell cycle progression after exposure to MNNG doses that lead to 90% killing.

Since a spontaneous mutator phenotype and resistance to the cytotoxic action of MNNG have been reported to be characteristics of *E. coli* mismatch repair mutants (44, 91, 92), the suggestion had been made that the MT1 phenotype is due to a defect in a strand-specific mismatch repair system analogous to the *E. coli* methyl-directed pathway. In this model, the cytotoxic effects of alkylation in mismatch proficient cells are attributed to futile attempts to correct mispairs that arise during replication of alkylated template strands (see chapter 3). Extracts of MT1 cells were found to be deficient in their ability to carry out strand-specific mismatch repair of all eight base-base mispairs (93). The block appears to be quite early, since MT1 nuclear extracts do not support the mismatch-provoked excision reaction described above (93). MT1 cells apparently lack a single component necessary for the reaction since in vitro complementation of MT1 nuclear extracts by partially purified fractions derived from HeLa cells has permitted the identification of a component that is lacking in the mutant line (93). In addition, traditional fractionation methods have led to the identification of four additional factors that are also necessary for bidirectional long-patch mismatch repair (49). Another set of O^6-methylguanine DNA methyltransferase-deficient, alkylation-resistant cells that display a weak spontaneous mutator phenotype have been shown to lack a G·T mismatch-binding protein (17), but the relationship of this G·T mismatch-binding protein to hMsh2 has not yet been established.

Destabilization of Tracts of Simple Repetitive DNA by Mutations Affecting Mismatch Repair: Implications for Human Diseases

In a particularly striking demonstration of how basic research can lead to breakthroughs in health-related or more-applied areas, evidence has recently been obtained indicating that defects in *hMSH2* or *hMLH1* are associated with hereditary nonpolyposis colorectal cancer (HNPCC) (also known as Lynch syndrome II and Muir-Torre syndrome). This is one of the most common genetic diseases of humans and affects as many as 1 in 200 individuals (128). HNPCC kindreds are commonly defined as those in which at least three relatives in two generations have colorectal cancer, with one of the relatives having been diagnosed at less

A

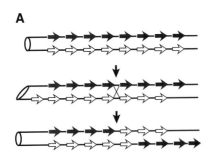

B

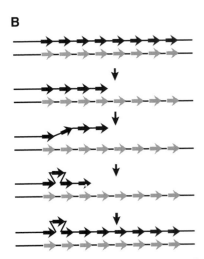

Figure 9–13 Mechanisms to account for the instability of tracts of short repeated sequences. (A) Unequal crossing over between two DNA duplexes (196). Crossovers between misaligned repeats (each repeat is indicated by an arrow) result in a deletion from one tract and an addition to another. Gene conversion events could cause nonreciprocal alterations in tract length. (B) DNA polymerase slippage during replication of DNA containing repeated sequences (200). During DNA replication (or repair synthesis), one DNA strand transiently dissociates from the other and then reanneals in a misaligned configuration. If the mismatched bases are located in the primer strand (as shown), continued elongation results in an increased tract length. If the unpaired bases are located in the template strand, one of the tracts would contain a deletion. (*Adapted from Strand et al. [199] with permission.*)

than 50 years of age. In spite of the name of the syndrome, affected individuals also develop tumors of the endometrium, ovary, and other organs. HNPCC is one type of colorectal cancer with a familial predisposition and accounts for 4 to 13% of all colorectal cancers in the industrial world.

A striking feature of HNPCC is that it has been found to be a class of cancer characterized by *instabilities* of simple repeated sequences (52, 111, 159). The genomes of all eukaryotes contain tracts of DNA in which a single base or a small number of bases is repeated, and the human genome in particular contains many sequences in which 1- to 6-nucleotide motifs are tandemly repeated numerous times (14). Mono-, di-, and trinucleotide repeats are unstable in HNPCC cells as well as in certain sporadic colorectal tumor cells (1, 83, 160, 204). Furthermore, expansions of trinucleotide tracts have been associated with several human disorders including the fragile X syndrome, myotonic dystrophy, spinal and bulbar muscular atrophy (Kennedy's disease), Huntington's disease, and spinocerebellar ataxia (14, 24, 98).

Tracts consisting of simple repetitive DNA change length at frequencies that are considerably higher than expected for base substitution mutations. Two possible mechanisms have been proposed to account for such alterations (Fig. 9–13). The first is that such alterations might occur by unequal crossing over (Fig. 9–13A). In other words, crossing over occurs with the tracts misaligned by an integral number of repeats (196). The second is that such alterations result from DNA polymerase slippage (200) (Fig. 9–13B). According to this model, during replication of the tract the primer and template strand transiently disassociate and then realign in a misaligned configuration. If the unpaired bases are in the primer strand, continued synthesis results in an elongation of the tract, whereas if the unpaired bases are in the template strand, continued synthesis results in a deletion.

Earlier in the chapter we discussed experiments (113) indicating that in *E. coli*, mismatch repair plays a major role in the stability of tracts of short repeated sequences. In the particular case examined of a (GT)$_n$ tract, the instabilities appeared to result from DNA polymerase slippage errors, and over 90% of them were corrected by a *mutS*$^+$ *mutL*$^+$-dependent process. A very similar situation has been observed in *S. cerevisiae*, in which mutations in genes implicated in DNA mismatch repair (*PMS1*, *MLH1*, and *MSH2*) result in 100- to 700-fold increases in the instabilities of (GT)$_n$ tracts (199). As in the case of *E. coli*, the instabilities of these tracts do not appear to arise from recombination, since they are unaffected by genes such as *RAD52*, which are required for recombination in *S. cerevisiae* (see chapter 12) (76). Furthermore, mutations that eliminate the proofreading function of replicative DNA polymerases were found to have little effect on the stabilities of such tracts, and the meiotic stability of the tracts was found to be similar to their mitotic stability (199). Taken together, these observations indicate that in *S. cerevisiae*, the replicative DNA polymerase has a very high rate of slippage on templates containing simple repeats but most of these errors are corrected by cellular mismatch repair systems.

These observations in *E. coli* and *S. cerevisiae* led to the prediction (52, 99, 111, 159, 199) that the phenotype of the mutation involved in HNPCC (1, 83, 160, 204) might result from a defect in a gene required for mismatch repair, such as a functional homolog of the *mutS*-like gene *MSH2* or the *mutL*-like genes *PMS1* and *MLH1*. Genetic linkage analysis of two large HNPCC kindreds had demonstrated the existence of a locus that caused a predisposition to colorectal cancer. This locus was located on chromosome 2, close to a microsatellite polymorphism marker termed D2S123 (160). One group, led by Richard Kolodner, obtained the *hMSH2* gene by using degenerate PCR primers as described above and then showed that it mapped to the same chromosome 2p locus (52). A second group, led by Bert Vogelstein and Albert de la Chapelle, first localized the *HNPCC* gene on chromosome 2p to a 0.8-Mb interval and then showed that the *hMSH2* gene mapped within this interval. Subsequent analysis then showed that affected individuals in the kindreds, for whom the *HNPCC* locus mapped to chromosome 2p, carried mutations in one of the two copies of their *hMSH2* gene (111). For example, affected members of one large HNPCC kindred had a C → T transition in one allele that changed a highly conserved Pro to Leu, while affected individuals in another large kindred had a C → T transition mutation in one allele that resulted in a nonsense mutation (111). Another mutation that was observed in some kindreds is located in an intron (52, 111) but appears to be simply a polymorphism since it was also observed in normal individuals (111). Taken together, these observations constitute compelling evidence that the *hMSH2* gene is the *HNPCC* gene located on chromosome 2p.

A second *HNPCC* locus was subsequently mapped to markers on chromosome 3p21 in several families (120, 153). The *hMLH1* gene (see above) was then shown to map to the same locus, and affected individuals from these families were found to have mutations in one of the two copies of their *hMLH1* genes (18, 157). Subsequent to this finding, *hPMS1* (chromosome 2) was found to be mutated in the germline of an HNPCC patient, as was *hPMS2* (chromosome 7) (152a). Thus four genes, one encoding a MutS homolog and three encoding MutL homologs, have been associated with HNPCC as of the writing of this book. It is possible that future analyses of other *HNPCC* genes will implicate other genes that are associated with mismatch repair.

The concept that HNPCC is associated with a deficiency in mismatch repair was strongly supported by the finding that a colorectal cancer cell line (termed H6) derived from a tumor with the characteristics of those in HNPCC patients, including instability of $(GT)_n$ tracts, was defective in strand-specific mismatch repair (159). Mismatch repair was assessed by using circular heteroduplex substrates, similar to those described above, that contain a single-strand break. The H6 cell line was found to be defective in the strand-specific repair of all eight possible base pair mismatches, and in the repair of two-, three-, and four-nucleotide insertion-deletion mismatches. In addition, the H6 cell line was defective in the repair of $(GT)_{20} \cdot (CA)_{19}$ and $(GT)_{19} \cdot (CA)_{20}$ heteroduplexes (Fig. 9–14). In contrast, a cell line (SO) derived from a different class of colorectal tumors that do not exhibit $(GT)_n$ tract instability was proficient in mismatch repair. The defect in the H6 cells was

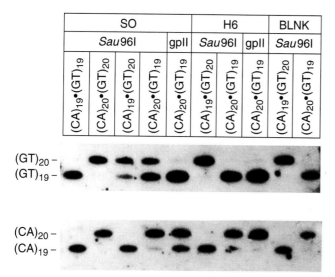

Figure 9–14 H6 cells are defective in the repair of $(GT)_{20} \cdot (CA)_{19}$ and $(GT)_{19} \cdot (CA)_{20}$ heteroduplexes. $(GT)_{20} \cdot (CA)_{19}$ and $(GT)_{19} \cdot (CA)_{20}$ heteroduplexes and $(GT)_{19} \cdot (CA)_{19}$ and $(GT)_{20} \cdot (CA)_{20}$ homoduplexes were subjected to mismatch repair by nuclear extracts of H6 cells (see text) and of SO cells (see text). BLNK indicates no extract. Strand specificity was provided by incision of the complementary or viral strands at *Sau*96I or gpII sites, respectively. After the reaction, the repeated sequence elements were excised from the reaction products by cleavage at flanking restriction enzyme sites and then separated by electrophoresis. The products were visualized by hybridization with strand-specific probes. (*Reproduced from Parsons et al. [159] with permission.*)

thought to be due to the absence of a single protein activity, since extracts of H6 cells could be complemented in vitro by a fraction from HeLa cells that chromatographs as a discrete species (159). It was subsequently found that the H6 cell line is devoid of a wild-type *hMLH1* allele (157), indicating that the deficiency in mismatch repair is due to the absence of a MutL-like function. Interestingly, the block in mismatch repair in H6 cells appears to be different from that in the MT1 cell line (see above), since nuclear extracts from the two cell lines complement each other in vitro (159).

It has not yet been established how the mutations in *hMSH2, hMLH1, hPMS1*, or *hPMS2* in afflicted individuals could give rise to an increased incidence of cancer. However, the genetic analyses discussed above have shown HNPCC individuals are heterozygous for these mutations, i.e., have a wild-type allele of these relevant genes in addition to the mutated one. Consistent with this, a lymphoblastoid cell line derived from an HNPCC individual, while proficient for mismatch repair, was found to be somewhat less proficient than normal cells (159). In contrast, the tumors that arise in HNPCC individuals seem to be largely defective in mismatch repair (111, 159). Since the transition from preneoplastic tissue to tumor tissue in HNPCC individuals is, in general, not accompanied by loss of heterozygosity as in many types of cancer (128), it would appear that the wild-type allele (of *hMSH2, hMLH1, hPMS1*, or *hPMS2*, depending on the kindred) must be lost by some other mechanism.

A plausible model (84, 111, 157) that could account for these observations is that heterozygous cells of an HNPCC individual might carry out mismatch repair more or less normally until some additional event occurred. It is possible that the somewhat reduced mismatch repair capability of heterozgygous HNPCC cells results in a modest increase in the mutation rate that would increase the probability of such an additional event occurring. For example, if a mutation occurred that brought about uncontrolled cellular growth, this might lead to a further increase in the mutation rate because of the reduced capacity of cells to carry out mismatch repair. This, in turn, might lead to inactivation of the wild-type allele of the relevant mismatch repair gene, an even higher mutation rate, and then the acquisition of other mutations that are responsible for the tumorigenic state. A model of this type, which postulates that the transition to the carcinogenic state is triggered by a event that is unrelated to the mismatch repair capability of the cell, is able to account for the fact that a strong mutator phenotype is observed only in tumor tissue from HNPCC individuals. Whatever the mechanism turns out to be, it seems likely that the analyses of HNPCC will serve as a useful paradigm for some other cancer predisposition syndromes and for genetic diseases involving instabilities of short repeated sequences.

Short-Patch Mismatch Repair in Prokaryotes and Eukaryotes

In addition to possessing a long-patch mismatch repair system that recognizes a variety of mismatches, both prokaryotic and eukaryotic cells possess mismatch repair systems that are characterized by *short excision repair tracts* (typically 10 nucleotides or less) (147). In general, these appear to have a relatively restricted specificity with respect to the mismatches that they repair. Some of these systems also appear to operate only within a restricted sequence context.

Very Short Patch Mismatch Correction in *E. coli*

The best characterized of any short-patch mismatch repair system is the very short patch (VSP) mismatch correction system of *E. coli* (58, 69, 139, 147, 171, 173). This system efficiently corrects the T in G·T mismatches that occur in sequences resembling CC(A/T)GG, the sequence at which the *dcm*-encoded methylase methylates the internal C's to generate 5-methylcytosine (132). Dcm recognition sites are hot spots for mutation. This effect is thought to be due to spontaneous deamination of 5-methylcytosine in the G·5-mC base pair, an event yielding a G·T mispair (32, 42) (Fig. 9–15). By restoring G·T mismatches occurring in sequences related to CC(A/T)GG to G·C, the VSP mismatch repair system plays a role in the avoidance of mutations resulting from the spontaneous deamination of 5-methylcytosine in DNA.

The phenomenon that first signaled the existence of the VSP repair pathway was the finding by Margaret Lieb that certain exceptional amber mutations in the *cI* gene of phage λ yielded unexpectedly high frequencies of recombination in multifactor crosses (114). These were identified as C → T transition mutations affecting the internal C's of CC(A/T)GG and several related sequences (115, 116, 119). These transition mutations were found to recombine at unexpectedly high frequency with markers as close as 10 bp away, with the excess recombinants recovered bearing the C·G genotype of the exceptional mutation (115, 116, 119). These findings led Lieb to suggest that the effect is due to a heteroduplex repair process that preferentially restores CC(A/T)GG and is characterized by excision tracts rarely exceeding 10 nucleotides, hence the name VSP repair (119). Additional evidence for the existence of the VSP repair system was also provided by experiments with heteroduplex DNA (87, 88, 175, 228). These experiments showed that of the two possible heteroduplex configurations, only the heteroduplex carrying the G·T mispair was subject to VSP repair and that the product was exclusively G·C.

Two of the genes required for methyl-directed mismatch repair, *mutS*+ and *mutL*+, are also required for VSP repair, but two others (*mutH*+ and *uvrD*+) are not (87, 88, 117, 228). In addition, the products of two other genes, *vsr*+ (197) and *polA*+ (43), are required. The *vsr*+ gene is located immediately downstream of the *dcm*+ gene and in fact partially overlaps with it (197). In fact, *dcm-6* mutants are deficient in both *dcm* and *vsr* function (197), and this accounts for the earlier erroneous conclusion that the *dcm*+ gene product is needed for VSP repair. The differing genetic requirements of methyl-directed mismatch repair and VSP repair allowed the performance of heteroduplex experiments that explored the relative roles of the two systems in processing G·T mispairs within the CC(A/T)GG sequence environment (87, 139). In heteroduplexes carrying fully methylated GATC sequences (and therefore representing nonreplicating DNA), G·T mispairs were corrected by the VSP system to G·C pairs. In contrast, in heteroduplexes carrying hemimethylated GATC sequences (and therefore representing newly synthesized DNA), the repair was carried out by the methyl-directed mismatch repair system so that the outcome of the repair (G·C versus A·T) was determined by the strand bearing the GATC methylation.

The *vsr*+ gene product has been shown to be an 18-kDa protein that serves as a *strand-specific mismatch endonuclease* (77). It recognizes G·T mismatches that occur in the following sequence contexts CT(A/T)GG and NT(A/T)GG and makes incisions 5′ of the underlined T that produce a 5′-phosphate terminus and a 3′-hydroxyl terminus. Since genetic studies have shown that DNA polymerase I is required for VSP, it appears that DNA polymerase I removes the T with its 5′ → 3′ exonucleolytic activity and then carries out repair synthesis, producing its usual short repair tracts (see chapter 4). Since the Vsr endonuclease can act in the absence of MutS and MutL proteins, it seems likely that these two proteins play a role in stimulating or regulating the activity of Vsr protein. This hypothesis is supported by the observation that the requirement for the *mutS*+ and *mutL*+ genes can be overcome by introduction of the *vsr*+ gene on a multicopy plasmid (197).

The hypothesis that VSP repair protects cells from mutations resulting from the deamination of 5-methylcytosine in CC(A/T)GG sequences has been directly tested and confirmed (118). A particular hot spot for spontaneous mutation that corresponds to a 5-methylcytosine was found to be absent in bacteria lacking both the Dcm methylase and VSP repair activity. Restoration of Dcm function in the absence of VSP repair resulted in a 10-fold higher spontaneous mutation frequency at that site than in wild-type bacteria. Finally, overproduction of the Vsr endonuclease in a wild-type strain resulted in a fourfold reduction in the spontaneous mutation frequency at that site. In addition, it has been suggested (15)

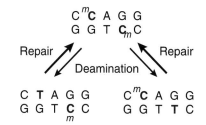

Figure 9–15 Production and repair of mismatches produced by the deamination of 5-methylcytosine in *E. coli*. Boldface letters signify sites of these two events. (*Adapted from Jones et al. [88] with permission.*)

that the capacity of the VSP system to repair G·T mismatches at sequences related to, but not recognized by, the Dcm methylase might play a role in shaping the sequence composition of the *E. coli* genome. For example, if the underlined T in the sequence C<u>T</u>AG were to end up in a G·T mismatch because of a replication error and were to have escaped long-patch mismatch repair, VSP repair would cause a T-to-C transition. This type of action would tend to deplete the genome of the T-containing sequences (e.g., CTAG) while enriching it for the corresponding C-containing sequences (e.g., CCAG) and might account for the scarcity of CTAG sequences in the genomes of enteric bacteria (15).

A second function for the VSP repair system appears to be to repair U·G mismatches in Dcm sequences that arise by deamination of cytosine (61). Cytosine deaminations at these sites might be promoted by the Dcm methylase since, in the absence of *S*-adenosylmethionine, DNA cytosine methylases promote the formation of a 5-dihydro intermediate that is much more susceptible to deamination than cytosine is (189, 225, 226). Normally, uracils arising by deamination of C are recognized by ura-DNA glycosylase and then repaired (see chapter 4). Although not as efficient as the ura-DNA glycosylase system, VSP repair of U·G mismatches has been shown to contribute to the maintenance of genetic fidelity at these sites (61).

MutY-Dependent Mismatch Repair in *E. coli*

In vitro analyses of repair by *E. coli* extracts of heteroduplexes containing A·G mismatches revealed that these mismatches could also be repaired by a short-patch repair system that was independent of the *mutH⁺L⁺S⁺*-dependent, ATP-dependent, methyl-dependent mismatch repair system discussed previously (124, 125, 201). The G·A mismatches were found to be corrected to G·C base pairs. In addition, in vivo analyses of heteroduplex repair indicated that A·G and A·C base pairs can be corrected independently of the methyl-directed mismatch repair system (170). Both in vitro and in vivo, the repair was dependent on DNA polymerase I and gave rise to repair patches in the range of 10 to 20 nucleotides (12, 169, 208). Furthermore, this repair was shown to require the function of the *mutY⁺* (*micA⁺*) gene both in vitro and in vivo (11, 170, 208). As discussed in chapter 4, the MutY protein plays an important role in the avoidance of mutations from 8-oxo-7,8-dihydrodeoxyguanine (GO) lesions by functioning as a DNA glycosylase that specifically removes A from a GO·A mispair (12, 141–143), the mispair that would be generated if A were inserted opposite a GO lesion during DNA replication. However, as mentioned in chapter 4, the MutY protein can also function as a DNA glycosylase with an associated 3′ AP lyase activity, which specifically removes A from G·A or C·A mispairs (12, 207). The nicking activity of highly purified MutY protein is about 20-fold lower on an A·C substrate than on an A·G substrate (207).

Several pieces of evidence have been obtained indicating that the G·C → T·A mutator phenotype of *mutY* mutants results from a failure to avoid mutagenesis resulting from 8-oxo-7,8-dihydrodeoxyguanine rather from a failure to correct A·G mismatches that arise as replication errors (141, 142). However, it appears that, along with VSP mismatch repair, MutY-dependent mismatch repair may play an important role in determining the outcome of recombinational events between closely linked markers (58, 170). Furthermore, there may also be other mismatch repair systems that have not yet been characterized. For example, a *mutY* mutant has 10% of the capacity of a wild-type strain to repair A·C mismatches, and this capacity is reduced still further by the additional introduction of a *fpg* (*mutM*) mutation (see chapter 4). This observation raises the possibility of the presence of a low-level *fpg⁺*-dependent system that can repair A·C mismatches (58). Also, in vitro experiments have shown that MutY-dependent repair of A·G mismatches is specific for correction to C·G (12, 207), yet in vivo experiments suggest that A·G mispairs can be corrected in a *mutY⁺*-dependent manner to either C·G (70%) or A·T (30%) (58). This discrepancy raises the possibility of the presence of a function which acts in a *mutY⁺*-dependent pathway to correct A·G mispairs to A·T.

Short-Patch Mismatch Repair in Mammalian Cells

In addition to being able to repair G·T mispairs by the long-patch system discussed above, mammalian cells have been shown to have the capacity to repair G·T to G·C base pairs by a short-patch mechanism. In vivo evidence for the existence of this system was obtained from experiments in which circularly closed simian virus 40 heteroduplexes were transfected in simian cells (20). Correction occurred predominantly to G·C base pairs regardless of the orientation of the mismatch within the DNA heteroduplex. Although these initial experiments failed to detect any sequence specificity of the repair, later work (67, 193) suggested that this repair occurs preferrentially at CpG sites. An earlier report (75) had inferred that correction of G·T mispairs in simian cells is directed by 5-methylcytosine. It is possible that the sequence preference of the system accounts for the differing reports concerning the mismatch repair of O^6-mG·T mispairs (91, 192, 193)

Although the biochemical mechanism of short-patch G·T mismatch repair in mammalian cells has not been completely established, considerable progress has been made. G·T → G·C mismatch repair had been demonstrated in nuclear extracts of HeLa cells (219, 220). Analyses of the reaction intermediates suggested that the excision repair involved the replacement of a single nucleotide. The polymerase required to fill the gap was inferred to be DNA polymerase β because anti-polymerase β antibodies block the reaction (220). Enzymatic activities capable of incising the phosphodiester bonds immediately 5′ and 3′ to the mismatched T were then identified in human cell extracts (192, 219, 220, 227). The observation that the mismatched thymine is released as the free base during the reaction suggested that a thymine-DNA glycosylase generates an intermediate abasic site that then serves as the substrate for the incisions (see chapter 4) (220). This activity was then purified from HeLa cells and shown to be a 55-kDa protein that functions as a *mismatch-specific thymine-DNA glycosylase* that is capable of removing the T from a G·T mispair (149). The purified glycosylase lacks an associated AP endonuclease. Although the purified glycosylase can weakly recognize T's in C·T and T·T mispairs in vitro, it is not clear that these activities are relevant to its role in vivo. The activity of the G·T mismatch-specific thymine-DNA glycosylase does not seem to require stimulation by the G·T mismatch binding protein, which has been found to be missing from certain O^6-methylguanine DNA-methyltransferase deficient, alkylation-resistant cells that have a weak mutator phenotype (67). As mentioned above, the relationship between this protein and hMsh2 has not yet been established.

The most common suggestion for the physiological role of the short patch G·T mismatch repair system has been that it repairs G·T mispairs that originate from deamination of 5-methylcytosine in CpG sequences. In this view, its role is analogous to that of the VSP repair system of *E. coli*, which repairs G·T mispairs that originate from cytosine deamination of 5-methylcytosine in Dcm sites (see above). However, two additional possibilities have been suggested by the subsequent finding that the same mismatch-specific thymine-DNA glycosylase can also remove uracil from G·U mispairs (150). The first is that the activity might play a role in the removal of U's that are generated at CpG sites by the action of a DNA cytosine methylase. This suggestion is conceptually similar to that discussed above for the possible second role of the VSP system in *E. coli* (61). The second additional possibility is that the activity plays a role in the repair of G·U mispairs that arise from C deamination occurring in G + C-rich regions of the genome. The major ura-DNA glycosylases may require local strand separation for efficient removal of the base from the DNA, and at least one purified mammalian DNA glycosylase is inefficient at removing uracil from G·U mispairs in G + C-rich regions of the genome (46).

It seems likely that other short-patch mismatch repair systems will be characterized in eukaryotic cells over the coming years. For example, the purification of an activity from HeLa cells that specifically nicks A·G mismatches has been reported (227). This enzyme simultaneously makes incisions at the first phosphodiester bond both 5′ and 3′ to the mispaired A base but not the G base. The existence of this activity suggests that it may play a role in a mismatch repair pathway that is similar to the *E. coli mutY*⁺-dependent pathway.

REFERENCES

1. **Aaltonen, L. A., P. Peltomaki, F. S. Leach, P. Sistonen, L. Pylkkanen, J.-P. Mecklin, H. Jarvinen, S. M. Powell, J. Jen, S. R. Hamilton, G. M. Petersen, K. W. Kinzler, B. Vogelstein, and A. de la Chapelle.** 1993. Clues to the pathogenesis of familial colorectal cancer. *Science* **260:**812–816.

2. **Abastado, J.-P., B. Cami, T. H. Dinh, J. Igolen, and P. Kourilsky.** 1984. Processing of complex heteroduplexes in *Escherichia coli* and Cos-1 monkey cells. *Proc. Natl. Acad. Sci. USA* **81:**5792–5796.

3. **Aboussekhra, A., R. Chanet, Z. Zgaga, C. Cassier-Chauvat, M. Heude, and F. Fabre.** 1989. *RADH*, a gene of *Saccharomyces cerevisiae* encoding a putative DNA helicase involved in DNA repair. *Nucleic Acids Res.* **17:**7211–7219.

4. **Adams, M. D., M. Dubnick, A. R. Kerlavage, R. Moreno, J. M. Kelley, T. R. Utterback, J. W. Nagle, C. Fields, and J. C. Venter.** 1992. Sequence identification of 2,375 human brain genes. *Nature* (London) **355:**632–634.

5. **Adams, M. D., J. M. Kelley, J. D. Gocayne, M. Dubnick, M. H. Polymeropoulos, H. Xiao, C. R. Merril, A. Wu, B. Olde, R. F. Moreno, A. R. Kerlavage, W. R. McCombie, and J. C. Venter.** 1991. Complementary DNA sequencing: expressed sequence tags and human genome project. *Science* **252:**1651–1656.

6. **Adams, M. D., M. B. Soares, A. R. Kerlavage, C. Fields, and J. C. Venter.** 1993. Rapid cDNA sequencing (expressed sequence tags) from a directionally cloned human infant brain cDNA library. *Nat. Genet.* **4:**373–380.

7. **Alani, E., R. A. G. Reenan, and R. D. Kolodner.** 1994. Interaction between mismatch repair and genetic recombination in *Saccharomyces cerevisiae. Genetics* **137:**19–39.

8. **Amati, P., and M. Meselson.** 1965. Localized negative interference in bacteriophage λ. *Genetics* **51:**369–379.

9. **Anderson, R. P., C. G. Miller, and J. R. Roth.** 1976. Tandem duplications of the histidine operon observed following generalized transduction in *Salmonella typhimurium. J. Mol. Biol.* **105:**201–218.

10. **Anderson, R. P., and J. R. Roth.** 1978. Gene duplication in bacteria: Alteration of gene dosage by sister-chromosome exchanges. *Cold Spring Harbor Symp. Quant. Biol.* **43:**1083–1087.

11. **Au, K. G., M. Cabrera, J. H. Miller, and P. Modrich.** 1988. *Escherichia coli mutY* gene product is required for specific A·G to C·G mismatch correction. *Proc. Natl. Acad. Sci. USA* **85:**9163–9166.

12. **Au, K. G., S. Clark, J. H. Miller, and P. Modrich.** 1989. The *Escherichia coli mutY* gene encodes an adenine glycosylase active on G-A mispairs. *Proc. Natl. Acad. Sci. USA* **86:**8877–8881.

13. **Au, K. G., K. Welsh, and P. Modrich.** 1992. Initiation of methyl-directed mismatch repair. *J. Biol. Chem.* **267:**12142–12148.

14. **Beckman, J. S., and J. L. Weber.** 1992. Survey of human and rat microsatellites. *Genomics* **12:**627–631.

15. **Bhagwat, A. S., and M. McClelland.** 1992. DNA mismatch correction by Very Short Patch repair may have altered the abundance of oligonucleotides in the *E. coli* genome. *Nucleic Acids Res.* **20:**1663–1668.

16. **Bishop, D. K., J. Andersen, and R. D. Kolodner.** 1989. Specificity of mismatch repair following transformation of *Saccharomyces cerevisiae* with heteroduplex plasmid DNA. *Proc. Natl. Acad. Sci. USA* **86:**3713–3717.

17. **Branch, P., G. Aquilina, M. Bignami and P. Karran.** 1993. Defective mismatch binding and a mutator phenotype in cells tolerant to DNA damage. *Nature* (London) **362:**652–654.

18. **Bronner, C. E., S. M. Baker, P. T. Morrison, G. Warren, L. G. Smith, M. K. Lescoe, M. Kane, C. Earabine, J. Lipford, A. Lindblom, P. Tannergard, R. J. Bollag, A. R. Godwin, D. C. Ward, M. Nordenskjold, R. Fishel, R. Kolodner, and R. M. Liskay.** 1994. Mutation in the DNA mismatch repair gene homologue *hMLH1* is associated with hereditary nonpolyposis colon cancer. *Nature* (London) **368:**258–261.

19. **Brooks, P., C. Dohet, G. Almouzni, M. Mechali, and M. Radman.** 1989. Mismatch repair involving localized DNA synthesis in extracts of *Xenopus* eggs. *Proc. Natl. Acad. Sci. USA* **86:**4425–4429.

20. **Brown, T. C., and J. Jiricny.** 1987. A specific mismatch repair event protects mammalian cells from loss of 5-methylcytosine. *Cell* **50:**945–950.

21. **Brown, T. C., and J. Jiricny.** 1988. Different base/base mispairs are corrected with different efficiencies and specificities in monkey kidney cells. *Cell* **54:**705–711.

22. **Bruni, R., D. Martin, and J. Jiricny.** 1988. d(GATC) sequences influence *Escherichia coli* mismatch repair in a distance-dependent manner from positions both upstream and downstream of the mismatch. *Nucleic Acids Res.* **16:**487 5–4890.

23. **Carraway, M., and M. G. Marinus.** 1993. Repair of heteroduplex DNA molecules with multibase loops in *Escherichia coli. J. Bacteriol.* **175:**3972–3980.

24. **Caskey, C. T., A. Pizzuti, Y.-H. Fu, R. G. Fenwick, and D. L. Nelson.** 1992. Triplet repeat mutations in human disease. *Science* **256:**784–788.

25. **Chang, D.-Y., and A.-L. Lu.** 1991. Base mismatch-specific endonuclease activity in extracts of *S. cerevisiae. Nucleic Acids Res.* **19:**4761–4766.

26. **Chase, M., and A. H. Doermann.** 1958. High negative interference over short segments of the genetic structure of bacteriophage T4. *Genetics* **43:**332–353.

27. **Chen, J.-D., and S. A. Lacks.** 1991. Role of uracil-DNA glycosylase in mutation avoidance by *Streptococcus pneumoniae. J. Bacteriol.* **173:**282–290.

28. **Claverys, J.-P., and S. A. Lacks.** 1986. Heteroduplex deoxyribonucleic acid base mismatch repair in bacteria. *Microbiol. Rev.* **50:**133–165.

29. **Claverys, J.-P., V. Mejean, A.-M. Gasc, and A. M. Sicard.** 1983. Mismatch repair in *Streptococcus pneumoniae*: relationship between base mismatches and transformation efficiencies. *Proc. Natl. Acad. Sci. USA* **80:**5956–5960.

30. **Claverys, J. P., M. Roger, and A. M. Sicard.** 1980. Excision and repair of mismatched base pairs in transformation of *Streptococcus pneumoniae. Mol. Gen. Genet.* **178:**191–201.

31. **Cooper, D. L., R. S. Lahue, and P. Modrich.** 1993. Methyl-directed mismatch repair is bidirectional. *J. Biol. Chem.* **268:**11823–11829.

32. **Coulondre, C., J. H. Miller, P. J. Farabaugh, and W. Gilbert.** 1978. Molecular basis of base substitution hot spots in *Escherichia coli. Nature* (London) **274:**775–780.

33. **Cox, E. C.** 1976. Bacterial mutator genes and the control of spontaneous mutation. *Annu. Rev. Genet.* **10:**135–156.

34. **DasGupta, C., and C. M. Radding.** 1982. Lower fidelity of RecA protein catalysed homologous pairing with a superhelical substrate. *Nature* (London) **295:**71–73.

35. **Deschavanne, P., and M. Radman.** 1991. Counterselection of GATC sequences in enterobacteriophages by the components of the methyl-directed mismatch repair system. *J. Mol. Evol.* **33:**125–132.

36. **Detloff, P., J. Sieber, and T. D. Petes.** 1991. Repair of specific base pair mismatches formed during meiotic recombination in the yeast *Saccharomyces cerevisiae. Mol. Cell. Biol.* **11:**737–745.

37. **Detloff, P., M. White, and T. D. Petes.** 1992. Analysis of a gene conversion gradient at the *HIS4* locus in *Saccharomyces cerevisiae. Genetics* **132:**113–123.

38. **Dohet, C., S. Dzidic, R. Wagner, and M. Radman.** 1987. Large nonhomology in heteroduplex DNA is processed differently than single base pair mismatches. *Mol. Gen. Genet.* **206:**181–184.

39. **Dohet, C., R. Wagner, and M. Radman.** 1985. Repair of defined single base-pair mismatches in *Escherichia coli. Proc. Natl. Acad. Sci. USA* **82:**503–505.

40. **Dohet, C., R. Wagner, and M. Radman.** 1986. Methyl-directed repair of frameshift mutations in heteroduplex DNA. *Proc. Natl. Acad. Sci. USA* **83:**3395–3397.

41. **Doutriaux, M. P., R. Wagner, and M. Radman.** 1986. Mismatch-stimulated killing. *Proc. Natl. Acad. Sci. USA* **83:**2576–2578.

42. **Duncan, B. K., and J. H. Miller.** 1980. Mutagenic deamination of cytosine residues in DNA. *Nature* (London) **287:**560–561.

43. **Dzidic, S., and M. Radman.** 1989. Genetic requirements for hyper-recombination by very short patch mismatch repair: involvement of *Escherichia coli* DNA polymerase I. *Mol. Gen. Genet.* **217:**254–256.

44. **Eadie, J. S., M. Conrad, D. Toorchen, and M. D. Topal.** 1984. Mechanism of mutagenesis by O^6-methylguanine. *Nature* (London) **308:**201–203.

45. **Echols, H., and M. F. Goodman.** 1991. Fidelity mechanisms in DNA replication. *Annu. Rev. Biochem.* **60:**477–511.

46. **Eftedal, I., P. H. Guddal, G. Slupphaug, G. Volden, and H. E. Krokan.** 1993. Consensus sequences for good and poor removal of uracil from double stranded DNA by uracil-DNA glycosylase. *Nucleic Acids Res.* **21:**2095–2101.

47. **Ephrussi-Taylor, H., and T. C. Gray.** 1966. Genetic studies of recombining DNA in pneumococcal transformation. *J. Gen. Physiol.* **49** (2):211–231.

48. **Ephrussi-Taylor, H., A. M. Sicard, and R. Kamen.** 1965. Genetic recombination in DNA-induced transformation of pneumococcus. I. The problem of relative efficiency of transforming factors. *Genetics* **51:**455–475.

49. **Fang, W. H., G. M. Li, M. Longley, J. Holmes, W. Thilly, and P. Modrich.** 1993. Mismatch repair and genetic stability in human cells. *Cold Spring Harbor Symp. Quant. Biol.* **58:**597–603.

50. **Fang, W.-H., and P. Modrich.** 1993. Human strand-specific mismatch repair occurs by a bidirectional mechanism similar to that of the bacterial reaction. *J. Biol. Chem.* **268:**11838–11844.

51. **Feinstein, S. I., and K. B. Low.** 1986. Hyper-recombining recipient strains in bacterial conjugation. *Genetics* **113:**13–33.

52. **Fishel, R., M. K. Lescoe, M. R. S. Rao, N. G. Copeland, N. A. Jenkins, J. Garber, M. Kane, and R. Kolodner.** 1993. The human mutator gene homolog *MSH2* and its association with hereditary nonpolyposis cancer. *Cell* **75:**1027–1038.

53. **Fishel, R. A., E. C. Siegel, and R. Kolodner.** 1986. Gene conversion in *Escherichia coli.* Resolution of heteroallelic mismatched nucleotides by corepair. *J. Mol. Biol.* **188:**147–157.

54. **Fleck, O., H. Michael, and L. Heim.** 1992. The *swi4$^+$* gene of *Schizosaccharomyces pombe* encodes a homologue of mismatch repair enzymes. *Nucleic Acids Res.* **20:**2271–2278.

55. **Folger, K. R., K. Thomas, and M. R. Capecchi.** 1985. Efficient correction of mismatched bases in plasmid heteroduplexes injected into cultured mammalian cell nuclei. *Mol. Cell. Biol.* **5:**70–74.

56. **Fox, M. S.** 1978. Some features of genetic recombination in procaryotes. *Annu. Rev. Genet.* **12:**47–68.

57. **Fox, M. S., and M. K. Allen.** 1964. On the mechanism of deoxyribonucleate integration in pneumococcal transformation. *Proc. Natl. Acad. Sci. USA* **52:**412–419.

58. **Fox, M. S., P. Radicella, and K. Yamamoto.** 1994. Some features of basepair mismatch repair and its role in the formation of genetic recombination. *Experientia* **50:**253–260.

59. **Friedberg, E. C.** 1985. *DNA Repair.* W. H. Freeman & Co., New York.

60. **Fujii, H., and T. Shimada.** 1989. Isolation and characterization of cDNA clones derived from the divergently transcribed gene in the region upstream from the human dihydrofolate reductase gene. *J. Biol. Chem.* **264:**10057–10064.

61. **Gabbara, S., M. Wyszynski, and A. S. Bhagwat.** 1994. A DNA repair process in *Escherichia coli* corrects U:G and T:G mismatches to C:G at sites of cytosine methylation. *Mol. Gen. Genet.* **243:**244–248.

62. **Glazer, P. M., S. N. Sarkar, G. E. Chisholm, and W. C. Summers.** 1987. DNA mismatch repair detected in human cell extracts. *Mol. Cell. Biol.* **7:**218–224.

63. **Glickman, B., P. van den Elsen, and M. Radman.** 1978. Induced mutagenesis in *dam⁻* mutants of *Escherichia coli:* a role for 6-methyladenine residues in mutation avoidance. *Mol. Gen. Genet.* **163:**307–312.

64. **Glickman, B. W., and M. Radman.** 1980. *Escherichia coli* mutator mutants deficient in methylation-instructed DNA mismatch correction. *Proc. Natl. Acad. Sci. USA* **77:**1063–1067.

65. **Goldmacher, V. S., R. A. Cuzick, and W. G. Thilly.** 1986. Isolation and partial characterization of human cell mutants differing in sensitivity to killing and mutation by methylnitrosourea and *N*-methyl-*N'*-nitro-nitrosoguanidine. *J. Biol. Chem.* **261:**12462–12471.

66. **Grafstrom, R. H., A. Amsterdam, and K. Zachariasewycz.** 1988. In vivo studies of repair of 2-aminopurine in *Escherichia coli. J. Bacteriol.* **170:**3485–3492.

67. **Griffin, S., and P. Karran.** 1993. Incision at DNA G.T mispairs by extracts of mammalian cells occurs preferentially at cytosine methylation sites and is not targeted by a separate G.T binding reaction. *Biochemistry* **32:**13032–13039.

68. **Grilley, M., J. Griffith, and P. Modrich.** 1993. Bidirectional excision in methyl-directed mismatch repair. *J. Biol. Chem.* **268:**11830–11837.

69. **Grilley, M., J. Holmes, B. Yashar, and P. Modrich.** 1990. Mechanisms of DNA-mismatch correction. *Mutat. Res.* **236:**253–267.

70. **Grilley, M., K. M. Welsh, S.-S. Su, and P. Modrich.** 1989. Isolation and characterization of the *Escherichia coli mutL* gene product. *J. Biol. Chem.* **264:**1000–1004.

71. **Guild, W. R., and N. B. Shoemaker.** 1976. Mismatch correction in pneumococcal transformation: donor length and *hex*-dependent marker efficiency. *J. Bacteriol.* **125:**125–135.

72. **Haber, J. E., B. L. Ray, J. M. Kolb, and C. I. White.** 1993. Rapid kinetics of mismatch repair of heteroduplex DNA that is formed during recombination in yeast. *Proc. Natl. Acad. Sci. USA* **90:**3363–3367.

73. **Haber, L. T., P. P. Pang, D. I. Sobell, J. A. Mankovich, and G. C. Walker.** 1988. Nucleotide sequence of the *Salmonella typhimurium mutS* gene required for mismatch repair: homology of MutS and HexA of *Streptococcus pneumoniae. J. Bacteriol.* **170:**197–202.

74. **Haber, L. T., and G. C. Walker.** 1991. Altering the conserved nucleotide binding motif in the *Salmonella typhimurium* MutS mismatch repair protein affects both its ATPase and mismatch binding activities. *EMBO J.* **10:**2707–2715.

75. **Hare, J. T., and J. H. Taylor.** 1988. Hemi-methylation dictates strand selection in repair of G/T and A/C mismatches in SV40. *Gene* **74:**159–161.

76. **Henderson, S. T., and T. D. Petes.** 1992. Instability of simple sequence DNA in *Saccharomyces cerevisiae. Mol. Cell. Biol.* **12:**2749–2757.

77. **Hennecke, F., H. Kolmar, K. Bründl, and H.-J. Fritz.** 1991. The *vsr* gene product of *E. coli* K-12 is a strand- and sequence-specific DNA mismatch endonuclease. *Nature* (London) **353:**776–778.

78. **Herman, G. E., and P. Modrich.** 1981. *Escherichia coli* K-12 clones that overproduce *dam* methylase are hypermutable. *J. Bacteriol.* **145:**644–646.

79. **Holliday, R. A.** 1964. A mechanism for gene conversion in fungi. *Genet. Res.* **5:**282–304.

80. **Holmes, J., S. Clark, and P. Modrich.** 1990. Strand-specific mismatch correction in nuclear extracts of human and *Drosophila melanogaster* cell lines. *Proc. Natl. Acad. Sci. USA* **87:**5837–5841.

81. **Hughes, M., and J. Jiricny.** 1992. The purification of a human mismatch-binding protein and identification of its associated ATPase sand helicase activities. *J. Biol. Chem.* **267:**8860–8864.

82. **Huisman, O., and M. S. Fox.** 1986. A genetic analysis of primary products of bacteriophage lambda recombination. *Genetics* **112:**409–420.

83. **Ionov, Y., M. A. Peinado, S. Malkhosyan, D. Shibata, and M. Perucho.** 1993. Ubiquitous somatic mutations in simple repeated sequences reveal a new mechanism for colonic carcinogenesis. Nature (London) **363:**558–561.

84. **Jiricny, J.** 1994. Colon cancer and DNA repair: have mismatches met their match? *Trends Genet.* **10:**164–168.

85. **Jiricny, J., M. Hughes, N. Corman, and B. B. Rudkin.** 1988. A human 200-kDa protein binds selectively to DNA fragments containing G·T mismatches. *Proc. Natl. Acad. Sci. USA* **85:**8860–8864.

86. Jiricny, J., S.-S. Su, S. G. Wood, and P. Modrich. 1988. Mismatch-containing oligonucleotide duplexes bound by the *E. coli mutS*-encoded protein. *Nucleic Acids Res.* **16:**7843–7854.

87. Jones, M., R. Wagner, and M. Radman. 1987. Mismatch repair and recombination in *E. coli. Cell* **50:**621–626.

88. Jones, M., R. Wagner, and M. Radman. 1987. Mismatch repair of deaminated 5-methylcytosine. *J. Mol. Biol.* **194:**155–159.

89. Jones, M., R. Wagner, and M. Radman. 1987. Repair of a mismatch is influenced by the base composition of the surrounding nucleotide sequence. *Genetics* **115:**605–610.

90. Karimova, G. S., P. S. Grigoriev, and V. N. Rybchin. 1985. Participation of genes for the system of mismatched bases correction in genetic recombination of *Escherichia coli. Mol. Genet. Mikrobiol. Virusol.* **10:**29–34.

91. Karran, P., and M. Bignami. 1992. Self-destruction and tolerence in resistance of mammalian cells to alkylation damage. *Nucleic Acids Res.* **20:**2933–2940.

92. Karran, P., and M. Marinus. 1982. Mismatch correction at O^6-methylguanine residues in *E. coli* DNA. *Nature* (London) **296:**868–869.

93. Kat, A., W. G. Thilly, W.-H. Fang, M. J. Longley, G.-M. Li, and P. Modrich. 1993. An alkylation-tolerant, mutator human cell line is deficient in strand-specific mismatch repair. *Proc. Natl. Acad. Sci. USA* **90:**6424–6428.

94. Kornberg, A., and T. A. Baker. 1991. *DNA Replication.* W. H. Freeman & Co., New York.

95. Kramer, B., W. Kramer, and H.-J. Fritz. 1984. Different base/base mismatches are corrected with different efficiencies by the methyl-directed DNA mismatch-repair system of *E. coli. Cell* **38:**879–887.

96. Kramer, B., W. Kramer, M. S. Williamson, and S. Fogel. 1989. Heteroduplex DNA correction in *Saccharomyces cerevisiae* is mismatch specific and requires functional *PMS* genes. *Mol. Cell. Biol.* **9:**4432–4440.

97. Kramer, W., B. Kramer, M. S. Williamson, and S. Fogel. 1989. Cloning and nucleotide sequence of DNA mismatch repair gene *PMS1* from *Saccharomyces cerevisiae*: homology of PMS1 to procaryotic MutL and HexB. *J. Bacteriol.* **171:**5339–5346.

98. Kuhl, D. P. A., and C. T. Caskey. 1993. Trinucleotide repeats and genome variation. *Curr. Opin. Genet. Dev.* **3:**404–407.

99. Kunkel, T. A. 1993. Slippery DNA and diseases. *Nature* (London) **365:**207–208.

100. Lacks, S. 1970. Mutants of *Diplococcus pneumoniae* that lack deoxyribonucleases and other activities possibly pertinent to genetic transformation. *J. Bacteriol.* **101:**373–383.

101. Lacks, S. A. 1962. Molecular fate of DNA in genetic transformation. *J. Mol. Biol.* **5:**119–131.

102. Lacks, S. A. 1966. Integration efficiency and genetic recombination in pneumococcal transformation. *Genetics* **53:**207–235.

103. Lacks, S. A. 1988. Mechanisms of genetic recombination in gram-positive bacteria, p. 43–85. *In* R. Kucherlapati and G. Smith (ed.), *Genetic Recombination.* American Society for Microbiology, Washington, D.C.

104. Lacks, S. A. 1989. Generalized DNA mismatch repair: its molecular basis in *Streptococcus pneumoniae* and other organisms., p. 325–339. *In* L. O. Butler, C. Harwood, and B. E. B. Mosley (ed.), *Genetic Transformation and Expression.* Intercept Ltd., Andover, United Kingdom.

105. Lacks, S. A., J. Dunn, and B. Greenberg. 1982. Identification of base mismatches recognized by the heteroduplex-DNA-repair system of *Streptococcus pneumoniae. Cell* **31:**327–336.

106. Laengle, R. F., M. G. Maenhaut, and M. Radman. 1986. GATC sequence and mismatch repair in *Escherichia coli. EMBO J.* **5:** 2009–2013.

107. Lahue, R. S., K. G. Au, and P. Modrich. 1989. DNA mismatch correction in a defined system. *Science* **245:**160–164.

108. Lahue, R. S., S. S. Su, and P. Modrich. 1987. Requirement for d(GATC) sequences in *Escherichia coli mutHLS* mismatch correction. *Proc. Natl. Acad. Sci. USA* **84:**1482–1486.

109. Lahue, R. S., S.-S. Su, K. Welsh, and P. Modrich. 1987. Analysis of methyl-directed mismatch repair in vitro, p. 125–134. *In* R. McMacken, and T. J. Kelly (ed.), *DNA Replication and Recombination.* Allen R. Liss, Inc., New York.

110. Längle-Rouault, F., M. G. Maenhaut, and M. Radman. 1987. GATC sequences, DNA nicks and the MutH function in *Escherichia coli* mismatch repair. *EMBO J.* **6:**1121–1127.

111. Leach, F. S., N. C. Nicolaides, N. Papadopoulos, B. Liu, J. Jen, R. Parsons, P. Peltomäki, P. Sistonen, L. A. Aaltonen, M. Nyström-Lahti, X.-Y. Guan, J. Zhang, P. S. Meltzer, J.-W. Yu, F.-T. Kao, D. J. Chen, K. M. Cerosaletti, R. E. K. Fournier, S. Todd, T. Lewis, R. J. Leach, S. L. Naylor, J. Weissenbach, J.-P. Mecklin, H. Järvinen, G. M. Petersen, S. R. Hamilton, J. Green, J. Jass, P. Watson, H. T. Lynch, J. M. Trent, A. de la Chapelle, K. W. Kinzler, and B. Vogelstein. 1993. Mutations of a *mutS* homolog in hereditary nonpolyposis colorectal cancer. *Cell* **75:**1215–1225.

112. Learn, B. A., and R. H. Grafstrom. 1989. Methyl-directed repair of frameshift heteroduplexes in cell extracts from *Escherichia coli. J. Bacteriol.* **171:**6473–6481.

113. Levinson, G., and G. A. Gutman. 1987. High frequencies of short frameshifts in poly-CA/TG tandem repeats borne by bacteriophage M13 in *Escherichia coli* K-12. *Nucleic Acids Res.* **15:**5323–5338.

114. Lieb, M. 1981. A fine structure map of spontaneous and induced mutations in the lambda repressor gene, including insertions of IS elements. *Mol. Gen. Genet.* **181:**364–371.

115. Lieb, M. 1983. Specific mismatch correction in bacteriophage lambda crosses by very short patch repair. *Mol. Gen. Genet.* **191:**118–125.

116. Lieb, M. 1985. Recombination in the lambda repressor gene: evidence that very short patch (vsp) mismatch correction restores a specific sequence. *Mol. Gen. Genet.* **199:**465–470.

117. Lieb, M. 1987. Bacterial genes *mutL, mutS,* and *dcm* participate in repair of mismatches at 5-methylcytosine sites. *J. Bacteriol.* **169:**5241–5246.

118. Lieb, M. 1991. Spontaneous mutation at a 5-methylcytosine hotspot is prevented by very short patch (VSP) mismatch repair. *Genetics* **128:**23–27.

119. Lieb, M., E. Allen, and D. Read. 1986. Very short patch mismatch repair in phage lambda: repair sites and length of repair tracts. *Genetics* **114:**1041–1060.

120. Lindblom, A., P. Tannergard, B. Werelius, and M. Nordenskjold. 1993. Genetic mapping of a second locus predisposing to hereditary nonpolyposis colon cancer. *Nat. Genet.* **5:**279–282.

121. Linton, J. P., J.-Y. J. Yen, E. Selby, Z. Chen, J. M. Chinsky, K. Liu, R. E. Kellems, and G. F. Crouse. 1989. Dual bidirectional promoters at the mouse *DHFR* locus: cloning and characterization of two mRNA classes of the divergently transcribed *Rep-1* gene. *Mol. Cell. Biol.* **9:**3058–3072.

122. Litman, R. M. 1961. Genetic and chemical alterations in the transforming DNA of pneumococcus caused by ultraviolet light and nitrous acid. *J. Chim. Phys.* **58:**997–1003.

123. Lu, A.-L. 1987. Influence of GATC sequences on *Escherichia coli* DNA mismatch repair in vitro. *J. Bacteriol.* **169:**1254–1259.

124. Lu, A.-L., and D. Y. Chang. 1988. A novel nucleotide excision repair for the conversion of an A/G mismatch to C/G base pair in *E. coli. Cell* **54:**805–812.

125. Lu, A.-L., and D. Y. Chang. 1988. Repair of single base-pair transversion mismatches of *Escherichia coli* in vitro: correction of certain A/G mismatches is independent of *dam* methylation and host *mutHLS* gene function. *Genetics* **118:**593–600.

126. Lu, A.-L., S. Clark, and P. Modrich. 1983. Methyl-directed repair of DNA base pair mismatches in vitro. *Proc. Natl. Acad. Sci. USA* **80:**4639–4643.

127. Lu, A.-L., K. Welsh, S. Clark, S.-S. Su, and P. Modrich. 1984. Repair of DNA base-pair mismatches in extracts of *Escherichia coli. Cold Spring Harbor Symp. Quant. Biol.* **49:**589–596.

128. Lynch, H. T., T. C. Smyrk, P. Watson, S. J. Lanspa, J. F. Lynch, P. M. Lynch, R. J. Cavalieri, and C. R. Boland. 1993. Genetics, natural history, tumor spectrum, and pathology of hereditary nonpolyposis colorectal cancer: an updated review. *Gasteroenterology* **104:**1535–1549.

129. **Lyons, S. M., and P. F. Schendel.** 1984. Kinetics of methylation in *Escherichia coli* K-12. *J. Bacteriol.* **159:**421–423.

130. **Mankovich, J. A., C. A. McIntyre, and G. C. Walker.** 1989. Nucleotide sequence of the *Salmonella typhimurium mutL* gene required for mismatch repair: homology of *mutL* to *hexB* of *Streptococcus pneumoniae* and to *PMS1* of the yeast *Saccharomyces cerevisiae. J. Bacteriol.* **171:** 5325–5331.

131. **Marinus, M. G.** 1976. Adenine methylation of Okazaki fragments in *Escherichia coli. J. Bacteriol.* **128:**853–854.

132. **Marinus, M. G.** 1984. Methylation of prokaryotic DNA, p. 81-109. *In* A. Razin, H. Cedar, and A. D. Riggs (ed.), *DNA Methylation.* Springer-Verlag, New York.

133. **Marinus, M. G., and N. R. Morris.** 1974. Biological function for 6-methyladenine residues in the DNA of *Escherichia coli* K12. *J. Mol. Biol.* **85:**309–322.

134. **Marinus, M. G., and N. R. Morris.** 1975. Pleiotropic effects of a DNA adenine methylation mutant (*dam-3*) in *Escherichia coli. Mutat. Res.* **28:**15–26.

135. **Marinus, M. G., A. Poteete, and J. A. Arraj.** 1984. Correlation of DNA adenine methylase activity with spontaneous mutability in *Escherichia coli* K-12. *Gene* **28:**123–125.

136. **Matic, I., M. Radman, and C. Rayssiguier.** 1994. Structure of recombinants from conjugational crosses between *Escherichia coli* donor and mismatch-repair deficient *Salmonella typhimurium* mutants. *Genetics* **136:**17–26.

137. **McGill, C., B. Shafer, and J. N. Strathern.** 1989. Coconversion of flanking sequences with homothallic switching. *Cell* **57:**459–467.

138. **McGraw, B. R., and M. G. Marinus.** 1980. Isolation and characterization of *dam*⁺ revertants and suppressor mutations that modify secondary phenotypes of *dam-3* strains of *Escherichia coli* K-12. *Mol. Gen. Genet.* **178:**309–315.

139. **Meselson, M.** 1988. Methyl-directed repair of DNA mismatches, p. 91–113. *In* K. B. Low (ed.), *Recombination of the Genetic Material.* Academic Press, Inc., San Diego, Calif.

140. **Meselson, M. S., and C. M. Radding.** 1975. A general model for genetic recombination. *Proc. Natl. Acad. Sci. USA* **72:**358–361.

141. **Michaels, M. L., C. Cruz, A. P. Grollman, and J. H. Miller.** 1992. Evidence that MutY and MutM combine to prevent mutations by an oxidatively damaged form of guanine in DNA. *Proc. Natl. Acad. Sci. USA* **89:**7022–7025.

142. **Michaels, M. L., and J. H. Miller.** 1992. The GO system protects organisms from the mutagenic effect of the spontaneous lesion 8-hydroxyguanine (7,8-dihydro-8-oxoguanine). *J. Bacteriol.* **174:**6321–6325.

143. **Michaels, M. L., J. Tchou, A. P. Grollman, and J. H. Miller.** 1992. A repair system for 8-oxo-7,8-dihydrodeoxyguanine. *Biochemistry* **31:**10964–10968.

144. **Miret, J. J., M. G. Milla, and R. S. Lahue.** 1993. Characterization of DNA mismatch-binding activity in yeast extracts. *J. Biol. Chem.* **208:**3507–3513.

145. **Modrich, P.** 1987. DNA mismatch correction. *Annu. Rev. Biochem.* **56:**435–466.

146. **Modrich, P.** 1989. Methyl-directed DNA mismatch correction. *J. Biol. Chem.* **264:**6597–6600.

147. **Modrich, P.** 1991. Mechanisms and biological effects of mismatch repair. *Annu. Rev. Genet.* **25:**229–253.

148. **Muster-Nassal, C., and R. Kolodner.** 1986. Mismatch correction catalyzed by cell-free extracts of *Saccharomyces cerevisiae. Proc. Natl. Acad. Sci. USA* **83:**7618–7622.

149. **Neddermann, P., and J. Jiricny.** 1993. The purification of a mismatch-specific thymine-DNA glycoslyase from HeLa cells. *J. Biol. Chem.* **268:**21218–21224.

150. **Neddermann, P., and J. Jiricny.** 1994. Efficient removal of uracil from G:U mispairs by the mismatch-specific thymine DNA glycosylase from HeLa cells. *Proc. Natl. Acad. Sci. USA* **91:**1642–1646.

151. **Nevers, P., and H. Spatz.** 1975. *Escherichia coli* mutants *uvrD uvrE* deficient in gene conversion of lambda heteroduplexes. *Mol. Gen. Genet.* **139:**233–243.

152. **New, L., K. Liu, and G. F. Crouse.** 1993. The yeast gene *MSH3* defines a new class of eukaryotic MutS homologs. *Mol. Gen. Genet.* **239:**97–108.

152a. **Nicolaides, N. C., N. Papadopoulos, B. Liu, Y.-F. Weil, K. C. Carter, S. M. Ruben, C. A. Rosen, W. A. Haseltine, R. D. Fleischmann, C. M. Fraser, M. D. Adams, J. C. Venter, M. G. Dunlop, S. R. Hamilton, G. M. Petersen, A. de la Chapelle, B. Vogelstein, and K. W. Kinzler.** 1994. Mutations of two *PMS* homologues in hereditary nonpolyposis colon cancer. *Nature* (London) **371:**75–80.

153. **Nyström-Lahti, M., P. Sistonen, J. P. Mecklin, L. Pylkkänen, L. Aaltonen, H. Järvinen, J. Weissenbach, A. de la Chapelle, and P. Peltomäki.** 1994. Close linkage to chromosome 3p and conservation of ancestral founding haplotype in hereditary nonpolyposis colorectal cancer cancer families. *Proc. Natl. Acad. Sci. USA* **91:**6054–6058.

154. **Palombo, F., M. Hughes, J. Jiricny, O. Truong, and J. Hsuan.** 1994. Mismatch repair and cancer. *Nature* (London) **367:**417.

155. **Pang, P. P., A. S. Lundberg, and G. C. Walker.** 1985. Identification and characterization of the *mutL* and *mutS* gene products of *Salmonella typhimurium* LT2. *J. Bacteriol.* **163:**1007–1015.

156. **Pang, P. P., S.-D. Tsen, A. S. Lundberg, and G. C. Walker.** 1984. The *mutH, mutL, mutS,* and *uvrD* genes of *Salmonella typhimurium* LT2. *Cold Spring Harbor Symp. Quant. Biol.* **49:**597–602.

157. **Papadopoulos, N., N. C. Nicolaides, Y.-F. Wei, S. M. Ruben, K. C. Carter, C. A. Rosen, W. A. Haseltine, R. D. Fleischmann, C. M. Fraser, M. D. Adams, J. C. Venter, S. R. Hamilton, G. M. Petersen, P. Watson, H. T. Lynch, P. Peltomäki, J.-P. Mecklin, A. de la Chapelle, K. W. Kinzler, and B. Vogelstein.** 1994. Mutation of a *mutL* homolog in hereditary colon cancer. *Science* **263:**1625–1629.

158. **Parker, B. O., and M. G. Marinus.** 1992. Repair of DNA heteroduplexes containing small heterologous sequences in *Escherichia coli. Proc. Natl. Acad. Sci. USA* **89:**1730–1734.

159. **Parsons, R., G.-M. Li, M. J. Longley, W.-H. Fang, N. Papadopoulos, J. Jen, A. de la Chapelle, K. W. Kinzler, B. Vogelstein, and P. Modrich.** 1993. Hypermutability and mismatch repair in RER⁺ tumor cells. *Cell* **75:**1227–1236.

160. **Peltomäki, P., L. A. Aaltonen, P. Sistonen, L. Pylkkäänen, J.-P. Mecklin, H. Järvinen, J. S. Green, J. R. Jass, J. L. Weber, F. S. Leach, G. M. Petersen, S. R. Hamilton, A. de la Chapelle, and B. Vogelstein.** 1993. Genetic mapping of a locus predisposing to human colorectal cancer. *Science* **260:**810–812.

161. **Petes, T. D., and C. W. Hill.** 1988. Recombination between repeated genes in microorganisms. *Annu. Rev. Genet.* **22:**147–168.

162. **Petit, M.-A., J. Dimpfl, M. Radman, and H. Echols.** 1991. Control of large chromosomal duplications in *Escherichia coli* by the mismatch repair system. *Genetics* **129:**327–332.

163. **Priebe, S. D., S. M. Hadi, B. Greenberg, and S. A. Lacks.** 1988. Nucleotide sequence of the *hexA* gene for DNA mismatch repair in *Streptococcus pneumoniae* and homology of *hexA* to *mutS* of *Escherichia coli* and *Salmonella typhimurium. J. Bacteriol.* **170:**190–196.

164. **Pritchard, R. H.** 1955. The linear arrangement of a series of alleles in *Aspergillus nidulans. Heredity* **9:**343.

165. **Prolla, T. A., D.-M. Christie, and R. M. Liskay.** 1994. Dual requirement in yeast DNA mismatch repair for *MLH1* and *PMS1,* two homologs of the bacterial *mutL* gene. *Mol. Cell. Biol.* **14:**407–415.

165a. **Prolla, T. A., Q. Pang, E. Alani, R. D. Kolodner, and R. M. Liskay.** 1994. MLH1, PMS1, and MSH2 interactions during the initiation of DNA mismatch repair in yeast. *Science* **265:**1091–1093.

166. **Prudhomme, M., B. Martin, V. Mejean, and J. P. Claverys.** 1989. Nucleotide sequence of the *Streptococcus pneumoniae hexB* mismatch repair gene: homology of *hexB* to *mutL* of *Salmonella typhi-*

murium and to *PMS1* of *Saccharomyces cerevisiae. J. Bacteriol.* **171:**5332–5338.

167. Prudhomme, M., V. Mejean, B. Martin, and J.-P. Claverys. 1991. Mismatch repair gene of *Streptococcus pneumoniae*: HexA confers a mutator phenotype in *Escherichia coli* by negative complementation. *J. Bacteriol.* **173:**7196–7203.

168. Pukkila, P. J., J. Peterson, G. Herman, P. Modrich, and M. Meselson. 1983. Effects of high levels of DNA adenine methylation on methyl-directed mismatch repair in *Escherichia coli. Genetics* **104:**571–582.

169. Radicella, J. P., E. A. Clark, S. Chen, and M. S. Fox. 1993. Patch length of localized repair events: the role of DNA polymerase I in *mutY*-dependent mismatch repair. *J. Bacteriol.* **174:**7732–7736.

170. Radicella, J. P., E. A. Clark, and M. S. Fox. 1988. Some mismatch repair activities in *Escherichia coli. Proc. Natl. Acad. Sci. USA* **85:**9674–9678.

171. Radman, M. 1988. Mismatch repair and genetic recombination, p. 169–192. *In* R. Kucherlapati and G. R. Smith (ed.), *Genetic Recombination.* American Society for Microbiology, Washington, D.C.

172. Radman, M. 1989. Mismatch repair and the fidelity of genetic recombination. *Genome* **31:**68–73.

173. Radman, M., and R. Wagner. 1986. Mismatch repair in *Escherichia coli. Annu. Rev. Genet.* **20:**523–538.

174. Radman, M., R. E. Wagner, B. W. Glickman, and M. Meselson. 1980. DNA methylation, mismatch correction and genetic stability, p. 121–130. *In* M. Alacevic (ed.), *Progress in Environmental Mutagenesis.* Elsevier/North-Holland Biomedical Press, Amsterdam.

175. Raposa, S., and M. S. Fox. 1987. Some features of base pair mismatch and heterology repair in *Escherichia coli. Genetics* **117:**381–390.

176. Ray, B., C. I. White, and J. E. Haber. 1991. Heteroduplex formation and mismatch repair of the "stuck" mutation during mating-type switching in *Saccharomyces cerevisiae. Mol. Cell. Biol.* **11:**5372–5380.

177. Rayssiguier, C., D. S. Thaler, and M. Radman. 1989. The barrier to recombination between *Escherichia coli* and *Salmonella typhimurium* is disrupted in mismatch-repair mutants. *Nature* (London) **342:**396–401.

178. Reenan, R. A. G., and R. D. Kolodner. 1992. Characterization of insertion mutations in the *Saccharomyces cerevisiae MSH1* and *MSH2* genes: evidence for separate mitochondrial nuclear functions. *Genetics* **132:**975–985.

179. Reenan, R. A. G., and R. D. Kolodner. 1992. Isolation and characterization of two *Saccharomyces cerevisiae* genes encoding homologs of the bacterial HexA and MutS mismatch repair proteins. *Genetics* **132:**963–973.

180. René, B., C. Auclair, and C. Paoletti. 1988. Frameshift lesions induced by oxazolopyridocarbazoles are recognized by the mismatch repair system in *Escherichia coli. Mutat. Res.* **193:**269–273.

181. Rong, L., F. Palladino, A. Aguilera, and H. L. Klein. 1991. The hyper-gene conversion *hpr5-1* mutation of *S. cerevisiae* is an allele of the *SRS2/RADH* gene. *Genetics* **127:**75–85.

182. Ross-MacDonald, P., and G. S. Roeder. 1994. Personal communication.

183. Rydberg, B. 1978. Bromouracil mutagenesis and mismatch repair in mutator strains of *Escherichia coli. Mutat. Res.* **5 2:**11–24.

184. Schaaper, R. M. 1988. Mechanisms of mutagenesis in the *Escherichia coli* mutator *mutD5*: role of DNA mismatch repair. *Proc. Natl. Acad. Sci. USA* **85:**8126–8130.

185. Schaaper, R. M. 1989. *Escherichia coli* mutator *mutD5* is defective in the *mutHLS* pathway of DNA mismatch repair. *Genetics* **121:**205–212.

186. Schaaper, R. M., and R. L. Dunn. 1987. Spectra of spontaneous mutations in *Escherichia coli* strains defective in mismatch correction: the nature of in vivo DNA replication errors. *Proc. Natl. Acad. Sci. USA* **84:**6220–6224.

187. Schaaper, R. M., and M. Radman. 1989. The extreme mutator effect of *Escherichia coli mutD5* results from saturation of mismatch repair by excessive DNA replication errors. *EMBO J.* **8:**3511–3516.

188. Service, R. F. 1994. Stalking the start of colon cancer. *Science* **263:**1559–1560.

189. Shen, J.-C., W. M. Rideout, and P. A. Jones. 1992. High mutation frequency mutagenesis by a DNA methyltransferase. *Cell* **71:**1073–1080.

190. Shen, P., and H. V. Huang. 1986. Homologous recombination in *Escherichia coli*: dependence on substrate length and homology. *Genetics* **112:**441–457.

191. Shen, P., and H. V. Huang. 1989. Effect of base pair mismatches on recombination via the RecBCD pathway. *Mol. Gen. Genet.* **218:**358–360.

192. Sibghat-Ullah, and R. S. Day. 1992. Incision at O^6-methylguanine:thymine mispairs in DNA by extracts of human cells. *Biochemistry* **31:**7998–8008.

193. Sibghat-Ullah, and R. S. Day. 1993. DNA-substrate sequence specificity of human G:T mismatch repair activity. *Nucleic Acids Res.* **21:**1281–1287.

194. Siddiqi, O., and M. S. Fox 1973. Integration of donor DNA in bacterial conjugation. *J. Mol. Biol.* **77:**101–123.

195. Skopek, T. R., and F. Hutchinson. 1984. Frameshift mutagenesis of lambda prophage by 9-aminoacridine, proflavin, and ICR-191. *Mol. Gen. Genet.* **195:**418–423.

196. Smith, G. P. 1973. Unequal crossover and the evolution of multigene families. *Cold Spring Harbor Symp. Quant. Biol.* **38:**507–513.

197. Sohail, A., M. Lieb, M. Dar, and A. S. Bhagwat. 1990. A gene required for very short patch repair in *Escherichia coli* is adjacent to the DNA cytosine methylase gene. *J. Bacteriol.* **172:**4214–4221.

198. Stephenson, C., and P. Karran. 1989. Selective binding to DNA base pair mismatches by proteins from human cells. *J. Biol. Chem.* **264:**21177–21182.

199. Strand, M., T. A. Prolla, R. M. Liskay, and T. D. Petes. 1993. Destabilization of tracts of simple repetitive DNA in yeast by mutations affecting DNA mismatch repair. *Nature* (London) **365:**274–276.

200. Streisinger, G., Y. Okada, J. Emrich, J. Newton, A. Tsugita, E. Terzaghi, and M. Inouye. 1966. Frameshift mutations and the genetic code. *Cold Spring Harbour Symp. Quant. Biol.* **31:**77–84.

201. Su, S.-S., R. S. Lahue, K. G. Au, and P. Modrich. 1988. Mispair specificity of methyl-directed DNA mismatch correction in vitro. *J. Biol. Chem.* **263:**6829–6835.

202. Su, S.-S., and P. Modrich. 1986. *Escherichia coli mutS*-encoded protein binds to mismatched DNA base pairs. *Proc. Natl. Acad. Sci. USA* **83:**5057–5061.

203. Szyf, M., K. Avraham-Haetzni, A. Reifman, J. Shlomai, F. Kaplan, A. Oppenheim, and A. Razin. 1984. DNA methylation pattern is determined by the intracellular level of the methylase. *Proc. Natl. Acad. Sci. USA* **81:**3278–3282.

204. Thibodeau, S. N., G. Bren, and D. Schaid. 1993. Microsatellite instability in cancer of the proximal colon. *Science* **260:**816–822.

205. Thomas, D. C., J. D. Roberts, and T. A. Kunkel. 1991. Heteroduplex repair in extracts of human HeLa cells. *J. Biol. Chem.* **266:**3744–3751.

206. Tiraby, J.-G., and M. S. Fox. 1973. Marker discrimination in transformation and mutation of pneumococcus. *Proc. Natl. Acad. Sci. USA* **70:**3541–3545.

207. Tsai-Wu, J.-J., H. L. Liu, and A.-L. Lu. 1992. *Escherichia coli* MutY protein has both N-glycosylase and apurinic-apyrimidinic endonuclease activities on A·C and A·G mispairs. *Proc. Natl. Acad. Sci. USA* **89:**8779–8783.

208. Tsai-Wu, J.-J., J. P. Radicella, and A.-L. Lu. 1991. Nucleotide sequence of the *Escherichia coli micA* gene required for A/G-specific mismatch repair: identity of MicA and MutY. *J. Bacteriol.* **173:**1902–1910.

209. Valle, G., E. Bergantino, G. Lanfranchi, and G. Carignani. 1991. The sequence of a 6.3 kb segment of yeast chromosome III reveals an open reading frame coding for a putative mismatch binding protein. *Yeast* **7:**981–988.

210. Varlet, I., M. Radman, and P. Brooks. 1990. DNA mismatch repair in *Xenopus* egg extracts: repair efficiency and DNA repair synthesis for all single base-pair mismatches. *Proc. Natl. Acad. Sci. USA* **87:**7883–7887.

211. Wagner, R., C. Dohet, M. Jones, M.-P. Doutriaux, F. Hutchinson, and M. Radman. 1984. Involvement of *Escherichia coli* mismatch repair in DNA replication and recombination. *Cold Spring Harbor Symp. Quant. Biol.* **49:**611–615.

212. Wagner, R., and M. Meselson. 1976. Repair tracts in mismatched DNA heteroduplexes. *Proc. Natl. Acad. Sci. USA* **73:**4135–4139.

213. Wang, T. C., and K. C. Smith. 1986. Inviability of *dam recA* and *dam recB* cells of *Escherichia coli* is correlated with their inability to repair DNA double-strand breaks produced by mismatch repair. *J. Bacteriol.* **165:**1023–1025.

214. Welsh, K. M., A.-L. Lu, S. Clark, and P. Modrich. 1987. Isolation and characterization of the *Escherichia coli mutH* gene product. *J. Biol. Chem.* **262:**15624–15629.

215. White, J. H., K. Lusnak, and S. Fogel. 1985. Mismatch-specific post-meiotic segregation frequency in yeast suggests a heteroduplex recombination intermediate. *Nature* (London) **315:**350–352.

216. White, R. L., and M. S. Fox. 1974. On the molecular basis of high negative interference. *Proc. Natl. Acad. Sci. USA* **71:**1544–1548.

217. White, R. L., and M. S. Fox. 1975. Genetic consequences of transfection with heteroduplex bacteriophage lambda DNA. *Mol. Gen. Genet.* **141:**163–171.

218. Whitehouse, H. L. K. 1963. A theory of crossing-over by means of hybrid deoxyribonucleic acid. *Nature* (London) **199:**1034–1040.

219. Wiebauer, K., and J. Jiricny. 1989. *In vitro* correction of G·T mispairs to G-C pairs in nuclear extracts from human cells. *Nature* (London) **339:**234–236.

220. Wiebauer, K., and J. Jiricny. 1990. Mismatch-specific thymine DNA glycosylase and DNA polymerase β mediate the correction of G·T mispairs in nuclear extracts from human cells. *Proc. Natl. Acad. Sci. USA* **87:**5842–5845.

221. Wildenberg, J., and M. Meselson. 1975. Mismatch repair in heteroduplex DNA. *Proc. Natl. Acad. Sci. USA* **72:**2202–2206.

222. Williamson, M. S., J. C. Game, and S. Fogel. 1985. Meiotic gene conversion mutants in *Saccharomyces cerevisiae*. I. Isolation and characterization of *pms1-1* and *pms1-2*. *Genetics* **110:**609–646.

223. Witkin, E. M. 1964. Pure clones of lactose-negative mutants obtained in *Escherichia coli* after treatment with 5-bromouracil. *J. Mol. Biol.* **8:**610–613.

224. Worth, L., S. Clark, M. Radman, and P. Modrich. 1994. Mismatch repair proteins MutS and MutL inhibit RecA-catalyzed strand transfer between diverged DNAs. *Proc. Natl. Acad. Sci. USA* **91:**3238–3241.

225. Wu, J. C., and D. V. Santi. 1987. Kinetic and catalytic mechanism of *Hha*I methyltransferase. *J. Biol. Chem.* **262:**4778–4786.

226. Wyszynski, M., S. Gabbara, and A. S. Bhagwat. 1994. Cytosine deamination catalyzed by DNA cytosine methyltransferases are unlikely to be the major cause of mutational hot-spots of cytosine methylation in *E. coli. Proc. Natl. Acad. Sci. USA* **91:**1574–1578.

227. Yeh, Y.-C., D.-Y. Chang, J. Masin, and A.-L. Lu. 1991. Two nicking enzyme systems specific for mismatch-containing DNA in nuclear extracts from human cells. *J. Biol. Chem.* **266:**6480–6484.

228. Zell, R., and H. J. Fritz. 1987. DNA mismatch-repair in *Escherichia coli* counteracting the hydrolytic deamination of 5-methylcytosine residues. *EMBO J.* **6:**1809–1815.

10

SOS Responses and DNA Damage Tolerance in Prokaryotes

U p to this point we have discussed the repair of damaged DNA by the direct reversal of damage or by excision repair of damage. However, cells have evolved a variety of other strategies for coping with damaged DNA. Some of these do not meet the strict biochemical definition of DNA repair used in this book since they do not result in the physical removal of the original lesion from the DNA and, in general, they are not understood at the same level of molecular detail as the systems that we have discussed previously. Hence, these cellular responses to DNA damage are referred to as *DNA damage tolerance mechanisms*. They play extremely important roles in allowing cells to deal with the consequences of unrepaired damage to their genomes. For example, prokaryotic cells have evolved mechanisms for repairing single-strand gaps and double-strand breaks in their DNA that have arisen either directly from DNA damage or indirectly as the result of processing of the initial DNA damage. These mechanisms involve proteins that also play roles in the homologous recombination of undamaged DNA. Neither of these processes appear to be particularly mutagenic. In addition, prokaryotic cells have evolved another class of mechanism for processing damaged DNA which, although not yet fully understood at a biochemical level, appears to involve the polymerization of DNA past a lesion and is often referred to as *translesion DNA synthesis*. In contrast to other systems that act on damaged DNA, this type of processing can be highly mutagenic and, in fact, is required for most UV radiation and chemical mutagenesis.

In the case of *Escherichia coli*, in which these alternative mechanisms for dealing with damaged DNA have been studied most closely, it has become clear that their regulation and operation is intimately related to the complex *SOS regulatory network*. The expression of the more than 20 genes in this network is induced by DNA damage and is regulated by the LexA and RecA proteins. The increased expression of these SOS genes results in the elaboration of a set of physiological responses that have been termed the *SOS responses*. At least some of the proteins that play roles in DNA damage tolerance mechanisms and in translesion synthesis are encoded by genes that are members of the SOS regulatory network. Furthermore, it seems likely that the RecA protein not only plays a central regulatory role with respect to these systems but also may play a mechanistic role in most, if not all, of them. It has also become clear that some genes required for excision repair are members of the SOS regulatory network.

In this chapter we will first discuss the development of the present model for SOS regulation and will focus particularly on the central role of the RecA protein in this process. We will then consider the present state of knowledge concerning DNA damage tolerance mechanisms and the dual regulatory and mechanistic roles of RecA protein in these processes. In the following chapter, we will focus specifically on events that are based on a translesion DNA synthesis mechanism and that can result in mutagenesis.

The SOS Responses

The idea that damage to DNA or that the physiological consequences of such damage might initiate a regulatory signal causing the simultaneous *recA+ lexA+*-regulated derepression of a number of genes in *E. coli*, the products of some or all of which enhance the survival of the cell or of its normal resident phages, was first suggested by Miroslav Radman in a memorandum which he circulated to various workers in the field (356). Encouraged by analyses in a paper from Jacqueline George's laboratory on the properties of the *tif-1* (now called *recA441*) mutant (33), Radman first formally promulgated his hypothesis at a conference on mutagenesis held in Rochester, N.Y., in 1973 (250). The *international distress signal (SOS)*, the term he appropriated to describe this phenomenon, is frequently misinterpreted to imply a last ditch attempt by the cell to survive the lethal effect of DNA damage after other cellular responses have failed. However, a direct quote from the written proceedings of the 1973 meeting (250) indicates that he coined the phrase SOS primarily to emphasize that this cellular response is to a distress signal, i.e., DNA damage.

> The principal idea is that *E. coli* possesses a DNA repair system which can be repressed under normal physiological conditions but which can be induced by a variety of DNA lesions. Because of its "response" to DNA-damaging agents we call this hypothetical repair "SOS repair." The 'danger' signal which induces SOS repair is probably just a temporary blockage of the normal DNA replication and possibly just the presence of DNA lesions in the cell.

Table 10–1 summarizes the SOS responses that have been recognized to date and, where known, the genes whose products are required for the various induced responses. It also includes genes whose products have known functions (for example, *polB* [formerly *dinA*], which encodes DNA polymerase II [see below]) but which have not yet been associated with a physiologically observable induced response. Also, Table 10–1 lists other SOS-regulated genes that have been identified solely on the basis of their regulatory characteristics by the use of gene and operon fusion technology but to which no function has yet been ascribed. In some cases, it is unclear whether the induction of these genes contributes to cellular survival under standard laboratory growth conditions.

Current Model for SOS Regulation

Development of the present model for SOS regulation in *E. coli* was made challenging by the diverse nature of the induced physiological responses, the complexity of genetic analyses of SOS regulation, and the dual regulatory and mechanistic roles played by the RecA protein. Therefore, before reviewing key experiments that led to its development, we will summarize the present model for SOS regulation.

The basic regulatory mechanism of the SOS system is now understood at a molecular level (175, 237, 330–332) and is diagrammed schematically in Fig. 10–1.

Table 10–1 SOS responses and genes of *E. coli*

Induced physiological responses or gene function	Induced genes[a]	References
E. coli		
Weigle reactivation of bacteriophage	*umuDC, recA, dinY*, uvrA, uvrB*	4, 51, 131, 132, 238
Weigle mutagenesis of bacteriophage	*umuDC, recA*	330
UV mutagenesis of bacterial chromosome	*umuDC, recA*	4, 330, 355
Filamentation (inhibition of cell division)	*sulA (sfiA)*	113
uvr⁺-dependent excision repair	*uvrA, uvrB, uvrD*	73, 134, 297
Long-patch repair	*uvrA, uvrB*	41, 42
Daughter-strand gap repair	*recA, ruvAB?*	183, 185
Double-strand break repair	*recA, recN*	151, 243, 267, 279
Tandem duplication	*recA*	55
*Nar*I frameshift mutagenesis	*recA, umuDC* independent	200
Increase in pBR322 plasmid copy number	?	12
Rifampin-resistant pBR322 replication	?	203
Alleviation of restriction	*umuDC*, ?*	49, 99, 323
Inducible stable DNA replication	*recA*	202
Inhibition of DNA degradation by *Exo*V	*recA, exi**	129, 241
Excision and transposition of Tn5	?	154
Induction of DNA polymerase II	*polB (dinA)*	15, 121
Induction of various SOS loci	*dinB, dinD, dinF, dinG, dinH, dinI*	134, 160–162
Induction of SOS loci apparently not repressed directly by LexA	*dinY*, recQ*, dnaA*, dnaN*, dinQ*, phr*, nrdAB**	160, 238
Phage and cryptic elements		
Prophage induction	Prophage genes	25, 96
Excision of element e14	?	88
Induction of defective retronphage(φR86)	?	141
φ186 induction	*tum*	155
Naturally occurring plasmids		
Induction of *umuDC* homologs (pKM101)	*mucAB*	70, 71
Induction of *umuDC* homologs (TP110)	*impAB*	187
Colicin production (ColE1)	*cea*	62, 63
Colicin production (ColA)	*caa*	181

> [a] Symbols: *, the gene or locus is induced as part of the SOS response but does not appear to be regulated by LexA (in some cases, it is possible that the conclusion that the gene is under SOS control is in error [see the text]; ?, the induced gene(s) has not yet been identified. The relationship between many of these induced processes and cell survival is not understood.

In an uninduced *E. coli* cell, the product of the *lexA⁺* gene acts as a *repressor* for more than 20 genes, including the *recA⁺* and *lexA⁺* genes, by binding to similar operator sequences at each gene or operon. The operator sequences bound by LexA protein are commonly referred to as *SOS boxes*. Many of these SOS genes, including the *recA⁺* and *lexA⁺* genes, are expressed at significant levels even in the repressed state. In particular, the RecA protein is expressed at approximately 7,200

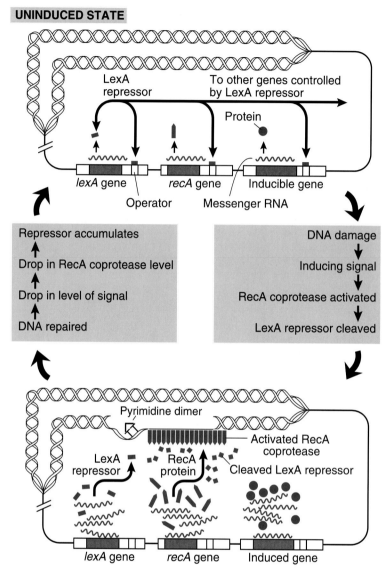

Figure 10–1 Diagrammatic representation of the mechanism by which the *lexA-recA* regulon is regulated. In the uninduced state (top), LexA repressor protein constitutively expressed in small amounts is bound to the *lexA* operator and to the operators of the *recA* gene and other genes under LexA control. These genes are still able to express small amounts of the proteins they encode; thus, there is some RecA protein constitutively present in uninduced cells. Following DNA damage (e.g., the presence of a pyrimidine dimer near a replication fork after induction by UV radiation), the coprotease activity of existing RecA protein is activated, probably by binding to the single-stranded DNA in the gaps created by discontinuous DNA synthesis past the dimers (bottom). The interaction between LexA and activated RecA results in the proteolytic cleavage of LexA. In the induced state (bottom), derepression of the *recA*⁺ gene results in the production of large amounts of RecA protein. Other genes under LexA control are also derepressed, although not necessarily with identical kinetics. When the inducing signal disappears (probably by repair of the single-strand gap), the level of active coprotease drops, LexA repressor accumulates, and genes under LexA control are once again repressed. (*Adapted from Howard-Flanders [107] and Little and Mount [175], and Friedberg [74] with permission.*)

molecules per cell in uninduced cells (280), which is evidently sufficient for its role in homologous recombination (130, 271).

When the genome of an *E. coli* cell is damaged or DNA replication is inhibited, an intracellular signal for SOS induction is generated. Both in vivo and in vitro evidence suggests that this intracellular inducing signal consists of regions of single-stranded DNA that are generated when the cell attempts to replicate a damaged template or when its normal process of DNA replication is interrupted. The binding of RecA protein to these regions of single-stranded DNA in the presence of a nucleoside triphosphate reversibly converts it to an activated form, often referred to as RecA*. LexA molecules then diffuse to the activated RecA protein and interact with this nucleoprotein complex in a way that results in LexA becoming *proteolytically cleaved* at a specific Ala-Gly bond near the middle of the protein. This proteolytic cleavage of LexA protein results from the ability of activated RecA to facilitate an otherwise latent capacity of LexA protein to autodigest. Activated RecA protein is also able to interact with several other proteins structurally related to LexA, including phage λ repressor and the UmuD protein (discussed in chapter 11), and similarly mediate their proteolytic cleavage at a specific bond.

RecA-mediated cleavage of LexA protein *inactivates* LexA as a repressor. Thus, as LexA cleavage proceeds after an SOS-inducing treatment, the pools of LexA protein begin to decrease so that various SOS genes, including the *recA*+ gene, are expressed at increased levels and SOS responses mediated by these genes can be observed. Genes with operators that bind LexA relatively weakly are the first to turn on fully. For example, at an intermediate SOS-inducing dose, the *lexA* gene would be fully induced and expressed at a five-fold higher level. In contrast, a gene such as *sulA*, which binds LexA more tightly, would be only partially derepressed (which for *sulA* would lead to a 10- to 20-fold increase in its level of expression). If the inducing treatment is sufficiently strong, more molecules of RecA are activated and more molecules of LexA are cleaved. As the pools of LexA decline to very low levels, even genes whose operators bind LexA very tightly are expressed at maximal levels. For *sulA*, full induction would lead to an approximately 100-fold increase in its level of expression. If the cell carries a λ prophage, the cleavage of λ repressor mediated by activated RecA results in *prophage induction*.

As the cell begins to recover from the SOS-inducing treatment, the regions of single-stranded DNA disappear as a consequence of various DNA repair processes and RecA molecules return to their nonactivated state. Continued synthesis of LexA protein then leads to an increase in LexA pools, which, in turn, leads to repression of the SOS genes and a return to the uninduced state.

Development of the Model for SOS Regulation

Since the SOS regulatory network of *E. coli* was the first regulatory network induced by DNA damage to be recognized in any organism, and since its regulation is understood in the greatest detail, it is instructive to review the key observations and insights that led to the development of the model just summarized.

HISTORICAL DEVELOPMENT OF THE SOS HYPOTHESIS

Certain induced responses, specifically the induction of λ prophage, UV radiation-induced filamentation of *E. coli* cells, induced mutagenesis of the *E. coli* chromosome, and Weigle reactivation and Weigle mutagenesis of UV-irradiated bacteriophage, played particularly important roles in the historical development of the model for SOS regulation. As early as 1950, it was recognized that exposure of bacteriophage λ lysogens to relatively small doses of UV light caused phage induction and lysis of cells (197, 198), and that this process required conditions in which protein synthesis can occur. It is now known, as originally postulated by Francois Jacob and Eli Wollman and their colleagues (95), that this induction of λ prophage involves the inactivation of a repressor protein that normally prevents the expression of the structural genes required for phage production. Other treatments were subsequently found to induce λ lysogens: exposure to DNA-damaging agents such as ionizing radiation, mitomycin, nitrogen mustard, hydrogen peroxide, and organic peroxides (125, 146); treatments with base analogs such as

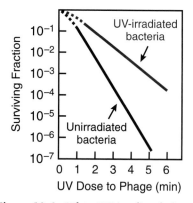

Figure 10–2 When UV-irradiated phage λ is plated on *E. coli* cells that were previously irradiated, the phage survival is greater than on unirradiated bacteria. This phenomenon was first described by Jean Weigle and is referred to as W (Weigle) reactivation of the phage. (*Adapted from Weigle [339].*)

6-azauracil and 5-fluorouracil; thymine starvation of thymine auxotrophs; and shifting of certain temperature-sensitive DNA replication mutants to the restrictive temperature (84, 125, 146, 365). All these physiological perturbations have in common the fact that they *arrest DNA replication*.

After exposure to a DNA-damaging treatment, *E. coli* cells continue to elongate but fail to septate, with the result that the cells grow as long filaments. This response is now known to be due to induction of an *inhibitor of septation* that is encoded by the *sulA⁺* (alternatively termed *sfiA⁺*) gene. It is now known that the SulA protein is unstable and is degraded by the *lon* protease. Hence, filamentation is particularly severe in *lon* mutants of *E. coli* K-12 and in wild-type *E. coli* B (which is a naturally occurring *lon* mutant) because of their failure to degrade the septum inhibitor protein (352). In 1967, Evelyn Witkin noted a number of similarities between the induction of λ prophage in *E. coli* K-12 and filament formation in *E. coli* B (352) (Table 10–2). On the basis of these similarities, she made the bold proposal that certain bacterial functions, including the synthesis of the septum inhibitor protein, might be governed by repressors similar enough to λ repressor to respond to the same inducer, an inducer produced only when DNA replication is interrupted. An independent report showed that an *E. coli* K-12 strain carrying the *tif-1* mutation (now known as *recA441*) not only induces λ prophage when shifted to 42°C but also grows in long filaments at 42°C without any of the normal inducing treatments (140). This report provided further evidence for the coordinate regulation of these two diverse physiological functions.

A particularly significant observation with respect to the development of the SOS hypothesis was made in 1953 by Jean Weigle (339). He found that UV irradiation of *E. coli* cells, prior to infection with UV-irradiated bacteriophage λ, resulted in increased survival of the UV-irradiated phage (Fig. 10–2). Photoreactivation of the host cells prior to phage infection eliminated these effects, suggesting that photoproducts such as pyrimidine dimers in host DNA were necessary for the induction of this phenomenon. This induced capacity to reactivate damaged bacteriophage is now generally referred to as *Weigle* (or *W*) *reactivation*, because the previously used terms, *UV restoration* (339) and *UV reactivation*, are readily confused with *host cell reactivation* and *multiplicity reactivation*, which describe entirely different phenomena (251). (Host cell reactivation describes the effect of constitutive host responses to DNA damage in phage genomes that do not encode their own DNA repair functions. Multiplicity reactivation describes the enhanced survival of phages exposed to DNA damage following infection of cells with *multiple* phage genomes which then reconstitute one or more undamaged genomes by recombinational events.) Weigle reactivation of UV-irradiated phage can be induced by the same spectrum of agents and conditions which elicit prophage induction or filamentation, so the designation *UV reactivation* is too limited (251). Furthermore, Weigle made the particularly interesting observation that a large proportion of the reactivated bacteriophage were mutant. The phenomenon of mutagenesis of damaged phage that infect cells that were exposed to a DNA-damaging agent prior to infection is now usually referred to as *Weigle* (or *W*) *mutagenesis*.

A particularly significant aspect of the development of the SOS hypothesis was the recognition of an additional striking similarity among these physiologically diverse responses. In addition to the common characteristic of being induced by agents and conditions that arrest DNA replication, they were all influenced by

Table 10–2 Similarities between filament formation and lysogenic induction in *E. coli*

1. Both are mass effects, occurring in virtually every member of a population
2. Both can be initiated by a variety of DNA damaging agents
3. Both occur occasionally in untreated cultures and in "old" cultures
4. Both are reduced by photoreactivation after exposure to UV light at 254 nm
5. Both are dependent on active protein synthesis

Source: Adapted from Witkin (352) and Friedberg (74) with permission.

mutations in the $recA^+$ gene. The first $recA$ mutant was isolated in the mid-1960s by John Clark and coworkers on the basis of its deficiency in homologous recombination (39). Clark and his colleagues observed that this mutant was unusually sensitive to UV radiation and that the induction of λ prophage by UV radiation was defective in this strain (25). It was subsequently shown that W reactivation (51, 142, 213, 224), W mutagenesis (51, 142, 213, 224), and UV radiation-induced filamentation of cells (86) were blocked by this class of $recA$ mutations. Furthermore, Evelyn Witkin showed that another phenomenon associated with DNA damage, i.e., UV radiation-induced mutagenesis of the bacterial chromosome, was also blocked by this class of $recA$ mutations (353). It was recognized that the induced responses were also affected by mutations in the $lexA^+$ gene. The first $lexA$ mutations (also originally called exr) were initially recognized as mutations that sensitized cells to killing by UV and ionizing radiation and abolished UV radiation-induced mutagenesis of the bacterial chromosome (217, 351). These $lexA$ mutations were subsequently found to also abolish W reactivation, W mutagenesis (51), and UV radiation-induced filamentation.

Since it had been established that UV radiation-induced mutagenesis of the *E. coli* genome was similarly dependent on the $recA^+$ $lexA^+$ genotype (see below), Miroslav Radman proposed that SOS-induced functions were also required for this process (250, 251, 356). Furthermore, he postulated that the induced cellular responses that led to both W reactivation and W mutagenesis were also responsible for mutagenesis of the bacterial chromosome. The implications of this suggestion will be considered more fully in the following chapter.

GENETIC STUDIES OF $recA^+$ AND $lexA^+$

The early recognition that induction of the various SOS responses had a common dependence on the products of the $recA^+$ and $lexA^+$ genes was complicated by the diverse phenotypes of different classes of $recA$ and $lexA$ mutants. In fact, in several cases mutations that turned out to be alleles of $recA$ or $lexA$ caused such diverse phenotypes that they were originally thought to define new genes. The properties of most of these mutations can now be explained relatively easily in terms of the model described above. Since it will be necessary to refer to a number of different classes of $recA$ and $lexA$ alleles in this and the following chapter, Table 10–3 summarizes the properties of the most important classes of these mutations.

recA Mutants

As mentioned above, the original $recA$ mutants that were characterized were virtually completely deficient in homologous recombination. (The roles of the RecA protein in homologous recombination are discussed below). The alleles of $recA$ present in these strains were recessive to $recA^+$ with respect to homologous recombination. Missense, amber, insertion, and deletion mutations (30, 214, 350) of this class have been isolated, and these will be referred to as $recA$(Def) alleles (330) to indicate that they result in defective RecA function. These $recA$(Def) mutations prevented the induction of the SOS response and were similarly recessive to $recA^+$ for this characteristic, suggesting that wild-type RecA protein functions as a positively acting control element in SOS induction (355). A second class of $recA$ alleles, designated $recA$(Cpt^c) (167), is exemplified by the $recA730$ allele (359) and causes expression of SOS-regulated genes in the absence of an inducing treatment. As discussed below, apparently the RecA proteins made by these mutants become activated in the cytoplasm of uninduced *E. coli* cells and mediate the cleavage of the LexA protein in the absence of an inducing signal. The $recA$(Cpt^{Ts}) class, exemplified by the $recA441$ (initially called $tif-1$ for thermal induction of filamentation) mutation, also causes constitutive expression of SOS-regulated genes, but only when the cells are shifted to 42°C; this effect is potentiated by the addition of adenine to the medium and is antagonized by the addition of guanosine and cytidine; the mechanisms of the potentiating and antagonizing effects are still not understood (33, 259, 330). Thus, the $recA441$ mutation can be thought of as a conditional mutant of the $recA$(Cpt^c) class. In fact, $recA441$ has recently been shown to carry two base pair changes. One is the same as the $recA730$ mutation, and the

Table 10–3 Properties of some important *recA* and *lexA* alleles

A. Allele	Recombinase	Coprotease
recA⁺	+	Inducible
Δ*recA*	−	Defective
recA430 (Cpt⁻) (formerly *lexB30*)	+	Defective (λ), partially defective (LexA), inducible (φ80)
recA441 (Cptᵀˢ)(formerly *tif-1*)	+ +	Constitutive (42°C), inducible (30°C)
recA730 (Cptᶜ)	+ +	Constitutive
recA718	+	Constitutive [*lexA*(Def)], inducible [*lexA*⁺]
recA1203	−	Constitutive (LexA), inducible (λ)
recA1730	+ [in *lexA*(Def)] − [in *lexA*⁺]	Deficient (LexA), inducible (λ)

B. Allele	Phenotypes	Biochemical change
lexA3(Ind⁻)	Defective in SOS induction, UV sensitive, dominant	Noncleavable LexA
lexA41(Tsl) (formerly *tsl*)	Partial expression of SOS functions at 30°C, higher expression at 42°C, recessive	Noncleavable but unstable LexA
lexA51(Def) (formerly *spr*)	Constitutive expression of LexA repressed genes	Defective LexA
lexA71::Tn5(Def)	Constitutive expression of LexA repressed genes	No LexA made

Source: Adapted from Walker (330) and Witkin (357) with permission.

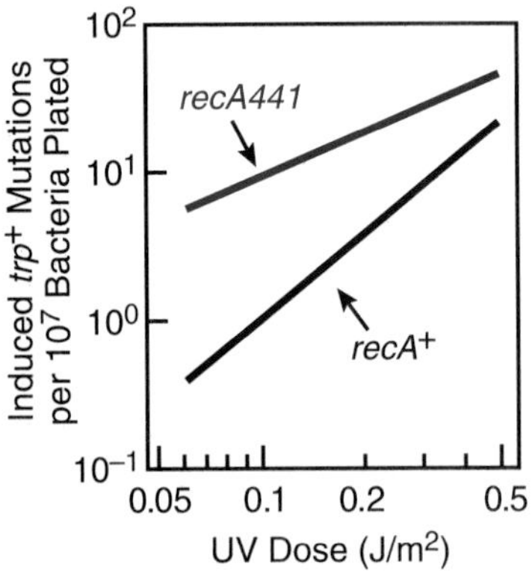

Figure 10–3 When wild-type (*recA*⁺) cells are UV irradiated and then incubated at either high or low temperatures, the slope of the curve relating mutation frequency to UV dose fits a theoretical two-hit curve. In contrast, *recA441* mutants generate mutations at the *trp* locus with a linear dependency on UV dose; i.e., the slope fits a theoretical one-hit curve. (*Adapted from Witkin [355] and Friedberg [74] with permission.*)

second is presumably responsible for the conditional characteristics of the RecA441 protein (143). Finally, the *recA*(Cpt⁻) class, exemplified by the *recA430* (formerly called *lexB30*) mutation, does not cause a major defect in homologous recombination. However, this class of mutations does cause a defect in the induction of some, but not all, of the SOS responses (54, 259, 330). Biochemical studies have revealed that the RecA430 protein is drastically defective in mediating the cleavage of λ repressor and UmuD protein (see chapter 11) and is partially defective in mediating the cleavage of LexA protein.

The *recA441* mutation played a particularly important role in the development of the SOS hypothesis. Although it was not found to be an allele of *recA*⁺ until later, analyses of the properties of strains carrying the *recA441* mutation helped build evidence for a common regulatory link between the physiologically diverse SOS functions. In a particularly influential set of experiments (356), Jacqueline George and her colleagues showed that, in addition to the induction of λ prophage and filamentation, W reactivation and W mutagenesis were heat inducible in the *recA441* mutant (33, 34). In fact, these responses were quantitatively indistinguishable from those observed when UV radiation was used as the inducing stimulus at noninducing temperatures (Table 10–4). Evelyn Witkin was able to exploit the properties of a *recA441* mutant to provide evidence that UV radiation mutagenesis required an inducible activity. It had previously been observed that the induction of mutations by UV irradiation of strains defective in nucleotide excision repair obeys *two-hit kinetics*; i.e., the slope of the mutation frequency curve shows a *quadratic* rather than a *linear* relationship to UV radiation dose (355) (Fig. 10–3). This had been interpreted as a requirement for two separate events for mutagenesis, both of which are dependent on UV irradiation of the cell (i.e., DNA damage as one event and induction of a mutagenic cellular response to that damage as a second event) (57, 358). Consistent with this interpretation, Witkin observed that induction of mutations at elevated temperatures following

Table 10–4 Mutagenesis of UV-irradiated phage λ when plated on *E. coli* *recA*⁺ and *recA441* strains

Strain	Preincubation temp (°C)	Frequency of clear-plaque mutants (× 10⁻⁴)	
		Unirradiated cells	UV-irradiated cells
recA⁺	30	3.6	25.0
	42	4.0	20.0
recA441	30	6.0	23.0
	42	20.0	36.0

Source: Adapted from Castellazzi et al. (33) with permission.

UV irradiation of the *recA441* mutant was increased and showed a linear rather than dose-squared relation (354) (Fig. 10–3). She and her colleagues therefore suggested that the requirement for one of the two UV radiation-dependent events (i.e., induction of the functions required for mutation) could be met by elevation of temperature in a *recA441* mutant. The later finding that the *tif-1* mutation was actually an allele of *recA* was important in the development of the concept that RecA had to be *activated* in order to play its role in SOS induction (35, 72, 89, 207).

lexA Mutants

The first *lexA* mutants to be characterized prevented the induction of SOS responses and were found to be *dominant* to *lexA*⁺ (87), i.e., cells carrying both the mutant and wild-type genes display the mutant phenotype (Fig. 10–4). The existence of these dominant negative mutations, which are referred to as *lexA*(Ind⁻) mutations (because they are defective in induction), suggested that LexA protein acts as a *negative control element* in SOS regulation (33, 217). It is now known that *lexA*(Ind⁻) mutations (110, 217) alter LexA protein in ways that interfere with RecA-mediated cleavage. In fact, the most widely used mutation of this class, *lexA3*(Ind⁻), changes the sequence of the Ala-Gly cleavage site to Ala-Asp (205). However, in contrast to *recA*(Def) mutants, *lexA*(Ind⁻) mutants were found to be recombination proficient, a property which indicated that a deficiency in SOS induction is not necessarily associated with a deficiency in homologous recombination.

The isolation of dominant *lexA*(Ind⁻) mutations that blocked SOS induction suggested that it should be possible to isolate *recessive* alleles of *lexA*⁺ [*lexA*(Def)], which constitutively express SOS responses. Such mutations were very difficult to isolate. In retrospect, it is clear that early studies of the *lexA*⁺ locus failed to yield mutants of the *lexA*(Def) class because constitutive expression of the LexA-repressed *sulA*⁺ (alternatively called *sfiA*⁺) gene (Table 10–1) leads to lethal filamentation. The first recessive *lexA* mutants were conditional mutants originally isolated as UV radiation-resistant, temperature-sensitive revertants of a *lexA*(Ind⁻) strain (218). These *lexA*(Tsl) (originally called *tsl*) mutants partially induced SOS responses when shifted to elevated temperature, owing to their synthesis of a thermolabile LexA protein which is unable to function as a repressor at elevated temperatures. The *lexA41*(Tsl) (formerly *tsl-1*) gene product is also sensitive to degradation by the *lon* protease, which is expressed at higher levels after a heat shock (236).

The recognition of the role of the *sulA*⁺ (*sfiA*⁺) gene led to the isolation of mutants carrying mutations (referred to as either *sulA* or *sfiA* [suppressor of filamentation]) which suppressed the filamentation associated with SOS induction (80). This in turn facilitated a search for derivatives of a *recA441 sfiA* strain which constitutively expressed the SOS responses (215). Such derivatives were identified by screening for mutants which could not be lysogenized by phage λ, because λ repressor is spontaneously inactivated. Hence, the mutations were originally called *spr* (spontaneous repressor inactivation). Mutations of this type are now referred to as *lexA*(Def) mutations and are known to prevent the synthesis of functional

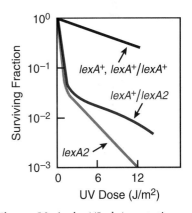

Figure 10–4 *lexA*(Ind⁻) mutations are dominant. A *lexA2*(Ind⁻) mutant is abnormally sensitive to UV radiation relative to *lexA*⁺ or *lexA*⁺/*lexA*⁺ strains. Following introduction of an episome carrying the *lexA*⁺ gene, the UV resistance of the *lexA2* strain is not enhanced. The shallower slope of the survival curve at higher UV doses suggests that a small fraction (about 5%) of the *lexA2* cells are UV resistant. These may be either *lexA*⁺/*lexA*⁺ homozygous segregants or haploid *lexA*⁺ segregants. (*Adapted from Mount et al. [217] with permission.*)

Table 10–5 Spontaneous mutator effect in *lexA* recessive mutants of *E. coli*

Genotype	Temp (°C)	No. of His⁺ colonies/ 10⁸ viable cells
lexA⁺ recA⁺ sulA +	30	12
	40	25
lexA⁺ recA441 sulA	30	17
	40	145
lexA3(Ind⁻) *recA441 sulA*	30	6
	40	12
lexA51(Def) *recA441 sulA*	30	145
	40	347

Source: Adapted from Mount (215) with permission.

LexA repressor protein. Missense mutants (215), amber mutants (229), Tn5-generated insertion mutants (152), and mutants with deletions of *lexA* (98) have now been isolated. These *lexA*(Def) *recA441 sfiA* strains also exhibit an elevated spontaneous mutation rate (Table 10–5), suggesting that the SOS responses were expressed constitutively. It is clear in retrospect that the ability of these strains to cleave λ repressor in the absence of an inducing treatment, as well as their elevated spontaneous mutation rate, is due not only to the loss of LexA function but also to the presence of the *recA441*(Cpt^Ts) mutation.

DEDUCTION OF THE ESSENTIAL ELEMENTS OF SOS REGULATION

The specific functions of the *recA⁺* and *lexA⁺* gene products in the regulation of the SOS system were initially deduced from studies of two particular SOS responses: λ induction and the induction of an approximately 40-kDa protein termed *protein X*, which later was shown to be the RecA protein. These responses were particularly amenable to study since the consequences of induction could be measured directly rather than by inference from more complicated physiological responses or from highly indirect end points such as mutagenesis.

PROTEOLYTIC CLEAVAGE OF λ REPRESSOR DURING SOS INDUCTION

The insight that *proteolytic cleavage of a repressor* could be involved in SOS regulation first came from studies of λ induction by Jeffrey Roberts and his colleagues (256).

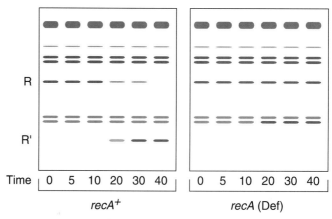

Figure 10–5 A *recA*(Def) mutation prevents cleavage of repressor. Cultures of λ lysogens of a *recA⁺* and a *recA*(Def) strain were grown, labeled with ³⁵SO₄²⁻, chased with cold SO₄²⁻ for 5 min, induced with 5 μg of mitomycin per ml, and sampled at the indicated intervals (in minutes). λ repressor was immunoprecipitated with antibodies to λ repressor and separated by SDS-polyacrylamide gel electrophoresis. R represents intact repressor, and R′ represents a cleavage fragment of repressor. (*Adapted from Roberts and Roberts [256] with permission.*)

These investigators showed that the treatment of a λ lysogen with UV radiation or mitomycin resulted in the proteolytic cleavage of the λ *cI* repressor and that this breakdown of the repressor correlated with the expression of the phage genes. As the induction proceeded, intact repressor protein disappeared and concurrently a protein fragment approximately half the size of the monomer appeared, suggesting a precursor-product relationship between the two (Fig. 10–5). The cleavage did not occur in a *recA*(Def) mutant. Importantly, a mutant repressor protein synthesized by a noninducible phage carrying the dominant *cI*(Ind⁻) mutation was not cleaved, suggesting that its dominance resulted from its resistance to cleavage thus allowing it to continue to function as a repressor when the wild-type protein had been inactivated by cleavage. Since both λ induction and the induced cleavage of λ repressor could be blocked by *recA*(Def) mutations, it was suggested that the RecA protein played a role in this process either by regulating a protease or by being a protease itself.

Extracts of *lexA*(Def) *recA441* cells were shown to contain an activity capable of mediating the proteolytic cleavage of λ repressor in vitro in an ATP-dependent reaction (258) (Fig. 10–6). Roberts and his collaborators then purified the protein responsible for the cleavage of λ repressor from these cells and found that it was RecA. Under conditions of in vitro cleavage of λ repressor (46, 47), two fragments of approximately equal size were recovered (Fig. 10–7), suggesting that λ repressor is inactivated by a single cleavage event near the middle of the repressor protein and that one of the fragments is degraded in the cell.

The RecA-mediated cleavage is now known to occur at the Ala-111–Gly-112 bond of λ repressor in a proteolytically sensitive hinge region which separates the two domains of the protein (Fig. 10–8) (105, 281). The N-terminal domain of λ repressor is the portion of the protein that recognizes the operator, whereas the C-terminal domain plays a key role in the dimerization of λ repressor ($K_d = 10^{-8}$ M) (175). The dimeric form of λ repressor is the form that binds operator sequences specifically and with high affinity. Since the intracellular concentration of λ repressor is about 10^{-6} M (280), most of the repressor molecules in a cell are in the form of dimers. Thus, RecA-mediated cleavage inactivates λ repressor by separating its two domains and thereby *preventing repressor dimerization* (175).

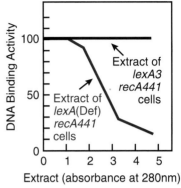

Figure 10–6 Extracts of *lexA3*(Ind⁻) *recA441* cells do not contain high levels of activated RecA coprotease activity, and, hence, when they are incubated with purified λ repressor in the presence of ATP, no inactivation of repressor (measured by binding to ³²P-labeled λ DNA) is observed. However, extracts of mutant *lexA*(Def) *recA441* cells contain high levels of activated coprotease, and these extracts degrade λ repressor in vitro. (*Adapted from Roberts et al. [258] with permission.*)

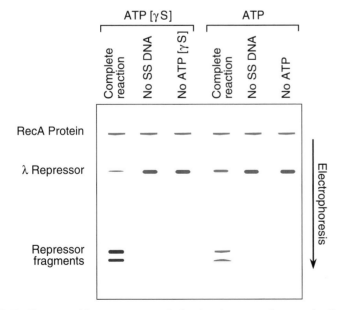

Figure 10–7 Cleavage of λ repressor protein in vitro has a requirement for RecA protein, single-stranded (SS) DNA, and ATP or ATPγS. Following gel electrophoresis of incubation mixtures, RecA protein, λ repressor, and degraded λ repressor can be observed in the appropriate lanes. (*Adapted from Craig and Roberts [46] with permission.*)

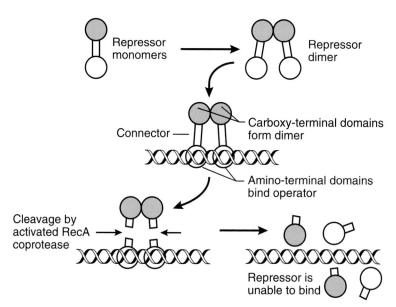

Figure 10–8 Mechanism of inactivation of phage λ repressor. The repressor protein exists in a monomeric form but dimerizes through interactions at its carboxyl-terminal domains. In the dimerized state, repressor binds to the operator in DNA through interactions involving the amino-terminal domains of the proteins. Cleavage mediated by the interaction of the repressor with activated RecA occurs in the connector regions of the proteins between the ends. (*Adapted from Lewin [159] and Friedberg [74] with permission.*)

Cleavage of λ repressor directed by highly purified RecA was found to be dependent on the presence of both ATP and polynucleotide (46, 47). The non-hydrolyzable ATP analog ATPγS substitutes effectively for ATP in vitro and, in fact, supports a severalfold-higher rate of cleavage than does ATP. Less than one ATP molecule is hydrolyzed for each molecule of repressor monomer cleaved, indicating that ATP hydrolysis is not directly coupled to the proteolytic cleavage event. A variety of single-stranded polynucleotides support the cleavage reaction, including oligonucleotides as short as 6 nucleotides, DNA restriction fragments, circular DNA molecules, polyribonucleotides, and polydeoxyribonucleotides (46). In contrast, native DNA is much less effective (46). The cleavage is catalytic, and its rate depends on the ratio of RecA monomer to nucleotide, reaching a maximum at a ratio of ~6 nucleotides to 1 monomer (46). Taken together, these observations suggested that RecA becomes activated for its role in λ repressor cleavage when it forms a ternary complex with single-stranded DNA and ATP. As mentioned above, the activated form of RecA is often referred to as RecA*. It was suggested that in vivo RecA might become activated by interaction with either single-stranded regions of DNA or short oligonucleotides generated as consequence of DNA damage.

INDUCTION OF RecA PROTEIN

At about the same time that Roberts and his colleagues reported their observation of the in vivo cleavage of λ repressor, Gudas and Pardee observed (90–92) that the induction of the synthesis of protein X could be blocked by *lexA*(Ind⁻) mutations, as well as by *recA*(Def) mutations (117). Protein X was a *ca.* 40-kDa, and its synthesis had been found to increase very strongly after DNA-damaging treatments. About 90% of protein X was present in the cytoplasm, but its induction could be particularly easily analyzed on sodium dodecyl sulfate (SDS)-gels of membrane fractions. The fact that induction of its synthesis was blocked by the same *recA*(Def) and *lexA*(Ind⁻) mutations that blocked the induction of the physiologically observable SOS responses suggested that induction of protein X could also be considered to be part of the SOS response. Furthermore, it was found that the synthesis of protein X could be induced by simply shifting a *lexA*(Ts) or *recA441* mutant or a *lexA*(Ts) *recA*(Def) double mutant to an elevated temperature (Fig. 10–9). [The *recA*(Def) alleles used in these studies were missense alleles so that the RecA protein was synthesized even though it was inactive]. On the basis of these observations, it was proposed that LexA *repressed* the gene coding for protein X and possibly other SOS genes and that the RecA protein was involved in the inactivation of LexA. Because of the observations (256) that λ repressor was proteolytically cleaved at the time of SOS induction, the possibility was raised that LexA was also inactivated by proteolytic cleavage. Shortly thereafter, several investigators (72, 89, 174, 207) demonstrated that protein X was actually the *recA*⁺ gene product and hence that RecA played a role in its own induction. It was later found that the association of a subfraction of RecA with the cell membrane occurred only if RecA is activated (78). The physiological significance of this association is not yet clear, but it is interesting that there are striking effects of SOS expression on the levels of the outer membrane proteins OmpA, OmpC, and OmpF (78).

The model that LexA protein functioned as a repressor of the SOS genes received additional support from the observation (89) that a *lexA*(Def) mutant, which, as discussed above, constitutively expresses the SOS responses, also synthesized high levels of RecA protein. It was noted that a double mutant carrying *lexA*(Def) and missense *recA*(Def) similarly synthesized high levels of RecA protein. The observation that the *recA*⁺ locus was expressed at high levels in the absence of functional LexA protein suggested that the LexA protein was acting as a repressor of this locus. Furthermore, these observations suggested that the role of the RecA protein in the regulation of *recA*⁺ expression was to *inactivate* LexA protein, since strains which lacked a functional LexA no longer required a functional RecA protein to express the *recA*⁺ locus at high levels.

LexA PROTEIN IS PROTEOLYTICALLY CLEAVED IN A RecA-DEPENDENT FASHION

Since RecA protein itself was capable of mediating the proteolytic cleavage of λ repressor and had been implicated in the inactivation of LexA protein by genetic studies, it seemed likely that LexA protein was cleaved in a RecA-mediated fashion at the time of SOS induction. This key element of the control circuit was established when the *lexA*⁺ gene product was identified as a 22.7-kDa protein (21, 172) that was at the time shown to undergo a RecA-mediated proteolytic cleavage in a fashion similar to λ repressor (105, 171). The RecA-mediated cleavage of LexA protein was first demonstrated in vitro by using maxicell extracts (74) containing

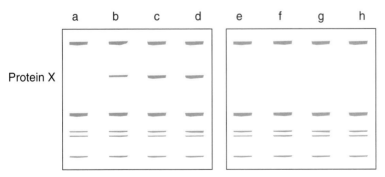

Figure 10–9 Induction of protein X (RecA). Nalidixic acid was added at a final concentration of 35 μg/ml to cultures of *E. coli lexA*⁺ and *lexA*(Ind⁻) strains. At various times after the addition of nalidixic acid, samples were removed and pulsed with [³⁵S] methionine for 8 min. Membrane proteins were prepared, separated by SDS-polyacrylamide gel electrophoresis, and autoradiographed. Lanes: a, *lexA*⁺, no nalidixic acid; b to d, *lexA*⁺ 10, 40, and 70 min, respectively, after nalidixic acid addition; e, *lexA*(Ind⁻), no nalidixic acid; f to h, *lexA*(Ind⁻) 10, 40, and 70 min, respectively, after nalidixic acid addition. A portion of the autoradiograph is depicted. (*Adapted from Gudas and Pardee [92] with permission.*)

radioactively labeled LexA protein. The LexA3(Ind⁻) protein was found to be largely resistant to cleavage, thus providing a satisfying biochemical demonstration of the dominance of the *lexA3*(Ind⁻) mutation. LexA protein was subsequently purified to homogeneity and cleavage was shown to occur at the Ala-84–Gly-85 bond, which is near the middle of the 202-amino-acid protein (105) (Fig. 10–10). As with λ repressor, RecA protein had to be activated by the addition of single-stranded DNA and ATP or ATP[γS] (105, 176). In vitro the cleavage of LexA protein was found to be far more rapid than that of λ repressor.

Subsequent pulse-chase studies (167, 168) patterned after those of Roberts and Roberts (256) and analyses of steady-state LexA levels by Western immunoblots (165, 280) showed that LexA is similarly cleaved in vivo after an SOS-inducing treatment. An uninduced *E. coli* cell contains approximately 1,300 LexA molecules and 7,200 RecA molecules (280), and, if it is a λ lysogen, it contains approximately 1,000 molecules of λ repressor (280). LexA protein is fairly stable in normally growing cells, with a half-life of ~1 h. Cleavage begins about 1 min after UV irradiation of cells and is largely complete by 4 to 5 min.

LexA PROTEIN REPRESSES BOTH THE *recA⁺* AND *lexA⁺* GENES

Purified LexA protein was shown to bind in vitro to regulatory sequences in both the cloned *recA⁺* and *lexA⁺* genes (20, 22, 176, 275). The sites where LexA binds were shown to be imperfect palindromic *operator sequences*, which are usually referred to as *SOS boxes*. The *recA⁺* gene has a single such box, and the *lexA⁺* gene has a tandem pair (Fig. 10–11). Related sequences were subsequently found in the 5′-noncoding regions of other SOS-regulated genes. The *binding* of LexA protein to the LexA boxes of these SOS-regulated genes was shown to *inhibit transcription* from the *recA⁺* and *lexA⁺* genes (22, 176), as well as from other SOS-regulated genes (275, 276), thus confirming the inference drawn from genetic studies that LexA protein functions as a *repressor*.

LexA protein binds to SOS boxes as a *dimer*. This concept was originally suggested by the twofold symmetry of the SOS boxes and has been supported by a variety of protection and interference studies, by measurement of the extent of retardation of a complex of LexA with a DNA fragment containing the *recA⁺* operator (285), and by analysis of the binding of a mixture of mutant and wild-type LexA proteins to a hybrid *recA⁺* operator consisting of mutant and wild-type half sites (322). The ability of LexA protein to dimerize is critical to its ability to repress SOS-regulated genes in vivo. LexA protein consists of two structurally defined domains. These domains are joined by a hinge region which appears to be relatively flexible, since deletions within this region do not strongly impair DNA binding of the protein (173). The N-terminal domain (amino acids 1 to 84) of LexA protein specifically recognizes SOS boxes but does so with lower affinity than the intact protein does (11, 114, 115, 137). The C-terminal domain of the protein contains the elements of the protein necessary for dimerization of LexA and thus indirectly increases DNA binding by allowing dimerization. Both the intact protein and the C-terminal domain form dimers in solution, with a rather low association constant $(2 \times 10^4 \text{ M}^{-1})$ (283, 285).

Recent experiments have indicated that the dimerization of LexA protein occurs after repressor monomers have become associated with the operator DNA (137). The Ala-84–Gly-85 bond is located within the hinge region between the two domains. Thus, cleavage of this bond during SOS induction separates the two domains (Fig. 10–10), thereby preventing the dimerization of LexA that is necessary for repression of SOS-regulated genes. The fact that dimerization occurs on the DNA leads to a high degree of cooperativity and therefore a steep binding curve so that less LexA must be cleaved to cause derepression of LexA-repressed genes (137).

It was noted that the N-terminal domain of LexA protein shares some homology with DNA-binding proteins containing a helix-turn-helix motif (228). However, subsequent two-dimensional nuclear magnetic resonance analysis of the structure and the sequence of mutations affecting LexA binding and LexA-DNA contacts suggested that it is not a canonical helix-turn-helix protein but is at

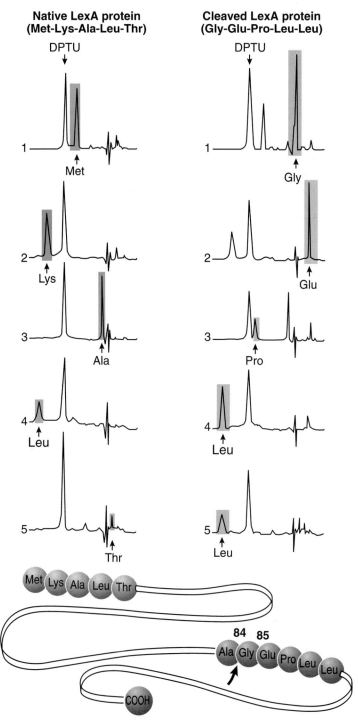

Figure 10–10 The site of cleavage of LexA protein by activated RecA coprotease is between Ala-84 and Gly-85 of the peptide (lower figure). LexA protein was incubated with RecA protein in the presence of ATPγS and single-stranded DNA. After incubation the reaction mixture was dialyzed, lyophilized, and subjected to partial degradation by the Edman procedure. Phenylthiohydantoin derivatives were determined by high-performance liquid chromatography and were detected by measuring the A_{269} (right panel). Diphenylthiourea (DPTU) is a by-product of the Edman degradation. In this way the amino acid sequence at the amino terminus of the cleavage product (and hence the site of cleavage) was determined. The left panel shows confirmation of the previously established amino-terminal sequence of native LexA protein. (*Adapted from Horii et al. [105] and Friedberg [74] with permission.*)

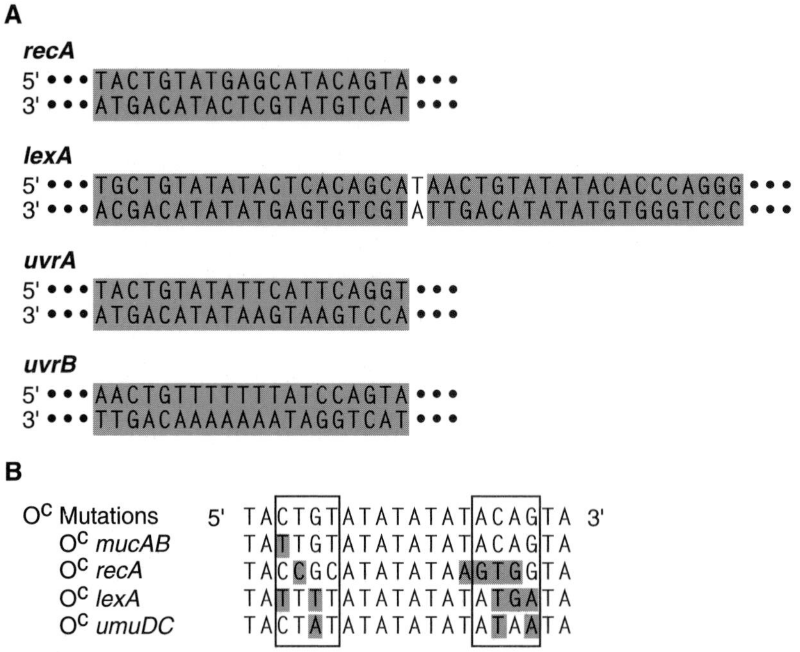

A

recA

```
5' ••• TACTGTATGAGCATACAGTA ••• 
3' ••• ATGACATACTCGTATGTCAT ••• 
```

lexA

```
5' ••• TGCTGTATATACTCACAGCATAACTGTATATACACCCAGGG ••• 
3' ••• ACGACATATATGAGTGTCGTATTGACATATATGTGGGTCCC ••• 
```

uvrA

```
5' ••• TACTGTATATTCATTCAGGT ••• 
3' ••• ATGACATATAAGTAAGTCCA ••• 
```

uvrB

```
5' ••• AACTGTTTTTTTATCCAGTA ••• 
3' ••• TTGACAAAAAAATAGGTCAT ••• 
```

B

O^C Mutations	5' TA⎡CTGT⎤ATATATAT⎡ACAG⎤TA 3'	
O^C *mucAB*	TA⎡TTGT⎤ATATATAT⎡ACAG⎤TA	
O^C *recA*	TA⎡CCGC⎤ATATATA⎡AGTGG⎤TA	
O^C *lexA*	TA⎡TTTT⎤ATATATAT⎡ATGA⎤TA	
O^C *umuDC*	TA⎡CTAT⎤ATATATAT⎡ATAA⎤TA	

Figure 10–11 (A) Operator regions near the beginning of the *recA*⁺, *lexA*⁺, *uvrA*⁺, and *uvrB*⁺ genes have similar base sequences, about 20 bp long, that are binding sites for the LexA repressor. The binding sites for LexA are often referred to as SOS boxes. The *lexA*⁺ gene has two nearly identical SOS boxes. (B) Operator-constitutive (O^c) mutations in SOS genes. Boxed nucleotides are the most highly conserved nucleotides in the consensus recognition sequence (341) of the LexA repressor (SOS box). The letters shaded in red indicate single-base-pair mutations that are O^c mutations. O^c mutations have been detected and characterized in the context of *mucA*⁺*B*⁺, *recA*⁺, *lexA*⁺, and *umuD*⁺*C*⁺ operators. (*Panel A adapted from Howard-Flanders [107] with permission; panel B adapted from Schnarr et al. [284] and Sommer et al. [309] with permission.*)

best a distant relative of this class of transcription factors (284). These results suggest that LexA protein interacts with one face of the DNA cylinder through contacts with the DNA backbone, while a protruding "reading head" probes the bottom of the major groove (where recognition takes place) (115). Not all of the bases shown in the consensus sequence for an SOS box (Fig. 10–11) are of equal importance. In fact, all the known SOS operators contain a consensus 5'-CTGT sequence, whereas the center of the different operators is rather variable, with some preference for an alternating (AT)₄ sequence. It therefore seems likely that the CTGT sequence contains most of the information read by the LexA repressor upon its interaction with operator DNA, with the central T and G bases being absolutely required for efficient interaction with wild-type LexA repressor. Furthermore, various operator-constitutive mutations that alter this sequence have been isolated (38, 208, 340, 341) (Fig. 10–11). In contrast, the central (AT)₄ stretch within the consensus sequence may favor LexA binding indirectly by providing the proper phasing for the two operator half sites (148, 284).

MECHANISM OF LexA REPRESSOR CLEAVAGE

The proteolytic cleavage of LexA when it interacts with activated RecA is central to SOS regulation. At first, it was assumed that RecA protein was acting as a classical protease. However, experiments by John Little and his colleagues have led to the view that activated RecA protein acts indirectly by stimulating an otherwise latent capacity of LexA to autodigest (169). According to Little (167),

> Autodigestion was discovered by accident in a control to an unrelated experiment. To test whether LexA is organized into domains like λ *cI* repressor, LexA was treated with trypsin, with the expectation that a domain structure might result in resistant

species, as with *cI*. Although this was observed, the control incubation lacking trypsin gave a far more interesting and provocative outcome. In the buffer supporting trypsin activity, part of the LexA broke down into 2 discrete fragments similar in size to those observed in RecA-mediated cleavage. The next experiments showed that the fragments were indeed of the same size and that breakdown was stimulated to a far greater extent at higher pH [approximately pH 10].

Furthermore, this cleavage reaction exhibited *first-order kinetics* and the rate constant was independent of protein concentration even when the protein concentration was varied widely, observations which indicated that it is an intramolecular reaction (169, 299). λ repressor was subsequently similarly found to exhibit self-cleavage under alkaline conditions (169). The elements of the protein required for both RecA-mediated cleavage and autodigestion are located in the carboxyl-terminal domain of the protein for both LexA and λ repressor (169, 281).

Both types of cleavage also occur in other homologous proteins. The C-terminal domain of LexA shares extensive homology with those of the repressors of bacteriophages λ, 434, and P22 (282). This homology has now been extended to the repressors of some other bacteriophages such as φ80 (68) and to UmuD protein of *E. coli* (6, 235) and its analogs MucA (6, 235) and ImpA (187) (see chapter 11) (Fig. 10–12). All of these proteins undergo RecA-mediated cleavage, and all that have been tested undergo autodigestion at alkaline pH, yet only a relatively small number of amino acids are conserved throughout the set of proteins. Consideration of the nature of these conserved amino acids and of possible mechanisms for the catalysis of peptide bond cleavage has led to a proposed mechanism for LexA autodigestion in which the uncharged form of Lys-156 helps to remove a proton from Ser-119, which then acts as a nucleophile to attack the Ala-84–Gly-85 bond (298). A second postulated role for Lys-156 could be to donate a proton to the α amino group when the peptide bond is broken. Similar roles have been postulated for a deprotonated lysine in the mechanism of action of a β-lactamase that has an active-site serine which forms a covalent ester with the substrate (170, 316) (Fig. 10–13). In this model, the titration of the Lys-156 is viewed as being crucial to the rate of cleavage. The interaction with RecA protein results in a titration of Lys-156, thereby affecting the rate of cleavage.

This model is supported by analysis of the properties of various classes of LexA mutants (165, 166, 298), including those that result in the alteration of Ser-119 and Lys-156, and by the demonstration that autodigestion can be inhibited by high concentrations of the serine protease inhibitor diisopropylfluorophosphate (260). The isolation of *lexA*(Ind^s) mutants, which greatly increase the rate of LexA cleavage (261, 307), has led to the further suggestion that LexA protein can exist in two conformations. In one of these, the pK$_a$ of Lys-156 has a normal value of about 10 and cleavage either does not occur or is extremely slow. In the other, Lys-156 has a pK$_a$ of about 5 to 6 as a consequence of being in a special environment and cleavage is much faster. The interaction with RecA protein is viewed as shifting the equilibrium in favor of the low-pK$_a$ conformation of LexA. It seems likely that the environment reducing the pK$_a$ is created by groups within LexA, not in RecA, since RecA appears to interact differently with different proteins. In a formal sense, LexA does not act as a catalyst in the cleavage reaction since it is changed during the reaction. It is therefore interesting that the C-terminal cleavage product of LexA can act as a relatively efficient enzyme to cleave other molecules of a truncated LexA protein, or intact LexA protein with a mutation (S119A) in its active site (138, 170).

Since it appears that activated RecA protein is not acting directly as a protease in LexA cleavage but, rather, is stimulating the autodigestion reaction and allowing the cleavage to proceed rapidly at physiological pH, we will use the term *RecA coprotease* (167) to refer to this activity. In this view, RecA protein is still acting as an enzyme since one molecule can mediate cleavage of many substrate molecules but the groups essential for the mechanism of catalysis are located in the LexA protein itself. The discovery of this coprotease action of RecA has not changed the basic logic of the SOS circuitry but only our view of the nature of the interactions between activated RecA protein and its substrates. Given the degree of divergence

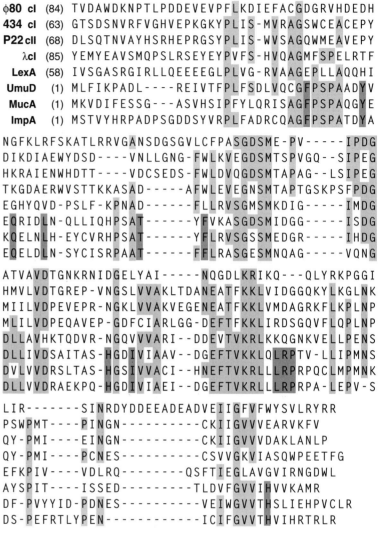

▼

φ80 cI	(84)	TVDAWDKNPTLPDDEVEVPFLKDIEFACGDGRVHDEDH
434 cI	(63)	GTSDSNVRFVGHVEPKGKYPLIS-MVRAGSWCEACEPY
P22 cII	(68)	DLSQTNVAYHSRHEPRGSYPLIS-WVSAGQWMEAVEPY
λcI	(85)	YEMYEAVSMQPSLRSEYEYPVFS-HVQAGMFSPELRTF
LexA	(58)	IVSGASRGIRLLQEEEEGLPLVG-RVAAGEPLLAQQHI
UmuD	(1)	MLFIKPADL----REIVTFPLFSDLVQCGFPSPAADYV
MucA	(1)	MKVDIFESSG---ASVHSIPFYLQRISAGFPSPAQGYE
ImpA	(1)	MSTVYHRPADPSGDDSYVRPLFADRCQAGFPSPATDYA

NGFKLRFSKATLRRVGANSDGSGVLCFPASGDSME-PV-----IPDG
DIKDIAEWYDSD----VNLLGNG-FWLKVEGDSMTSPVGQ--SIPEG
HKRAIENWHDTT----VDCSEDS-FWLDVQGDSMTAPAG--LSIPEG
TKGDAERWVSTTKKASAD-----AFWLEVEGNSMTAPTGSKPSFPDG
EGHYQVD-PSLF-KPNAD------FLLRVSGMSMKDIG-----IMDG
EQRIDLN-QLLIQHPSAT------YFVKASGDSMIDGG-----ISDG
KQELNLH-EYCVRHPSAT------YFLRVSGSSMEDGR-----IHDG
EQELDLN-SYCISRPAAT------FFLRASGESMNQAG-----VQNG

ATVAVDTGNKRNIDGELYAI-----NQGDLKRIKQ---QLYRKPGGI
HMVLVDTGREP-VNGSLVVAKLTDANEATFKKLVIDGGQKYLKGLNK
MIILVDPEVEPR-NGKLVVAKVEGENEATFKKLVMDAGRKFLKPLNP
MLILVDPEQAVEP-GDFCIARLGG-DEFTFKKLIRDSGQVFLQPLNP
DLLAVHKTQDVR-NGQVVVARI--DDEVTVKRLKKQGNKVELLPENS
DLLIVDSAITAS-HGDIVIAAV--DGEFTVKKLQLRPTV-LLIPMNS
DVLVVDRSLTAS-HGSIVVACI--HNEFTVKRLLLRPRPQCLMPMNK
DLLVVDRAEKPQ-HGDIVIAEI--DGEFTVKRLLLRPRPA-LEPV-S

LIR-------SINRDYDDEEADEADVEIIGFVFWYSVLRYRR
PSWPMT----PINGN----------CKIIGVVVEARVKFV
QY-PMI----EINGN----------CKIIGVVVDAKLANLP
QY-PMI----PCNES----------CSVVGKVIASQWPEETFG
EFKPIV----VDLRQ----------QSFTIEGLAVGVIRNGDWL
AYSPIT----ISSED----------TLDVFGVVIHVVKAMR
DF-PVYYID-PDNES----------VEIWGVVTHSLIEHPVCLR
DS-PEFRTLYPEN-----------ICIFGVVTHVIHRTRLR

Figure 10–12 Homology among the bacteriophages φ80, 434, and P22, λ repressors, LexA, and the mutagenesis proteins UmuD, MucA, and ImpA. The arrowhead indicates the site of RecA-mediated cleavage. Amino acids that are identical in four or more members of the set are shaded in grey. Amino acids that are identical in the three mutagenesis proteins but are not shared with LexA or the bacteriophage repressors are shaded in red (see chapter 11). (*Adapted from Battista et al. [6] with permission.*)

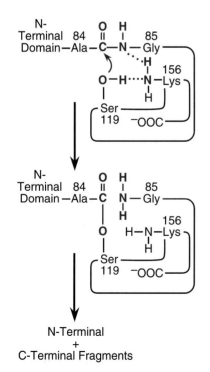

Figure 10–13 Proposed mechanism for LexA repressor cleavage. (*Adapted from Slilaty and Little [298] with permission.*)

between the proteins that undergo RecA-mediated cleavage, it seems likely that they differ with respect to the specific contacts they make with RecA protein. This view is supported by the observation that the RecA430 protein is deficient in the cleavage of LexA and is very deficient in the cleavage of λ repressor (257) but is proficient in the cleavage of φ80 repressor (68). Mutations in λ repressor which perturb RecA-mediated cleavage but not autodigestion at alkaline pH, and thus appear to affect the interaction between λ repressor and activated RecA protein, have been characterized (82). At least for λ repressor, it is clear that the repressor monomer is greatly preferred as a substrate over the repressor dimer (83).

IDENTIFICATION OF GENES IN THE SOS NETWORK

The demonstration that the RecA and LexA proteins regulate the expression of the *recA*[+] and *lexA*[+] genes suggested that other SOS responses were due to the increased expression of a set of genes controlled by RecA and LexA proteins. However, the physiological complexity of most of the SOS responses made it difficult to analyze the regulation of these responses directly. In an effort to dissociate the regulation of the SOS responses from their physiological complexity, experiments were carried out (134) which exploited a derivative of bacteriophage Mu termed Mu d1, which is capable of creating transcriptional fusions in a single step (29). This bacteriophage can integrate into the *E. coli* chromosome without any appreciable site specificity and carries the structural genes for the *lac* operon near one of its ends but no promoter capable of initiating their transcription (Fig. 10–14A). When the Mu d1 bacteriophage integrates into a host gene in the correct orientation, it creates a *transcriptional fusion* in which transcription initiates at the promoter of the host gene and continues into the *lac* genes on the phage. In such a fusion, β-galactosidase expression is controlled by regulatory elements which normally control the expression of the host gene.

With the use of this phage, it was possible to ask whether there is indeed a

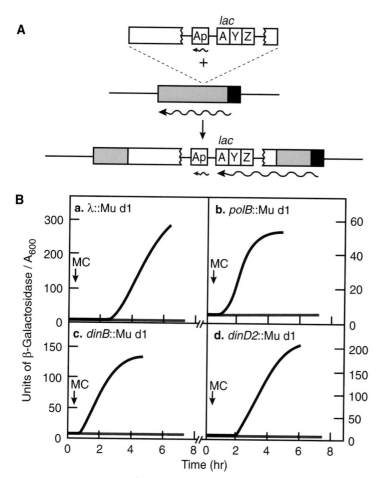

Figure 10–14 (A) Creation of an operon fusion to *lacZ* by the insertion of Mu d1 into a chromosomal gene. In such a fusion, the expression of β-galactosidase is now regulated by the control region of the chromosomal gene. (B) Kinetics of induction of β-galactosidase in strains carrying *lacZ* fusions to SOS genes. Mitomycin C (MC) at 1 μg/ml was added to exponentially growing cultures as indicated. Aliquots were removed periodically, and the total activity of β-galactosidase present in the culture was determined. Cell density was determined by measuring the A_{600}. For panels a to d: red line, untreated fusions; black line, fusion strains plus mitomycin; red line, *lexA*(Ind⁻) fusion strains plus mitomycin; red line, *recA*(Def) derivatives plus mitomycin. For panel a only: red line, λ::Mu d1/pKB280 (pKB280 is a plasmid that overproduces λ repressor). Ap = ampicillin resistance; A, Y, and Z = *lacA*, *lacY*, and *lacZ*, respectively (*Adapted from Walker et al. [333] and Kenyon and Walker [134] with permission.*)

set of genes whose expression was increased in response to SOS-inducing treatments (134). A set of random Mu d1-generated fusions in the *E. coli* chromosome was screened in a search for fusions which expressed β-galactosidase at higher levels in the presence of the DNA-damaging agent mitomycin than in its absence (Fig. 10–14B). By this procedure, a set of *din* (damage-inducible) loci whose expression was increased by exposure of the cells to a variety of SOS-inducing treatments were identified. The induction of these loci was blocked by *recA*(Def) and *lexA*(Ind⁻) mutations, indicating that there were in fact a set of inducible loci in the *E. coli* chromosome that were controlled by the *recA⁺ lexA⁺* regulatory circuit (134). Furthermore, these *din-lac* fusion strains expressed β-galactosidase constitutively at a high level whether they carried a *lexA*(Def) mutation or both a *lexA*(Def) and a *recA*(Def) mutation (133, 135). This genetic observation strongly suggested that LexA functioned as a repressor of the *din* genes and furthermore that RecA protein played a role in the inactivation of the LexA protein at the time of their induction.

One of the *din* mutants identified in this initial screen was as sensitive to UV radiation as were strains with mutations in the *uvr⁺* genes involved in nucleotide excision repair (see chapter 5), and it was subsequently shown that the Mu d1 insertion in this strain was in fact in the *uvrA⁺* gene (135). The observation that the *uvrA⁺* gene was inducible was consistent with various observations indicating that *uvr⁺*-dependent repair processes can be induced in *E. coli* (42, 289, 290) (see chapter 5). Another of the SOS-inducible loci isolated in this initial screen, *dinA⁺*, has been shown to be the *polB⁺* gene, which encodes DNA polymerase II (15, 120, 121) (see chapter 4). The *dinF⁺* locus has been sequenced (13), as has the *dinD* locus (196), but their functions are not known. The Mu d1 bacteriophage was subsequently used to identify a number of other SOS-regulated loci including *uvrB⁺* (73, 135), *sulA⁺* (113), *umuD⁺C⁺* (4), *himA⁺* (210), *uvrD⁺* (297), *ruvA⁺* and *ruvB⁺* (296), and *recN⁺* (184).

In addition, attempts have been made to identify SOS-regulated genes by creating recombinant plasmids which carry fusions of various genes to a reporter gene and then examining their induction. However, subsequent to a number of earlier studies it was discovered that the copy number of the commonly used cloning vector pBR322 increases in a *recA⁺-lexA⁺*-dependent fashion upon SOS induction (12) (in other words, an increase in the copy number of pBR322 is an SOS response) (Table 10–1). Thus, initial reports that certain genes, including *uvrC* (328), *phr* (116), and *ssb* (19), are SOS regulated are likely to have been incorrectly interpreted because of an SOS-induced increase in plasmid copy number. This probably accounts for the failure of other experimental approaches to confirm the SOS inducibility of these genes (85, 234, 272).

The *recA⁺-lexA⁺*-dependent inducibility of plasmid-borne fusions to the *recQ⁺* (128), *dinG⁺* (161), and *dinH⁺* (161) genes has also been reported, but to date LexA protein binding has been demonstrated directly only for *dinG⁺* (161). The expression of plasmid-borne fusions to *dnaN⁺* and *dnaQ⁺* has been reported to be inducible by DNA-damaging treatments and to be blocked by *recA* and *lexA* mutations, which block SOS induction (128, 246). However, these genes do not appear to be under normal SOS control, because the induction is abnormally slow, the fusions are not induced in a *lexA*(Def) *recA*(Cptᶜ) strain, and there is no SOS box in the promoter region of these genes (128). Similarly, plasmid-borne fusions to *dnaA⁺* have been reported to be inducible by DNA damage, but this induction does not appear to be a direct consequence of SOS control of *dnaA⁺* expression (245). Figure 10–15 summarizes the locations of the known SOS genes on the *E. coli* chromosome.

In addition to controlling the expression of genes located on the *E. coli* chromosome, the SOS system has been shown to regulate the expression of genes located on naturally occurring plasmids, bacteriophages, and transposons. Plasmid-encoded genes regulated by the SOS system include *mucAB* (an analog of *umuDC* [see chapter 11]) (70, 208), *cea* (which encodes colicin E1) (60, 61, 63, 273), and genes encoding other colicins (summarized in reference 274). In addi-

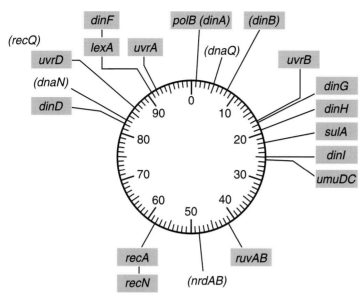

Figure 10–15 Locations of SOS genes on the *E. coli* chromosome. Genes which are components of the SOS regulon are shown in red boxes. In addition, *nrdA⁺B⁺* shows elevated synthesis in *lexA*(Def) hosts but does not contain a recognizable LexA-binding site in its sequenced regulatory region. Four other genes (*recQ⁺*, *dnaA⁺*, *dnaN⁺*, and *dnaQ⁺*) apparently require *recA* and *lexA* for transcriptional induction by DNA-damaging agents but may not be directly repressed by LexA (see the text). (*Adapted from Lewis et al. [161] with permission.*)

tion, the expression of a bacteriophage gene has been shown to be SOS regulated (155). Finally, LexA protein has been shown to weakly repress a transposon-encoded gene which encodes the transposase of Tn5 (153).

Differential Regulation of SOS-Controlled Genes

The various SOS genes differ with respect to the degree to which they are induced. As shown in Table 10–6, induction ranges from about 100-fold for *sulA⁺*, the most tightly repressed SOS gene identified so far, to only 4 to 5-fold for *uvrA⁺*, *uvrB⁺*,

Table 10–6 Induction ratios of the different SOS genes in vivo and relative binding affinity of the LexA repressor in vitro.

Gene	Induction ratio[a] at: 32° C	Induction ratio[a] at: 42°C	Relative binding affinity[b]
lexA⁺	6.7	4.8	15
recA⁺	11	13	3.8
sulA⁺	110	140	1
umuD⁺C⁺	28	17	1.1
uvrA⁺	3.4	6.2	14.6
uvrB⁺	3.6	3.7	8.8
uvrD⁺	4.6	7.2	17.9
polB⁺ (dinA⁺)	5.2	9.3	nd
dinB⁺	7.3	9.5	nd
dinD⁺	6.2	5.8	nd

[a] Induction ratios were determined at two different growth temperatures by determination of the β-galactosidase activity as follows: Miller units of *lexA71*(Def)/Miller units of *lexA⁺*, where *lexA71* is a mutation that prevents the synthesis of LexA.

[b] Relative LexA concentrations necessary to bind half of the DNA molecules in a gel retardation assay (pH 7.9, 200 mM KCl, 20°C) as compared with binding to the *sulA* operator. The higher the LexA concentration necessary for binding, the weaker the operator.

Source: Adapted from Schnarr (284) and Peterson and Mount (236) with permission.

uvrD$^+$, and *lexA$^+$*. The extent of repression may depend on at least four parameters: the operator strength, the localization of the operator relative to the promoter, the promoter strength, and the existence of additional, constitutive promoters. As shown in Fig. 10–16, the location of the SOS boxes varies with respect to the transcription start site. Some SOS boxes overlap with the −35 promoter region (*uvrA$^+$* and possibly *polB$^+$* and *ruv$^+$*), whereas others are situated between the −35 and −10 regions of the promoter (*recA$^+$* and *uvrB$^+$*). Some SOS boxes overlap with the −10 region of the promoter (*sulA$^+$*, *umuD$^+$C$^+$*, and *lexA$^+$*), whereas others are found downstream of the −10 region or even downstream of the +1 transcription start (*uvrD$^+$*, *cea$^+$*, and *caa$^+$*). Additionally, the number of operators ranges from a single operator observed with many SOS genes to three operators for *recN$^+$*. Individual SOS boxes also vary with respect to their ability to bind LexA protein (Table 10–6), and, as mentioned above, in some cases it has not been possible to demonstrate in vitro binding of LexA protein to a postulated SOS box (234). For the *uvrA$^+$* gene, where the SOS box overlaps with the −35 region of the promoter, LexA binding seems to interfere with RNA polymerase at an early stage, preventing the formation of a closed RNA polymerase-promoter complex (11). However, it is possible that LexA protein inhibits other stages of transcription initiation for other SOS genes. For example, in the *lac* operon, the operator lies downstream of the +1 position and the block appears to be at the step of promoter clearance; such a mechanism might apply to genes that have SOS boxes in a similar location.

What Is the SOS-Inducing Signal?

As discussed above, in vitro conditions that result in RecA-mediated cleavage of LexA protein and λ repressor have been clearly defined; the RecA protein, ATP or a nonhydrolyzable analog of ATP, and single-stranded DNA are required. Furthermore, measurements of the in vitro rate of LexA cleavage have correlated closely with estimates of the in vivo cleavage rate, suggesting that the important components of the in vivo reaction have been identified (280). However, it has been considerably more challenging to determine the nature of the in vivo-inducing signal that leads to RecA activation and subsequent LexA cleavage. Many of the agents that induce the SOS response are DNA-damaging agents, but this does not necessarily mean that the presence of the lesion in the DNA is the ultimate inducing signal, since DNA damage can have a number of *secondary consequences* including replication arrest, alterations in nucleoid structure, and altered superhelicity, to name just a few. Furthermore, the fact that the treatments and conditions that induce the SOS response have in common their ability to arrest DNA replication, as pointed out above, does not imply that simple arrest of replication is the inducing signal. Blockage of replication or the presence of an abnormal replication fork could have secondary consequences such as the generation of regions of single-stranded DNA or oligonucleotides, and these could serve as the ultimate inducing signal. Recent work has strongly supported the unifying view that, at least for many inducing treatments, the ultimate signal for SOS induction in vivo is the generation of regions of single-stranded DNA within the cell. Depending on the nature of the inducing agent, the generation of this single-stranded DNA may be independent of, or dependent on, DNA replication.

There is strong evidence that for SOS induction by nalidixic acid, the critical event is the generation of regions of exposed single-stranded DNA in the cell. Nalidixic acid inhibits DNA gyrase (79, 317) and causes double-strand breaks in DNA (58, 308). However, this is not sufficient for SOS induction, and subsequent processing of the DNA by the RecBCD nuclease is required (90, 130). Although this enzyme both degrades and unwinds DNA from double-strand breaks in vitro, only the *unwinding* activity of RecBCD is required for nalidixic acid to function as an SOS inducer (36).

Most SOS-inducing agents do not directly cause breaks in DNA but, rather, create lesions that alter the chemical structure of the bases thereby interfering with base pairing. Agents of this type include UV radiation and chemicals which react with DNA bases, such as activated derivatives of aflatoxin B$_1$ and dimethylbenz-

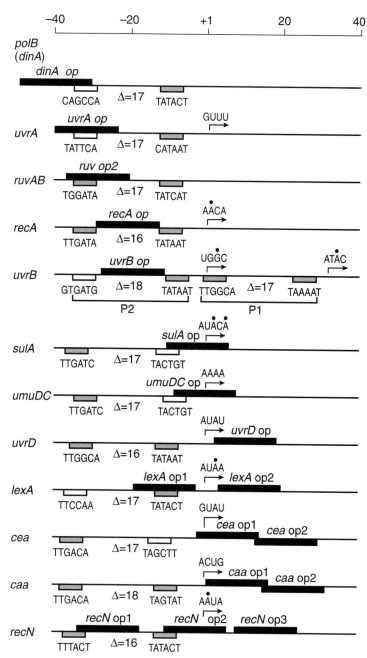

Figure 10–16 Disposition of the different SOS operators (black boxes) with respect to the promoters of the SOS genes. The promoters are characterized by their −35 regions (consensus sequence, TTGACA) and their −10 regions (consensus sequence, TATAAT). The corresponding regions are shown as red boxes if at least four base pairs are identical to the consensus sequence, otherwise these regions are shown as open boxes. The identification of these promoter motifs is reliable for genes whose transcriptional start site has been mapped but is tentative for *polB*⁺ and *ruvA*⁺*B*⁺. Black dots over some bases of the 5′ ends of the mRNA represent alternative transcription start sites. (*Adapted from Schnarr [284] with permission.*)

anthracene. As discussed in chapter 5, lesions of this type are usually removed by *uvr*$^+$-dependent nucleotide excision repair. It has been known for some time that UV radiation-induced lesions do not actually have to be present in the bacterial chromosome but will result in SOS induction if they are introduced into the cell on a DNA molecule such as an F or F' plasmid; P1, M13, or λ bacteriophage DNA; or Hfr DNA (17, 18, 48, 53, 81, 263). This phenomenon is known as *indirect induction*. An analysis of the efficiency of such damaged DNAs in inducing the expression of a *sulA-lacZ* fusion indicated that SOS induction was most efficient if the damaged DNAs were allowed to replicate after being introduced into the bacterial cell (48, 310).

A considerable body of evidence now indicates that the presence of UV radiation-induced lesions in DNA is not sufficient to cause SOS induction but, rather, that the SOS inducing signal is generated when the cell attempts to replicate the damaged DNA, leading to the generation of single-stranded regions. Initial evidence for this model was provided by using a *dnaC28*(Ts) *uvrB* double mutant (270). The *dnaC28*(Ts) mutation makes the strain temperature-sensitive for the initiation of DNA replication; if the cells are shifted to 42°C (the restrictive temperature), they complete the existing round of replication but are not able to initiate another. As discussed in chapter 5, a *uvrB* mutation inactivates the Uvr(A)BC endonuclease activity, which initiates nucleotide excision repair of UV radiation-induced lesions. It was demonstrated (270) that raising the temperature of the *dnaC28*(Ts) *uvrB* double mutant inhibited the induction of RecA protein by UV radiation, whereas some RecA induction occurred in a *dnaC28*(Ts) *uvrB*$^+$ strain following exposure to higher doses of UV light. These results suggested that the mere presence of UV radiation-induced lesions was not sufficient to cause SOS induction in a cell lacking Uvr(A)BC endonuclease activity and that DNA replication of the damaged template was required to generate the SOS-inducing signal. Furthermore, it was suggested that *uvr*$^+$-dependent SOS induction observed in the absence of replication resulted from the removal of pyrimidine dimers and the appearance of small gaps that could be processed into an SOS-inducing signal.

These results (270) were extended (280) in studies that directly measured LexA cleavage in the same strains. The latter studies found that if the *dnaC uvrB* cells were UV irradiated 70 min after the shift to the restrictive temperature, no LexA cleavage occurred, whereas 70% of the LexA protein was cleaved after UV irradiation within 10 min at the permissive temperature (Fig. 10–17). This result indicated that DNA replication of the damaged template was required to produce the inducing signal which activated RecA protein for LexA cleavage. Furthermore, under conditions where DNA replication can occur, the initial rates of LexA cleavage were the same in *uvr*$^+$ and *uvr* strains, implying that the action of the Uvr(A)BC endonuclease is not important for the generation of the SOS-inducing signal under ordinary physiological conditions. However, *uvr* mutants remained SOS induced for a much longer time than wild-type strains did, indicating the important role of nucleotide excision repair in the removal of these lesions. In a recent study, a persistent SOS-inducing signal observed in UV-irradiated *uvr*$^+$ *E. coli* cells was attributed to the excision of one member of a pair of closely spaced photoproducts to yield a gap opposite a lesion (24).

The suggestion that DNA replication is required for SOS induction may at first glance seem paradoxical since many temperature-sensitive *dna* mutants, in which the elongation phase of DNA replication is inhibited at the restrictive temperature, exhibit SOS induction when shifted to higher temperature (286, 355). It has been suggested (280) that the disintegration of the immobilized and hence unstable fork at the restrictive temperature gives rise to gaps and breaks where replicating DNA has been exposed to nuclease action. Similarly, it has been suggested (223) that *priA* mutants are induced for the SOS response because they lack the PriA protein, a component of the primosome, thereby resulting in a less-stable or less-efficient replication fork that is defective in its management of the single-stranded lagging template strand. The expression of the SOS response in certain *uvrD* mutants may be due to single-stranded regions generated at the rep-

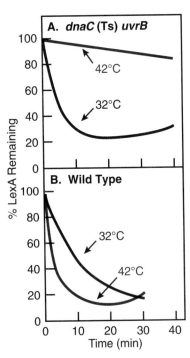

Figure 10–17 Inhibition of initiation of DNA replication blocks UV-induced LexA cleavage. *dnaC*(Ts) *uvrB* (A) or wild-type (B) cells were grown at 32°C. When the cell density of the cultures reached 10^8 cells per ml, the cells were diluted twofold with medium prewarmed to 55°C to establish a temperature of 42°C instantly. At 70 min after the temperature shift, the cells were UV irradiated with a dose of 5 J/m^2 and the rate of LexA decrease at 42°C (red line) was compared with that obtained with control cells kept at 32°C (black line). (*Adapted from Sassanfar and Roberts [280] with permission.*)

lication fork because of defects in the UvrD helicase protein (226). This view is supported by experiments involving a *dnaE486*(Ts) mutation which affects the catalytic α subunit of DNA polymerase III holoenzyme (280). When this mutant is shifted to 42°C, DNA replication stops abruptly and the amount of LexA in the cell rapidly declines until stabilizing at the 50% level, a level that was shown to represent a balance between LexA cleavage and resynthesis. However, once DNA replication had been arrested, UV irradiation of the cells was no longer capable of inducing further LexA cleavage. This experiment reinforces the conclusion that the SOS signal is generated when a cell attempts to replicate damaged DNA.

As discussed more fully below, DNA replication on damaged templates leads to the production of segments of single-stranded DNA which are generated when DNA synthesis reinitiates downstream of the lesion (268). It is the binding of RecA protein to these single-stranded regions that is postulated to result in its activation and hence the induction of the SOS response (270, 280). As also discussed more fully below, the binding of RecA protein to these single-stranded regions is additionally the event that initiates *daughter-strand gap repair*, a process that allows the cell to eliminate these gaps from its DNA by homologous recombination. At a UV radiation dose of 5 J/m², which introduces approximately 250 pyrimidine dimers per genome equivalent, a replication fork moving at a rate of 1,000 nucleotides per s at 30°C would be expected to encounter a pyrimidine dimer every 2 to 3 s (280). The lag of approximately 1 min after UV irradiation before LexA cleavage begins (Fig. 10–18) suggests that it takes some time for single-stranded DNA to be generated and for RecA protein to assemble on it (168). Neither photoreactivation nor nucleotide excision repair would be significant on this timescale. Even with lower doses of UV radiation exposure, the time required for the replication fork to travel between lesions may be negligible compared with the time required for reinitiation of the fork downstream from a lesion. According to this model, the fork would continuously encounter lesions, leaving gaps that are filled with activated RecA protein that mediates LexA cleavage. LexA cleavage would cease when the gaps are filled by daughter-strand gap repair (see below) and the lesions are removed by nucleotide excision repair.

If RecA protein is normally activated by binding to single-stranded gaps produced when DNA replication occurs on damaged templates, why is the protein not activated by single-stranded regions that are normally present on the lagging strand of the replication forks in undamaged cells? The reason may be a kinetic one; perhaps RecA protein cannot polymerize on this DNA or displace single-stranded-DNA-binding protein (SSB) from it before the DNA is covered again by replication (280). This view suggests that the ability of *recA*(Cpt^c) and *recA*(Cpt^{Ts}) mutants to express the SOS responses in the absence of exogenous DNA damage is due to the altered RecA protein, which is encoded by these mutants, becoming activated by polymerizing on these lagging-strand gaps during their relatively short half-life. The higher effective affinity of RecA441(Cpt^c) protein than wild-type RecA protein for DNA (239) is consistent with this view. The fact that the RecA730 protein displaces SSB from single-stranded DNA more efficiently than does RecA441, which in turn displaces SSB from single-stranded DNA more efficiently than does RecA$^+$ (156), also supports this model. An alternative suggestion is that some RecA mutants may exhibit constitutive coprotease activity because they can be activated by an expanded range of nucleotide and polynucleotide effectors such as tRNA and rRNA (337, 338). In view of the above results, it does not seem likely that the suggestion that RecA protein becomes activated directly by binding to damaged DNA (195) is physiologically significant. Although a relatively small extent of RecA activation can be detected by using double-stranded DNA carrying UV radiation-induced lesions (195), it seems likely that this is insufficient to promote LexA cleavage in vivo. The demonstration that the phenomenon of indirect induction requires replication of damaged DNA further supports this conclusion (310).

In the special case of φ80 induction, certain oligonucleotides such as d(GG) stimulate repressor cleavage. This finding was first made in a study in which the

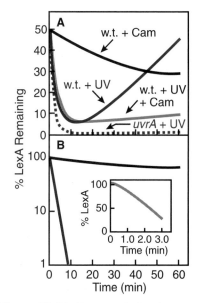

Figure 10–18 Decrease in LexA concentration following irradiation with UV. Exponentially growing wild-type (w.t.) and *uvrA* cells were UV irradiated (time zero) at 37°C for 20 s at 5 J/m² in the absence or presence of 60 µg of chloramphenicol (Cam) per ml added 90 s before irradiation with UV. As a control, chloramphenicol was added to unirradiated cells. At the indicated times, cells were analyzed for their LexA content by immunoblotting. (A) Quantification of LexA by densitometry. Red solid line, wild-type cells plus UV; black line, wild-type cells plus chloramphenicol; red dotted line, *uvrA* cells plus UV; grey line, wild-type cells plus UV plus chloramphenicol (B) Semilogarithmic plot of the disappearance of LexA in chloramphenicol-treated wild-type cells (same data as in panel A). Black line, wild-type cells plus chloramphenicol; red line, wild-type cells plus UV plus chloramphenicol. The inset shows LexA disappearance over the first 3 min after irradiation of wild-type cells with UV. (*Adapted from Sassanfar and Roberts [280] with permission.*)

derepression of a φ80 repressor-controlled reporter gene was measured in permeabilized cells exposed to potential inducing signals (118). For several years, this led to the speculation that oligonucleotides were the inducing signal for SOS induction, but no evidence was obtained to support this model. The demonstration that these oligonucleotides stimulate φ80 repressor cleavage in vitro but do not stimulate LexA or λ repressor cleavage has resolved this apparent inconsistency (69). The oligonucleotides interact with the φ80 repressor itself rather than with RecA protein and therefore exert a specific effect on φ80 repressor cleavage but not on LexA cleavage and hence not on SOS induction.

Physiological Considerations of the SOS Regulatory Circuit

Although the basic logic of the SOS circuitry is fairly simple—RecA protein becomes activated and mediates the cleavage of LexA repressor protein, thereby leading to the increased expression of LexA repressed genes—the detailed studies of SOS regulation carried out thus far indicate that there is considerably more subtlety to the regulation of the SOS response than simply the coordinated induction of a set of genes.

The SOS system can exist in two extreme states; *fully repressed* and *fully induced*. However, it can also exist in other states which are intermediate between these two extremes. As discussed above, cleavage of LexA protein occurs very rapidly after an SOS-inducing treatment, and the lowest concentration of LexA protein that results represents an equilibrium between LexA cleavage mediated by activated RecA protein and synthesis of intact LexA molecules (280). The various SOS-regulated genes differ with respect to the affinities with which their operators bind LexA protein, the number of operators, and other specific details of the regulation (Table 10–6; Fig. 10–16). Hence, the extent to which a given gene is expressed following a particular SOS-inducing treatment is dependent on the degree to which the LexA protein pool has been decreased in response to a given SOS-inducing treatment. As mentioned above, the fact that LexA dimerization occurs on the DNA leads to a high degree of cooperativity and therefore a steep binding curve so that less LexA must be cleaved to give derepression of LexA-repressed genes (137). Furthermore, cooperative binding of LexA protein to multiple SOS boxes, as in the case of *recN*⁺ (284) (see Fig. 10–16), leads to full induction within a narrower range of change of LexA protein concentration. The genes that bind LexA protein most weakly are fully turned on in response to even weak SOS-inducing treatments, whereas the genes that bind LexA protein most tightly are fully turned on only in response to stronger SOS-inducing treatments. As shown in Table 10–6, the SOS boxes in the *lexA*⁺, *uvrA*⁺, *uvrB*⁺, and *uvrD*⁺ genes bind LexA protein more weakly than does the operator in the *recA*⁺ gene, whereas the SOS boxes in *sulA*⁺ and *umuD*⁺*C*⁺ bind LexA protein more tightly. Thus, relatively weak SOS-inducing treatments lead to increased expression of *uvr*⁺ gene-dependent excision repair functions. However, other SOS responses, such as the accumulation of large amounts of RecA protein, *sulA*⁺-dependent filamentation, and the induction of *umuD*⁺*C*⁺-dependent translesion synthesis (see chapter 11), do not occur unless the cell receives a stronger SOS-inducing treatment. This feature of the regulatory system allows *E. coli* cells to utilize certain SOS-regulated functions, such as nucleotide excision repair, to recover from DNA damage without committing themselves to a full-fledged SOS response.

The repression of the *lexA*⁺ gene by LexA protein has three effects on SOS induction. First, it extends the range, in terms of inducing signal, over which the system can establish an intermediate state of induction and thus express only a subset of the SOS responses. Second, since the affinity of LexA protein for its operator is weak compared to its affinity for the operators of other genes, such as the *recA*⁺ gene, the system is buffered against significant levels of induction by a very limited inducing signal. Third, it speeds the return to the repressed state once the inducing signal begins to decrease (20–22, 168, 176).

The ability of the SOS system to be induced to an intermediate state may account, at least in part, for a variety of observations in the literature in which particular SOS-inducing treatments only induced subsets of the SOS responses, phenomena often referred to as *split phenotypes* (355). A particularly well-understood example is the split phenotype associated with the *lexA41*(Tsl) mutant. As mentioned above, this mutant was originally isolated as a UV radiation-resistant revertant of a *lexA3*(Ind⁻) mutant that was temperature sensitive for growth. At the permissive temperature (30°C), the *lexA41*(Tsl) mutant was nearly as resistant to UV radiation-induced killing as a wild-type strain but its UV mutability was greatly reduced (216, 218). At the restrictive temperature (42°C), the *lexA41*(Tsl) mutant constitutively expresses filamentous growth and the synthesis of RecA protein (30, 90, 218). The *lexA41*(Tsl) mutant gene was subsequently shown to have retained the original *lexA3* mutation (the Gly → Asp change at position 85, which prevents RecA-mediated cleavage) and to have a second mutation (Ala → Thr at position 131) (236). The LexA41 protein is susceptible to degradation by the *lon*-encoded protease, so that at 30°C the concentration of LexA protein is low enough that the *uvrA*⁺, *uvrB*⁺, and *uvrD*⁺ genes are expressed at an elevated level whereas the *umuD*⁺*C*⁺ genes are partially induced and the *sulA*⁺ gene is not induced (Table 10–7). The increased expression of these nucleotide excision repair genes accounts for the increased UV radiation resistance of the *lexA41*(Tsl) mutant, which apparently has an increased ability to remove pyrimidine dimers from its DNA (77), while the fact that the *umuD*⁺*C*⁺ genes are only partially induced accounts for the reduced UV mutability. After a shift to 42°C, the LexA41 protein is more unstable, mainly because of greater turnover by the *lon*-encoded protease (which is induced by a heat shock), but also less active, because of its intrinsic thermolability. This results in the additional induction of the tightly repressed *sulA*⁺ gene. The SulA protein then inhibits cell division, causing the lethal filamentation observed at 42°C.

The rate at which various proteins undergo RecA-mediated cleavage is also critical to the fine tuning of the SOS response. For example, λ repressor is cleaved much more slowly than LexA protein when it interacts with activated RecA protein. This may favor the survival of the phage by ensuring that lysogenic induction does not take place unless levels of DNA damage exceed the capacity of the other inducible responses. Furthermore, as discussed in chapter 11, the fact that UmuD protein is cleaved more slowly than LexA helps ensure that the active form of UmuD protein, which is required for UV radiation-induced mutagenesis, is not produced unless the cell has experienced a significant SOS-inducing treatment. The SOS system can also be induced to an intermediate level by the *indirect induction* procedures described above, and these provide examples of bacteriophages not being induced at intermediate states of SOS induction. For example, the introduction of a UV-irradiated F factor into *E. coli* leads to the induction of both Weigle reactivation and λ prophage, whereas the introduction of UV-irradiated

Table 10–7 β-Galactosidase activity in *lexA*⁺, *lexA41*, and *lexA71*(Def) derivatives of *E. coli* K-12 having *lacZ* transcriptional fusions to various SOS genes[a]

Fusion	Allele	β-Galactosidase activity (Miller units) at:	
		30°C	42°C
uvrA	*lexA*⁺	65	21
	lexA41(Tsl)	160	93
	lexA71(Def)	220	130
umuC	*lexA*⁺	6	14
	lexA41(Tsl)	26	32
	lexA71(Def)	170	240
sulA	*lexA*⁺	53	44
	lexA41(Tsl)	43	250
	lexA7(Def)	5,700	6,300

[a] Overnight cultures were diluted in fresh broth and were grown at 30°C to 5 × 10⁷ cells per ml, at which time they were split. One part was shifted to 42°C, and another part remained at 30°C. Incubation was continued for 4 h, at which time β-galactosidase was assayed.

Source: Adapted from Peterson and Mount (236) with permission.

Hfr DNA induces Weigle reactivation but not λ prophage (17, 18, 53, 81). Table 10–1 lists some genes that appear to be under SOS control but are not repressed by LexA. It is possible that a repressor for those genes that can be cleaved in a RecA-mediated fashion will be identified.

Finally, it is possible that various SOS-regulated genes are also subject to additional control by other regulatory systems. For example, the cea^+ gene, which encodes colicin E1, is regulated by both the SOS system and the cyclic AMP (cAMP)/cAMP receptor protein catabolite-repression system (62). This dual control system results in a delay during SOS induction that may achieve the physiological purpose of limiting the natural production of colicin E1 (which is lethal to the cells that produce it) to cells that had suffered a catastrophe and most probably would not survive anyway (274).

SOS Responses in Other Bacteria

Many prokaryotic organisms have been shown to induce physiological responses or new proteins in response to agents such as UV irradiation, which induce the SOS response in *E. coli*. It increasingly appears that most bacteria possess an SOS regulatory system that is significantly related to the *E. coli* system with respect to at least some of the induced physiological responses, the overall logic of its regulation, and, in some cases, even the specificity of key regulatory interactions. The list of such bacteria includes many gram-negative bacteria such as *Salmonella typhimurium* (225, 230, 288, 300), *Proteus mirabilis* (100, 101), *Haemophilus influenzae* (222), *Caulobacter crescentus* (7), and *Pseudomonas aeruginosa* (106, 211) and the gram-positive bacteria *Bacillus subtilis* (37, 188, 194) and *Bacillus thuringiensis* (3).

Recently, a broad-host-range plasmid carrying a fusion of the *E. coli recA*+ gene to *lacZ* was used to investigate the expression of the *E. coli recA*+ gene in these heterologous hosts after DNA damage (50). The plasmid was introduced into 30 different species belonging to 20 different genera of gram-negative bacteria (Table 10–8). With only a couple of exceptions, all of these bacteria were able to repress the expression of the *E. coli recA*+ gene to some extent in the absence of DNA

Table 10–8 Expression of a *recA-lacZ* fusion of *E. coli* in several gram-negative bacteria in the presence and absence of DNA damage

Species	Relative basal level[a]	Induction factor[b]
E. coli AB1157	1	6.8
E. coli AB1157 (*recA*)	0.25	1
Salmonella typhimurium LT2	1.15	5.8
Proteus rettgeri	1.25	7.17
Enterobacter aerogenes	0.59	8.71
Citrobacter freundii	1.13	5.73
Shigella flexneri	0.56	3.68
Klebsiella pneumoniae	1.16	8.36
Erwinia herbicola	1.1	5
Yersinia ruckeri	0.28	3.1
Alcaligenes faecalis	1.25	5.38
Acinetobacter lwofii	1.1	4.76
Azotobacter vinelandii	1.02	2.74
Agrobacterium tumefaciens	0.591	3.67
Rhizobium meliloti	0.56	1.95
Pseudomonas aeruginosa	1.86	3.2
Rhodobacter capsulatus	0.44	1

[a] The relative basal level is the ratio between the specific units of β-galactosidase of noninduced cultures of each strain and the specific units of β-galactosidase of noninduced cultures of *E. coli* AB1157. The basal level of *E. coli* AB1157(pUΔ89) was 85 specific units of β-galactosidase.

[b] The induction factor is the ratio between the specific units of β-galactosidase of DNA-damaged cells and the specific units of β-galactosidase of nontreated cells for each strain. Samples were taken at 120 min after mitomycin addition, which was used at a concentration producing 10% survival at this time for each of the species used.

Source: Adapted from de Henestrosa (50) with permission.

damage and to induce it after DNA damage. These observations suggest that these bacteria possess an SOS regulatory system and furthermore that the LexA-binding site is significantly conserved in these various systems. This interpretation was strengthened by the subsequent cloning of *lexA+* homologs from *S. typhimurium*, *Erwinia carotovora*, *P. aeruginosa*, and *P. putida* on the basis of their ability to suppress the lethal filamentation of an *E. coli lexA*(Def) *sulA+* strain (26). The LexA protein of *S. typhimurium* appeared to have the highest affinity for *E. coli* SOS boxes in vivo, while the LexA protein of *P. aeruginosa* had the lowest and the *Pseudomonas* LexA proteins were inferred to be less efficiently cleaved by activated *E. coli* RecA protein. Taken together, these results suggest that all of these systems have a LexA protein analog which is cleaved by activated RecA protein but that the more evolutionarily distant the organism from *E. coli*, the less likely it is that the LexA homolog will be able to bind to *E. coli* SOS boxes or to undergo cleavage upon interaction with activated *E. coli* RecA protein.

In the well-studied case of *B. subtilis* (361), the consensus SOS box has been postulated to be GAAC-N_4-GTTC (37), a sequence considerably diverged from the *E. coli* consensus SOS box (Fig. 10–11). Nevertheless, the purified *B. subtilis* RecA protein mediated cleavage of *E. coli* LexA protein in vitro (191) and the *E. coli recA+* gene was shown to suppress the deficiencies of a *B. subtilis recA* mutant in SOS induction (190). A 23-kDa protein that is functionally analogous to the *E. coli* LexA protein has been purified from *B. subtilis* (192). It binds to the consensus SOS box in the *B. subtilis dinA*, *dinB*, *dinC*, and *recA* genes, and its specific DNA-binding activity is abolished if the cells are treated with mitomycin. Furthermore, if the purified protein is incubated with *B. subtilis* RecA in the presence of single-stranded DNA and a nucleoside triphosphate, its binding activity is inactivated (192). It seems possible that this putative LexA analog is the product of the *B. subtilis dinR* gene (254), which has been shown to encode a 22.8-kDa protein that is 34% identical and 47% similar to *E. coli* LexA. Although it is possible that the regulation of the *B. subtilis* SOS system substantially resembles the regulation of the *E. coli* SOS system, alternative models for SOS regulation in *B. subtilis* have also been proposed (254, 361).

Homologs of the *E. coli recA+* gene have been cloned from more than 50 bacteria and in many cases have been shown to complement *E. coli recA* mutants for their deficiency in LexA cleavage (212, 259). These results imply that the RecA proteins encoded by these various bacteria are sufficiently related to the *E. coli* RecA protein to be able to interact with *E. coli* LexA protein and promote some degree of LexA cleavage. However, many of these RecA analogs are not sufficiently conserved with the *E. coli* RecA protein to be able to mediate the cleavage of λ repressor (212).

The specific SOS responses induced vary among different bacteria. For example, as discussed more fully in chapter 11, many bacteria do not exhibit an inducible capacity for UV radiation-induced mutagenesis but will do so if provided with analogs of the *umuD+C+* genes. In *E. coli* filamentation is an SOS response, whereas in *B. subtilis* it is not (189). Among the functions regulated by the SOS response in *Serratia marcescens* is the secretion of several extracellular proteins including an extracellular nuclease (5). SOS responses may be subject to additional controls. For example, in *B. subtilis* a class of SOS phenomena is also developmentally regulated and is spontaneously induced in the absence of externally generated DNA damage when the bacteria differentiate into the physiological state of natural competence (188, 193, 361).

RecA Protein: a Protein with Mechanistic Roles in Homologous Recombination and DNA Repair

In addition to its role in the SOS regulatory response, RecA protein plays roles in homologous recombination events and in a set of DNA repair and damage tolerance processes which help cells survive the effects of DNA damage. These processes include the repair of daughter-strand gaps generated when cells attempt to replicate a damaged DNA template (see below), the repair of double-strand breaks (see below), and translesion replicative bypass (see chapter 11). The molecular mechanisms of these processes are not yet understood at the same level of detail as nucleotide excision repair. However, it seems clear that the functions of RecA protein that are required for its roles in homologous recombination are important for its roles in these processes as well. In fact, it has been argued that the primary biological role of recombination is to restore the integrity of damaged genomes rather than to promote the formation of recombinant genotypes through the exchange of genetic markers (10, 28, 44, 259). Before discussing the possible mechanisms of RecA-dependent DNA repair and damage tolerance processes, we will summarize key features of RecA protein that are relevant to its roles in these processes. A number of excellent reviews on this topic provide references for most of the facts described in this section (45, 149, 249, 259, 311, 312, 343).

Recombination Reactions of RecA Protein

RecA is able to pair two homologous DNA molecules if one of them is single stranded or partially single stranded (Fig. 10–19) (149, 249, 259, 343). The pairing of single-stranded or partially single-stranded DNA with fully duplex DNA that has suitable ends leads to a further reaction: an exchange of strands that can proceed for several kilobases. When a partially single-stranded molecule reacts with a fully duplex molecule that has suitable ends, strand exchange is reciprocal and produces the classical recombination intermediate known as a *Holliday junction*, named after Robin Holliday, who first postulated its existence (102).

FORMATION OF A NUCLEOPROTEIN FILAMENT

The active species in RecA-mediated DNA strand exchange is a nucleoprotein filament composed of DNA and RecA protein that can be thought of as a scaffold to facilitate DNA pairing and strand exchange with a naked DNA molecule. RecA protein forms filaments on both single- and double-stranded DNA. The structure of the filaments has been determined by electron-microscopic studies combined with three-dimensional image reconstruction (66, 67, 312), and further details have been provided by X-ray crystallographic analyses of RecA protein (314, 315). The DNA in a nucleoprotein filament formed between RecA and double-stranded DNA in the presence of ATP is extended and underwound (Fig. 10–20). The filament is a right-handed helical structure with a diameter of 115 Å (11.5 nm) and a pitch of 95 Å (9.5 nm). There are 6.2 RecA monomers per helical turn, which corresponds to 3 bp per RecA monomer, and the helix has a deep groove. The DNA within the filament is stretched from 10.5 bp per turn (as it would be in B-form DNA) to 18.6 bp per turn. The nucleoprotein filament formed between RecA

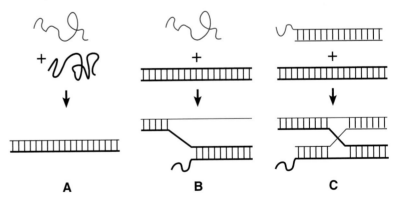

Figure 10–19 Examples of recombination reactions promoted by RecA protein. (A), Renaturation of complementary single strands; (B) asymmetric (nonreciprocal) strand exchange following the pairing of single-stranded and double-stranded DNA; (C) symmetric (reciprocal) strand exchange following the pairing of duplex DNA and partially single-stranded DNA. (*Adapted from Radding [249] with permission.*)

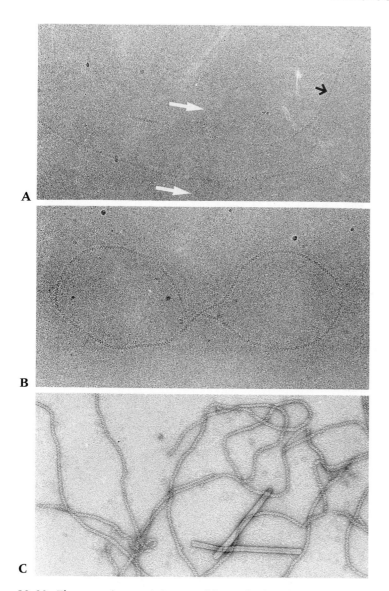

Figure 10–20 Electron-microscopic images of frozen-hydrated (A and B) and negatively stained (C) RecA-DNA-ATPγS filaments. The RecA filaments in panels A and C are on double-stranded DNA, while the RecA in panel B is on a circular single-stranded DNA molecule. The white arrows in panel A show the general direction of fluid flow that occurred on the grid during blotting, prior to rapid freezing, as judged by the preferred orientation of filaments in this direction on the grid. The pitch of the filament sections between the white arrows are 114 and 108 Å (11.4 and 10.8 nm), while the pitch of the filament section that runs perpendicular to the flow direction (black arrow) is 96 Å (9.6 nm). All panels are at the same magnification, and the tobacco mosaic virus particles in panel C are about 200 Å (20 nm) in diameter. (*Reproduced from Yu and Egelman [363] with permission.*)

protein and single-stranded DNA in the presence of ATP is somewhat less regular but otherwise seems very similar to the nucleoprotein filament formed between RecA protein and double-stranded DNA. The single-stranded DNA lies along the central axis of the filament and is bound by its sugar-phosphate backbone with its bases exposed in a wide helical groove. In the presence of ADP or the absence of ATP, RecA protein forms a compressed filament which appears not to be simply interconvertible with the extended active form (66, 67, 157, 158, 363).

Filament formation on naked single-stranded DNA is rapid, initiating any-where and extending in the $5' \rightarrow 3'$ direction at a rate that may approach 1,000 monomers per min. The presence of *E. coli* SSB can facilitate filament formation by removing secondary structure, but, as discussed below, it can also inhibit filament formation by inhibiting the initial nucleation event (325, 327). In contrast, RecA protein binds to duplex DNA very slowly. The rate-limiting step is nucleation, but, once it has been initiated, subsequent extension of the RecA filament on double-stranded DNA is rapid and the resulting nucleoprotein filament is quite stable. A major factor in the low rate of nucleation is the requirement for underwinding of the DNA. Nucleation is stimulated by any perturbation or sequence that facilitates underwinding. A single-strand gap provides an optimal nucleation site, and RecA filaments extend rapidly from the gap to incorporate adjacent duplex DNA.

The extension of the filament occurs in the 5′ → 3′ direction relative to the single-strand DNA in the gap. A physiologically relevant consequence of this may be that RecA binding is limited largely to regions of DNA containing suitable nucleation sites, especially single-strand gaps, and that this may target the protein to regions where repair is likely to be needed (44). A variety of DNA lesions also facilitate nucleation of RecA binding to duplex DNA, and this is the molecular basis for the observed preferential binding of RecA protein to DNA molecules containing such lesions (145, 262).

The crystal structure of the RecA protein in the absence of DNA (314, 315) (Fig. 10–21) showed that each monomer of RecA consists of a large central domain, forming nearly two-thirds of the protein, flanked by distinct N- and C-terminal domains. The ATP-binding site is located in the large central domain. Amino acids associated with the ATPase active site are highly conserved not only in RecA homologs encoded by various other bacteria but also in more distantly

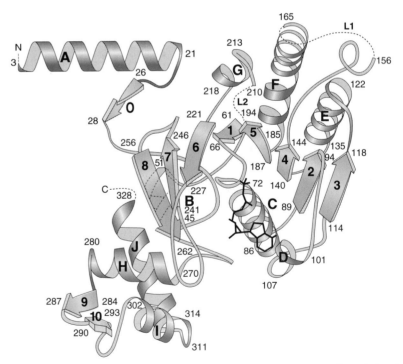

Figure 10–21 Structure of the RecA monomer. Schematic representation of the three-dimensional structure of RecA protein. β-Strands, shown as red arrows, are numbered 0 to 10 and α-helices are labeled A to J, according to their position in the primary sequence starting at the amino terminus. The sequence numbers of the approximate first and last residues of the secondary-structure elements are indicated. The position of a bound ADP that has been diffused into the crystal is shown in boldface. Two disordered loops (L1 and L2), proposed to be involved in DNA binding, and disordered chain termini are indicated by dashed lines. (*Adapted from Story et al. [315] with permission.*)

related RecA homologs such as the UvsX protein of bacteriophage T4 and the Dmc1, Rad51, and Rad57 proteins isolated from *Saccharomyces cerevisiae* and other eukaryotes (313) (see chapter 12). Residues known from genetic studies to affect RecA-mediated cleavage of LexA, λ repressor, and UmuD protein (see chapter 11) were found to be localized to a "notch" formed on the outer surface of the RecA polymer between adjacent monomers. This observation led to the suggestion that RecA-mediated cleavage of these repressors or of UmuD protein involves their binding to this intersubunit region (315). However, it appears that the crystal form of RecA protein corresponds more closely to the inactive than to the active RecA filament (67, 363). Three-dimensional reconstruction of electron micrographs has suggested instead that LexA protein is bound within the deep groove of the RecA filament (364). Also, within crystals of RecA, protein filaments are closely packed in parallel arrays that are stabilized by specific contacts between amino acids near the N terminus of a RecA monomer in one filament and amino acids near the C terminus of a RecA monomer in a neighboring filament. It is interesting that mutations leading to constitutively activated variants of RecA (i.e., Cpt^c) alter amino acids close to these contact points. This observation has led to the speculation that the effect of these mutations might be to disfavor the formation of bundles of RecA polymers in solution and thereby favor nucleoprotein filament formation (315).

HOMOLOGOUS PAIRING AND STRAND EXCHANGE

The nucleoprotein filament formed between RecA protein and single-stranded DNA is able to interact with naked duplex DNA via a second binding site on the filament and then to search for homology (45, 111, 149, 249, 259, 311, 312, 343). Homologous alignment produces a joint molecule in which the two partners are noncovalently joined by the hydrogen bonds of paired bases. The finding that RecA protein can pair a circular single-stranded DNA and a homologous circular duplex DNA revealed that no ends are required to form joints. Such molecules are topologically constrained, and the resulting joints, termed *paranemic joints*, are thought to represent the first paired intermediate.

Once the two molecules have become aligned, if a free end is available the reaction proceeds into its next phase, the formation of heteroduplex DNA. If a gapped DNA is paired with a duplex, as occurs during the repair of daughter-strand gaps (see below), the strand exchange is reciprocal and the heteroduplex molecules are thought to contain a Holliday junction (347). The strand exchange initiates within the single-stranded gap, which must be 40 to 100 bp long for maximal strand exchange efficiency and has a polarity of 5′ → 3′ relative to the single strand on which filament formation occurred. Thus, the single-stranded DNA in the gap serves a dual purpose: it serves as a nucleation site where RecA protein initiates binding, and it also serves to initiate strand exchange. Recent studies of RecA-mediated strand exchange between single-stranded DNA and linear duplexes have suggested that the two interacting DNA molecules become interwound within the RecA filament in the form of a triple-stranded DNA helix (111, 112, 209, 253) (Fig. 10–22). These triple helices were found to be surprisingly stable following deproteinization, suggesting the presence of hydrogen bonds between the single-stranded DNA and the duplex DNA. However, it has not yet been unambiguously established whether the search for homology involves a triple helix (27, 311).

Once the initial heteroduplex intermediate has been established, branch migration can occur. Although initial measurements of spontaneous branch migration with DNA alone indicated that it was fairly rapid, more-recent measurements have indicated that spontaneous branch migration is quite slow at physiological temperatures, occurs only over short distances, and is severely inhibited by nonhomologous sequences (127, 219). It can even be impeded by a single base mismatch (231). In contrast, strand exchange promoted by the RecA protein occurs at approximately 2 to 10 bases per s and can occur over several kilobases. During a typical strand exchange reaction, about 100 ATP molecules are hydrolyzed for every base pair of heteroduplex DNA generated, with ATP hydrolysis occurring throughout the filament. However, ATP hydrolysis is not required since a substantial three-strand exchange can be observed in the presence of ATPγS which is

RecA-DNA nucleoprotein filament

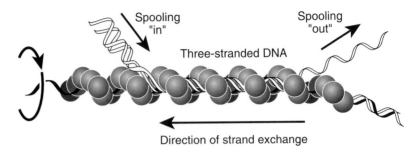

Figure 10–22 Model for homologous pairing within a RecA nucleoprotein filament. Courtesy of S. C. West.

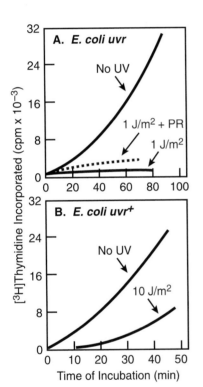

Figure 10–23 When *E. coli* cells are UV irradiated, semiconservative DNA synthesis is inhibited, as evidenced by the reduced rate of incorporation of [³H]thymidine (both panels). *uvr* cells, which are defective in excision repair of bulky base damage such as pyrimidine dimers (top panel), are much more sensitive to inhibition of DNA synthesis by UV radiation than are cells that are excision proficient (bottom panel). (Note the different doses of UV radiation used in the two panels.) If cells are exposed to enzymatic photoreactivation (PR) after irradiation, some recovery of the rate of thymidine incorporation is observed (top panel). (*Adapted from Setlow et al. [291] and Friedberg [74] with permission.*)

not appreciably hydrolyzed by RecA protein (209, 265). In contrast, the binding of ATP, which has been associated with a conformational change in the polypeptide, is required for RecA-mediated strand exchange. It is not too surprising that branch migration can occur in the absence of ATP hydrolysis, since the process of branch migration is isoenergetic and involves no net loss or gain of hydrogen bonds. As discussed in chapter 12, proteins believed to be involved in homologous pairing have been isolated from eukaryotic cells. These proteins are able to promote strand exchange in the absence of ATP hydrolysis.

The role of ATP hydrolysis by RecA protein has not been unambiguously established (43). One possibility is that it is required for dissociation of RecA from one or both DNA substrates, but it has been argued that this model provides an incomplete rationalization for ATP hydrolysis. Another possibility is to allow branch migration to occur on damaged DNA (45, 259). For example, RecA-mediated strand exchange can readily bypass lesions, mismatches, and even heterologous insertions ranging up to a few hundred base pairs in one or both DNA substrates (259). However, the bypass of a heterologous insert as small as 6 bp has been shown to have an absolute requirement for ATP hydrolysis (139, 266). Of particular interest is the observation that branch migration past a cyclobutane pyrimidine dimer is 50 times faster in the presence of RecA protein than in its absence (180). Yet a third possible role for ATP hydrolysis relevant to possible roles for RecA in DNA repair and damage tolerance is that ATP hydrolysis is required to render the strand exchange unidirectional (265).

DNA Replication and Damage Tolerance in Cells Exposed to DNA-Damaging Agents

Transient Inhibition of DNA Synthesis after DNA Damage

When *E. coli* cells are exposed to a DNA-damaging agent such as UV light, DNA synthesis is *transiently inhibited* (179). Figure 10–23 shows the incorporation of radiolabeled thymidine into the DNA of *E. coli* cells as a function of UV radiation dose. Note that a strain defective in the excision repair of pyrimidine dimers is much more sensitive to inhibition of DNA synthesis than is a wild-type strain and that photoreactivation of pyrimidine dimers partially reverses this effect (291).

Much of this inhibition is due to effects of the lesions on the progress of the replication fork. Analysis of the DNA synthesized in cells after UV irradiation showed that the DNA was present as short pieces, whose length corresponded approximately to the average interdimer distance (124, 268). The capacity of lesions such as cyclobutane pyrimidine dimers and abasic sites to inhibit the progress of DNA polymerases, including DNA polymerase III, is well established. As indicated in chapter 1, a particularly sensitive technique for measuring the effect of base damage on DNA synthesis in vitro involves the use of primed single-stranded

DNA templates that have been damaged in a variety of ways, e.g., following exposure to UV radiation or to various chemicals such N-acetoxy-N-2-acetyl-2-aminofluorene or the 7,8-hydroxy-9,10-epoxy derivative of benzo[a]- pyrene. DNA synthesis can be carried out in vitro with the DNA polymerase of interest. The products of the polymerization reactions can then be sequenced, and the sites of termination of DNA synthesis can be compared with the known nucleotide sequence of the template. With both N-acetoxy-N-2-acetyl-2-aminofluorene-treated and UV-irradiated DNA, synthesis arrests one nucleotide before the site of base damage. More recently, it has been possible to use a primed DNA template containing a single defined lesion, such as an abasic site (252), to demonstrate that particular lesions act as blocks to the progress of DNA polymerase III holoenzyme. In addition, studies of model in vitro DNA replication systems (147) have shown an inhibitory effect of UV radiation-induced lesions on elongation by DNA polymerase III holoenzyme (177, 178, 293).

In addition to inhibiting the progress of replication forks, damage to the *E. coli* genome may lead to an inhibition of new rounds of replication from the chromosomal origin of replication *oriC*. This inhibition is not due to the presence of DNA damage in *oriC* itself. This was demonstrated by showing that UV irradiation of *E. coli* inhibited *oriC*-dependent DNA replication of an undamaged λ phage carrying *oriC* (329). This observation was consistent with results of earlier work showing that DNA replication of undamaged coliphage 186, but not that of the related phages λ and P2, was transiently inhibited when unirradiated phages infected UV-irradiated *E. coli* cells (104). Unlike λ and P2, replication of coliphage 186 requires the DnaA and DnaC proteins used for *oriC* replication in *E. coli* (103). The nature of this inhibition has not been examined in detail, but it appears to require a higher dose of UV radiation than is required to induce the SOS responses and, on the basis of a limited characterization, appears to be independent of $recA^+$ (104). Taken together with results discussed above, these data indicate that DNA damage can affect two events in the *E. coli* cell cycle: *initiation of chromosome replication* and *cell division*. Effects of DNA damage on the eukaryotic cell cycle are discussed in chapter 13.

Recovery of DNA Replication after DNA Damage

REPLICATION RESTART

Following transient arrest after DNA damage, DNA synthesis recovers to its predamage rate in approximately 30–45 min (Fig. 10–24). This phenomenon has been termed *replication restart* (64, 65) or *inducible replisome reactivation* (136). It requires amplification of RecA protein plus the synthesis of at least one other protein that is not controlled by LexA (136, 360). In wild-type cells, replication restart does not require the $uvrA^+$, $recB^+$, or $umuC^+$ gene (136). The fact that it does not depend on $umuC^+$ function suggests that the major molecular mechanism responsible for replication restart is not translesion synthesis (see chapter 11). However, it has been observed that replication restart is dependent on $umuD^+C^+$ function in *recA718* (360) or *recA727* (318) mutants (Fig. 10–24). Strains carrying the *recA718* allele (Table 10–3) are recombination proficient, moderately UV radiation sensitive, and hypermutable at lower UV doses. In addition, RecA718 protein requires DNA damage to become activated for SOS induction and expression when present at basal levels. However, amplified levels of RecA718 protein are constitutively activated without DNA damage (206, 359). The $umuD^+C^+$ gene dependence of replication restart in these strains has been interpreted as being due to interaction of the UmuDC proteins with the RecA718 protein so that the the complex can perform the function in replication restart that is normally performed by wild-type RecA protein (357, 360).

The mechanistic basis of replication restart has not been clearly established, and it is possible that more than one molecular mechanism contributes to the recovery of DNA synthesis after DNA damage. Nevertheless, it appears that most of the recovery is due to effects on existing replication forks rather than to the initiation of new rounds of DNA replication. As discussed below, an additional

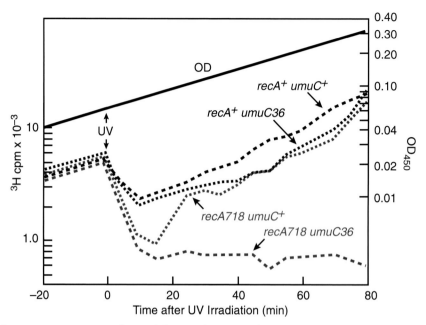

Figure 10–24 Recovery from inhibition of DNA synthesis after UV irradiation: synergism in double mutants combining *recA718* and *umuC36*. Black dashed line, *recA⁺ umuC⁺*; black dotted line, *recA⁺ umuC36*; red dotted line, *recA718 umuC⁺*; red dashed line, *recA718 umuC36*. The UV dose was 3 J/m². Cells growing exponentially at 37°C were pulse-labeled with [³H]thymidine for 2 min at various times before and after UV irradiation. The optical density (OD) curve defines the range of values obtained for all strains. (*Adapted from Witkin et al.* [360] *with permission.*)

consequence of exposure to DNA damage is that the mechanism of initiation of DNA replication is altered, a phenomenon referred to as inducible stable DNA replication (iSDR). The fact that the genetic and physiological requirements for iSDR differ from those for replication restart suggests that the initiation of new rounds of synthesis is not a major contributor to the resumption of DNA replication after DNA damage (2, 201, 202). This conclusion is also consistent with the fact that the transient arrest of DNA replication after a cell is exposed to a DNA-damaging agent occurs so quickly (Fig. 10–24) that it must be due mostly to inhibition of existing replication forks rather than to inhibition of new rounds of initiation.

There are at least two formal classes of alternatives for how cells could resume DNA synthesis on templates containing blocking lesions and, in so doing, enhance their potential for survival. One is by reinitiating DNA synthesis some distance downstream from the blocks (essentially circumventing them), thereby creating *gaps* or *discontinuities* in the daughter DNA strands, which subsequently become filled in by some mechanism. The second is by continuing DNA synthesis past the template lesion in a continuous mode.

MODEL FOR DAMAGE TOLERANCE INVOLVING THE REPAIR OF DAUGHTER-STRAND GAPS

A model for the former type of cellular response to DNA damage was first suggested by Paul Howard-Flanders and his colleagues (268, 269) and was subsequently modified to take into account the ability of RecA protein to catalyze pairing and strand exchange reactions between homologous DNA molecules (111, 344, 346). This model is frequently referred to as *postreplicational repair*. The model does not postulate biochemical events which lead to the removal of the blocking lesion from the DNA and therefore does not constitute "repair" as we have used the term up to this point in the book. The term *daughter-strand gap repair* (94) is sometimes used to make it clear that it is the gaps in the daughter strand, rather than the lesion itself, that are being repaired by this mechanism. Thus, this is a mechanism for *tolerance* of DNA damage rather than for *repair* of DNA damage. According to this model (Fig. 10–25), DNA synthesis resumes downstream of the lesion via a mechanism that remains uncertain but leaves a gap opposite the lesion in the template strand. RecA protein polymerizes at the gap to form a nucleoprotein filament which promotes homologous pairing and strand exchange with the undamaged sister molecule. Strand transfer past the lesion closes the gap. Further extension into duplex regions may result in reciprocal exchanges and the formation of classic Holliday junctions. Endonuclease cleavage across the point of strand exchange resolves the connection between the sister molecules, leaving a DNA polymerase(s) to fill in the gap formed in the parental strand.

MODEL FOR DAMAGE TOLERANCE INVOLVING A STRAND SWITCH MECHANISM

Two models have been proposed for how DNA synthesis could continue past the lesion in a continuous fashion. One is termed *translesion synthesis* and is a process in which the block to replication posed by the lesion is overcome by a polymerase inserting one or more nucleotides directly opposite the lesion and then extending the chain. As discussed in detail in chapter 11, this process is mutagenic and, for most lesions, requires the functions of the *umuD+* and *umuC+* genes. The fact that replication restart is not dependent on *umuC+* function in wild-type cells (136) suggests that translesion synthesis is not a major component of replication restart.

An alternative model for how DNA synthesis could continue past a lesion in a continuous fashion was proposed by Echols and Goodman (65) (Fig. 10–26).

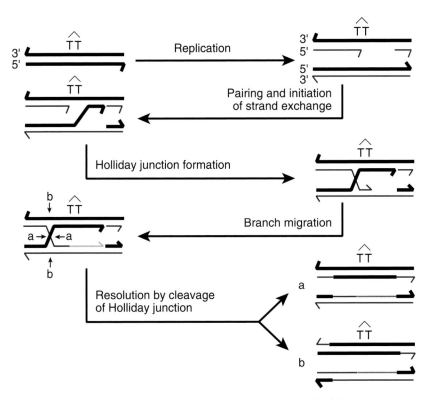

Figure 10–25 Model for daughter-strand gap repair. The thick black lines represent irradiated parental DNA containing a cyclobutane dimer. The thin red lines represent daughter DNA synthesized after irradiation. The Holliday junction can be resolved by either of two ways. The cleavages marked "a" result in an exchange event involving the parental strand without a dimer (269); the dimer stays in the parental strand. The cleavages marked "b" lead to an exchange event involving the parental strand with the dimer; the dimer is exchanged into the daughter strand (75). (*Adapted from Walker [331], Ganesan [75], and Tsaneva [326] with permission.*)

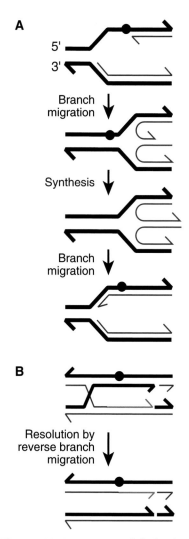

Figure 10–26 Two models for how a strand switch mechanism could permit accurate synthesis past a replication-blocking lesion. Such strand switch mechanisms might play a role in the phenomenon of replication restart. (A) In a model originally proposed by Higgins et al. (97) to account for repair replication in mammalian cells, strand displacement and branch migration create an alternate template allowing replication to bypass a lesion. This mechanism does not involve the formation of daughter-strand gaps. (B) This model shows how an intermediate in daughter-strand gap repair (see Fig. 10–25) could be resolved by reverse branch migration instead of by cleavage of the Holliday junction.

The same mechanism has been suggested as an explanation for events occurring in eukaryotic cells after DNA damage (97). This model suggests that the RecA-coated single-stranded DNA downstream from the blocked polymerase associates with the replicated duplex from the complementary strand just as in the formation of heteroduplex DNA in general recombination and in daughter-strand gap repair. In the resultant three-stranded structure, DNA polymerase III then bypasses the lesion by copying the newly synthesized daughter strand, which has the same polarity as the damaged parental strand, and then switching back after clearing the lesion. In this model, the recombination complex formed by RecA protein simply provides the means to switch strands. At present there is little direct evidence in support of this model. However, it could account for the observation that after infection with phage λ DNA carrying several pyrimidine dimers, efficient semiconservative replication was observed without evidence of multiple recombination exchanges (52). It could also account for the fact that W reactivation is more efficient with double-stranded phages than with single-stranded phages.

INDUCIBLE STABLE DNA REPLICATION

As mentioned above, exposure of cells to DNA-damaging agents that induce the SOS response leads to an altered form of initiation of DNA replication termed iSDR (1, 2, 201–203). Initiation of iSDR is independent of transcription, translation, and DnaA protein, which are essential for initiation of normal DNA replication from the *oriC* chromosomal origin of replication. Initiation of iSDR occurs primarily in the *oriC* and *ter* regions of the chromosome and leads to semiconservative replication of the entire *E. coli* genome for many hours. iSDR is considerably more resistant to UV radiation than is normal DNA replication and appears to be error prone. Except for DnaA protein, iSDR requires several gene products that are also necessary for normal DNA replication. They include DnaB, DnaC, DnaE, DnaG, and DnaT proteins. However, iSDR differs from normal replication in one important respect: it requires RecA function. Derepression of the *recA*⁺ gene and activation of amplified RecA protein are a necessary and sufficient condition for the induction of iSDR. No other LexA-repressed genes are necessary for iSDR. The recombination functions of RecA protein appear to be important since a *recA*(Cpt^c Rec⁻) mutant is defective in iSDR. Analyses of various mutations affecting the RecBCD enzyme suggest that its helicase activity is necessary for iSDR but that its exonuclease activity or chi-specific endonuclease activity is not necessary. Mutations that block the resolution of Holliday junctions or that block branch migration (*ruvC*, *recG*, or *ruvAB*) stimulate iSDR.

A model has been proposed for iSDR in which recombination intermediates (i.e., D-loop structures) created by the action of RecA recombinase and RecBC(D) helicase play a central role in the initiation of iSDR (1, 2). According to this model, D-loop formation provides the duplex opening necessary for iSDR initiation. This model for the initiation of DNA replication has similarities to the recombination-dependent initiation of replication that is observed during the late stage of bacteriophage T4 infection. In this system, a recombination event between homologous duplex and single-stranded DNA catalyzed by the UvsX protein, which is an analog of the RecA protein (see below), generates a primer-template structure that can serve as a substrate for replication by the T4 DNA polymerase holoenzyme.

MODEL FOR THE REPAIR OF DAUGHTER-STRAND GAPS

The experiments that led to the model shown in Fig. 10–25 provided the first indication that the survival of UV-irradiated cells can be affected by cellular responses other than excision repair of lesions such as pyrimidine dimers. The search for an alternative response to DNA damage was prompted by the observation that cells defective both in nucleotide excision repair and in *recA* function were killed by a dose of UV radiation that produced an average of 1 cyclobutane pyrimidine dimer per *E. coli* genome equivalent (108) (Fig. 10–27). One plausible interpretation of these results was that *recA*⁺ cells use a DNA damage tolerance mechanism which facilitates survival without excision repair of the dimers.

This putative excision-independent cellular response to DNA damage was investigated by examination of DNA synthesis in UV-irradiated *E. coli*. When excision repair-defective (*uvr*) cells were pulse-labeled with [³H]-thymidine shortly after exposure to UV radiation, the newly synthesized daughter strands were, on average, significantly smaller than those of unirradiated cells (109, 268) (Fig. 10–28). However, as the cells were incubated for increasing lengths of time after UV irradiation with nonlabeled thymidine, the ³H-labeled daughter strands became longer, ultimately reaching the size of those synthesized in unirradiated cells (109, 268). A number of explanations for this observation were considered. A reduced rate of DNA synthesis in UV-irradiated cells was ruled out, because when the radiolabeling periods for irradiated and unirradiated cells were adjusted so that equal amounts of isotope incorporation occurred in both cell cultures, the same result was obtained. A second possible explanation was that the shorter DNA strands reflected the onset of new rounds of DNA synthesis from the normal origin of replication. However, when UV-irradiated cells were incubated in unlabeled medium for a period long enough to ensure completion of any newly synthesized rounds of replication and then pulse-labeled, shorter DNA strands were still observed. It was therefore concluded that the DNA synthesized on damaged templates contained *discontinuities*.

This conclusion led to the specific suggestion that, after UV irradiation, DNA replication continues at a normal rate until a blocking lesion such as a pyrimidine dimer is reached, at which point replication is delayed before resuming at some point beyond the lesion, thereby generating a discontinuity or gap in the daughter strand (268) (Fig. 10–25). The increase in sedimentation rate of the daughter strands during subsequent incubation indicates that the defects (or gaps) in these strands ultimately disappear. A likely mechanism for the reconstitution of high-molecular-weight daughter-strand DNA was suggested by the fact that the process of replication generates two *identical sister molecules*, thus providing for the possibility of gap filling in daughter strands by recombinational events. As indicated in Fig. 10–25, a daughter-strand gap could be filled with a homologous region of undamaged DNA derived from a recombinational exchange from the isopolar parental strand present on the sister DNA duplex. The resulting discontinuity in the

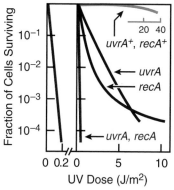

Figure 10–27 *E. coli* cells that are proficient in both excision repair (*uvrA*⁺) and discontinuous DNA synthesis with gap filling (*recA*⁺) are resistant to killing by UV radiation. Cells that are defective in both of these processes (*uvrA recA*) are extremely UV sensitive. The survival curve of the *uvrA recA* double mutant is reproduced on the left with an expanded UV dose scale. From this curve it can be estimated that one pyrimidine dimer per *E. coli* genome equivalent is lethal to a *uvrA recA* strain. (*Adapted from Howard-Flanders and Boyce [108] and Friedberg [74] with permission.*)

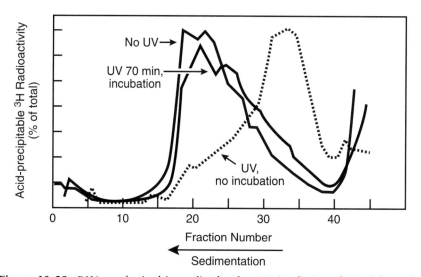

Figure 10–28 DNA synthesized immediately after UV irradiation of *E. coli* has a lower molecular weight than normal. *E. coli uvr* cells were exposed to UV radiation and pulse-labeled briefly with [³H]thymidine. Cells were analyzed by sedimentation in alkaline sucrose gradients either immediately after being labeled (no incubation; dotted red line) or following incubation for 70 min (red line). Immediately following the pulse-label, the newly synthesized DNA is of low molecular weight and sediments near the top of the gradient. However, with time the newly synthesized DNA approaches the size of the unirradiated control (no UV; black line). (*Adapted from Rupp and Howard-Flanders [268] with permission.*)

parental strand could then be eliminated by repair synthesis with the undamaged region of the complementary daughter strand as a template. The final products of these events would be two intact sister duplex DNA molecules, one of which still contained pyrimidine dimers in its parental strands.

Experiments with *E. coli* have yielded results that are consistent with the general model shown in Fig. 10–25. The predictions of the model that have been most extensively examined are the presence of gaps in newly replicated DNA, the specific location of these gaps with respect to pyrimidine dimers in template strands, and the reconstitution of high-molecular-weight DNA by recombinational and/or repair synthesis events (93, 94, 305, 306, 331).

EVIDENCE FOR GAPS IN NEWLY SYNTHESIZED DNA

The size of newly synthesized DNA fragments in UV-irradiated cells approximates the average interdimer distance in the template strands (287, 362), as expected if replication arrested at a dimer and then continued, leaving discontinuities. Similarly, if the number of "arresting" lesions (i.e., cyclobutane dimers) in the template DNA was reduced by prior nucleotide excision repair (i.e., when the cells are *uvr*⁺) or by enzymatic photoreactivation prior to replication, the formation of correspondingly longer daughter-strand fragments was observed during a brief labeling pulse. In addition, under such conditions, shorter post-UV-irradiation incubation times were required to convert newly synthesized low-molecular-weight DNA into high-molecular-weight products (304).

Further evidence in support of discontinuities or gaps in the DNA synthesized in UV-irradiated bacteria and an indication of their position relative to that of pyrimidine dimers in the template strands came from conjugal crosses between F*lac*⁺ donors and *lac* recipients (109). Such conjugations between unirradiated cells are normally accompanied by the transfer of one of the DNA strands of the F plasmid to the recipient cell and its replication in the recipient during transfer. The same process is presumed to occur when the F*lac*⁺ plasmid is UV irradiated. If discontinuities were generated in the daughter strands synthesized during the transfer of the plasmid, they could be opposite pyrimidine dimers or displaced between them. An attempt was made to distinguish between these two possible configurations by examination of the susceptibility of the transferred DNA to excision repair and to enzymatic photoreactivation. The former process depends on the presence of an intact complementary strand and presumably could not take place if dimers were opposite gaps. However, dimers opposite gaps should still be sensitive to enzymatic photoreactivation.

To minimize recombination events between the transferred plasmid and the recipient chromosome (which would have complicated the interpretation of the experiments), conjugations were carried out with *recA*(Def) recipient cells. Under these experimental conditions, the yield of Lac⁺ colonies decreased rapidly with increasing UV radiation dose (Fig. 10–29A). There was no significant difference in the yield of Lac⁺ colonies when recipient strains that were excision repair proficient or deficient were compared; i.e., excision repair was apparently not operative on the UV-irradiated F*lac*⁺ plasmid (Fig. 10–29A). The damaged plasmids were, however, sensitive to photoreactivation in the recipient by exposure to visible light (Fig. 10–29B). The lack of detectable excision repair of the transferred plasmid together with the apparent repair of dimers by photoreactivation is consistent with the presence of gaps in the strand synthesized during DNA transfer and the location of these gaps opposite pyrimidine dimers (109).

Estimates of average gap sizes of about 1,000 nucleotides have been obtained by the isolation of duplex molecules containing single-strand regions (presumed to be postreplicative gaps) by chromatography on benzoylated-naphthoylated DEAE-cellulose, a chromatographic matrix that discriminates between single- and double-stranded DNA (124). In addition, structures suggestive of postreplicative gaps in DNA have been visualized by electron microscopy and have yielded estimates of gap sizes ranging from 1,500 to 40,000 nucleotides (126). The mechanism by which DNA synthesis is primed when it resumes downstream of the blocking lesion is not known. However, the sizes of the gaps produced are consistent with

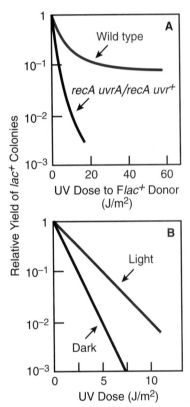

Figure 10–29 (A) The relative yield of *lac*⁺ colonies after an excision-defective F*lac*⁺ strain was mated to mutant *lac* recipients. When the recipient is defective in the *recA* gene, the yield of *lac*⁺ colonies is significantly reduced relative to that observed in a wild-type recipient. However, there is no detectable difference in the yield of *lac*⁺ colonies in recipients that are either excision repair proficient (*recA uvr*⁺) or excision repair deficient (*recA uvrA*). (B) F*lac*⁺ DNA transferred from a UV-irradiated donor to an unirradiated recipient shows enhanced survival if, following transfer, the cells are exposed to photoreactivating light. This result is consistent with (but certainly does not prove) the presence of replicative gaps opposite pyrimidine dimers. (*Adapted from Howard-Flanders et al. [109] with permission.*)

the notion that DNA synthesis initiates at the site for the next Okazaki fragment. The longer gaps might arise from gaps formed on the leading strand, where initiation sites are thought to be less frequent (306).

There is some limited evidence that levels of RNase H activity (which degrades RNA-DNA hybrids) may influence this priming of DNA synthesis downstream of blocking lesions. In the *E. coli* B/r strains used, overproduction of RNase H from a plasmid strongly sensitized *uvrA* cells to UV killing, moderately sensitized wild-type and *uvrA recF* cells, and did not sensitize *recA*(Def) cells (14). These observations suggested that the sensitization resulted from the inhibition of a RecA-dependent recombinational process. Biochemical studies showed that post-UV irradiation synthesis was sensitized and that the smaller amounts of DNA synthesized after UV irradiation, while of normal reduced size as indicated by sedimentation position in an alkaline sucrose density gradient, did not shift to larger size upon additional incubation. These results suggested that the excess RNase H activity was interfering with the initiation of synthesis downstream of lesions that created daughter-strand gaps, thereby sensitizing the cells to killing by UV radiation. This might occur by degradation of the RNA primers needed for this synthesis and might be of particular importance for priming on the leading strand (14). Furthermore, experiments measuring replication start (as in Fig. 10–24) showed that, in a double mutant lacking RNase H and having an amplified level of RecA protein, recovery of DNA synthesis occurred after a delay without the need for additional SOS-inducible functions. This observation would be consistent with a reduction in RNase H activity being one of the SOS responses. In fact, *rnh-lacZ* fusions have been used to show that the expression of the *rnh*+ gene, which encodes RNase H, is modestly inhibited upon SOS induction (31, 247), and it is formally possible that an inhibitor of RNase H is induced as well. However, direct measurements of RNase H activity in SOS-induced cells failed to detect a reduction in RNase H activity. Such measurements are complicated by the RNase H activities of other proteins, such as DNA polymerase I (31) and a recently discovered second RNase H activity encoded by *rnhB*+ (119).

EVIDENCE FOR RECOMBINATIONAL EVENTS

Gap filling in nascent DNA of UV-irradiated *E. coli* appears to be associated with recombinational events. Thus, cells carrying a mutation in the *recA*+ gene (and as a consequence severely deficient in generalized recombination) fail to convert short nascent DNA strands into high-molecular-weight products (304).

Direct evidence for exchanges between DNA strands during postreplicative gap filling comes from the use of appropriate density labels (269). *E. coli* cells were grown for several generations in medium containing the heavy isotopes [^{13}C]-thymine and [^{15}N]thymine as well as [^{14}C]thymine. The cells were then exposed to various doses of UV light and grown for less than one generation in a medium without density markers (light medium) containing [^{3}H]thymidine, which created hybrid-density DNA with the ^{3}H radiolabel uniquely in the light strands (Fig. 10–30). If exchanges were to occur between sister DNA duplexes, light (^{3}H-labeled) material should become covalently associated with heavy (^{14}C-labeled) DNA (Fig. 10–30). Heavy DNA from UV-irradiated cells replicated in the presence of [^{3}H]thymidine does indeed contain strands of intermediate density, as demonstrated by isopycnic sedimentation of denatured extracted DNA in CsCl (269). This DNA was resolved into light and heavy components after being sheared to a molecular weight of less than 5×10^5, suggesting that the exchanges involved segments of at least this size.

The hybrid molecules could not have arisen from degradation of DNA and reincorporation into progeny strands, since this should not result in *discrete segments* of heavy DNA becoming joined to ^{3}H-labeled light DNA (269). An estimate of the number of exchanges in these experiments showed that for every 1.7 pyrimidine dimers replicated in the ^{3}H-labeled light medium, one ^{3}H-labeled molecule containing one or more heavy segments was recovered; i.e., there was one genetic exchange for each dimer (269). This value correlates well with the bromouracil photolysis technique discussed in chapter 5 for measuring repair synthesis patches.

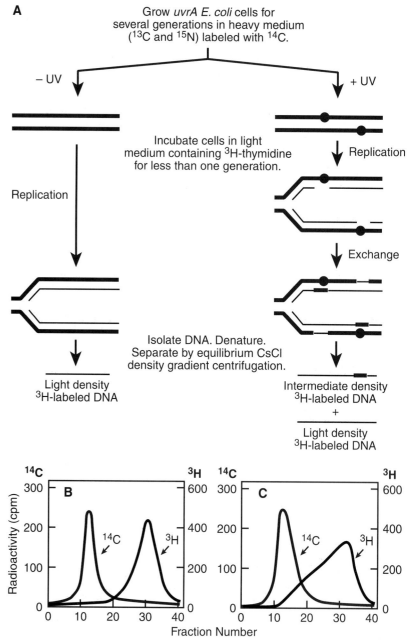

Figure 10–30 (A) Design and interpretation of an experiment (269) to detect exchanges between DNA strands during postreplicative gap filling. *E. coli uvrA* cells were grown for several generations in medium containing the heavy isotopes [^{13}C]thymine and [^{15}N]thymine as well as [^{14}C]thymine. The cells were then exposed to various doses of UV light and were grown for less than one generation in a medium without density markers (light medium) containing [^{3}H]thymidine. The newly replicated DNA is of hybrid density and has the ^{14}C label uniquely in the heavy strand and the ^{3}H label uniquely in the light strand. (B) If the cells are not irradiated and their DNA strands are separated by heat denaturation and equilibrium CsCl density gradient centrifugation, the strands separate cleanly as ^{14}C-labeled heavy strands and ^{3}H-labeled light strands. (C) If the cells are irradiated, strands of intermediate density are observed after heat denaturation and equilibrium CsCl density gradient centrifugation. This observation has been interpreted as indicating that exchanges have occurred between sister duplexes in the UV-irradiated cells so that light (^{3}H-labeled) DNA becomes covalently attached to heavy (^{14}C-labeled) DNA. This DNA of intermediate density could be resolved into heavy and light components after shearing to a molecular weight of less than 5×10^{5}, suggesting that the exchanges involved segments of at least this size.

The existence of 5-bromouracil in parental DNA has been observed after replication of UV-irradiated excision repair-defective cells in the presence of this thymine analog. Quantitative photolysis of the DNA by irradiation at 313 nm indicated that these regions were about 1.5×10^4 nucleotides long and that one such region was present for every 1.2 pyrimidine dimers in the parental DNA. These regions were presumed to have arisen either by repair synthesis across gaps created by recombinative transfer of parental DNA to fill postreplicative daughter-strand gaps or by the recombinational events themselves (163, 164) (Fig. 10–25).

Further evidence for DNA strand exchanges during the semiconservative replication of damaged DNA comes from analysis of the distribution of cyclobutane pyrimidine dimers in the progeny of UV-irradiated cells. Probing for the presence of dimers in DNA with T4 PD-DNA glycosylase/AP endonuclease (see chapter 4) demonstrated that these lesions were equally distributed between parental and progeny strands (75, 76). This observation supports the idea that strand exchanges occurred and indicates that dimers may be included in the exchanged regions. Furthermore, it reinforces the point that, by itself, the repair of daughter-strand gaps is a mechanism for *tolerating* DNA damage. In a nucleotide excision repair-defective strain, dimer-free DNA molecules are not generated. Rather, a slow dilution appears to occur as they become distributed among succeeding generations of daughter molecules.

POSSIBLE ROLES FOR VARIOUS PROTEINS IN THE REPAIR OF DAUGHTER-STRAND GAPS

A number of genes have been implicated in the tolerance of DNA damage associated with discontinuous synthesis and gap filling of daughter-strand DNA (94, 305, 306). However, the precise roles of most of these have not been clearly established. Furthermore, as discussed above, there are two formal classes of alternatives for how cells might resume synthesis on templates containing blocking lesions (discontinuous replication followed by daughter-strand gap repair and continuous DNA synthesis past template lesions). There is some uncertainty regarding whether particular functions are specifically required for one class of process or the other. Some of this uncertainty stems from the limitations of the techniques used to study these events. For example, replicative responses to DNA damage are frequently investigated by measuring changes in the molecular weight of newly synthesized DNA. Sedimentation of pulse-labeled DNA in alkaline sucrose density gradients is a particularly convenient procedure for this purpose (Fig. 10–28). However, demonstration of the conversion of small newly synthesized DNA fragments into larger ones does not necessarily imply that the conversion is associated with the formation of gaps that are subsequently filled by recombinational events. Thus, stalling of a replication complex at a lesion followed by resumption of fork progression, by either a strand switch mechanism (see above) or translesion synthesis (see chapter 11), cannot be distinguished from the repair of daughter-strand gaps by sedimentation analysis alone (Fig. 10–31). Furthermore, intermediate reactions and the formation of discrete reaction products such as gaps, gap filling by the exchange of DNA sequences between isologous strands of sister DNA molecules, and repair synthesis of DNA have not been specifically demonstrated in many studies (94).

These limitations notwithstanding, a large number of *E. coli* genes have been shown to be important for various aspects of homologous recombination, and there has been remarkable progress in identifying biochemical functions of the products of many of these genes. Table 10–9 summarizes the properties of a number of these (342). Knowledge of their roles in homologous recombination processes has made it possible to suggest roles for specific proteins in the repair of daughter-strand gaps. Figure 10–32 presents a model for daughter-strand gap repair which assigns particular proteins to particular steps on the basis of present knowledge of their biochemical properties.

HOMOLOGOUS PAIRING AND STRAND EXCHANGE (SYNAPSIS)

The properties of RecA protein described previously indicate that it would polymerize on the single-stranded DNA in the daughter-strand gap to form a right-

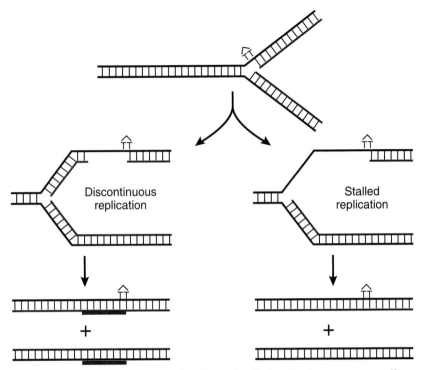

Figure 10–31 Both discontinuous replication and stalled replication generate small nascent DNA molecules that subsequently are converted into high-molecular-weight ones. Thus, it can be very difficult, if not impossible, to differentiate these processes by using the sedimentation velocity of radiolabeled newly synthesized DNA. The thicker lines in the lower left diagram indicate regions of gap filling by recombinational exchange and by non-semi-conservative DNA synthesis (see Fig. 10–25). (*Adapted from Friedberg [74] with permission.*)

handed helical filament. It is the generation of such a RecA filament after a cell has attempted to replicate its damaged DNA that is thought to activate RecA for inducing the SOS responses (280). This RecA filament could then promote a search of the sister duplex for homologous contacts via the formation of a three-stranded intermediate. Base pairing within the heteroduplex would then serve to properly align the homologous sequences and initiate strand transfer. RecA protein would drive strand exchange $5' \rightarrow 3'$ relative to the single-stranded gap to form a Holliday junction. As indicated in Fig. 10–32, evidence obtained both in vivo and in vitro suggests that an endonuclease introduces a nick into the sister duplex DNA opposite the gap. However, to date, no such putative activity has been purified, nor has any gene encoding the activity been identified (32, 264, 344, 345).

Recent results have indicated that the RecO, RecR, and possibly RecF proteins serve as RecA accessory proteins that help RecA overcome inhibition by SSB and utilize SSB–single-stranded DNA complexes as substrates (327). SSB is directly involved in recombination and stimulates strand exchange reactions promoted by RecA protein (45, 248). However, if SSB is bound to single-stranded DNA prior to the addition of RecA protein, it can inhibit the assembly of the RecA/single-stranded DNA nucleoprotein filament. Apparently it does so by inhibiting the initial binding of RecA protein to single-stranded DNA and hence the formation of a nucleation site (325) for filament formation (327). Thus, these proteins might also influence early stages of daughter-strand gap repair.

This hypothesis is supported by the observation that the *recA803* allele (which is able to partially suppress the deficiencies of a *recF* mutant in the repair of daughter-strand gaps) encodes a RecA derivative (336) that competes more effectively with SSB for binding to single-stranded DNA than does the wild type (199). This RecA derivative may be better able to participate in daughter-strand gap repair without the intervention of RecF protein. A role for RecF protein as a RecA ac-

Table 10–9 Classification of *E. coli* recombination genes and proteins

Gene	Gene product	Biochemical activities
Initiators (presynapsis)		
recB⁺		
recC⁺	RecBCD protein	ATPase, dsDNA exonuclease, DNA helicase, chi-specific endonuclease
recD⁺		
recE⁺	RecE protein	5′ → 3′ double-stranded DNA exonuclease (analogous to λ exonuclease)
recJ⁺	RecJ protein	5′ → 3′ single-stranded DNA exonuclease
recQ⁺	RecQ protein	DNA helicase
Homologous pairing and strand exchange (synapsis)		
recA⁺	RecA protein	Forms helical filaments, ATPase, catalyst of homologous pairing and strand exchange
recF⁺	RecF protein	single-stranded DNA binding protein
recO⁺	RecO protein	Stimulates binding of RecA to single-stranded DNA (with RecR)
recR⁺	RecR protein	Stimulates binding of RecA to single-stranded DNA (with RecO)
ssb⁺	SSB protein	SSB
Branch migration/resolution (postsynapsis)		
ruvA⁺	RuvA protein	Binds specifically to Holliday junctions; targets RuvB to DNA
ruvB⁺	RuvB protein	ATPase, promotes branch migration (with RuvA)
ruvC⁺	RuvC protein	Endonuclease, specifically resolves Holliday junctions
recG⁺	RecG protein	ATPase, binds Holliday junctions, promotes branch migration
rus⁺	?	Suppressor of *ruv*; may encode alternative resolvase

Source: Adapted from West (342) with permission.

cessory protein would also account for the impaired SOS induction observed in a *recF* mutant (324).

BRANCH MIGRATION AND RESOLUTION (POSTSYNAPSIS)

Once a Holliday junction has been established, branch migration can occur. As discussed above, RecA protein is able to promote branch migration even if the DNA contains lesions or substantial regions of nonhomology (44, 259). This latter property made it an attractive candidate for promoting branch migration on damaged DNA. However, recent work has shown that *E. coli* has two other means of promoting branch migration, one involving the *ruvA*⁺ and *ruvB*⁺ gene products and the other involving the *recG*⁺ gene product.

The *ruvA*⁺*B*⁺ operon is repressed by LexA protein and is regulated as part of the SOS response (9, 295, 296). RuvB protein is a DNA-dependent ATPase (122) and, when present at saturating amounts under high Mg^{2+} concentration, is able to promote branch migration without RuvA protein, indicating that it is the catalyst for branch migration (220). Visualization by electron microscopy has shown that RuvB binds circular duplex to form ring-like structures that associate in pairs (312a). RuvB binds to RuvA protein, which interacts with RuvB in solution (294), and the complex binds specifically to Holliday junctions (232) to form a RuvAB-Holliday junction complex (233). It appears that RuvA's primary function is to target the RuvB enzyme to the site of the junction, where it promotes ATP-dependent branch migration. The branch migration promoted by RuvA and RuvB proteins is significantly faster and much more energy efficient (326) than that promoted by RecA protein (259). Furthermore, RuvAB-mediated branch migra-

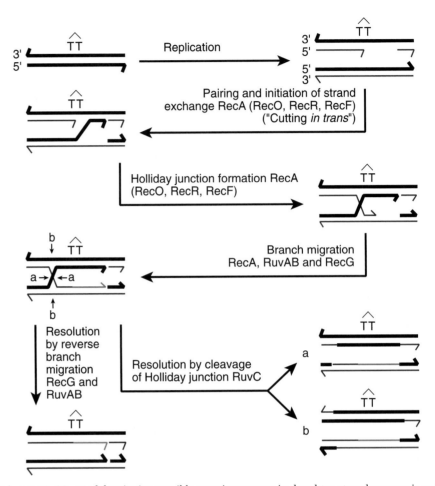

Figure 10–32 Model assigning possible proteins to steps in daughter-strand gap repair and to a strand-switching mechanism (see the text).

tion can bypass UV-induced lesions present in DNA at levels that inhibit RecA-mediated strand exchange (326). The fact that *ruv* mutants are sensitive to UV light, mitomycin, and ionizing radiation (182, 227), taken together with their known biochemical roles in homologous recombination, including the ability to promote branch migration on damaged DNA, makes RuvA and RuvB proteins attractive candidates for carrying out a portion of the branch migration that occurs during the repair of daughter-strand gaps. The fact that the *ruvAB* operon is induced as part of the SOS response is consistent with this point of view.

The product of a third gene at the *ruvC*+ locus, RuvC protein, is a 19-kDa protein which can resolve Holliday junctions into recombinant molecules by specific endonucleolytic cleavage across the point of strand exchange (8, 40, 59, 123). Although the *ruvC*+ gene is very close to the *ruvA*+*B*+ operon, it is in a separate operon that is not SOS regulated (292, 320). The phenotype of *ruvC* mutants is similar to that of *ruvA* and *ruvB* mutants, and it seems likely that RuvC resolvase activity is important for the repair of daughter-strand gaps. The cleavage of Holliday junctions by RuvC protein is symmetrical and, as shown in Fig. 10–32, produces recombinants of either the patch or splice type. The activity of purified RuvC is quite low, and a set of genetic arguments suggests that it may act in conjunction with the RuvA and RuvB proteins (204, 342).

RecG protein has also been shown to be a DNA-dependent ATPase that can bind specifically to a Holliday junction and promote branch migration (186). Furthermore, under certain experimental conditions it was shown to counter RecA-driven strand exchange by catalyzing branch migration in the Holliday junction in the reverse direction (349). The observation that both *recG* and *ruvAB* mutants have slight deficiencies in recombination but *recG ruvAB* double mutants were severely deficient in recombination suggested that the RecG and RuvAB proteins had overlapping functions but were not fully interchangeable (185). In a wild-type cell, RuvAB proteins appear to be considerably more important with respect to the processing of damaged DNA than is RecG protein. This inference is based on the fact that *ruvAB* mutants are far more sensitive to UV light than *recG* mutants are, although there is synergistic sensitization to killing by UV light when the mutations are combined (183, 185). These observations suggest that although RecG protein appears to play a role in the repair of daughter-strand gaps in wild-type cells, its role is not as prominent as that played by the RuvAB proteins. One possible role for the RuvA and RuvB proteins or for RecG protein in the repair of daughter-strand gaps would be to provide an alternative way to resolve the Holliday junction intermediate, without the intervention of a resolvase, by simply driving strand exchange in the reverse direction after the lesion has been removed from the heteroduplex by nucleotide excision repair (349). Another conceptually similar role for RuvA and RuvB proteins or for the RecG protein could conceivably be to resolve the intermediates in the postulated strand switch mechanism of replication restart (Fig. 10–26) by driving strand exchange in the reverse direction (349).

Although the functional overlap observed between RuvAB and RecG proteins suggests two alternative pathways for branch migration, this interpretation is complicated by the fact that *ruvC recG* double mutants are as recombination defective as *ruvAB recG* mutants, even though there is no evidence to suggest that RecG protein can cleave Holliday junctions. A consideration of these observations not only led to the suggestion mentioned above that RuvC interacts with RuvA and RuvB proteins but also raised the possibility that *E. coli* has a second Holliday junction resolvase besides RuvC protein (204, 342). A suppressor of *ruv* mutations that requires *recG*[+] function for its action has recently been isolated. It seems likely that this suppressor mutation leads to the increased expression of an activity (possibly another resolvase), which acts in conjunction with RecG protein to process Holliday junction intermediates (204).

Repair of Double-Strand Breaks

In addition to repairing daughter-strand gaps, bacteria have the capacity to repair double-strand breaks in their DNA. Double-strand breaks can be introduced directly by certain agents such as ionizing radiation (see chapter 1). They can also arise through the processing of DNA damaged by agents (such as UV radiation) which do not directly introduce double-strand breaks. In excision-proficient cells, one mechanism by which such double-strand breaks could arise in UV-irradiated cells would be by convergent excision of two pyrimidine dimers that occur in close proximity but on opposite strands of the DNA duplex (16). A second possible mechanism, which can operate even in excision-deficient cells, is that breaks occur in the parental DNA strand opposite unrepaired daughter strand gaps, possibly through the attack of an endonuclease specific for single-stranded DNA (335).

Pretreatment of *E. coli* with either X rays or UV radiation was shown to result in an induced resistance to killing by X rays or γ rays. This induced resistance was under SOS control and required new protein synthesis (242–244, 303). The discovery that the capacity to repair double-strand breaks caused by γ rays was an inducible SOS function (151) suggested that increased resistance resulted from an enhanced capacity to carry out double-strand break repair. It also indicated that the inducible inhibition of DNA degradation after X-irradiation (241) is probably caused by the inducible repair of double-strand breaks (151).

The inducible repair of double-strand breaks introduced by ionizing radiation or mitomycin requires the presence of another DNA duplex that has the same base sequence as that of the broken double helix (150). *E. coli* cells grown in medium that supports rapid cell growth initiate new rounds of replication prior to completing the first round and, as a consequence, have multiple replication forks and hence multiple copies of most of their genome. Such cells can carry out double-strand-break repair. In contrast, *E. coli* cells grown on a very poor medium, for example one in which aspartate serves as the carbon source, do not have multiple replication forks. They cannot carry out double-strand break repair, although they can efficiently repair single-strand breaks. These observations correlate well with the observation that *E. coli* cells that carry multiple genomes exhibit increased resistance to killing by X rays (243).

The basic elements of the mechanism by which double-strand break repair is thought to occur were first proposed by Michael Resnick, to account for the repair of radiation-induced double-chain breaks (255). Various amplifications and modifications of this model have been proposed by others (144, 302, 335; see reference 321 for a historical review of the development of the model), and essentially the same mechanism was postulated by Jack Szostak and his colleagues (319) to explain meiotic recombination in *S. cerevisiae*. The current model for double-strand break repair is summarized in Fig. 10–33. According to this model, the initial double-strand break is processed via exonuclease action to generate exposed 3' ends. One possibility is then for each 3' end to invade the homologous intact duplex and for DNA repair synthesis to initiate, primed by the invading 3' ends,

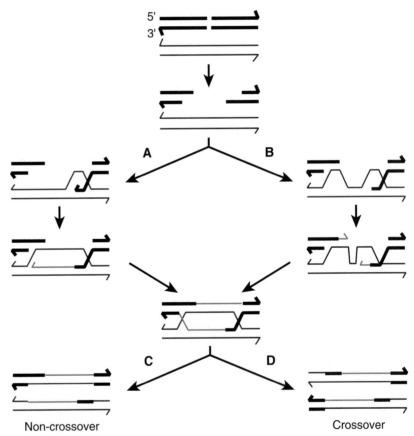

Figure 10–33 Model for the repair of double-strand breaks. A double-strand break is converted into a gap flanked by 3' single strands by the action of exonucleases. (A) One 3' end invades a homologous duplex, displacing a D-loop. The D-loop is enlarged by repair synthesis until the 3' end can anneal to complementary single-stranded sequences. Repair synthesis from the second 3' end completes the process of gap repair (319). (B) Alternatively, both 3' ends might invade (302). (C and D) Resolution of two junctions by cutting either the inner or outer strand leads to two possible noncrossover and two possible crossover configurations. In the illustrated resolutions, the right-hand junction was resolved by cutting the inner, crossed strands. (C) The left-hand junction was resolved by cutting the inner, crossed strands to generate a noncrossover product. (D) The left-hand junction was resolved by cutting the outer strands to generate a crossover product (319). (*Adapted from Szostak et al. [319] and Smith [302] with permission.*)

thus restoring the nucleotides lost from the broken ends. This would lead eventually to the formation of two Holliday junctions. An alternative would be for one 3′ end to invade and serve as a primer for DNA repair synthesis. The D-loop formed by this repair synthesis would continue to expand until the other 3′ end could anneal to complementary sequences. Repair synthesis would then initiate from the second 3′ end, completing the restoration of the nucleotides lost from the broken ends and also leading eventually to the formation of two Holliday junctions. Each Holliday junction could be resolved in two ways (cleavage of the inner strand or the outer strands in Fig. 10–33). Depending on how the Holliday junctions are resolved, the resulting DNA molecules could be nonrecombinant or recombinant, having exchanged flanking markers (319) (Fig. 10–33).

The repair of DNA double-strand breaks in *E. coli* exposed to ionizing radiation, mitomycin, or UV radiation requires a functional *recA*⁺ gene (150, 151, 277, 335). The known properties of RecA protein suggest that it could catalyze strand invasion by the 3′ ends and might play a role in branch migration (Fig. 10–33). In addition, in vitro experiments with model substrates have shown that RecA protein can promote strand exchanges past double-strand breaks (348). In wild-type *E. coli*, the repair of double-strand breaks is also dependent on *recB*⁺ and *recC*⁺ function (305, 306, 335). It seems likely that this reflects a role for the *recBCD*-encoded exonuclease V in the generation of the 3′ ends. On the basis of its known properties (301, 302), it seems likely that the mechanism for generation of these 3′ ends involves an initial entry of the RecBCD enzyme into the DNA at the double-strand break. It then travels along the DNA, unwinding and rewinding the DNA as it goes. When the enzyme encounters a properly oriented chi site (5′-GCTGGTGG-3′), it cuts one DNA strand to produce an invasive single-stranded tail with its 3′ end a few nucleotides 3′ of the chi site (upstream relative to the direction of travel of the RecBCD enzyme). These tails then serve to invade the homologous intact duplex in a reaction mediated by RecA (56).

The requirement for RecBCD function in the repair of double-strand breaks in wild-type *E. coli* appears absolute, since a single double-strand DNA break in the chromosome is lethal in the absence of RecA protein or the RecBCD enzyme but does not impair growth in the presence of these two proteins (221). However, the recombination and repair deficiencies of *recB recC* mutants can be suppressed by secondary mutations in either the *sbcA*⁺ or *sbcB*⁺ loci, and, in each case, the suppression can be rationalized in terms of an effect on the generation of 3′ ends (343). *sbcA* mutations result in the activation of exonuclease VIII, which digests double-stranded DNA to produce long 3′ tails and thus could provide an alternative means of producing the invasive 3′ ends needed for double-strand break repair. *sbcB* mutations inactivate exonuclease I, which digests single-stranded DNA from the 3′ end, so that its inactivation might leave 3′ ends available to initiate double-strand break repair. The additional requirement for RecF function in *sbcB* mutants (334) might reflect the requirement for RecF protein to serve as a RecA accessory protein in these cells, as discussed above (327).

The repair of double-strand breaks is obviously a complicated process. Furthermore, initial double-strand breaks might have a variety of structures at their ends depending on how they were generated. It is therefore perhaps not too surprising that mutations in a number of genes associated with recombinational processes (*recN*⁺, *recF*⁺, *recJ*⁺, *radA*⁺, and *uvrD*⁺) have been found to influence double-strand-break repair to some extent (277). The *recN*⁺ gene is of particular interest, since its expression is regulated by the SOS response (184, 267) and its product is required for the repair of double-strand breaks (240, 278, 279) but not for the repair of daughter-strand gaps (278, 279).

REFERENCES

1. **Asai, T., and T. Kogoma.** 1994. D-loops and R-loops: alternative mechanisms for the initiation of chromosome replication in *Escherichia coli. J. Bacteriol.* **176:**1807–1812.

2. **Asai, T., S. Sommer, A. Bailone, and T. Kogoma.** 1993. Homologous recombination-dependent initiation of DNA replication for DNA damage-inducible origins in *Escherichia coli. EMBO J.* **12:**3278–3295.

3. **Auffray, Y., and P. Boutibonnes.** 1987. Presence of inducible DNA repair in *Bacillus thuringiensis. Mutat. Res.* **183:**225–229.

4. **Bagg, A., C. J. Kenyon, and G. C. Walker.** 1981. Inducibility of a gene product required for UV and chemical mutagenesis in *Escherichia coli. Proc. Natl. Acad. Sci. USA* **78:**5749–5753.

5. **Ball, T. K., C. R. Wasmuth, S. C. Braunagel, and M. J. Benedik.** 1990. Expression of *Serratia marcescens* extracellular proteins requires *recA. J. Bacteriol.* **172:**342–349.

6. **Battista, J. R., T. Ohta, T. Nohmi, W. Sun, and G. C. Walker.** 1990. Dominant negative *umuD* mutations decreasing RecA-mediated cleavage suggest roles for intact UmuD in modulation of SOS mutagenesis. *Proc. Natl. Acad. Sci. USA* **87:**7190–7194.

7. **Bender, R. A.** 1984. Ultraviolet mutagenesis and inducible DNA repair in *Caulobacter crescentus. Mol. Gen. Genet.* **197:**399–402.

8. **Bennett, R. J., H. J. Dunderdale, and S. C. West.** 1993. Resolution of Holliday junctions by RuvC resolvase. *Cell* **74:**1021–1031.

9. **Benson, F. E., G. T. Illing, G. J. Sharples, and R. G. Lloyd.** 1988. Nucleotide sequencing of the *ruv* region of *Escherichia coli* K-12 reveals a LexA regulated operon encoding two genes. *Nucleic Acids Res.* **16:**1541–1549.

10. **Bernstein, H., H. C. Byerly, F. A. Hopf, and R. E. Michod.** 1985. Genetic damage, mutation and the evolution of sex. *Science* **229:**1277–1281.

11. **Bertrand-Burggraf, E., S. Hurstel, M. Daune, and M. Schnarr.** 1987. Promoter properties and negative regulation of the *uvrA* gene by the LexA repressor and its amino-terminal DNA binding domain. *J. Mol. Biol.* **193:**293–302.

12. **Bertrand-Burggraf, E., P. Oertel, M. Schnarr, M. Daune, and M. Granger-Schnarr.** 1989. Effect of induction of SOS response on expression of pBR322 genes and on plasmid copy number. *Plasmid* **22:**163–168.

13. **Blattner, F. R., V. Burland, G. Plunkett III, H. J. Sofia, and D. L. Daniels.** 1993. Analysis of the *Escherichia coli* genome. IV. DNA sequence of the region from 89.2 to 92.8 minutes. *Nucleic Acids Res.* **21:**5408–5417.

14. **Bockrath, R., L. Wolff, A. Farr, and R. J. Crouch.** 1987. Amplified RNase H activity in *Escherichia coli* B/r increases sensitivity to ultraviolet radiation. *Genetics* **115:**33–40.

15. **Bonner, C. A., S. Hays, K. McEntee, and M. F. Goodman.** 1990. DNA polymerase II is encoded by the DNA damage-inducible *dinA* gene of *Escherichia coli. Proc. Natl. Acad. Sci. USA* **87:**7663–7667.

16. **Bonura, T., and K. C. Smith.** 1975. Enzymatic production of deoxyribonucleic acid double-strand breaks after ultraviolet light irradiation of *Escherichia coli* K-12. *J. Bacteriol.* **121:**511–517.

17. **Borek, E., and A. Ryan.** 1958. The transfer of irradiation-elicited induction in a lysogenic organism. *Proc. Natl. Acad. Sci. USA* **44:**374–377.

18. **Borek, E., and A. Ryan.** 1973. Lysogenic induction. *Prog. Nucleic Acid Res. Mol. Biol.* **13:**249–300.

19. **Brandsma, J. A., D. Bosch, C. Backendorf, and P. van de Putte.** 1983. A common regulatory region shared by divergently transcribed genes of the *Escherichia coli* SOS system. *Nature* (London) **305:**243–245.

20. **Brent, R.** 1982. Regulation and autoregulation by *lexA* protein. *Biochimie* **64:**565–569.

21. **Brent, R., and M. Ptashne.** 1980. The *lexA* gene product represses its own promoter. *Proc. Natl. Acad. Sci. USA* **77:**1932–1936.

22. **Brent, R., and M. Ptashne.** 1981. Mechanism of action of the *lexA* gene product. *Proc. Natl. Acad. Sci. USA* **78:**4204–4208.

23. **Bridges, B. A.** 1990. Ultraviolet light mutagenesis in bacteria: the possible role of a DNA polymerase III complex lacking proofreading exonuclease, p. 27–35. *In* A. Kappas (ed.), *Mechanisms of Environmental Mutagenesis-Carcinogenesis.* Plenum Publishing Corp., New York.

24. **Bridges, B. A., and G. M. Brown.** 1992. Mutagenic DNA repair in *Escherichia coli.* XXI. A stable SOS-inducing signal persisting after excision repair of ultraviolet damage. *Mutat. Res.* **270:**135–144.

25. **Brooks, K., and A. J. Clark.** 1967. Behavior of lambda bacteriophage in a recombination-deficient strain of *Escherichia coli. J. Virol.* **1:**283–293.

26. **Calero, S., X. Garriga, and J. Barbe.** 1991. One-step cloning system for isolation of bacterial *lexA*-like genes. *J. Bacteriol.* **173:**7345–7350.

27. **Camerini, R. D., and P. Hsieh.** 1993. Parallel triple helices, homologous recombination, and homology-dependent DNA interactions. *Cell* **73:**217–223.

28. **Campbell, A.** 1984. Types of recombination: common problems and common strategies. *Cold Spring Harbor Symp. Quant. Biol.* **49:**839–844.

29. **Casadaban, M. J., and S. N. Cohen.** 1980. Lactose genes fused to exogenous promoters in one step using a Mu-*lac* bacteriophage: *in vivo* probe for transcriptional control sequences. *Proc. Natl. Acad. Sci. USA* **76:**4530–4533.

30. **Casaregola, S., R. D'Ari, and O. Huisman.** 1982. Quantitative evaluation of *recA* gene expression in *Escherichia coli. Mol. Gen. Genet.* **185:**430–439.

31. **Casaregola, S., M. Khidhir, and I. B. Holland.** 1987. Effects of modulation of RNase H production on the recovery of DNA synthesis following UV-irradiation in *Escherichia coli. Mol. Gen. Genet.* **209:**494–498.

32. **Cassuto, E., J. Musalim, and P. Howard-Flanders.** 1978. Homology-dependent cutting *in trans* in extracts of *Escherichia coli*: an approach to the enzymology of genetic recombination. *Proc. Natl. Acad. Sci. USA* **75:**620–624.

33. **Castellazzi, M., J. George, and G. Buttin.** 1972. Prophage induction and cell division in *E. coli.* I. Further characterization of the thermosensitive mutation *tif-1* whose expression mimics the effects of UV irradiation. *Mol. Gen. Genet.* **119:**139–152.

34. **Castellazzi, M., J. George, and G. Buttin.** 1972. Prophage induction and cell division in *E. coli.* II. Linked (*recA zab*) and unlinked (*lexA*) suppressors of *tif-1* mediated induction and filamentation. *Mol. Gen. Genet.* **119:**153–174.

35. **Castellazzi, M., P. Morand, J. George, and G. Buttin.** 1977. Prophage induction and cell division in *E. coli.* V. Dominance and complementation analysis in partial diploids with pleiotropic mutations (*tif, recA, zab,* and *lexB*) at the *recA* locus. *Mol. Gen. Genet.* **153:**297–310.

36. **Chaudhury, A. M., and G. R. Smith.** 1985. Role of *Escherichia coli* RecBC enzyme in SOS induction. *Mol. Gen. Genet.* **201:**525–528.

37. **Cheo, D., K. Bayles, and R. Yasbin.** 1991. Cloning and characterization of DNA damage-inducible promoter regions from *Bacillus subtilis. J. Bacteriol.* **173:**1696–1703.

38. **Clark, A. J.** 1982. RecA operator mutations and their usefulness. *Biochimie* **64:**669–675.

39. **Clark, A. J., and A. D. Margulies.** 1965. Isolation and characterization of recombination-deficient mutants of *Escherichia coli* K-12. *Proc. Natl. Acad. Sci. USA* **53:**451–459.

40. **Connelly, B., C. A. Parsons, F. E. Benson, H. J. Dunderdale, G. J. Sharples, R. G. Lloyd, and S. C. West.** 1991. Resolution of Holliday junctions *in vitro* requires the *Escherichia coli ruvC* gene product. *Proc. Natl. Acad. Sci. USA* **88:**6063–6067.

41. **Cooper, P.** 1981. Inducible excision repair in *Escherichia coli*, p. 139–146. *In* E. Seeberg and K. Kleppe (ed.), *Chromosome Damage and Repair.* Plenum Publishing Corp., New York.

42. **Cooper, P.** 1982. Characterization of long patch excision repair of DNA in ultraviolet-irradiated *Escherichia coli*: an inducible function under *rec-lex* control. *Mol. Gen. Genet.* **185:**189–197.

43. Cox, M. M. 1994. Why does RecA protein hydrolyze ATP. *Trends Biochem. Sci.* **19:**217–222.

44. Cox, M. M. 1993. Relating biochemistry to biology. *Bioessays* **15:**617–673.

45. Cox, M. M., and I. R. Lehman. 1987. Enzymes of general recombination. *Annu. Rev. Biochem.* **56:**229–262.

46. Craig, N. L., and J. W. Roberts. 1980. *E. coli recA* protein-directed cleavage of phage lambda repressor requires polynucleotide. *Nature* (London) **283:**26–30.

47. Craig, N. L., and J. W. Roberts. 1981. Function of nucleoside triphosphate and polynucleotide in *Escherichia coli recA* protein-directed cleavage of phage lambda repressor. *J. Biol. Chem.* **256:**8039–8044.

48. D'Ari, R., and O. Huisman. 1982. DNA replication and indirect induction of the SOS response in *Escherichia coli. Biochimie* **64:**623–627.

49. Day, R. S. 1977. UV-induced alleviation of K-specific restriction of bacteriophage lambda. *J. Virol.* **21:**1249–1251.

50. de Henestrosa, A. F., S. Calero, and J. Barbe. 1991. Expression of the *recA* gene of *Escherichia coli* in several species of gram-negative bacteria. *Mol. Gen. Genet.* **226:**503–506.

51. Defais, M., P. Fauquet, M. Radman, and M. Errera. 1971. Ultraviolet reactivation and ultraviolet mutagenesis of lambda in different genetic systems. *Virology* **43:**495–503.

52. Defais, M., C. Lesca, B. Monsarrat, and P. Hanawalt. 1989. Translesion synthesis is the main component of SOS repair in bacteriophage lambda DNA. *J. Bacteriol.* **171:**4938–4944.

53. Devoret, R., and J. George. 1967. Induction indirecte du prophage lambda par le rayonnement ultraviolet. *Mutat. Res.* **4:**713–714.

54. Devoret, R., M. Pierre, and P. L. Moreau. 1983. Prophage φ80 is induced in *Escherichia coli* K12 *recA430. Mol. Gen. Genet.* **189:**199–206.

55. Dimpfl, J., and H. Echols. 1989. Duplication mutation as an SOS response in *Escherichia coli*: enhanced duplication formation by a constitutively activated RecA. *Genetics* **123:**255–260.

56. Dixon, D. A., and S. C. Kowalczykowski. 1991. Homologous pairing in vitro stimulated by the recombination hotspot Chi. *Cell* **66:**361–371.

57. Doudney, C. O. 1975. The two-lesion hypothesis for UV-induced mutation in relation to recovery of capacity for DNA replication, p. 389–392. *In* P. C. Hanawalt and R. B. Setlow (ed.), *Molecular Mechanisms for the Repair of DNA.* Plenum Publishing Corp., New York.

58. Drlica, K., E. C. Engle, and S. H. Manes. 1980. DNA gyrase on the bacterial chromosome: possibility of two levels of action. *Proc. Natl. Acad. Sci. USA* **77:**6879–6883.

59. Dunderdale, H. J., F. E. Benson, C. A. Parsons, G. J. Sharples, R. G. Lloyd, and S. C. West. 1991. Formation and resolution of recombination intermediates by *E. coli* RecA and RuvC proteins. *Nature* (London) **354:**506–510.

60. Ebina, Y., F. Kishi, T. Miki, H. Kagamiyama, T. Nakazawa, and A. Nakazawa. 1981. The nucleotide sequence surrounding the promoter region of colicin E1 gene. *Gene* **15:**119–126.

61. Ebina, Y., F. Kishi, and A. Nakazawa. 1982. Direct participation of *lexA* protein in repression of colicin E1 synthesis. *J. Bacteriol.* **150:**1479–1481.

62. Ebina, Y., and A. Nakazawa. 1983. Cyclic AMP-dependent initiation and rho-dependent termination of colicin E1 gene transcription. *J. Biol. Chem.* **258:**7072–7078.

63. Ebina, Y., Y. Takahara, F. Kishi, A. Nakazawa, and R. Brent. 1983. LexA protein is a repressor of the colicin E1 gene. *J. Biol. Chem.* **258:**13258–13269.

64. Echols, H., and M. F. Goodman. 1990. Mutation induced by DNA damage: a many protein affair. *Mutat. Res.* **236:**301–311.

65. Echols, H., and M. F. Goodman. 1991. Fidelity mechanisms in DNA replication. *Annu. Rev. Biochem.* **60:**477–511.

66. Egelman, E. H. 1993. What do X-ray crystallographic and electron microscopic structural studies of the RecA protein tell us about recombination? *Curr. Opin. Struct. Biol.* **3:**189–197.

67. Egelman, E. H., and A. Stasiak. 1993. Electron microscopy of RecA-DNA complexes: two different states, their functional significance and relation to the solved crystal structure. *Micron* **24:**309–324.

68. Eguchi, Y., T. Ogawa, and H. Ogawa. 1988. Cleavage of bacteriophage φ80 cI repressor by RecA protein. *J. Mol. Biol.* **202:**565–573.

69. Eguchi, Y., T. Ogawa, and H. Ogawa. 1988. Stimulation of RecA-mediated cleavage of phage φ80 cI repressor by deoxydinucleotides. *J. Mol. Biol.* **204:**69–77.

70. Elledge, S. J., and G. C. Walker. 1983. The *muc* genes of pKM101 are induced by DNA damage. *J. Bacteriol.* **155:**1306–1315.

71. Elledge, S. J., and G. C. Walker. 1983. Proteins required for ultraviolet light and chemical mutagenesis: identification of the products of the *umuC* locus of *Escherichia coli. J. Mol. Biol.* **164:**175–192.

72. Emmerson, P. T., and S. C. West. 1977. Identification of protein X of *Escherichia coli* as the *recA⁺/tif⁺* gene product. *Mol. Gen. Genet.* **155:**77–85.

73. Fogliano, M., and P. F. Schendel. 1981. Evidence for the inducibility of the *uvrB* operon. *Nature* (London) **289:**196–198.

74. Friedberg, E. C. 1985. *DNA Repair.* W. H. Freeman & Co., New York.

75. Ganesan, A. K. 1974. Persistence of pyrimidine dimers during post-replication repair in ultraviolet light-irradiated *Escherichia coli* K-12. *J. Mol. Biol.* **87:**103–119.

76. Ganesan, A. K. 1975. Distribution of pyrimidine dimers during postreplication repair in UV-irradiated excision-deficient cells of *Escherichia coli* K12, p. 317–320. *In* P. C. Hanawalt, and R. B. Setlow (ed.), *Molecular Mechanisms for Repair of DNA.* Plenum Publishing Corp., New York.

77. Ganesan, A. K., and P. C. Hanawalt. 1985. Effect of a *lexA41* (Ts) mutation on DNA repair in *recA*(Def) derivatives of *Escherichia coli* K-12. *Mol. Gen. Genet.* **201:**387–392.

78. Garvey, N., A. St. John, and E. M. Witkin. 1985. Evidence for RecA protein association with the cell membrane and for changes in the levels of major outer membrane proteins in SOS-induced *Escherichia coli* cells. *J. Bacteriol.* **163:**870–876.

79. Gellert, M., K. Mizuuchi, M. O'Dea, T. Itoh, and J.-I. Tomizawa. 1977. Nalidixic acid resistance: A second genetic character involved in DNA gyrase activity. *Proc. Natl. Acad. Sci. USA* **74:**4772–4776.

80. George, J., M. Castellazzi, and G. Buttin. 1975. Prophage induction and cell division in *E. coli*. III. Mutations *sfiA* and *sfiB* restore division in *tif* and *lon* strains and permit the expression of mutator properties of *tif. Mol. Gen. Genet.* **140:**309–332.

81. George, J., R. Devoret, and M. Radman. 1974. Indirect ultraviolet-reactivation of phage lambda. *Proc. Natl. Acad. Sci. USA* **71:**144–147.

82. Gimble, F. S., and R. T. Sauer. 1986. λ repressor inactivation: properties of purified ind- proteins in the autodigestion and RecA-mediated cleavage reactions. *J. Mol. Biol.* **192:**39–47.

83. Gimble, F. S., and R. T. Sauer. 1989. Lambda repressor mutants that are better substrates for RecA-mediated cleavage. *J. Mol. Biol.* **206:**29–39.

84. Goldthwait, D., and F. Jacob. 1964. Sur le mechanisme de l'induction du developpement du prophage chez les bacteries lysogenes. *C. R. Acad. Sci.* (Paris) **259:**661–664.

85. Granger-Schnarr, M., M. Schnarr, and C. A. van Sluis. 1986. *In vitro* study of the interaction of the LexA repressor and the UvrC protein with a *uvrC* regulatory region. *FEBS Lett.* **198:**61–65.

86. Green, M. H., J. Greenberg, and J. Donch. 1969. Effects of a *recA* gene on cell division and capsular polysaccharide production in a *lon* strain of *Escherichia coli. Genet. Res.* **14:**159–162.

87. Greenberg, J., J. Donch, and L. Berends. 1975. The dominance of *exrB* over *exrB⁺. Genet. Res.* **25:**39–44.

88. Greener, A., and C. W. Hill. 1980. Identification of a novel genetic element in *Escherichia coli* K-12. *J. Bacteriol.* **144:**312–321.

89. Gudas, L. J., and D. W. Mount. 1977. Identification of the *recA(tif)* gene product of *Escherichia coli*. *Proc. Natl. Acad. Sci. USA* **74:**5280–5284.

90. Gudas, L. J., and A. B. Pardee. 1975. Model for regulation of *Escherichia coli* DNA repair functions. *Proc. Natl. Acad. Sci. USA* **72:**2330–2334.

91. Gudas, L. J., and A. B. Pardee. 1976. DNA synthesis inhibition and the induction of protein X in *Escherichia coli*. *J. Mol. Biol.* **101:**459–477.

92. Gudas, L. J., and A. B. Pardee. 1976. The induction of protein X in DNA repair and cell division mutants of *Escherichia coli*. *J. Mol. Biol.* **104:**567–587.

93. Hall, J. D., and D. Mount. 1981. Mechanisms of DNA replication and mutagenesis in ultraviolet-irradiated bacteria and mammalian cells. *Prog. Nucleic Acid Res. Mol. Biol.* **25:**53–126.

94. Hanawalt, P. C., P. K. Cooper, A. Ganesan, and C. A. Smith. 1979. DNA repair in bacteria and mammalian cells. *Annu. Rev. Biochem.* **48:**783–836.

95. Hayes, W. 1968. *The Genetics of Bacteria and Their Viruses*, 2nd ed. Blackwell Scientific Publications, Ltd., Oxford.

96. Hertman, I., and S. E. Luria. 1967. Transduction studies on the role of a *rec+* gene in ultraviolet induction of prophage lambda. *J. Mol. Biol.* **23:**117–133.

97. Higgins, N. P., K. Kato, and B. Strauss. 1976. A model for replication repair in mammalian cells. *J. Mol. Biol.* **101:**417–425.

98. Hill, S., and J. W. Little. 1988. Allele replacement in *Escherichia coli* by use of a selectable marker for resistance to spectinomycin: replacement of the *lexA* gene. *J. Bacteriol.* **170:**5913–5915.

99. Hiom, K. J., and S. G. Sedgwick. 1992. Alleviation of *EcoK* DNA restriction in *Escherichia coli* and involvement of *umuDC* activity. *Mol. Gen. Genet.* **231:**265–275.

100. Hofemeister, J. 1977. DNA repair in *Proteus mirabilis*. *Mol. Gen. Genet.* **154:**35–41.

101. Hofemeister, J., and G. Eitner. 1981. Repair and plasmid R46 mediated mutation requires inducible functions in *Proteus mirabilis*. *Mol. Gen. Genet.* **183:**369–375.

102. Holliday, R. 1964. A mechanism for gene conversion in fungi. *Genet. Res.* **5:**282–304.

103. Hooper, I., and J. B. Egan. 1981. Coliphage 186 replication requires host initiation functions DnaA and DnaC. *J. Virol.* **40:**599–601.

104. Hooper, I., W. H. Woods, and J. B. Egan. 1981. Coliphage 186 infection requires replication but is delayed when the host cell is irradiated before infection. *J. Virol.* **40:**341–349.

105. Horii, T., T. Ogawa, T. Nakatani, T. Hase, H. Matsubara, and H. Ogawa. 1981. Regulation of SOS functions: purification of *E. coli* LexA protein and determination of its specific site cleaved by the RecA protein. *Cell* **27:**515–522.

106. Horn, J. M., and D. E. Ohman. 1988. Autogenous regulation and kinetics of induction of *Pseudomonas aeruginosa recA* transcription as analyzed with operon fusions. *J. Bacteriol.* **170:**4699–4705.

107. Howard-Flanders, P. 1981. Inducible repair of DNA. *Sci. Am.* **245** (Nov.):56–64.

108. Howard-Flanders, P., and R. P. Boyce. 1966. DNA repair and genetic recombination: studies on mutants of *Escherichia coli* defective in these processes. *Radiat. Res.* **6:**156–184.

109. Howard-Flanders, P., W. D. Rupp, B. M. Wilkins, and R. S. Cole. 1968. DNA replication and recombination after UV-irradiation. *Cold Spring Harbor Symp. Quant. Biol.* **33:**195–205.

110. Howard-Flanders, P., and L. Theriot. 1966. Mutants of *Escherichia coli* K-12 defective in DNA repair and in genetic recombination. *Genetics* **53** (Suppl.):1137–1150.

111. Howard-Flanders, P., S. C. West, and A. J. Stasiak. 1984. Role of RecA spiral filaments in genetic recombination. *Nature* (London) **309:**215–220.

112. Hsieh, P., C. S. Camerini-Otero, and R. D. Camerini-Otero. 1990. Pairing of homologous DNA sequences by proteins: evidence for three-stranded DNA. *Genes Dev.* **4:**1951–1963.

113. Huisman, O., and R. D'Ari. 1981. An inducible DNA replication-cell division coupling mechanism in *E. coli*. *Nature* (London) **290:**797–799.

114. Hurstel, S., M. Granger-Schnarr, M. Daune, and M. Schnarr. 1986. *In vitro* binding of LexA repressor to DNA: evidence for the involvement of the amino-terminal domain. *EMBO J.* **5:**793–798.

115. Hurstel, S., M. Granger-Schnarr, and M. Schnarr. 1988. Contacts between the LexA repressor—or its DNA binding domain—and the backbone of the *recA* operator. *EMBO J.* **7:**269–275.

116. Ihara, M., K. Yamamato, and T. Ohnishi. 1987. Induction of *phr* gene expression by irradiation of ultraviolet light in *Escherichia coli*. *Mol. Gen. Genet.* **209:**200–202.

117. Inouye, M. 1971. Pleiotrophic effect of the *recA* gene of *Escherichia coli*: uncoupling of cell division from deoxyribonucleic acid replication. *J. Bacteriol.* **106:**539–542.

118. Irbe, R. M., L. M. E. Morin, and M. Oishi. 1981. Prophage (φ80) induction in *Escherichia coli* K-12 by specific deoxyoligonucleotides. *Proc. Natl. Acad. Sci. USA* **78:**138–142.

119. Itaya, M. 1990. Isolation and characterization of a second RNase H (RNase H II) encoded by the *rnhB* gene. *Proc. Natl. Acad. Sci. USA* **87:**8587–8591.

120. Iwasaki, H., Y. Ishino, H. Toh, A. Nakata, and H. Shinagawa. 1991. *Escherichia coli* DNA polymerase II is homologous to α-like DNA polymerases. *Mol. Gen. Genet.* **226:**24–33.

121. Iwasaki, H., A. Nakata, G. C. Walker, and H. Shinagawa. 1990. The *Escherichia coli polB* gene, which encodes DNA polymerase II, is regulated by the SOS system. *J. Bacteriol.* **172:**6268–6273.

122. Iwasaki, H., T. Shiba, K. Makino, A. Nakata, and H. Shingawa. 1989. Overproduction, purification, and ATPase activity of the *Escherichia coli* RuvB protein involved in DNA repair. *J. Bacteriol.* **171:**5276–5280.

123. Iwasaki, H., M. Takahagi, T. Shiba, A. Nakata, and H. Shinagawa. 1991. *Escherichia coli* RuvC protein is an endonuclease that resolves the Holliday structure. *EMBO J.* **10:**4381–4389.

124. Iyer, V. N., and W. D. Rupp. 1971. Usefulness of benzoylated naphthoylated DEAE-cellulose to distinguish and fractionate double-stranded DNA bearing different extents of single-stranded regions. *Biochim. Biophys. Acta* **228:**117.

125. Jacob, F., and E. L. Wollman. 1959. Lysogeny, p. 319. *In* F. N. Burnet and W. M. Stanley (ed.), *The Viruses*. Academic Press, Inc., New York.

126. Johnson, R. C., and W. F. McNeill. 1978. Electron microscopy of UV-induced postreplication repair of daughter strand gaps, p. 95–99. *In* P. C. Hanawalt, E. C. Friedberg, and C. F. Fox (ed.), *DNA Repair Mechanisms*. Academic Press, Inc., New York.

127. Johnson, R. D., and L. S. Symington. 1993. Crossed-stranded DNA structures for investigating the molecular dynamics of the Holliday junction. *J. Mol. Biol.* **229:**812–820.

128. Kaasch, M., J. Kaasch, and A. Quinones. 1989. Expression of the *dnaN* and *dnaQ* genes of *Escherichia coli* is inducible by mitomycin. *Mol. Gen. Genet.* **219:**187–192.

129. Kannan, P., and K. Dharmalingam. 1990. Induction of the inhibitor of the RecBCD enzyme in *Escherichia coli* is a *lexA*-independent SOS response. *Curr. Microbiol.* **21:**7–15.

130. Karu, A. E., and E. D. Belk. 1982. Induction of *E. coli recA* protein via *recBC* and alternate pathways: quantitation by enzyme-linked immunosorbent assay (ELISA). *Mol. Gen. Genet.* **185:**275–282.

131. Kato, T. 1977. Effects of chloramphenicol and caffeine on postreplication repair in *uvrA⁻ umuC⁻* and *uvrA⁻ recF⁻* strains of *Escherichia coli* K-12. *Mol. Gen. Genet.* **156:**115–120.

132. Kato, T., and Y. Shinoura. 1977. Isolation and characterization of mutants of *Escherichia coli* deficient in induction of mutations by ultraviolet light. *Mol. Gen. Genet.* **156:**121–131.

133. **Kenyon, C. J., R. Brent, M. Ptashne, and G. C. Walker.** 1982. Regulation of damage-inducible genes in *Escherichia coli. J. Mol. Biol.* **160:**445–457.

134. **Kenyon, C. J., and G. C. Walker.** 1980. DNA-damaging agents stimulate gene expression at specific loci in *Escherichia coli. Proc. Natl. Acad. Sci. USA* **77:**2819–2823.

135. **Kenyon, C. J., and G. C. Walker.** 1981. Expression of the *E. coli uvrA* gene is inducible. *Nature* (London) **289:**808–810.

136. **Khidhir, M. A., S. Casaregola, and I. B. Holland.** 1985. Mechanism of transient inhibition of DNA synthesis in ultraviolet-irradiated *E. coli*: inhibition is independent of *recA* whilst recovery requires RecA protein itself and an additional, inducible SOS function. *Mol. Gen. Genet.* **199:**133–140.

137. **Kim, B., and J. W. Little.** 1992. Dimerization of a specific DNA-binding protein on the DNA. *Science* **255:**203–206.

138. **Kim, B., and J. W. Little.** 1993. LexA and λ *cI* repressors as enzymes: specific cleavage in an intermolecular reaction. *J. Bacteriol.* **73:**1165–1173.

139. **Kim, J. I., M. M. Cox, and R. B. Inman.** 1992. On the role of ATP hydrolysis in RecA protein-mediated DNA strand exchange. *J. Biol. Chem.* **267:**16438–16443.

140. **Kirby, E. P., F. Jacob, and D. A. Goldthwait.** 1967. Prophage induction and filament formation in a mutant strain of *Escherichia coli. Proc. Natl. Acad. Sci. USA* **58:**1903–1910.

141. **Kirchner, J., D. Lim, E. M. Witkin, N. Garvey, and V. Roegner-Maniscalo.** 1992. An SOS-inducible defective retronphage (φR86) in *Escherichia coli* strain B. *Mol. Microbiol.* **6:**2815–2824.

142. **Kneser, H.** 1968. Relationship between K-reactivation and UV-reactivation of bacteriophage λ. *Virology* **36:**303–305.

143. **Knight, K. L., K. H. Aoki, E. L. Ujita, and K. McEntee.** 1984. Identification of the amino acid substitutions in two mutant forms of the RecA protein from *Escherichia coli* RecA441 and Rec629. *J. Biol. Chem.* **259:**11279–11283.

144. **Kobayashi, I., and H. Ikeda.** 1983. Double Holliday structure: a possible *in vivo* intermediate form of general recombination in *Escherichia coli. Mol. Gen. Genet.* **191:**213–220.

145. **Kojima, M., M. Suzuki, T. Morita, T. Ogawa, H. Ogawa, and M. Tada.** 1990. Interaction of RecA protein with pBR322 DNA modified by N-hydroxy-2-acetylaminofluorene and 4-hydroxyaminoquinoline 1-oxide. *Nucleic Acids Res.* **18:**2707–2714.

146. **Korn, D., and A. Weissbach.** 1962. Thymineless induction in *Escherichia coli* K-12 (λ). *Biochim. Biophys. Acta* **61:**775–790.

147. **Kornberg, A., and T. A. Baker.** 1992. *DNA Replication.* W. H. Freeman & Co., New York.

148. **Koudelka, G. B., P. Harbury, S. C. Harrison, and M. Ptashne.** 1988. DNA twisting and the affinity of the bacteriophage 434 operator for bacteriophage 434 repressor. *Proc. Natl. Acad. Sci. USA* **88:**4633–4637.

149. **Kowalczykowski, S. C.** 1991. Biochemistry of genetic recombination: energetics and mechanism of DNA strand exchange. *Annu. Rev. Biophys. Biophys. Chem.* **20:**539–575.

150. **Krasin, F., and F. Hutchinson.** 1977. Repair of DNA double-strand breaks in *Escherichia coli*, which requires RecA function and the presence of a duplicate genome. *J. Mol. Biol.* **116:**81–98.

151. **Krasin, F., and F. Hutchinson.** 1981. Repair of DNA double-strand breaks in *Escherichia coli* cells requires synthesis of proteins that can be induced by UV light. *Proc. Natl. Acad. Sci. USA* **78:**3450–3453.

152. **Krueger, J. H., S. J. Elledge, and G. C. Walker.** 1983. Isolation and characterization of Tn5 insertion mutations in the *lexA* gene of *Escherichia coli. J. Bacteriol.* **153:**1368–1378.

153. **Kuan, C.-T., and I. Tessman.** 1991. LexA protein of *Escherichia coli* represses expression of the Tn5 transposase gene. *J. Bacteriol.* **173:**6406–6410.

154. **Kuan, C.-T., and I. Tessman.** 1992. Further evidence that transposition of Tn5 in *Escherichia coli* is strongly enhanced by constitutively activated RecA proteins. *J. Bacteriol.* **174:**6872–6877.

155. **Lamont, I., A. M. Brumby, and J. B. Egan.** 1989. UV induction of coliphage 186: prophage induction as an SOS function. *Proc. Natl. Acad. Sci. USA* **86:**5492–5496.

156. **Lavery, P. E., and S. C. Kowalczykowski.** 1992. Biochemical basis of the constitutive repressor cleavage activity of recA730 protein. *J. Biol. Chem.* **29:**20648–20658.

157. **Lee, J. W., and M. M. Cox.** 1990. Inhibition of RecA protein promoted ATP hydrolysis. 1. ATP-γS and ADP are antagonistic inhibitors. *Biochemistry* **29:**7666–7676.

158. **Lee, J. W., and M. M. Cox.** 1990. Inhibition of RecA protein promoted ATP hydrolysis. 2. Longitudinal assembly and disassembly of RecA protein filaments mediated by ATP and ADP. *Biochemistry* **29:**7677–7683.

159. **Lewin, B.** 1983. *Genes.* John Wiley & Sons, Inc., New York.

160. **Lewis, L. K., G. R. Harlow, L. A. Gregg-Jolly, and D. W. Mount.** 1994. Identification of high affinity binding sites for LexA which define new DNA damage-inducible genes in *Escherichia coli. J. Mol. Biol.* **241:**507–523.

161. **Lewis, L. K., M. E. Jenkins, and D. W. Mount.** 1992. Isolation of DNA-damage inducible promoters in *E. coli*: regulation of *polB* (*dinA*), *dinG*, and *dinH* by LexA repressor. *J. Bacteriol.* **174:**3377–3385.

162. **Lewis, L. K., and D. W. Mount.** 1992. Interaction of LexA repressor with the asymmetric *dinG* operator and complete nucleotide sequence of the gene. *J. Bacteriol.* **174:**5110–5116.

163. **Ley, R. D.** 1973. Postreplication repair in an excision-defective mutant of *Escherichia coli*: ultraviolet light-induced incorporation of bromodeoxyuridine into parental DNA. *Photochem. Photobiol.* **18:**87–95.

164. **Ley, R. D.** 1975. Ultraviolet-light induced incorporation of bromodeoxyuridine into parental DNA of an excision-defective mutant of *Escherichia coli*, p. 313–316. *In* P. C. Hanawalt and R. B. Setlow (ed.), *Molecular Mechanisms for Repair of DNA.* Plenum Publishing Corp., New York.

165. **Lin, L., and J. W. Little.** 1989. Autodigestion and RecA-dependent cleavage of Ind⁻ mutant LexA proteins. *J. Mol. Biol.* **210:**439–452.

166. **Lin, L. L., and J. W. Little.** 1988. Isolation and characterization of noncleavable (Ind⁻) mutants of the LexA repressor of *Escherichia coli* K-12. *J. Bacteriol.* **170:**2163–2173.

167. **Little, J.** 1991. Mechanism of specific LexA cleavage: autodigestion and the role of RecA coprotease. *Biochimie* **73:**411–422.

168. **Little, J. W.** 1983. The SOS regulatory system: control of its state by the level of *recA* protease. *J. Mol. Biol.* **167:**791–808.

169. **Little, J. W.** 1984. Autodigestion of *lexA* and phage λ repressors. *Proc. Natl. Acad. Sci. USA* **81:**1375–1379.

170. **Little, J. W.** 1993. LexA cleavage and other self-processing reactions. *J. Bacteriol.* **175:**4943–4950.

171. **Little, J. W., S. H. Edmiston, Z. Pacelli, and D. W. Mount.** 1980. Cleavage of the *Escherichia coli lexA* protein by the *recA* protease. *Proc. Natl. Acad. Sci. USA* **77:**3225–3229.

172. **Little, J. W., and J. E. Harper.** 1979. Identification of the *lexA* gene product of *Escherichia coli. Proc. Natl. Acad. Sci. USA* **76:**6147–6151.

173. **Little, J. W., and S. A. Hill.** 1985. Deletions within a hinge region of a specific DNA-binding protein. *Proc. Natl. Acad. Sci. USA* **82:**2301–2305.

174. **Little, J. W., and D. G. Kleid.** 1977. *Escherichia coli* protein X is the *recA* gene product. *J. Biol. Chem.* **252:**6251–6252.

175. **Little, J. W., and D. W. Mount.** 1982. The SOS regulatory system of *Escherichia coli. Cell* **29:**11–22.

176. **Little, J. W., D. W. Mount, and C. R. Yanisch-Perron.** 1981. Purified *lexA* protein is a repressor of the *recA* and *lexA* genes. *Proc. Natl. Acad. Sci. USA* **78:**4199–4203.

177. **Livneh, Z.** 1986. Mechanism of replication of ultraviolet-irradiated single-stranded DNA by DNA polymerase III holoenzyme of *Escherichia coli*: implications for SOS mutagenesis. *J. Biol. Chem.* **261:**9526–9533.

178. Livneh, Z. 1986. Replication of UV-irradiated single-stranded DNA by DNA polymerase III holoenzyme of *Escherichia coli*: evidence for bypass of pyrimidine photodimers. *Proc. Natl. Acad. Sci. USA* **83**:4599–4603.

179. Livneh, Z., O. Cohen-Fix, R. Skaliter, and T. Elizur. 1993. Replication of damaged DNA and the molecular mechanism of ultraviolet light mutagenesis. *Crit. Rev. Biochem. Mol. Biol.* **28**:465–513.

180. Livneh, Z., and I. R. Lehman. 1982. Recombinational bypass of pyrimidine dimers promoted by the recA protein of *Escherichia coli*. *Proc. Natl. Acad. Sci. USA* **79**:3171–3175.

181. Lloubes, R., D. Baty, and C. Lazdunski. 1986. The promoters of the genes for colicin production, release and immunity in the ColA plasmid: effects of convergent transcription and LexA protein. *Nucleic Acids Res.* **14**:2621–2636.

182. Lloyd, R. G., F. E. Benson, and C. E. Shurvinton. 1984. Effect of *ruv* mutations on recombination and DNA repair in *Escherichia coli*. *Mol. Gen. Genet.* **194**:303–309.

183. Lloyd, R. G., and C. Buckman. 1991. Genetic analysis of the *recG* locus of *Escherichia coli* K-12 and of its role in recombination and DNA repair. *J. Bacteriol.* **173**:1004–1011.

184. Lloyd, R. G., S. M. Picksley, and C. Prescott. 1983. Inducible expression of a gene specific to the *recF* pathway for recombination in *Escherichia coli* K12. *Mol. Gen. Genet.* **190**:162–167.

185. Lloyd, R. G., and G. J. Sharples. 1991. Molecular organization and nucleotide sequence of the *recG* locus of *Escherichia coli* K-12. *J. Bacteriol.* **173**:6837–6843.

186. Lloyd, R. G., and G. J. Sharples. 1993. Dissociation of synthetic Holliday junctions by *E. coli* RecG protein. *EMBO J.* **12**:17–22.

187. Lodwick, D., D. Owen, and P. Strike. 1990. DNA sequence analysis of the *imp* UV protection and mutation operon of the plasmid TP110: identification of a third gene. *Nucleic Acids Res.* **18**:5045–5050.

188. Love, P. E., M. J. Lyle, and R. E. Yasbin. 1985. DNA-damage-inducible (*din*) loci are transcriptionally activated in competent *Bacillus subtilis*. *Proc. Natl. Acad. Sci. USA* **82**:6201–6205.

189. Love, P. E., and R. E. Yasbin. 1984. Genetic characterization of the inducible SOS-like system of *Bacillus subtilis*. *J. Bacteriol.* **160**:910–920.

190. Love, P. E., and R. E. Yasbin. 1986. Induction of the *Bacillus subtilis* SOS-like response by *Escherichia coli* RecA protein. *Proc. Natl. Acad. Sci. USA* **83**:5204–5208.

191. Lovett, C. M., and J. W. Roberts. 1985. Purification of a RecA analogue from *Bacillus subtilis*. *J. Biol. Chem.* **260**:3305–3313.

192. Lovett, C. M., Jr., K. C. Cho, and T. M. O'Gara. 1993. Purification of an SOS repressor from *Bacillus subtilis*. *J. Bacteriol.* **175**:6842–6849.

193. Lovett, C. M., Jr., P. E. Love, and R. E. Yasbin. 1989. Competence-specific induction of the *Bacillus subtilis* RecA protein analog: evidence for dual regulation of a recombination protein. *J. Bacteriol.* **171**:2318–2322.

194. Lovett, C. M., Jr., P. E. Love, R. E. Yasbin, and J. W. Roberts. 1988. SOS-like induction in *Bacillus subtilis*: induction of the RecA protein analog and a damage-inducible operon by DNA damage in Rec⁺ and DNA repair-deficient strains. *J. Bacteriol.* **170**:1467–1474.

195. Lu, C., and H. Echols. 1987. Correlation of mutagenesis phenotype with binding of mutant RecA proteins to duplex DNA and LexA cleavage. *J. Mol. Biol.* **196**:497–504.

196. Lundegaard, C., and K. F. Jensen. 1994. The DNA damage-inducible *dinD* gene of *Escherichia coli* is equivalent to *orfY* upstream of *pyrE*. *J. Bacteriol.* **176**:3383–3385.

197. Lwoff, A. 1966. The prophage and I, p. 88–99. *In* J. Cairns, G. S. Stent, and J. D. Watson (ed.), *Phage and the Origins of Molecular Biology*. Cold Spring Harbor Laboratory, Cold Spring Harbor, N.Y.

198. Lwoff, A., L. Siminovitch, and N. Kjelgaard. 1950. Induction de la production de bacteriophages chez une bacterie lysogene. *Ann. Inst. Pasteur* (Paris) **79**:815–859.

199. Madiraju, M. V., A. Templin, and A. J. Clark. 1988. Properties of a mutant *recA*-encoded protein reveal a possible role for *Esch-erichia coli recF*-encoded protein in genetic recombination. *Proc. Natl. Acad. Sci. USA* **85**:6592–6596.

200. Maenhaut-Michel, G., R. Janel-Bintz, and R. P. P. Fuchs. 1992. A *umuDC*-independent SOS pathway for frameshift mutagenesis. *Mol. Gen. Genet.* **235**:373–380.

201. Magee, T. R., T. Asai, D. Malka, and T. Kogoma. 1992. DNA damage-inducible origins of DNA replication in *Escherichia coli*. *EMBO J.* **11**:4219–4225.

202. Magee, T. R., and T. Kogoma. 1990. The requirement for RecBC enzyme and an elevated level of activated RecA for induced stable DNA replication in *Escherichia coli*. *J. Bacteriol.* **172**:1834–1839.

203. Magee, T. R., and T. Kogoma. 1991. Rifampin-resistant replication of pBR322 derivatives in *Escherichia coli* cells induced for the SOS response. *J. Bacteriol.* **173**:4736–4741.

204. Mandal, T. N., A. A. Mahdi, G. J. Sharples, and R. G. Lloyd. 1993. Resolution of Holliday intermediates in recombination and DNA repair: indirect suppression of *ruvA*, *ruvB*, and *ruvC* mutations. *J. Bacteriol.* **175**:4325–4334.

205. Markham, B. E., J. W. Little, and D. W. Mount. 1981. Nucleotide sequence of the *lexA* gene of *Escherichia coli* K-12. *Nucleic Acids Res.* **9**:4149–4161.

206. McCall, J. O., E. M. Witkin, T. Kogoma, and V. Roenger-Maniscalco. 1987. Constitutive expression of the SOS response in *recA718* mutants of *Escherichia coli* requires amplification of RecA718 protein. *J. Bacteriol.* **169**:728–734.

207. McEntee, K. 1977. Protein X is the product of the *recA* gene of *Escherichia coli*. *Proc. Natl. Acad. Sci. USA* **74**:5275–5279.

208. McNally, K. P., N. E. Freitag, and G. C. Walker. 1990. LexA-independent expression of a mutant *mucAB* operon. *J. Bacteriol.* **172**:6223–6231.

209. Menetski, J. P., D. G. Bear, and S. C. Kowalcyzkowski. 1990. Stable DNA heteroduplex formation catalyzed by the *Escherichia coli* RecA protein in the absence of ATP hydrolysis. *Proc. Natl. Acad. Sci. USA* **87**:21–25.

210. Miller, H. I., M. Kirk, and H. Echols. 1981. SOS induction and autoregulation of the *himA* gene for site-specific recombination in *E. coli*. *Proc. Natl. Acad. Sci. USA* **78**:6754–6758.

211. Miller, R. V., and T. A. Kokjohn. 1988. Expression of the *recA* gene of *Pseudomonas aeruginosa* PAO is inducible by DNA-damaging agents. *J. Bacteriol.* **170**:2384–2387.

212. Miller, R. V., and T. A. Kokjohn. 1990. General microbiology of *recA*: environmental and evolutionary significance. *Annu. Rev. Microbiol.* **44**:365–394.

213. Miura, A., and J. Tomizawa. 1968. Studies on radiation-sensitive mutants of *E. coli*. III. Participation of the Rec system in induction of mutation by ultraviolet irradiation. *Mol. Gen. Genet.* **103**:1–10.

214. Mount, D. W. 1971. Isolation and genetic analysis of a strain of *Escherichia coli* K-12 with an amber *recA* mutation. *J. Bacteriol.* **107**:388–389.

215. Mount, D. W. 1977. A mutant of *Escherichia coli* showing constitutive expression of the lysogenic induction and error-prone DNA repair pathways. *Proc. Natl. Acad. Sci. USA* **74**:300–304.

216. Mount, D. W., C. Kosel, and A. Walker. 1976. Inducible error-free DNA repair in *tsl recA* mutants of *E. coli*. *Mol. Gen. Genet.* **146**:37–42.

217. Mount, D. W., K. B. Low, and S. Edmiston. 1972. Dominant mutations (*lex*) in *Escherichia coli* K-12 which affect radiation sensitivity and frequency of ultraviolet light-induced mutations. *J. Bacteriol.* **112**:886–893.

218. Mount, D. W., A. C. Walker, and C. Kosel. 1973. Suppression of *lex* mutations affecting deoxyribonucleic acid repair in *Escherichia coli* K-12 by closely linked thermosensitive mutations. *J. Bacteriol.* **116**:950–956.

219. Müller, B., I. Burdett, and S. C. West. 1992. Unusual stability of recombination intermediates made by *E. coli* RecA protein. *EMBO J.* **11**:2685–2693.

220. **Müller, B., I. R. Tsaneva, and S. C. West.** 1993. Branch migration of Holliday junctions promoted by the *Escherichia coli* RuvA and RuvB proteins: comparison of the RuvAB- and RuvB-mediated reactions. *J. Biol. Chem.* **268:**17179–17184.

221. **Murialdo, H.** 1988. Lethal effect of λ terminase in recombination-deficient *Escherichia coli. Mol. Gen. Genet.* **213:**42–49.

222. **Notani, N. K., and J. E. Setlow.** 1980. Inducible repair system in *Haemophilus influenzae. J. Bacteriol.* **143:**516–519.

223. **Nurse, P., K. Zavitz, and K. Marians.** 1991. Inactivation of the *Escherichia coli* PriA DNA replication protein induces the SOS response. *J. Bacteriol.* **173:**6686–6693.

224. **Ogawa, H., K. Shimada, and J.-I. Tomizawa.** 1968. Studies on radiation-sensitive mutants of *E. coli*. I. Mutants defective in the repair synthesis. *Mol. Gen. Genet.* **101:**227–244.

225. **Orrego, C., and E. Eisenstadt.** 1987. An inducible pathway is required for mutagenesis in *Salmonella typhimurium* LT2. *J. Bacteriol.* **169:**2885–2888.

226. **Ossanna, N., and D. W. Mount.** 1989. Mutations in *uvrD* induce the SOS response in *Escherichia coli. J. Bacteriol.* **171:**303–307.

227. **Otsuji, N., H. Iyehara, and Y. Hideshima.** 1974. Isolation and characterization of an *Escherichia coli ruv* mutant which forms nonseptate filaments after low doses of ultraviolet light irradiation. *J. Bacteriol.* **117:**337–344.

228. **Pabo, C. O., and R. T. Sauer.** 1984. Protein-DNA recognition. *Annu. Rev. Biochem.* **53:**293–321.

229. **Pacelli, L. A., S. H. Edmiston, and D. W. Mount.** 1979. Isolation and characterization of amber mutations in the *lexA* gene of *Escherichia coli* K-12. *J. Bacteriol.* **137:**568–573.

230. **Pang, P. P., and G. C. Walker.** 1983. The *Salmonella typhimurium* LT2 *uvrD* gene is regulated by the *lexA* gene product. *J. Bacteriol.* **154:**1502–1504.

231. **Panyutin, I. G., and P. Hsieh.** 1993. Formation of single base mismatch impedes spontaneous DNA branch migration. *J. Mol. Biol.* **230:**413–424.

232. **Parsons, C. A., I. Tsaneva, R. G. Lloyd, and S. C. West.** 1992. Interaction of *Escherichia coli* RuvA and RuvB proteins with synthetic Holliday junctions. *Proc. Natl. Acad. Sci. USA* **89:**5452–5456.

233. **Parsons, C. A., and S. C. West.** 1993. Formation of a RuvAB-Holliday junction complex in vitro. *J. Mol. Biol.* **232:**397–405.

234. **Payne, N., and A. Sancar.** 1989. The LexA protein does not bind specifically to the two SOS box-like sequences immediately 5′ to the *phr* gene. *Mutat. Res.* **218:**207–210.

235. **Perry, K. L., S. J. Elledge, B. Mitchell, L. Marsh, and G. C. Walker.** 1985. *umuDC* and *mucAB* operons whose products are required for UV light- and chemical-induced mutagenesis: UmuD, MucA, and LexA proteins share homology. *Proc. Natl. Acad. Sci. USA* **82:**4331–4335.

236. **Peterson, K. R., and D. W. Mount.** 1987. Differential repression of SOS genes by unstable LexA41 (Tsl-1) protein causes a 'split-phenotype' in *Escherichia coli* K-12. *J. Mol. Biol.* **193:**27–40.

237. **Peterson, K. R., N. Ossanna, A. T. Thliveris, D. G. Ennis, and D. W. Mount.** 1988. Derepression of specific genes promotes DNA repair and mutagenesis in *Escherichia coli. J. Bacteriol* **170:**1–4.

238. **Petit, C., C. Cayrol, C. Lesca, P. Kaiser, C. Thompson, and M. Defais.** 1993. Characterization of *dinY*, a new *Escherichia coli* DNA repair gene whose products are damage inducible even in a *lexA*(Def) background. *J. Bacteriol.* **175:**642–646.

239. **Phizicky, E. M., and J. W. Roberts.** 1981. Induction of SOS functions: regulation of proteolytic activity of *E. coli* Rec protein by interaction with DNA and nucleoside triphosphate. *Cell* **25:**259–267.

240. **Picksley, S. M., P. V. Attfield, and R. G. Lloyd.** 1984. Repair of double-strand breaks in *Escherichia coli* requires a functional *recN* gene product. *Mol. Gen. Genet.* **195:**267–274.

241. **Pollard, E., and E. P. Randall.** 1973. Studies on the inducible inhibitor of radiation-induced DNA degradation of *E. coli. Radiat. Res.* **55:**265–279.

242. **Pollard, E. C., and P. M. Achey.** 1975. Induction of radioresistance in *Escherichia coli. Biophys. J.* **15:**1141–1154.

243. **Pollard, E. C., D. J. Fluke, and D. Kazanis.** 1981. Induced radioresistance: an aspect of induced repair. *Mol. Gen. Genet.* **184:**421–429.

244. **Pollard, E. C., S. Person, M. Rader, and D. J. Fluke.** 1977. Relationship of ultraviolet light mutagenesis to a radiation-damage inducible system in *Escherichia coli. Radiat. Res.* **72:**519–532.

245. **Quinones, A., W. Jüterbock, and W. Messer.** 1991. Expression of the *dnaA* gene of *Escherichia coli* is inducible by DNA damage. *Mol. Gen. Genet.* **227:**9–16.

246. **Quinones, A., J. Kaasch, M. Kaasch, and W. Messer.** 1989. Induction of *dnaN* and *dnaQ* gene expression in *Escherichia coli* by alkylation damage to DNA. *EMBO J.* **8:**587–593.

247. **Quinones, A., C. Kucherer, R. Piechocki, and W. Messer.** 1987. Reduced transcription of the *rnh* gene in *Escherichia coli* mutants expressing the SOS regulon constitutively. *Mol. Gen. Genet.* **206:**95–100.

248. **Radding, C. M.** 1988. Homologous pairing and strand exchange promoted by *Escherichia coli* RecA protein, p. 193–229. *In* R. Kucherlapati and G. R. Smith (ed.), *Genetic Recombination*. American Society for Microbiology, Washington, D.C.

249. **Radding, C. M.** 1991. Helical interactions in homologous pairing and strand exchange driven by RecA protein. *J. Biol. Chem.* **266:** 5355–5358.

250. **Radman, M.** 1974. Phenomenology of an inducible mutagenic DNA repair pathway in *Escherichia coli*: SOS repair hypothesis, p. 128–142. *In* L. Prakash, F. Sherman, M. Miller, C. Lawrence, and H. W. Tabor (ed.), *Molecular and Environmental Aspects of Mutagenesis*. Charles C Thomas, Springfield, Ill.

251. **Radman, M.** 1975. SOS repair hypothesis: phenomenology of an inducible DNA repair which is accompanied by mutagenesis, p. 355–367. *In* P. Hanawalt and R. B. Setlow (ed.), *Molecular Mechanisms for Repair of DNA*, part A. Plenum Publishing Corp., New York.

252. **Rajagopalan, M., C. Lu, R. Woodgate, M. O'Donnell, M. F. Goodman, and H. Echols.** 1992. Activity of the purified mutagenesis proteins UmuC, UmuD′, and RecA in replicative bypass of an abasic DNA lesion by DNA polymerase III. *Proc. Natl. Acad. Sci. USA* **89:**10777–10781.

253. **Rao, B. J., M. Dutreix, and C. M. Radding.** 1991. Stable three-stranded DNA made by RecA protein. *Proc. Natl. Acad. Sci. USA* **88:**2984–2988.

254. **Raymond-Denise, A., and N. Guillen.** 1991. Identification of *dinR*, a DNA damage-inducible regulator gene of *Bacillus subtilis. J. Bacteriol.* **173:**7084–7091.

255. **Resnick, M. A.** 1976. The repair of double-strand breaks in DNA: a model involving recombination. *J. Theor. Biol.* **59:**97–106.

256. **Roberts, J. W., and C. W. Roberts.** 1975. Proteolytic cleavage of bacteriophage lambda repressor in induction. *Proc. Natl. Acad. Sci. USA* **72:**147–151.

257. **Roberts, J. W., and C. W. Roberts.** 1981. Two mutations that alter the regulatory activity of *E. coli recA* protein. *Nature* (London) **290:**422–424.

258. **Roberts, J. W., C. W. Roberts, and D. W. Mount.** 1977. Inactivation and proteolytic cleavage of phage lambda repressor *in vitro* in an ATP-dependent reaction. *Proc. Natl. Acad. Sci. USA* **74:**2283–2287.

259. **Roca, A., and M. Cox.** 1990. The RecA protein: structure and function. *Crit. Rev. Biochem. Mol. Biol.* **25:**415–456.

260. **Roland, K. L., and J. W. Little.** 1990. Reaction of LexA repressor with diisopropylfluorophosphate: a test of the serine protease model. *J. Biol. Chem.* **265:**12828–12835.

261. **Roland, K. L., M. H. Smith, J. A. Rupley, and J. W. Little.** 1992. *In vitro* analysis of mutant LexA proteins with an increased rate of specific cleavage. *J. Mol. Biol.* **228:**395–408.

262. **Rosenberg, M., and H. Echols.** 1990. Differential recognition of ultraviolet lesions by RecA protein. *J. Biol. Chem.* **265:**20641–20645.

263. **Rosner, J. L., L. R. Kass, and M. B. Yarmolinksy.** 1968. Parallel behavior of F and Pl in causing indirect induction of lysogenic bacteria. *Cold Spring Harbor Symp. Quant. Biol.* **33:**785–789.

264. Ross, P., and P. Howard-Flanders. 1977. Initiation of *recA⁺*-dependent recombination in *Escherichia coli* (λ). *J. Mol. Biol.* **117:**137–158.

265. Rosselli, R., and A. Stasiak. 1990. Energetics of RecA-mediated recombination reactions—without ATP hydrolysis RecA can mediate polar exchange but is unable to recycle. *J. Mol. Biol.* **216:**335–352.

266. Rosselli, W., and A. Stasiak. 1991. The ATPase activity of RecA is needed to push the DNA strand exchange through heterologous regions. *EMBO J.* **10:**4391–4396.

267. Rostas, K., S. J. Morton, S. M. Picksley, and R. G. Lloyd. 1987. Nucleotide sequence and LexA regulation of the *Escherichia coli recN* gene. *Nucleic Acids Res.* **15:**5041–5050.

268. Rupp, W. D., and P. Howard-Flanders. 1968. Discontinuities in the DNA synthesized in an excision-defective strain of *Escherichia coli* following ultraviolet irradiation. *J. Mol. Biol.* **31:**291–304.

269. Rupp, W. D., C. E. I. Wilde, D. L. Reno, and P. Howard-Flanders. 1971. Exchanges between DNA strands in ultraviolet-irradiated *Escherichia coli*. *J. Mol. Biol.* **61:**25–44.

270. Salles, B., and M. Defais. 1984. Signal of induction of recA protein in *E. coli*. *Mutat. Res.* **131:**53–59.

271. Salles, B., and C. Paoletti. 1983. Control of UV induction of *recA* protein. *Proc. Natl. Acad. Sci. USA* **80:**65–69.

272. Salles, B., C. Paoletti, and G. Villani. 1983. Lack of single-strand DNA-binding protein amplification under conditions of SOS induction in *E. coli*. *Mol. Gen. Genet.* **189:**175–177.

273. Salles, B., and G. M. Weinstock. 1989. Mutation of the promoter and LexA binding sites of *cea*, the gene encoding colicin E1. *Mol. Gen. Genet.* **215:**483–489.

274. Salles, B., J. M. Weisemann, and G. Weinstock. 1987. Temporal control of colicin E1 induction. *J. Bacteriol.* **169:**5028–5034.

275. Sancar, A., G. B. Sancar, W. D. Rupp, J. W. Little, and D. W. Mount. 1982. *LexA* protein inhibits transcription of the *E. coli uvrA* gene *in vitro*. *Nature* (London) **298:**96–98.

276. Sancar, G. B., A. Sancar, J. W. Little, and W. D. Rupp. 1982. The *uvrB* gene of *Escherichia coli* has both *lexA*-repressed and *lexA*-independent promoters. *Cell* **28:**523–530.

277. Sargentini, N. J., and K. C. Smith. 1986. Quantitation of the involvement of the *recA, recB, recC, recF, recJ, recN, lexA, radA, radB, uvrD,* and *umuC* genes in the repair of X–ray-induced DNA double-strand breaks in *Escherichia coli*. *Radiat. Res.* **107:**58–72.

278. Sargentini, N. J., and K. C. Smith. 1986. Role of the *radB* gene in postreplicational repair in UV-irradiated *Escherichia coli uvrB*. *Radiat. Res.* **166:**17–22.

279. Sargentini, N. J., and K. C. Smith. 1988. Genetic and phenotypic analyses indicating occurrences of the *recN262* and *radB101* mutations at the same locus in *Escherichia coli*. *J. Bacteriol.* **170:**2392–2394.

280. Sassanfar, M., and J. W. Roberts. 1990. Nature of the SOS-inducing signal in *Escherichia coli*: the involvement of DNA replication. *J. Mol. Biol.* **212:**79–96.

281. Sauer, R. T., M. J. Ross, and M. Ptashne. 1982. Cleavage of the λ and P22 repressors by RecA protein. *J. Biol. Chem.* **257:**4458–4462.

282. Sauer, R. T., R. R. Yocum, R. F. Doolittle, M. Lewis, and C. O. Pabo. 1982. Homology among DNA-binding proteins suggests use of a conserved super-secondary structure. *Nature* (London) **298:**447–451.

283. Schnarr, M., M. Granger-Schnarr, S. Hurstel, and J. Pouyet. 1988. The carboxy-terminal domain of the LexA repressor oligomerises essentially as the entire protein. *FEBS Lett.* **234:**56–60.

284. Schnarr, M., P. Oertel-Buchheit, M. Kazmaier, and M. Granger-Schnarr. 1991. DNA binding properties of the LexA repressor. *Biochimie* **73:**423–431.

285. Schnarr, M., J. Pouyet, M. Granger-Schnarr, and M. Daune. 1985. Large-scale purification, oligomerization equilibria, and specific interaction of the LexA repressor of *Escherichia coli*. *Biochemistry* **24:**2812–2818.

286. Schuster, H., D. Beyersmann, M. Mikolajczyk, and M. Schlicht. 1973. Prophage induction by high temperature in thermo-sensitive *dna* mutants lysogenic for bacteriophage lambda. *J. Virol.* **11:**879–885.

287. Sedgwick, S. G. 1975. Genetic and kinetic evidence for different types of postreplication repair in *Escherichia coli* B. *J. Bacteriol.* **123:**154–161.

288. Sedgwick, S. G., and P. A. Goodwin. 1985. Interspecies regulation of the SOS response by the *E. coli lexA⁺* gene. *Mutat. Res.* **145:**103–106.

289. Sedliakova, M., V. Slezarikova, J. Brozmanova, F. Masek, and V. Bayerova. 1980. Role of UV-inducible proteins in repair of various wild-type *Escherichia coli* cells. *Mutat. Res.* **71:**15–23.

290. Sedliakova, M., V. Slezarikova, and M. Pirsel. 1978. UV-inducible repair. II. Its role in various defective mutants of *Escherichia coli* K-12. *Mol. Gen. Genet.* **167:**209–215.

291. Setlow, R. B., P. A. Swenson, and W. L. Carrier. 1963. Thymidine dimers and inhibition of DNA synthesis by ultraviolet irradiation of cells. *Science* **142:**1464–1466.

292. Sharples, G. J., and R. G. Lloyd. 1991. Resolution of Holliday junctions in *Escherichia coli*: identification of the *ruvC* gene product as a 19-kilodalton protein. *J. Bacteriol.* **173:**7711–7715.

293. Shavitt, O., and Z. Livneh. 1989. Rolling-circle replication of UV-irradiated duplex DNA in the φX174 replicative-form → single-strand replication system in vitro. *J. Bacteriol.* **171:**3530–3538.

294. Shiba, T., H. Iwasaki, A. Nakata, and H. Shinagawa. 1993. *Escherichia coli* RuvA and RuvB proteins involved in recombination repair: physical properties and interactions with DNA. *Mol. Gen. Genet.* **237:**395–399.

295. Shinagawa, H., K. Makino, M. Amemura, S. Kimura, H. Iwasaki, and A. Nakata. 1988. Structure and regulation of the *Escherichia coli ruv* operon involved in DNA repair and recombination. *J. Bacteriol.* **170:**4322–4329.

296. Shurvinton, C. E., and R. G. Lloyd. 1982. Damage to DNA induces expression of the *ruv* gene of *Escherichia coli*. *Mol. Gen. Genet.* **185:**352–355.

297. Siegel, E. C. 1983. The *Escherichia coli uvrD* gene is inducible by DNA damage. *Mol. Gen. Genet.* **191:**397–400.

298. Slilaty, S. N., and J. W. Little. 1987. Lysine-156 and serine-119 are required for LexA repressor cleavage: a possible mechanism. *Proc. Natl. Acad. Sci. USA* **84:**3987–3991.

299. Slilaty, S. N., J. A. Rupley, and J. W. Little. 1986. Intramolecular cleavage of LexA and phage lambda repressors: dependence of kinetics on repressor concentration, pH, temperature, and solvent. *Biochemistry* **25:**6866–6875.

300. Smith, C. M., Z. Arany, C. Orrego, and E. Eisenstadt. 1991. DNA damage-inducible loci in *Salmonella typhimurium*. *J. Bacteriol.* **173:**3587–3590.

301. Smith, G. R. 1987. Mechanism and control of homologous recombination in *Escherichia coli*. *Annu. Rev. Genet.* **21:**179–201.

302. Smith, G. R. 1991. Conjugational recombination in *E. coli*: myths and mechanisms. *Cell* **64:**19–27.

303. Smith, K. C., and K. D. Martignoni. 1976. Protection of *Escherichia coli* cells against the lethal effect of UV and X-irradiation by prior X-irradiation. *Photochem. Photobiol.* **24:**515–523.

304. Smith, K. C., and D. H. C. Meun. 1970. Repair of radiation-induced damage in *Escherichia coli*. I. Effect of *rec* mutations on post-replication repair of damage due to ultraviolet radiation. *J. Mol. Biol.* **51:**459–472.

305. Smith, K. C., and T.-C. V. Wang. 1987. *recA*-dependent DNA repair in UV-irradiated *Escherichia coli*. *J. Photochem. Photobiol.* **1:**1–11.

306. Smith, K. C., and T. V. Wang. 1989. *recA*-dependent DNA repair processes. *Bioessays* **10:**12–16.

307. Smith, M., M. M. Cavenaugh, and J. W. Little. 1991. Mutant LexA proteins with an increased rate of *in vivo* cleavage. *Proc. Natl. Acad. Sci. USA* **88:**7356–7360.

308. Snyder, M., and K. Drlica. 1979. DNA gyrase on the bacterial chromosome: DNA cleavage induced by oxolinic acid. *J. Mol. Biol.* **131:**287–302.

309. Sommer, S., J. Knezevic, A. Bailone, and R. Devoret. 1993. Induction of only one SOS operon, *umuDC*, is required for SOS mutagenesis in *E. coli. Mol. Gen. Genet.* **239:**137–144.

310. Sommer, S., A. Leitao, A. Bernardi, A. Bailone, and R. Devoret. 1991. Introduction of a UV-damaged replicon into a recipient cell is not a sufficient condition to produce an SOS-inducing signal. *Mutat. Res.* **254:**107–117.

311. Stasiak, A. 1992. Three-stranded DNA structure: is this the secret of DNA homologous recognition? *Mol. Microbiol.* **6:**3267–3276.

312. Stasiak, A., and E. H. Egelman. 1988. Visualization of recombination reactions, p. 265–308. *In* R. Kucherlapati and G. R. Smith (ed.), *Genetic Recombination.* American Society for Microbiology, Washington, D.C.

312a. Stasiak, A., I. R. Tsaneva, S. C. West, C. J B. Benson, X. Yu, and E. Egelman. 1994. The *Escherichia coli* RuvB branch migration protein forms double hexameric rings around DNA. *Proc. Natl. Acad. Sci. USA* **918:**7618–7622.

313. Story, R. M., D. K. Bishop, N. Kleckner, and T. A. Steitz. 1993. Structural relationship of bacterial RecA proteins to recombination proteins from bacteriophage T4 and yeast. *Science* **259:**1892–1896.

314. Story, R. M., and T. A. Steitz. 1992. Structure of the RecA protein-ADP complex. *Nature* (London) **355:**374–376.

315. Story, R. M., I. T. Weber, and T. A. Steitz. 1992. The structure of the *E. coli recA* protein monomer and polymer. *Nature* (London) **355:**318–325.

316. Strynadka, N. C. J., H. Adachi, S. E. Jensen, K. Johns, A. Sielecki, C. Betzel, K. Sutoh, and M. N. G. James. 1992. Molecular structure of the acyl-enzyme intermediate in β-lactam hydrolysis at 1.7 Angstrom resolution. *Nature* (London) **359:**700–705.

317. Sugino, A., C. L. Peebles, K. N. Kreuzer, and N. R. Cozzarelli. 1977. Mechanism of action of nalidixic acid: purification of *Escherichia coli nalA* gene product and its relationship to DNA gyrase and a novel nicking-closing enzyme. *Proc. Natl. Acad. Sci. USA* **74:**4767–4771.

318. Sweasy, J. B., and E. M. Witkin. 1991. Novel SOS phenotypes caused by second-site mutations in the *recA430* gene of *Escherichia coli. Biochimie* **73:**437–448.

319. Szostak, J. W., T. L. Orr-Weaver, and R. J. Rothstein. 1983. The double-strand-break repair model for recombination. *Cell* **33:**25–35.

320. Takahagi, M., H. Iwasaki, A. Nakata, and H. Shinagawa. 1991. Molecular analysis of the *Escherichia coli ruvC* gene, which encodes a Holliday junction-specific endonuclease. *J. Bacteriol.* **173:**5747–5753.

321. Thaler, D. S., and F. W. Stahl. 1988. DNA double-chain breaks in recombination of phage λ and of yeast. *Annu. Rev. Genet.* **22:**168–197.

322. Thliveris, A. T., J. W. Little, and D. W. Mount. 1991. Repression of the *E. coli recA* gene requires at least two LexA protein monomers. *Biochimie* **73:**449–455.

323. Thoms, B., and W. Wackernagel. 1984. Genetic control of damage-inducible restriction alleviation in *Escherichia coli* K12: an SOS function not repressed by *lexA. Mol. Gen. Genet.* **197:**297–303.

324. Thoms, B., and W. Wackernagel. 1987. Regulatory role of *recF* in the SOS response of *Escherichia coli*: impaired induction of SOS genes by UV irradiation and nalidixic acid in a *recF* mutant. *J. Bacteriol.* **169:**1731–1736.

325. Thresher, R., G. Christiansen, and J. D. Griffith. 1988. Assembly of presynaptic filaments. Factors affecting the assembly of RecA protein onto single-stranded DNA. *J. Mol. Biol.* **201:**101–113.

326. Tsaneva, I. R., B. Müller, and S. C. West. 1992. ATP-dependent branch migration of Holliday junctions promoted by the RuvA and RuvB proteins of *E. coli. Cell* **69:**1171–1180.

327. Umezu, K., N.-W. Chi, and R. D. Kolodner. 1993. Biochemical interactions of the *Escherichia coli* RecF, RecO, and RecR proteins with RecA protein and single-stranded binding protein. *Proc. Natl. Acad. Sci. USA* **90:**3875–3879.

328. van Sluis, C. A., G. F. Moolenaar, and C. Backendorf. 1983. Regulation of the *uvrC* gene of *Escherichia coli* K12: localization and characterization of a damage-inducible promoter. *EMBO J.* **2:**2313–2318.

329. Verma, M., K. G. Moffat, and J. B. Egan. 1989. UV irradiation inhibits initiation of DNA replication of *oriC* in *Escherichia coli. Mol. Gen. Genet.* **216:**446–454.

330. Walker, G. C. 1984. Mutagenesis and inducible responses to deoxyribonucleic acid damage in *Escherichia coli. Microbiol. Rev.* **48:**60–93.

331. Walker, G. C. 1985. Inducible DNA repair systems. *Annu. Rev. Biochem.* **54:**425–457.

332. Walker, G. C. 1987. The SOS response of *Escherichia coli*, p. 1346–1357. *In* F. C. Neidhardt, J. L. Ingraham, K. B. Low, B. Magasanik, M. Schaechter, and H. E. Umbarger (ed.), *Escherichia coli and Salmonella typhimurium: Cellular and Molecular Biology.* American Society for Microbiology, Washington, D.C.

333. Walker, G. C., C. J. Kenyon, A. Bagg, P. J. Langer, and W. G. Shanabruch. 1982. Mutagenesis and cellular responses to DNA damage. *J. Natl. Cancer Inst. Monogr.* **60:**257–267.

334. Wang, T.-C. V., and K. C. Smith. 1985. Mechanism of *sbcB*-suppression of the *recBC*-deficiency in postreplicational repair in UV-irradiated *Escherichia coli* K-12. *Mol. Gen. Genet.* **201:**186–191.

335. Wang, T.-C. V., and K. C. Smith. 1986. Postreplicational formation and repair of DNA double-strand breaks in UV-irradiated *Escherichia coli uvrB* cells. *Mutat. Res.* **165:**39–44.

336. Wang, T.-C. V., and K. C. Smith. 1986. *recA* (Srf) suppression of *recF* deficiency in the postreplication repair of UV-irradiated *Escherichia coli* K-12. *J. Bacteriol.* **168:**940–946.

337. Wang, W.-B., M. Sassanfar, I. Tessman, J. W. Roberts, and E. S. Tessman. 1988. Activation of protease-constitutive RecA proteins of *Escherichia coli* by all of the common nucleoside triphosphates. *J. Bacteriol.* **170:**4816–4822.

338. Wang, W.-B., I. Tessman, and E. S. Tessman. 1988. Activation of protease-constitutive RecA proteins of *Escherichia coli* by rRNA and tRNA. *J. Bacteriol.* **170:**4823–4827.

339. Weigle, J. J. 1953. Induction of mutation in a bacterial virus. *Proc. Natl. Acad. Sci. USA* **39:**628–636.

340. Wertman, K. F., J. W. Little, and D. W. Mount. 1984. Rapid mutational analysis of regulatory loci in *Escherichia coli* K-12 using bacteriophage M13. *Proc. Natl. Acad. Sci. USA* **81:**3801–3805.

341. Wertman, K. F., and D. W. Mount. 1985. Nucleotide sequence binding specificity of the LexA repressor of *Escherichia coli* K-12. *J. Bacteriol.* **163:**376–384.

342. West, S. C. 1994. The processing of recombination intermediates: mechanistic insights from studies of bacterial proteins. *Cell* **76:**9–15.

343. West, S. C. 1992. Enzymes and molecular mechanisms of genetic recombination. *Annu. Rev. Biochem.* **61:**603–640.

344. West, S. C., E. Cassuto, and P. Howard-Flanders. 1981. Mechanism of *E. coli* RecA protein-directed strand exchanges in postreplication repair of DNA. *Nature* (London) **294:**659–662.

345. West, S. C., E. Cassuto, and P. Howard-Flanders. 1982. Postreplication-repair in *E. coli*: strand exchange reactions of gapped DNA by RecA protein. *Mol. Gen. Genet.* **187:**209–217.

346. West, S. C., and B. Connolly. 1992. Biological roles of the *Escherichia coli* RuvA, RuvB and RuvC proteins revealed. *Mol. Microbiol.* **6:**2755–2759.

347. West, S. C., J. K. Countryman, and P. Howard-Flanders. 1983. Enzymatic formation of biparental figure-8 molecules from plasmid DNA and their resolution in *Escherichia coli. Cell* **32:**817–828.

348. West, S. C., and P. Howard-Flanders. 1984. Duplex-duplex interactions catalyzed by RecA protein allow strand exchanges to pass double-strand breaks in DNA. *Cell* **37:**683–691.

349. Whitby, M. C., L. Ryder, and R. G. Lloyd. 1993. Reverse branch migration of Holliday junctions by RecG protein: a new mechanism for resolution of Holliday junctions of intermediates in recombination. *Cell* **75:**341–350.

350. Winans, S. C., S. J. Elledge, J. H. Krueger, and G. C. Walker. 1985. Site-directed insertion and deletion mutagenesis with cloned fragments in *Escherichia coli. J. Bacteriol.* **161:**1219–1221.

351. Witkin, E. M. 1967. Mutation-proof and mutation-prone modes of survival in derivatives of *Escherichia coli* B differing in sensitivity to ultraviolet light. *Brookhaven Symp. Biol.* **20:**495–503.

352. Witkin, E. M. 1967. The radiation sensitivity of *Escherichia coli* B: a hypothesis relating filament formation and prophage induction. *Proc. Natl. Acad. Sci. USA* **57:**1275–1279.

353. Witkin, E. M. 1969. The mutability towards ultraviolet light of recombination-deficient strains of *Escherichia coli. Mutat. Res.* **8:**9–14.

354. Witkin, E. M. 1974. Thermal enhancement of ultraviolet mutability in a *tif-1 uvrA* derivative of *Escherichia coli* B/r: evidence that ultraviolet mutagenesis depends upon an inducible function. *Proc. Natl. Acad. Sci. USA* **71:**1930–1934.

355. Witkin, E. M. 1976. Ultraviolet mutagenesis and inducible DNA repair in *Escherichia coli.* Bacteriol. Rev. **40:**869–907.

356. Witkin, E. M. 1987. This week's citation classic. *Curr. Contents* **30:**17.

357. Witkin, E. M. 1991. RecA protein in the SOS response: milestones and mysteries. *Biochimie* **73:**133–141.

358. Witkin, E. M., and D. L. George. 1973. Ultraviolet mutagenesis in *polA* and *uvr polA* derivatives of *Escherichia coli* B/r: evidence for an inducible error-prone repair system. *Genetics* **73** (Suppl.) :91–108.

359. Witkin, E. M., J. O. McCall, M. R. Volkert, and I. E. Wermundsen. 1982. Constitutive expression of SOS functions and modulation of mutagenesis resulting from resolution of genetic instability at or near the *recA* locus of *Escherichia coli. Mol. Gen. Genet.* **185:**43–50.

360. Witkin, E. M., V. Roegner-Maniscalco, J. B. Sweasy, and J. O. McCall. 1987. Recovery from ultraviolet light-induced inhibition of DNA synthesis requires *umuDC* gene products in *recA718* mutant strains but not in *recA+* strains of *Escherichia coli. Proc. Natl. Acad. Sci. USA* **84:**6805–6809.

361. Yasbin, R. E., D. L. Cheo, and K. W. Bayles. 1992. Inducible DNA repair and differentiation in *Bacillus subtilis*: interactions between global regulons. *Mol. Microbiol.* **6:**1263–1270.

362. Youngs, D. A., and K. C. Smith. 1976. Genetic control of multiple pathways of post-replicational repair in *uvrB* strains of *Escherichia coli* K-12. *J. Bacteriol.* **125:**102–110.

363. Yu, X., and E. H. Egelman. 1992. Structural data suggest that active and inactive forms of the RecA filaments are not simply interconvertible. *J. Mol. Biol.* **227:**334–346.

364. Yu, X., and E. H. Egelman. 1993. The LexA repressor binds within the deep helical groove of the activated RecA filament. *J. Mol. Biol.* **231:**29–40.

365. Zgaga, V., and B. Militic. 1965. Superinfection with homologous phage of *Escherichia coli* K-12 (λ) treated with 6-azauracil. *Virology* **27:**205–212.

Mutagenesis in Prokaryotes

Most organisms are endowed with an impressive set of biochemical processes which are capable of accurately repairing DNA lesions. Nonetheless, some lesions either escape repair by these systems or for some reason cannot be accurately repaired. The presence of such lesions in DNA can lead to the *introduction of mutations*. There is now fairly widespread agreement that most mutagenesis that is induced as a consequence of DNA damage in prokaryotes occurs by a special DNA damage tolerance process termed *translesion DNA synthesis* in which a DNA polymerase inserts nucleotides opposite a misinstructional or noninstructional lesion (Fig. 11–1). It has been a challenging task to identify the proteins involved in this process, to determine their mechanistic roles, and to understand the influences of lesion structure on the resulting mutagenesis. In this chapter we will review the key approaches and experiments that have led to the current models for how such UV radiation and chemical mutagenesis occurs in *E. coli*. We will consider both mutagenesis that is dependent on the SOS system and mutagenesis that is independent of it (35, 87, 88, 195, 214, 228, 342–344, 352).

SOS-Dependent Mutagenesis: Requirements for Particular Gene Products

SOS Mutagenesis by UV Radiation and Most Chemicals Is Not a Passive Process

A fundamental concept that has emerged from studies of mutagenesis in *Escherichia coli* is that, at least in this organism, mutagenesis by UV and many chemicals is not a passive process. Rather, it requires the intervention of a cellular system to process damaged DNA efficiently in such a way that mutations result. This type of processing has often been referred to as *error-prone repair* or *SOS repair* (251, 352), terms proposed on the basis of a number of observations in *E. coli* which

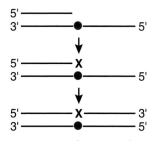

Figure 11–1 Translesion synthesis. A polymerase incorporates an incorrect nucleotide opposite a noninstructional or misinstructive lesion (represented by the red circle) and then continues synthesis.

suggested that cellular events involved in producing mutations from damaged DNA are closely associated with events increasing resistance to killing by chemicals and radiation. However, since it is not clear that its mechanism satisfies the definition of DNA repair used in this book, we shall refer to this specialized type of processing as *SOS mutagenesis*. Since it is involved with the introduction of mutations, this process contrasts with the accurate mechanisms of DNA damage tolerance discussed in chapter 10.

The concept that organisms can take an active role in mutating their DNA after exposure to UV radiation or many chemicals is somewhat counterintuitive. The idea that such a system might exist originally arose from studies of the genetic dependence of UV mutagenesis in *E. coli*, while the idea that such a system might be inducible came from experiments with bacteriophages as well as from studies of chromosomal mutagenesis.

W REACTIVATION AND W MUTAGENESIS: INDUCIBILITY OF SOS MUTAGENESIS AND EVIDENCE FOR TRANSLESION SYNTHESIS

The physiological experiments that first suggested that an inducible system is required for mutagenesis were those in which UV-irradiated bacteriophage infected UV-irradiated cells. As discussed in chapter 10, Jean Weigle (347) first noted that preirradiation of *E. coli* with a low dose of UV radiation prior to infection with UV-irradiated bacteriophage λ resulted in increased survival of the UV-irradiated phage (Weigle or W reactivation) (Fig. 10–2). Furthermore, and very strikingly, UV irradiation of the λ phage had very little effect on the phage mutation frequency unless the cells had been preirradiated with UV light before phage infection (Weigle or W mutagenesis). In other words, it seemed as though the UV-induced lesions in the phage DNA did not lead to the production of mutations unless the bacterial host had also been damaged. The subsequent demonstration (74) that neither W reactivation nor W mutagenesis was observed if the cells carried either a *recA*(Def) or *lexA*(Ind⁻) mutation (Table 11–1) suggested that the cellular functions responsible for increased phage survival and mutagenesis were induced in a *recA⁺-lexA⁺*-dependent fashion by UV irradiation of the bacteria and led to their subsequent classification as SOS responses.

Further evidence supporting the inducibility of the SOS mutagenesis system came from the observation that induction of W reactivation and W mutagenesis was inhibited by the presence of chloramphenicol during the period after the irradiation of the cells but before infection with the UV-irradiated bacteriophage (73). Additional support came from experiments with strains carrying the *recA441*(Cpt^Ts) mutation (see Table 10–4). As discussed in chapter 10, if a *recA441*(Cpt^Ts) strain was grown under conditions in which SOS responses were constitutively expressed, W reactivation and W mutagenesis were observed even though the bacterial DNA was not damaged. Since new cellular functions apparently had to be induced for UV radiation damage in the phage to lead to mutations, these results implied the existence of a specialized system required for SOS mutagenesis in *E. coli*. The further observations that the genetic requirements and dose requirements for W mutagenesis were extremely similar to those for W reactivation led to the suggestion (74) that the two processes might have some steps in common. Radman (251) then proposed that the mutations arose as a consequence of an error-prone bypass of lesions produced by UV radiation.

Initial evidence for translesion bypass during W reactivation came from an examination of the replication of UV-irradiated φX174 phage in SOS-induced cells (55). This single-stranded phage was used since it could not be repaired by excision repair and, at low multiplicities of infection, could not be repaired by recombination with another phage. The observation that a fraction of the phage molecules were replicated in SOS-induced cells supported the idea that translesion synthesis had occurred on the damaged single-stranded DNA molecule. Later work examining the replication of UV-irradiated λ in SOS-induced *E. coli uvrA* cells provided evidence consistent with the suggestion that translesion synthesis is the primary component of W reactivation of λ in the absence of excision repair (75). The

Table 11–1 W mutagenesis of UV-irradiated λ in *lexA*(Ind⁻) and *recA*(Def) strains

Relevant genotype	UV dose to bacteria (J/m²)	Frequency of λ clear plaque mutations ($\times 10^{-4}$) at UV dose (J/m²) to λ phage of:	
		0	50
recA⁺ lexA⁺	0	3.9	2.1
recA⁺ lexA⁺	30	4.9	22.6
recA⁺ lexA(Ind⁻)	0	2.5	3.7
recA⁺ lexA(Ind⁻)	30	1.7	1.9
recA(Def) *lexA⁺*	0	2.9	2.5
recA(Def) *lexA⁺*	2	6.5	4.1

Source: Data from Defais et al. (74).

greater efficiency of W reactivation of double-stranded DNA over single-stranded DNA in wild-type cells was attributed to some form of inducible excision repair that cannot occur on single-stranded phage DNA. However, as discussed in chapter 10, the observed results can also be interpreted (88) as reflecting a strand switch mechanism of DNA tolerance.

The power of the W mutagenesis experiments stemmed from the fact that they separated the role of DNA damage in mutagenesis from its role in the induction of the SOS mutagenesis system. Evidently, the basal level of expression of the SOS processing system is quite low in an uninduced cell, and, fortuitously, infection with the damaged λ DNA (indirect induction [see chapter 10]) does not cause a sufficient degree of induction of the functions required for SOS mutagenesis for the phenomenon to be observed. Thus, in these W mutagenesis experiments, an uninduced cell is acting in some senses as though it were a mutant defective in the SOS mutagenesis system.

BACTERIAL MUTAGENESIS

Studies of UV radiation-induced mutagenesis of the bacterial chromosome played a key role in the development of the notion that *recA⁺-lexA⁺*-dependent functions are required for the specialized processing of damaged DNA that gives rise to mutations and that this processing is inducible. Evelyn Witkin's observation that *lexA*(Ind⁻) mutants were not mutable by UV radiation led her to postulate that the *lexA⁺* gene might encode or control a new or modified DNA polymerase capable of inserting nucleotides opposite UV radiation lesions and that UV mutagenesis occurred by a mechanism of translesion replication (348). A similar suggestion was made by Bryn Bridges and coworkers during the same year (39). The findings that (i) *recA*(Def) mutants were also nonmutable by UV radiation (349) and (ii) UV mutagenesis of a *recA441*(Cpt^{Ts}) mutant grown at an elevated temperature displayed a linear rather than dosed-squared dose response (350) (see chapter 10) were important observations contributing to Radman's hypothesis (251, 252) that the same induced cellular processing system was responsible for W reactivation and W mutagenesis of UV-irradiated bacteriophage and also for UV mutagenesis of the bacterial chromosome.

The UmuD and UmuC Proteins Are Important for UV Radiation and Chemical Mutagenesis

SIGNIFICANCE OF *umuD* AND *umuC* MUTATIONS

For many years, genetic analysis of the molecular basis of the phenomenon of inducible SOS mutagenesis was complicated by the fact that the only mutations known to make *E. coli* nonmutable by UV radiation and various chemicals were alleles of either *recA⁺* or *lexA⁺*. In retrospect, it is obvious that the interpretation of the effects of these pleiotropic mutations on SOS mutagenesis was extremely complicated, since these mutations affect the regulation of the entire SOS system (see chapter 10). It was not coincidental that all of the mutations originally studied that affected the ability of *E. coli* to be mutated by UV radiation and other agents also affected SOS regulation. This was because all of these mutations had been originally identified on the basis of some other SOS phenotype and their effect on mutagenesis was noted subsequently. Thus, these studies had not led to the identification of mutations specifically affecting the process of SOS mutagenesis but, rather, to the isolation of SOS regulatory mutations. The isolation of *umuD* and *umuC* mutations, described in the following section, was of particular significance to the study of SOS mutagenesis because these were the first mutations that appeared to affect the process of SOS mutagenesis specifically.

ISOLATION AND CHARACTERIZATION OF *umuD* AND *umuC* MUTANTS

Two laboratories (147, 303, 304) independently screened directly for mutations which abolished the ability of *E. coli* cells to be mutagenized by agents such as UV radiation or 4-nitroquinoline 1-oxide (4-NQO); a conceptually similar screen for nonmutable mutants had been carried out previously with *Saccharomyces cerevisiae* (185). By screening directly for nonmutability, these investigators were able to search for mutants that were defective in SOS mutagenesis without necessarily being defective in SOS induction. One such screen (147) was based on the reversion of the *his-4* allele, yielding His[+] revertants that appeared as papillae on colonies of His[−] mutants grown on limiting histidine (Fig. 11–2) (also see chapter 2). Three classes of nonmutable mutations were obtained. Those in the first class appeared to be Ind[−] alleles of *lexA*[+]. Mutations in the second class mapped to the *recA*[+] locus and resembled the previously isolated *recA430*(Cpt[−]) mutation. Most interestingly, however, mutations in the third class mapped to a new locus called *umuC*[+] near 25 min on the *E. coli* genetic map (147). Mutations in this locus were later shown to lie in two adjacent genes, *umuD*[+] and *umuC*[+] (92, 291). Mutations in either gene cause the same nonmutable phenotype.

PHENOTYPE OF *umuD* AND *umuC* MUTANTS

E. coli strains carrying a *umuD* or *umuC* mutation are largely nonmutable by a variety of agents including UV radiation (144, 303), 4-NQO (144, 303), methyl methanesulfonate (MMS) (345), and neocarcinostatin (90) (Fig. 11–3A). *umuD* and *umuC* mutants are modestly sensitive to UV radiation but are much less sensitive than *uvr*, *recA*(Def), or *lexA*(Ind[−]) mutants (Fig. 11–3B). They are defective in W mutagenesis of UV-irradiated λ phage and show a reduced capacity to carry out W reactivation of UV-irradiated λ phage. However, *umuD* and *umuC* mutants differ from *recA* and *lexA* mutants in that they are still capable of expressing a variety of SOS responses such as induction of λ prophage (147, 304), filamentous

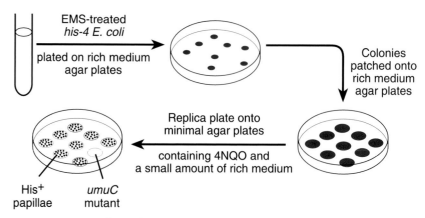

Figure 11–2 Screen used to isolate *umuD* and *umuC* mutants (147).

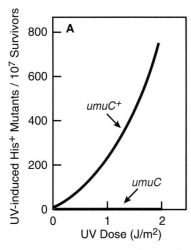
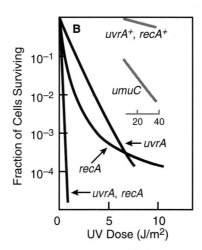

Figure 11–3 (A) Frequency of UV-induced His$^+$ mutations in umu^+ (black) and $umuC36$ (red) strains. Data from Kato and Shinoura (147). (B) UV survival curves of a $umuC$ strain compared with $uvrA$ and $recA$(Def). (*Adapted from Howard-Flanders and Boyce [131] and Bagg et al. [2], and Friedberg [110] with permission.*)

growth (147), and increased synthesis of RecA protein (345). The simplest interpretation of these observations is that the $umuD^+$ and $umuC^+$ genes code for products that are uniquely required for SOS mutagenesis.

Nevertheless, $umuD$ and $umuC$ mutants can still be mutated efficiently by chemical agents such as N-methyl-N'-nitro-N-nitrosoguanidine (MNNG) or ethyl methanesulfonate (EMS), which produce directly mispairing lesions (e.g., O^6-alkylguanine) that do not require the SOS processing system to cause a mutation (145, 273) (see below). Subsequent investigations have revealed that some agents, such as MNNG (70, 101), X rays (266, 303), and N-2-acetyl-2-aminofluorene (AAF) (204), have both $umuD^+$-$umuC^+$-dependent and $umuD^+$-$umuC^+$-independent components to their mutagenic effects and that the mutagenesis from these agents in $umuD$ or $umuC$ mutants reflects the $umuD^+$-$umuC^+$-independent component. Furthermore, in the case of UV radiation mutagenesis, it was found that up to a quarter of the normal yield of $lacI$ mutations could be induced by UV irradiation of a strain carrying a $umuC$ insertion mutation if no selection was applied so that all surviving cells could form full size colonies (64). In this case, the $lacI$ mutants were detected as sectors within the full-size colonies by including the chromogenic indicator X-Gal (5-bromo-4-chloro-3-indolyl-β-D-galactopyranoside) in the medium. This latter observation suggests that certain classes of UV radiation-induced lesions can give rise to mutations in the absence of $umuD^+C^+$ function. However, such $umuD^+C^+$-independent UV mutagenesis appears to be very inefficient compared with that in a wild-type cell because the experimental conditions under which it was observed permitted considerably more cell divisions than in those commonly used to measure UV mutagenesis. Furthermore, evidence discussed below suggests that there are qualitative as well as quantitative changes in the mutagenesis that occurs after SOS induction.

EXPRESSION OF THE $umuD^+C^+$ OPERON IS INDUCED BY DNA DAMAGE AND IS UNDER SOS CONTROL

An insertion mutation in $umuC^+$ (2) was generated by using the Mu d1 bacteriophage and, as discussed in chapter 10, was used to demonstrate that the $umuD^+C^+$ locus is induced by DNA damage and is a member of the SOS regulon. Subsequent sequencing of the locus revealed that the $umuD^+$ and $umuC^+$ genes are organized in an operon and encode products of ~15 and ~45 kDa, respectively. As expected, an SOS box is located upstream of the operon (151, 245) (see Fig. 10–16). Operator-constitutive mutations in the $umuD^+C^+$ operon have been isolated and change conserved nucleotides within this SOS box (301).

UmuD protein is present at ~180 copies per cell in uninduced $lexA^+$ strains but is present at ~2,400 copies per cell in a $lexA$(Def) strain (361). The $umuD^+$ and $umuC^+$ genes overlap slightly with the final A of the TGA stop codon of $umuD^+$, constituting the A of the ATG codon that initiates the $umuC^+$ open reading frame (245). The levels of UmuC protein in a $lexA$(Def) cell (200 molecules per cell) are about 12-fold lower than that of UmuD protein and were too low to be measured in uninduced $lexA^+$ cells (361).

PROPERTIES OF $umuD$ AND $umuC$ MUTANTS ARE DUE TO LOSS OF FUNCTION

The first $umuD$ and $umuC$ mutants isolated (130, 274) were derived by EMS and UV radiation mutagenesis, respectively, and all have been shown to be single-base-pair substitutions that resulted in missense mutations (153). Initially the possibility existed that the nonmutability of the strains was due to the alteration of some function rather than to the loss of some function. However, strains carrying insertion and deletion mutations of $umuC^+$ (2, 92), $umuD^+$ (3), and $umuD^+C^+$ (360) have been isolated, indicating that the phenotype of these mutants results from a loss of function and, furthermore, that the $umuD^+$ and $umuC^+$ genes are not essential for cell viability.

OTHER PHENOTYPES ASSOCIATED WITH $umuDC$ MUTANTS

Additional phenotypes associated with $umuD^+C^+$ deficiency have been reported. Although $umuDC$ mutants are only slightly sensitive to UV light compared with mutants defective in nucleotide excision repair and recombinational repair, they are very sensitive to angelicin, a monofunctional psoralen derivative (218). It seems likely that some important class of lethal lesion induced by angelicin is perhaps susceptible to a form of $UmuD^+$-$UmuC^+$-dependent processing but not to other forms of DNA repair or DNA damage tolerance mechanisms. A particularly strong sensitization to UV killing caused by the introduction of a $umuD$ or $umuC$ mutation into a $recA718$ mutant has been observed and was discussed in the section on replication restart in chapter 10. In addition, $umuDC$ mutants are defective in the SOS-induced inhibition of restriction of unmodified phages or plasmids in $E. coli$ K strains (128). This process requires cleavage of UmuD protein, but UV irradiation is still needed even if cleaved UmuD protein is provided. The mechanistic basis of the phenomenon is not yet clear. The $umuD^+C^+$ function has been reported to be required for the UV-stimulated precise excision of the transposon $Tn10$ (1), but this conclusion has not been supported by independent work (188).

Phenotypes associated with overexpression of $umuD^+C^+$ have been reported. For example, $lexA$(Def) cells carrying a multicopy $umuD^+C^+$ plasmid were cold sensitive for growth (207). This conditional lethality caused by overexpression appears to be due to a rapid and reversible inhibition of DNA synthesis at the nonpermissive temperature. The $umuC125$ mutation, which changes Ala-39 to Val, eliminates the ability of UmuC protein to cause the cold-sensitive growth phenotype but not its ability to participate in UV radiation mutagenesis (206). A high cellular level of UmuC seems to be particularly harmful if the cell also carries the $dnaQ49$ mutation, which affects the ε subunit of DNA polymerase III (237). In addition, more modest overexpression of $umuD^+C^+$ (by using a multicopy $umuD^+C^+$ plasmid in a $lexA^+$ strain) was found to inhibit the induction of SOS-controlled genes by UV radiation or mitomycin (206).

$umuD^+$ AND $umuC^+$ HOMOLOGS ON NATURALLY OCCURRING PLASMIDS

Analysis of the roles of the $umuD^+$ and $umuC^+$ genes in mutagenesis was aided by the discovery that various naturally occurring plasmids encode homologous genes. It was initially recognized that a subset of naturally occurring plasmids, for example R46 (81, 226), TP110 (117), and Col I (80, 132, 133), have the ability to make $E. coli$, $Salmonella typhimurium$, and other bacteria resistant to killing by UV radiation and a number of chemicals and also more susceptible to mutagenesis by these agents (49, 63, 222, 309, 332) (Fig. 11–4).

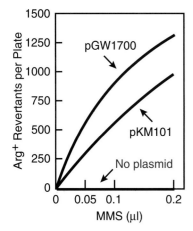

Figure 11–4 MMS-induced reversion of the $argE3$ mutation in a $umuC36$ strain mediated by pKM101 and pGW1700 ($mucA^+mucB^+$). To 2 ml of top agar supplemented with limiting (0.05 mM) arginine and nonlimiting concentrations of all other growth requirements were added 0.1 ml of a fresh stationary-phase culture and the appropriate amount of a freshly prepared solution of MMS in dimethyl sulfoxide. The top agar was then poured on minimal-glucose plates, which were incubated at 37°C for 4 days. Spontaneous revertants have been subtracted. (*Adapted from Perry and Walker [246] with permission.*)

pKM101: A MUTAGENESIS-ENHANCING PLASMID
IMPORTANT IN THE AMES TEST

A derivative of plasmid R46, termed pKM101, has been particularly intensively studied. pKM101 was derived from R46 (226) and apparently arose by an in vivo deletion of 14 kb of DNA which encoded several drug resistance genes (48, 172, 173). Because of its ability to enhance the susceptibility of cells to mutagenesis, pKM101 was introduced (210) into the Ames *S. typhimurium* strains used for the detection of mutagens and carcinogens (see chapter 2) and has played a major role in the success of this system (208, 209). In addition to enhancing the reversion of base pair and frameshift mutations with appropriate mutagens, it appears to increase the frequency of more-complex mutations (187) of a type discussed in chapter 2. In addition to increasing the resistance of cells to killing by various DNA-damaging agents and enhancing their susceptibility to mutagenesis, pKM101 increases both the survival and mutagenesis of UV-irradiated bacteriophages in unirradiated cells (337) and W reactivation and W mutagenesis of UV-irradiated phages in irradiated cells (337, 338). All of these pKM101-mediated effects were found to be blocked by *recA*(Def) and *lexA*(Ind⁻) mutations, suggesting that in some way the plasmid was enhancing the capacity of the cells to carry out SOS mutagenesis.

pKM101 Encodes Homologs of UmuD and UmuC Proteins
Which Suppress the Nonmutability of *umuD* and *umuC* Mutants

The key insight into the mode of action of pKM101 and other mutagenesis-enhancing plasmids stemmed from the observation that pKM101 and R46 suppress the nonmutability of *umuD* and *umuC* mutations (304, 345). The simplest interpretation of this observation was that pKM101 carried homologs of the *umuD⁺* and *umuC⁺* genes. Mutants of pKM101 that had completely lost their ability to influence mutagenesis and repair were then isolated (281, 339), and a ca. 2,000-bp DNA fragment that was responsible for these plasmid-mediated effects was identified (246). Sequencing of this region demonstrated that it encoded two genes called *mucA⁺* and *mucB⁺*, which are homologs of *umuD⁺* and *umuC⁺*, respectively (245). The deduced sequences of the MucA and UmuD proteins are 41% identical at the amino acid level (see Fig. 10–12), while the MucB and UmuC proteins are 55% identical. As with *umuD⁺C⁺*, the pKM101-encoded genes are organized in an operon, *mucA⁺B⁺* (245, 246), whose expression is induced by DNA damage (91). There is an SOS box located upstream of the operon that has been shown to bind the *E. coli* LexA protein (212, 245), and its expression is controlled by the *E. coli* SOS regulatory system (91). Since the *mucA⁺B⁺* operon is located on a broad-host-range plasmid that is capable of conjugal transfer to a wide variety of gram-negative bacteria, and since a number of gram-negative bacteria appear to have LexA homologs (see chapter 10), it seems likely that the *mucA⁺B⁺* operon is under SOS regulation when it is present in various other bacteria besides *E. coli*. Although a plasmid expressing the *mucA⁺B⁺* operon could suppress the nonmutability of either *umuD* or *umuC* mutants, a plasmid expressing only the *mucA⁺* gene or only the *mucB⁺* gene could not. This led to the suggestion that there might be a physical interaction between the two gene products of each operon and that the two pairs of proteins had diverged sufficiently that interactions between UmuD and MucB protein, or between MucA and UmuC protein, were excluded (245).

A number of other naturally occurring plasmids have also been shown to suppress the nonmutability of *umuD* or *umuC* mutants and are likely, therefore, to similarly encode homologs of UmuD and UmuC proteins (49, 80, 332). In the case of TP110, the plasmid has been directly shown to encode homologs of *umuD⁺* and *umuC⁺* called *impA⁺* and *impB⁺*, respectively (117, 196, 309) (see Fig. 10–12). As with *umuD⁺C⁺* and *mucA⁺B⁺*, the two genes are organized in an operon whose expression is controlled by the SOS regulatory system (80, 196, 309). However, an additional open reading frame (referred to as *impC⁺*) precedes the other two genes in the *impC⁺A⁺B⁺* operon (196). There is no equivalent of *impC⁺* in the *umuD⁺C⁺* or *mucA⁺B⁺* operons, and its function has not yet been determined.

Posttranslational Activation of UmuD Protein: a New Dimension to SOS Regulation

UmuD AND MucA PROTEINS SHARE HOMOLOGY WITH THE C-TERMINAL DOMAINS OF LexA AND λ REPRESSOR

When the $umuD^+$ and $mucA^+$ genes were sequenced, the surprising observation was made that the deduced amino acid sequences of the UmuD and MucA proteins shared significant homology with the C terminus of LexA protein and of λ cI repressor (245) (see Fig. 10–12). This led to the formulation of the hypothesis that the UmuD and MucA proteins might interact with activated RecA protein and that this interaction might result in a proteolytic cleavage of these proteins that would activate or unmask the function(s) required for mutagenesis (245). Furthermore, since there is very limited homology between the amino acids of UmuD and MucA on the N-terminal side of their putative cleavage sites, it seemed possible that this constituted a nonfunctional or expendable domain. The cleavage site was predicted to be Cys-24–Gly-25 (245). Although both λ repressor and LexA protein have Ala-Gly cleavage sites, φ80 repressor was subsequently found to have a Cys-Gly cleavage site (see Fig. 10–12).

A SECOND REQUIREMENT FOR RecA PROTEIN IN UV RADIATION MUTAGENESIS

The hypothesis that UmuD protein might become activated by RecA-mediated cleavage took on added significance in light of a number of observations which raised the possibility that RecA protein had an additional role in mutagenesis besides mediating the cleavage of LexA protein. This possibility was first suggested by the observation that lexA(Def) recA(Def) mutants are nonmutable by UV radiation even though the $umuD^+$ and $umuC^+$ genes are constitutively expressed at high levels in such strains (2, 19). However, as discussed previously (233, 313), an alternative explanation for such a result would be that all of the cells that underwent a mutation died because RecA protein might be required for their survival, in addition to be being required for the mutational event. More-convincing evidence came from the observation that lexA(Def) recA430(Cpt⁻) mutants were also very poorly mutated by UV radiation (19, 160). Cells carrying the recA430(Cpt⁻) mutation are proficient for homologous recombination (see below) and are much less sensitive to killing by UV radiation than are recA(Def) cells. As discussed in chapter 10, the RecA430 protein can catalyze strand exchange in vitro but is very defective in mediating the cleavage of λ repressor and is partially defective in mediating the cleavage of LexA protein.

A second requirement for RecA protein in UV radiation mutagenesis (besides mediating the cleavage of LexA protein) was clearly shown in a series of experiments that examined W mutagenesis of UV-irradiated phage λ (59, 93, 94). As shown in Table 11–2, unirradiated lexA(Def) recA⁺ cells showed no greater capacity for W mutagenesis of UV-irradiated bacteriophage λ than did unirradiated lexA⁺ recA⁺ cells, even though they constitutively express $umuD^+$, $umuC^+$, and other LexA-repressed genes at high levels. After irradiation, these lexA(Def) recA⁺ cells exhibited the same capacity for mutagenesis as lexA⁺ recA⁺ cells did. Consistent with these observations, W mutagenesis was not observed in lexA(Def) recA(Def) or lexA(Def) recA430(Cpt⁻) strains even when they were UV irradiated (Table 11–2). Taken together, these results indicated that RecA protein carries out at least one additional function in mutagenesis besides mediating LexA cleavage and that the RecA430 protein is not able to carry out this function. Furthermore, lexA(Def) recA730(Cptᶜ) cells exhibited higher than normal levels of mutagenesis of UV-irradiated phage, and there was little additional increase upon irradiation of the host cells (Table 11–2). This observation, in concert with the requirement for UV irradiation of lexA(Def) recA⁺ cells, indicated that RecA protein had to be activated to carry out its second role in mutagenesis and that the RecA730 protein was constitutively activated to carry out this second role.

Table 11–2 W mutagenesis of UV-irradiated λ requires RecA activation even in *lexA*(Def) strains

Relevant genotype		UV dose to cell (J/m²)	No. of mutant phage/10⁷ progeny	
recA	*lexA*		Without UV to phage	With UV to phage
+	+	0	5.8	34
		10	5.2	270
+	*51*(Def)	0	6.4	11
		10	5.6	310
Δ*306*(Def)	*51*(Def)	0	2.8	11
		10	5.0	8
430(Cpt⁻)	*51*(Def)	0	5.0	17
		10	7.0	40
730(Cptᶜ)	*51*(Def)	0	49	450
		10	76	510

Source: Data from Ennis et al. (93).

UmuD PROTEIN IS CLEAVED IN VIVO AND IN VITRO IN A RecA-MEDIATED FASHION

The demonstration that UmuD protein is cleaved in vivo in a RecA-dependent fashion after cells have been exposed to an SOS-inducing treatment (290) was made by using an experimental approach very similar to that initially used (257) to demonstrate the cleavage of λ repressor in vivo (see Fig. 10–5). Antibodies to UmuD protein were isolated and used in immunoblotting experiments to show that UV irradiation of the cells resulted in the conversion of UmuD protein to a smaller form (referred to as UmuD′). The molecular weight of UmuD′ was consistent with its being the C-terminal fragment of UmuD protein predicted from sequencing the gene (245). This in vivo cleavage of UmuD protein was not observed in *lexA*(Def) *recA*(Def) or *lexA*(Def) *recA430*(Cpt⁻) cells but was observed in the absence of an SOS-inducing treatment in *lexA*(Def) *recA730*(Cptᶜ) cells. This suggested that UmuD protein was cleaved as a consequence of an interaction with activated RecA protein, similar to the way in which LexA and λ *cI* repressor undergo RecA-mediated cleavage.

RecA-mediated cleavage of UmuD protein was also demonstrated in vitro (53). In these studies it was observed that, just as with λ repressor and LexA protein, purified UmuD protein was proteolytically cleaved to yield UmuD′ when incubated with RecA protein, single-stranded DNA, and ATPγS but that the cleavage of UmuD protein was less efficient than that of LexA protein. The bond cleaved was subsequently shown (362) to be the Cys-24–Gly-25 bond previously predicted (245) on the basis of the homology with LexA and λ repressor. Both UmuD and UmuD′ proteins are homodimers (362). UmuD protein can be cleaved by RecA or RecA441(CptᵀˢS) protein but not by RecA430 protein (Fig. 11–5), consistent with the in vivo observations (290). It was also shown that UmuD protein shares another characteristic with λ repressor and LexA, namely, autodigestion at alkaline pH (53).

RecA-MEDIATED CLEAVAGE ACTIVATES UmuD PROTEIN

To investigate the role of UmuD cleavage in UV radiation-induced mutagenesis, a special derivative of a plasmid carrying the *umuD*⁺ gene was constructed (233) in which overlapping termination (TGA) and initiation (ATG) codons were introduced at the site in the *umuD*⁺ sequence that corresponds to the cleavage site (Cys-24–Gly-25). The plasmid carrying this engineered form of UmuD protein thus encoded two polypeptides corresponding almost exactly to those that would normally be produced by RecA-mediated cleavage of UmuD at the Cys-24–Gly-25 bond. Even though UmuD′ protein synthesized from this plasmid differs from natural UmuD′ produced by RecA-mediated cleavage by having an N-terminal formylmethionine

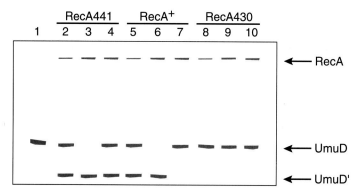

Figure 11–5 RecA-mediated cleavage of UmuD. UmuD (2 mg) was incubated with wild-type or mutant RecA at 37°C for 90 min in the presence of single-stranded DNA (40 ng) and ATPγS (1 mM) (or 2 mM dATP where indicated). To avoid depletion of the dATP by the dATPase activity of RecA, an additional 2 mM dATP was added after 45 min of incubation. The products of the reaction were separated by sodium dodecyl sulfate-polyacrylamide gel electrophoresis (19% polyacrylamide), and the proteins were visualized by staining with Coomassie blue. Migration positions are indicated for RecA, UmuD, and the largest cleavage fragment of UmuD (UmuD′). Lanes: 1, no RecA; 2, 0.5 mg of RecA441; 3, 1 mg of RecA441; 4, 1 mg of RecA441 plus dATP; 5, 0.5 mg of RecA+; 6, 1 mg of RecA+; 7, 1 mg of RecA+ plus dATP; 8, 0.5 mg of RecA430; 9, 1 mg of RecA430; 10, 1 mg of RecA430 plus dATP. (*Adapted from Burckhardt et al. [53] with permission.*)

residue, it is likely that this residue is removed with high efficiency by *E. coli* methionine aminopeptidase, since it is adjacent to a Gly residue (15). When the plasmid encoding the two polypeptides was introduced into a nonmutable *umuD44* strain, it restored the UV mutability of the cell to that of a *umuD*+ cell (Fig. 11–6). This observation ruled out the possibility that the purpose of UmuD cleavage was to inactivate UmuD protein. It was also observed that a plasmid encoding only the peptide corresponding to the small N-terminal fragment failed to complement the UV radiation nonmutability of a *umuD* strain, whereas a plasmid encoding only the large C-terminal fragment (UmuD′) did restore UV mutability. This observation strongly indicated that UmuD′ is both necessary and sufficient for the role of UmuD protein in UV radiation mutagenesis. Similar conclusions were later reached by using a lower-copy-number plasmid encoding the UmuD′ and UmuC proteins and strains carrying the *umuD24* insertion mutation (3).

To test the hypothesis that RecA-mediated cleavage activates UmuD protein for its role in mutagenesis, plasmids encoding either the two UmuD polypeptides or just the C-terminal polypeptide were introduced into a *lexA*(Def) *recA430*(Cpt⁻) strain (233). As discussed above, such strains are nonmutable by UV light, even though the *umuD*+ and *umuC*+ genes are constitutively expressed at high levels. Both plasmids made *lexA*(Def) *recA430*(Cpt⁻) cells UV mutable again (Fig. 11–7). Since the UmuD protein was not cleaved in *recA430* mutants (290), these results indicated that the nonmutability of *lexA*(Def) *recA430*(Cpt⁻) cells was solely due to their failure to cleave UmuD protein and hence that RecA-mediated cleavage of UmuD protein to yield UmuD′ activates UmuD protein for its role in mutagenesis.

Further evidence that the mechanism of RecA*-mediated cleavage of UmuD is mechanistically similar to that of the RecA*-mediated cleavage of LexA protein was provided by the observation that Ser-60 → Ala and Lys-97 → Ala mutations (233) block UmuD cleavage in vivo. These mutations correspond to the Ser-119 → Ala and Lys-156 → Ala mutations (see chapter 10 and Fig. 10–12), which block LexA cleavage (12). Interestingly, although the introduction of the Ser-60 → Ala mutation into a gene encoding intact UmuD protein greatly reduced its ability to participate in mutagenesis, introduction of the same mutation into a gene encoding UmuD′ only reduced the ability of UmuD′ protein to function in mutagenesis by 50%. This suggests that the primary requirement for Ser-60 in UmuD function

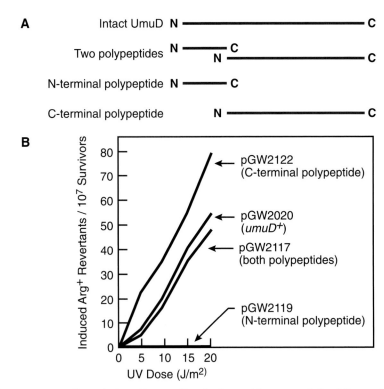

Figure 11–6 Effect of plasmids encoding both or either one of the NH$_2$- or C-terminal polypeptides of UmuD on UV radiation mutagenesis in a *umuD44* strain. Plasmids used were pGW2020 (*umuD*$^+$), pGW2117 (both polypeptides), and pGW2122 (C-terminal polypeptide), and pGW2119 (N-terminal polypeptide). The curve for a strain without a plasmid is the same as for the strain carrying pGW2119 (*Adapted from Nohmi et al. [233] with permission.*)

is for cleavage rather than for the subsequent role of UmuD′ in mutagenesis. In contrast, the Lys-97 residue appears to be important for both UmuD cleavage and for the subsequent role of UmuD′ in SOS mutagenesis (233).

SOS MUTAGENESIS IS REGULATED BOTH TRANSCRIPTIONALLY AND POSTTRANSLATIONALLY

Collectively, the results described above indicate that RecA protein carries out two mechanistically related roles in UV radiation and chemical mutagenesis: (i) transcriptional derepression of the *umuD*$^+$*C*$^+$ operon by mediating the cleavage of LexA protein, and (ii) posttranslational activation of UmuD protein by mediating its cleavage (Fig. 11–8). Since both of these events require the presence of activated RecA protein, which is the cell's internal indicator that it has suffered DNA damage, the biological purpose of this regulatory system is apparently to give the cell an extra measure of control over whether, and to what extent, it should express activities necessary for UV radiation and chemical mutagenesis.

It is worth noting that, by itself, the posttranslational activation of UmuD would be sufficient to account for the physiological observation of the "inducibility" of UV radiation and chemical mutagenesis in *E. coli*. Thus, although SOS mutagenesis in *E. coli* is widely known to be under transcriptional control, this system also provides a precedent for other organisms for how UV radiation and chemical mutagenesis could require "induction" at the physiological level without there being a requirement for the enhanced transcription of any genes (245, 346).

MucA PROTEIN IS CLEAVED MORE EFFICIENTLY THAN UmuD PROTEIN

SOS mutagenesis mediated by the plasmid pKM101-derived *mucA*$^+$*B*$^+$ genes results in a higher frequency of mutations for a given dose of UV radiation than does

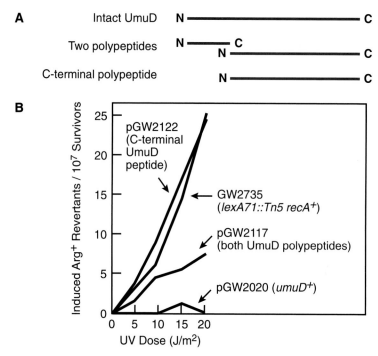

Figure 11–7 Plasmids encoding the C-terminal polypeptide of UmuD restore UV mutability to a *lexA71*::Tn5(Def) *recA430*(Cpt⁻) strain. Plasmids used were pGW2020 (*umuD⁺*), pGW2117 (encodes both UmuD polypeptides), and pGW2122 (encodes the C-terminal polypeptide of UmuD). GW2735 is a *lexA71*::Tn5 *recA⁺* strain carrying pGW2020 (*umuD⁺*). (*Adapted from Nohmi et al. [233] with permission.*)

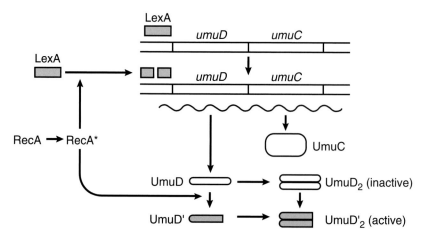

Figure 11–8 Posttranslational cleavage of UmuD activates it for its role in SOS mutagenesis. Once RecA has been activated in response to an SOS-inducing treatment, it mediates the proteolytic cleavage of LexA. This leads to increased expression of the SOS-regulated genes including the *umuDC* operon. RecA then mediates the proteolytic cleavage of UmuD at its Cys-24–Gly-25 bond in a process that is mechanistically and evolutionarily related to the cleavage of LexA. The carboxyl terminal domain of UmuD (termed UmuD') is the active form of UmuD that participates in SOS mutagenesis and is necessary and sufficient for the role in mutagenesis. Both UmuD and UmuD' form homodimers.

SOS mutagenesis mediated by the *umuD⁺C⁺* genes from the *E. coli* chromosome (18, 337). Furthermore, a *recA430* mutant carrying a plasmid encoding *mucA⁺B⁺* is UV mutable (20, 336). In contrast, as discussed above, a *recA430* mutant carrying a plasmid encoding *umuD⁺C⁺* is not UV mutable, but it is mutable if it carries a plasmid that encodes the UmuD′ and UmuC proteins (233). Analogously to UmuD, MucA protein has been shown to be cleaved at its Ala-26–Gly-27 bond (317) in a RecA*-mediated fashion (125, 288), and the C-terminal polypeptide MucA′ has been shown to be active in UV radiation mutagenesis (288). This cleavage was shown to be much more efficient than that of UmuD protein and to be almost as efficient as LexA cleavage. Although initial experiments failed to detect MucA protein cleavage in a *lexA*(Def) *recA430* strain (288), suggesting that both the intact and cleaved forms of MucA protein may be active in mutagenesis, the use of a more-sensitive detection assay demonstrated the presence of a significant amount of MucA′ (125). Taken together, these results suggest that the higher efficiency of *mucA⁺B⁺*-mediated mutagenesis compared to *umuD⁺C⁺*-mediated mutagenesis may be due, at least in part, to this more-efficient processing.

THE *umuD⁺C⁺* OPERON IS THE ONLY SOS LOCUS THAT MUST BE INDUCED FOR SOS MUTAGENESIS

Genetic studies have indicated that the *umuD⁺C⁺* operon is the only SOS locus that must be induced for SOS mutagenesis. This concept was first suggested by the observation that a plasmid carrying an operator-constitutive mutation for the *mucA⁺B⁺* genes made a *lexA*(Ind⁻) strain mutable (212). In such a *lexA*(Ind⁻) derivative, the genes in the SOS regulatory network are maintained in the repressed state, since LexA protein cannot be cleaved (see chapter 10) but the *mucA⁺B⁺* operon can be selectively expressed because of the operator-constitutive mutation. It was subsequently similarly demonstrated (301) that the *umuD⁺C⁺* operon is the only chromosomally encoded SOS locus that must be induced for SOS mutagenesis to occur. Operator-constitutive mutations of either *umuD⁺C⁺*, *umuD′C⁺*, or *recA⁺* were used to increase selectively the concentrations of the respective gene products. As shown in Fig. 11–9, *lexA*(Ind⁻) cells that expressed approximately twice the normal induced level of UmuD and UmuC proteins (5,000 and 500 molecules, respectively) became mutable by UV radiation. The unamplified level of RecA protein in these cells (8,000 molecules [223]) mediated sufficient UmuD cleavage for mutagenesis to occur. In contrast, *lexA*(Ind⁻) cells that instead expressed increased levels of RecA protein did not become more mutable, although they did become more resistant to killing by UV radiation.

POSSIBLE ROLE FOR INTACT UmuD PROTEIN IN THE MODULATION OF UV RADIATION AND CHEMICAL MUTAGENESIS: A SECOND LEVEL OF POSTTRANSLATIONAL CONTROL OF MUTAGENESIS?

The intact UmuD protein may be more than simply an inactive form that is converted to the activated form, UmuD′, upon RecA*-mediated cleavage. Rather, the intact form may play an active role in modulating the ability of *E. coli* to carry out SOS mutagenesis. This hypothesis was originally proposed (12) on the basis of experiments in which a number of missense mutants of a plasmid-borne *umuD⁺* gene were identified by screening for failure to complement the nonmutability of a *umuD44* strain. These missense *umuD* alleles encoded stable UmuD proteins that were deficient in RecA-mediated cleavage in vivo. Furthermore, most of these *umuD* alleles were dominant, and those with the strongest dominant negative effects caused the most severe deficiencies in RecA*-mediated cleavage. Since these *umuD* alleles did not interfere with SOS induction, it seemed likely that the mutant UmuD proteins were interfering with mutagenesis by titrating out essential components or by integrating into protein complexes necessary for mutagenesis. Since both UmuD and UmuD′ proteins form homodimers (362), the simplest model to account for these observations was that intact UmuD protein interacts

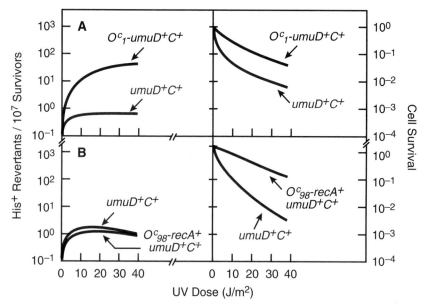

Figure 11–9 (A) Operator-constitutive (O^c) mutant of $umuD^+C^+$ enhances survival of and mutagenesis in *lexA1*(Ind⁻) cells. *lexA1* bacteria carrying $umuD^+C^+$ (red) and $O^cumuD^+C^+$ (black) plasmids were UV irradiated (doses are given on the abscissa). $hisG4 \rightarrow$ His⁺ reversions are shown in the left panel and cell survival is shown in the right panel. (B) No restoration of SOS mutagenesis by O^c mutant of $recA^+$ in a *lexA1*(Ind⁻) $umuD^+C^+$ strain. *lexA1* bacteria carrying $umuD^+C^+$ (red) or $umuD^+C^+$ and O^crecA^+ (black) were UV irradiated. $hisG4 \rightarrow$ His⁺ reversions are shown in the left panel, and cell survival is shown in the right panel. (*Adapted from Sommer et al. [301] with permission.*)

with UmuD′ to form a heterodimer that is inactive in mutagenesis. This model was supported by glutaraldehyde cross-linking experiments which demonstrated that UmuD-UmuD′ heterodimers were formed and also that they were more stable than either of the homodimers (12) (Fig. 11–10).

The preferential formation of UmuD-UmuD′ heterodimers may have interesting regulatory consequences. For example, assuming that the heterodimer is indeed inactive or weakly active in mutagenesis, a result supported by in vitro observations discussed below (254), substantial cleavage of UmuD protein would have to occur before the active UmuD′₂ homodimers could be produced in quantity. This would thus represent a third mechanism to ensure that cells would not express functions required for SOS mutagenesis unless their DNA had been damaged. In addition, a particularly interesting possibility is that formation of UmuD-UmuD′ heterodimers plays an important role in shutting off the capacity for SOS mutagenesis as the SOS response begins to diminish. Since UmuD protein is cleaved much less efficiently than LexA protein both in vivo (290) and in vitro (53), some intact UmuD protein would be expected to accumulate before the increase in intact LexA protein would return the expression of the $umuD^+C^+$ operon to its basal level. Each intact UmuD molecule would have the potential to inactivate a molecule of UmuD′ by heterodimer formation (Fig. 11–11). Such shutting off of the capability for SOS mutagenesis prior to shutoff of the entire set of SOS responses has been observed. For example, it has been shown (351) that SOS mutagenesis decays with a half-life of about 30 min in a strain defective in nucleotide excision repair. However, when such a *uvr* strain was irradiated at even lower doses, LexA protein continued to be cleaved for at least 60 min (268). Thus, it appears that inhibition of SOS mutagenesis by intact UmuD via UmuD-UmuD′ heterodimer formation may represent a third level of control of SOS mutagenesis beyond the transcriptional derepression and posttranslational activation mechanisms discussed above.

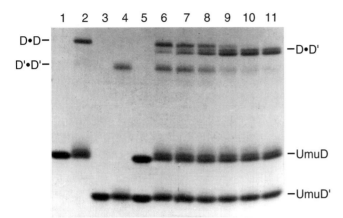

Figure 11–10 Formation of heterodimers of UmuD and UmuD'. UmuD and UmuD' were at 10 μM in 10 mM phosphate buffer (pH 6.8)–100 mM NaCl. Glutaraldehyde cross-linking was carried out by adding glutaraldehyde to 0.05%, incubating for 3 min at room temperature, and stopping the reaction by the addition of Tris-HCl and freezing quickly. Samples were subjected to electrophoresis on a polyacrylamide gel containing sodium dodecyl sulfate. Lanes: 1 and 2, UmuD with no treatment and with glutaraldehyde, respectively; 3 and 4, UmuD' with no treatment and with glutaraldehyde, respectively; 6 to 11, UmuD and UmuD' treated with glutaraldehyde after 1, 3, 5, 10, 20, and 30 min of incubation, respectively. (*Adapted from Battista et al. [12] with permission.*)

umuD⁺*C*⁺ Homologs in Other Organisms

A number of bacteria have been found to be naturally nonmutable by UV radiation. In this respect, their phenotype resembles that of *E. coli umuD* or *umuC* mutants (342). Others organisms, such as *S. typhimurium* LT2, are weakly mutable by UV radiation. It was found that *S. typhimurium* LT2 can be made highly mutable by the introduction of the *E. coli umuD*⁺*C*⁺ operon or homologs of the operon, suggesting that the weak mutability of this bacterium is due to a deficiency in *umuD*⁺*C*⁺ function (340, 341). Similarly, it has been found that some naturally nonmutable bacteria, for example *Proteus mirabilis* (129), become mutable if *umuD*⁺*C*⁺ homologs are introduced (342). On the basis of these early results, it seemed possible that the mutability or nonmutability of a bacterial species depended on whether it possessed a *umuD*⁺*C*⁺ homolog. Although this may be true in some cases, it is clear that the situation can be more complicated.

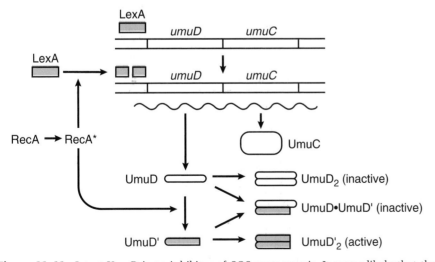

Figure 11–11 Intact UmuD is an inhibitor of SOS mutagenesis. It seems likely that the mechanism of this inhibition involves the formation of a heterodimer of UmuD and UmuD' since a UmuD·UmuD' heterodimer is more stable than either the UmuD₂ or the UmuD'₂ homodimer (see reference 12 and Fig. 11–10).

S. TYPHIMURIUM LT2 HAS TWO umuD⁺C⁺ OPERONS BUT IS WEAKLY MUTABLE BY UV RADIATION

Despite being very weakly UV mutable compared with *E. coli*, *S. typhimurium* LT2 has been found to possess two *umuD⁺C⁺* homologs. One, *umuD⁺C⁺*$_{ST}$ (297, 329), is located on the bacterial chromosome, while the other, *samA⁺B⁺* (234), is located on the 60-MDa cryptic plasmid present in *S. typhimurium* LT2. The deduced amino acid sequences of the *S. typhimurium* UmuC proteins are shown in Fig. 11–12. Although both *S. typhimurium umuD⁺C⁺* operons can complement the deficiencies of *E. coli umuD* or *umuC* mutants with respect to UV radiation-induced mutagenesis, genetic studies have shown that it is the chromosomally located *umuD⁺C⁺*$_{ST}$ operon that is responsible for the weak UV mutagenesis seen in wild-type *S. typhimurium* strains (152, 235).

The *umuD⁺C⁺* operon is located at 26 min on the map of the *E. coli* chromosome, but the *umuD⁺C⁺*$_{ST}$ operon is located at 40 min on the *S. typhimurium* chromosome near the end of the major chromosomal inversion that distinguishes the *E. coli* and *S. typhimurium* chromosomes (152, 297). Thus, it is likely that the *umuD⁺C⁺* operon was present in a common ancestor prior to the divergence of *E. coli* and *S. typhimurium* approximately 150 million years ago (297). The expression of the *umuD⁺C⁺*$_{ST}$ operon is induced by DNA damage, and UmuD$_{ST}$ protein appears to be cleaved even more efficiently than *E. coli* UmuD protein. Hence, the relatively weak mutability of *S. typhimurium* does not appear to be due to a limitation at either of those steps. The observation that a plasmid encoding the *E. coli umuD⁺* gene alone could increase the UV mutability of wild-type *S. typhimurium* originally

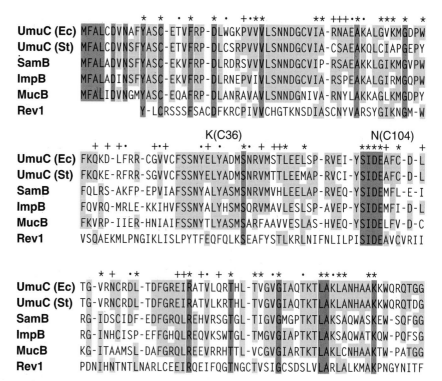

Figure 11–12 Conservation of protein sequence between the amino-terminal portions of UmuC-like proteins and the *Saccharomyces cerevisiae* Rev1 protein, residues 372 to 533. Red areas indicate residues that are identical in all proteins; grey areas indicate residues that are identical in three or more proteins; asterisks indicate identical matches of Rev1 to three or more bacterial sequences; plus signs indicate identical matches of Rev1 to one or two bacterial sequences; solid circles indicate that Rev1 is similar to one or more bacterial sequences. The positions and substitutions made by *E. coli umuC36* and *umuC104* mutations are also indicated. Ec, *E. coli*; St, *S. typhimurium*. (*Adapted from Woodgate and Sedgwick [363] with permission.*)

led to the suggestion that *S. typhimurium* was phenotypically UmuD$^-$ (126, 241). However, subsequent interspecies transfer experiments (274) indicated that the apparently poor activity of the *umuD$^+_{ST}$* gene in *S. typhimurium* is due to the simultaneous presence of the *S. typhimurium umuC$^+_{ST}$* gene. This led to the suggestion that the limitation of *umuD$^+_{ST}$* activity by *umuC$^+_{ST}$* contributes to the poor mutability of this organism.

OTHER ORGANISMS

A survey of 16 enterobacteria for UV radiation-induced mutability revealed that this capability is widely distributed among this group of bacteria (276). Furthermore, differences in UV-induced mutability of more than 200-fold were observed between different species of enteric bacteria and even between multiple natural isolates of *E. coli*. Interestingly, as in the case of *S. typhimurium*, some species that display a weakly mutable phenotype still have *umuD$^+$C$^+$*-like genes. In *E. coli* isolates the *umuD$^+$C$^+$* genes are at one end of a conserved 12-kb tract that is bracketed by restriction enzyme polymorphisms, and similar conserved units have been seen in other *Escherichia* species and in related bacteria (277). The presence of sequences resembling Tn*3* termini near where the polymorphisms begin has led to speculation that a transposon-like activity may have been responsible for rearrangements in this region of the chromosome (277). In this respect, it is interesting that the *mucA$^+$B$^+$* genes of pKM101 are flanked by inverted repeats (172).

A *umuD$^+$C$^+$* locus has been cloned from the SCP 1 plasmid of the gram-positive bacterium *Streptomyces coelicolor* (219). Thus, *umuD$^+$C$^+$* functions seem to be present in gram-positive as well as gram-negative bacteria. Interestingly, the N-terminal portion of the *Saccharomyces cerevisiae* Rev1 protein also shares homology with the set of UmuC homologs (Fig. 11–12) (174, 363). Mutants defective in the *REV1* gene are nonmutable by UV radiation (see chapter 12), so it is possible that Rev1 protein represents a eukaryotic version of UmuC protein.

Genetic and Physiological Evidence for the Involvement of RecA Protein, DNA Polymerase III, and GroEL/GroES Proteins in SOS Mutagenesis

One of the most challenging aspects of determining the actual biochemical mechanism of UV radiation and chemical mutagenesis has been how to establish whether particular reactions or conditions that lead to mutagenesis in vitro are relevant to the physiological processes that occur within living cells. Thus, genetic and physiological studies have played a critical role in identifying key proteins that participate in mutagenesis in vivo and in establishing criteria that allow the physiological relevance of biochemical models to be assessed. Genetic studies have revealed that, in addition to requiring the UmuD' and UmuC proteins, SOS mutagenesis requires an additional function of the RecA protein and the participation of some form of DNA polymerase III. The functions of the molecular chaperones GroEL and GroES are also normally needed for *umuD$^+$C$^+$*-dependent mutagenesis.

ANOTHER ROLE FOR RecA PROTEIN IN SOS MUTAGENESIS

As already indicated, a plasmid encoding the UmuD' and UmuC proteins will restore UV mutagenesis in a *lexA*(Def) *recA430*(Cpt$^-$) strain (233). However, this is not the case in a *lexA*(Def) *recA*(Def) strain even though LexA-repressed genes are fully expressed in this strain and both UmuD' and UmuC proteins are synthesized. This observation led to the suggestion (233) that there might be a third role for RecA protein in SOS mutagenesis beyond mediating the cleavage of LexA and UmuD proteins. However, there remained the formal possibility that mutagenesis was not detected because of the extreme UV radiation sensitivity of the *recA*(Def) strain. Subsequent experiments that circumvented the problem of detecting mutagenesis posed by the extreme radiation sensitivity of *recA*(Def) strains provided additional support for a third role for RecA protein in SOS mutagenesis. One approach was to measure mutagenesis of UV-irradiated bacteriophage λ instead of chromosomal mutagenesis. The observation that λ mutagenesis could not be observed in a *lexA*(Def) *recA56* strain expressing UmuD′ protein supported a third role for RecA and indicated that RecA56 protein was defective in this function (94). A second approach took advantage of an experimental system in which chromosomal His⁺ revertants arise from nonphotoreversible lesions so that almost all lethal photoproducts can be removed by photoreactivation immediately after UV irradiation. By using this system, it was possible to show (9) that *lexA*(Def) Δ*recA*(Def) strains were nonmutable by UV radiation under conditions where survival was not significantly affected.

Clues about the nature of this third role of RecA have been gleaned by analyses of various partial-loss-of-function *recA* mutants. The *recA1730* mutation was recognized as being defective in this third role of RecA in mutagenesis (4, 85). RecA1730 protein has a defect in nucleoprotein filament formation, but all of the phenotypes of a *recA1730* mutant, except for UV non-mutability, can be suppressed by overexpression of Rec1730 protein (84). Whether the deficiency of *recA1730* in mutagenesis is due to a problem in the formation of filaments or to some other defect is not yet clear. Interestingly, the amino acid change caused by the *recA1730* mutation affects the pocket between two RecA monomers postulated to form the binding site for LexA protein (306) (see chapter 10).

A particularly interesting discovery was that *recA* mutants [e.g., *recA1203* (Cptᶜ Rec⁻) (321, 322)] that are constitutive for coprotease function but defective in homologous recombination are proficient for SOS mutagenesis. This finding suggests that the third role of RecA in mutagenesis cannot require the full set of RecA functions necessary for homologous recombination. Analyses of the effect of various *recA* mutations on the spontaneous mutator activity seen in certain *recA* mutants (see below) led to speculation that the ability of RecA protein to perform its third role in this phenomenon correlates with its ability to bind to small single-stranded regions of DNA in undamaged cells (313). If this speculation were to be applied to UV mutagenesis as well, it would suggest that the third role of RecA might require it to bind to single-stranded DNA and form a nucleoprotein filament. Finally, the observation that *lexA*(Ind⁻) cells that express UmuD and UmuC proteins or MucA and MucB proteins are UV-mutable indicates that RecA protein does not have to be amplified to carry out its third role (212, 301). Specific possibilities for this third role are discussed below.

DNA POLYMERASE I IS NOT REQUIRED FOR SOS MUTAGENESIS

DNA polymerase I (157) is encoded by the *polA⁺* gene and, as discussed in chapters 4 and 5, plays an important role in excision repair. *E. coli polA1* mutants were found to be more mutable than *polA⁺* strains at low doses of UV radiation exposure (353), an effect attributed to induction of the SOS system at lower UV radiation doses in the *polA1* strain. These observations suggested that DNA polymerase I was not required for SOS mutagenesis. However, it was then reported (167, 168) that cells induced for the SOS responses have an altered form of DNA polymerase I, termed DNA polymerase I*, that exhibits reduced fidelity. This finding raised the possibility that small amounts of some form of DNA polymerase I were present in *polA1* mutants and led to an examination of the UV mutability of a *polA* deletion mutant (10). The finding that a strain lacking DNA polymerase I was indeed mutable by UV radiation ruled out the possibility that DNA polymerase I, or some altered form of DNA polymerase I, is required for SOS mutagenesis. However, this experiment leaves open the question whether some form of DNA polymerase I could participate if it is present.

DNA POLYMERASE II IS NOT REQUIRED FOR SOS MUTAGENESIS BUT MAY PLAY A ROLE IN PARTICULAR CIRCUMSTANCES

DNA polymerase II, encoded by the *polB* gene, has been characterized in detail (157). However, its physiological role remains ambiguous. Interest in the possibility that DNA polymerase II plays a role in SOS mutagenesis was sparked by the discovery (25) that DNA polymerase II activity is increased sevenfold in SOS-induced cells and that DNA polymerase II can bypass abasic lesions in vitro to a limited extent. The processivity of DNA polymerase II is stimulated by certain DNA polymerase auxiliary subunits (26, 134), and the β,γ complex and single-stranded-DNA-binding protein need to be present for the enzyme to bypass abasic sites. The SOS inducibility of DNA polymerase II activity was explained by the subsequent discovery (24, 61, 138) that the *polB* gene is identical to the *dinA* gene, one of the SOS-inducible loci identified by the use of operon fusions (149, 150) (see chapter 10). There is an SOS box upstream of the *polB* gene that binds LexA protein with relatively low affinity (137). Despite the attractiveness of the idea that DNA polymerase II might participate in SOS mutagenesis, the observation that *dinA*::Mu d1 mutants were normally mutable indicated that DNA polymerase II was not required for mutagenesis, although it did not rule out the possibility that the enzyme was able to participate in this process (138, 150). This conclusion is consistent with a more-extensive study (159) showing that several different *polB* mutants, including a *polB* deletion, were not affected in their ability to carry out W reactivation and W mutagenesis of single-stranded phage that carried either abasic sites, thymine glycols, or cyclobutane pyrimidine dimers.

A situation has been described in which mutagenesis appears to exhibit a dependence on *polB*⁺ (326). DNA of the single-stranded phage S13 was UV irradiated, incubated for 1 h at 37°C, exposed to photoreactivating light in the presence of photolyase, and treated with ura-DNA glycosylase. If this DNA was then transfected into an uninduced *recA*⁺ *polB*⁺ strain, an uninduced *recA*⁺ *polB* strain, or a *recA1202*(Cptᶜ) *polB* strain, phage survival and the specific mutation frequency were much lower than if the phage DNA was transfected into an unirradiated *recA1202*(Cptᶜ) *polB*⁺ strain. Thus, in this case, the S13 mutagenesis that occurred in the cells constitutively expressing the SOS response exhibited a dependence on DNA polymerase II. It was inferred (326) that the lesions responsible for the lethality and the mutagenesis were abasic sites that resulted from the deamination of cytosine-containing pyrimidine dimers, monomerization of the dimers to yield uracil, and removal of the uracil by ura-DNA glycosylase. Interestingly, this dependence on *polB*⁺ function was not observed if the SOS system was induced by UV irradiation of *recA*⁺ cells rather than by the *recA1202*(Cptᶜ) allele. Moreover, the dependence on *polB*⁺ function in *recA1202*(Cptᶜ) cells could be suppressed by increasing the gene dosage of the *groES*⁺*L*⁺ operon (see below), suggesting that the difference between the two cases might be related to the fact that the *groES*⁺*L*⁺ operon can be induced by UV radiation (161, 326). Further work is required to delineate whether there are differences in the mechanisms of mutagenesis depending on whether cells have been SOS induced by exposure to a DNA-damaging agent or as a consequence of the presence of a *recA*(Cptᶜ) mutation or whether the observed results are attributable to some other aspect of the experimental design.

DNA POLYMERASE III IS REQUIRED FOR SOS MUTAGENESIS

DNA polymerase III holoenzyme is the major replicative DNA polymerase of *E. coli* (see chapter 4) (Table 11–3). The extremely high processivity of DNA poly-

Table 11–3   Major subunits of DNA polymerase III holoenzyme

Subunit	Mass (kDa)	Gene
α	130	*dnaE*⁺
ε	27.5	*dnaQ*⁺ (*mutD*)⁺
θ	10	*holE*⁺
τ	71	*dnaX*⁺
γ	47.5	*dnaX*⁺
δ	35	*holE*⁺
δ′	33	*holB*⁺
χ	15	*holC*⁺
ψ	12	*holD*⁺
β	40.6	*dnaN*⁺

Source: Adapted from McHenry (211) and Kornberg and Baker (157) with permission.

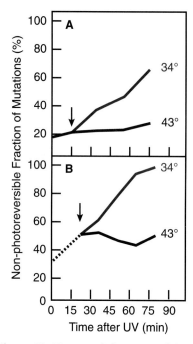

Figure 11–13 Loss of photoreversibility of UV-induced mutations in a *dnaE1026*(Ts) strain at 34°C (red) and 43°C (black). (A) Bacteria were incubated at 34°C after UV exposure for 15 min before the culture was divided (indicated by arrow) and half was incubated at 34°C and half was incubated at 43°C. (B) Similar experiment with a 22.5-min postirradiation incubation at 34°C before the temperature shift. (*Adapted from Bridges et al. [42] with permission.*)

merase III holoenzyme is due to the β subunit (157), which forms a ring-shaped sliding clamp that encircles the DNA (Color Plate 8) and tethers the polymerase to its substrate (156, 166, 239, 310). The γ complex subassembly of DNA polymerase III holoenzyme (which consists of the γ, δ, δ', χ, and ψ proteins) transfers the β subunit onto the DNA in a reaction that requires ATP hydrolysis.

A body of evidence, discussed here and in the following sections, that indicates that some form of DNA polymerase III is required for SOS mutagenesis has been accumulated. However, since DNA polymerase III is an essential enzyme for cell viability, it has been more difficult to determine whether DNA polymerase III is the physiologically relevant polymerase for SOS mutagenesis than in the cases of DNA polymerase I and II discussed above.

Experiments with Temperature-Sensitive *dnaE* Mutants

The first evidence that DNA polymerase III may be required for SOS mutagenesis came from experiments (42) with a nucleotide excision repair-defective strain carrying the temperature-sensitive *dnaE1026* (formerly *polC1026*) mutation. The *dnaE*⁺ gene encodes the α subunit of DNA polymerase III, which carries the polymerase active site. These experiments showed that, just as with a *dnaE*⁺ strain, UV-induced ochre suppressor mutations became nonphotoreversible as a function of time if the *dnaE1026* cells were incubated at the permissive temperature of 34°C after being UV irradiated (Fig. 11–13A). In contrast, shifting the cells to the nonpermissive temperature of 43°C after UV irradiation led to an immediate cessation of loss of photoreversibility (Fig. 11–13A), even though postreplicative strand joining still occurred and [³H]thymine incorporation into DNA continued for 20 to 30 min. Similar results were obtained if the cells were maintained at the permissive temperature for 22.5 min prior to the shift to 43°C (Fig. 11–13B), a time interval that should have been sufficient to allow induction of the SOS system (see chapter 10). In these experiments, the assumption was made (42) that photoreversibility reflected the enzymatic photoreactivation of a target photoproduct (pyrimidine dimer) and that the loss of photoreversibility reflected biochemical processing that gave rise to a mutation at that site. This led to the conclusion that some form of DNA polymerase III was required for SOS mutagenesis.

Experiments with Strains Carrying the *pcbA1* Mutation

Several of the subsequent efforts to investigate the requirement for DNA polymerase III in SOS mutagenesis have involved strains carrying the *pcbA1* mutation (*polC* bypass), which is thought to be an allele of *gyrB*⁺ (one of the subunits of DNA gyrase) (227), although this has not been unambiguously established. Cells carrying temperature-sensitive *dnaE* mutations (or amber *dnaE* mutations and a temperature-sensitive amber suppressor) can replicate at the restrictive temperature if they carry the *pcbA1* mutation but not if they lack it (205, 232). The replication that occurs under these conditions is DNA polymerase I dependent (232), although the Klenow fragment of DNA polymerase I will suffice (52), and it is sensitive to coumeromycin (51). The effect of the *pcbA1* mutation is not yet understood at the molecular level, but the DNA replication that occurs in a *pcbA1 dnaE*(Ts) mutant at the restrictive temperature has been attributed to the involvement of an alternative replication complex containing DNA polymerase I rather than the α subunit of DNA polymerase III (50, 205, 232). This interpretation was strengthened by the failure (205) to detect the α subunit of DNA polymerase when such cells were grown at 43°C and by the immunological detection of DNA polymerase I in holoenzyme preparations (50).

When grown at the restrictive temperature, *pcbA1 dnaE*(Ts) strains were found to be nonmutable by UV radiation even though their SOS system was induced (Fig. 11–14A) (37, 51, 122). Given the above model of the molecular basis of DNA replication in *pcbA1 dnaE*(Ts) strains, this finding was interpreted as reflecting a requirement for DNA polymerase III for SOS mutagenesis (37, 51, 122). Oddly, such cells were also found to be nonmutable with EMS, an SOS-independent mutagen (see below) (Fig. 11–14B). However, the finding that *pcbA1 dnaE*(Ts) strains are mutable by EMS at the restrictive temperature (Fig. 11–14C)

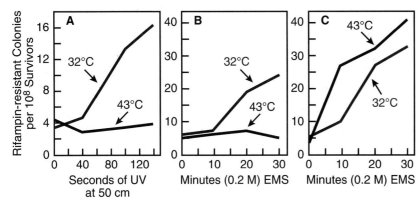

Figure 11–14 (A) Mutagenesis in CSM61 [*polA1 polB100 dnaE1026*(Ts), *pcbA1*] after UV exposure. Mutants were scored as rifampin-resistant colonies. D_{37} values at 32 and 43°C are 32 s of exposure in both. The UV flux was 1 J/m²/s. Cells held at 32 or 43°C in phosphate buffer did not show an increase in the number of rifampin-resistant mutants after 30 min. The survival rate was greater than 90% in buffer at both temperatures. (B) EMS mutagenesis in CSM61. CSM61 cells grown at 32 and 43°C were treated with 0.2 M EMS, and aliquots were removed at the indicated times, washed, and plated on L agar supplemented with 100 mg of rifampin per ml. The plates were incubated at 32 and 43°C, respectively, and the number of mutant (rifampin-resistant) colonies per 10⁸ survivors was calculated from the titer of surviving cells at each time point. (C) EMS mutagenesis in MH73 (CSM61 *uvrA*). MH73 cells were treated as described for panel B. (*Adapted from Hagensee et al [123] and Moses et al. [227] with permission.*)

if they also carry a *uvrA* mutation casts doubt on the interpretation that the non-mutability of these strains reflects a requirement for DNA polymerase III for mutagenesis. Rather, it suggests that the replication complex of *pcbA1 dnaE*(Ts) strains may pause longer at lesions than that of a wild-type cell, thereby allowing excision repair to occur more completely (227).

An alternative view of the mechanism of DNA replication in *pcbA1 dnaE*(Ts) strains can be gained from a consideration of the DNA polymerase I dependence of DNA replication in a *dnaQ* deletion mutant, which lacks the ε subunit of DNA polymerase III (189). Perhaps similarly in *pcbA1 dnaE*(Ts) strains, polymerase III holoenzyme containing the mutant α subunit is able to carry out a considerable amount of DNA replication, but there is a more stringent demand than normal for DNA polymerase I for filling single-strand gaps in these strains.

Overproduction of the ε or β Subunits of DNA Polymerase III: Inhibition of UV Radiation Mutagenesis and Partial Relief by Increased Expression of UmuDC Functions

An additional approach that has been used in attempts to provide genetic evidence for the involvement of some form of DNA polymerase III holoenzyme in SOS mutagenesis has been to examine UV mutagenesis when certain subunits of DNA polymerase III have been overproduced from recombinant plasmids. Overproduction of the α subunit (encoded by *dnaE*) had no effect on mutagenesis (315). However, overproduction of the ε subunit inhibited UV mutagenesis (66, 104, 140), an effect that could be partially alleviated by concomitant overexpression of the *umuD⁺C⁺* operon (104). Similarly, overproduction of the β subunit (encoded by *dnaN⁺*) caused a 5- to 10-fold inhibition of UV mutagenesis along with a slight increase in UV sensitivity (315). This inhibition of mutagenesis could be slightly alleviated by overexpression of the *umuD⁺C⁺* operon and partially alleviated by overexpression of the *mucA⁺B⁺* operon (315).

Both of these experiments indicate that an imbalance of certain subunits of DNA polymerase III interferes with SOS mutagenesis, but the mechanisms underlying these effects have not yet been clearly established. In the case of overproduction of the β subunit, it has been suggested (285, 315) that there are two forms of DNA polymerase III, the usual β-rich replicative enzyme and a β-poor form that performs translesion synthesis under SOS conditions and generates mutations. On the basis of this model, it has been proposed that overproduction of the β subunit affects the relative abundance of the normal and the SOS bypass polymerases by sequestering polymerase molecules in the β-rich form and blocking the SOS form. The alleviation of the inhibition of mutagenesis by overexpression of $umuD^+C^+$ and $mucA^+B^+$ raises the possibility that their gene products form an alternative sliding clamp adapted for replication on a damaged template. This possibility was suggested (11) on the basis of limited sequence similarities between UmuD protein and gp45 protein of bacteriophage T4 and between UmuC protein and gp44/62 proteins of bacteriophage T4. The gp45 protein appears to be a functional homolog of the β subunit of DNA polymerase III, which forms a sliding clamp on DNA, while gp44/62 proteins appear to form a functional analog of the γ complex (119, 255).

Models to account for the inhibition of UV mutagenesis by overproduction of the ε subunit are less straightforward since it has been shown that *dnaQ* deletion mutants of *S. typhimurium*, which completely lack the ε subunit, still require UmuD, UmuC, and RecA proteins for the bypass of lesions in single-stranded bacteriophage (296). This observation is inconsistent with the notion that the basis of SOS mutagenesis consists solely of an inhibition of the proofreading exonuclease of the polymerase carrying out the translesion synthesis (95, 200, 334). However, it does not exclude the possibility that inhibition of ε function, or the formation of a polymerase complex lacking ε (34), is one of the events necessary for SOS mutagenesis in wild-type cells. An analysis of the mutator phenotype of a *dnaQ49* strain (*dnaQ*⁺ encodes the ε subunit) led to the suggestions that the UmuD and UmuC proteins might interact with the replication complex and inhibit ε and, furthermore, that they might also interact with and stabilize the α subunit when ε activity is inhibited (103).

GENETIC EVIDENCE THAT THE MOLECULAR CHAPERONE PROTEINS GroEL AND GroES ARE REQUIRED FOR SOS MUTAGENESIS

The GroEL and GroES proteins are molecular chaperones that play roles in preventing the denaturation and aggregation of other proteins and also assist in the folding and oligomerization of proteins without being part of their final structure (111, 116, 367). The expression of the *E. coli groES⁺L⁺* operon is regulated as part of the heat shock response (367) and is increased by UV irradiation in an SOS-independent manner (161). The observation that *groEL* and *groES* mutations can suppress the cold-sensitive lethality caused by overexpression of the *umuD⁺C⁺* operon (discussed above) led to the discovery that *groEL* and *groES* mutants have a reduced ability to be mutated by UV radiation (76). These *groE* mutants were also shown to be defective in W mutagenesis of the single-stranded DNA phage S13 (193). SOS induction and UmuD protein cleavage occurred normally in these *groEL* and *groES* mutants, indicating that the *groE* mutations did not affect these processes (77). In addition, modest overexpression of UmuD′ and UmuC but not UmuD and UmuC proteins largely suppressed this deficiency in SOS mutagenesis, and UmuC protein was observed to coimmunoprecipitate with GroEL (77). Collectively, these observations led to the suggestion that the GroEL and GroES proteins play a role in folding UmuC protein into a conformation in which it is able to interact with UmuD and UmuD′ proteins (Fig. 11–15) (77).

Inferences about the Mechanism of SOS Mutagenesis Based on Mutational Spectra and Site-Directed Adduct Studies

Another approach that has led to insights about the nature of the molecular mechanisms of SOS mutagenesis has been the collection and analysis of spectra of mutations induced by various mutagens. Chapter 2 discusses the molecular basis of

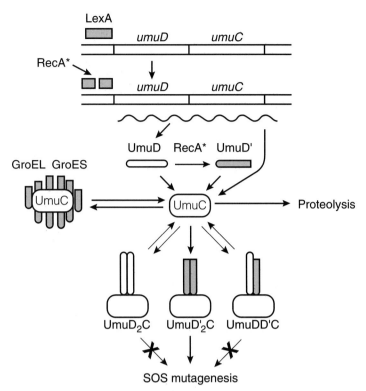

Figure 11–15 Model of the regulation and processing of the UmuD and UmuC proteins. After UV irradiation, the RecA protein is converted to its activated form (RecA*) and induces cleavage of the LexA repressor. Transcription of the *umuD*$^+$*C*$^+$ operon is increased, and UmuD and UmuC are synthesized. UmuD is activated by a RecA*-mediated cleavage, which results in UmuD′. UmuC undergoes an interaction with GroEL and GroES and then forms a complex with dimers of UmuD and UmuD′ or heterodimers of UmuD and UmuD′. The UmuC-UmuD′$_2$ complex is required for UV mutagenesis, but the UmuC-UmuD$_2$ and UmuC-UmuD·UmuD′ complexes are thought to be inactive in UV mutagenesis. The thick arrows indicate the major pathway of protein association, and the thin arrows indicate minor or reversible pathways. Arrows marked by an × indicate inactive pathways. (*Adapted from Donnelly and Walker [77] with permission.*)

representative systems that have been used to obtain such mutational spectra. As the years have passed, the ease with which such mutational spectra can be obtained has increased remarkably. Furthermore, the development of techniques for introducing a single defined DNA lesion into a specific site on a DNA molecule (chapter 2) has allowed various hypotheses to be tested and has led to further insights.

The Original *lacI* System: a Purely Genetic Means of Determining Mutational Spectra

MOST UV RADIATION AND CHEMICAL MUTAGENESIS IS TARGETED

The first system that allowed the nature of large numbers of mutations to be conveniently analyzed was the fine-structure *lacI* system of *E. coli* developed by Jeffrey Miller and his colleagues (69, 214, 216) (see chapter 2). With this system, a set of purely genetic techniques were used to analyze amber and ochre mutations occurring within the *lacI*$^+$ gene. These analyses allowed the sequence changes giving rise to these mutations to be inferred. All possible base pair changes could be detected in the original *lacI* system except for A·T → G·C. The spectra of *lacI* mutations caused by several SOS-dependent mutagens (e.g., UV [69, 331], 4-NQO [69], neocarcinostatin [100], (±)7a,8b-dihydroxy-9b,10b-epoxy-7,8,9,10-tetrahydrobenzo[*a*]pyrene [BPDE] [89], 3,4-epoxycyclopenta[*c,d*]pyrene [89, 102], activated aflatoxin B$_1$ [102], and cisplatin [47]) were obtained. As illustrated in Fig. 11–16, these studies clearly indicated that each agent causes its own characteristic spectrum of mutations. Thus, most of the base substitution mutations caused by these agents must have been targeted to different types of premutagenic lesions, for otherwise the same spectrum of mutations would have been generated by all of the mutagens. More specifically, this means that the particular chemical lesions generated in DNA as a consequence of exposure to these various agents must play a role in determining the outcome of the mutations that result. This conclusion was further strengthened by the finding that the mutational spectrum obtained with cells which were not exposed to mutagen but whose SOS system had been induced genetically [a *recA441*(Cpt^{Ts}) grown at 42°C] was very different from the spectrum obtained after exposure to a mutagen such as UV radiation (217). As discussed below, it is likely that a small fraction of the mutations observed in these experiments result from an untargeted process but it is clear that the majority of the mutational events in these spectra were targeted.

HOT SPOTS: NOT ALL SITES ARE EQUALLY MUTABLE

The data in Fig. 11–16 also clearly indicate a general aspect of mutational spectra—*mutational hot spots* (see chapter 2). Observed differences in mutability can result simply from different mutational events occurring at different sites. However, as shown in Fig. 11–16, the same mutational events can occur at different sites with widely different frequencies, indicating that the local DNA environment plays a role in determining the efficiency with which a given mutagen causes mutations at particular sites. The location of the hot spots observed depends on the mutagen used. Thus, hot spots are not due to the existence of regions of DNA that are simply intrinsically more mutable to all mutagens than are other regions of DNA. The DNA sequence surrounding a potentially mutable site could play a role in determining the frequency of mutation by affecting (i) the relative frequency and/or

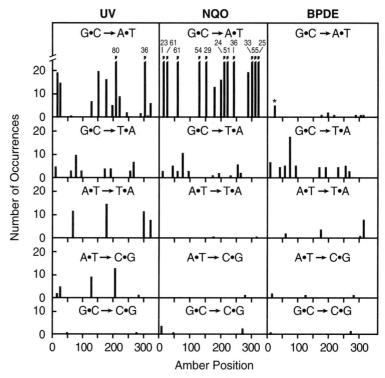

Figure 11–16 Amber mutations induced by UV (69), 4-NQO (69), and benzo[*a*]pyrene (89) in the *lacI* gene. (*Adapted from Miller [214] with permission.*)

nature of the lesions that occur at that site (i.e., by influencing the chemistry or photochemistry that occurs at that site), (ii) the efficiency of accurate repair systems in removing lesions from that site, and (iii) the efficiency with which SOS mutagenesis can occur at that site.

Mutational Spectra Obtained by Direct DNA Sequencing

Following the pioneering work in which the original *lacI* system was used to determine mutational spectra, several systems that utilized direct DNA sequencing to determine the nature of the mutation were developed. For example, substantial studies were performed with the *lacI⁺* gene carried on an F′ plasmid (182, 215, 269), the *cI⁺* gene of phage λ in both its lytic (359) and lysogenic (358) states, and the *lacZ′* gene segment carried on a recombinant M13 single-stranded DNA phage (183). Since mutations were determined directly by sequencing, other classes of changes besides base pair substitutions, particularly frameshift mutations, could readily be determined.

One of the most intensively studied mutagens has been UV light, and the results obtained after UV irradiation of these various systems are summarized (195) in Table 11–4. Although in most of these systems the target gene is carried on various extrachromosomal elements, the results obtained showed overall similarities. UV radiation caused primarily base substitutions (75%; mostly G·C → A·T) but in addition frameshifts (15%), tandem double-base substitutions (4%), and deletions (7%; but this may represent an overestimate since there is a specific deletion hot spot in *lacI* [195]). However, differences in the UV mutational spectra obtained from various systems were also observed. For example, nearest-neighbor analysis revealed that the preferred target site in phage M13 and lytic λ phage appeared to be T-T and T-C, whereas the target sites in the F′ episome appeared to be predominantly C-C and T-C (180). Such analyses alone do not lead to the identification of the premutagenic lesion but do suggest that the two major photoproducts that occur at dipyrimidine sequences, namely, cyclobutane dimers and pyrimidine-pyrimidone (6-4) photoproducts (see chapter 1), probably serve as premutagenic lesions.

Factors Influencing the Mutational Spectrum for a Given Mutagen

The intellectual driving force for the development of sophisticated systems for obtaining mutational spectra was a desire to identify premutagenic lesions and to learn to predict the mutagenic consequences of a given mutagen. However, this

Table 11–4 Specificity of UV mutagenesis[a]

| Assay system | Frequency (%) of following mutation type: | | | | |
	Transition	Transversion	Frameshift	Tandem double	Deletion
1. *lacI*/F′	40	23	30	1	5
2. *lacI*/F′	45	14	11	3	27
3. *lacI*/F′	51	22	8	5	14
4. *lacI*/F′	43	39	16	1	1
5. *lacI*/F′	62	23	11	4	0
6. *lacZ′*/M13	51	23	20	6	0
7. *cI*/λ[b]	66	13	11	5	0
Mean[b]	51	23	15	4	7
8. *cI*/λ lysogen[b]	43	38	7	5	2

[a] The spectra of UV mutations in the *lacI* gene were obtained under the following conditions: rows 1 (215) and 2 (269), an F′ in repair-proficient cells; 3 (269), an F′ in a *uvrB* strain; 4 (182), an F′ during vegetative growth; 5 (182), an F′ during conjugation.

[b] The mean values correspond to the mutagenesis data obtained with extrachromosomal replicons (rows 1 to 7). The λ *cI* data were obtained with lytic (359) or lysogenic (358) λ.

Source: Adapted from Livneh et al. (195) with permission.

endeavor is complicated by a number of factors (195). First, every selection or detection system is biased in some way. For example, if the detection system monitors the function of a protein, only a subset of mutational changes will give rise to a detectable phenotype (27). For both UV radiation and 4-NQO, most of the mutations detected in the 5′ end of the *lacI⁺* gene were base substitutions whereas most of the mutations detected in the remainder of the gene were frame shifts (215). A more-extensive analysis of 6,000 *lacI* mutations revealed that 40% of all sites and over one-half of all the substitutions found occur within the 59 N-terminal residues of the 360-amino-acid protein (120). Another source of bias might be the nature of the DNA replication system used in the detection system, since the replication of extrachromosomal elements differs in a variety of respects from chromosomal replication.

Second, a mutagen will generally introduce a variety of different DNA lesions that vary with respect to their overall frequency and with respect to their frequency at any specific site. The various lesions will also vary with respect to their intrinsic mutability. Thus, a relatively minor lesion which is intrinsically highly mutagenic may be responsible for much more of the observed mutagenesis than a more abundant lesion which is intrinsically less mutagenic. Third, the original lesion caused by the mutagen may not serve directly as a premutagenic lesion but may be processed to yield the actual premutagenic lesion. For example, a modified base might be removed by a DNA glycosylase to yield an AP site, which could then serve as the actual premutagenic lesion. Fourth, a complicating factor in the determination of mutagenic spectra can be the selective removal of particular lesions by various accurate repair systems so that their mutagenic consequences are not observed. For example, it was recognized early on (352) that *uvr* mutants, which are defective in nucleotide excision repair, are considerably more mutable by UV light than are their *uvr⁺* parents, implying that premutagenic lesions can be removed by nucleotide excision repair. Although it might seem possible to circumvent this problem by obtaining the mutagenic spectrum in a repair-deficient strain, this perturbation could introduce other problems. For example, some mutagenic processing might be initiated by the repair system, or else a partial action of the genetically inactivated repair system, such as the introduction of a nick in the vicinity of the lesion, might influence the mutagenic process (67).

Influence of Transcription-Coupled Excision Repair on Mutational Spectra

A more subtle example of the complicating effects of accurate DNA repair is illustrated by the recent realization that the potential of a lesion to cause a mutation can be influenced by whether the lesion is on the transcribed or nontranscribed strand of a gene and by the transcriptional status of the gene. This possibility was originally suggested (23) during studies on the phenomenon of mutation frequency decline (see chapter 5). However, the issue remained largely unexplored in prokaryotes until the initial report of selective repair of the transcribed strand of the *E. coli lac* operon (213). Later, re-analysis of previously published UV mutational spectra (154, 299) led to the realization that, in *uvr⁺* strains of *E. coli*, the frequency of UV-induced mutations presumed to have originated from dipyrimidine targets in the nontranscribed strand was considerably higher than the frequency of mutations presumed to have originated from dipyrimidine targets in the transcribed strand. This strong bias was not seen in mutational spectra of *uvr* mutants, suggesting that this bias was a consequence of the selective operation of UvrABC-dependent excision repair on the transcribed strand.

The discovery of the coupling between transcription and UvrABC-dependent excision repair mediated by the *mfd⁺* gene product, transcription-repair coupling factor (see chapter 5), led to a comparison of the spectrum of UV-induced mutations in *lacI* in a wild-type strain with that in an *mfd* mutant (240). As shown in Fig. 11–17, the observed mutational spectra were very different. In the wild-type strain there is a 3.2-fold bias in favor of mutations originating from dipyrimidine targets on the nontranscribed strand, whereas in *mfd* cells there was a 4.5-fold bias

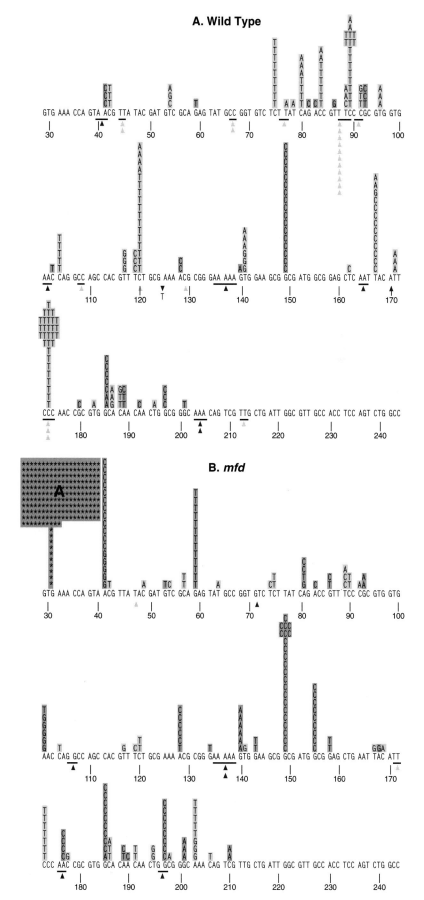

Figure 11–17 Distribution of dominant *lacI* base pair substitutions and frameshifts in the wild-type (A) and *mfd* (B) strains. The assumption is made that the mutations are originating from lesions at dipyrimidine targets. Thus, mutations originating from dipyrimidine targets in the nontranscribed strand appear at C or T in this figure (since the nontranscribed strand is the strand displayed). Conversely, mutations arising from dipyrimidine targets on the transcribed strand appear at G or A. Gray areas indicate mutations at C or T bases; red areas indicate mutations at G or A bases. Mutations above the DNA sequence are base pair substitutions; those below are single-base frame shifts. Underlined bases indicate ambiguity in the position of the missing base. The G → A transition at position 31 is included for completeness, although it is not a dominant mutation. (*Adapted from Oller et al. [240] with permission.*)

in favor of mutations originating from dipyrimidine targets on the transcribed strand. Thus, overall, there was a 14-fold switch in strand bias between the two strains. These observations are consistent with the requirement for transcription-repair coupling factor to direct UvrABC-dependent excision repair to the transcribed strand (279, 280) and with the observation that in the absence of transcription-repair coupling factor, transcription specifically inhibited repair of the transcribed strand (278). They are further consistent with direct measurements of the removal of cyclobutane dimers from each strand within the same system used to obtain the mutational spectrum (162).

Identification of Premutagenic Lesions

Determination of the spectrum of mutations caused by a given mutagen, information about the nature and sequence specificity of lesions introduced by the mutagen, and other experimental observations allowed the formulation of hypotheses concerning which DNA lesions might serve as premutagenic lesions for SOS mutagenesis. However, direct tests of these hypotheses did not become possible until technologies were developed that allowed the synthesis of DNA molecules containing a single adduct of known chemical structure at a defined position in the DNA molecule (see chapter 2). In this section, we summarize the evidence for a representative set of lesions that has led to their recognition as premutagenic lesions for SOS mutagenesis.

PYRIMIDINE-PYRIMIDONE (6-4) PHOTOPRODUCTS

All studies of the mutational spectra induced by UV light indicated that not only were single-base-pair transitions the most common mutagenic event but also they occurred far more frequently at sequences consisting of adjacent pyrimidines (Pyr-Pyr) than at sequences in which a pyrimidine was flanked by purines (Pur-Pyr-Pur). In an independent test of this correlation, it was shown that the conversion of a target A-C-A sequence to a T-C-A sequence resulted in a 15-fold increase in the frequency of C $\rightarrow$ T transitions at that site (215). This feature of UV-induced mutational spectra led to speculations that UV-induced lesions that occur at dipyrimidine sequences—in particular, cyclobutane dimers and pyrimidine-pyrimidone (6-4) adducts—might serve as premutagenic lesions.

The focus on pyrimidine-pyrimidone (6-4) lesions as possible premutagenic lesions had its origins in an effort (192) to correlate the observed spectrum of UV-induced mutations in the *lacI*+ gene with the frequency of UV-induced cyclobutane dimers within the gene. It was found that hot spots for UV mutagenesis did not correlate well with hot spots for dimer formation. This could have been simply due to sequence-specific effects on the SOS processing of cyclobutane dimers to mutations. However, a second possibility was that there was a minor UV-induced photoproduct that was more mutagenic that cyclobutane dimers. This latter possibility was supported by the observation that sites of strand breaks produced in UV-irradiated DNA by hot-alkali treatment (1 M piperidine at 90°C) correlated fairly well with the distribution of UV-induced mutations. The lesions responsible for this alkali lability were subsequently shown to be noncyclobutane dipyrimidine products referred to as pyrimidine-pyrimidone (6-4) lesions (30, 106–108, 221, 262) (see chapter 1) (Fig. 11–18).

These experiments raised the possibility that (6-4) photoproducts were premutagenic lesions, but they did not rule out the possibility that cyclobutane dimers were also premutagenic (30). The (6-4) photoproducts were reported to form preferentially at T-C dipyrimidine sequences and less well at C-C and C-T sites. Apparently, the presence of a methyl group at the 5 position of the 3′ pyrimidine in double-stranded DNA inhibits the formation of (6-4) products. This latter fact was used to provide additional support for the hypothesis that pyrimidine-pyrimidone (6-4) lesions could be premutagenic. Methylation of the 3′ C in the sequence C-C-A-G-G reduces the yield of (6-4) lesions but not of cyclobutane dimers at this sequence. Mutation of the 3′ C $\rightarrow$ T generates an amber mutation. In *dcm* mutants, which are defective in their ability to carry out this specific methylation, at three C-C-A-G-G sites in the *lacI*+ gene a specific increase in the formation of the (6-4)

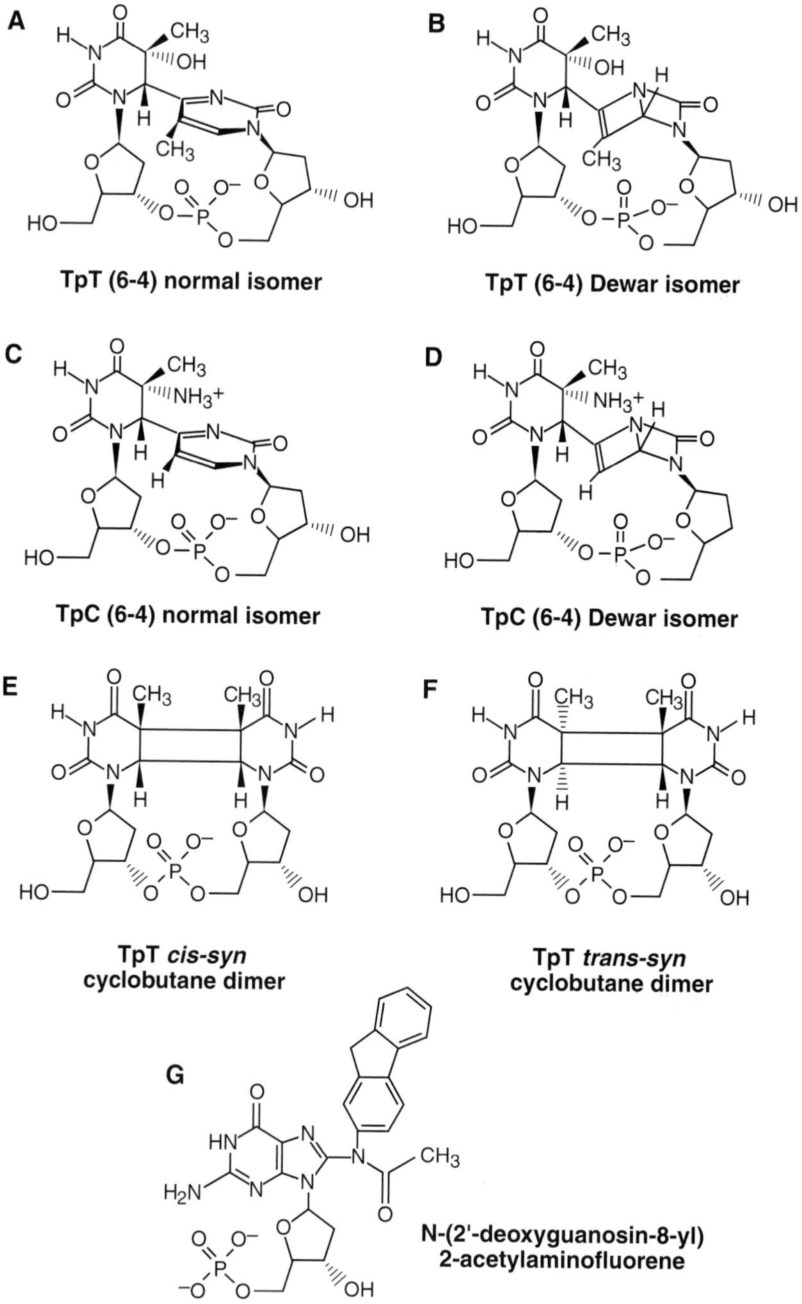

Figure 11–18 Premutagenic lesions for SOS mutagenesis. (A) T-T (6-4) normal isomer; (B) T-T (6-4) Dewar valence isomer; (C) T-C (6-4) normal isomer; (D) T-C (6-4) Dewar valence isomer; (E) *cis-syn*-T-T cyclobutane dimer; (F) *trans-syn*-T-T cyclobutane dimer; (G) N-(2'-deoxyguanosin-8-yl)-AAF.

photoproducts was accompanied by a concomitant increase in mutation frequency, to the extent that those sites became among the most common nonsense mutations recovered (118). Other results that supported roles for photoproducts other than cyclobutane dimers included the failure of photoreactivation of UV-induced *his-4* reversion in *E. coli* cells with constitutively induced SOS responses (260) and the reduced rate of photoreactivation of the reversion of *his* and *leu* mutations in UV-irradiated *E. coli* relative to colony-forming ability (316, 365).

It was subsequently discovered that the normal isomer of pyrimidine-pyrimidone (6-4) photoproducts could be converted to their Dewar valence isomers (Fig. 11–18) by irradiation with wavelengths of light between 280 and 360 nm (318, 320). Furthermore, it was found that both the normal isomer and the Dewar valence isomer can be produced by irradiation with 254-nm UV light (78, 79). However, the normal isomer could not be converted to the Dewar valence isomer by irradiation with 254-nm light (78), an observation suggesting that the normal isomer does not serve as an intermediate in the formation of the Dewar valence isomer under these irradiation conditions. It was also observed that the phosphodiester bond of the Dewar valence isomers of dTpC and TpT (6-4) photoproducts were sensitive to hot alkaline hydrolysis (1 M piperidine) but the phosphodiester bond of the normal isomers was not (78, 143). The latter observation raises the possibility that experiments in which strand breaks induced by hot-alkali treatment were used to monitor the frequency of (6-4) products in DNA irradiated with 254-nm UV light were, in fact, monitoring the frequency of formation of the Dewar valence isomers of those photoproducts.

A direct test of the ability of both the normal isomer (Fig. 11–18) and the Dewar valence isomer (Fig. 11–18) of the thymine-thymine pyrimidine-pyrimidone (6-4) photoproduct to function as premutagenic lesions was carried out by constructing single-stranded M13-based vectors containing these lesions at a specific site and introducing them into *E. coli uvrA* mutants (181). In the absence of SOS induction, vectors carrying the photoproducts were rarely replicated; relative to the lesion-free control, 1.9% of vectors carrying the normal (6-4) isomer produced plaques, while with the Dewar valence isomer the proportion was 0.4%. In SOS-induced cells, these frequencies were 22 and 12%, respectively, an observation consistent with the notion that SOS induction increases the capacity of the cell to carry out translesion synthesis across a lesion that normally blocks replication (Table 11–5). In SOS-induced cells, the normal (6-4) isomer was found to be extremely mutagenic (>91% of the phage that replicated contained targeted mutations) and to be highly specific with respect to the mutations induced (>93% of the mutations were 3′ T → C). Both the frequency and specificity of mutations were lower with the Dewar isomer (53% of the phage were mutants, and 46% of these were 3′ T → C mutants) (Table 11–5). Also, in the course of this study (181), it was found that T-T (6-4) photoproduct was considerably less alkali labile than were the other (6-4) photoproducts (T-C, C-T, and C-C), so that the yield of the T-T (6-4) photoproduct may have been substantially underestimated in earlier (30, 108) studies. This suggestion has received independent support from experiments involving antibodies specific for (6-4) photoproducts (220).

These results clearly demonstrated that both lesions were premutagenic lesions for SOS mutagenesis. A similar analysis (Table 11–5) of the normal isomer and the Dewar valence isomer of the thymine-cytosine pyrimidine-pyrimidone (6-4) photoproducts (Fig. 11–18) revealed that both of these lesions are also mu-

Table 11–5 SOS mutagenesis of single-stranded M13 vectors carrying single UV radiation-induced lesions at a unique site[a]

Lesion	% bypass in SOS-induced cells/% bypass in non-SOS-induced cells	% of mutagenic bypass events in SOS-induced cells
(6-4) T-T	22/1.9%	91
(6-4) T-T Dewar	12/0.4%	53
(6-4) T-C	25/0.7%	34
(6-4) T-C Dewar	13/0.5%	79
cis-syn-T-T dimer	27/0.5%	7
trans-syn-T-T dimer	29/14%	11

[a] Single-stranded M13mp7-based vectors carrying a single defined UV radiation-induced photoproduct at a unique site were transfected into an *E. coli uvrA6* strain that either had been induced for SOS (4-J/m² UV radiation) or were uninduced.

Source: Data from Banerjee et al. (5, 6), LeClerc et al. (181), and Horsfall and Lawrence (130).

tagenic lesions for SOS mutagenesis (130). In this case, the normal isomer was less intrinsically mutagenic than the Dewar isomer. However it was more specific, causing 80% 3' C → T transitions. As indicated above, C → T transitions are the most commonly observed mutations in UV radiation mutational spectra (195). For both the T-T and T-C (6-4) photoproducts, the differences in their mutagenicities clearly indicate that the chemical nature of the lesion influences the nature of the mutations that are produced.

CYCLOBUTANE PYRIMIDINE DIMERS

Cyclobutane pyrimidine dimers are the major photoproducts of UV irradiation of DNA (see chapter 1). As discussed below, they strongly inhibit the action of most polymerases and are particularly effective at inhibiting the action of *E. coli* DNA polymerase III holoenzyme. It was originally thought that cyclobutane dimers were unable to hydrogen bond, but nuclear magnetic resonance (NMR) studies (148, 319) have shown that they can participate in normal Watson-Crick base pairs, albeit with reduced bond strength. The reduced melting temperature associated with the dimer presumably reflects a premelting conformation over an extended region of the duplex. Pyrimidine dimers can be repaired by DNA photolyases that are specific for these lesions and not for pyrimidine-pyrimidone (6-4) lesions (see chapter 3) (29).

A number of experiments took advantage of the photoreversibility of pyrimidine dimers to provide evidence that these lesions were premutagenic, since treatment of cells with photoreactivating light reduced the number of mutants (352). However, the recognition that UV radiation-induced lesions play two different roles in mutagenesis, one in inducing the SOS response (see chapter 10) and another as targets for mutagenesis, complicated the interpretation of such experiments. For example, it was directly shown that photoreactivating treatments reduced the expression of SOS genes (including *umuC*+) in cells that had been previously exposed to UV radiation (31). Nevertheless, experiments that separated the induction of the host SOS system from the photoreversal of pyrimidine dimers in the DNA template provided support for the notion that dimers could serve as premutagenic lesions, since their removal from DNA was found to decrease the mutation frequency. Examples of such experiments included photoreactivating UV-irradiated cells containing an F' *lacI*+ plasmid prior to conjugation with an SOS-induced host (165, 179) and photoreactivating phages prior to introducing them into an SOS-induced host (135).

Another type of experimental approach that supported the concept that pyrimidine dimers functioned as premutagenic lesions involved the use of the photosensitizer acetophenone and 313-nm light, a regimen that results in the efficient formation of T-T dimers, some C-T dimers, but not other pyrimidine photoproducts (359). However, these experiments also indicated that the T-T dimer was not as mutagenic as some other UV-induced lesion that can form at a T-T site. The T-T (6-4) photoproduct would now seem to be a likely candidate.

The ability of a *cis-syn* thymine-thymine dimer (Fig. 11–18) to function as a premutagenic lesion for SOS mutagenesis in *E. coli* has been directly tested by constructing single-stranded M13-based vectors containing this lesion at a specific site and introducing them into *E. coli uvrA* mutants (6). If the host cells were not SOS induced, very few progeny phage were obtained, and the data indicate that the presence of the single dimer was sufficient to block the replication of at least 99.5% of the DNA molecules. In contrast, infection of SOS-induced cells resulted in a 50- to 100-fold increase in the number of phages replicated (to yield 27% of the number of plaques obtained with dimer-free phages) (Table 11–5). This observation also reinforces the concept that induction of the SOS system increases the capacity of cells to carry out translesion synthesis over a lesion that normally blocks DNA replication.

However, in marked contrast to the results described above with the T-T-(6-4) photoproducts, the *cis-syn* T-T cyclobutane dimer did not seem to be highly mutagenic, as only 7% of the plaques that formed after transformation into an SOS-induced cell contained mutations (Table 11–5). It thus appears that a *cis-syn* cyclobutane dimer acts as a major block to DNA replication in an uninduced cell but that once the SOS system is induced, it can be replicated and is able to direct the insertion of the correct nucleotides much of the time. The mutations that were obtained were targeted to the site of the lesion; of these, 57% were T → A transversions and 31% were T → C transitions. Interestingly, all the transversions, as well as 85% of the transitions, occurred at the 3′ thymine (the first base to be encountered during DNA replication).

Similar experiments were carried out to examine the ability of a *trans-syn*-T-T dimer (Fig. 11–18) to act as a premutagenic lesion for SOS mutagenesis (5). This is a minor UV radiation-induced lesion found in single-stranded but not double-stranded DNA (16). Interestingly, in contrast to phage carrying the *cis-syn*-T-T dimer, phage carrying the *trans-syn* dimer could be replicated with modest efficiency (14% of the plaque formation obtained with dimer-free phage) in uninduced cells. The replication that occurred in uninduced cells was surprisingly accurate, with only 4% of the replicated phage containing mutations, all of which were exclusively targeted single T deletions. When the phages carrying the *trans-syn* dimer were introduced into SOS-induced cells, not only did the proportion of replicated phages increase (to 29% of that obtained with dimer-free phage) but also the frequency of mutants increased and the specificity of the mutagenesis changed. By using SOS-induced cells, the proportion of replicated phages that carried mutations was found to be 11% (Table 11–5) and the resulting mutations included targeted substitutions (mostly T → A transversions and some T → C transitions) and near-targeted single-base additions, as well as the T deletion. In contrast to the results obtained with the *cis-syn* dimer, about 95% of the base substitutions occurred at the 5′ thymine, a result that again indicates that the chemical nature of the lesion influences the nature of the mutations which result. Furthermore, the difference between the mutations obtained in uninduced cells, as opposed to SOS-induced cells, provides evidence that SOS induction leads not only to an enhanced capacity to replicate damaged DNA, but also to a marked change in the mutation frequency and spectrum.

The classes of UV radiation-induced lesions discussed above clearly cause different mutagenic consequences in SOS-induced cells, indicating that they should be regarded as being *misinstructive* rather than *noninstructional* in the sense that the term is sometimes applied to apurinic/apyrimidinic (AP) sites (also see below) (130, 176, 177, 342). Although it is clear that these three UV-induced lesions can function as premutagenic lesions for SOS mutagenesis, at least in a single sequence context of a single-stranded phage, considerably more work must be done with respect to understanding UV radiation-induced premutagenic lesions. For example, more single-adduct studies of C-containing photoproducts are needed, as well as single-adduct studies of a number of minor UV radiation photoproducts (28) (see chapter 1). Furthermore, the ability of any of these single-adduct UV-induced photoproducts to function as premutagenic lesions in double-stranded DNA molecules has not been examined in detail either, in part because of the tendency for preferential replication of the undamaged strand (139). It has been proposed that the majority of UV-induced mutations in *uvr*⁺ cells occur early (40, 230, 231), presumably during excision repair and before replication, so that this type of mutagenesis cannot be analyzed by using single-stranded DNA molecules.

In wild-type cells, the contribution of specific premutagenic lesions to the final mutational spectrum is further influenced by their relative rates of repair. For example, studies with the purified Uvr(A)BC endonuclease have shown that the normal and Dewar valence isomers of pyrimidine-pyrimidone (6-4) photoproducts are removed 9 times faster that *cis-syn*-cyclobutane dimers (311). Furthermore, as discussed below, it is possible that at least some premutagenic lesions are of a more complex nature—for example, closely opposed photoproducts. Attempts to predict UV radiation-induced mutational spectra on the basis of our present knowledge of premutagenic lesions have been partially successful but have also highlighted the fact that specific details of the mutational spectrum obtained for a given mutagen depend on the nature of the detection system used (180).

AP SITES: IMPLICATIONS FOR THE "A RULE"

The lesion most often cited as an example of a noninstructional lesion is an AP site (197) (see chapter 4). Initial evidence of the mutagenic potential of AP sites as premutagenic lesions came from in vitro experiments which indicated that various DNA polymerases could misincorporate nucleotides opposite such sites with A being inserted preferentially (164, 197, 264, 286, 287). To test the hypothesis that AP sites serve as premutagenic lesions for SOS mutagenesis, *E. coli* spheroplasts were transformed with ɸX174 *am3* DNA molecules containing increasing numbers of AP sites. A major effect of the presence of these lesions was lethality: with single-stranded DNA, approximately 1 AP site per ɸX174 molecule (272) was lethal, while with double-stranded DNA, ~3.5 AP sites per DNA molecule produced death. In uninduced *E. coli*, the reversion frequency of phages produced by transfection with heat- and acid-treated DNA was at most two- to threefold greater than that seen with untreated DNA. However, in SOS-induced *E. coli*, mutagenesis of these phages was enhanced by 20- to 30-fold and was not observed in *recA*(Def) and *umuC* mutants (270).

Sequence analysis of revertants of ɸX174 DNA templates with amber mutations at different loci indicated that mutagenesis with heat- and acid-treated DNA (i.e., DNA with AP sites) occurs predominantly opposite purines and involves the preferential insertion of adenine. An experiment involving the use of the α segment of the *lacZ⁺* gene on an M13 derivative (163) provided further evidence that the premutagenic lesions introduced by heat and acid treatment were AP sites, since they could be abolished by preincubating the DNA with an AP endonuclease. Analyses of the mutations obtained in this experiment indicated that the frequency of misincorporation opposite AP sites was A (59%) > T (28%) > G (11%) > C (1%). The ability of an AP site to function as a premutagenic lesion was further tested by transfecting SOS-induced and uninduced cells with a single-stranded M13-based vector carrying a single AP site at two different locations (178). Consistent with previous results, very few of these phages (0.1 to 0.7%) could replicate in uninduced cells, whereas considerably more (5 to 7%) could replicate in SOS-induced cells (Table 11–6). In SOS-induced cells, 93 and 96% of the replicated phages had an insertion of a nucleotide opposite the AP site while the remainder resulted from the targeted omission of a nucleotide. At one of the AP sites the nucleotides inserted were A (54%) > T (25%) > G (20%) > C (1%), while at the other they were A (80%) > G (15%) > T (4%) > C (1%) (178) (Table 11–6). In

Table 11–6 SOS mutagenesis of single-stranded M13 vectors carrying a single AP lesion at a defined site[a]

% bypass in SOS-induced cells/ % bypass in non-SOS-induced cells	% of replicated phage in SOS-induced cells with a nucleotide inserted opposite the AP site[b]	Nucleotide inserted (frequency)
5–7/0.1–0.7	93	Site 1 A (54%) T (25%) G (20%) C (<1%)
	96	Site 2 A (80%) G (15%) T (4%) C (<1%)

[a] Single-stranded M13mp7-based vectors carrying a single AP lesion at a defined site were transfected into an *E. coli uvrA* strain that had been induced for SOS (4-J/m² UV irradiation) or were uninduced.
[b] The remaining replicated phage resulted from the targeted omission of a single nucleotide.
Source: Data from Lawrence et al. (178).

contrast to the above results, a set of experimental observations have been interpreted as indicating that there is a strong preference for a *polB*⁺-dependent insertion of A's opposite abasic sites in cells in which the SOS system is activated by a *recA1202*(Cptᶜ) mutation rather than by treatment with a DNA-damaging agent (326). In these latter experiments, the nature of the resulting mutations was inferred rather than being directly determined.

Thus, as in the case of the UV-induced premutagenic lesions discussed above, the presence of an AP site in a DNA molecule interferes with the replication of that molecule in uninduced *E. coli* and SOS induction increases the ability of the cell to replicate such molecules. Furthermore, in SOS-induced cells, AP sites are clearly premutagenic lesions. Studies based primarily on the behavior of purified DNA polymerase on DNA templates containing AP sites and other lesions have been interpreted as providing support for what is termed the *A rule* (307), which posits that DNA polymerases have an inherent preference for inserting A opposite a noninstructional site, with an AP site often taken as an example of such a site. As can be seen from the in vivo data discussed above, there is indeed a preference for the insertion of A opposite an AP site in *E. coli* cells whose SOS system has been induced by exposure to a DNA-damaging agent, but it varies from site to site and is not strongly pronounced. As mentioned above, it is possible that there is a strong preference for the insertion of A's opposite AP sites in *recA1202*(Cptᶜ) *polB*⁺ cells (326), but additional characterization of this phenomenon is required.

Although some polymerases may have a preference for insertion of an A opposite a noninstructional site, it is not clear that AP sites are truly noninstructional or that the preference for inserting an A opposite an AP is solely due to an inherent preference of the polymerase to insert an A opposite a noncoding lesion. The differences between the results obtained in *E. coli* with a single AP site in two difference sequence contexts indicate that other factors can influence the frequency with which different nucleotides are inserted. Factors that could result in the preference for A insertion include van der Waals and stacking interactions (342) and hydrogen binding to deoxyribose (197).

A two-dimensional NMR study of an oligonucleotide duplex containing an A opposite an AP site has indicated that the A is positioned within the helix as if it were paired with T, so that the helix retains a normal B-DNA structure (72). Furthermore, and surprisingly, the melting temperature (T_m) for the unpaired A was the same as that for other A·T base pairs in the same helix. In contrast, oligoduplexes containing G, T, or C opposite an AP site were less stable (71). When G was opposite the AP site, it was intrahelical and in the normal *anti* conformation at low temperatures, but with increasing temperature the G adopted a "melted" state in advance of the rest of the helix. T was found to be in both intrahelical and extrahelical states at low temperature, with the proportion of extrahelical state increasing with increasing temperature. If C was opposite the AP site, both the C and abasic sugar were extrahelical and the helix was collapsed with the adjacent base pairs stacking over each other. These results suggest a thermodynamic basis for the preference of A insertion opposite an AP site and also that suggest the "preference" for A insertion is not solely a property of the polymerase.

AP sites may play roles in the mutagenicity of a number of mutagens. For example, such sites have been postulated as intermediates in the mutagenicity of BPDE (89), 3,4-epoxycyclopenta[c,d]pyrene (89, 102), and activated aflatoxin B₁ (102), to account for the predominance of G·C → T·A transversions in the *lacI* mutational spectra obtained for these compounds. Data have also been reported which are consistent with the notion that mutagenesis by β-propiolactone involves AP sites generated by the depurination of alkylated bases (270). Another example is suggested by the report that *xthA* mutants (which lack the major AP endonuclease of *E. coli*) (see chapter 4) display an increased ability to be mutated by MNNG in an SOS-dependent fashion and that this increased component of mutagenesis can be eliminated by also introducing an *alkA* mutation which eliminates the major 3-methyladenine-DNA glycosylase activity. These observations have been interpreted to suggest that AP sites are generated by the action of 3-methylad-enine-DNA glycosylase on alkylated DNA and the resulting AP sites serve as pre-mutagenic lesions for SOS-dependent mutagenesis (99).

N-(2'-DEOXYGUANOSIN-8-YL)-2-AAF

The mutational spectrum induced by N-acetoxy-AAF (Fig. 11–18), an active me-tabolite of AAF (see chapter 1), revealed that more than 90% of the mutations were frame shifts clustered within two types of sequences: −1 deletions within runs of contiguous G residues, and −2 deletions within alternating GC sequences such as the NarI restriction site (113, 155). Since N-acetoxy-AAF reacts approxi-mately equally with all G residues (112), the presence of the mutational hot spots was attributed to specific mechanisms through which the lesions within the hot spot are mispaired or misreplicated. The −1 frame shifts were found to be recA⁺ and umuD⁺C⁺ dependent. The −2 frame shifts at alternating GC sequences arose from a different, as yet poorly characterized, process that required the induction of one or more LexA-repressed genes but was umuD⁺C⁺ independent and did not require RecA protein other than to mediate LexA cleavage (204).

N-Acetoxy-AAF reacts primarily at C-8 of guanine to form N-(2'-deoxygu-anosin-8-yl)-AAF adducts (see chapter 1). These adducts were shown to serve as premutagenic lesions for −1 frameshift mutations by constructing plasmids that contained a single AAF adduct within a run of 3 or 4 contiguous G residues. Mutagenesis was shown to be SOS dependent and to have a marked dependence on the position of the modification within the run of G residues (171).

In vitro, replication of model substrates carrying a single AAF adduct by DNA polymerase III holoenzyme and the Klenow fragment of DNA polymerase I stops 1 base prior to the adduct. In contrast, replication by a mutant Klenow fragment lacking the $3' \rightarrow 5'$ proofreading exonclease stops 1 nucleotide later (14). On the basis of these observations, it has been proposed that the reason that a polymerase with a proofreading activity stops 1 nucleotide before an AAF adduct is not because it is unable to insert a nucleotide opposite the adduct but because elongation is strongly impaired, thereby giving an increased amount of time for the proofread-ing exonuclease to remove the base inserted opposite the adduct (14). In the case of a T7 DNA polymerase lacking the $3' \rightarrow 5'$ proofreading exonuclease, single-turnover kinetics studies have shown that, although dCTP is incorporated opposite an AAF-dG adduct, its rate of incorporation is reduced $\sim 4 \times 10^6$ relative to its incorporation opposite an unmodified dG (191).

A mechanism to account for the observed specificity of −1 frameshift mu-tations induced by bulky lesions has been suggested (13, 170, 271), based on the strand slippage model originally proposed (308) to account for spontaneous −1 frameshift mutation hot spots (see chapter 2). This model (Fig. 11–19) suggests that strand slippage occurs after accurate incorporation of C opposite the lesion (171, 229). The misaligned intermediates are thought to be stabilized by the for-mation of normal G·C basepairs between the terminal C residue on the primer and a G residue 5' to the adduct on the template strand. According to this model, the AAF adduct would play at least two specific roles in promoting frameshift mutagenesis. First, the adduct would allow correct insertion of C opposite the lesion during DNA synthesis, probably because modification at the C-8 position of G does not interfere with G·C hydrogen-bonding groups. Second, the adduct hin-ders elongation so that the transient stalling of DNA synthesis increases the op-portunity for strand slippage to occur (171). Analyses of slipped intermediates of this type by using chemical probes and thermal denaturation have provided ad-ditional support for this model (115). In particular, it was shown that AAF mod-ification of a bulged G increases the stability of the bulged duplex to a level ap-proaching that of an unmodified homoduplex, probably by binding in the minor groove with the G residue in the syn conformation as previously suggested by NMR (236) and modeling studies (283).

These experiments have also revealed an unusual and, as yet, poorly char-acterized aspect of AAF-induced mutagenesis in SOS-induced cells. The presence of the same lesion within an alternating G-C sequence gives rises to −2 frameshift

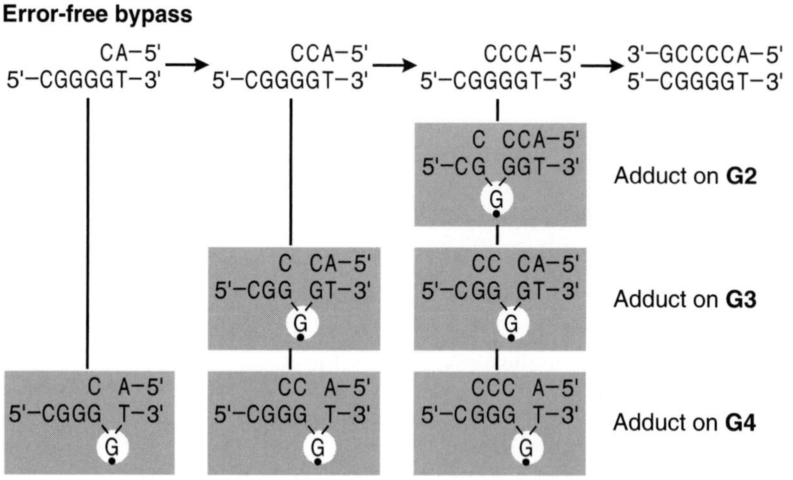

Error-free bypass

```
        CA-5'            CCA-5'            CCCA-5'      3'-GCCCCA-5'
5'-CGGGGT-3'  →  5'-CGGGGT-3'  →  5'-CGGGGT-3'  →  5'-CGGGGT-3'
```

```
                                          C CCA-5'
                                     5'-CG GGT-3'      Adduct on G2
                                            G
```

```
                       C CA-5'           CC CA-5'
                  5'-CGG GT-3'       5'-CGG GT-3'       Adduct on G3
                         G                  G
```

```
     C A-5'            CC A-5'           CCC A-5'
5'-CGGG T-3'      5'-CGGG T-3'       5'-CGGG T-3'       Adduct on G4
      G                  G                  G
```

Figure 11–19 Proposed model for AAF-induced frameshift mutagenesis. Error-free bypass of a sequence containing four guanines is shown at the top; this is by far the most frequent event (>97%). The potential slipped intermediates resulting from AAF adducts on G2, G3, or G4 are shown below in the red-shaded boxes. The observation that mutagenicity increases according to the number of guanine residues 5' to the AAF adduct is postulated to result from (i) the increased number of possible slipped intermediates resulting from adduction at 3' positions of the contiguous run as compared with 5' positions and (ii) the increased number of terminal C·G base pairs in the postulated intermediates. (*Adapted from Lambert et al. [171] with permission.*)

mutations in a fashion that requires SOS induction but is independent of *umuD⁺C⁺* function (54, 204).

The Predominance of C → T Transitions in UV Radiation Mutagenesis

As mentioned above, C → T transitions are a major class of UV radiation-induced mutation and there has been continued interest in understanding the molecular basis of these mutations. The original suggestion (323) that such mutations arise as a consequence of the A rule with a DNA polymerase inserting an A opposite a noninstructional lesion seems incompatible with the data discussed above. As discussed above, it seems likely that some of these arise as a consequence of (6-4) photoproducts at T-C sites.

Two other models have been proposed to account for the predominance of UV-induced C → T transitions in the absence of photoreactivation (Fig. 11–20). In the tautomer bypass model, the imino tautomer of C in a cyclobutane dimer directs the incorporation of A (21, 247). The feasibility of this model is supported by several arguments, including the fact that when the C-5–C-6 bond of cytosine becomes saturated, aromatic stabilization of the amino tautomer is no longer possible, resulting in an increase in the proportion of otherwise minor imino tautomers (139). In the deamination bypass mechanism (96, 139, 325, 327), deamination of a C in a cyclobutane dimer precedes replication. The saturation of the C-5,C-6 double bond in the dimer results in an increased rate of deamination (121), and the mechanism of deamination has been examined in considerable detail (184). The resulting U in the cyclobutane dimers then directs the incorporation of A during replicative bypass.

To explore further the feasibility of these models, a plasmid carrying a single *cis-syn* T-U dimer at a defined site was constructed and transformed into *uvrA phr* bacteria that were either uninduced or SOS induced (139). In uninduced cells the T-U cyclobutane dimer almost completely blocked replication, whereas in SOS-induced cells it directed the incorporation of two A's with a relatively high specificity (96%). This result has been interpreted as consistent with both the tautomer bypass model and the deamination bypass model, since the T-U dimer is isostruc-

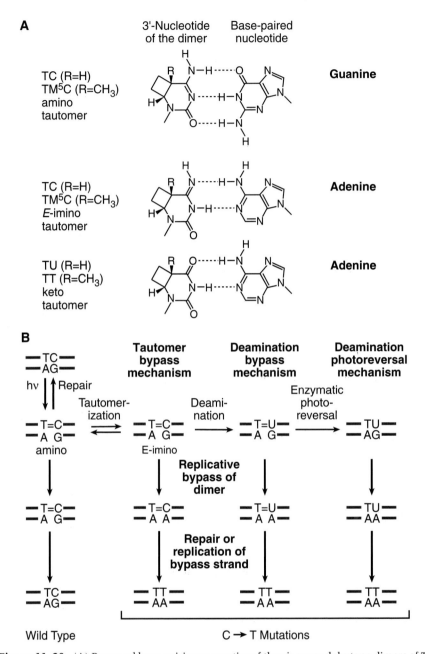

Figure 11–20 (A) Proposed base-pairing properties of the *cis-syn* cyclobutane dimers of TC, Tme⁵C, TU, and TT. Similar base pairing would apply to the dimers of CT and CC, and their me⁵C derivatives and deamination products, as well as to the 5'-pyrimidine of (6-4) and Dewar products. (B) Proposed mechanisms for the origin of UV-induced C → T mutation at TC sites. This scheme can also be applied to CT and CC sites and their me⁵C derivatives. (*Adapted from Jiang and Taylor [139] with permission.*)

tural with the unstable imino tautomer of a T-C dimer. The imino tautomer would be expected to be a prevalent tautomer of the T-C dimer, while the fraction of U-containing dimer would vary as a function of time. Thus, the extent to which tautomer bypass or deamination bypass or both play a significant role in a living system would be expected to depend on the delay time between UV irradiation and DNA replication. These experiments reinforce two inferences about the nature of the system carrying out translesion bypass during SOS mutagenesis. First, it is capable of replicating past lesions that normally block DNA replication. Second,

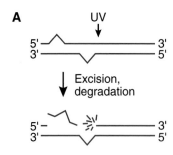

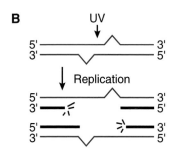

- Daughter-strand gaps overlap
- Gaps enlarged by degradation
- Recombinational repair impossible without translesion synthesis

Figure 11–21 Two ways in which closely opposed lesions could generate a substrate-requiring translesion bypass. (*Adapted from Witkin [352] with permission.*)

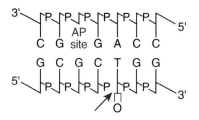

Figure 11–22 Proposed structure of the bivalent AP site/strand break lesion induced by neocarzinostatin. The arrow shows the strand break with 3'-phosphate and 5'-aldehyde termini. The AP site probably involves oxidation of C-1' to a lactone. (*Adapted from Povirk et al. [248] with permission.*)

the probability that a given nucleotide is inserted opposite the lesion is influenced by the chemical structure, conformation, and tautomeric state of the lesion.

More-Complex Lesions as Premutagenic Lesions

CLOSELY SPACED OPPOSING PHOTOPRODUCTS: POSSIBLE PREMUTAGENIC LESIONS FOR SOS MUTAGENESIS

The suggestion has been made that cells might be forced to use translesion bypass as a way of processing damaged double-stranded DNA if a lesion was situated opposite a single-stranded gap and if other mechanisms for repairing the lesion were precluded (32, 275, 352). In the case of UV radiation mutagenesis, this situation could arise if two closely spaced photoproducts on complementary strands were introduced by irradiation. As shown in Fig. 11–21, removal of one lesion by excision repair could lead to a lesion situated opposite a gap, either if the second lesion was opposite the 12- to 13-nucleotide gap created during nucleotide excision repair or if the initial gap was expanded. Such a substrate could not be repaired by the normal process of nucleotide excision repair because of the presence of the lesion in the template strand. If the DNA was replicated prior to the introduction of the lesion, repair involving recombination would not be an option.

The possibility of the formation of an excision gap when there are two extremely closely spaced opposing lesions has been supported by experiments showing that the Uvr(A)BC endonuclease can nick one strand when presented with a substrate having two closely opposed *cis-syn* dimers located 5' to each other (311). From a priori considerations, the probability of two closely spaced opposing lesions being generated by the doses of UV radiation used in biological experiments would appear to be low. However, direct measurements have shown that the induction of closely spaced opposed dimers can occur with greater probability than expected at specific sites in DNA sequences and that such sites are characterized by the presence of closely opposed pyrimidine runs (169). More work is required to establish the biological relevance of this model. However, one site in the i^{-d} portion of the *lacI*$^+$ gene that is a hot spot for mutations in a wild-type strain correlates well with high frequencies of closely opposed photoproducts (263), and a set of arguments has been presented suggesting that a component of the UV-induced mutations to streptomycin resistance in wild-type *E. coli* arises from closely spaced opposed lesions (38).

An alternative mechanism for creating a lesion opposite a gap would be by the generation of overlapping daughter-strand gaps (275). This mechanism could occur even in excision-repair-deficient strains, and neither excision repair nor daughter-strand gap repair would be an option for processing such a DNA substrate (Fig. 11–21). Furthermore, such overlapping daughter-strand gaps could arise from two opposing lesions that are farther apart than those considered in the model involving excision repair. The number of daughter-strand gaps determined to require some form of inducible repair was similar to the number calculated to be overlapping one another in opposite daughter chromosomes. This observation, together with an estimate that the survival with no repair of the gaps resembled the survival predicted with no mutagenesis, led to the proposal that the repair of these overlapping gaps by a process of translesion synthesis could result in mutagenesis (275).

COMPLEX LESIONS AFFECTING BOTH STRANDS OF DNA

An interesting class of lesions in the context of SOS mutagenesis is represented by a lesion introduced by neocarzinostatin (248). The lesion consists of an AP site plus a closely opposed break in the complementary strand (Fig. 11–22). The break occurs opposite the base 2 positions upstream from the AP site and has the 3'-phosphate and 5'-aldehyde termini of neocarzinostatin-induced breaks, suggesting that the neocarzinostatin simultaneously attacks two DNA sugars on opposite edges of the minor groove. The presence of the break almost opposite the AP site interferes with cleavage by AP endonucleases and would be expected to increase the likelihood that the AP site would be subject to translesion bypass under SOS-induced conditions. The hypothesis that these complex lesions serve as premutagenic lesions is supported by the observations that neocarcinostatin is an SOS-dependent mutagen (90) and that the sequence specificity for formation of AP sites with closely opposed breaks reflects that of neocarcinostatin-induced mutagenesis (248).

Model for the Molecular Mechanism of SOS Mutagenesis

The results described above concerning single-stranded DNA vectors carrying specific DNA adducts provide further support for the idea that SOS mutagenesis comes about via a mechanism of translesion DNA synthesis (55, 75, 251, 342, 352) and suggest that the nature of the lesion and the local DNA sequence play roles in determining the nature and frequency of the mutations which result. In this section we will summarize experiments that have led to the current model for SOS mutagenesis.

Experiments Involving UV Irradiation plus Photoreversal

UV RADIATION MUTAGENESIS OF *E. COLI* BY UV IRRADIATION PLUS DELAYED PHOTOREVERSAL

A set of experiments that have stimulated thinking about possible mechanisms of SOS mutagenesis (35) have resulted in the central observation that *umuC*, *umuD*, *lexA*(Ind⁻) (43–45) and Δ*recA*(Def) (33) *E. coli* strains, all of which are largely nonmutable by UV radiation (see above), can be mutated to a significant degree if, following exposure to UV radiation, the cells are incubated for a period (up to 4 h) and then exposed to photoreactivating light. By using this procedure, the frequency of UV mutations obtained was approximately 20% of that obtained with a wild-type *E. coli* strain. It was shown that most of the mutations obtained by using this "UV-plus-delayed-photoreversal" protocol were C → T transitions, suggesting that the phenomenon was confined to cyclobutane dimers containing cytosine (22, 330). These observations were interpreted in terms of a "two-step" model for UV radiation mutagenesis (Fig. 11–23). According to this model, in the first step of UV radiation mutagenesis, which was postulated to occur in *umuC*, *umuD*, *lexA*(Ind⁻), and Δ*recA* cells, an incorrect nucleotide is misincorporated opposite a dimer but continued chain elongation cannot occur because of the photoproduct. Photoreversal converts the dimer back to monomeric pyrimidines, thereby enabling the second step, chain elongation, to proceed.

This hypothesis thus suggests that the *umuC⁺* and *umuD⁺* gene products are required for the second step but not for the first (43–45). Initial experiments that addressed the role of RecA protein in this process involved the *recA430*(Cpt⁻) allele of *recA* (see chapter 10), and a reduced yield of mutants was observed, leading to the proposal that RecA protein does play a role in the misincorporation step (44). However, experiments with a Δ*recA* strain revealed that mutations could be obtained after UV plus delayed photoreversal, indicating that the RecA protein is not essential for the misincorporation step according to this model (33). Furthermore these experiments suggested that certain alleles of the *recA* gene (*recA1*, *recA56*, and *recA430*) (see chapter 10) could depress misincorporation to various extents, depending on the *recA* allele and the level of expression of the particular mutant RecA protein (33).

Physiological experiments involving the inhibition of DNA replication by chloramphenicol indicated that, in excision-repair-deficient bacteria, chromosomal replication was necessary for mutants to be obtained by the UV-plus-delayed-photoreversal procedures (43). This result was strengthened by the demonstration that at least the α subunit of DNA polymerase III appears to be required for mutagenesis by this procedure (284). There is as yet no evidence for the induction of frameshift mutations by the UV-radiation-plus-delayed photoreversal procedure, since no reversion of the *trpA9777* and *trpA9813* frameshift alleles in a *umuC122*::Tn5 mutant was observed (267). Both of these frameshifts occur in runs of A·T base pairs, where T-T photoproducts could occur, and evidence has been

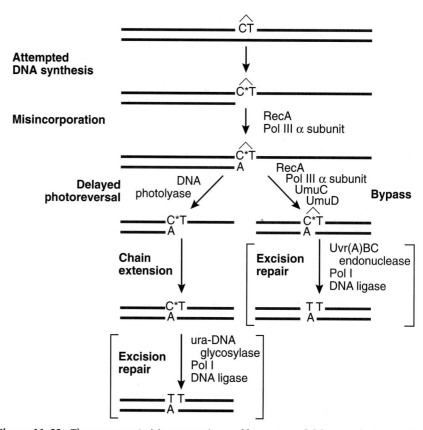

Figure 11–23 The two-step (misincorporation and bypass) model for translesion synthesis in UV mutagenesis in *E. coli*. C* indicates that a uracil may have formed by deamination of cytosine. Insertion of A opposite C or U is indicated as the most likely insertion, but others are not excluded. Likewise, misincorporations may occur more rarely opposite T in a cyclobutane-type dimer and more frequently opposite T in (6-4) photoproducts, but the latter will not be amenable to splitting by photoreversal. In this case, when insertion of A occurs, it would not lead to a mutation and would constitute part of an error-free UmuD-UmuC-dependent translesion synthesis. (*Adapted from Bridges [35] with permission.*)

presented that *trpA9777* is revertible by cyclobutane dimers (364). Interpretation of these latter observations in the context of this model has led to speculation that the roles of the *umuD⁺C⁺* gene products in frameshift mutagenesis might include a direct role in strand misalignment or in the initiation of replication on a misaligned primer (267).

EFFECT OF PHOTOREACTIVATION ON UV RADIATION MUTAGENESIS IN A SINGLE-STRANDED PHAGE

In a study with single-stranded DNA phage S13, it was reported (324) that UV radiation mutagenesis of the phage occurred in *umuC* or Δ*recA*(Def) bacteria when the bacteria were grown in the light instead of the dark. An important unstated feature of these experiments was the fact that the phage were stored for 4 h at 4°C in the dark after irradiation and before infection of the host cells and the exposure to light (325). If this period of storage was eliminated, no mutations were obtained under the same conditions (325). This finding was used to argue that, in phage S13 plated on *umuC* or Δ*recA*(Def) bacteria, no misincorporation occurs opposite a cyclobutane dimer. Instead, the suggestion was made that the mutations obtained upon storage prior to photoreactivation could be attributed entirely to deamination of cytosine in cyclobutane dimers. The kinetics of the postulated deamination reaction were highly unusual, exhibiting a step function (328), which the investigators have reported as being reproducible (327). It has been inferred that no uracil appears after 20 min of incubation of UV-irradiated single-stranded DNA at 37°C or after 40 min of incubation of double-stranded DNA. Most of the uracil was inferred to appear within a brief 14-min period centered at 29 min for single-stranded DNA or 55 min for double-stranded DNA (327). Other measurements of the deamination rates of cytosine-containing cyclobutane dimers, either in model dinucleotide monophosphates (78, 184) or in DNA in vivo (96, 258), did not reveal these unusual kinetics, so the molecular basis of this phenomenon in these experiments with S13 DNA remains unresolved (327).

The relationship of these S13 experiments to the UV-radiation-plus-delayed-photoreversal studies of chromosomal mutagenesis is not yet clear. However, a set of counterarguments against deamination of cytosine-containing dimers playing a role in the production of mutations in UV-radiation-plus-delayed-photoreversal experiments has been presented (35).

Premutagenic Lesions on the Lagging versus the Leading Strand May Influence the Probability of a Mutation

The chromosome of *E. coli* is thought to be replicated by an asymmetric dimer of DNA polymerase III (157, 238). The asymmetry arises because of the different requirements for replicating the leading strand of DNA as opposed to the lagging strand and is manifested as differences in the assembly of polymerase accessory subunits used to replicate the two strands. A preliminary study suggests that the probability of a premutagenic lesion giving rise to a mutation may be influenced by its presence on the leading or the lagging strand (333). The lesion used was the premutagenic AAF lesion at various locations within a run of three G's. The effect of asymmetry on mutagenesis was tested by using a pair of plasmids containing the unidirectional ColE1 origin of replication and a single lesion located in either the leading or lagging strand. The frequency of -1 frameshift mutations obtained in uninduced cells was approximately 20-fold higher when the adduct was located in the lagging strand than when it was located in the leading strand. SOS induction before transformation with damaged plasmid increased the frequency of mutations 50-fold for both strands but retained the bias. Various issues have been identified that could complicate the interpretation of this experiment (333). Nevertheless, it seems likely that this issue will be studied intensively in the coming years, and it will be interesting to see whether the phenomenon can be extended to other replicons and whether it is confined to frameshift mutations or applies also to base substitution mutations.

Biochemical Mechanism of SOS Mutagenesis

Determination of the biochemical mechanism of SOS mutagenesis has proved to be a very challenging problem. However, progress has been made recently by using two different experimental strategies. One has been to establish a system involving purified components that carries out translesion DNA synthesis and is dependent on proteins known from genetic experiments to be important for SOS mutagenesis. The other has been to establish a cell-free system that is capable of carrying out UV radiation-induced mutagenesis in vitro.

REPLICATIVE BYPASS BY PURIFIED COMPONENTS INCLUDING UmuD', UmuC, AND RecA PROTEINS

A direct approach toward establishing a biochemical system for SOS mutagenesis by using purified components was reported by Hatch Echols and his colleagues (254). On the basis of the genetic evidence implicating UmuD', UmuC, RecA, and DNA polymerase III (see above), these components were used in an assay designed to bypass a single AP lesion at a defined site in a primed template. The UmuC protein used in the assay was purified in 8 M urea and renatured by dilution and dialysis (362). When all of these proteins were present, a limited amount of bypass synthesis was observed (Fig. 11–24). The background level of bypass in the absence of UmuD', UmuC, or RecA proteins was ~0.5%, while the total amount of bypass was estimated to be 5% in the presence of all three proteins. This extent of bypass

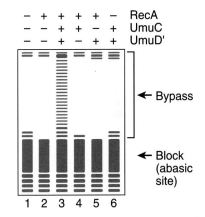

Figure 11–24 Bypass of abasic sites by DNA polymerase III in the presence of UmuC, UmuD', and RecA. A primed DNA substrate with a single abasic site located 65 nucleotides downstream of the radioactive primer was replicated by holoenzyme lacking the ε subunit in the presence or absence of UmuC, UmuD', and RecA. The replication products were separated in denaturing polyacrylamide gels and visualized by autoradiography. The heavy band resulted from the replication block before the abasic site. The bands above the abasic block derived from longer DNA chains were considered bypass events, as indicated. The misincorporation band cannot be seen because of the highly labeled abasic block. Additions to the replication reaction mixtures were as indicated. (*Adapted from Rajagopalan et al. [254] with permission.*)

is similar to that observed in vivo when a single-stranded vector carrying a single AP site is used (178) (see above). The ladder-like pattern of bypass bands suggests that the polymerase dissociates or ceases to replicate shortly after bypassing the lesion instead of continuing processively to the end of the template.

The requirement for UmuD′, UmuC, and RecA proteins was observed irrespective of whether the form of the DNA polymerase III holoenzyme contained ε, the proofreading subunit. DNA polymerase I did not exhibit bypass synthesis in the presence of UmuD′, UmuC, and RecA proteins, while DNA polymerase II, together with polymerase III processivity proteins, exhibited at best, marginal bypass synthesis. Interestingly, UmuD protein inhibited bypass, consistent with the hypothesis discussed above that the intact form of UmuD protein might function as an inhibitor of SOS mutagenesis (254).

CELL-FREE SYSTEM FOR MUTAGENESIS

The development of *cell-free systems* which mimic in vivo events played a pivotal role in the elucidation of the biochemical mechanism of DNA replication in *E. coli* (157). An in vitro system has recently been developed for UV radiation mutagenesis (67, 68). It consists of two stages. First, a UV-irradiated plasmid carrying the *cro*$^+$ gene is incubated with a soluble protein extract prepared from SOS-induced *E. coli* cells. Second, a bioassay in a *recA* deletion strain is used to detect mutations produced in the *cro* gene during the first stage. The use of this system has allowed the identification of two pathways for UV mutagenesis (67, 68). The first pathway depends on DNA replication and requires the *recA*$^+$ and *umuC*$^+$ gene products. The second pathway was revealed by enzymatically removing pyrimidine dimers from the plasmid DNA at the end of the first stage and prior to the bioassay. This treatment caused a large increase in the frequency of mutations detected in the bioassay. This photoreactivation-stimulated in vitro mutagenesis was dependent on the nucleotide excision repair genes *uvrA*$^+$, *uvrB*$^+$, and *uvrC*$^+$ and was partially dependent on *uvrD*$^+$. It did not, however, require the functions of the *umuC*$^+$ and *recA*$^+$ genes. The second mutagenic pathway occurred in the absence of plasmid DNA replication. Although not dependent on DNA polymerase I or II, it was dependent on DNA polymerase III but was not inhibited by antibodies against the β subunit of DNA polymerase III holoenzyme. The fact that the processivity subunit is not required is consistent with a mechanism for the second mutagenic pathway in which DNA polymerase III fills in short single-stranded DNA gaps (68). Sequencing of mutations arising via this second pathway revealed a spectrum similar to that of in vivo UV mutagenesis (68).

The relationship of this second pathway to UV mutagenesis in vivo is not yet clear, but it has been suggested that it may mimic the mutagenesis observed by using the UV-plus-delayed-photoreversal procedure described above. It is also possible that this in vitro phenomenon is related to the previous suggestions that many of the UV radiation-induced mutations that arise in wild-type bacteria are initiated by the process of nucleotide excision repair. The evidence for this is that certain classes of UV-induced mutations are not seen in wild-type bacteria if the cells are immediately exposed to visible light (photoreactivation) after UV irradiation but they rapidly lose their photoreactivability during subsequent incubation and can do so under conditions where there is no DNA replication (40, 41, 195, 230, 231).

POSSIBLE ROLES OF THE UmuD′, UmuC, AND RecA PROTEINS IN MUTAGENESIS

One suggestion for the third role of RecA in SOS mutagenesis (see above) is that it serves to inhibit the 3′ → 5′ proofreading exonuclease of DNA polymerase III (95, 200). If this does occur in vivo, it cannot solely account for the biochemical basis of SOS mutagenesis, since a *dnaQ* deletion mutant, which lacks the ε proofreading subunit, still displays a *recA*$^+$ dependence for UV radiation mutagenesis (296). This conclusion is also consistent with the observations that inhibition of proofreading does not affect the ability of DNA polymerase III to bypass photoproducts or AP sites (194, 293) and that bypass synthesis across an AP site requires RecA protein even when the ε subunit is missing from the DNA polymerase III holoenzyme (254).

An alternative suggestion for the third role of RecA protein is that it helps to direct the UmuD′ and UmuC proteins to the site of the lesion in DNA (4, 313) A more simple view is that UmuD′ and UmuC proteins may act only on DNA that is coated with RecA protein (363). A possible physical interaction between activated RecA and UmuC proteins was suggested by the demonstration that radiolabeled UmuC protein in a crude extract was retained on an activated RecA affinity column, but this experiment did not demonstrate a direct interaction (109). The possibility of an activated RecA-UmuD′ interaction has been suggested on the basis of the observations that RecA730 protein from a crude extract was specifically retained on UmuD and UmuD′ affinity columns and that UmuD and UmuD′ proteins could be shown to be targeted to single-stranded DNA only in the presence of RecA protein (105).

It has been observed that overproduction of UmuD′ and UmuC proteins in the recipient in a Hfr × F⁻ conjugal cross inhibits recombination but that this inhibition can be substantially suppressed by increasing the level of expression of RecA protein (300). On the basis of these observations, it has been suggested that the UmuD′ and UmuC proteins can inhibit recombination by interacting with the growing end of a RecA filament. If UmuD′ and UmuC proteins were to interact with the end of a RecA filament, this could play a role in positioning them at a lesion. Even if RecA protein were to play a role in directing the Umu proteins to the site of the lesions, this would not preclude its playing yet another role. It is even possible that RecA protein has to be released from the site of the lesion as a prerequisite for translesion synthesis.

The roles of the UmuD′ and UmuC proteins are not yet known. As discussed above, they might serve as alternative processivity elements that would facilitate synthesis on the damaged template, perhaps by preventing disassociation or by favoring reassociation. They might also play a role in modulating the action of ε (the proofreading subunit) or in stabilizing a form of the polymerase that lacks ε. It is also possible that they play some key role in modifying some other aspect of polymerase action that becomes limiting on damaged DNA templates, for example, a conformational shift to a catalytically active form (see chapter 2).

The relationship between the roles of RecA in translesion synthesis and replication restart is not yet clear. However, as discussed in chapter 10, it is possible that RecA plays a role in a strand-switch mechanism of DNA damage tolerance. Although replication restart does not normally require *umuD⁺C⁺* function, it is dependent on these functions the presence of certain *recA* alleles, raising the possibility that mechanisms of translesion synthesis and replication restart are related in some way (88).

SOS-Dependent Mutagenesis in Cells Not Exposed to Exogenous DNA-Damaging Agents

A number of experiments have clearly shown that induction of the SOS system can lead to an increase in the observed mutation frequency even in the absence of damage by an exogenous agent. The term *SOS mutator effect* has been used to describe this phenomenon (352), as have other terms such as *untargeted* or *nontargeted mutagenesis*. However, we will avoid the last two terms, since they impose an interpretative bias on the phenomenon. What has become increasingly clear is that the genetic dependence of SOS mutator effects is dependent on the system used to monitor such mutagenesis. With respect to its genetic dependence, the increased mutation frequency in SOS-induced cells that is observed by monitoring chromosomal loci bears a substantial resemblance to that for targeted mutagen-dependent SOS mutagenesis (see above), as does the increase observed when monitoring the mutation frequency by using loci carried by single-stranded DNA phages. In contrast, the genetic dependence of the SOS mutator effect observed when phage λ replicates in SOS-induced cells appears to differ in several respects from that of most events in targeted SOS mutagenesis.

The SOS Mutator Effect on Chromosomal Loci or on an F′ Plasmid

E. coli cells carrying the *recA730*(Cpt^c) allele (see chapter 10) have an elevated spontaneous mutation frequency, as do cells carrying the *recA441*(Cpt^Ts) allele (see chapter 10), when they are grown under conditions where RecA protein is activated (351, 352, 354). The elevated mutation frequency requires *umuD^+* and *umuC^+* functions (19, 65, 313). It also appears to be critically dependent on the presence of activated RecA protein in the cell, since depression of SOS functions by a *lexA*(Def) mutation does not result in elevation of the mutation rate in a *recA^+* cell (19), even if UmuD′ protein is provided directly from a plasmid (313). A study of the effect of various *recA* alleles on this SOS mutator effect has reinforced the conclusion that, as is the case with targeted mutagen-dependent SOS mutagenesis, RecA protein is required for some other function besides mediating the cleavage of LexA and UmuD proteins. Furthermore, in the case of the SOS mutator effect, it seems clear that RecA protein must be activated for the phenomenon to be observed. The elevated spontaneous mutation rate in a *polA* deletion mutant expressing the DNA polymerase I 5′ → 3′ exonuclease activity is also blocked by a *umuC* mutation and is likely to result from constitutive expression of SOS functions in this strain (10). The report that a *polB* mutation decreases the mutator phenotype of a *recA1202*(Cpt^c) strain (326) has raised the possibility that one significant difference between the SOS mutator effect observed in *recA*(Cpt^c) strains and targeted mutagen-dependent SOS mutagenesis is that a substantial proportion of the former involves DNA polymerase II. Further exploration of this issue is required.

An analysis of the spectrum of *lacI* nonsense mutations obtained in a *recA441*(Cpt^Ts) strain grown at 42°C revealed that the mutations resulting from the SOS mutator effect were mostly G·C → T·A transversions, along with some A·T → T·A transversions (217) (Fig. 11–25). Furthermore, this type of mutagenesis displayed a pronounced site specificity. However, the nature of the mutations obtained in this spectrum may have been influenced by the action of the *mutHLS*-dependent mismatch repair system, since the SOS mutator effect has been reported to increase in *mutS* or *mutL* strains (56), although a conflicting report has also been published (103). Thus, the spectrum of mutations initially introduced by the SOS mutator effect may differ from that experimentally observed in a wild-type cell. The fact that postreplicative *mutHLS*-dependent mismatch repair may reduce the magnitude of the SOS mutator effect has been used to argue that these mutations are introduced during a replicative rather than a repair process (56).

It is still unclear whether the chromosomal mutations that result from the SOS mutator effect are due to a decrease in replicative fidelity, to the processing of otherwise "cryptic" lesions by the SOS mutagenesis machinery, or both (342). Mutations arising from a decrease in replicative fidelity would be untargeted, in that they would not arise as a consequence of a lesion in the DNA, although their frequency and nature might be influenced by the DNA sequence context. However, if such a mechanism were to contribute to the spectrum of mutations observed after treatment of cells with a mutagen, they could still be what has been referred to as *regionally targeted* (342) or *hitchhiking* (259) *mutations*. That is to say, they might occur in the vicinity of lesions but not arise from events directly targeted by the lesion itself. Mutations arising from the processing of otherwise cryptic lesions could be targeted to the site of the cryptic lesion, just as in the case of mutations targeted to the site of lesions introduced by an exogenous mutagen. Such cryptic lesions would be spontaneously arising lesions which are innocuous or lethal unless the SOS system is induced (342, 352). Such cryptic lesions could arise from a variety of sources including fluorescent light (352), spontaneous depurinations, damage by endogenous alkylating agents, and damage by oxygen radicals generated during metabolism (190, 298). The fact that the specificity of the SOS mutator effect in the *lacI* gene resembled that of several bulky carcinogens

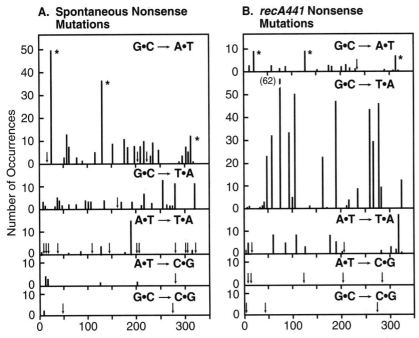

Figure 11–25 (A) Spontaneous amber (black bars) and ochre (red bars) mutations occurring in *recA⁺* strain. The height of each bar represents the number of independent occurrences in a collection of 306 nonsense mutations. (The ochre bar heights have been normalized to account for a small sample size.) Arrows indicate the position of nonsense sites at which there were zero occurrences in this collection. Asterisks indicate 5-methylcytosine residues. The positions of sites in the *lacI* gene are indicated on the horizontal axis by the number of the corresponding amino acid in the *lac* repressor. (B) Distribution of 586 spontaneous amber and ochre mutations occurring in a *recA441* strain at 42°C. Each mutation is of independent origin and corresponds to the actual number collected (no normalizations). (*Panel A redrawn from Coulondre and Miller [69] with permission; panel B adapted from Miller and Low [217] with permission.*)

(89, 102) led to the suggestion that the same lesion, for example an AP site, might serve as the premutagenic lesion in all these cases (217).

The SOS Mutator Effect on Single- and Double-Stranded Phages

The SOS mutator effect observed when a growing single-stranded phage (such as M13 or φX174) on SOS-induced cells resembles that observed when monitoring loci on the chromosome or on an F′ plasmid, in that it is *umuC⁺* dependent (203). It can be stimulated by irradiation of the host even when the cells carry the *recA730*(Cptᶜ) allele. This SOS mutator effect may be related to that discussed above, although it may be subject to specific influences because of the single-stranded nature of the initial DNA substrate.

In contrast, the majority of the SOS mutator effect observed when undamaged phage λ is grown in SOS-induced cells (136) appears to have a different mechanistic basis. In this situation, the SOS mutator effect is (i) independent of *umuC⁺* function (57, 357), (ii) dependent on UV irradiation of the host even if the cell carries a constitutively activated allele of *recA* (203, 357), (iii) dependent on *uvr⁺* function (203), (iv) dependent on *dinB⁺* function (46, 203), and (v) for certain classes of mutations, independent of *recA⁺* function if the cell carries a *lexA*(Def) mutation (46). Two reports have observed that *polA* mutations greatly decreased the SOS mutator effect observed with λ (136, 203), suggesting that the stimulation observed with the *polA480ex* allele (357) may not reflect the null phenotype. The SOS mutator effect observed with phage λ is higher in cells that are deficient in postreplicative *mutHLS*-dependent mismatch repair, an observation that in this case also has been used to suggest that the mutations are introduced by a replicative process. The fact that all the mutants detected by using this system were found in plaques containing both wild-type and mutant phage is also consistent with this view (202).

Sequencing of the mutations has revealed that three-quarters of them were frame shifts occurring at runs of identical bases (357). The relationship of the predominant SOS mutator effect observed in these experiments to phenomena affecting chromosomal loci is not clear, although it is possible that it is related to the SOS-dependent but $umuD^+C^+$-independent -2 frameshift mutagenesis induced by AAF at alternating G-C sequences (54, 204, 333). An SOS-dependent $umuD^+C^+$-dependent mutator effect has been reported for a certain class of λ mutants (250). However, mutations arising through this latter pathway appear to represent only a minor subset of mutations that are caused by the SOS mutator effect on λ.

What Is the Biological Purpose of SOS Mutagenesis?

Two rationalizations have been put forward to explain the physiological purpose of the SOS mutagenesis system. The first is that the ability to mutate in response to DNA damage confers an evolutionary advantage to the organism. The second is that the main purpose of the SOS mutagenesis system is to provide an additional mechanism for damage tolerance that helps organisms survive the lethal effect of DNA damage. In this view, the mutagenesis that results is a secondary consequence of the fact that translesion synthesis has occurred.

Inducible mutagenesis has been referred to as a form of inducible evolution for cells in stressful environments (86, 253). However, it has been argued (363) that this is not a particularly satisfactory explanation. Only a small fraction of the resulting mutations would be expected to be advantageous, with most of them being deleterious. In addition, SOS mutagenesis is not a particularly efficient system for introducing mutations compared, for example, with what could be achieved by eliminating proofreading by the replicative DNA polymerase (see above). Furthermore, as discussed above, the efficiency of mutation introduction, as well as the nature of the mutations that result, is highly influenced by the chemical nature of the lesions. If the primary purpose of the SOS mutagenesis system is to introduce mutations, it does not seem physiologically consistent that the functions for SOS mutagenesis are coinduced with the uvr^+ genes, since the operation of uvr^+-dependent excision repair reduces mutagenesis by removing premutagenic lesions.

In contrast, it is relatively easy to see how a system that provided extra resistance to killing by providing an additional mechanism of DNA damage tolerance would confer an evolutionary advantage. The effect of $umuDC$ mutations on the survival of *E. coli* exposed to UV light is smaller than the effect of uvr mutations, and this has led to speculation that translesion bypass might be important for surviving the lethal effect of a subclass of UV radiation-induced lesions that cannot be repaired accurately. In particular, as discussed above, cells might use translesion bypass to process damaged DNA in which a lesion is situated opposite a single-stranded gap (32, 275, 352).

An additional possibility is that SOS mutagenesis is important primarily for resistance to particular classes of environmental agents that introduce lesions that are not subject to repair by excision repair or other accurate repair systems (228, 363). An example of such an agent is neocarcinostatin (discussed above), which introduces a lesion consisting of an AP site plus a closely opposed break in the complementary strand Fig. 11–22 (248).

SOS-Independent Mutagenesis

Although most of the lesions induced by radiation and chemicals appear to require the participation of the SOS mutagenesis system to cause mutations, there are examples of lesions that can cause mutations in *E. coli* in an SOS-independent fashion. In addition, a number of base analogs (82) can cause mutagenesis in an SOS-independent fashion. Two common mutagens that introduce such lesions are the alkylating agents EMS and MNNG (83, 145, 146, 273, 292, 326). Analyses of the mutational spectra caused by these alkylating agents revealed that the predominant class of mutations were $G \cdot C \rightarrow A \cdot T$ transitions (69). These observations were consistent with the hypothesis that O^6-alkylguanine serves as the premutagenic lesion (199). Subsequently, the ability of O^6-methylguanine to serve as a premutagenic lesion in the absence of SOS induction was directly demonstrated by site-directed adduct studies (17, 127, 198).

Another lesion which contributes to SOS-independent mutagenesis by these agents is O^4-alkylthymine. A possible mutagenic role for this class of lesion was suggested by the observation of $A \cdot T \rightarrow G \cdot C$ transitions in the mutational spectra of these agents and by the observation that the fraction of mutations at $A \cdot T$ pairs caused by various alkylating agents correlates well with the amount of O^4-alkylthymine formed (7, 69, 256, 295). In this case as well, the ability of O^4-methylthymine to serve as a premutagenic lesion in *E. coli* was directly demonstrated by a site-directed adduct study (249). Both O^6-alkylguanine and O^4-alkylthymine lesions can be accurately repaired by DNA alkyltransferases (see chapter 3) and by nucleotide excision repair (265, 335). Hence, these processes serve to reduce the amount of mutagenesis that is observed when wild-type *E. coli* is treated with an alkylating agent. Alkylating agents also introduce 3-alkyladenine lesions, which appear to cause mutations in an SOS-dependent fashion (60), probably by first being converted to an AP site by the action of a DNA glycosylase (99) (see above), and they may contribute to the SOS-dependent component of alkylating mutagenesis (70, 101).

The high miscoding potential of O^6-alkylguanine has been demonstrated in in vitro studies with a variety of purified polymerases (294). For many years, the $G \rightarrow A$ transition mutations resulting from O^6-alkylguanine were thought to arise as a consequence of the ambiguous base-pairing properties of this lesion (199) (Fig. 11–26A). However, more recent studies have suggested that the mechanism is more complicated than a straightforward mispairing scheme (7, 312). For example, examination of a set of oligonucleotides duplexes in which O^6-methylguanine was situated opposite A, G, or C revealed that O^6-methylguanine forms the most stable base pair with C rather than with T (114). In fact, on a purely thermodynamic basis, T is a poorer pairing partner that either G or A. These results are also consistent with two-dimensional NMR studies which suggest that O^6-methylguanine and O^6-ethylguanine pair with C through two hydrogen bonds in a wobble arrangement (Fig. 11–26C) whereas, when O^6-methylguanine pairs with T, only one strong hydrogen bond between the 2-amino group of O^6-methylguanine and O^2 of T is evident (Fig. 11–26B) (7, 242–244).

The O^6-alkyl group is oriented *syn* to the N-1 of O^6-methylguanine, so there may be steric overlap of the O-alkyl group with the O^4 of T that weakens the hydrogen bonding between N-1 of O^6-methylguanine and the imino proton (N-3) of T. Given these observations, it is not clear which factors are responsible for the pairing of T with O^6-methylguanine (7, 312). However, a possible explanation is suggested by the observation that an O^6-methylguanine $\cdot$ T base pair maintains a geometry consistent with an unperturbed Watson-Crick pairing (141, 142, 244), and this may be a factor which favors the insertion of T by a DNA polymerase (see chapter 2). In the case of O^4-methylthymine, it had been postulated that mispairing between a O^4-methylthymine and G involves two hydrogen bonds (Fig. 11–26D). However, physical studies have predicted a situation similar to that with O^6-methylguanine. The pairing of O^4-methylthymine with A involves a wobble pairing (Fig. 11–26F), whereas its pairing with G involves a Watson-Crick-like structure (Fig. 11–26E).

Another lesion of particular interest that is mutagenic in an SOS-independent fashion is 8-oxo-7,8-dihydrodeoxyguanine (GO) (see chapter 4). Site-directed adduct studies were used to demonstrate that this lesion serves as an SOS-independent premutagenic lesion in wild-type *E. coli* and gives rise to targeted $G \rightarrow T$

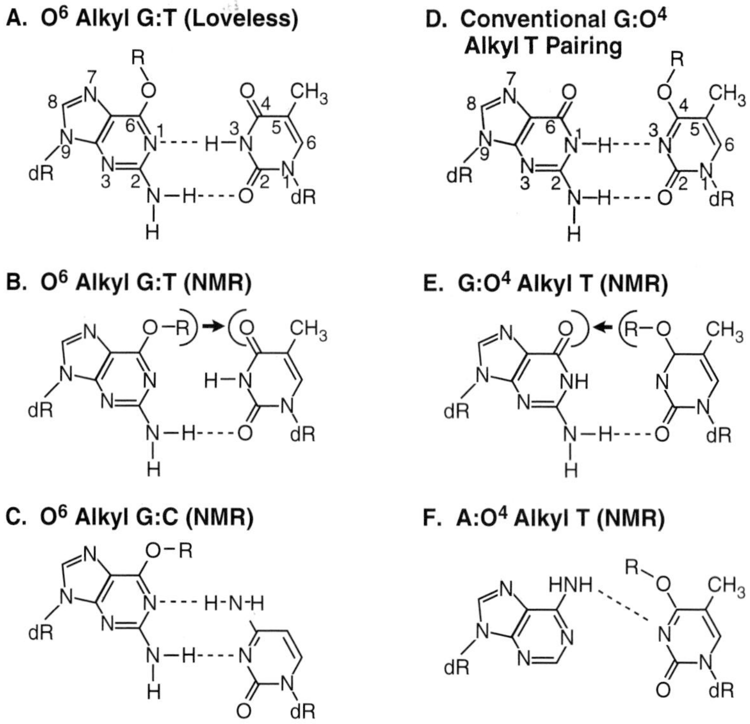

Figure 11–26 (A to C) Postulated base-pairing modes of O^6-alkylguanine (A) Scheme proposed on the basis of the original model of Loveless (199); (B and C) base pairing of O^6-alkylguanine with thymine and cytosine, respectively, predicted from two-dimensional NMR studies. (D to F) Postulated base-pairing models of O^4-alkylthymine. (D) Conventional scheme for mispairing of O^4-alkylthymine with guanine (175); (E and F) base pairing of O^4-alkylthymine with guanine and adenine, respectively, predicted from two-dimensional NMR studies. (*Adapted from Basu and Essigmann [7] with permission.*)

transversions (62, 225, 355, 356). When wild-type *E. coli* was used, the frequency of mutagenesis was about 1%, but in a *mutY* strain, which is defective in removing adenine from an A·GO pair in vitro, the frequency was ~25% (224) (see chapter 4). The conformation of the lesion at the replication fork is unknown, but the mutagenic intermediate, A·GO, forms a stable Hoogsteen pair in duplex DNA, with the modified base assuming the *syn* conformation (158). This base pair resists proofreading and is readily extended by DNA polymerases (289). Thymine glycol is another example of a lesion that has been shown to serve as a premutagenic lesion for SOS-independent mutagenesis, although it does not appear to be highly mutagenic (8).

Mutations Arising in Cell Populations That Are Not Actively Dividing

Although it is outside the scope of this book, an additional highly active area of investigation of mutagenesis in *E. coli* had focused on mutations that rise in cell populations that are not actively dividing. Previous evidence that mutations could arise in such cells had appeared in the literature (see, e.g., references 261 and 282), but much of the current interest was sparked by a controversial paper which suggested that such mutations might occur more readily if they were of benefit to the organism than if they were neutral or deleterious (58). At the time of writing of this book, it is still unresolved whether the mutations occurring under these conditions are caused by a single underlying mechanism or by a variety of processes. Several recent reviews and papers provide an entry to the literature of this active area of research (36, 97, 98, 124, 186, 201, 282, 302, 305, 314, 366).

REFERENCES

1. **Andreeva, I. V., O. Yu, E. E. Mirskaya, and A. G. Skavron-skaya.** 1990. The effect of plasmid pKM101 on *umuDC* gene function enhancing precise excision of transposons. *Mutat. Res.* **230:**55–60.

2. **Bagg, A., C. J. Kenyon, and G. C. Walker.** 1981. Inducibility of a gene product required for UV and chemical mutagenesis in *Escherichia coli. Proc. Natl. Acad. Sci. USA* **78:**5749–5753.

3. **Bailone, A., S. Sommer, J. Knezevic, and R. Devoret.** 1991. Substitution of UmuD' for UmuD does not affect SOS mutagenesis. *Biochimie* **73:**471–478.

4. **Bailone, A., S. Sommer, J. Knezevic, M. Dutreix, and R. Devoret.** 1991. A RecA protein deficient in its interaction with the UmuDC complex. *Biochimie* **73:**479–484.

5. **Banerjee, S. K., A. Borden, R. B. Christensen, J. E. LeClerc, and C. W. Lawrence.** 1990. SOS-dependent replication past a single *trans-syn* T-T cyclobutane dimer gives a different mutation spectrum and increased error rate compared with replication past this lesion in uninduced cells. *J. Bacteriol.* **172:**2105–2112.

6. **Banerjee, S. K., R. B. Christensen, C. W. Lawrence, and J. E. LeClerc.** 1988. Frequency and spectrum of mutations produced by a single *cis-syn* thymine-thymine cyclobutane dimer in a single-stranded vector. *Proc. Natl. Acad. Sci. USA* **85:**8141–8145.

7. **Basu, A. K., and J. M. Essigmann.** 1990. Site-specific alkylated oligodeoxynucleotides: probes for mutagenesis, DNA repair and the structural effects of DNA damage. *Mutat. Res.* **233:**189–201.

8. **Basu, A. K., E. L. Loechler, S. A. Leadon, and J. M. Essigmann.** 1989. Genetic effects of thymine glycol: site specific mutagenesis and molecular modeling studies. *Proc. Natl. Acad. Sci. USA* **86:**7677–7681.

9. **Bates, H., and B. A. Bridges.** 1991. Mutagenic DNA repair in *Escherichia coli.* XIX. On the roles of RecA protein in ultraviolet light mutagenesis. *Biochimie* **73:**485–489.

10. **Bates, H., S. K. Randall, C. Rayssiguier, B. A. Bridges, M. F. Goodman, and M. Radman.** 1989. Spontaneous and UV-induced mutations in *Escherichia coli* K-12 strains with altered or absent DNA polymerase I. *J. Bacteriol.* **171:**2480–2484.

11. **Battista, J. R., T. Nohmi, C. E. Donnelly, and G. C. Walker.** 1988. Role of UmuD and UmuC in UV and chemical mutagenesis, p. 455–459. *In* E. C. Friedberg and P. C. Hanawalt (ed.), *Mechanisms and Consequences of DNA Damage Processing.* Alan R. Liss, Inc., New York.

12. **Battista, J. R., T. Ohta, T. Nohmi, W. Sun, and G. C. Walker.** 1990. Dominant negative *umuD* mutations decreasing RecA-mediated cleavage suggest roles for intact UmuD in modulation of SOS mutagenesis. *Proc. Natl. Acad. Sci. USA* **87:**7190–7194.

13. **Bebenek, K., and T. A. Kunkel.** 1990. Frameshift errors initiated by nucleotide misincorporation. *Proc. Natl. Acad. Sci. USA* **87:**4946–4950.

14. **Belguise-Valladier, P., H. Maki, M. Sekiguichi, and R. P. P. Fuchs.** 1994. Effect of single DNA lesions on *in vitro* replication with DNA polymerase III holoenzyme. *J. Mol. Biol.* **236:**151–164.

15. **Ben-Bassat, A., K. Bauer, S.-Y. Chang, K. Myambo, A. Boosman, and S. Chang.** 1987. Processing of the initiation methionine from proteins: properties of the *Escherichia coli* methionine aminopeptidase and its gene structure. *J. Bacteriol.* **169:**751–757.

16. **Ben-Hur, E., and R. Ben-Ishai.** 1968. *Trans-syn* thymine dimers in ultraviolet-irradiated denatured DNA: identification and photoreactivability. *Biochim. Biophys. Acta* **166:**9–15.

17. **Bhanot, O. S., and A. Ray.** 1986. The in vivo mutagenic frequency and specificity of O⁶-methylguanine in φX174 replicative form DNA. *Proc. Natl. Acad. Sci. USA* **83:**7348–7352.

18. **Blanco, M., G. Herrera, and V. Aleixandre.** 1986. Different efficiency of UmuDC and MucAB proteins in UV light induced mutagenesis in *Escherichia coli. Mol. Gen. Genet.* **205:**234–239.

19. **Blanco, M., G. Herrera, P. Collado, J. E. Rebollo, and L. M. Botella.** 1982. Influence of RecA protein on induced mutagenesis. *Biochimie* **64:**633–636.

20. **Blanco, M., and J. E. Rebollo.** 1981. Plasmid pKM101-dependent repair and mutagenesis in *Escherichia coli* cells with mutations *lexB30 tif* and *zab* in the *recA* gene. *Mutat. Res.* **81:**265–275.

21. **Bockrath, R., and M. K. Cheung.** 1973. The role of nutrient broth supplementation in UV mutagenesis of *E. coli. Mutat. Res.* **19:**23–32.

22. **Bockrath, R., M. Ruiz-Rubio, and B. A. Bridges.** 1987. Specificity of mutation by ultraviolet light and delayed photoreversal in *umuC* defective *Escherichia coli* K-12: a targeting intermediate in pyrimidine dimers. *J. Bacteriol.* **169:**1410–1416.

23. **Bockrath, R. C., and J. E. Palmer.** 1977. Differential repair of premutational UV-lesions at tRNA genes in *E. coli. Mol. Gen. Genet.* **156:**133–140.

24. **Bonner, C. A., S. Hays, K. McEntee, and M. F. Goodman.** 1990. DNA polymerase II is encoded by the DNA damage-inducible *dinA* gene of *Escherichia coli. Proc. Natl. Acad. Sci. USA* **87:**7663–7667.

25. **Bonner, C. A., S. K. Randall, C. Rayssiguier, M. Radman, R. Eritja, B. E. Kaplan, K. McEntee, and M. F. Goodman.** 1988. Purification and characterization of an inducible *Escherichia coli* DNA polymerase capable of insertion and bypass at abasic lesions in DNA. *J. Biol. Chem.* **263:**18946–18952.

26. **Bonner, C. A., P. T. Stukenberg, M. Rajagopalan, R. Eritja, M. O'Donnell, K. McEntee, H. Echols, and M. F. Goodman.** 1992. Processive DNA synthesis by DNA polymerase II mediated by DNA polymerase III accessory proteins. *J. Biol. Chem.* **267:**11431–11438.

27. **Bowie, J. U., J. F. Reidhaar-Olson, W. A. Lim, and R. T. Sauer.** 1990. Deciphering the message in protein sequences: tolerance to amino acid substitutions. *Science* **247:**1306–1310.

28. **Brash, D. E.** 1988. UV mutagenic photoproducts in *Escherichia coli* and human cells: a molecular genetics perspective on human skin cancer. *Photochem. Photobiol.* **48:**59–66.

29. **Brash, D. E., W. A. Franklin, G. B. Sancar, A. Sancar, and W. A. Haseltine.** 1985. *Escherichia coli* DNA photolyase reverses cyclobutane pyrimidine dimers but not pyrimidine-pyrimidone (6-4) photoproducts. *J. Biol. Chem.* **260:**11438–11441.

30. **Brash, D. E., and W. A. Haseltine.** 1982. UV-induced mutation hotspots occur at DNA damage hotspots. *Nature* (London) **298:**189–192.

31. **Brash, D. E., and W. A. Haseltine.** 1985. Photoreactivation of *Escherichia coli* reverses *umuC* induction by UV light. *J. Bacteriol.* **163:**460–463.

32. **Bresler, S. E.** 1975. Theory of misrepair mutagenesis. *Mutat. Res.* **29:**467–474.

33. **Bridges, B. A.** 1988. Mutagenic DNA repair in *Escherichia coli.* XVI. Mutagenesis by ultraviolet light plus delayed photoreversal in *recA* strains. *Mutat. Res.* **198:**343–350.

34. **Bridges, B. A.** 1990. Ultraviolet light mutagenesis in bacteria: the possible role of a DNA polymerase III complex lacking proofreading exonuclease, p. 27–35. *In* A. Kappas (ed.), *Mechanisms of Environmental Mutagenesis-Carcinogenesis.* Plenum Publishing Corp., New York.

35. **Bridges, B. A.** 1992. Mutagenesis after exposure of bacteria to ultraviolet light and delayed photoreversal. *Mol. Gen. Genet.* **233:**331–336.

36. **Bridges, B. A.** 1994. Starvation-associated mutation in *Escherichia coli:* a spontaneous lesion hypothesis for "directed" mutation. *Mutat. Res.* **307:**149–156.

37. **Bridges, B. A., and H. Bates.** 1990. Mutagenic DNA repair in *Escherichia coli.* XVIII. Involvement of DNA polymerase III α-subunit (DnaE) in mutagenesis after exposure to UV light. *Mutagenesis* **5:**35–38.

38. **Bridges, B. A., and G. M. Brown.** 1992. Mutagenic DNA repair in *Escherichia coli.* XXI. A stable SOS-inducing signal persisting after excision repair of ultraviolet damage. *Mutat. Res.* **270:**135–144.

39. **Bridges, B. A., R. E. Dennis, and R. J. Munson.** 1967. Differential induction and repair of ultraviolet damage leading to true reversions and external suppressor mutations of an ochre codon in *Escherichia coli* B/r WP2. *Genetics* **57:**897–908.

40. Bridges, B. A., and R. P. Mottershead. 1971. RecA$^+$-dependent mutagenesis occurring before DNA replication in UV- and gamma-irradiated DNA. *Mutat. Res.* **13:**1–8.

41. Bridges, B. A., and R. P. Mottershead. 1978. Mutagenic repair in *E. coli*. VII. Constitutive and inducible manifestations. *Mutat. Res.* **52:**151–159.

42. Bridges, B. A., R. P. Mottershead, and S. G. Sedgwick. 1976. Mutagenic DNA repair in *Escherichia coli*. *Mol. Gen. Genet.* **144:**53–58.

43. Bridges, B. A., and R. Woodgate. 1984. Mutagenic repair in *Escherichia coli*: the *umuC* gene product may be required for replication past pyrimidine dimers but not for the coding error in UV-mutagenesis. *Mol. Gen. Genet.* **196:**364–366.

44. Bridges, B. A., and R. Woodgate. 1985. Mutagenic repair in *Escherichia coli*: products of the *recA* gene and of the *umuD* and *umuC* genes act at different steps in UV-induced mutagenesis. *Proc. Natl. Acad. Sci. USA* **82:**4193–4197.

45. Bridges, B. A., and R. Woodgate. 1985. The two-step model of bacterial UV mutagenesis. *Mutat. Res.* **150:**133–139.

46. Brotcorne-Lannoye, A., and G. Maenhaut-Michel. 1986. Role of RecA protein in untargeted UV mutagenesis of bacteriophage λ: evidence for the requirement for the *dinB* gene. *Proc. Natl. Acad. Sci. USA* **83:**3904–3908.

47. Brouwer, J., P. van de Putte, A. M. J. Fichtinger-Schepman, and J. Reedijk. 1981. Base-pair substitution hotspots in GAG and GCG nucleotide sequences in *Escherichia coli* K-12 induced by *cis*-diaminedichloroplatinum (II). *Proc. Natl. Acad. Sci. USA* **78:**7010–7014.

48. Brown, A. M. C., and N. S. Willetts. 1981. A physical and genetic map of the *incN* plasmid R46. *Plasmid* **5:**188–201.

49. Brunner, D. P., B. A. Traxler, S. M. Holt, and L. L. Crose. 1986. Enhancement of UV survival, UV- and MMS-mutability, precise excision of Tn*10* and complementation of *umuC* by plasmids of different incompatibility groups. *Mutat Res.* **166:**29–37.

50. Bryan, S., H. Chen, Y. Sun, and R. E. Moses. 1988. Alternative pathways of DNA replication in *Escherichia coli*. *Biochim. Biophys. Acta* **951:**249–254.

51. Bryan, S. K., M. Hagensee, and R. E. Moses. 1990. Holoenzyme DNA polymerase III fixes mutations. *Mutat. Res.* **243:**313–318.

52. Bryan, S. K., and R. E. Moses. 1988. Sufficiency of the Klenow fragment for survival of *polC*(Ts) *pcbA1 Escherichia coli* at 43°C. *J. Bacteriol.* **170:**456–458.

53. Burckhardt, S. E., R. Woodgate, R. H. Scheuermann, and H. Echols. 1988. UmuD mutagenesis protein of *Escherichia coli*: overproduction, purification, and cleavage by RecA. *Proc. Natl. Acad. Sci. USA* **85:**1811–1815.

54. Burnouf, D., P. Koehl, and R. P. P. Fuchs. 1989. Single adduct mutagenesis: strong effect of the position of a single acetylaminofluorene adduct with a mutation hotspot. *Proc. Natl. Acad. Sci. USA* **86:**4147–4151.

55. Caillet-Fauquet, P., M. Defais, and M. Radman. 1977. Molecular mechanism of induced mutagenesis. 1. *In vivo* replication of the single-stranded ultraviolet-irradiated φX174 phage DNA in irradiated host cells. *J. Mol. Biol.* **117:**95–112.

56. Caillet-Fauquet, P., and G. Maenhaut-Michel. 1988. Nature of the SOS mutator activity: genetic characterization of untargeted mutagenesis in *Escherichia coli*. *Mol. Gen. Genet.* **213:**491–498.

57. Caillet-Fauquet, P., G. Maenhaut-Michel, and M. Radman. 1984. SOS mutator effect in *E. coli* mutants deficient in mismatch correction. *EMBO J.* **3:**707–712.

58. Cairns, J., J. Overbaugh, and S. Miller. 1988. The origin of mutants. *Nature* (London) **335:**142–145.

59. Calsou, P., and M. Defais. 1985. Weigle reactivation and mutagenesis of bacteriophage λ in *lexA* (Def) mutants of *E. coli* K12. *Mol. Gen. Genet.* **201:**329–333.

60. Chaudhuri, I., and J. M. Essigmann. 1991. 3-Methyladenine mutagenesis under conditions of SOS induction in *Escherichia coli*. *Carcinogenesis* **12:**2283–2289.

61. Chen, H., Y. Sun, T. Stark, W. Beattie, and R. E. Moses. 1990. Nucleotide sequence and deletion analysis of the *polB* gene of *Escherichia coli*. *DNA Cell Biol.* **9:**631–635.

62. Cheng, K. C., D. S. Cahill, H. Kasai, S. Nishimura, and L. A. Loeb. 1992. 8-Hydroxyguanine, an abundant form of oxidative damage causes G to A and A to C substitutions. *J. Biol. Chem.* **267:**166–172.

63. Chernin, L. S., and V. S. Mikoyan. 1981. Effects of plasmids on chromosome metabolism in bacteria. *Plasmid* **6:**119–140.

64. Christensen, J. R., J. E. LeClerc, P. V. Tata, R. B. Christensen, and C. W. Lawrence. 1988. UmuC function is not essential for the production of all targeted *lacI* mutations induced by ultraviolet light. *J. Mol. Biol.* **203:**635–641.

65. Ciesla, Z. 1982. Plasmid pKM101-mediated mutagenesis in *Escherichia coli* is inducible. *Mol. Gen. Genet.* **186:**298–300.

66. Ciesla, Z., P. Jonczyk, and I. Fijalkowska. 1990. Effect of enhanced synthesis of the epsilon subunit of DNA polymerase III on spontaneous and UV-induced mutagenesis of the *Escherichia coli glyU* gene. *Mol. Gen. Genet.* **221:**251–255.

67. Cohen-Fix, O., and Z. Livneh. 1992. Biochemical analysis of UV mutagenesis in *Escherichia coli* by using a cell-free reaction coupled to a bioassay: identification of a DNA repair-dependent pathway. *Proc. Natl. Acad. Sci. USA* **89:**3300–3304.

68. Cohen-Fix, O., and Z. Livneh. 1994. *In vitro* UV mutagenesis associated with nucleotide excision-repair gaps in *Escherichia coli*. *J. Biol. Chem.* **269:**4953–4958.

69. Coulondre, C., and J. H. Miller. 1977. Genetic studies of the *lac* repressor. IV. Mutagenic specificity in the *lacI* gene of *Escherichia coli*. *J. Mol. Biol.* **117:**577–606.

70. Couto, L. B., I. Chaudhuri, B. A. Donahue, B. Demple, and J. M. Essigmann. 1989. Separation of the SOS-dependent and SOS-independent components of alkylating mutagenesis. *J. Bacteriol.* **171:**4170–4177.

71. Cuniasse, P., G. V. Fazakerley, W. Guschlbauer, B. E. Kaplan, and L. Sowers. 1990. The abasic site as a challenge to DNA polymerase. *J. Mol. Biol.* **213:**303–314.

72. Cuniasse, P., L. C. Sowers, R. Eritja, B. Kaplan, M. F. Goodman, J. A. H. Cognet, M. LeBret, W. Guschlbauer, and G. V. Fazakerly. 1987. An abasic site in DNA. Solution conformation determined by proton NMR and molecular mechanics calculations. *Nucleic Acids Res.* **15:**8003–8022.

73. Defais, M., P. Caillet-Fauquet, M. S. Fox, and M. Radman. 1976. Induction kinetics of mutagenic DNA repair activity in *E. coli* following ultraviolet irradiation. *Mol. Gen. Genet.* **148:**125–130.

74. Defais, M., P. Fauquet, M. Radman, and M. Errera. 1971. Ultraviolet reactivation and ultraviolet mutagenesis of lambda in different genetic systems. *Virology* **43:**495–503.

75. Defais, M., C. Lesca, B. Monsarrat, and P. Hanawalt. 1989. Translesion synthesis is the main component of SOS repair in bacteriophage lambda DNA. *J. Bacteriol.* **171:**4938–4944.

76. Donnelly, C. E., and G. C. Walker. 1989. *groE* mutants of *Escherichia coli* are defective in *umuDC*-dependent UV mutagenesis. *J. Bacteriol.* **171:**6117–6125.

77. Donnelly, C. E., and G. C. Walker. 1992. Coexpression of UmuD' with UmuC suppresses the UV mutagenesis deficiency of *groE* mutants. *J. Bacteriol.* **174:**3133–3139.

78. Douki, T., and J. Cadet. 1992. Far-UV photochemistry and photosensitization of 2'-deoxycytidylyl-(3'-5')-thymidine: isolation and characterization of the main photoproducts. *J. Photochem. Photobiol. Ser. B* **15:**199–213.

79. Douki, T., L. Voiturieux, and J. Cadet. 1991. Characterization of the (6-4) photoproducts of 2'-deoxycytidylyl-(3'-5')-thymidine and of its Dewar valence isomer. *Photochem. Photobiol.* **53:**293–297.

80. **Dowden, S. B., and P. Strike.** 1982. R46-derived recombinant plasmids affecting DNA repair and mutation in *E. coli. Mol. Gen. Genet.* **186:**140–144.

81. **Drabble, W. T., and B. A. D. Stocker.** 1968. R (transmissible drug-resistance) factors in *Salmonella typhimurium*: pattern of transduction of phage P22 and ultraviolet-protection effect. *J. Gen. Microbiol.* **53:**109–123.

82. **Drake, J. W.** 1970. *The Molecular Basis of Mutation.* Holden-Day, San Francisco.

83. **Drake, J. W., and R. H. Baltz.** 1976. The biochemistry of mutagenesis. *Annu. Rev. Biochem.* **45:**11–37.

84. **Dutreix, M., B. Burnett, A. Bailone, C. M. Radding, and R. Devoret.** 1992. A partially deficient mutant, *recA1730*, that fails to form normal nucleoprotein filaments. *Mol. Gen. Genet.* **232:**489–497.

85. **Dutreix, M., P. L. Moreau, A. Bailone, F. Galibert, J. R. Battista, G. C. Walker, and R. Devoret.** 1989. New *recA* mutations that dissociate the various RecA protein activities in *Escherichia coli* provide evidence for an additional role for RecA protein in UV mutagenesis. *J. Bacteriol.* **171:**2415–2423.

86. **Echols, H.** 1981. SOS functions, cancer and inducible evolution. *Cell* **25:**1–2.

87. **Echols, H., and M. F. Goodman.** 1990. Mutation induced by DNA damage: a many protein affair. *Mutat. Res.* **236:**301–311.

88. **Echols, H., and M. F. Goodman.** 1991. Fidelity mechanisms in DNA replication. *Annu. Rev. Biochem.* **60:**477–511.

89. **Eisenstadt, E., A. J. Warren, J. Porter, D. Atkins, and J. H. Miller.** 1982. Carcinogenic epoxides of benzo(*a*)pyrene and cyclopenta(*cd*)pyrene induce base substitutions via specific transversions. *Proc. Natl. Acad. Sci. USA* **79:**1945–1949.

90. **Eisenstadt, E., M. Wolf, and I. H. Goldberg.** 1980. Mutagenesis by neocarzinostatin in *Escherichia coli* and *Salmonella typhimurium*: requirement for *umuC*⁺ or plasmid pKM101. *J. Bacteriol.* **144:**656–660.

91. **Elledge, S. J., and G. C. Walker.** 1983. The *muc* genes of pKM101 are induced by DNA damage. *J. Bacteriol.* **155:**1306–1315.

92. **Elledge, S. J., and G. C. Walker.** 1983. Proteins required for ultraviolet light and chemical mutagenesis: identification of the products of the *umuC* locus of *Escherichia coli. J. Mol. Biol.* **164:**175–192.

93. **Ennis, D. G., B. Fisher, S. Edmiston, and D. W. Mount.** 1985. Dual role for *Escherichia coli* RecA protein in SOS mutagenesis. *Proc. Natl. Acad. Sci. USA* **82:**3325–3329.

94. **Ennis, D. G., N. Ossanna, and D. W. Mount.** 1989. Genetic separation of *Escherichia coli recA* functions for SOS mutagenesis and repressor cleavage. *J. Bacteriol.* **171:**2533–2541.

95. **Fersht, A. R., and J. W. Knill-Jones.** 1983. Contribution of 3′ → 5′ exonuclease activity of DNA polymerase from *Escherichia coli* to specificity. *J. Mol. Biol.* **165:**669–682.

96. **Fix, D.** 1986. Thermal resistance of UV-mutagenesis to photoreactivation in *E. coli* B/r *uvrA ung*: estimate of activation energy and further analysis. *Mol. Gen. Genet.* **204:**452–456.

97. **Foster, P. L.** 1992. Directed mutation: between unicorns and goats. *J. Bacteriol.* **174:**1711–1716.

98. **Foster, P. L.** 1993. Adaptive mutation: the uses of adversity. *Annu. Rev. Microbiol.* **47:**467–504.

99. **Foster, P. L., and E. F. Davis.** 1987. Loss of an apurinic/apyrimidinic site endonuclease increases mutagenicity of N-methyl-N′-nitro-N-nitrosoguanidine. *Proc. Natl. Acad. Sci. USA* **84:**2891–2895.

100. **Foster, P. L., and E. Eisenstadt.** 1983. Distribution and specificity of mutations induced by neocarzinostatin in the *lacI* gene of *Escherichia coli. J. Bacteriol.* **153:**379–383.

101. **Foster, P. L., and E. Eisenstadt.** 1985. Induction of transversion mutations in *Escherichia coli* by N-methyl-N′-nitro-N-nitrosoguanidine is SOS dependent. *J. Bacteriol.* **163:**213–220.

102. **Foster, P. L., E. Eisenstadt, and J. H. Miller.** 1983. Base substitution mutations induced by metabolically activated aflatoxin B₁. *Proc. Natl. Acad. Sci. USA* **80:**2695–2698.

103. **Foster, P. L., and A. D. Sullivan.** 1988. Interactions between epsilon, the proofreading subunit of DNA polymerase III, and proteins involved in the SOS response of *Escherichia coli. Mol. Gen. Genet.* **214:**467–473.

104. **Foster, P. L., A. D. Sullivan, and S. B. Franklin.** 1989. Presence of the *dnaQ-rnh* divergent transcriptional unit on a multicopy plasmid inhibits induced mutagenesis in *Escherichia coli. J. Bacteriol.* **171:**3144–3151.

105. **Frank, E. G., J. Hauser, A. S. Levine, and R. Woodgate.** 1993. Targeting of the UmuD, UmuD′ and MucA′ mutagenesis proteins to DNA by the RecA protein. *Proc. Natl. Acad. Sci. USA* **90:**8169–8173.

106. **Franklin, W. A., P. W. Doetsch, and W. A. Haseltine.** 1985. Structural determination of the ultraviolet light-induced thymine-cytosine pyrimidine-pyrimidone (6-4) photoproduct. *Nucleic Acids Res.* **13:**5317–5325.

107. **Franklin, W. A., and W. A. Haseltine.** 1986. The role of the (6-4) photoproduct in ultraviolet light-induced transition mutations in *E. coli. Mutat. Res.* **165:**1–7.

108. **Franklin, W. A., K. M. Low, and W. A. Haseltine.** 1982. Alkaline lability of novel fluorescent photoproducts produced in ultraviolet light irradiated DNA. *J. Biol. Chem.* **257:**13535–13543.

109. **Freitag, N., and K. McEntee.** 1989. "Activated"-RecA protein affinity chromatography of LexA repressor and other SOS-regulated proteins. *Proc. Natl. Acad. Sci. USA* **86:**8363–8367.

110. **Friedberg, E. C.** 1985. DNA Repair. W. H. Freeman & Co., New York.

111. **Frydman, J., and F.-U. Hartl.** 1994. Molecular chaperone functions of hsp70 and hsp60 in protein folding, p. 251–283. *In* R. I. Morimoto, A. Tissieres, and C. Georgopoulos (ed.), *The Biology of Heat-Shock Proteins and Molecular Chaperones.* Cold Spring Harbor Laboratory Press, Cold Spring Harbor, N.Y.

112. **Fuchs, R. P. P.** 1984. DNA binding spectrum of the carcinogen N-acetoxy-N-2-acetylaminofluorene differs significantly from the mutation spectrum. *J. Mol. Biol.* **177:**173–180.

113. **Fuchs, R. P. P., N. Schwartz, and M. P. Duane.** 1981. Hot spots of frameshift mutations induced by the ultimate carcinogen N-acetoxy-N-2-acetylaminofluorene. *Nature* (London) **294:**657–659.

114. **Gaffney, B. L., and R. A. Jones.** 1989. Thermodynamic comparison of the basepairs formed by the carcinogenic lesion O⁶-methylguanine with reference both to Watson-Crick pairs and to mismatched pairs. *Biochemistry* **28:**5881–5889.

115. **Garcia, A., I. B. Lambert, and R. P. P. Fuchs.** 1993. DNA adduct-induced stabilization of slipped frameshift intermediates within repetitive sequences: implications for mutagenesis. *Proc. Natl. Acad. Sci. USA* **90:**5989–5993.

116. **Georgopoulos, C., K. Liberek, M. Zylicz, and D. Ang.** 1994. Properties of the heat-shock proteins of *Escherichia coli* and the autoregulation of the heat-shock response, p. 209–249. *In* R. I. Morimoto, A. Tissieres, and C. Georgopoulos (ed.), *The Biology of Heat-Shock Proteins and Molecular Chaperones.* Cold Spring Harbor Laboratory Press, Cold Spring Harbor, N.Y.

117. **Glazebrook, J. A., K. K. Grewal, and P. Strike.** 1986. Molecular analysis of the UV protection and mutation genes carried by the I incompatibility group plasmid TP110. *J. Bacteriol.* **168:**251–256.

118. **Glickman, B. W., R. M. Schaaper, W. A. Haseltine, R. L. Dunn, and D. E. Brash.** 1986. The C-C (6-4) UV photoproduct is mutagenic in *Escherichia coli. Proc. Natl. Acad. Sci. USA* **83:**6945–6949.

119. **Gogol, E. P., M. C. Young, W. L. Kubasek, T. C. Jarvis, and P. H. von Hippel.** 1992. Cryoelectron microscopic visualization of functional subassemblies of the bacteriophage T4 DNA replication complex. *J. Mol. Biol.* **224:**395–412.

120. **Gordon, A. J. E., P. A. Burns, D. F. Fix, F. Yatagi, F. L. Allen, M. J. Horsfall, J. A. Halliday, J. Gray, C. Bernelot-Moens, and B. W. Glickman.** 1988. Missense mutation in the *lacI* gene of *Escherichia coli*: inferences on the structure of the repressor protein. *J. Mol. Biol.* **200:**239–251.

121. **Green, M., and S. S. Cohen.** 1958. Studies on the biosynthesis of bacterial and viral pyrimidines. III. Derivatives of dihydrocytosine. *J. Biol. Chem.* **228:**601–609.

122. **Hagensee, M. E., S. K. Bryan, and R. E. Moses.** 1987. DNA polymerase III requirement for repair of DNA damage caused by methyl methanesulfonate and hydrogen peroxide. *J. Bacteriol.* **169:**4608–4613.

123. **Hagensee, M. E., T. L. Timme, S. Bryan, and R. E. Moses.** 1987. DNA polymerase III of *Escherichia coli* is required for UV and ethyl methanesulfonate mutagenesis. *Proc. Natl. Acad. Sci. USA* **84:**4195–4199.

124. **Hall, B. G.** 1990. Spontaneous point mutations that occur more often when they are advantageous than when they are neutral. *Genetics* **126:**5–16.

125. **Hauser, J., A. S. Levine, D. G. Ennis, K. Chumakov, and R. Woodgate.** 1992. The enhanced mutagenic potential of the MucAB proteins correlates with the highly efficient processing of the MucA protein. *J. Bacteriol.* **174:**6844–6851.

126. **Herrera, G., A. Urios, V. Aleixandre, and M. Blanco.** 1988. UV light-induced mutability in *Salmonella* strains containing the *umuDC* or the *mucAB* operon: evidence for a *umuC* function. *Mutat. Res.* **198:**9–13.

127. **Hill-Perkins, M., M. D. Jones, and P. Karran.** 1986. Site-specific mutagenesis *in vivo* by single methylated or deaminated purine bases. *Mutat. Res.* **162:**153–163.

128. **Hiom, K. J., and S. G. Sedgwick.** 1992. Alleviation of *EcoK* DNA restriction in *Escherichia coli* and involvement of *umuDC* activity. *Mol. Gen. Genet.* **231:**265–275.

129. **Hofemeister, J., H. Kohler, and V. D. Filippov.** 1979. DNA repair in *Proteus mirabilis*. VI. Plasmid (R46)-mediated recovery and UV mutagenesis. *Mol. Gen. Genet.* **176:**265–273.

130. **Horsfall, M. J., and C. W. Lawrence.** 1994. Accuracy of replication past the T-C (6-4) adduct. *J. Mol. Biol.* **235:**465–471.

131. **Howard-Flanders, P., and R. P. Boyce.** 1966. DNA repair and genetic recombination: studies on mutants of *Escherichia coli* defective in these processes. *Radiat. Res.* **6:**156–184.

132. **Howarth, S.** 1965. Resistance to the bactericidal effect of ultraviolet radiation conferred on enterobacteria by the colicin factor Col. I. *J. Gen. Microbiol.* **40:**43–55.

133. **Howarth, S.** 1966. Increase in frequency of ultraviolet induced mutation brought about by the colicin factor Col I in *Salmonella typhimurium*. *Mutat. Res.* **3:**129–134.

134. **Hughes, A. J., Jr., S. K. Bryan, H. Chen, R. E. Moses, and C. S. McHenry.** 1991. *Escherichia coli* DNA polymerase II is stimulated by DNA polymerase III holoenzyme auxiliary subunits. *J. Biol. Chem.* **266:**4568–4573.

135. **Hutchinson, F., K. Yamamoto, J. Stein, and R. D. Wood.** 1988. Effect of photoreactivation on mutagenesis of lambda phage by ultraviolet light. *J. Mol. Biol.* **202:**593–601.

136. **Ichikawa-Ryo, H., and S. Kondo.** 1975. Indirect mutagenesis in phage lambda by ultraviolet preirradiation of host bacteria. *J. Mol. Biol.* **97:**77–92.

137. **Iwasaki, H., Y. Ishino, H. Toh, A. Nakata, and H. Shinagawa.** 1991. *Escherichia coli* DNA polymerase II is homologous to α-like DNA polymerases. *Mol. Gen. Genet.* **226:**24–33.

138. **Iwasaki, H., A. Nakata, G. C. Walker, and H. Shinagawa.** 1990. The *Escherichia coli polB* gene, which encodes DNA polymerase II, is regulated by the SOS system. *J. Bacteriol.* **172:**6268–6273.

139. **Jiang, N., and J.-S. Taylor.** 1993. In vivo evidence that UV-induced C → T mutations at dipyrimidine sites could result from the replicative bypass of *cis-syn* cyclobutane dimers or their deamination products. *Biochemistry* **32:**472–481.

140. **Jonczyk, P., I. Fijalkowski, and Z. Ciesla.** 1988. Overproduction of the epsilon subunit of DNA polymerase III counteracts the SOS mutagenic responses of *Escherichia coli*. *Proc. Natl. Acad. Sci. USA* **85:**9124–9127.

141. **Kalnik, M. W., B. F. L. Li, P. F. Swann, and D. J. Patel.** 1989. O⁶-Ethylguanine carcinogenic lesion in DNA: an NMR study of O⁶etG.C pairing in dodecanucleotide duplexes. *Biochemistry* **28:**6182–6192.

142. **Kalnik, M. W., B. F. L. Li, P. F. Swann, and D. J. Patel.** 1989. O⁶-Ethylguanine carcinogenic lesion in DNA: an NMR study of O⁶etG.T pairing in dodecanucleotide duplexes. *Biochemistry* **28:**6170–6181.

143. **Kan, L.-S., L. Voiturieux, and J. Cadet.** 1992. The Dewar valence isomer of the (6-4) photoadduct of thymidylyl-(3′-5′)-thymidine monophosphate formation, alkaline lability, and conformational properties. *J. Photochem. Photobiol. Ser. B* **12:**339–357.

144. **Kato, T.** 1977. Effects of chloramphenicol and caffeine on postreplication repair in *uvrA⁻ umuC⁻* and *uvrA⁻ recF⁻* strains of *Escherichia coli* K-12. *Mol. Gen. Genet.* **156:**115–120.

145. **Kato, T., T. Ise, and H. Shinagawa.** 1982. Mutational specificity of the *umuC* mediated mutagenesis in *Escherichia coli*. *Biochimie* **64:**731–733.

146. **Kato, T., and E. Nikano.** 1981. Effects of the *umuC36* mutation on ultraviolet-radiation-induced base change and frameshift mutagenesis in *Escherichia coli*. *Mutat. Res.* **83:**307–319.

147. **Kato, T., and Y. Shinoura.** 1977. Isolation and characterization of mutants of *Escherichia coli* deficient in induction of mutations by ultraviolet light. *Mol. Gen. Genet.* **156:**121–131.

148. **Kemmink, J., R. Boelens, T. Koning, G. A. van der Marel, J. H. van Boom, and R. Kaptein.** 1987. ¹H NMR study of the exchangeable protons of the duplex d(GCGT^TGCG)·d(CGCAACGC) containing a thymine photodimer. *Nucleic Acids Res.* **15:**4645–4653.

149. **Kenyon, C. J., R. Brent, M. Ptashne, and G. C. Walker.** 1982. Regulation of damage-inducible genes in *Escherichia coli*. *J. Mol. Biol.* **160:**445–457.

150. **Kenyon, C. J., and G. C. Walker.** 1980. DNA-damaging agents stimulate gene expression at specific loci in *Escherichia coli*. *Proc. Natl. Acad. Sci. USA* **77:**2819–2823.

151. **Kitagawa, Y., E. Akaboshi, H. Shinagawa, T. Horii, H. Ogawa, and T. Kato.** 1985. Structural analysis of the *umu* operon required for inducible mutagenesis in *Escherichia coli*. *Proc. Natl. Acad. Sci. USA* **82:**4336–4340.

152. **Koch, W. H., T. A. Cebula, P. L. Foster, and E. Eisenstadt.** 1992. UV mutagenesis in *Salmonella typhimurium* is *umuDC* dependent despite the presence of *samAB*. *J. Bacteriol.* **174:**2809–2815.

153. **Koch, W. H., D. G. Ennis, A. S. Levine, and R. Woodgate.** 1992. *Escherichia coli umuDC* mutants: DNA sequence alterations and UmuD cleavage. *Mol. Gen. Genet.* **233:**443–448.

154. **Koehler, D. R., S. S. Awadallah, and B. W. Glickman.** 1991. Sites of preferential induction of cyclobutane pyrimidine dimers in the nontranscribed strand of *lacI* correspond with sites of UV-induced mutation in *Escherichia coli*. *J. Biol. Chem.* **266:**11766–11773.

155. **Koffel-Schwartz, P., J. M. Verdier, J. F. Lefervre, M. Bichara, A. M. Fruend, M. P. Daune, and R. P. P. Fuchs.** 1984. Carcinogen-induced mutation spectrum in wild-type, *uvrA*, and *umuC* strains of *E. coli*: strain specificity and mutation-prone sequences. *J. Mol. Biol.* **177:**33–51.

156. **Kong, X.-P., R. Onrust, M. O'Donnell, and J. Kuriyan.** 1992. Three-dimensional structure of the β subunit of *E. coli* DNA polymerase III holoenzyme: a sliding DNA clamp. *Cell* **69:**425–437.

157. **Kornberg, A., and T. A. Baker.** 1992. *DNA Replication.* W. H. Freeman & Co., New York.

158. **Kouchakdjian, M., V. Bodepudi, S. Shibutani, M. Eisenberg, F. Johnson, A. P. Grollman, and D. J. Patel.** 1991. NMR structural studies of the ionizing radiation adduct 7-hydro-8-oxodeoxyguanosine (8-oxo-7H-dG) opposite deoxyadenosine in a DNA duplex. 8-oxo-7H-dG(syn).dA(anti) alignment at lesion site. *Biochemistry* **30:**1403–1412.

159. **Kow, Y. W., G. Fuandez, S. Hays, C. Bonner, M. F. Goodman, and S. S. Wallace.** 1993. Absence of a role for DNA polymerase II in SOS-induced translesion bypass of φX174. *J. Bacteriol.* **175:**561–564.

160. **Krueger, J. H., S. J. Elledge, and G. C. Walker.** 1983. Isolation and characterization of Tn*5* insertion mutations in the *lexA* gene of *Escherichia coli. J. Bacteriol.* **153:**1368–1378.

161. **Krueger, J. H., and G. C. Walker.** 1984. *groEL* and *dnaK* genes of *Escherichia coli* are induced by UV irradiation and nalidixic acid in an *htpR*-dependent manner. *Proc. Natl. Acad. Sci. USA* **81:**1499–1503.

162. **Kunala, S., and D. E. Brash.** 1992. Excision repair at individual bases of the *Escherichia coli lacI* gene: relation to mutation hot spots and transcription coupling activity. *Proc. Natl. Acad. Sci. USA* **89:**11031–11035.

163. **Kunkel, T. A., and L. A. Loeb.** 1984. Mutational specificity of depurination. *Proc. Natl. Acad. Sci. USA* **81:**1494–1498.

164. **Kunkel, T. A., R. M. Schaaper, and L. A. Loeb.** 1983. Depurination-induced infidelity of deoxyribonucleic acid synthesis with purified deoxyribonucleic acid replication proteins *in vitro. Biochemistry* **22:**2378–2384.

165. **Kunz, B. A., and B. W. Glickman.** 1984. The role of pyrimidine dimers as premutagenic lesions: a study of targeted *vs.* untargeted mutagenesis in the *lacI* gene of *Escherichia coli. Genetics* **106:**347–364.

166. **Kuriyan, J., and M. O'Donnell.** 1993. Sliding clamps of DNA polymerases. *J. Mol. Biol.* **234:**915–925.

167. **Lackey, D., S. W. Krauss, and S. Linn.** 1982. Isolation of an altered form of DNA polymerase I from *Escherichia coli* cells induced for *recA/lexA* functions. *Biochemistry* **79:**330–334.

168. **Lackey, D., S. W. Krauss, and S. Linn.** 1985. Characterization of DNA polymerase I*, a form of DNA polymerase I found in *Escherichia coli* expressing SOS functions. *J. Biol. Chem.* **260:**3178–3184.

169. **Lam, L. H., and R. J. Reynolds.** 1987. DNA sequence dependence of closely opposed cyclobutyl pyrimidine dimers induced by UV radiation. *Mutat. Res.* **178:**167–176.

170. **Lambert, I. B., A. J. E. Gordon, D. W. Bryant, and B. W. Glickman.** 1991. The action of 1-nitroso-8-nitropyrene in *Escherichia coli*: DNA adduct formation and mutational consequences in the absence of nucleotide excision repair. *Carcinogenesis* **12:**879–884.

171. **Lambert, I. B., R. L. Napolitano, and R. P. P. Fuchs.** 1992. Carcinogen-induced frameshift mutagenesis in repetitive sequences. *Proc. Natl. Acad. Sci. USA* **89:**1310–1314.

172. **Langer, P. J., W. G. Shanabruch, and G. C. Walker.** 1981. Functional organization of plasmid pKM101. *J. Bacteriol.* **145:**1310–1316.

173. **Langer, P. J., and G. C. Walker.** 1981. Restriction endonuclease cleavage map of pKM101: relationship to parental plasmid R46. *Mol. Gen. Genet.* **182:**268–272.

174. **Larimer, F., J. Perry, and A. Hardigree.** 1989. The *REV1* gene of *Saccharomyces cerevisiae*: isolation, sequence, and functional analysis. *J. Bacteriol.* **171:**230–237.

175. **Lawley, P. D.** 1984. Carcinogenesis by alkylating agents. *ACS Monogr.* **182:**325–484.

176. **Lawrence, C. W.** 1981. Are pyrimidine dimers non-instructive lesions? *Mol. Gen. Genet.* **182:**511–513.

177. **Lawrence, C. W., S. K. Banerjee, A. Borden, and J. E. LeClerc.** 1990. T-T cyclobutance dimers are misinstructive rather than non-instructive, mutagenic lesions. *Mol. Gen. Genet.* **222:**166–168.

178. **Lawrence, C. W., A. Borden, S. K. Banerjee, and J. E. LeClerc.** 1990. Mutation frequency and spectrum resulting from a single abasic site in a single-stranded vector. *Nucleic Acids Res.* **18:**2153–2157.

179. **Lawrence, C. W., R. B. Christensen, and J. R. Christensen.** 1985. Identity of the photoproduct that causes *lacI* mutations in UV-irradiated *Escherichia coli. J. Bacteriol.* **161:**767–768.

180. **Lawrence, C. W., P. E. M. Gibbs, A. Borden, M. J. Horsfall, and B. J. Kilbey.** 1993. Mutagenesis induced by single UV photoproducts in *E. coli* and yeast. *Mutat. Res.* **299:**157–163.

181. **LeClerc, J. E., A. Borden, and C. W. Lawrence.** 1991. The thymine-thymine pyrimidine-pyrimidone (6-4) ultraviolet light photoproduct is highly mutagenic and specifically induces 3' thymine-to-cytosine transitions in *Escherichia coli. Proc. Natl. Acad. Sci. USA* **88:**9685–9689.

182. **LeClerc, J. E., J. R. Christensen, P. V. Tata, R. B. Christensen, and C. W. Lawrence.** 1988. Ultraviolet light induces different spectra of *lacI* sequence changes in vegetative and conjugating cells of *Escherichia coli. J. Mol. Biol.* **203:**619–633.

183. **LeClerc, J. E., N. L. Istock, B. R. Saran, and R. Allen, Jr.** 1984. Sequence analysis of ultraviolet-induced mutations in M13*lacZ* hybrid phage DNA. *J. Mol. Biol.* **180:**217–237.

184. **Lemaire, D. G. E., and B. P. Ruzsicska.** 1993. Kinetic analysis of the deamination reactions of cyclobutane dimers of thymidylyl-3',5'-2'-deoxycytidine and 2'-deoxycytidylyl-3',5'-thymidine. *Biochemistry* **32:**2525–2533.

185. **Lemontt, J. F.** 1971. Mutants of yeast defective in mutation induced by ultraviolet light. *Genetics* **68:**21–33.

186. **Lenski, R. E., and J. E. Mittler.** 1993. Directed mutation. *Science* **260:**1222–1223.

187. **Levine, J. G., R. M. Schaaper, and D. M. DeMarini.** 1994. Complex frameshift mutations mediated by plasmid pKM101: mutational mechanisms deduced from 4-aminobiphenyl-induced mutational spectra in *Salmonella. Genetics* **136:**731–746.

188. **Levy, M. S., E. Balbinder, and R. Nagel.** 1993. The effect of mutations in SOS genes on UV-induced precise excision of Tn*10* in *Escherichia coli. Mutat. Res.* **293:**241–247.

189. **Lifsics, M. R., E. D. Lancy, Jr, and R. Maurer.** 1992. DNA replication defect in *Salmonella typhimurium* mutants lacking the editing (ε) subunit of DNA polymerase III. *J. Bacteriol.* **174:**6965–6973.

190. **Lindahl, T.** 1993. Instability and decay of the primary structure of DNA. *Nature* (London) **362:**709–715.

191. **Lindsley, J. E., and R. P. P. Fuchs.** 1994. Use of single-turnover kinetics to study bulky adduct bypass by T7 DNA polymerase. *Biochemistry* **33:**764–772.

192. **Lippke, J. A., L. K. Gordon, D. E. Brash, and W. A. Haseltine.** 1981. Distribution of UV light-induced damage in a defined sequence of human DNA: detection of alkaline-sensitive lesions at pyrimidine nucleoside-cytidine sequences. *Proc. Natl. Acad. Sci. USA* **78:**3388–3392.

193. **Liu, S.-K., and I. Tessman.** 1990. Error-prone SOS repair can be error-free. *J. Mol. Biol.* **216:**803–807.

194. **Livneh, Z.** 1986. Mechanism of replication of ultraviolet-irradiated single-stranded DNA by DNA polymerase III holoenzyme of *Escherichia coli*: implications for SOS mutagenesis. *J. Biol. Chem.* **261:**9526–9533.

195. **Livneh, Z., O. Cohen-Fix, R. Skaliter, and T. Elizur.** 1993. Replication of damaged DNA and the molecular mechanism of ultraviolet light mutagenesis. *Crit. Rev. Biochem. Mol. Biol.* **28:**465–513.

196. **Lodwick, D., D. Owen, and P. Strike.** 1990. DNA sequence analysis of the *imp* UV protection and mutation operon of the plasmid TP110: identification of a third gene. *Nucleic Acids Res.* **18:**5045–5050.

197. **Loeb, L. A., and B. D. Preston.** 1986. Mutagenesis by apurinic/apyrimidinic sites. *Annu. Rev. Genet.* **20:**201–230.

198. **Loechler, E. L., C. L. Green, and J. M. Essigmann.** 1984. *In vivo* mutagenesis by O-6-methylguanine built into a unique site in a viral genome. *Proc. Natl. Acad. Sci. USA* **81:**6271–6275.

199. **Loveless, A.** 1969. Possible relevance of O-6-alkylation of deoxyguanosine to the mutagenicity and carcinogenicity of nitrosamines and nitrosamides. *Nature* (London) **223:**206–207.

200. **Lu, C., R. H. Scheuermann, and H. Echols.** 1986. Capacity of RecA protein to bind preferentially to UV lesions and inhibit the editing subunit (ε) of DNA polymerase III: a possible mechanism for SOS-induced targeted mutagenesis. *Proc. Natl. Acad. Sci. USA* **83:**619–623.

201. **MacKay, W. J., S. Han, and L. D. Samson.** 1994. DNA alkylation repair limits spontaneous base substitution mutations in *Escherichia coli. J. Bacteriol.* **176:**3224–3230.

202. **Maenhaut-Michel, G., and P. Caillet-Fauquet.** 1984. Effect of *umuC* mutations on targeted and untargeted ultraviolet mutagenesis in bacteriophage λ. *J. Mol. Biol.* **177:**181–187.

203. Maenhaut-Michel, G., and P. Caillet-Fauquet. 1990. Genetic control of the UV-induced SOS mutator effect in single- and double-stranded DNA phages. *Mutat. Res.* **230:**241–254.

204. Maenhaut-Michel, G., R. Janel-Bintz, and R. P. P. Fuchs. 1992. A *umuDC*-independent SOS pathway for frameshift mutagenesis. *Mol. Gen. Genet.* **235:**373–380.

205. Maki, H., S. K. Bryan, T. Horiuchi, and R. E. Moses. 1989. Suppression of *dnaE* nonsense mutations by *pcbA1*. *J. Bacteriol.* **171:** 3139–3143.

206. Marsh, L., T. Nohmi, S. Hinton, and G. C. Walker. 1991. New mutations in cloned *Escherichia coli umuDC* genes: novel phenotypes of strains carrying a *umuC125* plasmid. *Mutat. Res.* **250:**183–197.

207. Marsh, L., and G. C. Walker. 1985. Cold sensitivity induced by overproduction of UmuDC in *Escherichia coli. J. Bacteriol.* **162:**155–161.

208. McCann, J., and B. N. Ames. 1976. Detection of carcinogens as mutagens in the *Salmonella*/microsome test: assay of 300 chemicals. Discussion. *Proc. Natl. Acad. Sci. USA* **73:**950–954.

209. McCann, J., E. Choi, E. Yamasaki, and B. N. Ames. 1975. Detection of carcinogens as mutagens in the *Salmonella*/microsome test: assay of 300 chemicals. *Proc. Natl. Acad. Sci. USA* **72:**5135–5139.

210. McCann, J., N. E. Spingarn, J. Kobori, and B. N. Ames. 1975. Detection of carcinogens as mutagens: bacterial tester strains with R factor plasmids. *Proc. Natl. Acad. Sci. USA* **72:**979–983.

211. McHenry, C. S. 1991. DNA polymerase III holoenzyme. *J. Biol. Chem.* **266:**19127–19130.

212. McNally, K. P., N. E. Freitag, and G. C. Walker. 1990. LexA independent expression of a mutant *mucAB* operon. *J. Bacteriol.* **172:** 6223–6231.

213. Mellon, I., and P. C. Hanawalt. 1989. Induction of the *Escherichia coli* lactose operon selectively increases repair of its transcribed DNA strand. *Nature* (London) **342:**95–98.

214. Miller, J. H. 1983. Mutational specificity in bacteria. *Annu. Rev. Genet.* **17:**215–238.

215. Miller, J. H. 1985. Mutagenic specificity of ultraviolet light. *J. Mol. Biol.* **182:**45–68.

216. Miller, J. H., C. Coulondre, and P. J. Farabaugh. 1978. Correlation of nonsense sites in the lacI gene with specific codons in the nucleotide sequence. *Nature* (London) **274:**770–775.

217. Miller, J. H., and K. B. Low. 1984. Specificity of mutagenesis resulting from the induction of the SOS system in the absence of mutagenic treatment. *Cell* **37:**675–682.

218. Miller, S. S., and E. Eisenstadt. 1985. Enhanced sensitivity of *Escherichia coli umuC* to photodynamic inactivation by angelicin (isopsoralen). *J. Bacteriol.* **162:**1307–1310.

219. Misuraca, F., D. Rampolla, and S. Grimaudo. 1991. Identification and cloning of a *umu* locus in *Streptomyces coelicolor* A3(2). *Mutat. Res.* **262:**183–188.

220. Mitchell, D. L. 1988. The relative cytotoxicity of (6-4) photoproducts and cyclobutane dimers in mammalian cells. *Photochem. Photobiol.* **48:**51–57.

221. Mitchell, D. L., and R. S. Nairn. 1989. The biology of the (6-4) photoproduct. *Photochem. Photobiol.* **49:**805–819.

222. Molina, A. M., N. Babudri, M. Tamaro, S. Venturini, and C. Monti-Bragadin. 1979. *Enterobacteriaceae* plasmids enhancing chemical mutagenesis and their distribution among incompatibility groups. *FEMS Microbiol. Lett.* **5:**33–37.

223. Moreau, P. 1987. Effects of overproduction of single-stranded DNA-binding protein on RecA protein-dependent processes in *Escherichia coli. J. Mol. Biol.* **194:**621–634.

224. Moriya, M., and A. P. Grollman. 1993. Mutations in the *mutY* gene of *Escherichia coli* enhance the frequency of targeted G:C → T:A transversions induced by a single-8-oxoguanine residue in single-stranded DNA. *Mol. Gen. Genet.* **239:**72–76.

225. Moriya, M., C. Ou, V. Bodepudi, F. Johnston, M. Takeshita, and A. P. Grollman. 1991. Site-specific mutagenesis using a gapped duplex vector: a study of translesion synthesis past 8-oxodeoxyguaninosine in *E. coli. Mutat. Res.* **254:**281–288.

226. Mortelmans, K. E., and B. A. D. Stocker. 1979. Segregation of the mutator property of plasmid R46 from its ultraviolet-protecting property. *Mol. Gen. Genet.* **167:**317–328.

227. Moses, R. E., A. Byford, and J. A. Hejna. 1992. Replisome pausing in mutagenesis. *Chromosoma* **102:**5157–5160.

228. Murli, S., and G. C. Walker. 1993. SOS mutagenesis. *Curr. Opin. Genet. Dev.* **3:**719–725.

229. Napolitano, R. L., I. B. Lambert, and R. P. P. Fuchs. 1994. DNA sequence determinants of carcinogen-induced frameshift mutagenesis. *Biochemistry* **33:**1311–1315.

230. Nishioka, H., and C. O. Doudney. 1969. Different modes of loss of photoreversibility of mutation and lethal damage in ultraviolet-light resistant and sensitive bacteria. *Mutat. Res.* **8:**215–228.

231. Nishioka, H., and C. O. Doudney. 1970. Different modes of loss of photoreversibility of ultraviolet light-induced true and suppressor mutations to tryptophan independence in an auxotrophic strain of *Escherichia coli. Mutat. Res.* **9:**349–358.

232. Niwa, O., S. K. Bryan, and R. E. Moses. 1981. Alternate pathways of DNA replication: DNA polymerase I-dependent replication. *Proc. Natl. Acad. Sci. USA* **78:**7024–7027.

233. Nohmi, T., J. R. Battista, L. A. Dodson, and G. C. Walker. 1988. RecA-mediated cleavage activates UmuD for mutagenesis: mechanistic relationship between transcriptional derepression and posttranslational activation. *Proc. Natl. Acad. Sci. USA* **85:**1816–1820.

234. Nohmi, T., A. Hakura, Y. Nakai, M. Watanabe, S. Y. Murayama, and T. Sofuni. 1991. *Salmonella typhimurium* has two homologous but different *umuDC* operons: cloning of a new *umuDC*-like operon (*samAB*) present in a 60-megadalton cryptic plasmid of *S. typhimurium. J. Bacteriol.* **173:**1051–1063.

235. Nohmi, T., M. Yamada, M. Watanabe, S. Murayama, and T. Sofuni. 1992. The roles of *umuDC*$_{ST}$ and *samAB* in UV mutagenesis and UV sensitivity of *Salmonella typhimurium. J. Bacteriol.* **174:**6948–6955.

236. Norman, D., P. Abuaf, B. E. Hingerty, D. Live, D. Grunberger, S. Broyde, and D. J. Patel. 1989. NMR and computational characteristics of the N-(deoxyguanosin-yl)aminofluorene adduct [(AF)G] opposite adenosine in DNA: (AF)G[syn]·A[anti] pair formation and its pH dependence. *Biochemistry* **28:**7462–7476.

237. Nowicka, A., M. Kanabus, E. Sledziewska-Gojska, and Z. Ciesla. 1994. Different UmuC requirements for generation of different kinds of UV-induced mutations in *Escherichia coli. Mol. Gen. Genet.* **243:**584–592.

238. O'Donnell, M. 1992. Accessory protein function in the DNA polymerase III holoenzyme from *Escherichia coli. Bioessays* **14:**105–111.

239. O'Donnell, M., J. Kuriyan, X.-P. Kong, P. T. Stukenberg, and R. Onrust. 1992. The sliding clamp of DNA polymerase III holoenzyme encircles DNA. *Mol. Biol. Cell* **3:**953–957.

240. Oller, A. R., I. J. Fijalkowska, R. L. Dunn, and R. M. Schaaper. 1992. Transcription-repair coupling determines the strandedness of ultraviolet mutagenesis in *Escherichia coli. Proc. Natl. Acad. Sci. USA* **89:**11036–11040.

241. Orrego, C., and E. Eisenstadt. 1987. An inducible pathway is required for mutagenesis in *Salmonella typhimurium* LT2. *J. Bacteriol.* **169:**2885–2888.

242. Patel, D. J., L. Shapiro, S. A. Kozlowski, B. L. Gaffney, S. Kuzmich, and R. A. Jones. 1986. Covalent carcinogenic O^6-methylguanosine lesions in DNA: structural studies of the O^6meG·A and O^6meG·G interactions in dodecanucleotide duplexes. *J. Mol. Biol.* **188:**677–692.

243. Patel, D. J., L. Shapiro, S. A. Kozlowski, B. L. Gaffney, S. Kuzmich, and R. A. Jones. 1986. Structural studies of the O^6-meG.C interaction in the d(C-G-C-G-A-A-T-T-C-O^6meG-C-G) duplex. *Biochemistry* **25:**1027–1036.

244. Patel, D. J., L. Shapiro, S. A. Kozlowski, B. L. Gaffney, S. Kuzmich, and R. A. Jones. 1986. Structural studies of the O^6-meG.T interaction in the d(C-G-T-G-A-A-T-T-C-O⁶meG-C-G) duplex. *Biochemistry* **25**:1036–1042.

245. Perry, K. L., S. J. Elledge, B. Mitchell, L. Marsh, and G. C. Walker. 1985. *umuDC* and *mucAB* operons whose products are required for UV light- and chemical-induced mutagenesis: UmuD, MucA, and LexA proteins share homology. *Proc. Natl. Acad. Sci. USA* **82**:4331–4335.

246. Perry, K. L., and G. C. Walker. 1982. Identification of plasmid (pKM101)-coded proteins involved in mutagenesis and UV resistance. *Nature* (London) **300**:278–281.

247. Person, S., J. A. McCloskey, W. Snipes, and R. C. Bockrath. 1974. Ultraviolet mutagenesis and its repair in an *Escherichia coli* strain containing a nonsense codon. *Genetics* **78**:1035–1049.

248. Povirk, L. F., C. W. Houlgrave, and Y.-H. Han. 1988. Neocarzinostatin-induced DNA base release accompanied by staggered oxidative cleavage of the complementary strand. *J. Biol. Chem.* **263**:19263–19266.

249. Preston, B. D., B. Singer, and L. A. Loeb. 1986. Mutagenic potential of O⁴-methylthymine *in vivo* determined by an enzymatic approach to site-specific mutagenesis. *Proc. Natl. Acad. Sci. USA* **83**:8501–8505.

250. Quillardet, P., and R. Devoret. 1982. Damaged-site independent mutagenesis of phage lambda produced by inducible error-prone repair. *Biochimie* **64**:789–796.

251. Radman, M. 1974. Phenomenology of an inducible mutagenic DNA repair pathway in *Escherichia coli*: SOS repair hypothesis, p. 128–142. *In* L. Prakash, F. Sherman, M. Miller, C. Lawrence, and H. W. Tabor (ed.), *Molecular and Environmental Aspects of Mutagenesis*. Charles C Thomas, Springfield, Ill.

252. Radman, M. 1975. SOS repair hypothesis: phenomenology of an inducible DNA repair which is accompanied by mutagenesis, p. 355–367. *In* P. Hanawalt and R. B. Setlow (ed.), *Molecular Mechanisms for Repair of DNA*, part A. Plenum Publishing Corp., New York.

253. Radman, M., G. Villani, S. Boiteux, R. Kinsella, B. W. Glickman, and S. Spadari. 1978. Replication fidelity: mechanisms of mutation avoidance and mutation fixation. *Cold Spring Harbor Symp. Quant. Biol.* **43**:937–946.

254. Rajagopalan, M., C. Lu, R. Woodgate, M. O'Donnell, M. F. Goodman, and H. Echols. 1992. Activity of the purified mutagenesis proteins UmuC, UmuD', and RecA in replicative bypass of an abasic DNA lesion by DNA polymerase III. *Proc. Natl. Acad. Sci. USA* **89**:10777–10781.

255. Reddy, M. K., S. E. Weitzel, and P. H. von Hippel. 1993. Assembly of a functional replication complex without ATP hydrolysis: a direct interaction of bacteriophage T4 gp45 with T4 DNA polymerase. *Proc. Natl. Acad. Sci. USA* **90**:3211–3215.

256. Richardson, K. K., F. C. Richardson, R. M. Crosby, J. A. Swenberg, and T. R. Skopek. 1987. DNA base changes and alkylation following *in vivo* exposure of *Escherichia coli* to N-methyl-N-nitrosourea or N-ethyl-N-nitrosourea. *Proc. Natl. Acad. Sci. USA* **84**:344–348.

257. Roberts, J. W., and C. W. Roberts. 1975. Proteolytic cleavage of bacteriophage lambda repressor in induction. *Proc. Natl. Acad. Sci. USA* **72**:147–151.

258. Ruiz-Rubio, M., and R. Bockrath. 1989. On the possible roles of cytosine deamination in delayed photoreversal mutagenesis targeted at thymine-cytosine dimers in *E. coli*. *Mutat. Res.* **210**:93–102.

259. Ruiz-Rubio, M., and B. A. Bridges. 1987. Mutagenic DNA repair in *Escherichia coli*. XIV. Influence of two DNA polymerase III mutator alleles on spontaneous and UV mutagenesis. *Mol. Gen. Genet.* **208**:542–548.

260. Ruiz-Rubio, M., R. Woodgate, B. A. Bridges, G. Herrera, and M. Blanco. 1986. New role for photoreversible dimers in the induction of prototrophic mutations in excision-deficient *Escherichia coli* by UV light. *J. Bacteriol.* **166**:1141–1143.

261. Ryan, F. J., D. Nakada, and M. J. Schneider. 1961. Is DNA replication a necessary condition for spontaneous mutation? *Z. Vererbungsl.* **92**:38–41.

262. Rycyna, R. E., and J. L. Alderfer. 1985. UV irradiation of nucleic acids: formation, purification, and solution conformational analysis of the (6-4) lesion of dTpdT. *Nucleic Acids Res.* **13**:5949–5963.

263. Sage, E., E. Cramb, and B. W. Glickman. 1992. The distribution of UV damage in the *lacI* gene of *Escherichia coli*: correlation with mutation spectrum. *Mutat. Res.* **269**:285–299.

264. Sagher, D., and B. Strauss. 1983. Insertion of nucleotides opposite purinic/apyrimidinic sites in deoxyribonucleic acid during *in vitro* synthesis: uniqueness of adenine nucleotides. *Biochemistry* **22**:4518–4526.

265. Samson, L., J. Thomale, and M. F. Rajewsky. 1988. Alternative pathways for the *in vivo* repair of O⁶-alkylguanine and O⁴-alkylthymine in *Escherichia coli*: the adaptive response and nucleotide excision repair. *EMBO J.* **7**:2261–2267.

266. Sargentini, N. J., and K. C. Smith. 1984. *umuC*-dependent and *umuC*-independent γ- and UV-radiation mutagenesis in *Escherichia coli*. *Mutat. Res.* **128**:1–9.

267. Sargentini, N. J., and K. C. Smith. 1987. Ionizing and ultraviolet radiation-induced reversion of sequenced frameshift mutations in *Escherichia coli*: a new role for *umuDC* suggested by delayed photoreactivation. *Mutat. Res.* **179**:55–63.

268. Sassanfar, M., and J. W. Roberts. 1990. Nature of the SOS-inducing signal in *Escherichia coli*: the involvement of DNA replication. *J. Mol. Biol.* **212**:79–96.

269. Schaaper, R. M., R. L. Dunn, and B. W. Glickman. 1987. Mechanisms of ultraviolet-induced mutation: mutational spectra in the *Escherichia coli lacI* gene for a wild type and excision-repair-deficient strain. *J. Mol. Biol.* **198**:187–202.

270. Schaaper, R. M., B. W. Glickman, and L. A. Loeb. 1982. Mutagenesis resulting from depurination is an SOS process. *Mutat. Res.* **106**:1–9.

271. Schaaper, R. M., N. Koffel-Schwartz, and R. P. P. Fuchs. 1990. N-acetoxy-N-acetylaminofluorene-induced mutagenesis in the *lacI* gene of *Escherichia coli*. *Carcinogenesis* **11**:1087–1095.

272. Schaaper, R. M., and L. A. Loeb. 1981. Depurination causes mutations in SOS-induced cells. *Proc. Natl. Acad. Sci. USA* **78**:1773–1777.

273. Schendel, P. F., and M. Defais. 1980. The role of *umuC* gene product in mutagenesis by simple alkylating agents. *Mol. Gen. Genet.* **177**:661–665.

274. Sedgwick, S., D. Lodwick, N. Doyle, H. Crowne, and P. Strike. 1991. Functional complementation between chromosomal and plasmid mutagenic DNA repair genes in bacteria. *Mol. Gen. Genet.* **229**:428–436.

275. Sedgwick, S. G. 1976. Misrepair of overlapping daughter strand gaps as a possible mechanism for UV-induced mutagenesis in *uvr* strains of *Escherichia coli*: a general model for induced mutagenesis by misrepair (SOS repair) of closely spaced DNA lesions. *Mutat. Res.* **41**:185–200.

276. Sedgwick, S. G., C. Ho, and R. Woodgate. 1991. Mutagenic DNA repair in enterobacteria. *J. Bacteriol.* **173**:5604–5611.

277. Sedgwick, S. G., M. Robson, and F. Malik. 1988. Polymorphisms in the *umuDC* region of *Escherichia* species. *J. Bacteriol.* **170**:1610–1616.

278. Selby, C. P., and A. Sancar. 1990. Transcription preferentially inhibits nucleotide excision repair of the template DNA strand *in vitro*. *J. Biol. Chem.* **265**:21330–21336.

279. Selby, C. P., and A. Sancar. 1993. Molecular mechanism of transcription-repair coupling. *Science* **260**:53–58.

280. Selby, C. P., E. M. Witkin, and A. Sancar. 1991. *Escherichia coli mfd* mutant deficient in "mutation frequency decline" lacks strand-specific repair: *in vitro* complementation with purified coupling factor. *Proc. Natl. Acad. Sci. USA* **88**:11574–11578.

281. Shanabruch, W. G., and G. C. Walker. 1980. Localization of the plasmid (pKM101) gene(s) involved in *recA⁺lexA⁺*-dependent mutagenesis. *Mol. Gen. Genet.* **179**:289–297.

282. Shapiro, J. A. 1984. Observations on the formation of clones containing *araB-lacZ* cistron fusions. *Mol. Gen. Genet.* **194**:79–90.

283. Shapiro, R., B. E. Hingerty, and S. Broyde. 1989. Minorgroove binding models for acetylaminofluorene modified DNA. *J. Biomol. Struct. Dyn.* **7:**493–513.

284. Sharif, F., and B. A. Bridges. 1990. Mutagenic DNA repair in *Escherichia coli*. XVII. Effect of temperature-sensitive DnaE proteins on the induction of streptomycin-resistant mutations by UV light. *Mutagenesis* **5:**31–34.

285. Shavitt, O., and Z. Livneh. 1989. The β subunit modulates bypass and termination at UV lesions during *in vitro* replication with DNA polymerase III holoenzyme of *Escherichia coli. J. Biol. Chem.* **264:**11275–11281.

286. Shearman, C. W., and L. A. Loeb. 1977. Depurination decreases fidelity of DNA synthesis *in vitro. Nature* (London) **270:**537–538.

287. Shearman, C. W., and L. A. Loeb. 1979. Effects of depurination on the fidelity of DNA synthesis. *J. Mol. Biol.* **128:**197–218.

288. Shiba, T., H. Iwasaki, A. Nakata, and H. Shinagawa. 1990. Proteolytic processing of MucA protein in SOS mutagenesis: both processed and unprocessed MucA may be active in the mutagenesis. *Mol. Gen. Genet.* **224:**169–176.

289. Shibutani, S., M. Takeshita, and A. P. Grollman. 1991. Insertion of specific bases during DNA synthesis past the oxidation-damaged base 8-oxodG. *Nature* (London) **349:**431–434.

290. Shinagawa, H., H. Iwasaki, T. Kato, and A. Nakata. 1988. RecA protein-dependent cleavage of UmuD protein and SOS mutagenesis. *Proc. Natl. Acad. Sci. USA* **85:**1806–1810.

291. Shinagawa, H., T. Kato, T. Ise, K. Makino, and A. Nakata. 1983. Cloning and characterization of the *umu* operon responsible for inducible mutagenesis in *Escherichia coli. Gene* **23:**167–174.

292. Shinoura, Y., T. Ise, T. Kato, and B. W. Glickman. 1983. *umuC*-mediated misrepair mutagenesis in *Echerichia coli*: extent and specificity of SOS mutagenesis. *Mutat. Res.* **111:**51–59.

293. Shwartz, H., O. Shavitt, and Z. Livneh. 1988. The role of exonucleolytic processing and polymerase-DNA association in bypass of lesions during replication *in vitro. J. Biol. Chem.* **263:**18227–18285.

294. Singer, B., and M. K. Dosanjh. 1990. Site-directed mutagenesis for quantitation of base-base interactions at defined sites. *Mutat. Res.* **233:**45–51.

295. Singer, B., and J. M. Essigmann. 1991. Site-specific mutagenesis: retrospective and prospective. *Carcinogenesis* **12:**949–955.

296. Slater, S. C., and R. Maurer. 1991. Requirement for bypass of UV-induced lesions in single-stranded DNA of bacteriophage φX174 in *Salmonella typhimurium. Proc. Natl. Acad. Sci. USA* **88:**1251–1255.

297. Smith, C. M., W. Koch, S. B. Franklin, P. L. Foster, T. A. Cebula, and E. Eisenstadt. 1990. Sequence analysis and mapping of the *Salmonella typhimurium* LT2 *umuDC* operon. *J. Bacteriol.* **172:**4979–4987.

298. Smith, K. C. 1992. Spontaneous mutagenesis: experimental, genetic and other factors. *Mutat. Res.* **277:**139–162.

299. Sockett, H., S. Romac, and F. Hutchinson. 1991. DNA sequence changes in mutations induced by ultraviolet light in the *gpt* gene on the chromosome of *Escherichia coli uvr*⁺ and *uvr* cells. *Mol. Gen. Genet.* **230:**295–301.

300. Sommer, S., A. Bailone, and R. Devoret. 1993. The appearance of the UmuD'C protein complex in *Escherichia coli* switches repair from homologous recombination to SOS mutagenesis. *Mol. Microbiol.* **10:**963–971.

301. Sommer, S., J. Knezevic, A. Bailone, and R. Devoret. 1993. Induction of only one SOS operon, *umuDC*, is required for SOS mutagenesis in *E. coli. Mol. Gen. Genet.* **239:**137–144.

302. Stahl, F. W. 1988. A unicorn in the garden. *Nature* (London) **335:**112–113.

303. Steinborn, G. 1978. Uvm mutants of *Escherichia coli* K12 deficient in UV mutagenesis. I. Isolation of *uvm* mutants and their phe-notypical characterization in DNA repair and mutagenesis. *Mol. Gen. Genet.* **165:**87–93.

304. Steinborn, G. 1979. Uvm mutants of *Escherichia coli* K12 deficient in UV mutagenesis. II. Further evidence for a novel function in error-prone repair. *Mol. Gen. Genet.* **175:**203–208.

305. Stewart, F. M., D. M. Gordon, and B. R. Levin. 1990. Fluctuation analysis: the probability distribution of the number of mutants under different conditions. *Genetics* **124:**175–185.

306. Story, R. M., I. T. Weber, and T. A. Steitz. 1992. The structure of the *E. coli recA* protein monomer and polymer. *Nature* (London) **355:**318–325.

307. Strauss, B. S. 1991. The "A rule" of mutagen specificity: a consequence of DNA polymerase bypass of non-instructional lesions. *Bioessays* **13:**79–84.

308. Streisinger, G., and J. Owen. 1985. Mechanisms of spontaneous and induced frameshift mutation in bacteriophage T4. *Genetics* **109:**633–659.

309. Strike, P., and D. Lodwick. 1987. Plasmid genes affecting DNA repair and mutation. *J. Cell Sci.* **6:**303–321.

310. Stukenberg, P. T., P. S. Studwell-Vaughn, and M. O'Donnell. 1991. Mechanism of the sliding clamp of DNA polymerase III holoenzyme. *J. Biol. Chem.* **266:**11328–11334.

311. Svoboda, D. L., C. A. Smith, J.-S. A. Taylor, and A. Sancar. 1993. Effect of sequence, adduct type, and opposing lesions on the binding and repair of ultraviolet photodamage by DNA photolyase and (A)BC excinuclease. *J. Biol. Chem.* **268:**10694–10700.

312. Swann, P. F. 1990. Why do O⁶-alkylguanine and O⁴-alkylthymine miscode? The relationship between the structure of DNA containing O⁶-alkylguanine and O⁴-alkylthymine and the mutagenic properties of these bases. *Mutat. Res.* **233:**81–94.

313. Sweasy, J. B., E. M. Witkin, N. Sinha, and V. Roegner-Maniscalco. 1990. RecA protein of *Escherichia coli* has a third essential role in SOS mutator activity. *J. Bacteriol.* **172:**3030–3036.

314. Symonds, N. 1989. Anticipatory mutagenesis? *Nature* (London) **337:**119–120.

315. Tadmor, Y., R. Ascarelli-Goell, R. Skaliter, and Z. Livneh. 1992. Overproduction of the β subunit of DNA polymerase III holoenzyme reduces UV mutagenesis in *Escherichia coli. J. Bacteriol.* **174:**2517–2524.

316. Tang, M.-S., J. Hrncir, D. Mitchell, J. Ross, and J. Clarkson. 1986. The relative toxicity and mutagenicity of cyclobutane pyrimidine dimers and (6-4) products in *Escherichia coli. Mutat. Res.* **161:**9–17.

317. Tanooka, H., K. Tanaka, and K. Shinozaki. 1991. Heterospecific expression of misrepair-enhancing activity of *mucAB* in *Escherichia coli* and *Bacillus subtilis. J. Bacteriol.* **173:**2906–2914.

318. Taylor, J.-S., and M. P. Cohrs. 1987. DNA, light, and Dewar pyrimidinones: the structure and biological significance of TpT3. *J. Am. Chem. Soc.* **109:**2834–2835.

319. Taylor, J.-S., D. S. Garrett, I. R. Brockie, D. L. Svoboda, and J. Telser. 1990. ¹NMR assignment and melting temperature of *cis-syn* and *trans-syn* thymine dimer containing duplexes of d(CGTATTATGC)·d(GCATAATACG). *Biochemistry* **29:**8858–8866.

320. Taylor, J.-S., D. S. Garrett, and M. P. Cohrs. 1988. Solution-state structure of the Dewar pyrimidinone photoproduct of thymidylyl-(3'-5')-thymine. *Biochemistry* **27:**7206–7215.

321. Tessman, E. S., and P. K. Peterson. 1985. Isolation of protease-proficient, recombinase-deficient *recA* mutants of *Escherichia coli* K-12. *J. Bacteriol.* **163:**688–695.

322. Tessman, E. S., I. Tessman, P. K. Peterson, and J. D. Forestal. 1986. Roles of RecA protease and recombinase activities of *Escherichia coli* in spontaneous and UV-induced mutagenesis and in Weigle repair. *J. Bacteriol.* **168:**1159–1164.

323. Tessman, I. 1976. A mechanism of UV reactivation, p. 87. *In* A. I. Bukhari and E. Ljungquist (ed.), *Abstracts of the Bacteriophage Meeting.* Cold Spring Harbor Laboratory, Cold Spring Harbor, N.Y.

324. Tessman, I. 1985. UV-induced mutagenesis of phage S13 can occur in the absence of the RecA and UmuC proteins of *Escherichia coli. Proc. Natl. Acad. Sci. USA* **82**:6614–6618.

325. Tessman, I., and M. A. Kennedy. 1991. The two-step model of UV mutagenesis reassessed: deamination of cytosine in cyclobutane dimers as the likely source of the mutations associated with photoreactivation. *Mol. Gen. Genet.* **227**:144–148.

326. Tessman, I., and M. A. Kennedy. 1994. DNA polymerase II of *Escherichia coli* in the bypass of apurinic sites *in vivo. Genetics* **136**:439–448.

327. Tessman, I., M. A. Kennedy, and S.-L. Liu. 1994. Unusual kinetics of uracil formation in single- and double-stranded DNA by deamination of cytosine in cyclobutane pyrimidine dimers. *J. Mol. Biol.* **235**:807–812.

328. Tessman, I., S.-K. Liu, and M. A. Kennedy. 1992. Mechanism of SOS mutagenesis of UV-irradiated DNA: mostly error-free processing of deaminated cytosine. *Proc. Natl. Acad. Sci. USA* **89**:1159–1163.

329. Thomas, S. M., H. M. Crowne, S. C. Pidsley, and S. G. Sedgwick. 1990. Structural characterization of the *Salmonella typhimurium* LT2 *umu* operon. *J. Bacteriol.* **172**:4979–4987.

330. Timms, A. R., H. Steingrimsdottir, A. R. Lehmann, and B. A. Bridges. 1992. Mutant sequences in the *rpsL* gene of *Escherichia coli* B/r: mechanistic implications for spontaneous and ultraviolet light mutagenesis. *Mol. Gen. Genet.* **232**:89–96.

331. Todd, P. A., and B. W. Glickman. 1982. Mutational specificity of UV light in *Escherichia coli*: indications for a role of DNA secondary structure. *Proc. Natl. Acad. Sci. USA* **79**:4123–4127.

332. Upton, C., and R. J. Pinney. 1983. Expression of eight unrelated Muc⁺ plasmids in eleven DNA repair-deficient *E. coli* strains. *Mutat. Res.* **112**:261–273.

333. Veaute, X., and R. P. P. Fuchs. 1993. Greater susceptibility to mutations in lagging strand of DNA replication in *Escherichia coli* than in leading strand. *Science* **261**:598–600.

334. Villani, G., S. Boiteux, and M. Radman. 1978. Mechanisms of ultraviolet induced mutagenesis: extent and fidelity of *in vitro* DNA synthesis on irradiated templates. *Proc. Natl. Acad. Sci. USA* **75**:3037–3041.

335. Voigt, J. M., B. Van Houten, and M. D. Topal. 1989. Repair of O⁶-methylguanine by ABC excinuclease of *Escherichia coli in vitro. J. Biol. Chem.* **264**:5172–5176.

336. Waleh, N. S., and B. A. D. Stocker. 1979. Effect of host *lex, recA, recF,* and *uvrD* genotypes on the ultraviolet light-protecting and related properties of plasmid R46 in *Escherichia coli. J. Bacteriol.* **137**:830–838.

337. Walker, G. C. 1977. Plasmid (pKM101)-mediated enhancement of repair and mutagenesis: dependence on chromosomal genes in *Escherichia coli* K-12. *Mol. Gen. Genet.* **152**:93–103.

338. Walker, G. C. 1978. Inducible reactivation and mutagenesis of UV-irradiated bacteriophage P22 in *Salmonella typhimurium* LT2 containing the plasmid pKM101. *J. Bacteriol.* **135**:415–421.

339. Walker, G. C. 1978. Isolation and characterization of mutants of the plasmid pKM101 deficient in their ability to enhance mutagenesis and repair. *J. Bacteriol.* **133**:1203–1211.

340. Walker, G. C. 1983. Molecular principles underlying the Ames *Salmonella*-microsome test: elements and design of short term tests, p. 15–39. *In* A. R. Kolber, T. K. Wang, L. D. Grant, R. S. DeWoskink, and T. J. Hughes (ed.), *In Vitro Toxicology Testing of Environmental Agents: Current and Future Possibilities.* Plenum Publishing Corp., New York.

341. Walker, G. C. 1983. Regulation and function of cellular gene products involved in UV and chemical mutagenesis in *E. coli*, p. 181–198. *In* C. W. Lawrence (ed.), *Induced Mutagenesis: Molecular Mechanisms and Their Implications for Environmental Protection.* Plenum Publishing Corp., New York.

342. Walker, G. C. 1984. Mutagenesis and inducible responses to deoxyribonucleic acid damage in *Escherichia coli. Microbiol. Rev.* **48**:60–93.

343. Walker, G. C. 1985. Inducible DNA repair systems. *Annu. Rev. Biochem.* **54**:425–457.

344. Walker, G. C. 1987. The SOS response of *Escherichia coli*, p. 1346–1357. *In* F. C. Neidhardt, J. L. Ingraham, K. B. Low, B. Magasanik, M. Schaechter, and H. E. Umbarger (ed.), *Escherichia coli and Salmonella typhimurium: Cellular and Molecular Biology.* American Society for Microbiology, Washington, D.C.

345. Walker, G. C., and P. P. Dobson. 1979. Mutagenesis and repair deficiencies of *Escherichia coli umuC* mutants are suppressed by the plasmid pKM101. *Mol. Gen. Genet.* **172**:17–24.

346. Walker, G. C., L. Marsh, and L. A. Dodson. 1985. Genetic analyses of DNA repair: inference and extrapolation. *Annu. Rev. Genet.* **19**:103–126.

347. Weigle, J. J. 1953. Induction of mutation in a bacterial virus. *Proc. Natl. Acad. Sci. USA* **39**:628–636.

348. Witkin, E. M. 1967. Mutation-proof and mutation-prone modes of survival in derivatives of *Escherichia coli* B differing in sensitivity to ultraviolet light. *Brookhaven Symp. Biol.* **20**:495–503.

349. Witkin, E. M. 1969. Ultraviolet-induced mutation and DNA repair. *Annu. Rev. Microbiol.* **23**:487–514.

350. Witkin, E. M. 1974. Thermal enhancement of ultraviolet mutability in a *tif-1 uvrA* derivative of *Escherichia coli* B/r: evidence that ultraviolet mutagenesis depends upon an inducible function. *Proc. Natl. Acad. Sci. USA* **71**:1930–1934.

351. Witkin, E. M. 1975. Persistence and decay of thermo-inducible error-prone repair activity in nonfilamentous derivatives of *tif-1 Escherichia coli* B/r: the timing of some critical events in ultraviolet mutagenesis. *Mol. Gen. Genet.* **142**:87–103.

352. Witkin, E. M. 1976. Ultraviolet mutagenesis and inducible DNA repair in *Escherichia coli. Bacteriol. Rev.* **40**:869–907.

353. Witkin, E. M., and D. L. George. 1973. Ultraviolet mutagenesis in *polA* and *uvrA polA* derivatives of *Escherichia coli* B/r: evidence for an inducible error-prone repair system. *Genetics* **73**(Suppl.):91–108.

354. Witkin, E. M., J. O. McCall, M. R. Volkert, and I. E. Wermundsen. 1982. Constitutive expression of SOS functions and modulation of mutagenesis resulting from resolution of genetic instability at or near the *recA* locus of *Escherichia coli. Mol. Gen. Genet.* **185**:43–50.

355. Wood, M. L., M. Dizdaroglu, E. Gajewski, and J. M. Essigmann. 1990. Mechanistic studies of ionizing radiation and oxidative mutagenesis: genetic effects of a single 8-hydroxyguanine (7-hydro-8-oxoguanine) residue inserted at a unique site in a viral genome. *Biochemistry* **29**:7024–7032.

356. Wood, M. L., A. Esteve, M. L. Morningstar, Kuziemko, and J. M. Essigmann. 1992. Genetic effects of oxidative DNA damage: comparative mutagenesis of 7,7-dihydro-8-oxoguanine and 7,8-dihydro-8-oxoadenine in *Escherichia coli. Nucleic Acids Res.* **20**:6023–6032.

357. Wood, R. D., and F. Hutchinson. 1984. Non-targeted mutagenesis in unirradiated lambda phage in *Escherichia coli* host cells irradiated with ultraviolet light. *J. Mol. Biol.* **173**:293–305.

358. Wood, R. D., and F. Hutchinson. 1987. Ultraviolet light-induced mutagenesis in the *Escherichia coli* chromosome: sequences of mutants in the *cI* gene of a lambda lysogen. *J. Mol. Biol.* **193**:637–641.

359. Wood, R. D., T. R. Skopek, and F. Hutchinson. 1984. Changes in DNA base sequence induced by targeted mutagenesis of lambda phage by ultraviolet light. *J. Mol. Biol.* **173**:273–291.

360. Woodgate, R. 1992. Construction of a *umuDC* operon substitution mutation in *Escherichia coli. Mutat. Res.* **281**:221–225.

361. Woodgate, R., and D. G. Ennis. 1991. Levels of chromosomally encoded Umu proteins and requirements for *in vivo* UmuD cleavage. *Mol. Gen. Genet.* **229**:10–16.

362. Woodgate, R., M. Rajagopalan, C. Lu, and H. Echols. 1989. UmuC mutagenesis protein of *Escherichia coli*: purification and interaction with UmuD and UmuD'. *Proc. Natl. Acad. Sci. USA* **86**:7301–7305.

363. Woodgate, R., and S. Sedgwick. 1992. Mutagenesis induced by bacterial UmuDC proteins and their plasmid homologues. *Mol. Microbiol.* **6:**2213–2218.

364. Yamamoto, K. 1985. Photoreactivation reverses ultraviolet-induced premutagenic lesions leading to frame-shift mutations in *Escherichia coli. Mol. Gen. Genet.* **201:**141–145.

365. Yamamoto, K., H. Shinagawa, and T. Ohnishi. 1985. Photoreactivation of UV damage in *Escherichia coli uvrA:* lethality is more effectively reversed than either premutagenic lesions of SOS induction. *Mutat. Res.* **146:**33–42.

366. Zambrano, M. M., D. A. Siegele, A. Marta, A. Tormo, and R. Kolter. 1993. Microbial competition: *Escherichia coli* mutants that take over stationary phase cultures. *Science* **259:**1757–1760.

367. Zeilstra-Ryalls, J., O. Fayet, and C. Georgopoulos. 1991. The universally conserved GroE (Hsp60) chaperonins. *Annu. Rev. Microbiol.* **45:**301–325.

12

DNA Damage Tolerance and Mutagenesis in Eukaryotic Cells

In this chapter, we will consider various processes that result in the tolerance of DNA damage in lower and higher eukaryotic cells, frequently at the cost of decreased genomic stability. As was the case when considering this topic in prokaryotes, we have divided this account into two primary sections that deal with *mutagenic responses* and with *recombinational processing*, the latter typically associated with the repair of DNA strand breaks. The fundamental concepts of mutagenesis and recombination are discussed in chapters 2, 10, and 11 in the context of prokaryotic cells. The utility of prokaryotic models notwithstanding, it must be emphasized that phenomena such as the *SOS response, SOS mutagenesis,* and *daughter-strand gap repair* cannot be simply extrapolated to considerations of the mechanisms of DNA damage tolerance in eukaryotic cells. As discussed in many of the previous chapters, *Escherichia coli* has served as an informative model for numerous aspects of DNA repair and DNA damage tolerance, and detailed insights into these processes have been gained as a result. However, a number of the more complex pathways affecting genomic stability in eukaryotes will almost certainly turn out to be mechanistically different in many respects.

The discussions in chapters 10 and 11 have illustrated how detailed information concerning genes and gene products in *E. coli* has led to testable models of the mechanisms by which mutations arise as a consequence of DNA damage and their biological significance. The situation in eukaryotic cells has not yet approached this level of understanding. Besides the limitations imposed by formidable technical limitations when working with many eukaryotes, especially cells from higher organisms, a major source of uncertainty stems from the lack of a sound genetic framework. In this necessarily incomplete accounting of DNA damage tolerance, we will focus on studies that at least hint at molecular mechanisms of mutagenesis rather than on more descriptive elements such as detailed analysis of mutational spectra. Additionally, rather than discussing the mutational specificity of a wide range of DNA-damaging agents, we have elected to focus primarily on the effects of UV radiation at 254 nm.

DNA Damage-Induced Mutagenesis in the Yeast *Saccharomyces cerevisiae*

The Genetic Framework

The best-understood eukaryotic model system for DNA damage-induced mutagenesis is the yeast *Saccharomyces cerevisiae*. Several mutants that are defective in genes required for damage-induced mutagenesis have been well characterized, and their corresponding gene products have been identified and in some cases purified. We will dwell on this organism to a considerable degree with the express intent of exploring the essential elements of damage-induced mutagenesis in a well-defined eukaryotic system. As we have seen in previous chapters, *S. cerevisiae* has served as an instructional model for several other aspects of cellular responses to DNA damage in eukaryotes in general.

The enhanced mutagenesis that results from the treatment of yeast cells with many DNA-damaging agents involves the active participation of multiple cellular functions (167, 311, 339, 471). The extent to which these functions depend on damage-inducible proteins (and hence at least superficially resemble the *E. coli* SOS system), is uncertain (see below) (548). We shall begin with a consideration of genes that are implicated in DNA damage-induced mutagenesis in *S. cerevisiae*, as evidenced by the fact that mutants defective in these genes have a reduced probability of mutation after treatment with DNA-damaging agents (167, 311, 339, 471). Almost all of the mutants studied fall into the heterogeneous *RAD6 epistasis group* (468). Some of these mutants carry the familiar *rad* designation because they were originally identified among those isolated in screens for enhanced sensitivity to radiation or chemical agents; others have turned out to be *cdc* mutants with defects in essential functions required for cell cycle progression. As with *E. coli* (see chapter 11), genetic screens have also been devised to specifically identify strains that are defective in mutation induction (323, 336, 338).

THE *RAD6* GENE

Mutants defective in the *RAD6* gene are abnormally sensitive to a variety of DNA-damaging agents including UV radiation, γ rays, and alkylating and cross-linking agents (99, 174, 220, 464, 465). Remarkably, *rad6* mutants show markedly *reduced* mutability after treatment with many DNA-damaging agents, regardless of the type of damage or type of mutation scored (311, 313, 325, 340, 382, 464, 468). However, this does not include induction of mitochondrial mutations by UV radiation (461). Comparable to the situation with *umuDC* mutants of *E. coli* (see chapter 11), the mutagenicity of base analogs such as 6-*N*-hydroxyaminopurine does not depend on a functional *RAD6* gene (72). On the other hand, *spontaneous mutability* is enhanced in *rad6* mutants, as are spontaneous recombination and damage-induced recombination (209, 238, 283). In particular, spontaneous mutation due to retrotransposition of the yeast transposable element Ty is significantly increased in a *rad6* mutant background (270, 452).

The characterization of different *rad6* alleles and of suppressors of *rad6* mutants indicates a multifunctional role for the *RAD6* gene product. For example, whereas haploid strains carrying the *rad6-1* or *rad6-3* alleles share all the characteristics listed above, *rad6-1/rad6-1* but not *rad6-3/rad6-3* diploids are additionally deficient in sporulation (398). Furthermore, suppressors which separate the role of *RAD6* in mutagenesis from its role in the repair of UV radiation-induced DNA damage have been isolated. In particular, suppressors of the sensitivity of *rad6* mutants to UV radiation and the antifolate drug trimethoprim have been identified which do not affect defective damage-induced mutagenesis, sporulation, or sensitivity to γ radiation (317, 606).

Experiments in a nucleotide excision repair-deficient background have indicated a defect in postreplication repair in *rad6* mutants (466), which probably accounts for the UV radiation sensitivity. However, given that the defect in DNA repair can be genetically separated from the defect in mutagenesis, this phenotype does not allow conclusions on the mechanism of *RAD6*-dependent mutational processing. Chromosomal *rad6* disruption mutants are viable but have a low plating efficiency and a prolonged S phase (305).

The *RAD6* transcript is UV radiation inducible and is characterized by enhanced steady-state levels early in meiosis (361). The *RAD6* open reading frame encodes a protein of 19.7 kDa containing a highly acidic C terminus of 13 consecutive Asp residues (467, 490). This protein has been shown to be one of five E2 enzymes in the yeast ubiquitin-ligase system, which catalyzes the transfer of activated ubiquitin to histones H2A and H2B (255). Ubiquitin is an evolutionarily conserved polypeptide of 76 amino acids. In general, conjugation of proteins to ubiquitin triggers their degradation, but the reaction may also play a role in modifying protein function (222, 254, 256). In the first step of the reaction, ubiquitin is activated by adenylation and transferred to a thiol site of the catalyzing enzyme, the so-called E1 enzyme, an essential gene in *S. cerevisiae* (380) (Fig. 12–1). In a subsequent step, ubiquitin is transferred to a cysteine residue of one of several E2 enzymes, which can then catalyze the direct ubiquitination of certain target proteins, such as histones (451) (Fig. 12–1). In other cases, an amino-end-recognizing E3 enzyme first binds to potential proteolytic substrates, and it is this complex which is recognized by the ubiquitin-E2 complex (83) (Fig. 12–1). This reaction leads to the sequential addition of several ubiquitin units. Thus, the E3 enzymes play an important role in selecting substrates for ubiquitination (32). The specificity of their binding sites constitutes the basis for the "amino-end rule" (22), which defines the structure of protein-destabilizing N termini. Indeed, in addition to ubiquitinating histones H2A and H2B in vitro (255, 577), purified Rad6 protein catalyzes the ubiquitination and degradation of protein substrates containing destabilizing N termini in an E3-dependent reaction (114, 576).

There is some similarity between the phenotypes of *rad6* mutants and those defective in other ubiquitin transactions. Additionally, the transcript of the polyubiquitin gene *UBI4* (encoding a single polypeptide consisting of multiple ubiquitin moieties) is inducible by DNA-damaging agents and by heat shock (605). Deletion

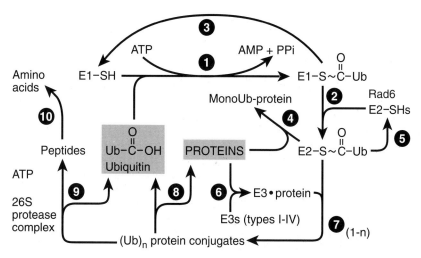

Figure 12–1 Modification and degradation of proteins by the ubiquitin system. The scheme depicts the following steps: activation of ubiquitin by an E1 enzyme (step 1), transfer of activated ubiquitin to an E2 enzyme such as Rad6 (step 2), recycling of E1 (step 3), conjugation of a protein to ubiquitin in an E2-mediated reaction (step 4), recycling of E2 (step 5), formation of an E3-protein complex (step 6), conjugation of multiple molecules of ubiquitin to the protein substrate in an E3-mediated reaction (step 7), recycling of intact substrates and ubiquitin (step 8), ATP-dependent degradation of conjugates into peptides (step 9), and release of amino acids from peptides (step 10). (*Adapted from Ciechanover [83] with permission.*)

of *UBI4* also results in defective sporulation and additionally causes abnormal sensitivity to heat stress and starvation (156). The yeast E3 enzyme is encoded by the *UBR1* gene, and a *ubr1* deletion mutant is characterized by defective sporulation and slow growth (32). Another E2 enzyme has been identified as the product of the *CDC34* gene, which plays an essential role during the G_1-S transition of the cell cycle and shares amino acid similarity with *RAD6* (190). These similarities notwithstanding, defective induction of mutations following exposure to DNA-damaging agents is apparently unique to *rad6* mutants.

Given the general conservation of the ubiquitin ligase system among eukaryotes, it is not surprising that homologs of the *RAD6* gene have been identified. Single-copy genes with extensive amino acid sequence identity have been characterized in *Schizosaccharomyces pombe* (*rhp6⁺*) (489) and *Drosophila melanogaster* (*Dhr6*) (292). In humans, two *RAD6* genes (*HHR6A* and *HHR6B*) (for human homolog of *RAD6*) which share 95% amino acid identity have been identified (293, 294). *HHR6B* is identical to *E214K*, the only E2 enzyme in rabbit reticulocyte lysates known to catalyze ubiquitination in an E3-dependent fashion (638). The acidic tail is unique to the *S. cerevisiae RAD6* gene but does not seem to be important for mutagenesis, since the *Drosophila* and human genes can complement the DNA repair and mutagenesis defects in a *rad6* deletion mutant. In contrast, these *RAD6* homologs weakly complement the sporulation defect of *rad6* mutants (293).

It was initially considered that *RAD6*-mediated ubiquitination of chromosomal proteins, specifically of histones, is essential for chromatin remodeling and that this might explain the various phenotypes of *rad6* mutants (539). However, deletion of the acidic tail, which is essential for the ubiquitination of histones, only inactivated sporulation. Other DNA repair functions and damage-induced mutagenesis were not affected by this deletion (404, 577). In contrast, a mutant with a point mutation of Cys-88 (the residue that forms a thioester conjugate with ubiquitin) resembles a *rad6* deletion mutant in all its phenotypes (578). These results suggest that Rad6-mediated modification of histone is essential only for sporulation and that the role of Rad6 protein in DNA repair and mutagenesis probably involves the modification of nonhistone proteins.

Is the E3-dependent ubiquitination of nonhistone target proteins essential for induced mutagenesis? Ubiquitination could conceivably change the biochemical activity of target molecules and, for example, lead to the alteration, dissociation, or degradation of a stalled DNA replication complex (471). This question was investigated in the *rad6Δ1–9* mutant, which contains a deletion of the highly conserved N terminus that results in a failure of Rad6 to interact with the *UBR1*-encoded E3 enzyme and a defect in amino-end rule-dependent protein degradation (629). Furthermore, engineered β-galactosidase proteins containing destabilizing N termini (Arg, Leu) are indeed more stable in a *rad6Δ1–9* mutant background (629) (Fig. 12–2). This mutant is also defective in sporulation and is moderately UV radiation sensitive. However, it is not defective in damage-induced mutagenesis (629), indicating that an interaction with the *UBR1*-encoded E3 enzyme is not necessary for the role of Rad6 protein in such mutagenesis.

The discovery of a biochemical function for Rad6 protein initially generated anticipation of major insights into the mechanism of DNA damage-induced mutagenesis in *S. cerevisiae*. However, the function of this protein in such mutagenesis is not correlated with its ability to ubiquitinate histones. Furthermore, unless one postulates some residual ability to ubiquitinate certain substrates in the *rad6Δ1–9* strain, no data directly link the ubiquitination of target proteins in E3-dependent reactions to induced mutagenesis. Hence, the role of this multifunctional protein in the mutagenic processing of DNA damage remains obscure. Its interaction with Rad18, another DNA repair protein considered in the next section, may be crucial for its mutagenic activity.

THE *RAD18* GENE

Like *rad6* mutants, *rad18* mutants are highly sensitive to UV light, γ rays, and a variety of chemical agents (174, 216). These mutants are also defective in postreplication repair after UV radiation treatment (108, 466) but are proficient in

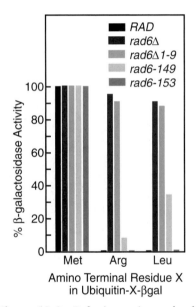

Figure 12–2 Defective amino-end rule-dependent degradation pathway in *rad6* deletion mutants. The steady-state levels of different ubiquitin–X–β-galactosidase constructs depend on the amino acid X to which ubiquitin is fused. Compared with Met (left), Arg and Leu are destabilizing residues in the wild-type strain (*RAD*) but not if Rad6 is completely deleted (*rad6Δ*) or deleted for the 9 most N-terminal residues (*rad6Δ1–9*) which are highly conserved among Rad6 homologs. *Rad6-149* and *rad6-153* are mutations that result in deletion of the last 23 or 19 amino acids of the acidic tail, respectively. Rad6-153 but not Rad6-149 retains normal levels of E3-dependent protein ubiquitination and degradation activity. Indeed, the destabilizing effect of Arg or Leu in *rad6-153* approaches that observed in the wild-type strain. (*Adapted from Watkins et al. [629] with permission.*)

nucleotide excision repair (492). Strand breaks induced by ionizing radiation are rejoined, but the reconstitution of high-molecular weight DNA during postirradiation incubation is defective (407). These mutants are also defective in the repair of clustered γ-ray-induced base damage (132). Spontaneous mutation and mitotic recombination are enhanced in *rad18* mutants (50, 474). The mutator effect specifically increases G·C → T·A transversions (303).

In contrast to *rad6* mutants, *rad18* mutants were initially reported not to be defective in sporulation, meiotic recombination, and UV radiation-induced mutability (118, 174, 311, 313, 326, 486). Hence, *RAD18* had been postulated to function in a nonmutagenic branch of the complex pathway(s) represented by the *RAD6* epistasis group. However, the phenotype of UV radiation-induced mutability has been reexamined with an *ochre* allele reversion system (73) which allows the distinction between locus-specific revertants of an auxotrophic marker and tRNA suppressor mutants (213). Defective DNA damage-induced locus-specific reversion was found to be comparable to that observed in *rad6* deletion mutants (73). A significant defect in UV radiation-induced mutagenesis in *rad18* mutants has also been reported when a plasmid-borne mutagenesis system was used (17).

Cloning of the *RAD18* gene revealed an open reading frame which can encode a protein of 55.5 kDa (77, 151, 264). Unlike the Rad6 polypeptide, no homology with ubiquitin-conjugating enzymes is evident. A consensus nucleotide-binding domain, as well as three regions resembling zinc fingers, has been identified (77, 264). The spacing of two acidic regions is similar to that of known transcriptional activator proteins of *S. cerevisiae*. These structural features hint at a role of Rad18 as a DNA-binding protein which either is directly involved in some aspect of DNA repair or functions indirectly as a transcription factor. Rad18 binds to single-stranded DNA (25). As is the case with the *RAD6* gene, the *RAD18* transcript is inducible following exposure of yeast cells to UV radiation but does not fluctuate during the mitotic cell cycle (263).

Immunoprecipitation experiments have demonstrated an interaction between Rad18 and Rad6 proteins (25). The biological significance of this interaction is suggested by several observations. For example, overexpression of the mutant protein Rad6/Ala-88 (which is devoid of ubiquitinating activity) confers a high degree of UV radiation sensitivity to wild-type cells. However, this protein retains the ability to interact with Rad18 protein and concomitant overexpression of Rad18 can alleviate the enhanced sensitivity to UV radiation. It has been suggested that the DNA-binding activity of Rad18 may be used to load Rad6 protein onto sites of DNA damage where its ubiquitin-conjugation activity could subsequently modify a protein or protein complex, e.g., a stalled DNA polymerization complex (25).

THE *SRS2* (*RADH*) GENE

The suppressors of UV radiation sensitivity of the *rad6* point and deletion mutants mentioned above have been mapped to a single gene originally named *SRS2* (for suppressor of Rad six) (531). Mutants with mutations in this gene have been independently isolated in two additional genetic screens. They were detected as suppressors of the UV radiation sensitivity of *rad18* deletion mutants (*radH*) (2) and as mutants characterized by an increased frequency of gene conversion (7, 502). In a *RAD6*⁺ background, all of these mutants manifest impaired mutation induction as well as slightly enhanced sensitivity to UV radiation (1). Interestingly, this enhanced sensitivity is confined to G₁ cells (1).

It has been suggested that the *SRS2* gene product plays a pivotal role in channeling single-strand gaps opposite DNA lesions that arise during replication into various repair pathways (1, 502, 531). This channeling function is presumed to initiate mutagenic processing by the *RAD6/RAD18* pathway and to compete with recombinational repair that is dependent on the *RAD52* epistasis group of genes (see below). This suggestion derives from the observation that the suppression of the UV radiation sensitivity of *rad6* mutants which is caused by the absence of a functional *SRS2* gene depends on a functional *RAD52* gene (531). Semidominant suppressors of the UV and γ radiation sensitivity of *SRS2* mutants were found to

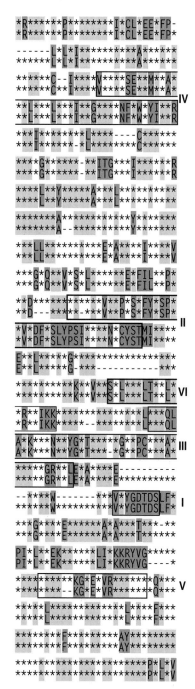

Figure 12–3 Alignment of the amino acid sequences of *S. cerevisiae* Rev3 (upper rows, residues 679 to 1315) and Epstein-Barr virus DNA polymerase (lower rows, residues 279 to 923). Identical residues are in red, and conservative amino acid substitutions are in grey. The numbered boxes indicate regions of homology that are conserved among several DNA polymerases (also see Fig. 8–20) (640). (*Adapted from Morrison et al. [403] with permission.*)

map to the *RAD51* gene (1), considered to be a yeast homolog of the *E. coli recA* gene (1, 547) (see chapter 10 and below). The single-stranded DNA-binding activity of the latter protein (547) may be essential for this effect.

The *SRS2* gene encodes a polypeptide with homology to the *E. coli* DNA helicases UvrD and Rep (2). The severity of the UV radiation-sensitive and hyperrecombination phenotypes of *srs2* mutants correlates with mutations in various helicase consensus amino acid motifs. Additionally, some alleles reduce spore viability (437). Srs2 protein has been purified and DNA helicase and ATPase activities have been demonstrated in vitro (501). Hence, as is the case in nucleotide excision repair in *S. cerevisiae*, DNA helicases have emerged as prominent catalytic activities in DNA damage tolerance pathways in *S. cerevisiae*.

THE *REV* GENES

Genetic screens for mutants characterized by reduced reverse mutation frequencies resulted in the isolation of a class of mutants designated *rev* (for defective mutation reversion) (323, 336). The related but less well characterized *umr* mutants were identified in a screen for defective forward mutability (338). Detailed characterization of the *rev* mutants has revealed important features of mutagenic processing in yeast cells. Some mutants (*rev3*, *rev6*, and *rev7* mutants) show a general defect in DNA damage-induced mutagenesis (311, 318, 321, 323–325, 336, 337, 339, 383, 465), whereas others are characterized by reduced mutation probability only for certain mutational systems (typically highly revertible nonsense alleles) (311, 314, 315, 323, 325, 336, 337, 339, 383). The results of reversion studies with certain mutant alleles of the *CYC1* gene (encoding iso-1-cytochrome C) (314–316, 339) indicate a general defect in mutation induction in *rev3* mutants, whereas *rev1* and *rev2* (= *rad5*) mutants show reduced mutation frequencies only for certain alleles. There is no correlation with the type of mutation scored, its position in the gene, or the immediate sequence context. It can be concluded from the few cases in which genes have been totally deleted that this phenomenon is not due to leakiness of the characterized *rev* point mutant alleles. Thus, no explanation for this surprising locus specificity of yeast mutagenic processes can yet be given.

The *REV3* gene is of special interest because of the dramatic and very general defect of *rev3* mutant strains in damage-induced and spontaneous mutability as measured in forward, frameshift, or reversion systems (311, 318, 325, 337, 339, 465, 474). However, these mutants are only slightly UV radiation sensitive and are normal with respect to overall postreplication repair (466). The cloned *REV3* gene encodes a predicted protein of ~173 kDa (403). The amino acid sequence has regions of homology with known DNA polymerases, in particular Epstein-Barr virus and herpes simplex virus DNA polymerases (403) (Figs. 6–32 and 12–3). Over a stretch of 673 residues there is 23.9% amino acid identity between the Rev3 protein and the Epstein-Barr virus DNA polymerase. The homology increases to >60% similarity if conservative exchanges are included (Fig. 12–3). Within the same segment, six motifs which are highly conserved among template-directed DNA polymerases (640) are also found in the correct order (Fig. 8–20 and 12–3). There are also general structural similarities between Rev3 protein and yeast and human DNA polymerases α, including a cysteine-rich C-terminal domain that may encode two zinc finger DNA-binding regions.

Direct evidence for a DNA polymerase activity of Rev3 protein has not been reported. If Rev3 protein is indeed a DNA polymerase, it must be considered a nonessential activity, since deletion of the gene is not lethal in haploid strains. Hence, this putative polymerase might conceivably function exclusively in some kind of translesion DNA synthesis mode. The contribution of the presumed Rev3 activity to overall repair synthesis remains unknown; it is apparently dispensable for postreplication repair of UV radiation damage (466), and base excision repair proceeds normally in cell extracts without functional Rev3 protein (628).

REV1 is another candidate for a role in translesion synthesis. This nonessential gene encodes a predicted protein of 112 kDa containing an internal stretch of 152 residues with 25% identity to the *E. coli* UmuC protein (309) (see chapter 11 and Fig. 11–12). It is interesting that overexpression of the $umuD^+C^+$ homolog $mucA^+B^+$ weakly enhances spontaneous and DNA damage-induced mutability of certain reverse mutation systems in *S. cerevisiae* (462). This result raises the possibility that a UmuC-like function plays a role in mutagenesis in *S. cerevisiae*. Rev1 protein is not induced by DNA-damaging agents or by heat shock (309).

The molecular cloning of the *REV2* gene (which is allelic to *RAD5*) revealed a third potential component of a replicational process (9, 261). *REV2* is a nonessential gene that can encode a predicted protein of 134 kDa containing the seven signature domains that characterize a subfamily of potential helicases which includes *ERCC6*, *RAD16*, *RAD54*, *SNF2*, *MOT1*, and the *Drosophila BRM* gene (see chapter 6 and Fig. 6–31). Additionally, the sequence contains a *RAD18*-like zinc finger motif and a potential leucine zipper preceded by a basic region. The *S. pombe rad8+* gene also shares amino acid sequence homology with *REV2*, indicating the evolutionary conservation of another protein involved in mutagenesis (113).

The cycloheximide sensitivity of mutation fixation in a thermoconditional *rev2* mutant suggested the UV radiation inducibility of the *REV2* gene (551, 552), but such induction has not been directly demonstrated. The stability of simple repetitive sequences is enhanced in *rev2* deletion mutants (261). Since the instability of these sequences is probably the result of template slippage during replication (see chapters 2 and 9), this observation hints at a role of *REV2* in a DNA replication complex. Deletion mutants are characterized by a greater sensitivity to UV radiation than that observed with *rev3* or *rev1* deletion mutants (261). The *REV2* gene product does not appear to act epistatically with Rev3 or Rev1 (261). It has therefore been suggested that *REV2* functions primarily in a *RAD18*-related, largely nonmutagenic postreplication repair mechanism.

THE *CDC7* GENE

CDC7 encodes an essential protein kinase which is associated with replication complexes (230, 248, 444). Thermoconditional *cdc7* mutant cells arrest during the $G_1/$S transition of the cell cycle when progression through the S phase no longer depends on de novo protein synthesis (207). Cdc7 kinase activity is confined to the G_1/S phase, although the protein level stays constant throughout the cell cycle (243). The activity is most probably posttranslationally regulated through an association with the Dbf4 protein (117, 243). Certain *cdc7* mutants have been reported to be slightly UV radiation sensitive and defective in induction of auxotrophic revertants at permissive temperatures (285, 422). Overexpression of Cdc7 leads to increased mutability (538). A systematic examination of a series of isogenic *cdc7* mutants indicated that some are hypomutable whereas others are hypermutable, showing up to 10-fold-higher damage-induced mutation frequencies (229). Hyper- as opposed to hypomutability does not correlate with kinase activity or stability of the mutant proteins. It has been suggested that differences in substrate affinities may account for these differential effects (229).

Timing and Regulation of UV Radiation-Induced Mutagenesis

As is generally the case, nucleotide excision repair in *S. cerevisiae* is typically an error avoidance process. Hence, mutants defective in this DNA repair mode are characterized by dramatically increased mutation frequencies (129, 216). However, nucleotide excision repair is not necessarily completely error free. Several lines of evidence suggest that this process has a profound effect on the timing and potentially the mechanism and regulation of mutation fixation in yeast cells. UV radiation-induced mutations arise with a high probability prereplicatively in excision-proficient cells and predominantly postreplicatively in the absence of nucleotide excision repair (a similar situation pertains in *E. coli* [see chapter 11]). In excision-deficient yeast cells, photoproducts can be a source of mutations (and possibly sister chromatid recombination [268]) during several rounds of DNA replication (246). Mutagenic replicative bypass of 8-methoxypsoralen monoadducts has also been demonstrated in excision-deficient yeast cells during at least one round of replication (76).

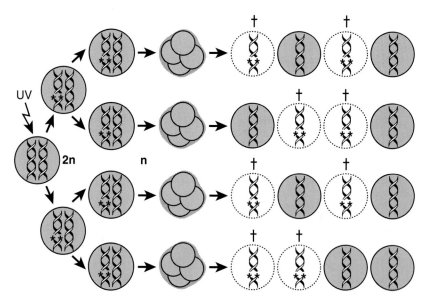

Figure 12–4 Pedigree analysis as a tool for determining the timing of fixation of UV radiation-induced mutations. Diploid *S. cerevisiae* cells are UV irradiated in stationary phase (i.e., with a G_1 DNA content). After two cell divisions, the four daughter cells of the microcolony are separated by micromanipulation and the derived clones are subjected to sporulation. The four meiosis products are analyzed for colony-forming ability. Inviable spores (indicated by unfilled circles with dotted outline) reflect the presence of recessive lethal mutations (red stars). The presence of inviable spores in the haploid progeny of all four diploid daughter cells (as shown in the diagram) demonstrates the fixation of a sequence alteration which caused recessive lethality in both DNA strands prior to the first cell division after UV irradiation (prereplicative mutation fixation). In postreplicative fixation, only one of the two daughter cells resulting from the first division of the diploid cell will inherit the sequence change. Consequently, two of the four daughter cells will not give rise to inviable spores.

These conclusions on the different timing of mutation fixation in UV treated excision-deficient versus excision-proficient cells have been inferred from several experimental systems.

1. A pedigree analysis was performed to detect the time of fixation of recessive lethal mutations in *S. cerevisiae* (245). Diploid cells were UV irradiated in the G_0 phase, the cells were separated after two rounds of replication, and the four derived clones were sporulated (Fig. 12–4). The existence of inviable spores in the meiosis products of all four clones was indicative of prereplicative fixation of a recessive lethal mutation in both DNA strands. This was true for two thirds of all pedigrees studied in excision-proficient cells after low doses of UV radiation (245).

2. A forward-mutation assay with a color marker allowed the distinction between pure and mosaic mutant clones (128, 131). An *ade2* mutant forms red-pigmented colonies owing to the accumulation of an intermediate of adenine metabolism. The presence of additional mutations in any gene in the same pathway upstream of *ADE2* prevents red coloration (Fig. 12–5). A substantial fraction of mutant clones recovered from UV-irradiated excision-proficient cells formed pure-white colonies, indicative of prereplicative fixation of a sequence alteration in both DNA strands, whereas mainly red-and-white sectored colonies were observed after treatment of excision-deficient cells (128, 131). The same was true for treatment with the alkylating agent methyl methanesulfonate (MMS); however, mainly sectored cloned were found after treatment with ethyl methanesulfonate (EMS) or nitrous acid (416).

A

B

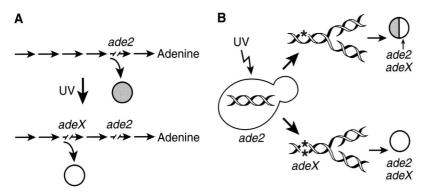

Figure 12–5   The *ade2 adeX* mutational system for determining the timing of mutation fixation. (A) *S. cerevisiae ade2* mutants form red colonies as a result of the accumulation of an intermediate of the adenine biosynthetic pathway. Inactivation of a metabolic step upstream of *ade2* by a second mutation (*adeX*) prevents the formation of the red pigment; hence, the colony is white. (B) If stationary-phase haploid *ade2* cells are UV irradiated and an *adeX* mutation has undergone mutational fixation in both DNA strands prereplicatively (lower section), a pure-white colony will result. If the sequence alteration affected only one DNA strand before replication, one daughter cell will show the *ade2* phenotype and (if this cell remains viable) a red-and-white sectored colony will result (upper section).

3. UV radiation-induced reversion of an auxotrophic marker was scored on medium containing traces of the required amino acid, thus allowing a limited number of divisions (286). This modification of the selection medium resulted in a dramatic enhancement of the mutant yield in excision-deficient strains, thus indicating the dependence of mutation fixation on replication in these cells. The effect was negligible for excision-proficient cells (286).

4. The timing of mutation fixation in a thermoconditional *rev2* mutant was determined by measuring the decreasing inhibitory effect of a temperature shift during postirradiation incubation at permissive temperature (551). *REV2*-dependent mutation fixation was substantially slower in an excision-deficient background (550). Again, this effect can be explained by a dependence of *REV2*-dependent mutation fixation on replication in the absence of nucleotide excision repair.

In contrast to what is observed in *E. coli* (639), dose-response curves for excision-deficient yeast cells show a linear increase of damage-induced mutation frequencies over a range of UV radiation doses (129, 216). Hence, mutation probability per dose increment remains constant and there is no evidence of an increase with increasing doses, an indication of regulated mutagenic processes. In haploid and diploid excision-proficient cells, however, this situation is different, at least for locus-specific reversion of auxotrophic markers. Mutation induction curves follow linear-quadratic kinetics, and mutation probability per dose increment increases with increasing dose (214–216) (Fig. 12–6). The quadratic component has been interpreted as the consequence of inducible mutagenic processes, while the linear component (which predominates at low doses) suggests the operation of a constitutive mutagenic mechanism (214, 215). This interpretation is supported by the observation that extended postirradiation incubation under nongrowth conditions in the presence of the protein synthesis inhibitor cycloheximide reduces the quadratic component (130) (Fig. 12–6). Additionally, a *rev2* mutant that is potentially deficient in an inducible (regulated) component of the mutagenic processing of UV radiation lesions abolishes the nonlinear component of the dose-response curve (549). Further support derives from UV radiation pretreatment studies. An "inducing" dose of UV radiation followed by the administration of a larger dose after a suitable time interval enhances survival and increases mutability in excision-proficient cells (130).

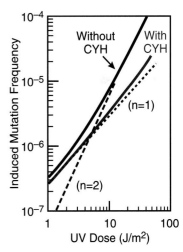

Figure 12–6   The dose dependency of reverse mutation induction in a UV-irradiated diploid wild-type strain of *S. cerevisiae* follows a biphasic linear-quadratic kinetics; i.e., with a double-logarithmic scale, the dose-response curve approaches a slope of $n = 1$ at low doses and a slope of $n = 2$ at higher doses (black line). Postirradiation incubation in the presence of the protein synthesis inhibitor cycloheximide (CYH) reduces the quadratic component (red line). (*Adapted from Eckardt et al. [130] with permission.*)

Contributions from Mutational Spectra

THE *SUP4*-o MUTATIONAL SYSTEM

In *S. cerevisiae*, plasmid-borne systems as well as chromosomally integrated gene probes have been used for mutational analyses (330, 453). Because of its technical ease of use, the *SUP4*-o-based mutational shuttle vector system is particularly popular. The *SUP4*-o gene encodes an ochre suppressor allele of a tyrosine-tRNA species of 89 residues and provides a convenient forward-mutation system for establishing mutational spectra (453). The gene has been placed in the vicinity of an M13 origin of replication on a yeast-*E. coli* shuttle vector which behaves as a single-copy centromeric plasmid in yeast cells. It is typically carried in a strain containing several ochre suppressible markers, e.g., *ade2-1*, *can1-100*, and *lys2-1*, which confer red pigmentation, canavanine sensitivity, and lysine auxotrophy, respectively, if not suppressed by *SUP4*-o. Mutant colonies showing loss of suppression of all three markers are indicative of the operation of a suppressor-inactivating mutation in the *SUP4*-o gene. These colonies are isolated, and the plasmids are rescued and sequenced.

DNA sequence analysis of mutants induced by UV light at 254 nm has led to the following conclusions (304) (Fig. 12–7).

1. The distribution of UV radiation-induced mutations differed from that of spontaneous mutations and was significantly more biased toward certain hot spots.

2. All kinds of mutations were recovered after UV radiation treatment, including single-base-pair deletions (~2%) and double-base-pair changes in the same molecule (~6%). However, the majority (84%) were transition mutations.

3. All transitions and the majority of transversions occurred at adjacent pyrimidine residues and were approximately equally frequent at TC, CC, and TT sites, with only a few events at CT sites.

4. In the majority of transitions the substitution took place opposite the 3′ residue of the pyrimidine dinucleotide.

As previously discussed for *E. coli* (see chapter 11), it can be concluded from these data that the majority of UV radiation-induced mutations are targeted opposite pyrimidine dimer photoproducts. Mechanisms that could account for the preference of transitions are discussed in detail in chapter 11. This bias may reflect the frequently observed preference of certain DNA polymerases to insert adenine opposite base adducts with limited or absent coding capacity in vitro (sometimes referred to as the *A-rule* [567]), the enhanced probability of deamination of cytosine in pyrimidine dimers (587), or a limited coding capacity of cyclobutane dimers (312). The tendency to insert the correct base opposite the 5′ photoproduct residue but not the 3′ residue can be explained by a limited base pairing capability of the 5′ residue.

It is difficult to determine the precise nature of the mutagenic photoproduct. So-called (6-4) photoproducts (see chapters 1 and 11) are formed at TC and CC sites almost exclusively (60), and these have indeed been found to be mutational hot spots in this system. However, TT sites accounted for about one-third of the transitions recovered. Indeed, a subsequent study on the effect of photoreactivation indicated that cyclobutane pyrimidine dimers are the predominant mutagenic lesions, since the UV radiation mutability of *SUP4*-o (including that attributed to TC hot spots) was efficiently reduced by photoreactivating light (19) (Fig. 12–7). As discussed in chapter 11, it has been argued that photoreactivation might indirectly affect the mutagenicity of (6-4) products by an overall reduction of the content of cyclobutane dimers in the genome, thereby interfering with the activation of inducible processes potentially required for UV radiation mutagenesis (59). If so, one would expect an *overall* reduction in the efficiency of mutation induction by photoreactivating light. However, the antimutagenic effect varied from site to site (Fig. 12–7) (19). Similar mutational studies performed after exposure to UV-B light (285 to 320 nm) yielded similar results, suggesting that the

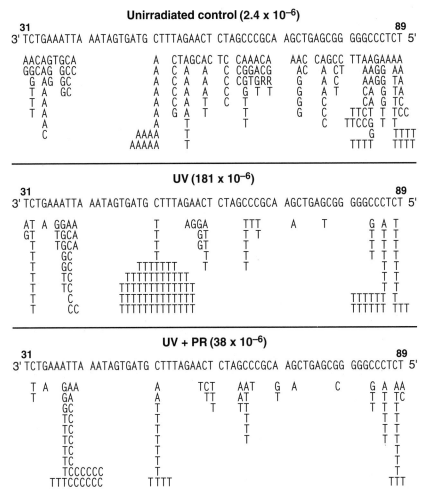

Figure 12–7 Distribution of single-base-pair substitutions in a portion of the *SUP4*-o tRNA gene (bp 31 to 89) of an *S. cerevisiae* wild-type strain treated with 254-nm UV radiation with and without subsequent photoreactivation (PR). Only the region of the transcribed strand encoding the tRNA is shown (the anticodon is at bp 36 to 38). The number of base substitutions shown reflects the number of analyzed events, not the total base pair substitution frequencies. These are indicated in red. (*Data from Armstrong and Kunz [19] and Kang et al. [270]. Courtesy of B. A. Kunz.*)

mutagenicity of this longer wavelength of radiation is also associated with cyclo-butane pyrimidine dimers (18).

STUDIES WITH DNA DAMAGE AT DEFINED SITES

Problems in establishing accurate mutational spectra resulting from biases introduced by any kind of selection were discussed in detail in chapters 2 and 11. In *S. cerevisiae* as in *E. coli*, a strategy used to avoid mutant selection bias has been to introduce defined photoproducts in fixed positions in DNA molecules and then to analyze all products of the presumed replicative bypass of the lesions (26, 27, 322). Additionally, such an analysis yields information on the overall bypass probability and on the frequency of error-free replication past a specific lesion.

Chemically synthesized TT cyclobutane dimers in the *cis-syn* and *trans-syn* configurations (see chapter 1 and Fig. 1–23) have been introduced into a single-stranded yeast-*E. coli* shuttle vector (322). A comparison of the transformation frequency with that achieved with the identical lesion-free plasmid provides a measure of the overall replicative bypass frequency (Table 12–1). In *E. coli*, such bypass is highly dependent on SOS induction by previous irradiation of the host

Table 12–1 Overall bypass and mutagenic bypass probability in *E. coli* and *S. cerevisiae* cells after transformation with a single-stranded plasmid containing a TT cyclobutane pyrimidine dimer in a defined position

Organism	Overall bypass frequency (%)		Mutagenic bypass frequency (%)	
	cis-syn-TT	*trans-syn*-TT	*cis-syn*-TT	*trans-syn*-TT
E. coli (SOS induced)	16	16	7.6	13
S. cerevisiae	80	17	0.4	36

Source: After Lawrence et al. (322).

(322). In contrast, maximum bypass frequencies in *S. cerevisiae* were achieved in unirradiated cells. The *trans-syn* dimer was bypassed with about equal probability in *E. coli* and *S. cerevisiae* (~17%) but was more mutagenic in *S. cerevisiae* than in SOS-induced *E. coli* (36 versus 16% targeted mutations) (Table 12–1).

The type of base substitutions found was similar in the two species. In contrast to the *S. cerevisiae SUP4-o* data discussed above, predominantly T·A → A·T transversions were detected, targeted mostly to the 5′ residue of the dimer. A significant fraction of single-base pair deletions was also recovered, less so in *S. cerevisiae* than in *E. coli*. The *cis-syn* dimer was bypassed in *S. cerevisiae* more frequently than in *E. coli* (80 versus 16%), and in both cases mutagenicity was less than that observed for the *trans-syn* dimer (7.6% in *E. coli* versus 0.4% in *S. cerevisiae*).

From this study, it was concluded that the type of lesion significantly influences the probability and type of mutagenic bypass. It was additionally concluded that in general, replicative bypass is frequently nonmutagenic and that this may be indicative of a limited capacity of cyclobutane dimers to form correct hydrogen bonds with appropriately paired bases (312). The spectrum of recovered mutants was quite different from that observed in the *SUP4-o* system, however, and it remains to be demonstrated that transformation with single-stranded DNA containing a single photoproduct in a unique sequence context accurately reflects the in vivo situation.

Untargeted Mutagenesis in Yeast Cells Exposed to DNA Damage

The sophistication of the genetic tools available in *S. cerevisiae* has facilitated studies of the contribution of sequence changes which are not opposite photoproducts (untargeted mutations; see chapters 2 and 11) to overall UV radiation-induced mutability (319). A haploid strain carrying a nonrevertable deletion mutation in the *CYC1* gene (*cyc1-363*) was UV irradiated and mated with an unirradiated partner carrying a point mutation (*cyc1-91*) in the region deleted in *cyc1-363*. Hence, *CYC⁺* revertants among the diploid progeny could result only from reversion of the point mutation *cyc1-91* of the unirradiated partner, i.e., in the absence of photoproducts in or in the vicinity of the mutated base pair (Fig. 12–8). By using this experimental procedure, a reversion frequency of ~40% of that obtained by direct irradiation of the *cyc1-91* parent alone was observed. Like any other UV radiation-induced mutation, this type of untargeted mutation was found to be dependent on the presence of functional *RAD6* and *REV3* genes (320). Untargeted mutagenesis was also shown to be dependent on nuclear fusion and was eliminated in a *kar1* mutant background in which only cytoplasmic fusion can occur (319).

The high percentage of untargeted mutations achieved in this system apparently contradicts the results from the *SUP4-o* system. However, a paucity of nondimer sites in the *SUP4-o* system has been noted and could have resulted in a potential underestimation of untargeted events (304). Additionally, it should also be pointed out that the *CYC* experiments were performed with nucleotide excision repair-deficient cells, since mutation fixation in excision-proficient cells is apparently faster than the time required for mating.

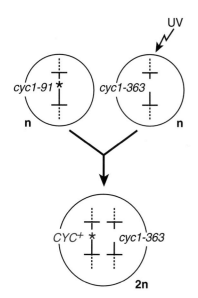

Figure 12–8 Detection of untargeted mutations in a yeast mating system. UV-irradiated haploid cells containing a *CYC1* deletion (*cyc1-363*) are fused with a nonirradiated partner carrying a point mutation (*cyc1-91*) in the region deleted in *cyc1-363*. Mutants with reversion of the point mutation (*cyc1-91* → *CYC⁺*) in the unirradiated genome are detected among the diploid progeny.

Summary and Conclusions

Although considerable information is now available on several of the genes (and their polypeptide products) involved, no single model satisfactorily explains all the features of UV radiation-induced mutagenesis in the yeast *S. cerevisiae*. The question of the regulation by gene induction of mutagenic processes in *S. cerevisiae* remains open and awaits a detailed study of the transcriptional regulation of the genes involved. However, as discussed above, it would appear that UV radiation mutagenesis in *S. cerevisiae* is not strictly dependent on the activation of inducible genes as is the case in *E. coli*. Inducible processes may play a role in special situations or in certain nucleotide sequence contexts. The inducible Rad51 protein of *S. cerevisiae* (the *E. coli* RecA homologue) has not been associated with mutagenic processing (see below).

The products of several genes such as *REV1*, *REV2*, *REV3*, *SRS2*, and *CDC7* may play roles as components of a specialized replication machine that is required for translesion DNA synthesis, and it is tempting to speculate that such a specialized replication complex is required to permit mutagenic bypass of noncoding lesions in *S. cerevisiae* (403). Whereas such a mechanism is tenable in mutants defective in nucleotide excision repair, any model of mutagenesis in *S. cerevisiae* must incorporate an explanation of the preferential prereplicative fixation of mutations in excision-proficient cells. One might consider that DNA polymerization processes may encounter photoproducts before or during the course of nucleotide excision repair (Fig. 12–9), leading to speculations about futile mismatch repair-like processes that might remove a stretch of DNA opposite an unexcised photoproduct (553). As discussed in chapter 11, another possible explanation can derive from the necessity to process closely spaced photoproducts on opposite DNA strands (64). During the course of nucleotide excision repair, a repair polymerase would encounter a photoproduct on the opposite strand, necessitating mutagenic bypass. A subsequent round of nucleotide excision repair would then lead to pre-replicative fixation of the sequence alteration in both strands (see Fig. 11–22). A sensitive method for detecting pyrimidine dimers on opposite strands with a maximum distance of ~15 bp has been developed (491). It has been shown that these

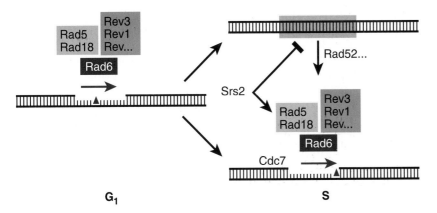

Figure 12–9 Hypothetical scheme for mutagenic processing of UV-induced lesions in *S. cerevisiae*. To account for the (mostly) targeted mutations, some form of replicative bypass of photoproducts is assumed. The bypass may occur before (G_1) (left) or during or after semiconservative DNA synthesis (S) (right). This reaction is dependent on wild-type proteins corresponding to mutants which define the *RAD6* epistasis group of genes. These have been subdivided into a group which is primarily responsible for mutagenic processing (e.g., Rev3), a group which mainly effects the repair of prelethal damage (e.g., Rad5), and Rad6 itself, which appears to be important for both processes. Biochemical correlates for these subgroups are unknown. A "channeling" function has been assumed for Srs2. When active, Srs2 protein diminishes the probability that gap filling opposite photoproducts by homologous recombination (mediated by the *RAD52* group of genes) will compete with mutagenic processing.

so-called *bifilar lesions* are indeed repairable in *S. cerevisiae* and that their removal depends on active nucleotide excision repair (491).

When one considers the relationship between postreplication repair and damage-induced mutagenesis in *S. cerevisiae*, key concepts to keep in mind are *channeling* and *redundancy*. Postreplication repair is operationally defined as the sum of processes that restore high-molecular weight DNA during or after replication of an irradiated template (usually studied in nucleotide excision repair-deficient cells by using pulse-labeled DNA). Mutants such as *rad6* are highly UV radiation sensitive and are deficient both in postreplication repair and in damage-induced mutagenesis. The phenotype conferred by the *srs2* mutant, which can suppress the UV radiation sensitivity of a *rad6* mutant, indicates that the restoration of high-molecular-weight DNA per se can occur by more than one mechanism (Fig. 12–9). One such mechanism is most probably *RAD52*-mediated homologous recombination, which is not associated with mutagenesis. Another may be mediated by genes such as *RAD6*, *RAD18*, and possibly *REV2* (*RAD5*) (Fig. 12–9). The latter mechanism may normally be associated with mutagenic events, but mutagenicity is apparently not a prerequisite for its function, since mutants such as *rev3* are severely impaired in damage-induced mutagenesis but are normal for postreplication repair.

DNA Damage-Induced Mutagenesis in Mammalian Cells

Genetic Control

Unlike in *S. cerevisiae*, there is no available collection of mammalian cell mutants which are characterized by reduced mutation frequencies following treatment with DNA-damaging agents. Nevertheless, some anecdotal information on this topic merits consideration. The Chinese hamster ovary (CHO) cell mutant UV-1 falls in the category of mutants defective in mutational processing (558, 559). The mutant is characterized by moderate sensitivity to UV radiation and alkylating agents and is less mutable at the *HPRT* locus (encoding hypoxanthine phosphoribosyltransferase) after exposure to these agents (233, 622). A defect in postreplication repair has been characterized, i.e., a slower reconstitution of high-molecular-weight DNA after UV irradiation compared with that in parental cells (558). However, nucleotide excision repair, as well as the capacity to exchange pyrimidine dimers between parental and daughter strands, appears to be normal (221).

The mouse lymphoblast line Ly-R (35) and a simian virus 40 (SV40)-transformed Indian muntjac cell line called SVM have also been found to be defective in postreplication repair. However, these cells are apparently not hypomutable (455, 456). After transfection with mouse DNA and cosmid rescue, a gene has been cloned which partially complements the UV radiation sensitivity and the defect in postreplication repair of SVM cells, but its presence increased the frequency of UV radiation-induced thioguanine-resistant mutants significantly above the wild-type level (52). It has been suggested that the cloned gene is not homologous with that defective in SVM but, nonetheless, represents a mammalian gene involved in mutation fixation.

Fanconi's anemia is an autosomal recessive disorder characterized by a predisposition to cancer and increased cellular sensitivity to cross-linking agents (see chapter 14). Lymphoblasts of genetic complementation groups A and B have been reported to be hypomutable after treatment with bifunctional psoralens (see chapter 1) (438, 438a, 439). A structural analysis of spontaneous and induced *hprt* mutants revealed a bias toward the recovery of small deletions in group B cells from patients with Fanconi's anemia. The correlation of hypomutability with the more frequent occurence of deletions among the induced mutants suggests the involvement of the *FAB* gene in some kind of mutagenic postreplicational gap-filling process. A defect in the processing of these gaps could explain the preferential appearance of small deletions among *hprt* mutants (438).

Replicational Bypass of UV Radiation Lesions

CORRELATION OF THE S PHASE WITH MUTATION FIXATION

There is no compelling imperative to invoke a process other than the misinsertion of bases during semiconservative DNA synthesis as the primary basis for the mutagenic processing of UV radiation lesions in mammalian cells (272, 504). Mutation frequencies are reduced (often to background levels) if irradiated cells are prevented from entering S phase for a sufficient period. This has been achieved by maintaining confluence (194, 367) or by synchronization with hydroxyurea (563) or low temperature (138).

An extended period in confluence does not, however, prevent induced mutagenesis in XP-A cells, which are defective in nucleotide excision repair (see chapters 8 and 14) (195, 367) and are hypermutable (297). Thus, it seems that removal of premutagenic lesions during the period of noncycling is due to the comparatively error-free process of nucleotide excision repair. This issue has been experimentally addressed in greater detail (297). Normal and XP-A fibroblasts were released from confluence by resuspension in fresh medium and plating at a lower density. The length and position of the subsequent S phase were then determined by autoradiography, bromouracil density shift analysis, and flow cytometry. Knowledge of these parameters allowed irradiation of cells at specific times before S phase (Fig. 12–10). The shorter the time interval between irradiation and S phase, the more mutants were recovered as a proportion of surviving cells, presumably because the time for removal of premutagenic photoproducts by nucleotide excision repair before mutational fixation during replication is shortened (297). Indeed, the time of irradiation before the onset of S phase does not make any difference for XP-A cells, which are of course deprived of functional nucleotide excision repair (297) (Fig. 12–10). For both cell lines the altered time window does not dramatically affect survival (Fig. 12–10), possibly because other repair pathways can substitute for nucleotide excision repair to some degree.

The amenability of S phase to mutational events appears to extend to the initiation of carcinogenesis. Carcinogenesis and transformation after treatment with chemical carcinogens or 254-nm UV radiation are most efficient if the target cells are treated just prior to or during S phase (272, 275, 276, 369, 393).

DNA REPLICATION IN UV-IRRADIATED CELLS

Before we discuss the available evidence that photoproducts are indeed bypassed during DNA replication in mammalian cells (and, in so doing, ignore the question of untargeted mutagenesis for the moment), it is useful to consider the known overall inhibitory effect of DNA damage on replication in mammalian cells. Such experiments have been notoriously difficult to interpret and are subject to controversy, mainly because of the small size of the multiple replicons in the mammalian genome and the variable sources of discontinuities. More-detailed discussions of this topic are presented elsewhere (166, 333, 334, 433).

It is important to discriminate between inhibition of replicon initiation and reduction in the rate of chain elongation, the latter presumably being due to arrest of replication at sites of damage in template strands. Inhibition of replicon initiation appears to be the predominant response to damage caused by ionizing radiation (310, 434, 435, 624) and low doses of UV light (273, 274, 432). This is exemplified in Fig. 12–11, where alkaline sucrose density gradient analysis of pulse-labeled DNA indicates a decrease in the number of small, newly synthesized DNA molecules relative to their number in untreated cells. The quantity of large, pulse-labeled molecules that reflect progression of replicons that were initiated prior to irradiation remains largely unaffected (435). Growing strands continue to be elongated, but new chains cannot be initiated. The inhibition of replicon initiation by ionizing radiation in a defined chromosomal replicon has also been demonstrated (310). Replication intermediates in an amplified dihydrofolate reductase (*DHFR*) domain can be visualized after hybridization of two-dimensional gels. Progressing initiation structures follow a distinctive pattern of electrophoretic

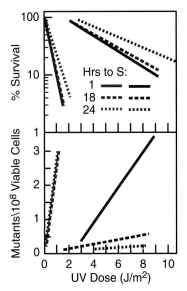

Figure 12–10 Dependence of UV radiation-induced cytotoxicity and mutagenicity in normal fibroblasts (black lines) and XP-A cells (red lines) on the length of time between irradiation and the onset of S phase. Cells were released from confluence and irradiated at the times indicated (1, 18, or 24 h before S phase). An increased time window reduces mutagenicity and, to a lesser extent cytotoxicity, in normal cells but not in the excision-deficient XP-A cells. (*Adapted from Konze-Thomas et al. [297] with permission.*)

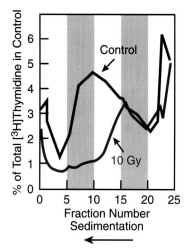

Figure 12–11 Alkaline sucrose density gradient profiles of DNA from unirradiated or X-irradiated (10 Gy) CHO cells. In the irradiated cells there is a decrease in the relative amount of small DNA molecules (fractions 5 to 10), reflecting regions in which replication was initiated soon after irradiation. However, the relative number of large molecules (fractions 15 to 20) which reflect the progression of replicons that were initiated prior to irradiation, is not significantly decreased. This suggests that growing DNA strands continue to be elongated but new chains cannot be initiated. (*Adapted from Painter and Young [434] with permission.*)

mobility ("bubble arcs") (Fig. 12–12), which can be separated from molecules representing replication intermediates that originated from initiation in adjacent regions.

Evidence is accumulating that the inhibition of replicon initiation is an actively regulated response and not the direct, *cis* effect of damage at replicational origins. (A similar mechanism of inhibition of replicon initiation appears to exist in *E. coli* [116, 613].) Low doses of X rays, UV light, or other DNA-damaging agents that result in low frequencies of DNA damage inhibit as much as 50% of the replicon initiation that is expected to occur within 1 h of exposure to damage. Hence, it has been suggested that the presence of a single lesion in a replicon domain (consisting of multiple replicons under coordinate regulation) might inhibit initiation of DNA synthesis in all the replicons in that domain (377, 435, 463). This view is confirmed by microbeam irradiation studies (105). More recent studies suggest that inhibition of replication of extrachromosomal plasmid DNA (but not of mitochondrial DNA [87]) can be detected at doses of radiation that produce no detectable damage in the plasmid molecules (89, 308). Finally, the existence of a human syndrome (ataxia telangiectasia) (see chapter 14) characterized by a reduced sensitivity to replicon initiation arrest (432, 436) is another strong argument in favor of a *trans*-acting regulatory mechanism. There are even arguments against a predominant *cis* effect of DNA damage. In *S. cerevisiae*, induction of a single localized double-strand break on a plasmid does not prevent the initiation of replication from a nearby origin (477). We will return to the issue of cell cycle regulation in the presence of DNA damage in chapter 13.

The inhibition of chain elongation in mammalian cells is more pronounced at higher doses of UV radiation (272, 384, 610). The existence of a damage-inducible protein (p21) which inhibits polymerase δ activity in vitro by binding to proliferating-cell nuclear antigen (Table 8–5) suggests that the inhibition of DNA chain elongation may also be subject to *trans*-acting regulatory mechanisms (621) (see chapter 13).

Rather than attempting a detailed discussion of the numerous studies available, we would like to offer a model for the different modes of DNA synthesis on damaged templates which has been put forward by several authors (387, 432, 556) (Fig. 12–13). DNA replication in mammalian cells progresses bidirectionally from

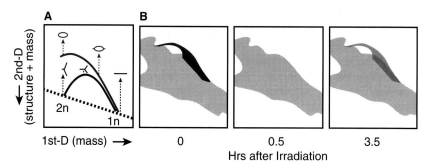

Figure 12–12 Detection of inhibition of replicon initiation in a defined gene (the *DHFR* cluster in asynchronous, methotrexate-resistant CHO cells) by a two-dimensional gel technique. (A) After restriction digest, in the first dimension (1st-D) DNA molecules are separated in an agarose gel according to their molecular mass along the *x* axis. The greater the extent of replication, the slower the fragments migrate in the gel. In the second dimension (2nd-D), separation is carried out under conditions in which the fragment shape contributes to the migration rate. The gel is then analyzed by Southern hybridization. Thus, the technique allows the distinction of replication intermediates originating in the immediate region of interest (bubble arc [red line]) from forks that enter from flanking regions (black line). The dashed arrows indicate the putative positions of replication intermediates (note the position of bubble versus branched structures). (B) After exposure of cells to γ radiation (9 Gy), a transient inhibition of replicon initiation is observed (as evidenced by loss and reappearance of the bubble arc). The red area in panel B corresponds to the red arc in panel A, and the grey area in panel B corresponds to the (idealized) black arc in panel A. (*Adapted from M. Larner et al. [310]*)

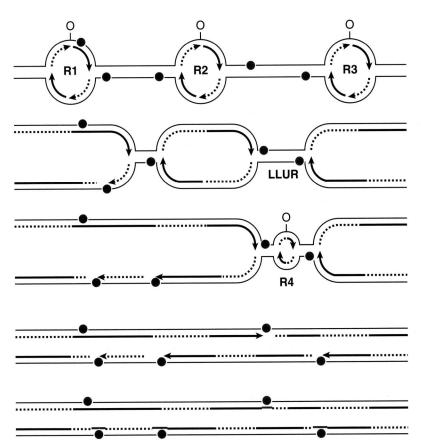

Figure 12–13 Model for replication of DNA containing photoproducts. Replication progresses bidirectionally from the three depicted origins (R1 to R3) in the presence of randomly located photoproducts (red dots). When replicon R1 encounters a photoproduct during lagging-strand DNA synthesis, replication continues with the next Okazaki fragment, and a gap is left. Progression of R2 is blocked on both sides of the replication bubble by photoproducts in the template strand for leading-strand synthesis. However, the adjacent replicon R1 is not blocked for leading-strand synthesis; hence, these replicons can fuse. On the other side, however, progression of the neighboring origin R3 is also arrested when the leading strand reaches a photoproduct. This situation results in a long-lived unreplicated region (LLUR), which may be replicated following the activation of an alternative replicational origin (R4). Finally, remaining gaps are filled by translesion synthesis (red lines). (*Adapted from Spivak and Hanawalt [556] with permission.*)

various origins. After primers have been synthesized by DNA polymerase α, DNA polymerase δ extends the leading strand whereas some other polymerase is responsible for lagging-strand synthesis (Table 8–5) (348). Damage in the leading strand can block the entire process of fork progression, possibly by inhibition of unwinding (Fig. 12–13) (36, 85, 100, 388, 521). Damage in the lagging strand, however, may not necessarily lead to replication arrest (85, 521). Synthesis can continue at the next Okasaki fragment, leaving a gap (332, 537, 544, 612, 633). When the replication fork of one replicon is arrested by damage in the leading strand, an adjacent replicon may not be inhibited, since there is no damage or damage only on the lagging strand (Fig. 12–13). Hence, the replicons can still be fused. However, if fork progression is blocked in adjacent replicons by damage in the leading strand, a long-lived unreplicated region may result (432, 440). This situation can be overcome by the activation of secondary origins in this region (193, 387) (Fig. 12–13).

Thus, a discontinuous mode of DNA replication can permit the completion of replication. Translesion synthesis could certainly occur at any stage but may be required only to fill gaps left in lagging strands opposite photoproducts. Alterna-

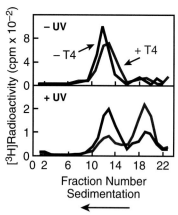

Figure 12–14 Pyrimidine dimers can be detected in completely replicated UV-irradiated SV40 DNA. After replication of the irradiated bromodeoxyuridine- and [³H]deoxythymidine-prelabeled SV40 genomes, DNA of intermediate density (indicating fully replicated DNA molecules) was isolated in neutral CsCl gradients, subjected to digestion with the pyrimidine dimer-specific T4 UV endonuclease (T4) (red lines), and analyzed in alkaline sucrose gradients. The peak at fractions 12 to 13 represents fully replicated covalently closed circular DNA molecules (form I DNA). The increase in the peak of relaxed circles (form II DNA around fraction 19) after digestion of the irradiated sample with the T4 enzyme indicates the presence of UV photoproducts in the replicated molecules. (*Adapted from Stacks et al. [557].*)

tively, gap filling by recombinational processes may occur at these sites, a topic that we will address below.

EVIDENCE FOR TRANSLESION DNA SYNTHESIS

Only a few studies have directly addressed translesion DNA synthesis at the level of chromosomal genes in mammalian cells. In a recent example (556), CHO cells were irradiated with UV light and incubated in the presence of bromodeoxyuridine. Unreplicated DNA of light density and replicated DNA of hybrid density were isolated, and the parental and daughter strands of the replicated DNA were separated. Unreplicated DNA and replicated DNA (separated as parental and daughter strands) were probed for the remaining presence of pyrimidine dimers. A 15-kb *Kpn*I fragment downstream of the *DHFR* gene (which contains an origin of replication) was found to be poorly repaired. Furthermore, after 24 h of postirradiation incubation, an almost unchanged number of dimers was detected in the parental strands in the fraction of replicated DNA. This result suggests that the photoproducts were efficiently bypassed.

Although a general inhibition of replication is evident (31, 36, 133), several studies indicate that the replication of pyrimidine dimer-containing viral genomes can be completed (85, 521, 557, 615, 636). This can be demonstrated by density labeling newly synthesized DNA (557). After irradiation of SV40-containing host cells, viral double-stranded DNA of hybrid density was isolated in neutral CsCl gradients. This DNA, which contains the light parental strand as well as a heavy newly synthesized strand, was isolated and subjected to nicking with the dimer-specific T4 DNA glycosylase/AP lyase (see chapter 4). Analysis in alkaline sucrose gradients indicated that these molecules still contained pyrimidine dimers (Fig. 12–14). Recombination with newly replicated intact molecules (which should have been detectable by a density shift of the parental heavy strand after denaturation) could not account for this observation.

There is no evidence for a decreased blockage of SV40 replication forks at UV radiation-induced lesions after preirradiation of host cells, which might be indicative of increased translesion synthesis or increased repair (524). However, pretreatment of host cells with *N*-acetoxy-*N*-2-acetyl-2-aminofluorene does enhance the overall replication of UV-treated SV40 (365). Pretreatment of host cells with UV radiation also alleviates replication inhibition at apurinic/apyrimidinic (AP) sites in single-stranded parvoviruses. The resulting increased mutagenesis is suggestive of increased replicative bypass (616).

By using human or monkey CV1 cell extracts and the addition of SV40 T antigen, an in vitro system has been established for replicating SV40 DNA and plasmids containing a SV40 origin (342). The relatively error-free complete replication of undamaged plasmids was confirmed with this system (211). Additionally, as is the case in vivo, inhibition of replication by the presence of UV radiation damage in the plasmid has been demonstrated (71, 192, 595). Inhibition of replication could be alleviated by photolyase treatment of the UV-irradiated plasmid prior to replication, thus proving that cyclobutane-type dimers primarily determine this effect (71, 595). Continued incubation of the damaged plasmid results in the appearance of double-stranded covalently closed circular DNA molecules which are sensitive to nicking by pyrimidine dimer DNA glycosylase/AP lyase (71, 595). Furthermore, resistance to the restriction endonuclease *Dpn*I indicates that these molecules contain a newly synthesized strand (71, 595), and density gradient analysis proves that these molecules originated by semiconservative replication (71). By including genes such as *supF* or *lacZ* (which allow detection of forward mutations after transformation of *E. coli*) in the plasmid, it was shown that mutations were introduced during in vitro replication of the damaged plasmid (Table 12–2) (71, 595). Removing pyrimidine dimers before replication abolished this effect (Table 12–2).

Collectively these data strongly suggest that cyclobutane pyrimidine dimers inhibit DNA replication but that they can also be bypassed in a mutation-prone manner. G·C → A·T transitions were predominant in the in vitro replication system. Thus, as we will discuss in greater detail below, the mutational specificity in

Table 12–2 Mutagenicity of SV40 origin-dependent replication of UV-irradiated DNA in HeLa cell extracts

Condition	UV dose (J/m²)	Mutant plaque frequency ($\times$ 10^{-4})
Control (SV40 DNA	0	5.0
incubated, no T antigen)	70	2.8
SV40 DNA replicated	0	3.7
(incubated, T antigen)	70	32
Pretreatment		
Pretreated with buffer	70	31
Pretreated with photolyase	70	5.1

Source: After Thomas and Kunkel (595).

vitro seems to closely mirror the in vivo situation. These observations correlate with the finding that calf thymus DNA polymerase δ (see Table 8–5) can synthesize past *cis-syn-* and *trans-syn*-thymine dimers in vitro, a reaction that is dependent on the presence of proliferating-cell nuclear antigen (424). Do the different modes of replication of the leading and lagging strands of DNA determine different probabilities for translesion synthesis? This question has been addressed in vitro by inserting the mutational target gene on opposite sides of the SV40 origin of replication (596). Changes in mutation probability at particular sites were indeed observed if the mode of replication changed. However, the overall mutation frequency remained unaltered.

The phenotypic characteristics of XP variant (XP-V) cells (see chapter 14) have been explained by a defect in translesion synthesis and thus provide arguments for the existence of such a process in normal mammalian cells (272). Patients with variant xeroderma pigmentosum (XP-V) exhibit many features of the majority of those with XP and share a predisposition to skin cancer. However, these cells are only marginally sensitive to UV radiation and are not notably defective in nucleotide excision repair (335). Undamaged DNA is replicated normally in XP-V cells, but abnormalities of DNA replication have been found in these cells after UV irradiation (57, 95, 273, 335, 440). An increased sensitivity to the inhibition of fork progression and elongation of daughter-strand DNA has been detected (57, 273, 610), implying that photoproducts are more efficiently bypassed in normal cells and raising the possibility that XP-V cells are defective in translesion DNA synthesis. We will return to this issue below (also see chapter 14).

Inducibility of Mutagenic Processes in Mammalian Cells?

As in yeast cells the search for phenomena in higher eukaryotic cells which can be correlated with SOS responses in *E. coli* has not provided definite answers (475, 503, 504, 515). The available studies have focused mainly on the regulatory induction of mutagenesis by pretreatment protocols in cells and viruses and on the detection of untargeted mutations.

REACTIVATION OF VIRAL PROBES AND SHUTTLE VECTORS

As described in chapter 10, one of the prominent SOS responses in *E. coli* is an increase in the proportion of mutant host cell-reactivated bacteriophages, phenomena termed *Weigle* (or *W) reactivation* and *Weigle* (or *W) mutagenesis*. Similar experiments have been attempted in mammalian cells by using a variety of pretreatments of the host and various double- or single-stranded viruses, as well as transfected viral DNA (reviewed in references 44, 104, 135, 284, 475, 504, 515). Reactivation of virus survival in various transformed and untransformed human, simian, and rodent cell lines has been observed in a number of such experiments, including the use of herpes simplex virus (4, 45, 46, 101, 217, 355–357, 359, 644),

SV40 (47, 96, 97, 516–519), simian adenovirus type 7 (47), adenovirus type 2 (249, 457–459), human cytomegalovirus (111), Kilham rat virus (354), parvovirus Lu III (196), minute virus of mice (500), and parvovirus H-1 (98).

Pretreatments of host cells which resulted in reactivation of UV radiation-treated viruses include not only UV radiation but also ionizing radiation (46, 249, 583), aflatoxin B₁ (356, 520), N-acetoxy-N-2-acetyl-2-aminofluorene (356, 518, 520), mitomycin (517), and monofunctional alkylating agents (520). Even hydroxyurea (159, 358), caffeine (159), bromouracil (159), ethanol (459), sodium arsenite (459), and heat (457, 458) have been used as mutation-inducing pretreatments. Additionally, reactivation in pretreated cells has been reported for viral genomes damaged by γ rays or ethylnitrosourea (568, 583). However, negative results have also been reported; e.g., UV-irradiated vaccinia virus or poliovirus propagated in CV1 monkey kidney cells pretreated with UV radiation or X rays (47) and N-acetoxy-N-2-acetyl-2-aminofluorene-treated SV40 in UV-irradiated cells (184).

When positive, the reactivation factor is highly variable and in general much lower than that observed for W reactivation in *E. coli*. A certain period of incubation of the treated host cells is required before enhanced virus reactivation can be detected. Frequently, the presence of the protein synthesis inhibitor cycloheximide during incubation can prevent enhanced viral reactivation (101, 358). These observations have been construed as arguments for the involvement of inducible processes. However, cycloheximide itself can inhibit mammalian DNA replication (562).

High levels of enhanced reactivation of herpes simplex virus have been observed in various cancer-prone syndromes, and it has been suggested that genes involved in this phenomenon may be controlled by tumor suppressor genes (3, 5). The phenomenon of enhanced viral reactivation appears to be independent of nucleotide excision repair, since reactivation of UV-irradiated herpes simplex virus has also been observed in UV-pretreated XP cells (4, 357). However, a different study that monitored the reactivation of a plasmid-borne chloramphenicol acetyltransferase (*cat*) gene in host cells pretreated with UV radiation or mitomycin suggests the opposite for UV-irradiated transfected plasmids (472).

Is enhanced reactivation correlated with an enhanced mutation probability of the viral progeny? Positive results have been reported for herpes simplex virus (Table 12–3) (4, 101). The use of herpes simplex virus as a model for endogenous mammalian processes has been criticized, since this virus encodes its own DNA polymerase (475). However, enhanced mutagenicity by host cell pretreatment has also been reported with SV40 (56, 517–519), parvovirus H-1 (98, 568), and human cytomegalovirus (111). UV radiation pretreatment also enhanced the mutagenicity of an AP site in a shuttle vector containing the SV40 origin of replication (185). Enhanced mutagenicity by pretreatment of host cells with mitomycin has also been observed in the UV-irradiated SV40-based shuttle vector pZ189 (see chapter 2) (112, 498). This effect was independent of nucleotide excision repair (498).

Several studies have failed to confirm many of these results. For example, no enhanced mutagenesis was detected with UV-treated herpes simplex virus (359, 582), SV40 (97, 585), or adenovirus (102). The multiplicity of infection seems to matter in certain cases. The detection of enhanced mutagenesis in herpes simplex virus appears to depend on a high multiplicity of infection (360), and that

Table 12–3 Mutagenesis in the thymidine kinase gene of herpes simplex virus

UV radiation to cells	Thymidine kinase mutant frequency (× 10⁻⁴) in:	
	UV-irradiated virus	Unirradiated virus
−	4.8	1.5
+	13.0	3.5

Source: After DasGupta and Summers (101).

in SV40 seems to depend on a low multiplicity of infection (517). The latter observation is understandable since a higher input of damaged viral DNA might give rise to increased recombination, which might result in a lower probability of mutagenic translesion DNA synthesis. Enhanced mutagenesis is also not observed when SV40 DNA, instead of intact virus, is transfected into host cells (182). This can also be explained by the uptake of multiple viral genomes per cell.

One of the few consistent observations that has emerged from such studies is that, in contrast to phage λ in *E. coli*, UV-irradiated mammalian viruses give rise to mutant progeny even without previous irradiation of the host cells. Hence, mutagenic processes operating in mammalian cells appear to have a significant constitutive activity.

UNTARGETED MUTATIONS IN MAMMALIAN CELLS

If pretreatment of cells were to result in the induction of a cellular SOS-like function which reduces replication fidelity, this would be expected to yield untargeted mutations (see chapter 2). Mutations have been detected in undamaged herpes simplex virus (4) and parvovirus H-1 (98) DNA following pretreatment of host cells, but conflicting results were encountered with SV40 DNA (97, 516, 518). A plasmid shuttle vector containing the SV40 origin of replication and an *E. coli* tyrosine suppressor tRNA (*supF*) was propagated in a COS-7 derivative of CV1 monkey kidney cells. DNA preparations from the cells were used to transform *E. coli*, and the frequency of *supF* mutations was measured (522). A 10-fold increase over the (rather high) spontaneous background frequency was observed after pretreatment of the host cells with EMS 40 h before transfection (522). However, mitomycin pretreatment failed to have such an effect in a similar host-vector system (498).

Another approach to the study of untargeted mutations involves the use of indirect induction protocols by essentially asking whether the introduction of a damaged DNA probe or viral genome enhances the mutability of endogenous host genes or undamaged extrachromosomal probes. Studies on cellular genes have not been reported. A mutator effect on unirradiated parvovirus H-1 was demonstrated following transfection of UV-irradiated calf thymus, SV40 DNA, or φX174 DNA into rat or human cells before parvovirus infection (109, 110). However, the simultaneous transfection of UV-irradiated SV40 DNA and an undamaged shuttle vector did not result in mutation induction in *trans* (473).

The emergence of mutants during several cellular divisions after DNA-damaging treatments may also be an indication of untargeted mutagenesis, if it can be assumed that all initial DNA lesions were repaired or eliminated from the growing population by cell death. Several studies have suggested such a possibility (49, 290, 349, 370). Fluctuation analysis was used to determine the time of mutation fixation during growth after introducing DNA damage by treatment with 8-methoxypsoralen plus UV-A (see chapter 1) (49) (Fig. 12–15). Treated mouse lymphoma cells were divided into subcultures, which were grown from a very small inoculum for several generations and then scored for *hprt* mutants. A subculture characterized by a mutational fixation event during early generations will contain a large fraction of mutant cells, whereas far fewer mutant cells will be detected in a subculture following a late mutational event. Mutations arose for up to 11 generations after treatment (Fig. 12–15), far beyond the time needed for the repair of psoralen-induced cross-links and monoadducts.

A final example of untargeted mutagenesis in mammalian cells concerns the excretion of *extracellular protein-inducing factor(s)* from UV-irradiated human fibroblasts. This factor(s) can apparently communicate the irradiation response to other unirradiated cells and induce the same pattern of proteins (see chapter 13) (370, 505, 534). Incubation in extracellular protein-inducing factor medium caused a moderate increase in mutation frequency of chromosomal genes (measured by 6-thioguanine resistance), whereas UV-induced mutation frequencies were unaffected (48, 370). Conditioned medium had a similar effect on the mutation frequency of the supF gene in an autonomously replicating shuttle vector (pZ189;

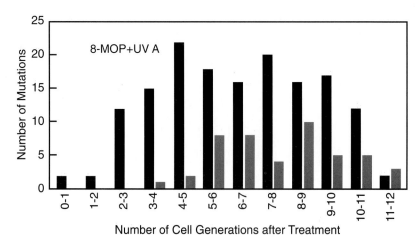

Figure 12–15 Delayed emergence of *hprt* mutants after treatment of mouse T lymphoma cells with 8-methxypsoralen (8-MOP) plus UV-A. The number of mutations arising during each cell generation after treatment was determined by fluctuation analysis. The number of mutational events in treated cells (red bars) approaches the spontaneous background level (grey bars) only after the 11th generation. (*Adapted from Boesen et al. [49] with permission.*)

see chapter 2). In this study, enhanced mutagenesis of the UV-damaged plasmid was also reported (112).

Mutational Specificity of UV-Induced Lesions

VIRAL PROBES AND SHUTTLE VECTORS

As in *E. coli*, it was anticipated that an analysis of the type and distribution of mutations in a defined target sequence in mammalian cells would allow insights into the molecular structure of the mutagenic lesions, the nature of the processes involved in mutational fixation, and the factors that influence their efficiency and specificity. Such factors include not only structural features of the target (e.g., the influence of neighboring bases, methylation, and secondary structure) but also the influence of the genetic background and of DNA repair, replication, and transcription. Initially such studies utilized shuttle vectors exclusively.

The development of mammalian shuttle vectors was reviewed in detail in chapter 2, and we have discussed the application of a particular *S. cerevisiae-E. coli* shuttle vector above. Suffice it to recapitulate that the use of viral probes and mutational systems was soon augmented by the use of plasmid vectors carrying a viral origin of replication and bacterial genes such as *supF*+ and *lacI*+ (54, 112). In these forward-mutation systems, mutation detection is performed in an appropriate *E. coli* host after rescue of the transfected and propagated plasmid from mammalian cells. The plasmid may replicate autonomously and independently of the host cell genome if it carries the SV40 origin of replication or in synchrony with the host cell S phase if the Epstein-Barr virus origin, *oriP*, is present (119, 123, 197).

Further technical refinements led to the use of bacterial or mammalian gene probes integrated into the mammalian genome and isolation of the target gene through packaging into phage λ (572) or following amplification induced by fusion with COS cells (see chapter 2) (20, 21, 122, 179). Here, the mutagenic treatment is performed in mammalian cells, and if a mammalian gene probe is used the mutant selection can be performed in the host cells.

Apart from the general problems associated with mutant selection and the unavoidable problems of selection bias (see chapters 2 and 11), there are numerous technical difficulties, whose severity depends to some extent on the particular system used. First, the undamaged vector must be characterized by a background level of spontaneous mutations which is sufficiently low to assign the mutagenic events unequivocally to the use of the DNA-damaging agent. This was initially a major problem, probably reflecting nuclease attack on transfected DNA (210, 328, 391, 482). Second, it must be ascertained that the mutation(s) has actually occurred in the mammalian cells under study and that a premutational lesion or its derivative was not fixed as a mutation after transformation of *E. coli*. Finally, a myriad of unknown variables pertinent to the genetic background of the cells used can influence mutational spectra (541).

With the advent of PCR technology direct sequencing of endogenous mutant genes has become feasible. However, the technical ease of the treatment of plasmid DNA probes in vitro still makes shuttle vector experiments appealing, especially for host pretreatment experiments of the type discussed above, and for the study of mutagenesis related to lesions at defined positions without the bias of selection (34, 136, 185, 395, 401) (see chapter 2).

Rather than discuss individual studies, we will highlight some of the significant observations that have emerged from the use of different experimental systems. As far as UV radiation mutagenesis is concerned, the general conclusions are similar to those dicussed for *S. cerevisiae*.

UV Light Induces Mostly Targeted Mutations

Most of the sequence alterations detected are single-base-pair substitutions at positions of potential pyrimidine dimers (Fig. 12–16) (53, 56, 58, 63, 122, 189, 212, 234, 327). The use of double-stranded DNA probes does not allow a formal conclusion about which strand the mutagenic lesion was originally located on. However, the assumption of targeted mutations opposite photoproducts formed at adjacent pyrimidines was confirmed by the use of single-stranded shuttle vectors (363). As in *S. cerevisiae*, single-base-pair substitutions are most frequently targeted to the 3′ residue of the pyrimidine dinucleotide (120, 234, 607). Tandem double-base pair changes at these positions have been observed at frequencies between 10 to 25% (of all mutational events) in the *supF*, *lacI*, and *aprt* genes (63, 122, 212, 234, 363). The distribution of base pair substitutions is nonrandom; hot spots are evident, and the positions of mutational hot spots after UV radiation treatment differ significantly from those in undamaged probes (53, 62, 63, 212, 234, 327, 473). Some of these hot spots are preserved if the same UV-irradiated vector is propagated in *E. coli* (234, 327).

Convincing arguments can be marshalled against a major role of processes that result in random, untargeted sequence alterations. Nonetheless, some data suggest that such processes do operate to some extent. For example, the occurrence of nonadjacent multiple substitutions in the same mutated plasmid has encouraged this interpretation. These types of mutations occur with a frequency of 20 to 30% among UV-induced mutants in the 160-bp *supF* target gene (543). One study on the spontaneous mutational spectrum in the endogenous *aprt* locus indicated that these types of events may be more frequent in tumor cell lines (208). However, there is no evidence that the underlying process is responsible for an inducible mutator phenotype and that its relative contribution to overall mutagenicity in a shuttle vector is enhanced by pretreatments of any kind. Although the overall mutation frequency is increased, no signifcant changes in the mutational spectrum are evident when cells were pretreated with mitomycin before transfection with a UV-irradiated plasmid (498).

UV Radiation Results Mainly in G·C → A·T Transitions

Independent of the system used, the productionof G·C → A·T transitions by UV radiation is a consistent feature of all studies on the mutational specificity of UV radiation in mammalian cells (Fig. 12–17) (53, 62, 63, 122, 189, 212, 234, 300, 327, 363, 473). Assuming mostly targeted mutations at pyrimidine dimers, these results imply that it is the C in TC, CC, or CT dimers and not the T in TC, TT, or CT dimers that is incorrectly bypassed.

An important question is whether the paucity of misinsertion events opposite T residues reflects a low frequency of mutagenic damage involving T's or a property of the mutational process itself. In favor of the former notion, one could

Spontaneous mutations

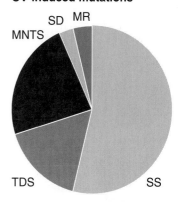

UV-induced mutations

SS Single substitutions

TDS Tandem double substitutions

MNTS Multiple non-tandem substitutions

SD Small deletions

MR Multiple rearrangements and large deletions

Figure 12–16 Probabilities of mutations of different molecular structure in the *supF* gene recovered in *E. coli* following passage of the shuttle vector pZ189 through CV1 monkey kidney cells. The mutational spectra detected with unirradiated (spontaneous mutations) and irradiated (500-J/m² UV radiation at 254 nm) plasmids are compared. The total area of the pie charts reflects the relative mutation frequencies (3.7 × 10⁻² in the unirradiated plasmid versus 7.6 × 10⁻¹ in the irradiated plasmid). (*Data from Hauser et al. [212].*)

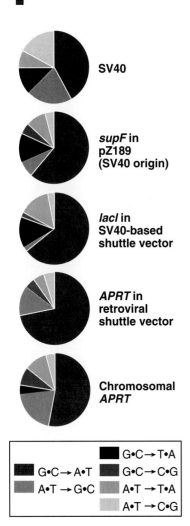

SV40

**supF in
pZ189
(SV40 origin)**

**lacI in
SV40-based
shuttle vector**

**APRT in
retroviral
shuttle vector**

**Chromosomal
APRT**

	■ G•C → T•A
■ G•C → A•T	■ G•C → C•G
▨ A•T → G•C	▨ A•T → T•A
	▨ A•T → C•G

Figure 12–17 Relative probabilities of UV-induced base pair substitutions detected in different target genes by using various viral and shuttle vectors and in a chromosomal gene. (*Data from Bourre et al. [53], Hauser et al. [212], Hsia et al. [234], and Drobetsky et al. [120, 122].*)

argue that (6-4) photoproducts are rarely formed at TT sites. However, the same is true for CT sites (see chapter 1) (60, 62). Cyclobutane dimers form most frequently at TT sites, but, as discussed below, extrapolation from other organisms that these lesions are mainly nonmutagenic has not been substantiated for mammalian cells (473). Additionally, at least in the *supF* gene, the formation of hairpin structures that can determine mutational hot spots in bacterial systems (493) cannot account for the observed hot spots in the mammalian shuttle vector pZ189 (410). In the final analysis, the observed mutational specificity is best explained by a mechanism which would render misinsertion opposite the T residue of photoproducts less likely and would lead preferentially to G·C → A·T transitions. Possible mechanisms have been discussed in chapter 11.

Pyrimidine Dimers Are a Major Source of Mutations, but (6-4) Photoproducts May Contribute to a Variable Extent

Selective removal of cyclobutane pyrimidine dimers by incubation with *E. coli* photolyase (59) before transfection with the UV-irradiated shuttle vector pZ189 or SV40 DNA was used to investigate the mutagenicity of these photoproducts (53, 62, 473). The removal of 90% of these photoproducts resulted in a 75%-higher transfection efficiency and a reduction of the mutation frequency by 80% (473). Although mutagenicity was in general substantially reduced, the position of mutational hot spots did not change dramatically, thus indicating that mutagenic photoproducts were still present at positions of adjacent pyrimidines.

A relative increase in the mutagenicity of photoproducts at CC and TC sites has been noted, concomitant with a relative decrease at CT and TT sites (473). This suggests an increased contribution of (6-4) photoproducts to the observed base substitution frequency, since CT and TT sites are poor substrates for the formation of (6-4) photoproducts (60, 62). Consequently, the selective removal of mutagenic pyrimidine dimers from these sites should be relatively more effective in reducing mutagenicity.

The inclusion of damaged SV40 DNA in the transfections did not affect the reduction in overall mutation frequency in the photoreactivated shuttle vector (473). This argues against the objection that these photoproducts by themselves are not the major mutagenic lesion but, rather, are required for the actual induction of an error-prone repair activity acting on (6-4) photoproducts. Such processes should have been triggered by the damaged cotransfected viral DNA.

Mutational Hot Spots Do Not Necessarily Correlate with Damage Hot Spots

The observation that mutational hot spots do not necessarily correlate with damage hot spots holds for both cyclobutane pyrimidine dimers and (6-4) photoproducts. The frequency of total photoproduct fomation in the *supF* gene has been determined by a DNA synthesis arrest assay (212). Additionally, levels of pyrimidine dimers and (6-4) photoproducts have been measured independently (62). The frequency of photoproduct formation varies considerably from site to site, but no correlation has been found with the occurrence of mutational hot spots, even in the absence of nucleotide excision repair (62). However, the immediate DNA sequence context may play a role in dermining mutational specificity (441).

A few concluding words of caution are in order. In analyzing mutational spectra following exposure to UV radiation, it is generally assumed that most, if not all, mutations arise from major photoproducts, i.e., pyrimidine dimers and (6-4) lesions. However, the possibility that "minor" photoproducts play a significant role and may even be responsible for individual mutational hot spots has not been excluded. In this regard it is important to at least mention the possible contribution of purine-pyrimidine dimers (170) and of alkali-sensitive photoproducts at ACA sites (55) and the (proposed) formation of pyrimidine dimers between nonadjacent nucleotides, mainly in single-stranded DNA, as possible progenitors of frameshift mutations (419).

Influence of Nucleotide Excision Repair

The frequency of base damage alone is not sufficient to explain mutational hot spots in mammalian cells. The mutational probability at a given site is the net result of damage frequency, error-free repair efficiency, and mutagenic bypass efficiency. The last-mentioned parameter reflects the frequency with which the lesion is successfully bypassed and, if bypass is indeed completed, how frequently a wrong base is inserted. Untargeted mutagenic processes add additional complexity. The availability of XP cell lines that are defective in nucleotide excision repair offers the potential of isolating the contribution of error-free DNA repair.

When SV40-transformed XP-A human fibroblasts (see chapter 14) were transfected with a UV-irradiated shuttle vector, the *plasmid survival* (measured as the relative number of bacterial transformants recovered from DNA preparations of the host cells) was dramatically lower than in a normal cell line but the frequency of mutant plasmids was higher (63). Mutational hot spots were not dramatically different from those found in normal cells. However, higher mutability at certain comparatively "silent" sites in normal cells was noted, and this might indicate the efficient removal of otherwise mutagenic lesions from these sites in excision-proficient cells (62).

A comparison of the probability of different types of base substitutions in XP-A and normal cells revealed an interesting feature. The spectrum appears to be more limited in XP-A cells, with a stronger bias toward G·C → A·T transitions at the expense of all other base substitutions (63, 300) (Fig. 12–18). This suggests that nucleotide excision repair is relatively more efficient in removing premutagenic lesions which result in these events relative to other lesions, e.g., nondimer products resulting in transversions. These observations also suggest that the process that results in G·C → A·T transitions is operative in cells which are defective in nucleotide excision repair.

Another class of mutations that is significantly reduced after passage of the vector through XP-A cells is multiple nontandem substitutions interpreted as untargeted events resulting from an error-prone DNA polymerase activity (63, 300, 543). These mutations reappear if the plasmid is nicked before transfection (543).

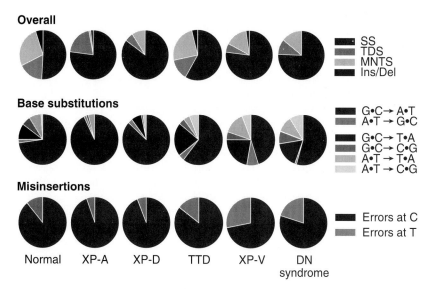

Figure 12–18 Overall spectrum of UV-induced mutations and relative probabilities of the different types of base pair substitutions detected in the *supF* gene after passage of the shuttle vector pZ189 through SV40-transformed normal, XP-A, XP-D, TTD, and XP-V fibroblasts and lymphoid cells of a patient with dysplastic nevus (DN) syndrome. Abbreviations are as in Fig. 12–16 (Ins/Del, insertions/deletions). In the third row, the respective base substitution frequencies have been converted to reflect the probabilities of misinsertion opposite T or C. (*Data from Bredberg et al. [63], Seetharam et al. [540, 542], Madzak et al. [362], and Wang et al. [626].*)

To explain the apparent dependence of the underlying process on incision activity, it has been suggested that these mutations are caused during the course of nucleotide excision repair by an error-prone DNA synthesis mode involved in gap filling.

A similarly biased mutational spectrum has been found in XP cells from other genetic complementation groups (Fig. 12–18) (540, 641, 642). It is of considerable interest to compare the mutational spectrum found in XP-D cells with that in cells from patients with trichothiodystrophy (TTD cells) from the same genetic complementation group (see chapter 14). Although the same gene is affected, patients with trichothiodystrophy are not characterized by the dramatically higher incidence of skin cancer typically observed in other XP-D patients. When tested with a UV-irradiated shuttle vector, the mutational spectrum in TTD cells was found to be much more similar to that observed in wild-type cells than in XP-D cells (Fig. 12–18) (362). However, considering that overall UV radiation-induced mutation frequencies are equally enhanced in the two mutant cell types, it seems unlikely that this difference in mutational specificity alone accounts for the higher cancer incidence in patients with XP-D than in those with trichothiodystrophy (362).

The mutational spectrum observed in XP-V cells (see chapter 14) is different from that in classical XP cells (626). A transfected, UV-irradiated plasmid suffered a decreased "survival" rate and enhanced mutation frequency in XP-V cells (626). Sequencing of the recovered base substitutions indicated a reduced prevalence of $G \cdot C \rightarrow A \cdot T$ transitions (626) (Fig. 12–18). In fact, the strongest hot spot involved a potential TT dimer position. In Fig. 12–18, we have pooled individual base substitution frequencies as events that indicate misinsertion opposite C or T. Apparently, XP-V cells are more likely to insert the wrong base opposite the T residue in a noninstructional photoproduct than are normal cells. This result has been confirmed by analysis of the mutational spectrum in an endogenous gene (627). Indeed, the known hypermutability of XP-V cells after exposure to UV radiation is specifically correlated with entry into S phase (368, 413). Thus, it appears that these cells are characterized by an abnormal, more mutagenic bypass of photoproducts during replicative DNA synthesis.

A similar bias is observed when the UV-irradiated vector pZ189 is passaged through cells from patients with another hereditary disease, termed *dysplastic nevus syndrome*, an autosomal dominant disease (542) (Fig. 12–18). Patients are characterized by multiple distinctive pigmented lesions (dysplastic nevi) and a predisposition to melanomas (299). Cultured cells are hypermutable (445), as are transfected plasmids (542), but some cell lines are also sensitive to killing by UV radiation or 4-nitroquinoline 1-oxide (231, 554, 555). As in XP-V cells, changes in the probabilities of different classes of base substitutions can be interpreted as indicative of a higher probability of misinsertion opposite T residues (Fig. 12–18).

The use of UV-irradiated single-stranded shuttle vector DNA is another way of excluding the template-dependent mode of nucleotide excision repair, at least before the first round of DNA replication. The replication origin of the single-stranded bacteriophage f1 was included in an SV40-based shuttle vector so that single-stranded plasmids could be propagated in *E. coli*. Thus, the outcome of transfection experiments involving single-stranded DNA and the same vector as a double-stranded plasmid could be compared (363, 364). The *E. coli supF* gene was once again the mutational target. Consistent with the absence of nucleotide excision repair, UV-irradiated single-stranded DNA was found to exhibit lower survival and a higher mutation frequency than was double-stranded DNA. The spontaneous mutation frequency after transfection of untreated DNA was also higher (363, 364). The positions of hot spots for UV radiation mutations were similar in single-stranded and double-stranded DNA, but some additional hot spots were observed in the former. This is in good agreement with the studies on XP-A cells discussed above. However, the absence of the class of multiple nontandem substitutions observed in the previously discussed study (63) was not reflected after transfection of single-stranded DNA (363, 364). These potentially untargeted mutations are usually coupled with base substitutions opposite dimer sites. Therefore, the dependence of an assumed error-prone DNA synthesis process on gap filling during nucleotide excision repair is not confirmed. However, it is possible that such a process operates after replication of the single-stranded dimer-containing probe. Alternatively, this mutagenic activity may generally use single-stranded DNA as a preferred substrate.

It was also noted that a higher percentage of mutations recovered after transfection with UV-irradiated single-stranded DNA were caused by frame shifts, usually at runs of three to eight pyrimidine residues, compared with the situation for double-stranded DNA transformation (363, 364). It has been suggested that this difference may reflect the higher proportion of *trans-syn*-cyclobutane pyrimidine dimers in single-stranded DNA (443). Indeed, this type of lesion also results in a high percentage of single-base-pair deletions or additions when it is introduced into *E. coli* as the exclusive lesion in DNA (26).

ANALYSIS OF MUTAGENESIS IN CHROMOSOMAL GENES

The analysis of mutational events in endogenous chromosomal genes is clearly preferable to the use of shuttle vectors. The availability of PCR technology has facilitated such an approach enormously (see chapter 2) (513). Following mutant selection, mRNA is prepared from mutant clones, cDNA is synthesized, and the coding region of the gene of interest is propagated with appropriate primers and sequenced. Such an approach typically precludes the detection of promoter mutations or mutations in introns, but deletion of an entire exon is indicative of a splice-site mutation. The sensitivity of the method makes it possible to compare in vitro and in vivo data and to determine the molecular nature of mutations recovered from the cells of individual patients in order to study the effect of occupational exposure to genotoxic agents or the mutational consequences of repair-deficient disorders (13, 483, 560, 617).

The first chromosomal systems used seemed to confirm results obtained with shuttle vectors. We have discussed these data in the comparison of UV radiation-induced base substitution frequencies observed in various systems (Fig. 12–17) (120, 122). In this case the mutant versions of the adenine phosphoribosyltransferase (*APRT*) gene were not isolated by PCR-mediated propagation but were rescued by recombination from a library constructed from size-fractionated chromosomal DNA into a phage containing *APRT* flanking sequences (121). Regrettably, further studies suggest that not all the characteristics of the mutagenic processing of UV radiation lesions inferred from shuttle vector studies hold for endogenous genes.

Influence of Transcription, Repair, and Replication on Mutagenic Processing of UV Radiation Lesions in the *HPRT* Gene

The *HPRT* locus is one of the earliest target genes used in mammalian systems. The human and mouse genes consist of nine exons contained in 44 or 34 kb of chromosomal DNA, respectively (386, 442). The gene resides on the X chromosome in all mammals studied, and therefore the locus is functionally hemizygous. The *HPRT* gene encodes the nonessential purine salvage enzyme hypoxanthine phosphoribosyltransferase, which catalyzes the formation of IMP and GMP from PP_i and hypoxanthine or guanine. This activity can also convert purine analogs such as 6-thioguanine into nucleotides which, if incorporated, result in cell death. Consequently, cells with impaired or absent *HPRT* activity can readily be selected by their resistance to 6-thioguanine. The locus has been used extensively for the screening of chemicals for mutagenic activity (235). Several studies have addressed the mutational specificity of 254-nm UV radiation in this system.

We have already discussed the influence of error-free (or mainly error-free) repair modes (such as nucleotide excision repair) as a potentially important factor in determining the relative mutagenicity of photoproduct sites. In recent years it has become evident that the transcriptional status of a gene can have a significant influence on the rate and extent of the removal of photoproducts. It is often observed that the transcribed strand is repaired more rapidly and efficiently than the nontranscribed strand (385). This observation holds true for the *HPRT* gene in Chinese hamster cells (619). These and related topics have been extensively discussed in chapters 7 and 8.

An experiment described above showed that the time window between UV irradiation and the beginning of S phase was crucial for the mutational yield in repair-proficient cells but had no significant influence in cells defective in nucleotide excision repair (297) (Fig. 12–10). By using the same method, the influence of excision repair and its bias toward the transcribed strand was investigated at the sequence level of UV radiation-induced *hprt* mutants recovered (381). Untransformed normal and XP-A human fibroblasts were treated with UV radiation at different times after release from confluence (G_1 phase), before the onset of S phase. Mutant clones were selected by resistance to thioguanine, and mutations in individual clones were determined by PCR. Assuming that most targeted mutations are opposite photoproducts, the strand containing premutagenic lesions can be determined.

If a strand-discriminating repair mode is functional and there is time to remove premutagenic photoproducts before DNA replication, mutational events in excision-proficient cells irradiated in the G_1 phase should occur preferentially opposite photoproducts in the nontranscribed strand (where repair is less efficient). If there is no time for repair before photoproducts can lead to misinsertion events during S phase, no strand bias should be apparent and only the initial photoproduct frequency should determine the distribution of mutations between the two strands. This is the expected result when both normal and excision-deficient cells are irradiated in S phase and when XP cells are irradiated in G_1, since the factor time for repair is irrelevant in such cells.

In agreement with related studies on endogenous or integrated genes (115, 120, 499, 607, 618), almost all recovered mutations can indeed be attributed to sites of potential pyrimidine dimers, and there is no evidence for untargeted events. However, it must be mentioned that these data are based on a much smaller number of mutant clones than are the data obtained in shuttle vector studies. Isolation of independent fibroblast mutant lines is a much more tedious task than the selection of forward mutations in a plasmid vector transformed into *E. coli*! Transitions are again the most frequent events, and a significant number of them (~20%) are tandem double substitutions (Fig. 12–19). The distribution among different classes of transitions and transversions does not seem to depend on functional nucleotide excision repair or on the time of irradiation in relation to S phase (Fig. 12–19).

Expectations regarding the strand specificity of mutations are, in general, confirmed (Fig. 12–19). The majority of substitution events detected in excision-repair-proficient cells irradiated in G_1 can be attributed to photoproducts in the nontranscribed strand. This is not the case if excision repair is inactive, i.e., in XP cells irrespective of the time of irradiation and in normal cells irradiated in S phase. Unexpectedly, a tendency toward mutations opposite photoproducts in the *transcribed* strand is also evident (an effect that has also been observed in *E. coli* [see chapter 11]).

As indicated previously (see chapter 7), this effect is even more dramatic in several studies involving Chinese hamster cell lines, including excision-defective mutant lines from different genetic complementation groups (389, 611, 618–620). Figure 12–19 shows the results of a study on the specificity and strand distribution of mutational events following treatment of a mutant lung fibroblast cell line (V-H1), which is defective in the rodent homolog of the human *XPD* gene (see chapter 8), and the parental cell line (V79) with a low dose of UV radiation (2 J/m²) delivered to asynchronously dividing cells (618, 619). The majority of recovered mutations are apparently targeted to photoproducts, but the mutational spectrum indicates differences compared with the human fibroblasts discussed above (Fig. 12–19). In the repair-proficient parental cells, transitions and transversions were recovered with approximately equal frequency. In the repair-deficient cell line, however, all the recovered base substitutions were $G \cdot C \rightarrow A \cdot T$ transitions. Addi-

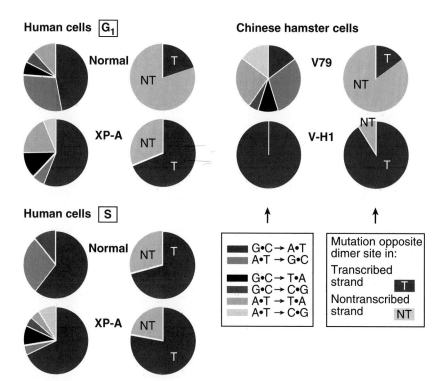

Figure 12–19 Relative probabilities of UV-induced base pair substitutions and strand assignment of *hprt* mutations recovered from diploid human fibroblasts and Chinese hamster cells. T and NT indicate a base change opposite a potential pyrimidine dimer site in the transcribed strand or nontranscribed strand, respectively. Normal and XP-A human fibroblasts synchronized in G_1 or S phase were treated with UV radiation fluences of 6.5 or 0.5 J/m², respectively. Unsynchronized normal hamster lung fibroblast cells (V79) and V-H1 cells deficient in nucleotide excision repair (the rodent homolog of human XP-D cells) were UV irradiated with 2 J/m². Note the predominance of G·C → A·T transitions (except in V79 cells), which becomes more pronounced in excision-deficient cells. Note also the strand bias in excision-competent cells and its reversal when nucleotide excision repair is inhibited (XP-A cells, V-H1 cells, normal human cells when irradiated in S). (*Data from McGregor et al. [381] and Vrieling et al. [381, 618, 619].*)

tionally, the proportion of tandem double substitutions and frameshift mutations was increased.

These differences notwithstanding, the reversed strand bias observed in human excision-deficient cells was confirmed in experiments with these rodent cells. Whereas 85% of the mutations occurred opposite photoproducts in the nontranscribed strand in repair-proficient cells (the bias was less pronounced at higher doses [618, 619]), only 10% did so in the V-H1 mutant. Essentially the same results were found in CHO cell lines AA8 and the AA8-derived excision-deficient mutants UV5 (*XPD*) (389) and UV24 (*XPB*) (620).

What is the explanation for this surprising, apparently general reversed strand bias of mutation induction in the *HPRT* gene as a consequence of defective nucleotide excision repair? As suggested in chapter 7, an unequal distribution of potential sites of pyrimidine dimer formation among the different strands might explain this result. However, examination of the sequence did not reveal a sufficiently strong bias of this type (618). Another possibility is that this reflects a selection bias, i.e., a pecularity of this particular gene is that mutations opposite C residues (G·C → A·T transitions were found in these excision-defective cells almost exclusively [Fig. 12–19]) in the transcribed strand have a higher probability of resulting in an inactive enzyme than does misinsertion opposite C residues on the nontranscribed strand (381). It has been argued that the observed bias is too pronounced for such an explanation (620). Finally, it has been suggested that the strand bias of misinsertion might be explained by different modes of replication of the two DNA strands (620), for example, if the polymerases which replicate the leading and lagging strands have different error rates or different processivities. However, as discussed previously, an overall enhanced mutational bypass frequency during leading strand DNA synthesis was not observed during SV40 origin-dependent replication of UV-irradiated templates in HeLa cell extracts (596).

As described above, inactivation of nucleotide excision repair in rodent cells also causes a dramatic shift in the frequency of different types of base substitutions; mutational events were due almost exclusively to $G \cdot C \rightarrow A \cdot T$ transitions and often tandem substitutions, with an overall significantly higher photoproduct mutagenicity (389, 618, 620). One can interpret this effect as the elimination of a distortion of the mutagenic spectrum introduced by excision repair, which removes mutagenic and nonmutagenic (or less-mutagenic) lesions with different efficiencies. Hence, true mutagenic lesions may be uncovered only in the absence of repair. Indeed, (6-4) photoproducts are removed significantly more quickly from the rodent genome than are cyclobutane pyrimidine dimers (394) (see chapter 7), and these lesions are therefore prime candidates for causing the increased mutagenicity with different specificity observed in the absence of nucleotide excision repair. The studies on the intrinsic mutagenicity of photoproducts in *E. coli*, without any bias introduced by nucleotide excision repair, support this view (see chapter 11). A (6-4) photoproduct in a defined position on a single-stranded DNA fragment is 10 to 20 times more mutagenic than is a pyrimidine dimer located in the same position (329).

Further support for this notion stems from reversion studies. A spontaneous revertant of the highly UV-sensitive and hypermutable V-H1 mutant strain has been isolated and manifests intermediate UV radiation sensitivity but an overall slightly lower mutability compared with the parental wild-type cell line (409, 648, 649). Removal of pyrimidine dimers as a whole from the genome was severely inhibited, apparently without major consequences for mutagenicity. However, removal of (6-4) photoproducts was normal. This leads to the conclusion that (6-4) photoproducts are the major mutagenic lesions in these Chinese hamster lines.

A similar correlation of (6-4) product removal and UV-induced mutability has been reported for a human XP-A revertant cell line (88). Mutability was normal, as was (6-4) photoproduct removal. However, pyrimidine dimer repair remained defective. In contrast to the Chinese hamster revertant line described above, survival was comparable to that of the wild-type line and was apparently not affected by impaired removal of pyrimidine dimers. These similarities notwithstanding, in general the nature and relative importance of the premutagenic lesions in UV-irradiated human cells remain to be elucidated. In contrast to the situation in rodent cells, defective nucleotide excision repair does not lead to differences in transition and transversion mutation frequencies in the endogenous *HPRT* gene (381) (Fig. 12–19). Additionally, the previously discussed studies on the effect of photoreactivation on UV-irradiated shuttle vectors in human and simian cells have convincingly shown that pyrimidine dimers can be a major source of mutations (62, 473).

Mutational Analysis of the *p53* Gene

One of the major advances in tumor biology in recent years has been the identification of distinct genes whose altered activity or enhanced expression (for oncogenes) or whose inactivation or diminished expression (for tumor suppressor genes) is required for, or at least correlated with, the process of neoplastic transformation. However, the causal relationship between genotoxic agents that enhance cancer risk and their specific effects on cellular target oncogenes or tumor suppressor genes is not straightforward (206). Skin cancer risk is enhanced in individuals with XP, and skin tumors appear preferentially in areas exposed to sunlight (see chapter 14). Together with the known enhanced frequency of mutations opposite UV-induced photoproducts due to defective repair, these observations suggest that skin cancer might be an ideal case for establishing such cause-and-effect relationships. The assumption that derives from the somatic mutation hypothesis is that skin cancer in patients with XP is caused by mutations in critical oncogenes or tumor suppressor genes and that these mutations are induced by photoproducts that have been characterized in cells in culture. If mutations in these genes are indeed recovered from skin tumors, the molecular nature of these mutations should be consistent with the known characteristics of photoproduct-induced mutations discussed above. These expectations have been confirmed in several such studies (61, 124).

Activation of the oncogene N-*ras*, together with amplification of the c-*myc* and c-H-*ras* oncogenes, has been observed in squamous cell carcinomas from patients with XP (569), and activating point mutations in N-*ras* map to potential pyrimidine dimer sites (278, 454, 569, 609). Similarly, insertion of a chemically synthesized pyrimidine dimer at a known activation site of N-*ras* led to transformation of cells after transfection (269).

The analysis of numerous inactivating mutations in the *p53* gene recovered from basal and squamous cell carcinomas in normal individuals and patients with XP or from murine skin cancers has substantiated the targeting of mutations to pyrimidine dimer sites (61, 124, 291, 397, 476, 523, 654). The *p53* tumor suppressor gene is the most frequently altered gene in a variety of human cancers; we will discuss its function and regulation in more detail in chapter 13. Point mutations have been detected in the *p53* gene in ~50% of skin tumors. Virtually all are targeted to dipyrimidine sites. Approximately 80% of all base substitutions are G·C → A·T transitions. Hot spots are present, some at identical positions in both non-XP- and XP-derived tumors (124). A high percentage of tandem substitutions (~60%) (exclusively CC → TT events) have been detected in tumors from patients with XP (124). With the exception of reactive oxygen species (484) and bisulfite (79), such tandem substitutions are rarely observed with agents other than UV radiation. These events are not detected in tumors in internal organs from the same patients. Additionally, dipyrimidine sites account for only ~60% of the *p53* mutations found in such tumors (61, 125).

In tumors derived from individuals with XP (Fig. 12–20A), a strand bias for *p53* mutations is also evident. The vast majority of mutations occur opposite photoproduct sites on the nontranscribed strand (124). Since this bias differs from that described above for XP-A cells in vitro, it had been suggested that most of the (largely unclassified) patients in this study belong to XP complementation group C, since competence for the preferential repair of pyrimidine dimers in the transcribed strand of the *p53* gene is retained in XP-C fibroblasts (146).

Mutant allele-specific PCR and ligase chain reactions have been designed to detect CC → TT substitution events at certain mutational hot spots in the *p53* gene (415). Such events are indeed detected after UV irradiation of cultured human skin cells and also in samples of Sun-exposed normal skin from a patient with skin cancer. In summary, the sequence specificity and molecular nature of mutational events in mutant *p53* genes recovered from skin tumors correlate well with in vitro data on mutations targeted to photoproducts. Thus, the significance of pyrimidine dimers for the molecular pathogenesis of skin cancer is now well established.

Ligation-mediated PCR (see chapter 7) allows the detection of initial damage incidence and repair rates in vivo at nucleotide resolution (408, 448, 449). A correlation of both factors with the distribution of mutational hot spots in the *p53* gene has been attempted (603, 604). The relative frequency of pyrimidine dimers and (6-4) photoproducts at different dipyrimidines known from bacterial studies has been confirmed to occur in diploid human fibroblasts over a range of UV radiation doses (604) (see chapter 1). Cytosine methylation apparently excludes (6-4) photoproduct formation, and this has been invoked to explain the absence of these lesion at certain sites in the *p53* gene (604). Some damage hot spots have been found to correlate with mutational hot spots in the *p53* gene from non-XP skin tumors. However, a very low photoproduct frequency was detected at other sites (e.g., codon 248, Fig. 12–20A).

A better correlation has been observed with repair rates of individual cyclo-

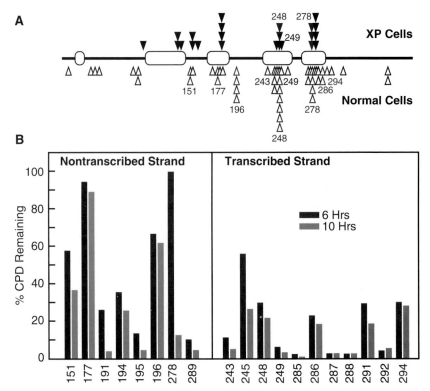

Figure 12–20 Correlation of positions of mutational hot spots in the *p53* gene derived from skin tumors with slow repair rates at individual pyrimidine dimer positions as measured by ligase-mediated PCR. (A) Open triangles indicate the number of mutations recovered from individuals without XP; solid triangles represent those found in skin tumors of patients with XP. The regions marked reflect domains that are conserved in *p53* genes from different organisms. (B) Relative repair rates for individual dimer sites on the transcribed and nontranscribed strands are expressed as the amount of cyclobutane dimers (CPD) remaining in normal skin fibroblasts after 6 and 10 h following UV radiation at a dose of 12 J/m². Note the overall faster removal of dimers from the transcribed strand but also the considerable site-specific variation. The numbers in panel B correspond to the map positions in panel A. Note the correlation between sites of mutational hot spots and those characterized by relatively slow dimer removal (indicated by red position numbers). (*Panel A adapted from Dumaz et al. [124]. Data for panel B from Tornaletti and Pfeiffer [603].*)

butane pyrimidine dimer sites (603). Repair rates overall are somewhat higher for the transcribed strand. However, individual rates are highly variable and are characterized by several "slow spots" (603). Sites of slow repair correlated with mutational hot spots in non-XP cells (Fig. 12–20B).

Summary and Conclusions

In the absence of a complete genetic framework, analyses of mutational mechanisms in mammalian cells have relied almost exclusively on studies of mutational specificity. We have limited our discussion to the effects of UV-C irradiation. There is an extensive literature on the mutagenic effects of other known DNA-damaging agents in mammalian cells, including UV-B radiation, oxidative damage, ionizing radiation, and various mutagenic chemicals. The use of these agents has extended our informational database on mutational spectra. However, in general, such studies have made limited contributions to our understanding of the molecular process(es) of mutagenesis in mammalian cells.

The initial hope that mutational processes would leave easily recognizable fingerprints and signature events has not been realized. Given our current incomplete understanding of the effects of chromatin structure, of the transcriptional status of genes, and of the secondary structure of DNA on the kinetics of error-free repair at a given site of premutagenic damage, this result is not unexpected. The efficiency of repair reflects one layer of complexity. Another layer is introduced by the even less well understood factors which determine replicative bypass efficiency and insertion preference, with possible contributions of untargeted processes. Elucidation of the mechanisms involved in mutational events induced by agents which can result in several different mutagenic adducts, which may be subject to different repair mechanisms (exemplified by certain alkylating agents [67]), provides a further daunting challenge.

Regardless of these current limitations, the study of UV-induced mutagenesis in mammalian cells has provided important insights, one of which is the unequivocal correlation of mutations opposite radiation-induced photoproducts with the inactivation of tumor suppressor genes in skin cancers. The majority of studies have come to the conclusion that most UV-induced mutations result from misinsertion opposite photoproducts during semiconservative DNA replication, and there is no reason to believe that other processes contribute significantly to overall mutagenicity. Inducible mechanisms do not seem to play the essential role that they play in *E. coli*. With the notable exception of the *HPRT* locus in rodent cells, the class of mutations with the highest incidence is that of $G \cdot C \rightarrow A \cdot T$ transitions, reflecting an insertion bias of the DNA polymerase(s) involved in translesion synthesis, a limited base pairing capacity of pyrimidine dimers, or the frequent deamination of C residues in photoproducts (see chapter 11). However, the nature of the mutagenic photoproduct(s) in mammalian cells is still unclear. Most probably, pyrimidine dimers and (6-4) photoproducts are mutagenic, with their individual contributions depending heavily on the particular system studied and the inherent relative kinetics of prereplicative repair of each species of damage.

DNA Strand Break Repair and Recombination in *S. cerevisiae*

In the second part of this chapter we will turn to eukaryotic *damage tolerance mechanisms associated with recombinational events*. As in *E. coli*, there is overwhelming evidence that these are initiated by strand breaks (590). Indeed, the occurrence of random strand breakage is in general regarded as the major source of spontaneous mitotic recombination events in eukaryotic cells (497). The two principal models for the initiation of recombination between homologous double-stranded DNA molecules, the *single-strand break invasion model* and the *double-strand break gap repair model,* were introduced in chapter 10 (see Fig. 10–25 and 10–33). The validity of these models in eukaryotic cells has been extensively debated, and the reader is referred to more specialized texts for detailed information. In the present discussion we will use the yeast *S. cerevisiae* as a model system. Genes involved in recombinational pathways are conserved among lower and higher eukaryotes. The same is apparently true for recombinational mechanisms. However, the probability that a recombinogenic substrate will be processed by a particular pathway differs between *S. cerevisiae* and higher eukaryotic cells.

DNA Strand Breaks and Their Repair after Treatment with Ionizing Radiation

Ionizing radiation induces a variety of DNA lesions (see chapter 1). These include single- and double-strand breaks, DNA-protein cross-links, S1 endonuclease-sensitive sites indicative of clustered base damage, and single-base damage such as 8-hydroxypurine, formamidopyrimidine, and thymine or cytosine glycol. Typically, single-strand breaks cannot be rejoined by one-step ligation reactions since ionizing radiation frequently results in damaged 3′ termini and often in a single-base gap rather than a simple nick. Hence, nonrecombinational repair requires the action of multiple enzymes involved in base excision repair (see chapter 4).

It is now widely accepted that the occurrence of double-strand breaks is the factor that determines whether yeast cells and other eukaryotic cells will die after treatment with ionizing radiation (162, 163, 240). In yeast cells, double-strand

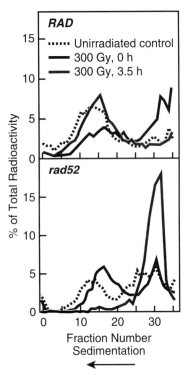

Figure 12–21 Detection of X-ray-induced double-strand breaks in yeast chromosomal DNA by neutral sucrose gradient centrifugation. Sedimentation profiles of labeled haploid yeast chromosomal DNA are shown before treatment, immediately after X-ray treatment (300 Gy), and following 3.5 h of postirradiation incubation (red lines). A shift to a lower average molecular weight is evident following X-ray treatment. The restoration of high-molecular-weight DNA after incubation is observed in wild-type cells (*RAD*, top panel). The latter is not detected in the *rad52* mutant (bottom panel), which is defective in double-strand break repair and shows excessive degradation of nuclear DNA. The presence of a higher fraction of low-molecular-weight DNA than in the wild-type strain even in unirradiated cells can be attributed to the presence of degraded DNA in nondividing cells in *rad52* cultures. (*Adapted from Ho [225] with permission.*)

breaks increase linearly with radiation dose under various irradiation conditions, and mutants that are defective in double-strand break repair show pure exponential survival curves (162, 164). For these cases it has been calculated that one double-strand break per cell corresponds to one lethal hit (161). When differences in DNA content are taken into account, the survival curves of yeast and mammalian cells are quite similar, and the higher resistance of yeast cells can be explained by their significantly lower DNA content (smaller target) (163).

S. cerevisiae is an excellent eukaryotic system for studying strand break repair. A major reason is that the small size of the yeast chromosomes allows their isolation intact. The repair of double-strand breaks can then be detected by analyzing the molecular weight distribution of chromosomal DNA in neutral sucrose gradients (Fig. 12–21) (225, 487). More recently, pulsed-field gel electrophoresis techniques have been adapted to the analysis of strand break repair in individual chromosomes (Fig. 12–22) and to the analysis of chromosome abnormalities following treatment with ionizing radiation (94, 178).

Aneuploidy is a frequent consequence of double-strand breaks in yeast cells. The significance of repair processes involving homologous chromosomes is well demonstrated by the observation that chromosomes are lost with high frequency at sublethal γ-ray doses if the only counterpart to a particular chromosome in a diploid G₁ *S. cerevisiae* cell is a functionally equivalent chromosome of divergent sequence (e.g., of *S. carlsbergensis*) (488). The absence of precise homology apparently limits recombinational repair. If cells are irradiated in G₂, the repair of double-strand breaks by sister chromatid recombination is preferred over recombination with the homologous chromosome (267).

X-ray-induced double-strand breaks also constitute dominant lethal events.

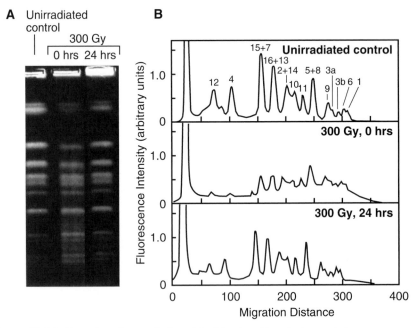

Figure 12–22 Detection of γ-ray-induced double-strand breaks in yeast chromosomes separated by orthogonal-field-alternation gel electrophoresis. The original agarose gel (A) and a quantitative analysis of band fluorescence (B) are shown. Note the decrease in signal intensities of individual bands in γ-irradiated cells (300 Gy) concomitant with an increase of DNA fluorescence interspersed between bands. Numbers at peaks indicate a putative chromosome assignment, some chromosomes cannot be separated under these conditions. Additionally, this diploid wild-type strain is heterozygous for the length of chromosome 3 (note peaks 3a and 3b). The more slowly migrating, larger chromosomes are more susceptible to breakage. Restoration of the original profile after 24 h of incubation indicates repair of chromosomal double strand breaks. (*Courtesy of A. Friedl and F. Eckardt-Schupp.*)

Although inactivated by X rays, a mutant defective in double-strand break repair can still undergo mating with an unirradiated cell. However, the diploid fusion product is unable to grow (226). We shall reexamine this lethal effect of double-strand breaks in *trans* in chapter 13.

There are indications that yeast cells can repair low levels of staggered double-strand breaks with intact ends (such as those introduced by restriction enzymes) by direct ligation (527). However, mitotic recombinational mechanisms appear to be the only way of overcoming DNA double-strand breaks induced by ionizing radiation. In *S. cerevisiae*, most of the genes required for homologous recombination fall into the *RAD52* epistasis group mentioned above. Mutants with mutations in the genes *RAD50* through *RAD57* are characterized by sensitivity to ionizing radiation and to chemicals known to induce strand breakage and typically do not show significantly enhanced UV radiation sensitivity or altered UV-induced mutability (171, 172, 216). The recombinational nature of the underlying repair process, requiring the presence of two DNA copies, is indicated by the relative X-ray resistance of haploid G_2 and diploid wild-type cells, which is completely abolished in mutants in this group (172, 406). Most of these mutants also show defects in meiosis and frequently produce inviable spores (144). There is good evidence that at least certain kinds of meiotic and mitotic recombination events are initiated by double-strand breaks (70, 175, 295, 379, 574) which require processing by gene products encoded by the *RAD52* group of genes. Mutations in genes in this group of mutants also frequently inactivate or delay mating-type switching, which involves the processing of targeted double-strand breaks and the transfer of sequence information from a silent mating-type locus (*HML* or *HMR*) to the transcriptionally active *MAT* locus (see chapter 6) (200, 298, 565).

The recombinogenic effects of single-strand breaks or gaps have been less well studied. Agents such as UV radiation, which are not typically associated with the induction of double-strand breaks, are known to induce various types of recombination (142, 154, 302). This effect is dramatically enhanced in mutants defective in nucleotide excision repair, presumably because unexcised lesions create a higher frequency of single-strand gaps during replication (216, 268, 302). The accumulation of single-strand breaks and enhanced recombination frequencies (30, 173) have been observed when strains containing a *cdc9* mutant gene (which encodes a thermoconditional DNA ligase) are incubated at the restrictive temperature. Furthermore, expression of the gene II protein of bacteriophage f1 (which results in nicking at a specific recognition site engineered between two selectable yeast marker genes) induces homologous recombination events (566). In this case, the location of the single-strand break acts as the recipient site for gene conversion.

The Genetic Framework

In this section we shall describe the known characteristics of genes in the *RAD52* epistasis group and then move on to a few selected experimental systems that have provided insights into mechanisms of recombinational repair. This account is focused on mitotic events. Although some mutant phenotypes indicate overlap between pathways of mitotic and meiotic recombination, the characteristics of meiotic recombination are not necessarily relevant for repair events in mitotically dividing cells.

THE *RAD50* GENE

The *RAD50* gene encodes a predicted 153-kDa protein containing a nucleotide-binding domain in the N-terminal region and two regions of heptad repeat sequence separated by a large repeat-free spacer (12). Such a motif can form an α-helical structure in which the two repeat units can coil around one another and form a regular helical structure termed an α-helical coiled coil, as observed in myosin or keratin. The purified protein binds to double-stranded DNA in an ATP-dependent reaction (481), but ATP hydrolysis may not be required for this. No evidence for preferential binding to specific sequences or to DNA ends has been observed. Rad50 protein forms a homodimer at high ionic strength (481). Gradual lowering of the ionic strength induces a pattern of changes in its physical properties reminiscent of those detected in other coiled-coil proteins. Bending of the heptad-free hinge region may be an important determinant of the observed transitions. A transient increase in *RAD50* mRNA levels has been observed during meiosis. However, the protein level stays constant (480).

Strains carrying point mutations in the *RAD50* gene are highly sensitive to X rays and MMS (171, 174, 216) and are impaired in double-strand break repair (485, 570). However, in contrast to mutants with mutations in other genes of the same epistasis group, spontaneous mitotic recombination (242, 375) and the chromosomal integration of linearized plasmids by homologous recombination (11, 527) are not reduced. The mutants are capable of mating-type switching, albeit with slower kinetics (242).

The meiotic defect conferred by mutations in *RAD50* has been studied in detail (11, 70, 371, 372, 485). A deletion or mutation of a conserved residue within the ATP-binding consensus domain results in marked sensitivity to MMS and blockage of meiosis at an early stage (11). There is also a failure to undergo so-called synaptonemal complex formation, a process that yields a tripartite structure which connects condensed homologous paired chromosomes during meiotic prophase (11). It was initially suggested, on the basis of the dimensions of the coiled-coil domain, that Rad50 itself might be part of the synaptonemal complex (12). It seems more likely that the inability to form this structure is a consequence rather than the cause of the failure to undergo meiotic recombination in *rad50* cells (431). Recombinogenic double-strand breaks are not formed in a *rad50* deletion mutant during meiosis (70) (Fig. 12–23).

A point mutation in *RAD50* (*rad50S*) has been characterized in detail. This mutation does not confer sensitivity to MMS and allows progression of meiosis to an intermediate stage (11). In contrast to *RAD50* deletion mutants, in these mutants persistent strand breaks at recombinational hot spots are generated but are not processed appropriately (Fig. 12–23) (70). Apparently these breaks do not undergo processing that results in single-stranded tails typically found in wild-type cells (574). The ability of these *rad50* mutants to carry out meiotic recombination has been further analyzed in "return-to-growth" experiments (141). Meiosis in a

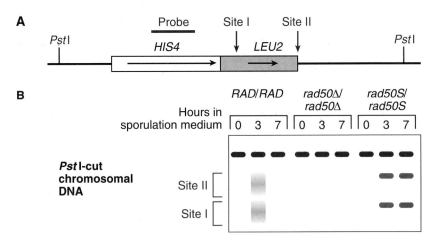

Figure 12–23 The *RAD50* gene is required for the induction and processing of meiotic double-strand breaks in *S. cerevisiae*. (A) Two sites of double-strand breaks in the region of the *LEU2* gene during meiosis I can be physically mapped by Southern analysis with the depicted probe. (B) Concomitant with a commitment to recombination, signals corresponding to these break sites can be detected in DNA prepared at different times after induction of synchronous meiosis by incubation in sporulation medium. The signals are diffuse and transient in the wild type (*RAD/RAD*), suggesting some kind of degradative processing. No signals are detected in a *rad50* deletion mutant (*rad50Δ/rad50Δ*). However, certain point mutants of *RAD50* (*rad50S/rad50S*) show distinct and persistent bands, suggesting the accumulation of unprocessed double-strand breaks. (*Adapted from Cao et al. [70] with permission.*)

sporulating culture was interrupted at different times by plating on vegetative growth medium, and gene conversion events among heteroalleles (indicating a commitment to meiotic recombination) were scored. Neither *rad50* point mutants nor null mutants show meiosis-specific enhancement of recombination during incubation in sporulation medium (11).

How might a protein function both in the *repair* of double-strand breaks in mitotic cells and in the *induction* and correct processing of double-strand breaks in meiotic cells? It has been suggested that Rad50 protein binds to certain hypersensitive sites in chromatin which are destined to be double-strand break sites during meiosis (481). After these strand breaks have formed, the continued presence of Rad50 protein at positions of double-strand breaks may provide a specific signal or a nucleation point for the recombinational machinery. Hence, the latter function would explain its repair function in mitotic cells.

The phenotype of *xrs2* mutants is similar to that of *rad50* mutants; the strains are viable and sensitive to ionizing radiation, produce inviable spores, and are blocked in an early step of meiotic recombination (241). A deletion mutant is viable and capable of mating-type switching but only with the delayed kinetics observed in *rad50* mutants (242). The *XRS2* gene encodes a predicted protein of 96.3 kDa with no apparent homology to other known proteins (242).

THE *RAD51* GENE AND ITS HOMOLOGS

The highly X-ray- and MMS-sensitive *rad51* mutants are defective in double-strand break repair (94), in spontaneous and radiation-induced mitotic recombination, and in the formation of viable spores during meiosis (171, 405, 447, 512). Cloning of the gene revealed a predicted 43-kDa protein with significant homology to the bacterial RecA protein (see chapter 10) (1, 547) (Fig. 12–24). The two proteins share a region of 30% amino acid identity corresponding to amino acid 154

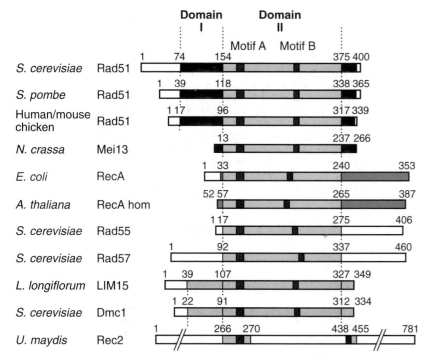

Figure 12–24 Sequence comparison of Rad51 protein and homologs isolated from fungal, plant, chicken, mouse, and human cells with *E. coli* RecA protein. The various proteins are aligned relative to a homologous core (domain II [pink]) containing motifs A and B of the ATP-binding consensus sequence of RecA (red bars). The different grey-shaded areas indicate regions that are conserved only in a subfamily of these proteins. Open boxes indicate regions without apparent homology. (*Adapted from Heyer [224] with permission.*)

to 374 of Rad51 protein and 33 to 240 of RecA protein. This region of RecA protein is responsible for oligomer formation and recombination (426). It includes the Walker A- and B-type nucleotide-binding domains (623). The importance of this region for the repair and recombination function of Rad51 has been verified by site-specific mutational studies (547).

A *rad51* deletion mutant is viable, but spore formation during meiosis is reduced to ~10% of the wild-type level and hardly any of the recovered spores are viable (547). Meiosis is blocked at a later step than in a *rad50* deletion mutant. Double-strand breaks at recombinational hot spots are formed but are apparently not processed or are incorrectly processed (547). In spite of the dramatically lower level of spontaneous or damage-induced mitotic gene conversion, commitment to meiotic recombination (to a level of about 20 to 70% of the wild-type) has been detected in return-to-growth experiments (547). Mitotic intrachromosomal recombination between repeated sequences accompanied by a deletion of the intervening sequence ("pop-out" events) is not reduced in a *rad51* deletion mutant (547).

RAD51 mRNA levels are significantly increased during meiosis and are also regulated during the mitotic cell cycle, with higher levels found at the G_1/S boundary (33, 547). The *RAD51* promoter region contains two restriction sites for *MluI*, indicating the presence of sequences known to regulate genes involved in DNA metabolism that are coordinately expressed at higher levels at the G_1/S boundary (262). Additionally, the gene is inducible by DNA-damaging agents in mitotic cells (33, 547).

The *DMC1* gene encodes a predicted protein with 46% amino acid identity and 68% overall homology to *RAD51* (Fig. 12–24). It was isolated in a screen for genes specifically expressed during meiosis (41). The meiotic phenotypes of *rad51* and *dmc1* deletion mutants are indeed quite similar (41, 547). However, the mitotic levels of *DMC1* mRNA are almost undetectable, and deletion mutants do not confer a mitotic phenotype, e.g., resistance to MMS is not affected, and the frequency of spontaneous or damage-induced mitotic recombination is not reduced (41).

Other yeast proteins involved in recombinational repair, Rad55 and Rad57, also share homology with Rad51 (Fig. 12–24) (271, 350). The phenotype of mutants with point mutations and deletion mutations in the *RAD55* and *RAD57* genes indicates that these proteins are specifically required for strand break repair at low temperatures, since these mutants are X-ray sensitive at 23 but not 36°C (172, 227, 351). This may be explained by different repair mechanisms at different temperatures or by an additional requirement for Rad55 and Rad57 to stabilize a multimeric repair complex at low temperatures (172, 351).

Several homologs of Rad51 have been isolated from lower and higher eukaryotes (224), including *S. pombe* (247, 411, 546), *Neurospora crassa* (80), *Ustilago maydis* (510), *Arabidopsis thaliana* (75), *Lilium longiflorum* (546), and chicken (37), mouse (400, 546), and human (546) cells (Fig. 12–24). The predicted mouse and human proteins differ by only four amino acids. These proteins share 83% homology with the *S. cerevisiae* and *S. pombe* Rad51 proteins, 64% homology with Dmc1, and 55% homology with *E. coli* RecA protein (Fig. 12–24). The overexpressed mouse gene can complement the MMS sensitivity of an *S. cerevisiae rad51* deletion mutant (400).

Studies on the expression of these homologs have revealed an interesting tissue specificity. The mouse and chicken genes are highly expressed in spleen and thymus cells as well as in testis and ovary cells (37, 400, 546). This pattern of expression is consistent with an assumed role of the protein in double-strand break repair during meiosis or V(D)J recombination during lymphocyte development (see below).

Purified yeast Rad51 protein is endowed with ATP-dependent double- and single-stranded-DNA-binding activities (224, 427, 547). The protein also has an ATP-independent single-stranded-DNA-binding activity and a single-stranded DNA-dependent ATPase activity (427, 547). The protein binds to Rad52 protein as observed by its retention on Rad52-Sepharose columns (547) and as observed in the two-hybrid genetic system (392) (see chapter 6). The phenotypes of *rad51* and *rad52* mutants are quite similar (171, 172, 447) (see below).

In the presence of ATP, Rad51 protein forms a helical filament with double-stranded DNA, which is almost identical to that formed with RecA protein (see chapter 11) (427). Common features include extensive stretching and unwinding of B-DNA. Further comparison of the primary and three-dimensional structures of RecA and its homologs, including the bacteriophage T4 UvsX protein, indicates conservation of residues not only in or near the nucleotide-binding domains but also in other identically located and presumably functionally equivalent regions, e.g., in the hydrophobic core and in the monomer interface (564). This analysis has strengthened the assumption of a family of RecA homologs which have diversified and evolved from a common ancestral protein.

Indeed, like RecA, Rad51 catalyzes complete strand exchange between single-stranded and double-stranded homologous DNA in an ATP- and Mg^{2+}-dependent reaction in vitro (575). The single-strand DNA-binding protein replication protein A was found to be an important accessory factor for this reaction. Thus, Rad51 must be regarded as a true functional homolog of RecA. Whether all of the eukaryotic RecA homologs are endowed with similar activities remains to be demonstrated.

THE *RAD52* GENE

Mutants with mutations in *RAD52* are characterized by a general defect in meiotic (176, 373, 470) and mitotic (176, 244, 374, 470, 637) recombination, in the repair of double-strand breaks induced by ionizing radiation (171, 172, 174, 225, 259, 487), and in postreplicative repair following UV irradiation of cells defective in nucleotide excision repair (466). The gene is absolutely required for integration of linearized plasmids (527) (see below) and for mating-type switching (374); attempts to perform mating-type switching in a *rad52* background result in cell death (631). However, unequal sister chromatid exchange within the cluster of rDNA repeats and certain intrachromosomal events (discussed in more detail below) are not affected in these mutants (244, 390, 428, 469, 645).

Some data suggest a direct or indirect role of Rad52 protein in regulatory phenomena. A yeast DNA endonuclease has been identified by using antibodies raised against a single-stranded DNA-dependent endonuclease of *N. crassa*. The activity (which is not encoded by *RAD52* itself) is absent in a *rad52-1* mutant (81, 82). Certain transcripts that are inducible by DNA-damaging agents are also abnormally regulated in a *rad52* mutant (366, 545).

RAD52 has been cloned (6), and we have already mentioned the physical interaction of Rad52 with the RecA-like strand exchange protein Rad51 (392, 547; also see above). As with Rad51, there are indications that Rad52 is conserved among lower and higher eukaryotes. Like *RAD52*, the *rad22⁺* gene of *S. pombe* is involved in mating-type switching and mutants are highly γ-ray sensitive (429). The *rad22⁺* gene encodes a protein with significant homology to Rad52; a region of 126 amino acids in the N-terminal half shows amino acid identity (429). Degenerate oligonucleotide primers derived from this region were used to isolate a homologous single-copy chicken gene which is also characterized by conservation of this region (38). The region may be required for interaction with Rad51. Like *RAD51*, the chicken *RAD52* homolog is highly expressed in testis, ovary, and thymus cells (38). The *S. cerevisiae* gene is characterized by higher mRNA levels during meiosis (91). Unlike *RAD51*, however, the yeast *RAD52* gene is not inducible by DNA-damaging agents (90).

OTHER GENES IN THE *RAD52* EPISTASIS GROUP

The *RAD53* gene encodes an essential serine/threonine protein kinase (14, 651). The X-ray sensitivity of available mutants is best explained by a failure to arrest cell cycle progression correctly in response to DNA damage (14, 632). The role of such *checkpoint functions* is discussed in detail in chapter 13.

The amino acid sequence of the damage-inducible *RAD54* gene suggests a product with DNA helicase activity (137). This has already been discussed in a comparison with other potential helicases from the other two yeast epistasis groups for DNA repair (28, 113, 261, 532) (see chapter 6 and Fig. 6–31). Despite the presence of a severe defect in the repair of X-ray-induced double-strand breaks, meiotic recombination is less strongly affected in *rad54* mutants (172). A thermoconditional mutant (*rad54-3*) which is characterized by a defect in double-strand break repair at 36°C but undergoes nearly normal repair at 23°C has been informative in analyses of the lethal and recombinogenic effects of double-strand breaks in *S. cerevisiae* (65, 165).

Several other mutants which are defective in meiotic or mitotic recombination have been selected (10, 139, 447, 460, 496). These may or may not belong in the *RAD52* epistasis group. Some of these are X-ray sensitive (such as the thermoconditional mutant *rec1-1* mutant [139]), but others are not (460).

We have already mentioned the *S. pombe* and *U. maydis* homologs of *RAD51* and *RAD52*. Two other genes, designated *rad9*[+], and *rad21*[+] that are involved in the repair of X-ray damage have been isolated from *S. pombe* but have no identified counterparts in *S. cerevisiae* (40, 412). The *rad21*[+] gene is essential for mitotic growth (40). The *REC1* gene of *U. maydis* is another example of a gene that is involved in the repair of X-ray-induced damage but has no homology to other known genes (591).

Strand Exchange Proteins

Apart from Rad51, at least three more strand exchange proteins have been purified from *S. cerevisiae* (127, 203, 224, 296, 571). Sep1 (Dst2) is a large (175-kDa) protein that was purified from mitotic cells and can catalyze strand exchange between linear and circular double-stranded DNA (127, 260). However, unlike *RAD51* deletion mutants, *SEP1* deletion mutants are not sensitive to ionizing radiation and are only slightly UV-sensitive (601). Another strand exchange protein (Dst1) has been isolated from meiotic cells and, like Sep1 (Dst2), catalyzes strand exchange in an ATP-independent reaction (84). The phenotype conferred by a chromosomal deletion of the *DST1* gene is restricted to meiotic recombination.

Several other strand exchange proteins have been isolated from lower and higher eukaryotic cells, such as *U. maydis* (289), *D. melanogaster* (352), and human (399) cells. Their significance for repair and DNA damage tolerance phenomena is unknown.

Intrachromosomal Recombination and the Role of the *RAD1/RAD10* Nuclease

Numerous studies have indicated the existence of multiple, competing mitotic pathways of recombination in *S. cerevisiae*. The predominance of a particular process appears to depend on the experimental system studied (158, 201, 228, 288, 497, 529, 530, 650). Whereas mutants involved in general homologous recombination contain the *RAD52* group of genes, more-specialized intrachromosomal recombination events may also require the action of certain nucleotide excision repair proteins, notably the Rad1/Rad10 endonuclease (579, 602) (see chapter 6).

Nonfunctional heteroalleles of the same gene located on the same chromosome can recombine by various mechanisms to generate a functional copy. These mechanisms include gene conversion, pop-out events, and, apparently much less frequently, unequal sister chromatid exchanges (288). An illustrative example is the spontaneous reconstitution of a functional *HIS3* gene from a duplication contained on the same chromosome in which one copy contains a 3' deletion and the other contains a 5' deletion, with an overlapping region of sequence homology. Elimination of the *RAD1* or *RAD10* gene results in an approximately sixfold-lower frequency of these intrachromosomal recombination events. Deletion of *RAD52* has a similar effect (Table 12–4) (529, 530) (other studies have indicated an even more dramatic reduction [244]). The *rad1 rad10* double deletion does not show any further decrease of recombination frequencies with respect to the effect of deletion of either of these genes alone (i.e., the mutants behave epistatically). However, additional deletion of *RAD52* results in dramatically lower

Table 12–4 Effect of deletions in the yeast *RAD52*, *RAD1*, and *RAD10* genes on the frequency of spontaneous mitotic recomination between duplicated *HIS3* genes, one carrying a 3' deletion and the other carrying a 5' deletion

Genotype	Mean recombination rate (no. of *HIS3*⁺ recombinants/ 10⁴ viable cells)
RAD	5.0
rad1Δ	0.52
rad10Δ	0.46
rad52Δ	0.56
rad10Δ rad1Δ	0.84
rad10Δ rad52Δ	0.018

Source: Adapted from Schiestl and Prakash (530).

frequencies (i.e., the double mutants behave synergistically) (Table 12–4) (529, 530). This indicates that the *RAD52* gene and the *RAD1/RAD10* genes operate in different, competing pathways of intrachromosomal recombination.

RAD1 and *RAD10* are also required for efficient integration of circular and linearized plasmids (529, 530). However, they are not involved in meiotic gene conversion and intrachromosomal recombination between direct-repeat *his4* heteroalleles differing only by single base pairs (529, 530). This is in contrast to the direct-repeat arrangement of the *his3* deletion alleles discussed above. As noted below, it may be important that in the former case the region of homology is only 400 bp long. An influence of *RAD1* has been noted in other studies in which nondeletion heteroalleles of *HIS3* and *LEU2* were used as direct and inverted repeats (8, 288). The amount of intervening sequence may be critical in these experiments as well. *RAD1* affects primarily events which result in a deletion of a marker inserted between the recombined alleles (288).

Other intrachromosomal events influenced by *RAD1* (and presumably also by *RAD10*) are mediated by transcriptional enhancer sequences (594, 650). When *HOT1* (which constitutes an initiation and enhancer site for RNA polymerase I [614]), is placed upstream of a duplication of *his4* heteroalleles separated by a marker gene (*URA3*), the frequency of *HIS*⁺ recombinants is dramatically increased (650). This transcription-dependent increase is due mainly to an increase in events leading to a concomitant loss of the intervening marker and depends on the presence of functional *RAD1* (and presumably *RAD10*) and *RAD52* genes (650). Again, a synergistic interaction is observed in double mutants (594, 650).

The influence of *RAD1/RAD10* is apparently not restricted to intrachromosomal events. The increase in ectopic recombination between *homeologous* genes (i.e., recombination between genes of similar but not identical sequence located on different chromosomes [see chapter 9]) in a mutant defective in mismatch repair or in a topoisomerase mutant (*top3*) also depends on a functional *RAD1* (*RAD10*) gene (23, 24).

We have already mentioned that *S. cerevisiae* is capable of switching its mating type by unidirectional transfer of information from a transcriptionally silent gene to the transcriptionally active *MAT* locus (200). This reaction is initiated by a double-strand break targeted to a specific sequence in the *MAT* region. The responsible enzyme is an endonuclease encoded by the *HO* gene (298, 565). Cell cycle stage-specific expression of the *HO* gene normally restricts mating-type switching to the G₁ phase (417, 418). However, this system has been exploited to induce site-specific double-strand breaks synchronously at any given time in vivo and to permit investigators to study recombinational products and intermediates by Southern blotting (93). This is achieved by regulated expression of the *HO* gene controlled by the galactose-inducible *GAL10* promoter. Induction of double-strand breaks at the HO recognition site inserted into heterologous DNA stimulates recombination mitotically (420) and meiotically (295). If the target sequence is inserted between surrounding regions of homology, double-strand break induction results primarily in intrachromosomal recombination accompanied by a deletion of the intervening sequences (430, 511).

These processes have been studied in more detail in a centromeric plasmid system carrying a selectable marker (*URA3*) as well as two nonfunctional *lacZ* copies (Fig. 12–25) (157, 158). The HO recognition site was introduced into one *lacZ* copy, which was consequently rendered non-functional. The adjacent *lacZ* copy was not expressed because of the absence of an adequate promoter. Thus, the *lacZ* color marker could be used to detect successful repair events. A double-strand break at the HO site resulted in the restoration of a functional *lacZ*⁺ gene by the transfer of sequence information following gene conversion or deletion of the intervening sequences (Fig. 12–25) (158). Single-strand annealing was suggested as the mechanism for deletion. This process involves bidirectional single-strand degradation and homology search until regions of sufficient homology are exposed (Fig. 12–26). Indeed, deletion formation was observed to be decelerated when sequences of increasing length were inserted between the two *lacZ* copies (158). Furthermore, exposure of single-stranded DNA during deletion formation was detected by hybridizing single-strand-specific RNA probes to nondenatured recombination intermediates. Only 3′ overhanging DNA ends were observed.

Consistent with the general role of *RAD52* in recombination, a *rad52* mutant was completely blocked in the gene conversion pathway. However, deletion by single-strand annealing could still occur (158). On the other hand, both *rad1* and *rad10* mutants were severely inhibited in both processes, resulting in drastically reduced plasmid survival after introduction of the double-strand break (157, 198). Remarkably, this phenotype was alleviated if a sequence identical to the HO site (except for a single-base-pair change to prevent cutting at this second site) was

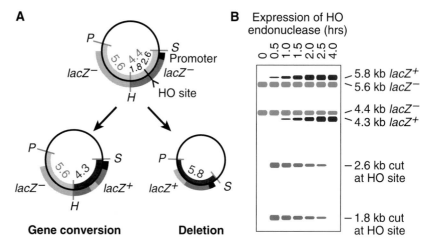

Figure 12–25 Plasmid system to monitor the repair of a double-strand break in a defined position. (A) The depicted centromeric plasmid contains two inactive copies of *lacZ* as direct repeats. One copy is inactivated by insertion of the recognition site for the HO endonuclease, and the other lacks a functional promoter. After induction of HO endonuclease expression, *lacZ*⁺ colonies can arise by gap repair and transfer of information from the second *lacZ* copy, leading to a gene conversion event without crossover. Alternatively, the deletion of the intervening sequences results in a single expressed intact *lacZ*⁺ gene copy. A mechanism for deletion formation is shown in Fig. 12–26. Probing DNA preparations digested with *Pst*I (P), *Hin*dIII (H), and *Sma*I (S) with an internal *lacZ* fragment in Southern blots allows the detection of certain diagnostic fragments (compare panels A and B). The presence of a 5.6-kb *Hin*dIII-*Pst*I and a 4.4-kb *Sma*I-*Hin*IdII fragment indicates unaltered *lacZ* copies present in the uncut plasmid. Fragments of 1.8 and 2.6 kb indicate the presence of a double-strand break in the formerly 4.4-kb fragment containing the *lacZ* copy with the HO site. Deletion results in the formation of a 5.8-kb *Pst*I-*Sma*I fragment. The presence of a 4.3-kb *Hin*dIII-*Sma*I fragment indicates a gene conversion event. (B) The schematized gel shows the different kinetics for deletion formation and gene conversion; the 5.8-kb fragment is detectable at earlier times after HO endonuclease induction than is the 4.3-kb fragment. (*Adapted from Fishman-Lobell et al. [158].*)

introduced into the second *lacZ* copy (157). The original double-strand break was in an immediate sequence context (i.e., the HO site), which did not have a homologous counterpart in the second *lacZ* copy. Thus, apparently the Rad1/Rad10 endonuclease (579, 602) is required to remove nonhomologous short 3' ends (Fig. 12–26) in order to complete the recombinational process. As discussed in chapter 6, the enzyme has been shown to be junction specific with specificity for duplex/3'-single-stranded regions in DNA (29). Recombination events which require such an endonuclease activity should be less frequent in a *rad1* or *rad10* background. This could also explain why this activity plays only a minor role, if any, in the repair of double-strand breaks introduced by ionizing radiation or during meiosis. In both cases, double-strand breaks are repaired mainly by recombination between homologous sequences and there is no need for the removal of nonhomologous stretches (157).

Illegitimate Recombination

We have noted that intrachromosomal recombination between repeated homologous elements can give rise to deletions and that these events contribute to genomic instability. However, even in cases when homologous or homeologous recombination is impossible or not favored, yeast cells have the ability to tolerate double-strand breaks. These "illegitimate" events have been investigated by the integration of plasmids linearized in a region which does not have any homology with the yeast genome (528). Analysis of the target sites indicates the presence of short stretches of homology (microhomologies) corresponding to the terminal base pairs of the plasmid and the occurrence of short deletions or duplications during the integration process.

The probability of homologous versus nonhomologous recombination in various repair mutants has been studied (Table 12–5) (527). Whereas all mutants of the *RAD52* epistasis group exhibit reduced integration frequencies in general, the effect is most dramatic in a *rad52* deletion mutant. *RAD50* seems to be involved almost exclusively in illegitimate events. *RAD52* seems to represent the opposite case; some integrants that arise by illegitimate recombination can be recovered in a *rad52* deletion mutant, but no integration due to homologous recombination has been detected.

The processing of double-strand breaks by events involving non-homologous sequences has also been studied with replicating plasmids (390) and in a selection system which allows conditional introduction of lethal strand breaks in a yeast chromosome (301). This can be achieved by regulated expression of the HO endonuclease and also by functional activation of the second centromere of a dicentric chromosome, resulting in strand breaks during mitosis. The latter can be achieved by silencing a nearby *GAL1* upstream activating sequence, which, if it is

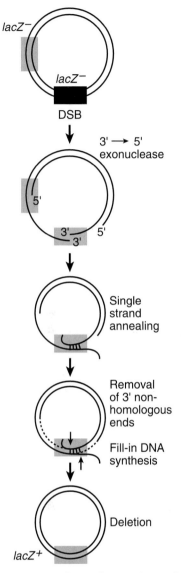

Figure 12–26 The single-strand annealing model of double-strand-break-initiated recombination. 5' → 3' degradation occurs bidirectionally at sites of breakage until homologous single-stranded regions are exposed. After annealing, nonhomologous 3' overhanging sequences are excised and remaining gaps are filled by repair synthesis and ligation. This mechanism leads to a deletion product with removal of the intervening DNA between the direct repeats. The Rad1/Rad10 endonuclease is required for this process (see chapter 6). (*Adapted from Fishman-Lobell et al. [158].*)

Table 12–5 Frequency of illegitimate recombination versus homologous integration after transformation of various yeast mutants with linearized plasmids

	Frequency[a] (transformants/μg of DNA) of:		Ratio of illegitimate/ homologous events
Genotype	Illegitimate recombination	Homologous integration	
RAD	18.2	408	0.05
rad1Δ	6.5	164	0.04
rad50Δ	0.07	264	0.0003
rad51Δ	1.0	60	0.02
rad52Δ	1.2	<0.03	>7

[a] The frequency of homologous integration is determined by using a linearized DNA fragment containing a marker gene (*URA3*) and flanking sequences homologous to another yeast gene (*LYS2*). The chromosomal *LYS2* gene is inactivated by this integration event. Illegitimate recombination was detected by use of a linearized plasmid carrying the same marker gene but no homology to yeast genes in the vicinity of the ends.

Source: Adapted from Schiestl et al. (527).

activating transcription, renders this "conditional" centromere nonfunctional. Since homologous recombination in yeast cells is so efficient, illegitimate events can be successfully studied only in a *rad52* background. In this background, induced double-strand breakage is highly lethal and clones that survive strand break conditions contain deletions of the HO target sequence or the conditional centromere. Deletions which span a significantly larger region than that required for inactivation of the centromere or the HO cut site can be recovered. Sequence analysis of the deletion junctions revealed very short regions of sequence identity at the deletion junctions (1 to 7 bp), consistent with single-strand annealing after $5' \rightarrow 3'$ exonuclease degradation (Fig. 12–27). Events at the HO cut site also include small duplications, which can be explained by annealing of the terminal residue of the 3' overhang and filling in by DNA synthesis from the 3' ends (Fig. 12–27).

Recombination in the Cellular Context

The outcome of recombinational repair cannot be fully explained by considering only the type of DNA damage, the availability of sequence homology, and the integrity of the recombinational pathways. In this section, we will introduce an additional layer of complexity by addressing aspects of other cellular processes that influence recombination in *S. cerevisiae*.

One area that we have already touched upon is the role of transcription. Hot spots for mitotic recombination have been found in promoter regions, e.g., in *GAL10* (593). Perhaps the best-studied example is *HOT1*, a region encompassing the enhancer and transcriptional initiation site of the rDNA repeat unit mentioned above (279, 347, 561, 614, 650). Yeast rDNA is organized as tandem repeat units. *HOT1* not only enhances mitotic recombination between these units (347) but also stimulates recombination when inserted into novel locations (279). The correlation between activated transcription and enhanced recombination is well established in this case (561, 614). A transcriptional terminator inserted between *HOT1* and recombining sequences eliminates the effect of *HOT1* (614). However, the molecular mechanism of this correlation remains to be determined. It is also unknown whether these findings can be generalized, and it is certainly premature to extend the concept of "transcription-coupled repair" to "transcription-coupled recombination."

The effect of transcription on meiotic recombination is less clear. In several instances, hot spots for meiotic double-strand breakage and recombination have

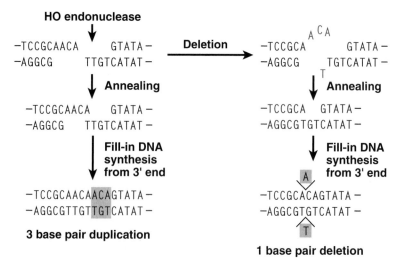

Figure 12–27 Proposed mechanism for the formation of small duplications and deletions at the HO recognition site. After the terminal base pairs of the HO cleavage site anneal, the gaps are filled, and thus a 3-bp (ACA) insertion is created (left). Loss of nucleotides from both 3' ends before annealing can result in small deletions (right). (*Adapted from Kramer et al. [301].*)

been found to coincide with transcriptional promoters (143, 199, 573), and certain transcriptional motifs such as poly(A) stretches are apparently important for this effect (536). However, active transcription counteracts rather than enhances the hot-spot character of a region (495). Mutational analysis of the *HIS4* promoter indicates that the binding of transcription factors and presumably the conformational changes associated with such binding correlate better with enhanced meiotic recombination than the transcriptional activity conferred by altered promoters (634).

Another aspect of the interplay between recombination and cellular controls concerns the position of a cell in the mitotic cell cycle at the time when a recombinogenic substrate originates. The issue of whether a particular cell cycle stage is a precondition for homologous mitotic recombination has been addressed for several systems (140, 147, 150, 191). In one system, diploid yeast cells carrying heteroallelic mutations that confer a thermoconditional block of an essential cell cycle step (e.g., *cdc4-1/cdc4-3*) were used (147). If incubated at the restrictive temperature, *cdc* mutant cells will arrest synchronously at a specific stage of the cycle (often termed the execution point). When cells are irradiated under these conditions, colonies recovered at the restrictive temperature must have a restored wild-type *CDC* gene as a result of intragenic recombination between the *cdc* heteroalleles, and this event must have occurred at the specific arrest stage. When a mutant that arrests in G_1 (*cdc4*) is used, the frequency of radiation-induced recombinants recovered is the same as that after irradiation of exponentially growing cells (147). This implies that intragenic mitotic recombination can occur in G_1. Analysis of further heteroallelic *cdc* strains indicates that mitotic intragenic recombination can occur at any stage of the cell cycle (150). However, recombination at restrictive temperatures is inhibited in strains heteroallelic for *cdc2*, indicating a role of the *CDC2*-encoded polymerase δ (Table 8-5) in mitotic gene conversion (149).

Finally, other cellular regulation phenomena can influence the probability of recombination and most probably the outcome at a specific site. The cytoplasmic transmission of enhanced recombination competence has been detected in a system conceptually similar to that described above for investigating untargeted mutations (Fig. 12–8) (152). A diploid strain homozygous for mating type and heterozygous for *ade6* alleles (a/a *ade6-21/ade6-45*) was mated to a UV- or γ-irradiated haploid strain containing the identical *ade6* alleles as a double mutation of opposite mating type (α *ade6-21,45*). The irradiated genome itself does not contribute to the ADE^+ recombinants recovered among the triploid progeny. Recombination levels above background and a dose-dependent increase in the frequency of such ''untargeted'' recombinants are observed, even if only cytoplasmic fusion and not nuclear fusion is allowed (*kar1* mutant background) (152). Thus, radiation damage can apparently induce a state of enhanced recombination competence that can be cytoplasmically transmitted to an untreated cell. Further studies indicate that this recombination-competent state can persist for up to five generations (148). This phenomenon has been known for long time, but a molecular explanation remains elusive. We have mentioned that certain genes of the *RAD52* group are DNA damage inducible. However, the increase in the level of transcripts studied is too short-lived to account for such a long-lived enhancement of recombination probability.

DNA Strand Break Repair and Recombination in Mammalian Cells

The Genetic Framework

At the outset it must be appreciated that the genetic framework for the repair of DNA strand breaks in mammalian cells is not nearly as extensive as that in yeast cells. Additionally, the assumption that strand break repair is effected mainly by recombinational mechanisms is less well founded for mammalian cells. Some of the mutants included in the following discussion are clearly defective in processes that are not necessarily recombinational, such as the ligation of DNA strand breaks. In fact, it has been suggested that the fast component of strand break rejoining in mammalian cells (in which the kinetics of double-strand break repair are typically biphasic) reflects direct DNA strand rejoining, whereas the slower component reflects recombinational processing (630).

X-ray-sensitive rodent cell lines fall into at least nine genetic complementation groups (92). Several cross-complementing human genes (*XRCC* genes) have been mapped to specific chromosomes (78, 188, 250, 600). The phenotype of these mutant cell lines is highly variable, and various degrees of cross-sensitivity to alkylating agents, topoisomerase inhibitors, and alkylating agents have been observed. Mutants in three complementation groups show a pronounced defect in double-strand break repair (187, 253, 281, 589, 635, 647). Mutants in these three groups (representative cell lines are XR-1, xrs1 to xrs7 [*XRCC5*], and V3) are among the most X-ray-sensitive mutant rodent cell lines. Other groups (such as EM9 [*XRCC1*] and irs1SF [*XRCC3*]) are characterized by a moderate defect in the repair of single-strand breaks (168, 597, 598), while mutants in yet another group display somewhat reduced rates of repair of both single- and double-strand breaks (BLM2) (494). Finally, mutants in groups represented by the Chinese hamster V79 cell lines irs1, irs2, and irs3 are normal with regard to single- and double-strand break repair (265, 266, 646), but at least one these (irs1 [*XRCC2*]) shows reduced fidelity of double-strand break rejoining in a plasmid assay (103). Mutants in the irs2 group are characterized by radioresistant DNA synthesis comparable to that of human AT cells (646).

THE *XRCC1* GENE

The human *XRCC1* gene has been cloned by complementation of the repair defect of the CHO mutant cell line EM9 (598, 599). These cells are characterized by enhanced sensitivity to EMS or ionizing radiation, a high level of sister chromatid exchange, and reduced homologous recombination following plasmid transfection (232, 597). The underlying defect appears to be a delayed repair of single-strand breaks in DNA (597). The *XRCC1* gene has been localized to the long arm of chromosome 19 (396). The 33-kb gene encodes a single transcript of 2.2 kb (598). The gene is not DNA damage inducible. However, high transcript levels are detected in testis, ovary, and brain cells (643). The 69.5-kDa predicted polypeptide is not homologous with any known protein, with the exception of two short regions of homology with the product of *S. pombe* checkpoint control gene *rad4⁺* (*cut5⁺*) (155, 514) (see chapter 13).

XRCC1 cDNA has been fused to a hexahistidine tag, thus facilitating purification of XRCC1 protein from cell extracts by binding to Ni^{2+}-nitrilotriacetic acid agarose (a technique termed *immobilized metal affinity chromatography*) (69). As already indicated in chapter 8, mammalian DNA ligase III copurifies with XRCC1 protein (69). DNA ligase III activity can be detected by formation of the adenylate-ligase intermediate in the presence of [α-³²P]ATP (69). No such intermediate is formed in extracts from EM9 mutant cells, but the activity can be restored after transformation with an *XRCC1* expression plasmid. Hence, complex formation of XRCC1 protein with mammalian DNA ligase III may be necessary for efficient ligase activity. This activity may be required for the repair of X-ray-induced single-strand breaks and those generated during base excision repair.

THE *XRCC5* GENE

A different line of study has provided interesting information on the molecular defect in the Chinese hamster complementation group represented by cell lines such as xrs5, XR-V15B, and XR-V9B, and on the function of the human homolog *XRCC5*. By using electrophoretic mobility shift assays, a DNA end-binding activity has been detected in nuclear extracts of CHO wild-type cells (Fig. 12–28) (186, 478). The protein(s) binds specifically to the ends of double-stranded DNA, but an affinity for closed-circular single-stranded molecules has been observed. The activity is not enhanced by pretreatment of cells with X rays. The potential involvement of this factor in DNA repair is indicated by the observation that the activity is absent in extracts of mutants of the *XRCC5* group (such as xrs5) (186, 478), which show a pronounced defect in the repair of double-strand breaks (280, 281,

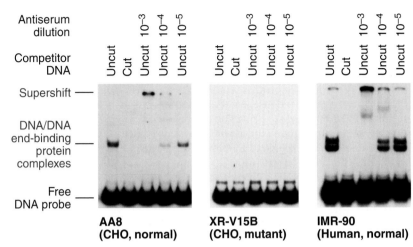

Figure 12–28 Detection of a DNA end-binding activity that is antigenically related (or identical) to the 70-kDa autoantigen Ku subunit, which is part of a DNA-activated protein kinase. With labeled linearized DNA as a probe, protein-DNA complexes can be detected as retarded bands in gel shift assays. An excess of unlabeled linearized DNA but not of uncut circular DNA can successfully compete for this DNA end-binding activity (compare lanes with cut and uncut competitor DNA). Addition of antiserum raised against the 70-kDa subunit of (human) autoantigen Ku results in a band of even lower mobility ("supershift") at the expense of the original retarded band. This indicates binding of the complex by the antibody. The DNA-end-binding activity is absent in an X-ray sensitive mutant CHO cell line (XR-V15B, a mutant of the XRCC5 group [middle panel]). A double band is found if extracts of certain normal human cells (such as the IMR-90 line, right panel) are probed. (*Adapted from Rathmell and Chu [479]. Courtesy of G. Chu.*)

647). These mutants are also characterized by a decreased frequency of chromosomal integration of transfected DNA (252). Treatment with azacytidine (a methyltransferase inhibitor) induces X-ray-resistant revertants of xrs5 cells with a high frequency, most probably by permitting the expression of an otherwise silent wild-type allele (251). Reversion to X-ray resistance is correlated with the reappearance of end-binding activity. Thus, this factor may participate in double-strand break repair by protecting free ends from further degradation.

The binding factor(s) is identical to the autoantigen Ku (186, 479). Ku consists of two subunits of 70 and 80 kDa and forms a complex with a 450-kDa polypeptide. This complex represents a well-characterized DNA-dependent protein kinase which is activated by free DNA ends (16). It has been suggested that this activity participates in a variety of regulatory responses, most notably cell cycle regulation in the presence of DNA damage (see chapter 13). It is interesting that the phenotype of xrs5 cells does indeed include defective transcriptional activation; several transcripts have been identified as being underexpressed in mutants of this rodent genetic complementation group (652, 653). Antibodies raised against the 70- and 80-kDa subunits bind to the complex containing the DNA-end-binding protein(s), indicated by a supershift detected in gel shift assays (Fig. 12–28) (186, 479). *XRCC5* maps to the same region of human chromosome 2 (78, 202, 250, 580a) as does the gene encoding the 80 kDa subunit of autoantigen Ku (68). Indeed, expression of the human 80-kDa Ku subunit in xrs5 hamster cells results in partial restoration of X-ray resistance and V(D)J recombination (553a, 580a). These results confirm the identity of *XRCC5* with the 80-kDa Ku subunit and suggest a role of DNA-dependent protein kinase in double-strand break repair.

Repair of Localized Double-Strand Breaks: V(D)J Recombination

The formation of the enormous repertoire of antigen-binding proteins, T-cell receptors, and immunoglobulins during the development of lymphoid cells is a remarkable example of a programmed DNA rearrangement with the purpose of generating diversity (15, 42, 180, 181, 343, 526). To this end, germ line V, J, and in some cases D segments are fused in a site-specific recombinational joining process to give rise to a functional antigen-binding molecule (Fig. 12–29A). Diversity is generated because several germ line V, J, and D segments can result in multiple fusion products and because specific mechanisms which generate inaccuracies are operative during the joining process. These segments are flanked by identical signal sequences that constitute a conserved heptamer and nonamer motif separated by a spacer of 12 or 23 bp (Fig. 12–29B). The recombination events occur at the border between the signal heptamer and coding sequences; the actual sequence at the specific break point is not relevant. Stable broken molecules reflecting these events have been directly detected (507, 533).

A double-strand break is introduced by a recombinase, and there is good circumstantial evidence that this recombinase is activated or encoded by the genes

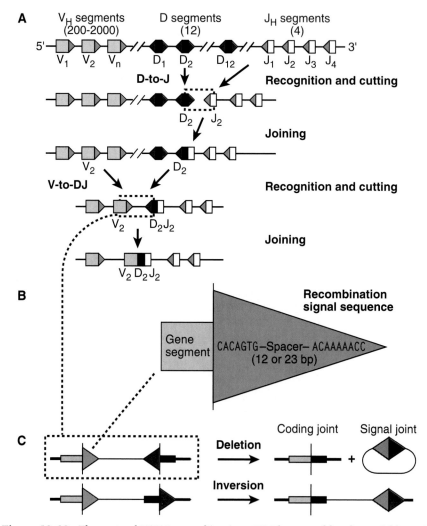

Figure 12–29 Elements of V(D)J recombination. (A) The assembly of a variable region gene by joining of segments from a large repertory of germ line V, D, and J segments is depicted. (B) The coding segments are flanked by recombination signal sequences (represented by arrowheads). These consist of conserved heptamer and nonamer motifs, separated by 12- or 23-bp spacers of variable sequence. Joining of the noncoding regions leads to a heptamer-heptamer fusion: the *signal joint*. As oulined in more detail in Fig. 12–30, the (usually imprecise) fusion of the ends of coding sequences results in the formation the *coding joint*. (C) Depending on the relative orientation of the recombination signal sequences, a deletion or inversion of the intervening region is created. (*Adapted from Schatz et al. [526] and Blackwell and Alt [42] with permission.*)

RAG-1 and *RAG-2* (for recombination activation genes) (15). These genes are expressed only in primary lymphoid tissues and in precursor lymphocytes, but their simultaneous overexpression in any cell type is sufficient to generate V(D)J recombinase activity (425, 525). Following strand breakage at two participating recombination sites, two kinds of joints are formed: a *coding joint* and a *signal joint* (Fig. 12–29C) (425, 525). Depending on the relative orientation of the signal sequences, this can lead to circularization and deletion of the intervening sequence or to an inversion (Fig. 12–29C). The signal joint is usually a precise heptamer-heptamer fusion. The coding joint, however, is imprecise, and there are apparently two mechanisms that can lead to variable insertions between the two coding ends. One of these is template independent, and the other is template dependent. The template-independent process involves random extension by *terminal deoxynucleotidyltransferase*, an enzyme which is expressed only in early lymphoid cells. For the template-dependent process, a model has been proposed which takes into account the unique molecular structures of the coding ends (181, 343). These are covalently linked, giving rise to hairpin structures (Fig. 12–30) (506). Subsequently, this hairpin can be randomly cut by a single-strand endonuclease. Such a process can explain the occurrence of insertions of short inverted repeats at coding joints (307). Further proposed events include elongation by terminal deoxynucleotidyl-transferase, homology search by single-strand annealing, gap filling by DNA synthesis, and trimming of single-stranded DNA ends which protrude from an otherwise double-stranded fused joint ("flaps") (Fig. 12–30).

The second recombination event distinct from V(D)J recombination, which occurs after antigen-induced differentiation of B lymphocytes, concerns a switch in the constant region of the heavy chain (42, 145, 204). This is accomplished by a deletion involving switch regions, i.e., 2 to 10-kb segments of DNA located upstream of heavy-chain constant region genes. A compilation of sequences of switch recombination sites does not reveal a conserved recognition motif. These events are apparently mediated by an as yet uncharacterized mechanism of illegitimate recombination (126).

V(D)J RECOMBINATION IN REPAIR MUTANTS

Tissue-specific elements are clearly required for V(D)J recombination. However, it is likely that components of the ubiquitous repair system(s) for double-strand breaks are also recruited for the processing of these site-specific breaks. Yeast mating-type switching provides such an example. It has long been suspected that defects in these common steps are responsible for immunodeficiencies frequently associated with human syndromes characterized by defective repair of exogenous DNA damage. V(D)J recombination has therefore been examined in these human cell lines and in the available rodent mutants defective in strand break repair (236, 446, 580, 581).

A convenient test system is facilitated by the availibity of the *RAG-1* and *RAG-2* genes. Following cotransfection on an expression plasmid, these genes can initiate V(D)J-like recombination in extrachromosomal substrates, regardless of the cell type. The recombination test plasmid contains two antibiotic resistance markers detectable in *E. coli* (Fig. 12–31) (223). Expression of one of these is possible only if a prokaryotic transcriptional termination sequence is removed or inactivated by recombinational excision or inversion. This sequence is flanked by V(D)J recombination signal sequences in different orientations to detect signal joint or coding-joint formation (Fig. 12–31). After transformation of *E. coli*, the fraction of chloramphenicol-resistant transformants among all ampicillin-resistant transformants is a measure of the V(D)J recombination frequency.

Cell lines from the three CHO cell complementation groups defective in double-strand break repair are defective in signal joint and coding-joint formation (446, 580, 581). In addition to a decrease in recombination frequencies in xrs5/xrs6 and XR-1 cells, recombinant plasmids have been shown to have suffered inaccurate joint formation and frequently contain large deletions (446, 581). These observations are consistent with a defect in a presumed DNA end-binding protein which prevents degradation and which is defective in the group of mutants dis-

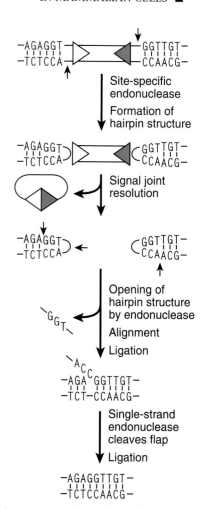

Figure 12–30 Model for coding-joint formation and mechanisms of template-dependent variability. After induction of double-strand breaks, coding-segment ends are sealed by the formation of hairpin structures. These are opened by random cleavage by a single-strand endonuclease, which may lead to variable gains or loss of base pairs. Single-strand ends are then aligned and ligated. Protruding single strands are trimmed, and gaps are filled if necessary. (*Adapted from Lieber [343] with permission.*)

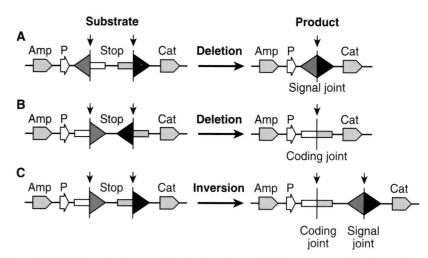

Figure 12–31 Plasmid substrates for the detection of V(D)J-type recombination. After passage through mammalian cells, the ampicillin resistance marker (Amp) is used to recover transformants in *E. coli*. The second antibiotic resistance marker (Cat), conferring resistance to chloramphenicol, is not expressed in unaltered plasmids since a transcriptional termination sequence has been inserted downstream of its promotor. This region is flanked by V(D)J recombination signal sequences. A deletion (A and B) or inversion (C) of the intervening terminator sequence resulting from V(D)J-type recombination in mammalian host cells enables the expression of chloramphenicol resistance. Depending on the orientation of the recombination signal sequences, coding (A and C) or signal (B) joint formation can be detected. (*Adapted from Lieber [344] with permission.*)

cussed above that are complemented by *XRCC5*. V3 cells were found to have near normal levels of V(D)J recombination, however, coding joint formation was reduced and frequently inaccurate (446, 580). Complementation of the radiosensitivity of these mutant cell lines by a human chromosome also restores V(D)J recombination in all cases (446, 580, 581).

X-ray-sensitive mutants proficient in double-strand break repair and UV radiation-sensitive CHO mutants (see chapter 8) are not impaired in V(D)J recombination (446, 581, 588). Additionally, cells from patients with ataxia telangiectasia (AT cells) or Bloom's syndrome or from a patient afflicted by an immunodeficiency disorder associated with a DNA ligase I deficiency (see chapter 14) are all normal for signal and coding-joint formation (236). These results indicate that the immundeficiencies in these syndromes are not caused by a general inability to perform V(D)J recombination. Furthermore, this process appears to be independent of DNA ligase I activity.

The possibility must be entertained that some of these human cell lines are defective in a particular aspect(s) of V(D)J recombination that cannot be simulated in vitro. For instance, there is evidence that efficient V(D)J recombination is correlated with the transcriptional activity of the V gene segments (15, 42), a condition which may not be adequately reproduced with a plasmid substrate. Furthermore, *RAG* gene activity is cell cycle regulated by phosphorylation under normal circumstances and is confined to the G_1/G_0 stage of the cell cycle (345, 346, 533). AT cells do not delay cell cycle progression in G_1 in the presence of strand breakage (see chapters 13 and 14). Hence, they may convert attempted V(D)J recombinations to lethal events. Continuous overexpression of plasmid-borne *RAG* genes may override such a cell cycle dependency of V(D)J recombination.

THE *scid* MUTATION

Mice that are homozygous for the *scid* mutation (severe combined immune deficiency) are characterized by the absence of functional B and T lymphocytes because of a severe deficiency in V(D)J recombination (51, 535). However, cells from these animals also manifest a pronounced sensitivity to DNA strand-breaking agents, such as X rays or bleomycin, that is associated with a defect in double-strand break repair (39, 169, 218). Enhanced sensitivity is not limited to lymphocytes; the *scid* phenotype represents a general repair defect (169). *scid* fibroblasts are also more sensitive to treatment with certain (but not all) cross-linking agents, but they are not sensitive to monofunctional alkylating agents or to UV light (584). A human gene that restores resistance toward ionizing radiation in *scid* mouse cells has been mapped to the centromeric region of chromosome 8 (306).

scid cells are capable of forming correct signal joints with near-normal efficiency (43, 344), but can only rarely form coding joints (219, 581). This occasional joining is not necessarily accompanied by major deletions, as found for two complementation groups involved in double-strand break repair. In this respect (and also regarding the degree of sensitivity toward DNA-damaging agents), *scid* cells resemble the V3 hamster cells mentioned above, and cell fusion data indicate that both mutations affect the same genetic locus (580). Coding ends with hairpin structure are detected as the major product of attempted V(D)J recombination in *scid* cells (506). A defect in the "opening" of hairpin structures is not a likely explanation, since typical inverted-repeat insertions can be found (205). The defective repair of random strand breakage introduced by ionizing radiation in *scid* cells also argues against a failure of such a specialized reaction.

Otherwise-lethal strand breakage can be rescued by illegitimate recombination in *scid* cells (376). Irregular fusion products with coding ends have been detected, and these have also been observed at a low frequency in wild-type cells (205, 343, 344). Such products involve indiscriminate fusions between coding and signal ends (*hybrid joints*) (341) or the insertion of short oligonucleotides cleaved from a coding end (*oligonucleotide capture*) (344). *scid* cells are also defective in the integration of exogenous DNA (205). To date, the various phenotypes of *scid* cells can probably best be accommodated by the assumption of a (possibly leaky) mutation affecting recognition and/or the joining of ends: perhaps a failure to resolve unusual DNA end structures (205).

Mammalian Proteins Possibly Involved in Recombination

There have been several reports of the partial purification of DNA strand-pairing activities from human cell extracts (74, 237, 277, 282). One such activity (designated HPP-1) has been purified to apparent homogeneity and is a 120-kDa polypeptide (399). HPP-1 has a 3' → 5' exonuclease activity and is stimulated by the trimeric human single-strand-binding factor RP-A. The significance of all of these activities for DNA repair is unclear.

Partial purification of an activity analogous to the *E. coli* RuvC resolvase (see chapter 11) from calf thymus and CHO cells has also been reported (134, 239). This activity is capable of symmetrically nicking synthetic Holliday junctions and recombinational intermediates formed in vitro by *E. coli* RecA protein. Its level is normal in hamster cells defective in the *ERCC1* gene, in xrs5/xrs6 mutant lines, and in murine *scid* cells (see below).

Plasmids containing double-strand deletions or gaps (i.e., linearized molecules) in an antibiotic resistance gene (*neo*) can be recombined in mammalian cell extracts with donor plasmids containing a homologous region spanning the deletion or gap in the recipient plasmid (257). Thereafter, homologous recombination events can be detected by transformation of *E. coli* and scoring of kanamycin-resistant colonies. This in vitro DNA transfer assay has been further developed to measure the exchange activity more quantitatively (257) and has been used to purify a recombinationally active multiprotein complex (258). In this assay, the recipient plasmid DNA containing gaps or deletions (70 to 390 bp) was labeled with biotin or digoxigenin and incubated with nuclear extract or extract fractions together with donor [³H] DNA (257). The recipient DNA was then bound to streptavidin-agarose or antidigoxigenin-Sepharose beads, and the physical association of ³H label with the washed beads was used as a measure of the extent of strand exchange. Completed recombination was verified by PCR performed with Sepharose-bound DNA (257). On the basis of elution characteristics on gel filtration columns, two purified recombination complexes (RC-1 and RC-2) have molecular masses of 550 to 600 and 250 kDa, respectively (258). RC-1 contains DNA polymerase ε, DNA ligase (most probably DNA ligase III), and low levels of 5' → 3' as well as 3' → 5' exonuclease activities (258). The most highly purified fractions lack helicase, topoisomerase, and terminal deoxynucleotidyltransferase activities. The recombinational activity of RC-1 is stimulated by bovine single-strand-binding protein.

Illegitimate Recombination and DNA End Joining

On the basis of experiments involving the integration of transfected DNA, it has long been assumed that homologous recombination is not favored in mammalian cells and that the majority of the integration events are the result of illegitimate recombination involving microhomologies (508, 509), not unlike the mechanisms described for *S. cerevisiae*. Recently, it was demonstrated that targeting frequencies of up to 78% can be achieved, and the use of strictly isogenic DNA in the regions of homology has been identified as a key factor (586). However, illegitimate events are likely to play a role in double-strand break repair. Sequence analysis of break points of deletions found in the *HPRT* gene after X-ray irradiation did not indicate sequence homologies, and it has been suggested that illegitimate recombination may be the underlying mechanism for the formation of chromosomal deletions in cells treated with ionizing radiation (402).

Joining of nonhomologous ends can be readily detected in cell extracts (177, 423), but this reaction may not always be faithful. A higher proportion of misjoining has been found in AT cell extracts (177, 423). Misjoining is caused by deletions involving regions of short repeats in the vicinity of the break site. Formally, this mechanism appears to be similar to the single-strand-annealing mechanism discussed above for yeast cells (Figure 12–27).

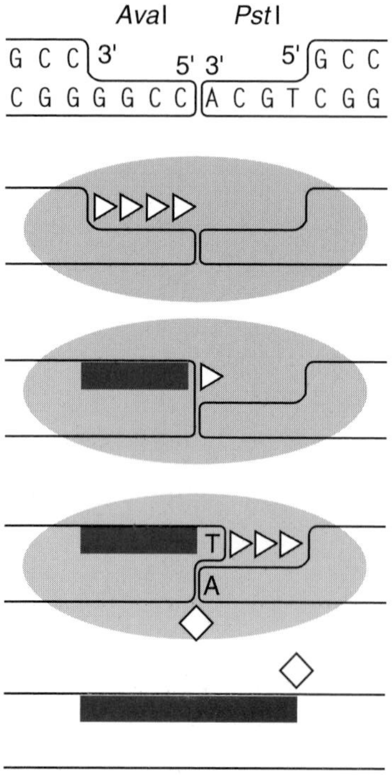

Blunt ends, as well as complementary and noncomplementary protruding ends, can be joined, although with different efficiencies. In this respect, *Xenopus* egg extracts are characterized by extremely promiscuous activities (331, 450, 592). Overhanging ends with noncomplementary sequence but with the same polarity are fused following alignment of single strands in such a way that the maximum number of base pairs is formed (one base pair is indeed sufficient). Exonucleolytic degradation or gap filling may be needed. The same type of reactions have been found after transformation of human cells with linearized plasmids (287, 353). One reaction is especially puzzling: *Xenopus* extracts are capable of fusing overhanging 5' ends with overhanging 3' ends without base loss or alteration, the gap apparently being filled in by error-free template-dependent DNA synthesis (Fig. 12–32) (592). Since no single-stranded DNA ligases are known, the existence has been postulated of an alignment protein that would provide enough stability to enable the priming of DNA synthesis from the recessed 3' end, DNA synthesis across the interrupted template, and, finally, ligation (Fig. 12–32) (592). Recent studies have indicated that the DNA polymerase itself can function as an alignment protein (286a).

Double-strand-break-joining activities which do not correspond to known DNA ligases have been purified from human cells (107, 153). One such activity results in the formation of circular monomers from linear molecules (153). Another results in the formation of concatemeric products from the same species of linear molecules with nonrandom head-to-head and tail-to-tail ligations (as opposed to random fusions including head-to-tail fusions) (107). During this reaction, 3' nucleotides are apparently modified. The activity resides in a high-molecular-weight complex and is associated with a double-stranded 3' → 5' exonuclease and with the homologous pairing protein HPP-1 (also capable of joining heterologous molecules). This has led to a hypothesis suggesting that covalent head-to-head or tail-to-tail linkage of linear molecules occurs after pairing of homologs and 3' → 5' end resection (Fig. 12–33) (107). A correlate for this reaction may be the cytological observation of the so-called breakage-fusion-bridge cycle of a broken chromosome (378). This cycle is characterized by head-to-head or tail-to-tail fusions of homologous broken sister chromatids and breakage at a second site of

Figure 12–32 Hypothetical mechanism for joining of 5' and 3' single-stranded ends of broken DNA molecules. The ends are held together by an alignment protein (oval in the background), allowing continuation of fill-in DNA synthesis primed at the recessed 3' end in order to bridge the unligated overhangs. Establishing the first base pair at the 3' overhang creates a substrate for DNA ligase (diamond). The fill-in synthesis is continued, and a second ligation completes the joining. (*Adapted from Thode et al. [592] with permission.*)

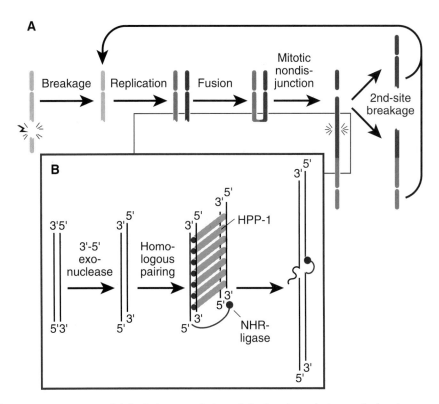

Figure 12–33 (A) Model depicting correlation of the breakage-fusion cycle for the perpetuation of double-strand-break-induced chromosome damage with the properties of an activity that can join nonhomologous ends purified from human cells (NHR ligase). (B) NHR ligase activity is stimulated by protein HPP-1, which promotes side-by-side pairing of homologous double-stranded DNA and thus leads to nonrandom, head-to-head and tail-to-tail joining by NHR ligase. The associated $3' \rightarrow 5'$ exonuclease activity creates single-stranded DNA ends facilitating covalent ligation of $5'$ overhangs and $3'$ recessed ends. The underlying mechanism may resemble the single-strand annealing process discussed previously (see Fig. 12–26) and may be accompanied by nucleotide loss. This reaction scheme can explain the cytologically observed end-to-end fusion of sister chromatids after induction of broken chromosomes, resulting in nondisjunction during mitosis and the occurrence of secondary break sites (A). (*Adapted from Derbyshire et al. [107].*)

the dicentric chromosome during mitosis, which can initiate further such cycles (Fig. 12–33).

UV Radiation-Stimulated Recombination

The recombinational filling of gaps opposite pyrimidine dimers was the earliest suggestion of such a damage tolerance mechanism in *E. coli*. The evidence for the existence of such a process in eukaryotic cells is inconclusive. Earlier in this chapter we discussed evidence that replication in mammalian cells can be completed in the presence of unexcised photoproducts. However, the bypass mechanism(s) seems to be replicative rather than recombinational. On the other hand, transfer of photodimers into daughter strand DNA does occur to some extent in mammalian cells. If cells are irradiated in G_1, sites sensitive to dimer-specific enzyme probes (corresponding to 1 to 3% of the dimers present in the parental DNA) can be detected in newly synthesized daughter strands (160). It is likely that the underlying mechanism is recombinational. The appearance of double-strand breaks in S phase following UV irradiation (of excision-deficient cells) is consistent with such a model (625).

Lesions caused by UV radiation are recombinogenic in mammalian cells. Irradiation of one or both genomes dramatically increases recombination between two SV40 genomes containing different *ts* mutations, upon transfection (183). The effect is not enhanced in preirradiated cells (608). Furthermore, integration of homologous transfected DNA in mammalian and yeast cells is typically enhanced by the presence of UV radiation damage in the extrachromosomal DNA (414, 421). Finally, transcription-stimulated repair of photoproducts correlates with a diminished frequency of intrachromosomal recombination (106).

Numerous aspects of the interaction of damaged DNA with cellular processes that stimulate recombination should be elucidated further. We have indicated earlier in this chapter that radiation-induced recombination in yeast cells can take place in G_1 without the need for formation of replication-dependent single strand gaps. Experiments on viral recombination in mammalian cells indicate that these events are also independent of replication (66) and possibly also of strand breaks introduced by nucleotide excision repair (4).

Summary and Conclusions

Space constraints limit our discussion to a few selected aspects of recombination in eukaryotes, which may be of special relevance to cellular responses to DNA damage, in particular the presence of DNA strand breaks. *RAD52*-dependent homologous recombination is the major repair pathway for double-strand breaks in *S. cerevisiae*. However, mutational inactivation of homologous recombination uncovers an array of alternative mechanisms, which may often be accompanied by deletions or other genetic alterations. Examples include single-strand annealing during homology search after strand resection and the joining of nonhomologous ends after the maximum possible base pairing has been established. So-called illegitimate events are in fact frequently directed by very short stretches of homology. Characterized gene products known to participate in these events include the Rad1/Rad10 nuclease, which is essential for nucleotide excision repair.

There is no indication of fundamental mechanistic differences in the various pathways of recombinational repair in higher eukaryotic cells, yeast cells, or even *E. coli*. RecA homologs from various eukaryotic organisms have been isolated, and for one of them, Rad51 of *S. cerevisiae*, strand exchange between homologous single- and double-stranded DNA has been demonstrated in vitro. However, the relative probability of a given operational mechanism seems to have changed during evolution. Illegitimate mechanisms may play a more important role in higher eukaryotes than in *S. cerevisiae*, and direct end joining may be an important mechanism for double-strand break repair in human cells. The frequent consequence of chromosomal translocation as the result of the fusion of broken heterologous chromosome ends is apparently critical for the activation of oncogenes (86).

We have discussed examples of targeted strand breakage in yeast (mating type switching) and human [V(D)J recombination] cells, which are accessible to experimental manipulation. Here the lessons are that these processes require event- or tissue-specific functions and that more-general strand break repair or recombination functions are also recruited. The same seems to hold for meiotic recombination.

The homologous pairing and strand exchange activities of Rad51 and the identity of a subunit of the DNA end-binding autoantigen Ku with the *XRCC5* gene are examples in which known biochemical activities are absent in double-strand-break-repair-deficient cells. However, the biochemical details of recombination reactions that are essential for repair are unknown. We have mentioned the significance of multimeric complexes on several occasions, and it is likely that, as with several other areas of DNA repair, a deeper understanding of the underlying molecular reactions involved in recombinational repair will require the characterization of more than individual gene products.

REFERENCES

1. **Aboussekhra, A., R. Chanet, A. Adjiri, and F. Fabre.** 1992. Semidominant suppressors of Srs2 helicase mutation of *Saccharomyces cerevisiae* map in the *RAD51* gene, whose sequence predicts a protein with similarities to procaryotic RecA proteins. *Mol. Cell. Biol.* **12:**3224–3234.

2. **Aboussekhra, A., R. Chanet, Z. Zgaga, C. Cassier-Chauvat, M. Heude, and F. Fabre.** 1989. *RADH*, a gene of *Saccharomyces cerevisiae* encoding a putative DNA helicase involved in DNA repair. Characteristics of *radH* mutants and sequence of the gene. *Nucleic Acids Res.* **17:**7211–7220.

3. **Abrahams, P. J., A. Houweling, and A. J. van der Eb.** 1992. High levels of enhanced reactivation of herpes simplex virus in skin fibroblasts from various hereditary cancer-prone syndromes. *Cancer Res.* **52:**53–57.

4. **Abrahams, P. J., B. A. Huitema, and A. J. van der Eb.** 1984. Enhanced reactivation and enhanced mutagenesis of herpes simplex virus in normal human and xeroderma pigmentosum cells. *Mol. Cell. Biol.* **4:**2341–2346.

5. **Abrahams, P. J., A. A. M. van der Kleij, R. Schouten, and A. J. van der Eb.** 1988. Absence of induction of enhanced recombination of herpes simplex virus in cells from xeroderma pigmentosum patients without skin cancer. *Cancer Res.* **48:**6054–6057.

6. **Adzuma, K., T. Ogawa, and H. Ogawa.** 1984. Primary structure of the *RAD52* gene in *Saccharomyces cerevisiae*. *Mol. Cell. Biol.* **4:**2735–2744.

7. **Aguilera, A., and H. L. Klein.** 1988. Genetic control of intrachromosomal recombination in *Saccharomyces cerevisiae*. I. Isolation and genetic characterization of hyper-recombination mutations. *Genetics* **119:**779–790.

8. **Aguilera, A., and H. L. Klein.** 1989. Yeast intrachromosomal recombination: long gene conversion tracts are preferentially associated with reciprocal exchange and require the *RAD1* and *RAD3* gene products. *Genetics* **123:**683–694.

9. **Ahne, F., M. Baur, and F. Eckardt-Schupp.** 1992. The *REV2* gene of *Saccharomyces cerevisiae*: cloning and DNA sequence. *Curr. Genet.* **22:**277–282.

10. **Ajimura, M., S.-H. Leem, and H. Ogawa.** 1992. Identification of new genes required for meiotic recombination in *Saccharomyces cerevisiae*. *Genetics* **133:**51–66.

11. **Alani, E., R. Padmore, and N. Kleckner.** 1990. Analysis of wild-type and *rad50* mutants of yeast suggests an intimate relationship between meiotic chromosome synapsis and recombination. *Cell* **61:**419–436.

12. **Alani, E., S. Subbiah, and N. Kleckner.** 1989. The yeast *RAD50* gene encodes a predicted 153-kD protein containing a purine-nucleotide binding domain and two large heptad-repeat regions. *Genetics* **122:**47–57.

13. **Albertini, R. J., L. M. Sullivan, J. K. Berman, C. J. Greene, J. A. Stewart, J. M. Silveira, and J. P. O'Neill.** 1988. Mutagenicity monitoring in humans by autoradiographic assay for mutant T lymphocytes. *Mutat. Res.* **204:**481–492.

14. **Allen, J. B., Z. Zhou, W. Siede, E. C. Friedberg, and S. J. Elledge.** 1994. The *SAD1/RAD53* protein kinase controls multiple checkpoints and DNA damage-induced transcription in yeast. *Genes Dev.* **8:**2416–2428.

15. **Alt, F. W., E. M. Oltz, F. Young, J. Gorman, G. Taccioli, and J. Chen.** 1992. VDJ recombination. *Immunol. Today* **13:**306–314.

16. **Anderson, C. W.** 1993. DNA damage and the DNA-activated protein kinase. *Trends Biochem. Sci.* **18:**433–437.

17. **Armstrong, J. D., D. N. Chadee, and B. A. Kunz.** 1994. Roles for the yeast *RAD18* and *RAD52* DNA repair genes in UV mutagenesis. *Mutat. Res.*, **315:**281–293.

18. **Armstrong, J. D., and B. A. Kunz.** 1990. Site and strand specificity of UVB mutagenesis in the *SUP4-o* gene of yeast. *Proc. Natl. Acad. Sci. USA* **87:**9005–9009.

19. **Armstrong, J. D., and B. A. Kunz.** 1992. Photoreactivation implicates cyclobutane dimers as the major UVB lesions in yeast. *Mutat. Res.* **268:**83–94.

20. **Ashman, C. R.** 1989. Retroviral shuttle vectors as a tool for the study of mutational specificity (base substitution/deletion/mutational hotspot). *Mutat. Res.* **220:**143–149.

21. **Ashman, C. R., P. Jagadeeswaran, and R. L. Davidson.** 1986. Efficient recovery and sequencing of mutant genes from mammalian chromosomal DNA. *Proc. Natl. Acad. Sci. USA* **83:**3356–3360.

22. **Bachmair, A., D. Finley, and A. Varshavsky.** 1986. *In vivo* halflife of a protein is a function of its amino-terminal residues. *Science* **234:**179–186.

23. **Bailis, A. M., L. Arthur, and R. Rothstein.** 1992. Genome rearrangement in *top3* mutants in *Saccharomyces cerevisiae* requires a functional *RAD1* excision repair gene. *Mol. Cell. Biol.* **12:**4988–4993.

24. **Bailis, A. M., and R. Rothstein.** 1990. A defect in mismatch repair in *Saccharomyces cerevisiae* stimulates ectopic recombination between homeologous genes by an excision repair dependent process. *Genetics* **126:**535–547.

25. **Bailly, V., J. Lamb, P. Sung, S. Prakash, and L. Prakash.** 1994. Specific complex formation between yeast RAD6 and RAD18 proteins: a potential mechanism for targeting RAD6 ubiquitin-conjugating activity to DNA damage sites. *Genes Dev.* **8:**811–820.

26. **Banerjee, S. K., A. Borden, R. B. Christensen, J. E. LeClerc, and C. W. Lawrence.** 1990. SOS-dependent replication past a single *trans-syn* T-T cyclobutane dimer gives a different mutation spectrum and increased error rate compared with replication past this lesion in uninduced cells. *J. Bacteriol.* **172:**2105–2112.

27. **Banerjee, S. K., R. B. Christensen, C. W. Lawrence, and J. E. LeClerc.** 1988. Frequency and spectrum of mutations produced by a single *cis-syn* thymine-thymine cyclobutane dimer in a single-stranded vector. *Proc. Natl. Acad. Sci. USA* **85:**8141–8145.

28. **Bang, D. D., R. Verhage, N. Goosen, J. Brouwer, and P. van de Putte.** 1992. Molecular cloning of *RAD16*, a gene involved in differential repair in *Saccharomyces cerevisiae*. *Nucleic Acids Res.* **20:**3925–3931.

29. **Bardwell, A. J., L. Bardwell, A. E. Tomkinson, and E. C. Friedberg.** 1994. Specific cleavage of model recombination and repair intermediates by the yeast Rad1/Rad10 endonuclease. *Science* **265:**2082–2085.

30. **Barker, D. G., A. L. Johnson, and L. H. Johnston.** 1985. An improved assay for DNA ligase reveals temperature-sensitive activity in *cdc9* mutants of *Saccharomyces cerevisiae*. *Mol. Gen. Genet.* **200:**458–462.

31. **Barnett, S. W., E. M. Landaw, and K. Dixon.** 1984. Test of models for replication of simian virus 40 DNA following ultraviolet irradiation. *Biophys. J.* **46:**307–321.

32. **Bartel, B., I. Wünning, and A. Varshavsky.** 1990. The recognition component of the N-end rule pathway. *EMBO J.* **9:**3179–3189.

33. **Basile, G., M. Aker, and R. K. Mortimer.** 1992. Nucleotide sequence and transcriptional regulation of the yeast recombinational repair gene *RAD51*. *Mol. Cell. Biol.* **12:**3235–3246.

34. **Basu, A. K., and J. M. Essigmann.** 1990. Site-specifically alkylated oligodeoxynucleotides: probes for mutagenesis, DNA repair and the structural effects of DNA damage. *Mutat. Res.* **233:**189–201.

35. **Beer, J. Z., E. D. Jacobson, H. H. Evans, and I. Szumiel.** 1984. X-ray and UV mutagenesis in two L51784 cell lines differing in tumorigenicity, radiosensitivity and DNA repair. *Br. J. Cancer* **49**(Suppl. VI):107–111.

36. **Berger, C. A., and H. J. Edenberg.** 1986. Pyrimidine dimers block simian virus 40 replication forks. *Mol. Cell. Biol.* **6:**3443–3450.

37. **Bezzubova, O., A. Shinohara, R. G. Mueller, H. Ogawa, and J.-M. Buerstedde.** 1993. A chicken *RAD51* homologue is expressed at high levels in lymphoid and reproductive organs. *Nucleic Acids Res.* **21:**1577–1580.

38. **Bezzubova, O. Y., H. Schmidt, K. Ostermann, W.-D. Heyer, and J.-M. Buerstedde.** 1993. Identification of a chicken *RAD52* homologue suggests conservation of the *RAD52* recombination pathway

throughout the evolution of higher eukaryotes. *Nucleic Acids Res.* **21:**5945–5949.

39. **Biedermann, K. A., J. R. San, A. J. Giaccia, L. M. Tosto, and J. M. Brown.** 1991. Scid mutation in mice confers hypersensitivity to ionizing radiation and a deficiency in DNA double-strand break repair. *Proc. Natl. Acad. Sci. USA* **88:**1394–1397.

40. **Birkenbihl, R. P., and S. Subramani.** 1992. Cloning and characterization of *rad21*, an essential gene of *Schizosaccharomyces pombe* involved in DNA double-strand-break repair. *Nucleic Acids Res.* **20:**6605–6611.

41. **Bishop, D., D. Park, L. Xu, and N. Kleckner.** 1992. *DCM1*: a meiosis-specific yeast homolog of *E. coli recA* required for recombination, synaptonemal complex formation, and cell cycle progression. *Cell* **69:**439–456.

42. **Blackwell, T. K., and F. W. Alt.** 1989. Mechanism and developmental program of immunoglobulin gene rearrangement in mammals. *Annu. Rev. Genet.* **23:**605–636.

43. **Blackwell, T. K., B. A. Malynn, R. R. Pollock, P. Ferrier, L. R. Covey, G. M. Fulop, R. A. Phillips, G. D. Yancopoulos, and F. W. Alt.** 1989. Isolation of *scid* pre-B cells that rearrange kappa light chain genes: formation of normal signal and abnormal coding joints. *EMBO J.* **8:**735–742.

44. **Bockstahler, L. E.** 1981. Induction and enhanced reactivation of mammalian viruses by light. *Prog. Nucleic Acid Res. Mol. Biol.* **26:**303–313.

45. **Bockstahler, L. E., and C. D. Lytle.** 1970. Ultraviolet light enhanced reactivation of a mammalian virus. *Biochem. Biophys. Res. Commun.* **41:**184–189.

46. **Bockstahler, L. E., and C. D. Lytle.** 1971. X-ray-enhanced reactivation of ultraviolet-irradiated human virus. *J. Virol.* **8:**601–602.

47. **Bockstahler, L. E., and C. D. Lytle.** 1977. Radiation enhanced reactivation of nuclear replicating mammalian viruses. *Photochem. Photobiol.* **25:**477–482.

48. **Boesen, J. J. B., N. Dietersen, E. Bal, P. H. M. Lohman, and J. W. I. M. Simons.** 1992. A possible factor in genetic instability of cancer cells: stress-induced proteins lead to decrease in replication fidelity. *Carcinogenesis* **13:**2407–2413.

49. **Boesen, J. J. B., S. Stuivenberg, C. H. M. Thyssens, H. Panneman, F. Darroudi, P. H. M. Lohman, and J. W. I. M. Simons.** 1992. Stress response induced by DNA damage leads to specific, delayed and untargeted mutations. *Mol. Gen. Genet.* **234:**217–227.

50. **Boram, R., and H. Roman.** 1976. Recombination in *Saccharomyces cerevisiae*: a DNA repair mutation associated with elevated mitotic gene conversion. *Proc. Natl. Acad. Sci. USA* **73:**2828–2832.

51. **Bosma, G. C., R. P. Custer, and M. J. Bosma.** 1983. A severe combined immunodeficiency mutation in the mouse. *Nature* (London) **301:**527–530.

52. **Bouffler, S. D., D. Godfrey, M. J. Raman, S. R. R. Musk, and R. T. Johnson.** 1990. Molecular cloning of a mammalian gene involved in the fixation of UV-induced mutations. *Somatic Cell Mol. Genet.* **16:**507–516.

53. **Bourre, F., A. Benoit, and A. Sarasin.** 1989. Respective roles of pyrimidine dimer and pyrimidine (6-4) pyrimidone photoproducts in UV mutagenesis of simian virus 40 DNA in mammalian cells. *J. Virol.* **63:**4520–4524.

54. **Bourre, F., G. Renault, and A. Gentil.** 1989. From simian virus 40 to transient shuttle vectors in mutagenesis studies. *Mutat. Res.* **220:**107–113.

55. **Bourre, F., G. Renault, and A. Sarasin.** 1987. Sequence effect on alkali-sensitive sites in UV-irradiated SV40 DNA. *Nucleic Acids Res.* **15:**8861–8875.

56. **Bourre, F., and A. Sarasin.** 1983. Targeted mutagenesis of SV40 DNA induced by UV light. *Nature* (London) **305:**68–70.

57. **Boyer, J. C., W. K. Kaufmann, B. P. Brylawski, and M. Cordeiro-Stone.** 1990. Defective postreplication repair in xeroderma pigmentosum variant fibroblasts. *Cancer Res.* **50:**2593–2598.

58. **Braggar, H., J. J. Cornelis, J. L. M. van der Lubbe, and A. J. van der Eb.** 1985. Mutagenesis in UV-irradiated simian virus 40 occurs predominantly at pyrimidine doublets. *Mutat. Res.* **142:**75–81.

59. **Brash, D. E., W. A. Franklin, G. B. Sancar, A. Sancar, and W. A. Haseltine.** 1985. *Escherichia coli* DNA photolyase reverses cyclobutane dimers but not pyrimidine-pyrimidone (6-4) photoproducts. *J. Biol. Chem.* **260:**11438–11441.

60. **Brash, D. E., and W. A. Haseltine.** 1982. UV-induced mutation hotspots occur at DNA damage hotspots. *Nature* (London) **298:**189–192.

61. **Brash, D. E., J. A. Rudolph, J. A. Simon, A. Lin, G. J. McKenna, H. P. Baden, A. J. Halperin, and J. Pontén.** 1991. A role for sunlight in skin cancer: UV-induced p53 mutations in squamous cell carcinoma. *Proc. Natl. Acad. Sci. USA* **88:**10124–10128.

62. **Brash, D. E., S. Seetharam, K. H. Kraemer, M. M. Seidman, and A. Bredberg.** 1987. Photoproduct frequency is not the major determinant of UV base substitution hot spots or cold spots in human cells. *Proc. Natl. Acad. Sci. USA* **84:**3782–3786.

63. **Bredberg, A., K. H. Kraemer, and M. M. Seidman.** 1986. Restricted ultraviolet mutational spectrum in a shuttle vector propagated in xeroderma pigmentosum cells. *Proc. Natl. Acad. Sci. USA* **83:**8273–8277.

64. **Bresler, S. E.** 1975. Theory of misrepair mutagenesis. *Mutat. Res.* **29:**467–474.

65. **Budd, M., and R. K. Mortimer.** 1982. Repair of double-strand breaks in a temperature conditional radiation-sensitive mutant of *Saccharomyces cerevisiae. Mutat. Res.* **103:**19–24.

66. **Burke, J. F., and A. E. Mogg.** 1993. UV-stimulated recombination in mammalian cells is not dependent upon DNA replication. *Mutat. Res.* **294:**309–315.

67. **Burkhart, J. G., and H. V. Malling.** 1993. Mutagenesis and transgenic systems: perspective from the mutagen, *N*-ethyl-*N*-nitrosourea. *Environ. Mol. Mutagen.* **22:**1–6.

68. **Cai, Q.-Q., A. Plet, J. Imbert, M. Lafage-Pochitaloff, C. Cerdan, and J.-M. Blanchard.** 1994. Chromosomal location and expression of the genes coding for Ku p70 and p80 in human cell lines and normal tissues. *Cytogenet. Cell Genet.* **65:**221–227.

69. **Caldecott, K. W., C. K. McKeown, J. D. Tucker, S. Ljungquist, and L. H. Thompson.** 1994. An interaction between the mammalian DNA repair protein XRCC1 and DNA ligase III. *Mol. Cell. Biol.* **14:**68–76.

70. **Cao, L., E. Alani, and N. Kleckner.** 1990. A pathway for generation and processing of double-strand breaks during meiotic recombination in *S. cerevisiae. Cell* **61:**1089–1101.

71. **Carty, M. P., J. Hauser, A. S. Levine, and K. Dixon.** 1993. Replication and mutagenesis of UV-damaged DNA templates in human and monkey cell extracts. *Mol. Cell. Biol.* **13:**533–542.

72. **Cassier, C., R. Chanet, J. A. P. Henriques, and E. Moustacchi.** 1980. The effects of three *pso*-genes on induced mutagenesis: a novel class of mutationally defective yeast. *Genetics* **96:**841–857.

73. **Cassier-Chauvat, C., and F. Fabre.** 1991. A similar defect in UV-induced mutagenesis conferred by the *rad6* and *rad18* mutations of *Saccharomyces cerevisiae. Mutat. Res.* **254:**247–253.

74. **Cassuto, E., L.-A. Lightfoot, and P. Howard-Flanders.** 1987. Partial purification of an activity from human cells that promotes homologous pairing and the formation of heteroduplex DNA in the presence of ATP. *Mol. Gen. Genet.* **208:**10–14.

75. **Cerutti, H., M. Osman, P. Grandoni, and A. T. Jagendorf.** 1992. A homolog of *Escherichia coli* RecA protein in plastids of higher plants. *Proc. Natl. Acad. Sci. USA* **89:**8068–8072.

76. **Chanet, R., C. Cassier, N. Magaña-Schwencke, and E. Moustacchi.** 1983. Fate of photo-induced 8-methoxypsoralen monoadducts in yeast. Evidence for bypass of these lesions in the absence of excision repair. *Mutat. Res.* **112:**201–214.

77. **Chanet, R., N. Magaña-Schwencke, and F. Fabre.** 1988. Potential DNA-binding domains in the *RAD18* gene product of *Saccharomyces cerevisiae. Gene* **74:**543–547.

78. Chen, D. J., M. S. Park, E. Campbell, M. Oshimura, P. Liu, Y. Zhao, B. F. White, and M. J. Siciliano. 1992. Assignment of a human DNA double-strand break repair gene (*XRCC5*) to chromosome 2. *Genomics* **13**:1088–1094.

79. Chen, H., and B. R. Shaw. 1994. Bisulfite induces tandem double CC → TT mutations in double-stranded DNA. 2. Kinetics of cytosine deamination. *Biochemistry* **33**:4121–4129.

80. Cheng, R., T. I. Baker, C. E. Cords, and R. J. Radloff. 1993. *mei-3*, a recombination and repair gene of *Neurospora crassa*, encodes a RecA-like protein. *Mutat. Res.* **294**:223–234.

81. Chow, T. Y.-K., and M. A. Resnick. 1987. Purification and characterization of an endo-exonuclease from *Saccharomyces cerevisiae* that is influenced by the *RAD52* gene. *J. Biol. Chem.* **262**:17659–17667.

82. Chow, T. Y.-K., and M. A. Resnick. 1988. An endo-exonuclease activity of yeast that requires a functional *RAD52* gene. *Mol. Gen. Genet.* **211**:41–48.

83. Ciechanover, A., and A. L. Schwartz. 1989. How are substrates recognized by the ubiquitin-mediated proteolytic system? *Trends Biochem. Sci.* **14**:483–488.

84. Clark, A. B., C. C. Dykstra, and A. Sugino. 1991. Isolation, DNA sequence, and regulation of a *Saccharomyces cerevisiae* gene that encodes DNA strand transfer protein α. *Mol. Cell. Biol.* **11**:2576–2582.

85. Clark, J. M., and P. C. Hanawalt. 1984. Replicative intermediates in UV-irradiated simian virus 40. *Mutat. Res.* **132**:1–14.

86. Cleary, M. L. 1991. Oncogenic conversion of transcription factors by chromosomal translocations. *Cell* **66**:619–622.

87. Cleaver, J. E. 1992. Replication of nuclear and mitochondrial DNA in X-ray-damaged cells: evidence for a nuclear-specific mechanism that down-regulates replication. *Radiat. Res.* **131**:338–344.

88. Cleaver, J. E., F. Cortés, L. H. Lutze, W. F. Morgan, A. N. Player, and D. L. Mitchell. 1987. Unique DNA repair properties of a xeroderma pigmentosum revertant. *Mol. Cell. Biol.* **7**:3353–3357.

89. Cleaver, J. E., R. Rose, and D. L. Mitchell. 1990. Replication of chromosomal and episomal DNA in X-ray-damaged human cells: a *cis*- or *trans*-acting mechanism? *Radiat. Res.* **124**:294–299.

90. Cole, G. M., D. Schild, S. T. Lovett, and R. K. Mortimer. 1987. Regulation of *RAD54*- and *RAD52-lacZ* gene fusions in *Saccharomyces cerevisiae* in response to DNA damage. *Mol. Cell. Biol.* **7**:1078–1084.

91. Cole, G. M., D. Schild, and R. K. Mortimer. 1989. Two DNA repair and recombination genes in *Saccharomyces cerevisiae*, *RAD52* and *RAD54*, are induced during meiosis. *Mol. Cell. Biol.* **9**:3101–3104.

92. Collins, A. R. 1993. Mutant rodent cell lines sensitive to ultraviolet light, ionizing radiation and cross-linking agents: a comprehensive survey of genetic and biochemical characteristics. *Mutat. Res.* **293**:99–118.

93. Connolly, B., C. I. White, and J. E. Haber. 1988. Physical monitoring of mating type switching in *Saccharomyces cerevisiae*. *Mol. Cell. Biol.* **8**:2342–2349.

94. Contopoulou, C. R., V. E. Cook, and R. K. Mortimer. 1987. Analysis of DNA double strand breakage and repair using orthogonal field alternation gel electrophoresis. *Yeast* **3**:71–76.

95. Cordeiro-Stone, M., J. C. Boyer, B. A. Smith, and W. K. Kaufmann. 1986. Xeroderma pigmentosum variant and normal fibroblasts show the same response to the inhibition of DNA replication by benzo[*a*]pyrene-diol-epoxide. *Carcinogenesis* **7**:1783–1786.

96. Cornelis, J., Z.-Z. Su, C. Dinsart, and J. Rommelaere. 1982. Ultraviolet-irradiated simian virus 40 activates a mutator function in rat cells under conditions preventing viral DNA replication. *Biochimie* **64**:677–680.

97. Cornelis, J. J., J. H. Lupker, and A. J. van der Eb. 1980. UV-reactivation, virus production and mutagenesis of SV40 in UV-irradiated monkey kidney cells. *Mutat. Res.* **71**:139–146.

98. Cornelis, J. J., Z. Z. Su, and J. Rommelaere. 1982. Direct and indirect effects of ultraviolet light on the mutagenesis of parvovirus H-1. *EMBO J.* **1**:693–699.

99. Cox, B. S., and J. M. Parry. 1968. The isolation, genetics and survival characteristics of ultraviolet light-sensitive mutants in yeast. *Mutat. Res.* **6**:37–55.

100. Dahle, D., T. D. Griffith, and J. G. Carpenter. 1980. Inhibition and recovery of DNA synthesis in UV irradiated Chinese hamster V-79 cells. *Photochem. Photobiol.* **32**:157–165.

101. DasGupta, U. B., and W. C. Summers. 1978. Ultraviolet reactivation of herpes simplex virus is mutagenic and inducible in mammalian cells. *Proc. Natl. Acad. Sci. USA* **75**:2378–2381.

102. Day, R. S., and C. H. J. Ziolkowski. 1981. UV-induced reversion of adenovirus 5 ts2 infecting human cells. *Photochem. Photobiol.* **34**:403–406.

103. Debenham, P. G., N. J. Jones, and M. B. T. Webb. 1988. Vector-mediated DNA double-strand break analysis in normal and radiation-sensitive Chinese hamster wild-type cells. *Mutat. Res.* **199**:1–9.

104. Defais, M. J., P. C. Hanawalt, and A. Sarasin. 1983. Viral probes for DNA repair. *Adv. Radiat. Biol.* **10**:1–37.

105. Dendy, P. P., and C. L. Smith. 1964. Effects on DNA synthesis of localized irradiation of cells in tissue culture by (i) a U.V. microbeam and (ii) an α-particle microbeam. *Proc. R. Soc. London Ser. B* **160**:328–344.

106. Deng, W. P., and J. A. Nickoloff. 1994. Preferential repair of UV damage in highly transcribed DNA diminishes UV-induced intrachromosomal recombination in mammalian cells. *Mol. Cell. Biol.* **14**:391–399.

107. Derbyshire, M. K., L. H. Epstein, C. S. H. Young, P. L. Munz, and R. Fishel. 1994. Nonhomologous recombination in human cells. *Mol. Cell. Biol.* **14**:156–169.

108. di Caprio, L., and B. S. Cox. 1981. DNA synthesis in UV-irradiated yeast. *Mutat. Res.* **82**:69–85.

109. Dinsart, C., J. J. Cornelis, M. Decaesstecker, J. van der Lubbe, A. J. van der Eb, and J. Rommelaere. 1985. Differential effect of ultraviolet light on the induction of simian virus 40 and a cellular mutator phenotype in transformed mammalian cells. *Mutat. Res.* **151**:9–14.

110. Dinsart, C., J. J. Cornelis, B. Klein, A. J. van der Eb, and J. Rommelaere. 1984. Transfection of extracellular UV-damaged DNA induces human and rat cells to express a mutator phenotype towards parvovirus H1. *Mol. Cell. Biol.* **4**:324–328.

111. Dion, M., and C. Hamelin. 1987. Relationship between enhanced reactivation and mutagenesis of UV-irradiated human cytomegalovirus in normal human cells. *EMBO J.* **6**:397–399.

112. Dixon, K., E. Roilides, J. Hauser, and A. S. Levine. 1989. Studies on direct and indirect effects of DNA damage on mutagenesis in monkey cells using an SV40-based shuttle vector. *Mutat. Res.* **220**:73–82.

113. Doe, C. L., J. M. Murray, M. Shayegi, M. Hoskins, A. R. Lehmann, A. M. Carr, and F. Z. Watts. 1993. Cloning and characterisation of the *Schizosaccharomyces pombe rad8* gene, a member of the *SNF2* helicase family. *Nucleic Acids Res.* **21**:5964–5971.

114. Dohmen, R. J., K. Madura, B. Bartel, and A. Varshavsky. 1991. The N-end rule is mediated by the *UBC2*(*RAD6*) ubiquitin-conjugating enzyme. *Proc. Natl. Acad. Sci. USA* **88**:7351–7355.

115. Dorado, G., H. Steingrimsdottir, C. F. Arlett, and A. R. Lehmann. 1991. Molecular analysis of ultraviolet-induced mutations in a xeroderma pigmentosum cell line. *J. Mol. Biol.* **217**:217–222.

116. Doudney, C. A. 1990. DNA-replication recovery inhibition and subsequent reinitiation in UV-radiation damaged *E. coli*: a strategy for survival. *Mutat. Res.* **243**:179–186.

117. Dowell, S. J., P. Romanowski, and J. F. X. Diffley. 1994. Interaction of Dbf4, the Cdc7 protein kinase regulatory subunit, with yeast replication origins in vivo. *Science* **265**:1243–1246.

118. Dowling, E. L., D. H. Maloney, and S. Fogel. 1985. Meiotic recombination and sporulation in repair-deficient strains of yeast. *Genetics* **109**:283–302.

119. Drinkwater, N. R., and D. K. Klinedinst. 1986. Chemically induced mutagenesis in a shuttle vector with a low-background mutant frequency. *Proc. Natl. Acad. Sci. USA* **83**:3402–3406.

120. Drobetsky, E. A., A. J. Grosovsky, and B. W. Glickman. 1987. The specificity of UV-induced mutations at an endogenous locus in mammalian cells. *Proc. Natl. Acad. Sci. USA* **84:**9103–9107.

121. Drobetsky, E. A., A. J. Grosovsky, and B. W. Glickman. 1989. Perspectives on the use of an endogenous gene target in studies of mutational specificity. *Mutat. Res.* **220:**235–240.

122. Drobetsky, E. A., A. J. Grosovsky, A. Skandalis, and B. W. Glickman. 1989. Perspectives on UV light mutagenesis: investigation of the CHO *aprt* gene carried on a retroviral shuttle vector. *Somatic Cell Mol. Genet.* **15:**401–409.

123. DuBridge, R. B., P. Tang, H.-C. Hsia, P.-M. Leong, J. H. Miller, and M. P. Calos. 1987. Analysis of mutation in human cells by using an Epstein-Barr virus shuttle vector. *Mol. Cell. Biol.* **7:**379–387.

124. Dumaz, N., C. Drougard, A. Sarasin, and L. Daya-Grosjean. 1993. Specific UV-induced mutation spectrum in the p53 gene of skin tumors from DNA-repair-deficient xeroderma pigmentosum patients. *Proc. Natl. Acad. Sci. USA* **90:**10529–10533.

125. Dumaz, N., A. Stary, T. Soussi, L. Daya-Grosjean, and A. Sarasin. 1994. Can we predict solar ultraviolet radiation as the causal event in human tumours by analysing the mutation spectra of the p53 gene? *Mutat. Res.* **307:**375–386.

126. Dunnick, W., G. Z. Hertz, L. Scappino, and C. Gritzmacher. 1993. DNA sequences at immunoglobulin switch region recombination sites. *Nucleic Acids Res.* **21:**365–372.

127. Dykstra, C. C., R. K. Hamatake, and A. Sugino. 1990. DNA strand transfer protein β from yeast mitotic cells differs from strand transfer protein α from meiotic cells. *J. Biol. Chem.* **265:**10968–10973.

128. Eckardt, F., and R. H. Haynes. 1977. Induction of pure and sectored mutant clones in excision-proficient and deficient strains of yeast. *Mutat. Res.* **43:**327–338.

129. Eckardt, F., and R. H. Haynes. 1977. Kinetics of mutation induction by ultraviolet light in excision-deficient yeast. *Genetics* **85:**225–247.

130. Eckardt, F., E. Moustacchi, and R. H. Haynes. 1978. On the inducibility of error-prone repair in yeast. *ICN-UCLA Symp. Mol. Cell. Biol.* **9:**421–423.

131. Eckardt, F., S.-J. Teh, and R. H. Haynes. 1980. Heteroduplex repair as an intermediate step of UV-mutagenesis in yeast. *Genetics* **95:**63–80.

132. Eckardt-Schupp, F., F. Ahne, E.-M. Geigl, and W. Siede. 1988. Is mismatch repair involved in UV-induced mutagenesis in *Saccharomyces cerevisiae*? p. 355–361. *In* R. E. Moses and W. C. Summers (ed.), *DNA Replication and Mutagenesis*. American Society for Microbiology, Washington, D.C.

133. Edenberg, H. J. 1983. Inhibition of simian virus 40 DNA replication by ultraviolet light. *Virology* **128:**298–309.

134. Elborough, K. M., and S. C. West. 1990. Resolution of synthetic Holliday junctions in DNA by an endonuclease activity from calf thymus. *EMBO J.* **9:**2931–2936.

135. Elespuru, R. K. 1987. Responses to DNA damage in bacteria and mammalian cells. *Environ. Mol. Mutagen.* **10:**97–116.

136. Ellison, K. S., E. Dogliotti, T. D. Connors, A. K. Basu, and J. M. Essigmann. 1989. Site-specific mutagenesis by O⁶-alkylguanines located in the chromosomes of mammalian cells: influence of the mammalian O⁶-alkylguanine-DNA alkyltransferase. *Proc. Natl. Acad. Sci. USA* **86:**8620–8624.

137. Emery, H. S., D. Schild, D. E. Kellogg, and R. K. Mortimer. 1991. Sequence of *RAD54*, a *Saccharomyces cerevisiae* gene involved in recombination and repair. *Gene* **104:**103–106.

138. Ennings, I., R. T. L. Groenendijk, A. A. van Zeeland, and J. W. I. M. Simons. 1985. Differential response of human fibroblasts to the cytotoxic and mutagenic effects of UV radiation in different phases of the cell cycle. *Mutat. Res.* **148:**119–128.

139. Esposito, M. S., and J. T. Brown. 1990. Conditional hyporecombination mutants of three *REC* genes of *Saccharomyces cerevisiae*. *Curr. Genet.* **17:**7–12.

140. Esposito, M. S. 1978. Evidence that spontaneous mitotic recombination occurs at the 2-strand stage. *Proc. Natl. Acad. Sci. USA* **75:**4436–4440.

141. Esposito, M. S., and R. E. Esposito. 1974. Genetic recombination and commitment to meiosis in *Saccharomyces*. *Proc. Natl. Acad. Sci. USA* **71:**3172–3176.

142. Esposito, M. S., D. T. Maleas, K. A. Bjornstad, and C. V. Bruschi. 1982. Simultaneous detection of changes in chromosome number, gene conversion and intergenic recombination during mitosis of *Saccharomyces cerevisiae*: spontaneous and ultraviolet light induced events. *Curr. Genet.* **6:**5–11.

143. Esposito, M. S., R. M. Ramirez, and C. V. Bruschi. 1994. Recombinators, recombinases and recombination genes of yeasts. *Curr. Genet.* **25:**1–11.

144. Esposito, R. E., and S. Klapholz. 1981. Meiosis and ascospore development, p. 211–287. *In* J. N. Strathern, E. W. Jones, and J. R. Broach (ed.), *The Molecular Biology of the Yeast Saccharomyces*. Cold Spring Harbor Laboratory, Cold Spring Harbor, N.Y.

145. Esser, C., and A. Radbruch. 1990. Immunoglobulin class switching: molecular and cellular analysis. *Annu. Rev. Immunol.* **8:**717–735.

146. Evans, M. K., B. G. Taffe, C. C. Harris, and V. A. Bohr. 1993. DNA strand bias in the repair of the *p53* gene in normal human and xeroderma pigmentosum group C fibroblasts. *Cancer Res.* **53:**5377–5381.

147. Fabre, F. 1978. Induced intragenic recombination in yeast can occur during the G1 mitotic phase. *Nature* (London) **272:**795–798.

148. Fabre, F. 1983. Mitotic transmission of induced recombinational ability in yeast, p. 379–384. *In* E. C. Friedberg and B. A. Bridges (ed.), *Cellular Responses to DNA Damage*, vol. II. Alan R. Liss, Inc., New York.

149. Fabre, F., A. Boulet, and G. Faye. 1991. Possible involvement of the yeast POLIII DNA polymerase in induced gene conversion. *Mol. Gen. Genet.* **229:**353–356.

150. Fabre, F., A. Boulet, and H. Roman. 1984. Gene conversion at different points in the mitotic cycle of *Saccharomyces cerevisiae*. *Mol. Gen. Genet.* **195:**139–143.

151. Fabre, F., N. Magaña-Schwencke, and R. Chanet. 1989. Isolation of the *RAD18* gene of *Saccharomyces cerevisiae* and construction of *rad18* deletion mutants. *Mol. Gen. Genet.* **215:**425–430.

152. Fabre, F., and H. Roman. 1977. Genetic evidence for inducibility of recombination competence in yeast. *Proc. Natl. Acad. Sci. USA* **74:**1667–1671.

153. Fairman, M. P., A. P. Johnson, and J. Thacker. 1992. Multiple components are involved in the efficient joining of double stranded DNA breaks in human extracts. *Nucleic Acids Res.* **20:**4145–4152.

154. Fasullo, M., P. Dave, and R. Rothstein. 1994. DNA-damaging agents stimulate the formation of directed reciprocal translocations in *Saccharomyces cerevisiae*. *Mutat. Res.* **314:**121–133.

155. Fenech, M., A. M. Carr, J. Murray, F. Z. Watts, and A. R. Lehmann. 1991. Cloning and characterization of the *rad4* gene of *Schizosaccharomyces pombe*: a gene showing short regions of sequence similarity to the human *XRCC1* gene. *Nucleic Acids Res.* **19:**6737–6741.

156. Finley, D., E. Özkaynak, and A. Varshavsky. 1987. The yeast polyubiquitin gene is essential for resistance to high temperatures, starvation and other stresses. *Cell* **48:**1035–1046.

157. Fishman-Lobell, J., and J. Haber. 1992. Removal of nonhomologous DNA ends in double-strand break recombination—the role of the yeast ultraviolet repair gene *RAD1*. *Science* **258:**480–484.

158. Fishman-Lobell, J., N. Rudin, and J. E. Haber. 1992. Two alternative pathways of double-strand break repair that are kinetically separable and independently modulated. *Mol. Cell. Biol.* **12:**1292–1303.

159. Fogel, M., K. Yamanishi, and F. Rapp. 1979. Enhancement of host cell reactivation of ultraviolet-irradiated herpes simplex virus by caffeine, hydroxyurea and 5-bromodeoxyuridine. *Int. J. Cancer* **23:**657–662.

160. Fornace, A. J., Jr. 1983. Recombination of parent and daughter strand DNA after UV-irradiation in mammalian cells. *Nature* (London) **304:**552–554.

161. Frankenberg, D., M. Frankenberg-Schwager, D. Blöcher, and R. Harbich. 1981. Evidence for DNA double-strand breaks as the critical lesions in yeast cells irradiated with sparsely or densely ionizing radiation under oxic and anoxic conditions. *Radiat. Res.* **88:**524–532.

162. Frankenberg-Schwager, M. 1990. Induction, repair and biological relevance of radiation-induced DNA lesions in eukaryotic cells. *Radiat. Environ. Biophys.* **29:**273–292.

163. Frankenberg-Schwager, M., and D. Frankenberg. 1990. DNA double-strand breaks: their repair and relationship to cell killing in yeast. *Int. J. Radiat. Biol.* **58:**569–575.

164. Frankenberg-Schwager, M., D. Frankenberg, D. Blöcher, and C. Adamczyk. 1979. The influence of oxygen on the survival and yield of DNA double-strand breaks in irradiated yeast cells. *Int. J. Radiat. Biol.* **36:**261–270.

165. Frankenberg-Schwager, M., D. Frankenberg, and R. Harbich. 1994. Radiation-induced mitotic gene conversion frequency in yeast is modulated by the conditions allowing DNA double-strand break repair. *Mutat. Res.* **314:**57–66.

166. Friedberg, E. C. 1985. *DNA Repair.* W. H. Freeman & Co., New York.

167. Friedberg, E. C., W. Siede, and A. J. Cooper. 1991. Cellular responses to DNA damage in yeast, p. 147–192. *In* E. Jones, J. R. Pringle, and J. Broach (ed.), *The Molecular Biology of the Yeast Saccharomyces: Genome Dynamics, Protein Synthesis, and Energetics,* vol. I. Cold Spring Harbor Laboratory, Cold Spring Harbor, N.Y.

168. Fuller, L. F., and R. B. Painter. 1988. A Chinese hamster ovary cell line hypersensitive to ionizing radiation and deficient in repair replication. *Mutat. Res.* **193:**109–121.

169. Fulop, G. M., and R. A. Phillips. 1990. The *scid* mutation in mice causes a general defect in DNA repair. *Nature* (London) **347:**479–482.

170. Gallagher, P. E., and N. J. Duker. 1986. Detection of purine photoproducts in a defined sequence of human DNA. *Mol. Cell. Biol.* **6:**707–709.

171. Game, J. C. 1983. Radiation-sensitive mutants and repair in yeast, p. 109–137. *In* J. F. T. Spencer, D. M. Spencer, and A. R. W. Smith (ed.), *Yeast Genetics: Fundamental and Applied Aspects.* Springer-Verlag, New York.

172. Game, J. C. 1993. DNA double-strand breaks and the *RAD50-RAD57* genes in *Saccharomyces. Semin. Cancer Biol.* **4:**73–83.

173. Game, J. C., L. H. Johnston, and R. C. von Borstel. 1979. Enhanced mitotic recombination in a ligase deficient mutant of the yeast *Saccharomyces cerevisiae. Proc. Natl. Acad. Sci. USA* **76:**4589–4592.

174. Game, J. C., and R. K. Mortimer. 1974. A genetic study of X-ray sensitive mutants in yeast. *Mutat. Res.* **24:**281–292.

175. Game, J. C., K. C. Sitney, V. E. Cook, and R. K. Mortimer. 1989. Use of a ring chromosome and pulsed-field gels to study interhomolog recombination, double-strand breaks and sister chromatid exchange in yeast. *Genetics* **123:**695–713.

176. Game, J. C., T. J. Zamb, R. J. Braun, M. A. Resnick, and R. M. Roth. 1980. The role of radiation (*rad*) genes in meiotic recombination in yeast. *Genetics* **94:**51–68.

177. Ganesh, A., P. North, and J. Thacker. 1993. Repair and misrepair of site-specific DNA double-strand breaks by human cell extracts. *Mutat. Res.* **299:**251–259.

178. Geigl, E. M., and F. Eckardt-Schupp. 1991. The repair of double-strand breaks and S1 nuclease-sensitive sites can be monitored chromosome-specifically in *Saccharomyces cerevisiae* using pulsed-field gel electrophoresis. *Mol. Microbiol.* **5:**1615–1620.

179. Gelbert, L. M., and R. L. Davidson. 1988. A sensitive molecular assay for mutagenesis in mammalian cells: reversion analysis in cells with a mutant shuttle vector gene integrated into chromosomal DNA. *Proc. Natl. Acad. Sci. USA* **85:**9143–9147.

180. Gellert, M. 1992. Molecular analysis of V(D)J recombination. *Annu. Rev. Genet.* **22:**425–426.

181. Gellert, M. 1992. V(D)J recombination gets a break. *Trends Genet.* **8:**408–412.

182. Gentil, A., A. Margot, and A. Sarasin. 1982. Enhanced reactivation and mutagenesis after transfection of carcinogen-treated monkey kidney cells with UV-irradiated simian virus 40 (SV40) DNA. *Biochimie* **64:**693–696.

183. Gentil, A., A. Margot, and A. Sarasin. 1983. Effect of UV-irradiation on genetic recombination of Simian virus 40 mutants, p. 385–396. *In* E. C. Friedberg and P. C. Hanawalt (ed.), *Cellular Responses to DNA Damage.* Alan R. Liss, Inc., New York.

184. Gentil, A., A. Margot, and A. Sarasin. 1986. 2-(*N*-Acetoxy-*N*-acetylamino)fluorene mutagenesis in mammalian cells: sequence specific hot spot. *Proc. Natl. Acad. Sci. USA* **83:**9556–9560.

185. Gentil, A., G. Renault, C. Madzak, A. Margot, J. B. Cabral-Neto, J. J. Vasseur, B. Rayner, J. L. Imbach, and A. Sarasin. 1990. Mutagenic properties of a unique abasic site in mammalian cells. *Biochem. Biophys. Res. Commun.* **173:**704–710.

186. Getts, R. C., and T. D. Stamato. 1994. Absence of a Ku-like DNA end binding activity in the *xrs* double-strand DNA repair-deficient mutant. *J. Biol. Chem.* **269:**15981–15984.

187. Giaccia, A., R. Weinstein, J. Hu, and T. D. Stamato. 1985. Cell cycle-dependent repair of double-strand DNA breaks in a gamma-ray-sensitive Chinese hamster cell. *Somatic Cell Mol. Genet.* **11:**485–491.

188. Giaccia, A. J., N. Denko, R. MacLaren, D. Mirman, C. Waldren, I. Hart, and T. D. Stamato. 1990. Human chromosome 5 complements the DNA double-strand break-repair deficiency and gamma-ray sensitivity of the XR-1 hamster variant. *Am. J. Hum. Genet.* **47:**359–469.

189. Glazer, P. M., S. N. Sarkar, and W. C. Summers. 1986. Detection and analysis of UV-induced mutations in mammalian cell DNA using a lambda phage shuttle vector. *Proc. Natl. Acad. Sci. USA* **83:**1041–1044.

190. Goebl, M. G., J. Yochem, S. Jentsch, J. P. McGrath, A. Varshavsky, and B. Byers. 1988. The yeast cell cycle gene *CDC34* encodes a ubiquitin-conjugating enzyme. *Science* **241:**1331–1335.

191. Golin, J. E., and M. S. Esposito. 1981. Mitotic recombination: mismatch correction and replicational resolution of Holliday structures formed at the two strand stage in *Saccharomyces. Mol. Gen. Genet.* **183:**252–263.

192. Gough, G., and R. D. Wood. 1989. Inhibition of in vitro SV40 DNA replication by ultraviolet light. *Mutat. Res.* **227:**193–197.

193. Griffiths, T. D., and S. Y. Ling. 1989. Activation of alternative sites of replicon initiation in Chinese hamster cells exposed to ultraviolet light. *Mutat. Res.* **184:**39–46.

194. Grosovsky, A. J., and J. B. Little. 1983. Influence of confluent holding times on UV light mutagenesis in human diploid fibroblasts. *Mutat. Res.* **110:**401–412.

195. Grosovsky, A. J., and J. B. Little. 1985. Confluent holding leads to a transient enhancement in mutagenesis in UV-light-irradiated xeroderma pigmentosum, Gardner's syndrome and normal human diploid fibroblasts. *Mutat. Res.* **149:**147–155.

196. Gunther, M., R. Wicker, S. Tiravy, and J. Coppey. 1981. Enhanced survival of ultraviolet-damaged parvovirus LuIII and herpes virus in carcinogen pretreated transformed human cells, p. 605–609. *In* E. Seeberg and K. Kleppe (ed.), *Chromosome Damage and Repair.* Plenum Publishing Corp., New York.

197. Haase, S. B., S. S. Heinzel, P. J. Krysan, and M. P. Calos. 1989. Improved EBV shuttle vectors. *Mutat. Res.* **220:**125–132.

198. Haber, J. E. Personal communication.

199. Haber, J. E. 1992. Exploring the pathways of homologous recombination. *Curr. Opin. Cell Biol.* **4:**401–412.

200. Haber, J. E. 1992. Mating-type gene switching in *Saccharomyces cerevisiae. Trends Genet.* **8:**446–452.

201. Haber, J. E., and M. Hearn. 1985. *RAD52*-independent mitotic gene conversion in *Saccharomyces cerevisiae* frequently results in chromosome loss. *Genetics* **111:**7–22.

202. Hafezparast, M., G. P. Kaur, M. Zdzienicka, R. S. Athwal, A. R. Lehmann, and P. A. Jeggo. 1993. Subchromosomal localization of a gene (*XRCC5*) involved in double strand break repair to the region 2q34–36. *Somatic Cell Mol. Genet.* **19:**413–421.

203. Halbrook, J., and K. McEntee. 1989. Purification and characterization of a DNA-pairing and strand transfer activity from mitotic *Saccharomyces cerevisiae. J. Biol. Chem.* **264:**21403–21412.

204. Harriman, W., H. Völk, N. Defranoux, and M. Wabl. 1993. Immunoglobulin class switch recombination. *Annu. Rev. Immunol.* **11:** 361–384.

205. Harrington, J., C.-L. Hsieh, J. Gerton, G. Bosma, and M. R. Lieber. 1992. Analysis of the defect in DNA end joining in the murine *scid* mutation. *Mol. Cell. Biol.* **12:**4758–4768.

206. Harris, C. C. 1991. Chemical and physical carcinogenesis: advances and perspectives for the 1990s. *Cancer Res.* **51**(Suppl.):5023s–5044s.

207. Hartwell, L. H. 1974. *Saccharomyces cerevisiae* cell cycle. *Bacteriol. Rev.* **38:**164–198.

208. Harwood, J., A. Tachinaba, and M. Meuth. 1991. Multiple dispersed spontaneous mutations: a novel pathway of mutation in a malignant human cell line. *Mol. Cell. Biol.* **11:**3163–3170.

209. Hastings, P. J., S.-K. Quah, and R. C. von Borstel. 1976. Spontaneous mutation by mutagenic repair of spontaneous lesions in DNA. *Nature* (London) **264:**719–722.

210. Hauser, J., A. S. Levine, and K. Dixon. 1987. Unique pattern of point mutations arising after gene transfer into mammalian cells. *EMBO J.* **6:**63–67.

211. Hauser, J., A. S. Levine, and K. Dixon. 1988. Fidelity of DNA synthesis in a mammalian in vitro replication system. *Mol. Cell. Biol.* **8:**3267–3271.

212. Hauser, J., M. M. Seidman, K. Sidur, and K. Dixon. 1986. Sequence specificity of point mutations induced during passage of a UV-irradiated shuttle vector plasmid in monkey cells. *Mol. Cell. Biol.* **6:**277–285.

213. Hawthorne, D. C., and U. Leupold. 1974. Suppressor mutation in yeast. *Curr. Top. Microbiol. Immunol.* **64:**1–47.

214. Haynes, R. H., and F. Eckardt. 1980. Mathematical analysis of mutation induction kinetics. *Chem. Mutagens* **6:**271–307.

215. Haynes, R. H., F. Eckardt, and B. A. Kunz. 1985. Analysis of nonlinearities in mutation frequency curves. *Mutat. Res.* **150:**51–59.

216. Haynes, R. H., and B. A. Kunz. 1981. DNA repair and mutagenesis in yeast, p. 371–414. *In* J. N. Strathern, E. W. Jones, and J. R. Broach (ed.), *The Molecular Biology of the Yeast Saccharomyces*, vol. I. Cold Spring Harbor Laboratory, Cold Spring Harbor, N.Y.

217. Hellman, K. B., and K. F. Haynes. 1974. Radiation-enhanced survival of a human virus in normal and malignant rat cells. *Proc. Soc. Exp. Biol. Med.* **145:**255–262.

218. Hendrickson, E. A., X.-Q. Qin, E. A. Bump, D. G. Schatz, M. Oettinger, and D. T. Weaver. 1991. A link between double-strand break-related repair and V(D)J recombination: the *scid* mutation. *Proc. Natl. Acad. Sci. USA* **88:**4061–4065.

219. Hendrickson, E. A., M. S. Schlissel, and D. T. Weaver. 1990. Wild-type V(D)J recombination in *scid* pre-B cells. *Mol. Cell. Biol.* **10:**5397–5407.

220. Henriques, J. A. P., and E. Moustacchi. 1981. Interactions between mutations for sensitivity to psoralen photoaddition (*pso*) and radiations (*rad*) in *Saccharomyces cerevisiae. J. Bacteriol.* **148:**248–256.

221. Hentosh, P., A. R. S. Collins, L. Correll, A. J. Fornace Jr., A. Giaccia, and C. A. Waldren. 1990. Genetic and biochemical characterization of the CHO-UV-1 mutant defective in postreplication recovery of DNA. *Cancer Res.* **50:**2356–2362.

222. Hershko, A. 1991. The ubiquitin pathway for protein degradation. *Trends Biochem. Sci.* **16:**265–268.

223. Hesse, J. E., M. R. Lieber, M. Gellert, and K. Mizuuchi. 1987. Extrachromosomal substrates undergo inversion or deletion at immunoglobulin VDJ joining signals. *Cell* **49:**775–783.

224. Heyer, W.-D. 1994. The search for the right partner: homologous pairing and DNA strand exchange proteins in eukaryotes. *Experientia* **50:**223–233.

225. Ho, K. S. Y. 1975. Induction of DNA double-strand breaks by X-rays in a radiosensitive strain of the yeast *Saccharomyces cerevisiae. Mutat. Res.* **30:**327–334.

226. Ho, K. S. Y., and R. K. Mortimer. 1975. Induction of dominant lethality by X-rays in a radiosensitive strain of yeast. *Mutat. Res.* **20:**45–51.

227. Ho, K. S. Y., and R. K. Mortimer. 1975. Two mutations which confer temperature-sensitive radiation sensitivity in the yeast *Saccharomyces cerevisiae. Mutat. Res.* **33:**157–164.

228. Hoekstra, M. F., T. Naughton, and R. E. Malone. 1986. Properties of spontaneous mitotic recombination occuring in the presence of the *rad52-1* mutation of *Saccharomyces cerevisiae. Genet. Res.* **48:**9–17.

229. Hollingsworth, R. E., Jr., R. M. Ostroff, M. B. Klein, L. A. Niswander, and R. A. Sclafani. 1992. Molecular genetic studies of the Cdc7 protein kinase and induced mutagenesis in yeast. *Genetics* **132:**53–62.

230. Hollingsworth, R. E., Jr., and R. A. Sclafani. 1990. DNA metabolism gene *CDC7* from yeast encodes a serine (threonine) protein kinase. *Proc. Natl. Acad. Sci. USA* **87:**6272–6276.

231. Howell, J. N., M. H. Greene, R. C. Corner, V. M. Maher, and J. J. McCormick. 1984. Fibroblasts from patients with hereditary cutaneous malignant melanoma are abnormally sensitive to the mutagenic effect of simulated sunlight and 4-nitroquinoline 1-oxide. *Proc. Natl. Acad. Sci. USA* **81:**1179–1183.

232. Hoy, C. A., J. C. Fuscoe, and L. H. Thompson. 1987. Recombination and ligation of transfected DNA in CHO mutant EM9, which has high levels of sister chromatid exchange. *Mol. Cell. Biol.* **7:**2007–2011.

233. Hoy, C. A., L. H. Thompson, E. P. Salazar, and S. A. Stewart. 1985. Different genetic alterations underlie dual hypersensitivity of CHO mutant UV-1 to DNA methylating and cross-linking agents. *Somatic Cell Mol. Genet.* **11:**523–532.

234. Hsia, H. C., J. S. Lebkowski, P.-M. Leong, M. P. Calos, and J. H. Miller. 1989. Comparison of ultraviolet irradiation-induced mutagenesis of the *lacI* gene in *Escherichia coli* and in human 293 cells. *J. Mol. Biol.* **205:**103–113.

235. Hsie, A. W. 1987. The use of the *hgprt* versus *gpt* locus for quantitative mammalian cell mutagenesis, p. 37–52. *In* M. M. Moore, D. M. DeMarini, F. J. de Serres, and K. R. Tindall (ed.), *Mammalian Cell Mutagenesis.* Cold Spring Harbor Laboratory Press, Cold Spring Harbor, N.Y.

236. Hsieh, C.-L., C. F. Arlett, and M. R. Lieber. 1993. V(D)J recombination in ataxia telangiectasia, Bloom's syndrome and a DNA ligase I-associated immunodeficiency disorder. *J. Biol. Chem.* **268:**20105–20109.

237. Hsieh, P., M. S. Meyn, and R. D. Camerini-Otero. 1986. Partial purification and characterization of a recombinase from human cells. *Cell* **44:**885–894.

238. Hunnable, E. G., and B. S. Cox. 1971. The genetic control of dark recombination in yeast. *Mutat. Res.* **13:**297–309.

239. Hyde, H., A. A. Davies, F. E. Benson, and S. C. West. 1994. Resolution of recombination intermediates by a mammalian activity functionally analogous to *Escherichia coli* RuvC resolvase. *J. Biol. Chem.* **269:**5202–5209.

240. Iliakis, G. 1991. The role of DNA double strand breaks in ionizing radiation-induced killing of eukaryotic cells. *Bioessays* **13:**641–648.

241. Ivanov, E. L., V. G. Korolev, and F. Fabre. 1992. *XRS2*, a DNA repair gene of *Saccharomyces cerevisiae*, is needed for meiotic recombination. *Genetics* **132:**651–664.

242. **Ivanov, E. L., N. Sugawara, C. I. White, F. Fabre, and J. E. Haber.** 1994. Mutations in *XRS2* and *RAD50* delay but do not prevent mating-type switching in *Saccharomyces cerevisiae*. *Mol. Cell. Biol.* **14**:3414–3425.

243. **Jackson, A. L., P. M. B. Pahl, K. Harrison, J. Rosamond, and R. A. Sclafani.** 1993. Cell cycle regulation of the yeast Ccd7 protein kinase by association with the Dbf4 protein. *Mol. Cell. Biol.* **13**:2899–2908.

244. **Jackson, J. A., and G. R. Fink.** 1981. Gene conversion between duplicated genetic elements in yeast. *Nature* (London) **292**:306–311.

245. **James, A. P., and B. J. Kilbey.** 1977. The timing of UV mutagenesis in yeast: a pedigree analysis of induced recessive mutation. *Genetics* **87**:237–248.

246. **James, A. P., B. J. Kilbey, and G. J. Prefontaine.** 1978. The timing of UV mutagenesis in yeast: continuing mutation in an excision-defective (*rad1-1*) strain. *Mol. Gen. Genet.* **165**:207–212.

247. **Jang, Y. K., Y. H. Jin, E. M. Kim, F. Fabre, S. H. Hong, and S. D. Park.** 1994. Cloning and sequence analysis of *rhp51*+, a *Schizosaccharomyces pombe* homolog of the *Saccharomyces cerevisiae RAD51* gene. *Gene* **142**:207–211.

248. **Jazwinski, S. M.** 1988. *CDC7*-dependent protein kinase activity in yeast replicative-complex preparations. *Proc. Natl. Acad. Sci. USA* **85**:2101–2105.

249. **Jeeves, W. P., and A. J. Rainbow.** 1979. γ-ray enhanced reactivation of UV-irradiated adenovirus in human cells. *Mutat. Res.* **60**:33–41.

250. **Jeggo, P. A., M. Hafezparast, A. F. Thompson, B. C. Broughton, G. P. Kaur, M. Z. Zdzienicka, and R. S. Athwal.** 1992. Localization of a DNA repair gene (*XRCC5*) involved in double strand break rejoining to human chromosome 2. *Proc. Natl. Acad. Sci. USA* **89**:6423–6427.

251. **Jeggo, P. A., and R. Holliday.** 1986. Azacytidine-induced reactivation of a DNA repair gene in Chinese hamster ovary cells. *Mol. Cell. Biol.* **6**:2944–2949.

252. **Jeggo, P. A., and J. Smith-Ravin.** 1989. Decreased stable transfection frequencies of six X-ray-sensitive CHO strains, all members of the *xrs* complementing group. *Mutat. Res.* **218**:75–86.

253. **Jeggo, P. A., J. Tesmer, and D. J. Chen.** 1991. Genetic analysis of ionising radiation sensitive mutants of cultured mammalian cell lines. *Mutat. Res.* **254**:125–133.

254. **Jentsch, S.** 1992. The ubiquitin-conjugation system. *Annu. Rev. Genet.* **26**:179–207.

255. **Jentsch, S., J. P. McGrath, and A. Varshavsky.** 1987. The yeast DNA repair gene *RAD6* encodes a ubiquitin-conjugating enzyme. *Nature* (London) **329**:131–134.

256. **Jentsch, S., W. Seufert, T. Sommer, and H.-A. Reins.** 1990. Ubiquitin-conjugating enzymes: novel regulators of eukaryotic cells. *Trends Biochem. Sci.* **15**:195–198.

257. **Jessberger, R., and P. Berg.** 1991. Repair of deletions and double-strand gaps by homologous recombination in a mammalian in vitro system. *Mol. Cell. Biol.* **11**:445–457.

258. **Jessberger, R., V. Podust, U. Hübscher, and P. Berg.** 1993. A mammalian protein complex that repairs double-strand breaks and deletions by recombination. *J. Biol. Chem.* **268**:15070–15079.

259. **Jha, B., F. Ahne, and F. Eckardt-Schupp.** 1993. The use of a double-marker shuttle vector to study DNA double-strand break repair in wild-type and radiation-sensitive mutants of the yeast *Saccharomyces cerevisiae*. *Curr. Genet.* **23**:402–407.

260. **Johnson, A. W., and R. D. Kolodner.** 1991. Strand exchange protein 1 from *Saccharomyces cerevisiae*—a novel multifunctional protein that contains DNA strand exchange and exonuclease activity. *J. Biol. Chem.* **266**:14046–14054.

261. **Johnson, R. E., S. T. Henderson, T. D. Petes, S. Prakash, M. Bankmann, and L. Prakash.** 1992. *Saccharomyces cerevisiae RAD5*-encoded DNA repair protein contains DNA helicase and zinc-binding sequence motifs and affects the stability of simple repetitive sequences in the genome. *Mol. Cell. Biol.* **12**:3807–3818.

262. **Johnston, L. H.** 1990. Periodic events in the cell cycle. *Curr. Opin. Cell Biol.* **2**:274–279.

263. **Jones, J. S., L. Prakash, and S. Prakash.** 1990. Regulated expression of the *Saccharomyces cerevisiae* DNA repair gene *RAD7* in response to DNA damage and during sporulation. *Nucleic Acids Res.* **18**:3281–3285.

264. **Jones, J. S., S. Weber, and L. Prakash.** 1988. The *Saccharomyces cerevisiae RAD18* gene encodes a protein that contains potential zinc finger domains for nucleic acid binding and a putative nucleotide binding sequence. *Nucleic Acids Res.* **16**:7119–7131.

265. **Jones, N. J., R. Cox, and J. Thacker.** 1988. Six complementation groups for ionizing-radiation sensitivity in Chinese hamster cells. *Mutat. Res.* **193**:139–144.

266. **Jones, N. J., S. A. Stewart, and L. H. Thompson.** 1990. Biochemical and genetic analysis of the Chinese hamster mutants *irs1* and *irs2* and their comparison to cultured ataxia telangiectasia cells. *Mutagenesis* **5**:15–23.

267. **Kadyk, L. C., and L. H. Hartwell.** 1992. Sister chromatids are preferred over homologs as substrates for recombinational repair in *Saccharomyces cerevisiae*. *Genetics* **132**:387–402.

268. **Kadyk, L. C., and L. H. Hartwell.** 1993. Replication-dependent sister chromatid recombination in *rad1* mutants of *Saccharomyces cerevisiae*. *Genetics* **133**:469–487.

269. **Kamiya, H., N. Murata, T. Murata, S. Iwai, A. Matsukage, C. Masutani, F. Hanaoka, and E. Ohtsuka.** 1993. Cyclobutane thymine dimers in a *ras*-protooncogene hot spot activate the gene by point mutation. *Nucleic Acids Res.* **21**:2355–2361.

270. **Kang, X., F. Yadao, R. D. Gietz, and B. A. Kunz.** 1992. Elimination of the yeast *RAD6* ubiquitin conjugase enhances base-pair transitions and G·C → T·A transversions as well as transposition of the Ty element: implications for the control of spontaneous mutagenesis. *Genetics* **130**:285–294.

271. **Kans, J. A., and R. K. Mortimer.** 1991. Nucleotide sequence of the *RAD57* gene of *Saccharomyces cerevisiae*. *Gene* **105**:139–140.

272. **Kaufmann, W. K.** 1989. Pathways of human cell post-replication repair. *Carcinogenesis* **10**:1–11.

273. **Kaufmann, W. K., and J. E. Cleaver.** 1981. Mechanisms of inhibition of DNA replication by ultraviolet radiation in normal human and xeroderma pigmentosum fibroblasts. *J. Mol. Biol.* **149**:171–187.

274. **Kaufmann, W. K., J. E. Cleaver, and R. B. Painter.** 1980. Ultraviolet radiation inhibits replicon initiation in S phase human cells. *Biochim. Biophys. Acta* **608**:191–195.

275. **Kaufmann, W. K., R. J. Rahija, S. A. MacKenzie, and D. G. Kaufman.** 1987. Cell cycle-dependent initiation of hepatocarcinogenesis in rats by (±)-7r,8t-dihydroxy-9t,10t-epoxy-7,8,9,10-tetrahydrobenzo[a]pyrene. *Cancer Res.* **47**:3771–3775.

276. **Kaufmann, W. K., J. M. Rice, M. L. Wenk, and D. G. Kaufman.** 1981. Reversible inhibition of rat hepatocyte proliferation by hydrocortisone and its effect on cell cycle-dependent hepatocarcinogenesis by *N*-methyl-*N*-nitrosourea. *Cancer Res.* **41**:4653–4660.

277. **Kawasaki, I., S. Sugano, and H. Ikeda.** 1989. Calf thymus histone H1 is a recombinase that catalyzes ATP-independent DNA strand transfer. *Proc. Natl. Acad. Sci. USA* **86**:5281–5285.

278. **Keijzer, W., M. P. Mulder, J. C. M. Langeveld, E. M. E. Smit, J. L. Bos, D. Bootsma, and J. H. J. Hoeijmakers.** 1989. Establishment and characterization of a melanoma cell line from a xeroderma pigmentosum patient: activation of N-*ras* at a potential pyrimidine dimer site. *Cancer Res.* **49**:1229–1235.

279. **Keil, R. L., and G. S. Roeder.** 1984. *Cis*-acting, recombination stimulating activity in a fragment of the ribosomal DNA of *S. cerevisiae*. *Cell* **39**:377–386.

280. **Kemp, L. M., and P. A. Jeggo.** 1986. Radiation-induced chromosome damage in X-ray-sensitive mutants (xrs) of the Chinese hamster ovary cell line. *Mutat. Res.* **166**:255–263.

281. Kemp, L. M., S. G. Sedgwick, and P. A. Jeggo. 1984. X-ray sensitive mutants of Chinese hamster ovary cells defective in double-strand break rejoining. *Mutat. Res.* **132:**189–196.

282. Kenne, K., and S. Ljungquist. 1984. A DNA-recombinogenic activity in human cells. *Nucleic Acids Res.* **12:**3057–3068.

283. Kern, R., and F. K. Zimmermann. 1978. The influence of defects in excision and error prone repair on spontaneous and induced mitotic recombination and mutation in *Saccharomyces cerevisiae*. *Mol. Gen. Genet.* **161:**81–88.

284. Keyse, S. M. 1993. The induction of gene expression in mammalian cells by radiation. *Semin. Cancer Biol.* **4:**119–128.

285. Kilbey, B. J. 1986. *cdc7* alleles and the control of induced mutagenesis in yeast. *Mutagenesis* **1:**29–31.

286. Kilbey, B. J., T. Brychcy, and A. Nasim. 1978. Initiation of UV mutagenesis in *Saccharomyces cerevisiae*. *Nature* (London) **274:**889–891.

286a. King, J. S., C. F. Fairley, and W. F. Morgan. 1994. Bridging the gap. Joining of non-homologous ends by DNA polymerases. *J. Biol. Chem.* **269:**13061–13064.

287. King, J. S., E. R. Valcarcel, J. T. Rufer, J. W. Phillips, and W. F. Morgan. 1993. Noncomplementary DNA double-strand-break rejoining in bacterial and human cells. *Nucleic Acids Res.* **21:**1055–1059.

288. Klein, H. L. 1988. Different types of recombination events are controlled by the *RAD1* and *RAD52* genes of *Saccharomyces cerevisiae*. *Genetics* **120:**367–377.

289. Kmiec, E. B., and W. K. Holloman. 1984. Synapsis promoted by *Ustilago* Rec1 protein. *Cell* **36:**593–598.

290. Knaap, A. G. A. C., and J. W. I. M. Simons. 1983. Ascertainment of the effect of differential growth of mutants on observed mutant frequencies in X-irradiated mammalian cells. An indication for the occurrence of X-ray-induced untargeted mutagenesis. *Mutat. Res.* **110:**413–422.

291. Knajilal, S., W. E. Pierceall, K. K. Cummings, M. L. Kripke, and H. N. Ananthaswamy. 1993. High frequency of p53 mutations in ultraviolet radiation-induced murine skin tumors: evidence for strand bias and tumor heterogeneity. *Cancer Res.* **53:**2961–2964.

292. Koken, M., P. Reynolds, D. Bootsma, J. H. J. Hoeijmakers, S. Prakash, and L. Prakash. 1991. *Dhr6*, a *Drosophila* homolog of the yeast DNA repair gene *RAD6*. *Proc. Natl. Acad. Sci. USA* **88:**3832–3836.

293. Koken, M. H. M., P. Reynolds, I. Jaspers-Dekker, L. Prakash, S. Prakash, D. Bootsma, and J. H. J. Hoeijmakers. 1991. Structural and functional conservation of two human homologs of the yeast DNA repair gene *RAD6*. *Proc. Natl. Acad. Sci. USA* **88:**8865–8869.

294. Koken, M. H. M., E. M. E. Smit, I. Jaspers-Dekker, B. A. Oostra, A. Hagemeijer, D. Bootsma, and J. H. J. Hoeijmakers. 1992. Localization of two human homologs, HHR6A and HHR6B, of the yeast DNA repair gene *RAD6* to chromosomes Xq24-q25 and 5q23-q32. *Genomics* **12:**447–453.

295. Kolodkin, A. L., A. J. S. Klar, and F. Stahl. 1986. Double-strand breaks can initiate meiotic recombination in *Saccharomyces cerevisiae*. *Cell* **46:**733–740.

296. Kolodner, R., D. H. Evans, and P. T. Morrison. 1987. Purification and characterization of an activity from *Saccharomyces cerevisiae* that catalyzes homologous pairing and strand exchange. *Proc. Natl. Acad. Sci. USA* **84:**5560–5564.

297. Konze-Thomas, B., R. M. Hazard, V. M. Maher, and J. J. McCormick. 1982. Extent of excision repair before DNA synthesis determines the mutagenic but not the lethal effect of UV radiation. *Mutat. Res.* **94:**421–434.

298. Kostriken, R., J. N. Strathern, A. J. S. Klar, J. B. Hicks, and F. Heffron. 1983. A site-specific endonuclease essential for mating-type switching in *Saccharomyces cerevisiae*. *Cell* **35:**167–174.

299. Kraemer, K. H., and M. H. Greene. 1985. Dysplastic nevus syndrome: familial and sporadic precursors of cutaneous melanoma. *Dermatol. Clin.* **3:**225–237.

300. Kraemer, K. H., S. Seetharam, M. Protic-Sabljic, D. E. Brash, A. Bredberg, and M. M. Seidman. 1988. Defective DNA repair and mutagenesis by dimer and non-dimer photoproducts in xeroderma pigmentosum measured with plasmid vectors, p. 325–335. *In* E. C. Friedberg and P. C. Hanawalt (ed.), *Mechanisms and Consequences of DNA Damage Processing*. Alan R. Liss, Inc., New York.

301. Kramer, K. M., J. A. Brock, K. Bloom, J. K. Moore, and J. E. Haber. 1994. Two different types of double-strand breaks in *Saccharomyces cerevisiae* are repaired by similar *RAD52*-independent, nonhomologous recombination events. *Mol. Cell. Biol.* **14:**1293–1301.

302. Kunz, B. A., and R. H. Haynes. 1981. Phenomenology and genetic control of mitotic recombination in yeast. *Annu. Rev. Genet.* **15:**57–89.

303. Kunz, B. A., X. Kang, and L. Kohalmi. 1990. The yeast *rad18* mutator specifically increases G·C-T·A transversions without reducing correction of G·A or C·T mismatches to G·C pairs. *Mol. Cell. Biol.* **11:**218–225.

304. Kunz, B. A., M. K. Pierce, J. R. A. Mis, and C. N. Giroux. 1987. DNA sequence analysis of the mutational specificity of u.v. light in the *SUP4*-o gene of yeast. *Mutagenesis* **2:**445–453.

305. Kupiec, M., and G. Simchen. 1984. Cloning and integrative deletion of the *RAD6* gene of *Saccharomyces cerevisiae*. *Curr. Genet.* **8:**559–566.

306. Kurimasa, A., Y. Nagata, M. Shimizu, M. Emi, Y. Nakamura, and M. Oshimura. 1994. A human gene that restores the DNA-repair defect in SCID mice is located on 8p11.1 → q11.1. *Hum. Genet.* **93:**21–26.

307. Lafaille, J. J., A. DeCloux, M. Bonneville, Y. Takagaki, and S. Tonegawa. 1989. Junctional sequences of T cell receptor γδ genes: implications for γδ T cell lineages and for a novel intermediate of V-(D)-J joining. *Cell* **59:**859–870.

308. Lamb, J. R., C. Petit-Frère, B. C. Broughton, A. R. Lehmann, and M. H. L. Green. 1989. Inhibition of DNA replication by ionizing radiation is mediated by a *trans*-acting factor. *Int. J. Radiat. Biol.* **56:**125–130.

309. Larimer, F. W., J. R. Perry, and A. A. Hardigree. 1989. The *REV1* gene of *Saccharomyces cerevisiae*: isolation, sequence and functional analysis. *J. Bacteriol.* **171:**230–237.

310. Larner, J. M., H. Lee, and J. L. Hamlin. 1994. Radiation effects on DNA synthesis in a defined chromosomal replicon. *Mol. Cell. Biol.* **14:**1901–1908.

311. Lawrence, C. W. 1982. Mutagenesis in *Saccharomyces cerevisiae*. *Adv. Genet.* **21:**173–254.

312. Lawrence, C. W., S. K. Banerjee, A. Borden, and J. E. LeClerk. 1990. T-T cyclobutane dimers are misinstructive, rather than non-instructive, mutagenic lesions. *Mol. Gen. Genet.* **222:**166–169.

313. Lawrence, C. W., and R. Christensen. 1976. UV mutagenesis in radiation-sensitive strains of yeast. *Genetics* **82:**207–232.

314. Lawrence, C. W., and R. B. Christensen. 1978. Ultraviolet-induced reversion of *cyc1* alleles in radiation sensitive strains of yeast. I. *rev1* mutant strains. *J. Mol. Biol.* **122:**1–22.

315. Lawrence, C. W., and R. B. Christensen. 1978. Ultraviolet-induced reversion of *cyc1* alleles in radiation sensitive strains of yeast. II. *rev2* mutant strains. *Genetics* **90:**213–226.

316. Lawrence, C. W., and R. B. Christensen. 1979. Absence of relationship between UV-induced reversion frequency and nucleotide sequence at the *CYC1* locus of yeast. *Mol. Gen. Genet.* **177:**31–38.

317. Lawrence, C. W., and R. B. Christensen. 1979. Metabolic suppressors of trimethoprim and ultraviolet light sensitivities of *Saccharomyces cerevisiae rad6* mutants. *J. Bacteriol.* **139:**866–876.

318. Lawrence, C. W., and R. B. Christensen. 1979. Ultraviolet-induced reversion of *cyc1* alleles in radiation-sensitive strains of yeast. III. *rev3* mutant strains. *Genetics* **92:**397–408.

319. Lawrence, C. W., and R. B. Christensen. 1982. The mechanism of untargeted mutagenesis in UV-irradiated yeast. *Mol. Gen. Genet.* **186:**1–9.

320. Lawrence, C. W., R. B. Christensen, and A. Schwartz. 1982. Mechanisms of UV mutagenesis in yeast, p. 109–120. *In* J. F. Lemontt and W. M. Generoso (ed.); *Molecular and Cellular Mechanisms of Mutagenesis*. Basic Life Sciences. Plenum Publishing Corp., New York.

321. Lawrence, C. W., G. Das, and R. B. Christensen. 1985. *REV7*, a new gene concerned with UV mutagenesis in yeast. *Mol. Gen. Genet.* **200:**80–85.

322. Lawrence, C. W., P. E. M. Gibbs, A. Borden, M. J. Horsfall, and B. J. Kilbey. 1993. Mutagenesis induced by single UV photoproducts in *E. coli* and yeast. *Mutat. Res.* **299:**157–164.

323. Lawrence, C. W., B. R. Krauss, and R. B. Christensen. 1985. New mutations affecting induced mutagenesis in yeast. *Mutat. Res.* **150:**211–216.

324. Lawrence, C. W., P. E. Nisson, and R. B. Christensen. 1985. UV and chemical mutagenesis in *rev7* mutants of yeast. *Mol. Gen. Genet.* **200:**86–91.

325. Lawrence, C. W., T. O'Brien, and J. Bond. 1984. UV-induced reversion of *his4* frameshift mutations in *rad6*, *rev1*, and *rev3* mutants of yeast. *Mol. Gen. Genet.* **195:**487–490.

326. Lawrence, C. W., J. W. Stewart, F. Sherman, and R. B. Christensen. 1974. Specificity and frequency of UV-induced reversion of an iso-1-cytochrome-C ochre mutant in radiation-sensitive strains of yeast. *J. Mol. Biol.* **85:**137–162.

327. Lebkowski, J. S., S. Clancy, J. H. Miller, and M. P. Calos. 1985. The *lacI* shuttle: rapid analysis of the mutagenic specificity of ultraviolet light in human cells. *Proc. Natl. Acad. Sci. USA* **82:**8606–8610.

328. Lebkowski, J. S., R. B. DuBridge, E. A. Antell, K. S. Greisen, and M. P. Calos. 1984. Transfected DNA is mutated in monkey, mouse, and human cells. *Mol. Cell. Biol.* **4:**1951–1960.

329. LeClerc, J. E., A. Borden, and C. W. Lawrence. 1991. The thymine-thymine pyrimidine-pyrimidone(6-4) photoproduct is highly mutagenic and specifically introduces 3′ thymine-to-cytosine transitions in *Escherichia coli*. *Proc. Natl. Acad. Sci. USA* **88:**9685–9689.

330. Lee, G. S.-F., E. A. Savage, R. G. Ritzel, and R. C. von Borstel. 1988. The base-alteration spectrum of spontaneous and ultraviolet radiation-induced forward mutations in the *URA3* locus of *Saccharomyces cerevisiae*. *Mol. Gen. Genet.* **214:**396–404.

331. Lehman, C. W., J. K. Trautman, and D. Carroll. 1994. Illegitimate recombination in *Xenopus*: characterization of end-joined junctions. *Nucleic Acids Res.* **22:**434–442.

332. Lehmann, A. R. 1972. Postreplication repair of DNA in ultraviolet-irradiated mammalian cells. *J. Mol. Biol.* **66:**319–337.

333. Lehmann, A. R. 1975. Postreplication repair of DNA in mammalian cells. *Life Sci.* **15:**2005–2016.

334. Lehmann, A. R. 1981. DNA replication in mammalian cells damaged by mutagens, p. 383–388. *In* E. Seeberg and K. Kleppe (ed.), *Chromosome Damage and Repair*. Plenum Publishing Corp., New York.

335. Lehmann, A. R., S. Kirk-Bell, C. F. Arlett, M. C. Paterson, P. H. M. Lohman, E. A. de Weerd-Kastelein, and D. Bootsma. 1975. Xeroderma pigmentosum cells with normal levels of excision repair have a defect in DNA synthesis after UV-irradiation. *Proc. Natl. Acad. Sci. USA* **72:**219–223.

336. Lemontt, J. F. 1971. Mutants of yeast defective in mutation induction by ultraviolet light. *Genetics* **68:**21–33.

337. Lemontt, J. F. 1972. Induction of forward mutations in mutationally defective yeast. *Mol. Gen. Genet.* **119:**27–42.

338. Lemontt, J. F. 1977. Pathways of ultraviolet mutability in *Saccharomyces cerevisiae*. III. Genetic analysis and properties of mutants resistant to ultraviolet-induced forward mutation. *Mutat. Res.* **43:**179–204.

339. Lemontt, J. F. 1980. Genetic and physiological factors affecting repair and mutagenesis in yeast, p. 85–120. *In* W. M. Generoso, M. D. Shelby, and F. J. de Serres (ed.), *DNA Repair and Mutagenesis in Eukaryotes*. Plenum Publishing Corp., New York.

340. Lemontt, J. F., and S. V. Lair. 1982. Plate assay for chemical and radiation-induced mutagenesis of *CAN1* in yeast as a function of posttreatment DNA replication: effect of *rad6-1*. *Mutat. Res.* **93:**339–352.

341. Lewis, S. M., J. E. Hesse, K. Mizuuchi, and M. Gellert. 1988. Novel strand exchanges in V(D)J recombination. *Cell* **55:**1099–1107.

342. Li, J. J., and T. J. Kelly. 1984. Simian virus 40 DNA replication in vitro. *Proc. Natl. Acad. Sci. USA* **81:**6973–6977.

343. Lieber, M. R. 1992. The mechanism of V(D)J recombination: a balance of diversity, specificity, and stability. *Cell* **70:**873–876.

344. Lieber, M. R., J. E. Hesse, S. Lewis, G. C. Bosma, N. Rosenberg, K. Mizuuchi, M. J. Bosma, and M. Gellert. 1988. The defect in murine severe combined immune deficiency: joining of signal sequences but not coding segments in V(D)J recombination. *Cell* **55:**7–16.

345. Lin, W.-C., and S. Desiderio. 1993. Regulation of V(D)J recombination activator protein RAG-2 by phosphorylation. *Science* **260:**953–959.

346. Lin, W.-C., and S. Desiderio. 1994. Cell cycle regulation of V(D)J recombination-activating protein RAG-2. *Proc. Natl. Acad. Sci. USA* **91:**2733–2737.

347. Lin, Y.-H., and R. L. Keil. 1991. Mutations affecting RNA polymerase I-stimulated exchange and rDNA recombination in yeast. *Genetics* **127:**31–38.

348. Linn, S. 1991. How many pols does it take to replicate nuclear DNA? *Cell* **66:**185–187.

349. Little, J. B., L. Gorgojo, and H. Veltroos. 1990. Delayed appearance of lethal and specific gene mutations in irradiated mammalian cells. *Int. J. Radiat. Oncol. Biol. Phys.* **19:**1425–1429.

350. Lovett, S. T. 1994. Sequence of the *RAD55* gene of *Saccharomyces cerevisiae*: similarity of *RAD55* to prokaryotic RecA and other RecA-like proteins. *Gene* **142:**103–106.

351. Lovett, S. T., and R. K. Mortimer. 1987. Characterization of null mutants of the *RAD55* gene of *Saccharomyces cerevisiae*: effects of temperature, osmotic strength and mating type. *Genetics* **116:**547–553.

352. Lowenhaupt, K., M. Sander, C. Hauser, and A. Rich. 1989. *Drosophila melanogaster* strand transferase: a protein that forms heteroduplex DNA in the absence of both ATP and single-strand DNA binding protein. *J. Biol. Chem.* **264:**20568–20575.

353. Lutze, L. H., J. E. Cleaver, W. F. Morgan, and R. A. Winegar. 1993. Mechanisms involved in rejoining DNA double-strand breaks induced by ionizing radiation and restriction enzymes. *Mutat. Res.* **299:**225–232.

354. Lytle, C. D. 1978. Radiation-enhanced virus reactivation in mammalian cells. *Natl. Cancer Inst. Monogr.* **50:**145–149.

355. Lytle, C. D., S. G. Benane, and L. E. Bockstahler. 1974. Ultraviolet enhanced reactivation of herpes virus in human tumor cells. *Photochem. Photobiol.* **19:**91–94.

356. Lytle, C. D., J. Coppey, and W. D. Taylor. 1978. Enhanced survival of ultraviolet-irradiated herpes simplex virus in carcinogen-pretreated cells. *Nature* (London) **272:**60–62.

357. Lytle, C. D., R. S. Day III, K. B. Hellman, and L. E. Bockstahler. 1976. Infection of UV-irradiated xeroderma pigmentosum fibroblasts by herpes simplex virus: study of capacity and Weigle reactivation. *Mutat. Res.* **36:**257–264.

358. Lytle, C. D., and J. G. Goddard. 1979. Enhanced virus reactivation in mammalian cells: effects of metabolic inhibitors. *Photochem. Photobiol.* **29:**959–962.

359. Lytle, C. D., J. G. Goddard, and G. Lin. 1980. Repair and mutagenesis of herpes simplex in UV irradiated monkey cells. *Mutat. Res.* **70:**139–149.

360. Lytle, C. D., and D. C. Knott. 1982. Enhanced mutagenesis parallels enhanced reactivation of herpes simplex virus in a human cell line. *EMBO J.* **1:**701–703.

361. Madura, K., S. Prakash, and L. Prakash. 1990. Expression of the *Saccharomyces cerevisiae* DNA repair gene *RAD6* that encodes a ubiquitin conjugating enzyme, increases in response to DNA damage and in meiosis but remains constant during the mitotic cell cycle. *Nucleic Acids Res.* **18:**771–778.

362. Madzak, C., J. Armier, A. Stary, L. Daya-Grosjean, and A. Sarasin. 1993. UV-induced mutations in a shuttle vector replicated in repair deficient trichothiodystrophy cells differ with those in genetically-related cancer prone xeroderma pigmentosum. *Carcinogenesis* **14**:1255–1260.

363. Madzak, C., J. B. Cabral-Neto, C. F. M. Menck, and A. Sarasin. 1992. Spontaneous and ultraviolet-induced mutations on a single-stranded shuttle vector transfected into monkey cells. *Mutat. Res.* **274**:135–145.

364. Madzak, C., and A. Sarasin. 1991. Mutation spectrum following transfection of ultraviolet-irradiated single-stranded or double-stranded shuttle vector DNA into monkey cells. *J. Mol. Biol.* **218**:667–673.

365. Maga, J. A., and K. Dixon. 1984. Enhanced replication of damaged SV40 DNA in carcinogen-treated monkey cells. *Photochem. Photobiol.* **40**:473–478.

366. Maga, J. A., T. A. McClanahan, and K. McEntee. 1986. Transcriptional regulation of DNA damage responsive (*DDR*) genes in different *rad* mutant strains of *Saccharomyces cerevisiae. Mol. Gen. Genet.* **205**:276–284.

367. Maher, V. M., D. J. Dorney, A. L. Mendrala, B. Konze-Thomas, and J. J. McCormick. 1979. DNA excision repair processes in human cells can eliminate the cytotoxic and mutagenic consequences of ultraviolet radiation. *Mutat. Res.* **62**:311–323.

368. Maher, V. M., L. M. Ouellete, R. D. Curren, and J. J. McCormick. 1976. Frequency of ultraviolet light-induced mutations is higher in xeroderma pigmentosum variant cells than in normal cells. *Nature* (London) **261**:593–595.

369. Maher, V. M., L. A. Rowan, K. C. Silinskas, S. A. Kateley, and J. J. McCormick. 1982. Frequency of UV-induced neoplastic transformation of diploid human fibroblasts is higher in xeroderma pigmentosum cells than in normal cells. *Proc. Natl. Acad. Sci. USA* **79**:2613–2617.

370. Maher, V. M., K. Sato, S. Kateley-Kohler, H. Thomas, S. Michaud, J. J. McCormick, M. Kraemer, and H. J. Rahmsdorf. 1988. Evidence of inducible error-prone repair mechanisms in diploid human fibroblasts, p. 465–471. *In* R. E. Moses and W. C. Summers (ed.), *DNA Replication and Mutagenesis.* American Society for Microbiology, Washington, D.C.

371. Malone, R., and R. E. Esposito. 1981. Recombinationless meiosis in *Saccharomyces cerevisiae. Mol. Cell. Biol.* **1**:891–901.

372. Malone, R., K. Jordan, and W. Wardman. 1985. Extragenic revertants of *rad50*, a yeast mutation causing defects in recombination and repair. *Curr. Genet.* **9**:453–461.

373. Malone, R. E. 1983. Multiple mutant analysis of recombination in yeast. *Mol. Gen. Genet.* **189**:405–412.

374. Malone, R. E., and R. E. Esposito. 1980. The *RAD52* gene is required for homothallic interconversion of mating types and spontaneous mitotic recombination in yeast. *Proc. Natl. Acad. Sci. USA* **77**:503–507.

375. Malone, R. E., T. Ward, S. Lin, and J. Waring. 1990. The *RAD50* gene, a member of the doublestrand break repair epistasis group, is not required for spontaneous recombination in yeast. *Curr. Genet.* **18**:111–116.

376. Malynn, B. A., T. K. Blackwell, G. M. Fulop, G. A. Rathburn, A. J. Furley, P. Ferrier, L. B. Heinke, R. A. Phillips, G. D. Yancopoulos, and F. W. Alt. 1988. The *scid* defect affects the final step of the immunoglobulin VDJ recombinase mechanism. *Cell* **54**:453–460.

377. Mattern, M. R., and R. B. Painter. 1977. The organization of repeated nucleotide sequences in the replicons of mammalian DNA. *Biophys. J.* **19**:117–123.

378. McClintock, B. 1951. Chromosome organization and genic expression. *Cold Spring Harbor Symp. Quant. Biol.* **16**:13–47.

379. McGill, C. B., B. K. Shafer, L. K. Derr, and J. N. Strathern. 1993. Recombination initiated by double-strand breaks. *Curr. Genet.* **23**:305–314.

380. McGrath, J. P., S. Jentsch, and A. Varshavsky. 1991. *UBA1*: an essential yeast gene encoding ubiquitin-activating enzyme. *EMBO J.* **10**:227–236.

381. McGregor, W. G., R.-H. Chen, L. Lukash, V. M. Maher, and J. J. McCormick. 1991. Cell cycle-dependent strand bias for UV-induced mutations in the transcribed strand of excision repair-proficient human fibroblasts but not in repair-deficient cells. *Mol. Cell. Biol.* **11**:1927–1934.

382. McKee, R. H., and C. W. Lawrence. 1979. Genetic analysis of gamma ray mutagenesis in yeast. I. Reversion in radiation sensitive strains. *Genetics* **93**:361–373.

383. McKee, R. H., and C. W. Lawrence. 1979. Genetic analysis of gamma ray mutagenesis in yeast. II. Allele specific control of mutagenesis. *Genetics* **93**:375–381.

384. Meechan, P. J., J. G. Carpenter, and T. D. Griffiths. 1986. Recovery of subchromosomal DNA synthesis in synchronous V-79 Chinese hamster cells after ultraviolet light exposure. *Photochem. Photobiol.* **43**:149–156.

385. Mellon, I., G. Spivak, and P. C. Hanawalt. 1987. Selective removal of transcription-blocking DNA damage from the transcribed strand of the mammalian DHFR gene. *Cell* **51**:241–249.

386. Melton, D. W., D. S. Konecki, J. Brennand, and C. T. Caskey. 1984. Structure, expression and mutation of the hypoxanthine phosphoribosyltransferase gene. *Proc. Natl. Acad. Sci. USA* **81**:2147–2151.

387. Meneghini, R., and P. Hanawalt. 1976. T4-endonuclease V-sensitive sites in DNA from ultraviolet-irradiated human cells. *Biochim. Biophys. Acta* **425**:428–437.

388. Meneghini, R., C. F. M. Menck, and R. I. Schumacher. 1981. Mechanism of tolerance to DNA lesions in mammalian cells. *Q. Rev. Biophys.* **14**:381–432.

389. Menichini, P., H. Vrieling, and A. A. van Zeeland. 1991. Strand-specific mutation spectra in repair-proficient and repair-deficient hamster cells. *Mutat. Res.* **251**:143–155.

390. Mezard, C., and A. Nicolas. 1994. Homologous, homeologous, and illegitimate repair of double-strand breaks during transformation of a wild-type strain and a *rad52* mutant strain of *Saccharomyces cerevisiae. Mol. Cell. Biol.* **14**:1278–1292.

391. Miller, J. H., J. S. Lebkowski, K. S. Greisen, and M. P. Calos. 1984. Specificity of mutations induced in transfected DNA by mammalian cells. *EMBO J.* **3**:3117–3121.

392. Milne, G. T., and D. T. Weaver. 1993. Dominant negative alleles of *RAD52* reveal a DNA repair/recombination complex including Rad51 and Rad52. *Genes Dev.* **7**:1755–1765.

393. Milo, G. E., S. A. Weisbrode, R. Zimmerman, and J. A. McCloskey. 1981. Ultraviolet radiation-induced neoplastic transformation of normal human cells *in vitro. Chem.-Biol. Interact.* **36**:45–49.

394. Mitchell, D. L., C. A. Haipek, and J. M. Clarkson. 1985. (6-4) photoproducts are removed from the DNA of UV-irradiated mammalian cells more efficiently than cyclobutane dimers. *Mutat. Res.* **143**:109–112.

395. Mitra, G., G. T. Pauly, R. Kumar, G. K. Pei, S. H. Hughes, R. C. Moschel, and M. Barbacid. 1989. Molecular analysis of O^6-substituted guanine-induced mutagenesis of *ras* oncogenes. *Proc. Natl. Acad. Sci. USA* **86**:8650–8654.

396. Mohrenweiser, H. W., A. V. Carrano, A. Fertitta, B. Perry, L. H. Thompson, J. D. Tucker, and C. A. Weber. 1989. Refined mapping of the three DNA repair genes, *ERCC1*, *ERCC2*, and *XRCC1*, on human chromosome 19. *Cytogenet. Cell. Genet.* **52**:11–14.

397. Molès, J.-P., C. Moyret, B. Guillot, P. Jeanteur, J.-J. Guilhou, C. Theillet, and N. Basset-Sèguin. 1993. p53 gene mutations in human epithelial skin cancers. *Oncogene* **8**:583–588.

398. Montelone, B. A., S. Prakash, and L. Prakash. 1981. Recombination and mutagenesis in *rad6* mutants of *Saccharomyces cerevisiae*: evidence for multiple functions of the *RAD6* gene. *Mol. Gen. Genet.* **184**:410–415.

399. Moore, S. P., and R. Fishel. 1990. Purification and characterization of a protein from human cells which promotes homologous pairing of DNA. *J. Biol. Chem.* **265**:11108–11117.

400. Morita, T., Y. Yoshimura, A. Yamamoto, K. Murata, M. Mori, H. Yamamoto, and A. Matsushiro. 1993. A mouse homolog of the *Escherichia coli recA* and *Saccharomyces cerevisiae RAD51* genes. *Proc. Natl. Acad. Sci. USA* **90**:6577–6580.

401. Moriya, T., M. Takeshita, F. Johnson, K. Peden, S. Will, and A. P. Grollman. 1988. Targeted mutations induced by a single acetylaminofluorene DNA adduct in mammalian cells and bacteria. *Proc. Natl. Acad. Sci. USA* **85**:1586–1589.

402. Morris, T., and J. Thacker. 1993. Formation of large deletions by illegitimate recombination in the *HPRT* gene of primary human fibroblasts. *Proc. Natl. Acad. Sci. USA* **90**:1392–1396.

403. Morrison, A., R. B. Christensen, J. Alley, A. K. Beck, E. G. Bernstine, J. F. Lemontt, and C. W. Lawrence. 1989. *REV3*, a yeast gene whose function is required for induced mutagenesis, is predicted to encode a nonessential DNA polymerase. *J. Bacteriol.* **171**:5659–5667.

404. Morrison, A., E. J. Miller, and L. Prakash. 1988. Domain structure and functional analysis of the carboxyl-terminal polyacidic sequence of the *RAD6* protein of *Saccharomyces cerevisiae*. *Mol. Cell. Biol.* **8**:1179–1185.

405. Morrison, D. P., and P. J. Hastings. 1979. Characterization of the mutator mutation *mut5-1*. *Mol. Gen. Genet.* **175**:57–65.

406. Mortimer, R. K. 1958. Radiobiological and genetic studies on a polyploid series (haploid to hexaploid) of *Saccharomyces cerevisiae*. *Radiat. Res.* **9**:312–316.

407. Mowat, M. R. A., W. J. Jachymczyk, P. J. Hastings, and R. C. von Borstel. 1983. Repair of gamma-ray induced DNA strand breaks in the radiation-sensitive mutant *rad18-2* of *Saccharomyces cerevisiae*. *Mol. Gen. Genet.* **189**:256–262.

408. Mueller, P. R., and B. Wold. 1989. In vivo footprinting of a muscle specific enhancer by ligation mediated PCR. *Science* **246**:780–786.

409. Mullenders, L. H. F., A.-M. Hazekamp-van Dokkum, W. H. J. Kalle, H. Vrieling, M. Z. Zdzienicka, and A. A. van Zeeland. 1993. UV-induced photolesions, their repair and mutations. *Mutat. Res.* **299**:271–276.

410. Munson, P. J., J. Hauser, A. S. Levine, and K. Dixon. 1987. Test of models for the sequence specificity of UV-induced mutations in mammalian cells. *Mutat. Res.* **179**:103–114.

411. Muris, D. F. R., K. Vreeken, A. M. Carr, B. C. Broughton, A. R. Lehmann, P. H. M. Lohman, and A. Pastink. 1993. Cloning the *RAD51* homologue of *Schizosaccharomyces pombe*. *Nucleic Acids Res.* **21**:4586–4591.

412. Murray, J. M., A. M. Carr, A. R. Lehmann, and F. Z. Watts. 1991. Cloning and characterization of the *rad9* DNA repair gene from *Schizosaccharomyces pombe*. *Nucleic Acids Res.* **19**:3525–3531.

413. Myhr, B. C., D. Turnbull, and J. A. DiPaolo. 1979. Ultraviolet mutagenesis of normal and xeroderma pigmentosum fibroblasts. *Mutat. Res.* **62**:341–353.

414. Nairn, R. S., R. M. Humphrey, and G. M. Adair. 1988. Transformation depending on intermolecular homologous recombination is stimulated by UV damage in transfected DNA. *Mutat. Res.* **208**:137–141.

415. Nakazawa, H., D. English, P. L. Randell, K. Nakazawa, N. Martel, B. K. Armstrong, and H. Yamasaki. 1994. UV and skin cancer: specific p35 gene mutation in normal skin as a biologically relevant exposure measurement. *Proc. Natl. Acad. Sci. USA* **91**:360–364.

416. Nasim, A., M. A. Hannan, and E. R. Nestmann. 1981. Pure and mosaic clones—a reflection of differences in mechanisms of mutagenesis by different agents in *Saccharomyces cerevisiae*. *Can. J. Genet. Cytol.* **23**:73–79.

417. Nasmyth, K. 1983. Molecular analysis of a cell lineage. *Nature* (London) **302**:670–676.

418. Nasmyth, K., and D. Shore. 1987. Transcriptional regulation in the yeast cell cycle. *Science* **237**:1162–1170.

419. Nguyen, H. T., and K. W. Minton. 1988. Ultraviolet-induced dimerization of non-adjacent pyrimidines. A potential mechanism for the targeted −1 frameshift mutation. *J. Mol. Biol.* **200**:681–693.

420. Nickoloff, J. A., E. Y. Chen, and F. Heffron. 1986. A 24-base-pair DNA sequence from the *MAT* locus stimulates intergenic recombination in yeast. *Proc. Natl. Acad. Sci. USA* **83**:7831–7835.

421. Ninkovic, M., M. Alacevic, F. Fabre, and Z. Zgaga. 1994. Efficient UV stimulation of yeast integrative transformation requires damage to both plasmid strands. *Mol. Gen. Genet.* **243**:308–314.

422. Njagi, G. D. E., and B. J. Kilbey. 1982. *cdc7-1*, a temperature sensitive cell-cycle mutant which interferes with induced mutagenesis in *Saccharomyces cerevisiae*. *Mol. Gen. Genet.* **186**:478–481.

423. North, P., A. Ganesh, and J. Thacker. 1990. The rejoining of double-strand breaks in DNA by human cell extracts. *Nucleic Acids Res.* **18**:6205–6210.

424. O'Day, C. L., P. M. J. Burgers, and J.-S. Taylor. 1992. PCNA-induced DNA synthesis past *cis-syn* and *trans-syn-I* thymine dimers by calf thymus DNA polymerase δ in vitro. *Nucleic Acids Res.* **20**:5403–5406.

425. Oettinger, M. A., D. G. Schatz, C. Gorka, and D. Baltimore. 1990. RAG-1 and RAG-2, adjacent genes that synergistically activate V(D)J recombination. *Science* **248**:1517–1523.

426. Ogawa, T., A. Shinohara, H. Ogawa, and J. Tomizawa. 1992. Functional structure of the RecA protein found by the chimera analysis. *J. Mol. Biol.* **226**:651–660.

427. Ogawa, T., X. Yu, A. Shinohara, and E. H. Egelman. 1993. Similarity of the yeast RAD51 filament to the bacterial RecA filament. *Science* **259**:1896–1899.

428. Orr-Weaver, T. L., J. W. Szostak, and R. J. Rothstein. 1981. Yeast transformation: a model system for the study of recombination. *Proc. Natl. Acad. Sci. USA* **78**:6354–6358.

429. Ostermann, K., A. Lorentz, and H. Schmidt. 1993. The fission yeast *rad22* gene, having a function in mating-type switching and repair of DNA damages, encodes a protein homolog to Rad52 of *Saccharomyces cerevisiae*. *Nucleic Acids Res.* **21**:5940–5944.

430. Ozenberger, B. A., and G. S. Roeder. 1991. A unique pathway of double-strand break repair operates in tandemly repeated genes. *Mol. Cell. Biol.* **11**:1222–1231.

431. Padmore, R., L. Cao, and N. Kleckner. 1991. Temporal comparison of recombination and synaptonemal complex formation during meiosis in *S. cerevisiae*. *Cell* **66**:1239–1256.

432. Painter, R. B. 1985. Inhibition and recovery of DNA synthesis in human cells after exposure to ultraviolet light. *Mutat. Res.* **145**:63–69.

433. Painter, R. B. 1986. Inhibition of mammalian cell DNA synthesis by ionizing radiation. *Int. J. Radiat. Biol.* **49**:771–781.

434. Painter, R. B., and B. R. Young. 1975. X-ray-induced inhibition of DNA synthesis in Chinese hamster, human HeLa, and mouse L cells. *Radiat. Res.* **64**:648–656.

435. Painter, R. B., and B. R. Young. 1976. Formation of nascent DNA molecules during inhibition of replicon initiation in mammalian cells. *Biochim. Biophys. Acta* **418**:146–153.

436. Painter, R. B., and B. R. Young. 1980. Radiosensitivity in ataxia telangiectasia: a new explanation. *Proc. Natl. Acad. Sci. USA* **77**:7315–7317.

437. Palladino, F., and H. L. Klein. 1992. Analysis of mitotic and meiotic defects in *Saccharomyces cerevisiae*. *Genetics* **132**:23–37.

438. Papadopoulo, D., C. Guillouf, H. Mohrenweiser, and E. Moustacchi. 1990. Hypomutability in Fanconi anemia cells is associated with increased deletion frequency at the *HPRT* locus. *Proc. Natl. Acad. Sci. USA* **87**:8383–8387.

438a. Papadopoulo, D., A. Laquerbe, C. Guillouf, and E. Moustacchi. 1993. Molecular spectrum of mutations induced at the *HPRT* locus by a cross-linking agent in human cell lines with different repair capacities. *Mutat. Res.* **294**:167–177.

439. Papadopoulo, D., B. Porfirio, and E. Moustacchi. 1990. Mutagenic response of Fanconi's anemia cells from a defined complementation group after treatment with photoactivated bifunctional psoralens. *Cancer Res.* **50**:3289–3294.

440. Park, S. D., and J. E. Cleaver. 1979. Postreplication repair: questions of its definition and possible alteration in xeroderma pigmentosum strains. *Proc. Natl. Acad. Sci. USA* **76**:3927–3931.

441. Parris, C. N., D. D. Levy, J. Jessee, and M. M. Seidman. 1994. Proximal and distal effects of sequence context on ultraviolet mutational hotspots in a shuttle vector replicated in xeroderma cells. *J. Mol. Biol.* **236:**491–502.

442. Patel, P. I., P. E. Famson, C. T. Caskey, and A. C. Chinault. 1986. Fine structure of the human hypoxanthine phosphoribosyltransferase gene. *Mol. Cell. Biol.* **6:**393–403.

443. Patrick, M. H., and R. O. Rahn. 1976. Photochemistry of DNA and polynucleotides: photoproducts, p. 35–95. *In* S. Y. Wang (ed.), *Photochemistry and Photobiology of Nucleic Acids.*, vol. II. Academic Press, Inc., New York.

444. Patterson, M., R. A. Sclafani, W. L. Fangman, and J. Rosamond. 1986. Molecular characterization of cell cycle gene *CDC7* from *Saccharomyces cerevisiae. Mol. Cell. Biol.* **6:**1590–1598.

445. Perera, M. I. R., K. I. Um, M. H. Greene, H. L. Waters, A. Bredberg, and K. H. Kraemer. 1986. Hereditary dysplastic nevus syndrome: lymphoid cell ultraviolet hypermutability in association with increased melanoma susceptibility. *Cancer Res.* **46:**1005–1009.

446. Pergola, F., M. Z. Zdzienicka, and M. A. Lieber. 1993. V(D)J recombination in mammalian cell mutants defective in DNA double-strand break repair. *Mol. Cell. Biol.* **13:**3464–3471.

447. Petes, T. D., R. E. Malone, and L. S. Symington. 1991. Recombination in yeast, p. 407–521. *In* J. R. Broach, J. R. Pringle, and E. W. Jones (ed.), *The Molecular and Cellular Biology of the Yeast Saccharomyces: Genome Dynamics, Protein Synthesis, and Energetics.* Cold Spring Harbor Laboratory, Cold Spring Harbor, N.Y.

448. Pfeifer, G. P., R. Drouin, A. D. Riggs, and G. P. Holmquist. 1991. *In vivo* mapping of a DNA adduct at nucleotide resolution: detection of pyrimidine (6-4) pyrimidone photoproducts by ligation-mediated polymerase chain reaction. *Proc. Natl. Acad. Sci. USA* **88:**1374–1378.

449. Pfeifer, G. P., R. Drouin, A. D. Riggs, and G. P. Holmquist. 1992. Binding of transcription factors creates hot spots for UV photoproducts in vivo. *Mol. Cell. Biol.* **12:**1798–1804.

450. Pfeiffer, P., and W. Vielmetter. 1988. Joining of nonhomologous DNA double strand breaks *in vitro. Nucleic Acids Res.* **16:**907–924.

451. Pickart, C. M., and I. A. Rose. 1985. Functional heterogeneity of ubiquitin carrier proteins. *J. Biol. Chem.* **260:**1573–1581.

452. Picologlou, S., N. Brown, and S. W. Liebman. 1990. Mutations in *RAD6*, a yeast gene encoding a ubiquitin-conjugating enzyme, stimulate retrotransposition. *Mol. Cell. Biol.* **10:**1017–1022.

453. Pierce, M. H., C. N. Giroux, and B. A. Kunz. 1987. Development of a yeast system to assay mutational specificity. *Mutat. Res.* **182:**65–74.

454. Pierceall, W. E., L. H. Goldberg, M. A. Tainsky, T. Mukhopadhyay, and H. N. Ananthaswamy. 1991. *Ras* gene mutation and amplification in human nonmelanoma skin cancers. *Mol. Carcinog.* **4:**196–202.

455. Pillidge, L., C. S. Downes, and R. T. Johnson. 1986. Defective post-replication recovery and u.v. sensitivity in a simian virus 40-transformed Indian muntjac cell line. *Int. J. Radiat. Biol.* **50:**119–136.

456. Pillidge, L., S. R. R. Musk, R. T. Johnson, and C. A. Waldren. 1986. Excessive chromosome fragility and abundance of sister-chromatid exchanges by UV in an Indian muntjac cell line defective in postreplication (daughter strand) repair. *Mutat. Res.* **166:**265–273.

457. Piperakis, S. M., and A. G. McLennan. 1984. Heat-enhanced reactivation of UV-irradiated adenovirus 2 is not associated with enhanced mutagenesis in HeLa cells. *Mutat. Res.* **139:**173–176.

458. Piperakis, S. M., and A. G. McLennan. 1984. Hyperthermia enhances the reactivation of irradiated adenovirus in HeLa cells. *Br. J. Cancer* **49:**199–205.

459. Piperakis, S. M., and A. G. McLennan. 1985. Enhanced reactivation of UV-irradiated adenovirus 2 in HeLa cells treated with nonmutagenic chemical agents. *Mutat. Res.* **142:**83–85.

460. Pittman, D., W. Lu, and R. E. Malone. 1993. Genetic and molecular analysis of *REC114*, an early meiotic recombination gene in yeast. *Curr. Genet.* **23:**295–304.

461. Polakowska, R., L. Prakash, and S. Prakash. 1983. Ultraviolet light induced mutagenesis of mitochondrial genes in the *rad6, rev3*

and *cdc8* mutants of *Saccharomyces cerevisiae. Mol. Gen. Genet.* **189:**513–515.

462. Potter, A. A., E. R. Nestmann, and V. N. Iyer. 1984. Introduction of the plasmid pKM101-associated *muc* genes in *Saccharomyces cerevisiae. Mutat. Res.* **131:**197–204.

463. Povirk, L. F., and R. B. Painter. 1976. The effect of 313 nanometer light on initiation of replicons in mammalian cell DNA containing bromodeoxyuridine. *Biochim. Biophys. Acta* **432:**267–272.

464. Prakash, L. 1974. Lack of chemically induced mutation in repair deficient mutants of yeast. *Genetics* **78:**1101–1118.

465. Prakash, L. 1976. Effect of genes controlling radiation sensitivity on chemically induced mutations in *Saccharomyces cerevisiae. Genetics* **83:**285–301.

466. Prakash, L. 1981. Characterization of postreplication repair in *Saccharomyces cerevisiae* and effects of *rad6, rad18, rev3* and *rad52* mutations. *Mol. Gen. Genet.* **184:**471–478.

467. Prakash, L., and A. Morrison. 1988. The *RAD6* gene and protein of *Saccharomyces cerevisiae*, p. 237–243. *In* E. C. Friedberg and P. C. Hanawalt (ed.), *Mechanisms and Consequences of DNA Damage Processing.* Alan R. Liss, Inc., New York.

468. Prakash, L., and S. Prakash. 1980. Genetic analysis of error-prone repair systems in *Saccharomyces cerevisiae*, p. 141-158. *In* W. M. Generoso, M. D. Shelby, and F. J. de Serres (ed.), *DNA Repair and Mutagenesis in Eukaryotes.* Plenum Publishing Corp., New York.

469. Prakash, L., and P. Taillon-Miller. 1981. Effects of the *RAD52* gene on sister chromatid recombination in *Saccharomyces cerevisiae. Curr. Genet.* **3:**247–250.

470. Prakash, S., L. Prakash, W. Burke, and B. A. Montelone. 1980. Effects of the *RAD52* gene on recombination in *Saccharomyces cerevisiae. Genetics* **94:**31–50.

471. Prakash, S., P. Sung, and L. Prakash. 1993. DNA repair genes and proteins of *Saccharomyces cerevisiae. Annu. Rev. Genet.* **27:**33–70.

472. Protić, M., E. Roilides, A. S. Levine, and K. Dixon. 1988. Enhancement of DNA repair capacity of mammalian cells by carcinogen treatment. *Somatic Cell Mol. Genet.* **14:**351–357.

473. Protić-Sabljić, M., N. Tuteja, P. J. Munson, J. Hauser, K. H. Kraemer, and K. Dixon. 1986. UV light-induced cyclobutane pyrimidine dimers are mutagenic in mammalian cells. *Mol. Cell. Biol.* **6:**3349–3356.

474. Quah, S.-K., R. C. von Borstel, and P. J. Hastings. 1980. The origin of spontaneous mutation in *Saccharomyces cerevisiae. Genetics* **96:**819–839.

475. Radman, M. 1980. Is there SOS induction in mammalian cells? *Photochem. Photobiol.* **32:**823–830.

476. Rady, P., F. Scinicariello, R. F. Wagner, and S. K. Tyring. 1992. *p53* mutations in basal cell carcinomas. *Cancer Res.* **52:**3804–3806.

477. Raghuraman, M. K., B. J. Brewer, and W. L. Fangman. 1994. Activation of a yeast replication origin near a double-stranded DNA break. *Genes Dev.* **8:**554–562.

478. Rathmell, W. K., and G. Chu. 1994. A DNA end-binding factor involved in double-strand break repair and V(D)J recombination. *Mol. Cell. Biol.* **14:**4741–4748.

479. Rathmell, W. K., and G. Chu. 1994. Involvement of the Ku autoantigen in the cellular response to DNA double-strand breaks. *Proc. Natl. Acad. Sci. USA* **91:**7623–7627.

480. Raymond, W. E., and N. Kleckner. 1993. Expression of the *Saccharomyces cerevisiae RAD50* gene during meiosis: steady-state transcript levels rise and fall while steady-state protein levels remain constant. *Mol. Gen. Genet.* **238:**390–400.

481. Raymond, W. E., and N. Kleckner. 1993. *RAD50* protein of *S. cerevisiae* exhibits ATP-dependent DNA binding. *Nucleic Acids Res.* **21:**3851–3856.

482. Razzaque, A., S. Chakrabarti, S. Joffe, and M. M. Seidman. 1984. Mutagenesis of a shuttle vector plasmid in mammalian cells. *Mol. Cell. Biol.* **4:**435–441.

483. Recio, L., J. Cochrane, D. Simpson, T. R. Skopek, J. P. O'Neill, and R. J. Albertini. 1987. DNA sequence analysis of in vivo *hprt* mutation in human T lymphocytes. *Mutagenesis* **5:**505–510.

484. Reid, T. M., and L. A. Loeb. 1993. Tandem double CC → TT mutations are produced by reactive oxygen species. *Proc. Natl. Acad. Sci. USA* **90:**3904–3907.

485. Resnick, M. A. 1987. Investigating the genetic control of biochemical events in meiotic recombination, p. 157–210. *In* P. B. Moens (ed.), *Meiosis.* Academic Press, Inc., New York.

486. Resnick, M. A. 1969. Genetic control of radiation sensitivity in *Saccharomyces cerevisiae. Genetics* **62:**519–531.

487. Resnick, M. A., and P. Martin. 1976. The repair of double-strand breaks in the nuclear DNA of *Saccharomyces cerevisiae* and its genetic control. *Mol. Gen. Genet.* **143:**119–129.

488. Resnick, M. A., M. Skaanild, and T. Nilsson-Tillgren. 1989. Lack of DNA homology in a pair of divergent chromosomes greatly sensitizes them to loss by DNA damage. *Proc. Natl. Acad. Sci. USA* **86:** 2276–2280.

489. Reynolds, P., M. H. M. Koken, J. H. J. Hoeijmakers, S. Prakash, and L. Prakash. 1990. The *rhp6+* gene of *Schizosaccharomyces pombe*: a structural and functional homolog of the *RAD6* gene from the distantly related yeast *Saccharomyces cerevisiae. EMBO J.* **9:**1423–1430.

490. Reynolds, P., S. Weber, and L. Prakash. 1985. *RAD6* gene of *Saccharomyces cerevisiae* encodes a protein containing a tract of 13 consecutive aspartates. *Proc. Natl. Acad. Sci. USA* **82:**168–172.

491. Reynolds, R. J. 1987. Induction and repair of closely opposed pyrimidine dimers in *Saccharomyces cerevisiae. Mutat. Res.* **184:**197–207.

492. Reynolds, R. J., and E. C. Friedberg. 1981. Molecular mechanism of pyrimidine dimer excision in *Saccharomyces cerevisiae*: incision of ultraviolet-irradiated deoxyribonucleic acid in vivo. *J. Bacteriol.* **146:**692–704.

493. Ripley, L. S., and B. W. Glickman. 1983. DNA secondary structure and mutation. *UCLA Symp. Mol. Cell. Biol. New Ser.* **11:**521–540.

494. Robson, C. N., A. L. Harris, and I. D. Hickson. 1989. Defective repair of DNA single- and double-strand breaks in the bleomycin- and X-ray-sensitive Chinese hamster ovary cell mutant BLM-2. *Mutat. Res.* **217:**93–100.

495. Rocco, V., B. de Masy, and A. Nicolas. 1992. The *Saccharomyces cerevisiae ARG4* initiator of meiotic gene conversion and its associated double-strand DNA breaks can be inhibited by transcriptional interference. *Proc. Natl. Acad. Sci. USA* **89:**12068–12072.

496. Rodarte-Ramón, U. S., and R. K. Mortimer. 1972. Radiation induced recombination in *Saccharomyces*: isolation and genetic study of recombination-deficient mutants. *Radiat. Res.* **49:**133–147.

497. Roeder, G. S., and S. E. Stewart. 1988. Mitotic recombination in yeast. *Trends Genet.* **4:**263–267.

498. Roilides, E., P. J. Munson, A. S. Levine, and K. Dixon. 1988. Use of a simian virus 40-based shuttle vector to analyze enhanced mutagenesis in mitomycin C-treated monkey cells. *Mol. Cell. Biol.* **8:** 3943–3946.

499. Romac, S., P. Leong, H. Sockett, and F. Hutchinson. 1989. DNA base sequence changes induced by ultraviolet light mutagenesis of a gene on a chromosome in Chinese hamster ovary cells. *J. Mol. Biol.* **209:**195–204.

500. Rommelaere, J., J.-M. Vos, J. J. Cornelis, and D. C. Ward. 1981. UV-enhanced reactivation of minute-virus-of-mice: stimulation of a late step in the viral life cycle. *Photochem. Photobiol.* **33:**845–854.

501. Rong, L., and H. Klein. 1993. Purification and characterization of the *SRS2* DNA helicase of the yeast *Saccharomyces cerevisiae. J. Biol. Chem.* **268:**1252–1259.

502. Rong, L., F. Palladino, A. Aguilera, and H. Klein. 1991. The hyper-gene conversion *hpr5-1* mutation of *Saccharomyces cerevisiae* is an allele of the *SRS2/RADH* gene. *Genetics* **127:**75–85.

503. Rossman, T. G., and C. B. Klein. 1985. Mammalian SOS system: a case of misplaced analogies. *Cancer Invest.* **3:**175–187.

504. Rossman, T. G., and C. B. Klein. 1988. From DNA damage to mutation in mammalian cells: a review. *Environ. Mol. Mutagen.* **11:**119–133.

505. Rotem, N., J. H. Axelrod, and R. Miskin. 1987. Induction of urokinase-type plasminogen activator by UV light in human fetal fibroblasts is mediated through a UV-induced secreted protein. *Mol. Cell. Biol.* **7:**622–631.

506. Roth, D. B., J. P. Menetski, P. B. Nakajima, M. J. Bosma, and M. Gellert. 1992. V(D)J recombination: broken DNA molecules with covalently sealed (hairpin) ends in scid mouse thymocytes. *Cell* **70:**983–991.

507. Roth, D. B., P. B. Nakajima, J. P. Menetski, M. J. Bosma, and M. Gellert. 1992. V(D)J recombination in mouse thymocytes: double-strand breaks near T cell receptor δ rearrangements. *Cell* **69:**41–53.

508. Roth, D. B., and J. H. Wilson. 1986. Nonhomologous recombination in mammalian cells: role for short sequence homologies in the joining reaction. *Mol. Cell. Biol.* **6:**4295–4304.

509. Roth, D. B., and J. H. Wilson. 1988. Illegitimate recombination in mammalian cells, p. 621–653. *In* R. Kucherlapati and G. R. Smith (ed.), *Genetic Recombination.* American Society for Microbiology, Washington, D.C.

510. Rubin, B. P., D. Ferguson, and W. K. Holloman. 1994. Structure of *REC2*, a recombination repair gene of *Ustilago maydis*, and its function in homologous recombination between plasmid and chromosomal sequences. *Mol. Cell. Biol.* **14:**6287–6296.

511. Rudin, N., and J. E. Haber. 1988. Efficient repair of HO-induced chromosomal breaks in *Saccharomyces cerevisiae* by recombination between flanking homologous sequences. *Mol. Cell. Biol.* **8:**3918–3928.

512. Saeki, T., I. Machida, and S. Nakai. 1980. Genetic control of diploid recovery after γ-irradiation in the yeast *Saccharomyces cerevisiae. Mutat. Res.* **73:**251–265.

513. Saiki, R. K., S. Scharf, F. Faloona, K. B. Mullis, G. T. Horn, H. A. Erlich, and N. Arnheim. 1985. Enzymatic amplification of beta-globin genomic sequences and restriction site analysis for diagnosis of sickle cell anemia. *Science* **230:**1350–1354.

514. Saka, Y., and M. Yanagida. 1993. Fission yeast *cut5+*, required for S phase onset and M phase restraint, is identical to the radiation-damage repair gene *rad4+. Cell* **74:**383–393.

515. Sarasin, A. 1985. SOS response in mammalian cells. *Cancer Invest.* **3:**163–174.

516. Sarasin, A., and A. Benoit. 1980. Induction of an error-prone mode of DNA repair in UV-irradiated monkey kidney cells. *Mutat. Res.* **70:**71–81.

517. Sarasin, A., and A. Benoit. 1986. Enhanced mutagenesis of UV-irradiated simian virus 40 occurs in mitomycin C-treated host cells only at a low multiplicity of infection. *Mol. Cell. Biol.* **6:**1102–1107.

518. Sarasin, A., F. Bourre, and A. Benoit. 1982. Error-prone replication of ultraviolet-irradiated simian virus 40 in carcinogen-treated monkey kidney cells. *Biochimie* **64:**815–821.

519. Sarasin, A., C. Gaillard, and J. Feunteun. 1983. Induced mutagenesis of simian virus 40 in carcinogen-treated monkey cells, p. 311–334. *In* C. W. Lawrence (ed.), *Induced Mutagenesis.* Plenum Publishing Corp., New York.

520. Sarasin, A., and P. C. Hanawalt. 1978. Carcinogens enhance survival of UV-irradiated simian virus 40 in treated monkey kidney cells: induction of a recovery pathway? *Proc. Natl. Acad. Sci. USA* **75:**346–350.

521. Sarasin, A. R., and P. C. Hanawalt. 1980. Replication of ultraviolet-irradiated simian virus 40 in monkey kidney cells. *J. Mol. Biol.* **138:**299–319.

522. Sarkar, S., U. B. DasGupta, and W. C. Summers. 1984. Error-prone mutagenesis detected in mammalian cells by a shuttle vector containing the *supF* gene of *Escherichia coli. Mol. Cell. Biol.* **4:**2227–2230.

523. Sato, M., C. Nishigori, M. Zghal, T. Yagi, and H. Takebe. 1993. Ultraviolet-specific mutations in *p53* gene in skin tumors in xeroderma pigmentosum patients. *Cancer Res.* **53:**2944–2946.

524. Scaria, A., and H. J. Edenberg. 1988. Preirradiation of host cells does not alter blockage of simian virus 40 replication forks by pyrimidine dimers. *Mutat. Res.* **193:**11–20.

525. Schatz, D. G., M. A. Oettinger, and D. Baltimore. 1989. The V(D)J recombination activating gene (RAG-1). *Cell* **59:**1035–1048.

526. Schatz, D. G., M. A. Oettinger, and M. S. Schlissel. 1992. V(D)J recombination: molecular biology and regulation. *Annu. Rev. Immunol.* **10:**359–383.

527. Schiestl, R., J. Zhu, and T. D. Petes. 1994. Effect of mutations in genes affecting homologous recombination on restriction enzyme-mediated and illegitimate recombination in *Saccharomyces cerevisiae. Mol. Cell. Biol.* **14:**4493–4500.

528. Schiestl, R. H., M. Dominska, and T. D. Petes. 1993. Transformation of *Saccharomyces cerevisiae* with non-homologous DNA: illegitimate integration of transforming DNA into yeast chromosomes and in vivo ligation of transforming DNA to mitochondrial DNA sequences. *Mol. Cell. Biol.* **13:**2697–2705.

529. Schiestl, R. H., and S. Prakash. 1988. *RAD1*, an excision repair gene of *Saccharomyces cerevisiae*, is also involved in recombination. *Mol. Cell. Biol.* **8:**3619–3626.

530. Schiestl, R. H., and S. Prakash. 1990. *RAD10*, an excision repair gene of *Saccharomyces cerevisiae*, is also involved in the *RAD1* pathway of mitotic recombination. *Mol. Cell. Biol.* **10:**2485–2491.

531. Schiestl, R. H., S. Prakash, and L. Prakash. 1990. The *SRS2* suppressor of *rad6* mutations of *Saccharomyces cerevisiae* acts by channeling DNA lesions into the *RAD52* DNA repair pathway. *Genetics* **124:**817–831.

532. Schild, D., B. J. Glassner, R. K. Mortimer, M. Carlson, and B. C. Laurent. 1992. Identification of *RAD16*, a yeast excision repair gene homologous to the recombinational repair gene *RAD54* and to the *SNF2* gene involved in transcriptional activation. *Yeast* **8:**385–396.

533. Schlissel, M., A. Constantinescu, T. Morrow, M. Baxter, and A. Peng. 1993. Double-strand signal sequence breaks in V(D)J recombination are blunt, 5'-phosphorylated, RAG-dependent, and cell cycle regulated. *Genes Dev.* **7:**2520–2532.

534. Schorpp, M., U. Mallick, H. J. Rahmsdorf, and P. Herrlich. 1984. UV-induced extracellular factor from human fibroblasts communicates the UV response to nonirradiated cells. *Cell* **37:**861–868.

535. Schuler, W., I. J. Weiler, A. Schuler, R. A. Phillips, N. Rosenberg, T. W. Mak, J. F. Kearney, R. P. Perry, and M. J. Bosma. 1986. Rearrangement of antigen receptor genes is defective in mice with severe combined immune deficiency. *Cell* **46:**963–972.

536. Schultes, N. P., and J. W. Szostak. 1991. A poly (dA·dT) tract is a component of the recombination initiation site at the *ARG4* locus in *Saccharomyces cerevisiae. Mol. Cell. Biol.* **11:**322–338.

537. Schumacher, R. I., C. F. Menck, and R. Meneghini. 1983. Sites sensitive to S1 nuclease and discontinuities in DNA nascent strands of ultraviolet irradiated mouse cells. *Photochem. Photobiol.* **37:**605–610.

538. Sclafani, R. A., M. Patterson, J. Rosamond, and W. L. Fangman. 1988. Differential regulation of the yeast *CDC7* gene during mitosis and meiosis. *Mol. Cell. Biol.* **8:**293–300.

539. Sedgwick, S., and L. Johnston. 1987. Ubiquitous cycles of repair. *Nature* (London) **329:**109.

540. Seetharam, S., M. Protić-Sabljić, M. M. Seidman, and K. H. Kraemer. 1987. Abnormal ultraviolet mutagenic spectrum in plasmid DNA replicated in cultured fibroblasts from a patient with the skin cancer-prone disease, xeroderma pigmentosum. *J. Clin. Invest.* **80:**1613–1617.

541. Seetharam, S., and M. M. Seidman. 1992. Modulation of ultraviolet light mutational hotspots by cellular stress. *J. Mol. Biol.* **228:**1031–1036.

542. Seetharam, S., H. L. Waters, M. M. Seidman, and K. H. Kraemer. 1989. Ultraviolet mutagenesis in a plasmid vector replicated in lymphoid cells from a patient with the melanoma-prone disorder dysplastic nevus syndrome. *Cancer Res.* **49:**5918–5921.

543. Seidman, M. M., A. Bredberg, S. Seetharam, and K. H. Kraemer. 1987. Multiple point mutations in a shuttle vector propa-

gated in human cells: evidence for an error-prone DNA polymerase activity. *Proc. Natl. Acad. Sci. USA* **84:**4944–4948.

544. Setlow, R. B., and S. M. D'Ambrosio. 1978. Size of putative gaps in post replication repair. *Biophys. J.* **21:**94a.

545. Sheng, S., and S. M. Schuster. 1993. Purification and characterization of *Saccharomyces cerevisiae* DNA damage-responsive protein 48 (*DDRP 48*). *J. Biol. Chem.* **268:**4752–4758.

546. Shinohara, A., H. Ogawa, Y. Matsuda, N. Ushio, K. Ikeo, and T. Ogawa. 1993. Cloning of human, mouse and fission yeast recombination genes homologous to *RAD51* and *recA. Nat. Genet.* **4:**239–243.

547. Shinohara, A., H. Ogawa, and T. Ogawa. 1992. Rad51 protein involved in repair and recombination in *S. cerevisiae* is a RecA-like protein. *Cell* **69:**457–470.

548. Siede, W., and F. Eckardt. 1984. Inducibility of error-prone DNA repair in yeast? *Mutat. Res.* **129:**3–11.

549. Siede, W., and F. Eckardt. 1986. Analysis of mutagenic DNA repair in a thermoconditional mutant of *Saccharomyces cerevisiae*. III. Dose-response pattern of mutation induction in UV-irradiated *rev2^{ts}* cells. *Mol. Gen. Genet.* **202:**68–74.

550. Siede, W., and F. Eckardt. 1986. Analysis of mutagenic DNA repair in a thermoconditional mutant of *Saccharomyces cerevisiae*. IV. Influence of DNA replication and excision repair on *REV2* dependent UV-mutagenesis and repair. *Curr. Genet.* **10:**871–878.

551. Siede, W., F. Eckardt, and M. Brendel. 1983. Analysis of mutagenic DNA repair in a thermoconditional mutant of *Saccharomyces cerevisiae*. I. Influence of cycloheximide on UV-irradiated *rev2^{ts}* cells. *Mol. Gen. Genet.* **190:**406–412.

552. Siede, W., F. Eckardt, and M. Brendel. 1983. Analysis of mutagenic DNA repair in a thermoconditional mutant of *Saccharomyces cerevisiae*. II. Influence of cycloheximide on UV-irradiated exponentially growing *rev2^{ts}* cells. *Mol. Gen. Genet.* **190:**413–416.

553. Siede, W., and F. Eckardt-Schupp. 1986. A mismatch repair-based model can explain some features of u.v. mutagenesis in yeast. *Mutagenesis* **1:**471–474.

553a. Smider, V., W. K. Rathmell, M. Lieber, and G. Chu. 1994. Restoration of X-ray resistance and V(D)J recombination in mutant cells by Ku cDNA. *Science* **266:**288–291.

554. Smith, P. J., M. H. Greene, D. Adams, and M. Paterson. 1983. Abnormal responses to the carcinogen 4-nitroquinoline 1-oxide of cultured fibroblasts from patients with hereditary cutaneous malignant melanoma to killing and mutations by 4-nitroquinoline 1-oxide and adduct formation. *Cancer Res.* **46:**1005–1009.

555. Smith, P. J., M. H. Greene, D. A. Devlin, E. A. McKeen, and M. Paterson. 1982. Abnormal sensitivity to UV-radiation in cultured skin fibroblasts from patients with hereditary cutaneous malignant melanoma and dysplastic nevus syndrome. *Int. J. Cancer* **30:**39–45.

556. Spivak, G., and P. C. Hanawalt. 1992. Translesion DNA synthesis in the dihydrofolate reductase domain of UV-irradiated CHO cells. *Biochemistry* **31:**6794–6800.

557. Stacks, P. C., J. H. White, and K. Dixon. 1983. Accommodation of pyrimidine dimers during replication of UV-damaged simian virus 40 DNA. *Mol. Cell. Biol.* **3:**1403–1411.

558. Stamato, T. D., L. Hinkle, A. R. S. Collins, and C. A. Waldren. 1981. Chinese hamster ovary mutant UV-1 is hypomutable and defective in a postreplication recovery process. *Somatic Cell Genet.* **7:**307–320.

559. Stamato, T. D., and C. A. Waldren. 1977. Isolation of UV-sensitive variants of CHO-K1 by nylon cloth replica plating. *Somatic Cell Genet.* **3:**431–440.

560. Steingrimsdottir, H., G. Rowley, A. Waugh, D. Beare, I. Ceccherini, J. Cole, and A. R. Lehmann. 1993. Molecular analysis of mutations in the *hprt* gene in circulating lymphocytes from normal and DNA-repair-deficient donors. *Mutat. Res.* **294:**29–41.

561. Stewart, S. E., and G. S. Roeder. 1989. Transcription by RNA polymerase I stimulates mitotic recombination in *Saccharomyces cerevisiae. Mol. Cell. Biol.* **9:**3464–3472.

562. Stone-Wolff, D. S., and T. G. Rossman. 1981. Effects of inhibitors of *de novo* protein synthesis on UV-mutagenesis via inducible, error-prone DNA repair. *Mutat. Res.* **82:**147–157.

563. Stone-Wolff, D. S., and T. G. Rossman. 1982. Demonstration of recovery from the potentially mutagenic effects of ultra-violet light by replication-inhibited Chinese hamster V79 cells. *Mutat. Res.* **95:**493–503.

564. Story, R. M., D. Bishop, N. Kleckner, and T. A. Steitz. 1993. Structural relationships of bacterial RecA proteins to recombination proteins from bacteriophage T4 and yeast. *Science* **259:**1892–1896.

565. Strathern, J. N., A. J. S. Klar, J. B. Hicks, J. A. Abraham, J. M. Ivy, K. A. Nasmyth, and C. McGill. 1982. Homothallic switching of yeast mating type cassettes is initiated by a double-stranded cut in the *MAT* locus. *Cell* **31:**183–192.

566. Strathern, J. N., K. G. Weinstock, D. R. Higgins, and C. B. McGill. 1991. A novel recombinator in yeast based on gene II protein from bacteriophage f1. *Genetics* **127:**61–73.

567. Strauss, B. S. 1991. The 'A' rule of mutagen specificity: a consequence of DNA polymerase bypass of non-instructional lesions? *Bioessays* **13:**1–6.

568. Su, Z. Z., B. Avalosse, J.-M. Vos, J. J. Cornelis, and J. Rommelaere. 1985. Mutagenesis at putative apurinic sites in alkylated single-stranded DNA of parvovirus H-1 propagated in human cells. *Mutat. Res.* **149:**1–8.

569. Suarez, H. G., L. Daya-Grosjean, D. Schlaifer, P. Nardeux, G. Renault, J. L. Bos, and A. Sarasin. 1989. Activated oncogenes in human skin tumors from a repair-deficient syndrome, xeroderma pigmentosum. *Cancer Res.* **49:**1223–1228.

570. Sugawara, N., and J. E. Haber. 1992. Characterization of double-strand break-induced recombination: homology requirements and single-stranded DNA formation. *Mol. Cell. Biol.* **12:**563–575.

571. Sugino, A., J. Nitiss, and M. A. Resnick. 1988. ATP-independent DNA strand transfer catalyzed by protein(s) from meiotic cells of the yeast *Saccharomyces cerevisiae. Proc. Natl. Acad. Sci. USA* **85:**3683–3687.

572. Summers, W. C., P. M. Glazer, and D. Malkevich. 1989. Lambda phage shuttle vectors for analysis of mutations in mammalian cells in culture and in transgenic mice. *Mutat. Res.* **220:**263–268.

573. Sun, H., D. Dawson, and J. W. Szostak. 1991. Genetic and physical analysis of sister chromatid exchange in yeast meiosis. *Mol. Cell. Biol.* **11:**6328–6336.

574. Sun, H., D. Treco, N. P. Schultes, and J. W. Szostak. 1989. Double-strand breaks at an initiation site for meiotic gene conversion. *Nature* (London) **338:**87–90.

575. Sung, P. 1994. Catalysis of ATP-dependent homologous DNA pairing and strand exchange by yeast RAD51 protein. *Science* **265:**1241–1243.

576. Sung, P., E. Berleth, C. Pickart, S. Prakash, and L. Prakash. 1991. Yeast *RAD6* encoded ubiquitin conjugating enzyme mediates protein degradation dependent on the N-end-recognizing E3 enzyme. *EMBO J.* **10:**2187–2193.

577. Sung, P., S. Prakash, and L. Prakash. 1988. The *RAD6* protein of *Saccharomyces cerevisiae* polyubiquitinates histones and its acidic domain mediates this activity. *Genes Dev.* **2:**1476–1485.

578. Sung, P., S. Prakash, and L. Prakash. 1990. Mutation of cysteine-88 in the *Saccharomyces cerevisiae RAD6* protein abolishes its ubiquitin-conjugating activity and its various biological functions. *Proc. Natl. Acad. Sci. USA* **87:**2695–2699.

579. Sung, P., P. Reynolds, L. Prakash, and S. Prakash. 1993. Purification and characterization of the *Saccharomyces cerevisiae RAD1/RAD10* endonuclease. *J. Biol. Chem.* **268:**26391–26399.

580. Taccioli, G. E., H.-L. Cheng, A. J. Varghese, G. Whitmore, and F. W. Alt. 1994. A DNA repair defect in Chinese hamster ovary cells affects V(D)J recombination similarly to the murine *scid* mutation. *J. Biol. Chem.* **269:**7439–7442.

580a. Taccioli, G. E., T. M. Gottlieb, T. Blunt, A. Priestley, J. Demengeot, R. Mizuta, A. R. Lehmann, F. W. Alt, S. P. Jackson, and **P. A. Jeggo.** 1994. Ku80: Product of the *XRCC5* gene and its role in DNA repair and V(D)J recombination. *Science* **265:**1442–1445.

581. Taccioli, G. E., G. Rathbun, E. Oltz, T. Stamato, P. A. Jeggo, and F. W. Alt. 1993. Impairment of V(D)J recombination in double-strand break repair mutants. *Science* **260:**207–210.

582. Takimoto, K. 1983. Lack of enhanced mutation of UV- and γ-irradiated herpes virus in UV-irradiated CV-1 monkey cells. *Mutat. Res.* **121:**159–166.

583. Takimoto, K., O. Niwa, and T. Sugahara. 1982. Reactivation of UV- and γ-irradiated herpes virus in UV- and X-irradiated CV-1 cells. *Photochem. Photobiol.* **35:**495–499.

584. Tanaka, T., T. Yamagami, Y. Oka, T. Nomura, and H. Sugiyama. 1993. The scid mutation in mice causes defects in the repair system for both double-strand DNA breaks and DNA cross-links. *Mutat. Res.* **288:**277–280.

585. Taylor, W. D., L. E. Bockstahler, J. Montes, M. A. Babich, and C. D. Lytle. 1982. Further evidence that ultraviolet radiation-enhanced reactivation of SV40 in monkey cells is not accompanied by mutagenesis. *Mutat. Res.* **105:**291–298.

586. te Riele, H., E. R. Maandag, and A. Berns. 1992. Highly efficient gene targeting in embryonic stem cells through homologous recombination with isogenic DNA constructs. *Proc. Natl. Acad. Sci. USA* **89:**5128–5132.

587. Tessman, I., and M. A. Kennedy. 1991. The two-step model of UV mutagenesis reassessed: deamination of cytosine as the likely source of the mutations associated with deamination. *Mol. Gen. Genet.* **227:**144–148.

588. Thacker, J., A. N. Ganesh, A. Stretch, D. M. Benjamin, A. J. Zahalski, and E. A. Hendrickson. 1994. Gene mutation and V(D)J recombination in the radiosensitive *irs* lines. *Mutagenesis* **9:**163–168.

589. Thacker, J., and R. E. Wilkinson. 1991. The genetic basis of resistance to ionizing radiation damage in cultured mammalian cells. *Mutat. Res.* **254:**135–142.

590. Thaler, D. S., and F. W. Stahl. 1988. DNA double-chain breaks in recombination of phage λ and of yeast. *Annu. Rev. Genet.* **22:**169–197.

591. Thelen, M. P., K. Onel, and W. K. Holloman. 1994. The *REC1* gene of *Ustilago maydis* involved in the cellular response to DNA damage encodes an exonuclease. *J. Biol. Chem.* **269:**747–754.

592. Thode, S., A. Schäfer, P. Pfeiffer, and W. Vielmetter. 1990. A novel pathway of DNA end-to-end joining. *Cell* **60:**921–928.

593. Thomas, B. J., and R. Rothstein. 1989. Elevated recombination rates in transcriptionally active DNA. *Cell* **56:**619–630.

594. Thomas, B. J., and R. Rothstein. 1989. The genetic control of direct-repeat recombination in *Saccharomyces*: the effect of *rad52* and *rad1* on mitotic recombination at *GAL10*, a transcriptionally regulated gene. *Genetics* **123:**725–738.

595. Thomas, D. C., and T. A. Kunkel. 1993. Replication of UV-irradiated DNA in human cell extracts—evidence for mutagenic bypass of pyrimidine dimers. *Proc. Natl. Acad. Sci. USA* **90:**7744–7753.

596. Thomas, D. C., D. C. Nguyen, W. W. Piegorsch, and T. A. Kunkel. 1993. Relative probabilities of mutagenic translesion synthesis on the leading and lagging strands during replication of UV-irradiated DNA in a human cell extract. *Biochemistry* **32:**11476–11482.

597. Thompson, L. H., K. W. Brookman, L. E. Dillehay, A. V. Carrano, J. A. Mazrimas, C. L. Mooney, and J. L. Minkler. 1982. A CHO-cell strain having hypersensitivity to mutagens, a defect in strand-break repair, and an extraordinarily baseline frequency of sister chromatid exchange. *Mutat. Res.* **95:**427–440.

598. Thompson, L. H., K. W. Brookman, N. J. Jones, S. A. Allen, and A. V. Carrano. 1990. Molecular cloning of the human XRCC1 gene, which corrects defective DNA strand break repair and sister chromatid exchange. *Mol. Cell. Biol.* **10:**6160–6171.

599. Thompson, L. H., K. W. Brookman, J. L. Minkler, J. C. Fuscoe, K. A. Henning, and A. V. Carrano. 1985. DNA-mediated transfer of a human DNA repair gene that controls sister chromatid exchange. *Mol. Cell. Biol.* **5:**881–884.

600. Thompson, L. H., K. W. Brookman, and M. J. Siciliano. 1990. Mammalian genes involved in the repair of damage from ionizing radiation, p. 271–279. *In* S. S. Wallace and R. B. Painter (ed.), *Ionizing Radiation Damage to DNA: Molecular Aspects.* Wiley-Liss, New York.

601. Tishkoff, D. X., A. W. Johnson, and R. D. Kolodner. 1991. Molecular and genetic analysis of the gene encoding the *Saccharomyces cerevisiae* strand exchange protein Sep1. *Mol. Cell. Biol.* **11:**2593–2608.

602. Tomkinson, A., A. J. Bardwell, L. Bardwell, N. J. Tappe, and E. C. Friedberg. 1993. Yeast DNA repair and recombination proteins Rad1 and Rad10 constitute a single-stranded DNA endonuclease. *Nature* (London) **362:**860–862.

603. Tornaletti, S., and G. P. Pfeifer. 1994. Slow repair of pyrimidine dimers at *p53* mutation hotspots in skin cancer. *Science* **263:**1436–1438.

604. Tornaletti, S., D. Rozek, and G. P. Pfeifer. 1993. The distribution of UV photoproducts along the human p53 gene and its relation to mutations in skin cancer. *Oncogene* **8:**2051–2057.

605. Treger, J. M., K. A. Heichman, and K. McEntee. 1988. Expression of the yeast *UBI4* gene increases in response to DNA-damaging agents and in meiosis. *Mol. Cell. Biol.* **8:**1132–1136.

606. Tuite, M. F., and B. S. Cox. 1981. *RAD6*⁺ gene of *Saccharomyces cerevisiae* codes for two mutationally separable deoxyribonucleic acid repair functions. *Mol. Cell. Biol.* **1:**153–157.

607. Urlaub, G., P. J. Mitchell, C. J. Ciudad, and L. A. Chasin. 1989. Nonsense mutations in the dihydrofolate reductase gene after RNA processing. *Mol. Cell. Biol.* **9:**2868–2880.

608. van der Lubbe, J. L. M., H. J. M. Rosdorff, and A. J. van der Eb. 1989. Homologous recombination is not enhanced in UV-irradiated normal and repair-deficient human fibroblasts. *Mutat. Res.* **217:**153–161.

609. van't Veer, L. J., B. M. T. Burgering, R. Versteeg, A. J. M. Boot, D. J. Ruiter, S. Osanto, P. I. Schrier, and J. L. Bos. 1989. N-*ras* mutations in human cutaneous melanoma from Sun-exposed body sites. *Mol. Cell. Biol.* **9:**3114–3116.

610. van Zeeland, A. A., and A. R. Filon. 1982. Post-replication repair: elongation of daughter strand DNA in UV-irradiated mammalian cells in culture. *Prog. Mutat. Res.* **4:**375–384.

611. van Zeeland, A. A., L. Mullenders, M. Z. Zdzienicka, and H. Vrieling. 1992. Influence of strand-specific repair of DNA damage on molecular mutation spectra, p. 117–122. *In* V. A. Bohr, K. Wassermann, and K. H. Kraemer (ed.), *DNA Repair Mechanisms,* vol. 35. *Alfred Benzon Symposium.* Munksgaard, Copenhagen.

612. Ventura, A. M., and R. Meneghini. 1984. Inhibition and recovery of the rate of DNA synthesis in V79 Chinese hamster cells following ultraviolet light irradiation. *Mutat. Res.* **131:**81–88.

613. Verma, M., K. G. Moffat, and J. B. Egan. 1989. UV irradiation inhibits initiation of DNA replication from *oriC* in *Escherichia coli. Mol. Gen. Genet.* **216:**446–454.

614. Voelkel-Meiman, K., R. L. Keil, and G. S. Roeder. 1987. Recombination-stimulating sequences in yeast ribosomal DNA correspond to sequences regulating transcription by RNA polymerase I. *Cell* **48:**1071–1079.

615. Vos, J.-M., and J. Rommelaere. 1982. Are pyrimidine dimers tolerated during DNA replication of UV-irradiated parvovirus minute-virus-of-mice in mouse fibroblasts? *Biochimie* **64:**839–844.

616. Vos, J.-M., and J. Rommelaere. 1987. Replication of DNA containing apurinic sites in human and mouse cells probed with parvoviruses MVM and H-1. *Mol. Cell. Biol.* **7:**2620–2624.

617. Vrieling, H., J. C. P. Thijssen, A. M. Rossi, F. J. van Dam, A. T. Natarajan, A. D. Tates, and A. A. van Zeeland. 1992. Enhanced *hprt* mutant frequency but no significant difference in mutation spectrum between a smoking and a non-smoking human population. *Carcinogenesis* **13:**1625–1632.

618. Vrieling, H., M. L. van Rooijen, N. A. Groen, M. Z. Zdzienicka, J. W. I. M. Simons, P. H. M. Lohman, and A. A. van Zeeland. 1989. DNA strand specificity for UV-induced mutations in mammalian cells. *Mol. Cell. Biol.* **9:**1277–1283.

619. Vrieling, H., J. Venema, M. L. van Rooyen, A. van Hoffen, P. Menichini, M. Z. Zdzienicka, J. W. I. M. Simons, L. H. F. Mullenders, and A. A. van Zeeland. 1991. Strand specificity for UV-induced DNA repair and mutations in the Chinese hamster HPRT gene. *Nucleic Acids Res.* **19:**2411–2415.

620. Vrieling, H., L.-H. Zhang, A. A. van Zeeland, and M. Z. Zdzienicka. 1992. UV-induced *hprt* mutations in a UV-sensitive hamster cell line from complementation group 3 are biased towards the transcribed strand. *Mutat. Res.* **274:**147–155.

621. Waga, S., G. J. Hannon, D. Beach, and B. Stillman. 1994. The p21 inhibitor of cyclin-dependent kinases controls DNA replication by interaction with PCNA. *Nature* (London) **369:**574–578.

622. Waldren, C., D. Snead, and T. Stamato. 1983. Restoration of normal resistance to killing and of post-replication recovery (PRR) in CHO-UV-1 cells by transformation with hamster or human DNA, p. 637–646. *In* E. C. Friedberg and B. A. Bridges (ed.), *Cellular Responses to DNA Damage.* Alan R. Liss, Inc., New York.

623. Walker, J. E., M. Saraste, M. J. Runswick, and N. J. Gay. 1982. Distantly related sequences in the α- and β-subunits of ATP synthase, myosin, kinases and other ATP-requiring enzymes and a common nucleotide binding fold. *EMBO J.* **1:**945–951.

624. Walters, R. A., and C. E. Hildebrand. 1975. Evidence that X-irradiation inhibits DNA replicon initiation in Chinese hamster cells. *Biochem. Biophys. Res. Commun.* **65:**265–271.

625. Wang, T., and K. C. Smith. 1986. Postreplication repair in ultraviolet-irradiated human fibroblasts: formation and repair of DNA double-strand breaks. *Carcinogenesis* **7:**389–392.

626. Wang, Y.-C., V. M. Maher, and J. J. McCormick. 1991. Xeroderma pigmentosum variant cells are less likely than normal cells to incorporate dAMP opposite photoproducts during replication of UV-irradiated plasmids. *Proc. Natl. Acad. Sci. USA* **88:**7810–7814.

627. Wang, Y.-C., V. M. Maher, D. L. Mitchell, and J. J. McCormick. 1993. Evidence from mutation spectra that the UV hypermutability of xeroderma pigmentosum variant cells reflects abnormal, error-prone replication on a template containing photoproducts. *Mol. Cell. Biol.* **13:**4276–4283.

628. Wang, Z., X. Wu, and E. C. Friedberg. 1993. DNA repair synthesis during base excision repair in vitro is catalyzed by DNA polymerase ε and is influenced by DNA polymerases α and δ in *Saccharomyces cerevisiae. Mol. Cell. Biol.* **13:**1051–1058.

629. Watkins, J. F., P. Sung, S. Prakash, and L. Prakash. 1993. The extremely conserved amino terminus of RAD6 ubiquitin-conjugating enzyme is essential for amino-end rule-dependent protein degradation. *Genes Dev.* **7:**250–261.

630. Weibezahn, K. F., H. Lohrer, and P. Herrlich. 1985. Double-strand break repair and G2 block in Chinese hamster ovary cells and their radiosensitive mutants. *Mutat. Res.* **145:**177–183.

631. Weiffenbach, B., and J. E. Haber. 1981. Homothallic mating type switching generates lethal chromosome breaks in *rad52* strains of *Saccharomyces cerevisiae. Mol. Cell. Biol.* **1:**522–534.

632. Weinert, T. A., G. L. Kiser, and L. H. Hartwell. 1994. Mitotic checkpoint genes in budding yeast and the dependence of mitosis on DNA replication and repair. *Genes Dev.* **8:**652–665.

633. White, J. H., and K. Dixon. 1984. Gap filling and not replication fork progression is the rate-limiting step in the replication of UV-damaged simian virus 40 DNA. *Mol. Cell. Biol.* **4:**1286–1292.

634. White, M., M. Wierdl, P. Detloff, and T. Petes. 1991. DNA-binding protein RAP1 stimulates meiotic recombination at the *HIS4* locus in yeast. *Proc. Natl. Acad. Sci. USA* **88:**9755–9759.

635. Whitmore, G. F., A. J. Varghese, and S. Gulyas. 1989. Cell cycle responses of two X-ray sensitive mutants defective in DNA repair. *Int. J. Radiat. Biol.* **56:**657–665.

636. Williams, J. I., and J. E. Cleaver. 1978. Perturbations in simian virus 40 DNA synthesis by ultraviolet light. *Mutat. Res.* **52:**301–311.

637. Willis, K. K., and H. L. Klein. 1987. Intrachromosomal recombination in *Saccharomyces cerevisiae*: reciprocal exchange in an inverted repeat and associated gene conversion. *Genetics* **117**:633–643.

638. Wing, S. S., F. Dumas, and D. Banville. 1992. A rabbit reticulocyte ubiquitin carrier protein that supports ubiquitin dependent proteolysis (E2 14K) is homologous to the yeast DNA repair gene *RAD6*. *J. Biol. Chem.* **267**:6495–6501.

639. Witkin, E. M., and D. L. George. 1973. Ultraviolet mutagenesis and inducible DNA repair in *polA* and *uvrA polA* derivatives of *Escherichia coli* B/r: evidence for an inducible error-prone repair system. *Genetics* **73**(Suppl.):91–108.

640. Wong, S. W., A. F. Wahl, P.-M. Yuan, N. A. Arai, B. E. Pearson, K.-I. Arai, D. Korn, M. W. Hunkapiller, and T. S.-F. Wang. 1988. Human polymerase α gene expression is cell proliferation dependent and its primary structure is similar to both prokaryotic and eukaryotic replicative DNA polymerases. *EMBO J.* **7**:37–47.

641. Yagi, T., M. Sato, J. Tatsumi-Miyajima, and H. Takebe. 1992. UV-induced base substitution mutations in a shuttle vector plasmid propagated in group C xeroderma pigmentosum cells. *Mutat. Res.* **273**:213–220.

642. Yagi, T., J. Tatsumi-Miyajima, M. Sato, K. H. Kraemer, and H. Takebe. 1991. Analysis of point mutations in an ultraviolet-irradiated shuttle vector plasmid propagated in cells from Japanese xeroderma pigmentosum patients in complementation groups A and F. *Cancer Res.* **51**:3177–3182.

643. Yoo, H., L. Li, P. G. Sacks, L. H. Thompson, F. F. Becker, and J. Y.-H. Chan. 1992. Alterations in expression and structure of the DNA repair gene *XRCC1*. *Biochem. Biophys. Res. Commun.* **186**:900–910.

644. Zamansky, G. B., L. F. Kleinman, P. H. Black, and J. C. Kaplan. 1980. Reactivation of herpes simplex virus in a cell line inducible for simian virus 40 synthesis. *Mutat. Res.* **70**:1–9.

645. Zamb, T. J., and T. D. Petes. 1981. Unequal sister-strand recombination within yeast ribosomal DNA does not require the *RAD52* gene product. *Curr. Genet.* **3**:125–132.

646. Zdzienicka, M. Z., N. G. J. Jaspers, G. P. van der Schans, A. T. Natarajan, and J. W. I. M. Simons. 1989. Ataxia-telangiectasia-like Chinese hamster V79 cell mutants with radioresistant DNA synthesis, chromosomal instability, and normal DNA strand break repair. *Cancer Res.* **49**:1481–1485.

647. Zdzienicka, M. Z., Q. Tran, G. P. van der Schans, and J. W. I. M. Simons. 1988. Characterization of an X-ray sensitive mutant of V79 Chinese hamster cells. *Mutat. Res.* **194**:239–249.

648. Zdzienicka, M. Z., G. P. van der Schans, A. Westerveld, A. A. van Zeeland, and J. W. I. M. Simons. 1988. Phenotypic heterogeneity within the first complementation group of UV-sensitive mutants of Chinese hamster cell lines. *Mutat. Res.* **193**:31–41.

649. Zdzienicka, M. Z., J. Venema, D. L. Mitchell, A. van Hoffen, A. A. van Zeeland, H. Vrieling, L. H. F. Mullenders, P. H. M. Lohman, and J. W. I. M. Simons. 1992. (6-4) photoproducts and not cyclobutane pyrimidine dimers are the main UV-induced mutagenic lesions in Chinese hamster cells. *Mutat. Res.* **273**:73–83.

650. Zehfus, B. R., A. D. McWilliams, Y. H. Lin, M. F. Hoekstra, and R. L. Keil. 1990. Genetic control of RNA polymerase I-stimulated recombination in yeast. *Genetics* **126**:41–52.

651. Zheng, P., D. S. Fay, J. Burton, H. Xiao, J. L. Pinkham, and D. F. Stern. 1993. *SPK1* is an essential S-phase-specific gene of *Saccharomyces cerevisiae* that encodes a nuclear serine/threonine/tyrosine kinase. *Mol. Cell. Biol.* **13**:5829–5842.

652. Zhu, W., P. C. Keng, and W.-G. Chou. 1992. Differential gene expression in wild-type and X-ray sensitive Chinese hamster ovary cells. *Mutat. Res.* **274**:237–245.

653. Zhu, W., M. Kriajevskaia, P. C. Keng, and W.-G. Chou. 1993. A "trans-acting" factor for activation of transcription in the xrs-5 mutant of the Chinese hamster ovary cell line. *Mutat. Res.* **294**:101–108.

654. Ziegler, A., D. J. Leffell, S. Kunala, H. W. Sharma, M. Gailani, J. A. Simon, A. J. Halperin, H. P. Baden, P. E. Shapiro, A. E. Bale, and D. E. Brash. 1993. Mutation hotspots due to sunlight in the p53 gene of nonmelanoma skin cancers. *Proc. Natl. Acad. Sci. USA* **90**:4216–4220.

13

Regulatory Responses to DNA-Damaging Agents in Eukaryotic Cells

In the previous chapter we touched on the inducibility of some mutagenic processes in eukaryotic cells. We also mentioned the importance of cell cycle regulation for cellular resistance to treatment with ionizing radiation. In this chapter we will examine observations that are indicative of actively regulated responses to DNA damage or other forms of cell injury invoked by DNA-damaging agents in both lower and higher eukaryotic cells. The physiological significance of these phenomena for DNA repair and mutagenesis is frequently unknown. Even the assumption that DNA damage or a consequence of DNA damage constitutes the triggering signal for the regulatory cascades involved has not been uniformly tested and does not appear to be true in all cases.

Gene Activation following Exposure to DNA-Damaging Agents

Induction of Genes in *Saccharomyces cerevisiae*

The yeast *Saccharomyces cerevisiae* has been extensively used as a model organism to explore whether eukaryotes possess a regulated system(s) of gene expression which responds to DNA damage and whether such a system(s) is formally analogous to the prokaryotic SOS system discussed in chapter 10. Selected genes in all three DNA repair epistasis groups in this yeast (see chapter 6) have indeed been found to be regulated in response to treatment with DNA-damaging agents (Table 13–1). However, there is no evidence of a coordinated regulon that can be recognized as the formal eukaryotic equivalent of the SOS regulon in *Escherichia coli*.

INDUCIBLE GENES KNOWN TO BE INVOLVED IN DNA REPAIR

The genes among the *RAD3* epistasis group whose expression is induced in the presence of DNA damage have already been discussed in considerable detail in

Table 13–1 Yeast genes inducible by DNA-damaging agents

Gene	Function	Heat shock inducible[a]	Cell cycle regulated[a]	Increased transcript levels in meiosis[a]	Reference(s)
PHR1	Photoreactivating enzyme	−			273, 274
RAD2	Nucleotide excision repair; required for incision	−	−	+	188, 189, 247, 287, 288
RAD7	Nucleotide excision repair	−	−	+	136
RAD16	Nucleotide excision repair				38
RAD23	Nucleotide excision repair	−	−	+	190
RAD5 (= REV2)	Helicase-like protein; required for mutagenesis, possibly postreplication repair				23
RAD6	Ubiquitin conjugation; required for mutagenesis, sporulation, postreplication repair	−	−	+	191
RAD18	Postreplication repair	−	−	+	135
RAD51	Double-strand break repair, recombination		+		21
RAD54	Double-strand break repair, recombination	−		+	53–55
SNM1 (= PSO2)	Cross-link repair				338
CDC8	Thymidylate kinase		+	+	73, 133, 331
CDC9	DNA ligase		+	+	20, 132, 236, 332
CDC17 (= POL1)	DNA polymerase I		+	+	134
UBI4	Polyubiquitin	+		+	309
MAG1	3-Methyladenine DNA glycosylase	−			47, 341
RNR1	Ribonucleotide reductase, large subunit		+		76, 77, 355
RNR2	Ribonucleotide reductase, small subunit	−	(+)		5, 15, 73–75, 77, 355
RNR3 (= DIN1)	Ribonucleotide reductase, large subunit	−			5, 76, 77, 193, 258, 259, 344, 354, 355
DIN2	Unknown				258, 259
DIN3	Unknown				258, 259
DIN4	Unknown				258, 259
DIN5	Unknown				258, 259
DIN6	Unknown				258, 259
DDRA2	Unknown	+			192, 193, 199, 200
DDR48	ATP, GTPase of unknown function; role in spontaneous mutagenesis	+			192, 193, 199, 200, 280, 344
Ty (= DDR78A)	Yeast transposon				36, 199, 201, 210, 211, 251, 252
HIS3	Histidine metabolism				79
HIS4	Histidine metabolism				79
	Various unknown proteins				14, 100, 250
	UV-inducible proteins on the basis of activity measurements				
	Proteinase B				271, 272
	Cytochrome P-450				212
	Catalase				212

[a] + indicates inducibility by the treatment or condition indicated, (+) indicates a low level of regulation, and − indicates unregulated expression. No symbol is given if a particular agent or condition has not been studied.

chapter 6. These include the nucleotide excision repair group of genes *RAD2* (188, 247), *RAD7* (136), *RAD16* (38), *RAD23* (190), and *SNM1* (*PSO2*) (338). Regulation in response to DNA damage has also been observed in members of the *RAD6* epistasis group, i.e., *RAD18* (135) and *RAD6* itself (191), as well as in *RAD51* (21) and *RAD54* (54), members of the *RAD52* recombinational repair epistasis group.

Two different methods of screening have been used to identify additional genes which are inducible by DNA-damaging agents, regardless of their relationship to the processing of such damage. One such screen involved differential hybridization (199), and the other utilized the power of transcriptional fusions, a general experimental strategy extensively discussed in chapter 10 with respect to inducible genes in *E. coli* (258, 259). On the basis of the efficiency of the latter approach, it has been estimated that there are at least 50 DNA damage-inducible genes in *S. cerevisiae* (259). This is probably a conservative estimate since these screens were biased for the detection of genes with a high induction ratio. As discussed in chapter 6, the inducible yeast genes that are known to participate in nucleotide excision repair are characterized by only a moderate degree of enhanced transcription (up to about sixfold).

Several other general observations are worth mentioning about inducible functions in yeast cells associated with exposure to DNA-damaging agents. First, induction generally is not agent specific but occurs after treatment with a variety of different agents, including some that result in types of DNA damage that are probably not substrates for the activated protein. A noteworthy example previously discussed is the gene for the yeast DNA photolyase (see chapter 3), which is activated by various chemical agents in addition to UV radiation (273). Second, among the DNA damage-inducible genes of yeast cells there is some overlap with heat shock-inducible genes (as there is in *E. coli*). However, none of the well-characterized DNA repair genes apparently responds to heat shock. Third, transcripts for many yeast genes that are inducible by DNA-damaging agents are known to accumulate in one or another stage of meiosis. However, the presence of enhanced levels of mRNA during meiosis is a feature of many yeast genes and in most cases does not reflect a true meiotic regulation phenomenon (138). The physiological significance of increased gene expression during meiosis is particularly questionable when the corresponding mutant does not have a meiotic phenotype. Finally, with the notable exception of the *E. coli recA* homolog *RAD51* (21) (see chapter 12), none of the known inducible repair genes are cell cycle regulated in untreated mitotically dividing cells. In contrast, genes with a possible dual function in DNA repair and replication, such as *CDC9* (20, 236), *CDC8*, *CDC17* (*POL1*), and *RNR1*, are cell cycle regulated (77, 134, 331). Promoter elements that confer enhanced expression of *CDC9* during S phase in untreated cells have been characterized and appear to be distinct from those that are responsive to DNA-damaging agents (332).

The *RAD6* gene is required for a poorly understood postreplicational repair response to damaged DNA and for mutagenesis and sporulation. *RAD6* encodes a ubiquitin-conjugating (E2) enzyme (131). The properties of this gene and its function were discussed in chapter 12. *RAD6* is a DNA damage-inducible gene but is not regulated in a cell cycle stage-specific fashion (191). Increased levels of *RAD6* mRNA are also observed during meiosis, coincident with the time of recombination (191).

The *RAD54* gene is required for the repair of double-strand breaks in DNA and for meiotic and mitotic recombination in *S. cerevisiae* (95, 96, 101) (see chapter 12). The precise function of Rad54 protein is unknown. *RAD54* is transcriptionally induced by γ rays and by UV radiation. It is also induced following the introduction of strand breaks in DNA by overexpression of *Eco*RI endonuclease (54). Increased levels of *RAD54* transcripts are also found during meiosis (55).

The *MAG1* gene encodes a 3-methyladenine DNA glycosylase (see chapter 4) and is inducible not only by alkylating agents but also by exposure to UV radiation (46, 47). In this respect, its regulation differs from that of the *alkA* gene of *E. coli*, whose expression is induced by alkylating agents but not by UV radiation (see chapter 4).

INDUCIBLE GENES POSSIBLY INVOLVED IN ASPECTS OF DNA REPAIR

Certain other yeast genes that participate in cellular responses to DNA damage (some of which may play auxiliary roles in DNA repair) are inducible by DNA-damaging treatments. These include the *RNR1*, *RNR2*, and *RNR3* genes, all of which encode subunits of the enzyme ribonucleotide reductase (73–77), and *CDC17* (*POL1*), the gene for yeast DNA polymerase α (134). A series of mutations that result in constitutively high expression of the *RNR3* gene have been isolated. These are designated *crt* (for constitutive *RNR3* transcription) (354). Five *CRT* genes have been identified as *TUP1*, *POL1*, *RNR2*, *RNR1*, and *SSN6*. These genes are almost certainly not directly involved in regulation. It is more likely that mutations in these genes result in endogenous DNA damage or a metabolic stress situation (e.g., nucleotide depletion) which results in up-regulation of the *RNR* genes. A second screen devised for mutations that prevented the induction of *RNR3* transcription in response to DNA damage yielded mutations in genes designated *DUN1* to *DUN5* (for damage uninducible)(355). The *dun1* mutants are sensitive to UV radiation and methyl methanesulfonate (MMS) and are also defective in induction of *RNR1* and *RNR2*. However, they do not confer a general defect in the regulation of damage-inducible transcripts, and MMS inducibility of *UBI4* and *DDR48* (see below) is normal (355). Thus, the *RNR* genes may represent a special subclass of DNA damage-inducible yeast genes.

DUN1 has been shown to encode a nonessential serine/threonine protein kinase (355), and a sixfold increase in phosphorylation of Dun1 protein has been observed in response to DNA damage. This increase depends on another serine/threonine protein kinase encoded by the *RAD53* gene (*MEC2*/*SPK1*/*SAD1*) (5). Additionally, the *RAD53* gene is required for DNA damage inducibility of *RNR2* and *RNR3* (but once again not for *UBI4*) and also for arrest in various stages of the cell cycle following treatment with DNA-damaging agents or hydroxyurea (see below) (5, 328).

INDUCIBLE GENES OF UNKNOWN FUNCTION

In addition to the handful of inducible genes of known function discussed above, a number of inducible genes of unknown function have been isolated by screening yeast promoter fusion libraries. In these experiments, genomic libraries randomly fused to the *E. coli lacZ+* gene were screened for enhanced expression of β-galactosidase activity after treatment with the UV radiation-mimetic chemical 4-nitro-quinoline 1-oxide (4-NQO). These genes are designated *DIN* (for damage inducible). Other investigators have isolated damage-inducible genes designated *DDR* (for DNA damage-responsive) by differential plaque hybridization with cDNA isolated from treated and untreated cells (199). Finally, the examination of protein synthesis by two-dimensional gel electrophoresis (100) in yeast cells exposed to UV radiation has identified several inducible proteins.

The *DIN1* gene is in fact the *RNR3* gene discussed above (344). The product of the *DDR48* gene has been purified and found to be a negatively charged hydrophilic glycoprotein of 65 kDa that has ATPase and GTPase activity (280). However, the nature and function of other genes in this group are not known. Nonetheless, these studies have yielded several interesting general insights. Multiple and diverse DNA-damaging agents can induce the expression of genes in yeast cells, but the spectrum of inducing agents varies considerably for different genes (199, 258). Despite some overlap, heat shock treatment induces a set of genes different from that induced by DNA-damaging agents (200). Furthermore, the constitutive expression and inducibility of genes can be influenced by the repair capacity of cells. For example, the *DDR48*-encoded product is uninducible in *rad52* mutant strains following ethyl methanesulfonate or heat shock treatments (192, 280), and induction of *DDRA2* by 4-NQO or nitrosoguanidine is dependent on a functional *RAD3* gene (192). Since *RAD3* encodes a component of the yeast RNA polymerase II basal transcription factor TFIIH (see chapter 6), this observation may perhaps be explained by a transcription defect conferred by certain *rad3* alleles.

Among the *DDR* genes, one is present in multiple copies in the yeast genome and is homologous with Ty (Transposable yeast) elements (199). Indeed, Ty transcription has been shown to be inducible by UV radiation and other DNA-damaging agents (36, 251, 252). A corresponding increase in the frequency of insertion mutations in the *ADH1* promoter mediated by Ty transposition has been detected (36, 210). It is interesting to note that transposition of Tn*5* in *E. coli* has also been reported to be increased in an SOS-regulated fashion in the presence of DNA damage (see chapter 10).

The *UBI4* gene, which codes for polyubiquitin (see chapter 12), is required for protein degradation and resistance to starvation conditions (85). This gene is induced by treatment with 4-NQO and also during meiosis (309). These observations suggest that some genes might be induced in response to aberrant proteins generated as the result of DNA- and/or protein-damaging treatment (284).

Increased activity of catalase and cytochrome P-450 has been observed after UV irradiation of logarithmically growing diploid yeast cells (212). These observations provide the first indication of an inducible response to oxidative stress which might overlap with the network(s) of DNA damage-inducible genes. Another possible area of overlap concerns recombinational repair (see chapter 12) and heat shock responsiveness. This notion is supported by the observation that resistance to ionizing radiation can be enhanced by previous heat shock treatment (33, 208).

Although several yeast genes are induced following treatment with alkylating agents, there is no evidence for an adaptive response that is specifically induced by alkylating agents and that is analogous to the one in *E. coli* (see chapters 3 and 4) (193, 238).

PROMOTER ELEMENTS AND COREGULATION OF DAMAGE-INDUCIBLE GENES

Not much is known about the promoter sequences involved in the regulation of the DNA damage-inducible genes discussed above. Deletion analysis has been performed with only a few of these genes. The well-studied example *RAD2* was discussed in chapter 6. In addition to A + T-rich regions, two positive-acting elements of related sequence termed DRE1 and DRE2 at positions − 93 and −168 (relative to the translational start site), respectively, have been shown to be necessary for induction of the *RAD2* gene and, in the case of DRE1, for efficient constitutive expression as well (288). Gel shift experiments have revealed the formation of specific protein-DNA complexes, as well as enhanced binding of one or more (presumed) regulatory proteins to both sequences after treatment of cells with UV radiation or 4-NQO (Fig. 13–1) (287). This increased protein binding is dependent on postirradiation protein synthesis and hence is probably due to increased synthesis of the binding factors themselves rather than posttranslational modifications that enhance the affinity of these proteins for promoter sequences.

Similar sequences have been identified in the promoters of several other inducible repair genes in *S. cerevisiae*, including *RAD7*, *RAD23*, *PHR1*, *SNM1* (287, 338), and *MAG1* (341) (Fig. 13–2). Protein-DNA complexes of apparently similar mobility have also been detected in several cases. However, competition experiments do not support the notion of a single transcription factor which coregulates all these genes (287).

In the case of *PHR1*, a repressor protein has been found to bind a 39-bp region at position − 40, which includes a palindromic sequence (274). Binding activity was not detected in fractionated extracts of UV-irradiated cells. Furthermore, deletion of the repressor-binding region resulted in enhanced constitutive expression and reduced induction (Fig. 13–3). Insertion of this sequence into a heterologous unregulated promoter resulted in repression of constitutive expression and inducibility by UV radiation (274) (Fig. 13–3). A positive-acting element has been identified in the *RAD54* promoter 229 to 258 bp upstream of the translational start site (53). Similarities to both of these sequences have been observed in the *RAD51* promoter region, but their significance remains to be determined

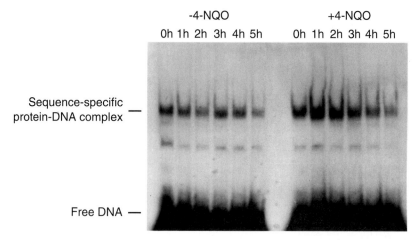

Figure 13–1   Increased protein binding to a radioactively labeled fragment containing the regulatory sequence DRE1 of the *RAD2* promoter (see Fig. 13–2) is observed in electrophoretic mobility shift assays with cell extracts prepared at different times after treatment of the cells with 4-NQO. Complexes formed are compared with extracts from untreated cells. Equal amounts of total protein were loaded per lane. (*Adapted from Siede and Friedberg [287] with permission.*)

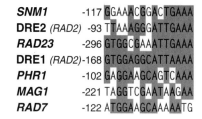

Figure 13–2   Similarities among sequences in the upstream regions of DNA damage-inducible *S. cerevisiae* DNA repair genes. So far, a role in gene regulation has been demonstrated for the DRE1$_{RAD2}$, DRE2$_{RAD2}$, *RAD23*, and *MAG1* sequences only (135, 287, 288, 341). (*Adapted from Siede and Friedberg [287] and Xiao et al. [341] with permission.*)

(283). As is the case with *RAD2*, A + T-rich regions are required for efficient constitutive expression of *RAD54* (53).

As indicated above, the *MAG1* gene is inducible not only by alkylating agents but also by UV light (46, 47). Deletion analysis of the promoter has revealed an activating sequence between nucleotide − 376 and − 330, and an upstream repressing sequence between nucleotides − 221 and − 190 (341). A related upstream repressing sequence has been identified in the promoter of the *MGT1* gene, which encodes a DNA methyltransferase specific for O^6-methylguanine and O^4-methylthymine (see chapter 3) (340). Protein-DNA complexes of similar mobility have been identified by using these repressing elements of the *MAG1* and *MGT1* genes, and competition studies are consistent with the involvement of identical proteins in the two complexes (341). Indeed, DNA-protein (Southwestern) blotting indicated that proteins of 39 and 26 kDa bind to both sequences. These data suggest coregulation of the two alkylation repair genes. It is interesting that there is also some homology between the negatively regulated *MAG1* upstream repressing sequence and the positively regulated DRE1 element of the *RAD2* promoter (Fig. 13–2) (341). Thus, identical or related *trans*-acting factors may function as either activating or repressing factors, a surprising but not an unprecedented phenomenon in gene regulation.

The physiological significance of DNA damage-induced gene regulation in *S. cerevisiae* is difficult to assess. To address this question more completely, it is necessary to measure the effect on one or more genetic end points of specific promoter mutations which result in noninducibility of the gene in response to DNA damage (and which do not influence constitutive expression). For *RAD54*, rendering its expression uninducible to DNA- damaging agents has no detectable effect on sensitivity to X rays or MMS, or on spontaneous or X-ray-induced recombination, sporulation, or mating-type switching (53). As already indicated (see chapter 6), deletion of the DRE promoter elements in *RAD2* (which eliminates induction by UV radiation) has a modest effect on survival, which is detectable only when cells are synchronized and UV irradiated in the G$_1$ phase of the cell cycle (287).

These phenotypic uncertainties notwithstanding, there is indirect evidence for inducible components of nucleotide excision repair in *S. cerevisiae* which influence genetic end points. For example, the removal of pyrimidine dimers from DNA in both transcriptionally active and inactive genes (*MATα* and *HMLα*, respectively) exposed to UV radiation is enhanced after pretreatment of the cells with a lower radiation dose (Fig. 13–4) (320). Protein synthesis is required during the interval

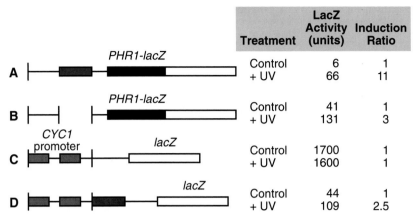

	Treatment	LacZ Activity (units)	Induction Ratio
A	Control	6	1
	+ UV	66	11
B	Control	41	1
	+ UV	131	3
C	Control	1700	1
	+ UV	1600	1
D	Control	44	1
	+ UV	109	2.5

Figure 13–3 Deletion of a 39-bp fragment containing a potential repressor sequence (red box) from the *PHR1* promoter results in higher constitutive expression of a *PHR1-lacZ* fusion gene, measured by β-galactosidase activity, and in lower induction ratios after treatment with UV radiation or MMS (compare rows A and B). However, the gene is still inducible, which indicates the existence of further elements which regulate *PHR1* expression. Insertion of the same 39-bp fragment into a heterologous promoter (*CYC1*) distal to the normal *CYC1* upstream activating sequences (grey boxes) results in repression of constitutive expression and UV-inducibility (compare rows C and D). (*Adapted from Sebastian and Sancar [274] with permission.*)

of incubation between the two doses for this effect to be apparent. This enhancement is abolished for the *MATa* locus in *rad16* mutants but not in *rad7* mutants (both mutants are deficient in the repair of the *HMLa* locus [see chapter 6]). In addition to its role in the repair of inactive genes, *RAD16* might therefore play an important role in the regulation of overall nucleotide excision repair in yeast cells.

Regulation of base excision repair in *Schizosaccharomyces pombe* has been studied (129) by using UV-damaged plasmids in an in vitro system originally developed for *S. cerevisiae* (318). When extracts were prepared from irradiated *S. pombe* cells at different times after UV radiation treatment, a transiently enhanced capacity to carry out base excision repair was detected, which returned to the level for untreated cells after about 2 h. UV radiation inducibility of certain components of base excision repair is consistent with these results. However, the inducibility of individual *S. pombe* genes in response to DNA damage is a largely unexplored area.

Gene Activation in Mammalian Cells following Exposure to DNA-Damaging Agents

Mammalian genes and gene products known to be activated at the transcriptional or posttranscriptional level in response to treatment with DNA-damaging agents include not only those involved in DNA repair and repair-associated processes but also transcription factors, secreted growth factors and growth factor receptors, protective cytoplasmic enzymes, and proteins normally associated with tissue injury and inflammation (Table 13–2) (88, 118, 119, 125, 149, 322). Several are known or suspected protooncogenes. There is extensive overlap between mammalian genes induced by UV radiation and those induced by the phorbol ester tumor promoters and by growth factors which are known activators of protein kinase C (118, 125, 161). Among these genes are those that encode metalloproteases (11) and metallothionein II (13, 92), transcription factor genes such as c-*fos* (40, 61) and c-*jun* (61), various growth factor genes (78, 109), and the viral genomes of simian virus 40 (SV40) (161) and human immunodeficiency virus type 1 (HIV-I) (298, 311, 356). Growth factor secretion accounts for the observation that conditioned medium from UV-irradiated cells can elicit a pattern of inducible proteins in nonirradiated cells (256, 269). Interleukin-1α and basic fibroblast growth factors have been identified as constituents responsible for this communicated response (160).

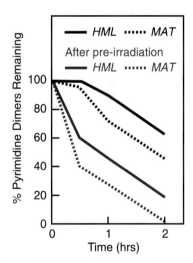

Figure 13–4 Preirradiation accelerates the removal of UV-induced pyrimidine dimers from transcriptionally active and silent *S. cerevisiae* genes. Removal of pyrimidine dimers from the transcriptionally active mating-type locus *MAT* and the silent mating-type locus *HML* was measured by using the dimer-specific T4 endonuclease and gene-specific probes. The kinetics of dimer removal after irradiation with a single dose of 70 J/m² of UV radiation (black lines) and after irradiation with 25 J/m², followed by 70 J/m² after 1 h of incubation in growth medium (red lines), are compared. (*Adapted from Waters et al. [320] with permission.*)

Table 13–2 Mammalian genes inducible at the mRNA or protein level by radiation or DNA-damaging agents

Gene or gene product	Inducible by[a]:				Reference(s)
	UV-C	Ionizing radiation	Other DNA-damaging agents	UV-A and/or reactive oxygen intermediates	
c-*fos*	+	+	+	+	40, 57, 126, 278, 281
c-*jun*	+	+		+	61, 281
jun-B	+	+		+	61, 281
NF-κB		+			34
EGR-1		+			112
c-*myc*	+	+	+	+	57, 254, 301
c-H-*ras*	+	+			8, 254
c-*src*		+			8
C-5 (Trk-2 homolog)	+				25
PKC (protein kinase C)	−	+		+	234, 335
CL100 (tyrosine phosphatase)	−	−	−	+	150
EGF receptor				+	346
α interferon		+			334
Interleukin-1	+	+			99, 160, 336
Tumor necrosis factor α		+			111
Basic fibroblast growth factor	+	+			109, 160
Transforming growth factor α	+				78
p53	+	+	+		97, 110, 143, 186, 195, 221, 351
MDM-2	+				235, 239
WAF1/CIP1	+				70
MHC (CL I) (major histocompatibility complex)	+		+	+	163
MHC (CL II)-associated invariant-chain protein		+	+	+	242–244
uPA (urokinase plasminogen activator)	+				207
Tissue type plasminogen activator	+	+			31, 32
Nmo-1 (= DT diaphorase)		+	+		31, 91, 219
Pgp (P-glycoprotein involved in multiple-drug resistance)		+			227
Collagenase	+	+	+		140, 161, 298
Stromelysin	+				140
α-Tubulin		+			337
γ-Actin		+			337
Keratins	+		+		141
Metallothioneins	+		+		13, 92, 176
Heme oxygenase				+	151
Ornithine decarboxylase	+	+		+	196, 255
Thymidine kinase	+	+			31
Ribosomal protein L7A	+		+		25
p-Glycoprotein		+			121
O⁶-Methylguanine-DNA methyltransferase	+	+	+		98, 172, 220
N³-Methyladenine-DNA glycosylase		+	+		168
DNA polymerase β	−	−	+	+	93
DNA ligase	+				205
DDI (class 1)	+	−			89
DDI (class 2) (= *GADD*)	+	−	+		89, 91, 231
spr-I, spr-II (small proline-rich proteins)	+		+		142

[a] See footnote to Table 13–1 for use of symbols.

Source: Adapted from Fornace (88) and Keyse (149) with permission.

(continued)

Table 13–2   *Continued*

Gene or gene product	UV-C	Ionizing radiation	Other DNA-damaging agents	UV-A and/or reactive oxygen intermediates	Reference(s)
Ubiquitin			+		90
HSP27, HSP70	+	+	+	+	39, 90
Various X-ray-induced transcripts	+ (some)	+			30, 31
DNA-binding proteins	+	+	+		1, 49, 50, 102, 122, 145, 233, 290
RNA-binding proteins	+	+	+		42
RP-2, RP-8 (involved in apoptosis)		+			228
HIV-1	+	−	+	+	270, 296, 311, 356
SV40 enhancer	+		+		108, 154, 161
Moloney murine sarcoma virus	+	+			177
Virus-like 30S element		+			17
RaLV (rat leukemia virus-related sequence)	+				254

There is also an extensive overlap between genes induced by reactive oxygen species and by UV-A radiation (i.e., radiation at wavelengths between 320 and 380 nm). Two genes that are highly inducible by these agents and only moderately inducible by other DNA-damaging agents have been identified. Since UV-A light causes oxidative stress (310), this response may be a specific response to oxygen radicals. One of these genes encodes heme oxygenase (151), and the other encodes a novel tyrosine phosphatase (150). Additionally, the induction of metallothioneins may protect cells from oxidative stress by sequestration of metal ions like Fe^{2+} (see chapter 1), thereby affording protection against genotoxic agents (139). Some studies also indicate an influence of enhanced metallothionein expression on transcriptional regulation and cell cycle parameters (249).

A third group of genes which respond to more than one inducing condition is represented by the *GADD* (growth arrest and DNA damage-inducible) genes (91). These genes were originally isolated as UV radiation-inducible genes in CHO cells by subtractive hybridization (89). They are divided into two classes: class I genes are responsive only to UV radiation and UV-mimetic agents, whereas class II genes are induced by UV radiation and non-UV-mimetic agents such as MMS and by growth arrest. Inhibition of DNA synthesis, however, is not sufficient for inducing *GADD* transcripts (88).

Several DNA-binding proteins that may be important for DNA repair in mammalian cells are activated by DNA-damaging agents such as UV radiation (102) and ionizing radiation (290, 303). In both cases there is no specific requirement for de novo protein synthesis. Another protein that specifically binds UV radiation damage [most probably (6-4) photoproducts in double-stranded DNA] has been isolated from monkey kidney cells, and its binding activity has been shown to be enhanced after irradiation, but in this case protein synthesis is required (122). The purified native protein is about ~210 kDa in size. However, sodium dodecyl sulfate-gel electrophoresis indicates a size of ~126 kDa. Hence, the native protein may exist as a homodimer (1). Microsequence information facilitated the molecular cloning of monkey and human cDNAs which encode this protein. The predicted amino acid sequences of these cDNAs show homology to proteins of unknown function from several organisms, most significantly 44% identity with an uncharacterized protein from the slime mold *Dictyostelium discoideum* (302). This protein may be homologous to a similar DNA-binding activity which is absent in extracts of cells from some patients with xeroderma pigmentosum group E (see chapters 8 and 14) and is increased in tumor cell lines selected for resistance to cisplatin (49, 50).

Hybridization of an RNA probe with stem-loop structure to blotted, filter-bound mammalian nuclear proteins (Northwestern [RNA-protein] blotting) was used to identify RNA-binding proteins in various cell lines (42). The probe was derived from a site in nascent HIV-1 transcripts which binds a transcriptional activator protein called Tat (262). This activator is dispensable in UV-irradiated cells, indicating the existence of UV radiation-activated cellular factors which can replace Tat protein (262). Studies of UV-irradiated CHO cells indicated increased RNA-binding activity of five proteins present in unirradiated cells (42). Additionally, eight new RNA-binding proteins were identified after irradiation. These proteins may play a role in radiation-induced transcriptional activation or even in transcription-coupled repair.

Some studies suggest a direct role of gene activation during DNA repair in mammalian cells. As indicated in chapter 3, transcription of mammalian O^6-methylguanine-DNA methyltransferase protein is inducible in Mex$^+$ cell lines by treatment with the alkylating agents N-methyl-N'-nitro-N-nitrosoguanidine (MNNG), N-methyl-N-nitrosourea, and MMS and also by UV light and X rays (98). However, the extent of the transcriptional activation of the gene which encodes this transferase is considerably smaller than that observed in *E. coli* and does not correlate with a protective effect from alkylation-induced mutagenesis by a challenging dose of MNNG or X rays (98). Furthermore, inducibility of O^6-methylguanine-DNA methyltransferase is apparently confined to mouse and rat liver cell lines (91, 98, 168). DNA polymerase β mRNA has been reported to be inducible by MNNG, MMS, or N-acetoxy-N-2-acetyl-2-aminofluorene but not in protein kinase A-deficient mutant cell lines (80). However, increased transcript levels are not reflected by increased enzyme activity (93).

It is remarkable that the list of activated mammalian genes (Table 13–2) includes only a few bona fide DNA repair enzymes. Many well-characterized genes like the human nucleotide excision repair genes involved in xeroderma pigmentosum (see chapters 8 and 14) have not emerged in screens for DNA damage-inducible genes, suggesting that they are not significantly induced in response to DNA damage. The regulation of these nucleotide excision repair genes and other known repair genes has not been studied in detail, but it would appear that overall chromosomal DNA repair in mammalian cells results from constitutive levels of repair enzymes. The situation may differ for viral DNA probes as substrates, since preirradiation experiments have frequently been interpreted as indicative of the presence of inducible repair components (see chapter 12).

The levels of p53 protein increase in mammalian cells exposed to DNA damage. Consequently, the expression of genes regulated by p53 is altered. These observations have attracted considerable interest, and this topic is discussed in some detail below. The following section highlights the features of another group of genes which have provided interesting insights into molecular events associated with gene activation following exposure of mammlian cells to DNA-damaging agents.

GENES REGULATED BY AP-1 AND NF-κB

Heterologous expression of reporter gene constructs coupled with mutational analyses have identified promoter elements responsive to DNA-damaging treatments in a number of genes that are inducible by UV radiation or phorbol esters. Some of these encode transcription factors. In general, positive regulation of gene expression by enhancer elements has been the rule. In cases when the enhanced binding of proteins to promoter elements has been observed in gel retardation assays, this effect was independent of protein synthesis, suggesting that posttranslational modification of preexisting transcription factors is a primary operational mechanism (118, 119, 161, 261, 298).

AP-1 is a heterodimeric complex consisting of c-Fos and c-Jun proteins (12). All proteins of the Jun and Fos family can form DNA-binding heterodimers with similar sequence specificity. Jun proteins but not Fos proteins can also form homodimers (12). AP-1-type binding sites have been identified as the UV radiation responsive elements in the promoters of the collagenase gene and in c-*jun* itself (61, 161, 298). Although the two characterized AP-1-type binding sites in the c-*jun* promoter bind AP-1 (consisting of Jun/Jun or Jun/Fos) in vitro, it was subsequently determined that a heterodimer of Jun and ATF-2 (a member of the ATF/CREB family of transcription factors) functions in vivo as the transcriptional activator of c-*jun* following treatment with UV radiation or tetradecanoyl phorbol acetate (TPA) (117, 297, 312). c-*jun* activation in particular has been detected as an early activation event after exposure of cells to UV radiation, and it has been suggested that AP-1 activation may be critical for the expression of numerous genes which are characterized by slower induction kinetics after UV radiation exposure. Rapid AP-1 activation correlates with posttranslational phosphorylation of c-Jun after UV irradiation of cells (Fig. 13–5), involving Ser-63 and Ser-73 residues in the transactivation domain of the protein (62, 241).

Coexpression of *trans*-dominant negative mutant proteins that block the UV radiation inducibility of reporter constructs has demonstrated that the serine/threonine kinase Raf-1 and the GTP-binding protein Ha-Ras are essential for AP-1 activation (Fig. 13–6) (62, 241). The MAP2 (mitogen-activated protein) kinase has been proposed as a substrate for Raf-1 (29). MAP2 kinase does indeed phosphorylate c-Jun in vitro, and transient modification of MAP2 kinase has been observed in vivo after treatment of cells with UV radiation or phorbol ester (29, 241). However, it has been pointed out that the MAP2-induced modification of c-Jun involves Ser-243, which is different from the activating reaction observed after UV irradiation (60). A novel distant relative of the MAP kinases, a c-Jun N-terminal kinase termed JNK, has been shown to mediate UV radiation-induced activation of c-Jun by phosphorylation of Ser-63 and Ser-73 (60, 120).

A family of membrane-associated tyrosine kinases, the *Src kinases*, have been shown to be rapidly activated by exposure to UV-C radiation (62). Additionally, tyrosine kinase inhibitors interfere with the induction of c-Jun mRNA by UV-C radiation (but not by TPA) and also increase the sensitivity of cells to UV-C radiation (62). The involvement of two cell surface membrane-bound proteins (Src and Ha-Ras) in AP-1 activation suggests that the primary inducing signal is generated outside the nucleus, possibly as a response to the generation of free radicals through membrane lipid peroxidation after UV irradiation (Fig. 13–7). However, it should be noted that the generation of free oxygen radicals is well established only for long-wave UV radiation (28, 45, 310). The intracellular level of reduced glutathione (a free radical scavenger [see chapter 1]) can be increased by the addition of *N*-acetylcysteine (202), which inhibits Src activation and blocks the UV-C induction of c-*jun* mRNA (62).

DNA binding of AP-1 is also subject to redox regulation through a conserved cysteine residue in the DNA-binding domain of Fos and Jun (339). Ref-1 is a ubiquitous nuclear redox factor that stimulates DNA binding of AP-1 and of NF-κB through cysteine reduction (316). Interestingly, this protein also possesses AP endonuclease activity (see chapter 4) (339) and is identical to a previously characterized repair enzyme designated HAP-1 or APEX, the human homolog of *E. coli* exonuclease III (248, 275). It remains to be determined whether this reaction is somehow involved in coupling transcriptional activation, DNA repair, and oxidative signaling.

Involvement of membrane-bound signaling elements and free oxygen radicals appears to hold for the activation of the transcription factor NF-κB, which is responsible for UV-stimulated transcription of HIV-1 (60, 296, 298, 311). Activation of NF-κB-binding activity involves dissociation of the inactive cytoplasmic NF-κB–I-κB complex (Fig. 13–7). Again, coexpression of dominant negative alleles of v-*src*, Ha-*ras* and *raf1*, as well as the examination of stable transfectants of these mutant alleles, has established the dependence of HIV-1 promoter activation on these proteins (63). The nucleus is dispensable for activation by UV light, as evidenced by the observation that NF-κB-binding activity is efficiently induced in HeLa cells enucleated by cytochalasin B treatment (63) or (most probably by a different mechanism) by irradiation of cytosolic cell extracts of keratinocytes (289).

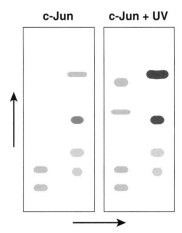

Figure 13–5 c-Jun labeled with ^{32}P in vivo was immunoprecipitated and analyzed by high-voltage electrophoresis (horizontal axis) and thin-layer chromatography (vertical axis) after complete digestion with trypsin. Enhanced phosphorylation of specific tryptic fragments is evident after exposure of cells to UV radiation. (*Adapted from Devary et al. [62] with permission.*)

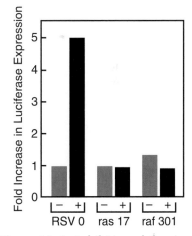

Figure 13–6 Inhibition of the signal-transduction chain, resulting in c-*jun* promoter activation by dominant-negative Ha-Ras and Raf-1 mutant alleles. HeLa cells were cotransfected with a c-Jun–luciferase reporter gene and an "empty" expression vector (RSV 0), or the same vector expressing dominant-negative alleles of Ha-Ras (ras 17) or Raf (raf 301). After transfection, cells were exposed to UV radiation at 0 or 40 J/m^2 (−/+), and luciferase activity was measured after 21 h. No UV-inducible expression of c-Jun–luciferase is evident if ras 17 or raf 301 is coexpressed with the reporter gene. (*Adapted from Devary et al. [62] with permission.*)

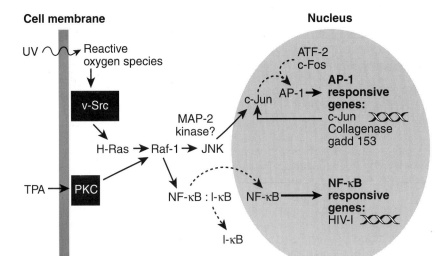

Figure 13–7 Hypothetical scheme of a cascade of protein kinases and other activators
resulting in activation of the transcription factors AP-1 and NF-κB following UV radiation
or TPA treatment of cells. Note that transcription of c-*jun* is activated by an AP-1-like transcription factor containing c-Jun itself. See the text for more details on UV-induced gene
activation. Gene activation mediated by TPA and protein kinase C (PKC) most probably
feeds into the signal transduction chain at Raf-1 (156), since expression of dominant-negative alleles of H-Ras do not interfere (261). This scheme is almost certainly an oversimpli-
fication (e.g., it does not reflect the known influence of radicals on TPA-induced gene activation [270]). It is not clear whether all intermediate steps are known. Steps in this system
might also be activated independently by different, converging pathways.

From the observations summarized above, it has been suggested that mem-
brane-generated oxygen radicals constitute the main signal for UV-induced expression of mammalian genes which are coregulated by phorbol esters and growth
factors (Fig. 13–7) (62). Such a growth response may be responsible for the tumor
promotion effect of UV light (198, 253). This response seems to counteract mechanisms that induce cell cycle arrest in the presence of DNA damage, a topic extensively discussed below. It has been speculated that such a growth response may
play a role in replacing damaged cellular components when the capacity for DNA
repair is not saturated, so that growth and cell cycle progression can be safely
resumed without delay (62).

Apparently, some elements of these systems are conserved among lower and
higher eukaryotes. In *S. cerevisiae* the transcription factor Gcn4 is homologous to
AP-1 factors like Fos and Jun, and these factors are indeed functionally interchangeable (299). Gcn4 is involved in regulation of amino acid biosynthesis upon
amino acid or purine starvation, but Gcn4-mediated transcription is also activated
by UV irradiation of cells (79). This pathway of activation is apparently different
from the starvation signaling system, and, as in mammalian cells, it depends on
the yeast homolog of Ras protein (79).

Data have been reported that suggest a direct role of DNA damage in triggering the UV radiation responses discussed above. For example, the action spectrum for UV-induced gene activation correlates almost exactly with that for the
induction of DNA damage and not with that for oxygen radical production (261,
298). Additionally, it has been reported for several genes that a much lower UV
radiation dose is required for induction in repair-deficient cell lines, e.g., XP group
A lines, than in repair-proficient cells (194, 206, 261, 269, 298). This observation
hints at the significance of unexcised photoproducts in eliciting an inducing signal.
One study has actually shown that a UV-responsive construct (SV40 enhancer)
can be activated by cotransfection with UV-irradiated DNA (161). However, similar experiments with c-Jun were unsuccessful (62). Theoretical considerations
have also suggested that damage to a non-DNA target is unlikely to be the critical

signal for gene activation following γ irradiation (319). It remains to be seen whether radiation of mammalian cells is followed by several different, redundant mechanisms of gene induction, which feed into common or overlapping signal transduction pathways.

Cell Cycle Arrest as a Response to DNA Damage

Regulation of Cell Cycle Progression

Recent years have witnessed great progress in our understanding of the molecular participants and mechanisms involved in the regulation of the eukaryotic cell cycle (214). Perhaps the primary reason for this startling progress is the recognition of the tremendous conservation of the genetic components of cell cycle regulation and the fact that they are often functionally interchangeable, even between human and yeast cells (224). Consequently, the genetic frameworks established in the fission yeast *S. pombe* and in the budding yeast *S. cerevisiae* have been extremely useful for our understanding of the general mechanisms of cell cycle progression in eukaryotes.

Simply stated, progression through the various defined stages of the eukaryotic cell cycle is driven by a family of protein kinases termed *cyclin-dependent kinases* (Cdks). The simplest case is represented by *S. cerevisiae*, in which a single kinase of this kind (Cdc28) is sufficient (218). Cdk activity depends on the association of the kinase with regulatory subunits (66). These are the *cyclins*, proteins whose abundance fluctuates dramatically as a function of the position in the cell cycle. G_1- and G_2-specific cyclins have been characterized in various organisms (237, 282). Association with cyclins not only activates Cdks but also determines their substrate specificity. Depending on the cyclin partner and consequently on the cell cycle stage, different (largely unknown) key target molecules are phosphorylated and thus activated or inactivated at the appropriate time (Fig. 13–8). Cdks themselves are also subject to regulation by phosphorylation.

Given this organization of a central "cell cycle engine" (215) that drives different forces at different times, the critical importance of efficient regulation is obvious. The cell requires feedback controls to monitor the completion of one stage of the cell cycle before initiating the next. In the absence of such controls, dire consequences ensue for cells, such as the initiation of mitosis in the presence of unreplicated DNA or chromosome segregation in the face of an incompletely assembled mitotic spindle. Such phenotypes are in fact observed in mutants that are defective in cell cycle control processes. Hence, eukaryotic cells can arrest cell cycle progression transiently at discrete stages to allow completion of the biosynthetic and mechanical processes associated with each stage, thereby averting potentially deleterious consequences during subsequent cell cycle events (175, 215). Lee Hartwell and his colleagues first formalized this important concept and termed the general phenomenon *checkpoint control* (115, 323, 325).

Checkpoint control can be viewed as a signal transduction pathway, i.e., a chain of events by which certain cell cycle conditions are recognized and transmitted to effector molecules which drive the cycle. The action of these effectors is consequently inhibited, or they are converted to an inactive state. Hence, mutations in participating checkpoint control genes may affect the recognition or transmission processes or the response elements themselves, and can therefore be highly pleiotropic. Research in this area has attracted considerable interest in recent years and the prevailing view of checkpoint controls is becoming increasingly complex. In the following discussion we will limit ourselves specifically to a consideration of the interaction of cell cycle checkpoint controls with DNA damage.

It is not surprising that DNA damage can trigger cell cycle checkpoint controls. This signal has presumably been an important factor in eukaryotic evolution, since DNA damage can interfere significantly with cell cycle processes and diminish their accuracy. For instance, one can imagine that in the presence of DNA double-strand breaks, mitosis would almost certainly result in irreversible chromosome aberrations, and that DNA replication of a damaged template would pose

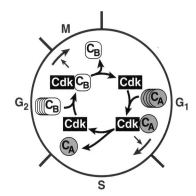

Figure 13–8 Schematic view of a cyclin-dependent kinase activity driving the cell cycle. Its association with different cyclins present during different stages of the cell cycle determines its activity and substrate specificity. Only one Cdk species and two cyclins specific for G_1 and G_2 (C_A and C_B, respectively) are depicted.

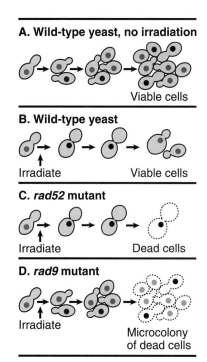

A. Wild-type yeast, no irradiation

Viable cells

B. Wild-type yeast

Irradiate   Viable cells

C. rad52 mutant

Irradiate   Dead cells

D. rad9 mutant

Irradiate

Microcolony
of dead cells

Figure 13–9 (A and B) When irradiated with X rays, wild-type *S. cerevisiae* cells arrest as a large-budded cell in G₂ to allow time for DNA repair before resuming cell cycle progression. (C) A repair-deficient mutant such as a *rad52* mutant stays arrested in G₂ and cannot resume cell division. (D) An arrest-deficient mutant such as a *rad9* mutant does not arrest in G₂ but continues cell cycle progression in the presence of unrepaired damage, resulting in the formation of microcolonies of dead cells. (*Adapted from Murray and Hunt [214] with permission.*)

a high risk of mutation. One might therefore intuitively expect that controls would have evolved in cells to delay entry into the S and M phases in response to DNA damage, in order to increase the kinetic window for DNA repair. This is indeed the case.

G₂ Arrest in the Presence of DNA Damage: the *S. cerevisiae RAD9* Gene as a Paradigm

It has long been known that cells treated with DNA-damaging agents, in particular ionizing radiation, arrest cell cycle progression in the G₂ phase of the cycle (187, 307, 345). However, an important distinction must be made here. Initially it was not known whether this phenomenon simply reflects the passive inhibition of cell cycle progression due to direct inactivation of components of the mitotic machinery by radiation damage or whether it is a *regulated cellular response* with potential benefit to the cell. If the latter is true, one would expect to be able to isolate mutants which are sensitive to DNA-damaging agents as a consequence of defective radiation-induced cell cycle delay.

Screening of the available collection of radiation-sensitive *S. cerevisiae* mutants led indeed to the recognition that *rad9* (and subsequently *rad17* and *rad24*) mutants are defective in G₂ arrest after treatment with γ rays or 254-nm UV radiation (323–325, 327, 328) (Fig. 13–9). This established the actively regulated nature of the arrest phenomenon. Artificial imposition of mitotic delay by treatment of cells with an agent that inhibits normal assembly of the mitotic spindle can in fact restore the γ-ray resistance of a *rad9* mutant to wild-type levels (Fig. 13–10) (325). This observation strongly argues that the *RAD9* gene is not directly involved in the DNA repair response but, rather, operates to generate an increased time window for DNA repair in G₂ by actively delaying entry into mitosis.

RAD9-ASSOCIATED ACTIVITIES

The *RAD9* gene has been cloned and sequenced (268, 326). It encodes a protein with a predicted molecular mass of 148 kDa, with no apparent amino acid homology to previously sequenced proteins or genes. *RAD9* transcript levels are constant during cell cycle progression. The gene is not essential for the growth of unirradiated cells, since deletion mutants are viable and show unaltered growth parameters (326). In fact, the only known phenotype conferred by deletion of the gene to cells not exposed to DNA-damaging agents is an increased rate of spontaneous chromosome loss (326). This is consistent with a checkpoint function for Rad9 protein, which somehow regulates entry into mitosis. Breakage of a dicentric chromosome or loss of a single telomere is sufficient to trigger *RAD9*-dependent cell cycle arrest (37, 264).

DNA alterations can also be generated by mutational inactivation of proteins involved in DNA metabolism. *cdc* (for cell division cycle) mutants of *S. cerevisiae* are thermoconditionally defective in various stages of mitotic cell division and have historically constituted valuable tools for mapping the sequence of cell cycle events. Some *cdc* mutants undergo regulated checkpoint arrest that is dependent on a functional *RAD9* gene when the cells are incubated under restrictive conditions. An example is *CDC9*, which encodes DNA ligase I (see chapter 6). Inactivation of DNA ligase I activity by incubation of the corresponding temperature-sensitive (*ts*) mutant at 36°C leads to cell cycle arrest in late S/G₂. However, in the background of a *rad9* mutation, cells do not arrest but continue cell cycle progression (268). This has fatal consequences: a *cdc9* mutant without a functional *RAD9* gene loses viability at 36°C much more rapidly than does a corresponding *RAD9 cdc9* mutant. These observations have been interpreted as the result of alterations in the structure of DNA, most probably strand breaks, which accumulate following inactivation of DNA ligase and hence trigger the *RAD9* checkpoint pathway. This phenomenon has been observed with several other *cdc* mutations which inactivate replicative enzymes (173, 327). In the case of *cdc13*, the "rapid death" phenotype of the mutant at the restrictive temperature in the absence of G₂ checkpoint control has been used as a selection system to isolate additional mutants with a *rad9*-like phenotype (323, 324, 328).

Some of these mutants (designated *mec* [for mitotic entry checkpoint]) have another interesting phenotype. They are unable to arrest in late S phase after exposure of cells to the chemical hydroxyurea and die rather quickly after such treatment (5, 324, 328). Hydroxyurea is an inhibitor of ribonucleotide reductase, and hydroxyurea-induced arrest in wild-type cells is interpreted as a checkpoint control triggered by the presence of unreplicated DNA. *rad9* mutants are perfectly normal in this respect, as are *rad17*, *rad24*, and *mec3* (324, 328). It remains to be seen whether this reflects the existence of multiple discrete checkpoints and hence discrete but overlapping regulatory pathways triggered independently by unreplicated DNA and by DNA damage or whether there is simply an additional requirement for *RAD9*, *RAD17*, *RAD24*, and *MEC3* in the recognition of damaged DNA in an otherwise identical signal transduction mechanism (Fig. 13–11).

The mechanism of DNA damage-induced G_2 cell cycle arrest is not understood. Dephosphorylation of Tyr-15 in the ATP-binding site of p34^{cdc2} protein of *S. pombe* (the homolog of the *S. cerevisiae* Cdc28 protein) accelerates entry into mitosis. Phosphorylation of the equivalent residue in Cdc28 (Tyr-19) has been examined as a potential mechanism by which the cell cycle is arrested in the presence of unreplicated DNA. However, a mutation of this residue that prevents phosphorylation does not interfere with feedback controls, and cells arrest normally after treatment with hydroxyurea or X rays (7, 293). Isolated Cdc28 protein remains active during the mitotic block imposed by hydroxyurea, as determined by its ability to phosphorylate histone H1 in vitro. However, its association as a complex appears to have changed, and this might hint at the mechanism of checkpoint arrest in *S. cerevisiae* (300).

G_2 Checkpoint Control in *S. pombe*

Careful examination of previously isolated radiation-sensitive mutants of *S. pombe* also led to the identification of strains that are defective in G_2 arrest (Fig. 13–12) (3, 257, 279). As in *S. cerevisiae*, the response to unreplicated DNA among these mutants is not uniform. Comparable to the *S. cerevisiae* mutants *rad9*, *rad17*, and *rad24*, late S/G_2 arrest after hydroxyurea treatment is unaffected in an *S. pombe* *rad21* mutant (Table 13–3). *rad21*$^+$ is an essential gene involved in double-strand break repair (27). Its (not undisputed [27]) partial arrest-defective phenotype is specific for DNA damage (3). Like the *S. cerevisiae* mutants *rad53* (*mec2*) and *mec1*, the *S. pombe* mutants *rad1*, *rad3* *rad9*, *rad17* and *rad26* are defective in the checkpoint response to both damaged and unreplicated DNA (Table 13–3) (3, 4, 257). The same is true for the mutant *hus1* (for hydroxyurea-sensitive) and possibly other *hus* mutants isolated in a screen for mutants sensitive to hydroxyurea (81).

S. pombe mutants with mutations in *rad24* and *rad25* are partially defective in G_2 arrest following γ or UV irradiation (87). These two gene products are 71% identical in amino acid sequence; in fact overexpression of *rad25*$^+$ can suppress the UV radiation sensitivity of *rad24* mutants. The *rad24 rad25* double null mutant is inviable. Both gene products show homology to the so-called 14-3-3 family of eukaryotic gene products (87), which has been implicated in several functions, including the regulation of protein kinase C (2).

The *S. pombe* checkpoint gene *chk1*$^+$ (*rad27*$^+$) was identified by complementation of the radiation sensitivity of a *rad27* mutant (4) and independently as a gene that allows colony formation of a cold-sensitive *cdc2* mutant at the restrictive temperature when overexpressed (317). Overexpression of *chk1*$^+$ can also rescue the UV radiation sensitivity of the checkpoint mutant *rad1*. Indeed, *chk1* disruptants are defective in G_2 arrest (Table 13–3) (4, 317). As is the case with *S. cerevisiae* *rad9* mutants, the UV radiation sensitivity of *S. pombe* *chk1* mutants can be largely rescued by the artificial imposition of a G_2 delay, and rapid loss of viability is observed after temperature shift of a *chk* mutant containing thermosensitive DNA ligase I. The cell cycle response to hydroxyurea is normal (Table 13–3) (4, 317).

These interesting parallels notwithstanding, the *chk1*$^+$ gene product has no amino acid sequence homology with Rad9 protein. The *chk1*$^+$ gene encodes a protein with 30 to 40% sequence identity to serine/threonine protein kinases. The

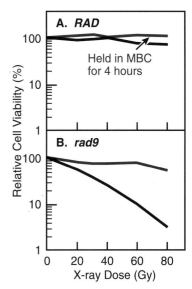

Figure 13–10 Influence of inhibition of M phase on X-ray sensitivity of *S. cerevisiae*. Cells were synchronized in G_2/M by incubation in the presence of the tubulin-destabilizing drug methyl benzimidazole-2-yl-carbamate (MBC), treated with X rays, and plated immediately (black lines) or after an additional 4 h of incubation in the presence of MBC (red lines). In contrast to the *RAD* wild-type strain (A), survival of colony-forming cells is significantly enhanced in the *rad9* mutant strain defective in G_2 arrest by imposition of such an artificial G_2 block after irradiation (B). (*Adapted from Weinert and Hartwell [325] with permission.*)

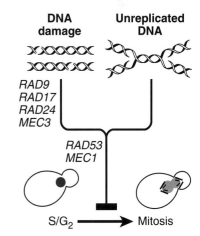

Figure 13–11 S/G_2 cell cycle arrest of *S. cerevisiae* in response to DNA damage (shown here as double-strand breaks) or unreplicated DNA is interpreted as a converging pathway with common steps (dependent on *RAD53* and *MEC1*) and steps specific for damaged DNA (dependent on *RAD9*, *RAD17*, *RAD24*, and *MEC3*). (*Adapted from Murray and Hunt [214] and Weinert et al. [328] with permission.*)

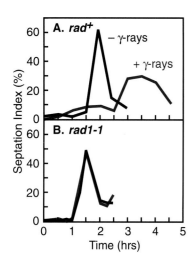

Figure 13–12 Defective mitotic delay after irradiation of the *S. pombe* mutant *rad1-1*. Cells were synchronized at the G_2/M border and treated with γ-irradiation. The percentage of cells undergoing nuclear division ("septation index") was measured during postirradiation incubation. The mitotic delay observed in the wild-type strain (A) is absent in the *rad1-1* mutant (B). (*Adapted from Rowley et al. [257] with permission.*)

gene is not essential for mitotic viability. The isolation of *chk1*⁺ as a multicopy suppressor of a cold-sensitive *cdc2* mutant suggests that this kinase links the DNA damage surveillance pathway to the cell cycle by regulating p34*cdc2* kinase activity. However, it remains to be determined how an overexpressed gene that is presumably required for restricting mitosis can overcome the defect of a mutant defective in cell cycle progression.

In contrast to *S. cerevisiae*, cell cycle arrest in the presence of unreplicated DNA in *S. pombe* appears to be linked to phosphorylation at Tyr-15 of p34*cdc2* (279). A mutant that overproduces the *cdc25*⁺-encoded *phosphatase*, which activates p34*cdc2* by dephosphorylation of Tyr-15, initiates mitosis in the presence of unreplicated DNA (Table 13–3) (82) . The same is true for *wee1 mik1* double mutants defective in both corresponding *kinase* activities (279) and for certain *cdc2-w* ("wee like") mutants which are impaired in regulation of p34*cdc2* by *wee1/mik1*-dependent tyrosine phosphorylation (104). Both types of mutants are also severely crippled for growth in the presence of several other checkpoint mutations, but only *wee1* mutants are sensitive to UV or γ radiation (3, 279). Checkpoint arrest following treatment with UV and ionizing radiation is normal, however, in *wee1* mutants, which argues against an involvement of Tyr-15 phosphorylation of Cdc2 in the DNA damage checkpoint pathway in *S. pombe* and hints at a different function of *wee1*⁺ contributing to radiation resistance (18, 279).

A somewhat different type of checkpoint control is represented by the *cut5*⁺ (*rad4*⁺) gene of *S. pombe*. A mutation in this essential gene causes uncoupling of the completion of DNA replication from mitosis, but the gene is also required for entry into the S phase (Table 13–3) (263). Mutants enhance sensitivity to UV and ionizing radiation (84, 263). Amino acid sequence comparison has revealed two short regions of homology with the human *XRCC1* gene, which is involved in strand break repair following treatment of cells with ionizing radiation or alkylating agents (84, 304) (see chapter 12).

Table 13–3 Phenotype of *S. pombe* checkpoint mutants[a]

Defective S/G_2 arrest in the presence of unreplicated DNA and DNA damage, nonessential genes
  rad1
  rad3
  rad9
  rad17
  rad24 (partial defect)
  rad26
  hus1

Defective G_2 arrest in the presence of DNA damage
  chk1 (*rad27*)
  rad21 (partial defect)
  rad25 (partial defect)

Defective S arrest in the presence of unreplicated DNA
  hus6

Defective G_2 arrest in the presence of unreplicated/damaged DNA and defective initiation of S phase
  cut5 (*rad4*)

Defective S/G_2 arrest in the presence of unreplicated DNA due to impaired responsiveness of *cdc2*
  cdc2.3w
  cdc25 (overexpression)
  wee1 mik1

[a] The given phenotypes apply to deletion mutants unless a specific allele is listed. See reference 4 for more details.

A dual function similar to that of the *cut5⁺* (*rad4⁺*) gene product, i.e., regulated entry into the S phase and restrained entry into mitosis in the presence of unreplicated DNA, has been suggested for the mammalian RCC1 (for regulator of chromatin condensation) protein (59, 222). Curiously, the *S. pombe* homolog *pim1⁺/dcd1⁺* is not required for entry into S phase (266). RCC1 appears to be required for normal chromatin structure and may render the genome "unreadable" for certain checkpoint controls (94).

G₂ Checkpoint Control in Mammalian Cells

Homologs of the yeast DNA damage-responsive checkpoint control genes have not yet been characterized in mammalian cells. Higher eukaryotic cells also respond to the presence of incompletely replicated DNA by delayed mitosis, as has been convincingly shown by the ability of cycling *Xenopus* oocyte extracts to support mitosis in added sperm nuclei (58) (Fig. 13–13). This response is mediated through tyrosine phosphorylation of p34*cdc2* (292). The mechanism of G₂ arrest following DNA damage is less clear. In CHO cells treated with γ rays or the topoisomerase-interacting drug etoposide (both of which induce DNA strand breaks), G₂ arrest is correlated with diminished histone H1 kinase activity and increased tyrosine phosphorylation of p34*cdc2* (182, 183). In human lymphoma cells treated with nitrogen mustard or γ rays, hyperphosphorylation of the cyclin B1-p34*cdc2* complex has been observed in G₂-arrested cells, suggesting a similar mechanism (153, 155, 225). The level of cyclin B1 remains unchanged, in contrast to results with γ-irradiated HeLa cells, in which inhibition of cyclin B1 expression has been observed (216).

G₁ and G₁/S Arrest

As suggested in chapter 12, the active cessation of DNA replication in the presence of DNA damage is presumably an important determinant for the preservation of genomic stability. Noninstructional base damage or other structural alterations such as cross-links or strand breaks could potentially lead to the irreversible stalling of replication forks. In the absence of other cellular responses, such lesions may necessitate an enhanced frequency of translesion DNA synthesis, thereby enhancing the probability of mutations and other chromosome aberrations. Inhibition of the initiation of S phase after treatment of cells with DNA-damaging agents has now been shown to be due to an actively regulated checkpoint arrest mechanism in several biological systems.

Once again, the *rad9* mutant of *S. cerevisiae* has been informative (286). As is the case for G₂ arrest, *rad9* mutants are also defective in G₁ arrest. As shown in Fig. 13–14, a *rad9* mutant culture synchronized early in G₁ enters the S phase essentially without delay following exposure to UV radiation, in contrast to the significantly arrested isogenic wild-type strain. At higher doses of UV radiation, arrest is observed, but to a much lesser extent than in wild-type cells. This delay is probably a passive, unregulated phenomenon caused by the inhibition of DNA replication and/or transcription on damaged templates.

It would therefore appear that the *RAD9* gene is involved in the control of multiple checkpoints in the mitotic cycle in *S. cerevisiae*. The same is apparently true for meiosis (305, 321). START defines the point in the mitotic cell cycle when yeast cells become committed to entering S phase and no longer respond to nutritional factors or other environmental signals (114). An analysis of the precise stage of *RAD9*-dependent G₁ arrest indicates that cells irradiated at START also arrest at this point (285). However, if cells are irradiated later in the cycle, they can still actively delay entry into S phase. A second *RAD9*-dependent checkpoint has been mapped downstream of START between the cell cycle steps represented by the thermoconditional mutant genes *cdc4* and *cdc7* (285). This stage is characterized by S phase levels of all replication-related transcripts except histone mRNA. Other yeast genes involved in G₂ as well as G₁ checkpoint control include *RAD24* (285) and *RAD53* (= *MEC2/SPK1/SAD1*) (5). *RAD53* encodes an essential serine/threonine protein kinase (353), and, as discussed above, this gene is also required for DNA damage induction of ribonucleotide reductase genes (5) (see chapter 12).

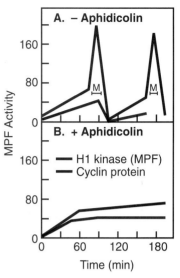

Figure 13–13 (A) *Xenopus* oocyte extracts are characterized by periodic oscillations of Cdk activity (referred to as MPF, for mitosis promoting factor), measured by histone H1 kinase activity, and correlated with G₂ cyclin accumulation. These extracts support DNA synthesis of added sperm nuclei. A feedback mechanism couples DNA synthesis to Cdk activity: the more nuclei are added, the longer the time between oscillations. (B) This feedback mechanism results in cessation of MPF oscillations when an inhibitor of DNA synthesis (aphidicolin) is added together with the identical number of nuclei. (*Adapted from Dasso and Newport [58] with permission.*)

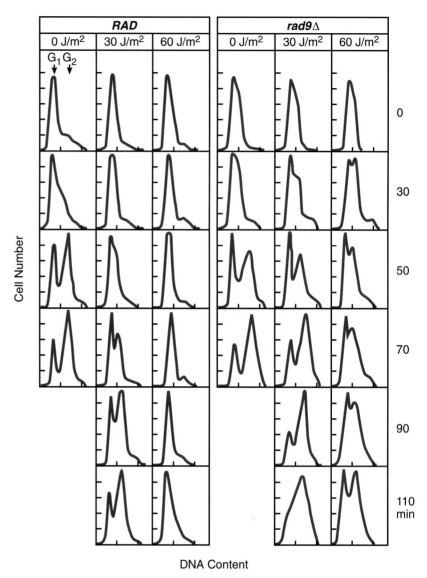

DNA Content

Figure 13–14   Defective UV-induced G_1 arrest in the *S. cerevisiae* mutant *rad9*. Cells of a
wild-type strain and an isogenic *rad9* deletion mutant were synchronized in early G_1 by use
of the yeast pheromone α factor; treated with UV radiation at 0, 30, or 60 J/m²; and immediately released from α-factor arrest in fresh medium. Cell cycle progression was monitored as a function of time (0 to 110 min) by flow-cytometric analysis of DNA content. The
left-hand peak in every panel corresponds to cells in G_1, and the right-hand peak corresponds to cells in G_2. Compared with the significant G_1 arrest in wild-type cells, no arrest is
observed in *rad9* cells treated with the lower UV dose (30 J/m²). In *rad9* cells, entry into S
phase is also accelerated following treatment with the higher UV dose (60 J/m²). (*Adapted
from Siede et al. [286].*)

The existence of multiple checkpoint controls is not an unprecedented situation. Cells from patients with the human heritable disorder *ataxia telangiectasia*
(AT) (see chapter 14) are characterized by increased sensitivity to ionizing radiation and increased cancer incidence. A prominent phenotype of AT cells is a failure
to delay DNA replication in the presence of DNA damage resulting from exposure
to ionizing radiation (*radioresistant DNA synthesis*). This phenotype may be a consequence of a loss of G_1 checkpoint control (230). However, there are indications
for defects in multiple checkpoints for damaged DNA in AT cells.

There are apparently also at least two checkpoints operative in the G_1 phase of the mammalian cell cycle. One appears to equate with the restriction point (the equivalent of START in yeast cells) (232). The other checkpoint seems to coincide with replicon initiation (see chapter 12) and is sometimes referred to simply as the G_1/S checkpoint. AT cells are apparently defective in both G_1 checkpoints (217, 229, 230) as well as in G_2 arrest (260, 350). The defect in G_2 arrest may not be detectable under all conditions, and some discrepancies in the literature (22) may be explained by the position in the cycle of the majority of cells at the time of irradiation. In a recent study, a reduced G_2 arrest in AT cells of several complementation groups was observed only when cells were irradiated in G_2 and not when they were treated in earlier stages of the cell cycle (24).

The *p53* Gene and Its Role in the Cell Cycle

The *p53* gene has been shown to be required for G_1 arrest in human cells following treatment with ionizing radiation (143, 144, 162). Inactivation of this tumor suppressor gene by deletion or point mutation appears to be a precondition for neoplastic transformation in many types of human cancer (127). At the time of writing, at least 51 types of human cancer that carry *p53* mutations have been documented. *p53* mutations have been linked to 70% of colorectal cancers, 50% of lung cancers, and 40% of breast cancers. A heterozygous *p53* defect is found in the germ cells of patients affected by a hereditary predisposition to cancer, the Li-Fraumeni syndrome (174), whereas tumors in these patients are homozygous for mutant *p53* alleles (294, 295).

The p53 protein binds sequence specifically to DNA as a tetramer and acts as a transcriptional activator (313, 349). Its overexpression leads to G_1 arrest prior to or near the restriction point (178, 203, 204). Thus, its activity as a growth-controlling gene can presumably be attributed at least in part to transcriptional activation of growth-inhibiting genes. Propagation of certain viruses depends on p53 inactivation through the binding of viral gene products, e.g., T antigen of SV40 (165, 265, 329).

The level of p53 protein is increased by posttranslational stabilization of the protein following UV irradiation of cells (195) (Fig. 13–15). An increase in p53 levels has also been observed in situ following irradiation of human skin with sunlight (110). Interestingly, this increase correlates with an increase in the levels of proliferating-cell nuclear antigen (110). Enhanced levels of p53 are found after treatment with ionizing irradiation or MMS and also after serum starvation (351). Heat shock, however, does not lead to accumulation of p53 protein (186). Hence, this regulation cannot be regarded as a general stress response. In primary mouse prostate cells induction of transcriptionally active p53 after UV irradiation follows quite different kinetics from induction after ionizing radiation treatment (186). In contrast to the transient short-lived response to X rays, a delayed induction which reaches much higher levels is observed after UV irradiation. This seems to correlate with the persistence of DNA damage, since DNA repair is detectable for a longer interval after UV radiation exposure than after X-ray exposure. Indeed, if DNA double-strand break repair after X-ray treatment is inhibited by 3-aminobenzamide [an inhibitor of poly(ADP-ribose) polymerase] (see chapter 7), prolonged p53 induction is observed (186). The exposure of permeabilized cells to restriction enzymes also causes accumulation of p53 protein, which again hints at the significance of DNA double-strand breaks (186, 221). Other agents that cause strand breaks and also induce p53 rapidly and efficiently include bleomycin and topoisomerase-targeted drugs such as camptothecin. However, base-modifying, intercalating, and cross-linking agents are less effective, at least at early time points (221). Exposure of cells to antimetabolites such as methotrexate or *N*-(phosphonoacetyl)-L-aspartate (PALA) resulted in a late increase in p53 levels after continuous exposure, which has also been temporally correlated with the appearance of strand breaks (221).

The recently discovered role of p53 as a G_1 checkpoint determinant has provided an explanation for this regulation. Normal asynchronously dividing human cells treated with ionizing radiation arrest cell cycle progression in the G_1 and G_2 phases of the cycle. If the *p53* gene is inactivated, no accumulation of cells in G_1

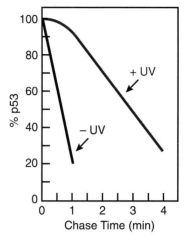

Figure 13–15 UV irradiation enhances the half-life of p53 protein. Proteins were labeled with [³⁵S] methionine in mouse cells before treatment of the cells with UV radiation at 0 (−UV) or 10 (+UV) J/m². Relative steady-state levels of p53 were detected by immunprecipitation during postirradiation incubation in the presence of nonradioactive methionine. These chase periods are indicated on the *x* axis. (*Adapted from Maltzman and Czyzyk [195].*)

is observed (Fig. 13–16). However, G$_2$ arrest is normal (143, 144, 162). G$_1$ arrest is also sensitive to caffeine treatment, and no increase in p53 levels is found after irradiation of caffeine-treated cells (143). Because of the known defect of AT cells in cell cycle arrest following irradiation, it was of interest to determine the regulation of p53 in these cells. A defect would place a potential checkpoint control function conferred by the AT gene(s) upstream of p53. Conflicting data have been reported. Absent or reduced p53 induction was found after γ irradiation of AT cells (41a, 144, 152). However, other studies indicate only a delayed increase of p53 protein but normal induced levels (29, 186). Induction of p53 after UV irradiation was normal (152).

Several cell lines from patients with Bloom's syndrome (see chapter 14) have been found to be defective in the induction of p53 (186), whereas XP-A cell lines are normal for p53 accumulation after UV radiation or X-ray treatment, arguing against the requirement of functional nucleotide excision repair in the p53-activating mechanism (186). However, when normal and XP-A lymphoblastoid cells were compared under conditions of inhibition of DNA replication (mediated by hydroxyurea or cytosine arabinoside), only the excision-proficient cells showed p53 induction after UV radiation treatment (221). Both cell lines were normal for p53 induction under these conditions following treatment with ionizing radiation (221). These results suggest that p53 induction depends on the presence of DNA

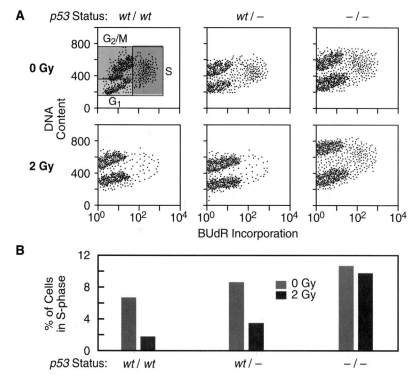

Figure 13–16 Defective G$_1$ arrest in *p53*-deficient cells following exposure to ionizing radiation. (A) The dot diagram visualizes the distribution of cells among different cycle stages in an unsynchronized culture of embryonic mouse fibroblasts before and after X-ray treatment. Cells were pulse-labeled with bromodeoxyuridine (BUdR). The flow-cytometric analysis of the distribution of cells among different cell cycle stages uses two criteria. The *y* axis indicates total DNA content (detected by propidium iodide staining), whereas the *x* axis shows the amount of incorporated BUdR (detected by binding of a BUdR-specific antibody conjugated to a fluorescent dye). Cells in S phase are characterized by high BUdR levels and an intermediate total DNA content. A depletion of cells in S phase is evident 16 h after irradiation with 2 Gy of normal cells (*wt/wt*) and those carrying a heterozygous p53 defect (*wt/*−). However, there is no indication of G$_1$ arrest in cells homozygous for a p53 defect (−/−). (B) Quantitation of this type of analysis as the percentage of cells remaining in S phase 16 h after irradiation with 0 or 2 Gy. (*Adapted from Kastan et al. [144] with permission.*)

strand breaks, which can be induced either directly (e.g., by ionizing radiation) or indirectly during replication or through the action of nucleotide excision repair (e.g., after UV irradiation). Hence, the induction of p53 after UV irradiation of cells requires functional nucleotide excision repair in nonreplicating cells.

GENE REGULATION BY p53 INDUCTION

Several downstream effectors of p53 regulation which may be important in DNA repair or in mediating arrest of the cell cycle have been identified. The promoter of *GADD45* (a gene already mentioned) that was isolated during a screen for genes that are inducible by DNA-damaging agents and growth arrest (91) contains a p53-binding sequence. Furthermore, inducibility by ionizing radiation is indeed absent in cells without a functional p53 gene (144) and is reduced in AT cells (41a, 231).

The *MDM-2* gene has been identified as a negative regulator of transcriptional activation by p53 (209). Both gene products appear to function in a feedback loop (349), since *MDM-2* is itself activated by p53 from an internal promoter (137) and is UV radiation inducible (235). The kinetics of delayed induction of *MDM-2* suggest a function of the gene product in recovery from p53-imposed G_1 arrest (235).

WAF1 has been identified as a gene encoding the most abundant p53-inducible transcript (72), and a p53-binding site in the *WAF1* promoter that confers inducibility on the gene has been identified. Random fragments of the genomic *WAF1* clone were inserted upstream of a truncated *GAL1* promoter driving a *HIS3* reporter gene. Yeast clones were scored for histidine prototrophy in the presence or absence of expression of functional p53. In this manner, the consensus sequence (CRRRCATGYYYRRRCATGYYY, where R is purine and Y is pyrimidine) (71) was identified as a p53-reponsive element 2.4 kb upstream of the *WAF1* coding sequence (72). In several tumor cell lines, *WAF1* expression caused growth suppression similar to that observed following p53 expression.

WAF1 encodes a protein with a predicted size of 18.1 kDa. The amino acid sequence indicates a potential zinc-binding motif as well as a nuclear localization signal. No significant homology with known proteins was found initially. However, the identical gene was independently cloned by a functional assay that immediately suggested its physiological role (113). The protein was identified as one that forms a tight complex with the mammalian cyclin-dependent kinase Cdk2. This kinase associates with cyclins A, D, and E and has been implicated in the G_1-to-S transition. The isolated protein (independently termed Cip1, for Cdk-interacting protein) has turned out to be a potent inhibitor of phosphorylation mediated by Cdk2-cyclin A, Cdk2-cyclin E, Cdk4-cyclin D1, and Cdk4-cyclin D2 (Fig. 13–17). The protein is identical to a 21-kDa protein (p21) previously found to be associated with most cyclin-Cdk complexes in normal cells (106, 342, 343, 352). The *WAF1/CIP1* gene was also identified in a screen for transcripts, which are preferentially expressed in senescent human cells, that can prevent growing cells from entering the S phase when overexpressed by transient transfection (128, 223).

Several lines of evidence suggest that p53 protein executes its function in delaying S phase in the presence of DNA damage through the transcriptional activation of growth-inhibiting genes, in particular the Cdk inhibiting protein Waf1/Cip1.

1. Among different cell lines, there is an excellent correlation between p53 status, capability for G_1 arrest and *WAF1/CIP1* induction after treatment with DNA-damaging agents (70). Waf1/Cip1 protein is inducible only in cell lines containing a normal p53 gene and is undetectable in cells containing mutant p53. Imposing p53-independent G_1 arrest (e.g., by treatment of cells with mimosine) does not result in *WAF1/CIP1* induction (70).

2. The association of cyclin E with Cdk2 is believed to be required for the G1-to-S transition (246). Immunprecipitation of Cdk2-cyclin E complexes after γ-irradiation of synchronized cells indicates that the kinase activities of the isolated complexes are reduced following irradiation. However, the amount of cyclin E and

Figure 13–17 The kinase activity of a Cdk2-cyclin A complex is inhibited by increasing amounts of Cip1. Kinase activity is measured by histone H1 phosphorylation with ^{32}P. (*Adapted from Harper et al. [113] with permission.*)

the level of complex formation between Cdc2 and cyclin E remain unchanged compared with those in untreated cells (67). However, if extracts from irradiated cells containing functional p53 prepared at the time of G_1 arrest are added to isolated Cdk2-cyclin E complexes, kinase activity (as measured by phosphorylation of histone H1) is inhibited (67) (Fig. 13–18A). Passage of these extracts through anti-p21$^{WAF1/CIP1}$ immunobeads efficiently removes the inhibitor (Fig. 13–18B).

3. Waf1/Cip1 can be recovered from immunprecipitated Cdk2-cylin E complexes (70). Induced Waf1/Cip1 expression in the presence of DNA damage results in increased binding to Cdk2-cyclin E complexes and reduced kinase activity of the complexes.

In summary, Cdk proteins are regulated by their association with cyclins, by phosphorylation, and by binding to proteins of the growing family of Cdk inhibitors (116, 277). Apparently, the latter is the mechanism of p53- and Waf1/Cip1-dependent G_1 arrest in mammalian cells treated with DNA-damaging agents and in senescent cells (Fig. 13–19) (67).

A ROLE OF p53 IN G_1/S ARREST?

Other mechanisms of p53-mediated cell cycle arrest cannot be excluded. The protein is also capable of suppressing the activity of promoters containing TATA elements, apparently by associating with TATA-binding protein or other transcription factors (180). Furthermore, the observations that p53 binds replicational origins (19, 240) and binds and inactivates a cellular replication initiation factor, the single-stranded-DNA-binding complex replication protein A (68), suggest the possibility of cell cycle arrest by direct interaction with the replicative machinery as an additional or alternative mechanism of action.

Additionally, proteins encoded by genes that are regulated by p53 can apparently arrest DNA synthesis by binding to replication components. Waf1/Cip1 has been shown to inhibit the activity of polymerase δ by binding to proliferating-cell nuclear antigen in a reconstituted in vitro DNA replication system (Fig. 13–19) (314). Proliferating-cell nuclear antigen interacts also with Gadd45 (Fig. 13–19) (291a).

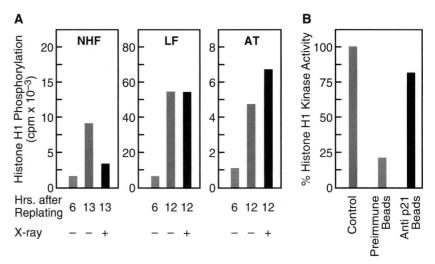

Figure 13–18 (A) Extracts from irradiated, G_1-arrested normal cells (NHF) inhibit the phosphorylation activity of Cdk2-cyclin E complexes. Extracts were prepared at 6 or 13 h after replating of G_0 cells (13 h = S phase in unirradiated cells) and tested (grey bars). Inhibition of kinase activity is observed if the added extracts were prepared 13 h after X-ray treatment and replating, i.e., during the period of G_1 arrest (red bars). No such inhibition is evident if extracts from irradiated Li-Fraumeni (LF) cells (without functional p53) or AT cells were used. (B) Incubation of the inhibiting extract with protein A beads coupled to WAF1/CIP1 antibodies removes the inhibiting activity (but not if beads are coupled to preimmune serum). (*Adapted from Dulic et al. [67] with permission.*)

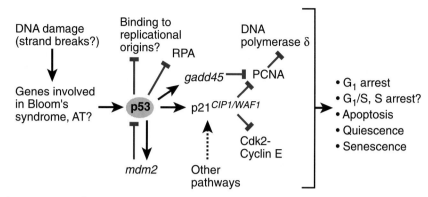

Figure 13–19 Elements potentially involved in the multiple mechanisms of p53-mediated growth inhibition, such as DNA damage-induced G_1 arrest. PCNA, proliferating-cell nuclear antigen. See the text for details.

Such mechanisms most probably affect replicon initiation or elongation and would result in G_1/S or even S phase arrest, downstream of the restriction point. In chapter 12 we discussed evidence that the observed inhibition of replicon initiation after irradiation of cells is indeed caused by an actively regulated mechanism. However, a study of the inhibition of replicon initiation within the *DHFR* domain after ionizing radiation (discussed in chapter 12) indicates that this response does not necessarily depend on a functional *p53* gene (166).

Phosphorylation of replication components may also be involved in the delay of S phase. Extracts from HeLa cells have reduced replication activity in vitro following UV irradiation (44). This effect has been correlated with the appearance of a hyperphosphorylated form of human single-stranded DNA binding protein, and the addition of purified (dephosphorylated) human single-stranded-DNA-binding protein restores the replicative activity (44). Phosphorylation of RPA following UV or γ irradiation may also play a role in cell cycle regulation and in DNA repair (179).

THE CHECKPOINT FUNCTION OF p53 AND PATHOGENESIS OF CANCER

The involvement of a general tumor suppressor gene in G_1 checkpoint control mechanisms responsive to DNA-damaging treatments has potentially important implications for the understanding of neoplastic transformation (146, 164). Regulated delay of S phase in the presence of DNA damage conceivably provides increased time for error-free removal of damage. On the basis of this premise, it is not unreasonable to expect that the loss of G_1 checkpoint control would enhance radiation sensitivity and genetic variability through the potentially lethal, mutagenic, or clastogenic consequences of interactions of unrepaired DNA damage with the DNA replication machinery. Loss of this checkpoint control would thus provide a reasonable explanation for the enhanced genetic instability that is a general feature of cancer cells (306). Such a concept would also be consistent with the known multistep nature of carcinogenesis. Hence, inactivation of a G_1 checkpoint control gene is an attractive hypothetical mechanism for an initiation step in carcinogenesis that could provide a growth advantage to preneoplastic cells. Such inactivation might also potentiate the probability of subsequent genetic alterations induced by exogenous or spontaneous DNA damage, ultimately establishing the full-blown neoplastic phenotype.

Inactivated checkpoint control might also provide the basis for rational anticancer therapy (146, 155, 164, 226) by further enhancing the typically increased sensitivity of tumor cells to radiation regimens and to chemotherapeutic treatments with DNA-damaging agents. Improvement of intact checkpoint control in untransformed cells and exploitation of the defective checkpoint control in transformed cells are likely to be exciting new areas of chemotherapeutic research in the future. There are already encouraging experimental indications in this direction. A defect in p53 has been demonstrated as one factor responsible for the enhanced rate of spontaneous gene amplification in tumor cells, an event which is undetectable in untransformed cells (181, 347). Furthermore, the development of cervical carcinoma after integration of certain types of human papillomavirus into the host genome is correlated with loss of p53 activity (through binding to the virus-encoded E6 protein and promotion of ubiquitin-mediated degradation [267]), and loss of G_1 arrest (107, 148).

Other aspects of the concept outlined above await further experimental confirmation. In fact, it has been argued that G_2 arrest is of greater importance for resistance to radiation damage and the maintenance of genetic stability (213). Certainly, the significance of G_1 arrest for survival after DNA-damaging treatments is challenged by several disquieting observations. For example, SV40-transformed cells (in which p53 protein is inactivated by binding to T antigen) are not more sensitive to ionizing radiation than are untransformed cells (16). Additionally, there are indications that the time interval between induction of DNA damage and the onset of S phase is more important for reducing UV radiation-induced mutagenicity than for the repair of prelethal damage (see chapter 12) (157). There is also no correlation of radiosensitivity with inactivation of *p53* in cancer cell lines (35). Finally, the availability of a mouse model homozygously deleted for *p53* (65) has facilitated a comparison of the radiation sensitivity of isogenic primary cell lines. Surprisingly, fibroblasts carrying a homozygous *p53* deletion are not more UV sensitive and even more resistant to ionizing radiation than are their normal counterparts (128a, 169, 291). It remains to be seen whether this enhanced resistance coincides with enhanced recovery of genetically altered clones. It has been argued that the observed enhanced probability of survival is in fact due to the abrogation of another cancer-preventing control function of p53, possibly associated with G_1 checkpoint control: the initiation of apoptosis, or programmed cell death.

Programmed Cell Death (Apoptosis)

It has long been known that cells of multicellular organisms have the capacity for invoking a suicide program leading to programmed cell death, a phenomenon termed *apoptosis* (pronounced with a silent second "p") (56, 197, 308, 330). The most prominent hallmarks of apoptosis include cell shrinkage, membrane blebbing, chromosome condensation, influx of Ca^{2+}, and early onset of endonucleolytic attack of DNA in chromatin at internucleosomal regions (Fig. 13–20). Conditions that induce apoptosis include DNA damage, hyperthermia, glucocorticoids, infections, and withdrawal of growth factors (308). However, apoptosis is also an important mechanism for tissue remodeling during normal developmental processes (147).

Apoptosis following exposure to DNA-damaging agents can be regarded as a mechanism for preventing the propagation of genetically aberrant cells which have sustained a high level of DNA damage. Several pathways of apoptosis have been postulated. *p53* activates a DNA damage-responsive pathway which also includes c-*myc* among other genes (51, 184, 185, 333, 348). *p53*-mediated apoptosis is correlated with enhanced levels of Waf1/Cip1 protein (70). However, the function of p53 as a transcriptional activator may not be required for apoptosis. In a cell line harboring a thermoconditionally active p53 protein, UV-induced apoptosis, as measured by intranucleosomal DNA cleavage, was strictly dependent on a functional p53, i.e., on incubation at the permissive temperature following treatment (41). However, p53-dependent apoptosis could readily be detected in the presence of inhibitors of transcription and protein synthesis (41). This result indicates that another function of p53 (Fig. 13–19), not its role in transcriptional activation, may be critical for apoptosis.

The *BCL2* proto-oncogene is an antagonist of this apoptotic program in multiple contexts (48, 158). *BCL2* activity also counteracts γ radiation-induced cell death (276) and protects cells from the apoptotic consequences of oxidative dam-

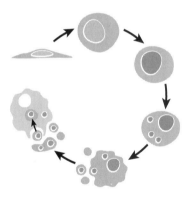

24 hrs   72 hrs

Figure 13–20 Scheme for apoptosis. Activation of an apoptotic pathway results in rounding of an adherent cell, DNA condensation, and fragmentation. Finally, disintegration of the entire cell into membrane-bound vesicles (apoptotic bodies) occurs. These can be phagocytosed by macrophages. The schematized agarose gel electropherogram demonstrates the appearance of a ladder of DNA fragments during progressing apoptosis, indicating nuclease attack in internucleosomal regions of chromatin. (*Adapted from Martin et al. [197] with permission.*)

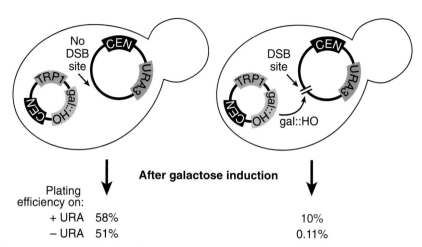

Figure 13–21 Cell death induced by a double-strand break in a nonessential yeast gene. Upon galactose induction, the plasmid-encoded site-specific HO endonuclease is overexpressed and cleaves an HO target site on a second plasmid (right). The otherwise identical control plasmid lacks the HO cleavage site (left). If the cells are plated on uracil-free media, survival is very low since the linearized plasmid carrying the *URA3* gene is almost quantitatively lost (right, last line). However, even in uracil-containing medium (where the *URA3*-containing plasmid is dispensable) survival of colony-forming cells after induction of the double-strand break is only 17% of that of the control. (*Data from Bennett et al. [26].*)

age (124). Most notably, its overexpression blocks lipid peroxidation that accompanies apoptotic cell death (124). These observations, together with its localization as a mitochondrial membrane protein (123), suggest a primary function of Bcl-2 protein as a free radical scavenger molecule in an antioxidant pathway.

While apoptosis appears to be a mechanism of programmed cell death specifically adapted to multicellular organisms, it is interesting that DNA damage even in unicellular eukaryotes can apparently elicit signals leading to cell death. A system has been developed to study the consequences of a localized double-strand break in a nonessential gene in *S. cerevisiae* (26). A cell deleted of the endogenous chromosomal site-specific mating type HO endonuclease was transformed with two plasmids carrying different selectable markers. One plasmid encodes the HO endonuclease (159) under the control of the *GAL10* promoter. The other contains the endonuclease target site in a nonessential region of the plasmid (Fig. 13–21). When cells were plated on galactose-containing medium, HO endonuclease expression was induced and a double-strand break was created in the target plasmid. Although the conditions do not specifically select for the presence of the target plasmid (it is in fact completely dispensable), the survival of macrocolony-forming cells was only 17% of that of an appropriate control,, i.e., cells containing identical constructs but no HO site on the second plasmid (Fig. 13–21). This effect is partly dependent on a functional *RAD9* gene, since the fraction of survivors increased to 30% in a *rad9* deletion mutant (26). Thus, it would appear that DNA damage-induced death in a unicellular eukaryote such as *S. cerevisiae* is actively regulated by *trans*-acting components. Such a phenomenon could explain the known dominant-lethal effect of X-ray-induced damage in *S. cerevisiae* (see chapter 12). However, a lot more remains to be learned about the mechanisms of cell death after DNA damage in eukaryotic cells.

Potential Signaling Enzymes

If DNA damage is indeed the crucial underlying factor for some of the regulatory responses discussed here, there are several possible ways that DNA alterations may be perceived and converted to a signal which can be transduced to specific transcriptional elements or to the cell cycle machinery. Such signals could result from

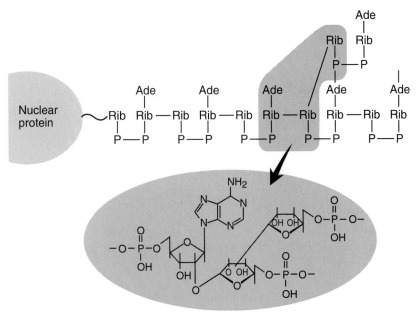

Figure 13–22 Molecular structure of the branched protein-poly(ADP) adduct resulting from poly(ADP-ribosylation). (*Adapted from Althaus and Richter [6] with permission.*)

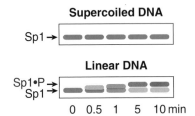

Figure 13–23 DNA-dependent protein kinase activity in crude HeLa cell extracts is detected by using unphosphorylated ^{35}S-labeled transcription factor Sp1 as a substrate. The presence of linear plasmid DNA results in the conversion of Sp1 into a phosphorylated form (Sp1·P) of lower electrophoretic mobility within 10 min of incubation. The addition of supercoiled DNA is ineffective. (*Adapted from Gottlieb and Jackson [103] with permission.*)

indirect consequences of the interaction of cellular components with unrepaired DNA damage, i.e., signals that emanate from stalled or idling DNA or RNA polymerase complexes, or from the complete assembly of a DNA repair complex(es). On the other hand, there may be more-direct signaling mechanisms, which include a dedicated surveillance system for monitoring the integrity of DNA, in which specialized proteins may respond to an initial DNA damage load. Candidate proteins have indeed been identified with properties that one might expect from such hypothetical damage-signaling moieties.

Poly(ADP-ribose) polymerase is an abundant nuclear enzyme which binds to nuclear proteins or DNA strand breaks and catalyzes the synthesis of poly(ADP-ribose) from NAD$^+$ precursors in the form of branched homopolymer chains of 200 to 300 residues (Fig. 13–22) (see chapter 7) (52, 167). The role of this enzyme in protecting DNA strand breaks and hence delaying certain modes of base excsion repair was discussed in chapter 7. Conceivably, poly(ADP-ribosylation) could also be a potent unique signal for genomic injury and might trigger certain regulatory responses. In particular, the regulation of c-Fos expression by poly(ADP-ribosylation) has been suggested (45).

The DNA-dependent protein kinases are also candidates for a signaling enzyme linking the presence of DNA damage to a regulatory cascade(s) (see chapter 12) (9, 10). DNA-dependent phosphorylation has been observed in cell-free lysates of different sources (315). An activity has been purified from HeLa cells, and a 450-kDa ATP-binding polypeptide has been identified as the catalytic subunit of a DNA-regulated serine/threonine kinase (43, 170). Its activation depends on free ends of duplex DNA; closed-circular plasmids are poor activators unless mechanically broken or cleaved with restriction enzymes (Fig. 13–23) (43, 103). Blunt ends are as efficient as 3′ or 5′ overhangs. Activation is largely sequence independent. However, double-stranded DNA of >25 bp is required. RNA-DNA hybrids are ineffective.

The human *Ku autoantigen* has been identified as the regulatory and targeting subunit for DNA-activated protein kinase (69, 103). Ku binds in vitro to the ends of naked double-stranded DNA and can translocate in an ATP-independent reaction (64). Ku consists of two subunits (Fig. 13–24). The 80-kDa subunit is encoded by a unique gene, whereas the 70-kDa subunit is likely to be member of a

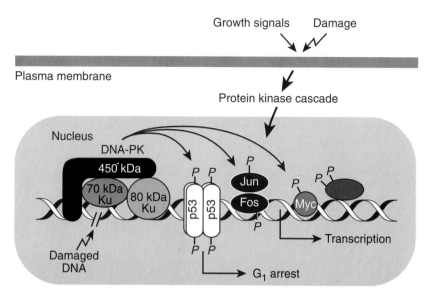

Figure 13–24 Possible roles of DNA-activated protein kinase (DNA-PK). The DNA-PK complex consists of the two DNA-end binding Ku subunits (of 70 and 80 kDa) and the catalytic subunit (of 450 kDa). It is hypothesized that activation of DNA-PK activity by strand breaks can result in the phosphorylation of transcriptional regulators such as p53, AP-1, Myc, and others. Their activation leads to cell cycle arrest and enhanced expression of genes inducible by treatment with DNA-damaging agents. The latter group of genes can also respond to cytoplasmically originated protein kinase cascades (see Fig. 13–7). This scheme is hypothetical and is based primarily on in vitro data. (*Adapted from Anderson [9] with permission.*)

gene family (105). Different family members may have different affinities for different DNA structures. An activity similar to human Ku has been characterized in yeast cells, and a putative yeast homolog of the 70-kDa subunit has been cloned. Deletion mutants were found to be viable but thermosensitive (83).

A variety of substrates for DNA-activated protein kinase have been identified in vitro (Table 13–4) (9, 10). These include transcription factors such as c-Jun, c-Fos, c-Myc, and p53, the C-terminal domain of RNA polymerase II, replication factor A, and SV40 T antigen. Heat shock protein 90 (hsp90) is among the few substrates which are not DNA-binding proteins. The 450-kDa catalytic subunit of DNA-activated protein kinase and both Ku components are also subject to autophosphorylation, which inhibits kinase activity. Human p53 protein is phosphorylated by DNA-activated protein kinase at Ser-15 and Ser-37 (171). A change of Ser-15 to Ala in the transactivation domain enhances the half-life of p53, and it has been suggested that DNA-activated kinase phosphorylation might have the same effect (86).

A paucity of in vivo data precludes firm conclusions about the precise role(s) of DNA-dependent kinases. It is, however, tempting to speculate that DNA-dependent kinase is a major player in sensing DNA damage and converting a signal through a cascade of kinase activation to modifications of transcription, replication, and cell cycle regulation (9) (Fig. 13–24). A defect in the gene encoding the 80-kDa subunit of Ku autoantigen in CHO mutants that are defective in double-strand break repair and that are complemented by the human gene *XRCC5* (see chapter 12) hints at a direct role in DNA repair (245, 301a). However, this finding does not suggest a regulatory role (at least not for G_1 or G_1/S arrest), since, unlike AT cells or other CHO mutants, normal inhibition of DNA synthesis is found after X ray treatment of these mutants (130).

Summary and Conclusions

We have discussed various examples of regulatory responses that are observed after treatment of eukaryotic cells with DNA-damaging agents. Their significance

Table 13–4 In vitro substrates of DNA-activated protein kinase

DNA-binding proteins
  SV40 large tumor antigen
  p53
  Ku autoantigen (p70 and p80)
  RNA polymerase II CTD domain
  Serum response factor
  Transcription factor c-Jun
  Transcription factor c-Fos
  Transcription factor Oct1
  Transcription factor Sp1
  Transcription factor c-Myc
  Transcription factor CTF/NF-1
  Transcription factor TFIID
  Chicken progesterone receptor
  Replication factor A
  Topoisomerases I and II
  Xenopus histone 2A.X
  Adenovirus type 2 72-kDa DNA-binding protein
  Bovine papillomavirus E2 protein
  Polyomavirus VP1
  Unidentified HeLa 110-kDa polypeptide
  Unidentified HeLa 52-kDa polypeptide

Non-DNA-binding proteins
  Heat shock protein 90 (hsp90)
  Microtubule-associated protein τ
  Casein
  Phosvitin

Source: Adapted from Anderson (9) with permission.

for DNA repair and mutagenesis is in general not well understood. It is tempting to assume the existence of a single signal transduction pathway that regulates gene activation, cell cycle arrest, and eventually programmed cell death. However, the available data hint at a complicated, flexible network of overlapping systems. It is safe to say that the idea of a unified regulatory circuit responding to DNA damage, analogous to the *E. coli* SOS system, should be abandoned for eukaryotic cells. For cell cycle arrest and gene activation in mammalian cells, it would appear that signal transduction mechanisms which are not DNA damage specific can be triggered. However, the notion that DNA damage was an important evolutionary driving force in shaping their emergence in eukaryotic cells merits consideration. Some of these pathways seem to transmit apparently opposing signals (e.g., cell cycle arrest versus general growth stimulation). How a cell integrates these different inputs is completely unknown.

Signals for these regulatory circuits are also largely unknown. In several instances (e.g., p53 induction, G_2 arrest, and "regulated death" in yeast cells), the available data hint at the significance of DNA double-strand breaks. In other cases the involvement of cytoplasmic oxidative damage seems likely (e.g., AP-1- and NF-κB-dependent gene activation and Bcl-2-regulated apoptosis). DNA-dependent kinases and poly(ADP-ribose) polymerase are interesting candidates for signaling enzymes that can detect DNA damage. All the above described phenomena have important implications for tumor biology, and it is certain that this will be an area of active research in years to come.

REFERENCES

1. **Abramić, M., A. S. Levine, and M. Protić.** 1991. Purification of an ultraviolet-inducible, damage-specific DNA-binding protein from primate cells. *J. Biol. Chem.* **266:**22493–22500.

2. **Aitken, A., D. B. Collinge, B. P. H. van Heusden, T. Isobe, P. H. Roseboom, G. Rosenfeld, and J. Soll.** 1992. 14-3-3 proteins: a highly conserved, widespread family of eukaryotic proteins. *Trends Biochem. Sci.* **17:**498–501.

3. **Al-Khodairy, F., and A. M. Carr.** 1992. DNA repair mutants defining G2 checkpoint pathways in *Schizosaccharomyces pombe*. *EMBO J.* **11:**1343–1350.

4. **Al-Khodairy, F., E. Fotou, K. S. Sheldrick, D. J. F. Griffiths, A. R. Lehmann, and A. M. Carr.** 1994. Identification and characterization of new elements involved in checkpoint and feedback controls in fission yeast. *Mol. Biol. Cell* **5:**147–160.

5. **Allen, J. B., Z. Zhou, W. Siede, E. C. Friedberg, and S. J. Elledge.** 1994. The *SAD1/RAD53* protein kinase controls multiple checkpoints and DNA-damage induced transcription in yeast. *Genes Dev.* **8:** 2416–2428.

6. **Althaus, F. R., and C. Richter.** 1987. *ADP-Ribosylation of Proteins, Enzymology and Biological Significance.* Springer-Verlag KG, Berlin.

7. **Amon, A., U. Surana, I. Muroff, and K. Nasmyth.** 1992. Regulation of p34^{cdc28} tyrosine phosphorylation is not required for entry into mitosis in *S. cerevisiae*. *Nature* (London) **355:**365–368.

8. **Anderson, A., and G. E. Woloschak.** 1992. Cellular proto-oncogene expression following exposure of mice to gamma rays. *Radiat. Res.* **130:**340–344.

9. **Anderson, C. W.** 1993. DNA damage and the DNA-activated protein kinase. *Trends Biochem. Sci.* **18:**433–437.

10. **Anderson, C. W., and S. P. Lees-Miller.** 1992. The nuclear serine/threonine protein kinase DNA-PK. *Crit. Rev. Eukaryotic Gene Expression* **2:**283–314.

11. **Angel, P., M. Imagawa, R. Chiu, B. Stein, R. J. Imbra, H. J. Rahmsdorf, C. Jonat, P. Herrlich, and M. Karin.** 1987. Phorbol ester-inducible genes contain a common cis element recognized by a TPA-modulated trans-acting factor. *Cell* **49:**729–739.

12. **Angel, P., and M. Karin.** 1991. The role of Jun, Fos and the AP-1 complex in cell-proliferation and transformation. *Biochim. Biophys. Acta* **1072:**129–157.

13. **Angel, P., A. Pöting, U. Mallick, H. J. Rahmsdorf, H. Schorpp, and P. Herrlich.** 1986. Induction of metallothionein and other mRNA species by carcinogens and tumor promoters in primary human skin fibroblasts. *Mol. Cell. Biol.* **6:**1760–1766.

14. **Angulo, J. F., J. Schwencke, I. Fernandez, and E. Moustacchi.** 1986. Induction of polypeptides in *Saccharomyces cerevisiae* after ultraviolet irradiation. *Biochem. Biophys. Res. Commun.* **138:**679–686.

15. **Angulo, J. F., J. Schwencke, P. L. Moreau, E. Moustacchi, and R. Devoret.** 1985. A yeast protein analogous to *Escherichia coli REC A* protein whose cellular level is enhanced after UV irradiation. *Mol. Gen. Genet.* **201:**20–24.

16. **Arlett, C. F., M. H. L. Green, A. Priestley, S. A. Harcourt, and L. V. Mayne.** 1988. Comparative human cellular radiosensitivity. I. The effect of SV40 transformation and immortalisation on the gamma-irradiation survival of skin derived fibroblasts from normal individuals and from ataxia-telangiectasia patients and heterozygotes. *Int. J. Radiat. Biol.* **54:**911–928.

17. **Banozzo, J., D. Bertoncini, D. Miller, C. R. Libertin, and G. E. Woloschak.** 1991. Modulation of expression of virus-like elements following exposure of mice to high- and low-LET radiations. *Carcinogenesis* **12:**801–804.

18. **Barbet, N. C., and A. M. Carr.** 1993. Fission yeast *wee1* protein kinase is not required for DNA damage-dependent mitotic arrest. *Nature* (London) **364:**824–827.

19. **Bargonetti, J., P. N. Friedman, S. E. Kern, B. Vogelstein, and C. Prives.** 1991. Wild-type but not mutant p53 immunopurified proteins bind to sequences adjacent to the SV40 origin of replication. *Cell* **65:**1083–1091.

20. **Barker, D. G., J. H. M. White, and L. H. Johnston.** 1985. The nucleotide sequence of the DNA ligase gene (*CDC9*) from *Saccharomyces cerevisiae*, a gene which is cell-cycle regulated and induced in response to DNA damage. *Nucleic Acids Res.* **13:**8323–8338.

21. **Basile, G., M. Aker, and R. K. Mortimer.** 1992. Nucleotide sequence and transcriptional regulation of the yeast recombinational repair gene *RAD51*. *Mol. Cell. Biol.* **12:**3235–3246.

22. **Bates, P. R., and M. F. Lavin.** 1989. Comparison of γ-radiation-induced accumulation of ataxia telangiectasia and control cells in G2 phase. *Mutat. Res.* **218:**165–170.

23. **Baur, M., and F. Eckardt-Schupp.** Personal communication.

24. **Beamish, H., and M. F. Lavin.** 1994. Radiosensitivity in ataxia-telangiectasia: anomalies in radiation-induced cell cycle delay. *Int. J. Radiat. Biol.* **65:**175–184.

25. **Ben-Ishai, R., R. Scharf, R. Sharon, and I. Kapten.** 1990. A human cellular sequence implicated in *trk* oncogene activation is DNA damage inducible. *Proc. Natl. Acad. Sci. USA* **87:**6039–6043.

26. **Bennett, C. B., A. L. Lewis, K. K. Baldwin, and M. A. Resnick.** 1993. Lethality induced by a single site-specific double-strand break in a dispensable yeast plasmid. *Proc. Natl. Acad. Sci. USA* **90:**5613–5617.

27. **Birkenbihl, R. P., and S. Subramani.** 1992. Cloning and characterization of *rad21* an essential gene of *Schizosaccharomyces pombe* involved in DNA double-strand-break repair. *Nucleic Acids Res.* **20:**6605–6611.

28. **Black, H. S.** 1987. Potential involvement of free radical reactions in ultraviolet light-mediated cutaneous damage. *Photochem. Photobiol.* **46:**213–221.

29. **Blattner, C., A. Knebel, A. Radler-Pohl, C. Sachsenmaier, P. Herrlich, and H. J. Rahmsdorf.** 1994. DNA damaging agents and growth factors induce changes in the program of expressed gene products through common routes. *Environ. Mol. Mutagen* **24:**3–10.

30. **Boothman, D. A., I. Bouvard, and E. N. Hughes.** 1989. Identification and characterisation of X-ray-induced proteins in human cells. *Cancer Res.* **49:**2871–2878.

31. **Boothman, D. A., M. Meyers, N. Fukunaga, and S. W. Lee.** 1993. Isolation of x-ray-inducible transcripts from radioresistant human melanoma cells. *Proc. Natl. Acad. Sci. USA* **90:**7200–7204.

32. **Boothman, D. A., M. Wang, and S. W. Lee.** 1991. Induction of tissue type plasminogen activator by ionizing radiation in human malignant melanoma cells. *Cancer Res.* **51:**5587–5595.

33. **Boreham, D. R., and R. E. J. Mitchel.** 1994. Regulation of heat and radiation stress responses in yeast by hsp-104. *Radiat. Res.* **137:**190–195.

34. **Brach, M. A., R. Hass, M. L. Sherman, H. Gunji, R. Weichselbaum, and D. Kufe.** 1991. Ionizing radiation induces expression and binding activity of the nuclear factor κB. *J. Clin. Invest.* **88:**691–695.

35. **Brachman, D. G., M. Beckett, D. Graves, D. Haraf, E. Vokes, and R. R. Weichselbaum.** 1993. p53 mutation does not correlate with radiosensitivity in 24 head and neck cancer cell lines. *Cancer Res.* **53:**3667–3669.

36. **Bradshaw, V. A., and K. McEntee.** 1989. DNA damage activates transcription and transposition of yeast Ty retrotransposons. *Mol. Gen. Genet.* **218:**465–474.

37. **Brock, J.-A. K., and K. Bloom.** 1994. A chromosome breakage assay to monitor mitotic forces in budding yeast. *J. Cell Sci.* **107:**891–902.

38. **Brouwer, J.** Personal communication.

39. **Brunet, S., and P. U. Giacomini.** 1989. Heat shock mRNA in mouse epidermis after UV irradiation. *Mutat. Res.* **219:**217–224.

40. **Büscher, M., H. J. Rahmsdorf, M. Litfin, M. Karin, and P. Herrlich.** 1988. Activation of the *c-fos* gene by UV and phorbol ester: different signal transduction pathways converge to the same enhancer element. *Oncogene* **3:**301–311.

41. **Caelles, C., A. Helmberg, and M. Karin.** 1994. p53-dependent apoptosis in the absence of transcriptional activation of p53-target genes. *Nature* (London) **370:**220–223.

41a. Canman, C. E., A. C. Wolff, C.-Y. Chen, A. J. Fornace, Jr., and M. B. Kastan. 1994. The p53-dependent G₁ cell cycle checkpoint pathway and ataxia-telangiectasia. *Cancer Res.* **54:**5054–5058.

42. Carrier, F., A. Gatignol, M. C. Hollander, K. T. Jeang, and A. J. Fornace, Jr. 1994. Induction of RNA-binding proteins in mammalian cells by DNA-damaging agents. *Proc. Natl. Acad. Sci. USA* **91:**1554–1558.

43. Carter, T., I. Vancurová, I. Sun, W. Lou, and S. DeLeon. 1990. A DNA-activated protein kinase from HeLa cell nuclei. *Mol. Cell. Biol.* **10:**6460–6471.

44. Carty, M. P., M. Zernik-Kobak, S. McGrath, and K. Dixon. 1994. UV light-induced DNA synthesis arrest in HeLa cells is associated with changes in phosphorylation of human single-stranded DNA-binding protein. *EMBO J.* **13:**2114–2123.

45. Cerutti, P. A., and B. F. Trump. 1991. Inflammation and oxidative stress in carcinogenesis. *Cancer Cells* **3:**1–7.

46. Chen, J., B. Derfler, and L. Samson. 1989. *Saccharomyces cerevisiae* 3-methyladenine DNA glycosylase has homology to the *alkA* glycosylase of *E. coli* and is induced in response to DNA alkylation agents. *EMBO J.* **9:**4569–4575.

47. Chen, J., and L. Samson. 1991. Induction of *S. cerevisiae MAG* 3-methyladenine DNA glycosylase transcript levels in response to DNA damage. *Nucleic Acids Res.* **19:**6427–6432.

48. Chiou, S.-K., L. Rao, and E. White. 1994. Bcl-2 blocks p53-dependent apoptosis. *Mol. Cell. Biol.* **14:**2556–2563.

49. Chu, G., and E. Chang. 1988. Xeroderma pigmentosum group E cells lack a nuclear factor that binds to damaged DNA. *Science* **242:**564–567.

5(Chu, G., and E. Chang. 1990. Cisplatin-resistant cells express increased levels of a factor that recognizes damaged DNA. *Proc. Natl. Acad. Sci. USA* **87:**3324–3327.

51. Clarke, A. R., C. A. Purdie, D. J. Harrison, R. G. Morris, C. C. Bird, M. L. Hooper, and A. H. Wyllie. 1993. Thymocyte apoptosis induced by p53-dependent and independent pathways. *Nature* (London) **362:**849–852.

52. Cleaver, J. E., and W. F. Morgan. 1991. Poly(ADP-ribose) polymerase: a perplexing participant in cellular responses to DNA breakage. *Mutat. Res.* **257:**1–18.

53. Cole, G. M., and R. K. Mortimer. 1989. Failure to induce a DNA repair gene, *RAD54*, in *Saccharomyces cerevisiae* does not affect DNA repair or recombination pathways. *Mol. Cell. Biol.* **9:**3314–3322.

54. Cole, G. M., D. Schild, S. T. Lovett, and R. K. Mortimer. 1987. Regulation of *RAD54*- and *RAD52-lacZ* gene fusions in *Saccharomyces cerevisiae* in response to DNA damage. *Mol. Cell. Biol.* **7:**1078–1084.

55. Cole, G. M., D. Schild, and R. K. Mortimer. 1989. Two DNA repair and recombination genes in *Saccharomyces cerevisiae*, *RAD52* and *RAD54*, are induced during meiosis. *Mol. Cell. Biol.* **9:**3101–3104.

56. Collins, M. K. L., and A. L. Rivas. 1993. The control of apoptosis in mammalian cells. *Trends Biochem. Sci.* **18:**307–309.

57. Crawford, D., I. Zbinden, P. Amstad, and P. Cerutti. 1988. Oxidant stress induces the proto-oncogenes c-*fos* and c-*myc* in mouse epidermal cells. *Oncogene* **3:**27–32.

58. Dasso, M., and J. W. Newport. 1990. Completion of DNA replication is monitored by a feedback system that controls the initiation of mitosis *in vitro*: studies in *Xenopus*. *Cell* **61:**811–823.

59. Dasso, M., H. Nishitani, S. Kornbluth, T. Nishimoto, and J. W. Newport. 1992. RCC1, a regulator of mitosis, is essential for DNA replication. *Mol. Cell. Biol.* **12:**3337–3345.

60. Dérijard, B., M. Hibi, I.-H. Wu, T. Barrett, B. Su, T. Deng, M. Karin, and R. J. Davis. 1994. JNK1: a protein kinase stimulated by UV light and Ha-Ras that binds and phosphorylates the c-Jun activation domain. *Cell* **76:**1025–1037.

61. Devary, Y., R. A. Gottlieb, L. F. Lau, and M. Karin. 1991. Rapid and preferential activation of the c-*jun* gene during the mammalian UV response. *Mol. Cell. Biol.* **11:**2804–2811.

62. Devary, Y., R. A. Gottlieb, T. Smeal, and M. Karin. 1992. The mammalian ultraviolet response is triggered by activation of Src tyrosine kinase. *Cell* **71:**1081–1091.

63. Devary, Y., C. Rosette, J. A. DiDonato, and M. Karin. 1993. NF-κB activation by ultraviolet light not dependent on a nuclear signal. *Science* **261:**1442–1445.

64. de Vries, E., W. van Driel, W. G. Bergsma, A. C. Arnberg, and P. C. van der Vliet. 1989. HeLa nuclear protein recognizing DNA termini and translocating on DNA forming a regular DNA-multimeric protein complex. *J. Mol. Biol.* **208:**65–78.

65. Donehower, L. A., M. Harvey, B. L. Slagle, M. J. McArthur, C. A. Montgomery Jr., J. S. Butel, and A. Bradley. 1992. Mice deficient for p53 are developmentally normal but susceptible to spontaneous tumors. *Nature* (London) **356:**215–221.

66. Draetta, G. 1990. Cell cycle control in eukaryotes: molecular mechanisms of cdc2 activation. *Trends Biochem. Sci.* **15:**378–383.

67. Dulić, V., W. K. Kaufmann, S. J. Wilson, T. D. Tlsty, E. Lees, J. W. Harper, S. J. Elledge, and S. I. Reed. 1994. p53-dependent inhibition of cyclin-dependent kinase activities in human fibroblasts during radiation-induced G1 arrest. *Cell* **76:**1013–1023.

68. Dutta, A., J. M. Ruppert, J. C. Aster, and E. Winchester. 1993. Inhibition of DNA replication factor RPA by p53. *Nature* (London) **365:**79–82.

69. Dvir, A., S. R. Peterson, M. W. Knuth, H. Lu, and W. S. Dynan. 1992. Ku autoantigen is the regulatory component of a template-associated protein kinase that phosphorylates RNA polymerase II. *Proc. Natl. Acad. Sci. USA* **89:**11920–11924.

70. El-Deiry, W. S., J. W. Harper, P. M. O'Connor, V. E. Velculescu, C. E. Canman, J. Jackman, J. A. Pietenpol, M. Burrell, D. E. Hill, Y. Wang, K. G. Wiman, W. E. Mercer, M. B. Kastan, K. W. Kohn, S. J. Elledge, K. W. Kinzler, and B. Vogelstein. 1994. *WAF1/CIP1* is induced in *p53*-mediated G₁ arrest and apoptosis. *Cancer Res.* **54:**1169–1174.

71. El-Deiry, W. S., S. E. Kern, J. A. Pietenpol, K. W. Kinzler, and B. Vogelstein. 1992. Human genomic DNA sequences define a consensus binding site for human p53 protein complexes. *Nat. Genet.* **1:**44–49.

72. El-Deiry, W. S., T. Tokino, V. E. Velculescu, D. B. Levy, R. Parsons, J. M. Trent, D. Lin, W. E. Mercer, K. W. Kinzler, and B. Vogelstein. 1993. *WAF1*, a potential mediator of p53 tumor suppression. *Cell* **75:**817–825.

73. Elledge, S. J., and R. W. Davis. 1987. Identification and isolation of the gene encoding the small subunit of ribonucleotide reductase from *Saccharomyces cerevisiae*: DNA damage-inducible gene required for mitotic viability. *Mol. Cell. Biol.* **7:**2783–2793.

74. Elledge, S. J., and R. W. Davis. 1989. DNA damage induction of ribonucleotide reductase. *Mol. Cell. Biol.* **9:**4932–4940.

75. Elledge, S. J., and R. W. Davis. 1989. Identification of the DNA damage-responsive element of *RNR2* and evidence that four distinct cellular factors bind. *Mol. Cell. Biol.* **9:**5373–5386.

76. Elledge, S. J., and R. W. Davis. 1990. Two genes differentially regulated in the cell cycle and by DNA-damaging agents encode alternative regulatory subunits of ribonucleotide reductase. *Genes Dev.* **4:**740–751.

77. Elledge, S. J., Z. Zhou, J. B. Allen, and T. A. Navas. 1993. DNA damage and cell cycle regulation of ribonucleotide reductase. *Bioessays* **15:**333–339.

78. Ellem, K. A. O., M. Cullinan, K. C. Baumann, and A. Dunstan. 1988. UVR induction of TGFα: a possible autocrine mechanism for the epidermal melanocyte response and for promotion of epidermal carcinogenesis. *Carcinogenesis* **9:**797–801.

79. Engelberg, D., C. Klain, H. Martinetto, K. Struhl, and M. Karin. 1994. The UV response involving the Ras signaling pathway and AP-1 transcription factors is conserved between yeast and mammals. *Cell* **77:**381–390.

80. Englander, E. W., and S. H. Wilson. 1992. DNA damage response of cloned DNA β-polymerase promoter is blocked in mutant cell lines deficient in protein kinase A. *Nucleic Acids Res.* **20:**5527–5531.

81. **Enoch, T., A. M. Carr, and P. Nurse.** 1992. Fission yeast genes involved in coupling mitosis to completion of DNA replication. *Genes Dev.* **6:**2035–2046.

82. **Enoch, T., and P. Nurse.** 1990. Mutation of fission yeast cell cycle control genes abolishes dependence of mitosis on DNA replication. *Cell* **60:**665–673.

83. **Feldmann, H., and E. L. Winnacker.** 1993. A putative homologue of the human autoantigen Ku from *Saccharomyces cerevisiae. J. Biol. Chem.* **268:**12895–12900.

84. **Fenech, M., A. M. Carr, J. Murray, F. Z. Watts, and A. R. Lehmann.** 1991. Cloning and characterization of the *rad4* gene of *Schizosaccharomyces pombe*: a gene showing short regions of sequence similarity to the human *XRCC1* gene. *Nucleic Acids Res.* **19:**6737–6741.

85. **Finley, D., E. Özkaynak, and A. Varshavsky.** 1987. The yeast polyubiquitin gene is essential for resistance to high temperatures, starvation and other stresses. *Cell* **48:**1035–1046.

86. **Fiscella, M., S. J. Ullrich, N. Zambrano, M. T. Shields, D. Lin, S. P. Lees-Miller, C. W. Anderson, W. E. Mercer, and E. Appella.** 1993. Mutation of the serine 15 phosphorylation site of human p53 reduces the ability of p53 to inhibit cell cycle progression. *Oncogene* **8:**1519–1528.

87. **Ford, J. C., F. Al-Khodairy, E. Fotou, K. S. Sheldrick, D. J. F. Griffiths, and A. M. Carr.** 1994. 14-3-3 protein homologs required for the DNA damage checkpoint in fission yeast. *Science* **265:**533–535.

88. **Fornace, A. J., Jr.** 1992. Mammalian genes induced by radiation; activation of genes associated with growth control. *Annu. Rev. Genet.* **26:**507–526.

89. **Fornace, A. J., Jr., I. Alamo, Jr., and M. C. Hollander.** 1988. DNA damage-inducible transcripts in mammalian cells. *Proc. Natl. Acad. Sci. USA* **85:**8800–8804.

90. **Fornace, A. J., Jr., I. Alamo, Jr., M. C. Hollander, and E. Lamoreaux.** 1989. Ubiquitin mRNA is a major stress-induced transcript in mammalian cells. *Nucleic Acids Res.* **17:**1215–1230.

91. **Fornace, A. J., Jr., D. W. Nebert, M. C. Hollander, J. D. Luethy, M. Papathanasiou, J. Fargnoli, and N. J. Holbrook.** 1989. Mammalian genes coordinately regulated by growth arrest signals and DNA-damaging agents. *Mol. Cell. Biol.* **9:**4196–4203.

92. **Fornace, A. J., Jr., H. Schalch, and I. Alamo, Jr.** 1988. Coordinate induction of metallothioneins I and II in rodent cells by UV irradiation. *Mol. Cell. Biol.* **8:**4716–4720.

93. **Fornace, A. J., Jr., B. Zmudzka, M. C. Hollander, and S. H. Wilson.** 1989. Induction of β-polymerase mRNA by DNA-damaging agents in Chinese hamster ovary cells. *Mol. Cell. Biol.* **9:**851–853.

94. **Frasch, M.** 1991. The maternally expressed *Drosophila* gene encoding the chromatin-binding BJ1 is a homolog of the vertebrate gene regulator of chromatin condensation RCC1. *EMBO J.* **5:**1225–1236.

95. **Friedberg, E. C.** 1988. Deoxyribonucleic acid repair in the yeast *Saccharomyces cerevisiae. Microbiol. Rev.* **52:**70–102.

96. **Friedberg, E. C., W. Siede, and A. J. Cooper.** 1991. Cellular responses to DNA damage in yeast, p. 147–192. *In* E. Jones, J. R. Pringle, and J. Broach (ed.), *The Molecular Biology of the Yeast Saccharomyces: Genome Dynamics, Protein Synthesis, and Energetics*, vol. I. Cold Spring Harbor Laboratory, Cold Spring Harbor, N.Y.

97. **Fritsche, M., C. Haessler, and G. Brandner.** 1993. Induction of nuclear accumulation of the tumor-suppressor protein p53 by DNA-damaging agents. *Oncogene* **8:**307–318.

98. **Fritz, G., K. Tano, S. Mitra, and B. Kaina.** 1991. Inducibility of the DNA repair gene encoding O^6-methylguanine-DNA methyltransferase in mammalian cells by DNA-damaging treatments. *Mol. Cell. Biol.* **11:**4660–4668.

99. **Gahring, L. B., M. B. Pepys, and R. Daynes.** 1984. Effect of UV radiation on production of epidermal cell thymocyte-activating factor/interleukin1 *in vivo* and *in vitro. Proc. Natl. Acad. Sci. USA* **81:**1198–1202.

100. **Gailit, J.** 1990. Identification of proteins whose synthesis in *Saccharomyces cerevisiae* is induced by DNA damage and heat shock. *Int. J. Radiat. Biol.* **57:**981–992.

101. **Game, J. C.** 1983. Radiation-sensitive mutants and repair in yeast, p. 109–137. *In* J. F. T. Spencer, D. M. Spencer, and A. R. W. Smith (ed.), *Yeast Genetics: Fundamental and Applied Aspects.* Springer-Verlag, New York.

102. **Glazer, P. M., N. A. Greggio, J. E. Metherall, and W. C. Summers.** 1989. UV-induced DNA-binding proteins in human cells. *Proc. Natl. Acad. Sci. USA* **86:**1163–1167.

103. **Gottlieb, T. M., and S. P. Jackson.** 1993. The DNA-dependent protein kinase: requirement for DNA ends and association with Ku antigen. *Cell* **72:**131–142.

104. **Gould, K., and P. Nurse.** 1989. Tyrosine phosphorylation of the fission yeast *cdc2⁺* protein kinase regulates entry into mitosis. *Nature* (London) **342:**39–45.

105. **Griffith, A. J., J. Craft, J. Evans, T. Mimori, and J. A. Hardin.** 1992. Nucleotide sequence and genomic structure analyses of the p70 subunit of the human Ku autoantigen: evidence for a family of genes encoding Ku(p70)-related polypeptides. *Mol. Biol. Rep.* **16:**91–97.

106. **Gu, Y., C. W. Turck, and D. O. Morgan.** 1993. Inhibition of Cdk2 activity *in vivo* by an associated 20K regulatory subunit. *Nature* (London) **366:**707–710.

107. **Gu, Z., D. Pim, S. Labrecque, L. Banks, and G. Matlashewski.** 1994. DNA damage induced p53 mediated transcription is inhibited by human papillomavirus type 18 E6. *Oncogene* **9:**629–633.

108. **Hagedorn, R., H. W. Thielmann, H. Fischer, and C. H. Schroeder.** 1983. SV40-induced transformation and T-antigen production is enhanced in normal and repair deficient human fibroblasts after pretreatment of cells with UV light. *J. Cancer Res. Clin. Oncol.* **106:**93–96.

109. **Haimovitz-Friedman, A., I. Vlodavsky, A. Chaudhuri, L. Witte, and Z. Fuks.** 1991. Autocrine effects of fibroblast growth factor in repair of radiation damage in endothelial cells. *Cancer Res.* **51:**2552–2558.

110. **Hall, P. E., P. H. McKee, H. D. Menage, R. Dover, and D. P. Lane.** 1993. High levels of p53 protein in UV-irradiated normal human skin. *Oncogene* **8:**203–207.

111. **Hallahan, D. E., D. R. Spriggs, M. A. Beckett, D. W. Kufe, and R. R. Weichselbaum.** 1989. Increased tumor necrosis factor-α mRNA after cellular exposure to ionizing radiation. *Proc. Natl. Acad. Sci. USA* **86:**10104–10107.

112. **Hallahan, D. E., V. P. Sukhatme, M. L. Sherman, S. Virudachalam, D. Kufe, and R. R. Weichselbaum.** 1991. Protein kinase C mediates X-ray inducibility of nuclear signal transducers EGR1 and JUN. *Proc. Natl. Acad. Sci. USA* **88:**2156–2160.

113. **Harper, J. W., G. R. Adami, N. Wei, K. Keyomarsi, and S. J. Elledge.** 1993. The p21 Cdk-interacting protein Cip1 is a potent inhibitor of G1 cyclin-dependent kinases. *Cell* **75:**805–816.

114. **Hartwell, L. H.** 1974. *Saccharomyces cerevisiae* cell cycle. *Bacteriol. Rev.* **38:**164–198.

115. **Hartwell, L. H., and T. A. Weinert.** 1989. Checkpoints: controls that ensure the order of cell cycle events. *Science* **246:**629–634.

116. **Hengst, L., V. Dulić, J. M. Slingerland, E. Lees, and S. I. Reed.** 1994. A cell cycle-regulated inhibitor of cyclin-dependent kinases. *Proc. Natl. Acad. Sci. USA* **91:**5291–5295.

117. **Herr, I., H. van Dam, and P. Angel.** 1994. Binding of promoter-associated AP-1 is not altered during induction and subsequent repression of the *c-jun* promoter by TPA and UV irradiation. *Carcinogenesis* **15:**1105–1113.

118. **Herrlich, P., H. Ponta, and H. J. Rahmsdorf.** 1992. DNA damage induced gene expression: signal transduction and relation to growth factor signalling. *Rev. Physiol. Biochem. Pharmacol.* **119:**187–223.

119. **Herrlich, P., and H. J. Rahmsdorf.** 1994. Transcriptional and posttranscriptional responses to DNA-damaging agents. *Curr. Opin. Cell Biol.* **6:**425–431.

120. **Hibi, M., A. Lin, T. Smeal, A. Minden, and M. Karin.** 1993. Identification of an oncoprotein- and UV-responsive protein kinase that

binds and potentiates the c-Jun activation domain. *Genes Dev.* **7:**2135–2148.

121. Hill, B. T., K. Deuchars, L. K. Hosking, V. Ling, and R. D. H. Whelan. 1990. Overexpression of P-glycoprotein in mammalian tumor cell lines after fractionated X-irradiation *in vitro. J. Natl. Cancer Inst.* **82:**607–612.

122. Hirschfeld, S., A. S. Levine, K. Ozato, and M. Protić. 1990. A constitutive damage-specific DNA-binding protein is synthesized at higher levels in UV-irradiated primate cells. *Mol. Cell. Biol.* **10:**2041–2048.

123. Hockenbery, D. M., G. Nuñez, C. Milliman, R. D. Schreiber, and S. J. Korsmeyer. 1990. Bcl-2 is an inner mitochondrial membrane protein that blocks programmed cell death. *Nature* (London) **348:**334–336.

124. Hockenbery, D. M., Z. N. Oltvai, X. M. Yin, C. L. Milliman, and S. J. Korsmeyer. 1993. Bcl-2 functions in an antioxidant pathway to prevent apoptosis. *Cell* **75:**241–251.

125. Holbrook, N. J., and A. J. Fornace, Jr. 1991. Response to adversity: molecular control of gene activation following genotoxic stress. *New Biol.* **3:**825–833.

126. Hollander, M. C., and A. J. Fornace, Jr. 1989. Induction of fos RNA by DNA damaging agents. *Cancer Res.* **49:**1687–1692.

127. Hollstein, M., D. Sidransky, B. Vogelstein, and C. C. Harris. 1991. p53 Mutations in human cancers. *Science* **253:**49–53.

128. Hunter, T. 1993. Braking the cycle. *Cell* **75:**839–841.

128a. Ishizaki, K., Y. Ejima, T. Matsunage, R. Hara, A. Sakamoto, M. Ikenaga, Y. Ikawa, and S. Aizawa. 1994. Increased UV-induced SCEs but normal repair of DNA damage in *p53*-deficient mouse cells. *Int. J. Cancer* **58:**254–257.

129. Jaeg, J.-P., K. Bouayadi, P. Calsou, and B. Salles. 1994. UV induction of excision repair enzymes detected in protein extracts from *Schizosaccharomyces pombe. Biochem. Biophys. Res. Commun.* **198:**770–779.

130. Jeggo, P. A. 1985. X-ray sensitive mutants of Chinese hamster ovary cell line: radio-sensitivity of DNA synthesis. *Mutat. Res.* **145:**171–176.

131. Jentsch, S., J. P. McGrath, and A. Varshavsky. 1987. The yeast DNA repair gene *RAD6* encodes a ubiquitin-conjugating enzyme. *Nature* (London) **329:**131–134.

132. Johnson, A. L., D. G. Barker, and L. H. Johnston. 1986. Induction of yeast DNA ligase genes in exponential and stationary phase cultures in response to DNA damaging agents. *Curr. Genet.* **11:**107–112.

133. Johnston, L. H., A. L. Johnson, and D. G. Barker. 1986. The expression in meiosis of genes which are transcribed periodically in the mitotic cell cycle of budding yeast. *Exp. Cell Res.* **165:**541–549.

134. Johnston, L. H., J. H. M. White, A. L. Johnson, G. Lucchini, and P. Plevani. 1987. The yeast DNA polymerase I transcript is regulated in both the mitotic cell cycle and in meiosis and is also induced after DNA damage. *Nucleic Acids Res.* **15:**5017–5030.

135. Jones, J. S., and L. Prakash. 1991. Transcript levels of the *Saccharomyces cerevisiae* DNA repair gene *RAD18* increase in UV irradiated cells and during meiosis but not during the mitotic cell cycle. *Nucleic Acids Res.* **19:**893–898.

136. Jones, J. S., L. Prakash, and S. Prakash. 1990. Regulated expression of the *Saccharomyces cerevisiae* DNA repair gene *RAD7* in response to DNA damage and during sporulation. *Nucleic Acids Res.* **18:**3281–3285.

137. Juven, T., Y. Barak, A. Zauberman, D. L. George, and M. Oren. 1993. Wild type p53 can mediate sequence-specific transactivation of an internal promoter within the *mdm2* gene. *Oncogene* **8:**3411–3416.

138. Kaback, D. B., and L. R. Feldberg. 1985. *Saccharomyces cerevisiae* exhibits a sporulation-specific temporal pattern of transcript accumulation. *Mol. Cell. Biol.* **5:**751–761.

139. Kaina, B., H. Lohrer, M. Karin, and P. Herrlich. 1990. Overexpressed human metallothionein IIA gene protects Chinese hamster ovary cells from killing by alkylating agents. *Proc. Natl. Acad. Sci. USA* **87:**2710–2714.

140. Kaina, B., B. Stein, A. Schönthal, H. J. Rahmsdorf, H. Ponta, and P. Herrlich. 1989. An update of the mammalian UV response: gene regulation and induction of a protective function, p. 149–156. *In* M. W. Lambert and J. Laval (ed.), *DNA Repair Mechanisms and Their Biological Implications in Mammalian Cells.* Plenum Publishing Corp., New York.

141. Kartasova, T., B. J. C. Cornelissen, P. Belt, and P. van de Putte. 1987. Effects of UV, 4-NQO and TPA on gene expression in cultured human epidermal keratinocytes. *Nucleic Acids Res.* **15:**5945–5962.

142. Kartasova, T., and P. van de Putte. 1988. Isolation, characterization, and UV-stimulated expression of two families of genes encoding polypeptides of related structure in human epidermal keratinocytes. *Mol. Cell. Biol.* **8:**2195–2203.

143. Kastan, M. B., O. Onyekwere, D. Sidransky, B. Vogelstein, and R. W. Craig. 1991. Participation of p53 protein in the cellular response to DNA damage. *Cancer Res.* **51:**6304–6311.

144. Kastan, M. B., Q. Zhan, W. S. El-Deiry, F. Carrier, T. Jacks, W. V. Walsh, B. S. Plunkett, B. Vogelstein, and A. J. Fornace, Jr. 1992. A mammalian cell cycle checkpoint pathway utilizing p53 and GADD45 is defective in ataxia-telangiectasia. *Cell* **71:**587–597.

145. Kataoka, H., and Y. Fujiwara. 1991. UV damage specific protein in xeroderma pigmentosum complementation group E. *Biochem. Biophys. Res. Commun.* **175:**1139–1143.

146. Kaufmann, W. K., and D. G. Kaufman. 1993. Cell cycle control, DNA repair and initiation of carcinogenesis. *FASEB J.* **7:**1188–1191.

147. Kerr, J. F. R., J. Searle, B. V. Harmon, and C. J. Bishop. 1987. Apoptosis, p. 93–128. *In* C. S. Potten (ed.), *Perspectives on Mammalian Cell Death.* Oxford University Press, Oxford.

148. Kessis, T. D., R. J. Slebos, W. G. Nelson, M. B. Kastan, B. S. Plunkett, S. M. Han, A. T. Lorincz, L. Hedrick, and K. R. Cho. 1993. Human papillomavirus 16 E6 expression disrupts the p53-mediated cellular response to DNA damage. *Proc. Natl. Acad. Sci. USA* **90:**3988–3992.

149. Keyse, S. M. 1993. The induction of gene expression in mammalian cells by radiation. *Semin. Cancer Biol.* **4:**119–128.

150. Keyse, S. M., and E. A. Emslie. 1992. Oxidative stress and heat shock induce a human gene encoding a protein tyrosine phosphatase. *Nature* (London) **359:**644–646.

151. Keyse, S. M., and R. M. Tyrell. 1989. Heme oxygenase is the major 32-kDa stress protein induced in human skin fibroblasts by UVA radiation, hydrogen peroxide, and sodium arsenite. *Proc. Natl. Acad. Sci. USA* **86:**99–103.

152. Khanna, K. K., and M. F. Lavin. 1993. Ionizing radiation and UV induction of p53 protein by different pathways in ataxia-telangiectasia cells. *Oncogene* **8:**3411–3416.

153. Kharbanda, S., A. Saleem, R. Datta, Z.-M. Yuan, R. Weichselbaum, and D. Kufe. 1994. Ionizing radiation induces rapid tyrosine phosphorylation of p34^cdc2. *Cancer Res.* **54:**1412–1414.

154. Kleinberger, T., Y. B. Flint, M. Blank, S. Etkin, and S. Lavi. 1988. Carcinogen-induced *trans*-activation of gene expression. *Mol. Cell. Biol.* **8:**1366–1370.

155. Kohn, K. W., J. Jackman, and P. M. O'Connor. 1994. Cell cycle control and cancer chemotherapy. *J. Cell. Biochem.* **54:**440–452.

156. Kolch, W., G. Heidecker, G. Kochs, R. Hummel, H. Vahidi, H. Mischak, G. Finkenzeller, D. Marmé, and U. R. Rapp. 1993. Protein kinase Cα activates RAF-1 by direct phosphorylation. *Nature* (London) **364:**249–252.

157. Konze-Thomas, B., R. M. Hazard, V. M. Maher, and J. J. McCormick. 1982. Extent of excision repair before DNA synthesis determines the mutagenic but not the lethal effect of UV radiation. *Mutat. Res.* **94:**421–434.

158. Korsmeyer, S. J. 1992. Bcl-2 initiates a new category of oncogenes: regulators of cell death. *Blood* **80:**879–886.

159. Kostriken, R., J. N. Strathern, A. J. S. Klar, J. B. Hicks, and F. Heffron. 1983. A site-specific endonuclease essential for mating-type switching in *Saccharomyces cerevisiae. Cell* **35:**167–174.

160. **Krämer, M., C. Sachsenmaier, P. Herrlich, and H. J. Rahmsdorf.** 1993. UV irradiation-induced interleukin-1 and basic fibroblast growth factor synthesis and release mediate part of the UV response. *J. Biol. Chem.* **268:**6734–6741.

161. **Krämer, M., B. Stein, S. Mai, E. Kunz, H. König, H. Loferer, H. H. Grunicke, H. Ponta, P. Herrlich, and H. J. Rahmsdorf.** 1990. Radiation-induced activation of transcription factors in mammalian cells. *Radiat. Environ. Biophys.* **29:**303–313.

162. **Kuerbitz, S. J., B. S. Plunkett, W. V. Walsh, and M. B. Kastan.** 1992. Wild-type p53 is a cell cycle checkpoint determinant following irradiation. *Proc. Natl. Acad. Sci. USA* **89:**7491–7495.

163. **Lambert, M. E., Z. A. Ronai, I. B. Weinstein, and J. I. Garrels.** 1989. Enhancement of major histocompatibility class 1 protein synthesis by DNA damage in cultured human fibroblasts and keratinocytes. *Mol. Cell. Biol.* **9:**847–850.

164. **Lane, D. P.** 1992. p53, guardian of the genome. *Nature* (London) **358:**15–16.

165. **Lane, D. P., and L. V. Crawford.** 1979. T antigen is bound to a host protein in SV40 transformed cells. *Nature* (London) **278:**261–263.

166. **Larner, J. M., H. Lee, and J. L. Hamlin.** 1994. Radiation effects on DNA synthesis in a defined chromosomal replicon. *Mol. Cell. Biol.* **14:**1901–1908.

167. **Lautier, D., J. Lagueux, J. Thibodeau, L. Ménard, and G. G. Porier.** 1993. Molecular and biochemical features of poly(ADP-ribose) metabolism. *Mol. Cell. Biochem.* **122:**171–193.

168. **Laval, F.** 1991. Increase of O^6-methylguanine-DNA-methyltransferase and N^3-methyladenine glycosylase in rat hepatoma cells treated with DNA-damaging agents. *Biochem. Biophys. Res. Commun.* **176:**1086–1092.

169. **Lee, J. M., and A. Bernstein.** 1993. p53 mutations increase resistance to ionizing radiation. *Proc. Natl. Acad. Sci. USA* **90:**5742–5756.

170. **Lees-Miller, S. P., Y.-R. Chen, and C. W. Anderson.** 1990. Human cells contain a DNA-activated protein kinase that phosphorylates simian virus 40 T antigen, mouse p53, and the human Ku autoantigen. *Mol. Cell. Biol.* **10:**6472–6481.

171. **Lees-Miller, S. P., K. Sakaguchi, S. J. Ullrich, E. Appella, and C. W. Anderson.** 1992. Human DNA-activated protein kinase phosphorylates serines 15 and 37 in the amino-terminal transactivation domain of human p53. *Mol. Cell. Biol.* **12:**5041–5049.

172. **Lefebvre, P., and F. Laval.** 1986. Enhancement of O^6-methylguanine-DNA methyltransferase activity induced by various treatments in mammalian cells. *Cancer Res.* **46:**5701–5705.

173. **Levin, N., M.-A. Bjornsti, and G. R. Fink.** 1993. A novel mutation in DNA topoisomerase I of yeast causes DNA damage and *RAD9*-dependent cell cycle arrest. *Genetics* **133:**799–814.

174. **Li, F., and J. Fraumeni.** 1969. Soft-tissue sarcomas, breast cancer and other neoplasms: a familial syndrome? *Ann. Intern. Med.* **71:**747–753.

175. **Li, J. J., and R. J. Deshaies.** 1993. Exercising self-restraint: discouraging illicit acts of S and M in eukaryotes. *Cell* **74:**223–226.

176. **Lieberman, M. W., L. R. Beach, and R. D. Palmiter.** 1983. Ultraviolet radiation-induced metallothionein-1 gene activation is associated with extensive DNA demethylation. *Cell* **35:**207–214.

177. **Lin, C. S., D. A. Goldthwait, and D. Samols.** 1990. Induction of transcription from the long terminal repeat of Moloney murine sarcoma provirus by UV-irradiation, X-irradiation and phorbol ester. *Proc. Natl. Acad. Sci. USA* **87:**36–40.

178. **Lin, D., M. T. Shields, S. J. Ullrich, E. Appella, and W. E. Mercer.** 1992. Growth arrest induced by wild-type p53 protein blocks cells prior to or near the restriction point in late G1-phase. *Proc. Natl. Acad. Sci. USA* **89:**9210–9214.

179. **Liu, V. F., and D. T. Weaver.** 1993. The ionizing radiation-induced replication protein A phosphorylation response differs between ataxia telangiectasia and normal human cells. *Mol. Cell. Biol.* **13:**7222–7231.

180. **Liu, X., C. W. Miller, P. H. Koeffler, and A. J. Berk.** 1993. The p53 activation domain binds the TATA box-binding polypeptide in holo-TFIID, and a neighboring p53 domain inhibits transcription. *Mol. Cell. Biol.* **13:**3291–3300.

181. **Livingstone, L. R., A. White, J. Sprouse, E. Livanos, T. Jacks, and T. D. Tlsty.** 1992. Altered cell cycle arrest and gene amplification potential accompany loss of wild-type p53. *Cell* **70:**923–935.

182. **Lock, R. B.** 1992. Inhibition of $p34^{cdc2}$ kinase activation, $p34^{cdc2}$ dephosphorylation, and mitotic progression in Chinese hamster ovary cells exposed to etoposide. *Cancer Res.* **52:**1817–1822.

183. **Lock, R. B., and W. E. Ross.** 1990. Inhibition of $p34^{cdc2}$ kinase activity by etoposide or irradiation as a mechanism of G2 arrest in Chinese hamster ovary cells. *Cancer Res.* **50:**3761–3766.

184. **Lowe, S. W., H. E. Ruley, T. Jacks, and D. E. Housman.** 1993. p53-dependent apoptosis modulates the cytotoxicity of anticancer agents. *Cell* **74:**957–967.

185. **Lowe, S. W., E. M. Schmitt, S. W. Smith, B. A. Osborne, and T. Jacks.** 1993. p53 is required for radiation-induced apoptosis in mouse thymocytes. *Nature* (London) **362:**847–849.

186. **Lu, X., and D. P. Lane.** 1993. Differential induction of transcriptionally active p53 following UV or ionizing radiation: defects in chromosome instability syndromes. *Cell* **75:**765–778.

187. **Lücke-Huhle, C.** 1982. Alpha-irradiation-induced G2 delay: a period of cell recovery. *Radiat. Res.* **89:**298–308.

188. **Madura, K., and S. Prakash.** 1986. Nucleotide sequence, transcript mapping, and regulation of the *RAD2* gene of *Saccharomyces cerevisiae*. *J. Bacteriol.* **166:**914–923.

189. **Madura, K., and S. Prakash.** 1990. The *Saccharomyces cerevisiae* DNA repair gene *RAD2* is regulated in meiosis but not during the mitotic cell cycle. *Mol. Cell. Biol.* **10:**3256–3257.

190. **Madura, K., and S. Prakash.** 1990. Transcript levels of the *Saccharomyces cerevisiae* DNA repair gene *RAD23* increase in response to UV light and in meiosis but remain constant in the mitotic cell cycle. *Nucleic Acids Res.* **18:**4737–4742.

191. **Madura, K., S. Prakash, and L. Prakash.** 1990. Expression of the *Saccharomyces cerevisiae* DNA repair gene *RAD6* that encodes a ubiquitin conjugating enzyme, increases in response to DNA damage and in meiosis but remains constant during the mitotic cell cycle. *Nucleic Acids Res.* **18:**771–778.

192. **Maga, J. A., T. A. McClanahan, and K. McEntee.** 1986. Transcriptional regulation of DNA damage responsive (*DDR*) genes in different *rad* mutant strains of *Saccharomyces cerevisiae*. *Mol. Gen. Genet.* **205:**276–284.

193. **Maga, J. A., and K. McEntee.** 1985. Response of *S. cerevisiae* to N-methyl-N'-nitro-N-nitrosoguanidine: mutagenesis, survival and *DDR* gene expression. *Mol. Gen. Genet.* **200:**313–321.

194. **Maher, V. M., K. Sato, S. Kateley-Kohler, H. Thomas, S. Michaud, J. J. McCormick, M. Kraemer, and H. J. Rahmsdorf.** 1988. Evidence of inducible error-prone repair mechanisms in diploid human fibroblasts, p. 465–471. *In* R. E. Moses and W. C. Summers (ed.), *DNA Replication and Mutagenesis*. American Society for Microbiology, Washington, D.C.

195. **Maltzman, W., and L. Czyzyk.** 1984. UV irradiation stimulates levels of p53 cellular tumor antigen in nontransformed mouse cells. *Mol. Cell. Biol.* **4:**1689–1694.

196. **Marsh, J. P., and B. T. Mossman.** 1991. Role of asbestos and active oxygen species in activation and expression of ornithine decarboxylase in hamster tracheal epithelial cells. *Cancer Res.* **51:**167–173.

197. **Martin, S. J., D. R. Green, and T. G. Cotter.** 1994. Dicing with death: dissecting the components of the apoptosis machinery. *Trends Biochem. Sci.* **19:**26–30.

198. **Matsui, M. S., and V. A. DeLeo.** 1991. Longwave ultraviolet radiation and promotion of skin cancer. *Cancer Cells* **3:**8–12.

199. **McClanahan, T., and K. McEntee.** 1984. Specific transcripts are elevated in *Saccharomyces cerevisiae* in response to DNA damage. *Mol. Cell. Biol.* **4:**2356–2363.

200. McClanahan, T., and K. McEntee. 1986. DNA damage and heat shock dually regulate genes in *Saccharomyces cerevisiae. Mol. Cell. Biol.* **6**:90–96.

201. McEntee, K., and V. A. Bradshaw. 1988. Effects of DNA damage on transcription and transposition of Ty retrotransposons in yeast. *Banbury Rep.* **30**:245–253.

202. Meister, A. 1991. Glutathione deficiency produced by inhibition of its synthesis and its reversal, applications in research and therapy. *Pharmacol. Ther.* **51**:155–194.

203. Mercer, W. E., M. T. Shields, M. Amin, G. J. Suave, E. Appella, A. J. Ullrich, and J. W. Romano. 1990. Antiproliferative effects of wild-type human p53. *J. Cell. Biochem.* **14C**:285.

204. Mercer, W. E., M. T. Shields, D. Lin, E. Appella, and S. J. Ullrich. 1991. Growth suppression induced by wild-type p53 protein is accompanied by selective down-regulation of proliferating-cell antigen expression. *Proc. Natl. Acad. Sci. USA* **88**:1958–1962.

205. Mezzina, M., and S. Nocentini. 1978. DNA ligase activity in UV-irradiated monkey kidney cells. *Nucleic Acids Res.* **5**:4317–4328.

206. Miskin, R., and R. Ben-Ishai. 1981. Induction of plasminogen activator by UV light in normal and xeroderma pigmentosum fibroblasts. *Proc. Natl. Acad. Sci. USA* **78**:6236–6240.

207. Miskin, R., and E. Reich. 1980. Plasminogen activator: induction of synthesis by DNA damage. *Cell* **19**:217–224.

208. Mitchel, R. E. J., and D. P. Morrison. 1982. Heat shock induction of ionizing radiation resistance in *Saccharomyces cerevisiae*. Transient changes in growth cycle distribution and recombinational ability. *Radiat. Res.* **92**:182–187.

209. Momand, J., G. P. Zambetti, D. C. Olson, D. George, and A. J. Levine. 1992. The *mdm-2* oncogene product forms a complex with the p53 protein and inhibits p53 mediated transactivation. *Cell* **69**:1237–1245.

210. Morawetz, C. 1987. Effect of irradiation and mutagenic chemicals on the generation of *ADH2*-constitutive mutants in yeast. Significance for the inducibility of Ty transposition. *Mutat. Res.* **177**:53–60.

211. Morawetz, C., and U. Hagen. 1990. Effect of irradiation and mutagenic chemicals on the generation of *ADH2*- and *ADH4*-constitutive mutants in yeast: the inducibility of Ty transposition by UV and ethyl methansulfonate. *Mutat. Res.* **229**:69–77.

212. Morichetti, E., E. Cundari, R. Del Carratore, and G. Bronzetti. 1989. Induction of cytochrome P-450 and catalase activity in *Saccharomyces cerevisiae* by UV and X-ray irradiation. Possible role for cytochrome P-450 in cell protection against oxidative damage. *Yeast* **5**:141–148.

213. Murnane, J. P., and J. L. Schwartz. 1993. Cell checkpoint and radiosensitivity. *Nature* (London) **365**:22.

214. Murray, A., and T. Hunt. 1993. *The Cell Cycle—An Introduction*. W. H. Freeman & Co., New York.

215. Murray, A. W. 1992. Creative blocks: cell-cycle checkpoints and feedback controls. *Nature* (London) **359**:599–604.

216. Muschel, R. J., H. B. Zhang, and W. G. McKenna. 1992. Differential effect of ionizing radiation on the expression of cyclin A and cyclin B in HeLa cells. *Cancer Res.* **53**:1128–1135.

217. Nagasawa, H., K. H. Kraemer, Y. Shiloh, and J. B. Little. 1987. Detection of ataxia telangiectasia heterozygous cell lines by postirradiation cumulative labeling index: measurements with coded samples. *Cancer Res.* **47**:398–402.

218. Nasmyth, K. 1993. Control of the yeast cell cycle by the Cdc28 protein kinase. *Curr. Opin. Cell Biol.* **5**:166–179.

219. Nebert, D. W., D. D. Petersen, and A. J. Fornace, Jr. 1990. Cellular responses to oxidative stress: the [Ah] gene battery as a paradigm. *Environ. Health Perspect.* **88**:13–25.

220. Nehls, P., van Beuningen, D., Karwowski, M. 1991. After X-irradiation a transient arrest of L929 cells in G2-phase coincides with a rapid elevation of the level of O⁶-methylguanine-DNA-transferase. *Radiat. Environ. Biophys.* **30**:21–31.

221. Nelson, W. G., and M. B. Kastan. 1994. DNA strand breaks: the DNA template alterations that trigger p53-dependent DNA damage response pathways. *Mol. Cell. Biol.* **14**:1815–1823.

222. Nishitani, H., M. Ohtsubo, K. Yamashita, H. Iida, J. Pines, H. Yasudo, Y. Shibata, T. Hunter, and T. Nishimoto. 1991. Loss of RCC1, a nuclear DNA-binding protein, uncouples the completion of DNA replication from the activation of cdc2 protein kinase and mitosis. *EMBO J.* **10**:1555–1564.

223. Noda, A., Y. Ning, S. F. Venable, O. M. Pereira-Smith, and J. R. Smith. 1994. Cloning of senescent cell-derived inhibitors of DNA synthesis using an expression screen. *Exp. Cell Res.* **211**:90–98.

224. Nurse, P. 1990. Universal control mechanism regulating onset of M-phase. *Nature* (London) **344**:503–507.

225. O'Connor, P. M., D. K. Ferris, M. Pagano, G. Draetta, J. Pines, T. Hunter, D. L. Longo, and K. W. Kohn. 1993. G2 delay induced by nitrogen mustard in human cells affects cyclin A/cdk2 and cyclin B1/cdc2-kinase complexes differently. *J. Biol. Chem.* **268**:8298–8308.

226. O'Connor, P. M., and K. W. Kohn. 1992. A fundamental role for cell cycle regulation in the chemosensitivity of cancer cells? *Semin. Cancer Biol.* **3**:409–416.

227. Osmak, M., S. Miljanic, and S. Kapitanovic. 1994. Low doses of γ-rays can induce the expression of *mdr* gene. *Mutat. Res.* **324**:35–41.

228. Owens, G. P., W. E. Hahn, and J. J. Cohen. 1991. Identification of mRNAs associated with programmed cell death in immature thymocytes. *Mol. Cell. Biol.* **11**:4177–4188.

229. Painter, R. B. 1986. Inhibition of mammalian cell DNA synthesis by ionizing radiation. *Int. J. Radiat. Biol.* **49**:771–781.

230. Painter, R. B., and B. R. Young. 1980. Radiosensitivity in ataxia telangiectasia: a new explanation. *Proc. Natl. Acad. Sci. USA* **77**:7315–7317.

231. Papathanasiou, M. A., N. C. K. Kerr, J. H. Robbins, O. W. McBride, I. Alamo, Jr., S. F. Barrett, I. D. Hickson, and A. J. Fornace, Jr. 1991. Induction by ionizing radiation of the *gadd45* gene in cultured human cells: lack of mediation by protein kinase C. *Mol. Cell. Biol.* **11**:1009–1016.

232. Pardee, A. B. 1989. G1 events and regulation of cell proliferation. *Science* **246**:603–608.

233. Patterson, M., and G. Chu. 1989. Evidence that xeroderma pigmentosum cells from complementation group E are deficient in a homolog of yeast photolyase. *Mol. Cell. Biol.* **9**:5105–5112.

234. Peak, J. G., G. E. Woloschak, and M. J. Peak. 1991. Enhanced expression of protein kinase C gene caused by solar radiation. *Photochem. Photobiol.* **53**:395–398.

235. Perry, M. E., J. Piette, J. A. Zawadzki, D. Harvey, and A. J. Levine. 1993. The mdm-2 gene is induced in response to UV light in a p53-dependent manner. *Proc. Natl. Acad. Sci. USA* **90**:11623–11627.

236. Peterson, T. A., L. Prakash, S. Prakash, M. A. Osley, and S. I. Reed. 1985. Regulation of *CDC9*, the *Saccharomyces cerevisiae* gene that encodes DNA ligase. *Mol. Cell. Biol.* **5**:226–235.

237. Pines, J. 1991. Cyclins: wheels within wheels. *Cell Growth Differ.* **2**:305–310.

238. Polakowska, R., G. Perozzi, and L. Prakash. 1986. Alkylation mutagenesis in *Saccharomyces cerevisiae*: lack of evidence for an adaptive reponse. *Curr. Genet.* **10**:647–655.

239. Price, B. D., and S. J. Park. 1994. DNA damage increases the levels of *MDM2* messenger RNA in wtp53 human cells. *Cancer Res.* **54**:896–899.

240. Prives, C. 1993. Doing the right thing: feedback control and p53. *Curr. Opin. Cell Biol.* **5**:214–218.

241. Radler-Pohl, A., C. Sachsenmaier, S. Gebel, H. P. Auer, J. T. Bruder, U. Rapp, P. Angel, H. J. Rahmsdorf, and P. Herrlich. 1993. UV-induced activation of AP-1 involves obligatory extranuclear steps including Raf-1 kinase. *EMBO J.* **12**:1005–1012.

242. Rahmsdorf, H. J., N. Harth, A. M. Eades, M. Litfin, M. Steinmetz, L. Forni, and P. Herrlich. 1986. Interferon-γ, mitomycin-

C and cycloheximide as regulatory agents of MHC class II-associated gene expression. *J. Immunol.* **136:**2293–2299.

243. Rahmsdorf, H. J., N. Koch, U. Mallick, and P. Herrlich. 1983. Regulation of MHC class II invariant chain expression: induction of synthesis in human and murine plasmocytoma cells by arresting replication. *EMBO J.* **2:**811–816.

244. Rahmsdorf, H. J., U. Mallick, H. Ponta, and P. Herrlich. 1982. A B-lymphocyte-specific high turnover protein: constitutive expression in resting B-cells and induction of synthesis in proliferating cells. *Cell* **29:**459–468.

245. Rathmell, W. K., and G. Chu. 1994. Involvement of the Ku autoantigen in the cellular response to DNA double-strand breaks. *Proc. Natl. Acad. Sci. USA* **91:**7623–7627.

246. Reed, S. I. 1992. The role of p34 kinases in the G1- to S-phase transition. *Annu. Rev. Cell Biol.* **8:**529–561.

247. Robinson, G. W., C. M. Nicolet, D. Kalainov, and E. C. Friedberg. 1986. A yeast excision repair gene is inducible by DNA damaging agents. *Proc. Natl. Acad. Sci. USA* **83:**1842–1846.

248. Robson, C. N., and I. D. Hickson. 1991. Isolation of cDNA clones encoding a human apurinic/apyrimidinic endonuclease that corrects DNA repair and mutagenesis defects in *E. coli xth* (exonuclease III) mutants. *Nucleic Acids Res.* **19:**5519–5523.

249. Robson, T., H. Grindley, A. Hall, J. Vormoor, and H. Lohrer. 1994. Increased DNA-repair capacity and the modulation of 2 proteins in an metallothionein overexpressing Chinese hamster cell line. *Mutat. Res.* **314:**143–157.

250. Rolfe, M. 1985. UV-inducible proteins in *Saccharomyces cerevisiae*. *Curr. Genet.* **9:**529–532.

251. Rolfe, M. 1985. UV-inducible transcripts in *Saccharomyces cerevisiae*. *Curr. Genet.* **9:**533–538.

252. Rolfe, M., A. Spanos, and G. Banks. 1986. Induction of yeast Ty element transcription by ultraviolet light. *Nature* (London) **319:**339–340.

253. Romerdahl, C. A., L. C. Stephens, C. Bucarra, and M. L. Kripke. 1989. The role of ultraviolet radiation in the induction of melanocytic skin tumors in inbred mice. *Cancer Commun.* **1:**209–216.

254. Ronai, Z. A., E. Okin, and I. B. Weinstein. 1988. Ultraviolet light induces the expression of oncogenes in rat fibroblasts and human keratinocyte cells. *Oncogene* **2:**201–204.

255. Rosen, C. F., D. Gajic, and D. J. Drucker. 1990. UV radiation induction of ornithine decarboxylase in rat keratinocytes. *Cancer Res.* **50:**2631–2635.

256. Rotem, N., J. H. Axelrod, and R. Miskin. 1987. Induction of urokinase-type plasminogen activator by UV light in human fetal fibroblasts is mediated through a UV-induced secreted protein. *Mol. Cell. Biol.* **7:**622–631.

257. Rowley, R., S. Subramani, and P. G. Young. 1992. Checkpoint controls in *Schizosaccharomyces pombe: rad1. EMBO J.* **11:**1335–1342.

258. Ruby, S. W., and J. W. Szostak. 1985. Specific *Saccharomyces cerevisiae* genes are expressed in response to DNA-damaging agents. *Mol. Cell. Biol.* **5:**75–84.

259. Ruby, S. W., J. W. Szostak, and A. W. Murray. 1983. Cloning regulated yeast genes from a pool of *lacZ* fusions. *Methods Enzymol.* **101:**253–269.

260. Rudolph, N. S., and S. A. Latt. 1989. Flow cytometric analysis of X-ray sensitivity in ataxia telangiectasia. *Mutat. Res.* **211:**31–41.

261. Sachsenmaier, C., A. Radler-Pohl, A. Müller, P. Herrlich, and H. J. Rahmsdorf. 1994. Damage to DNA by UV light and activation of transcription factors. *Biochem. Pharmacol.* **47:**129–136.

262. Sadaie, M. R., E. Tschachler, K. Valerie, M. Rosenberg, B. K. Felber, G. N. Pavlakis, M. E. Klotman, and F. Wong-Staal. 1990. Activation of *tat*-defective human immunodeficiency virus by ultraviolet light. *New Biol.* **2:**479–486.

263. Saka, Y., and M. Yanagida. 1993. Fission yeast *cut5*⁺, required for S phase onset and M phase restraint, is identical to the radiation-damage repair gene *rad4*⁺. *Cell* **74:**383–393.

264. Sandell, L. L., and V. A. Zakian. 1993. Loss of a yeast telomere: arrest, recovery, and chromosome loss. *Cell* **75:**729–739.

265. Sarnow, P., Y. S. Ho, J. Williams, and A. J. Levine. 1982. Adenovirus E1B-58Kd tumor antigen and SV40 large tumor antigen are physically associated with the same 54Kd cellular protein in transformed cells. *Cell* **28:**387–394.

266. Sazer, S., and P. Nurse. 1994. The fission yeast RCC1-related protein is required for the mitosis to interphase transition. *EMBO J.* **13:**606–615.

267. Scheffner, M., J. M. Hulbregtse, R. D. Vierstra, and P. M. Howley. 1993. The HPV-16 E6 and E6-AP complex functions as a ubiquitin-protein ligase in the ubiquitination of p53. *Cell* **75:**495–505.

268. Schiestl, R. H., P. Reynolds, S. Prakash, and L. Prakash. 1989. Cloning and sequence analysis of the *Saccharomyces cerevisiae RAD9* gene and further evidence that its product is required for cell cycle arrest induced by DNA damage. *Mol. Cell. Biol.* **9:**1882–1896.

269. Schorpp, M., U. Mallick, H. J. Rahmsdorf, and P. Herrlich. 1984. UV-induced extracellular factor from human fibroblasts communicates the UV response to nonirradiated cells. *Cell* **37:**861–868.

270. Schreck, R., P. Rieber, and P. A. Baeuerle. 1991. Reactive oxygen intermediates as apparently widely used messenger in the activation of the NFκB transcription factor and HIV-1. *EMBO J.* **10:**2247–2258.

271. Schwencke, J., and E. Moustacchi. 1982. Proteolytic activities in yeast after UV irradiation. I. Variation in proteinase levels in repair proficient *RAD*⁺ strains. *Mol. Gen. Genet.* **185:**290–295.

272. Schwencke, J., and E. Moustacchi. 1982. Proteolytic activities in yeast after UV irradiation. II. Variation in proteinase levels in mutants blocked in DNA-repair pathways. *Mol. Gen. Genet.* **185:**296–301.

273. Sebastian, J., B. Kraus, and G. B. Sancar. 1990. Expression of the yeast *PHR1* gene is induced by DNA-damaging agents. *Mol. Cell. Biol.* **10:**4630–4637.

274. Sebastian, J., and G. Sancar. 1991. A damage-responsive DNA binding protein regulates transcription of the yeast DNA repair gene *PHR1. Proc. Natl. Acad. Sci. USA* **88:**11251–11255.

275. Seki, S., M. Hatsushika, S. Watanabe, K. Akiyama, K. Nagao, and K. Tsutsui. 1992. cDNA cloning, sequencing, expression and possible domain structure of human APEX nuclease homologous to *Escherichia coli* exonuclease III. *Biochim. Biophys. Acta* **1131:**287–299.

276. Sentman, C. L., J. R. Shutter, D. Hockenbery, O. Kanagawa, and S. J. Korsmeyer. 1991. bcl-2 inhibits multiple forms of apoptosis but not negative selection in thymocytes. *Cell* **67:**879–888.

277. Serrano, M., G. J. Hannon, and D. Beach. 1993. A new regulatory motif in cell-cycle control causing specific inhibition of cyclin D/CDK4. *Nature* (London) **366:**704–707.

278. Shah, G., R. Ghosh, P. A. Amstad, and P. A. Cerutti. 1993. Mechanism of induction of c-fos by ultraviolet B (290-320 nm) in mouse JB6 epidermal cells. *Cancer Res.* **53:**38–45.

279. Sheldrick, K. S., and A. M. Carr. 1993. Feedback controls and G2 checkpoints: fission yeast as a model system. *Bioessays* **15:**775–782.

280. Sheng, S., and S. M. Schuster. 1993. Purification and characterization of *Saccharomyces cerevisiae* DNA damage-responsive protein 48 (DDRP 48). *J. Biol. Chem.* **268:**4752–4758.

281. Sherman, M. L., R. Datta, D. E. Hallahan, R. R. Weichselbaum, and D. W. Kufe. 1990. Ionizing radiation regulates expression of the c-*jun* protooncogene. *Proc. Natl. Acad. Sci. USA* **87:**5663–5666.

282. Sherr, C. J. 1993. Mammalian G1 cyclins. *Cell* **73:**1059–1065.

283. Shinohara, A., H. Ogawa, and T. Ogawa. 1992. Rad51 protein involved in repair and recombination in *S. cerevisiae* is a RecA-like protein. *Cell* **69:**457–470.

284. Siede, W. 1988. The *RAD6* gene of yeast: a link between DNA repair, chromosome structure and protein degradation. *Radiat. Environ. Biophys.* **27:**277–286.

285. Siede, W., A. S. Friedberg, I. Dianova, and E. C. Friedberg. 1994. Characterization of G₁ checkpoint control in the yeast *Sac-*

charomyces cerevisiae following exposure to DNA-damaging agents. *Genetics* **138**:271–281.

286. Siede, W., A. S. Friedberg, and E. C. Friedberg. 1993. *RAD9*-dependent G1 arrest defines a second checkpoint for damaged DNA in the cell cycle of *Saccharomyces cerevisiae*. *Proc. Natl. Acad. Sci. USA* **90**:7985–7989.

287. Siede, W., and E. C. Friedberg. 1992. Regulation of the yeast *RAD2* gene: DNA damage-dependent induction correlates with protein binding to regulatory sequences and their deletion influences survival. *Mol. Gen. Genet.* **232**:247–256.

288. Siede, W., G. W. Robinson, D. Kalainov, T. Malley, and E. C. Friedberg. 1989. Regulation of the *RAD2* gene of *Saccharomyces cerevisiae*. *Mol. Microbiol.* **3**:1697–1707.

289. Simon, M. M., Y. Aragane, A. Schwarz, T. A. Luger, and T. Schwarz. 1994. UVB light induces nuclear factor κB (NFκB) activity independently from chromosomal DNA damage in cell-free cytosolic extracts. *J. Invest. Dermatol.* **102**:422–427.

290. Singh, P., and M. F. Lavin. 1990. DNA-binding protein activated by gamma radiation in human cells. *Mol. Cell. Biol.* **10**:5279–5285.

291. Slichenmyer, W. J., W. G. Nelson, R. J. Slebos, and M. B. Kastan. 1993. Loss of a *p53*-associated G1 checkpoint does not decrease cell survival following DNA damage. *Cancer Res.* **53**:4164–4168.

291a. Smith, M. L., I.-T. Chen, Q. Zhan, I. Bae, C.-Y. Chen, T. M. Gilmer, M. B. Kastan, P. M. O'Connor, and A. J. Fornace, Jr. Interaction of the p53-regulated protein Gadd45 with proliferating cell nuclear antigen. *Science*, in press.

292. Smythe, C., and J. W. Newport. 1992. Coupling of mitosis to the completion of S phase in *Xenopus* occurs via modulation of the tyrosine kinase that phosphorylates p34*cdc2*. *Cell* **68**:787–797.

293. Sorger, P. K., and A. W. Murray. 1992. S-phase feedback control in budding yeast independent of tyrosine phosphorylation of p34*cdc28*. *Nature* (London) **355**:365–368.

294. Srivastava, S., Y. Tong, K. Devadas, Z.-Q. Zou, V. Sykes, Y. Chen, W. Blattner, K. Pirollo, and E. Chang. 1992. Detection of both mutant and wild-type p53 proteins in normal skin fibroblasts and demonstration of a shared "second hit" on p53 in diverse tumors from a cancer-prone family with Li-Fraumeni syndrome. *Oncogene* **7**:987–991.

295. Srivastava, S., Z. Zou, K. Pirollo, W. Blattner, and E. H. Chang. 1990. Germ-line transmission of a mutated *p53* in a cancer-prone family with Li-Fraumeni syndrome. *Nature* (London) **348**:747–749.

296. Staal, F. J. T., M. Roederer, L. A. Herzenberg, and L. A. Herzenberg. 1990. Intracellular thiols regulate activation of nuclear factor κB and transcription of human immunodeficiency virus. *Proc. Natl. Acad. Sci. USA* **87**:9943–9947.

297. Stein, B., P. Angel, H. van Dam, H. Ponta, P. Herrlich, A. van der Eb, and H. J. Rahmsdorf. 1992. Ultraviolet-radiation induced *c-jun* gene transcription: two AP-1 like binding sites mediate the response. *Photochem. Photobiol.* **55**:409–415.

298. Stein, B., H. J. Rahmsdorf, A. Steffen, M. Litfin, and P. Herrlich. 1989. UV-induced DNA damage is an intermediate step in UV-induced expression of human immunodeficiency virus type 1, collagenase, c-fos, and metallothionein. *Mol. Cell. Biol.* **9**:5169–5181.

299. Struhl, K. 1987. The DNA-binding domains of the jun oncoprotein and the yeast *GCN4* transcriptional activator protein are functionally homologous. *Cell* **50**:841–846.

300. Stueland, C. S., D. J. Lew, M. J. Cismowski, and S. I. Reed. 1993. Full activation of p34*cdc28* histone H1 kinase activity is unable to promote entry into mitosis in checkpoint-arrested cells of the yeast *Saccharomyces cerevisiae*. *Mol. Cell. Biol.* **13**:3744–3755.

301. Sullivan, N. F., and A. E. Willis. 1989. Elevation of c-myc protein by DNA strand breakage. *Oncogene* **4**:1497–1502.

301a. Taccioli, G. E., T. M. Gottlieb, T. Blunt, A. Priestley, J. Demengeot, R. Mizuta, A. R. Lehmann, F. W. Alt, S. P. Jackson, and P. A. Jeggo. 1994. Ku80: product of the *XRCC5* gene and its role in DNA repair and V(D)J recombination. *Science* **265**:1442–1445.

302. Takao, M., M. Abramić, M. Moos, Jr., V. R. Otrin, J. C. Wootton, M. McLenigan, A. S. Levine, and M. Protić. 1993. A 127 kDa component of a UV-damaged DNA-binding complex, which is defective in some xeroderma pigmentosum group E patients, is homologous to a slime mold protein. *Nucleic Acids Res.* **21**:4111–4118.

303. Teale, B., and M. Lavin. 1994. A specific DNA-binding protein activated by ionizing radiation in normal cells and constitutively present in ataxia telangiectasia cells. *Radiat. Res.* **138**:S52–S55.

304. Thompson, L. H., K. W. Brookman, J. J. Jones, S. A. Allen, and A. V. Carrano. 1990. Molecular cloning of the human *XRCC1* gene, which corrects defective DNA repair and sister chromatid exchange. *Mol. Cell. Biol.* **10**:6160–6171.

305. Thorne, L. W., and B. Byers. 1993. Stage-specific effects of X-irradiation on yeast meiosis. *Genetics* **134**:29–42.

306. Tlsty, T. D., B. Margolin, and K. Lum. 1989. Differences in the rates of gene amplification in nontumorigenic and tumorigenic cell lines as measured by Luria-Delbrück fluctuation analysis. *Proc. Natl. Acad. Sci. USA* **86**:9441–9445.

307. Tobey, R. A. 1975. Different drugs arrest cells at a number of distinct stages in G2. *Nature* (London) **254**:245–247.

308. Tomei, L. D., and F. O. Cope (ed). 1991. Apoptosis: the molecular basis of cell death. *Curr. Commun. Cell. Mol. Biol.* **3**:1–321.

309. Treger, J. M., K. A. Heichman, and K. McEntee. 1988. Expression of the yeast *UBI4* gene increases in response to DNA-damaging agents and in meiosis. *Mol. Cell. Biol.* **8**:1132–1136.

310. Tyrrell, R. M. 1991. UVA (320-380 nm) radiation as an oxidative stress, p. 57–83. *In* H. Sies (ed.), *Oxidative Stress: Oxidants and Antioxidants*. Academic Press, Ltd., London.

311. Valerie, K., A. Delers, C. Bruck, C. Thiriart, H. Rosenberg, C. Debouck, and M. Rosenberg. 1988. Activation of human immunodeficiency virus type 1 by DNA damage in human cells. *Nature* (London) **333**:78–81.

312. van Dam, H., M. Duyndam, R. Rottier, A. Bosch, L. de Vries-Smits, P. Herrlich, and A. J. van der Eb. 1993. Heterodimer formation of cJun and ATF-2 is responsible for induction of *c-jun* by the 243 amino acid adenovirus E1A protein. *EMBO J.* **12**:479–487.

313. Vogelstein, B., and K. W. Kinzler. 1992. p53 function and dysfunction. *Cell* **70**:523–526.

314. Waga, S., G. J. Hannon, D. Beach, and B. Stillman. 1994. The p21 inhibitor of cyclin-dependent kinases controls DNA replication by interaction with PCNA. *Nature* (London) **369**:574–578.

315. Walker, A. I., T. Hunt, R. J. Jackson, and C. W. Anderson. 1985. Double-stranded DNA induces the phosphorylation of several proteins including the 90 000 mol. wt. heat-shock protein in animal cell extracts. *EMBO J.* **4**:139–145.

316. Walker, L. J., C. N. Robson, E. Black, D. Gillespie, and I. D. Hickson. 1993. Identification of residues in the human DNA repair enzyme HAP1 (Ref-1) that are essential for redox regulation of Jun DNA binding. *Mol. Cell. Biol.* **13**:5370–5376.

317. Walworth, N., S. Davey, and D. Beach. 1993. Fission yeast *chk1* protein kinase links the *rad* checkpoint pathway to *cdc2*. *Nature* (London) **363**:368–371.

318. Wang, Z., X. Wu, and E. C. Friedberg. 1992. Excision repair of DNA in nuclear extracts from the yeast *Saccharomyces cerevisiae*. *Biochemistry* **31**:3694–3702.

319. Ward, J. F. 1994. DNA damage as the cause of ionizing radiation-induced gene activation. *Radiat. Res.* **138**:S85–S88.

320. Waters, R., R. Zhang, and N. J. Jones. 1993. Inducible removal of UV-induced pyrimidine dimers from transcriptionally active and inactive genes of *Saccharomyces cerevisiae*. *Mol. Gen. Genet.* **239**:28–32.

321. Weber, L., and B. Byers. 1992. A *RAD9*-dependent checkpoint blocks meiosis of *cdc13* yeast cells. *Genetics* **131**:55–63.

322. Weichselbaum, R. R., D. E. Hallahan, V. Sukhatme, A. Dritschilo, M. L. Sherman, and D. W. Kufe. 1991. Biological consequences of gene regulation after ionizing radiation exposure. *J. Natl. Cancer Inst.* **83**:480–484.

323. **Weinert, T., and D. Lydall.** 1993. Cell cycle checkpoints, genetic instability and cancer. *Semin. Cancer Biol.* **4:**129–140.

324. **Weinert, T. A.** 1992. Dual cell cycle checkpoints sensitive to chromosome replication and DNA damage in the budding yeast *Saccharomyces cerevisiae. Radiat. Res.* **132:**141–143.

325. **Weinert, T. A., and L. H. Hartwell.** 1988. The *RAD9* gene controls the cell cycle response to DNA damage in *Saccharomyces cerevisiae. Science* **241:**317–322.

326. **Weinert, T. A., and L. H. Hartwell.** 1990. Characterization of the *RAD9* gene of *Saccharomyces cerevisiae* and evidence that it acts posttranslationally in cell cycle arrest after DNA damage. *Mol. Cell. Biol.* **10:**6554–6564.

327. **Weinert, T. A., and L. H. Hartwell.** 1993. Cell cycle arrest of *cdc* mutants and specificity of the *RAD9* checkpoint. *Genetics* **134:**63–80.

328. **Weinert, T. A., G. L. Kiser, and L. H. Hartwell.** 1994. Mitotic checkpoint genes in budding yeast and the dependence of mitosis on DNA replication and repair. *Genes Dev.* **8:**652–665.

329. **Werness, B. A., A. J. Levine, and P. M. Howley.** 1990. Association of human papillomavirus types 16 and 18 E6 proteins with p53. *Science* **248:**76–79.

330. **White, E.** 1993. Death-defying acts: a meeting review on apoptosis. *Genes Dev.* **7:**2277–2284.

331. **White, J. H. M., S. R. Green, D. G. Barker, L. B. Dumas, and L. H. Johnston.** 1987. The *CDC8* transcript is cell cycle regulated in yeast and is expressed coordinately with *CDC9* and *CDC21* at a point preceding histone transcription. *Exp. Cell Res.* **171:**223–231.

332. **White, J. M. H., A. L. Johnson, N. F. Lowndes, and L. H. Johnston.** 1991. The yeast DNA ligase gene *CDC9* is controlled by six orientation specific upstream activating sequences that respond to cellular proliferation but which alone cannot mediate cell cycle regulation. *Nucleic Acids Res.* **19:**359–364.

333. **Williams, G. T., and C. A. Smith.** 1993. Molecular regulation of apoptosis: genetic controls on cell death. *Cell* **74:**777–779.

334. **Woloschak, G. E., and C. M. Chang-Lui.** 1990. Differential modulation of specific gene expression following high- and low-LET radiations. *Radiat. Res.* **124:**183–187.

335. **Woloschak, G. E., C. M. Chang-Lui, and P. Shearin-Jones.** 1990. Regulation of protein kinase C by ionizing radiation. *Cancer Res.* **50:**3963–3967.

336. **Woloschak, G. E., C. M. Chang-Lui, P. Shearin-Jones, and C. A. Jones.** 1990. Modulation of gene expression in Syrian hamster embryo cells following ionizing radiation. *Cancer Res.* **50:**339–344.

337. **Woloschak, G. E., P. Shearin-Jones, and C. M. Chang-Lui.** 1990. Effects of ionizing radiation on expression of genes encoding cytoskeletal elements: kinetics and dose effects. *Mol. Carcinog.* **3:**374–378.

338. **Wolter, R., W. Siede, and M. Brendel.** Personal communication.

339. **Xanthoudakis, S., G. Miao, F. Wang, Y. E. Pan, and T. Curran.** 1992. Redox activation of Fos-Jun DNA binding activity is mediated by a DNA repair enzyme. *EMBO J.* **11:**3323–3335.

340. **Xiao, W., and L. Samson.** 1992. The *Saccharomyces cerevisiae MGT1* DNA repair methyltransferase gene—its promoter and entire coding sequence, regulation and *in vivo* biological functions. *Nucleic Acids Res.* **20:**3599–3606.

341. **Xiao, W., K. K. Singh, B. Chen, and L. Samson.** 1993. A common element involved in transcriptional regulation of two DNA alkylation repair genes (*MAG* and *MGT1*) of *Saccharomyces cerevisiae. Mol. Cell. Biol.* **13:**7213–7221.

342. **Xiong, Y., G. J. Hannon, H. Zhang, D. Casso, R. Kobayashi, and D. Beach.** 1993. p21 is a universal inhibitor of cyclin kinases. *Nature* (London) **366:**701–704.

343. **Xiong, Y., H. Zhang, and D. Beach.** 1992. D type cyclins associate with multiple protein kinases and the DNA replication and repair factor PCNA. *Cell* **71:**505–514.

344. **Yagle, K., and K. McEntee.** 1990. The DNA damage-inducible gene *DIN1* of *Saccharomyces cerevisiae* encodes a regulatory subunit of ribonucleotide reductase and is identical to *RNR3. Mol. Cell. Biol.* **10:**5553–5557.

345. **Yamada, M., and T. T. Puck.** 1961. Action of radiation on mammalian cells, IV. Reversible mitotic lag in the S3 HeLa cells produced by low doses of X-rays. *Proc. Natl. Acad. Sci. USA* **47:**1181–1191.

346. **Yang, X. Y., Z. A. Ronai, R. M. Santella, and I. B. Weinstein.** 1988. Effects of 8-methoxypsoralen and ultraviolet light A on EGF receptor (HER-1) expression. *Biochem. Biophys. Res. Commun.* **157:**590–596.

347. **Yin, Y., M. A. Tainsky, F. Z. Bischoff, L. C. Strong, and G. M. Wahl.** 1992. Wild-type p53 restores cell cycle control and inhibits gene amplification in cells with mutant p53 alleles. *Cell* **70:**937–948.

348. **Yonish-Rouach, E., D. Resnitzky, J. Lotem, L. Sachs, A. Kimichi, and M. Oren.** 1991. Wild type p53 induces apoptosis of myeloid leukemic cells that is inhibited by interleukin-6. *Nature* (London) **352:**345–347.

349. **Zambetti, G. P., and A. J. Levine.** 1993. A comparison of the biological activities of wild-type and mutant p53. *FASEB J.* **7:**855–865.

350. **Zambetti-Bosseler, F., and D. Scott.** 1981. Cell death, chromosome damage and mitotic delay in normal human, ataxia telangiectasia, and retinoblastoma fibroblasts after X-irradiation. *Int. J. Radiat. Biol.* **39:**547–558.

351. **Zhan, Q., F. Carrier, and A. J. Fornace, Jr.** 1993. Induction of cellular p53 activity by DNA-damaging agents and growth arrest. *Mol. Cell. Biol.* **13:**4242–4250.

352. **Zhang, H., Y. Xiong, and D. Beach.** 1993. Proliferating cell nuclear antigen and p21 are components of multiple cell cycle kinase complexes. *Mol. Biol. Cell* **4:**897–906.

353. **Zheng, P., D. S. Fay, J. Burton, H. Xiao, J. L. Pinkham, and D. F. Stern.** 1993. *SPK1* is an essential S-phase-specific gene of *Saccharomyces cerevisiae* that encodes a nuclear serine/threonine/tyrosine kinase. *Mol. Cell. Biol.* **13:**5829–5842.

354. **Zhou, Z., and S. J. Elledge.** 1992. Isolation of *crt* mutants constitutive for transcription of the DNA damage inducible *RNR* in *Saccharomyces cerevisiae. Genetics* **131:**851–866.

355. **Zhou, Z., and S. J. Elledge.** 1993. *DUN1* encodes a protein kinase that controls the DNA damage response in yeast. *Cell* **75:**1119–1127.

356. **Zmudzka, B. Z., and J. Z. Beer.** 1990. Activation of human immunodeficiency virus by ultraviolet radiation. *Photochem. Photobiol.* **52:**1153–1162.

14

Human Hereditary Diseases with Defective Processing of DNA Damage

The study of DNA repair in human cells was limited for many years by the lack of available mutant cells defective in their response to DNA damage. In 1968 James Cleaver reported that fibroblasts in culture derived from the skin of a human patient with the disease *xeroderma pigmentosum* (XP) were unable to carry out nucleotide excision repair following exposure to UV radiation (55, 56). These observations were confirmed by Richard Setlow and his colleagues (364). This was the first indication of a DNA repair defect associated with a hereditary human disease. Since then, an enormous amount of attention has been focused on the phenotypic responses of individuals with XP, and particularly of XP cells in culture, to DNA damage by a variety of chemical and physical agents.

Cleaver's observations on XP cells provided an impetus to examine the response to DNA-damaging agents of cells from a number of other hereditary human diseases, particularly those associated with chromosomal abnormalities or with an abnormally raised incidence of neoplasia. In addition, the observation that human subjects suffering from XP are highly susceptible to malignant neoplasms caused by well-characterized mutagens and carcinogens prompted further exploration of the relation between DNA damage, mutagenesis, and neoplastic transformation in a more general sense. This chapter deals with a number of selected hereditary diseases in which there is evidence for defects in some aspect of DNA metabolism following DNA damage by physical and/or chemical agents, although not necessarily in the repair of such damage.

There is little doubt that the disease XP is characterized primarily by defective or deficient nucleotide excision repair of DNA (see below). Defective DNA repair as the primary basis for the other human diseases considered here is much less certain, however. In this regard it should be noted that in general, the demonstration of abnormal sensitivity of human (or other) cells to killing by physical and/or chemical agents known to interact with DNA does not necessarily imply a defect in DNA repair. The persistence of damage to DNA may, for example, result from abnormal DNA replication. Indeed, as will be seen below, this has been suggested as a possible basis for the abnormal sensitivity of ataxia telangiectasia (AT) cells to ionizing radiation. In addition, some cells may sustain greater levels of DNA damage than normal cells following an equivalent insult. A general mechanism by which this could occur is by defects in active transport that allow access of higher levels of certain chemicals to the nucleus. Another possible mechanism could involve qualitative and/or quantitative disturbances in the metabolic activation of nonpolar compounds to forms that are more reactive with DNA (see chapter 1). For example, some agents have a particular avidity for mitochondria because of their marked lipophilicity (18). Such agents might cause selective or preferential damage to mitochondrial rather than nuclear DNA, interfering with normal mitochondrial functions and thereby altering cell survival. Finally, there are indications that some human diseases may be characterized by the abnormal production of metabolic products that damage DNA in the absence of exogenous damaging agents and also create heightened sensitivity to DNA damage in their presence (121).

Hereditary Diseases Characterized by Defective DNA Repair

Xeroderma Pigmentosum

This chapter presents the primary discussion of XP as a clinical and biological entity. The reader is particularly referred to chapter 8 and also to chapter 7, which present several aspects of the biochemistry and molecular pathology of XP in the context of the process of nucleotide excision repair in human cells.

CLINICAL FEATURES

XP has been extensively reviewed in the literature. Pertinent citations prior to 1984 are provided in chapter 9 of *DNA Repair*, the predecessor to this book (102). More-recent reviews are included here (31, 32, 58, 62, 109, 130, 142, 194, 198, 199, 202, 299, 317, 418, 451). XP is clinically characterized chiefly by the early onset of severe photosensitivity of the exposed regions of the skin (Color Plate 9), eyes, and tongue; a very high incidence of skin cancers; and frequent neurological abnormalities. The term *xeroderma*, meaning parchment skin, was coined by Hebra and Kaposi in 1870 (148) on the basis of clinical observations of a patient with the disease. The term *pigmentosum* was added later to emphasize the dramatic pigmentary disturbances that characterize this disease (178). An association of the cutaneous manifestations of XP with neurological features was first noted by Albert Neisser (272). A particular syndrome of neurological defects in association with the dermatological features of XP was first described by De Sanctis and Cacchione and is referred to as the *De Sanctis-Cacchione syndrome* (79). However, it should be noted that neurological abnormalities frequently accompany the dermatological manifestations of XP (see below) and not all such abnormalities qualify as examples of this particular syndrome.

Among the ocular tissues, the eyelids, conjunctiva, and cornea are the most frequently affected (Color Plate 9), because they are subjected to substantial UV radiation exposure (198). Of a total of 337 patients investigated, only 9 were found to be free of ocular complications (198). Oral abnormalities are also frequently encountered for similar reasons. Indeed, the incidence of squamous cell carcinoma of the tip of tongue in individuals with XP is more than 20,000 times that in the general population (198). There are sporadic reports of internal neoplasms and occasional reports of abnormalities of the immune system in XP (198). It has also been suggested that a case of XP associated with bone marrow failure may reflect an increased susceptibility of hematopoietic stem cells to endogenous DNA-damaging agents (331).

XP has a worldwide distribution, but the incidence varies from about 1 in 250,000 in Europe and the United States to as high as 1 in 40,000 in Japan (318, 399). The reported incidence in Egypt is also very high (62). In a comprehensive review of 830 cases of XP culled from the literature between 1874 and 1982, the mean patient age at the time of diagnosis was 12 years, with approximately equal numbers of affected males and females (198). Cutaneous symptoms such as Sun sensitivity or freckling were noted as early as 1 to 2 years after birth (Color Plate 9) (198). Consanguinity in the parents of affected individuals was associated with ~30% of cases, and nearly 20% of patients had neurological abnormalities. In all known cases the disease is inherited through an autosomal recessive mode.

Patients with XP have a very high incidence of premalignant actinic keratoses, as well as benign and malignant skin tumors. The median age of onset of skin cancer is 8 years, nearly 50 years younger than in the general U.S. population (Fig. 14–1). It has been noted that this might be the largest reduction in age of onset of neoplasia documented for any recessive human hereditary disease (62). The neoplasms are predominantly basal cell and squamous cell carcinomas, but they also include melanomas, keratoacanthomas, angiomas, and sarcomas (198, 418). Other Sun-exposed areas of the body are also severely predisposed to neoplastic change, especially the conjunctivae and, as noted above, even the tip of the tongue.

XP is typically uncomplicated by other hereditary diseases. However, the disease is sometimes accompanied by the clinical features of another genetic disease called Cockayne's syndrome (CS). The clinical features of this disease and the genetic and molecular basis for its association with XP are discussed below. An additional aspect of the genetic complexity of XP derives from the recognition that some individuals carrying mutations in at least two genes that are known to be involved in XP can present with the clinical features of a third disease called trichothiodystrophy (TTD), which appears to be quite distinct from XP at the clinical level. This conundrum will also be addressed below.

PHENOTYPES OF XP CELLS

A vast array of biological and biochemical phenotypes have been found in XP cells exposed to chemical or radiation damage (Table 14–1). Cultured cells from most patients with XP have a normal karyotype and a normal frequency of spontaneous *sister chromatid exchanges* (SCEs). However, the frequency of these and of other chromosomal abnormalities is increased in most XP cells exposed to UV radiation or UV-mimetic chemicals (2, 45, 82, 278, 293, 334, 458). Indeed, one study has reported increased levels of SCEs and micronucleation in XP heterozygotes compared with normal individuals, suggesting a possible phenotype for diagnosing heterozygous individuals (27). Chromosomal abnormalities are not common to all forms of XP, however. In the so-called variant form of the disease and in the XP-E genetic complementation group, both of which are characterized at the cellular level by undetectable or mild defects in nucleotide excision repair (see below), no chromosomal abnormalities were detected in one study (359). These observations suggest that the marked predisposition to cancer is not causally related to chromosomal aberrations per se (359). Nonetheless, it has been suggested that cells from all individuals with defective cellular responses to DNA damage that are associated with cancer proneness, including but not limited to XP, manifest a persistence of chromatid breaks and gaps after irradiation with X rays or near-UV visible light, specifically in the G_2 phase of the cell cycle (310). Cells from patients in XP genetic complementation groups A and D (see below) apparently do not manifest these abnormalities.

One of the most consistent cellular phenotypes of XP cells is an increased sensitivity to killing following exposure to a wide variety of DNA-damaging agents, including UV radiation, the agent most frequently used experimentally (418) (Fig. 14–2). As noted below, not all XP cells are equally sensitive to killing by these agents (Fig. 14–2), reflecting differences between cells from different genetic complementation groups and within individual complementation groups. This phenotype of increased UV radiation sensitivity led to the hypothesis that XP cells are defective in nucleotide excision repair of DNA, a hypothesis that has since been extensively substantiated by studies at the biochemical level (see below).

Consistent with their sensitivity to killing resulting from defective DNA re-

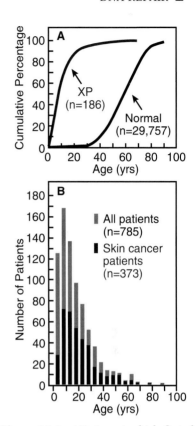

Figure 14–1 (A) Age at which first skin cancer was reported for 186 individuals with XP compared with the age distribution for over 29,000 patients without XP but with either basal or squamous cell carcinoma in the U.S. general population. (B) Age distribution of patients with XP. The age at last clinical observation is shown for 785 patients, 373 of whom were also reported to have skin cancer. (*Adapted from Cleaver and Kraemer [62] and Kraemer et al. [197] with permission.*)

Table 14–1 Clinical features observed in 830 cases of xeroderma pigmentosum

Feature	Value or frequency (%)[a]
Cutaneous abnormalities	
Median age of onset of symptoms	1.5 yr
Median age of onset of freckling	1.5 yr
Photosensitivity	19%
Cutaneous atrophy	23%
Cutaneous telangiectasia	17%
Actinic keratoses	19%
Malignant skin neoplasms	45%
Median age of first cutaneous neoplasm	8 yr
Ocular abnormalities	
Frequency	40%
Median age of onset	4 yr
Conjunctival injection	18%
Corneal abnormalities	17%
Impaired vision	12%
Photophobia	2%
Ocular neoplasms	11%
Median age of first ocular neoplasm	11 yr
Neurological abnormalities	
Median age of onset	6 mo
Association with skin problems	33%
Association with ocular abnormalities	36%
Low intelligence	80%
Abnormal motor activity	30%
Areflexia	20%
Impaired hearing	18%
Abnormal speech	13%
Abnormal EEG	11%
Microcephaly	24%
Other abnormalities in association with neurological defects	
Slow growth	23%
Delayed secondary sexual development	12%

[a] Frequencies are given as percentages; values are accompanied by units.
[b] EEG, electroencephalogram.

Source: Adapted from Kraemer et al. (198) with permission.

pair, XP cells have an increased frequency of mutations following exposure to DNA damage (78, 133, 242, 243, 302) (Fig. 14–3). Additionally, there are indications that XP cells are more readily transformed to anchorage independence following exposure to carcinogens than are normal cells (249). Mutagenesis in XP cells has been explored at the molecular level by using shuttle vector plasmids that are capable of autonomous replication in both mammalian cells and *E. coli* and that carry suitable target genes for measuring both the frequency and spectrum of mutations following exposure of transfected cells to DNA damage (Table 14–2). Following replication in XP cells, there are significantly fewer plasmids with multiple base substitutions and with single or tandem transversion mutations (Table 14–2). In both cell types the major base substitution mutation is the $G \cdot C \rightarrow A \cdot T$ transition. Thus, the major photoproduct in DNA, the T<>T dimer, is not the major source of mutations (62). This observation is consistent with the observation that replicational polymerases apparently prefer to insert adenine opposite sites of base damage during translesion DNA synthesis (62).

The capacity of XP cells to repair DNA damage can also be assessed by using transfected damaged plasmid DNA molecules as substrates, recovering the plasmid molecules from the cells after some designated time, and then monitoring the

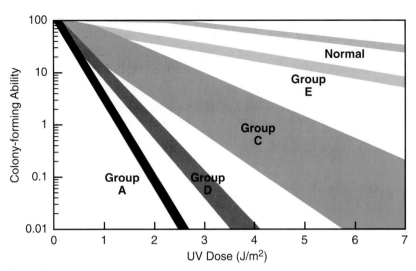

Figure 14–2 XP cells in culture from most genetic complementation groups are hypersensitive to UV radiation. However, the precise level of sensitivity varies somewhat from cell line to cell line within a given genetic complementation group and particularly between genetic complementation groups. (*Adapted from Cleaver and Kraemer [62] with permission.*)

extent of their biological recovery, e.g., by their ability to propagate in bacterial hosts (37, 38, 84, 146, 200, 201, 313, 314, 357, 361, 462). In general, such studies have supported the notion that XP cells are defective in the repair of UV radiation damage and that such damage causes mutagenic lesions during semiconservative DNA synthesis.

The demonstration of differential repair of DNA strands in normal mammalian cells, i.e., the preferential repair of the transcribed strand (see chapter 7), has prompted an analysis of point mutations in the two DNA strands in XP cells. If nucleotide excision repair preferentially removes base damage from the transcribed DNA strand, one would expect that mutations would be observed in this strand more frequently in cells defective (or deficient) in repair relative to normal

Table 14–2 Mutations in the plasmid vector pZ189 replicated in xeroderma pigmentosum or normal cells

Mutation	Value in:	
	XP cells	Normal cells
Independent plasmids sequenced	61	89
Point mutations		
Single base substitutions	71%	53%
Tandem base substitutions	20%	18%
Multiple base substitutions	2%	28%
Base insertions and deletions		
Single base insertion	0	2
Single or tandem base deletions	1	3
Types of base substitutions		
$G \cdot C \rightarrow A \cdot T$ transitions	93%	73%
$A \cdot T \rightarrow G \cdot C$ transitions	66%	59%
$G \cdot C \rightarrow T \cdot A$ transversions	0%	10%
$G \cdot C \rightarrow C \cdot G$ transversions	1%	6%
$A \cdot T \rightarrow T \cdot A$ transversions	4%	8%
$A \cdot T \rightarrow C \cdot G$ transversions	0%	1%

Source: Adapted from Cleaver and Kraemer (62) with permission.

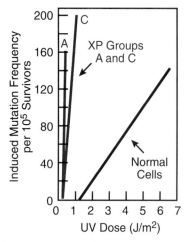

Figure 14–3 XP cells (genetic complementation groups A and C are shown) have an increased frequency of mutations at various genetic loci (such as the locus for azaguanine resistance) when exposed to DNA-damaging agents such as UV radiation. (*Adapted from Friedberg [102] with permission.*)

cells. Several studies suggest that this indeed is the case. For example, an analysis of the *HPRT* gene in XP cells revealed that virtually all point mutations were on the transcribed strand and presumably arose from unrepaired pyrimidine dimers or (6-4) photoproducts (84, 241, 253, 440). Curiously, in cells from genetic complementation group F (see below), less residual DNA repair was detected in transcriptionally active genes than in the genome overall, suggesting that this class of XP cells may be selectively defective in gene-specific repair, a phenomenon referred to as *dyspreferential repair* (94). As will be seen below, this general phenotype is also observed in cells from patients with CS.

GENETIC COMPLEXITY OF XP

Cells from patients with the first cases of XP documented in terms of their defect in nucleotide excision repair were exposed to UV radiation and examined by measurement of repair synthesis and/or the ability to excise pyrimidine dimers as a function of post-UV incubation time (55, 364). Both parameters of nucleotide excision repair were found to be defective (Fig. 14–4 to 14–6). However, as increasing numbers of cases were studied, a large variability in the repair defect as measured by unscheduled DNA synthesis (UDS) was noted (35). This suggested genetic heterogeneity, a postulate that was examined by fusing cells from different patients with XP and comparing the levels of UDS in heterodikaryons and unfused monokaryons (81). It was observed that cells from different patients with XP can complement one another, restoring levels of UDS to normal (Fig. 14–7 and 14–8).

Systematic complementation analysis by cell fusion has led to the identification of seven complementation groups, designated A through G (62) (Table 14–3). (Complementation groups initially designated H and I were subsequently

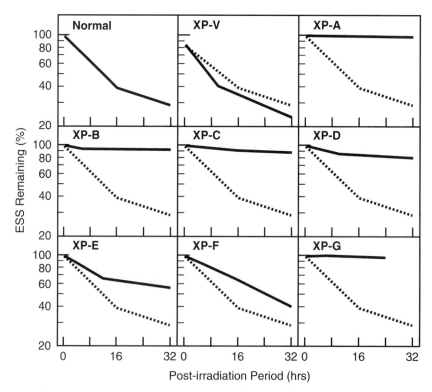

Figure 14–4 The kinetics of the disappearance of sites in DNA (pyrimidine dimers) that are sensitive to attack by a pyrimidine dimer-DNA glycosylase (see chapter 5 for details of this technique). The percentages shown are relative to the enzyme-sensitive sites (ESS) detected in unirradiated cells. The dotted line in each panel is reproduced from the kinetics observed with normal cells in the first panel. (*Adapted from Cleaver and Kraemer [62] with permission.*)

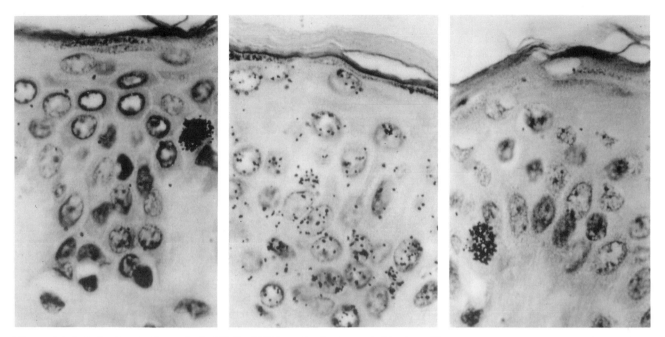

Figure 14–5 Defective repair synthesis (UDS) of DNA can be demonstrated in the epidermal cells of living individuals with XP following the injection of [³H]thymidine into an area of skin previously exposed to UV radiation. The panel on the left is from an unirradiated normal subject, and that in the middle is from a UV-irradiated normal subject. The panel on the right is from a UV-irradiated individual with XP. (*Reproduced from Friedberg [102] with permission.*)

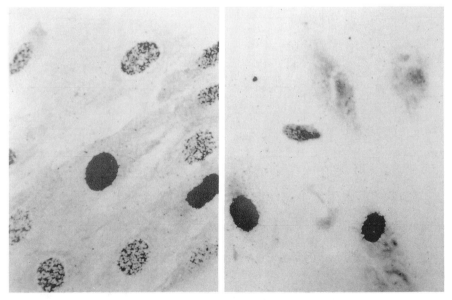

Figure 14–6 Defective repair synthesis (UDS) in XP cells in culture. Normal cells (left panel) show autoradiographic labeling of the great majority of the nuclei that are not in S phase (intensely labeled cells), reflecting repair synthesis of DNA. XP cells (right panel) show normal S phase (scheduled) DNA synthesis but are defective in UDS. (*Courtesy of J. E. Cleaver.*)

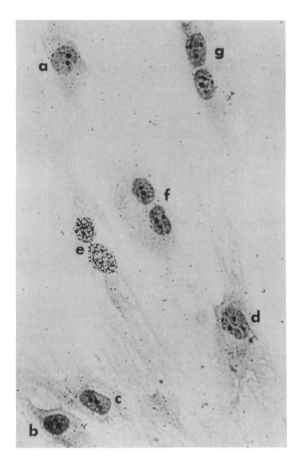

Figure 14–7 Complementation of defective repair synthesis (UDS) in heterodikaryons derived by fusing cells from individuals with XP from different genetic complementation groups. The cells labeled a to d are monokaryons which are defective in UDS. The cells labeled f and g are homodikaryons resulting from fusion of cells from the same individual and hence are also defective in UDS. The heterodikaryon labeled e shows restoration of normal levels of UDS in both nuclei. (*Courtesy of J. E. Cleaver.*)

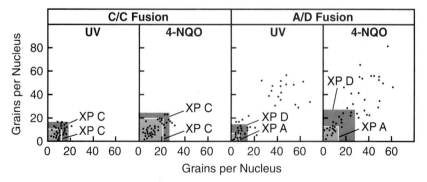

Figure 14–8 Quantitative complementation analysis by monitoring UDS after the treatment of cells with either UV radiation or the UV-radiomimetic chemical 4-NQO. When cells from two group C individuals are fused (C/C Fusion), the levels of UDS in the heterodikaryons are no greater than in each of the monokaryons. However, when XP-A and XP-D cells are fused (A/D Fusion) the levels of UDS in the heterodikaryons are greater than in each of the monokaryons. (*Adapted from Friedberg [102] with permission.*)

Table 14–3 Relative frequencies of various XP complementation groups among different populations[a]

Population	No. of patients in complementation group:							
	A	B	C	D	E	F	G	Variant
North America	9	1	11	8	0	0	0	5
Europe	12	0	28	17	2	0	2	19
Japan	30	0	5	4	3	11	1	21
Egypt	7	0	12	0	0	0	0	5
Other	5	0	6	0	0	0	0	4
Total	63	1	62	29	5	11	3	54

[a] Circa 1987.

Source: Adapted from Cleaver and Kraemer (62) with permission.

shown to be erroneous and to belong to complementation groups D and C, respectively [34, 436].) In addition to these so-called classical forms of XP, ~10% of all cases of clinically typical XP are not associated with an obvious defect in nucleotide excision repair in vivo (42, 57). These cases are referred to as the *XP variant* form (57, 61). However, cell-free assays which measure nucleotide excision repair in vitro (see chapter 8) have shown some XP variant cells to be indistinguishable from those of the classical XP form (460). This result has not been consistently observed with extracts of XP variant cells, suggesting that the putative XP variant protein is unusually labile (459).

A worldwide compilation of XP cases and the complementation groups into which they fall (62, 198, 399, 418) is shown in Table 14–3. Patients from genetic complementation groups A, C, and D and the variant group (which we will refer to as XP-V) constitute the majority of cases. There appears to be a preponderance of XP-E and XP-F cases in Japan, with relatively fewer cases of XP-C and XP-D. Since the time that this table was compiled, more cases have been identified, including a second case of XP group B; this one was found in association with CS (see below). As discussed in chapter 8, the genes for XP genetic complementation groups A, B, C, D, and G (and probably F) have been cloned. Hence, there is now unequivocal evidence in support of the notion that many, and probably all, of the complementation groups in XP reflect mutations at different genetic loci. XP must therefore be thought of as a single gene disorder with extensive *locus heterogeneity* (105, 421).

There have been suggestions that XP may be a multigene disease, based on the possibility that the expression of the disease requires the corecessive inheritance of more than one mutant gene (211, 212). Such suggestions have been prompted by the observation of the uncoupling of the repair of different photoproducts in an XP-A revertant cell line (59, 60). Specifically, in this revertant line (which is no longer abnormally sensitive to UV radiation), (6-4) photoproducts are apparently excised normally but cyclobutyl pyrimidine dimers are not. This has invited the notion that reversion at the *XPA* locus might have restored the ability of the cells to repair (6-4) photoproducts, but a persistent mutation in a second, unidentified gene accounts for the continued failure to repair thymine dimers. As indicated in chapter 8, the fact that different types of DNA damage are repaired with different efficiencies very probably reflects idiosyncrasies in the biochemistry of nucleotide excision repair rather than intrinsic genetic complexity. The stoichiometry of the assembly of a multiprotein excision repair complex may vary according to the differential affinity of one or more components of the human nucleotide excision repair complex for different types of damage. Additionally, complex formation may be influenced by the level of expression of nucleotide excision repair genes. Variations in the efficiency of the repair of different types of base damage might therefore result when XP cells are transfected with complementing genes that are driven by different exogenous promoters and that integrate randomly into the genome at different copy numbers.

The spurious correction of the sensitivity of XP cells to various DNA-damaging agents by various genes has added further confusion to this genetic debate. It should be appreciated that the correction of mutant phenotypes, particularly relatively nonspecific phenotypes such as the sensitivity to killing by DNA-damaging agents, may also be quite general. For example, the sensitivity of cells to DNA damage can vary as a function of cell cycle progression and, as already suggested, as a function of differential uptake and metabolism of chemicals. Additionally, cell survival may reflect perturbations of cellular physiology associated with the overexpression of certain genes. To quote a specific example, the observation that the gene that encodes the β subunit of casein kinase II partially restores the UV sensitivity of certain XP cells (413) might reflect the phosphorylation-dependent activation of one or more proteins which directly or indirectly affect the sensitivity of cells to killing by UV radiation.

OTHER ASPECTS OF GENETIC COMPLEXITY IN XP

An intriguing feature of XP alluded to several times already is the observation that individuals with this disorder sometimes manifest the clinical features of CS. CS is considered in detail below. For the purposes of the present discussion, it is relevant to point out that the association of XP with CS has thus far been observed to be confined to the three XP genetic complementation groups B, D, and G. The genes for each of these complementation groups (*XPB*, *XPD*, and *XPG*, respectively) have all been cloned (see chapter 8). A cardinal question is whether all mutations in these genes result in the combined phenotype of XP and CS. If so, this would suggest that the biochemistry of nucleotide excision repair in human cells involves a branched pathway with both common and independent gene products. Mutations in the genes which encode common products could then result in both XP and CS, while mutations which inactivate genes that are specific to each pathway would result in one or the other disease. An alternative possibility is that the expression of two different genes may be coordinately affected by a mutation in the *XPB*, *XPD*, or *XPG* gene. At present there are insufficient numbers of cases from genetic complementation groups B and G to evaluate these models. However, it is clear that not all *XPD* mutant alleles result in CS. Hence, a more likely hypothesis for the relationship between XP group D and CS is that only certain mutations in these genes result in the combined phenotype, i.e., that in addition to *locus heterogeneity*, XP is a disease that is characterized by *allelic heterogeneity*.

Allelic heterogeneity is an important determinant of phenotypic and hence clinical variation in other hereditary disease. In some cases different mutations in the same gene have been shown to result in clinically indistinguishable or closely similar disorders (105, 421). However, in other cases, different mutant alleles at the same locus result in very different clinical presentations (393, 421). The clinical complexity that accompanies allelic heterogeneity can result from relatively small variations in residual enzyme activity. Partial defects typically present with clinical abnormalities which represent a subset of those associated with the complete deficiency. However, the etiologic relationship between diseases caused by complete and partial defects in a single gene product may not be immediately obvious (421). Complex clinical phenotypes can also arise from mutations that result in the altered association of proteins, e.g., mutations that disrupt subunit assembly or that disrupt subunit interactions in multimeric proteins. Finally, increased expression of a mutant protein may exert a dominant-negative effect if it interacts with the product of the normal allele or in a complex with other proteins to impair their function (421). In summary, the relationship between the molecular pathology of proteins and the clinical phenotype can be very complex. The biochemical and clinical consequences of different mutations in a single gene are often unpredictable and can produce very different clinical phenotypes. A specific model of allelic heterogeneity to account for the association of XP, TTD, and CS is presented below.

Now that multiple XP genes have been cloned and mutations have been identified in the alleles of affected individuals, the stage is set to examine this hypothesis in detail. Certainly it will be of great interest to determine whether the association of XP and CS correlates with particular mutations in certain XP loci. The phenotypic characterization of mouse strains carrying genetically engineered mutations at defined locations in single XP genes is expected to be singularly informative in this regard. However, it must be borne in mind that mouse cells in culture have historically shown significant differences from human cells in the way in which they respond to DNA damage. In particular, as discussed in chapters 7 and 8, rodent cells manifest significantly reduced rates of nucleotide excision repair of UV radiation damage.

In the context of the present discussion, it is germane to point out that similar considerations apply to the relationship between XP and TTD (see below for a more complete discussion of TTD). At the clinical level, individuals with TTD do not manifest the characteristic features of XP and, in particular, do not appear to be unduly prone to skin cancer. Nonetheless, at the cellular level, some cases of TTD manifest clear sensitivity to UV radiation and are defective in nucleotide excision repair (see below). Heterokaryon analysis has placed many of these TTD cell lines in genetic complementation group D of XP and at least one case in genetic complementation group B, once again prompting the question of how mutations in the same gene, which cause apparently identical defects in cellular responses to DNA damage, can result in clinically distinct syndromes. One possible explanation is that XP is a multigene disease and hence that the complete clinical syndrome of XP, including its tendency to cause cancer, requires independent mutations in another gene(s), which has not yet been identified (218). While this remains a viable hypothesis, there is no direct experimental evidence in support of the involvement of multiple independent genes in XP. As discussed below, a more compelling alternative hypothesis derives from the observations (see chapter 8) that nucleotide excision repair genes implicated in both XP and TTD are also known to be involved in RNA polymerase II basal transcription.

XP HETEROZYGOTES

Heterozygous carriers of a gene for a rare autosomal syndrome are common in the general population (149). For example, in a disease such as XP, even though the true incidence may be as low as 1 in 250,000, the heterozygote frequency has been calculated at <1 in 300. Clinical and laboratory investigations on obligate XP heterozygotes have failed to uncover consistent clinical or cellular abnormalities. Hence, detection of the carrier state of this disease has not been possible to date. However, at least one study suggests that in metaphase preparations of T lymphocytes or skin fibroblasts from XP heterozygotes, there is an increased frequency of chromatid breaks or gaps some hours after exposure of cells to ionizing radiation (295). Another study has reported increased levels of SCEs and other chromosomal abnormalities in the cells of XP heterozygotes after exposure to UV radiation (27). An increase in the frequency of spontaneous chromosome aberrations has also been observed in two obligate XP heterozygotes (45).

As more and more XP genes are cloned and sequenced, the identification, by techniques such as PCR, of sites at which mutations commonly occur may facilitate the detection of carriers. The identification of a known common mutation in a specific gene simultaneously verifies the carrier state and identifies the genetic complementation group of the disease, as has already been demonstrated in Japan, where a particular mutation accounts for the majority of XP group A cases (192).

BIOCHEMICAL DEFECTS IN XP CELLS

The evidence that XP cells are defective in nucleotide excision repair of DNA was extensively discussed in chapter 8, which deals with the molecular biology and biochemistry of this repair mode in mammalian cells. As indicated in those discussions, the available evidence suggests that much, if not all, of the genetic complexity in XP concerns the requirement for multiple genes for damage-specific recognition and/or damage-specific incision of DNA during nucleotide excision repair. In general, the various parameters by which nucleotide excision repair capacity is measured in human cells correlate well within any given genetic complementation group, but there are significant exceptions (62, 194, 198, 202, 317).

For example, most cells in group A have very low rates of loss of sites sensitive to pyrimidine dimer-specific enzyme probes (Fig. 14–4) and of thymine dimers (measured directly) from their DNA. Additionally, these cells have low levels of UDS (Table 14–4) and are very sensitive to killing by UV radiation (62, 194, 198, 202, 317) (Fig. 14–2). On the other hand, many group D cells are as UV sensitive as most group A cells (3) (Fig. 14–2) and show a rate of loss of specific enzyme-sensitive sites (pyrimidine dimers) no higher than that of group C cells (471) (Fig. 14–4). However, group D cells often manifest significant residual levels of UDS (195, 203) (Table 14–4). Although group F cells show a pronounced residual capacity for loss of enzyme-sensitive sites (Fig. 14–4), UDS is disproportionately low (470) (Table 14–4).

Of the various parameters mentioned, UDS is perhaps the least reliable indicator of nucleotide excision repair, since it is sensitive to physiological perturbations in the pool size of thymidine in the cells. In addition, repair synthesis reflects not only the number of sites in DNA repaired but also the size of the repair patch at any given site. Thus, longer average repair patch sizes at a reduced number of sites of repair may not be reflected by a reduction in total repair synthesis. Furthermore, recent studies (111) suggest that the standard UDS experimental protocol with cells in culture reflects the repair of (6-4) photoproducts more than it does the repair of cyclobutane pyrimidine dimers. This correlates well with observations in a cell-free system for nucleotide excision repair (see chapter 8). As indicated in chapter 7, the kinetics of the excision of (6-4) lesions are different from those of excision of cyclobutane pyrimidine dimers. Hence, these caveats may also complicate the direct correlation of UDS and dimer excision.

MOLECULAR DEFECTS IN XP

As indicated in chapter 8, the human *XPA*, *XPB*, *XPC*, *XPD*, and *XPG* genes have been isolated by molecular cloning, and at the time of writing there are indications that the recently cloned *ERCC4* gene may be *XPF*. Molecular analysis of these genes and in some cases the proteins they encode is fully consistent with the notion that classical XP (i.e., all known forms except the XP variant form) is a disease that arises from defective nucleotide excision repair of DNA. Furthermore, the majority of these genes have been mapped to specific different chromosomal locations in the human genome (see chapter 8 and Tables 8–2 and 8–3), directly confirming the locus heterogeneity of XP. Several aspects of the discussion in chapter 8 are reiterated here in the interests of presenting a coherent description of the molecular pathology of XP.

XP Group A

A reduction in *XPA* mRNA levels in several Japanese patients has been demonstrated and probably results from defective processing of mRNA as a result of a G → C transition mutation that alters a consensus AG dinucleotide sequence in the 3' splice acceptor site of intron 3 of the *XPA* gene (403). More-extensive studies on patients from XP genetic complementation group A showed this mutation to

Table 14–4   Complementation group, clinical type, and defective repair in XP

Complementation group	Predominant clinical type	UDS (% of normal)
A	Neurological	2–5
B	Neurological (plus CS)	3–7
C	Classical	5–20
D	Neurological (plus CS) (plus TTD)	25–50
E	Classical	40–50
F	Classical	20
G	Neurological (plus CS)	2

Source: Adapted from Cleaver and Kraemer (62) with permission.

be present in 16 of 21 unrelated Japanese homozygous individuals. However, 11 Caucasians and 2 blacks with XP group A did not have this mutation (338). More-extensive mutational analyses have demonstrated that the majority of all Japanese cases of XP group A are caused by one or more of three mutations: a splicing mutation in intron 3 or a nonsense mutation at either codon 116 or codon 228 (340). There are suggestions that the homozygous state for the splicing mutation in intron 3 may produce more-severe clinical symptoms than does the compound heterozygous state (276).

In general these mutations correlate well with the clinical severity of the disease (339, 340). For example, the nonsense mutation at codon 116 is expected to result in a truncated polypeptide missing the C-terminal 157 amino acids of the 273-amino-acid XPA polypeptide. Patients with this mutation have severe clinical symptoms. On the other hand, the nonsense mutation at codon 228 is expected to result in a polypeptide that is deleted of only the C-terminal 45 amino acids. A patient carrying this mutation manifests only mild skin symptoms and minimal neurological involvement (340). Patients from Tunisia with XP group A have a high prevalence of a nonsense mutation in exon 6 of the *XPA* gene (277). This mutation was also found in several Japanese patients, who had notably clinically milder disease than did most patients in this racial group. The Tunisian cases are also characterized by less severe clinical disease (277).

The polypeptide product of the *XPA* gene has been purified and partially characterized (see chapter 8). It is a DNA-binding protein that preferentially binds to UV-irradiated compared with native DNA. However, exactly how this DNA-binding activity contributes to the repair of damaged DNA remains to be elucidated in more-extensive biochemical studies, when all of the multiple proteins required for nucleotide excision repair have been isolated. The purified protein specifically corrects defective nucleotide excision repair in cell extracts of XP-A cells (291, 320). A second mutation in the originally affected amino acid in the XPA protein from an XP-A cell line defective in nucleotide excision repair has been shown to cause the inactive protein to revert to a different mutant form that now expresses XPA activity (170).

Parenthetically, it is interesting that the revertant form of XPA protein is able to effect the efficient repair of (6-4) photoproducts but not of pyrimidine dimers (170). This provides a direct demonstration of phenotypic variability in nucleotide excision repair for different types of DNA damage and supports the contention expressed above that such variability does not require explanations that invoke the existence of multiple pathways for nucleotide excision repair or hidden genetic complexity in XP. As might be expected from the severe defect in nucleotide excision repair observed in XP group A cells, such cells show no retention of gene-specific or strand-specific nucleotide excision repair (94).

XP Group B

Analysis of the *XPB* gene in an XP-B proband (who also had clinical features of CS) revealed that her maternal allele harbored a C → A transversion mutation in the last intron, generating a splicing acceptor sequence 4 bp upstream of the normal 3' splice site (452). Sequence analysis of *XPB* cDNA from this individual has confirmed that the vast majority (and possibly all) of the transcripts carry the intron-derived 4-bp insert (452). Two further patients with XP have been placed in genetic complementation group B by the demonstration that defective nucleotide excision repair in cells from a sibling pair with the disease is uniquely corrected by microinjection of the *XPB* (*ERCC3*) gene (435). Analysis by somatic cell hybridization confirmed this genetic assignment. Molecular analysis of the *XPB* gene in cells from both individuals revealed a single missense mutation in an evolutionarily highly conserved region of the putative XPB polypeptide (435). The second (presumably mutated) *XPB* allele is apparently not expressed.

Like the first XP patient with XP group B, the two new patients demonstrate clinical features of both XP and CS (351). Curiously, however, although these two individuals have a profound defect in nucleotide excision repair at the cellular level (435), the clinical features of XP are not as severe and at age 40 neither patient has shown a predisposition to skin cancer despite a history of no unusual protective measures from Sun exposure (351) (see below). Interestingly, in striking contrast to what is observed in other XP cells, measurements of the spontaneous mutation frequency in the *HPRT* gene in T lymphocytes from these individuals with XP group B did not show an increase relative to that in normal individuals (435).

XPB protein is a DNA helicase which is a component of the human basal transcription factor TFIIH (342), and microinjection of this transcription factor corrects defective nucleotide excision repair specifically in XP-B cells (429). The precise role of TFIIH in nucleotide excision repair is unknown at present. The relationship between some mutant *XPB* alleles and CS is discussed below.

XP Group C

Mutations in the *XPC* gene have been demonstrated in cell lines from several individuals with XP group C (234). It has been suggested that the product of the *XPC* gene is specifically required for the repair of transcriptionally inactive bulk DNA as well the nontranscribed strand in transcriptionally active genes (176, 177, 433). The *XPC* gene encodes a protein with a molecular mass of ~130 kDa, which has an unknown function in nucleotide excision repair.

XP Group D

The *XPD* gene (448) is conserved in eukaryotes, including the yeasts *Saccharomyces cerevisiae* and *Schizosaccharomyces pombe* (see chapter 8). Mutations have been identified in the *XPD* gene which inactivate its ability to correct the UV radiation sensitivity of XP-D cells (101). The *XPD* gene product is a protein of ~87 kDa with a *DNA helicase activity* (392). The role of this DNA helicase in nucleotide excision repair is unknown. Like XPB protein, XPD protein is a component of the RNA polymerase II-dependent basal transcription factor BTF2 (TFIIH) and hence, like XPB protein, is required for both DNA repair and transcription (85, 341).

XP Group E

The binding of cellular proteins to damaged DNA has been examined by gel retardation assays. These studies identified a defect in a DNA-binding protein in XP-E cells, designated *XP-E-binding factor* (54, 301). However, defective protein binding is not a universal biochemical phenotype of XP-E cells. At least four fibroblast strains from unrelated Japanese patients with this type of XP do not show this defect (182). In fact only 3 of 12 XP-E cell lines examined do (185). Nonetheless, microinjection of purified XPE protein into XP-E cells specifically corrects deficient nucleotide excision repair (184). Correction of defective repair synthesis has also been observed following the microinjection of poly(A)$^+$ RNA into these cells (280). The size of the correcting mRNA species has been estimated at ~1.5 kb (280). Several explanations are consistent with the notion that the XPE protein is functionally defective in all XP-E cell lines. Of these, the most compelling is the suggestion that the protein in the majority of cell lines is altered at some domain other than that required for binding to DNA (185).

XP Group G

The *XPG* gene has also been cloned. The size of the open reading frame predicts a polypeptide of ~134 kDa (344). The amino acid sequence of the *XPG* gene indicates that it is the human homolog of the yeast *RAD2* gene (344, 369) and strongly suggests that, like the yeast Rad2 protein (see chapter 6), XPG protein is an endonuclease involved in the damage-specific incision step of nucleotide excision repair. XPG protein has been purified ~3,000-fold on the basis of its ability to correct defective nucleotide excision repair in human cell extracts (279).

SUMMARY OF THE MOLECULAR PATHOLOGY OF XP

It is obvious from the preceding discussion that the understanding of the molecular basis for the defect(s) in nucleotide excision repair in XP cells has advanced rapidly. The remaining gaps in our knowledge notwithstanding, the biochemical and cellular defects in XP cells discussed above are consistent with the marked sensitivity to sunlight in individuals with XP, and the prevalence of skin cancer in patients with this disease provides important indirect evidence that DNA damage can result in mutation and neoplastic transformation of human cells.

Since defective nucleotide excision repair of DNA is common to all cell types in patients with XP, one might expect that these individuals would also manifest an increased frequency of neoplasms in other organs, particularly organs such as the lungs and gastrointestinal tract, which are (presumably) constantly subjected to DNA damage by environmental agents. However, XP is a rare disease, and most individuals succumb to skin cancers at a relatively early age. This, combined with the relative paucity of detailed autopsy information on deceased patients with XP, has precluded the obvious statistical weight that large numbers of cases could bring to such an issue. Nonetheless, one study suggests that patients with XP do indeed suffer a 10- to 20-fold-increased frequency of neoplasms affecting internal organs. Notable among the cases investigated in this study was a patient who died of lung cancer at age 34 after smoking a pack of cigarettes a day for 16 years (62, 197). Several studies have identified mutations in specific proto-oncogenes, including c-*ras*, *p53*, and c-*met*, in cells from individuals with XP (75, 87, 257). In several of these studies (75, 87), mutations were identified at dipyrimidine sites, consistent with the notion that photoproducts induced by actinic radiation were the primary source of DNA damage associated with activation of these proto-oncogenes. One might anticipate that XP heterozygotes would be at increased risk for skin cancers, since such individuals have no reason to take special precautions against exposure to sunlight. It would be of interest to determine whether skin cancers in younger individuals in the population turn out to show loss of heterozygosity for specific loci associated with XP. A 3-year controlled prospective study of five individuals with XP who had a history of multiple skin cancers showed a statistically significant protective effect of ingestion of high levels of retinoids (196).

The pathogenesis of the neurological complications that accompany many cases of XP remains a mystery. The most compelling and obvious explanation is that the neurological symptoms and signs reflect defective or deficient repair of endogenous base damage in neurons and/or other cells in the central nervous system. It has been pointed out that the brain is certainly a prominent site for oxidative metabolism (58). It has also been shown that plasmid DNA exposed to oxidative damage carries a minor component of base damage that is repaired in vitro by cell extracts of normal human cells but not those of XP cells (337) (Fig. 14–9). The apparent heterogeneity of the neurological complications among different XP complementation groups may reflect the fact that not all the polypeptide components of the nucleotide excision repair apparatus are required to the same extent for the repair of such damage. In addition to the potential direct relationship between defective nucleotide excision repair and persisting oxidative base damage, there are reports in the literature suggesting that XP cells may be defective in specific enzymes required for normal oxidative metabolism (275, 441).

The extensive literature on XP that has accumulated in the past 25 years is filled with descriptions of a host of intriguing and as yet unexplained cellular and biochemical phenotypes that are not obviously reconciled with defects in genes involved in nucleotide excision repair. These will not be considered here, but the interested reader is referred to some of the recent reviews cited at the end of this chapter (58, 62, 102, 198, 202, 317). The availability of mutant mice (and conceivably other vertebrates) in which specific mutations have been engineered into selected genes can be reasonably expected to shed significant insights on how multiple cellular and clinical phenotypes might relate to defects in nucleotide excision repair proteins. At the time of writing, mice carrying subtle and deletion mutations in several XP genes, including *XPA* (401) and *XPC* (36, 52) are already available. Additionally, although a direct relationship between the human *ERCC1* nucleotide excision repair gene (see chapter 8) and XP remains to be established, *ERCC1* mutant mice are also now available for study (157, 254). Since it is now firmly established that several nucleotide excision repair proteins are also involved

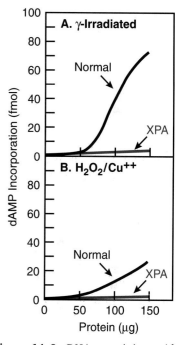

Figure 14–9 DNA containing oxidative base damage produced by exposure to γ rays (A) or H_2O_2 plus Cu^{2+} (B) is repaired by extracts of normal cells but not by extracts of cells from XP complementation group A. (*Adapted from Satoh et al. [337].*)

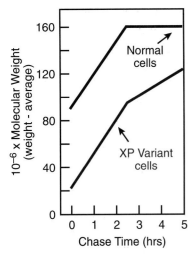

Figure 14–10 Semiconservative DNA synthesis in UV-irradiated XP variant cells. Normal and XP-variant fibroblasts were irradiated, pulse-labeled for 1 h with [³H]thymidine, and then chased for various times in medium with unlabeled thymidine. Weight-average molecular weights were calculated from molecular weight distributions obtained from alkaline sucrose sedimentation profiles. Immediately after irradiation the weight-average molecular weight of the DNA from the XP-variant cells is lower than that of normal cells and takes longer to reach normal high-molecular-weight values. (*Adapted from Friedberg [102] with permission.*)

in transcription, the possibility that some cellular and clinical phenotypes reflect subtle defects in transcription is a compelling notion. We will revisit this notion in greater detail below, when we discuss the relationship of XP to TTD and CS.

THE XP VARIANT FORM

As indicated previously, shortly after the discovery of defective DNA repair in classical XP cell lines, a group of clinically typical patients with XP were found to have normal levels of UDS and of repair synthesis (42, 57, 61). Cells from these patients have only a slightly increased sensitivity to UV radiation and a modest decrease in the ability to reactivate UV-irradiated adenovirus (73, 74). The assumption that pyrimidine dimer excision is normal in these cells is supported by a normal rate of loss of sites sensitive to pyrimidine dimer-specific enzyme probes in studies with intact cells (225). Also, the kinetics of the loss of thymine dimers from the acid-insoluble fraction of UV-irradiated cells is normal in intact XP variant cells (92). The term *XP variant* was coined to describe this form of the disease, characterized by typical clinical features but with no apparent defect in nucleotide excision repair following exposure to UV radiation. Studies with human cell extracts have identified defective nucleotide excision repair in some cases (460). Hence, as indicated in chapter 8, it is possible that the gene product of the as yet uncloned putative *XPV* gene will turn out to be a protein involved in the process of nucleotide excision repair.

XP variant cells also have a clear defect in the replication of damaged DNA. The weight-average molecular weight of newly synthesized DNA (labeled with a pulse of [³H]thymidine) in unirradiated or UV-irradiated normal skin fibroblasts is about 8×10^7. Following a chase in nonradioactive thymidine, the molecular weight reaches about 1.5×10^8 (the maximal weight-average molecular weight measurable under the experimental conditions generally used) (225). The molecular weight distribution of the DNA of unirradiated XP variant cells is identical to that of normal cells. However, following irradiation, XP variant DNA pulse-labeled for the same time as normal cells has a lower molecular weight (about 3×10^7) and consequently requires a longer chase time to reach the maximal detectable molecular weight (225) (Fig. 14–10). In the presence of caffeine, the restoration of high-molecular-weight DNA is completely inhibited in XP variant cells within the time of the usual chase, whereas normal cells are hardly affected at all by this compound (Fig. 14–11). Since the process of conversion of low-molecular-weight DNA synthesized on damaged templates into high-molecular-weight DNA constitutes a particular cellular response to DNA damage whereby lesions that are not repaired may be tolerated (so-called postreplication repair) (see chapter 12), it has been suggested that XP variant cells are defective in this response (224–226).

The type of pulse-chase experiments just described are open to multiple interpretations. For example, as indicated above, when semiconservative DNA synthesis is examined in UV-irradiated XP variant and normal fibroblasts, the absolute size of the DNA fragments synthesized during brief pulses of [³H]thymidine given 1 to 2 h after irradiation is decreased relative to that in unirradiated cells, with the XP variant showing a greater decrease than normal cells (Fig. 14–10). It has been proposed that pyrimidine dimers (and presumably other sites of base damage in DNA) act as all-or-nothing blocks to the progress of replication forks and that XP variant cells have an increased probability (relative to that of normal cells) of replicative arrest at sites of DNA damage (57a, 182a). XP variant cells have been reported to be defective in the ability to bypass psoralen adducts in the DNA of rRNA genes during semiconservative DNA synthesis (258).

XP variant cells, like classical XP cells, are hypermutable by a variety of DNA-damaging agents (243, 270). A comparison of the mutational spectrum in the *HPRT* gene of synchronized populations of variant and normal cells irradiated with UV light in early S and in G_1 revealed base substitutions as the majority type of mutation in both cell types (446). However, in the XP variant cells, these substitutions were mainly transversions, which were rare in normal fibroblasts. Similar results have been obtained by measuring mutagenesis in a plasmid vector in XP variant cells (447). A detailed analysis of these transversions suggests that variant cells are much less likely than normal cells to incorporate either A or G opposite pyrimidines that are potential sites for dipyrimidine photoproducts. Hence, for reasons that are currently obscure, XP variant cells apparently do not preferentially incorporate adenine opposite pyrimidine adducts during semiconservative DNA synthesis (445, 446) (see chapter 12).

Classical XP cells also show a reduced ability to convert low molecular weight DNA into a higher-molecular-weight form following exposure to UV radiation, although quantitatively the effect is not as marked as that observed in XP variant cells (224, 226) (Fig. 14–11). This raises the interesting question whether XP variant and classical XP cells share a common defect in the response to base damage at replication forks. Some studies suggest that they may. When normal human cells are exposed to UV radiation, there is an initial decrease in the rate of semiconservative DNA synthesis followed by a dose-dependent recovery toward normal rates (Fig. 14–12) (325). However, when the UV dose is fractionated over a 2-h period, that is, when UV radiation at 1 J/m² is followed 2 h later by irradiation at 9 J/m² instead of being given as a single dose of 10 J/m², the rate of recovery from maximal inhibition of DNA synthesis is increased in normal cells (265) (Fig. 14–12). Both classical XP and XP variant cells fail to show this split-dose-dependent enhanced recovery phenomenon (265) (Fig. 14–12).

In the final analysis, our current understanding of the molecular pathology of the XP variant form points to a defect in the tolerance of base damage in DNA rather than a defect in its repair. The cloning of the putative *XPV* gene and the detailed biochemical characterization of its protein product are eagerly awaited, with the optimistic expectation that new and more definitive insights into this puzzling aspect of XP will be provided.

Trichothiodystrophy

A syndrome of sulfur-deficient brittle hair, ichthyosis (fish-like scales on the skin), and mental and physical retardation was first described by Pollitt and his colleagues in 1968 (308) (Color Plate 10). The term *trichothiodystrophy* (TTD) was coined in subsequent reports (311). The disease is an autosomal recessive disorder (218, 308, 311). The hair is brittle and weathers badly, and when exposed to trauma the hair shafts may break (*trichoschisis*) (Fig. 14–13) or form brush-like ends or nodes (*trichorrhexis nodosa*). The defining characteristic of TTD hair is a "tiger tail" appearance of alternating bright and dark bands of the hair shafts when viewed under the polarizing microscope. In the scanning electron microscope, hairs are flattened and the shafts are irregular with ridging and fluting. Biochemical analyses have demonstrated a low sulfur content of the hair in patients with TTD, which is believed to be related to a decrease in low-molecular-weight sulfur-rich proteins (132). In TTD hair, constituent proteins with a low sulfur content were found to be normal with respect to the range of molecular weights, isoelectric points, and subunit polypeptides. On the other hand, the sulfur-rich proteins were found to be substantially decreased in amount, had a lower cysteine content, and had a greatly altered amino acid composition. It has been suggested that these differences reflect a sulfur deficiency state (132). Patients with TTD also frequently have a distinctive facies with protruding ears and a receding chin (Color Plate 10). As of 1988, about 20 cases of this hereditary disorder were reported in the literature. This number has since increased substantially.

Various syndromes that include brittle hair have been associated with TTD (71). One of these has been christened with the acronym BIDS, and includes the principal clinical features of brittle hair, intellectual impairment, decreased fertility, and short stature. The acronym IBIDS refers to the additional presence of ichthyosis, and PIBIDS refers to the yet further presence of photosensitivity (71). PIBIDS and the photosensitive form of TTD may be the same (or closely related) clinical entities. However, it should be borne in mind that all of these syndromes are extremely rare and there is as yet no consensus as to their clinical relationship. As discussed below, there are suggestions that these and possibly other syndromes, including CS, represent a spectrum of diseases with overlapping clinical features

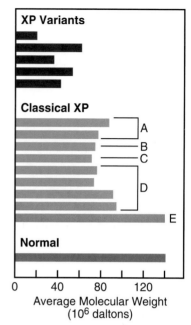

Figure 14–11 Semiconservative DNA synthesis in UV irradiated XP variant, classical XP, and normal cells. Cells were pulse-labeled with [³H]thymidine and then chased for various times in medium with unlabeled thymidine in the presence of caffeine. The cells were lysed and sedimented on alkaline sucrose gradients, and the weight-average molecular weights of the DNA were determined. The letters in the middle panel (classical XP) refer to XP genetic complementation groups. Note that caffeine inhibits the restoration of high-molecular-weight DNA in both variant and (to a lesser extent) classical XP cells. (*Adapted from Friedberg [102] with permission.*)

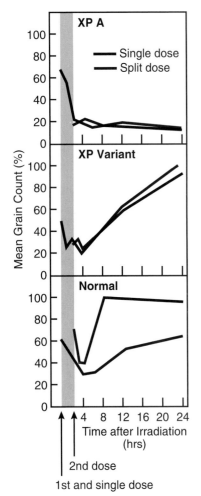

Figure 14–12 When normal human fibroblasts are exposed to a split dose of UV radiation (bottom panel), the rate and extent of the recovery of semiconservative DNA synthesis measured by quantitative autoradiography are enhanced compared with the rate and extent observed in cells that received an equivalent amount of UV radiation as a single dose. However, both XP variant (middle panel) and XP-A (top panel) cells are defective in this enhanced recovery. (*Adapted from Friedberg [102] with permission.*)

linked by a common pathogenetic mechanism, i.e., defective transcription of a particular subset of genes.

Photosensitivity is apparent in about half of the reported cases of TTD. This phenotype attracted the attention of Miria Stefanini and her colleagues, who were the first to investigate the DNA repair capacity of cell lines from Italian patients from three apparently unrelated families and demonstrate defective nucleotide excision repair of UV-induced damage in these (384) (Fig. 14–14). Complementation studies showed that normal levels of UDS could be restored in heterokaryons obtained by fusion of TTD cells with normal fibroblasts and with cells from XP genetic complementation group A (Table 14–5). However fusion with XP cells from group D failed to complement TTD cells (Table 14–5), suggesting that the DNA repair defect in these particular cases was somehow related to XP group D (384). Consistent with this conclusion, a patient with TTD but without accompanying photosensitivity did not show defective nucleotide excision repair (386).

More-extensive studies have shown that the relationship between TTD and defective DNA repair is more complex. For example, different degrees of responsiveness to UV radiation have been defined at the cellular level (40, 218, 219). In some individuals the response is completely normal. In others there is a clear defect in nucleotide excision repair. In many members of the latter group, the defect is not corrected (complemented) by fusion with cells from XP complementation group D. As will be seen below, one other XP gene involved in nucleotide excision repair has been implicated in this group of TTD cases. A third group of individuals with TTD have normal cell survival following exposure to UV radiation, and their cells manifest normal rates of removal of pyrimidine dimers. However, these cells show some reduction in repair synthesis of damaged DNA and a marked reduction in the loss of (6-4) photoproducts from DNA (40).

Despite clear evidence of photosensitivity and defective nucleotide excision repair in some cases (which cannot be corrected following fusion with XP-D cells), these patients with TTD do not manifest clinical features of typical XP, such as xerosis and pigmentary disturbances, and do not appear to be particularly prone to skin cancer (217, 218). Similarly, classical XP group D patients do not have

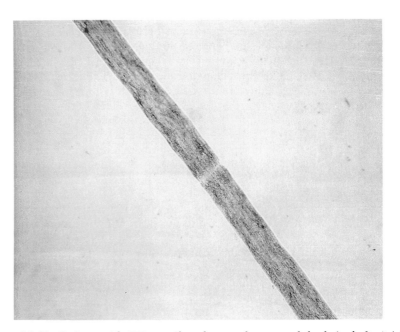

Figure 14–13 Patients with TTD manifest cleavage fractures of the hair shafts (trichoschisis). Also note the slightly undulating contour of the hair shaft and the wavy distribution pattern of melanin granules in the hair cortex. The latter feature reflects the varying orientation of the internal fiber structure in TTD. (*Reproduced from Price et al. [311] with permission. With special thanks to Vera Price.*)

Table 14–5 Complementation analysis in heterokaryons obtained by fusion of TTD and other human fibroblasts

| Fusions | No. of grains/nucleus | |
	Homokaryons	Heterokaryons
TTD1 × TTD2	2.67, 2.37	2.67
TTD2 × TTD4	3.15, 2.70	4.05
TTD1 × Normal	3.00, 30.13	21.56
TTD2 × Normal	1.02, 31.40	22.00
XP-A × TTD1	2.00, 2.22	20.26
XP-A × TTD2	1.42, 1.90	21.72
XP-D × TTD1	4.60, 2.65	5.57
XP-D × TTD2	5.12, 3.50	4.07

Source: Adapted from Stefanini et al. (384) with permission.

brittle hair or other clinical features typical of TTD. Hence, the simple model that the *TTD* and *XPD* genes are tightly linked and that in some cases of TTD a chromosomal deletion includes the *XPD* gene is not tenable. Indeed, Southern analyses of the *XPD* gene in normal individuals, individuals with classical XP, and those with the photosensitive form of TTD due to a defect in the *XPD* gene (based on lack of complementation by XP-D cells) have not revealed any major modifications (260). However, a more-detailed analysis of the *XPD* gene in individuals with TTD has identified mutations in this gene in several of the patients (41) (Fig. 14–15). All of these individuals were shown to be compound heterozygotes with different mutations in the two *XPD* alleles (41).

THE "TRANSCRIPTION SYNDROME" HYPOTHESIS

As indicated above and in other discussions concerning the *XPD* gene (see chapter 8) and its highly conserved yeast homolog *RAD3* (see chapter 6), this gene is known to be multifunctional, with defined roles both in nucleotide excision repair and in RNA polymerase II-dependent basal transcription. Hence, an intriguing model that has been advanced to explain TTD (as well as other diseases associated with defective nucleotide excision repair [see below]) is that the disease is caused by defective transcription (33, 41, 103, 437). The model essentially postulates that certain mutations in *XPD* (and possibly other genes with dual functional roles in transcription and DNA repair), which do not inactivate its essential transcription function, may nonetheless perturb the efficiency of transcription of a particular subset of genes and in some cases also the ability of the *XPD* gene to support normal nucleotide excision repair. This subset of genes may be specifically defined by a particular requirement for very high levels of expression or highly sensitive

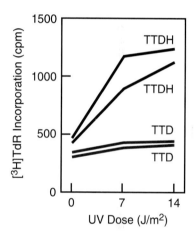

Figure 14–14 Quantitative expression of repair synthesis levels in UV-irradiated TTD homozygous (TTD) and heterozygous (TTDH) G_0 lymphocytes. (*Adapted from Stefanini et al. [384] with permission.*)

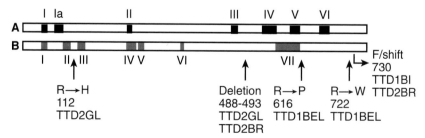

Figure 14–15 Diagrammatic representation of the linear human XPD polypeptide showing DNA helicase motifs (A) and RNA helicase motifs (B), and the position and nature of mutations mapped in the *XPD* gene of patients with TTD with defective nucleotide excision repair. Note that patients TTD2GL, TTD2BR, and TTD1BEL are compound heterozygotes. Also note that the codon 488 to 493 deletion mutation is present in both TTD2GL and TTD2BR cells. (*Adapted from Broughton et al. [41] with permission.*)

fine tuning of expression (437) (Fig. 14–16). Gene size and sequence-specific repair efficiency may also influence the transcription rate of some genes (112, 425).

Several lines of evidence are consistent with this hypothesis. One might predict that TTD (and related syndromes such as BIDS, IBIDS, PIBIDS, and even CS)
may arise as a consequence of defects in genes other than *XPD*, which also have
dual roles in transcription and nucleotide excision repair. Such is indeed the case.
An analysis of 10 cases of TTD with defective nucleotide excision repair at the
cellular level demonstrated the *XPD* defect in 7 of them (385). The remaining three
cases showed correction of defective repair synthesis after fusion with cells from
XP group D, suggesting that a different gene(s) is affected. One of these cell lines
(designated TTD1BR) is corrected by fusion with cell lines from all known XP
genetic complementation groups, suggesting that this gene is not one of the known
XP genes (387). Furthermore, microinjection of the cloned *ERCC1*, *XPB*, or *CSB*
(for Cockayne's syndrome group B [see chapter 8 and below]) gene into the nucleus of TTD1BR cells failed to correct defective nucleotide excision repair, indicating that this putative TTD gene is not one of these three genes either (387).
The TTD1BR cell line has therefore been placed in TTD complementation group
A (437).

The remaining two cell lines characterized as non-*XPD* TTD cells are designated TTD6VI and TTD4VI (385). Phenotypic correction of TTD6VI cells was observed following fusion with XP cells from genetic complementation groups A, C,
D, E, F, and G but not following fusion with XP cells from group genetic complementation group B, derived from a patient with clinical features of both TTD and
CS (see below) (437). Furthermore, expression of the *XPB* gene following its microinjection into the nuclei of TTD6VI cells resulted in specific correction of defective UDS in these cells (437) (Fig. 14–17). Hence, the TTD6VI (and presumably

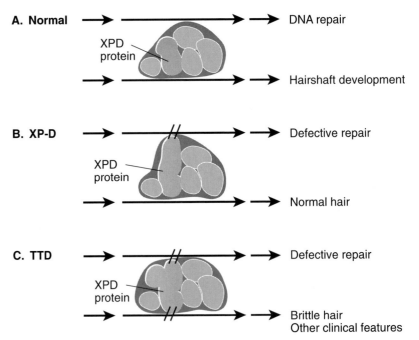

Figure 14–16 The transcription hypothesis. (A) The multisubunit human TFIIH RNA polymerase II basal transcription complex containing XPD protein is shown between the sets
of arrows. (B) Some mutations inactivate just the nucleotide excision repair function of the
protein and result in typical XP accompanied by skin cancer. (C) Other mutations may
inactivate the excision repair function and/or alter the level of transcription of one or more
specific genes, resulting in the clinical and cellular phenotypes of TTD, including the lack of
a predisposition to skin cancer. Depending on which particular genes are affected in this
way, the clinical complexity of TTD may vary. (*Adapted from Broughton et al. [41] with permission.*)

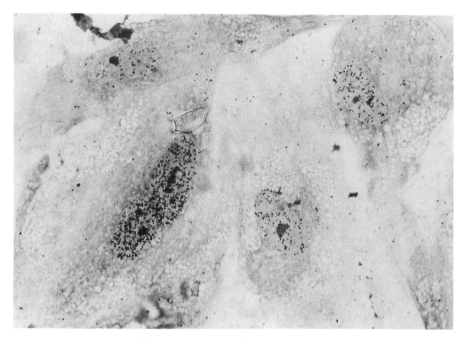

Figure 14–17 Autoradiograph showing that expression of the *XPB* gene following its microinjection into the nuclei of TTD6VI cells results in the correction of defective UDS in these cells. (*With thanks to W. Vermeulen and Jan H. J. Hoeijmakers.*)

the TTD4VI) cell line represents an example of TTD resulting from a mutation in the *XPB* gene, although this remains to be proven definitively.

The observation that a known component of the RNA polymerase II basal transcription complex TFIIH (BTF2), notably XPB protein, corrects the cellular phenotype of TTD6VI cells prompted an examination of the effect of microinjection of this protein complex into TTD-A (TTD1BR) cells. Correction of defective repair synthesis was indeed observed (437) (Table 14–6). Additionally, such correction was observed in cell extracts of TTD-A cells (as well as similarly prepared extracts of rodent cells corresponding to the human XP-D and XP-B complementation groups [437]). Hence, in addition to XPD and XPB proteins, another (as yet undetermined) component of TFIIH is implicated in TTD and in the transcription syndrome complex of diseases. These experiments therefore define the existence of at least three distinct genetic complementation groups for the photosensitive form of TTD (PIBIDS). One complementation group (TTD-A) defines a form of TTD associated with defects in a gene that encodes an identified basal transcription factor. A second complementation group is defined by a form of TTD associated

Table 14–6 Complementation of defective nucleotide excision repair by human TFIIH (BTF2)

Human mutant	Correction by microinjection of TFIIH
XP-A	−
XP-B	+
XP-C	−
XP-D	+
XP-E	−
XP-F	−
XP-G	−
TTD-A	+

Source: Adapted from Vermeulen et al. [437] with permission.

with defects in the *XPB* gene, and a third group is defined by a form of the disease with mutations in the *XPD* gene. TTD that is not associated with photosensitivity, i.e., defective nucleotide excision repair, may define the existence of yet further genetic complementation groups. Alternatively, these cases may represent defects in the XPB or XPD proteins that alter their transcription function but not their nucleotide excision repair function. As discussed more fully below, both the *XPB* and *XPD* genes are also implicated in XP associated with CS.

On the basis of these experimental observations, it seems reasonable to suggest that when the genes that encode all of the polypeptide components of the basal transcription complex are defined, they will all be found to have dual functional roles in nucleotide excision repair and transcription, and all will be found to be targets for mutations that result in clinical phenotypes which may include various combinations of the following:

defective nucleotide excision repair in cells

brittle hair or related abnormalities of the hair and/or nails

skin abnormalities

impaired growth

neurological defects

impaired sexual development

dental problems

Particular combinations of these clinical features may account for diseases and syndromes now referred to as CS, TTD, BIDS, IBIDS, and PIBIDS, as well as Sjögren-Larsson syndrome and conditions designated as RUD syndrome (ichthyosis and male hypogonadism), ichthyosis-cheek-eyebrow (ICE) syndrome, oculotrichodysplasia (OTD), ichthyosis-follicularis-atrichia-photophobia (IFAP) syndrome, cataract-microcephaly-failure-to-thrive-kyphoscoliosis (CAMFAK) syndrome, Rothmund-Thompson syndrome, and keratosis-ichthyosis-deafness (KID) syndrome (250, 437).

TTD AND CANCER

A curious and unresolved conundrum is the observation that classical XP, including the XP-D form, is clearly associated with an increased risk for skin cancer (and possibly other cancers as well). On the other hand, even though most, if not all, patients with TTD who have associated photosensitivity are defective in nucleotide excsion repair at the cellular level and have a pronounced increased mutation frequency (219, 240), these individuals are not especially prone to skin cancer (217, 218). This paradox seriously challenges the central dogma that defective nucleotide excision repair, hypermutability, and cancer proneness are pathogenetically related events. The fact that some individuals with TTD also do not manifest the typical skin symptoms of XP, i.e., dry scaly skin (xerosis) and severe freckling (pigmentary problems), has led to the further suggestion (229) (discussed above) that the repair deficiency in XP is necessary but not sufficient for the full-blown clinical picture, including cancer proneness, and that mutational inactivation in a second as yet undetected gene is required. As already indicated above, the notion that the genetics of XP is not simple recessive inheritance of a single mutated locus but rather involves corecessive inheritance of mutations at two or more loci has been independently postulated (211, 212). The nature of such a putative second gene(s) is unknown. Since some individuals with XP manifest immune deficiencies of various sorts (68, 222, 262) and since immune surveillance is a prominent component of cancer progression, a gene(s) that affects the immune function has been suggested as a possible candidate (222, 229).

Other phenotypic differences between XP and TTD cells have been observed. An examination of 21 different XP cell lines showed reduced levels of catalase activity in cell extracts, whereas catalase levels in nucleotide excision repair-defective TTD cells were normal (442). As a corollary of these observations, it was found that exposure of these XP and TTD cells to UV radiation led to a significant increase in the level of hydrogen peroxide in the former cells (442). It remains unexplained why XP cells have reduced levels of catalase activity. However, the consequent increase in the levels of compounds that might contribute oxidative damage to DNA provides an intriguing explanation for the cancer proneness of these cells (442).

The transcription syndrome hypothesis provides the basis for considering alternative explanations for the failure to observe an obvious cancer predisposition in individuals with defective nucleotide excision repair associated with other cellular phenotypes, such as in TTD. Conceivably, the subset of genes whose expression is perturbed by defects in basal transcription factors contains one or more genes that encode polypeptides which are required at particular levels for the process of neoplastic transformation to transpire. Hence, altered levels of expression of these genes could have an antitumorogenic effect (41, 103).

Cockayne's Syndrome

We have alluded to the disease CS in several contexts already, both in this chapter and in chapter 8. As discussed above, there are indications that some features of CS may represent part of the continuum of so-called transcription syndromes, especially when considered in terms of the relationship of this disease to XP. However, CS often occurs without any apparent relationship to XP and merits discussion here as an independent entity. This rare syndrome was originally described by Cockayne in the mid-1930s (63, 64). It is believed to be transmitted genetically in an autosomal recessive mode, as evidenced by the familial incidence and the history of consanguinity in some parents of affected children (137).

CLINICAL AND CELLULAR PHENOTYPES

Individuals with CS are characteristically dwarfed (Fig. 14–18), with growth and development arrested at ages from several months to a few years, although levels of growth hormone are normal (106). The other prevalent symptoms of the disease include photosensitivity, deafness, optic atrophy, intracranial calcifications, mental deficiency, large ears and nose, sunken eyes, long arms and legs proportional to body size, type 11 lipoproteinemia, and a general appearance of premature aging (Fig. 14–18) (63, 64, 106, 137, 271). In a recent comprehensive review of 140 cases of CS (271), it was concluded that the criteria of poor growth and neurological abnormalities are cardinal requirements for the diagnosis of the disease. Other common clinical manifestations include sensorineural hearing loss, cataracts, pigmentary retinopathy, cutaneous photosensitivity, and dental caries (271). The mean age of death of reported patients is about 12 years. Prenatal growth failure, severe neurological dysfunction, congenital structural anomalies affecting the eyes, and the presence of cataracts before the age of 3 have very poor prognostic significance (271). The occasional patients with CS who have survived to early adulthood do not manifest skin cancer, unlike the patients with XP (217).

The unusual sensitivity of individuals with CS to sunlight led to the notion that they may be defective in some form of repair of UV radiation damage to DNA. Skin fibroblasts from these individuals are indeed distinctly more sensitive to killing by UV radiation at 254 nm than are normal cells (4, 7, 80, 245, 345, 443, 444) (Fig. 14–19). The sensitivity of CS cells to killing by UV light is paralleled by their sensitivity to certain UV-mimetic chemicals. Thus, the survival of CS cells treated with either N-acetoxy-N-2-acetyl-2-aminofluorene or with 4-nitroquinoline 1-oxide (4-NQO) is significantly lower than that of normal cells so treated (7, 443, 444). However, according to most studies, CS cells survive as well as normal cells when treated with agents that are X-ray mimetic (e.g., monofunctional alkylating agents) (7, 345, 443, 444).

A number of CS cell lines show an increased number of SCEs induced by UV radiation; however, there is no close correlation between the levels of exchange and of cell killing by UV radiation (245). Several parameters of DNA repair in CS cells have been studied. Cells show a normal decrease in molecular weight of DNA during postirradiation incubation, suggesting that DNA incisions are made

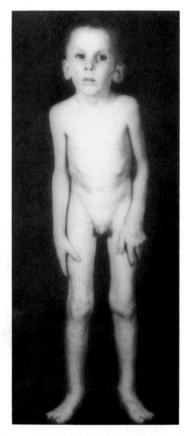

Figure 14–18 Individuals with CS have arrested growth and development, resulting in a dwarfed appearance. The sunken eyes and disproportionately long arms and legs are also very characteristic. (*Courtesy of James German.*)

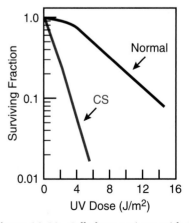

Figure 14–19 Cells from patients with CS are hypersensitive to killing by UV radiation. (*Adapted from Friedberg [102] with permission.*)

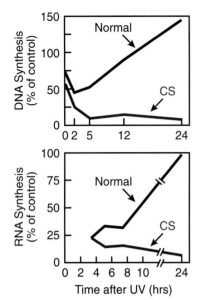

Figure 14–20 When cells from individuals with CS are exposed to UV radiation, the recovery of semiconservative DNA synthesis, measured by the incorporation of [³H]thymidine (top panel), and of total RNA synthesis, measured by the incorporation of [³H]uridine (bottom panel), is delayed relative to that observed in normal cells. (*Adapted from Lehmann [216] with permission.*)

at pyrimidine dimer and (6-4) photoproduct sites. Excision of thymine-containing pyrimidine dimers also occurs at normal levels in CS cells, and UDS is normal (1, 227). However, more-sensitive assays for measuring nucleotide excision repair of DNA do show a defect in CS cells. Thus, reactivation of the chloramphenicol acetyltransferase (*cat*) gene expression from a nonreplicating plasmid, following its inactivation by UV irradiation, is significantly reduced in CS cells relative to normal fibroblasts (21, 349). Additionally, the ability of CS cells to reactivate UV-irradiated adenoviruses is reduced (315). Several studies have demonstrated hypermutability of CS cells following exposure to UV radiation (6, 9, 199).

As mentioned above in the discussion of the XP variant form, when normal mammalian cells are exposed to UV radiation, there is an immediate and dose-dependent depression in the overall rate of DNA synthesis in the culture, with a progressive recovery to normal rates of DNA synthesis in the succeeding 5 to 8 h (Fig. 14–12). CS cells, like cells from patients with XP, show a pronounced defect in this recovery phenomenon (Fig. 14–20) (161, 227, 228, 245, 402). In growing CS cells, RNA synthesis is also inhibited by UV radiation (Fig. 14–20), but this aspect of nucleic acid metabolism recovers more rapidly after irradiation than DNA synthesis does (216, 228, 232, 247, 391). Indeed, the phenotype of defective recovery of RNA synthesis following the exposure of CS cells to UV radiation has proven to be a useful diagnostic criterion. When 52 patients with a possible clinical diagnosis of CS were tested, 29 (55.7%) showed a characteristic defect in RNA synthesis. Further analysis indicated that defective recovery of RNA synthesis correlated best with the presence of dwarfism, mental retardation, and Sun sensitivity (232).

These observations on the cellular phenotypes of CS cells led to the early suggestion that immediately following UV irradiation of normal cells there is rapid nucleotide excision repair of photoproducts from actively transcribing regions, thereby preventing significant inhibition of RNA synthesis. This preferential DNA repair was considered to be defective in CS cells but was undetectable by conventional measurements because quantitatively (as was suggested for XP variant cells) it represents a negligible proportion of total nucleotide excision repair (228, 247).

In recent studies, CS cells in culture have been reported to be hypersensitive to the toxic effects of deoxyguanosine (382) and to the DNA topoisomerase I inhibitor camptothecin (383). Additionally, it has been shown that whereas in normal cells treatment with deoxyguanosine results in a threefold increase in the pool sizes of purine deoxyribonucleotides (dATP and dGTP), these pool sizes are unaffected in CS cells (382). Exposure to UV radiation also results in different effects on the pool sizes of deoxyribonucleotides in CS and normal cells (382).

CS CELLS ARE DEFECTIVE IN THE PREFERENTIAL REPAIR OF TRANSCRIPTIONALLY ACTIVE DNA

Studies in which DNA probes for specific genes were used have demonstrated a defect in the preferential repair of actively transcribed genes in CS cells (140, 432). Additionally, such cells are defective in the repair of nuclear matrix-associated genes (248). By using plasmids carrying the actively transcribed *cat* gene as a probe for excision repair in CS and XP cells, it was observed that both cell types were defective in the repair of photoreactivable damage, i.e., pyrimidine dimers. When the UV-irradiated *cat* plasmid was pretreated with photoreactivating enzyme, no defect in *cat* gene reactivation was observed in CS cells, suggesting that pyrimidine dimers are the only substrate recognized (21).

Fusion of CS cells from different individuals to form heterodikaryons results in the recovery of normal rates of defective RNA synthesis after UV irradiation. On the basis of the results of these types of experiments, three genetic complementation groups have been identified in the disease (216, 402). Two of these are apparently uncomplicated by XP and are designated CS groups A and B. The third complementation group overlaps with XP group B and is associated with mutational inactivation of the *XPB* gene (318, 351, 435) (see above and chapter 8). Since CS can also be associated with XP groups D (207) and G (434) (see below), at least five complementation groups are apparently represented in this disease.

However, in keeping with recent nomenclature recommendations (221), only two CS genetic complementation groups are formally recognized (CS-A and CS-B). CS associated with XP is referred to as XP-B/CS, XP-D/CS, or XP-G/CS.

As indicated in chapter 8, the *ERCC6* gene has been shown to correct the cellular phenotypes of CS group B cells, suggesting that the *ERCC6* and the *CSB* genes are identical. At the molecular level, the primary, if not the exclusive, phenotype of CS-B cells is an inability to carry out preferential nucleotide excision repair of template strands (267, 428). However, the defect in CS cells is not strictly confined to the repair of the template strand of transcriptionally active genes. Whereas in CS cells the rate of repair of cyclobutane pyrimidine dimers in transcriptionally active genes is higher than that in transcriptionally silent loci, this rate is lower than that observed in the nontranscribed strand of transcriptionally active genes in normal cells (53, 428). Hence, the following kinetic hierarchy has been proposed for nucleotide excision repair of pyrimidine dimers in human cells. Repair is fastest in the template strand of transcriptionally active genes in normal cells. However, the rate of repair of the coding (nontranscribed) strand of transcriptionally active genes in normal cells is faster than that of either strand in CS cells, which in turn are still repaired more rapidly than transcriptionally silent genes in either cell type.

Cells from a second CS complementation group (designated CS-A) have also been shown to be defective in the preferential repair of the template strand in actively transcribed DNA (431). In addition to defective preferential strand-specific repair of bulky base damage, CS cells have been shown to be defective in such repair of DNA damage produced by exposure of cells to ionizing radiation (215). Consistent with the notion that CS cells are also defective in the repair of the nontranscribed strand in transcriptionally active genes, it has been shown that both CS-A and CS-B cells repair rDNA (which is considered to be transcriptionally silent with respect to RNA polymerase II [see chapter 7]) more slowly than do normal cells (53).

The *CSB* gene is expected to encode a polypeptide of 168 kDa (426). A notable feature of the predicted amino acid sequence of this polypeptide is the presence of a helicase signature that is highly conserved in a number of other eukaryotic polypeptides, some of which are known transcriptional activators (426). Although these motifs do not perfectly accommodate the amino acid sequence of the *E. coli* transcription-repair coupling factor (see chapter 5), there are remarkable similarities in the sequence of the putative CSB and transcription-repair coupling factor polypeptides (362). Thus, the possibility must be entertained that CSB protein is the functional homolog of transcription-repair coupling factor and that it (and presumably the polypeptide encoded by the putative *CSA* gene) plays a specific role in the recruitment of nucleotide excision repair proteins to sites of stalled transcription, and/or interacts with the stalled RNA polymerase complex (141).

Proteins such as transcription-repair coupling factor, which are apparently involved in the coupling of transcription to excision repair, are not expected to be absolutely required for transcription itself. Assuming this expectation to be true, it is not obvious how to correlate defects in the *CSA* and *CSB* genes with the transcription syndrome model discussed above, as may be the case with the forms of CS that are associated with defects in the *XPB* and *XPD* genes which encode proteins known to be required for normal transcription. A similar problem arises when considering the molecular pathogenesis of CS associated with XP complementation group G. As discussed above and in chapter 8, the *XPG* gene encodes a protein that is homologous to the *S. cerevisiae* Rad2 protein, a nonessential protein with endonuclease activity. Hence, as is suspected for the CSA and CSB proteins, it is unlikely that human XPG protein is required for transcription. Although they are not absolutely required, it is nevertheless possible that the products of the *CSA*, *CSB*, and *XPG* genes do in fact participate in the process of transcription. If so, functional defects in these proteins might indeed manifest with phenotypes consistent with altered transcription efficiency, as suggested for the other transcription syndromes. In this regard it is of interest that an analysis of the yeast Rad2 protein in the two-hybrid protein-protein interaction system (see chapter 6) revealed an ability of Rad2 protein to activate transcription of a *GAL1/lacZ* reporter gene when fused to the DNA-binding domain of Gal4 protein (19). To date it has not been possible to distinguish whether this is a completely spurious effect, not uncommonly encountered in this experimental system, or represents a true transcriptional activation function of Rad2 protein (19).

ASSOCIATION BETWEEN CS AND XP

The association of XP with CS has been extensively discussed. Before leaving this topic, a final word of diagnostic caution is appropriate. There is no single cellular or biochemical phenotype that categorically distinguishes XP from CS when both diseases are present in the same individual (except for the possible hypersensitivity to deoxyguanosine and to camptothecin mentioned above [382, 383]). Hence, it is imperative that when the diagnosis of both diseases is made it be unequivocal, since the association has profound biological and pathogenetic implications that are clearly confusing if the diagnosis is inaccurate.

The conundrum addressed above, of why a predisposition to skin cancer is not observed in TTD patients with proven defective nucleotide excision repair at the cellular level but without the clinical features of XP, emerges again when we consider CS uncomplicated by XP. Considerable attention has been focused on the observation that such individuals do not typically present with skin cancer. Indeed, some studies have even reported that two siblings with CS with clinically proven XP from complementation group B (including the presence of Sun sensitivity, freckling, and xerosis) are also less cancer prone (351, 435). Aside from the implication of the presence of other as yet unidentified genes to satisfy the cancer proneness in XP discussed above when we addressed this enigma in relation to TTD, other models have been advanced. As noted above, CS cells are apparently defective in the repair of cyclobutane pyrimidine dimers but not in the repair of other photoproducts in a plasmid-borne *cat* gene (21, 294). It has been suggested that skin cancer might be determined by the inability to repair other UV radiation-induced lesions such as (6-4) photoproducts or possibly even minor nondimer thymine adducts (21, 294). Cells from individuals with CS have the same abnormally increased number of UV radiation-induced chromosomal aberrations as is observed in cells from individuals with XP (360).

CS/XP-LIKE STATES

Several reports have documented the clinical and cellular phenotypes of cases that are neither typical XP nor CS but have features reminiscent of both. One such report concerns two Japanese siblings with slight cutaneous photosensitivity and hyperpigmentation but with increased cellular UV radiation sensitivity and profoundly decreased rates of recovery of RNA synthesis after exposure to UV radiation (166). Cells from these individuals are complemented by all known XP complementation groups and by CS group A and B cells (166). A second report has described a pair of Hispanic siblings with a clinical presentation of cutaneous photosensitivity and central nervous system dysfunction strongly reminiscent of the DeSanctis-Cacchione syndrome (see above). Cellular studies demonstrated hypersensitivity to UV radiation. However, there was no evidence of defective repair synthesis as is typically observed in XP cells. Rather, the cells manifested a defect in the rate of recovery of RNA synthesis, as is observed in CS (117).

Fanconi's Anemia

CLINICAL AND CELLULAR PHENOTYPES

Fanconi's anemia (FA) is characterized clinically by pancytopenia (depression of all cellular elements in the blood, i.e., erythrocytes, leukocytes, and platelets) and diverse congenital abnormalities, including short stature, hyperpigmentation, and radial aplasia (95) (Table 14–7). Initial studies on the inheritance of FA were uncertain because of the limited number of documented cases and the high ratio of affected to unaffected siblings (316, 397). However, subsequent segregation analy-

Table 14–7 Features of Fanconi's anemia

Type	Feature
Genetics	Autosomal recessive
	Recurrence risk of 25%
	Equal male/female sex ratio
	Occurs in all ethnic groups
	Carrier frequency approximately 1:200
	Elevated spontaneous chromosome breakage
	Cellular hypersensitivity to DNA-cross-linking agents
	Prenatal diagnosis available
	At least four complementation groups
Physical anomalies	Growth retardation and failure to thrive
	Hyperpigmentation
	Thumb and radial ray
	Other skeletal abnormalities
	Urogenital abnormalities
	Microcephaly, CNS[a] malformations
	Microphthalmia
	Ear malformations, hearing loss
	Cardiac abnormalities
	Gastrointestinal abnormalities
Developmental	Learning disabilities
Hematology	Bone marrow failure
	Macrocytic anemia, thrombocytopenia, or pancytopenia
	Elevated fetal hemoglobin level
	Highly increased risk of leukemia
	Frequent occurrence of cytogenetically abnormal bone marrow clones
Cancer risk	Increased risk of leukemia and solid tumors

[a] CNS, central nervous system.

Source: Reproduced from Auerbach (12) with permission.

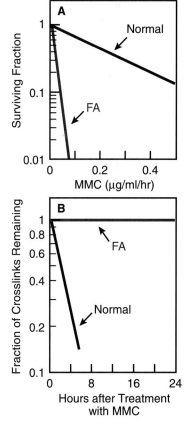

Figure 14–21 (A) Cells from patients with FA are typically hypersensitive to treatment with agents such as mitomycin (MMC) that result in the formation of interstrand cross-links in DNA. (B) Cells from certain FA genetic complementation groups also show a defect in the removal of cross-links from DNA (*Adapted from Poll et al. [307] and Fujiwara [107] with permission.*)

sis on a total of 90 families led to the firm conclusion that the disease is inherited through an autosomal recessive mode (346a). Several reports have noted a high incidence of leukemia and other malignant neoplasms in individuals with the disease and in asymptomatic blood relatives (113, 120, 346, 394, 397, 466a). A review of all cases in the International Fanconi Anemia Registry as of 1992 indicated that at least 15% manifested acute myelogenous leukemia or preleukemia, an incidence ~15,000-fold greater than that observed in children in the general population (12, 13).

Cells from individuals with FA have an increased number of spontaneous chromosomal aberrations and aberrations in response to treatment with a variety of mutagenic and carcinogenic agents, particularly those that cause interstrand DNA cross-links, such as nitrogen mustard, mitomycin, diepoxybutane, and photoactivated psoralen (14, 15, 181, 336, 346). FA cells are also abnormally sensitive to killing by these agents (Fig. 14–21) but not to DNA damage produced by UV or ionizing radiation (335). This sensitivity is mirrored by defective host cell reactivation of a plasmid-borne *cat* gene following transfection of nitrogen mustard-treated plasmid DNA into FA cells (76). Cultured FA lymphocytes have also been reported to exhibit a G_2 phase of the cell cycle that is about twice as long as in normal cells (88, 156). This parameter is potentially useful for diagnostic purposes. However, its biological relevance is unclear since FA lymphoblasts transformed with Epstein-Barr virus, and SV40-transformed fibroblasts do not show this cell-cycle abnormality. This defect is oxygen sensitive in FA fibroblasts and can be alleviated by exposure of the cells to low levels of oxygen (156). Subsequent studies have confirmed a marked increase in the percentage of FA cells in the G_2/M phase of the cycle after treatment with nitrogen mustard. However, the results were ambiguous in some cases (26).

A variety of other miscellaneous and poorly understood cellular phenotypes have been described in FA, including abnormal lymphokine (interleukin 6) production (18a, 18b, 323, 461) defective (210) or altered (330) endonuclease activities, a defective damage recognition protein for cross-links in DNA (144), overproduction of tumor necrosis factor α (322b, 347), increased formation of 8-hydroxydeoxyadenosine possibly associated with catalase deficiency (400), impaired metabolism of NAD^+ accompanied by a decreased ability to transfer ADP-ribose to acid-precipitable material (186), and cellular hypersensitivity to oxygen (329). Additionally, it has been observed that when FA cells are exposed to γ radiation they suffer a markedly reduced level of apoptosis compared to normal cells (322a). Consistent with the observation that radiation-induced apoptosis in normal cells is dependent on the *p53* gene, p53 protein is not induced following exposure of FA cells to γ radiation (322a). It remains to be determined how these diverse phenotypes relate to the primary molecular pathology of this disease. Overproduction of tumor necrosis factor α has also been observed in heterozygous individuals (322b), and hence may be of diagnostic significance in the detection of such people in the population.

DNA REPAIR IN FA CELLS

The demonstration of the persistence of reversibly denaturable DNA (an expression of DNA cross-links) after treatment with mitomycin first suggested that FA cells are deficient or defective in the repair of interstrand DNA cross-links (107, 108, 110) (Fig. 14–21). However, these results have not been unambiguously reproduced when a variety of cross-linking agents were used (99, 183, 307). A possible explanation for this ambiguity is the presence of genetic heterogeneity in this disease (288). Two genetic complementation groups (designated A and B) have been identified on the basis of complementation of various phenotypes in somatic cell hybrids (86). Recovery of the inhibition of semiconservative DNA synthesis following exposure to cross-linking agents is defective in FA cells of only one of these groups (group A) (264, 266) (Fig. 14–22). Consistent with this observation, the incision of DNA from FA cells containing psoralen cross-links is slower in FA group A cells than in normal or FA group B cells and the removal of psoralen (324) and of mitomycin-induced cross-links (246) is reduced in these cells relative to group B cells. The emergence of techniques for examining gene-specific excision repair in human cells (see chapter 7) has demonstrated that FA cells from genetic complementation group A repair only 50% of the interstrand cross-links present

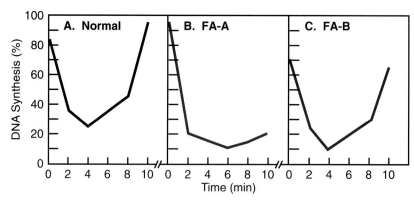

Figure 14–22 FA cells from genetic complementation group A (middle panel) are defective in the recovery of normal rates of semiconservative DNA synthesis after exposure to psoralen plus 365-nm radiation. Normal cells and FA cells from genetic complementation group B are shown in the left and right panels, respectively. (*Adapted from Moustacchi and Diatloff-Zito [264] with permission.*)

in the *DHFR* gene by 24 h after exposure to cisplatin (472). These FA cells are also deficient in the removal of cisplatin-induced cross-links in ribosomal DNA (472).

GENETIC COMPLEXITY OF FA

Following the identification of FA complementation groups A and B, an analysis of available FA cell lines provided evidence for the existence of two other complementation groups, designated C and D. These assignments are based on the observation of complementation of FA cellular phenotypes (such as sensitivity to cross-linking agents) in hybrids derived by fusion of different cell lines (389). Hence, like XP, CS, and TTD, FA is characterized by locus heterogeneity. One of the FA genes (*FAC*) (for Fanconi's anemia group C) has been mapped to chromosome 9q22.3 (389), not far from the region (9q34) where the *XPA* gene maps (see chapter 8). (This gene was originally designated as *FACC*, but is changed here to *FAC* to be consistent with the altered nomenclature for human repair-related genes discussed earlier in the chapter [221]). A second FA gene has been tentatively localized to chromosome 20q by linkage analysis (244).

FA cells have a decreased frequency of point mutations induced by psoralen treatment relative to that observed in normal cells (289, 290). However, they have an increased frequency of deletions (289). These observations have led to the suggestion that FA cells may be defective in an error-prone pathway of DNA repair that involves the replicative bypass of lesions in normal cells (289) (see chapter 12).

Mouse lymphoma cell line mutants from two different genetic complementation groups have been shown to be abnormally sensitive to killing by cross-linking agents (138, 139, 321). However, there is no evidence of defective excision repair of cross-links in these cells (139). A Chinese hamster cell line designated V-H4 is sensitive to killing by mitomycin and does not correct the sensitivity to this agent in FA group A cells in interspecies hybrids, suggesting that this cell line represents the rodent equivalent of FA genetic complementation group A (17). Cocultivation of human FA cells with ''FA-like'' rodent cell lines designated MCN-151 and MCE-50, assigned to rodent complementation groups I and II, respectively, results in partial correction of mitomycin-induced chromosomal aberrations in the FA cells (322).

MOLECULAR CLONING OF FA GENES

Functional complementation of the sensitivity of FA group C cells to mitomycin and diepoxybutane facilitated the molecular cloning of an *FAC* cDNA (390). The cDNA contains an open reading frame of 557 amino acids that could encode a protein with a calculated size of ~63 kDa (390). The predicted amino acid sequence of the *FAC* open reading frame does not reveal any clues to the function of the protein. Northern (RNA) analysis of RNA from lymphoblasts and from several other cell types indicates the presence of three distinct transcripts of 2.3, 3.2, and 4.6 kb, which differ exclusively in the length of their untranslated regions (390). Surprisingly, immunoprecipitation and immunofluorescence studies with an antibody to the *FAC* gene product suggest that the protein is localized in the cytoplasm of lymphoblast cell lines (464, 466b).

Confirmation that the *FAC* gene is indeed related to FA group C was provided by the identification of a mutation which changes codon 553 from Leu to Pro in an FA group C cell line (390). This mutation has been shown to abolish the functional activity of the *FAC* gene (116a). Analysis of the *FAC* gene detected mutations in 4 of 17 (23.5%) individuals with FA. An identical splice site mutation was detected in two Ashkenazi Jewish individuals, and this allele was also found in three other Ashkenazi Jews following the screening of a further 21 families (455). These observations suggest that this splice site mutation accounts for the majority of Ashkenazi Jewish patients with FA.

More extensive mutational analysis of 174 racially and ethnically diverse FA families has identified disease-associated mutations in the *FAC* gene in 14.4% of the families (430). At present there is no indication of how mutations in genes that confer an apparent inability to repair interstrand DNA cross-links result in the complex phenotype of FA, in particular the pancytopenia that is so characteristic of this disease. One way out of this problem is to consider the possibility that defective DNA repair is not the primary phenotype on FA cells. This possibility is strengthened by the observation mentioned above that the *FAC* gene product is localized in the cytoplasm rather than the nucleus. Additionally, antibodies against tumor necrosis factor α result in the correction of multiple phenotypes of FA cells in vitro, including defective repair of mitomycin. Indeed, these and other observations have led to the suggestion that FA genes control the expression of a network of other genes involved in the development of the hematopoietic system, genomic stability, and the repair of DNA cross-links (322b).

FA cells from genetic complementation group D are partially corrected for their sensitivity to mitomycin following transfection with a fragment of cloned mouse DNA (83a). This fragment is conserved in the human genome and maps to human chromosome 11q23 (83a), a region of the genome that is of particular interest with respect to ataxia telangiectasia, another hereditary disease with abnormal cellular responses to DNA damage (see below).

Hereditary Diseases Characterized by Defective Cellular Responses to DNA Damage

The diseases discussed thus far are all characterized by persuasive evidence for primary (or possibly secondary) defects in nucleotide excision repair or processes believed to be related to this DNA repair mode (i.e., the excision repair of interstrand cross-links). Several other inherited human diseases are phenotypically characterized by defective cellular responses to DNA-damaging agents. Among these diseases are *ataxia telangiectasia* (AT), *Bloom's syndrome* (BS), and an as yet unnamed clinical syndrome involving a single patient thus far, from whom a cell line designated *46BR* was isolated.

Ataxia Telangiectasia (Louis-Bar Syndrome)

AT is another genetic disorder with an autosomal recessive mode of inheritance (395, 398a). The disease occurs with an incidence of about 1 per 40,000 live births and affects many systems of the body, particularly the nervous system, the immune system, and the skin (10, 29, 30, 39, 104, 115, 116, 193, 237, 252, 296, 356, 363). The most characteristic neurological disorder is a profound *cerebellar ataxia*, resulting in a staggering gait, severe muscular uncoordination, and progressive mental retardation (10, 29, 30, 39, 104, 115, 116, 193, 237, 252, 296, 356, 363) (Color Plate 11). The term *telangiectasia* describes the marked dilation of small blood vessels, most easily observed in the eye and the skin (Color Plate 11). A significant feature of the disease is an *immune dysfunction* that may render affected individuals susceptible to intercurrent infection (10, 29, 30, 39, 104, 115, 116, 193, 237, 252, 296, 356, 363).

This dysfunction also manifests with an increased incidence of *neoplasms of the lymphoreticular system*. Approximately 10% of individuals with AT develop neoplasms, 88% of which are lymphoreticular, including Hodgkin's disease and non-Hodgkin lymphomas (305, 395, 405). Almost all of the neoplasms occur before individuals with AT are 20 years of age, in contrast to the median age in the general population of about 55 years. Overall cancer incidence was measured retrospectively in 263 AT homozygotes and was found to be increased 61- and 184-fold in whites and blacks, respectively. This cancer excess was most pronounced for lymphomas (263). Heterozygous relatives of patients with AT are more likely to develop neoplasms than are persons unrelated to these patients, and it has been estimated that the presence of the AT gene in the population may be responsible for about 5% of all persons who die of cancer before the age of 45 years (193). One retrospective study of adult blood relatives of patients with AT in 110 white families showed a significant elevation of cancer incidence rates compared with those in spouse controls (396). In this study, breast cancer was most clearly associated with heterozygosity for the disease, leading to the suggestion that as many as 8.8% of patients with breast cancer in the U.S. white population might be heterozygous for AT (396).

CELLULAR PHENOTYPES IN AT

Sensitivity to Killing by Ionizing Radiation

Cells from individuals with AT present a number of phenotypes. The most consistent of these is the phenomenon of increased sensitivity to killing by ionizing radiation and X-ray-mimetic chemicals (67, 92, 226, 231, 268, 406, 407, 410). This phenotype was first suspected when several patients with AT who had tumors demonstrated severe or fatal reactions to ionizing radiation therapy (134, 261). This prompted an examination of the response of AT fibroblasts in culture to ionizing radiation and the startling observation by Malcolm Taylor and his colleagues that cells from such individuals are indeed radiosensitive (406). There is virtually no overlap in the survival of normal and AT cells exposed to ionizing radiation (8, 51, 297, 298), and this extreme radiosensitivity is one of the most outstanding features common to all AT cell lines so far examined (70) (Fig. 14–23).

AT cells are not abnormally sensitive to UV radiation or to chemicals classified as UV radiation mimetic, e.g., N-acetoxy-N-2-acetyl-2-aminofluorene or 4-NQO. Cells from obligatory heterozygous individuals have been reported to show less-pronounced but nonetheless detectably increased sensitivity to ionizing radiation, and it has been suggested that this might constitute a means for detecting heterozygotes in the population (11, 297). However, this difference is small and of insufficient sensitivity to constitute a reliable means of detecting heterozygotes. Cells from heterozygous individuals with AT also manifest more-pronounced chromatid damage in the G_2 phase of the cell cycle after exposure to ionizing radiation, and this test offers greater potential for detecting heterozygous AT carriers (366).

Abnormal DNA Synthesis

Semiconservative DNA synthesis in cells from individuals with AT differs from that in normal cells in two principal ways (285). First, in unirradiated cultures some AT cells have a lower inherent rate of DNA synthesis and hence a longer S phase of the cell cycle (66, 269, 285, 350). Second, when normal human fibroblasts are exposed to ionizing radiation, there is a dose-dependent inhibition of the rate of DNA synthesis, as measured by the uptake of radiolabeled thymidine (83, 97, 159, 214, 286) (Fig. 14–24). On the other hand, most, if not all, AT cells are significantly more resistant to radiation- and bleomycin-induced DNA synthesis inhibition (83, 89, 91, 97, 159, 214, 282, 286) (Fig. 14–24). This phenomenon of radioresistant DNA synthesis was detected in all 27 cell lines examined from patients with AT (466).

On the basis of the results of experiments in which the molecular weight distribution of newly synthesized DNA in alkaline sucrose gradients was examined, it appears that replicon initiation is inhibited by X irradiation or by bleomycin to a much smaller extent in AT cells than in normal cells (286). It has been suggested that in normal cells, DNA damage induces a change in clusters of replicons (the domain of a cluster having a molecular weight of about 10^9) that precludes the initiation of DNA synthesis in any replicon in that cluster until the damage is repaired (281, 309). In addition to the partial defect in replicon initiation, it has been suggested that AT cells show complete radioresistance with respect to DNA chain elongation (285). However, studies with AT lymphoblastoid lines have shown that the rate of DNA chain elongation is similar to that in normal cells (259).

The biochemical basis for the radioresistant DNA synthesis in AT cells is not known. Since poly(ADP-ribose) synthesis is stimulated by strand breaks in DNA (see chapter 7), it has been suggested that AT cells fail to inhibit DNA synthesis because of lower rates of poly(ADP-ribose) synthesis relative to normal cells (89). However, when cells were exposed to 3-aminobenzamide, a known inhibitor of poly(ADP-ribose) synthesis, the dose-response for ionizing radiation-induced DNA synthesis was unaffected in normal cells (284). Exposure of normal fibroblasts carrying an extrachromosomally replicating plasmid to γ radiation results in inhibition of DNA synthesis of both chromosomal and plasmid DNA at radiation

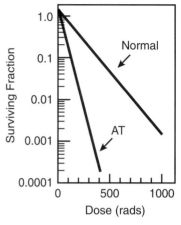

Figure 14–23 Cells from individuals with AT are abnormally sensitive to killing by ionizing radiation. (*Adapted from Friedberg [102] with permission.*)

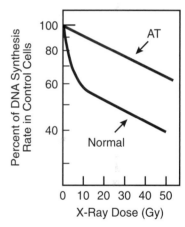

Figure 14–24 Radioresistant DNA synthesis in AT cells exposed to ionizing radiation. Note that normal cells manifest an immediate rapid inhibition component in the presence of low doses of ionizing radiation, followed by a less dramatic inhibition in the presence of higher doses. The former component has been suggested to reflect the inhibition of replicon initiation and is sensitive to low doses of radiation because the target (a large replicon cluster) is relatively large. The less-severe component of replication inhibition has been suggested to reflect the inhibition of DNA replication fork progression, which is a smaller target for damage. The observation that the (presumed) inhibition of fork progression occurs to about the same extent in AT cells and normal cells (indicated by the parallel slopes of the two curves) suggests that AT cells are primarily insensitive (resistant) to inhibition of the initiation of replication following exposure to ionizing radiation. (*Adapted from Friedberg [102] with permission.*)

doses which produce insignificant levels of damage in the plasmid DNA. These results suggest that inhibition of DNA synthesis is mediated by a *trans*-acting factor rather than by a *cis*-acting signal (208).

Radioresistant DNA synthesis can be partly mimicked by treating normal cells with caffeine or by incubating them in hypertonic medium after irradiation (283). Because both of these perturbations are known to alter chromatin structure, it has been suggested that the synthesis of radioresistant DNA in AT cells may be due to an intrinsic difference in chromatin structure between AT and normal cells (283). This suggestion has prompted several studies on enzymes that alter the topology of DNA and hence might affect chromatin structure. Reduced DNA topoisomerase II activity has been observed in extracts of Epstein-Barr virus-transformed lymphocytes from some individuals with AT relative to normal controls (373). Additionally, some AT cell lines have been reported to be abnormally sensitive to killing by treatment with novobiocin (a coumarin which blocks the ATPase activity of the β subunit of *E. coli* DNA gyrase, the prokaryotic equivalent of topoisomerase II) (77) and with the epipodophyllotoxin VP16, an inhibitor of topoisomerase II (151, 376). Reduced levels of DNA topoisomerase II would be expected to reduce the amount of target enzyme available for interaction with these drugs. Hence, the observation of reduced topoisomerase II activity in AT cells seems contradictory to that of increased sensitivity to topoisomerase inhibitors. At least one cell line has been shown to overproduce topoisomerase II in all phases of the cell cycle but retains typical X-ray sensitivity and radioresistant-DNA synthesis. Thus, altered levels of this enzyme do not explain the fundamental biochemical defect in this disease.

Differences in the levels of topoisomerase I activity have not been detected in AT cells (377). Nonetheless, consistent with the abnormal sensitivity of AT cells to agents which induce DNA strand breaks, these cells are also abnormally sensitive to camptothecin, an agent which traps covalently bound topoisomerase I-DNA intermediates, thereby leaving strand breaks in DNA (377).

It is not clear that the phenotype of radioresistant-DNA synthesis explains the sensitivity of AT cells to ionizing radiation and to radiomimetic chemicals (214). Following transfection of AT cells from complementation group D (see below) with total human genomic DNA, a cell line with normal levels of radiation resistance was isolated. However, this cell line retained a DNA synthetic pattern which closely resembles that observed in the parental AT line (220). Similarly, when AT cells from the same genetic complementation group were fused to HeLa cells, the resulting hybrids showed normal levels of radiation sensitivity but retained the radioresistant-DNA synthesis phenotype (191). In a related study, partial correction of radiosensitivity was observed following gene transfer in AT group D cells, but radiosensitive DNA synthesis was not complemented (179).

Cell Cycle Defects in AT Cells

Perhaps no single phenotype of AT cells is more difficult to understand than the plethora of conflicting cell cycle perturbations after exposure to ionizing radiation. Several studies have documented an accumulation of AT cells in the G_2 phase of the cell cycle following exposure to ionizing radiation or radiomimetic chemicals (22, 23, 98, 162, 326, 375, 404). Some of these studies indicate a delay in the G_2 phase of the cycle (23). However, other studies suggest that the accumulation of AT cells in G_2 reflects an abnormally hurried passage of cells through G_1 and S (404). Yet other studies suggest a delay in both G_1 and in S (24). Perhaps some of this confusion stems from differences in different AT cell types and from the particular cell cycle phase during which cells are irradiated. For example, it has been suggested that the G_2 arrest following ionizing radiation is diminished in untransformed AT fibroblasts (467) but is enhanced in transformed fibroblast and lymphoblast cell lines (98, 162, 375).

A series of diploid fibroblast cell strains obtained from normal, AT heterozygous, and AT homozygous individuals were monitored for DNA synthesis by incorporation of bromodeoxyuridine and were then examined by flow cytometry after staining with a fluoresceinated antibody to bromodeoxyuridine, thereby providing a direct measure of the number of functional S phase cells as well as the cell cycle distribution, based on DNA content. These studies showed that after exposure to ionizing radiation, normal and heterozygous cells underwent a transient G_2/M block which delayed their passage into G_1 by 6 to 8 h. Additionally, the passage of these cells from G_1 to S was blocked. On the other hand, AT homozygous cells failed to arrest in G_2/M during the first cell cycle and were also not delayed in their passage from G_1 to S (326) (Fig. 14–25). Most recently it has been reported that at early times after irradiation, AT cells show a reduced G_2 block, but at later times (possibly after DNA damage has been incorrectly processed), the G_2 block is prolonged (25).

The prominent phenotype of the failure of AT cells to delay DNA replication in the presence of DNA damage resulting from exposure to ionizing radiation may be a consequence of a loss of G_1 checkpoint control. However, there are indications for defects in multiple checkpoints for damaged DNA in AT cells (see chapter 13). It has been suggested that AT cells may suffer delayed induction of p53 protein. A different study has reported reduced levels of p53 induction in AT cells after treatment with ionizing radiation. However, induction after UV irradiation was normal (see chapter 13).

An interesting insight into the possible molecular basis for defective cell cycle control in AT stems from observations on the phosphorylation of one of the subunits of replication protein A (236). Replication protein A is a trimeric single-stranded-DNA-binding protein that is required for DNA replication and for nucleotide excision repair (see chapter 8). Phosphorylation of the p34 subunit of this protein normally occurs during the S and G_2 phases of the cell cycle but not during G_1. When normal cells are exposed to ionizing radiation phosphorylation of p34 is induced; however, this response is significantly delayed in AT cells (236). The phosphorylation response to UV radiation in normal and AT cells was found to be identical, however.

Chromosomal Abnormalities

Cells from individuals with AT have a high frequency of spontaneous chromosomal abnormalities, and X irradiation markedly increases the number of chromosome breaks in such cells (65, 190, 287). AT lymphocytes frequently manifest spontaneous rearrangements involving chromosomes 7 and 14, particularly at bands 7p14, 7q35, 14q12, and 14q32 (16, 189, 190, 343, 409), which correspond to the locations of the T-cell α-, β-, and γ-chain gene complexes. Indeed, detailed study of a single AT T-cell line with inversion of chromosome 7 demonstrated that the inversion arose by site-specific recombination between the T-cell receptor γ variable segment and the T-cell receptor β joining segment (388). Despite the increased general chromosomal instability in AT fibroblasts, the spontaneous rearrangements observed in these cells are more random.

Integrating plasmid vectors carrying selectable genes have provided interesting insights into possible defects in genetic recombination in AT cells. The vectors were cut at unique restriction sites in the coding regions of the selectable genes to model strand breaks produced by ionizing radiation or by radiomimetic chemicals. Both DNA rejoining and recombination were assessed by the ability of cells to survive growth under selective conditions. An AT cell line showed normal levels of homologous recombination under these experimental conditions. However, the rejoining of double strand breaks was reduced relative to that in wild-type cells (416).

It has been suggested (44, 190, 304) that the fundamental defect in AT may be defective recombination, resulting in an inability to carry out normal rearrangement and repair of genes. This hypothesis could explain the observed immune defects, the numerous karyotypic abnormalities, and the radiation sensitivity of AT cells (255). This model of defective recombination has been examined by measuring the rates of spontaneous mitotic recombination between directly repeated genes carried on vectors integrated into the genome of AT cells. Surprisingly, spontaneous recombination rates were found to be 30 to 200 times greater in AT cells than in normal or XP cells (255). These observations are not obviously consistent

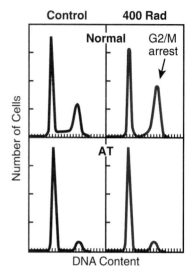

Figure 14–25 Normal (and AT heterozygote cells [not shown]) show an arrest in the G_2/M phase of the cell cycle when exposed to ionizing radiation. AT cells fail to show this cell cycle arrest. (*Adapted from Rudolph and Latt [326] with permission.*)

with the inability of AT cells to rearrange DNA productively. However, they are certainly consistent with the notion of defective cell cycle control stated above (255).

Defective DNA Repair

There is no definitive evidence that the profound radiosensitivity of AT is the direct result of a defect in the repair of ionizing-radiation damage, although in early studies this was a highly favored hypothesis of the molecular pathology of this disease. Investigations of the biochemistry of DNA repair in irradiated AT cells have not revealed detectable defects in the kinetics of rejoining of single- or double-strand breaks in DNA (230, 365, 408, 438). However, some (but not all) AT cell lines show a decreased rate and extent of repair synthesis of DNA after exposure to ionizing radiation (296). In addition, when DNA from irradiated normal and AT cells is incubated with an extract of *Micrococcus luteus* containing an enzyme activity that apparently recognizes an as yet unidentified form of base damage caused by ionizing radiation, sites sensitive to the enzyme disappear from the DNA of some AT cells more slowly than from the DNA of normal cells, suggesting that AT cells are defective in the excision of some form of ionizing-radiation damage (300).

Other experiments have provided provocative suggestions of a DNA repair defect in AT cells (90, 163, 164, 367). For example, crude extracts of normal human cells apparently contain an enzyme(s) that enhances the priming activity of γ-irradiated ColE1 plasmid DNA for purified DNA polymerase I of *E. coli*. Exactly what effects this enhancement of DNA synthesis is not known, but it has been speculated that sites of base damage in γ-irradiated DNA may be attacked by endonucleases which leave 3' moieties that are ineffective primers for DNA synthesis. This is reminiscent of the effect of 3'-terminal apurinic or apyrimidinic sites on DNA synthesis catalyzed by *E. coli* DNA polymerase I (see chapter 4). The removal of these moieties by activities present in human cell extracts may then facilitate enhanced DNA synthesis. Extracts of a number of AT cell lines contain smaller amounts of this putative priming-enhancer activity, whereas the activity of AT heterozygotes is indistinguishable from that of normal cells (90, 163, 164). Studies designed to elucidate this phenomenon led to the observation that AT cell extracts modify deoxyribose-phosphate (dRp) residues by converting them to a form (dRp-X) with altered chromatographic properties (180). This modification requires both an enzymatic activity and a low-molecular-weight cofactor. Extracts of normal cells contain a dialyzable inhibitor which suppresses the reaction catalyzed by extracts of AT cells.

The more-recent literature continues to allude to possible biochemical defects in AT cells that are consistent with at least a secondary, if not primary, DNA repair or processing defect. For example, to assess the ability of AT cells to join the ends of linear plasmid molecules, such cells were transfected with plasmids which require joining of ends in order to survive (328). AT cells were found to circularize these plasmids less efficiently than normal cells did (328). In different studies, evidence has been presented for a defect in a specific nuclear DNA-binding protein in AT cells which is observed when normal cells are exposed to ionizing radiation damage (372). The protein is apparently present in the cytoplasm of unperturbed normal cells and is translocated to the nucleus when cells are irradiated (412). The protein has been purified and characterized as three DNA-binding species (412). A similar activity has been reported to be constitutively present in the nuclei of AT (but not normal) cells (411).

Genetic Heterogeneity in AT

The radioresistant DNA synthesis that characterizes AT cells has provided the basis for an assay to detect complementation of the defect in cell hybrids (Fig. 14–26). A comprehensive study of 50 different cell strains identified four genetic complementation groups, designated groups A/B, C, D, and E, of which group A/B is the largest (167, 168). In addition to these 50 patients, additional patients with clinical features characteristic of AT but with weak cellular phenotypes have been

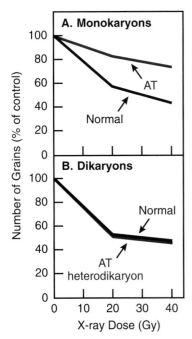

Figure 14–26 AT cells can be assigned to different genetic complementation groups on the basis of the ability of heterodikaryons to support the normal inhibition of DNA synthesis (radiosensitive DNA synthesis) after exposure to ionizing radiation. (A) Monokaryons from AT cells show reduced inhibition of (radioresistant) DNA synthesis relative to control cells. (B) Heterodikaryons (dikaryons) formed by fusing AT cells from different complementation groups show a normal inhibition response. (*Adapted from Friedberg [102] with permission.*)

identified (96, 167, 473). These cases typically follow a relatively mild clinical course. They may constitute a separate genetic class(es) from the four genetic complementation groups described above. Alternatively, they may represent leaky alleles of the four major groups. Regardless, it is evident that like XP, TTD, and CS, AT is a genetically complex disease, and it is likely that its total complexity is not yet fully defined.

The Nijmegen Breakage Syndrome and Other Possible AT Variants

Some patients with clinical and cellular phenotypes that resemble AT in some respects have been identified. In particular, these individuals manifest immunological deficiencies, chromosomal instability, microcephaly, and developmental delay but do not have ataxia and telangiectasia. This collection of clinical features is referred to as the *Nijmegen breakage syndrome* (453). Fibroblasts from such individuals show increased sensitivity to killing by ionizing radiation and also manifest radioresistant-DNA synthesis (169, 398). Complementation of the radioresistant-DNA synthesis phenotype has identified genetic heterogeneity in the disease. Cell lines from individuals with Nijmegen breakage syndrome complement AT cells (72, 169). However, cells from twin girls with typical clinical and cellular phenotypes of AT, who additionally presented with microcephaly and mental retardation, failed to complement radioresistant-DNA synthesis in cells from an individual with the Nijmegen breakage syndrome (72). These observations suggest that the Nijmegen breakage syndrome and AT represent a spectrum of a single clinical disorder. This concept is supported by various studies in which other "AT-like" syndromes have been identified, including syndromes involving typical clinical and cellular features of the disease but without evidence of radioresistant-DNA synthesis (473).

Concluding Comments on AT Phenotypes

Consultation with one of our knowledgeable colleagues in the field, whose opinion we respect highly, led to the suggestion that the AT field "is a mess" and that the preceding discussion reflects little more insight than that! If so, we hope that enthusiastic young (and old) investigators who are undaunted by messy fields will recognize this as a challenge. Further phenotypic characterization of AT cells, especially those immortalized with viruses like simian virus 40, is unlikely to clarify this mess to any significant extent. The next frontier is surely the isolation and characterization of the AT gene(s), with the hope that the translated structure of this gene will be informative and with the expectation that it will prove to be a powerful probe for exploring interactions between the AT proteins and other cellular proteins whose properties will be at least as informative. As discussed below, however, progress toward the cloning of the AT gene(s) has been slow and holds no promise for immediate reward at the time of this writing.

PURSUING AT GENES

The isolation of different AT genes will provide a definitive description of the genetic complexity of this disease. Unlike the situation with XP, progress toward this holy grail has been slow, despite the presence of an apparently selectable cellular phenotype for phenotypic complementation. At least one report has documented the problems that have plagued numerous attempts to isolate stable transfectants following DNA-mediated gene transfer into AT cells (238). A more serious and recently appreciated problem is the limited specificity for correction of AT cellular phenotypes, in particular the phenotype of radiation or radiomimetic chemical sensitivity. In one study, the screening of an episomal vector-based human cDNA library for complementation of hypersensitivity to the radiomimetic agent streptonigrin yielded plasmids carrying nine unrelated cDNAs (256). Some of these partially or nearly completely corrected streptonigrin sensitivity, the hyperrecombination phenotype discussed above, and even radioresistant-DNA synthesis. However, none of these cDNAs were from the chromosome 11q23 region of the genome where the genes implicated in AT groups A/B, C, and D have been mapped either by linkage analysis and/or independently by chromosome transfer studies (100, 114, 175, 188, 209, 332, 380, 474). In independent investigations, another cDNA which partially complemented the radiosensitivity and sensitivity to streptonigrin of AT group D cells failed to correct the radioresistant DNA synthesis phenotype of these cells and also did not map to human chromosome 11 (152). Hence, the inviting complementable phenotypes of AT cells notwithstanding, the high background of nonspecific partial and even complete correction of these may preclude the cloning of the AT gene(s) by a straightforward functional approach, at least as attempted thus far, and the AT gene(s) remains elusive.

While genetic complementation studies lead to the strong suggestion of multiple AT genes, the observation that these genes all map to the same region of the human genome begs the question whether there is more than a single AT locus. Assuming that phenotypic complementation of radioresistant DNA synthesis in heterodikaryons from different individuals with AT reflects the existence of distinct genetic loci, as is the case with XP and CS, the mapping data suggest that human cells contain multiple closely linked AT genes. Alternatively, functional complementation in AT heterodikaryons may be intragenic, in which case a single AT gene may be represented in the population by multiple mutant alleles.

X-RAY-SENSITIVE RODENT CELL LINES AS MODELS FOR AT

A number of X-ray-sensitive rodent cell lines have been isolated (69, 171, 173, 174, 379, 417, 469) and represent potential in vitro systems for cloning human AT genes, in the same way that rodent mutants defective in nucleotide excision repair facilitated the isolation of *ERCC* genes, several of which have been implicated in XP and CS. There are indications for at least six distinct genetic complementation groups among these mutants (172). However, to date all of these have been shown to complement AT cells in fusion hybrids, and none are complemented by the introduction of human chromosome 11, suggesting that they are not defective in homologous loci.

Bloom's Syndrome

CLINICAL AND CELLULAR PHENOTYPES

The syndrome originally described by Bloom (28) is characterized by a strikingly low birth weight and by stunted growth (28, 123, 124). Light-induced telangiectasia develops on the skin of the face in typical *"butterfly" lesions* (Color Plate 12), the head is elongated, and affected individuals are highly susceptible to infections (28, 123, 124). The incidence of BS is very low in the general population but occurs in about 1 in 58,000 Ashkenazi Jews (126). Evidence points to an autosomal recessive mode of inheritance, and there is no evidence of locus heterogeneity in this disease (454). Several notable features distinguish BS cells in culture. Generation times for these cells are longer than for normal fibroblasts, and plating efficiencies are lower (131). Cells undergoing DNA synthesis incorporate about one-half as much labeled DNA precursor per unit time as normal cells, and newly synthesized DNA strands elongate more slowly in some BS cells than in normal cells (131, 143, 153). However, the replication intermediates which accumulate in vivo are ~20 kb in size, significantly larger than Okazaki fragments (235, 239). Levels of DNA polymerases α, β, and γ are normal in these cells, however (292).

BS cells in culture also have large numbers of spontaneous chromosomal breaks and rearrangements (125, 129). The distinguishing chromosome rearrangements in this syndrome are SCEs, which occur with a frequency 10 to 15 times greater than in normal cells (46, 129). Disproportionate numbers of particular chromosomal abnormalities referred to as *quadriradial figures*, so called because the chromosome is apparently extended in four directions (Fig. 14–27), are also present in BS cells (46, 127).

The aforementioned characteristics of sunlight sensitivity and large numbers of chromosomal aberrations, together with the observation that one in nine BS patients develops malignancies at an average age of 20 years, initially prompted speculation that the primary defect in BS may be in DNA repair (reviewed in

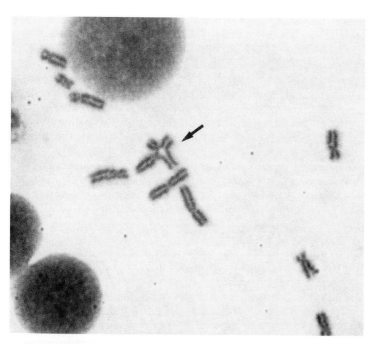

Figure 14–27 A quadriradial chromosome typically observed in the cells of individuals with BS. (*Courtesy of Roger A. Schultz.*)

reference 104). To date, direct evidence for this assertion is marginal. One study has reported that two BS fibroblast lines are abnormally sensitive to UV radiation at 254 nm (131) and other studies have reported increased sensitivity to killing by mitomycin (158, 165). However, several studies have found no such differences (reviewed in reference 104). A number of BS cell lines showed increased numbers of DNA strand breaks following exposure to near-UV light (313 nm) (155, 468). The sensitivity of BS cells to chemical mutagens has also been tested. Treatment of cells with ethyl methanesulfonate (7) or *N*-ethyl-*N*-nitrosourea (205) has been reported to result in increased killing relative to normal cells. Additionally, when peripheral blood lymphocytes from patients with BS are treated with ethyl methanesulfonate, the number of SCEs is increased relative to that in cells from normal donors (204). However, in general most other measures of DNA repair in UV-irradiated BS cells have proven to be normal.

An intriguing and long-standing observation in the literature is that the co-cultivation of BS cells with other cell types affects the frequency of chromosomal abnormalities in all the cells (93, 422). Furthermore, the propagation of normal lymphocytes in conditioned medium derived from BS fibroblast cultures results in a significant increase in the frequency of SCEs in the normal lymphocytes, while cocultivation of BS skin fibroblasts with Chinese hamster ovary (CHO) cells or with normal human cells reduces the frequency of SCEs in the former population. Addition of bovine superoxide dismutase to cultures of phytohemagglutinin-stimulated lymphocytes from normal individuals has been reported to suppress the clastogenic activity present in concentrated ultrafiltrates of BS cell media (93). This observation has led to the suggestion that BS cells may be deficient in the detoxification of reactive oxygen species which lead to DNA damage (93). It has also been reported that whereas normal human cells in culture show a defined temporal relationship of the activity of several DNA repair modes relative to DNA replication (135, 136, 374), including the time of expression of the base excision repair enzyme uracil-DNA glycosylase, this temporal regulation of repair activity is disturbed in BS cells (136, 463).

There is no direct evidence that BS cells are defective in uracil-DNA glycosylase activity. However, this enzyme purified from five different BS fibroblast cell

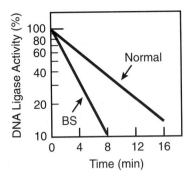

Figure 14–28 DNA ligase activity from extracts of BS cells is more heat labile than that from normal cells. (*Adapted from Willis and Lindahl [456] with permission.*)

lines is characterized by a unique immunological response to a series of human monoclonal antibodies raised against a uracil-DNA glycosylase activity purified from human placenta (354, 439). Specifically, a particular monoclonal antibody designated 40.10.09 fails to recognize the native enzyme from all BS lines. Similar results have been obtained with the enzyme purified from a BS lymphoblastoid line (355). The protein with uracil-DNA glycosylase activity recognized by this antibody has been shown to be the 37-kDa subunit of human glyceraldehyde-3-phosphate dehydrogenase (see chapter 4). While the molecular basis and significance of this observation remain obscure, it has potential for the development of a diagnostic test for BS.

Finally, among the diverse repertoire of abnormal cellular phenotypes that characterize BS, mention should be made of elevated levels of superoxide dismutase (274) and active oxygen species (273), altered topoisomerase II activity (147), and altered thymidylate synthetase activity (370).

ABNORMAL DNA LIGASE ACTIVITY IN BS CELLS

BS cells have a high spontaneous mutation rate (213) and a reduced ability to ligate linear plasmid DNA, with an increased frequency of mutations at the ligation site (327). These characteristics, together with the features of abnormal DNA synthesis and chromosomal abnormalities mentioned above, have sparked interest in the possibility that BS represents a defect in one or more DNA ligases. Several studies do indeed suggest a defect in DNA ligase activity in BS cells. As indicated in chapter 8, extracts of human cells contain three distinct DNA ligases, designated I, II, and III. Extracts of cells from individuals with BS show either reduced levels or abnormal activity of DNA ligase I (48, 49, 206, 456, 457) (Fig. 14–28). However, PCR amplification and DNA sequencing of DNA ligase I cDNA from several BS cell lines have not revealed mutations in the coding region of the *LIG1* gene (235, 306). Hence, the molecular basis for the reduced DNA ligase I activity in extracts of BS cells is unclear.

It has been suggested that the activity of the enzyme may be dependent on its phosphorylation state and that the phosphorylation pattern and/or other post-translational modifications may be altered in BS cells (235). It is now recognized that the fractions analyzed for DNA ligase I activity contain substantial amounts of DNA ligase III. Hence, an alternative explanation is that BS may represent a defect in the activity of DNA ligase III. However, direct measurements of DNA ligase III activity have not detected any defect in BS cells (424). Several cellular proteins that modulate the activity of DNA ligase I have been identified, including an inhibitor of 55 to 75 kDa (465). One or more of these factors could conceivably be altered in BS cells (235). Finally, alteration in DNA ligase I may be a more general secondary consequence of an as yet unidentified single gene defect in BS.

A CHO cell line designated EM9 is defective in the rejoining of DNA strand breaks and has a high frequency of SCEs (419). These phenotypes have been complemented by transfection of EM9 cells with human DNA (420). The complementing gene (designated *XRCC1*) has been cloned and maps to human chromosome 19, the same chromosome on which the excision repair genes *ERCC1* and *ERCC2* reside (371). Despite the similarities in the phenotypes of EM9 and BS cells, BS cells are not defective in the *XRCC1* gene. Indeed, the introduction of human chromosome 15 into BS by microcell-mediated chromosome transfer results in the correction of the high frequency of SCEs, suggesting that the BS gene maps to that chromosome rather than to chromosome 19 (251). The latter observation offers the promise of the isolation of the BS gene by positional cloning. This chromosomal assignment has been confirmed and refined by an independent linkage analysis that mapped the BS locus close to *FES* in the chromosomal region 15q26.1 (128).

A Clinical Syndrome That Yielded the 46BR Cell Line

In the early 1980s, a cell line designated 46BR was found to be markedly sensitive to killing by a variety of DNA-damaging agents, including UV radiation, γ radiation, and a number of alkylating agents (450). The individual from whom these cells were derived was a woman who manifested a complex clinical syndrome of dwarfism, the absence of secondary sexual characteristics, dilated venous capillaries, and severe photosensitivity. The patient also had severe immunodeficiency, characterized by a deficiency in immunoglobulins A and G, and a failure of her lymphocytes to respond to mitogenic stimuli in vitro (223). The patient suffered recurrent infections and died at the age of 19, apparently as the result of a slowly growing lymphoma and an acute respiratory infection (235, 449). No definitive diagnosis of this patient's disease was made; hence, this clinical picture has no formal designation.

Further biochemical and cellular biological characterization of the 46BR cell line as well as a simian virus 40-transformed immortalized derivative designated 46BR.1G1 (150, 233, 414, 415), demonstrated defective joining of Okazaki fragments during semiconservative DNA synthesis. Additionally, the rate of joining of DNA strand breaks following the exposure of cells to DNA-damaging agents was markedly reduced. These observations led to the suggestion that 46BR cells might be defective in DNA ligase activity (Fig. 14–29) (150, 233, 414, 415). This hypothesis was directly corroborated by the demonstration of a marked reduction of enzymatic activity (312) and the ability of DNA ligase I from these cells to form a DNA ligase-AMP complex, despite residual DNA-joining activity (20).

Sequencing of PCR-amplified cDNA from the *LIG1* gene revealed the presence of two mutations in 46BR cells. One of these is predicted to result in the substitution of a highly conserved Glu residue with Lys at position 566 of the polypeptide chain (20). Site-specific mutagenesis of codon 566 has shown that this change indeed leads to a severe loss of DNA ligase I activity (187). The second mutation is predicted to result in the substitution of Arg-771 with Trp in the DNA ligase I polypeptide and is also located in a highly conserved region of the gene (235). The two mutations are carried on different alleles. Transfection of 46BR cells with the cloned *LIG1* gene has been shown to rescue the hypersensitivity of these cells to ethyl methanesulfonate and to the poly(ADP-ribose) inhibitor 3-aminobenzamide (381).

The defect in DNA ligase I activity provides a reasonable explanation for the immunodeficiency in this patient, since this enzyme is presumably required for the recombinational events associated with normal antibody diversity. Similarly, the hypersensitivity of 46BR cells to DNA damage is presumably the result of defective excision repair. Remarkably, this single cell line, representative of a poorly defined clinical entity, currently represents the sole example among the spectrum of human diseases characterized by defective processing of damaged DNA in which a homozygous mutation in a single gene correlates with defective activity of an enzyme with a clearly defined role in DNA metabolism (235). If BS is indeed the result of a defect that specifically affects a DNA ligase activity, the 46BR syndrome might appropriately be considered a variant of this disease. On the other hand, if defective DNA ligase activity in BS is a relatively nonspecific secondary effect of a mutation that affects multiple genes, it might be more appropriate to consider 46BR a novel human syndrome (235).

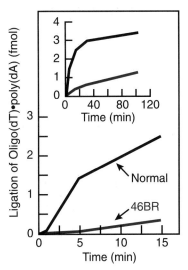

Figure 14–29 DNA I ligase activity measured by the joining of oligo(dT) fragments annealed to a poly(dA) template is reduced in 46BR cells. The inset shows later kinetics of enzyme activity (*Adapted from Prigent et al. [312].*)

Miscellaneous Diseases with Possible Defects in the Processing of Damaged DNA

The correlation between sensitivity to sunlight, predisposition to malignant neoplasms of the skin, and defective DNA repair in XP cells has stimulated a broad search for DNA repair defects in other diseases that are characterized by light sensitivity or by a cancer predisposition. Early studies yielded negative results for diseases such as Rothmund-Thomson disease, lupus erythematosus, psoriasis, basal cell nevus syndrome, and malignant melanoma (reviewed in reference 104). However, more-recent studies have demonstrated increased sensitivity to killing of cells from individuals with the dysplastic nevus syndrome following treatment with 4-NQO (160), and a patient with Rothmund-Thomson syndrome has been reported to have reduced repair synthesis of DNA in cells following exposure to

UV radiation (368). Additionally, two strains of dysplastic nevus syndrome cells were found to be hypermutable after 4-NQO treatment (160), and in independent studies, hypermutability was observed in three lymphoblastoid lines from individuals with this disease following exposure to UV radiation (303, 358). It has also been reported that individuals with nevoid basal cell carcinoma syndrome are more sensitive to killing by ionizing-radiation treatment (47) and to UV-B (but not UV-C) radiation (5).

The association of neurological defects in AT with increased sensitivity to ionizing radiation and to chemicals such as N-methyl-N'-nitro-N-nitrosoguanidine (MNNG) has prompted investigators to examine cells from individuals with other primary neurological disorders. Initial case studies suggested that individuals with Huntington's disease, Friedreich's ataxia, familial dysautonomia, and oligoponto-cerebellar atrophy are abnormally sensitive to ionizing radiation (352). In a more comprehensive study it has been shown that cell strains from six patients with Huntington's disease and from four patients with familial dysautonomia are abnormally sensitive to killing by MNNG (352), and that cells from three individuals with Alzheimer's disease are more sensitive to killing by MNNG (353). However, no defects in the repair of cells from patients with Alzheimer's disease were detected after exposure to ionizing radiation (378). When chromosome aberrations were used as an indicator of abnormal sensitivity to ionizing radiation, cells from patients with Alzheimer's disease and Down's syndrome were found to be abnormal (50, 333). Hypersensitivity to ionizing radiation in patients with neurofibromatosis has also been reported (145). This potpourri is completed by a consideration of a rare clinical entity called Roberts syndrome. This disorder is recognized clinically by mental and growth retardation, tetraphocomelia, and a variety of craniofacial abnormalities (154, 319). Cell lines derived from patients with Roberts syndrome exhibit cytogenetic abnormalities that include random chromosomal loss or gain, as well as premature repulsion of highly heterochromatic chromosomal regions at metaphase (122, 423). These cell lines also demonstrate abnormal sensitivity to killing by γ radiation, mitomycin, G418, hygromycin B, the topoisomerase inhibitor VM-26, and, to a lesser degree, UV radiation and N-methyl-N-nitrosourea (43, 118, 119, 348, 423, 427).

In general, it is fair to state that while isolated reports of abnormal cellular responses to DNA damage associated with various rare syndromes (some of which predispose to cancer) continue to emerge in the literature, there is as yet no clear-cut evidence of defective DNA repair in any of these syndromes. A feature that merits specific comment in this regard is the consistent emergence of neurological, and, to a lesser extent, developmental defects in hereditary diseases linked to defective processing of DNA damage. Finally, the reader is referred to chapter 9 on mismatch repair of DNA for a discussion of the relationship of defective mismatch repair to human colon (and other) cancer.

REFERENCES

1. Ahmed, F. E., and R. B. Setlow. 1978. Excision repair in ataxia telangiectasia, Fanconi's anemia, Cockayne syndrome, and Bloom's syndrome after treatment with ultraviolet radiation and N-acetoxy-2-acetylaminofluorene. *Biochim. Biophys. Acta* **521:**805–817.

2. Aledo, R., G. Renault, M. Prieur, M. F. Avril, B. Chretein, B. Dutrillaux, and A. Aurias. 1989. Increase of sister chromatid exchanges in excision repair deficient xeroderma pigmentosum. *Hum. Genet.* **81:**221–225.

3. Andrews, A. D., S. F. Barrett, and J. H. Robbins. 1978. Xeroderma pigmentosum neurological abnormalities correlate with colony-forming ability after ultraviolet radiation. *Proc. Natl. Acad. Sci. USA* **75:**1984–1988.

4. Andrews, A. D., S. F. Barrett, F. W. Yoder, and J. H. Robbins. 1978. Cockayne's syndrome fibroblasts have increased sensitivity to ultraviolet light but normal rates of unscheduled DNA synthesis. *J. Invest. Dermatol.* **70:**237–239.

5. Applegate, L. E., L. H. Goldberg, R. D. Ley, and H. N. Ananthaswamy. 1990. Hypersensitivity of skin fibroblasts from basal cell nevus syndrome patients to killing by ultraviolet B but not by ultraviolet C radiation. *Cancer Res.* **50:**637–641.

6. Arlett, C. F. 1980. Mutagenesis in repair-deficient human cell strains, p. 161–174. *In* M. Alačević (ed.), *Progress in Environmental Mutagenesis.* Elsevier Biomedical Press, Amsterdam.

7. Arlett, C. F., and S. A. Harcourt. 1978. Cell killing and mutagenesis in repair-defective human cells, p. 633–636. *In* P. C. Hanawalt, E. C. Friedberg, and C. F. Fox (ed.), *DNA Repair Mechanisms.* Academic Press, Inc., New York.

8. Arlett, C. F., and S. A. Harcourt. 1980. Survey of radiosensitivity in a variety of human cell strains. *Cancer Res.* **40:**926–932.

9. Arlett, C. F., and S. A. Harcourt. 1982. Variation in response to mutagens amongst normal and repair-defective human cells, p. 249–266. *In* C. W. Lawrence (ed.), *Induced Mutagenesis. Molecular Mechanisms and Their Implications for Environmental Protection.* Plenum Publishing Corp., New York.

10. Arlett, C. F., and A. R. Lehmann. 1978. Human disorders showing increased sensitivity to the induction of genetic damage. *Annu. Rev. Genet.* **12:**95–115.

11. Arlett, C. F., and A. Priestly. 1985. As assessment of the radiosensitivity of ataxia-telangiectasia heterozygotes, p. 1–63. *In* R. A. Gatti and M. Swift (ed.), *Ataxia-Telangiectasia: Genetics, Neuropathology, and Immunology of a Degenerative Disease of Childhood.* Alan R. Liss, Inc., New York.

12. Auerbach, A. D. 1992. Fanconi anemia and leukemia: tracking the genes. *Leukemia* **6**(Suppl. 1)**:**1–4.

13. Auerbach, A. D., and R. G. Allen. 1991. Leukemia and preleukemia in Fanconi anemia patients: a review of the literature and report of the International Fanconi Anemia Registry. *Cancer Genet. Cytogenet.* **51:**1–12.

14. Auerbach, A. D., and S. R. Wolman. 1976. Susceptibility of Fanconi's anemia fibroblasts to chromosome damage by carcinogens. *Nature* (London) **261:**494–496.

15. Auerbach, A. D., and S. R. Wolman. 1978. Carcinogen-induced chromosome breakage in Fanconi's anemia heterozygous cells. *Nature* (London) **271:**69–71.

16. Aurias, A., B. Dutrillaux, D. Buriot, and J. Lejeune. 1980. High frequencies of inversions and translocations of chromosome 7 and 14 in AT. *Mutat. Res.* **69:**369–374.

17. Awert, F., M. A. Rooimans, A. Westerfeld, J. W. I. M. Simons, and M. Z. Zdziencka. 1991. The Chinese hamster cell mutant V-H4 is homologous to Fanconi anemia (complementation group A). *Cytogenet. Cell Genet.* **56:**23–26.

18. Backer, J. M., and I. B. Weinstein. 1982. Interaction of benzo[a]pyrene and its dihydrodiol-epoxide derivative with nuclear and mitochondrial DNA in C3H10T1/2 cell cultures. *Cancer Res.* **42:**2764–2769.

18a. Bagby, G. C., G. M. Segal, A. D. Auerbach, T. Onega, W. Keeble, and M. C. Heinrich. 1993. Constitutive and induced expression of hemopoietic growth factors by fibroblasts from children with Fanconi anemia. *Exp. Hematol.* **21:**1419–1426.

18b. Bagnara, G. P., L. Bonsi, P. Strippoli, U. Ramenghi, F. Timeus, F. Bonifazi, M. Bonafe, R. Tonelli, G. Bubola, M. F. Brizzi, L. Vitale, G. Paolucci, L. Pegoraro, and V. Gabutti. 1993. Production of interleukin 6, leukocyte inhibitory factor and granulocyte-macrophage colony stimulating factor by peripheral blood mononuclear cells in Fanconi's anemia. *Stem Cells* **11**(Suppl. 2)**:**137–143.

19. Bardwell, J., L. Bardwell, N. Iyer, J. Q. Svejstrup, W. J. Feaver, R. D. Kornberg, and E. C. Friedberg. 1994. Yeast nucleotide excision repair proteins Rad2 and Rad4 interact with RNA polymerase II basal transcription factor b (TFIIH). *Mol. Cell. Biol.* **14:**3569–3576.

20. Barnes, D. E., A. E. Tomkinson, A. R. Lehmann, D. B. Webster, and T. Lindahl. 1992. Mutations in the DNA ligase I gene of an individual with immunodeficiencies and cellular hypersensitivity to DNA-damaging agents. *Cell* **69:**495–503.

21. Barrett, S. F., J. H. Robbins, R. E. Tarone, and K. H. Kraemer. 1991. Evidence for defective repair of cyclobutane pyrimidine dimers with normal repair of other DNA photoproducts in a transcriptionally active gene transfected into Cockayne's syndrome cells. *Mutat. Res.* **255:**281–291.

22. Bates, P. R., F. P. Imray, and M. F. Lavin. 1985. Effect of caffeine on gamma-ray-induced G2 delay in ataxia-telangiectasia. *Int. J. Radiat. Biol.* **47:**713–722.

23. Bates, P. R., and M. F. Lavin. 1989. Comparison of gamma-radiation-induced accumulation of ataxia telangiectasia and control cells in G2 phase. *Mutat. Res.* **218:**165–170.

24. Beamish, H., K. K. Khana, and M. F. Lavin. 1994. Ionizing radiation and cell cycle progression in ataxia telangiectasia. *Radiat. Res.* **138:**S130–S133.

25. Beamish, H., and M. F. Lavin. 1994. Radiosensitivity in ataxia-telangiectasia: anomalies in radiation-induced cell cycle delay. *Int. J. Radiat. Biol.* **65:**175–184.

26. Berger, R., M. Le Coniant, and M.-C. Gendron. 1993. Fanconi anemia. Chromosome breakage and cell cycle studies. *Cancer Genet. Cytogenet.* **69:**13–16.

27. Bielfeld, V., M. Weichenthal, M. Roser, E. Breitbart, J. Berger, E. Seemanova, and H. W. Rudiger. 1989. Ultraviolet-induced chromosomal instability in cultured fibroblasts of heterozygous carriers for xeroderma pigmentosum. *Cancer Genet. Cytogenet.* **43:**219–226.

28. Bloom, D. 1954. Congenital telangiectatic erythema resembling lupus erythematosis in dwarfs. *Am. J. Dis. Child.* **88:**754–758.

29. Boder, E. 1975. Ataxia telangiectasia. Some historic, clinical and pathologic observations. *Birth Defects Orig. Artic. Ser.* **11:**255–270.

30. Boder, E., and R. P. Sedgewick. 1958. Ataxia telangiectasia. A familial syndrome of progressive cerebellar ataxia, oculocutaneous telangiectasia and frequent pulmonary infection. *Pediatrics* **21:**526–554.

31. Bohr, V. A., M. K. Evans, and A. J. Fornace, Jr. 1989. Biology of disease. DNA repair and its pathogenetic implications. *Lab. Invest.* **61:**143–161.

32. Bootsma, D. 1993. The genetic defect in DNA repair deficiency syndromes. *Eur. J. Cancer* **29A:**1482–1488.

33. Bootsma, D., and J. H. J. Hoeijmakers. 1993. Engagement with transcription. *Nature* (London) **363:**114–115.

34. Bootsma, D., W. Keijzer, E. G. Jung, and E. Bohnert. 1989. Xeroderma pigmentosum complementation group XP-I withdrawn. *Mutat. Res.* **218:**149–151.

35. Bootsma, D., M. P. Mulder, E. Pot, and J. A. Cohen. 1970. Different inherited levels of DNA repair replication in xeroderma pigmentosum cell strains after exposure to ultraviolet irradiation. *Mutat. Res.* **9:**507–516.

36. Bradley, A. 1994. Personal communication.

37. Brash, D. E., S. Seetheram, K. H. Kraemer, M. M. Seidman, and A. Bredberg. 1987. Photoproduct frequency is not the major determinant of UV base substitution hot spots or cold spots in human cells. *Proc. Natl. Acad. Sci. USA* **84**:3782–3786.

38. Bredberg, A., K. H. Kraemer, and M. M. Seidman. 1986. Restricted ultraviolet mutational spectrum in a shuttle vector propagated in xeroderma pigmentosum cells. *Proc. Natl. Acad. Sci. USA* **83:** 8273–8277.

39. Bridges, B. A., and D. G. Harnden (ed.). 1982. *Ataxia Telangiectasia—a Cellular and Molecular Link between Cancer, Neuropathology, and Immune Disease.* John Wiley & Sons, Inc., New York.

40. Broughton, B. C., A. R. Lehmann, S. A. Harcourt, C. F. Arlett, A. Sarasin, W. J. Kleijer, F. A. Breemer, R. Nairn, and D. L. Mitchell. 1990. Relationship between pyrimidine dimers, 6-4 photoproducts, repair synthesis and cell survival: studies using cells from patients with trichothiodystrophy. *Mutat. Res.* **235**:33–40.

41. Broughton, B. C., H. Steingrimsdottir, C. Weber, and A. R. Lehmann. 1994. Mutations in the xeroderma pigmentosum group D DNA repair/transcription gene in patients with trichothiodystrophy. *Nat. Genet.* **7**:189–194.

42. Burk, P. G., M. A. Lutzner, D. D. Clarke, and J. H. Robbins. 1971. Ultraviolet-stimulated thymidine incorporation in xeroderma pigmentosum lymphocytes. *J. Lab. Clin. Med.* **77**:759–767.

43. Burns, M. A., and D. J. Tomkins. 1989. Hypersensitivity to mitomycin C cell killing in Roberts syndrome fibroblasts with, but not without, the heterochromatin abnormality. *Mutat. Res.* **261**:243–249.

44. Carbonari, M., R. Paganelli, G. Giannini, E. Galli, C. Gaetano, C. Papetti, and M. Fiorilli. 1990. Relative increase of T cells expressing the gamma/delta rather than the alpha/beta receptor in ataxia-telangiectasia. *N. Engl. J. Med.* **322**:73–76.

45. Casati, A., M. Stefanini, R. Giorgi, P. Ghetti, and F. Nuzzo. 1991. Chromosome rearrangements in normal fibroblasts from xeroderma pigmentosum homozygotes and heterozygotes. *Cancer Genet. Cytogenet.* **51**:89–101.

46. Chaganti, R. S. K., S. Schonberg, and J. German. 1974. A manyfold increase in sister chromatid exchanges in Bloom's syndrome lymphocytes. *Proc. Natl. Acad. Sci. USA* **71**:4508–4512.

47. Chan, G. L., and J. B. Little. 1983. Cultured diploid fibroblasts from patients with the nevoid basal cell carcinoma syndrome are hypersensitive to killing by ionizing radiation. *Am. J. Pathol.* **111**:50–55.

48. Chan, J. Y., F. F. Becker, J. German, and J. H. Ray. 1987. Altered DNA ligase I activity in Bloom's syndrome cells. *Nature* (London) **325**:357–359.

49. Chan, Y.-H., and F. F. Becker. 1988. Defective DNA ligase I in Bloom's syndrome cells. *J. Biol. Chem.* **263**:18231–18235.

50. Chen, P., C. Kidson, and M. Lavin. 1993. Evidence of different complementation groups amongst human genetic disorders characterized by radiosensitivity. *Mutat. Res.* **285**:669–677.

51. Chen, P. C., M. F. Lavin, C. Kidson, and D. Moss. 1978. Identification of ataxia telangiectasia heterozygotes, a cancer prone population. *Nature* (London) **274**:484–486.

52. Cheo, D. L., and E. C. Friedberg. 1994. Unpublished observations.

53. Christinas, F. C., and P. C. Hanawalt. 1994. Repair of ribosomal RNA gene is deficient in xeroderma pigmentosum group C and in Cockayne's syndrome cells. *Mutat. Res.* **323**:179–187.

54. Chu, G., and E. Chang. 1988. Xeroderma pigmentosum group E cells lack a nuclear factor that binds to damaged DNA. *Science* **242**:564–567.

55. Cleaver, J. E. 1968. Defective repair replication of DNA in xeroderma pigmentosum. *Nature* (London) **218**:652–656.

56. Cleaver, J. E. 1969. Xeroderma pigmentosum: a human disease in which an initial stage of DNA repair is defective. *Proc. Natl. Acad. Sci. USA* **63**:428–435.

57. Cleaver, J. E. 1972. Xeroderma pigmentosum: variants with normal DNA repair and normal sensitivity to ultraviolet light. *J. Invest. Dermatol.* **58**:124–128.

57a. Cleaver, J. E. 1981. Inhibition of DNA replication by hydroxyurea and caffeine in an ultraviolet-irradiated human fibroblast line. *Mutat. Res.* **82**:159–171.

58. Cleaver, J. E. 1990. Do we know the cause of xeroderma pigmentosum? *Carcinogenesis* **11**:875–882.

59. Cleaver, J. E., F. Cortes, D. Karentz, L. H. Lutze, W. F. Morgan, A. N. Player, L. Vuksanovic, and D. L. Mitchell. 1988. The relative biological importance of cyclobutane and (6-4) pyrimidine-pyrimidone dimer photoproducts in human cells: evidence from a xeroderma pigmentosum revertant. *Photochem. Photobiol.* **48**:41–49.

60. Cleaver, J. E., F. Cortes, L. H. Lutze, W. F. Morgan, A. N. Player, and D. L. Mitchell. 1987. Unique DNA repair properties of a xeroderma pigmentosum revertant. *Mol. Cell. Biol.* **7**:3353–3357.

61. Cleaver, J. E., A. E. Greene, L. L. Coriell, and R. A. Mulivor. 1981. Xeroderma pigmentosum variants. *Cytogenet. Cell Genet.* **31**:188–192.

62. Cleaver, J. E., and K. H. Kraemer. 1989. Xeroderma pigmentosum, p. 2949–2971. *In* C. R. Scriver, A. L. Beaudet, W. S. Sly, and D. Valle (ed.), *The Metabolic Basis of Inherited Disease.* McGraw-Hill Book Co., New York.

63. Cockayne, E. A. 1936. Dwarfism with retinal atrophy and deafness. *Arch. Dis. Child.* **11**:1–8.

64. Cockayne, E. A. 1946. Dwarfism with retinal atrophy and deafness. *Arch. Dis. Child.* **21**:52–54.

65. Cohen, M. M., M. Shaham, J. Dagan, E. Shmueli, and G. Kohn. 1975. Cytogenetic investigations in families with ataxia telangiectasia. *Cytogenet. Cell Genet.* **15**:338–356.

66. Cohen, M. M., and S. J. Simpson. 1980. Growth kinetics of ataxia-telangiectasia lymphoblastoid cells. Evidence for a prolonged S period. *Cytogenet. Cell Genet.* **28**:24–33.

67. Cohen, M. M., S. J. Simpson, and L. Pazos. 1981. Specificity of bleomycin-induced cytotoxic effects on ataxia telangiectasia lymphoid cell lines. *Cancer Res.* **41**:1817–1823.

68. Cole, J., C. F. Arlett, P. G. Norris, G. Stephens, A. P. W. Waugh, D. M. Beare, and M. H. L. Green. 1992. Elevated *hprt* mutant frequency in circulating T-lymphocytes of xeroderma pigmentosum patients. *Mutat. Res.* **273**:171–178.

69. Collins, A. 1993. Mutant rodent cell lines sensitive to ultraviolet light, ionizing radiation, and cross-linking agents: a comprehensive survey of genetic and biochemical properties. *Mutat. Res.* **293**:99–118.

70. Cox, R., G. P. Hosking, and J. Wilson. 1978. Ataxia telangiectasia. Evaluation of radiosensitivity in cultured skin fibroblasts as a diagnostic test. *Arch. Dis. Child.* **53**:386–390.

71. Crovato, F., C. Barrone, and A. Rebora. 1983. Trichothiodystrophy, BIDS, IBIDS and PIBIDS? *Br. J. Dermatol.* **108**:247–253.

72. Curry, C. J. R., P. O'Lague, J. Tsai, H. T. Hutchison, N. G. J. Jaspers, D. Wara, and R. A. Gatti. 1989. AT(Fresno): a phenotype linking ataxia-telangiectasia with the Nijmegen breakage syndrome. *Am. J. Hum. Genet.* **45**:270–275.

73. Day, R. S., III. 1974. Studies on repair of adenovirus 2 by human fibroblasts using normal, xeroderma pigmentosum, and xeroderma pigmentosum heterozygous strains. *Cancer Res.* **34**:1965–1970.

74. Day, R. S., III. 1975. Xeroderma pigmentosum variants have decreased repair of ultra-violet damaged DNA. *Nature* (London) **253**:748–749.

75. Daya-Grosjean, L., C. Robert, C. Drougart, H. Suarez, and A. Sarasin. 1993. High mutation frequency in *ras* genes of skin tumors isolated from DNA repair deficient xeroderma pigmentosum patients. *Cancer Res.* **53**:1625–1629.

76. Dean, S. W., H. R. Sykes, and A. R. Lehmann. 1988. Inactivation by nitogen mustard of plasmids introduced into normal and Fanconi's anemia cells. *Mutat. Res.* **194**:57–63.

77. Debenham, P. G., M. B. T. Webb, N. J. Jones, and R. Cox. 1987. Molecular studies on the nature of the repair defect in ataxia telangiectasia and their implications for radiobiology. *J. Cell Sci. Suppl.* **6**:179–189.

78. Deluca, J. G., D. A. Kaden, E. A. Komives, and W. G. Thilly. 1984. Mutation of xeroderma pigmentosum lymphoblasts by far-ultraviolet light. *Mutat. Res.* **128:**47–57.

79. DeSanctis, C., and A. Cacchione. 1932. L'idiozia xerodermica. *Riv. Sper. Frentiatr. Med. Leg. Alienazioni Ment.* **56:**269–292.

80. Deschavanne, P. J., N. Chavaudra, B. Fertil, and E. P. Malaise. 1984. Abnormal sensitivity of some Cockayne's syndrome cell strains to UV and gamma rays. Association with a reduced ability to repair potentially lethal damage. *Mutat. Res.* **131:**61–70.

81. De Weerd-Kastelein, E. A., W. Keijzer, and D. Bootsma. 1972. Genetic heterogeneity of xeroderma pigmentosum demostrated by somatic cell hybridization. *Nature* (London) *New Biol.* **238:**80–83.

82. De Weerd-Kastelein, E. A., W. Keijzer, G. Rinaldi, and D. Bootsma. 1977. Induction of sister chromatid exchanges in xeroderma pigmentosum cells after exposure to ultraviolet light. *Mutat. Res.* **45:**253–261.

83. deWit, J., N. G. J. Jaspers, and D. Bootsma. 1981. The rate of DNA synthesis in normal human and ataxia telangiectasia cells after exposure to X-irradiation. *Mutat. Res.* **80:**221–226.

83a. Diatloff-Zito, C., E. Duchaud, E. Viegas-Pequignot, D. Fraser, and E. Moustacchi. 1994. Identification and chromosomal localization of a DNA fragment implicated in the partial correction of the Fanconi anemia group D cellular defect. *Mutat. Res.* **307:**33–42.

84. Dorado, G., H. Steingrimsdottir, C. F. Arlett, and A. R. Lehmann. 1991. Molecular analysis of ultraviolet-induced mutations in a xeroderma pigmentosum cell line. *J. Mol. Biol.* **217:**217–222.

85. Drabkin, R., J. T. Reardon, A. Ansari, J.-C. Huang, L. Zawel, K.-J. Ahn, A. Sancar, and D. Reinberg. 1994. Dual role of TFIIH in DNA excision repair and transcription by RNA polymerase II. *Nature* (London) **368:**769–772.

86. Duckworth-Rysiecki, G., K. Cornish, C. A. Clarke, and M. Buchwald. 1985. Identification of two complementation groups in Fanconi's anemia. *Somatic Cell Mol. Genet.* **11:**35–43.

87. Dumaz, N., C. Drougard, A. Sarasin, and L. Daya-Grosjean. 1993. Specific UV-induced mutation spectrum in the p53 gene of skin tumors from DNA-repair-deficient xeroderma pigmentosum patients. *Proc. Natl. Acad. Sci. USA* **90:**10529–10533.

88. Dutrillaux, B., A. Aurias, A.-M. Fosse, D. Buriot, and M. Prieur. 1982. The cell cycle of lymphocytes in Fanconi anemia. *Hum. Genet.* **62:**327–332.

89. Edwards, M. J., and A. M. R. Taylor. 1980. Unusual levels of (ADP-ribose) and DNA synthesis in ataxia-telangiectasia cells following γ-ray-irradiation. *Nature* (London) **287:**745–747.

90. Edwards, M. J., A. M. R. Taylor, and G. Duckworth. 1980. An enzyme activity in normal and ataxia telangiectasia cell lines which is involved in the repair of gamma-irradiation-induced DNA damage. *Biochem. J.* **188:**677–682.

91. Edwards, M. J., A. M. R. Taylor, and E. J. Flude. 1981. Bleomycin induced inhibition of DNA synthesis in ataxia-telangiectasia cell lines. *Biochem. Biophys. Res. Commun.* **102:**610–616.

92. Ehmann, U. K., K. H. Cook, and E. C. Friedberg. 1978. The kinetics of thymine dimer excision in ultraviolet irradiated human cells. *Biophys. J.* **22:**249–264.

93. Emerit, I., and P. Cerutti. 1981. Clastogenic activity from Bloom syndrome fibroblasts cultures. *Proc. Natl. Acad. Sci. USA* **78:**1868–1872.

94. Evans, M. K., J. H. Robbins, M. B. Ganges, R. E. Tarone, R. S. Nairn, and V. A. Bohr. 1993. Gene-specific DNA repair in xeroderma pigmentosum complementation groups A, C, D, and F. Relation to cellular survival and clinical features. *J. Biol. Chem.* **268:**4839–4847.

95. Fanconi, G. 1967. Familial constitution panmyelocytopathy, Fanconi's anemia (F.A.). I. Clinical aspects. *Semin. Hematol.* **4:**233–240.

96. Fiorilli, M., A. Antonelli, G. Russo, M. Crescenzin, M. Carbonari, and P. Petrinelli. 1985. Variant of ataxia-telangiectasia with low-level radiosensitivity. *Hum. Genet.* **70:**274–277.

97. Ford, M. D., and M. F. Lavin. 1981. Ataxia telangiectasia: an anomaly in DNA replication after irradiation. *Nucleic Acids Res.* **9:**1395–1404.

98. Ford, M. D., L. Martin, and M. F. Lavin. 1984. The effects of ionizing radiation on cell cycle progression in ataxia telangiectasia. *Mutat. Res.* **125:**115–122.

99. Fornace, A. J., J. B. Little, and R. R. Weichselbaum. 1979. DNA repair in a Fanconi's anemia fibroblast cell strain. *Biochim. Biophys. Acta* **561:**99–109.

100. Foroud, T., S. Wei, Y. Ziv, E. Sobel, E. Lange, A. Chao, T. Goradia, Y. Hou, A. Tolun, L. Chessa et al. 1991. Localization of an ataxia-telangiectasia locus to a 3-cM interval on chromosome 11q23: linkage analysis of 111 families by an international consortium. *Am. J. Hum. Genet.* **49:**1263–1279.

101. Frederick, G. D., R. Amirkhan, R. A. Schultz, and E. C. Friedberg. 1994. Structural and mutational analysis of the xeroderma pigmentosum group D (*XPD*) gene. *Hum. Mol. Genet.* **3:**1783–1788.

102. Friedberg, E. C. (ed.). 1985. *DNA Repair.* W. H. Freeman & Co., New York.

103. Friedberg, E. C., A. J. Bardwell, L. Bardwell, Z. Wang, and G. Dianov. 1994. Transcription and nucleotide excision repair-reflections, considerations and recent biochemical insights. *Mutat. Res.* **307:**5–14.

104. Friedberg, E. C., U. K. Ehmann, and J. I. Williams. 1979. Human diseases associated with defective DNA repair. *Adv. Radiat. Biol.* **8:**86–174.

105. Friedberg, E. C., and K. A. Henning. 1993. The conundrum of xeroderma pigmentosum—a rare disease with frequent complexities. *Mutat. Res.* **289:**47–53.

106. Fujimoto, W. Y., M. L. Greene, and J. E. Seegmiller. 1969. Cockayne's syndrome: report of a case with hyperlipoproteinemia, hyperinsulinemia, renal disease, and normal growth hormone. *J. Pediatr.* **75:**881–884.

107. Fujiwara, Y. 1982. Defective repair of mitomycin C crosslinks in Fanconi's anemia and loss in confluent normal human and xeroderma pigmentosum cells. *Biochim. Biophys. Acta* **699:**217–225.

108. Fujiwara, Y., Y. Kano, and Y. Yamamoto. 1984. DNA interstrand cross-linking, repair and SCE mechanism in human cells in special reference to Fanconi anemia, p. 787–800. *In* R. Tice and A. Hollaender (ed.), *Sister Chromatid Exchanges.* Plenum Publishing Corp., New York.

109. Fujiwara, Y., A. Matsumoto, M. Itchihashi, and Y. Satoh. 1987. Heritable disorders of DNA repair: xeroderma pigmentosum and Fanconi's anemia. *Curr. Probl. Dermatol.* **17:**182–198.

110. Fujiwara, Y., M. Tatsumi, and M. S. Sasaki. 1977. Crosslink repair in human cells with its possible defects in Fanconi's anemia cells. *J. Mol. Biol.* **113:**635–649.

111. Galloway, A. M., M. Liuzzi, and M. C. Paterson. 1994. Metabolic processing of cyclobutyl pyrimidine dimers and (6-4) photoproducts in UV-treated human cells. Evidence for distinct excision-repair pathways. *J. Biol. Chem.* **269:**974–980.

112. Gao, S., R. Drouin, and G. P. Holmquist. 1994. DNA repair rates mapped along the human *PGK1* gene at nucleotide resolution. *Science* **263:**1438–1440.

113. Garriga, S., and W. H. Crosby. 1959. The incidence of leukemia in families of patients with hypoplasia of the marrow. *Blood* **14:**1008–1014.

114. Gatti, R. A., I. Berkel, E. Boder, G. Braedt, P. Charmley, P. Concannon, F. Ersoy, T. Faroud, N. G. J. Jaspers, K. Lange et al. 1988. Localization of an ataxia-telangiectasia gene to chromosome 11q22-23. *Nature* (London) **336:**557–580.

115. Gatti, R. A., E. Boder, H. V. Vinters, R. S. Sparkes, A. Norman, and K. Lange. 1991. Ataxia-telangiectasia: an interdisciplinary approach to pathogenesis. *Medicine* (Baltimore) **70:**99–117.

116. Gatti, R. A., and M. Swift (ed.), 1989. *Ataxia-Telangiectasia: Genetics, Neuropathology, and Immunology of a Degenerative Disease of Childhood.* Alan R. Liss, Inc., New York.

116a. Gavish, H., C. C. dos Santos, and M. Buchwald. 1993. A Leu$_{554}$-to-Pro substitution completely abolishes the functional complementing activity of the Fanconi anemia (FACC) protein. *Hum. Mol. Genet.* **2:**123–126.

117. Geernhaw, G. A., A. Hebert, M. E. Duke-Woodsie, I. J. Butler, J. T. Hecht, J. E. Cleaver, G. H. Thomas, and W. A. Horton. 1992. Xeroderma pigmentosum and Cockayne syndrome: overlapping clinical and biochemical defects. *Am. J. Hum. Genet.* **50:**677–689.

118. Gentner, N. E., B. P. Smith, G. M. Norton, L. Courchesne, L. Moeck, and D. J. Tomkins. 1986. Carcinogen hypersensitivity in cultured fibroblast strains from Roberts syndrome patients. *Proc. Can. Fed. Biol. Soc.* **29:**144.

119. Gentner, N. E., D. J. Tomkins, and M. C. Paterson. 1985. Roberts syndrome fibroblasts with heterochromatin abnormality show hypersensitivity to carcinogen-induced cytoxicity. *Am. J. Hum. Genet.* **37:**A231.

120. German, J. 1972. Genes which increase chromosomal instability in somatic cells and predispose to cancer. *Prog. Med. Genet.* **VIII:**61–101.

121. German, J. 1978. DNA repair defects and human diseases, p. 625–631. *In* P. C. Hanawalt, E. C. Friedberg, and C. F. Fox (ed.), *DNA Repair Mechanisms.* Academic Press, Inc., New York.

122. German, J. 1979. Roberts syndrome. I. Cytological evidence for a disturbance in chromatid pairing. *Clin. Genet.* **16:**441–447.

123. German, J. 1990. The chromosome-breakage syndromes: rare disorders that provide models for studying somatic mutation. *Birth Defects Orig. Art. Ser.* **26:**85–111.

124. German, J. 1993. Bloom syndrome: a mendalian prototype of somatic mutational disease. *Medicine* (Baltimore) **72:**393–406.

125. German, J., R. Archibald, and D. Bloom. 1965. Chromosome breakage in a rare and probably genetically determined syndrome of man. *Science* **148:**506–507.

126. German, J., D. Bloom, E. Passarge, K. Fried, R. M. Goodman, I. Katzenellenbogen, Z. Larcon, C. Legum, S. Levin, and J. Wahrman. 1977. Bloom's syndrome. VI. The disorder in Israel, and an estimation of the gene frequency in the Ashkenazim. *Am. J. Hum. Genet.* **29:**553–562.

127. German, J., and E. Passarge. 1987. Bloom's syndrome. XII. Report from the registry for 1987. *Clin. Genet.* **35:**57–69.

128. German, J., A. M. Roe, M. F. Leppert, and N. A. Ellis. 1994. Bloom syndrome: an analysis of consanguineous families assigns the locus to chromosome band 15q26.1. *Proc. Natl. Acad. Sci. USA* **91:**6669–6673.

129. German, J., S. Schonberg, E. Louis, and R. S. K. Chaganti. 1977. Bloom's syndrome. VI. Sister-chromatid exchanges in lymphocytes. *Am. J. Hum. Genet.* **29:**248–255.

130. Gianelli, F. 1986. DNA maintenance and its relation to human pathology. *J. Cell Sci.* **4:**383–416.

131. Gianelli, F., P. F. Benson, S. A. Pawsey, and P. E. Polani. 1977. Ultraviolet light sensitivity and delayed DNA-chain maturation in Bloom's syndrome. *Nature* (London) **265:**466–469.

132. Gillespie, J. M., and R. C. Marshall. 1983. Comparison of the proteins of normal and trichothiodystrophy human hair. *J. Invest. Dermatol.* **80:**195–205.

133. Glover, T. W., C. C. Chang, J. E. Trosko, and S. S. Li. 1979. Ultraviolet light induction of diphtheria toxin-resistant mutants of normal and xeroderma pigmentosum human fibroblasts. *Proc. Natl. Acad. Sci. USA* **76:**3982–3986.

134. Gotoff, S. P., E. Amirmokri, and E. Leibner. 1967. Ataxia telangiectasia, untoward response to X-irradiation and tuberous sclerosis. *Am. J. Dis. Child.* **114:**617–625.

135. Gupta, P. K., and M. A. Sirover. 1980. Sequential stimulation of DNA repair and DNA replication in normal human cells. *Mutat. Res.* **72:**273–284.

136. Gupta, P. K., and M. A. Sirover. 1984. Altered temporal expression of DNA repair in hypermutable Bloom's syndrome cells. *Proc. Natl. Acad. Sci. USA* **81:**757–761.

137. Guzzetta, F. 1972. Cockayne-Neill-Dingwall syndrome. *Handb. Clin. Neurol.* **13:**431–440.

138. Hama-Inaba, H., N. Hieda-Shiomi, T. Shiomi, and K. Sato. 1983. Isolation and characterization of mitomycin C-sensitive mouse lymphoma cell mutants. *Mutat. Res.* **108:**405–416.

139. Hama-Inaba, H., K. Sato, and E. Moustacchi. 1988. Survival and mutagenic responses of mitomycin C-sensitive mouse lymphoma cell mutants to other DNA cross-linking agents. *Mutat. Res.* **194:**121–129.

140. Hanawalt, P. C. 1991. Heterogeneity of DNA repair at the gene level. *Mutat. Res.* **247:**203–211.

141. Hanawalt, P. C. 1992. Transcription-dependent and transcription coupled DNA repair responses, p. 231–246. *In* V. A. Bohr, K. Wasserman, and K. H. Kraemer (ed.), *DNA Repair Mechanisms.* Munksgaard, Copenhagen.

142. Hanawalt, P. C., and A. Sarasin. 1986. Cancer-prone hereditary diseases with DNA processing abnormalities. *Trends Genet.* **2:**124–129.

143. Hand, R., and J. German. 1975. A retarded rate of DNA chain growth in Bloom's syndrome. *Proc. Natl. Acad. Sci. USA* **72:**758–762.

144. Hang, B., A. T. Yeung, and M. W. Lambert. 1993. A damage-recognition protein which binds to DNA containing interstrand cross-links is absent or defective in Fanconi anemia, complementation group A, cells. *Nucleic Acids Res.* **21:**4187–4192.

145. Hannan, M. A., K. Sackey, and D. Sigut. 1993. Cellular radiosensitivity of patients with different types of neurofibromatosis. *Cancer Genet. Cytogenet.* **66:**120–125.

146. Hauser, J., A. S. Levine, and K. Dixon. 1987. Unique pattern of point mutations arising after gene transfer into mammalian cells. *EMBO J.* **6:**63–67.

147. Heartlein, M. W., H. Tsuji, and S. A. Latt. 1987. 5-Bromodeoxyuridine-dependent increase in sister chromatid exchange formation in Bloom's syndrome is associated with reduction in topoisomerase II activity. *Exp. Cell Res.* **169:**245–254.

148. Hebra, F., and M. Kaposi. 1874. On diseases of the skin, including the exanthemata. *New Sydenham Soc.* **61:**252–258. (Translated by W. Tay, London.)

149. Heim, R. A., N. J. Lench, and M. Swift. 1992. Heterozygous manifestations in four recessive human cancer-prone syndromes: ataxia telangiectasia, xeroderma pigmentosum, Fanconi anemia, and Bloom syndrome. *Mutat. Res.* **284:**25–36.

150. Henderson, L. M., C. F. Arlett, S. A. Harcourt, A. R. Lehmann, and B. C. Broughton. 1985. Cells from an immunodeficient patient (46BR) with a defect in DNA ligation are hypomutable but hypersensitive to the induction of sister chromatid exchanges. *Proc. Natl. Acad. Sci. USA* **82:**2044–2048.

151. Henner, W. D., and M. E. Blazka. 1986. Hypersensitivity of cultured ataxia-telangiectasia cells to etoposide. *JNCI* **76:**1007–1011.

152. Henning, K. A. 1993. The molecular genetics of human diseases with defective DNA damage processing. Ph.D. thesis. Stanford University, Stanford, Calif.

153. Henson, P., C. A. Selsky, and J. B. Little. 1981. Excision of ultraviolet damage and the effect of irradiation on DNA synthesis in a strain of Bloom's syndrome fibroblasts. *Cancer Res.* **42:**760–766.

154. Herrman, J., and J. M. Opitz. 1977. The SC phocomelia and the Roberts syndromes: nosologic aspects. *Eur. J. Pediatr.* **125:**117–134.

155. Hirschi, M., M. S. Netrawali, J. F. Remsen, and P. A. Cerutti. 1981. Formation of DNA single-strand breaks by near-ultraviolet and gamma-rays in normal and Bloom's syndrome skin fibroblasts. *Cancer Res.* **41:**2003–2007.

156. Hoehn, H., M. Kubbies, D. Schindler, M. Poot, and P. S. Rabinovitch. 1987. BrdU-Hoechst flow cytometry links the cell kinetic defect of Fanconi's anemia to oxygen hypersensitivity, p. 161–173. *In* T. M. Schroeder-Kurth, A. D. Auerbach, and G. Obe (ed.), *Fanconi Anemia*: *Clinical, Cytogenic and Experimental Aspects.* Springer-Verlag KG, Heidelberg, Germany.

157. Hoeijmakers, J. H. J. 1994. Personal communication.

158. Hook, G. J., E. Kwok, and J. A. Heddle. 1984. Sensitivity of Bloom syndrome fibroblasts to mitomycin C. *Mutat. Res.* **131:**223–230.

159. Houldsworth, J., and M. F. Lavin. 1980. Effect of ionizing radiation on DNA synthesis in ataxia telangiectasia cells. *Nucleic Acids Res.* **8:**3709–3730.

160. Howell, J. N., M. H. Greene, R. C. Corner, V. M. Maher, and J. J. McCormick. 1984. Fibroblasts from patients with hereditary cutaneous malignant melanoma are abnormally sensitive to the mutagenic effect of simulated sunlight and 4-nitroquinoline-1-oxide. *Proc. Natl. Acad. Sci. USA* **81:**1179–1183.

161. Ikenage, M., M. Inoue, T. Kozuka, and T. Sugita. 1981. The recovery of colony-forming ability and the rate of semi-conservative DNA synthesis in ultraviolet-irradiated Cockayne and normal human cells. *Mutat. Res.* **91:**87–91.

162. Imray, F. P., and C. Kidson. 1983. Perturbations of cell-cycle progression in gamma-irradiated ataxia telangiectasia and Huntington's disease cells detected by flow cytometry analysis. *Mutat. Res.* **112:**369–382.

163. Inoue, T., K. Hirano, A. Yokoiyama, and H. C. Kato. 1977. DNA repair enzymes in ataxia telangiectasia and Bloom's syndrome fibroblasts. *Biochim. Biophys. Acta* **479:**497–500.

164. Inoue, T., K. A. Yokoiyama, and T. Kada. 1981. DNA repair enzyme deficiency and in vitro complementation of the enzyme activity in cell free extracts from ataxia telangiectasia fibroblasts. *Biochim. Biophys. Acta* **655:**49–53.

165. Ishisaki, K., T. Yagi, M. Inoue, O. Nakaido, and H. Takebe. 1981. DNA repair in Bloom's syndrome fibroblasts after UV-irradiation or treatment with mitomycin C. *Mutat. Res.* **80:**213–219.

166. Itoh, T., T. Ono, and M. Yamaizumi. 1994. A new UV-sensitive syndrome not belonging to any complementation groups of xeroderma pigmentosum or Cockayne syndrome: siblings showing biochemical characteristics of Cockayne syndrome without typical clinical manifestations. *Mutat. Res.* **314:**233–248.

167. Jaspers, N. G. J., R. A. Gatti, C. Baan, P. C. M. L. Linssen, and D. Bootsma. 1988. Genetic complementation analysis of ataxia-telangiectasia and Nijmegen breakage syndrome: a survey of 50 patients. *Cytogenet. Cell Genet.* **49:**259–263.

168. Jaspers, N. G. J., R. B. Painter, M. C. Paterson, C. Kidson, and T. Inoue. 1985. Complementation analysis of ataxia telangiectasia, p. 147–162. *In* R. A. Gatti and M. Swift (ed.), *Ataxia-Telangiectasia: Genetics, Neuropathology, and Immunology of a Degenerative Disease of Childhood.* Alan R. Liss, Inc., New York.

169. Jaspers, N. G. J., R. D. F. M. Taalman, and C. Baan. 1988. Patients with an inherited syndrome characterized by immunodeficiency, microcephaly, and chromosomal instability: genetic relationship to ataxia telangiectasia. *Am. J. Hum. Genet.* **42:**66–73.

170. Jones, C. J., J. E. Cleaver, and R. D. Wood. 1992. Repair of damaged DNA by extracts from xeroderma pigmentosum complementation group A revertant and expression of a protein absent in its parental cell line. *Nucleic Acids Res.* **20:**991–995.

171. Jones, N. J., R. Cox, and J. Thacker. 1987. Isolation and cross-sensitivity of X-ray-sensitive mutants of V79-4 hamster cells. *Mutat. Res.* **183:**279–286.

172. Jones, N. J., R. Cox, and J. Thacker. 1988. Six complementation groups for ionizing-radiation sensitivity in Chinese hamster cells. *Mutat. Res.* **193:**139–144.

173. Jones, N. J., S. A. Stewart, and L. H. Thompson. 1990. Biochemical and genetic analysis of the Chinese hamster mutants *irs1* and *irs2* and their comparison to cultured ataxia-telangiectasia. *Mutagenesis* **5:**15–23.

174. Jongmans, W., G. W. C. T. Verhaegh, K. Sankaranarayanan, P. H. M. Lohman, and M. Z. Zdzienicka. 1993. Cellular characteristics of Chinese hamster cell mutants resembling ataxia telangiectasia. *Mutat. Res.* **294:**207–214.

175. Jongmans, W., J. Wiegant, M. Oshimura, M. R. James, P. M. H. Lohman, and M. Z. Zdzienicka. 1993. Human chromosome 11 complements ataxia-telangiectasia cells but does not complement the defect in AT-like Chinese hamster cell mutants. *Hum. Genet.* **92:**259–264.

176. Kantor, G. J., L. S. Barsalou, and P. C. Hanawalt. 1990. Selective repair of specific chromatin domains in UV-irradiated cells from xeroderma pigmentosum complementation group C. *Mutat. Res.* **235:**171–180.

177. Kantor, G. J., and C. F. Elking. 1988. Biological significance of domain-oriented DNA repair in xeroderma pigmentosum. *Cancer Res.* **48:**844–849.

178. Kaposi, M. 1883. Xeroderma pigmentosum. *Ann. Dermatol. Venereol.* **4:**29–38. (In French.)

179. Kapp, L. N., and R. B. Painter. 1989. Stable radioresistance in ataxia-telangiectasia cells containing DNA from normal human cells. *Int. J. Radiat. Biol.* **56:**667–675.

180. Karam, L. R., P. Calsou, W. A. Franklin, M. Olsson, and T. Lindahl. 1990. Modification of deoxyribose-phosphate residues by extracts of ataxia telangiectasia cells. *Mutat. Res.* **236:**19–26.

181. Kato, H., and H. F. Stich. 1976. Sister chromatid exchanges in ageing and repair-deficient human fibroblasts. *Nature* (London) **260:**447–448.

182. Katoaka, H., and Y. Fujiwara. 1991. UV damage-specific DNA-binding protein in xeroderma pigmentosum complementation group E. *Biochem. Biophys. Res. Commun.* **175:**1139–1143.

182a. Kaufmann, W. K., and J. E. Cleaver. 1981. Mechanisms of inhibition of DNA replication by ultraviolet light in normal human and xeroderma pigmentosum fibroblasts. *J. Mol. Biol.* **149:**171–187.

183. Kaye, J., C. A. Smith, and P. C. Hanawalt. 1980. DNA repair in human cells containing photoadducts of 8-methoxypsoralen or angelicin. *Cancer Res.* **40:**696–702.

184. Keeney, S., A. P. M. Eker, T. Brody, W. Vermeulen, D. Bootsma, J. H. J. Hoeijmakers, and S. Linn. 1994. Correction of the DNA repair defect in xeroderma pigmentosum group E by injection of a DNA damage binding protein. *Proc. Natl. Acad. Sci. USA* **91:**4053–4056.

185. Keeney, S., H. Wein, and S. Linn. 1992. Biochemical heterogeneity in xeroderma pigmentosum complementation group E. *Mutat. Res.* **273:**49–56.

186. Klocker, H., B. Auer, M. Hirsch-Kauffmann, H. Altmann, H. J. Burtscher, and M. Schweiger. 1983. DNA repair dependent NAD$^+$ metabolism is impaired in patients with Fanconi anemia. *EMBO J.* **2:**303–307.

187. Kodama, K., D. E. Barnes, and T. Lindahl. 1992. In vitro mutagenesis and functional expression in *Escherichia coli* of a cDNA encoding the catalytic domain of human DNA ligase I. *Nucleic Acids Res.* **19:**6093–6099.

188. Kodama, S., K. Komatsu, Y. Okumura, and M. Oshimura. 1992. Suppression of X-ray-induced chromosome aberrations in ataxia telangiectasia cells by introduction of a normal human chromosome 11. *Mutat. Res.* **293:**31–37.

189. Kohn, P. H., J. Whang-Peng, and W. R. Levis. 1982. Chromosomal instability in ataxia telangiectasia. *Cancer Genet. Cytogenet.* **6:**289–302.

190. Kojis, T. L., R. A. Gatti, and R. S. Sparks. 1991. The cytogenetics of ataxia telangiectasia. *Cancer Genet. Cytogenet.* **56:**143–156.

191. Komatsu, K., Y. Okumura, S. Kodama, M. Yoshida, and R. C. Muller. 1989. Lack of correlation between radiosensitivity and inhibition of DNA synthesis in hybrids (A-T × HeLa). *Int. J. Radiat. Biol.* **56:**863–867.

192. Kore-eda, S., T. Tanaka, S.-I. Morikawa, C. Nishiguri, and S. Imamura. 1992. A case of xeroderma pigmentosum group A diagnosed with a polymerase chain reaction (PCR) technique. *Arch. Dermatol.* **128:**971–974.

193. Kraemer, K. H. 1977. Progressive degenerative diseases associated with defective DNA repair: xeroderma pigmentosum and ataxia telangiectasia, p. 37–71. *In* W. W. Nichols and D. G. Murphy (ed.), *DNA Repair Processes.* Symposia Specialists, Miami.

194. Kraemer, K. H. 1991. Twenty years of research on xeroderma pigmentosum at the National Institutes of Health, p. 211–221. *In* E. Riklis (ed.), *Photobiology.* Plenum Publishing Corp., New York.

195. Kraemer, K. H., H. G. Coon, R. A. Petinga, S. F. Barrett, A. E. Rahe, and J. H. Robbins. 1975. Genetic heterogeneity in xeroderma pigmentosum: complementation groups and their relationship to DNA repair rates. *Proc. Natl. Acad. Sci. USA* **72:**59–63.

196. Kraemer, K. H., J. J. DiGiovana, A. N. Moshell, R. E. Tarone, and G. L. Peck. 1988. Prevention of skin cancer in xeroderma pigmentosum with the use of oral isotretinoin. *N. Engl. J. Med.* **318:** 1633–1637.

197. Kraemer, K. H., M. M. Lee, and J. Scotto. 1984. DNA repair protects against cutaneous and internal neoplasia; evidence from xeroderma pigmentosum. *Carcinogenesis* **5:**511–514.

198. Kraemer, K. H., M. L. Myung, and J. Scotto. 1987. Xeroderma pigmentosum. Cutaneous, ocular, and neurologic abnormalities in 830 published cases. *Arch. Dermatol.* **123:**241–250.

199. Kraemer, K. H., C. N. Parris, E. M. Gozukara, D. D. Levy, S. Adelberg, and M. M. Seidman. 1992. Human DNA repair-deficient diseases: clinical disorders and molecular defects, p. 15–22. *In* V. A. Bohr, K. Wassermann, and K. H. Kraemer (ed.), *DNA Repair Mechanisms.* Munksgaard, Copenhagen.

200. Kraemer, K. H., S. Seetheram, M. Protic-Sabljic, A. Bredberg, D. E. Brash, and M. M. Seidman. 1989. DNA repair and mutagenesis induced by dimer and non-dimer photoproducts measured with plasmid vectors in xeroderma pigmentosum cells, p. 169–181. *In* A. Castellani (ed.), *DNA Damage and Repair.* Plenum Publishing Corp., New York.

201. Kraemer, K. H., and M. M. Seidman. 1989. Use of *supF,* an *Escherichia coli* tyrosine suppressor tRNA gene, as a mutagenic target in shuttle-vector plasmids. *Mutat. Res.* **220:**61–72.

202. Kraemer, K. H., and H. Slor. 1985. Xeroderma pigmentosum, p. 33–69. *In* D. J. Demis, R. L. Dobson, and J. McGuire (ed.), *Clinics in Dermatology.* Harper & Row, New York.

203. Kraemer, K. H., D. Weerd-Kastelein, J. H. Robbins, W. Keijzer, S. F. Barrett, R. A. Petinga, and D. Bootsma. 1975. Five complementation groups in xeroderma pigmentosum. *Mutat. Res.* **33:**327–340.

204. Krepinsky, A. B., J. A. Heddle, and J. German. 1979. Sensitivity of Bloom's syndrome lymphocytes to ethyl methanesulfonate. *Hum. Genet.* **50:**151–156.

205. Kurihara, T., M. Inoue, and K. Tatsumi. 1987. Hypersensitivity of Bloom's syndrome fibroblasts to N-ethyl-N-nitrosourea. *Mutat. Res.* **184:**147–151.

206. Kurihara, T., H. Teraoka, M. Inoue, H. Takebe, and K. Tatsumi. 1991. Two types of DNA ligase I activity in lymphoblastoid cells from patients with Bloom's syndrome. *Jpn. J. Cancer Res.* **82:**51–57.

207. Lafforet, D., and J. M. Dupuy. 1978. Photosensibilité et réparation de l'ADN: possibilité d'une parenté nosologique entre xeroderma pigmentosum et syndrome de Cockayne. *Arch. Fr. Pediatr.* **35:**65–74.

208. Lamb, J. R., C. Petit-Frere, B. C. Broughton, A. R. Lehmann, and M. H. L. Green. 1989. Inhibition of DNA replication by ionizing radiation is mediated by a trans-acting factor. *Int. J. Radiat. Biol.* **56:**125–130.

209. Lambert, C., R. A. Schultz, M. Smith, C. Wagner-McPherson, L. D. McDaniel, T. Donlon, E. J. Stambridge, and E. C. Friedberg. 1991. Functional complementation of ataxia-telangiectasia group D (AT-D) cells by microcell-mediated chromosome transfer and mapping of the AT-D locus to the region 11q22-23. *Proc. Natl. Acad. Sci. USA* **88:**5907–5911.

210. Lambert, M., G. J. Tsongalis, W. C. Lambert, B. Hang, and D. D. Parrish. 1992. Defective endonuclease activities in Fanconi's anemia cells, complementation groups A and B. *Mutat. Res.* **273:**57–71.

211. Lambert, W. C., and M. L. Lambert. 1985. Co-recessive inheritance: a model for DNA repair, genetic disease and carcinogenesis. *Mutat. Res.* **145:**227–234.

212. Lambert, W. C., and M. L. Lambert. 1989. Co-recessive inheritance: a model for diseases associated with defective DNA repair, p. 399–428. *In* M. W. Lambert and J. Laval (ed.), *DNA Repair Mechanisms and Their Biological Implications in Mammalian Cells.* Plenum Publishing Corp., New York.

213. Langlois, R. G., W. L. Bigbee, R. H. Jensen, and J. German. 1989. Evidence for increased in vivo mutation and somatic recombination in Bloom's syndrome. *Proc. Natl. Acad. Sci. USA* **86:**670–674.

214. Lavin, M., and A. L. Schroeder. 1988. Damage-resistant DNA synthesis in eukaryotes. *Mutat. Res.* **193:**193–206.

215. Leadon, S. A., and P. K. Cooper. 1993. Preferential repair of ionizing radiation-induced damage in the transcribed strand of an active gene is defective in Cockayne syndrome. *Proc. Natl. Acad. Sci. USA* **90:**10499–10503.

216. Lehmann, A. R. 1982. Three complementation groups in Cockayne syndrome. *Mutat. Res.* **106:**347–356.

217. Lehmann, A. R. 1987. Cockayne's syndrome and trichothiodystrophy: defective repair without cancer. *Cancer Rev.* **7:**82–103.

218. Lehmann, A. R. 1989. Trichothiodystrophy and the relationship between DNA repair and cancer. *Bioessays* **11:**168–170.

219. Lehmann, A. R., C. F. Arlett, B. C. Broughton, S. A. Harcourt, H. Steingrimsdottir, M. Stefanini, M. R. Taylor, A. T. Natarajan, S. Green, M. D. King, R. M. MacKie, J. B. P. Stephenson, and J. L. Tolmie. 1988. Trichothiodystrophy, a human DNA repair disorder with heterogeneity in the cellular response to ultraviolet light. *Cancer Res.* **48:**6090–6096.

220. Lehmann, A. R., C. F. Arlett, J. F. Burke, M. H. L. Green, M. R. James, and J. E. Lowe. 1986. A derivative of an ataxia-telangiectasia (A-T) cell line with normal radiosensitivity but an A-T-like inhibition of DNA synthesis. *Int. J. Radiat. Biol.* **49:**639–643.

221. Lehmann, A. R., D. Bootsma, S. G. Clarkson, J. E. Cleaver, P. J. McAlpine, K. Tanaka, L. H. Thompson, and R. D. Wood. 1994. Nomenclature of human DNA repair genes. *Mutat. Res.* **315:**41–42.

222. Lehmann, A. R., and B. A. Bridges. 1990. Sunlight-induced cancer: some new aspects and implications of the xeroderma pigmentosum model. *Br. J. Dermatol.* **122**Suppl. 35:115–119.

223. Lehmann, A. R., B. C. Broughton, M. R. James, C. F. Arlett, and L. Henderson. 1983. Cells from an immunodeficient patient may have a defective DNA ligase, p. 700–719. *In* E. C. Friedberg and B. A. Bridges (ed.), *Cellular Responses to DNA Damage.* Alan R. Liss, Inc., New York.

224. Lehmann, A. R., S. Kirk-Bell, and C. F. Arlett. 1977. Post-replication repair in human fibroblasts, p. 293–298. *In* A. Castellani (ed.), *Research in Photobiology.* Plenum Publishing Corp., New York.

225. Lehmann, A. R., S. Kirk-Bell, C. F. Arlett, M. C. Paterson, P. M. H. Lohman, E. A. De Weerd-Kastelein, and D. Bootsma. 1975. Xeroderma pigmentosum cells with normal levels of excision repair have a defect in DNA synthesis after UV-irradiation. *Proc. Natl. Acad. Sci. USA* **72:**219–223.

226. Lehmann, A. R., S. Kirk-Bell, and N. G. J. Jaspers. 1977. Post-replication repair in normal and abnormal human fibroblasts, p. 203–215. *In* W. W. Nichols and D. G. Murphy (ed.), *DNA Repair Processes.* Symposia Specialists, Miami.

227. Lehmann, A. R., S. Kirk-Bell, and L. Mayne. 1979. Abnormal kinetics of DNA synthesis in ultraviolet light-irradiated cells from patients with Cockayne's syndrome. *Cancer Res.* **39:**4237–4241.

228. Lehmann, A. R., and L. V. Mayne. 1981. The response of Cockayne syndrome cells to UV-irradiation, p. 367–371. *In* E. Seeberg and K. Kleppe (ed.), *Chromosome Damage and Repair.* Plenum Publishing Corp., New York.

229. Lehmann, A. R., and P. G. Norris. 1989. DNA repair and cancer: speculations based on studies with xeroderma pigmentosum, Cockayne's syndrome and trichothiodystrophy. *Carcinogenesis* **10:**1353–1356.

230. Lehmann, A. R., and S. Stevens. 1977. The production and repair of double strand breaks in cells from normal humans and from patients with ataxia telangiectasia. *Biochim. Biophys. Acta* **474:**49–60.

231. Lehmann, A. R., and S. Stevens. 1979. The response of ataxia telangiectasia cells to bleomycin. *Nucleic Acids Res.* **6:**1953–1960.

232. Lehmann, A. R., A. F. Thompson, S. A. Harcourt, M. Stefanini, and P. G. Norris. 1993. Cockayne's syndrome: correlation of clinical features with cellular sensitivity of RNA synthesis to UV irradiation. *J. Med. Genet.* **30:**679–682.

233. Lehmann, A. R., A. E. Willis, B. C. Broughton, M. R. James, H. Steingrimsdottir, S. A. Harcourt, C. F. Arlett, and T. Lindahl. 1988. Relation between the human fibroblast strain 46BR and cell lines representative of Bloom's syndrome. *Cancer Res.* **48:**6343–6347.

234. Li, L., E. S. Bales, C. A. Peterson, and R. J. Legerski. 1993. Characterization of molecular defects in xeroderma pigmentosum group C. *Nat. Genet.* **5:**413–417.

235. Lindahl, T., and D. Barnes. 1992. Mammalian DNA ligases. *Annu. Rev. Biochem.* **61:**251–281.

236. Liu, V. F., and D. T. Weaver. 1993. The ionizing radiation-induced replication protein A phosphorylation reponse differs between ataxia telangiectasia and normal human cells. *Mol. Cell. Biol.* **13:**7222–7231.

237. Llerena, J. C., Jr., and M. Murer-Orlando. 1991. Bloom syndrome and ataxia telangiectasia. *Semin. Hematol.* **28:**95–103.

238. Lohrer, H., M. Blum, and P. Herrlich. 1988. Ataxia telangiectasia resists gene cloning: an account of parameters determining gene transfer into human recipient cells. *Mol. Gen. Genet.* **212:**474–480.

239. Lonn, U., S. Lonn, U. Nylen, G. Winblad, and J. German. 1990. An abnormal profile of DNA replication intermediates in Bloom's syndrome. *Cancer Res.* **50:**3141–3145.

240. Madzak, C., J. Armier, A. Stary, L. Daya-Grosjean, and A. Sarasin. 1993. UV-induced mutations in a shuttle vector replicated in repair deficient trichothiodystrophy cells differ with those in genetically-related cancer prone xeroderma pigmentosum. *Carcinogenesis* **14:**1255–1260.

241. Maher, V. M., R.-H. Chen, W. G. McGregor, and J. J. McCormick. 1992. Biological evidence of strand-specific repair of carcinogen-induced DNA damage in diploid human cells, p. 126–137. *In* V. A. Bohr, K. Wassermann, and K. H. Kraemer (ed.), *DNA Repair Mechanisms.* Munksgaard, Copenhagen.

242. Maher, V. M., D. J. Dorney, A. L. Mendrala, B. Konze-Thomas, and J. J. McCormick. 1979. DNA excision-repair processes in human cells can eliminate the cytotoxic and mutagenic consequences of ultraviolet irradiation. *Mutat. Res.* **62:**311–323.

243. Maher, V. M., L. M. Oulettte, R. D. Curren, and J. J. McCormick. 1976. Frequency of ultraviolet light-induced mutations is higher in xeroderma pigmentosum variant cells. *Nature* (London) **261:**593–595.

244. Mann, W. R., V. S. Venkatraj, R. G. Allen, Q. Liu, D. A. Olsen, B. Adler-Brecher, J.-I. Mao, B. Weiffenbach, S. L. Sherman, and A. D. Auerbach. 1991. Fanconi anemia: evidence for linkage heterogeneity on chromosome 20q. *Genomics* **9:**329–337.

245. Marshall, R. R., C. F. Arlett, S. A. Harcourt, and B. A. Broughton. 1980. Increased sensitivity of cell strains from Cockayne's syndrome to sister chromatid exchange induction and cell killing by UV light. *Mutat. Res.* **69:**107–112.

246. Matsumoto, A., J.-M. H. Vos, and P. C. Hanawalt. 1989. Repair analysis of mitomycin C-induced DNA crosslinking in ribosomal RNA genes in lymphoblastoid cells from Fanconi's anemia patients. *Mutat. Res.* **217:**185–192.

247. Mayne, L. V., and A. R. Lehmann. 1982. Failure of RNA synthesis to recover after UV-irradiation: an early defect in cells from individuals with Cockayne's syndrome and xeroderma pigmentosum. *Cancer Res.* **42:**1473–1478.

248. Mayne, L. V., L. H. F. Mullenders, and A. A. van Zeeland. 1988. Cockayne's syndrome: a UV sensitive disorder with a defect in the repair of transcribing DNA but normal overall excision repair, p. 349–353. *In* E. C. Friedberg and P. C. Hanawalt (ed.), *Mechanisms and Consequences of DNA Damage Processing.* Alan R. Liss, Inc., New York.

249. McCormick, J. J., S. Kateley-Kohler, M. Watanabe, and V. M. Maher. 1986. Abnormal sensitivity of human fibroblasts from xeroderma pigmentosum variants to transformation to anchorage independence by ultraviolet radiation. *Cancer Res.* **46:**489–492.

250. McCusick, V. A. (ed.), 1992. *Mendelian Inheritance in Man. Catalogs of Autosomal Dominant, Recessive and X-Linked Phenotypes.* Johns Hopkins University Press, Baltimore.

251. McDaniel, L. D., and R. A. Schultz. 1992. Elevated sister chromatid exchange phenotype on Bloom syndrome cells is complemented by human chromosome 15. *Proc. Natl. Acad. Sci. USA* **89:**7968–7972.

252. McFarlin, D. E., W. Strober, and T. A. Waldmann. 1972. Ataxia-telangiectasia. *Medicine* **51:**281–314.

253. McGregor, W. G., R.-H. Chen, L. Lukash, V. M. Maher, and J. J. McCormick. 1991. Cell cycle-dependent strand bias for UV-induced mutations in the transcribed strand of excision repair-proficient human fibroblasts but not in repair-deficient cells. *Mol. Cell. Biol.* **11:** 1927–1934.

254. McWhir, J., J. Selfridge, D. J. Harrison, S. Squires, and D. W. Melton. 1993. Mice with DNA repair gene (*ERCC-1*) deficiency have elevated levels of p53, liver nuclear abnormalities and die before weaning. *Nat. Genet.* **5:**217–224.

255. Meyn, S. M. 1993. High spontaneous intrachromosomal recombination rates in ataxia-telangiectasia. *Science* **260:**1327–1330.

256. Meyn, S. M., J. M. Lu-Kuo, and L. B. K. Herzing. 1993. Expression cloning of multiple human cDNAs that complement the phenotypic defects of ataxia-telangiectasia group D fibroblasts. *Am. J. Hum. Genet.* **53:**1206–1216.

257. Michelin, S., L. Daya-Grosjean, F. Sureau, S. Said, A. Sarasin, and H. G. Suarez. 1993. Characterization of a *c-met* proto-oncogene activated in human xeroderma pigmentosum cells after treatment with N-methyl-N'-nitro-N-nitrosoguanidine (MNNG). *Oncogene* **8:**1983–1991.

258. Mistra, R. R., and J.-M. H. Vos. 1993. Defective replication of psoralen adducts detected at the gene-specific level in xeroderma pigmentosum variant cells. *Mol. Cell. Biol.* **13:**1002–1012.

259. Mohamed, R., M. Ford, and M. F. Lavin. 1986. Ionizing radiation and DNA-chain elongation in ataxia telangiectasia lymphoblastoid cells. *Mutat. Res.* **165:**117–122.

260. Mondello, C., T. Nardo, S. Giliani, J. E. Arrand, C. A. Weber, A. R. Lehmann, F. Nuzzo, and M. Stefanini. 1994. Molecular analysis of the *XP-D* gene in Italian families with patients affected by trichothiodystrophy and xeroderma pigmentosum group D. *Mutat. Res.* **314:**159–165.

261. Morgan, J. L., T. M. Holcomb, and R. W. Morrissey. 1968. Radiation reactions in ataxia telangiectasia. *Am. J. Dis. Child.* **116:**557–558.

262. Morison, W. L., C. Bucana, N. Hashem, M. L. Kripke, J. E. Cleaver, and J. L. German. 1985. Impaired immune function in patients with xeroderma pigmentosum. *Cancer Res.* **45:**3929–3931.

263. Morrell, D., E. Cromartie, and M. Swift. 1986. Mortality and cancer incidence in 263 patients with ataxia-telangiectasia. *JNCI* **77:**89–92.

264. Moustacchi, E., and C. Diatloff-Zito. 1985. DNA semi-conservative synthesis in normal and Fanconi anemia fibroblasts following treatment with 8-methoxypsoralen and near ultraviolet light or with X-rays. *Hum. Genet.* **70:**236–242.

265. Moustacchi, E., U. K. Ehmann, and E. C. Friedberg. 1979. Defective recovery of semi-conservative DNA synthesis in xeroderma pigmentosum cells following split-dose ultraviolet irradiation. *Mutat. Res.* **62:**159–171.

266. Moustacchi, E., D. Papadopoulo, C. Diatloff-Zito, and M. Buchwald. 1987. Two complementation groups of Fanconi's anemia differ in their phenotypic response to a DNA-crosslinking treatment. *Hum. Genet.* **75:**45–47.

267. Mullenders, L. H. F., R. J. Sakker, A. van Hoffen, J. Venema, A. T. Natarajan, and A. A. van Zeeland. 1992. Genomic heterogeneity of UV-induced repair: relationship to chromatin structure and transcriptional activity, p. 247–254. *In* V. A. Bohr, K. Wasserman, and K. H. Kraemer (ed.), *DNA Repair Mechanisms.* Munksgaard, Copenhagen.

268. Murnane, J. P., and L. N. Kapp. 1993. A critical look at the association of human genetic syndromes with sensitivity to ionizing radiation. *Cancer Biol.* **4:**93–104.

269. Murnane, J. P., and R. B. Painter. 1982. Complementation of the defects in DNA synthesis in irradiated and unirradiated ataxia-telangiectasia cells. *Proc. Natl. Acad. Sci. USA* **79:**1960–1963.

270. Myhr, B. C., D. Turnbull, and J. A. DiPaolo. 1979. Ultraviolet mutagenesis of normal and xeroderma pigmentosum variant human fibroblasts. *Mutat. Res.* **63:**341–353.

271. Nance, M. A., and S. A. Berry. 1992. Cockayne syndrome: review of 140 cases. *Am. J. Med. Genet.* **42:**68–84.

272. Neisser, A. 1883. Ueber das xeroderma pigmentosum. Lioderma essentialis cum melanosi et telangiectasia. *Jahrschr. Dermatol. Syphil.* 1883. 47-62.

273. Nicotera, T., K. Thusu, and P. Dandona. 1993. Elevated production of active oxygen in Bloom's syndrome cell lines. *Cancer Res.* **53:**5104–5107.

274. Nicotera, T. M., J. Notaro, S. Notaro, J. Schumer, and A. Sandberg. 1989. Elevated superoxide dismutase activity in Bloom's syndrome: a genetic condition of oxidative stress. *Cancer Res.* **49:**5239–5243.

275. Nishigori, C., Y. Miyachi, S. Imamura, and H. Takebe. 1989. Reduced superoxide dismutase in xeroderma pigmentosum fibroblasts. *J. Invest. Dermatol.* **93:**506–510.

276. Nishigori, C., S.-I. Moriwaki, H. Takebe, T. Tanaka, and S. Imamura. 1994. Gene alterations and clinical characteristics of xeroderma pigmentosum group A patients in Japan. *Arch. Dermatol.* **130:**191–197.

277. Nishigori, C., M. Zghal, T. Yagi, S. Imamura, M. R. Komoun, and H. Takebe. 1993. High prevalence of the point mutation in exon 6 of the xeroderma pigmentosum group A-complementing (*XPAC*) gene in xeroderma pigmentosum group A patients in Tunisia. *Am. J. Hum. Genet.* **53:**1001–1006.

278. Nuzzo, F., P. Lagomarsini, A. Casati, R. Giorgi, E. Berardesca, and M. Stefanini. 1989. Clonal chromosome rearrangements in a fibroblast strain from a patient affected by xeroderma pigmentosum (complementation group C). *Mutat. Res.* **219:**209–215.

279. O'Donovan, A., and R. D. Wood. 1993. Identical defects in DNA repair in xeroderma pigmentosum group G and rodent ERCC group 5. *Nature* (London) **363:**185–188.

280. Okuna, Y., S. Tateishi, and M. Yamaizumi. 1994. Complementation of xeroderma pigmentosum cells by microinjection of mRNA fractionated under denaturing conditions: an estimation of sizes of XP-E and XP-G mRNA. *Mutat. Res.* **314:**11–19.

281. Painter, R. B. 1978. Inhibition of DNA replicon initiation by 4-nitroquinoline-1-oxide, adriamycin, and ethyleneimine. *Cancer Res.* **38:**4445–4449.

282. Painter, R. B. 1981. Radioresistant DNA synthesis: an intrinsic feature of ataxia telangiectasia. *Mutat. Res.* **84:**183–190.

283. Painter, R. B. 1982. Structural changes in chromatin as the basis for radiosensitivity in ataxia telangiectasia. *Cytogenet. Cell Genet.* **33:**139–144.

284. Painter, R. B. 1985. 3-Aminobenzamide does not affect radiation-induced inhibition of DNA synthesis in human cells. *Mutat. Res.* **143:**113–115.

285. Painter, R. B. 1985. Altered DNA synthesis in irradiated and unirradiated ataxia-telangiectasia cells, p. 89–100. *In* R. A. Gatti and M. Swift (ed.), *Ataxia-Telangiectasia: Genetics, Neuropathology, and Immunology of a Degenerative Disease of Childhood.* Alan R. Liss, Inc., New York.

286. Painter, R. B., and B. R. Young. 1980. Radiosensitivity in ataxia-telangiectasia, a new explanation. *Proc. Natl. Acad. Sci. USA* **77:**7315–7317.

287. Pandita, T. K., and W. N. Hittleman. 1992. Initial chromosome damage but not DNA damage is greater in ataxia telangiectasia cells. *Radiat. Res.* **130:**94–103.

288. Papadopoulo, D., D. Averbeck, and E. Moustacchi. 1987. The fate of 8-methoxypsoralen-photoinduced DNA interstrand cross-links in Fanconi's anemia cells of defined genetic complementation groups. *Mutat. Res.* **184:**271–280.

289. Papadopoulo, D., C. Guillouf, H. Mohrenweiser, and E. Moustacchi. 1990. Hypomutability in Fanconi anemia cells is associated with increased deletion frequency at the *HPRT* locus. *Proc. Natl. Acad. Sci. USA* **87:**8383–8387.

290. Papadopoulo, D., B. Porfirio, and E. Moustacchi. 1990. Mutagenic response of Fanconi's anemia cells from a defined complementation group after treatment with photoactivated bifunctional psoralens. *Cancer Res.* **50:**3289–3294.

291. Park, C.-H., and A. Sancar. 1993. Reconstitution of mammalian excision repair ability with mutant cell-free extracts and XPAC and ERCC1 proteins expressed in *Escherichia coli. Nucleic Acids Res.* **21:**5110–5116.

292. Parker, V. P., and M. W. Lieberman. 1977. Levels of DNA polymerases α, β, and γ in control and repair-deficient human diploid fibroblasts. *Nucleic Acids Res.* **4:**2029–2037.

293. Parrington, J. M., J. D. A. Delhanty, and H. P. Baden. 1971. Unscheduled DNA synthesis, UV-induced chromosome aberrations and SV40 transformation in cultured cells from xeroderma pigmentosum. *Ann. Hum. Genet.* **35:**149–160.

294. Parris, C. N., and K. H. Kraemer. 1993. Ultraviolet-induced mutations in Cockayne syndrome cells are primarily caused by cyclobutane dimer photoproducts while repair of other photoproducts is normal. *Proc. Natl. Acad. Sci. USA* **90:**7260–7264.

295. Parshad, R., K. K. Sanford, K. H. Kraemer, G. M. Jones, and R. E. Tarone. 1990. Carrier detection in xeroderma pigmentosum. *J. Clin. Invest.* **85:**135–138.

296. Paterson, M. C. 1979. Ataxia telangiectasia: an inherited human disorder involving hypersensitivity to ionizing radiation and related DNA-damaging chemicals. *Annu. Rev. Genet.* **13:**291–318.

297. Paterson, M. C., A. K. Anderson, B. P. Smith, and P. J. Smith. 1979. Enhanced radiosensitivity of cultured fibroblasts from ataxia telangiectasia heterozygotes manifested by defective colony-forming ability and reduced DNA repair replication after hypoxic γ-irradiation. *Cancer Res.* **39:**3725–3734.

298. Paterson, M. C., N. T. Bech-Hansen, and P. J. Smith. 1981. Heritable radiosensitivity and DNA repair deficient disorders in man, p. 335–354. *In* E. Seeberg and K. Kleppe (ed.), *Chromosome Damage and Repair.* Plenum Publishing Corp., New York.

299. Paterson, M. C., N. E. Gentner, M. V. Middlestadt, and M. Weinfeld. 1984. Cancer predisposition, carcinogen hypersensitivity, and aberrant DNA metabolism. *J. Cell. Physiol.* **3:**45–62.

300. Paterson, M. C., B. P. Smith, P. H. M. Lohman, A. K. Anderson, and L. Fishman. 1976. Defective excision repair of gamma-ray-damaged DNA in human [ataxia telangiectasia] fibroblasts. *Nature* (London) **260:**444–447.

301. Patterson, M., and G. Chu. 1989. Evidence that XP cells from complementation group E are deficient in a homolog of yeast photolyase. *Mol. Cell. Biol.* **9:**5105–5112.

302. Patton, J. D., L. A. Rowan, A. L. Mendrala, J. N. Howell, V. A. Maher, and J. J. McCormick. 1984. Xeroderma pigmentosum fibroblasts including cells from XP variants are abnormally sensitive to the mutagenic and cytotoxic action of broad spectrum simulated light. *Photochem. Photobiol.* **39:**37–42.

303. Perera, M. I. R., K. I. Um, M. G. Greene, H. L. Waters, A. Bredberg, and K. H. Kraemer. 1986. Hereditary dysplastic nevus syndrome: lymphoid cell ultraviolet hypermutability is association with increased melanoma susceptibility. *Cancer Res.* **46:**1005–1009.

304. Peterson, R. D. A., and J. D. Funkhouser. 1989. Speculations on ataxia-telangiectasia: defective regulation of the immunoglobin gene superfamily. *Immunol. Today* **10:**313–314.

305. Peterson, R. D. A., J. D. Funkhouser, C. M. Tuck-Miller, and R. A. Gatti. 1992. Cancer susceptibility in ataxia-telangiectasia. *Leukemia* **6**(Suppl. 1):8–13.

306. Petrini, J. H. J., K. G. Huwiler, and D. T. Weaver. 1991. A wild-type DNA ligase I gene is expressed in Bloom's syndrome cells. *Proc. Natl. Acad. Sci. USA* **88:**7615–7619.

307. Poll, E. H. A., F. Arwert, H. T. Kortbeek, and A. W. Eriksson. 1984. Fanconi anemia cells are not uniformly deficient in unhooking of DNA interstrand crosslinks, induced by mitomycin C or 8-methoxypsoralen plus UVA. *Hum. Genet.* **68:**228–234.

308. Pollitt, R. J., F. A. Fenner, and M. Davies. 1968. Sibs with mental and physical retardation and trichorrhexis nodosa with abnormal amino acid composition of the hair. *Arch. Dis. Child.* **43:**211–216.

309. Povrik, L. F., and R. B. Painter. 1976. The effect of 313 nanometer light on initiation of replicons in mammalian cell DNA containing bromodeoxyuridine. *Biochim. Biophys. Acta* **432:**267–272.

310. Price, F. M., R. Parshad, R. T. Tarone, and K. K. Sanford. 1991. Radiation-induced chromatid aberrations in Cockayne syndrome and xeroderma pigmentosum group C fibroblasts in relation to cancer predisposition. *Cancer Genet. Cytogenet.* **57:**1–10.

311. Price, V. H., R. B. Odom, W. H. Ward, and F. T. Jones. 1980. Trichothiodystrophy: sulfur-deficient brittle hair as a marker for a neuroectodermal symptom complex. *Arch. Dermatol.* **116:**1375–1384.

312. Prigent, C., M. S. Satoh, G. Daly, D. E. Barnes, and T. Lindahl. 1994. Aberrant DNA repair and DNA replication due to an inherited enzymatic defect in DNA ligase I. *Mol. Cell. Biol.* **14:**310–317.

313. Protic-Sabljic, M., T. Narendra, P. J. Munson, J. Hauser, K. H. Kraemer, and K. Dixon. 1986. UV light-induced cyclobutane pyrimidine dimers are mutagenic in mammalian cells. *Mol. Cell. Biol.* **6:**3349–3356.

314. Protic-Sabljic, M., S. Seetharam, M. M. Seidman, and K. H. Kraemer. 1986. An SV40-transformed xeroderma pigmentosum group D cell line: establishment, ultraviolet sensitivity, transfection efficiency and plasmid mutation induction. *Mutat. Res.* **166:**287–294.

315. Rainbow, A. J., and M. Howes. 1982. A deficiency in the repair of UV and gamma-ray damaged DNA in fibroblasts from Cockayne's syndrome. *Mutat. Res.* **93:**235–247.

316. Reinhold, J. D. L., E. Neumark, R. Lightwood, and C. O. Carter. 1952. Familial hypoplastic anemia with congenital abnormalities (Fanconi's syndrome). *Blood* **7:**915–926.

317. Robbins, J. H. 1988. Xeroderma pigmentosum. Defective DNA repair causes skin cancer and neurodegeneration. *JAMA* **260:**384–388.

318. Robbins, J. H., K. H. Kraemer, M. A. Lutzner, B. W. Festoff, and H. G. Coon. 1974. Xeroderma pigmentosum. An inherited disease with sun sensitivity, multiple cutaneous neoplasms and abnormal DNA repair. *Ann. Intern. Med.* **80:**221–248.

319. Roberts, J. B. 1917. A child with double cleft lip and palate, protrusion of the intermaxillary portion of the upper jaw and imperfect development of the bones of the four extremities. *Ann. Surg.* **70:**252–254.

320. Robins, P., C. J. Jones, M. Biggerstaff, T. Lindahl, and R. D. Wood. 1991. Complementation of DNA repair in xeroderma pigmentosum group A cell extracts by a protein with affinity for damaged DNA. *EMBO J.* **10:**3913–3921.

321. Rosselli, F., and E. Moustacchi. 1989. Chromosomal hypersensitivity in mutant MCN-151 mouse cells exposed to mitomycin C. *Mutat. Res.* **225:**115–119.

322. Rosselli, F., and E. Moustacchi. 1990. Cocultivation of Fanconi anemia cells and of mouse lymphoma mutants leads to interspecies complementation of chromosomal hypersensitivity to DNA cross-linking agents. *Hum. Genet.* **84:**517–521.

322a. Roselli, F., A. Ridet, T. Soussi, E. Duchaud, C. Alapetite, and E. Moustacchi. 1995. p53-Dependent pathway of radio-induced apoptosis is altered in Fanconi anemia. *Oncogene,* in press.

322b. Roselli, F., J. Sanceau, E. Gluckman, J. Wietzerbin, and E. Moustacchi. 1994. Abnormal lymphokine production: a novel feature of the genetic disease Fanconi anemia. II. In vitro and in vivo spontaneous overproduction of tumor necrosis factor α. *Blood* **83:**1216–1225.

323. Rosselli, F., J. Sanceau, J. Wietzerbin, and E. Moustacchi. 1992. Abnormal lymphokine production: a novel feature of the genetic disease Fanconi anemia. I. Involvement of interleukin-6. *Hum. Genet.* **89:**42–48.

324. Rousset, S., S. Nocentini, B. Revet, and E. Moustacchi. 1990. Molecular analysis by electron microscopy of the removal of psoralen-photoinduced DNA cross-links in normal and Fanconi's anemia fibroblasts. *Cancer Res.* **50:**2443–2448.

325. Rudé, J., and E. C. Friedberg. 1977. Semiconservative deoxyribonucleic acid synthesis in unirradiated and ultraviolet irradiated xeroderma pigmentosum and normal human skin fibroblasts. *Mutat. Res.* **42:**433–442.

326. Rudolph, N. S., and S. A. Latt. 1989. Flow cytometric analysis of X-ray sensitivity in ataxia telangiectasia. *Mutat. Res.* **211:**31–41.

327. Rünger, T. M., and K. H. Kraemer. 1989. Joining of linear plasmid DNA is reduced and error-prone in Bloom's syndrome cells. *EMBO J.* **8:**1419–1425.

328. Rünger, T. M., M. Poot, and K. H. Kraemer. 1992. Abnormal processing of transfected plasmid DNA in cells from patients with ataxia telangiectasia. *Mutat. Res.* **293:**47–54.

329. Saito, H., A. T. Hammond, and R. E. Moses. 1993. Hypersensitivity to oxygen is a uniform and secondary defect in Fanconi anemia cells. *Mutat. Res.* **294:**255–262.

330. Sakaguchi, K., P. V. Harris, C. Ryan, M. Buchwald, and J. B. Boyd. 1991. Alteration of a nuclease in Fanconi anemia. *Mutat. Res.* **255:**31–38.

331. Salob, S. P., D. K. H. Webb, and D. J. Atherton. 1992. A child with xeroderma pigmentosum and bone marrow failure. *Br. J. Dermatol.* **126:**372–374.

332. Sanal, O., S. Wei, T. Faroud, U. Malhotra, P. Concannon, P. Charmley, W. Salser, K. Lange, and R. A. Gatti. 1990. Further mapping of an ataxia-telangiectasia locus to the chromosome 11q23 region. *Am. J. Hum. Genet.* **47:**860–866.

333. Sanford, K. K., R. Parshad, F. M. Price, R. E. Tarone, and M. B. Schapiro. 1993. X-ray-induced chromatid damage in cells from Down syndrome and Alzheimer disease patients in relation to DNA repair and cancer proneness. *Cancer Genet. Cytogenet.* **70:**25–30.

334. Sasaki, M. S. 1973. DNA repair capacity and susceptibility to chromosome breakage in xeroderma pigmentosum cells. *Mutat. Res.* **20:**291–293.

335. Sasaki, M. S. 1978. Fanconi's anemia. A condition possibly associated with defective DNA repair, p. 675–684. *In* P. C. Hanawalt, E. C. Friedberg and C. F. Fox (ed.), *DNA Repair Mechanisms.* Academic Press, Inc., New York.

336. Sasaki, M. S., and A. Tonomura. 1973. A high susceptibility of Fanconi's anemia to chromosome breakage by DNA cross-linking agents. *Cancer Res.* **33:**1829–1836.

337. Satoh, M. S., C. J. Jones, R. D. Wood, and T. Lindahl. 1993. DNA excision-repair defect of xeroderma pigmentosum prevents removal of a class of oxygen free radical-induced base lesions. *Proc. Natl. Acad. Sci. USA* **90:**6335–6339.

338. Satokata, I., K. Tanaka, N. Miura, I. Miyamoto, Y. Sato, S. Kondo, and Y. Okada. 1990. Characterization of a splice mutation in group A xeroderma pigmentosum. *Proc. Natl. Acad. Sci. USA* **87:**9908–9912.

339. Satokata, I., K. Tanaka, S. Yuba, and Y. Okada. 1992. Identification of splicing mutations of the last nucleotides of exons, a nonsense mutation, and a missense mutation of the *XPAC* gene as causes of group A xeroderma pigmentosum. *Mutat. Res.* **273:**203–212.

340. Satokata, I., K. Tanaka, S. Yuba, and Y. Okada. 1992. Three non-sense mutations responsible for group A xeroderma pigmentosum. *Mutat. Res.* **273:**193–202.

341. Schaeffer, L., V. Moncolin, R. Roy, A. Staub, M. Mezzina, A. Sarasin, G. Weeda, J. H. J. Hoeijmakers, and J.-M. Egly. 1994. The ERCC2/DNA repair protein is associated with the class II BTF2/TFIIH transcription factor. *EMBO J.* **13:**2388–2392.

342. Schaeffer, L., R. Roy, S. Humbert, V. Moncollin, W. Vermeulen, J. H. J. Hoeijmakers, P. Chambon, and J.-M. Egly. 1993. DNA repair helicase: a component of BTF2(TFIIH) basic transcription factor. *Science* **260:**58–63.

343. Scheres, J. M. J. C., T. W. J. Hustinx, and C. M. P. Weemaes. 1980. Chromsome 7 in ataxia telangiectasia. *J. Pediatr.* **97:**440–441.

344. Scherly, D., T. Nouspikel, J. Corlet, C. Ucla, A. Bairoch, and S. G. Clarkson. 1993. Complementation of the DNA repair defect in xeroderma pigmentosum group G cells by a human cDNA related to yeast *RAD2. Nature* (London) **363:**182–185.

345. Schmickel, R. D., E. H. Y. Chu, J. E. Trosko, and C. C. Chang. 1977. Cockayne syndrome: a cellular sensitivity to ultraviolet light. *Pediatrics* **60:**135–139.

346. Schroeder, T. M., and R. Kurth. 1971. Spontaneous chromosomal breakage and high incidence of leukemia in inherited disease. *Blood* **37:**96–112.

346a. Schroeder, T. M., D. Tigen, J. Kruger, and F. Vogen. 1976. Formal genetics of Fanconi's anemia. *Hum. Genet.* **32:**257–288.

347. Schultz, J. C., and N. T. Shahidi. 1993. Tumor necrosis factor-α overproduction in Fanconi's anemia. *Am. J. Hematol.* **42:**196–201.

348. Schultz, R. A. 1994. Personal communication.

349. Schweiger, M., B. Auer, H. J. Burtscher, M. Hirsch-Kauffmann, H. Klocker, and R. Schneider. 1987. DNA repair in human cells. Biochemistry of the hereditary diseases Fanconi's anemia and Cockayne syndrome. *Eur. J. Biochem.* **165:**235–242.

350. Scott, D., and F. Zampetti-Bosseler. 1982. Cell cycle dependence of mitotic delay in X-irradiated normal and ataxia-telangiectasia fibroblasts. *Int. J. Radiat. Biol. Relat. Stud. Phys. Chem. Med.* **42:**679–683.

351. Scott, R. J., P. Itin, W. J. Kleijer, K. Kolb, C. Arlett, and H. Muller. 1993. Xeroderma pigmentosum-Cockayne syndrome complex in two patients: absence of skin tumors despite severe deficiency of DNA excision repair. *J. Am. Acad. Dermatol.* **29:**883–889.

352. Scudiero, D. A., S. A. Meyer, B. E. Clatterbuck, R. E. Tarone, and J. H. Robbins. 1981. Hypersensitivity to N′-methyl-N′-nitro-N-nitrosoguanidine in fibroblasts from patients with Huntington's disease, familial dysautonomia, and other primary neuronal degenerations. *Proc. Natl. Acad. Sci. USA* **78:**6451–6455.

353. Scudiero, D. A., R. J. Polinsky, R. A. Brumback, R. E. Tarone, L. E. Nee, and J. H. Robbins. 1986. Alzheimer disease fibroblasts are hypersensitive to the lethal effects of a DNA-damaging chemical. *Mutat. Res.* **159:**125–131.

354. Seal, G., K. Brech, S. J. Karp, B. J. Cool, and M. A. Sirover. 1988. Immunological lesions in human uracil DNA glycosylase: association with Bloom's syndrome. *Proc. Natl. Acad. Sci. USA* **85:**2339–2343.

355. Seal, G., E. E. Henderson, and M. A. Sirover. 1990. Immunological alteration of the Bloom's syndrome uracil DNA glycosylase in Epstein-Barr virus-transformed human lymphoblastoid cells. *Mutat. Res.* **243:**241–248.

356. Sedgewick, R. P., and E. Boder. 1972. Ataxia telangiectasia, p. 267–339. *In* P. J. Vinken and G. W. Bruyn (ed.), *Handbook of Clinical Neurology.* North-Holland, Amsterdam.

357. Seetharam, S., M. Protic-Sabljic, M. M. Seidman, and K. H. Kraemer. 1987. Abnormal ultraviolet mutagenic spectrum in plasmid DNA replicated in cultured fibroblasts from a patient with the skin cancer-prone disease xeroderma pigmentosum. *J. Clin. Invest.* **80:**1613–1617.

358. Seetharam, S., H. L. Waters, M. M. Seidman, and K. H. Kraemer. 1989. Ultraviolet mutagenesis in a plasmid vector replicated in lyphoid cells from a patient with the melanoma-prone disorder dysplastic nevus syndrome. *Cancer Res.* **49:**5918–5921.

359. Seguin, L. R., M. B. Ganges, R. E. Tarone, and J. H. Robbins. 1992. Skin cancer and chromosomal aberrations induced by ultraviolet radiation. Evidence for lack of correlation in xeroderma pigmentosum variant and group E patients. *Cancer Genet. Cytogenet.* **60:**111–116.

360. Seguin, L. R., R. E. Tarone, K. Liao, and J. H. Robbins. 1988. Ultraviolet light-induced chromosomal aberrations in cultured cells from Cockayne syndrome and complementation group C xeroderma pigmentosum patients: lack of correlation with cancer susceptibility. *Am. J. Hum. Genet.* **42:**468–475.

361. Seidman, M. M., A. Bredberg, S. Seetheram, and K. H. Kraemer. 1987. Multiple point mutations in a shuttle vector propagated in human cells: evidence for an error-prone DNA polymerase activity. *Proc. Natl. Acad. Sci. USA* **84:**4944–4948.

362. Selby, C. P., and A. Sancar. 1993. Molecular mechanism of transcription-repair coupling. *Science* **260:**53–58.

363. Setlow, R. B. 1978. Repair deficient human disorders and cancer. *Nature* (London) **271:**713–717.

364. Setlow, R. B., J. D. Regan, J. German, and W. L. Carrier. 1969. Evidence that xeroderma pigmentosum cells do not perform the first step in the repair of ultraviolet damage to their DNA. *Proc. Natl. Acad. Sci. USA* **64:**1035–1041.

365. Sheridan, R. B., III, and P. C. Haung. 1977. Single strand breakage and repair in eukaryotic cells as assayed by S1 nuclease. *Nucleic Acids Res.* **4:**299–318.

366. Shiloh, Y., R. Pashad, M. Frydman, K. K. Sanford, S. Portnoi, Y. Ziv, and G. M. Jones. 1989. G2 chromosomal radiosensitivity in families with ataxia-telangiectasia. *Hum. Genet.* **84:**15–18.

367. Shiloh, Y., E. Tabor, and Y. Becker. 1985. In vitro phenotype of ataxia-telangiectasia (AT) fibroblast strains: clues to the nature of the ''AT DNA lesion'' and the molecular defect in AT, p. 111–121. *In* R. A. Gatti and M. Swift (ed.), *Ataxia-Telangiectasia: Genetics, Neuropathology, and Immunology of a Degenerative Disease of Childhood.* Alan R. Liss, Inc., New York.

368. Shinya, A., C. Nishigori, S.-I. Moriwaki, H. Takebe, M. Kubota, A. Ogino, and S. Imamura. 1993. A case of Rothmund-Thomson syndrome with reduced DNA repair capacity. *Arch. Dermatol.* **129:**332–336.

369. Shiomi, T., Y.-N. Harada, T. Saito, N. Shiomi, Y. Okuno, and M. Yamaizumi. 1994. An *ERCC5* gene with homology to yeast *RAD2* is involved in group G xeroderma pigmentosum. *Mutat. Res.* **314:**167–175.

370. Shiraishi, Y., T. Taguchi, M. Ozawa, and R. Bamezai. 1989. Different mutations responsible for the elevated sister-chromatid exchange frequencies in Bloom syndrome and X-irradiated B-lymphoblastoid cell lines originating from acute leukemia. *Mutat. Res.* **211:**273–278.

371. Siciliano, M. J., A. V. Carrano, and L. H. Thompson. 1986. Assignment of a human DNA repair-gene associated with sister-chromatid exchange to chromosome 19. *Mutat. Res.* **174:**303–308.

372. Singh, S. P., and M. F. Lavin. 1990. DNA-binding protein activated by gamma radiation in human cells. *Mol. Cell. Biol.* **10:**5279–5285.

373. Singh, S. P., R. Mohamed, C. Salmond, and M. F. Lavin. 1988. Reduced topoisomerase II activity in ataxia-telangiectasia cells. *Nucleic Acids Res.* **16:**3919–3929.

374. Sirover, M. A. 1979. Induction of the DNA repair enzyme uracil DNA glycosylase in stimulated human lymphocytes. *Cancer Res.* **39:**2090–2095.

375. Smith, P. J., C. O. Anderson, and J. V. Watson. 1985. Abnormal retention of X-irradiated ataxia-telangiectasia fibroblasts in G2 phase of the cell cycle: cellular RNA content, chromatin stability and the effect of 3-aminobenzamide. *Int. J. Radiat. Biol.* **47:**701–712.

376. Smith, P. J., C. O. Anderson, and J. V. Watson. 1986. Predominant role for DNA damage in etoposide-induced cytotoxicity and cell cycle perturbation in human SV40-transformed fibroblasts. *Cancer Res.* **46:**5641–5645.

377. Smith, P. J., T. A. Makinson, and J. V. Watson. 1989. Enhanced sensitivity to camptothecin in ataxia-telangiectasia cells and its relationship with the expression of DNA topoisomerase I. *Int. J. Radiat. Biol.* **55:**217–231.

378. Smith, T. A., D. Neary, and R. F. Itzhaki. 1987. DNA repair in lymphocytes from young and old individuals and from patients with Alzheimer's disease. *Mutat. Res.* **184:**107–112.

379. Smith-Ravin, J., and P. A. Jeggo. 1989. Use of damaged plasmid to study DNA repair in X-ray sensitive (*xrs*) strains of Chinese hamster ovary (CHO) cells. *Int. J. Radiat. Biol.* **56:**951–961.

380. Sobel, E., E. Lange, N. G. J. Jaspers, L. Chessa, O. Sanal, Y. Shiloh, A. M. R. Taylor et al. 1992. Ataxia-telangiectasia: linkage evidence for genetic heterogeneity. *Am. J. Hum. Genet.* **50:**1343–1348.

381. Somia, N. V., J. K. Jessop, and D. W. Melton. 1993. Phenotypic correction of a human cell line (46BR) with aberrant DNA ligase I activity. *Mutat. Res.* **294:**51–58.

382. Squires, S., D. J. Oates, S. D. Bouffler, and R. T. Johnson. 1992. Cockayne's syndrome fibroblasts are characterized by hypersensitivity to deoxyguanosine and abnormal DNA precursor pool metabolism in response to deoxyguanosine or ultraviolet light. *Somatic Cell Mol. Genet.* **18:**387–401.

383. Squires, S., A. J. Ryan, H. L. Strutt, and R. T. Johnson. 1993. Hypersensitivity of Cockayne's syndrome cells to camptothecin is associated with the generation of abnormally high levels of double strand breaks in nascent DNA. *Cancer Res.* **53:**2012–2019.

384. Stefanini, M., P. Lagomarsini, C. F. Arlett, S. Marioni, C. Barrone, F. Crovato, G. Trevisan, G. Cordone, and F. Nuzzo. 1986. Xeroderma pigmentosum (complementation group D) mutation is present in patients affected by trichothiodystrophy with photosensitivity. *Hum. Genet.* **74:**107–112.

385. Stefanini, M., P. Lagomarsini, S. Giliani, T. Nardo, E. Botta, A. Peserico, W. J. Kleijer, A. R. Lehmann, and A. Sarasin. 1993. Genetic heterogeneity of the excision repair defect associated with trichothiodystrophy. *Carcinogenesis* **14:**1101–1105.

386. Stefanini, M., P. Lagomarsini, P. Giorgi, and F. Nuzzo. 1987. Complementation studies in cells from patients affected by trichothiodystrophy with normal or enhanced photosensitivity. *Mutat. Res.* **191:**117–119.

387. Stefanini, M., W. Vermeulen, G. Weeda, S. Giliani, T. Nardo, M. Mezzina, A. Sarasin, J. I. Harper, C. F. Arlett, J. H. J. Hoeijmakers, and A. R. Lehmann. 1993. A new nucleotide-excision-repair gene associated with the disorder trichothiodystrophy. *Am. J. Hum. Genet.* **53:**817–821.

388. Stern, M.-H., S. Lipkowitz, A. Aurias, C. Griscelli, G. Thomas, and I. R. Kirsch. 1989. Inversion of chromosome 7 in ataxia telangiectasia is generated by a rearrangement between T-cell receptor β and T-cell receptor γ genes. *Blood* **74:**2076–2080.

389. Strathdee, C. A., A. M. V. Duncan, and M. Buchwald. 1992. Evidence for at least four Fanconi anemia genes including *FACC* on chromosome 9. *Nat. Genet.* **1:**196–198.

390. Strathdee, C. A., H. Gavish, W. R. Shannon, and M. Buchwald. 1992. Cloning of cDNAs for Fanconi's anemia by functional complementation. *Nature* (London) **356:**763–767.

391. Sugita, K., N. Suzuki, T. Kojima, Y. Tanabe, H. Nakajima, A. Hayashi, and M. Arima. 1987. Cockayne syndrome with delayed recovery of RNA synthesis after ultraviolet irradiation but normal ultraviolet survival. *Pediatr. Res.* **21:**34–37.

392. Sung, P., V. Bailly, C. Weber, L. H. Thompson, L. Prakash, and S. Prakash. 1993. Human xeroderma pigmentosum group D gene encodes a DNA helicase. *Nature* (London) **365:**852–855.

393. Suthers, G. K., and K. E. Davies. 1992. Phenotypic heterogeneity and the single gene. *Am. J. Hum. Genet.* **50:**887–891.

394. Swift, M. 1971. Fanconi's anemia in the genetics of neoplasia. *Nature* (London) **230:**370–373.

395. Swift, M., R. A. Heim, and N. J. Lench. 1993. Genetic aspects of ataxia telangiectasia. *Adv. Neurol.* **61:**115–125.

396. Swift, M., P. J. Reitnauer, D. Morrell, and C. L. Chase. 1987. Breast and other cancers in families with ataxia telangiectasia. *N. Engl. J. Med.* **316:**1289–1294.

397. Swift, M. R., and K. Hirschhorn. 1966. Fanconi's anemia. Inherited susceptibility to chromosome breakage in various tissues. *Ann. Intern. Med.* **65:**496–503.

398. Taalman, R. D. F. M., N. G. J. Jaspers, J. M. J. C. Scheres, J. de Wit, and T. W. J. Hustinx. 1983. Hypersensitivity to ionizing radiation, in vitro, in a new chromosomal breakage disorder, the Nijmegen breakage syndrome. *Mutat. Res.* **112:**23–32.

398a. Tadjoedin, M. K., and E. C. Fraser. 1965. Heredity of ataxia telangiectasia (Louis-Barr syndrome). *Am. J. Dis. Child.* **110:**64–68.

399. Takebe, H., Y. Miki, T. Kozuka, J.-I. Fujiwara, K. Tanaka, M. S. Sasaki, and H. Akiba. 1977. DNA repair characteristics and skin cancers of xeroderma pigmentosum patients in Japan. *Cancer Res.* **367:**490–495.

400. Takeuchi, T., and K. Morimoto. 1993. Increased formation of 8-hydroxyguanosine, an oxidative DNA damage, in lymphoblasts from Fanconi's anemia patients due to possible catalase deficiency. *Carcinogenesis* **14:**1115–1120.

401. Tanaka, K. 1994. Personal communication.

402. Tanaka, K., K. Y. Kawai, Y. Kumahara, and M. Ikenaga. 1981. Genetic complementation groups in Cockayne syndrome. *Somatic Cell Genet.* **7:**445–456.

403. Tanaka, K., N. Miura, I. Satokata, I. Miyamoto, M. C. Yoshida, Y. Sato, S. Kondo, A. Yasui, H. Okayama, and Y. Okada. 1990. Analysis of a human DNA excision repair gene involved in group A xeroderma pigmentosum and a zinc-finger domain. *Nature* (London) **348:**73–76.

404. Tatsuka, M., O. Nikaido, K. Tatsumi, and H. Takebe. 1989. X-ray-induced G2 arrest in ataxia telangiectasia lymphoblastoid cells. *Mutat. Res.* **214:**321–328.

405. Taylor, A. M. R. 1992. Ataxia telangiectasia genes and predisposition to leukemia, lymphoma and breast cancer. *Br. J. Cancer* **66:**5–9.

406. Taylor, A. M. R., D. G. Harnden, C. F. Arlett, S. A. Harcourt, A. R. Lehmann, S. Stevens, and B. A. Bridges. 1975. Ataxia-telangiectasia: a human mutation with abnormal radiation sensitivity. *Nature* (London) **258:**427–429.

407. Taylor, A. M. R., J. A. Metcalfe, and C. McConville. 1989. Increased radiosensitivity and the basic defect in ataxia telangiectasia. *Int. J. Radiat. Biol.* **56:**677–684.

408. Taylor, A. M. R., J. A. Metcalfe, J. M. Oxford, and D. G. Harnden. 1976. Is chromatid-type damage in ataxia telangiectasia after irradiation at G0 a consequence of defective repair? *Nature* (London) **260:**441–443.

409. Taylor, A. M. R., J. M. Oxford, and J. A. Metcalfe. 1981. Spontaneous cytogenetic abnormalities in lymphocytes from thirteen patients with ataxia telangiectasia. *Int. J. Cancer* **27:**311–319.

410. Taylor, A. M. R., C. M. Rosney, and J. B. Campbell. 1979. Unusual sensitivity of ataxia telangiectasia cells to bleomycin. *Cancer Res.* **39:**1046–1050.

411. Teale, B., K. K. Khanna, S. Singh, and M. F. Lavin. 1993. Radiation-activated DNA-binding protein constitutively present in ataxia telangiectasia nuclei. *J. Biol. Chem.* **268:**22450–22455.

412. Teale, B., S. Singh, K. K. Khanna, D. Findik, and M. F. Lavin. 1992. Purification and characterization of a DNA-binding protein activated by ionizing radiation. *J. Biol. Chem.* **267:**10295–10301.

413. Teitz, T., D. Eli, M. Penner, M. Bakhanashvili, T. Naiman, T. Timme, C. M. Wood, R. E. Moses, and D. Canaani. 1990. Expression of the cDNA for the β subunit of human casein kinase II confers partial UV resistance on xeroderma pigmentosum cells. *Mutat. Res.* **236:**85–97.

414. Teo, I. A., C. F. Arlett, S. A. Harcourt, A. Priestly, and B. C. Broughton. 1983. Multiple hypersensitivity to mutagens in a cell strain (46BR) derived from a patient with immuno-deficiencies. *Mutat. Res.* **107:**371–386.

415. Teo, I. A., B. C. Broughton, R. S. Day, M. R. James, P. Karran, L. V. Mayne, and A. R. Lehmann. 1983. A biochemical defect in the repair of alkylated DNA in cells from an immunodeficient patient (46BR). *Carcinogenesis* **4**:559–564.

416. Thacker, J. 1989. The use of integrating DNA vectors to analyze the molecular defects in ionizing radiation-sensitive mutants of mammalian cells including ataxia telangiectasia. *Mutat. Res.* **220**:187–204.

417. Thacker, J., and A. N. Ganesh. 1990. DNA-break repair, radioresistance of DNA synthesis, and camptothecin sensitivity in the radiation-sensitive *irs* mutants: comparisons to ataxia-telangiectasia cells. *Mutat. Res.* **235**:49–58.

418. Thielmann, H. W., O. Popanda, L. Edler, and E. G. Jung. 1991. Clinical symptoms and DNA repair characteristics of xeroderma pigmentosum patients from Germany. *Cancer Res.* **51**:3456–3470.

419. Thompson, L. H., K. W. Brookman, L. E. Dillehay, C. L. Mooney, and A. V. Carrano. 1982. Hypersensitivity to mutation and sister-chromatid exchange induction in CHO cell mutants defective in incising DNA containing UV lesions. *Somatic Cell Genet.* **8**:759–773.

420. Thompson, L. H., K. W. Brookman, J. L. Minkler, J. C. Fuscoe, K. A. Henning, and A. V. Carrano. 1985. DNA-mediated transfer of a human DNA repair gene that controls sister chromatid exchange. *Mol. Cell. Biol.* **5**:881–884.

421. Thompson, M. W., R. D. McInnes, and H. Willard (ed.). 1991. *Thompson and Thompson—Genetics in Medicine*, 5th ed. The W. B. Saunders Co., Philadelphia.

422. Tice, R., G. Windler, and J. M. Bary. 1978. Effect of cocultivation on sister chromatid exchange frequencies in Bloom's syndrome and normal fibroblast cells. *Nature* (London) **273**:538–540.

423. Tomkins, D., A. Hunter, and M. Roberts. 1979. Cytogenetic findings in Roberts-SC phocomelia syndrome(s). *Am. J. Hum. Genet.* **4**:17–26.

424. Tomkinson, A. E., R. Starr, and R. A. Schultz. 1993. DNA ligase III is the major high molecular weight DNA joining activity in SV40-transformed human fibroblasts: normal levels of DNA ligase III activity in Bloom syndrome cells. *Nucleic Acids Res.* **21**:5425–5430.

425. Tornaletti, S., and G. P. Pfeifer. 1994. Slow repair of pyrimidine dimers at p53 mutation hotspots in skin cancer. *Science* **263**:1436–1438.

426. Troelstra, C., A. van Gool, J. de Wit, W. Vermeulen, D. Bootsma, and J. H. J. Hoeijmakers. 1992. ERCC6, a member of a subfamily of putative helicases, is involved in Cockayne's syndrome and preferential repair of active genes. *Cell* **71**:939–953.

427. Van Den Berg, D. J., and U. Francke. 1993. Sensitivity of Roberts syndrome cells to gamma radiation, mitomycin C, and protein synthesis inhibitors. *Somatic Cell Mol. Genet.* **19**:377–392.

428. van Hoffen, A., A. T. Natarajan, L. V. Mayne, A. A. van Zeeland, L. H. F. Mullenders, and J. Venema. 1993. Deficient repair of the transcribed stand of active genes in Cockayne's syndrome. *Nucleic Acids Res.* **21**:5890–5895.

429. van Vuuren, A. J., W. Vermeulen, L. Ma, G. Weeda, E. Appeldoorn, N. G. J. Jaspers, A. J. van der Eb, D. Bootsma, J. H. J. Hoeijmakers, S. S. Humbert, L. Schaeffer, and J.-M. Egly. 1994. Correction of xeroderma pigmentosum repair defect by basal transcription factor BTF2(TFIIH). *EMBO J.* **13**:1345–1353.

430. Varlander, P. C., J. D. Lin, M. U. Udono, Q. Zang, R. A. Gibson, C. G. Mathew, and A. D. Auerbach. 1994. Mutation analysis of the Fanconi anemia gene *FACC. Am. J. Hum. Genet.* **54**:595–601.

431. Venema, J. 1991. Ph.D. thesis. Leiden University, Leiden, The Netherlands.

432. Venema, J., L. H. F. Mullenders, A. T. Natarajan, A. A. van Zeeland, and L. V. Mayne. 1990. The genetic defect in Cockayne's syndrome is associated with a defect in repair of UV-induced DNA damage in transcriptionally active DNA. *Proc. Natl. Acad. Sci. USA* **87**:4704–4711.

433. Venema, J., A. van Hoffen, V. Karcagi, A. T. Natarajan, A. A. van Zeeland, and L. H. F. Mullenders. 1991. Xeroderma pigmentosum complementation group C cells remove pyrimidine dimers selectively from the transcribed strand of active genes. *Mol. Cell. Biol.* **11**:4128–4134.

434. Vermeulen, W., J. Jaeken, N. G. J. Jaspers, D. Bootsma, and J. H. J. Hoeijmakers. 1993. Xeroderma pigmentosum complementation group G associated with Cockayne syndrome. *Am. J. Hum. Genet.* **53**:185–192.

435. Vermeulen, W., R. J. Scott, S. Rodgers, H. J. Muller, J. Cole, C. F. Arlett, W. J. Kleijer, D. Bootsma, J. H. J. Hoeijmakers, and G. Weeda. 1994. Clinical heterogeneity within xeroderma pigmentosum associated with mutations in the DNA repair and transcription gene *ERCC3. Am. J. Hum. Genet.* **54**:191–200.

436. Vermeulen, W., M. Stefanini, S. Giliani, J. H. J. Hoeijmakers, and D. Bootsma. 1991. Xeroderma pigmentosum complementation group H falls into complementation group D. *Mutat. Res.* **255**:201–208.

437. Vermeulen, W., A. J. van Vuuren, M. Chipoulet, L. Schaeffer, E. Appeldoorn, G. Weeda, N. G. J. Jaspers, A. Priestly, C. F. Arlett, A. R. Lehmann, M. Stefanini, M. Mezzina, A. Sarasin, D. Bootsma, J.-M. Egly, and J. H. J. Hoeijmakers. 1994. Three unusual repair deficiencies associated with transcription factor BTF2(TFIIH). Evidence for the existence of a transcription syndrome. *Cold Spring Harbor. Symp. Quant. Biol.* **59**:(in press).

438. Vincent, R. A., R. B. Sheridan, and P. C. Haung. 1975. DNA strand breakage repair in ataxia telangiectasia fibroblast-like cells. *Mutat. Res.* **33**:357–366.

439. Volberg, T. M., G. Seal, and M. A. Sirover. 1987. Monoclonal antibodies detect conformational abnormality of uracil DNA glycosylase in Bloom's syndrome cells. *Carcinogenesis* **8**:1725–1729.

440. Vrieling, H., M. L. van Rooijen, N. A. Groen, M. Z. Zdzienicka, J. W. I. M. Simons, P. H. M. Lohman, and A. A. van Zeeland. 1991. DNA strand specificity for UV-induced mutations in mammalian cells. *Mol. Cell. Biol.* **9**:1277–1283.

441. Vuillaume, M., R. Calvayrac, M. Best-Belpomme, P. Tarroux, M. Hubert, Y. Decroix, and A. Sarasin. 1986. Deficiency in the catalase activity of xeroderma pigmentosum cells and simian virus 40-transformed human cell extracts. *Cancer Res.* **46**:538–544.

442. Vuillaume, M., L. Daya-Grosjean, P. Vincens, J. L. Pennetier, P. Tarroux, A. Baret, R. Calvayrac, A. Taieb, and A. Sarasin. 1992. Striking differences in cellular catalase activity between two DNA repair-deficient diseases: xeroderma pigmentosum and trichothiodystrophy. *Carcinogenesis* **13**:321–328.

443. Wade, M. H., and E. H. Y. Chu. 1978. Effects of DNA damaging agents on cultured fibroblasts derived from patients with Cockayne syndrome, p. 667–670. *In* P. C. Hanawalt, E. C. Friedberg, and C. F. Fox (ed.), *DNA Repair Mechanisms*. Academic Press, Inc., New York.

444. Wade, M. H., and E. H. Y. Chu. 1979. Effects of DNA damaging agents on cultured fibroblasts derived from patients with Cockayne syndrome. *Mutat. Res.* **59**:49–60.

445. Wang, Y.-C., V. M. Maher, and J. J. McCormick. 1991. Xeroderma pigmentosum variant cells are less likely than normal cells to incorporate dAMP opposite photoproducts during replication of UV-irradiated plasmids. *Proc. Natl. Acad. Sci. USA* **88**:7810–7814.

446. Wang, Y.-C., V. M. Maher, D. L. Mitchell, and J. J. McCormick. 1993. Evidence from mutation spectra that the UV hypermutability of xeroderma pigmentosum variant cells reflects abnormal, error-prone replication on a template containing photoproducts. *Mol. Cell. Biol.* **13**:4276–4283.

447. Waters, H. L., S. Seetharam, M. M. Seidman, and K. H. Kraemer. 1993. Ultraviolet hypermutability of a shuttle vector propagated in xeroderma pigmentosum variant cells. *J. Invest. Dermatol.* **101**:744–748.

448. Weber, C. A., E. P. Salazar, S. A. Stewart, and L. H. Thompson. 1990. *ERCC2*: cDNA cloning and molecular characterization of a human nucleotide excision repair gene with high homology to yeast *RAD3. EMBO J.* **9**:1437–1447.

449. Webster, A. D. B., D. E. Barnes, C. F. Arlett, A. R. Lehmann, and T. Lindahl. 1992. Growth retardation and immunodeficiency in a patient with mutations in the DNA ligase I gene. *Lancet* **339**:1508–1509.

450. Webster, D., C. F. Arlett, S. A. Harcourt, I. A. Teo, and L. Henderson. 1982. A new syndrome of immunodeficiency and increased cellular sensitivity to DNA damaging agents, p. 379–386. *In* B. A. Bridges and D. G. Harnden (ed.), *Ataxia-Telangiectasia: Cellular and Molecular Link between Cancer, Neuropathology, and Immune Deficiency.* John Wiley & Sons, London.

451. Weeda, G., and J. H. J. Hoeijmakers. 1993. Genetic analysis of nucleotide excision repair in mammalian cells. *Semin. in Cancer Biol.* **4:**105–117.

452. Weeda, G., R. C. A. van Ham, W. Vermeulen, D. Bootsma, A. J. van der Eb, and J. H. J. Hoeijmakers. 1990. A presumed DNA helicase encoded by *ERCC-3* is involved is involved in the human repair disorders xeroderma pigmentosum and Cockayne's syndrome. *Cell* **62:**777–791.

453. Weemaes, C. M. R., T. W. J. Hustinx, J. M. J. C. Scheres, P. J. J. van Munster, J. A. J. M. Bakkeren, and R. D. F. M. Taalman. 1981. A new chromosomal instability disorder: the Nijmegen breakage syndrome. *Acta Paediat. Scand.* **70:**557–564.

454. Weksberg, R., C. Smith, L. Anson-Cartwright, and K. Maloney. 1988. Bloom syndrome: a single complementation group defines patients of diverse ethnic origin. *Am. J. Hum. Genet.* **42:**816–824.

455. Whitney, M. A., H. Sato, P. M. Jakobs, R. A. Gibson, R. A. Moses, and M. Grompe. 1993. A common mutation in the *FACC* gene causes Fanconi anemia in Ashkenazi Jews. *Nat. Genet.* **4:**202–205.

456. Willis, A. E., and T. Lindahl. 1987. DNA ligase I deficiency in Bloom's syndrome. *Nature* (London) **325:**355–357.

457. Willis, A. E., R. Weksberg, S. Tomlinson, and T. Lindahl. 1987. Structual alterations of DNA ligase I in Bloom syndrome. *Proc. Natl. Acad. Sci. USA* **84:**8016–8020.

458. Wolff, S., J. Bodycote, G. H. Thomas, and J. E. Cleaver. 1975. Sister chromatid exchanges in xeroderma pigmentosum cells that are defective in DNA excision repair or post-replication repair. *Genetics* **81:**349–355.

459. Wood, R. D., and D. Coverley. 1991. DNA excision repair in mammalian cell extracts. *Bioessays* **13:**447–453.

460. Wood, R. D., P. Robins, and T. Lindahl. 1988. Complementation of the xeroderma pigmentosum DNA repair defect in cell-free extracts. *Cell* **53:**97–106.

461. Wunder, E., B. T. Mortensen, F. Schilling, and P. R. Henon. 1993. Anomalous plasma concentrations and impaired secretion of growth factors in Fanconi's anemia. *Stem Cells* **11**(Suppl. 2):144–149.

462. Yagi, T., J. Tatsumi-Miyajima, M. Sato, K. H. Kraemer, and H. Takebe. 1991. Analysis of point mutations in an ultraviolet-irradiated shuttle vector plasmid propagated in cells from Japanese xeroderma pigmentosum patients in complementation groups A and F. *Cancer Res.* **31:**3177–3182.

463. Yamamoto, Y., and Y. Fujiwara. 1986. Abnormal regulation of uracil-DNA glycosylase induction during cell cycle and cell passage in Bloom's syndrome fibroblasts. *Carcinogenesis* **7:**305–310.

464. Yamashita, T., D. L. Barber, Y. Zhu, N. Wu, and A. D. D'Andrea. 1994. The Fanconi anemia polypeptide FACC is localized to the cytoplasm. *Proc. Natl. Acad. Sci. USA* **91:**6712–6716.

465. Yang, S.-W., F. F. Becker, and J. Y.-H. Chan. 1992. Identification of a specific inhibitor for DNA ligase I in human cells. *Proc. Natl. Acad. Sci. USA* **89:**2227–2231.

466. Young, B. R., and R. B. Painter. 1989. Radioresistant DNA synthesis and human genetic disease. *Hum. Genet.* **82:**113–117.

466a. Young, N. S., and B. P. Alter. 1993. *Aplastic Anemia-Acquired and Inherited.* W. B. Saunders, Philadelphia.

466b. Youssoufian, H. 1994. Localization of Fanconi anemia C protein to the cytoplasm of mammallian cells. *Proc. Natl. Acad. Sci. USA* **91:**7975–7979.

467. Zampetti-Bosseler, F., and D. Scott. 1981. Cell death, chromosome damage and mitotic delay in normal human ataxia telangiectasia and retinoblastoma fibroblasts after X-irradiation. *Int. J. Radiat. Biol.* **39:**547–558.

468. Zbinden, I., and P. Cerutti. 1981. Near-ultraviolet sensitivity of skin fibroblasts of patients with Bloom's syndrome. *Biochem. Biophys. Res. Commun.* **98:**579–587.

469. Zdzienicka, M. Z., and J. W. I. M. Simons. 1987. Mutagen-sensitive cell lines are obtained with a high frequency in V79 Chinese hamster cells. *Mutat. Res.* **178:**235–244.

470. Zelle, B., F. Berends, and P. M. H. Lohman. 1980. Repair of damage by ultraviolet readiation in xeroderma pigmentosum cell strains of complementation groups E and F. *Mutat. Res.* **73:**157–169.

471. Zelle, B., and P. M. H. Lohman. 1979. Repair of UV-endonuclease-susceptible sites in the 7 complementation groups of xeroderma pigmentosum A through G. *Mutat. Res.* **62:**363–368.

472. Zhen, W., M. K. Evans, C. M. Haggerty, and V. A. Bohr. 1993. Deficient gene specific repair of cisplatin-induced lesions in xeroderma pigmentosum and Fanconi's anemia cell lines. *Carcinogenesis* **14:**919–924.

473. Ziv, Y., N. G. J. Jaspers, A. I. Berkel, and Y. Shiloh. 1989. Ataxia-telangiectasia: a variant with altered in vitro phenotype of fibroblast cells. *Mutat. Res.* **210:**211–219.

474. Ziv, Y., G. Rotman, M. Frydman, J. Dagan, T. Cohen, T. Faroud, R. A. Gatti, and Y. Shiloh. 1991. The ATC (ataxia-telangiectasia complementation group C) locus localizes to 11q22-23. *Genomics* **9:**373–375.

Index